PRINCIPLES OF MATERIALS SCIENCE AND ENGINEERING

THIRD EDITION

PRINCIPLES OF MATERIALS SCIENCE AND ENGINEERING

William F. Smith

Professor of Engineering
University of Central Florida

McGraw-Hill, Inc.

New York   St. Louis   San Francisco   Auckland   Bogotá
Caracas   Lisbon   London   Madrid
Mexico City   Milan   Montreal   New Delhi
San Juan   Singapore   Sydney   Tokyo   Toronto

PRINCIPLES OF MATERIALS SCIENCE AND ENGINEERING

This book is printed on acid-free paper.

2 3 4 5 6 7 8 9 0   DOC   DOC   9 0 9 8 7 6

ISBN 0-07-059241-1

This book was set in Cheltenham Light by York Graphic Services, Inc.
The editors were B. J. Clark and Jack Maisel;
the production supervisor was Denise L. Puryear.
The cover was designed by Jo Jones.
R. R. Donnelley & Sons Company was printer and binder.

On the Cover:
Polymer-matrix composite tube being filament-braided with aramid fibers and an epoxy matrix. Computerized braiding machines fabricate these tubes for critical applications in the medical, electrical, automotive, and aerospace industries. (*Courtesy of Polygon Company, Walkerton, Ind.*)

Library of Congress Cataloging-in-Publication Data

Smith, William Fortune, (date).
    Principles of materials science and engineering / William F. Smith.—3d ed.
        p.   cm.—(McGraw-Hill series in materials science and engineering)
    Includes bibliographical references (p.   ) and index.
    ISBN 0-07-059241-1
    1. Materials.   I. Title.   II. Series.
TA403.S596   1996
620.1′1—dc20   94-48015

ABOUT THE AUTHOR

William F. Smith is professor of engineering in the Mechanical and Aerospace Engineering Department of the University of Central Florida at Orlando, Florida. He was awarded an M.S. degree in metallurgical engineering from Purdue University and an Sc.D. degree in metallurgy from Massachusetts Institute of Technology. Dr. Smith, who is a registered professional engineer in the states of California and Florida, has been teaching undergraduate and graduate materials science and engineering courses and actively writing textbooks for many years. He is also the author of *Structure and Properties of Engineering Alloys,* Second Edition (McGraw-Hill, 1993) and *Foundations of Materials Science and Engineering,* Second Edition (McGraw-Hill, 1993).

*It is difficult to say what is impossible,
for the dream of yesterday is the hope of
today and the reality of tomorrow.*
Robert H. Goddard

CONTENTS

CHAPTER 7 POLYMERIC MATERIALS

CHAPTER 8 PHASE DIAGRAMS

PREFACE

Designed to serve as a text for a first course in materials science and engineering for engineering students, this third edition of *Principles of Materials Science and Engineering* is an enlarged, updated, and improved version of the second edition. In this edition there are over 280 new or modified problems with numerical answers. Some of the excellent problems from Professor Lew Rabenberg, University of Texas, Austin, have been retained as well as some of the excellent suggestions of Professor Patricia Shamamy of Lawrence Technological University. New subsections on polyphthalamides and thermoplastic polyurethane elastomers have been added. Enlarged subsections on silicon nitride and silicon carbide structural ceramics and ion implantation in microelectronic materials have been included. A subsection on new trends and technologies in microelectronic microprocessors has also been added because of the major changes that have been taking place in this field in the past years. Other improvements include additional information on the mechanical properties of carbon fibers, some new technical photos and diagrams, and updated data on polymeric materials. New sections on materials selection of plastic and metallic materials have also been added because of their importance in engineering designs.

Again, it is a pleasure to acknowledge the help and support of everyone concerned with this ongoing project. For those professors who have used or are using this text, your support is gratefully acknowledged and needed since without it, there would be no book. So, many thanks to all of you for your *sustained* backing. The comments and helpful suggestions of the following reviewers are also appreciated: Chung-Jen Hsu, Feng Chia University, Taichung, Taiwan; Jose Pires, University of California–Irvine; and Lew Rabenberg, University of Texas at Austin.

Finally, I would like to acknowledge the sustained help and encouragement from my engineering editor B. J. Clark of McGraw-Hill, chairman David Nicholson of the Mechanical and Aerospace Engineering Department, and Dean Marty Wanielista of the College of Engineering of the University of Central Florida.

William F. Smith

PRINCIPLES OF MATERIALS SCIENCE AND ENGINEERING

Introduction to Materials Science and Engineering

1.1 MATERIALS AND ENGINEERING

Materials are substances of which something is composed or made. Since civilization began, materials along with energy have been used by people to improve their standard of living. Materials are everywhere about us since products are made of materials. Some of the commonly encountered materials are wood (timber), concrete, brick, steel, plastic, glass, rubber, aluminum, copper, and paper. There are many more kinds of materials, and one only has to look around oneself to realize that. Because of constant research and development, new materials are frequently being created.

The production and processing of materials into finished goods constitutes a large part of our present economy. Engineers design most manufactured products and the processing systems required for their production. Since products require materials, engineers should be knowledgeable about the internal structure and properties of materials so that they will be able to select the most suitable ones for each application and be able to develop the best processing methods.

Research and development engineers work to create new materials or to modify the properties of existing ones. Design engineers use existing, modified,

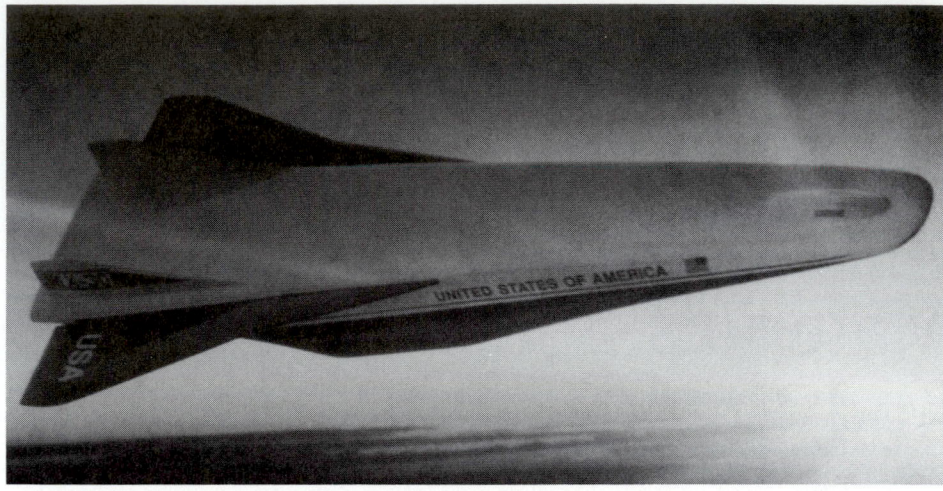

FIGURE 1.1 The hydrogen-powered national aerospace plane will be capable of using standard runways for takeoff and landing and will be able to travel to the space station, but only if materials can be developed that are strong, lightweight, and capable of resisting extreme temperature variations. (*After Advanced Materials & Processes, September 1993, p. 24.*)

or new materials to design and create new products and systems. Sometimes the reverse is the case, and design engineers have a problem in their design which requires a new material to be created by research scientists and engineers. For example, engineers designing a hypersonic-speed transport such as the X-30 (Fig. 1.1) will have to develop new high-temperature advanced materials able to withstand temperatures as high as 1800°C (3250°F) so that airspeeds as high as Mach 12 to 25* can be attained. Materials currently (1993) under investigation for the X-30 include silicon carbide–fiber-reinforced Timetal 21S titanium alloy for load bearing structures, carbon-carbon composites for the mechanically attached thermal protection system, and carbon-fiber-epoxy composites for the fuel tank. Another example of a challenge for engineers is the permanently manned space station (Fig. 1.2a). One proposal involves fabricating the space station's main structure in space using in-orbit-produced I beams and channels made of polyetherimide and polyetheretherketone composite materials (Fig. 1.2b and c).

The search for new materials goes on continuously. For example, mechanical engineers search for higher-temperature materials so that jet engines can operate more efficiently. Electrical engineers search for new materials so that electronic devices can operate faster and at higher temperatures. Aerospace engineers search for materials with higher strength-to-weight ratios for aircraft and space vehicles. Chemical engineers look for more highly corrosion-

---

*Mach 1 equals the speed of sound in air.

(a)

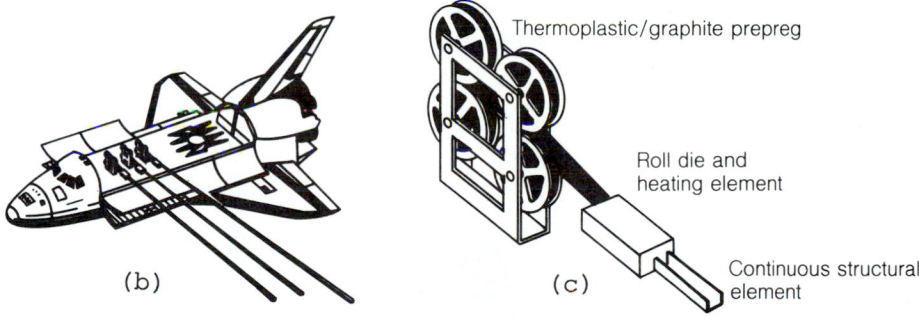

Thermoplastic/graphite prepreg

Roll die and
heating element

(b)

(c)

Continuous structural
element

FIGURE 1.2 (*a*) An artist's concept of the enhanced configuration of the permanently manned space station produced by Martin Marietta. [*After Journal of Metals, **40**(5):12(1988).*] (*b*) and (*c*) One proposal is to construct the station frame from in-orbit-fabricated I beams and channels made from advanced thermoplastic composites using the pultrusion process. (*After Modern Plastics, August 1988, p. 102.*)

resistant materials. These are only a few examples of the search by engineers for new and improved materials for applications. In many cases what was impossible yesterday is a reality today!

Engineers in all disciplines should have some basic and applied knowledge of engineering materials so that they will be able to do their work more effectively when using materials. The purpose of this book is to serve as an introduction to the internal structure, properties, processing, and applications of engineering materials. Because of the enormous amount of information available about engineering materials and due to the limitations of this book, the material presented has had to be selective.

1.2   MATERIALS SCIENCE AND ENGINEERING

Materials science is primarily concerned with the search for basic knowledge about the internal structure, properties, and processing of materials. *Materials engineering* is mainly concerned with the use of fundamental and applied

FIGURE 1.3 Materials knowledge spectrum. Using the combined knowledge of materials from materials science and materials engineering enables engineers to convert materials into the products needed by society.

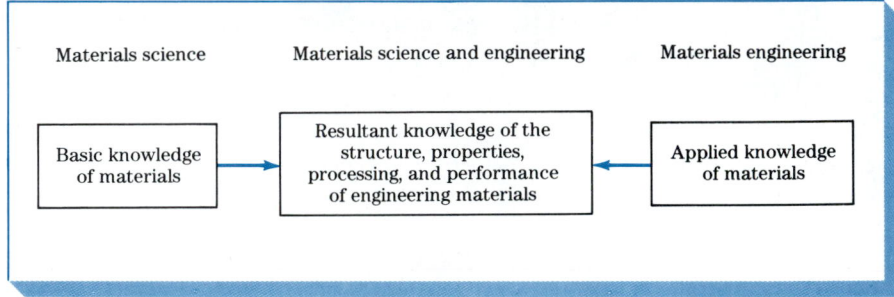

FIGURE 1.4 This diagram illustrates how materials science and engineering form a bridge of knowledge from the basic sciences to the engineering disciplines. (*Courtesy of the National Academy of Science.*)

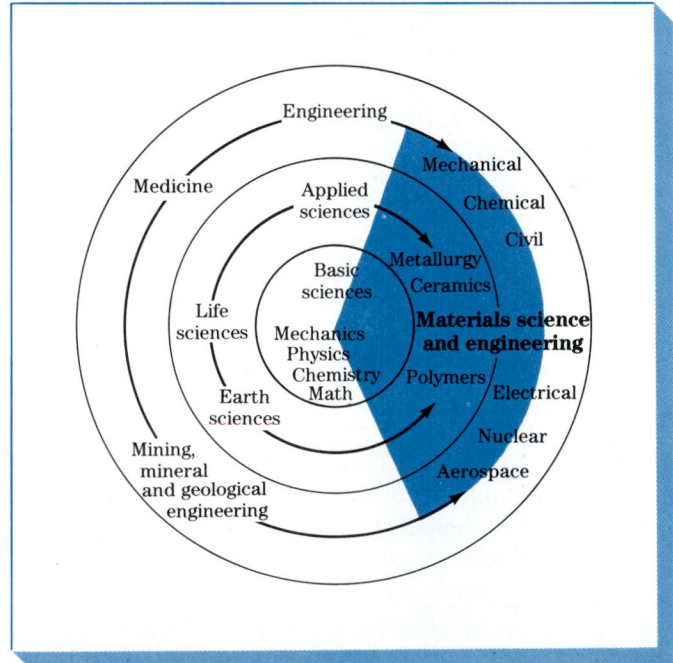

knowledge of materials so that the materials can be converted into products necessary or desired by society. The name *materials science and engineering* combines both materials science and materials engineering and is the subject of this book. Materials science is at the basic knowledge end of the materials knowledge spectrum and materials engineering is at the applied knowledge end, and there is no demarcation line between the two (Fig. 1.3).

Figure 1.4 shows a three-ringed diagram which indicates the relationship among the basic sciences (and mathematics), materials science and engineering, and the other engineering disciplines. The basic sciences are located within the first ring or core of the diagram, while the various engineering disciplines (mechanical, electrical, civil, chemical, etc.) are located in the outermost third ring. The applied sciences, metallurgy, ceramics, and polymer science are located in the middle or second ring. Materials science and engineering is shown to form a bridge of materials knowledge from the basic sciences (and mathematics) to the engineering disciplines.

1.3   TYPES OF MATERIALS

For convenience most engineering materials are divided into three main classes: *metallic, polymeric (plastic)*, and *ceramic materials*. In this chapter we shall distinguish among them on the basis of some of their important mechanical, electrical, and physical properties. In subsequent chapters we shall study the internal structural differences of these types of materials. In addition to the three main classes of materials, we shall consider two more types, *composite materials* and *electronic materials*, because of their great engineering importance.

Metallic Materials   These materials are inorganic substances which are composed of one or more metallic elements and may also contain some nonmetallic elements. Examples of metallic elements are iron, copper, aluminum, nickel, and titanium. Nonmetallic elements such as carbon, nitrogen, and oxygen may also be contained in metallic materials. Metals have a crystalline structure in which the atoms are arranged in an orderly manner. Metals in general are good thermal and electrical conductors. Many metals are relatively strong and ductile at room temperature, and many maintain good strength even at high temperatures.

Metals and alloys[1] are commonly divided into two classes: *ferrous metals and alloys* that contain a large percentage of iron such as the steels and cast irons and *nonferrous metals and alloys* that do not contain iron or only a relatively small amount of iron. Examples of nonferrous metals are aluminum, copper, zinc, titanium, and nickel.

---

[1] A metal alloy is a combination of two or more metals or a metal (metals) and a nonmetal (nonmetals).

Figure 1.5 is a photograph of a commercial aircraft jet engine which is made primarily of metal alloys. The metal alloys used inside the engine must be able to withstand the high temperatures and pressures generated during its operation. Many years of research and development work by scientists and engineers were required to perfect this advanced-performance engine. Figure

FIGURE 1.5 The aircraft turbine engine (PW2037) shown is made principally of metal alloys. The latest high-temperature-resistant, high-strength nickel-base alloys are used in this engine. The engine features a new solid-state digital electronic control system that eliminates frequent movement of the throttle by the flight crew. More efficient engine cooling also improves engine efficiency and fuel consumption. Engines of this type are constantly being improved by intensive engineering research and developmental work. (*Courtesy of the Pratt and Whitney Co.*)

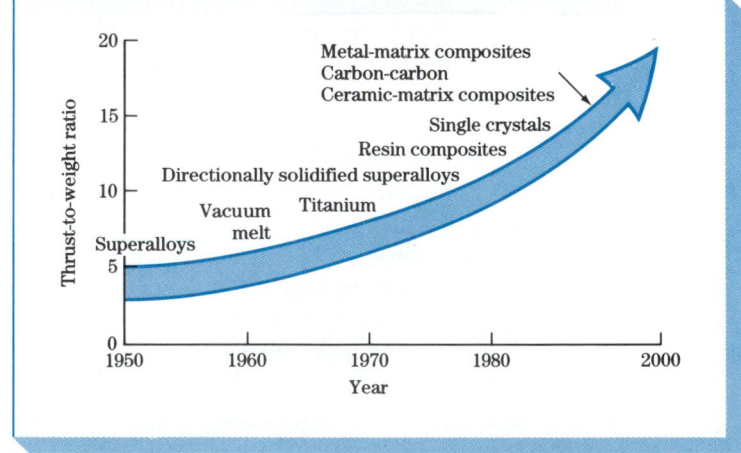

FIGURE 1.6 Materials and materials processes over the past years have been associated with the increase in performance of gas-turbine engines. [*After Adv. Mat. & Proc., **133**(1):88(1988).*]

1.6 shows how materials and materials processes have been associated with more efficient gas-turbine engine propulsion performance over the past years. Metal-matrix and ceramic-matrix materials may lead to even further increased performance in the future

Polymeric (Plastic) Materials

Most polymeric materials consist of organic (carbon-containing) long molecular chains or networks. Structurally, most polymeric materials are noncrystalline but some consist of mixtures of crystalline and noncrystalline regions. The strength and ductility of polymeric materials vary greatly. Because of the nature of their internal structure, most polymeric materials are poor conductors of electricity. Some of these materials are good insulators and are used for electrical insulative applications (Fig. 1.7). In general, polymeric materials have low densities and relatively low softening or decomposition temperatures.

Ceramic Materials

Ceramic materials are inorganic materials which consist of metallic and nonmetallic elements chemically bonded together. Ceramic materials can be crystalline, noncrystalline, or mixtures of both. Most ceramic materials have high hardness and high-temperature strength but tend to have mechanical brittleness. Lately, new ceramic materials have been developed for engine applications (Fig. 1.8). Advantages of ceramic materials for engine applications are light weight, high strength and hardness, good heat and wear resistance, reduced friction, and insulative properties.

The insulative property along with high heat and wear resistance of many ceramics make them useful for furnace linings for high-temperature liquid

FIGURE 1.7 The circuit board and connectors shown here utilize the engineering thermoplastic polyetheretherketone to meet stringent high-temperature resistance and dimensional stability requirements and to assure material integrity under soldering conditions. (*Courtesy of ICI Americas.*)

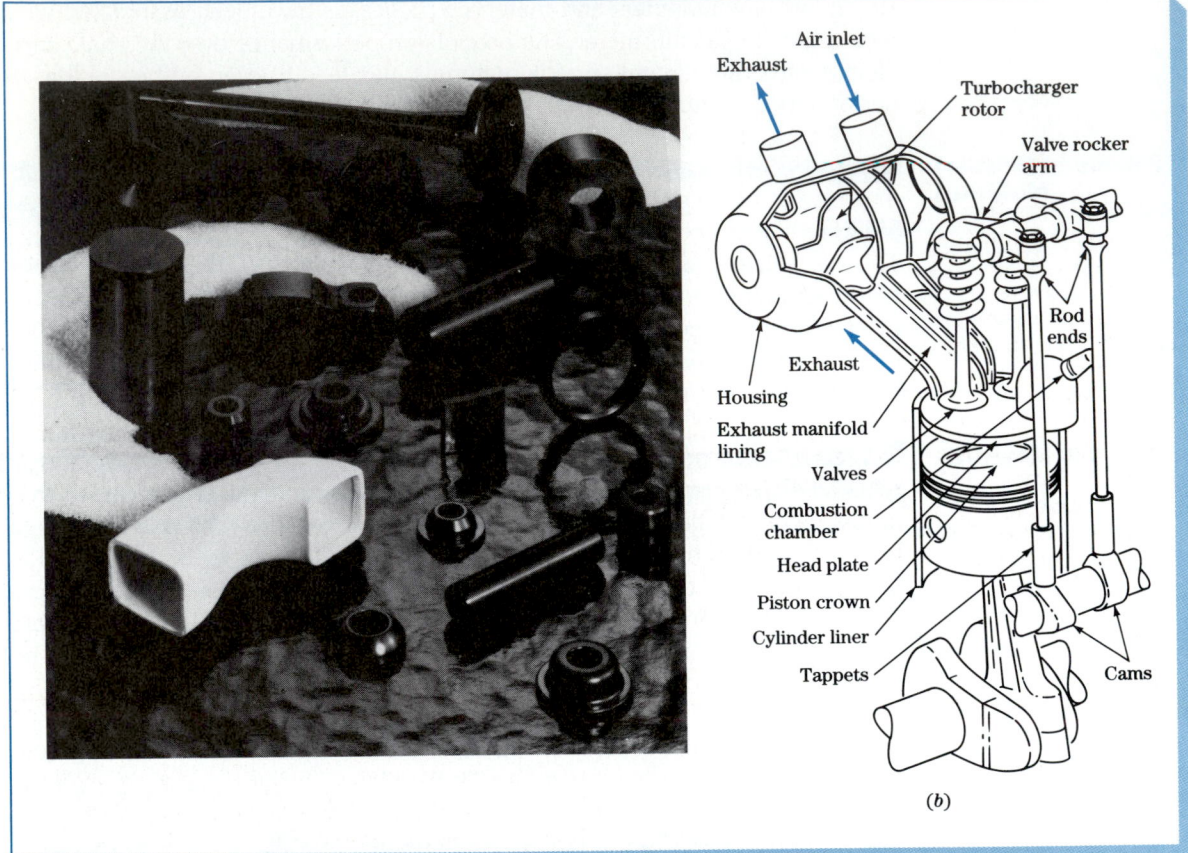

Air inlet

Exhaust

Turbocharger rotor

Valve rocker arm

Rod ends

Exhaust

Housing

Exhaust manifold lining

Valves

Combustion chamber

Head plate

Piston crown

Cylinder liner

Tappets

Cams

(b)

FIGURE 1.8 (*a*) Examples of a newly developed generation of engineered ceramic materials for advanced engine applications. The black items include engine valves, valve seat inserts, and piston pins made of silicon nitride. The white item is a port-manifold liner made of an alumina ceramic material. (*Courtesy of Norton/TRW Ceramics.*) (*b*) Potential ceramic component applications in a turbocharged diesel engine. (*After Metals and Materials, December 1988.*)

metals such as steel. An important space application for ceramics are the ceramic tiles for the space shuttle. These ceramic materials thermally protect the aluminum internal structure of the space shuttle during ascent out of and reentry into the earth's atmosphere (Figs. 10.50–10.51).

Composite Materials   Composite materials are mixtures of two or more materials. Most composite materials consist of a selected filler or reinforcing material and a compatible resin binder to obtain the specific characteristics and properties desired. Usually, the components do not dissolve in each other and can be physically

FIGURE 1.9 Glass-fiber-reinforced (20 percent chopped-fiber or 40 percent continuous-strand) polyphenylene sulfide (PPS) composites have good resistance to corrosive chemicals encountered in oil-field environments. Shown here are high-strength, thermally stable, compression-molded PPS pipe fittings undergoing field tests. (*Courtesy of Phillips 66 Co.*)

identified by an interface between the components. Composites can be of many types. Some of the predominant types are fibrous (composed of fibers in a matrix) and particulate (composed of particles in a matrix). There are many different combinations of reinforcements and matrices used to produce composite materials. Two outstanding types of *modern composite materials* used for engineering applications are fiberglass-reinforcing material in a polyester or epoxy matrix and carbon fibers in an epoxy matrix. Figure 1.15 shows schematically where carbon-fiber-epoxy composite material will be used for the wings and engines of the new 777 airliner. Another example of the use of composites is glass-reinforced polyphenylene sulfide (PPS) for oil-field fittings. This application utilizes the excellent corrosion resistance of this material (Fig. 1.9).

Electronic Materials   Electronic materials are not a major type of material by volume but are an extremely important type of material for advanced engineering technology. The most important electronic material is pure silicon which is modified in various ways to change its electrical characteristics. A multitude of complex electronic circuits can be miniaturized on a silicon chip which is about $\frac{1}{4}$ in square (0.635 cm square) (Fig. 1.10). Microelectronic devices have made possible such new products as communication satellites, advanced computers, hand-held calculators, digital watches, and welding robots (Fig. 1.11).

FIGURE 1.10 A microprocessor, which is the central processing element of a microcomputer, is shown with part of its case removed to expose the silicon chip. The pins below the case connect it to the external circuitry (magnification ½X). (*Courtesy of Intel Corporation.*)

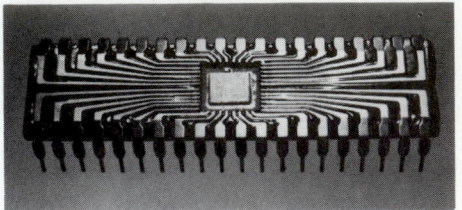

FIGURE 1.11 Electronic materials play a major role in the construction of computer-controlled robots. Shown above are four computer-controlled robots welding a 1985 full-size automobile to achieve high structural soundness and economy. (*Courtesy of General Motors Co.*)

1.4 COMPETITION AMONG MATERIALS

Materials compete with each other for existing and new markets. Over a period of time many factors arise which make it possible for one material to replace another for certain applications. Certainly cost is a factor. If a breakthrough is made in the processing of a certain type of material so that its cost is decreased substantially, this material may replace another for some applications. Another factor which causes material replacement changes is the development of a new material with special properties for some applications. As a result, over a period of time, the usage of different materials changes.

Figure 1.12 shows graphically how the production of six materials in the United States on a weight basis varied over the past years. Aluminum and polymers show an outstanding increase in production since 1930. On a volume basis the production increases for aluminum, and polymers are even more accentuated since these are light materials.

The competition among materials is evident in the composition of the U.S. auto. In 1978 the average U.S. auto weighed 4000 lb (1800 kg) and consisted of about 60 percent iron and steel, 10 to 20 percent plastics and rubber, and 3 to 5 percent aluminum. The 1993 U.S. auto by comparison weighed an average of 3150 lb (1430 kg) and consisted of 50 to 60 percent iron and steel, 10 to 20 percent plastics and rubber, and 5 to 10 percent aluminum (Fig. 1.13 and Table 1.1). Thus in the period 1978 to 1993 the percentage of steel and iron declined while that of aluminum and plastics increased. No major changes appear in the short-term future.

For some applications only certain materials are able to meet the engineering requirements for a design, and these materials may be relatively expensive. For example, the modern jet engine (Fig. 1.5) requires high-temperature nickel-base superalloys to function. These materials are expensive, and

FIGURE 1.12 Competition of six major materials produced in the United States on a weight (pound) basis. The rapid rise in the production of aluminum and polymers (plastics) is evident.

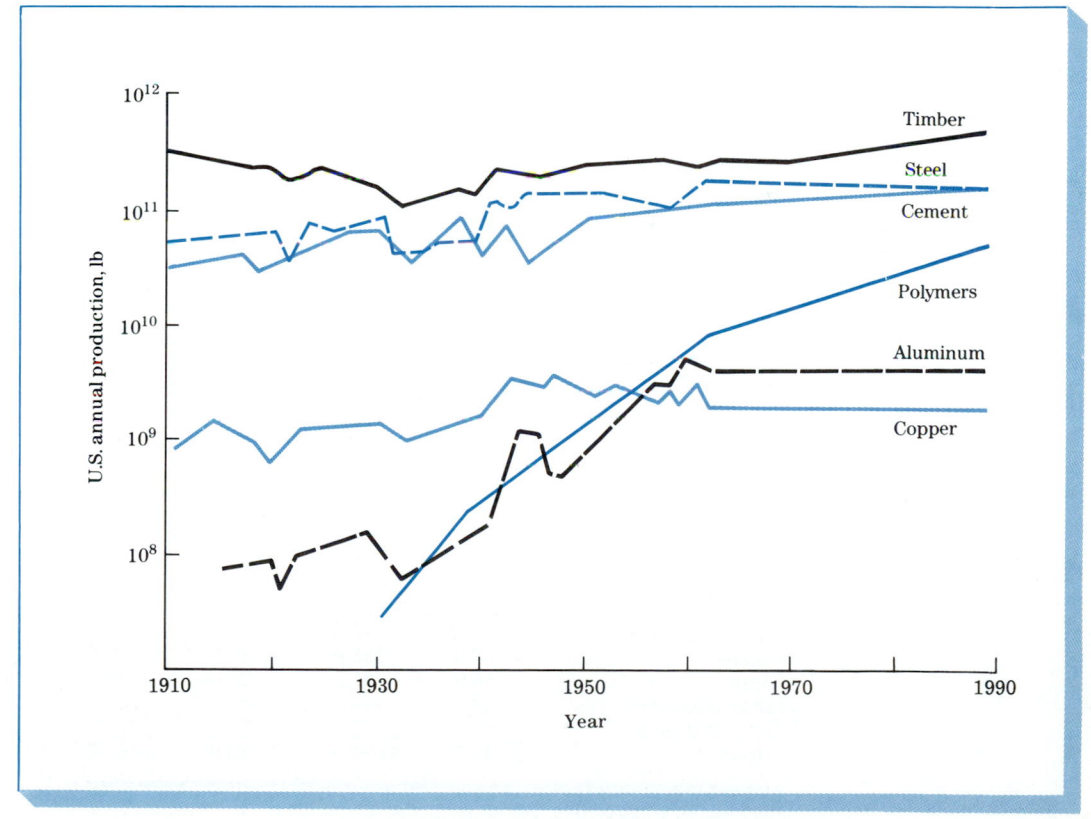

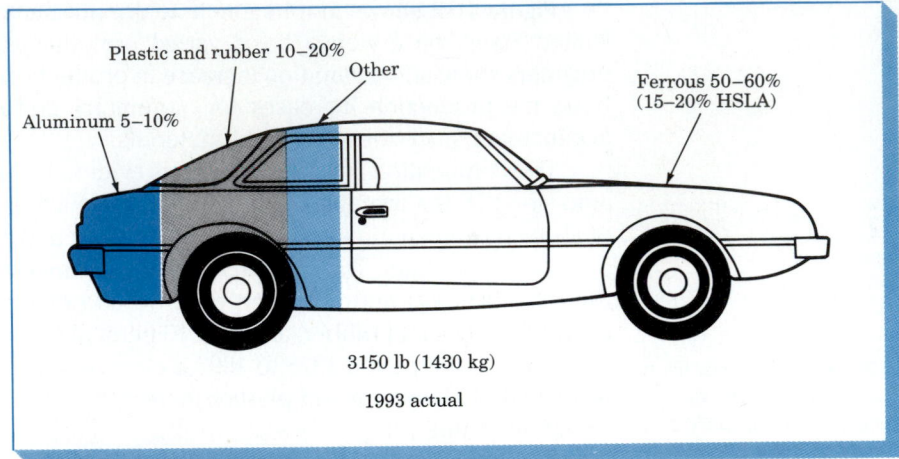

FIGURE 1.13 Breakdown of weight percentages of major materials used in the average 1993 U.S. automobile. Some predictions made in the past decade were off-base because the forecasters failed to account for downsizing, aerodynamics, and use of computer control of engine functions. (*After J. Bittence, Mater. Eng., with modified data.*)

TABLE 1.1 **Comparison of the Average Weight (in Pounds*) of Materials in U.S. Autos in Period 1977–1993**

Material	1993	1991	1987	1977
Regular steel sheet, strip, bar	1,376.0	1,341.0	1,459.0	1,995.0
High- and medium-strength steel	259.0	240.5	228.0	125.0
Stainless steel	43.5	37.0	32.0	26.0
Other steels	48.0	41.5	55.5	56.0
Iron	411.5	431.0	460.0	540.0
Plastic and plastic composites	245.0	238.0	221.5	168.0
Aluminum	177.0	166.0	146.0	97.0
Copper, brass	43.5	46.0	46.0	38.5
Magnesium castings	6.5	5.0	2.5	1.0
Powder metal parts	26.0	23.5	19.5	15.5
Zinc die-castings	16.0	17.5	18.0	38.0
Rubber	134.5	135.5	135.5	150.0
Glass	88.5	86.0	86.0	87.5
Fluids and lubricants	188.5	174.0	183.0	200.0
Other materials	86.0	76.5	85.5	128.0
Total	**3,149.5**	**3,059.0**	**3,178.0**	**3,665.5**

*To convert to kilograms, divide by 2.204.
Source: American Metal Market from industry reports.

no cheap substitute has been found to replace them. Thus, although cost is an important factor in engineering design, the materials used must also meet performance specifications. Replacement of one material by another will continue in the future since new materials are being discovered and new processes are being developed.

1.5 FUTURE TRENDS IN MATERIALS USAGE

Metallic Materials
The U.S. production of basic metals such as iron, steel, aluminum, copper, zinc, and magnesium are expected to follow the U.S. economy fairly closely. However, existing alloys may be improved by better chemistry, composition control, and processing techniques.

New and improved aerospace alloys such as the nickel-base high-temperature superalloys are constantly being researched for increased high-temperature strength and corrosion resistance. These alloys are used for jet engines, and increased engine efficiency can be obtained by higher operating temperatures. New processing techniques such as hot isostatic pressing and isothermal forging can lead to improved fatigue life of aircraft alloys. Also powder metallurgy techniques will continue to be important since improved properties can be obtained for some alloys with lower finished-product costs. Rapid solidification technology can now produce metal alloy powders which have been rapidly cooled from the melt at rates as high as 1 million degrees Celsius per second. These powders are then consolidated into bars by various processes including hot isostatic pressing. For example, new high-temperature nickel-based superalloys, aluminum alloys, and titanium alloys have been produced by this new process.

Polymeric (Plastic) Materials
Historically, plastic materials have been the fastest growing basic material in the United States over the past years, with a growth rate of 9 percent/year on a weight basis (Fig. 1.12). However, the growth rate for plastics through 1995 is expected to average below 5 percent, a significant decrease. This drop is expected because plastics have already substituted for metals, glass, and paper in most of the main volume markets such as packaging and construction for which plastics are suitable.

Engineering plastics such as nylon are expected to be competitive with metals at least to the year 2000, according to some predictions. Figure 1.14 shows expected costs for engineering plastic resins vs. some common metals. With the exception of hot-rolled steel, engineering plastics are expected to be the least expensive material. An important trend in the development of engineering plastics is to blend or alloy different polymeric materials together to produce new synergistic plastic alloys.[1] For example, in the period from June 1987 to June 1988, approximately 1000 new thermoplastic resins, alloys, and

---

[1]Synergism means that the cooperative action of discrete agencies is such that the total effect is greater than the sum of the effects taken independently.

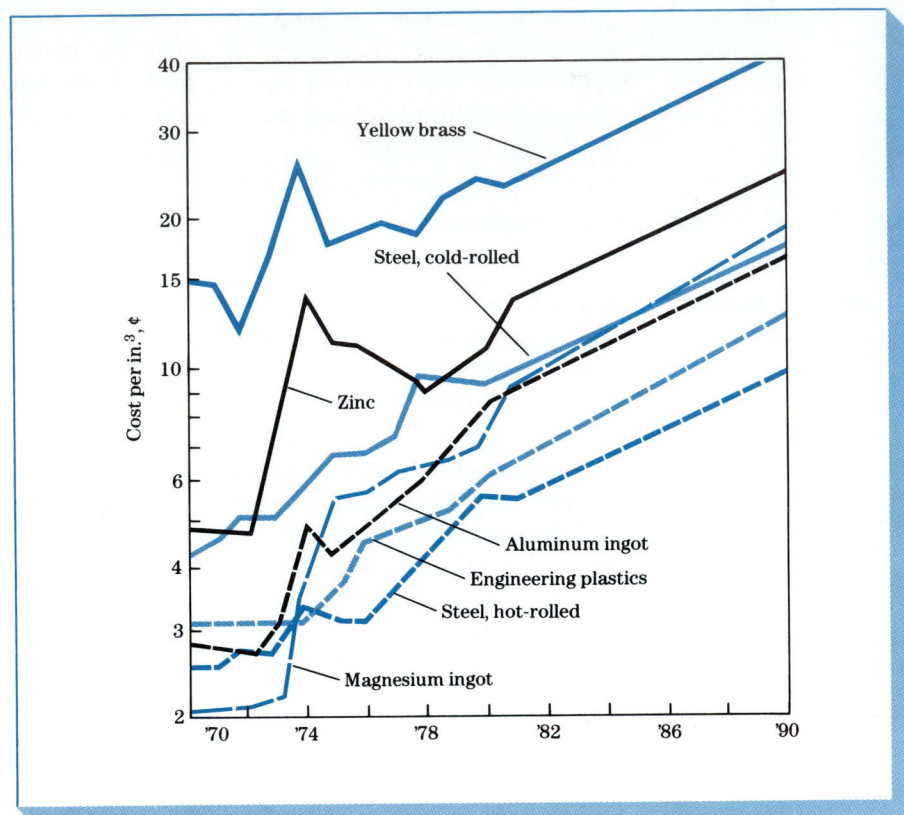

FIGURE 1.14 Historical and expected competitive costs of engineering plastic resins vs. some common metals from 1970 to 1990. Engineering plastics are expected to remain competitive with cold-rolled steel and other metals. (*After Modern Plastics, August 1982, p. 12.*)

compounds were introduced worldwide. New plastic alloys and blends accounted for about 10 percent of these.

Ceramic Materials The historical growth of traditional ceramic materials such as clay, glass, and stone in the United States has been 3.6 percent (1966 to 1980). The expected growth rate of these materials from 1982 to 1995 is expected to be about 2 percent.

In the past decade an entire new family of *engineering ceramics* of nitrides (Fig. 1.8), carbides, and oxides has been produced. New applications for these materials are being found constantly, particularly for high-temperature uses and for electronic ceramics.

Ceramic materials are low in cost, but their processing into finished products is usually slow and costly. Also, most ceramic materials are easily damaged by impact because of their low or nil ductility. If new techniques for developing high-impact ceramics could be found, these materials could show an upsurge for engineering applications where high-temperature and high-wear environments exist.

Composite Materials

Fiber-reinforced plastics are the main type of composite material used by industry, with glass being the dominant fiber. Unsaturated polyester, which is the main matrix resin used for many glass-reinforced plastics, increased in usage from 1407 million lb (640 Mkg) in 1991 to 1467 million lb (667 MKg) in 1992, which is an increase of about 4 percent.

Advanced composite materials such as fiberglass-epoxy and graphite-epoxy combinations are becoming more important all the time for high-performance and critical structural applications. An average annual gain of about 7.5 percent is predicted for the future usage of these materials. Modern commercial aircraft are using more composite materials. For example, the new 777 aircraft (Fig. 1.15) scheduled for 1995 utilizes about 10 percent by weight of composite materials.

Electronic Materials

The use of silicon and other semiconductor materials in solid-state and microelectronics has shown a tremendous growth since 1970, and this growth pattern is expected to continue beyond the year 2000. The impact of computers and other industrial types of equipment using integrated circuits made from silicon chips has been spectacular. The full effect of computerized robots in modern manufacturing is yet to be determined. Electronic materials will undoubtedly play a vital role in the "factories of the future" where almost all manufacturing may be done by robots assisted by computer-controlled machines.

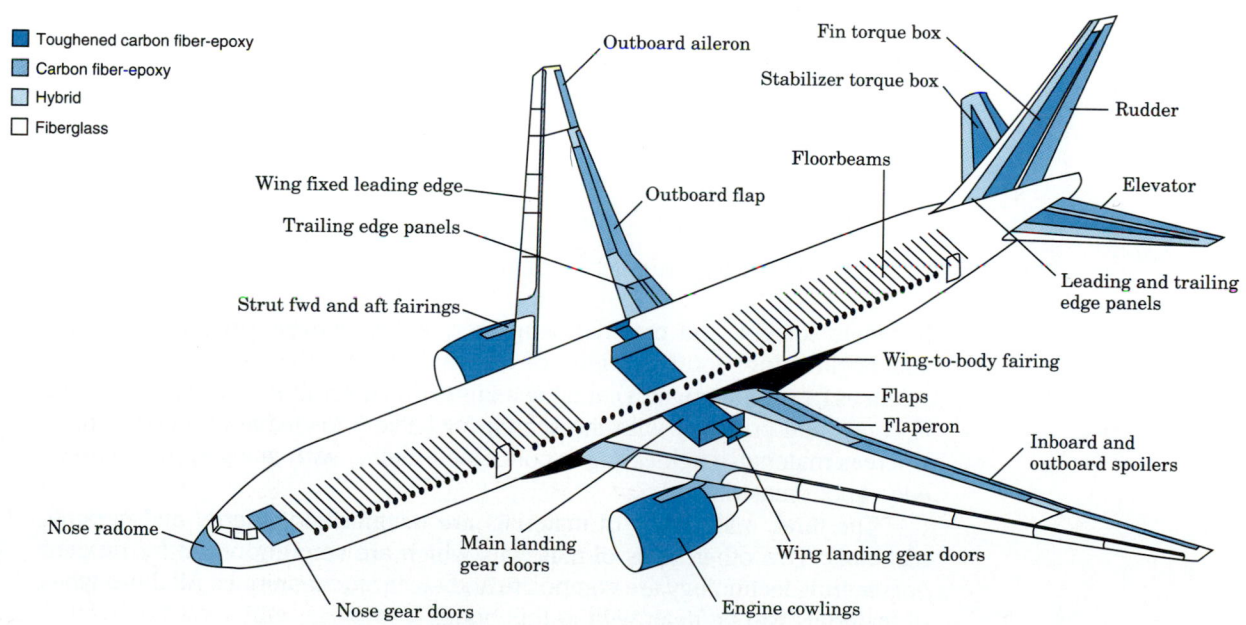

FIGURE 1.15 Types and distribution of composite material parts for the new 777 airliner. (*Courtesy of Boeing Aircraft Co.*)

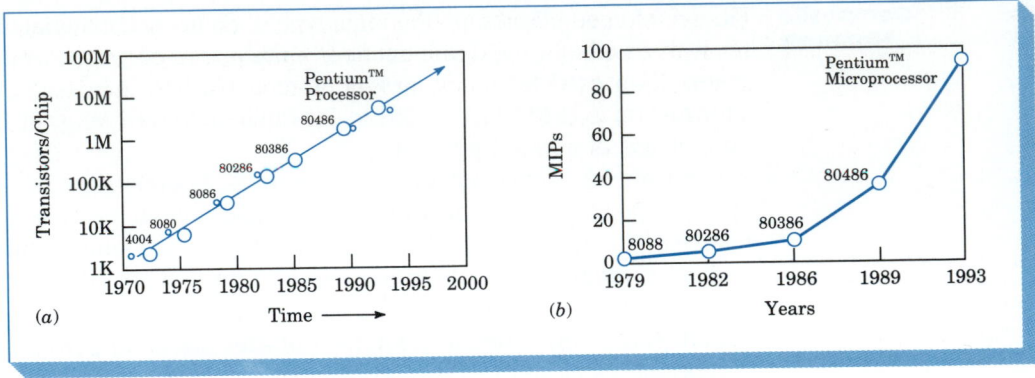

FIGURE 1.16 Trends in electronic integrated circuits: (*a*) transistors per chip vs. time; (*b*) processing power in millions of instructions per second (MIPs) vs. time. (*Intel Corp. data.*)

Over the years integrated circuits are being made that have a greater and greater density of transistors located on a single silicon chip with a corresponding decrease in transistor width. As indicated in Fig. 1.16*a*, the number of transistors in an integrated circuit chip has about doubled every 18 months for the past 30 years. Figure 1.16*b* shows that the processing power for a microprocessor has greatly increased over the 1989–1993 period.

Welcome to the fascinating and exceedingly interesting world of materials science and engineering!

SUMMARY

Materials science and materials engineering (collectively, materials science and engineering) form a bridge of materials knowledge between the basic sciences (and mathematics) and the engineering disciplines. Materials science is concerned primarily with the search for basic knowledge about materials, whereas materials engineering is concerned mainly with using applied knowledge about materials.

The three main types of materials are metallic, polymeric, and ceramic materials. Two other types of materials which are very important for modern engineering technology are composite and electronic materials. All these types of materials will be dealt with in this book.

Materials compete with each other for existing and new markets, and so the replacement of one material by another for some applications occurs. The availability of raw materials, cost of manufacturing, and the development of new materials and processes for products are major factors which cause changes in materials usage.

DEFINITIONS

Materials: substances of which something is composed or made. The term *engineering materials* is sometimes used to refer specifically to materials used to produce technical products. However, there is no clear demarcation line between the two terms and they are used interchangeably.

Materials science: a scientific discipline which is primarily concerned with the search for basic knowledge about the internal structure, properties, and processing of materials.

Materials engineering: an engineering discipline which is primarily concerned with use of fundamental and applied knowledge of materials so that they can be converted into products needed or desired by society.

Metallic materials (metals and metal alloys): materials which are characterized by high thermal and electrical conductivities. Examples are iron, steel, aluminum, and copper.

Ferrous metals and alloys: metals and alloys which contain a large percentage of iron such as steels and cast irons.

Nonferrous metals and alloys: metals and alloys which do not contain iron or if they do contain iron, it is only in a relatively small percentage. Examples of nonferrous metals are aluminum, copper, zinc, titanium, and nickel.

Ceramic materials: materials consisting of compounds of metals and nonmetals. Ceramic materials are usually hard and brittle. Examples are clay products, glass, and pure aluminum oxide which has been compacted and densified.

Polymeric materials: materials consisting of long molecular chains or networks of low-weight elements such as carbon, hydrogen, oxygen, and nitrogen. Most polymeric materials have low electrical conductivities. Examples are polyethylene and polyvinyl chloride (PVC).

Composite materials: materials which are mixtures of two or more materials. Examples are fiberglass-reinforcing material in a polyester or epoxy matrix.

Electronic materials: materials used in electronics, especially microelectronics. Examples are silicon and gallium arsenide.

PROBLEMS

1.1.1 What are materials? List eight commonly encountered engineering materials.
1.2.1 Define materials science and materials engineering.

1.3.1 What are the main classes of engineering materials?

1.3.2 What are some of the important properties of each of these engineering materials?

1.3.3 Define a composite material. Give an example of a composite material.

1.4.1 List some materials usage changes which you have observed over a period of time in some manufactured products. What reasons can you give for the changes that have occurred?

1.4.2 What factors might cause materials usage predictions to be incorrect?

Atomic Structure and Bonding

2.1 THE STRUCTURE OF ATOMS

Let us now review some of the fundamentals of atomic structure since atoms are the basic structural unit of all engineering materials.

Atoms consist primarily of three basic subatomic particles: *protons*, *neutrons*, and *electrons*. The current simple model of an atom envisions a very small nucleus of about 10^{-14} m in diameter surrounded by a relatively thinly dispersed electron cloud of varying density so that the diameter of the atom is of the order of 10^{-10} m. The nucleus accounts for almost all the mass of the atom and contains the protons and neutrons. A proton has a mass of 1.673×10^{-24} g and a unit charge of $+1.602 \times 10^{-19}$ coulombs (C). The neutron is slightly heavier than the proton and has a mass of 1.675×10^{-24} g but no charge. The electron has a relatively small mass of 9.109×10^{-28} g ($\frac{1}{1836}$ that of the proton) and a unit charge of -1.602×10^{-19} C (equal in charge but opposite in sign from the proton). Table 2.1 summarizes these properties of subatomic particles.

The electron charge cloud thus constitutes almost all the volume of the atom but accounts for only a very small part of its mass. The electrons, particularly the outer ones, determine most of the electrical, mechanical, chemical,

TABLE 2.1 **The Mass and Charge of the Proton, Neutron, and Electron**

	Mass, grams (g)	Charge, coulombs (C)
Proton	1.673×10^{-24}	$+ 1.602 \times 10^{-19}$
Neutron	1.675×10^{-24}	0
Electron	9.109×10^{-28}	$- 1.602 \times 10^{-19}$

and thermal properties of the atoms, and thus a basic knowledge of atomic structure is important in the study of engineering materials.

2.2 ATOMIC NUMBERS AND ATOMIC MASSES

Atomic Numbers   The *atomic number* of an atom indicates the number of protons (positively charged particles) which are in its nucleus, and in a neutral atom the atomic number is also equal to the number of electrons in its charge cloud. Each element has its own characteristic atomic number, and thus the atomic number identifies the element. Atomic numbers of the elements from hydrogen, which has an atomic number of 1 to hahnium, which has an atomic number of 105, are located above the atomic symbols of the elements in the periodic table of Fig. 2.1.

Atomic Masses   The *relative atomic mass* of an element is the mass in grams of 6.023×10^{23} atoms (Avogadro's number N_A) of that element. The relative atomic masses of the elements from 1 to 105 are located below the atomic symbols in the periodic table of the elements (Fig. 2.1). The carbon atom with 6 protons and 6 neutrons is the carbon 12 atom and is the reference mass for atomic masses. One *atomic mass unit* (u) is defined as exactly one-twelfth of the mass of a carbon atom which has a mass of 12 u. One molar relative atomic mass of carbon 12 has a mass of 12 g on this scale. *One gram-mole* or *mole* (abbreviated mol) of an element is defined as having the mass in grams of the relative molar atomic mass of that element. Thus, for example, 1 gram-mole of aluminum has a mass of 26.98 g and contains 6.023×10^{23} atoms.

Example Problem 2.1

(*a*) What is the mass in grams of one atom of copper?
(*b*) How many copper atoms are in 1 g of copper?

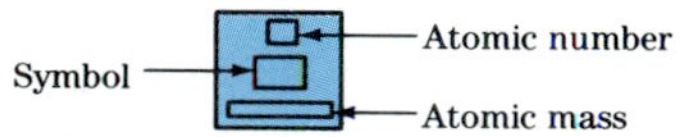

FIGURE 2.1 The periodic table of the elements. (*After F. M. Miller, "Chemistry: Structure and Dynamics," McGraw-Hill, 1984, p. 170.*)

Legend:
- Atomic number
- Atomic mass
- Symbol

	1A	2A	3B	4B	5B	6B	7B	8B			1B	2B	3A	4A	5A	6A	7A	8A
1 — 1s	1 **H** 1.00797																	2 **He** 4.0026
2 — 2s2p	3 **Li** 6.941	4 **Be** 9.0122											5 **B** 10.811	6 **C** 12.01115	7 **N** 14.0067	8 **O** 15.9994	9 **F** 18.9984	10 **Ne** 20.179
3 — 3s3p	11 **Na** 22.9898	12 **Mg** 24.305											13 **Al** 26.9815	14 **Si** 28.086	15 **P** 30.9738	16 **S** 32.064	17 **Cl** 35.453	18 **Ar** 39.948
4 — 4s3d4p	19 **K** 39.098	20 **Ca** 40.08	21 **Sc** 44.956	22 **Ti** 47.88	23 **V** 50.942	24 **Cr** 51.996	25 **Mn** 54.9380	26 **Fe** 55.847	27 **Co** 58.9332	28 **Ni** 58.69	29 **Cu** 63.54	30 **Zn** 65.38	31 **Ga** 69.72	32 **Ge** 72.59	33 **As** 74.9216	34 **Se** 78.96	35 **Br** 79.904	36 **Kr** 83.80
5 — 5s4d5p	37 **Rb** 85.47	38 **Sr** 87.62	39 **Y** 88.906	40 **Zr** 91.22	41 **Nb** 92.906	42 **Mo** 95.94	43 **Tc** (98)	44 **Ru** 101.07	45 **Rh** 102.906	46 **Pd** 106.4	47 **Ag** 107.870	48 **Cd** 112.41	49 **In** 114.82	50 **Sn** 118.69	51 **Sb** 121.75	52 **Te** 127.60	53 **I** 126.905	54 **Xe** 131.29
6 — 6s(4f)5d6p	55 **Cs** 132.905	56 **Ba** 137.33	71 **Lu** 174.97	72 **Hf** 178.49	73 **Ta** 180.948	74 **W** 183.85	75 **Re** 186.2	76 **Os** 190.2	77 **Ir** 192.2	78 **Pt** 195.08	79 **Au** 196.967	80 **Hg** 200.59	81 **Tl** 204.38	82 **Pb** 207.19	83 **Bi** 208.980	84 **Po** (209)	85 **At** (210)	86 **Rn** (222)
7 — 7s(5f)6d7p	87 **Fr** (223)	88 **Ra** 226	103 **Lr** (260)	104 **Rf** (257)	105 **Ha** (260)													

*Lanthanide series 4f

57 **La** 138.91	58 **Ce** 140.12	59 **Pr** 140.907	60 **Nd** 144.24	61 **Pm** (145)	62 **Sm** 150.36	63 **Eu** 151.96	64 **Gd** 157.25	65 **Tb** 158.924	66 **Dy** 162.50	67 **Ho** 164.930	68 **Er** 167.26	69 **Tm** 168.934	70 **Yb** 173.04

†Actinide series 5f

89 **Ac** 227.03	90 **Th** 232.038	91 **Pa** 231.04	92 **U** 238.03	93 **Np** 237.05	94 **Pu** (244)	95 **Am** (243)	96 **Cm** (247)	97 **Bk** (247)	98 **Cf** (251)	99 **Es** (252)	100 **Fm** (257)	101 **Md** (258)	102 **No** (259)

(Atomic masses are relative to $^{12}C = 12.0000$. Numbers in parentheses are for the most stable isotope.)

Solution:

(a) The atomic mass of copper is 63.54 g/mol. Since in 63.54 g of copper there are 6.02×10^{23} atoms, the number of grams in one atom of copper is

$$\frac{63.54 \text{ g/mol Cu}}{6.02 \times 10^{23} \text{ atoms/mol}} = \frac{x \text{ g Cu}}{1 \text{ atom}}$$

x = mass of 1 Cu atom

$$= \frac{63.54 \text{ g/mol}}{6.02 \times 10^{23} \text{ atoms/mol}} \times 1 \text{ atom} = 1.05 \times 10^{-22} \text{ g} \blacktriangleleft$$

(b) The number of copper atoms in 1 g of copper is

$$\frac{6.02 \times 10^{23} \text{ atoms/mol}}{63.54 \text{ g/mol Cu}} = \frac{x \text{ atoms of Cu}}{1 \text{ g Cu}}$$

$$x = \text{no. of Cu atoms} = \frac{(6.02 \times 10^{23} \text{ atoms/mol})(1 \text{ g Cu})}{63.54 \text{ g/mol Cu}}$$

$$= 9.47 \times 10^{21} \text{ atoms!} \blacktriangleleft$$

Example Problem 2.2

The cladding (outside layers) of the U.S. quarter coin consists of an alloy[1] of 75 wt % copper and 25 wt % nickel. What are the atomic percent Cu and atomic percent Ni contents of this material?

Solution:

Using a basis of 100 g of the 75 wt % Cu and 25 wt % Ni alloy, there are 75 g Cu and 25 g Ni. Thus the number of gram-moles of copper and nickel is

$$\text{No. of gram-moles of Cu} = \frac{75 \text{ g}}{63.54 \text{ g/mol}} = 1.1803 \text{ mol}$$

$$\text{No. of gram-moles of Ni} = \frac{25 \text{ g}}{58.69 \text{ g/mol}} = \underline{0.4260 \text{ mol}}$$

$$\text{Total gram-moles} = 1.6063 \text{ mol}$$

Thus the atomic percentages of Cu and Ni are

---

[1] An alloy is a combination of two or more metals or a metal (metals) and a nonmetal (nonmetals).

$$\text{Atomic \% Cu} = \left(\frac{1.1803 \text{ mol}}{1.6063 \text{ mol}}\right)(100\%) = 73.5 \text{ at \%} \blacktriangleleft$$

$$\text{Atomic \% Ni} = \left(\frac{0.4260 \text{ mol}}{1.6063 \text{ mol}}\right)(100\%) = 26.5 \text{ at \%} \blacktriangleleft$$

Example Problem 2.3

An intermetallic compound has the general chemical formula Ni_xAl_y, where x and y are simple integers, and consists of 42.04 wt % nickel and 57.96 wt % aluminum. What is the simplest formula of this nickel aluminide?

Solution:

We first determine the gram-mole fractions of nickel and aluminum in this compound. Using a basis of 100 g of the compound, we have 42.04 g of Ni and 57.96 g of Al. Thus,

$$\text{No. of gram-moles of Ni} = \frac{42.04 \text{ g}}{58.71 \text{ g/mol}} = 0.7160 \text{ mol}$$

$$\text{No. of gram-moles of Al} = \frac{57.96 \text{ g}}{26.98 \text{ g/mol}} = \underline{2.1483 \text{ mol}}$$

$$\text{Total gram-moles} = 2.8643 \text{ mol}$$

Thus,

$$\text{Gram-mole fraction of Ni} = \frac{0.7160 \text{ mol}}{2.8643 \text{ mol}} = 0.25$$

$$\text{Gram-mole fraction of Al} = \frac{2.1483 \text{ mol}}{2.8643 \text{ mol}} = 0.75$$

Next, we replace the x and y in the Ni_xAl_y compound with 0.25 and 0.75, respectively, to give $Ni_{0.25}Al_{0.75}$, which is the simplest chemical formula on a mole-fraction basis. The simplest chemical formula on an integral basis is obtained by multiplying both the 0.25 and 0.75 by 4 to give $NiAl_3$ for the simplest chemical formula of this nickel aluminide.

2.3   THE ELECTRONIC STRUCTURE OF ATOMS

The Hydrogen Atom   The hydrogen atom is the simplest atom and consists of one electron surrounding a nucleus of one proton. If we consider the orbital motion of the hydrogen electron around its nucleus, only certain definite orbits (energy levels)

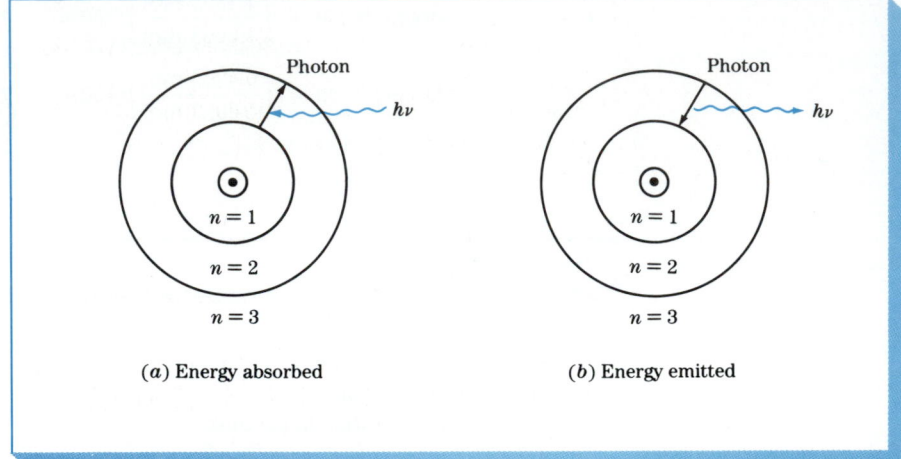

FIGURE 2.2 The hydrogen electron being (a) excited into a higher orbit, and (b) an electron in a higher energy orbit dropping to a lower orbit, and as a result, a photon of energy $h\nu$ is emitted.

are allowed. The reason for the restricted energy values is that electrons obey the laws of quantum mechanics which only allow certain energy values and not any arbitrary ones. Thus, if the hydrogen electron is excited to a higher orbit (energy level), energy is absorbed by a discrete amount (Fig. 2.2a). Similarly, if the electron drops to a lower orbit (energy level), energy is emitted by a discrete amount (Fig. 2.2b)

During the transition to a lower energy, the hydrogen electron will emit a discrete amount (quantum) of energy in the form of electromagnetic radiation called a *photon*. The energy change ΔE associated with the transition of the electron from one level to another is related to the frequency ν (nu) of the photon by Planck's[1] equation

$$\Delta E = h\nu \tag{2.1}$$

where h = Planck's constant = 6.63×10^{-34} joule-second (J·s). Since for electromagnetic radiation $c = \lambda\nu$, where c is the velocity of light equal to 3.00×10^{8} meters/second (m/s) and λ (lambda) is its wavelength, the energy change ΔE associated with a photon can be expressed as

$$\Delta E = \frac{hc}{\lambda} \tag{2.2}$$

[1] Max Ernst Planck (1858–1947). German physicist who received the Nobel Prize in Physics in 1918 for his quantum theory.

Example Problem 2.4

Calculate the energy in joules (J) and electron volts (eV) of the photon whose wavelength λ is 121.6 nanometers (nm). (1.00 eV $= 1.60 \times 10^{-19}$ J; $h = 6.63 \times 10^{-34}$ J·s; 1 nm $= 10^{-9}$ m.)

Solution:

$$\Delta E = \frac{hc}{\lambda} \qquad (2.2)$$

$$\Delta E = \frac{(6.63 \times 10^{-34} \text{ J·s})(3.00 \times 10^8 \text{ m/s})}{(121.6 \text{ nm})(10^{-9} \text{ m/nm})}$$

$$= 1.63 \times 10^{-18} \text{ J} \blacktriangleleft$$

$$= 1.63 \times 10^{-18} \text{ J} \times \frac{1 \text{ eV}}{1.60 \times 10^{-19} \text{ J}} = 10.2 \text{ eV} \blacktriangleleft$$

Experimental verification of the energies associated with electrons being excited to discrete higher energy levels or losing energy and dropping to lower discrete levels is obtained mainly by the analyses of the wavelengths and intensities of spectral lines. Using hydrogen spectral data, Niels Bohr[1] in 1913

---

[1] Niels Henrik Bohr (1885–1962). Danish physicist who was one of the founders of modern physics. In 1922 he received the Nobel Prize in Physics for his theory of the hydrogen atom.

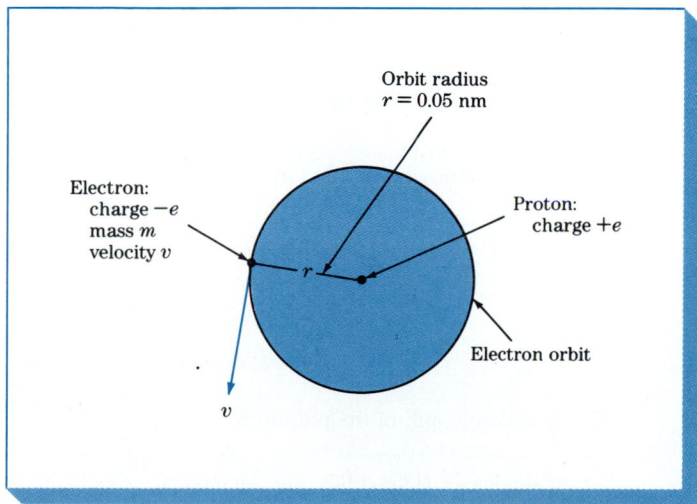

FIGURE 2.3 The Bohr hydrogen atom. In the Bohr hydrogen atom an electron revolves in a circular orbit of 0.05 nm radius around a central proton.

developed a model for the hydrogen atom which consisted of a single electron orbiting a proton at a fixed radius (Fig. 2.3). A good approximation of the energy of the hydrogen electron for allowed energy levels is the Bohr equation

$$E = -\frac{2\pi^2 m e^{4*}}{n^2 h^2} = -\frac{13.6}{n^2}\,eV \qquad (n = 1, 2, 3, 4, 5, \ldots) \qquad (2.3)$$

where e = electron charge
m = electron mass
n = an integer called the *principal quantum number*

(in cgs units)

Example Problem 2.5

A hydrogen atom exists with its electron in the $n = 3$ state. The electron undergoes a transition to the $n = 2$ state. Calculate (*a*) the energy of the photon emitted, (*b*) its frequency, and (*c*) its wavelength. (*d*) Is energy emitted or absorbed in the transition?

Solution:

(*a*) Energy of the photon emitted is

$$E = \frac{-13.6\ eV}{n^2} \qquad (2.3)$$

$$\Delta E = E_3 - E_2$$

$$= \frac{-13.6}{3^2} - \frac{-13.6}{2^2} = 1.89\ eV \blacktriangleleft$$

$$= 1.89\ eV \times \frac{1.60 \times 10^{-19}\ J}{eV} = 3.02 \times 10^{-19}\ J \blacktriangleleft$$

(*b*) The frequency of the photon is

$$\Delta E = h\nu \qquad (2.1)$$

$$\nu = \frac{\Delta E}{h} = \frac{3.02 \times 10^{-19}\ J}{6.63 \times 10^{-34}\ J \cdot s}$$

$$= 4.55 \times 10^{14}\ s^{-1} = 4.55 \times 10^{14}\ Hz \blacktriangleleft$$

(*c*) The wavelength of the photon is

$$\Delta E = \frac{hc}{\lambda} \qquad (2.2)$$

or

$$\lambda = \frac{hc}{\Delta E} = \frac{(6.63 \times 10^{-34} \text{ J·s})(3.00 \times 10^{8} \text{ m/s})}{3.02 \times 10^{-19} \text{ J}}$$

$$= 6.59 \times 10^{-7} \text{ m}$$

$$= 6.59 \times 10^{-7} \text{ m} \times \frac{1 \text{ nm}}{10^{-9} \text{ m}} = 659 \text{ nm} \blacktriangleleft$$

(*d*) Energy is emitted by this transition.

In modern atomic theory, the *n* in Bohr's equation is designated the *principal quantum number* and represents the principal energy levels for electrons in atoms. From Bohr's equation (2.3) the energy level of the hydrogen electron in its ground state is -13.6 eV and corresponds to the line where $n = 1$ on the hydrogen energy-level diagram (Fig. 2.4). When the hydrogen electron is excited to higher energy levels, its energy is raised but its numerical value is less. For example, when the hydrogen electron is excited to the second principal quantum level, its energy is -3.4 eV, and if the hydrogen electron is excited to the free state where $n = \infty$, the electron has a zero value of energy. The energy required to remove the electron completely from the hydrogen atom is 13.6 eV, which is the *ionization energy* of the hydrogen electron.

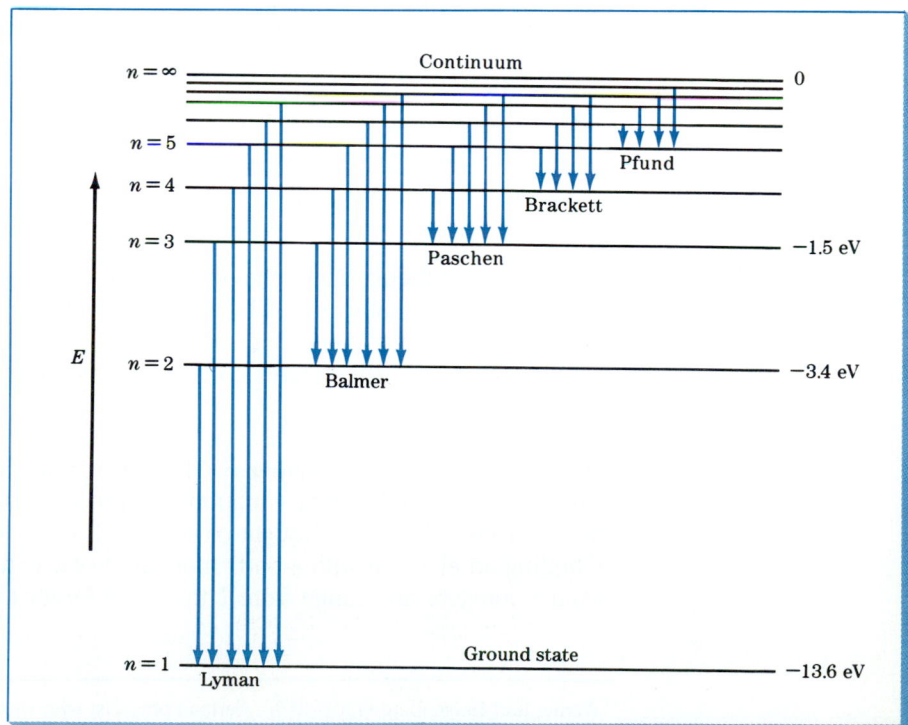

FIGURE 2.4 Energy-level diagram for the line spectrum of hydrogen. (*After F. M. Miller, "Chemistry: Structure and Dynamics," McGraw-Hill, 1984, p. 141.*)

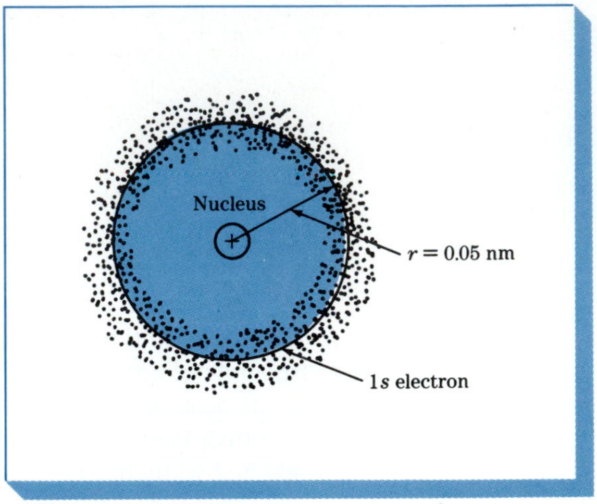

FIGURE 2.5 Electron charge cloud (schematic) surrounding the nucleus of a hydrogen atom in the ground state. The outer circle of $r = 0.05$ nm corresponds to the radius of the first Bohr orbit (i.e., for $n = 1$) and indicates the most probable region for finding the electron.

The motion of electrons in atoms is more complicated than that presented by the simple Bohr atomic model. Electrons can have noncircular (elliptical) orbits around the nucleus, and according to Heisenberg's[1] uncertainty principle the position and momentum (mass × velocity) of a small particle such as an electron cannot be determined simultaneously. Thus, the exact position of the electron at any instant cannot be determined because the electron is so small a particle. Since the position of the hydrogen electron cannot be precisely determined, an electron charge cloud density distribution is sometimes used to represent the position of the hydrogen electron in its orbital motion about its nucleus (Fig. 2.5). The highest electron density is at a radius of about 0.05 nm, which corresponds to the Bohr radius of the hydrogen atom.

Quantum Numbers of Electron of Atoms

Modern atomic theory states that the motion of an electron about its nucleus and its energy is characterized by not just one principal quantum number but by four quantum numbers: the principal quantum number n, the subsidiary quantum number l, the magnetic quantum number m_l, and the electron spin quantum number m_s.

The principal quantum number n   The principal quantum number n corresponds to the n in the Bohr equation. It represents the main energy levels for the electron and can be thought of as a shell in space where the probability of finding an electron with a particular value of n is high. The values for n are positive integers and range from 1 to 7. The larger the value of n, the farther

---

[1] Werner Karl Heisenberg (1901–1976). German physicist who was one of the founders of the modern quantum theory. In 1932 he received the Nobel Prize in Physics.

the shell is from the nucleus, and hence the larger the value of the principal quantum number of an electron, the farther the electron is (on an average time basis) away from the nucleus. Also, in general, the higher the principal quantum number of an electron, the higher its energy.

The subsidiary quantum number l The second quantum number is the subsidiary quantum number l. This quantum number specifies subenergy levels within the main energy levels and also specifies a subshell where the probability of finding an electron is high if that energy level is occupied. The allowed values of l are $l = 0, 1, 2, 3, \ldots, n - 1$. The letters s, p, d, and f are used[1] to designate the l subenergy levels as follows:

$$\text{Number designation } l = 0 \quad 1 \quad 2 \quad 3$$
$$\text{Letter designation } l = s \quad p \quad d \quad f$$

The s, p, d, and f subenergy levels of an electron are termed *orbitals*, and so one speaks, for example, of an s or p subenergy level. The term orbital also refers to a subshell of an atom where the density of a particular electron or pair of electrons is high. Thus, one may speak of an s or p subshell of a particular atom.

The magnetic quantum number m_l The third quantum number, the magnetic quantum number m_l, specifies the spatial orientation of a single atomic orbital and has little effect on the energy of an electron. The number of different permissible orientations of an orbital depends on the l value of a particular orbital. The m_l quantum number has permissible values of $-l$ to $+l$, including zero. When $l = 0$, there is only one allowed value for m_l, which is zero. When $l = 1$, there are three permissible values for m_l, which are -1, 0, and $+1$. In general, there are $2l + 1$ allowed values for m_l. In terms of s, p, d, and f orbital notation, there is a maximum of one s orbital, three p orbitals, five d orbitals, and seven f orbitals for each allowed s, p, d, and f subenergy level.

Electron spin quantum number m_s The fourth quantum number, the electron spin quantum number m_s, specifies two allowed spin directions for an electron spinning on its own axis. The directions are clockwise and counterclockwise rotation, and their allowed values are $+\frac{1}{2}$ and $-\frac{1}{2}$. The spin quantum number has only a very minor effect on the energy of an electron. It should be pointed out that two electrons may occupy the same orbital, and if they do, they must have opposed spins.

Table 2.2 summarizes the allowed values for the four quantum numbers

---

[1] The letters s, p, d, and f were adopted from the first letters of the old spectral-line intensity designations: *s*harp, *p*rincipal, *d*iffuse, and *f*undamental.

TABLE 2.2   **Allowed Values for the Quantum Numbers of Electrons**

n	Principal quantum number	$n = 1, 2, 3, 4, \ldots$	All positive integers
l	Subsidiary quantum number	$l = 0, 1, 2, 3, \ldots, n - 1$	n allowed values of l
m_l	Magnetic quantum number	Integral values from $-l$ to $+l$, including zero	$2l + 1$
m_s	Spin quantum number	$+\frac{1}{2}, -\frac{1}{2}$	2

of electrons. According to the *Pauli*[1] *exclusion principle* of atomic theory, *no two electrons can have the same set of four quantum numbers.*

Electronic Structure of Multielectron Atoms

Maximum number of electrons for each principal atomic shell   Atoms consist of principal shells of high electron densities as dictated by the laws of quantum mechanics. There are seven of these principal shells when the atomic number of the atom reaches 87 for the element francium (Fr). Each shell can only contain a maximum number of electrons, which is again dictated by the laws of quantum mechanics. The maximum number of electrons which can be contained in each shell in an atom is defined by different sets of the four quantum numbers (Pauli principle) and is $2n^2$, where n is the principal quantum number. Thus, there can be only a maximum of 2 electrons for the first principal shell, 8 for the second, 18 for the third, 32 for the fourth, etc., as indicated in Table 2.3.

Atomic size   Each atom can be considered to a first approximation as a sphere with a definite radius. The radius of the atomic sphere is not a constant but depends to some extent on its environment. Figure 2.6 shows the relative atomic sizes of many of the elements along with their atomic radii. Many of the values of the atomic radii are still not fully agreed upon and vary to some extent depending on the reference source.

From Fig. 2.6, some trends in atomic size are evident. In general, as each successive shell of increasing principal quantum number is added to the atom of an element, the size of the atom increases. There are, however, a few exceptions where the atomic size actually gets smaller. The alkali elements of group 1A of the periodic table (Fig. 2.1) are a good example of atoms whose size increases with each added electron shell. For example, lithium ($n = 2$) has an atomic radius of 0.157 nm, whereas cesium ($n = 6$) has an atomic radius of 0.270 nm. In progressing across the period table from an alkali group 1A element to a noble gas of group 8A, the atomic size, in general, decreases.

---

[1] Wolfgang Pauli (1900–1958). Austrian physicist who was one of the founders of the modern quantum theory. In 1945 he received the Nobel Prize in Physics.

TABLE 2.3   Maximum Number of Electrons for Each Principal Atomic Shell

Shell number, n (principal quantum number)	Maximum number of electrons in each shell $(2n^2)$	Maximum number of electrons in orbitals
1	$2(1^2) = 2$	s^2
2	$2(2^2) = 8$	s^2p^6
3	$2(3^2) = 18$	$s^2p^6d^{10}$
4	$2(4^2) = 32$	$s^2p^6d^{10}f^{14}$
5	$2(5^2) = 50$	$s^2p^6d^{10}f^{14}\cdots$
6	$2(6^2) = 72$	$s^2p^6\cdots$
7	$2(7^2) = 98$	$s^2\cdots$

However, again there are some small exceptions. Atomic size will be important to us in the study of atomic diffusion in metal alloys.

Electron configurations of the elements   The electron configuration of an atom describes the way in which electrons are arranged in orbitals in an atom. Electron configurations are written in a conventional notation that lists the principal quantum number first, which is then followed by an orbital letter s, p, d, or f. A superscript above the orbital letter indicates the number of electrons it contains. The order by which the electrons fill up the orbitals is as follows:[1]

$$1s^22s^22p^63s^23p^64s^23d^{10}4p^65s^24d^{10}5p^66s^24f^{14}5d^{10}6p^67s^25f^{14}6d^{10}7p^6$$

The order for writing the orbitals for the electron configurations (for this book) will be by increasing principal quantum number, as

$$1s^22s^22p^63s^23p^63d^{10}4s^24p^64d^{10}4f^{14}5s^25p^65d^{10}5f^{14}6s^26p^66d^{10}7s^2$$

---

[1] A convenient memory device for this order is to list the orbitals as shown below and to use a series of arrows as drawn on the orbitals. By following the arrows from tail to head, the order is indicated.

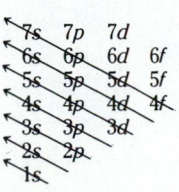

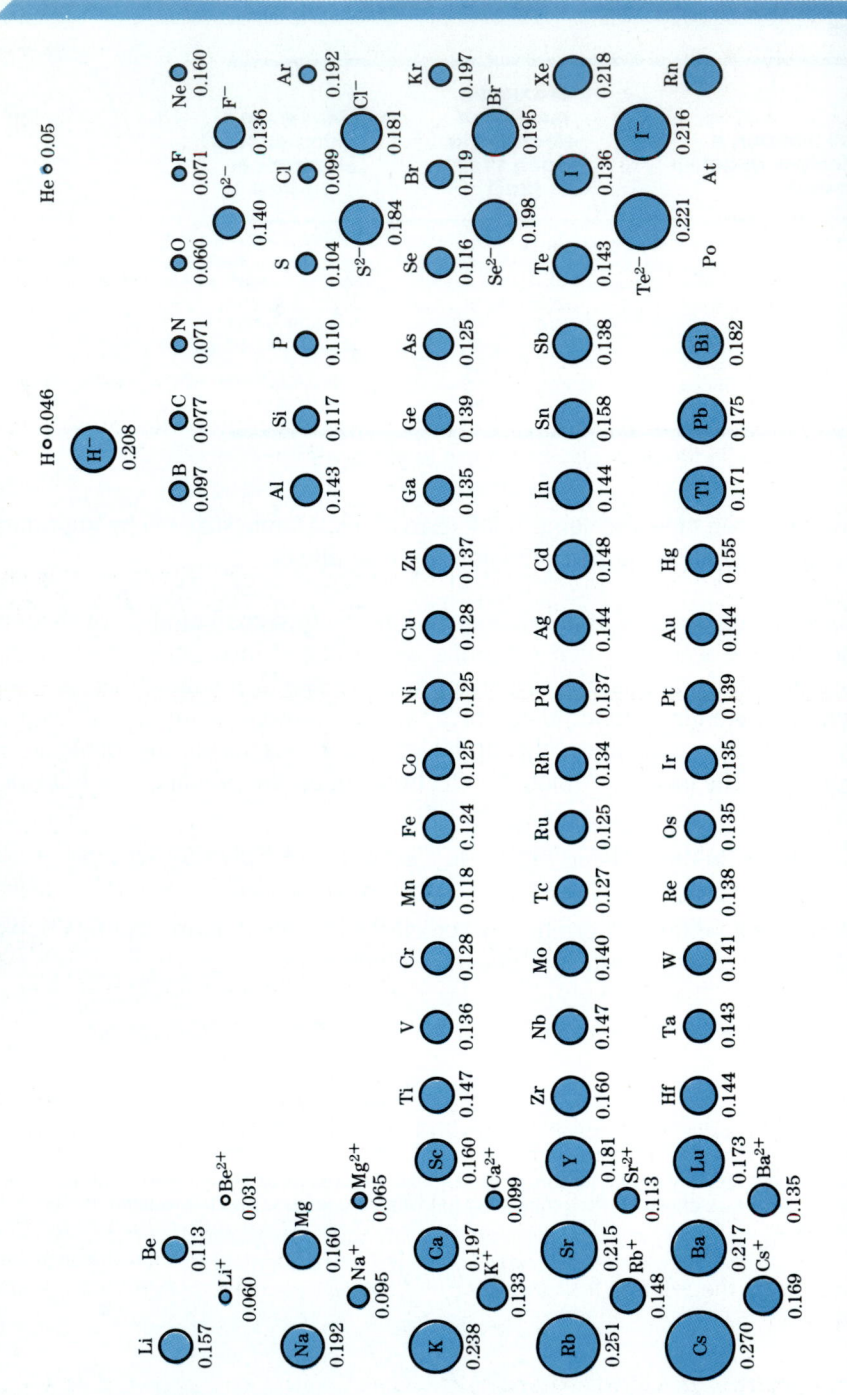

FIGURE 2.6 Relative sizes of some atoms and ions. Values are given in nanometers for the radii of the atoms and ions. Metallic radii are given for atoms where applicable. (*Adapted from F. M. Miller, "Chemistry: Structure and Dynamics," McGraw-Hill, 1984, p. 176.*)

Example Problem 2.6

Write the electron configurations of the elements (*a*) iron, $Z = 26$, and (*b*) samarium, $Z = 62$.

Solution:

Using the order of filling the orbitals listed previously, the following electron configurations for these elements are

Iron ($Z = 26$): $1s^2 2s^2 2p^6 3s^2 3p^6 3d^6 4s^2$

Note the incomplete $3d$ orbitals. The electron arrangement of the $3d^6$ electrons in the iron atom will be important in the study of ferromagnetism.

Samarium ($Z = 62$): $1s^2 2s^2 2p^6 3s^2 3p^6 3d^{10} 4s^2 4p^6 4d^{10} 4f^6 5s^2 5p^6 6s^2$

Note that for this rare-earth element the $4f$ orbitals are incomplete. The close similarity of chemical properties of the rare-earth elements is due to the filling in of the $4f$ orbitals, which are two shells below the outer shell containing two $6s$ electrons.

Table 2.4 lists the electron configurations of the elements as determined experimentally. It is noted that there are some irregularities inconsistent with the above-listed system. For example, copper ($Z = 29$) has the outer electron configuration of $3d^{10} 4s^1$. One would expect the outer configuration according to the above-indicated system to be $3d^9 4s^2$. The reason for these irregularities is not precisely known.

Experimental evidence also shows that electrons with the same subsidiary quantum number have as many parallel spins as possible. Thus, if there are five electrons in d orbitals, there will be one electron in each of the d orbitals and the spin directions of all the electrons will be parallel, as shown in Fig. 2.7.

Electronic Structure and Chemical Reactivity

Noble gases   The chemical properties of the atoms of the elements depend principally on the reactivity of their outermost electrons. The most stable and least reactive of all the elements are the noble gases. With the exception of helium, which has a $1s^2$ electron configuration, the outermost shell of all the other noble gases (Ne, Ar, Kr, Xe, and Rn) has an $s^2 p^6$ electron configuration. The $s^2 p^6$ configuration for the outermost shell has high chemical stability as indicated by the relative inactivity of the noble gases to react chemically with other atoms.

Electropositive and electronegative elements   Electropositive elements are metallic in nature and give up electrons in chemical reactions to produce

TABLE 2.4 **Electron Configurations of the Elements**

Z	Element	Electron configuration	Z	Element	Electron configuration
1	H	$1s$	53	I	$[Kr]4d^{10}5s^25p^5$
2	He	$1s^2$	54	Xe	$[Kr]4d^{10}5s^25p^6$
3	Li	$[He]2s$	55	Cs	$[Xe]6s$
4	Be	$[He]2s^2$	56	Ba	$[Xe]6s^2$
5	B	$[He]2s^22p$	57	La	$[Xe]5d6s^2$
6	C	$[He]2s^22p^2$	58	Ce	$[Xe]4f5d6s^2$
7	N	$[He]2s^22p^3$	59	Pr	$[Xe]4f^26s^2$
8	O	$[He]2s^22p^4$	60	Nd	$[Xe]4f^46s^2$
9	F	$[He]2s^22p^5$	61	Pm	$[Xe]4f^56s^2$
10	Ne	$[He]2s^22p^6$	62	Sm	$[Xe]4f^66s^2$
11	Na	$[Ne]3s$	63	Eu	$[Xe]4f^76s^2$
12	Mg	$[Ne]3s^2$	64	Gd	$[Xe]4f^75d6s^2$
13	Al	$[Ne]3s^23p$	65	Tb	$[Xe]4f^96s^2$
14	Si	$[Ne]3s^23p^2$	66	Dy	$[Xe]4f^{10}6s^2$
15	P	$[Ne]3s^23p^3$	67	Ho	$[Xe]4f^{11}6s^2$
16	S	$[Ne]3s^23p^4$	68	Er	$[Xe]4f^{12}6s^2$
17	Cl	$[Ne]3s^23p^5$	69	Tm	$[Xe]4f^{13}6s^2$
18	Ar	$[Ne]3s^23p^6$	70	Yb	$[Xe]4f^{14}6s^2$
19	K	$[Ar]4s$	71	Lu	$[Xe]4f^{14}5d6s^2$
20	Ca	$[Ar]4s^2$	72	Hf	$[Xe]4f^{14}5d^26s^2$
21	Sc	$[Ar]3d4s^2$	73	Ta	$[Xe]4f^{14}5d^36s^2$
22	Ti	$[Ar]3d^24s^2$	74	W	$[Xe]4f^{14}5d^46s^2$
23	V	$[Ar]3d^34s^2$	75	Re	$[Xe]4f^{14}5d^56s^2$
24	Cr	$[Ar]3d^54s$	76	Os	$[Xe]4f^{14}5d^66s^2$
25	Mn	$[Ar]3d^54s^2$	77	Ir	$[Xe]4f^{14}5d^76s^2$
26	Fe	$[Ar]3d^64s^2$	78	Pt	$[Xe]4f^{14}5d^96s$
27	Co	$[Ar]3d^74s^2$	79	Au	$[Xe]4f^{14}5d^{10}6s$
28	Ni	$[Ar]3d^84s^2$	80	Hg	$[Xe]4f^{14}5d^{10}6s^2$
29	Cu	$[Ar]3d^{10}4s$	81	Tl	$[Xe]4f^{14}5d^{10}6s^26p$
30	Zn	$[Ar]3d^{10}4s^2$	82	Pb	$[Xe]4f^{14}5d^{10}6s^26p^2$
31	Ga	$[Ar]3d^{10}4s^24p$	83	Bi	$[Xe]4f^{14}5d^{10}6s^26p^3$
32	Ge	$[Ar]3d^{10}4s^24p^2$	84	Po	$[Xe]4f^{14}5d^{10}6s^26p^4$
33	As	$[Ar]3d^{10}4s^24p^3$	85	At	$[Xe]4f^{14}5d^{10}6s^26p^5$
34	Se	$[Ar]3d^{10}4s^24p^4$	86	Rn	$[Xe]4f^{14}5d^{10}6s^26p^6$
35	Br	$[Ar]3d^{10}4s^24p^5$	87	Fr	$[Rn]7s$
36	Kr	$[Ar]3d^{10}4s^24p^6$	88	Ra	$[Rn]7s^2$
37	Rb	$[Kr]5s$	89	Ac	$[Rn]6d7s^2$
38	Sr	$[Kr]5s^2$	90	Th	$[Rn]6d^27s^2$
39	Y	$[Kr]4d5s^2$	91	Pa	$[Rn]5f^26d7s^2$
40	Zr	$[Kr]4d^25s^2$	92	U	$[Rn]5f^36d7s^2$
41	Nb	$[Kr]4d^45s$	93	Np	$[Rn]5f^46d7s^2$
42	Mo	$[Kr]4d^55s$	94	Pu	$[Rn]5f^67s^2$
43	Tc	$[Kr]4d^55s^2$	95	Am	$[Rn]5f^77s^2$
44	Ru	$[Kr]4d^75s$	96	Cm	$[Rn]5f^76d7s^2$
45	Rh	$[Kr]4d^85s$	97	Bk	$[Rn]5f^97s^2$
46	Pd	$[Kr]4d^{10}$	98	Cf	$[Rn]5f^{10}7s^2$
47	Ag	$[Kr]4d^{10}5s$	99	Es	$[Rn]5f^{11}7s^2$
48	Cd	$[Kr]4d^{10}5s^2$	100	Fm	$[Rn]5f^{12}7s^2$
49	In	$[Kr]4d^{10}5s^25p$	101	Md	$[Rn]5f^{13}7s^2$
50	Sn	$[Kr]4d^{10}5s^25p^2$	102	No	$[Rn]5f^{14}7s^2$
51	Sb	$[Kr]4d^{10}5s^25p^3$	103	Lr	$[Rn]5f^{14}6d7s^2$
52	Te	$[Kr]4d^{10}5s^25p^4$			

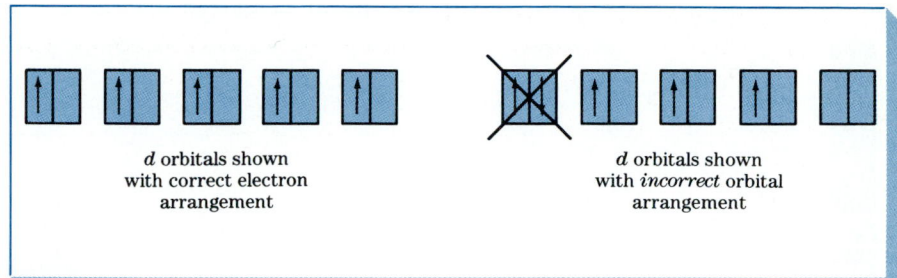

d orbitals shown
with correct electron
arrangement

d orbitals shown
with *incorrect* orbital
arrangement

FIGURE 2.7 Spin directions of unpaired electrons in *d* orbitals

positive ions, or *cations*. The number of electrons given up by an electropositive atom of an element is indicated by a *positive oxidation number*. Figure 2.8 lists the oxidation numbers for the elements. Note that some elements have more than one oxidation number. The most electropositive elements are in groups 1A and 2A of the periodic table.

Electronegative elements are nonmetallic in nature and accept electrons in chemical reactions to produce *negative ions*, or *anions*. The number of electrons accepted by an electronegative atom of an element is indicated by a *negative oxidation number* (Fig. 2.8). The most electronegative elements are in groups 6A and 7A of the periodic table in Fig. 2.1.

Some elements in groups 4A to 7A of the periodic table can behave in either an electronegative or an electropositive manner. This dual behavior is shown by elements such as carbon, silicon, germanium, arsenic, antimony, and phosphorus. Thus, in some reactions they have positive oxidation numbers, where they show electropositive behavior, and in others they have negative oxidation numbers, where they show electronegative behavior.

Example Problem 2.7

Write the electron configuration for the Fe atom ($Z = 26$) and the Fe^{2+} and Fe^{3+} ions by using conventional *spdf* notation.

Solution:

$$Fe: \quad 1s^2 2s^2 2p^6 3s^2 3p^6 3d^6 4s^2$$
$$Fe^{2+}: 1s^2 2s^2 2p^6 3s^2 3p^6 3d^6$$
$$Fe^{3+}: 1s^2 2s^2 2p^6 3s^2 3p^6 3d^5$$

Note that the outer-two $4s$ electrons are lost first since they have the highest energy and are the easiest to remove.

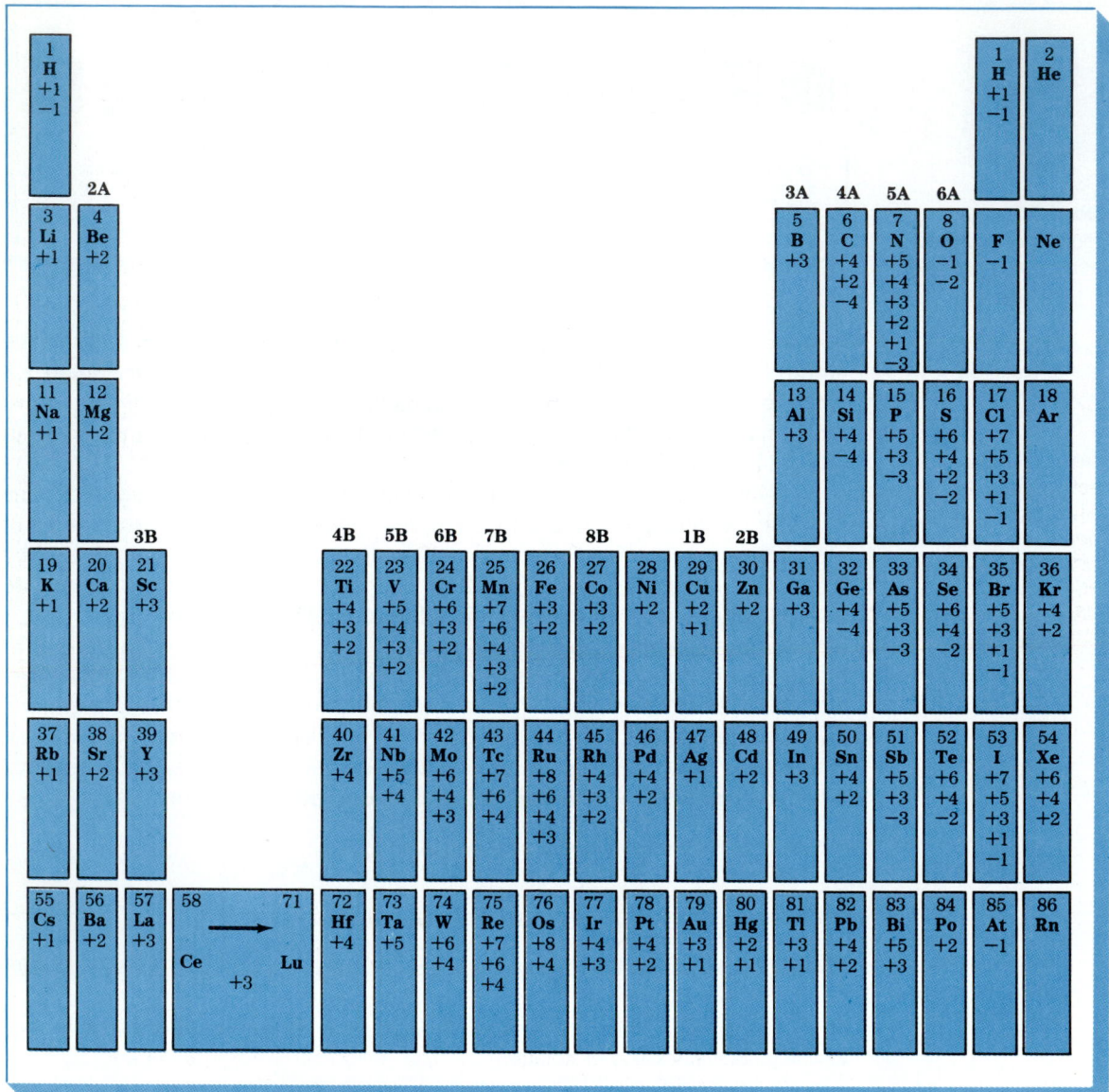

FIGURE 2.8 Oxidation numbers of the elements with respect to their positions in the periodic table. *(After R. E. Davis, K. D. Gailey, and K. W. Whitten, "Principles of Chemistry," CBS College Publishing, 1984, p. 299.)*

Electronegativity *Electronegativity* is defined as the degree to which an atom attracts electrons to itself. The comparative tendency of an atom to show electropositive or electronegative behavior can be made quantitative by assigning each element an electronegativity number. Electronegativity is measured on a scale from 0 to 4.1, and each element is assigned a value on this scale, as shown in Fig. 2.9. The most electropositive elements are the alkali metals, which have electronegativities ranging from 0.9 for cesium, rubidium,

													H 2.1			
Li 1.0	Be 1.5											B 2.0	C 2.5	N 3.1	O 3.5	F 4.1
Na 1.0	Mg 1.3											Al 1.5	Si 1.8	P 2.1	S 2.4	Cl 2.9
K 0.9	Ca 1.1	Sc 1.2	Ti 1.3	V 1.5	Cr 1.6	Mn 1.6	Fe 1.7	Co 1.7	Ni 1.8	Cu 1.8	Zn 1.7	Ga 1.8	Ge 2.0	As 2.2	Se 2.5	Br 2.8
Rb 0.9	Sr 1.0	Y 1.1	Zr 1.2	Nb 1.3	Mo 1.3	Tc 1.4	Ru 1.4	Rh 1.5	Pd 1.4	Ag 1.4	Cd 1.5	In 1.5	Sn 1.7	Sb 1.8	Te 2.0	I 2.2
Cs 0.9	Ba 0.9	La 1.1	Hf 1.2	Ta 1.4	W 1.4	Re 1.5	Os 1.5	Ir 1.6	Pt 1.5	Au 1.4	Hg 1.5	Tl 1.5	Pb 1.6	Bi 1.7	Po 1.8	At 2.0
Fr 0.9	Ra 0.9	Ac 1.0														

Lanthanides: 1.0–1.2
Actinides: 1.0–1.2

FIGURE 2.9 The electronegativities of the elements. (*After F. M. Miller, "Chemistry: Structure and Dynamics," McGraw-Hill, 1984, p. 185.*)

and potassium to 1.0 for sodium and lithium. The most electronegative elements are fluorine, oxygen, and nitrogen, which have electronegativities of 4.1, 3.5, and 3.1, respectively. The electronegativity concept helps in understanding the bonding behavior of the elements.

Summary of Some of the Electronic Structure–Chemical Property Relationships for Metals and Nonmetals

Metals	Nonmetals
1. Have few electrons in outer shells, usually three or less	1. Have four or more electrons in outer shells
2. Form cations by losing electrons	2. Form anions by gaining electrons
3. Have low electronegativities	3. Have high electronegativities

2.4 TYPES OF ATOMIC AND MOLECULAR BONDS

Chemical bonding between atoms occurs since there is a net decrease in the potential energy of atoms in the bonded state. That is, atoms in the bonded state are in a more stable energy condition than when they are unbonded. In general, chemical bonds between atoms can be divided into two groups: primary or strong bonds and secondary or weak bonds.

Primary Atomic Bonds

Primary atomic bonds in which relatively large interatomic forces develop can be subdivided into the following three classes:

1. *Ionic bonds.* Relatively large interatomic forces are set up in this type of bonding by an electron transfer from one atom to another to produce ions which are bonded together by coulombic forces (attraction of positively and negatively charged ions). The ionic bond is a relatively strong nondirectional bond.
2. *Covalent bonds.* Relatively large interatomic forces are created by the sharing of electrons to form a bond with a localized direction.
3. *Metallic bonds.* Relatively large interatomic forces are created by the sharing of electrons in a delocalized manner to form strong nondirectional bonding between atoms.

Secondary Atomic and Molecular Bonds

1. *Permanent dipole bonds.* Relatively weak intermolecular bonds are formed between molecules which possess permanent dipoles. A dipole in a molecule exists due to asymmetry in its electron density distribution.
2. *Fluctuating dipole bonds.* Very weak electric dipole bonding can take place among atoms due to the asymmetrical distribution of electron densities around their nuclei. This type of bonding is termed *fluctuating* since the electron density is continuously changing with time.

2.5   IONIC BONDING

Ionic Bonding in General

Ionic bonds can form between highly electropositive (metallic) elements and highly electronegative (nonmetallic) elements. In the ionization process electrons are transferred from atoms of electropositive elements to atoms of electronegative elements, producing positively charged cations and negatively charged anions. The ionic bonding forces are due to the electrostatic or coulombic force attraction of oppositely charged ions. Ionic bonds form between oppositely charged ions because there is a net decrease in potential energy of the ions after bonding.

An example of a solid which has a high degree of ionic bonding is sodium chloride (NaCl). In the ionization process to form a Na^+Cl^- ion pair, a sodium atom gives up its outer $3s^1$ electron and transfers it to a half-filled $3p$ orbital of a chlorine atom, producing a pair of Na^+ and Cl^- ions (Fig. 2.10).

In the ionization process the sodium atom which originally had a 0.192-

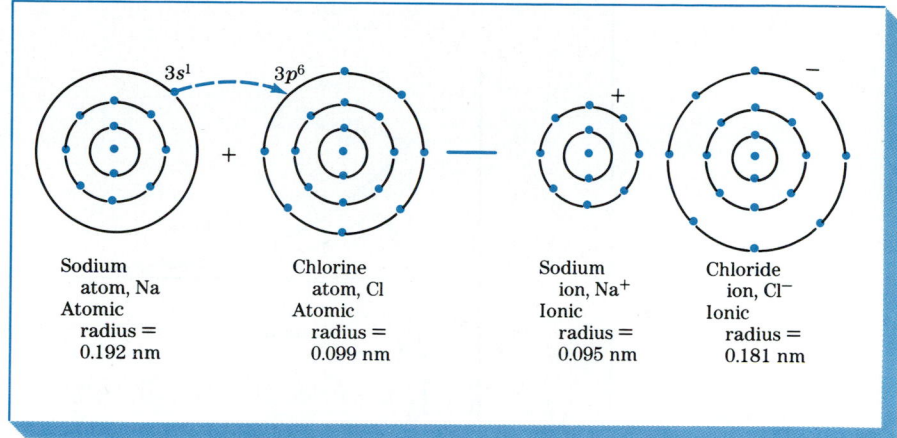

FIGURE 2.10 Formation of a sodium chloride ion pair from sodium and chlorine atoms. In the ionization process a $3s^1$ electron from the sodium atom is *transferred* to a half-empty $3p$ orbital of the chlorine atom.

Sodium
atom, Na
Atomic
radius =
0.192 nm

Chlorine
atom, Cl
Atomic
radius =
0.099 nm

Sodium
ion, Na$^+$
Ionic
radius =
0.095 nm

Chloride
ion, Cl$^-$
Ionic
radius =
0.181 nm

nm radius is reduced in size to a sodium cation with a radius of 0.095 nm, and the chlorine atom which originally had a radius of 0.099 nm is expanded into a chloride anion with a radius of 0.181 nm.

The sodium atom is reduced in size when the ion is formed because of the loss of its outer-shell $3s^1$ electron and because of the decrease in the electron-to-proton ratio. The higher positively charged nucleus of the sodium ion attracts the electron charge cloud closer to itself, causing the size of the atom to decrease during ionization. In contrast, during ionization the chlorine atom expands due to an increase in the electron-to-proton ratio. During ionization, atoms decrease in size when they form cations and increase in size when they form anions, as shown for some atoms in Fig. 2.6.

Interionic Forces for an Ion Pair

Consider a pair of oppositely charged ions (for example, a Na$^+$Cl$^-$ ion pair) approaching each other from a large distance of separation a. As the ions come closer together, they will be attracted to each other by coulombic forces. That is, the nucleus of one ion will attract the electron charge cloud of the other, and vice versa. When the ions come still closer together, eventually their two electron charge clouds will interact and repulsive forces will arise. When the attractive forces equal the repulsive forces, there will be no net force between the ions and they will remain at an equilibrium separation distance, the interionic distance a_0. Figure 2.11 schematically shows force vs. separation distance curves for an ion pair.

The net force between a pair of oppositely charged ions is equal to the sum of the attractive and repulsive forces. Thus

$$F_{net} = F_{attractive} + F_{repulsive} \qquad (2.4)$$

The attractive force between the ion pair is the coulombic force which

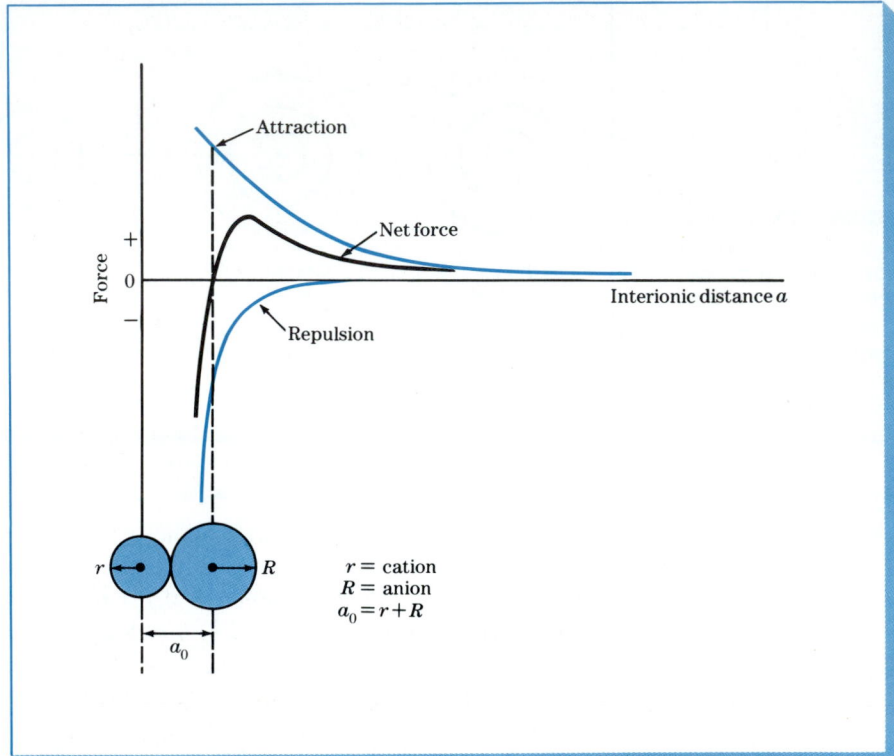

FIGURE 2.11 Force vs. separation distance for a pair of oppositely charged ions. The equilibrium interionic separation distance a_0 is reached when the force between the ions is zero.

results when the ions are considered as point charges. Using Coulomb's[1] law with SI units, the following equation can be written:

$$F_{\text{attractive}} = -\frac{(Z_1 e)(Z_2 e)}{4\pi\epsilon_0 a^2} = -\frac{Z_1 Z_2 e^2}{4\pi\epsilon_0 a^2} \tag{2.5}$$

where Z_1, Z_2 = number of electrons removed or added from the atoms during the ion formation

e = electron charge

a = interionic separation distance

ϵ_0 = permittivity of free space = 8.85×10^{-12} C^2/(N·m^2)

The repulsive force between an ion pair is found by experiment to be

---

[1] Charles Augustin Coulomb (1736–1806). French physicist who demonstrated experimentally that the force between two charged bodies varied inversely as the square of the distance between them.

inversely proportional to the interionic separation distance a and can be described by the equation

$$F_{repulsive} = -\frac{nb}{a^{n+1}} \tag{2.6}$$

where a is the interionic separation distance and b and n are constants; n usually ranges from 7 to 9 and is 9 for NaCl.

Substituting Eqs. (2.5) and (2.6) into (2.4) gives the net force between an ion pair:

$$F_{net} = -\frac{Z_1 Z_2 e^2}{4\pi\epsilon_0 a^2} - \frac{nb}{a^{n+1}} \tag{2.7}$$

Example Problem 2.8

Calculate the coulombic attractive force (●➤ ◄●) between a pair of Na^+ and Cl^- ions that just touch each other. Assume the ionic radius of the Na^+ ion to be 0.095 nm and that of the Cl^- ion to be 0.181 nm.

Solution:

The attractive force between the Na^+ and Cl^- ions can be calculated by substituting the appropriate values into the Coulomb's law equation (2.5).

$$Z_1 = +1 \text{ for } Na^+ \qquad Z_2 = -1 \text{ for } Cl^-$$

$$e = 1.60 \times 10^{-19} \text{ C} \qquad \epsilon_0 = 8.85 \times 10^{-12} \text{ C}^2/(\text{N·m}^2)$$

$$a_0 = \text{sum of the radii of } Na^+ \text{ and } Cl^- \text{ ions}$$
$$= 0.095 \text{ nm} + 0.181 \text{ nm}$$
$$= 0.276 \text{ nm} \times 10^{-9} \text{ m/nm} = 2.76 \times 10^{-10} \text{ m}$$

$$F_{attractive} = -\frac{Z_1 Z_2 e^2}{4\pi\epsilon_0 a_0^2}$$

$$= -\frac{(+1)(-1)(1.60 \times 10^{-19} \text{ C})^2}{4\pi[8.85 \times 10^{-12} \text{ C}^2/(\text{N·m}^2)](2.76 \times 10^{-10}\text{m})^2}$$

$$= +3.02 \times 10^{-9} \text{ N} \blacktriangleleft$$

Thus, the attractive force (●➤ ◄●) between the ions is $+3.02 \times 10^{-9}$ N. The repulsive force (◄● ●➤) will be equal and opposite in sign and thus will be -3.02×10^{-9} N.

Example Problem 2.9

If the attractive force between a pair of Mg^{2+} and S^{2-} ions is 1.49×10^{-8} N and if the S^{2-} ion has a radius of 0.184 nm, calculate a value for the ionic radius of the Mg^{2+} ion in nanometers.

Solution:

The value of a_0, the sum of the Mg^{2+} and S^{2-} ionic radii, can be calculated from a rearranged form of Coulomb's law equation (2.5):

$$a_0 = \sqrt{\frac{-Z_1 Z_2 e^2}{4\pi\epsilon_0 F_{\text{attractive}}}}$$

$$Z_1 = +2 \text{ for } Mg^{2+} \qquad Z_2 = -2 \text{ for } S^{2-}$$

$$|e| = 1.60 \times 10^{-19} \text{ C} \qquad \epsilon_0 = 8.85 \times 10^{-12} \text{ C}^2/(\text{N} \cdot \text{m}^2)$$

$$F_{\text{attractive}} = 1.49 \times 10^{-9} \text{ N}$$

Thus,

$$a_0 = \sqrt{\frac{-(2)(-2)(1.60 \times 10^{-19} \text{ C})^2}{4\pi[8.85 \times 10^{-12} \text{ C}^2/(\text{N} \cdot \text{m}^2)](1.49 \times 10^{-8} \text{ N})}}$$

$$= 2.49 \times 10^{-10} \text{ m} = 0.249 \text{ nm}$$

$$a_0 = r_{Mg^{2+}} + r_{S^{2-}}$$

or

$$0.249 \text{ nm} = r_{Mg^{2+}} + 0.184 \text{ nm}$$

$$r_{Mg^{2+}} = 0.065 \text{ nm} \blacktriangleleft$$

Interionic Energies for an Ion Pair

The net potential energy E between a pair of oppositely charged ions, for example, $Na^+ Cl^-$, which are brought closer together is equal to the sum of the energies associated with the attraction and repulsion of the ions and can be written in equation form as

$$E_{\text{net}} = +\underbrace{\frac{Z_1 Z_2 e^2}{4\pi\epsilon_0 a}}_{\substack{\text{Attractive} \\ \text{energy}}} + \underbrace{\frac{b}{a^n}}_{\substack{\text{Repulsive} \\ \text{energy}}} \qquad (2.8)$$

The attractive-energy term of Eq. (2.8) represents the energy released when the ions come close together and is negative because the product of $(+Z_1)(-Z_2)$

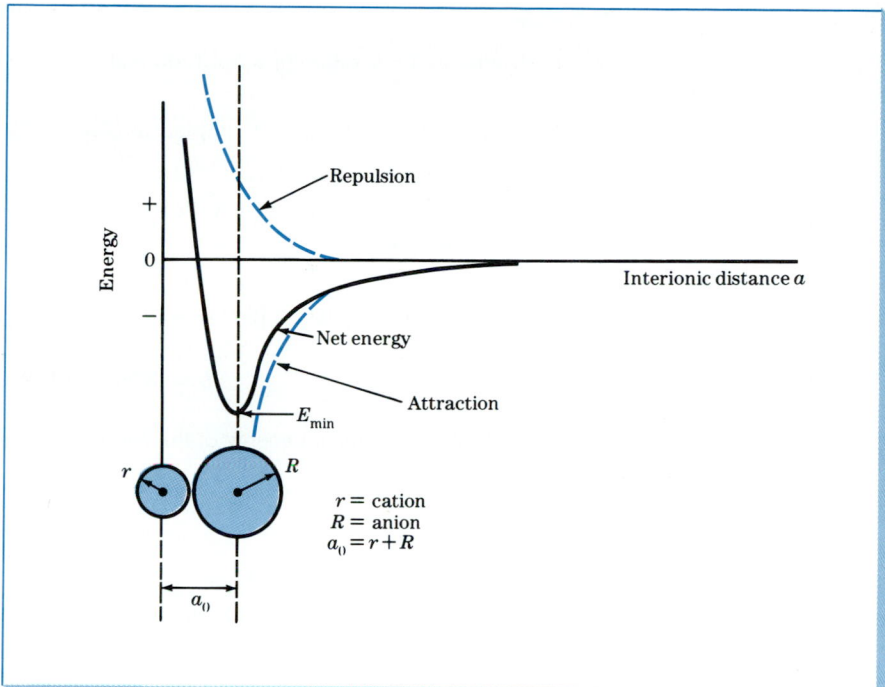

FIGURE 2.12 Energy vs. separation distance for a pair of oppositely charged ions. The equilibrium interionic separation distance a_0 is reached when the net potential energy is a minimum.

is negative. The repulsive-energy term of Eq. (2.8) represents the energy absorbed as the ions come close together and is positive. The sum of the energies associated with the attraction and repulsion of the ions equals the net energy, which is a minimum when the ions are at their equilibrium separation distance a_0. Figure 2.12 shows the relationship among these three energies and indicates the minimum energy E_{min}. At the minimum energy, the force between the ions is zero.

Example Problem 2.10

Calculate the net potential energy of a simple Na^+Cl^- ion pair by using the equation

$$E_{net} = \frac{+Z_1Z_2e^2}{4\pi\epsilon_0 a} + \frac{b}{a^n}$$

and by using the b value obtained from the repulsive force calculated for the Na^+Cl^- ion pair in Example Problem 2.8. Assume $n = 9$ for NaCl.

Solution:

(a) To determine the b value for a NaCl ion pair,

$$F = -\frac{nb}{a^{n+1}} \tag{2.6}$$

The repulsive force between a Na^+Cl^- ion pair from Example Problem 2.8 is -3.02×10^{-9} N. Thus

$$-3.02 \times 10^{-9}\ N = \frac{-9b}{(2.76 \times 10^{-10}\ m)^{10}}$$

$$b = 8.59 \times 10^{-106}\ N \cdot m^{10}\ \blacktriangleleft$$

(b) To calculate the potential energy of the Na^+Cl^- ion pair,

$$E_{Na^+Cl^-} = \frac{+Z_1Z_2e^2}{4\pi\epsilon_0 a} + \frac{b}{a^n}$$

$$= \frac{(+1)(-1)(1.60 \times 10^{-19}\ C)^2}{4\pi[8.85 \times 10^{-12}\ C^2/(N \cdot m^2)](2.76 \times 10^{-10}\ m)}$$

$$+ \frac{8.59 \times 10^{-106}\ N \cdot m^{10}}{(2.76 \times 10^{-10}\ m)^9}$$

$$= -8.34 \times 10^{-19}\ J* + 0.92 \times 10^{-19}\ J*$$

$$= -7.42 \times 10^{-19}\ J\ \blacktriangleleft$$

Ion Arrangements in Ionic Solids

Until now our discussion of ionic bonding has been for a pair of ions. We shall now briefly discuss ionic bonding in three-dimensional ionic solids. Since elemental ions have approximately spherical symmetrical charge distributions, they can be considered spherical with a characteristic radius. Ionic radii of some selected elemental cations and anions are listed in Table 2.5. As previously stated, elemental cations are smaller than their atoms and anions are larger. Also, as in the case of atoms, elemental ions, in general, increase in size as their principal quantum number increases (i.e., as the number of electron shells increases).

When ions pack together in a solid, they do so with no preferred orientation since electrostatic attraction of symmetrical charges is independent of the orientation of the charges. Thus the ionic bond is *nondirectional* in character. However, the packing of ions in an ionic solid is governed by the geometric arrangement of the ions that is possible and the necessity to maintain electrical neutrality of the solid.

---

*1 N·m = 1 J.

TABLE 2.5   Ionic Radii of Selected Elements

Ion	Ionic radius, nm	Ion	Ionic radius, nm
Li^+	0.060	F^-	0.136
Na^+	0.095	Cl^-	0.181
K^+	0.133	Br^-	0.195
Rb^+	0.148	I^-	0.216
Cs^+	0.169		

Geometric arrangement of ions in an ionic solid   Ionic crystals may have extremely complex structures. At this point we shall only consider the *geometric arrangement* of the ions in two simple ionic structures, CsCl and NaCl. In the case of CsCl, *eight* Cl^- ions ($r = 0.181$ nm) pack around a central Cs^+ ion ($r = 0.169$ nm), as shown in Fig. 2.13a. However, in NaCl, only *six* Cl^- ions ($r = 0.181$ nm) can pack around a central Na^+ ion ($r = 0.095$ nm), as shown in Fig. 2.13b. For CsCl, the ratio of the cation-to-anion radius is $0.169/0.181 = 0.934$, whereas for NaCl the ratio is $0.095/0.181 = 0.525$. Thus, as the ratio of the cation-to-anion radius decreases, fewer anions can surround a central cation in this type of structure. Ionic crystal structures will be discussed in more detail in Chap. 10 on Ceramic Materials.

Electrical neutrality of ionic solids   The ions in an ionic solid must be arranged in a structure so that local charge neutrality is maintained. Thus, in an ionic solid such as CaF_2, the ionic arrangement will be partly governed by the fact that there has to be two fluoride ions for every calcium ion.

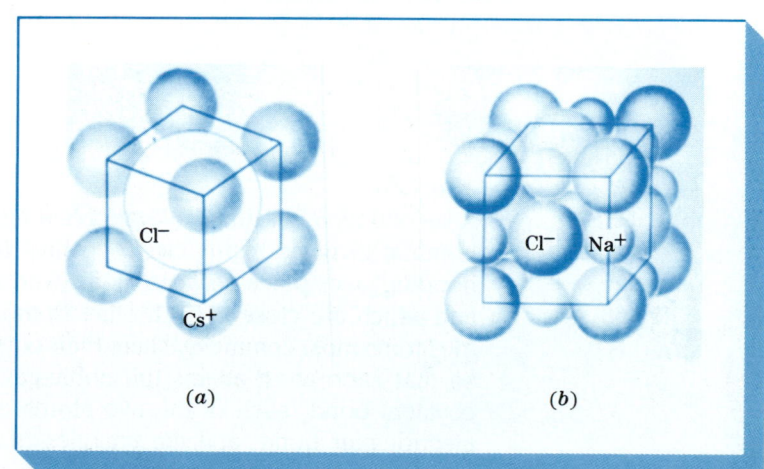

FIGURE 2.13 Ionic packing arrangements in (*a*) CsCl and (*b*) NaCl. Eight Cl^- ions can pack around a Cs^+ ion, but only six Cl^- ions can pack around a Na^+ ion. (*After C. R. Barrett, W. D. Nix, and A. S. Tetelman, "The Principles of Engineering Materials," Prentice-Hall, 1973, p. 27.*)

(*a*)   (*b*)

TABLE 2.6   Lattice Energies and Melting Points of Selected Ionic Solids

Ionic solid	Lattice energy*		Melting point, °C
	kJ/mol	kcal/mol	
LiCl	829	198	613
NaCl	766	183	801
KCl	686	164	776
RbCl	670	160	715
CsCl	649	155	646
MgO	3932	940	2800
CaO	3583	846	2580
SrO	3311	791	2430
BaO	3127	747	1923

* All values are negative for bond formation (energy is released).

Bonding Energies of Ionic Solids

The lattice energies and melting points of ionically bonded solids are relatively high, as indicated in Table 2.6 which lists the lattice energies and melting points of selected ionic solids. As the size of the ion increases in a group in the periodic table, the lattice energy decreases. For example, the lattice energy for LiCl is 829 kJ/mol (198 kcal/mol), whereas that of CsCl is only 649 kJ/mol (155 kcal/mol). The reason for this decrease in lattice energy is that the bonding electrons in the larger ions are farther away from the attractive influence of the positive nucleus. Also, multiple bonding electrons in an ionic solid increase the lattice energy, as exemplified by MgO which has a lattice energy of 3932 kJ/mol (940 kcal/mol).

2.6   COVALENT BONDING

A second type of primary atomic bonding is *covalent bonding.* Whereas ionic bonding involves highly electropositive and electronegative atoms, covalent bonding takes place between atoms with small differences in electronegativity and which are close to each other in the periodic table. In covalent bonding the atoms most commonly share their outer *s* and *p* electrons with other atoms so that each atom attains the noble-gas electron configuration. In a single covalent bond, each of the two atoms contributes one electron to form an electron-pair bond, and the energies of the two atoms associated with the covalent bond are lowered (made more stable) because of the electron interaction. In covalent bonding, mutliple electron-pair bonds can be formed by one atom with itself or other atoms.

Covalent Bonding in the Hydrogen Molecule

The simplest case of covalent bonding occurs in the hydrogen molecule in which two hydrogen atoms contribute their $1s^1$ electrons to form an electron-pair covalent bond, as indicated by the electron-dot notation reaction

$1s^1$ electron

Electron-pair covalent bond

$$\text{H} \cdot + \text{H} \colon \longrightarrow \text{H} \colon \text{H}$$

Hydrogen   Hydrogen   Hydrogen
atom   atom   molecule

Although the electron-dot notation is useful to represent the covalent bond, it does not consider the valence-electron density distribution. In Fig. 2.14 the two hydrogen atoms are shown as forming an electron-pair covalent bond with a high electron charge cloud density between the two nuclei of the hydrogen atoms. As the two hydrogen atoms come together to form the hydrogen molecule, their electron charge clouds interact and overlap, creating a high probability of finding the $1s^1$ electrons of the atoms between the two nuclei in the molecule (Fig. 2.14). In the bonding process of forming the hydrogen molecule, the potential energy of the hydrogen atoms is lowered, as indicated in Fig. 2.15, and energy is released. If the hydrogen atoms in the hydrogen molecule are separated, energy is required since the atoms would then be in a higher energy state.

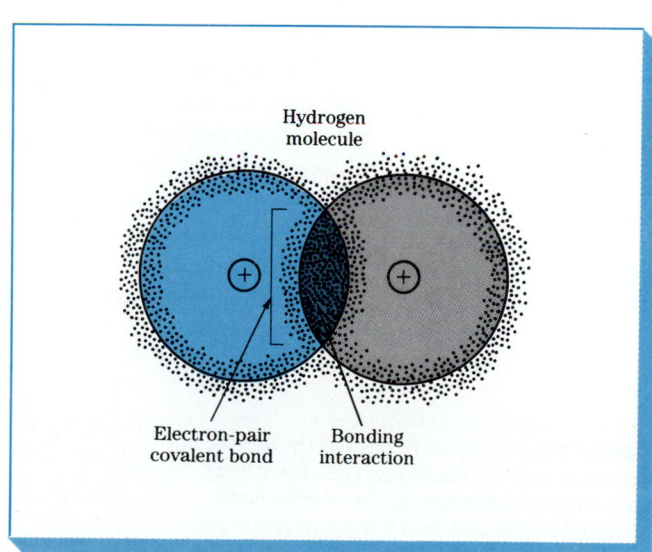

Hydrogen
molecule

Electron-pair   Bonding
covalent bond   interaction

FIGURE 2.14 Covalent bonding in the hydrogen molecule. The highest electron charge cloud density is in the region of overlap between the hydrogen atom nuclei.

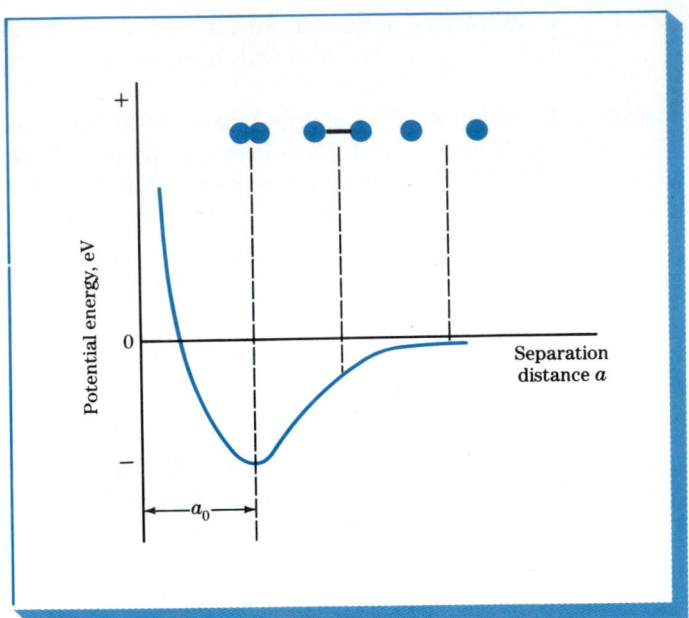

FIGURE 2.15 Potential energy vs. separation distance for two hydrogen atoms. The equilibrium interatomic distance a_0 in the hydrogen molecule occurs at the minimum potential energy E_{min}.

Covalent Bonding in Other Diatomic Molecules

Electron-pair covalent bonds are also formed in other diatomic molecules such as F_2, O_2, and N_2. In these cases p electrons are shared between the atoms. The fluorine atom with its outer seven electrons $(2s^22p^5)$ can attain the neon noble-gas electron configuration by sharing one $2p$ electron with another fluorine atom, as shown in the electron-dot notation reaction in Fig. 2.16a. Similarly, the oxygen atom with its outer six electrons $(2s^22p^4)$ can attain the neon noble-gas electron configuration $(2s^22p^6)$ by sharing *two* $2p$ electrons with another oxygen atom to form the diatomic O_2 molecule (Fig. 2.16b). Nitrogen also with its five outer valence electrons $(2s^22p^3)$ can attain the neon noble-gas electron configuration of $(2s^22p^6)$ by sharing three $2p$ electrons with another nitrogen atom to form the diatomic nitrogen molecule (Fig. 2.16c).

Chemical reactions involving covalent bonds are sometimes written by using electron-dot notation for the covalent bonds, as shown in Fig. 2.16. However, more commonly the straight-line notation for covalent bonds (Fig. 2.16) is used. We shall encounter both types of these notations in Chap. 7 on Polymeric Materials.

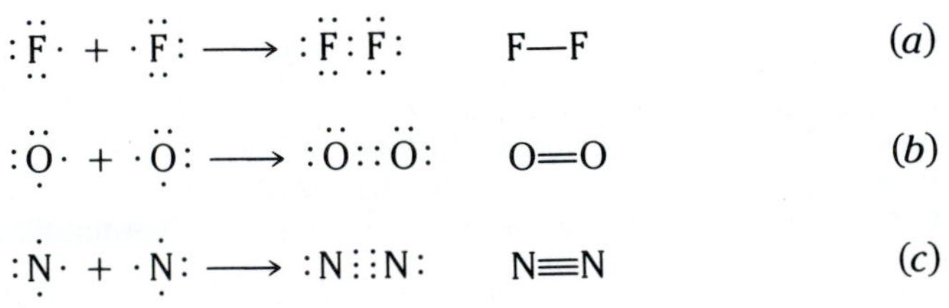

FIGURE 2.16 Covalent bonding in molecules of fluorine (single bond), oxygen (double bond), and nitrogen (triple bond). The electron-pair covalent bonds between the atoms are shown on the left by electron-dot notation and on the right by straight-line notation.

$$:\overset{..}{\underset{..}{F}}\cdot \ + \ \cdot \overset{..}{\underset{..}{F}}: \ \longrightarrow \ :\overset{..}{\underset{..}{F}}:\overset{..}{\underset{..}{F}}: \qquad F{-}F \qquad\qquad (a)$$

$$:\overset{..}{O}\cdot \ + \ \cdot \overset{..}{O}: \ \longrightarrow \ :\overset{..}{O}::\overset{..}{O}: \qquad O{=}O \qquad\qquad (b)$$

$$:\overset{.}{N}\cdot \ + \ \cdot \overset{.}{N}: \ \longrightarrow \ :N::N: \qquad N{\equiv}N \qquad\qquad (c)$$

TABLE 2.7   **Bond Energies and Lengths of Selected Covalent Bonds**

Bond	Bond energy*		Bond length, nm
	kcal/mol	kJ/mol	
C—C	88	370	0.154
C=C	162	680	0.13
C≡C	213	890	0.12
C—H	104	435	0.11
C—N	73	305	0.15
C—O	86	360	0.14
C=O	128	535	0.12
C—F	108	450	0.14
C—Cl	81	340	0.18
O—H	119	500	0.10
O—O	52	220	0.15
O—Si	90	375	0.16
N—O	60	250	0.12
N—H	103	430	0.10
F—F	38	160	0.14
H—H	104	435	0.074

*Approximate values since environment changes energy. All values are negative for bond formation (energy is released).
Source: L. H. Van Vlack, "Elements of Materials Science," 4th ed., Addison-Wesley, 1980.

Table 2.7 lists approximate bond energies and lengths of selected covalent bonds. Note that higher bond energies are associated with multiple bonds. For example, the C—C bond has an energy of 370 kJ/mol (88 kcal/mol), whereas the C=C bond has a much higher energy of 680 kJ/mol (162 kcal/mol).

Covalent Bonding by Carbon

In the study of engineering materials carbon is very important since it is the basic element in most polymeric materials. The carbon atom in the ground state has the electron configuration $1s^2 2s^2 2p^2$. This electron arrangement indicates that carbon should form *two covalent bonds* with its two half-filled $2p$ orbitals. However, in many cases carbon forms *four covalent bonds* of equal strength. The explanation for the four carbon covalent bonds is provided by the concept of *hybridization* whereby upon bonding, one of the $2s$ orbitals is promoted to a $2p$ orbital so that *four equivalent sp^3 hybrid orbitals* are produced, as indicated in the orbital diagrams of Fig. 2.17. Even though energy is required to promote the $2s$ electron to the $2p$ state in the hybridization process, the energy necessary for the promotion is more than compensated for by the decrease in energy accompanying the bonding process.

Carbon in the form of diamond exhibits sp^3 tetrahedral covalent bonding.

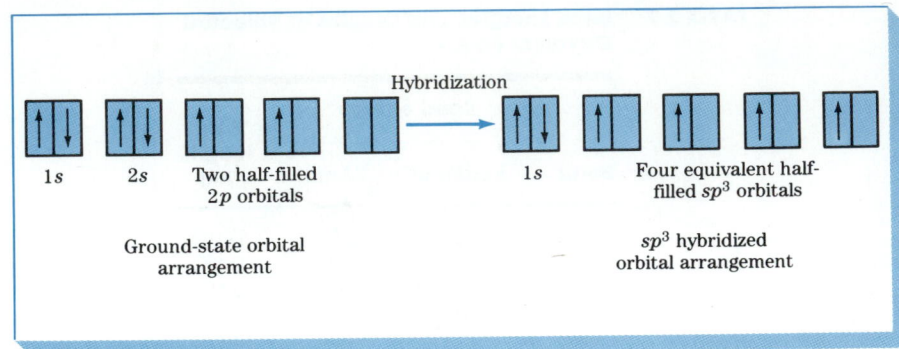

FIGURE 2.17
Hybridization of carbon
orbitals for the formation
of single covalent bonds.

The four sp^3 hybrid orbitals are directed symmetrically toward the corners of a regular tetrahedron, as shown in Fig. 2.18. The structure of diamond consists of a massive molecule with sp^3 tetrahedral covalent bonding, as shown in Fig. 2.19. This structure accounts for the extremely high hardness of diamond and its high bond strength and melting temperature. Diamond has a bond energy of 711 kJ/mol (170 kcal/mol) and a melting temperature of 3550°C.

**Covalent Bonding in
Carbon-Containing
Molecules**

Covalently bonded molecules containing only carbon and hydrogen are hydrocarbons. The simplest hydrocarbon is methane in which carbon forms four sp^3 tetrahedral covalent bonds with hydrogen atoms, as shown in Fig. 2.20. The intramolecular bonding energy of methane is relatively high at 1650 kJ/mol (396 kcal/mol), but the intermolecular bond energy is very low at about 8 kJ/mol (2 kcal/mol). Thus the methane molecules are very weakly bonded together, resulting in a low melting point of −183°C.

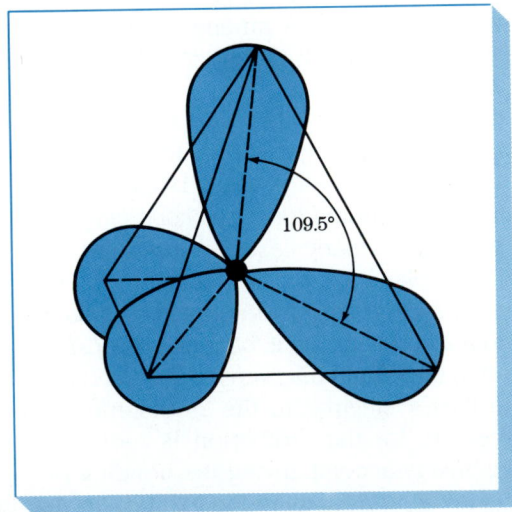

FIGURE 2.18 A carbon
atom with four equivalent
sp^3 orbitals directed
symmetrically toward the
corners of a tetrahedron.
The angle between the
orbitals is 109.5°.

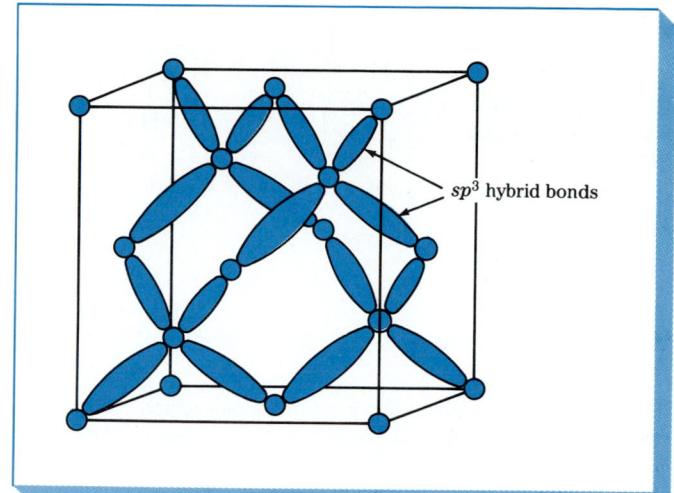

FIGURE 2.19 Tetrahedral *sp³* covalent bonds in the diamond structure. Each shaded region represents an electron-pair covalent bond.

sp³ hybrid bonds

Figure 2.21 shows structural formulas for methane, ethane, and normal (*n*-) butane, which are single covalently bonded hydrocarbons. As the molecular mass of the molecule increases, so does its stability and melting point.

Carbon can also bond to itself to form double and triple bonds in molecules, as indicated in the structural formulas for ethylene and acetylene shown in Fig. 2.22. Double and triple carbon-carbon bonds are chemically more reactive than single carbon-carbon bonds. Multiple carbon-carbon bonds in carbon-containing molecules are referred to as *unsaturated bonds*.

Benzene   An important molecular structure for some polymeric materials is the benzene structure. The benzene molecule has the chemical composition C_6H_6, with the

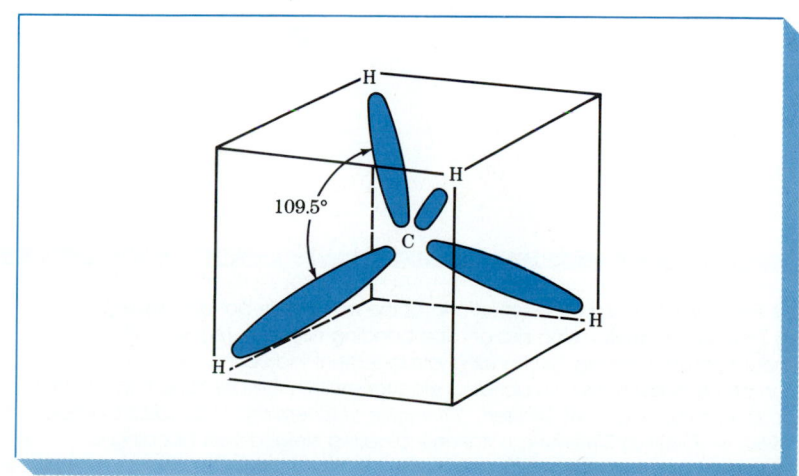

FIGURE 2.20 The methane molecule, CH_4, contains four tetrahedral *sp³* covalent bonds. Each shaded region represents an electron-pair covalent bond.

109.5°

H

H

C

H

H

$$
\begin{array}{c}
\text{H} \\
| \\
\text{H}-\text{C}-\text{H} \\
| \\
\text{H}
\end{array}
\qquad
\begin{array}{c}
\text{H}\quad\text{H} \\
|\quad| \\
\text{H}-\text{C}-\text{C}-\text{H} \\
|\quad| \\
\text{H}\quad\text{H}
\end{array}
\qquad
\begin{array}{c}
\text{H}\quad\text{H}\quad\text{H}\quad\text{H} \\
|\quad|\quad|\quad| \\
\text{H}-\text{C}-\text{C}-\text{C}-\text{C}-\text{H} \\
|\quad|\quad|\quad| \\
\text{H}\quad\text{H}\quad\text{H}\quad\text{H}
\end{array}
$$

FIGURE 2.21 Structural formulas for simple covalently bonded hydrocarbons. Note that as the molecular mass increases, the melting point increases.

Methane
mp = −183°C

Ethane
mp = −172°C

n-Butane
mp = −135°C

$$
\begin{array}{c}
\text{H}\quad\text{H} \\
|\quad| \\
\text{C}=\text{C} \\
|\quad| \\
\text{H}\quad\text{H}
\end{array}
\qquad
\text{H}-\text{C}\equiv\text{C}-\text{H}
$$

FIGURE 2.22 Structural formulas for two covalently bonded molecules which have multiple carbon-carbon covalent bonds.

Ethylene
mp = −169.4°C

Acetylene
mp = −81.8°C

carbon atoms forming a hexagonal-shaped ring referred to sometimes as the *benzene ring* (Fig. 2.23). The six hydrogen atoms of benzene are covalently bonded with single bonds to the six carbon atoms of the ring. However, the bonding arrangement between the carbon atoms in the ring is complex. The simplest way of satisfying the requirement that each carbon atom have four covalent bonds is to assign alternating single and double bonds to the carbon atoms in the ring itself (Fig. 2.23*a*). This structure can be represented more

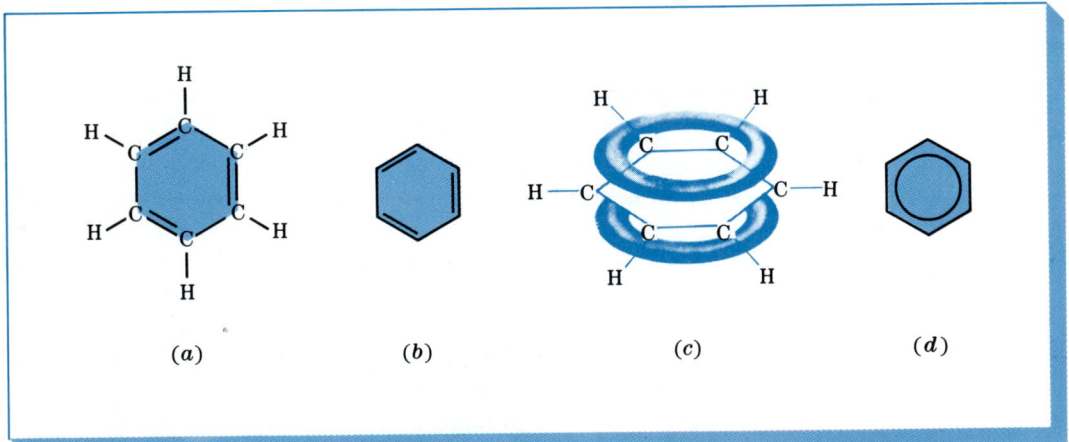

(*a*)   (*b*)   (*c*)   (*d*)

FIGURE 2.23 Structural formula representations for the covalently bonded molecule benzene. (*a*) Structural formula using straight-line bonding notation. (*b*) Simplified notation without hydrogen atoms. (*c*) Bonding arrangement indicating the delocalization of the carbon-carbon bonding electrons within the benzene ring. (*After R. E. Davis, K. D. Gailey, and K. W. Whitten, "Principles of Chemistry," Saunders College Publishing, 1984, p. 830.*) (*d*) Simplified notation indicating delocalized bonding of electrons within the benzene ring.

simply by omitting the external hydrogen atoms (Fig. 2.23b). This structural formula for benzene will be used in this book since it more clearly indicates the bonding arrangement in benzene.

However, experimental evidence indicates that a normal reactive double carbon-carbon bond does not exist in benzene and that the bonding electrons within the benzene ring are delocalized, forming an overall bonding structure intermediate in chemical reactivity to that between single and double carbon-carbon bonds (Fig. 2.23c). Thus, most chemistry books are written using a circle inside a hexagon to represent the structure of benzene (Fig. 2.23d).

2.7 METALLIC BONDING

A *third* primary type of atomic bonding is metallic bonding which occurs in solid metals. In metals in the solid state, atoms are packed relatively close together in a systematic pattern or crystal structure. For example, the arrangement of copper atoms in crystalline copper is shown in Fig. 2.24a. In this structure the atoms are so close together that their outer valence electrons are attracted to the nuclei of their numerous neighbors. In the case of solid copper, each atom is surrounded by *12* nearest neighbors. The valence electrons are therefore not closely associated with any particular nucleus and are thus spread out among the atoms in the form of a low-density electron charge cloud, or "electron gas."

Solid metals are therefore visualized as consisting of *positive-ion cores* (atoms without their valence electrons) and of *valence electrons* dispersed in the form of an electron cloud which covers a large extent of space (Fig. 2.24b). The valence electrons are weakly bonded to the positive-ion cores and can readily move in the metal crystal and are frequently referred to as *free electrons*. The high thermal and electrical conductivities of metals support the theory that some electrons are free to move through the metal crystal lattice. Most metals can be deformed a considerable amount without fracturing because the metal atoms can slide past each other without completely disrupting the metallically bonded structure.

Atoms in a solid metal bond themselves together by metallic bonding to achieve a lower energy (or more stable) state. For metallic bonding there are *no electron-pair restrictions* as for the covalent bond or *charge neutrality restrictions* as for the ionic bond. In metallic bonding the outer valence electrons of the atoms are shared by many surrounding atoms, and so, in general, metallic bonding is *nondirectional*.

When metallic atoms bond together by valence-electron sharing to form a solid crystal, the overall energy of the individual atoms is lowered by the bonding process. As is the case for ionic and covalent bonding, a minimum in energy between a pair of atoms is attained when the equilibrium atomic

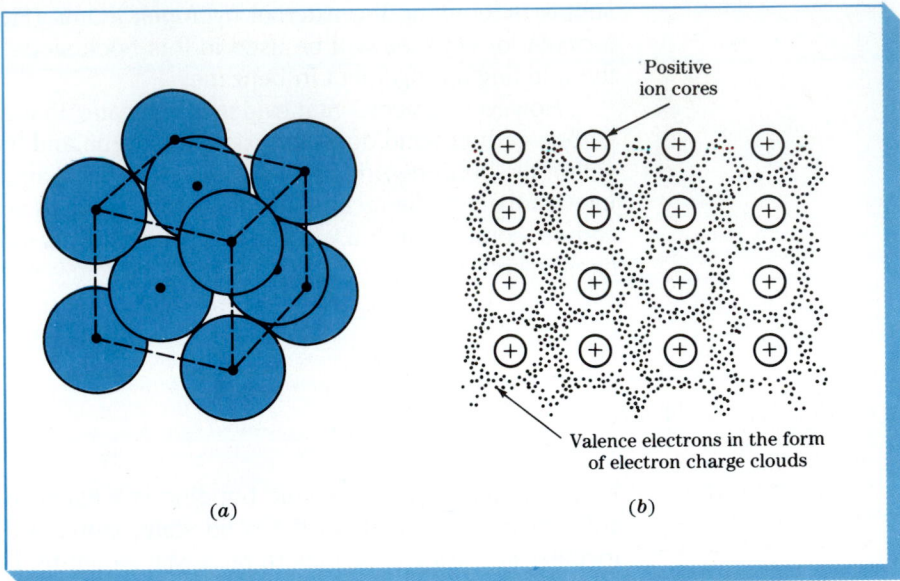

FIGURE 2.24 (a) Atomic arrangement in a metallic copper crystal. Each copper atom is coordinated with 12 other copper atoms, producing a crystal structure called the *face-centered-cubic structure*. The atoms are bonded together by an "electron gas" of delocalized valence electrons. (b) Two-dimensional schematic diagram of metallically bonded atoms. The circles with the inner positive signs represent positive-ion cores, and the charge clouds around the ion cores represent the dispersed valence electrons.

separation distance a_0 is reached, as shown in Fig. 2.25. The magnitude of $E_0 - E_{min}$ in Fig. 2.25 is a measure of the bonding energy between the atoms of a particular metal.

The energy levels in multiatom metal crystals differ from those of single atoms. When metal atoms bond together to form a metal crystal, their energies are lowered but to slightly different levels. The valence electrons in a metal crystal therefore form an energy "band." The energy-band theory of metals will be discussed further in Sec. 5.2.

The bond energies and melting points of different metals vary greatly. In general, the fewer valence electrons per atom involved in metallic bonding, the more metallic the bonding is. That is, the valence electrons are freer to move. The highest degree of metallic bonding is found in the alkali metals, which have just one bonding electron outside a noble-gas electron configuration. Thus, the bonding energies and melting points of the alkali metals are relatively low. For example, the bonding energy for sodium is 108 kJ/mol (25.9 kcal/mol) and that for potassium is 89.6 kJ/mol (21.4 kcal/mol). The melting points of sodium (97.9°C) and potassium (63.5°C) are also relatively low.

However, as the number of bonding electrons increases, the bonding energies and melting points of the metals also increase, as indicated in Table 2.8 for the metals of the fourth period. Calcium with two valence electrons per

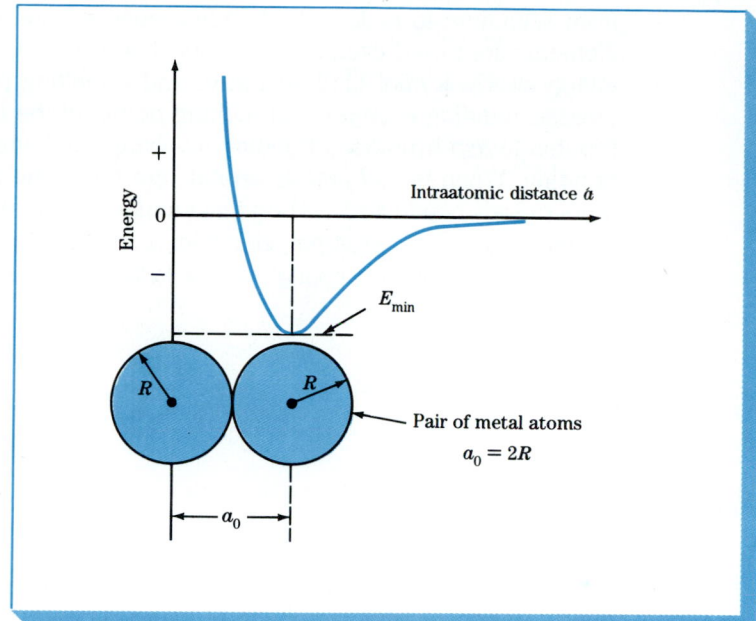

FIGURE 2.25 Energy vs. separation distance for a pair of metal atoms. The equilibrium atomic separation distance a_0 is reached when the net potential energy is a minimum.

atom has its bonding electrons more tightly bound than potassium, and as a result calcium's bonding energy of 177 kJ/mol (42.2 kcal/mol) and melting point (851°C) are both considerably higher than those of potassium. With the introduction of the $3d$ electrons in the transition metals of the fourth period

TABLE 2.8   Bonding Energies, Melting Points, and Electron Configurations of the Fourth Period Metals of the Periodic Table

Element	Electron configuration	Bonding energy		Melting point, °C
		kJ/mol	kcal/mol	
K	$4s^1$	89.6	21.4	63.5
Ca	$4s^2$	177	42.2	851
Sc	$3d^14s^2$	342	82	1397
Ti	$3d^24s^2$	473	113	1812
V	$3d^34s^2$	515	123	1730
Cr	$3d^54s^1$	398	95	1903
Mn	$3d^54s^2$	279	66.7	1244
Fe	$3d^64s^2$	418	99.8	1535
Co	$3d^74s^2$	383	91.4	1490
Ni	$3d^84s^2$	423	101	1455
Cu	$3d^{10}4s^1$	339	81.1	1083
Zn	$4s^2$	131	31.2	419
Ga	$4s^24p^1$	272	65	29.8
Ge	$4s^24p^2$	377	90	960

from scandium to nickel, the bonding energies and melting points of these elements are raised even much higher. For example, titanium has a bonding energy of 473 kJ/mol (113 kcal/mol) and a melting point of 1812°C. The increased bonding energies and melting points of the transition metals are attributed to *dsp* hybridized bonding involving a significant fraction of covalent bonding. When the 3*d* and 4*s* orbitals are filled, the outer electrons become more loosely bound and the bonding energies and melting points of the metals decrease again. For example, zinc with a $3d^{10}4s^2$ electron configuration has a relatively low bonding energy of 131 kJ/mol (31.2 kcal/mol) and a melting temperature of 419°C.

2.8   SECONDARY BONDING

Until now we have considered only primary bonding between atoms and showed that it depends on the interaction of their valence electrons. The driving force for primary atomic bonding is the lowering of the energy of the bonding electrons. Secondary bonds are relatively weak in contrast to primary bonds and have energies of only about 4 to 42 kJ/mol (1 to 10 kcal/mol). The driving force for secondary bonding is the attraction of the electric dipoles contained in atoms or molecules.

An electric dipole moment is created when two equal and opposite charges are separated, as shown in Fig. 2.26*a*. Electric dipoles are created in atoms or molecules when positive and negative charge centers exist (Fig. 2.26*b*).

Dipoles in atoms or molecules create dipole moments. A *dipole moment* is defined as the charge value multiplied by the separation distance between

FIGURE 2.26 (*a*) An electric dipole. The dipole moment is *qd*. (*b*) An electric dipole moment in a covalently bonded molecule. Note the separation of the positive and negative charge centers.

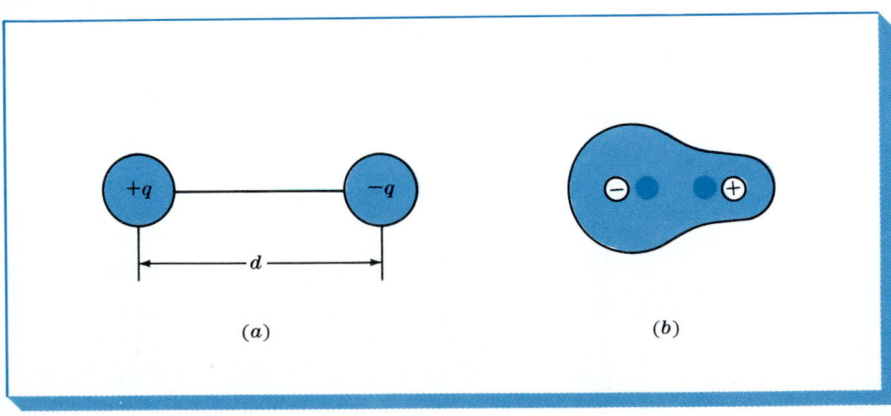

(*a*)

(*b*)

positive and negative charges, or

$$\mu = qd \qquad (2.9)$$

where μ = dipole moment
q = magnitude of electric charge
d = separation distance between the charge centers

Dipole moments in atoms and molecules are measured in coulomb-meters (C·m) or in debye units, where 1 debye = 3.34×10^{-30} C·m.

Electric dipoles interact with each other by electrostatic (coulombic) forces, and thus atoms or molecules containing dipoles are attracted to each other by these forces. Even though the bonding energies of secondary bonds are weak, they become important when they are the only bonds available to bond atoms or molecules together.

In general, there are two main kinds of secondary bonds between atoms or molecules involving electric dipoles: fluctuating dipoles and permanent dipoles. Collectively, these secondary dipole bonds are sometimes called *van der Waals bonds (forces)*.

Fluctuating Dipoles   Very weak secondary bonding forces can develop between the atoms of noble-gas elements which have complete outer-valence-electron shells (s^2 for helium and s^2p^6 for Ne, Ar, Kr, Xe, and Rn). These bonding forces arise because the asymmetrical distribution of electron charge distribution in these atoms creates electric dipoles. At any instant there is a high probability that there will be more electron charge on one side of an atom than on the other (Fig. 2.27).

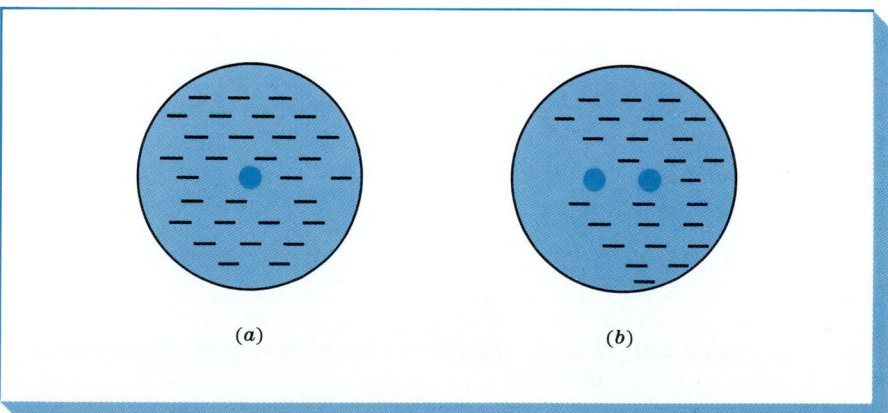

(a)   (b)

FIGURE 2.27 Electron charge cloud distribution in a noble-gas atom. (*a*) Idealized symmetrical electron charge cloud distribution. (*b*) Real case with unsymmetrical electron charge cloud distribution which changes with time, creating a "fluctuating electric dipole."

TABLE 2.9   **Melting and Boiling Points of Noble Gases at Atmospheric Pressure**

Noble gas	Melting point, °C	Boiling point, °C
Helium	-272.2	-268.9
Neon	-248.7	-245.9
Argon	-189.2	-185.7
Krypton	-157.0	-152.9
Xenon	-112.0	-107.1
Radon	-71.0	-61.8

Thus, in a particular atom, the electron charge cloud will change with time, creating a "fluctuating dipole." Fluctuating dipoles of nearby atoms can attract each other, creating weak interatomic nondirectional bonds. The liquefaction and solidification of the noble gases at low temperatures and high pressures are attributed to fluctuating dipole bonds. The melting and boiling points of the noble gases at atmospheric pressure are listed in Table 2.9. Note that as the atomic size of the noble gases increases, the melting and boiling points also increase due to stronger bonding forces since the electrons have more freedom to create stronger dipole moments.

Permanent Dipoles   Weak bonding forces among covalently bonded molecules can be created if the molecules contain *permanent dipoles*. For example, the methane molecule, CH_4, with its four C—H bonds arranged in a tetrahedral structure (Fig. 2.20), has a zero dipole moment because of its symmetrical arrangement of four C—H bonds. That is, the vectorial addition of its four dipole moments is zero. The chloromethane molecule, CH_3Cl, in contrast, has an unsymmetrical tetrahedral arrangement of three C—H bonds and one C—Cl bond, resulting in a net dipole moment of 2.0 debyes. The replacement of one hydrogen atom in methane with one chlorine atom raises the boiling point of methane of $-128°C$ to $-14°C$ for chloromethane. The much higher boiling point of chloromethane is due to the permanent dipole bonding forces among the chloromethane molecules. Table 2.10 lists some experimental dipole moments of some compounds.

TABLE 2.10   **Experimental Dipole Moments of Some Compounds (Debyes)**

H_2O	1.84	CH_3Cl	2.00
H_2	0.00	$CHCl_3$	1.10
CO_2	0.00	HCl	1.03
CCl_4	0.00	NH_3	1.46

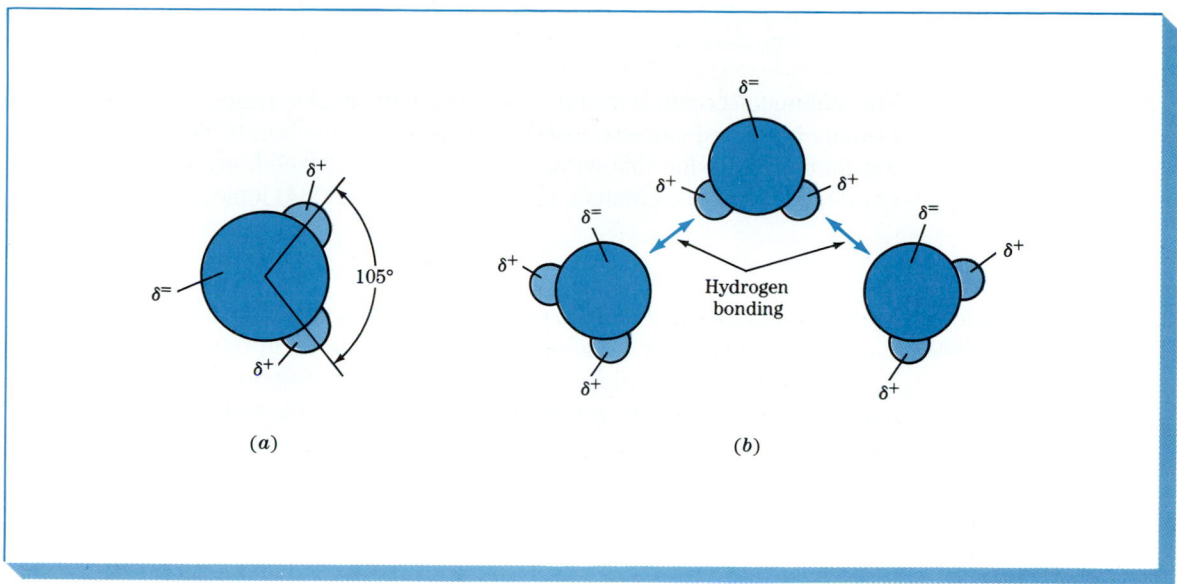

FIGURE 2.28 (a) Permanent dipole nature of the water molecule. (b) Hydrogen bonding among water molecules due to permanent dipole attraction.

The *hydrogen bond* is a special case of a permanent dipole-dipole interaction between polar molecules. Hydrogen bonding occurs when a polar bond containing the hydrogen atom, O—H or N—H, interacts with the electronegative atoms O, N, F, or Cl. For example, the water molecule, H_2O, has a permanent dipole moment of 1.84 debyes due to its asymmetrical structure with its two hydrogen atoms at an angle of 105° with respect to its oxygen atom (Fig. 2.28a).

The hydrogen atomic regions of the water molecule have positively charged centers, and the opposite end region of the oxygen atom has a negatively charged center (Fig. 2.28a). In hydrogen bonding between water molecules the negatively charged region of one molecule is attracted by coulombic forces to the positively charged region of another molecule (Fig. 2.28b).

In liquid and solid water relatively strong intermolecular permanent dipole forces (hydrogen bonding) are formed among the water molecules. The energy associated with the hydrogen bond is about 29 kJ/mol (7 kcal/mol) as compared to about 2 to 8 kJ/mol (0.5 to 2 kcal/mol) for fluctuating dipole forces in the noble gases. The exceptionally high boiling point of water (100°C) for its molecular mass is attributed to the effect of hydrogen bonding. Hydrogen bonding is also very important for strengthening the bonding between molecular chains of some types of polymeric materials.

2.9 MIXED BONDING

The chemical bonding of atoms or ions can involve more than one type of primary bond and can also involve secondary dipole bonds. For primary bonding there can be the following combinations of mixed-bond types: (1) ionic-covalent, (2) metallic-covalent, (3) metallic-ionic, and (4) ionic-covalent-metallic.

Ionic-Covalent Mixed Bonding   Most covalent-bonded molecules have some ionic binding, and vice versa. The partial ionic character of covalent bonds can be interpreted in terms of the electronegativity scale of Fig. 2.9. The greater the difference in the electronegativities of the elements involved in a mixed ionic-covalent bond, the greater the degree of ionic character of the bond. Pauling[1] proposed the following equation to determine the percentage ionic character of bonding in a compound AB:

$$\% \text{ ionic character} = (1 - e^{(-1/4)(X_A - X_B)^2})(100\%) \qquad (2.10)$$

where X_A and X_B are the electronegativities of the atoms A and B in the compound.

Many semiconducting compounds have mixed ionic-covalent bonding. For example, GaAs is a 3–5 compound (Ga is in group 3A and As is in group 5A of the periodic table) and ZnS is a 2–6 compound. The degree of ionic character in the bonding of these compounds increases as the electronegativity between the atoms in the compounds increases. Thus one would expect a 2–6 compound to have more ionic character than a 3–5 compound because of the greater electronegativity difference in the 2–6 compound. Example Problem 2.11 illustrates this.

Example Problem 2.11

Calculate the percentage ionic character of the semiconducting compounds GaAs (3–5) and ZnSe (2–6) by using Pauling's equation

$$\% \text{ ionic character} = (1 - e^{(-1/4)(X_A - X_B)^2})(100\%)$$

[1]Linus Carl Pauling (1901–). American chemist whose pioneer work led to a greater understanding of chemical bonding. He received the Nobel Prize in Chemistry in 1954.

(*a*) For GaAs, electronegativities from Fig. 2.9 are $X_{Ga} = 1.8$ and $X_{As} = 2.2$. Thus

$$\% \text{ ionic character} = (1 - e^{(-1/4)(1.8-2.2)^2})(100\%)$$
$$= (1 - e^{(-1/4)(-0.4)^2})(100\%)$$
$$= (1 - 0.96)(100\%) = 4\%$$

(*b*) For ZnSe, electronegativities from Fig. 2.9 are $X_{Zn} = 1.7$ and $X_{Se} = 2.5$. Thus

$$\% \text{ ionic character} = (1 - e^{(-1/4)(1.7 - 2.5)^2})(100\%)$$
$$= (1 - e^{(-1/4)(-0.8)^2})(100\%)$$
$$= (1 - 0.85)(100\%) = 15\%$$

Note that as the electronegativities differ more for the 2–6 compound, the percentage ionic character increases.

Metallic-Covalent Mixed Bonding   Mixed metallic-covalent bonding occurs commonly. For example, the transition metals have mixed metallic-covalent bonding involving *dsp* bonding orbitals. The high melting points of the transition metals are attributed to mixed metallic-covalent bonding. Also in group 4A of the periodic table there is a gradual transition from pure covalent bonding in carbon (diamond) to some metallic character in silicon and germanium. Tin and lead are primarily metallically bonded.

Metallic-Ionic Mixed Bonding   If there is a significant difference in electronegativity in the elements that form an intermetallic compound, there may be a significant amount of electron transfer (ionic binding) in the compound. Thus, some intermetallic compounds are good examples for mixed metallic-ionic bonding. Electron transfer is especially important for intermetallic compounds such as $NaZn_{13}$ and less important for compounds Al_9Co_3 and Fe_5Zn_{21} since the electronegativity differences for the latter two compounds are much less.

SUMMARY

Atoms consist of mainly three basic subatomic particles: *protons, neutrons,* and *electrons.* The electrons are envisaged as forming a thinly dispersed electron cloud of varying density around a denser atomic nucleus containing almost all the mass of the atom. The outer electrons are the valence electrons, and it is mainly their behavior that determines the chemical reactivity of each individual atom.

Electrons obey the laws of quantum mechanics, and as a result the energies of electrons are *quantized.* That is, an electron can have only certain allowed values of energies. If an electron changes its energy, it must change to a new allowed energy level. During an energy change, an electron emits or absorbs a photon of energy according to Planck's equation $\Delta E = h\nu$, where ν is the frequency of the radiation. Each electron is associated with four quantum numbers: the principal quantum number n, the subsidiary quantum number l, the magnetic quantum number m_l, and the spin quantum number m_s. According to Pauli's exclusion principle, *no two electrons can have all four quantum numbers the same.* Electrons also obey Heisenberg's uncertainty principle, which states that it is impossible to determine the momentum and position of an electron simultaneously. Thus the location of electrons in atoms must be considered in terms of electron density distributions.

There are two main types of atomic bonds: (1) *strong primary bonds* and (2) *weak secondary bonds.* Primary bonds can be subdivided into (1) *ionic,* (2) *covalent,* and (3) *metallic bonds,* and secondary bonds can be subdivided into (1) *fluctuating dipoles* and (2) *permanent dipoles.*

Ionic bonds are formed by the transfer of one or more electrons from an electropositive atom to an electronegative one. The ions are bonded together in a solid crystal by electrostatic (coulombic) forces and are *nondirectional.* The size of the ions (geometric factor) and electrical neutrality are the two main factors which determine the ion packing arrangement. *Covalent bonds* are formed by the sharing of electrons in pairs by half-filled orbitals. The more the bonding orbitals overlap, the stronger the bond. Covalent bonds are *directional. Metallic bonds* are formed by metal atoms by a mutual sharing of valence electrons in the form of delocalized electron charge clouds. In general the fewer the valence electrons, the more delocalized they are and the more metallic the bonding. *Metallic bonding* only occurs among an aggregate of atoms and is *nondirectional.*

Secondary bonds are formed by the electrostatic attraction of electric dipoles within atoms or molecules. *Fluctuating dipoles* bond atoms together due to an asymmetrical distribution of electron charge within atoms. These bonding forces are important for the liquefaction and solidification of noble gases. *Permanent dipole* bonds are important in the bonding of polar covalently bonded molecules such as water and hydrocarbons.

Mixed bonding commonly occurs between atoms and in molecules. For example, metals such as titanium and iron have mixed metallic-covalent bonds; covalently bonded compounds such as GaAs and ZnSe have a certain amount of ionic character; some intermetallic compounds such as $NaZn_{13}$ have some ionic bonding mixed with metallic bonding. In general, bonding occurs between atoms or molecules because their energies are lowered by the bonding process.

DEFINITIONS

Sec. 2.1

Atom: the basic unit of an element that can undergo chemical change.

Sec. 2.2

Atomic number: the number of protons in the nucleus of an atom of an element.
Atomic mass unit (u): mass unit based on the mass of exactly 12 for $^{12}_{6}C$.
Avogadro's number: 6.023×10^{23} atoms/mol; the number of atoms in one relative gram-mole or mole of an element.

Sec. 2.3

Quantum mechanics: a branch of physics in which systems under investigation can have only discrete allowed energy values which are separated by forbidden regions.
Ground state: the quantum state with the lowest energy.
Heisenberg's uncertainty principle: the statement that it is impossible to determine accurately at the same time the position and momentum of a small particle such as an electron.
Photon: a particle of radiation with an associated wavelength and frequency. Also referred to as a *quantum* of radiation.
Quantum numbers: the set of four numbers necessary to characterize each electron in an atom. These are the principal quantum number n, the orbital quantum number l, the magnetic quantum number m_l, and the spin quantum number m_s.
Pauli exclusion principle: the statement that no two electrons can have the same four quantum numbers.
Electron shell: a group of electrons with the same principal quantum number n.
Electron configuration: the distribution of all the electrons in an atom according to their atomic orbitals.
Atomic orbital: the region in space about the nucleus of an atom in which an electron with a given set of quantum numbers is most likely to be found. An atomic orbital is also associated with a certain energy level.
Ionization energy: the energy required to remove an electron from its ground state in an atom to infinity.

Sec. 2.5

Anion: an ion with a negative charge.
Cation: an ion with a positive charge.
Ionic bond: a primary bond resulting from the electrostatic attraction of oppositely charged ions. It is a nondirectional bond. An example of an ionically bonded material is a NaCl crystal.

Sec. 2.6

Covalent bond: a primary bond resulting from the sharing of electrons. In most cases the covalent bond involves the overlapping of half-filled orbitals of two atoms. It is a directional bond. An example of a covalently bonded material is diamond.

Hybrid orbital: an atomic orbital obtained when two or more nonequivalent orbitals of an atom combine. The process of the rearrangement of the orbitals is called *hybridization*.

Sec. 2.7

Valence electrons: electrons in the outermost shells which are most often involved in bonding.

Positive-ion core: an atom without its valence electrons.

Metallic bond: a primary bond resulting from the sharing of delocalized outer electrons in the form of an electron charge cloud by an aggregate of metal atoms. It is a nondirectional bond. An example of a metallically bonded material is elemental sodium.

Sec. 2.8

Permanent dipole bond: a secondary bond created by the attraction of molecules which have permanent dipoles. That is, each molecule has positive and negative charge centers separated by a distance.

Hydrogen bond: a special type of intermolecular permanent dipole attraction that occurs between a hydrogen atom bonded to a highly electronegative element (F,O,N, or Cl) and another atom of a highly electronegative element.

PROBLEMS

2.1.1 What is the mass in grams of (*a*) a proton, (*b*) a neutron, and (*c*) an electron?

2.1.2 What is the electric charge in coulombs of (*a*) a proton, (*b*) a neutron, and (*c*) an electron?

2.2.1 Define (*a*) atomic number, (*b*) atomic mass unit, (*c*) Avogadro's number, and (*d*) relative gram atomic mass.

2.2.2 What is the mass in grams of one atom of silver?

2.2.3 How many atoms are there in 1 g of silver?

2.2.4 A nickel wire is 0.90 mm in diameter and 10 cm in length. How many atoms does it contain? The density of nickel is 8.90 g/cm^3.

2.2.5 What is the mass in grams of one atom of tin?

2.2.6 How many atoms are there in 1 g of tin?

2.2.7 A solder contains 55 wt % tin and 45 wt % lead. What are the atomic percentages of Sn and Pb in the solder?

2.2.8 A monel alloy consists of 65 wt % Ni and 35 wt % Cu. What are the atomic percentages of Ni and Cu in this alloy?

2.2.9 A brass consists of 80 wt % Cu and 20 wt % Zn. What are the atomic percentages of Cu and Zn in the alloy?

2.2.10 What is the chemical formula of an intermetallic compound which consists of 54.07 wt % Cu and 45.93 wt % Al?

2.2.11 What is the chemical formula of an intermetallic compound which consists of 45.3 wt % Mg and 54.7 wt % Ni?

2.3.1 Define a photon.

2.3.2 Calculate the energy in joules and electron volts of the photon whose wavelength is 286.3 nm.

2.3.3 Calculate the energy in joules and electron volts of the photon whose wavelength is 425.2 nm.

2.3.4 A hydrogen atom exists with its electron in the $n = 5$ state. The electron undergoes a transition to the $n = 3$ state. Calculate (a) the energy of the photon emitted, (b) its frequency, and (c) its wavelength in nanometers (nm).

2.3.5 A hydrogen atom exists with its electron in the $n = 6$ state. The electron undergoes a transition to the $n = 4$ state. Calculate (a) the energy of the photon emitted, (b) its frequency, and (c) its wavelength in nanometers.

2.3.6 In a commercial x-ray generator, a stable metal such as Cu or Fe is exposed to an intense beam of high-energy electrons. These electrons cause ionization events in the metal atoms. When the metal atoms regain their ground state, they emit x-rays of characteristic energy and wavelength. For example, an Fe atom struck by a high-energy electron may lose one of its K shell electrons. When this happens, another electron, probably from the Fe L shell, will "fall" into the vacant site in the K shell. If such a $2p \rightarrow 1s$ transition in Fe occurs, an Fe K_α x-ray is emitted. An Fe K_α x-ray has a wavelength λ of 0.1937 nm. What is its energy? What is its frequency?

2.3.7 Most modern scanning electron microscopes (SEM) are equipped with energy-dispersive x-ray detectors for the purpose of chemical analysis of the specimens. This x-ray analysis is a natural extension of the capability of the SEM because the electrons that are used to form the image are also capable of creating characteristic x-rays in the sample. When the electron beam hits the specimen, x-rays specific to the elements in the specimen are created. These can be detected and used to deduce the composition of the specimen from the well-known wavelengths of the characteristic x-rays of the elements. For example:

Element	Wavelength of K_α x-rays
Cr	0.2291 nm
Mn	0.2103 nm
Fe	0.1937 nm
Co	0.1790 nm
Ni	0.1659 nm
Cu	0.1542 nm
Zn	0.1436 nm

Suppose a metallic alloy is examined in an SEM and three different x-ray energies are detected. If the three energies are 5911, 6417, and 6994 eV, what elements are present in the sample? What would you call such an alloy?

2.3.8 Describe the Bohr model of the hydrogen atom. What are the major shortcomings of the Bohr model?

2.3.9 What is the ionization energy in electron volts for a hydrogen electron in the ground state?

2.3.10 Describe the four quantum numbers of an electron and give their allowed values.

2.3.11 Write the electron configurations of the following elements by using *spdf* notion: (*a*) yttrium, (*b*) hafnium, (*c*) samarium, (*d*) rhenium.

2.3.12 What is the outer electron configuration for all the noble gases except helium?

2.3.13 Of the noble gases Ne, Ar, Kr, and Xe, which should be the most chemically reactive?

2.3.14 Define the term *electronegativity*.

2.3.15 Which five elements are the most electropositive according to the electronegativity scale?

2.3.16 Which five elements are the most electronegative according to the electronegativity scale?

2.3.17 Write the electron configuration of the following ions by using *spdf* notation: (*a*) Cr^{2+}, Cr^{3+}, Cr^{6+}; (*b*) Mo^{3+}, Mo^{4+}, Mo^{6+}; (*c*) Se^{4+}, Se^{6+}, Se^{2-}

2.4.1 Briefly describe the following types of primary bonding: (*a*) ionic, (*b*) covalent, and (*c*) metallic.

2.4.2 Briefly describe the following types of secondary bonding: (*a*) fluctuating dipole, and (*b*) permanent dipole.

2.4.3 In general, why does bonding between atoms occur?

2.5.1 Describe the ionic bonding process between a pair of Na and Cl atoms. Which electrons are involved in the bonding process?

2.5.2 After ionization, why is the sodium ion smaller than the sodium atom?

2.5.3 After ionization, why is the chloride ion larger than the chlorine atom?

2.5.4 Calculate the attractive force ($\bullet\!\!\rightarrow \leftarrow\!\!\bullet$) between a pair of K^+ and Cl^- ions that just touch each other. Assume the ionic radius of the K^+ ion to be 0.133 nm and that of the Cl^- ion to be 0.181 nm.

2.5.5 Calculate the attractive force ($\bullet\!\!\rightarrow \leftarrow\!\!\bullet$) between a pair of Ca^{2+} and S^{2-} ions that just touch each other. Assume the ionic radius of the Ca^{2+} ion to be 0.106 nm and that of the S^{2-} ion to be 0.184 nm.

2.5.6 Calculate the net potential energy for a K^+Cl^- pair by using the *b* constant calculated from Prob. 2.5.4. Assume $n = 9.0$.

2.5.7 Calculate the net potential energy for a $Ca^{2+}S^{2-}$ ion pair by using the *b* constant calculated from Prob. 2.5.5. Assume $n = .9.0$.

2.5.8 If the attractive force between a pair of Na^+ and F^- ions is 4.439×10^{-9} N and the ionic radius of the Na^+ ion is 0.095 nm, calculate the ionic radius of the F^- ion in nanometers.

2.5.9 If the attractive force between a pair of Sr^{2+} and S^{2-} ions is 9.544×10^{-9} N and the ionic radius of the S^{2-} ion is 0.184 nm, calculate the ionic radius of the Sr^{2+} ion in nanometers.

2.5.10 Describe the two major factors that must be taken into account in the packing of ions in an ionic crystal.

2.6.1 Describe the covalent-bonding process between a pair of hydrogen atoms. What is the driving energy for the formation of a diatomic molecule?

2.6.2 Describe the covalent-bonding electron arrangement in the following diatomic molecules: (*a*) fluorine, (*b*) oxygen, (*c*) nitrogen.

2.6.3 Describe the hybridization process for the formation of four equivalent sp^3 hybrid orbitals in carbon during covalent bonding. Use orbital diagrams.

2.6.4 List the number of atoms bonded to a C atom that exhibits sp^3, sp^2, and sp hybridization. For each, give the geometrical arrangement of the atoms in the molecule.

2.6.5 Why is diamond such a hard material?

2.7.1 Describe the metallic bonding process among an aggregate of copper atoms.

2.7.2 How can the high electrical and thermal conductivities of metals be explained by the "electron gas" model of metallic bonding? Ductility?

2.7.3 The melting point of the metal potassium is 63.5°C, while that of titanium is 1812°C. What explanation can be given for this great difference in melting temperatures?

2.7.4 Is there a correlation between the electron configurations of the elements potassium ($Z = 19$) through copper ($Z = 29$) and their melting points? (See Table 2.8).

2.8.1 Define an electric dipole moment.

2.8.2 Describe fluctuating dipole bonding among the atoms of the noble gas neon. Of a choice between the noble gases krypton and xenon, which noble gas would be expected to have the strongest dipole bonding and why?

2.8.3 Describe permanent dipole bonding among polar covalent molecules.

2.8.4 Carbon tetrachloride (CCl_4) has a zero dipole moment. What does this tell us about the C—Cl bonding arrangement in this molecule?

2.8.5 Describe the hydrogen bond. Among what elements is this bond restricted?

2.8.6 Describe hydrogen bonding among water molecules.

2.8.7 Methane (CH_4) has a much lower boiling temperature than does water (H_2O). Explain why this is true in terms of the bonding between molecules in each of these two substances.

2.8.8 Sketch a tetrahedron and make a small model of one from cardboard and tape with a 2-in side. Do the same for an octahedron.

2.9.1 What is Pauling's equation for determining the percentage ionic character in a mixed ionic-covalently bonded compound?

2.9.2 Compare the percentage ionic character in the semiconducting compounds ZnS and GaP.

2.9.3 Compare the percentage ionic character in the semiconducting compounds CdS and InAs.

2.9.4 For each of the following compounds, state whether the bonding is essentially metallic, covalent, ionic, van der Waals, or hydrogen: (*a*) Ni, (*b*) ZrO_2, (*c*) graphite, (*d*) solid Kr, (*e*) Si, (*f*) BN, (*g*) SiC, (*h*) Fe_2O_3, (*i*) MgO, (*j*) W, (*k*) H_2O within the molecules, (*l*) H_2O between the molecules. If ionic and covalent bonding are involved in the bonding of any of the compounds above, calculate the percentage ionic character in the compound.

3

Crystal Structures and Crystal Geometry

3.1 THE SPACE LATTICE AND UNIT CELLS

The physical structure of solid materials of engineering importance depends mainly on the arrangements of the atoms, ions, or molecules which make up the solid and the bonding forces between them. If the atoms or ions of a solid are arranged in a pattern that repeats itself in three dimensions, they form a solid which is said to have a *crystal structure* and is referred to as a *crystalline solid* or *crystalline material.* Examples of crystalline materials are metals, alloys, and some ceramic materials.

Atomic arrangements in crystalline solids can be described by referring the atoms to the points of intersection of a network of lines in three dimensions. Such a network is called a *space lattice* (Fig. 3.1a) and can be described as an infinite three-dimensional array of points. Each point in the space lattice has identical surroundings. In an ideal crystal the grouping of lattice points about any given point are identical with the grouping about any other lattice point in the crystal lattice. Each space lattice can thus be described by specifying the atom positions in a repeating *unit cell*, such as the one heavily

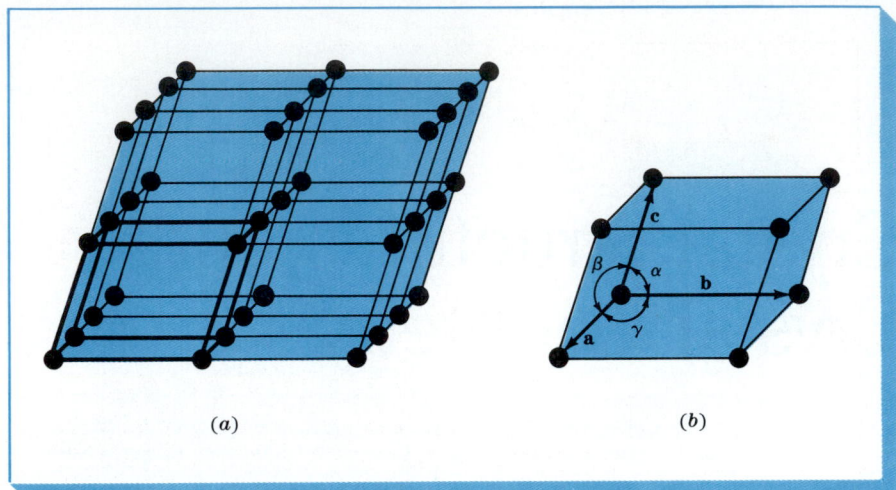

FIGURE 3.1 (*a*) Space lattice of ideal crystalline solid. (*b*) Unit cell showing lattice constants.

(*a*)

(*b*)

outlined in Fig. 3.1*a*. The size and shape of the unit cell can be described by three lattice vectors **a, b,** and **c,** originating from one corner of the unit cell (Fig. 3.1*b*). The axial lengths *a, b,* and *c* and the interaxial angles α, β, and γ are the *lattice constants* of the unit cell.

3.2 CRYSTAL SYSTEMS AND BRAVAIS LATTICES

By assigning specific values for axial lengths and interaxial angles, unit cells of different types can be constructed. Crystallographers have shown that only seven different types of unit cells are necessary to create all point lattices. These crystal systems are listed in Table 3.1.

Many of the seven crystal systems have variations of the basic unit cell. A. J. Bravais[1] showed that 14 standard unit cells could describe all possible lattice networks. These Bravais lattices are illustrated in Fig. 3.2. There are four basic types of unit cells: (1) simple, (2) body-centered, (3) face-centered, and (4) base-centered.

In the cubic system there are three types of unit cells: simple cubic, body-centered cubic, and face-centered cubic. In the orthorhombic system all four types are represented. In the tetragonal system there are only two: simple and body-centered. The face-centered tetragonal unit cell appears to be missing but can be constructed from four body-centered tetragonal unit cells. The

[1] August Bravais (1811–1863). French crystallographer who derived the 14 possible arrangements of points in space.

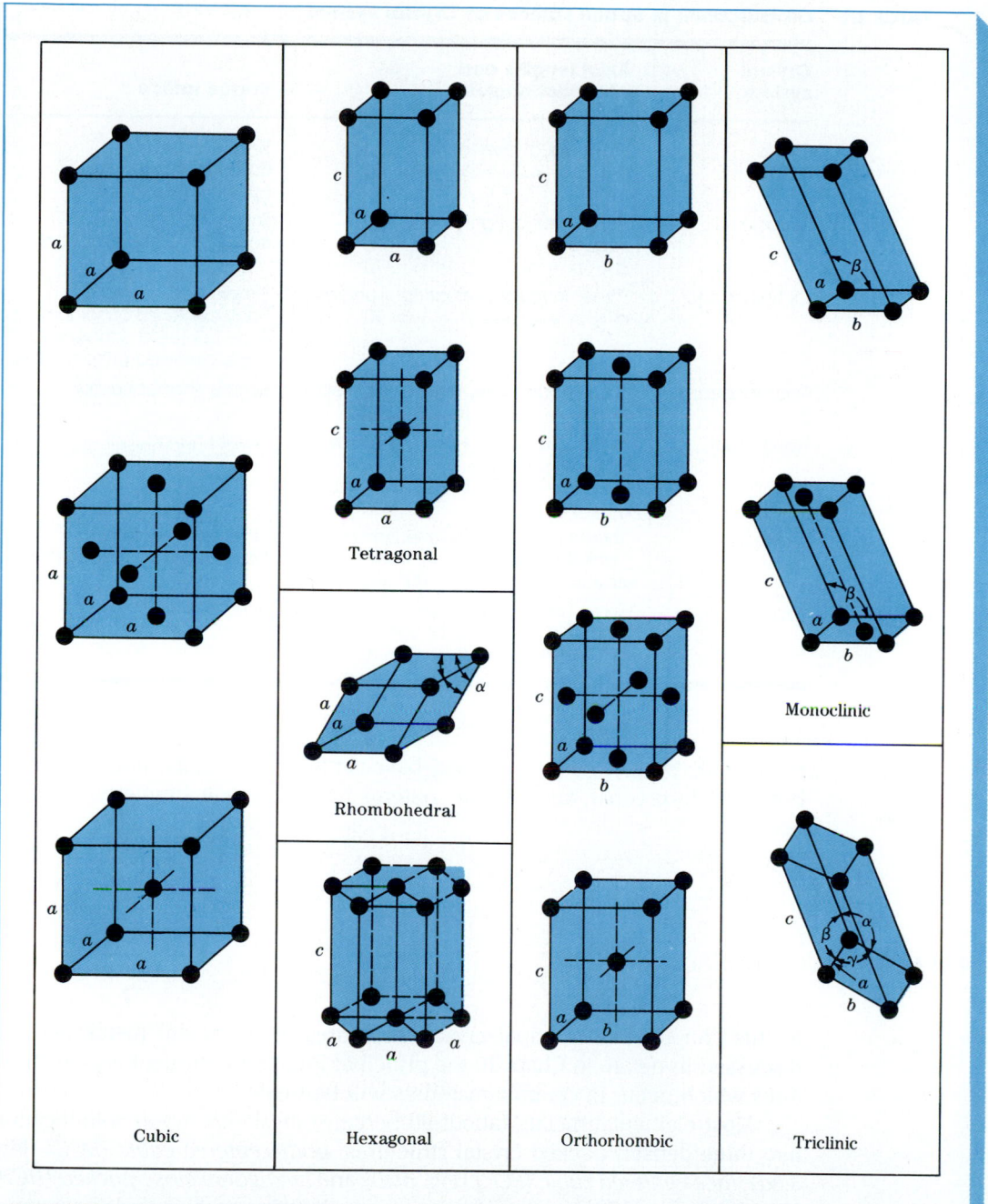

FIGURE 3.2 The 14 Bravais conventional unit cells grouped according to crystal system. The dots indicate lattice points that, when located on faces or at corners, are shared by other identical lattice unit cells. (*After W. G. Moffatt, G. W. Pearsall, and J. Wulff, "The Structure and Properties of Materials," vol. I: "Structure," Wiley, 1964, p. 47.*)

TABLE 3.1   Classification of Space Lattices by Crystal System

Crystal system	Axial lengths and interaxial angles	Space lattice
Cubic	Three equal axes at right angles $a = b = c, \alpha = \beta = \gamma = 90°$	Simple cubic Body-centered cubic Face-centered cubic
Tetragonal	Three axes at right angles, two equal $a = b \neq c, \alpha = \beta = \gamma = 90°$	Simple tetragonal Body-centered tetragonal
Orthorhombic	Three unequal axes at right angles $a \neq b \neq c, \alpha = \beta = \gamma = 90°$	Simple orthorhombic Body-centered orthorhombic Base-centered orthorhombic Face-centered orthorhombic
Rhombohedral	Three equal axes, equally inclined $a = b = c, \alpha = \beta = \gamma \neq 90°$	Simple rhombohedral
Hexagonal	Two equal axes at 120°, third axis at right angles $a = b \neq c, \alpha = \beta = 90°,$ $\gamma = 120°$	Simple hexagonal
Monoclinic	Three unequal axes, one pair not at right angles $a \neq b \neq c, \alpha = \gamma = 90° \neq \beta$	Simple monoclinic Base-centered monoclinic
Triclinic	Three unequal axes, unequally inclined and none at right angles $a \neq b \neq c, \alpha \neq \beta \neq \gamma \neq 90°$	Simple triclinic

monoclinic system has simple and base-centered unit cells, and the rhombohedral, hexagonal, and triclinic systems have only one simple type of unit cell.

3.3   PRINCIPAL METALLIC CRYSTAL STRUCTURES

In this chapter the principal crystal structures of elemental metals will be discussed in detail. In Chap. 10 the principal ionic and covalent crystal structures which occur in ceramic materials will be treated.

Most elemental metals (about 90 percent) crystallize upon solidification into three densely packed crystal structures: *body-centered cubic (BCC)* (Fig. 3.3*a*), *face-centered cubic (FCC)* (Fig. 3.3*b*) and *hexagonal close-packed (HCP)* (Fig. 3.3*c*). The HCP structure is a denser modification of the simple hexagonal crystal structure shown in Fig. 3.2. Most metals crystallize in these dense-packed structures because energy is released as the atoms come closer together and bond more tightly with each other. Thus, the densely packed structures are in lower and more stable energy arrangements.

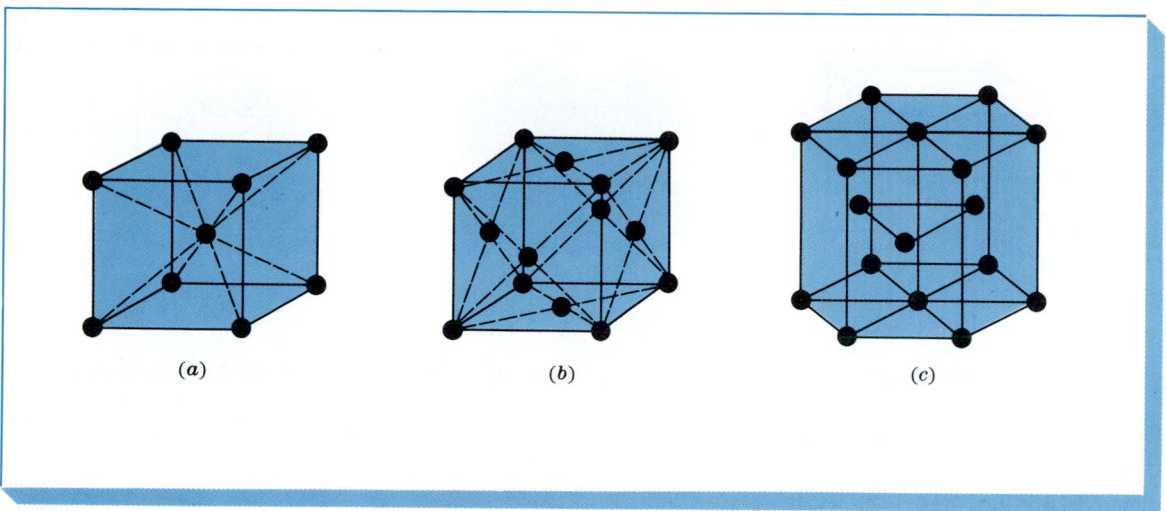

FIGURE 3.3 Principal metal crystal structure unit cells: (*a*) body-centered cubic, (*b*) face-centered cubic, (*c*) hexagonal close-packed.

The extremely small size of the unit cells of crystalline metals which are shown in Fig. 3.3 should be emphasized. The cube side of the unit cell of body-centered cubic iron, for example, at room temperature is equal to 0.287×10^{-9} m, or 0.287 nanometer (nm).[1] Therefore, if unit cells of pure iron are lined up side by side, in 1 mm there will be

$$1 \text{ mm} \times \frac{1 \text{ unit cell}}{0.287 \text{ nm} \times 10^{-6} \text{ mm/nm}} = 3.48 \times 10^6 \text{ unit cells!}$$

Let us now examine in detail the arrangement of the atoms in the three principal crystal structure unit cells. Although an approximation, we shall consider atoms in these crystal structures to be hard spheres. The distance between the atoms (interatomic distance) in crystal structures can be determined experimentally by x-ray diffraction analysis.[2] For example, the interatomic distance between two aluminum atoms in a piece of pure aluminum at 20°C is 0.2862 nm. The radius of the aluminum atom in the aluminum metal is assumed to be half the interatomic distance, or 0.143 nm. The atomic radii of selected metals are listed in Tables 3.2 to 3.4.

Body-Centered Cubic (BCC) Crystal Structure

First, consider the atomic-site unit cell for the BCC crystal structure shown in Fig. 3.4*a*. In this unit cell the solid spheres represent the centers where atoms are located and clearly indicate their relative positions. If we represent the

---

[1] 1 nanometer = 10^{-9} meter.
[2] Some of the principles of x-ray diffraction analysis will be studied in Sec. 3.11.

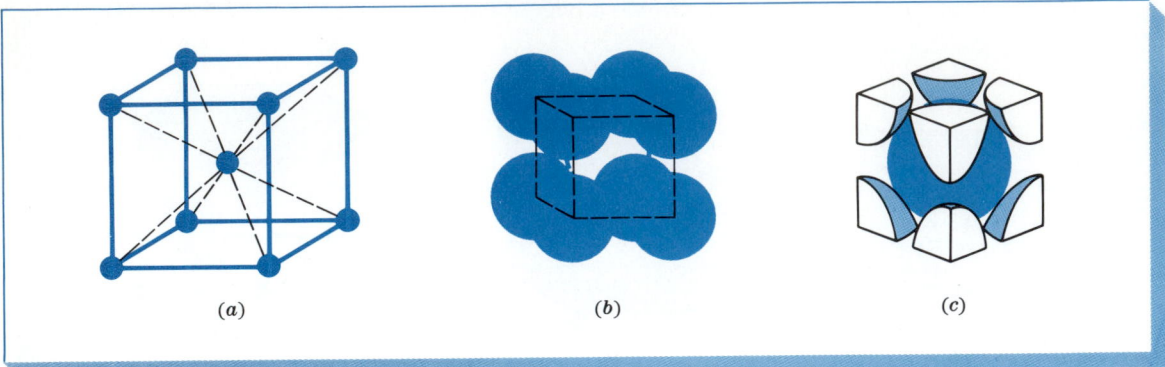

(a)   (b)   (c)

FIGURE 3.4 BCC unit cells: (*a*) atomic-site unit cell, (*b*) hard-sphere unit cell, and (*c*) isolated unit cell.

atoms in this cell as hard spheres, then the unit cell appears as shown in Fig. 3.4*b*. In this unit cell we see that the central atom is surrounded by eight nearest neighbors and is said to have a coordination number of 8.

If we isolate a single hard-sphere unit cell, we obtain the model shown in Fig. 3.4*c*. Each of these cells has the equivalent of two atoms per unit cell. One complete atom is located at the center of the unit cell, and an eighth of a sphere is located at each corner of the cell, making the equivalent of another atom. Thus there is a total of 1 (at the center) $+ 8 \times \frac{1}{8}$ (at the corners) $= 2$

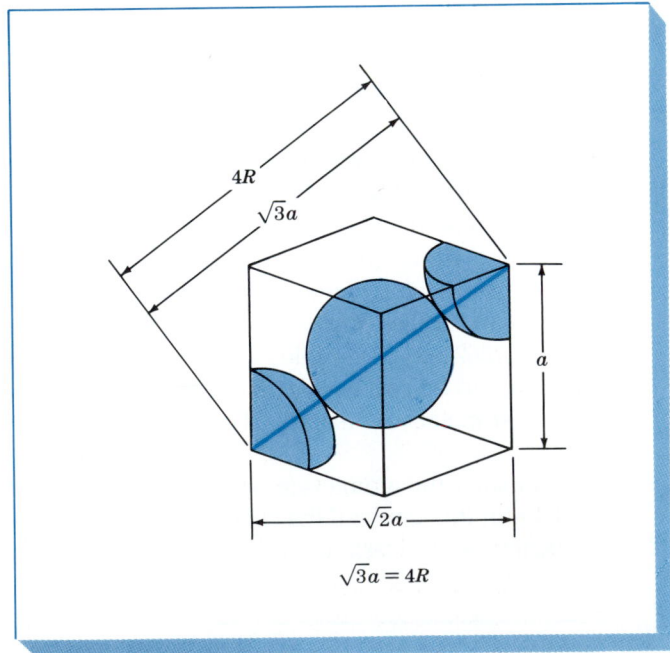

$4R$

$\sqrt{3}a$

a

$\sqrt{2}a$

$\sqrt{3}a = 4R$

FIGURE 3.5 BCC unit cell showing relationship between the lattice constant *a* and the atomic radius *R*.

atoms per unit cell. The atoms in the BCC unit cell contact each other across the cube diagonal, as indicated in Fig. 3.5, so that the relationship between the length of the cube side a and the atomic radius R is

$$\sqrt{3}a = 4R \quad \text{or} \quad a = \frac{4R}{\sqrt{3}} \tag{3.1}$$

Example Problem 3.1

Iron at 20°C is BCC with atoms of atomic radius 0.124 nm. Calculate the lattice constant a for the cube edge of the iron unit cell.

Solution:

From Fig. 3.5 it is seen that the atoms in the BCC unit cell touch across the cube diagonals. Thus, if a is the length of the cube edge, then

$$\sqrt{3}a = 4R \tag{3.1}$$

where R is the radius of the iron atom. Therefore

$$a = \frac{4R}{\sqrt{3}} = \frac{4(0.124 \text{ nm})}{\sqrt{3}} = 0.2864 \text{ nm} \blacktriangleleft$$

TABLE 3.2   Selected Metals Which Have the BCC Crystal Structure at Room Temperature (20°C) and Their Lattice Constants and Atomic Radii

Metal	Lattice constant a, nm	Atomic radius R,* nm
Chromium	0.289	0.125
Iron	0.287	0.124
Molybdenum	0.315	0.136
Potassium	0.533	0.231
Sodium	0.429	0.186
Tantalum	0.330	0.143
Tungsten	0.316	0.137
Vanadium	0.304	0.132

*Calculated from lattice constants by using Eq. (3.1), $R = \sqrt{3}a/4$.

If the atoms in the BCC unit cell are considered to be spherical, an atomic packing factor (APF) can be calculated by using the equation

$$\text{Atomic packing factor (APF)} = \frac{\text{volume of atoms in unit cell}}{\text{volume of unit cell}} \qquad (3.2)$$

Using this equation, the APF for the BCC unit cell (Fig. 3.3a) is calculated to be 68 percent (see Example Problem 3.2). That is, 68 percent of the volume of the BCC unit cell is occupied by atoms and the remaining 32 percent is empty space. The BCC crystal structure is *not* a close-packed structure since the atoms could be packed closer together. Many metals such as iron, chromium, tungsten, molybdenum, and vanadium have the BCC crystal structure at room temperature. Table 3.2 lists the lattice constants and atomic radii of selected BCC metals.

Example Problem 3.2

Calculate the atomic packing factor (APF) for the BCC unit cell, assuming the atoms to be hard spheres.

Solution:

$$\text{APF} = \frac{\text{volume of atoms in BCC unit cell}}{\text{volume of BCC unit cell}} \qquad (3.2)$$

Since there are two atoms per BCC unit cell, the volume of atoms in the unit cell of radius R is

$$V_{\text{atoms}} = (2)(\tfrac{4}{3}\pi R^3) = 8.373R^3$$

The volume of the BCC unit cell is

$$V_{\text{unit cell}} = a^3$$

where a is the lattice constant. The relationship between a and R is obtained from Fig. 3.5, which shows that the atoms in the BCC unit cell touch each other across the cubic diagonal. Thus

$$\sqrt{3}a = 4R \qquad \text{or} \qquad a = \frac{4R}{\sqrt{3}} \qquad (3.1)$$

Thus

$$V_{\text{unit cell}} = a^3 = 12.32R^3$$

The atomic packing factor for the BCC unit cell is, therefore,

$$\text{APF} = \frac{V_{atoms}/\text{unit cell}}{V_{unit\ cell}} = \frac{8.373R^3}{12.32R^3} = 0.68 ◄$$

Face-Centered Cubic (FCC) Crystal Structure

Consider next the FCC lattice-point unit cell of Fig. 3.6a. In this unit cell there is one lattice point at each corner of the cube and one at the center of each cube face. The hard-sphere model of Fig. 3.6b indicates that the atoms in the FCC crystal structure are packed as close together as possible. The APF for this close-packed structure is 0.74 as compared to 0.68 for the BCC structure which is not close-packed.

The FCC unit cell as shown in Fig. 3.6c has the equivalent of four atoms per unit cell. The eight corner octants account for one atom ($8 \times \frac{1}{8} = 1$), and the six half-atoms on the cube faces contribute another three atoms, making a total of four atoms per unit cell. The atoms in the FCC unit cell contact each other across the cubic face diagonal, as indicated in Fig. 3.7, so that the relationship between the length of the cube side a and the atomic radius R is

$$\sqrt{2}a = 4R \qquad \text{or} \qquad a = \frac{4R}{\sqrt{2}} \tag{3.3}$$

The APF for the FCC crystal structure is 0.74, which is greater than the 0.68 factor for the BCC structure. The APF of 0.74 is for the closest packing possible of "spherical atoms." Many metals such as aluminum, copper, lead, nickel, and iron at elevated temperatures (912 to 1394°C) crystallize with the FCC crystal structure. Table 3.3 lists the lattice constants and atomic radii for some selected FCC metals.

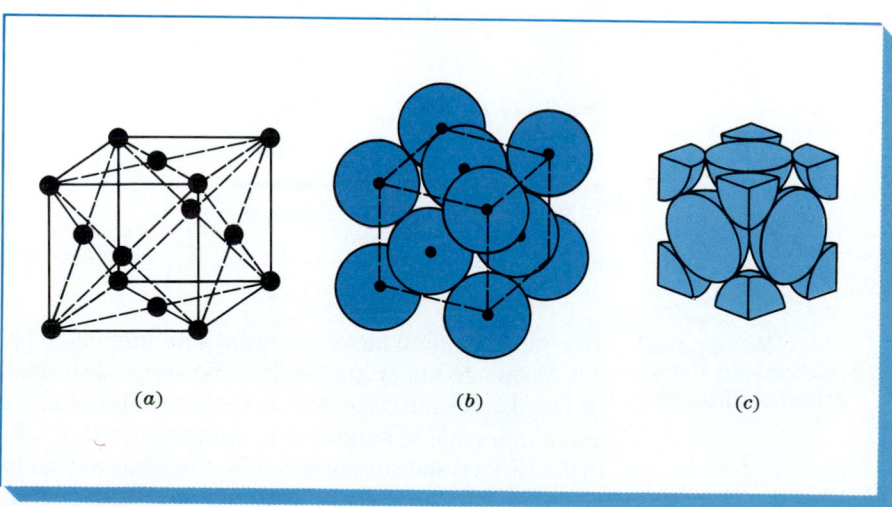

(a)   (b)   (c)

FIGURE 3.6 FCC unit cells: (a) atomic-site unit cell, (b) hard-sphere unit cell, and (c) isolated unit cell.

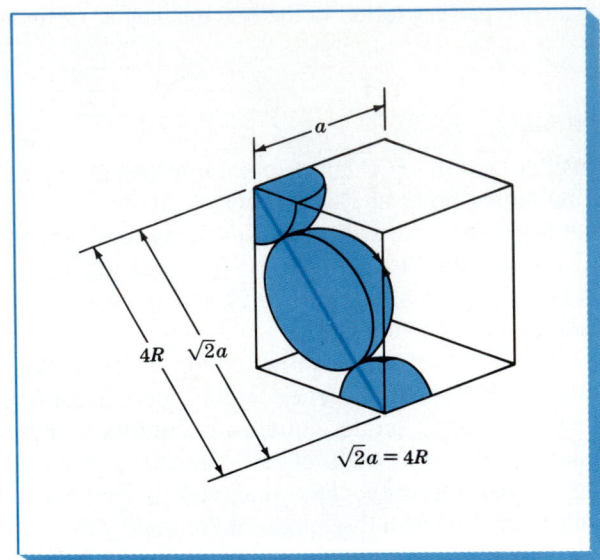

FIGURE 3.7 FCC unit cell showing relationship between the lattice constant a and atomic radius R. Since the atoms touch across the face diagonals $\sqrt{2}a = 4R$.

TABLE 3.3 Selected Metals Which Have the FCC Crystal Structure at Room Temperature (20°C) and Their Lattice Constants and Atomic Radii

Metal	Lattice constant a, nm	Atomic radius R,* nm
Aluminum	0.405	0.143
Copper	0.3615	0.128
Gold	0.408	0.144
Lead	0.495	0.175
Nickel	0.352	0.125
Platinum	0.393	0.139
Silver	0.409	0.144

*Calculated from lattice constants by using Eq. (3.3), $R = \sqrt{2}a/4$.

Hexagonal Close-Packed (HCP) Crystal Structure

The third common metallic crystal structure is the HCP structure shown in Fig. 3.8. Metals do not crystallize into the simple hexagonal crystal structure shown in Fig. 3.2 because the APF is too low. The atoms can attain a lower energy and a more stable condition by forming the HCP structure of Fig. 3.8. The APF of the HCP crystal structure is 0.74, the same as that for the FCC crystal structure since in both structures the atoms are packed as tightly as possible. In both

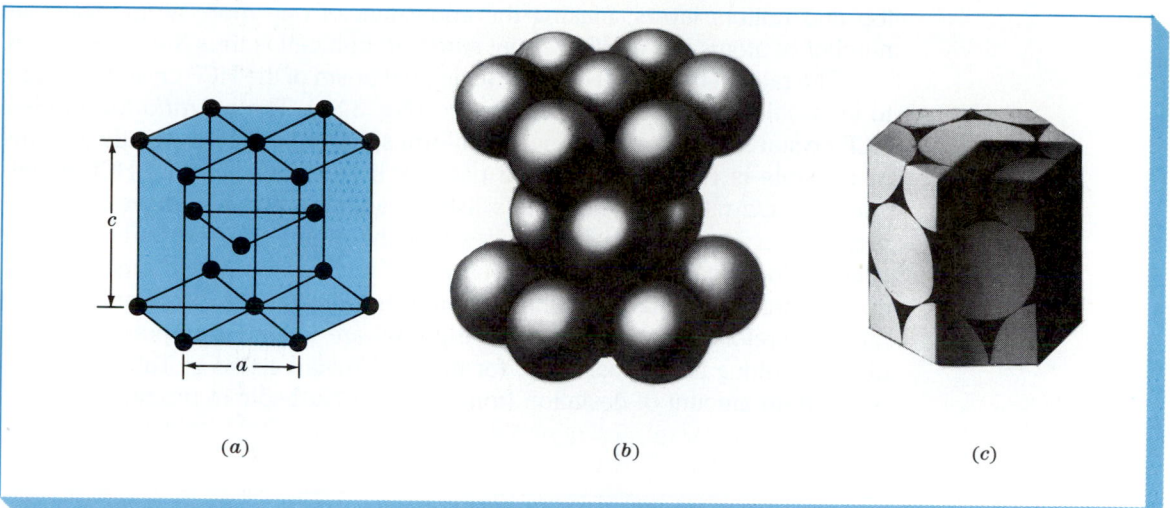

(a)   (b)   (c)

FIGURE 3.8 HCP unit cells: (a) atomic-site unit cell, (b) hard-sphere unit cell, and (c) isolated unit cell. [(b) and (c) After F. M. Miller, "Chemistry: Structure and Dynamics," McGraw-Hill, 1984, p. 296.]

the HCP and FCC crystal structures each atom is surrounded by 12 other atoms, and thus both structures have a coordination number of 12. The differences in the atomic packing in FCC and HCP crystal structures will be discussed in Sec. 3.8.

The isolated HCP unit cell is shown in Fig. 3.8c and has the equivalent of six atoms per unit cell. Three atoms form a triangle in the middle layer, as indicated by the atomic sites in Fig. 3.8a. There are six $\frac{1}{6}$-atom sections on both the top and bottom layers, making an equivalent of two more atoms ($2 \times 6 \times \frac{1}{6} = 2$). Finally, there is one-half an atom in the center of both the

TABLE 3.4 **Selected Metals Which Have the HCP Crystal Structure at Room Temperature (20°C) and Their Lattice Constants, Atomic Radii, and c/a Ratios**

Metal	Lattice constants, nm		Atomic radius R, nm	c/a ratio	% deviation from ideality
	a	c			
Cadmium	0.2973	0.5618	0.149	1.890	+15.7
Zinc	0.2665	0.4947	0.133	1.856	+13.6
Ideal HCP				1.633	0
Magnesium	0.3209	0.5209	0.160	1.623	−0.66
Cobalt	0.2507	0.4069	0.125	1.623	−0.66
Zirconium	0.3231	0.5148	0.160	1.593	−2.45
Titanium	0.2950	0.4683	0.147	1.587	−2.81
Beryllium	0.2286	0.3584	0.113	1.568	−3.98

top and bottom layers, making the equivalent of one more atom. The total number of atoms in the HCP crystal structure unit cell is thus $3 + 2 + 1 = 6$.

The ratio of the height c of the hexagonal prism of the HCP crystal structure to its basal side a is called the *c/a ratio* (Fig. 3.8a). The *c/a ratio* for an ideal HCP crystal structure consisting of uniform spheres packed as tightly together as possible is 1.633. Table 3.4 on page 79 lists some important HCP metals and their *c/a* ratios. Of the metals listed, cadmium and zinc have *c/a* ratios higher than ideality, which indicates that the atoms in these structures are slightly elongated along the *c* axis of the HCP unit cell. The metals magnesium, cobalt, zirconium, titanium, and beryllium have *c/a* ratios less than the ideal ratio. Therefore, in these metals the atoms are slightly compressed in the direction along the *c* axis. Thus, for the HCP metals listed in Table 3.4 there is a certain amount of deviation from the ideal hard-sphere model.

Example Problem 3.3

Calculate the volume of the zinc crystal structure unit cell by using the following data: pure zinc has the HCP crystal structure with lattice constants $a = 0.2665$ nm and $c = 0.4947$ nm.

Solution:

FIGURE 3.9 Diagrams for calculating the volume of an HCP unit cell. (*a*) HCP unit cell. (*b*) Base of HCP unit cell. (*c*) Triangle *ABC* removed from base of unit cell.

The volume of the zinc HCP unit cell can be obtained by determining the area of the base of the unit cell and then multiplying this by its height (Fig. 3.9).

The area of the base of the unit cell is area *ABDEFG* of Fig. 3.9a and b. This total area consists of the areas of six equilateral triangles of area *ABC* of Fig. 3.9b. From Fig. 3.9c,

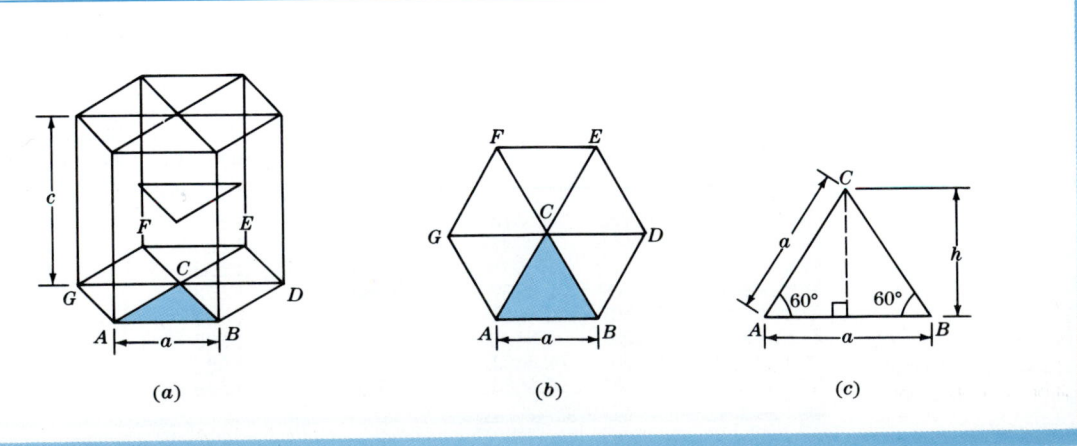

(*a*)   (*b*)   (*c*)

$$\text{Area of triangle } ABC = \tfrac{1}{2}(\text{base})(\text{height})$$

$$= \tfrac{1}{2}(a)(a \sin 60°) = \tfrac{1}{2}a^2 \sin 60°$$

From Fig. 3.9b,

$$\text{Total area of HCP base} = (6)(\tfrac{1}{2}a^2 \sin 60°)$$

$$= 3a^2 \sin 60°$$

From Fig. 3.9a,

$$\text{Volume of zinc HCP unit cell} = (3a^2 \sin 60°)(c)$$

$$= (3)(0.2665 \text{ nm})^2(0.8660)(0.4947 \text{ nm})$$

$$= 0.0913 \text{ nm}^3 \blacktriangleleft$$

3.4 ATOM POSITIONS IN CUBIC UNIT CELLS

To locate atom positions in cubic unit cells, we use rectangular x, y, and z axes. In crystallography the positive x axis is usually the direction coming out of the paper, the positive y axis is the direction to the right of the paper, and the positive z axis is the direction to the top (Fig. 3.10). Negative directions are opposite to those just described.

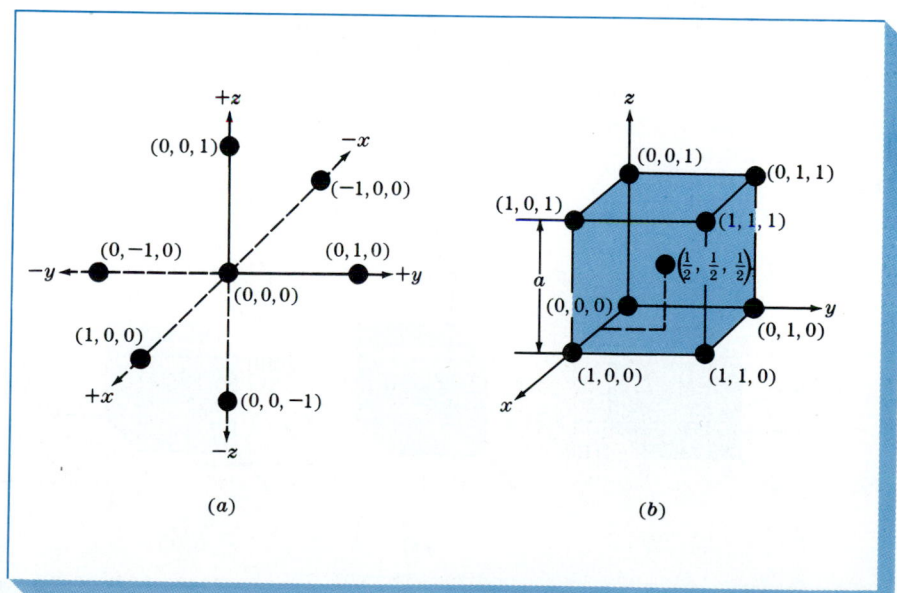

FIGURE 3.10 (*a*) Rectangular x, y, and z axes for locating atom positions in cubic unit cells. (*b*) Atom positions in a BCC unit cell.

Atom positions in unit cells are located by using unit distances along the x, y, and z axes, as indicated in Fig. 3.10a. For example, the position coordinates for the atoms in the BCC unit cell are shown in Fig. 3.10b. The atom positions for the eight corner atoms of the BCC unit cell are

$$
\begin{array}{cccc}
(0, 0, 0) & (1, 0, 0) & (0, 1, 0) & (0, 0, 1) \\
(1, 1, 1) & (1, 1, 0) & (1, 0, 1) & (0, 1, 1)
\end{array}
$$

The center atom in the BCC unit cell has the position coordinates $(\frac{1}{2}, \frac{1}{2}, \frac{1}{2})$. For simplicity sometimes only two atom positions in the BCC unit cell are specified which are $(0, 0, 0)$ and $(\frac{1}{2}, \frac{1}{2}, \frac{1}{2})$. The remaining atom positions of the BCC unit cell are assumed to be understood. In the same way the atom positions in the FCC unit cell can be located.

3.5 DIRECTIONS IN CUBIC UNIT CELLS

Frequently it is necessary to refer to specific directions in crystal lattices. This is especially important for metals and alloys that have properties which vary with crystallographic orientation. *For cubic crystals the crystallographic direction indices are the vector components of the direction resolved along each of the coordinate axes and reduced to the smallest integers.*

To diagrammatically indicate a direction in a cubic unit cell, we draw a direction vector from an origin which is usually a corner of the cubic cell until it emerges from the cube surface (Fig. 3.11). The position coordinates of the unit cell where the direction vector emerges from the cube surface after being converted to integers are the direction indices. The direction indices are enclosed by square brackets with no separating commas.

For example, the position coordinates of the direction vector *OR* in Fig.

FIGURE 3.11 Some directions in cubic unit cells.

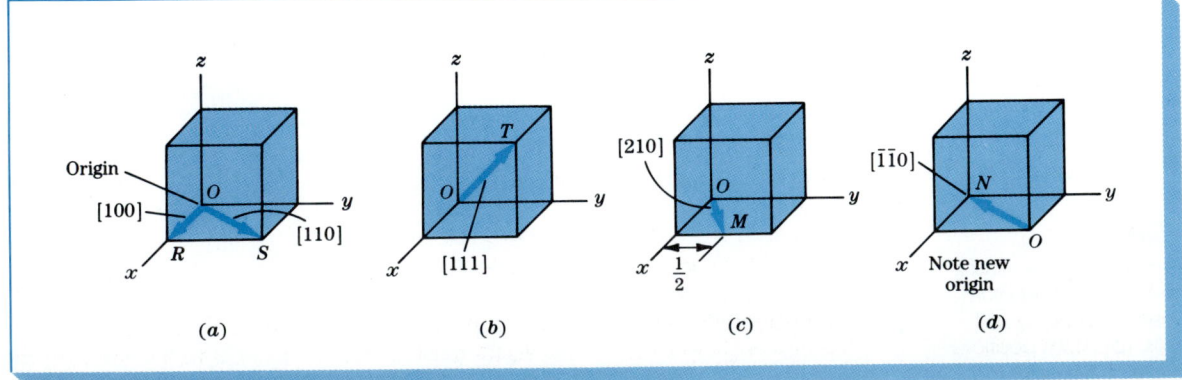

(a)   (b)   (c)   (d)

3.11*a* where it emerges from the cube surface are $(1, 0, 0)$, and so the direction indices for the direction vector *OR* are [100]. The position coordinates of the direction vector *OS* (Fig. 3.11*a*) are $(1, 1, 0)$, and so the direction indices for *OS* are [110]. The position coordinates for the direction vector *OT* (Fig. 3.11*b*) are $(1, 1, 1)$, and so the direction indices of *OT* are [111].

The position coordinates of the direction vector *OM* (Fig. 3.11*c*) are $(1, \frac{1}{2}, 0)$, and since the direction vectors must be integers, these position coordinates must be multiplied by 2 to obtain integers. Thus, the direction indices of *OM* become $2(1, \frac{1}{2}, 0) = [210]$. The position coordinates of the vector *ON* (Fig. 3.11*d*) are $(-1, -1, 0)$. A negative direction index is written with a bar over the index. Thus, the direction indices for the vector *ON* are [$\bar{1}\bar{1}0$]. Note that to draw the direction *ON* inside the cube the origin of the direction vector had to be moved to the front lower-right corner of the unit cube (Fig. 3.11*d*). Further examples of cubic direction vectors are given in Example Problem 3.4.

The letters u, v, w are used in a general sense for the direction indices in the x, y, and z directions, respectively, and are written as [uvw]. It is also important to note that *all parallel direction vectors have the same direction indices.*

Directions are said to be *crystallographically equivalent* if the atom spacing along each direction is the same. For example, the following cubic edge directions are crystallographic equivalent directions:

$$[100], [010], [001], [0\bar{1}0], [00\bar{1}], [\bar{1}00] \equiv \langle 100 \rangle$$

Equivalent directions are called *indices of a family or form.* The notation $\langle 100 \rangle$ is used to indicate cubic edge directions collectively. Other directions of a form are the cubic body diagonals $\langle 111 \rangle$ and the cubic face diagonals $\langle 110 \rangle$.

Example Problem 3.4

Draw the following direction vectors in cubic unit cells:
(*a*) [100] and [110]   (*b*) [112]   (*c*) [$\bar{1}$10]   (*d*) [$\bar{3}2\bar{1}$]

Solution:

(*a*) The position coordinates for the [100] direction are $(1, 0, 0)$ (Fig. 3.12*a*). The position coordinates for the [110] direction are $(1, 1, 0)$ (Fig. 3.12*a*).

(*b*) The position coordinates for the [112] direction are obtained by dividing the direction indices by 2 so that they will lie within the unit cube. Thus they are $(\frac{1}{2}, \frac{1}{2}, 1)$ (Fig. 3.12*b*).

(*c*) The position coordinates for the [$\bar{1}$10] direction are $(-1, 1, 0)$ (Fig. 3.12*c*). Note that the origin for the direction vector must be moved to the lower-left front corner of the cube.

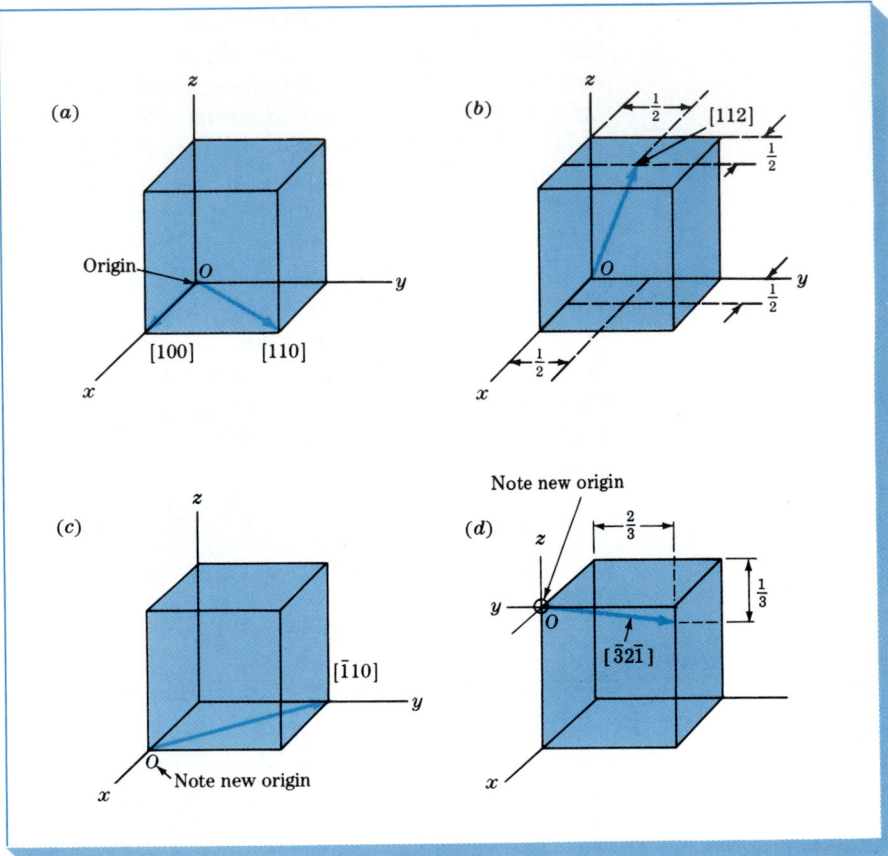

FIGURE 3.12 Direction vectors in cubic unit cells.

(*d*) The position coordinates for the [$\bar{3}2\bar{1}$] direction are obtained by first dividing all the indices by 3, the largest index. This gives $-1, \frac{2}{3}, -\frac{1}{3}$ for the position coordinates of the exit point of the direction [$\bar{3}2\bar{1}$] which are shown in Fig. 3.12*d*.

Example Problem 3.5

Determine the direction indices of the cubic direction shown in Fig. EP3.5*a*.

Solution:

Parallel directions have the same direction indices, and so we move the direction vector in a parallel manner until its tail reaches the nearest corner of the cube, still keeping the vector within the cube. Thus, in this case, the upper-left front corner becomes the new origin for the direction vector (Fig. EP3.5*b*). We can now determine the position coordinates where the direction vector leaves the unit cube. These are $x = -1$, $y =$

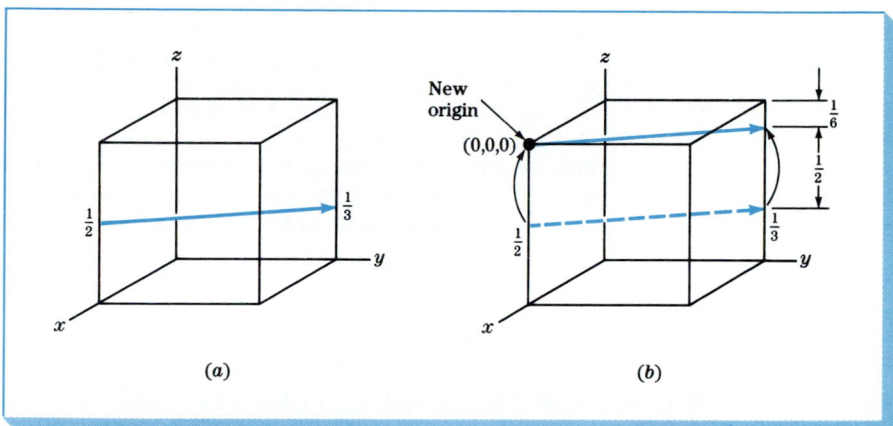

FIGURE EP3.5

$+1$, and $z = -\frac{1}{6}$. The position coordinates of the direction where it leaves the unit cube are thus $(-1, +1, -\frac{1}{6})$. The direction indices for this direction are, after clearing the fraction $6x$, $(-1, +1, -\frac{1}{6})$, or $[\bar{6}6\bar{1}]$.

Example Problem 3.6

Determine the direction indices of the cubic direction between the position coordinates $(\frac{3}{4}, 0, \frac{1}{4})$ and $(\frac{1}{4}, \frac{1}{2}, \frac{1}{2})$.

Solution:

First we locate the origin and termination points of the direction vector in a unit cube, as shown in Fig. EP3.6. The fraction vector components for this direction are

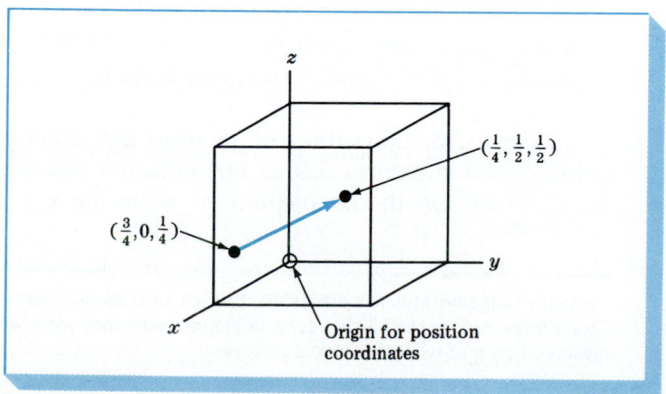

FIGURE EP3.6

$$x = -(\tfrac{3}{4} - \tfrac{1}{4}) = -\tfrac{1}{2}$$
$$y = (\tfrac{1}{2} - 0) = \tfrac{1}{2}$$
$$z = (\tfrac{1}{2} - \tfrac{1}{4}) = \tfrac{1}{4}$$

Thus, the vector direction has fractional vector components of $-\tfrac{1}{2}, \tfrac{1}{2}, \tfrac{1}{4}$. The direction indices will be in the same ratio as their fractional components. By multiplying the fraction vector components by 4, we obtain $[\bar{2}21]$ for the direction indices of this vector direction.

3.6   MILLER INDICES FOR CRYSTALLOGRAPHIC PLANES IN CUBIC UNIT CELLS

Sometimes it is necessary to refer to specific lattice planes of atoms within a crystal structure, or it may be of interest to know the crystallographic orientation of a plane or group of planes in a crystal lattice. To identify crystal planes in cubic crystal structures, the *Miller*[1] *notation system* is used. The *Miller indices of a crystal plane* are defined as the *reciprocals of the fractional intercepts (with fractions cleared) which the plane makes with the crystallographic x, y, and z axes of the three nonparallel edges of the cubic unit cell.* The cube edges of the unit cell represent unit lengths, and the intercepts of the lattice planes are measured in terms of these unit lengths.

The procedure for determining the Miller indices for a cubic crystal plane is as follows:

1. Choose a plane that does *not* pass through the origin at $(0, 0, 0)$.
2. Determine the intercepts of the plane in terms of the crystallographic *x, y,* and *z* axes for a unit cube. These intercepts may be fractions.
3. Form the reciprocals of these intercepts.
4. Clear fractions and determine the *smallest* set of whole numbers which are in the same ratio as the intercepts. These whole numbers are the Miller indices of the crystallographic plane and are enclosed in parentheses without the use of commas. The notation *(hkl)* is used to indicate Miller indices in a general sense, where *h, k,* and *l* are the Miller indices of a cubic crystal plane for the *x, y,* and *z* axes, respectively.

Figure 3.13 shows three of the most important crystallographic planes of cubic crystal structures. Let us first consider the shaded crystal plane in Fig. 3.13*a*, which has the intercepts 1, ∞, ∞ for the *x, y,* and *z* axes, respectively.

---

[1] William Hallowes Miller (1801–1880). English crystallographer who published a "Treatise on Crystallography" in 1839, using crystallographic reference axes which were parallel to the crystal edges and using reciprocal indices.

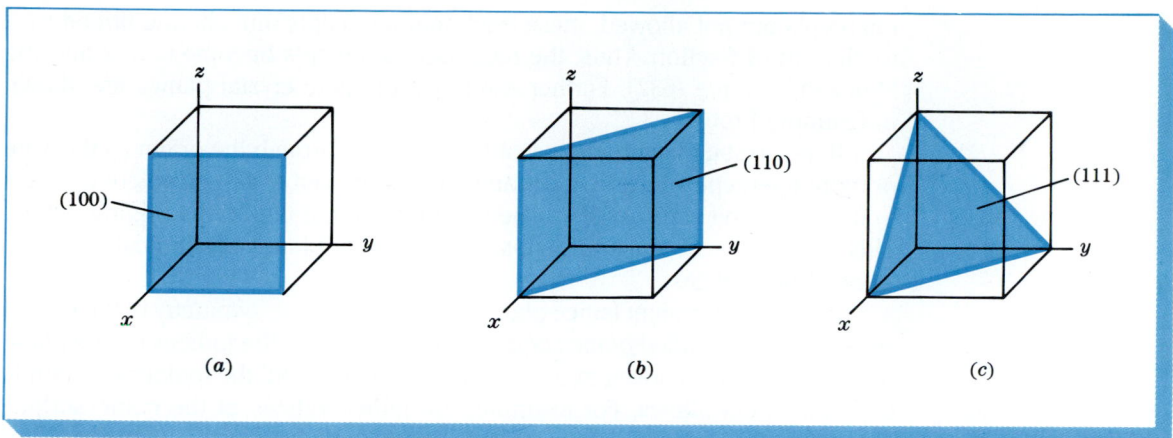

FIGURE 3.13 Miller indices of some important cubic crystal planes: (*a*) (100), (*b*) (110), and (*c*) (111).

We take the reciprocals of these intercepts to obtain the Miller indices, which are therefore 1, 0, 0. Since these numbers do not involve fractions, the Miller indices for this plane are (100), which is read as the one-zero-zero plane. Next let us consider the second plane shown in Fig. 3.13*b*. The intercepts of this plane are 1, 1, ∞. Since the reciprocals of these numbers are 1, 1, 0, which do not involve fractions, the Miller indices of this plane are (110). Finally, the third plane (Fig. 3.13*c*) has the intercepts 1, 1, 1, which give the Miller indices (111) for this plane.

Consider now the cubic crystal plane shown in Fig. 3.14 which has the intercepts $\frac{1}{3}$, $\frac{2}{3}$, 1. The reciprocals of these intercepts are 3, $\frac{3}{2}$, 1. Since fractional

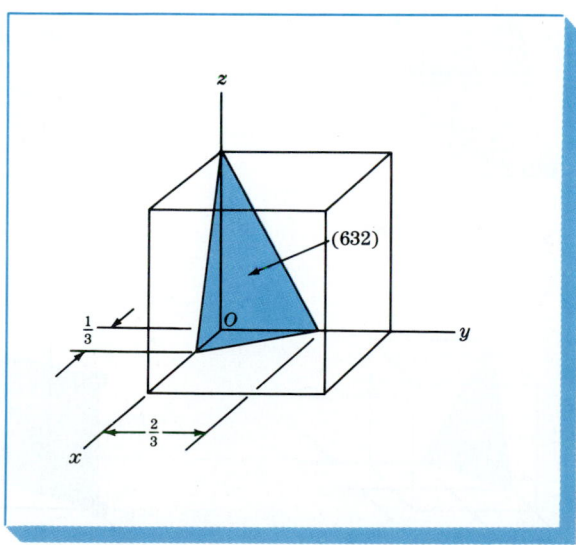

FIGURE 3.14 Cubic crystal plane (632) which has fractional intercepts.

intercepts are not allowed, these fractional intercepts must be multiplied by 2 to clear the $\frac{3}{2}$ fraction. Thus, the reciprocal intercepts become 6, 3, 2 and the Miller indices are (632). Further examples of cubic crystal planes are shown in Example Problem 3.7.

If the crystal plane being considered passes through the origin so that one or more intercepts are zero, the plane must be moved to an equivalent position in the same unit cell and the plane must remain parallel to the original plane. This is possible because all equispaced parallel planes are indicated by the same Miller indices.

If sets of equivalent lattice planes are related by the symmetry of the crystal system, they are called *planes of a family or form,* and the indices of one plane of the family are enclosed in braces as {*hkl*} to represent the indices of a family of symmetrical planes. For example, the Miller indices of the cubic surface planes (100), (010), and (001) are designated collectively as a family or form by the notation {100}.

Example Problem 3.7

Draw the following crystallographic planes in cubic unit cells:
(*a*) (101)   (*b*) (1$\bar{1}$0)   (*c*) (221)
(*d*) Draw a (110) plane in a BCC atomic-site unit cell, and list the position coordinates of the atoms whose centers are intersected by this plane.

Solutions:

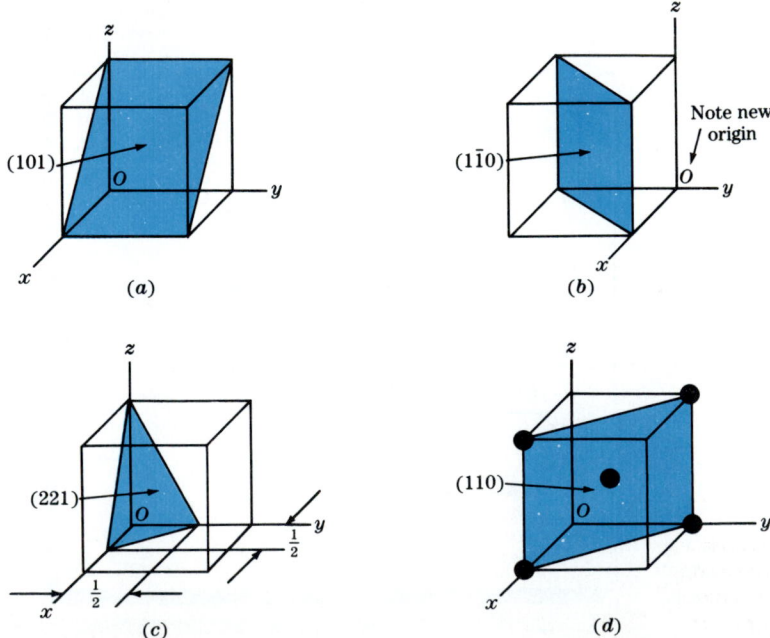

(a) First determine the reciprocals of the Miller indices of the (101) plane. These are 1, ∞, 1. The (101) plane must pass through a unit cube at intercepts $x = 1$ and $z = 1$ and be parallel to the y axis.

(b) First determine the reciprocals of the Miller indices of the ($1\bar{1}0$) plane. These are 1, -1, ∞. The ($1\bar{1}0$) plane must pass through a unit cube at intercepts $x = 1$ and $y = -1$ and be parallel to the z axis. Note that the origin of axes must be moved to the lower-right back side of the cube.

(c) First determine the reciprocals of the Miller indices of the (221) plane. These are $\frac{1}{2}, \frac{1}{2}, 1$. The (221) plane must pass through a unit cube at intercepts $x = \frac{1}{2}$, $y = \frac{1}{2}$, and $z = 1$.

(d) Atom positions whose centers are intersected by the (110) plane are (1, 0, 0), (0, 1, 0), (1, 0, 1), (0, 1, 1), and ($\frac{1}{2}, \frac{1}{2}, \frac{1}{2}$). These positions are indicated by the solid circles.

An important relationship for the cubic system, and *only the cubic system*, is that the direction indices of a direction *perpendicular* to a crystal plane are the same as the Miller indices of that plane. For example, the [100] direction is perpendicular to the (100) crystal plane.

In cubic crystal structures the *interplanar spacing* between two closest parallel planes with the same Miller indices is designated d_{hkl}, where h, k, and l are the Miller indices of the planes. This spacing represents the distance from a selected origin containing one plane and another parallel plane with the

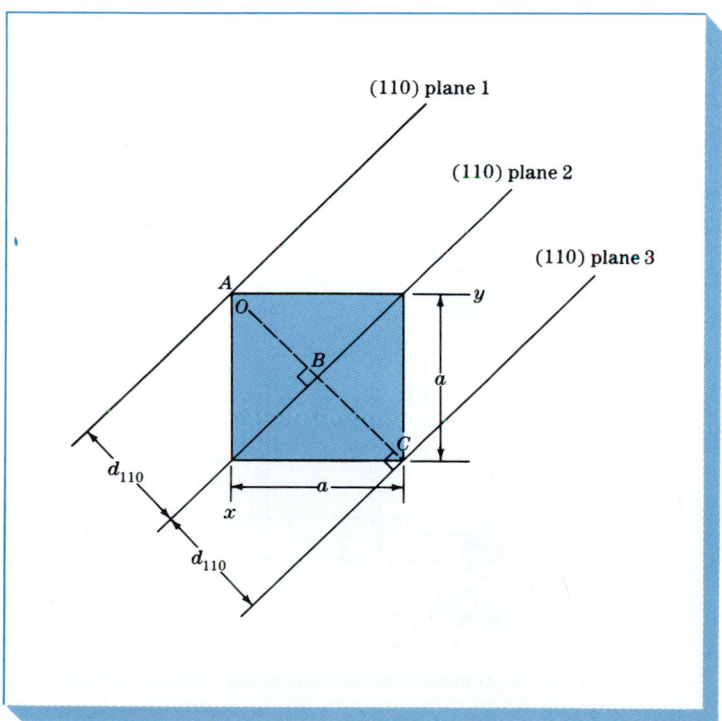

FIGURE 3.15 Top view of cubic unit cell showing the distance between (110) crystal planes, d_{110}.

same indices which is closest to it. For example, the distance between (110) planes 1 and 2, d_{110}, in Fig. 3.15 on page 89 is AB. Also, the distance between (110) planes 2 and 3 is d_{110} and is length BC in Fig. 3.15. From simple geometry, it can be shown that for cubic crystal structures

$$d_{hkl} = \frac{a}{\sqrt{h^2 + k^2 + l^2}}$$

(3.4)

where d_{hkl} = interplanar spacing between parallel closest planes with Miller indices h, k, and l

a = lattice constant (edge of unit cube)

h, k, l = Miller indices of cubic planes being considered

Example Problem 3.8

Determine the Miller indices of the cubic crystallographic plane shown in Fig. EP3.8a.

Solution:

First, transpose the plane parallel to the z axis $\frac{1}{4}$ unit to the right along the y axis as shown in Fig. EP3.8b so that the plane intersects the x axis at a unit distance from the new origin located at the lower-right back corner of the cube. The new intercepts of the transposed plane with the coordinate axes are now $(+1, -\frac{5}{12}, \infty)$. Next, we take the reciprocals of these intercepts to give $(1, -\frac{12}{5}, 0)$. Finally, we clear the $\frac{12}{5}$ fraction to obtain $(5\overline{12}0)$ for the Miller indices of this plane.

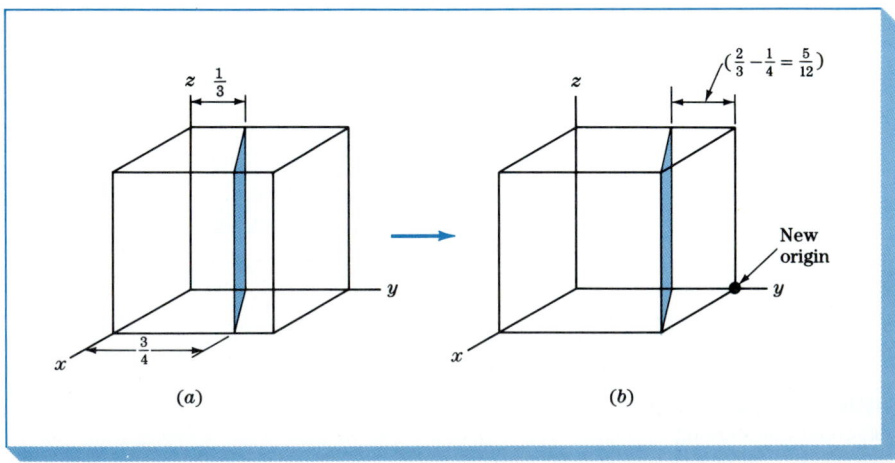

(a)   (b)

FIGURE EP3.8

Example Problem 3.9

Determine the Miller indices of the cubic crystal plane which intersects the position coordinates $(1, \frac{1}{4}, 0)$, $(1, 1, \frac{1}{2})$, $(\frac{3}{4}, 1, \frac{1}{4})$, and all coordinate axes.

Solution:

First, we locate the three position coordinates as indicated in Fig. EP3.9 at A, B, and C. Next, we join A and B and extend AB to D and then join A and C. Finally, we join A to C to complete plane ACD. The origin for this plane in the cube can be chosen at E, which gives axial intercepts for plane ACD at $x = -\frac{1}{2}$, $y = -\frac{3}{4}$, and $z = \frac{1}{2}$. The reciprocals of these axial intercepts are -2, $-\frac{4}{3}$, and 2. Multiplying these intercepts by 3 clears the fraction, giving Miller indices for the plane of $(\bar{6}\bar{4}6)$.

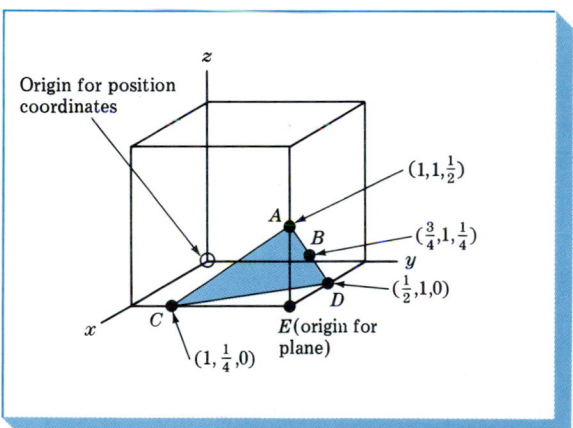

FIGURE EP3.9

Example Problem 3.10

Copper has an FCC crystal structure and a unit cell with a lattice constant of 0.361 nm. What is its interplanar spacing d_{220}?

Solution:

$$d_{hkl} = \frac{a}{\sqrt{h^2 + k^2 + l^2}} = \frac{0.361 \text{ nm}}{\sqrt{(2)^2 + (2)^2 + (0)^2}} = 0.128 \text{ nm} \blacktriangleleft$$

3.7 CRYSTALLOGRAPHIC PLANES AND DIRECTIONS IN HEXAGONAL UNIT CELLS

Indices for Crystal Planes in HCP Unit Cells

Crystal planes in HCP unit cells are commonly identified by using four indices instead of three. The HCP crystal plane indices, called *Miller-Bravais indices,* are denoted by the letters h, k, i, and l and are enclosed in parentheses as $(hkil)$. These four-digit hexagonal indices are based on a coordinate system with four axes, as shown in Fig. 3.16 in an HCP unit cell. There are three basal axes, a_1, a_2, and a_3, which make 120° with each other. The fourth axis or c axis is the vertical axis located at the center of the unit cell. The a unit of measurement along the a_1, a_2, and a_3 axes is the distance between the atoms along these axes and is indicated in Fig. 3.16. The unit of measurement along the c axis is the height of the unit cell. The reciprocals of the intercepts that a crystal plane makes with the a_1, a_2, and a_3 axes give the h, k, and i indices, while the reciprocal of the intercept with the c axis gives the l index.

Basal planes   The basal planes of the HCP unit cell are very important planes for this unit cell and are indicated in Fig. 3.17a. Since the basal plane on the top of the HCP unit cell in Fig. 3.17a is parallel to the a_1, a_2, and a_3 axes, the intercepts of this plane with these axes will all be infinite. Thus, $a_1 = \infty$, $a_2 = \infty$, and $a_3 = \infty$. The c axis, however, is unity since the top basal plane intersects the c axis at unit distance. Taking the reciprocals of these intercepts gives the Miller-Bravais indices for the HCP basal plane. Thus $h = 0$, $k = 0$, $i = 0$, and $l = 1$. The HCP basal plane is, therefore, a zero-zero-zero-one or (0001) plane.

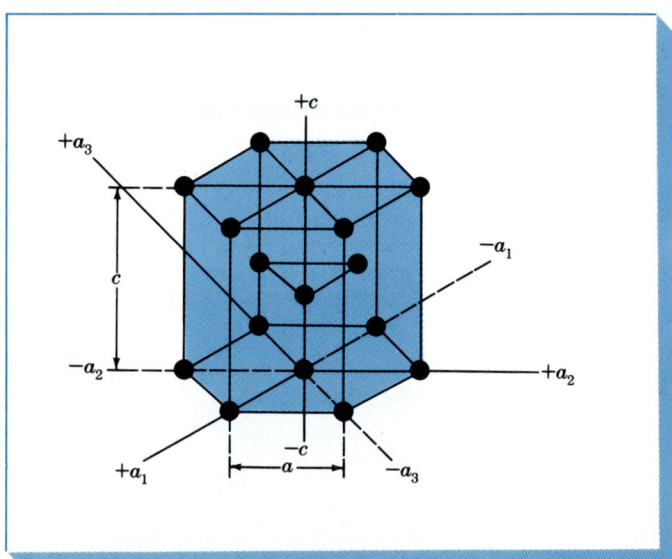

FIGURE 3.16 The four coordinate axes (a_1, a_2, a_3, and c) of the HCP crystal structure unit cell.

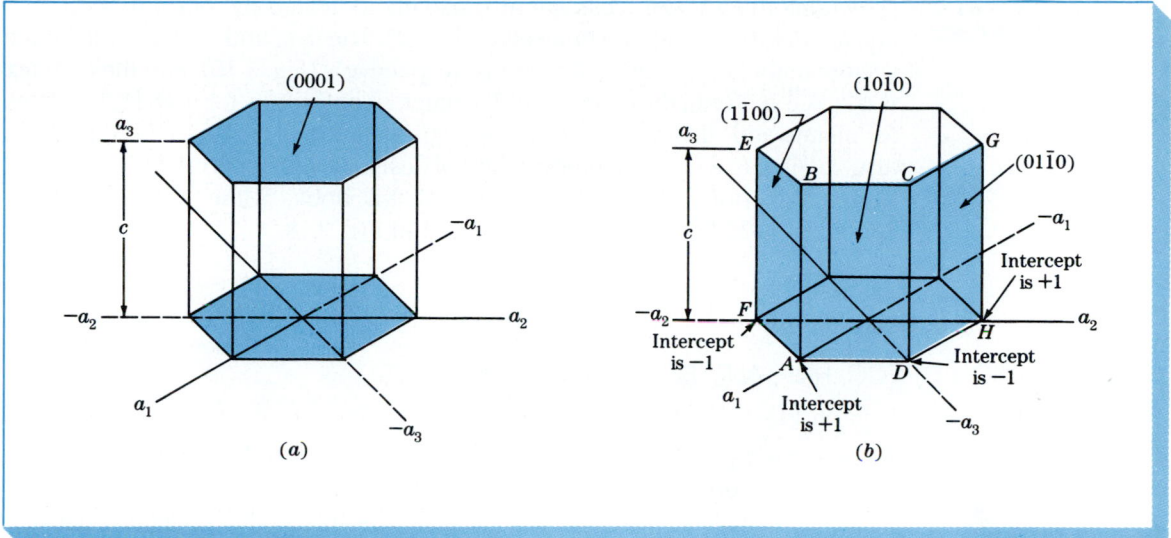

FIGURE 3.17 Miller-Bravais indices of hexagonal crystal planes: (a) basal planes, and (b) prism planes.

Prism planes   Using the same method, the intercepts of the front prism plane ($ABCD$) of Fig. 3.17b are $a_1 = +1$, $a_2 = \infty$, $a_3 = -1$, and $c = \infty$. Taking the reciprocals of these intercepts gives $h = 1$, $k = 0$, $i = -1$, and $l = 0$, or the ($10\bar{1}0$) plane. Similarly, the $ABEF$ prism plane of Fig. 3.17b has the indices ($1\bar{1}00$) and the $DCGH$ plane the indices ($01\bar{1}0$). All HCP prism planes can be identified collectively as the $\{10\bar{1}0\}$ family of planes.

Sometimes HCP planes are identified only by three indices (hkl) since $h + k = -i$. However, the ($hkil$) indices are used more commonly because they reveal the hexagonal symmetry of the HCP unit cell.

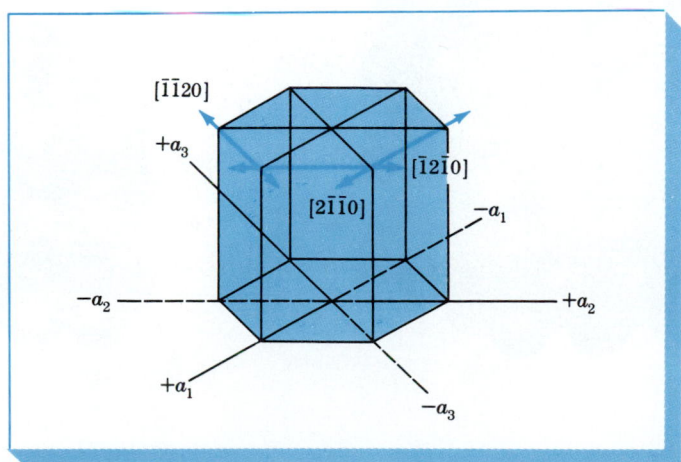

FIGURE 3.18 Some important directions in HCP unit cells.

Direction Indices in HCP Unit Cells

Directions in HCP unit cells are also usually indicated by four indices u, v, t, and w enclosed by square brackets as $[uvtw]$. The u, v, and t indices are lattice vectors in the a_1, a_2, and a_3 directions, respectively (Fig. 3.16), and the w index is a lattice vector in the c direction. To maintain uniformity for both HCP indices for planes and directions, it has been agreed that $u + v = -t$ for directions also, leading to a cumbersome method of designating directions whose method of determination is beyond the scope of this book. Some of the important directions in HCP unit cells are indicated in Fig. 3.18.

3.8   COMPARISON OF FCC, HCP, AND BCC CRYSTAL STRUCTURES

Face-Centered Cubic and Hexagonal Close-Packed Crystal Structures

As previously pointed out, both the HCP and FCC crystal structures are close-packed structures. That is, their atoms which are considered approximate "spheres" are packed together as closely as possible so that an atomic packing factor of 0.74 is attained.[1] The (111) planes of the FCC crystal structure shown

---

[1] As pointed out in Sec. 3.3, the atoms in the HCP structure deviate to varying degrees from ideality. In some HCP metals the atoms are elongated along the c axis, and in other cases they are compressed along the c axis (see Table 3.4).

FIGURE 3.19 Comparison of the (a) FCC crystal structure showing the close-packed (111) planes, and (b) the HCP crystal structure showing the close-packed (0001) planes. (*After W. G. Moffatt, G. W. Pearsall, and J. Wulff, "The Structure and Properties of Materials," vol. I: "Structure," Wiley, 1964, p. 51.*)

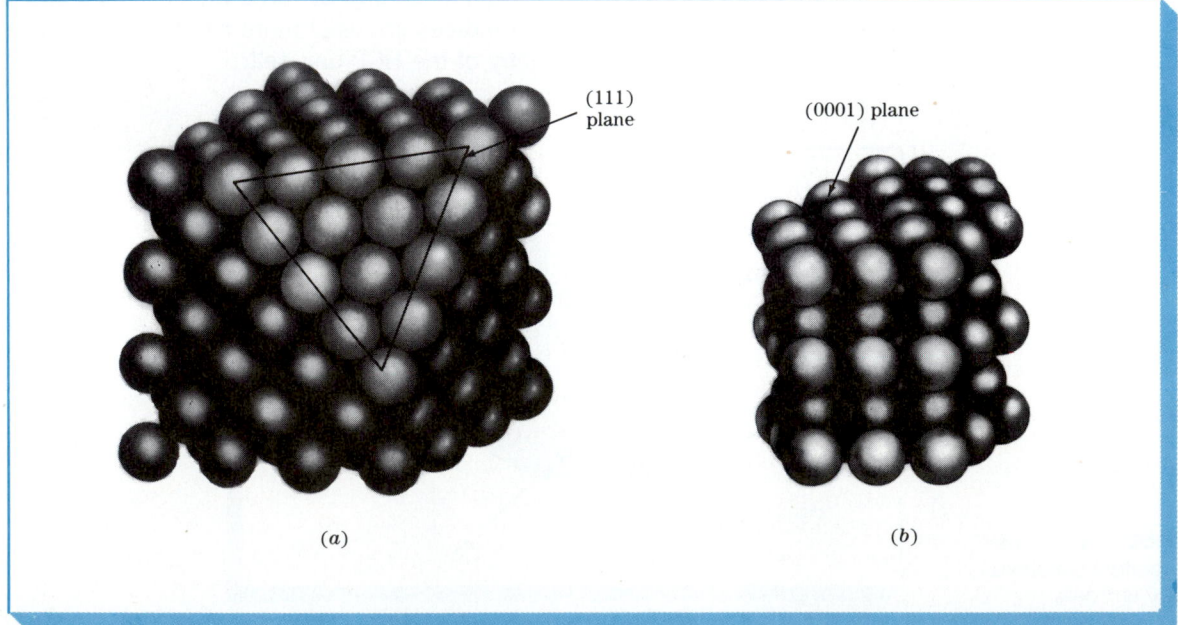

(111) plane

(0001) plane

(a)   (b)

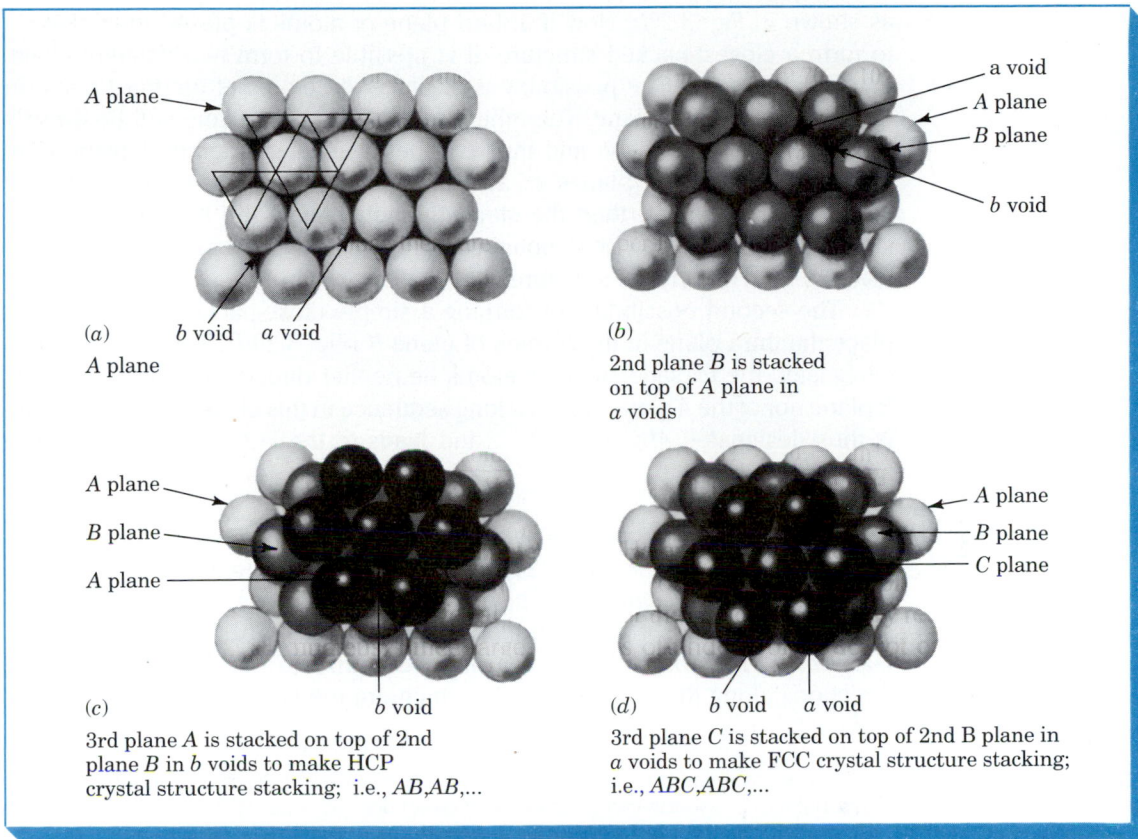

(a)   b void   a void

A plane

(b)

2nd plane B is stacked
on top of A plane in
a voids

A plane

B plane

A plane

(c)   b void

3rd plane A is stacked on top of 2nd
plane B in b voids to make HCP
crystal structure stacking; i.e., AB,AB,...

(d)   b void   a void

3rd plane C is stacked on top of 2nd B plane in
a voids to make FCC crystal structure stacking;
i.e., ABC,ABC,...

FIGURE 3.20 Formation of HCP and FCC crystal structures by changing the stacking of close-packed planes of atoms. (*a*) A plane containing *a*- and *b*-type voids between the atoms. (*b*) B plane is stacked on top of A plane in *a* voids. (*c*) Third plane: Another A plane is stacked on top of the *b* voids of B plane to make HCP crystal structure stacking. i.e., *ABAB*. . . . (*d*) Third plane (alternate): a C plane is stacked on the *a* voids of B plane to make an FCC stacked crystal structure: ABCABC. . . .

in Fig. 3.19*a* have the identical packing arrangement as the (0001) planes of the HCP crystal structure shown in Fig. 3.19*b*. However, the three-dimensional FCC and HCP crystal structures are not identical because there is a difference in the stacking arrangement of their atomic planes, which can best be described by considering the stacking of hard spheres representing atoms. As a useful analogy, one can imagine the stacking of planes of equal-sized marbles on top of each other, minimizing the space between the marbles.

Consider first a plane of close-packed atoms designated the *A* plane, as shown in Fig. 3.20*a*. Note that there are two different types of empty spaces or voids between the atoms. The voids pointing to the top of the page are designated *a* voids and those pointing to the bottom of the page, *b* voids. A second plane of atoms can be placed over the *a* or *b* voids and the same three-dimensional structure will be produced. Let us place plane *B* over the *a* voids, as shown in Fig. 3.20*b*. Now if a third plane of atoms is placed over plane *B* to form a closest-packed structure, it is possible to form two different close-packed structures. One possibility is to place the atoms of the third plane in the *b* voids of the *B* plane. Then the atoms of this third plane will lie directly over those of the *A* plane and thus can be designated another *A* plane (Fig. 3.20*c*). If subsequent planes of atoms are placed in this same alternating stacking arrangement, then the stacking sequence of the three-dimensional structure produced can be denoted by *ABABAB*.... Such a stacking sequence leads to the HCP crystal structure (Fig. 3.19*b*).

The second possibility for forming a simple close-packed structure is to place the third plane in the *a* voids of plane *B* (Fig. 3.20*d*). This third plane is designated the *C* plane since its atoms lie neither directly above those of the *B* plane nor of the *A* plane. The stacking sequence in this close-packed structure is thus designated *ABCABCABC* ... and leads to the FCC structure shown in Fig. 3.19*a*.

Body-Centered Cubic Crystal Structure

The BCC structure is not a close-packed structure and hence does not have close-packed planes like the {111} planes in the FCC structure and the {0001} planes in the HCP structure. The most densely packed planes in the BCC structure are the {110} family of planes of which the (110) plane is shown in Fig. 3.21*b*. The atoms, however, in the BCC structure do have close-packed directions along the cube diagonals, which are the ⟨111⟩ directions.

FIGURE 3.21 BCC crystal structure showing (*a*) the (100) plane and (*b*) a section of the (110) plane. Note that this is not a close-packed structure but that diagonals are close-packed directions. [(*a*) After W. G. Moffatt, G. W. Pearsall, and J. Wulff, "The Structure and Properties of Materials," vol. I: "Structure," Wiley, 1964, p. 51.]

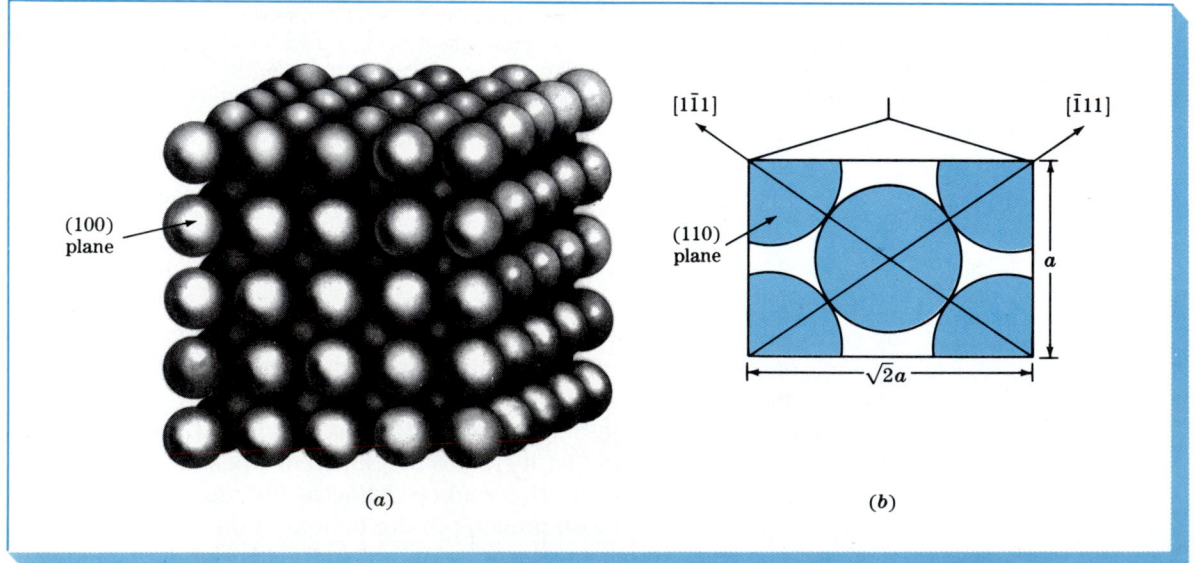

3.9   VOLUME, PLANAR, AND LINEAR DENSITY UNIT-CELL CALCULATIONS

Volume Density   Using the hard-sphere atomic model for the crystal structure unit cell of a metal and a value for the atomic radius of the metal obtained from x-ray diffraction analysis, a value for the volume density of a metal can be obtained by using the equation

$$\text{Volume density of metal} = \rho_v = \frac{\text{mass/unit cell}}{\text{volume/unit cell}} \qquad (3.5)$$

In Example Problem 3.11 a value of 8.98 Mg/m^3 (8.98 g/cm^3) is obtained for the density of copper. The handbook experimental value for the density of copper is 8.96 Mg/m^3 (8.96 g/cm^3). The slightly lower density of the experimental value could be attributed to the absence of atoms at some atomic sites (vacancies), line defects, and mismatch where grains meet (grain boundaries). These crystalline defects are discussed in Chap. 4. Another cause of the discrepancy could also be due to the atoms not being perfect spheres.

Example Problem 3.11

Copper has an FCC crystal structure and an atomic radius of 0.1278 nm. Assuming the atoms to be hard spheres which touch each other along the face diagonals of the FCC unit cell as shown in Fig. 3.7, calculate a theoretical value for the density of copper in megagrams per cubic meter. The atomic mass of copper is 63.54 g/mol.

Solution:

For the FCC unit cell, $\sqrt{2}a = 4R$, where a is the lattice constant of the unit cell and R is the atomic radius of the copper atom. Thus

$$a = \frac{4R}{\sqrt{2}} = \frac{(4)(0.1278 \text{ nm})}{\sqrt{2}} = 0.361 \text{ nm}$$

$$\text{Volume density of copper} = \rho_v = \frac{\text{mass/unit cell}}{\text{volume/unit cell}} \qquad (3.5)$$

In the FCC unit cell there are four atoms/unit cell. Each copper atom has a mass of (63.54 g/mol)/(6.02 $\times$ 10^{23} atoms/mol). Thus the mass m of Cu atoms in the FCC unit cell is

$$m = \frac{(4 \text{ atoms})(63.54 \text{ g/mol})}{6.02 \times 10^{23} \text{ atoms/mol}} \left(\frac{10^{-6} \text{ Mg}}{\text{g}} \right) = 4.22 \times 10^{-28} \text{ Mg}$$

The volume V of the Cu unit cell is

$$V = a^3 = \left(0.361 \text{ nm} \times \frac{10^{-9} \text{ m}}{\text{nm}}\right)^3 = 4.70 \times 10^{-29} \text{ m}^3$$

Thus the density of copper is

$$\rho_v = \frac{m}{V} = \frac{4.22 \times 10^{-28} \text{ Mg}}{4.70 \times 10^{-29} \text{ m}^3} = 8.98 \text{ Mg/m}^3 \qquad (8.98 \text{ g/cm}^3) \blacktriangleleft$$

Planar Atomic Density   Sometimes it is important to determine the atomic densities on various crystal planes. To do this a quantity called the *planar atomic density* is calculated by using the relationship

$$\text{Planar atomic density} = \rho_p$$
$$= \frac{\text{equiv. no. of atoms whose centers are intersected by selected area}}{\text{selected area}}$$

(3.6)

For convenience the area of a plane which intersects a unit cell is usually used in these calculations, as shown, for example, in Fig. 3.22 for the (110) plane in a BCC unit cell. In order for an atom area to be counted in this calculation, the plane of interest must intersect the center of an atom. In Example Problem 3.12 the (110) plane intersects the centers of five atoms, but the equivalent of only two atoms is counted since only one-quarter of each of the four corner atoms is included in the area inside the unit cell. Figure 3.22A shows area occupied by atoms on a planar section of an (111) plane in the FCC lattice.

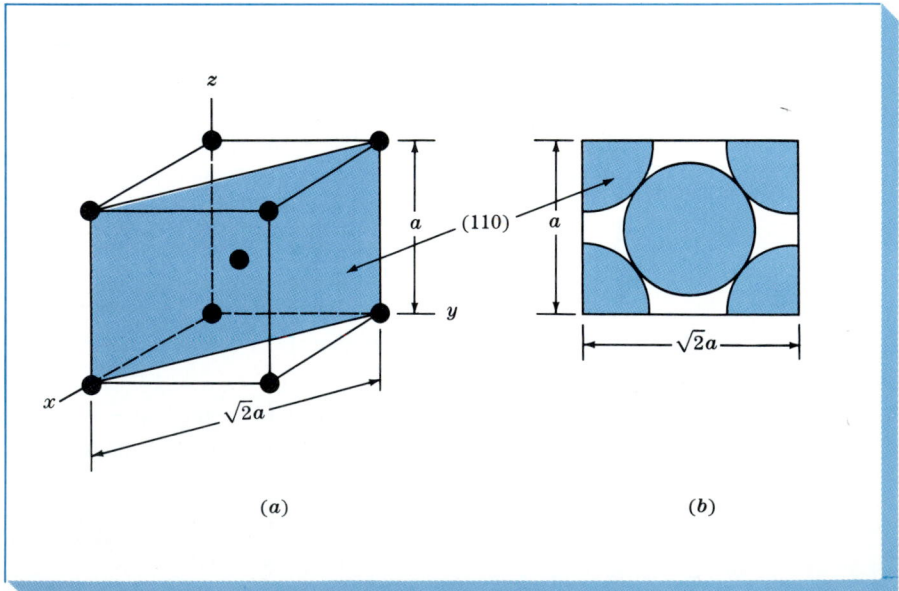

FIGURE 3.22 (*a*) A BCC atomic-site unit cell showing a shaded (110) plane. (*b*) Areas of atoms in BCC unit cell cut by the (110) plane.

(*a*)

(*b*)

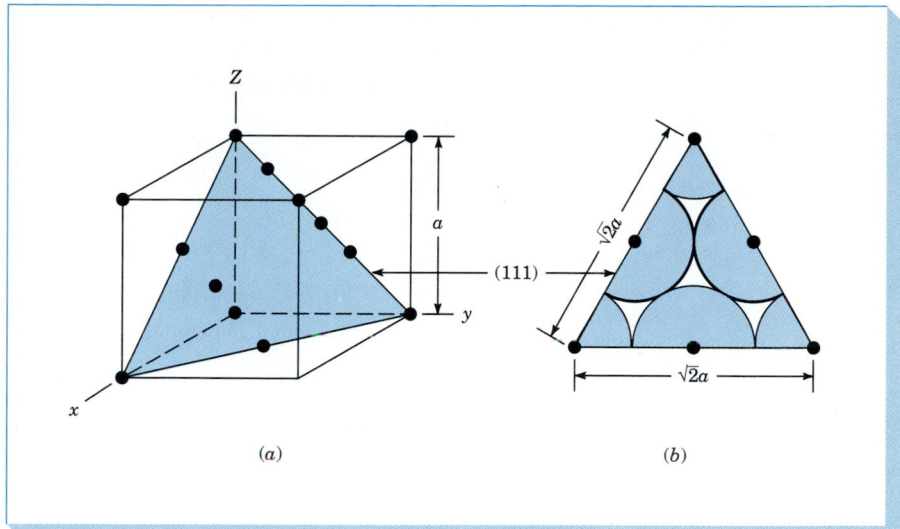

FIGURE 3.22A (*a*) An FCC atomic-site unit cell showing a shaded (111) plane. (*b*) Areas of atoms in FCC unit cell cut by the (111) plane.

Example Problem 3.12

Calculate the planar atomic density ρ_p on the (110) plane of the α iron BCC lattice in atoms per square millimeter. The lattice constant of α iron is 0.287 nm.

Solution:

$$\rho_p = \frac{\text{equiv. no. of atoms whose centers are intersected by selected area}}{\text{selected area}} \qquad (3.6)$$

The equivalent number of atoms intersected by the (110) plane in terms of the surface area inside the BCC unit cell is shown in Fig. 3.22 and is

$$1 \text{ atom at center} + 4 \times \tfrac{1}{4} \text{ atoms at four corners of plane} = 2 \text{ atoms}$$

The area intersected by the (110) plane inside the unit cell (selected area) is

$$(\sqrt{2}a)(a) = \sqrt{2}a^2$$

Thus the planar atomic density is

$$\rho_p = \frac{2 \text{ atoms}}{\sqrt{2}(0.287 \text{ nm})^2} = \frac{17.2 \text{ atoms}}{\text{nm}^2}$$

$$= \frac{17.2 \text{ atoms}}{\text{nm}^2} \times \frac{10^{12} \text{ nm}^2}{\text{mm}^2}$$

$$= 1.72 \times 10^{13} \text{ atoms/mm}^2 \blacktriangleleft$$

Linear Atomic Density   Sometimes it is important to determine the atomic densities in various directions in crystal structures. To do this a quantity called the *linear atomic density* is calculated by using the relationship

$$\text{Linear atomic density} = \rho_l$$

$$= \frac{\begin{array}{c}\text{no. of atomic diam. intersected by selected} \\ \text{length of line in direction of interest}\end{array}}{\text{selected length of line}} \quad (3.7)$$

Example Problem 3.13 shows how the linear atomic density can be calculated in the [110] direction in a pure copper crystal lattice.

Example Problem 3.13

Calculate the linear atomic density ρ_l in the [110] direction in the copper crystal lattice in atoms per millimeter. Copper is FCC and has a lattice constant of 0.361 nm.

Solution:

The atoms whose centers the [110] direction intersects are shown in Fig. 3.23. We shall select the length of the line to be the length of the face diagonal of the FCC unit cell, which is $\sqrt{2}a$. The number of atomic diameters intersected by this length of line are $\frac{1}{2} + 1 + \frac{1}{2} = 2$ atoms. Thus using Eq. (3.7), the linear atomic density is

$$\rho_l = \frac{2 \text{ atoms}}{\sqrt{2}a} = \frac{2 \text{ atoms}}{\sqrt{2}(0.361 \text{ nm})} = \frac{3.92 \text{ atoms}}{\text{nm}}$$

$$= \frac{3.92 \text{ atoms}}{\text{nm}} \times \frac{10^6 \text{ nm}}{\text{mm}}$$

$$= 3.92 \times 10^6 \text{ atoms/mm} \blacktriangleleft$$

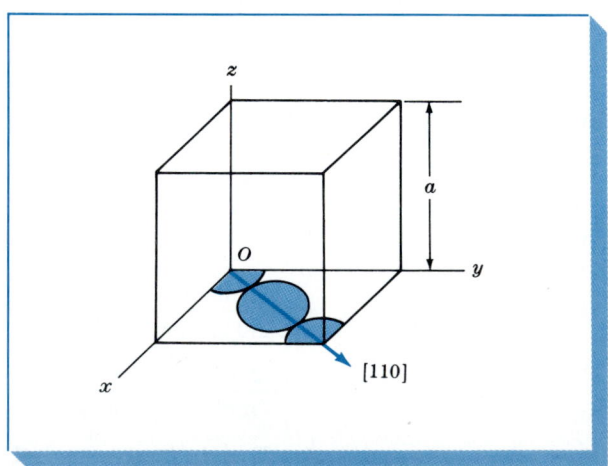

FIGURE 3.23 Diagram for calculating the atomic linear density in the [110] direction in an FCC unit cell.

3.10 POLYMORPHISM OR ALLOTROPY

Many elements and compounds exist in more than one crystalline form under different conditions of temperature and pressure. This phenomenon is termed *polymorphism,* or *allotropy.* Many industrially important metals such as iron, titanium, and cobalt undergo allotropic transformations at elevated temperatures at atmospheric pressure. Table 3.5 lists some selected metals which show allotropic transformations and the structure changes which occur.

Iron exists in both BCC and FCC crystal structures over the temperature range from room temperature to its melting point at 1539°C, as shown in Fig. 3.24. Alpha (α) iron exists from -273 to 912°C and has the BCC crystal structure.

TABLE 3.5   Allotropic Crystalline Forms of Some Metals

Metal	Crystal structure at room temperature	At other temperatures
Ca	FCC	BCC (>447°C)
Co	HCP	FCC (>427°C)
Hf	HCP	BCC (>1742°C)
Fe	BCC	FCC (912–1394°C)
		BCC (>1394°C)
Li	BCC	HCP (<−193°C)
Na	BCC	HCP (<−233°C)
Tl	HCP	BCC (>234°C)
Ti	HCP	BCC (>883°C)
Y	HCP	BCC (>1481°C)
Zr	HCP	BCC (>872°C)

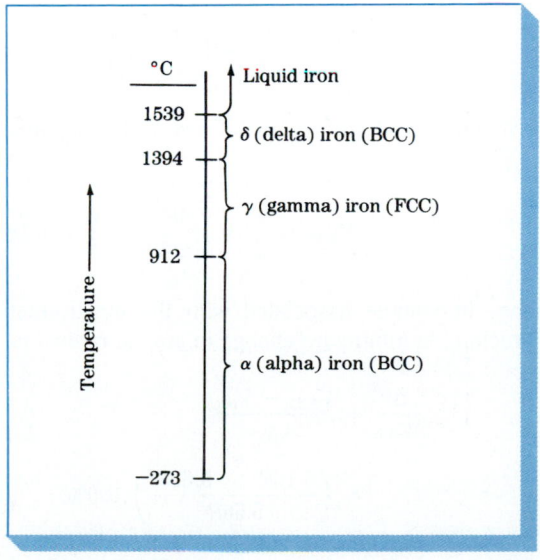

FIGURE 3.24 Allotropic crystalline forms of iron over temperature ranges at atmospheric pressure.

Gamma (γ) iron exists from 912 to 1394°C and has the FCC crystal structure. Delta (δ) iron exists from 1394 to 1539°C, which is the melting point of iron. The crystal structure of δ iron is also BCC but with a larger lattice constant than α iron.

Example Problem 3.14

Calculate the theoretical volume change accompanying a polymorphic transformation in a pure metal from the FCC to BCC crystal structure. Assume the hard-sphere atomic model and that there is no change in atomic volume before and after the transformation.

Solution:

In the FCC crystal structure unit cell, the atoms are in contact along the face diagonal of the unit cell, as shown in Fig. 3.7. Hence

$$\sqrt{2}a = 4R \quad \text{or} \quad a = \frac{4R}{\sqrt{2}} \tag{3.3}$$

In the BCC crystal structure unit cell, the atoms are in contact along the body diagonal of the unit cell as shown in Fig. 3.5. Hence

$$\sqrt{3}a = 4R \quad \text{or} \quad a = \frac{4R}{\sqrt{3}} \tag{3.1}$$

The volume per atom for the FCC crystal lattice, since it has four atoms per unit cell, is

$$V_{\text{FCC}} = \frac{a^3}{4} = \left(\frac{4R}{\sqrt{2}}\right)^3 \left(\frac{1}{4}\right) = 5.66R^3$$

The volume per atom.for the BCC crystal lattice, since it has two atoms per unit cell, is

$$V_{\text{BCC}} = \frac{a^3}{2} = \left(\frac{4R}{\sqrt{3}}\right)^3 \left(\frac{1}{2}\right) = 6.16R^3$$

The change in volume associated with the transformation from the FCC to BCC crystal structure, assuming no change in atomic radius, is

$$\frac{\Delta V}{V_{\text{FCC}}} = \frac{V_{\text{BCC}} - V_{\text{FCC}}}{V_{\text{FCC}}}$$

$$= \left(\frac{6.16R^3 - 5.66R^3}{5.66R^3}\right) 100\% = +8.8\% \blacktriangleleft$$

3.11   CRYSTAL STRUCTURE ANALYSIS

Our present knowledge of crystal structures has been obtained mainly by x-ray diffraction techniques which utilize x-rays about the same wavelength as the distance between crystal lattice planes. However, before discussing the manner in which x-rays are diffracted in crystals, let us consider how x-rays are produced for experimental purposes.

X-Ray Sources

X-rays used for diffraction are electromagnetic waves with wavelengths in the range 0.05 to 0.25 nm (0.5 to 2.5 Å). By comparison the wavelength of visible light is of the order of 600 nm (6000 Å). In order to produce x-rays for diffraction purposes, a voltage of about 35 kV is necessary and is applied between a cathode and an anode target metal, both of which are contained in a vacuum, as shown in Fig. 3.25. When the tungsten filament of the cathode is heated, electrons are released by thermionic emission and accelerated through the vacuum by the large voltage difference between the cathode and anode, thereby gaining kinetic energy. When the electrons strike the target metal (e.g., molybdenum), x-rays are given off. However, most of the kinetic energy (about 98 percent) is converted into heat, so the target metal must be cooled externally.

The x-ray spectrum emitted at 35 kV using a molybdenum target is shown in Fig. 3.26. The spectrum shows continuous x-ray radiation in the wavelength range from about 0.2 to 1.4 Å (0.02 to 0.14 nm) and two spikes of characteristic radiation which are designated the K_α and K_β lines. The wavelengths of the K_α and K_β lines are characteristic for an element. For molybdenum, the K_α line occurs at a wavelength of about 0.7 Å (0.07 nm). The origin of the characteristic

FIGURE 3.25 Schematic diagram of the cross section of a sealed-off filament x-ray tube. (*After B. D. Cullity, "Elements of X-Ray Diffraction," 2d ed., Addison-Wesley, 1978, p. 23.*)

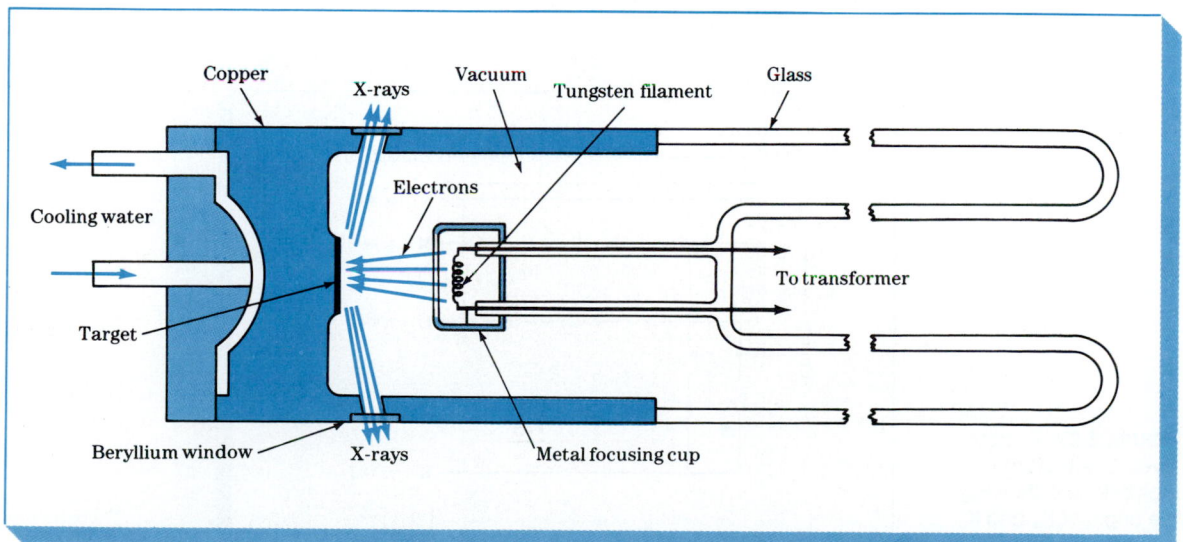

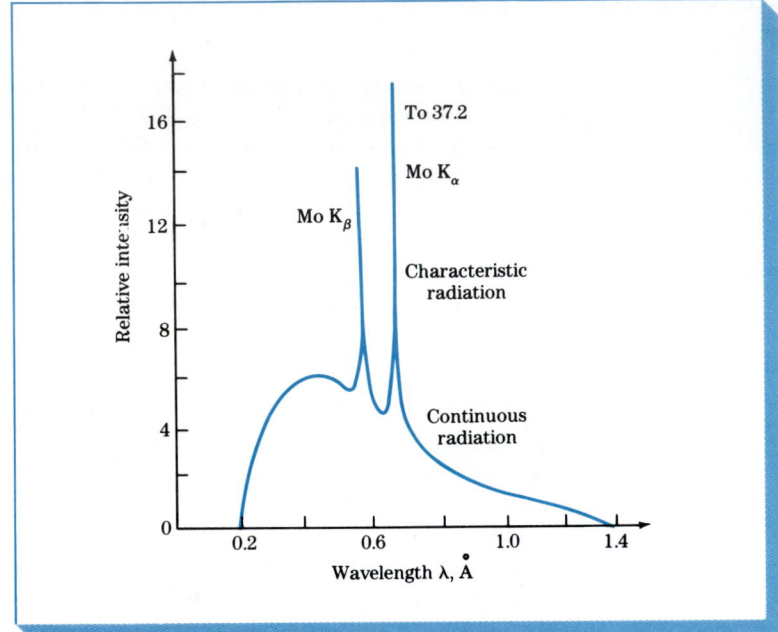

FIGURE 3.26 X-ray emission spectrum produced when molybdenum metal is used as the target metal in an x-ray tube operating at 35 kV.

radiation is explained as follows. First, K electrons (electrons in the $n = 1$ shell) are knocked out of the atom by highly energetic electrons bombarding the target, leaving excited atoms. Next, some electrons in higher shells (that is, $n = 2$ or 3) drop down to lower energy levels to replace the lost K electrons, emitting energy of a characteristic wavelength. The transition of electrons from

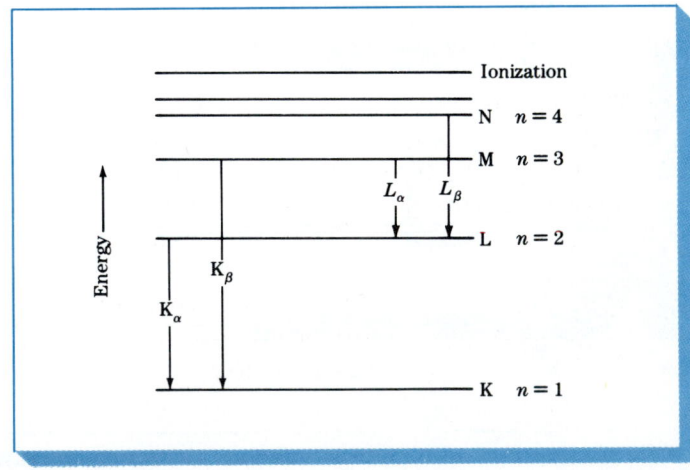

FIGURE 3.27 Energy levels of electrons in molybdenum showing the origin of K_α and K_β radiation.

the L ($n = 2$) shell to the K ($n = 1$) shell creates energy of the wavelength of the K_α line, as indicated in Fig. 3.27.

X-Ray Diffraction Since the wavelengths of some x-rays are about equal to the distance between planes of atoms in crystalline solids, reinforced diffraction peaks of radiation of varying intensities can be produced when a beam of x-rays strikes a crystalline solid. However, before considering the application of x-ray diffraction techniques to crystal structure analysis, let us examine the geometric conditions necessary to produce diffracted or reinforced beams of reflected x-rays.

Consider a monochromatic (single-wavelength) beam of x-rays to be incident on a crystal, as shown in Fig. 3.28. For simplification let us allow the crystal planes of atomic scattering centers to be replaced by crystal planes which act as mirrors in reflecting the incident x-ray beam. In Fig. 3.28 the horizontal lines represent a set of parallel crystal planes with Miller indices (hkl). When an incident beam of monochromatic x-rays of wavelength λ strikes this set of planes at an angle such that the wave patterns of the beam leaving the various planes are *not in phase, no reinforced beam will be produced* (Fig. 3.28a). Thus destructive interference occurs. If the reflected wave patterns of the beam leaving the various planes are in phase, then reinforcement of the beam or constructive interference occurs (Fig. 3.28b).

Let us now consider incident x-rays 1 and 2 as indicated in Fig. 3.28c. For these rays to be in phase, the extra distance of travel of ray 2 is equal to $MP + PN$, which must be an integral number of wavelengths λ. Thus

$$n\lambda = MP + PN \tag{3.8}$$

where $n = 1, 2, 3, \ldots$ and is called the *order of the diffraction*. Since both MP and PN equal $d_{hkl} \sin \theta$, where d_{hkl} is the interplanar spacing of the crystal planes of indices (hkl), the condition for constructive interference (i.e., the production of a diffraction peak of intense radiation) must be

$$n\lambda = 2d_{hkl} \sin \theta \tag{3.9}$$

This equation, known as *Bragg's law*,[1] gives the relationship among the angular positions of the reinforced diffracted beams in terms of the wavelength λ of the incoming x-ray radiation and of the interplanar spacings d_{hkl} of the crystal planes. In most cases, the first order of diffraction where $n = 1$ is used, and so for this case Bragg's law takes the form

$$\lambda = 2d_{hkl} \sin \theta \tag{3.10}$$

---

[1] William Henry Bragg (1862–1942). English physicist who worked in x-ray crystallography.

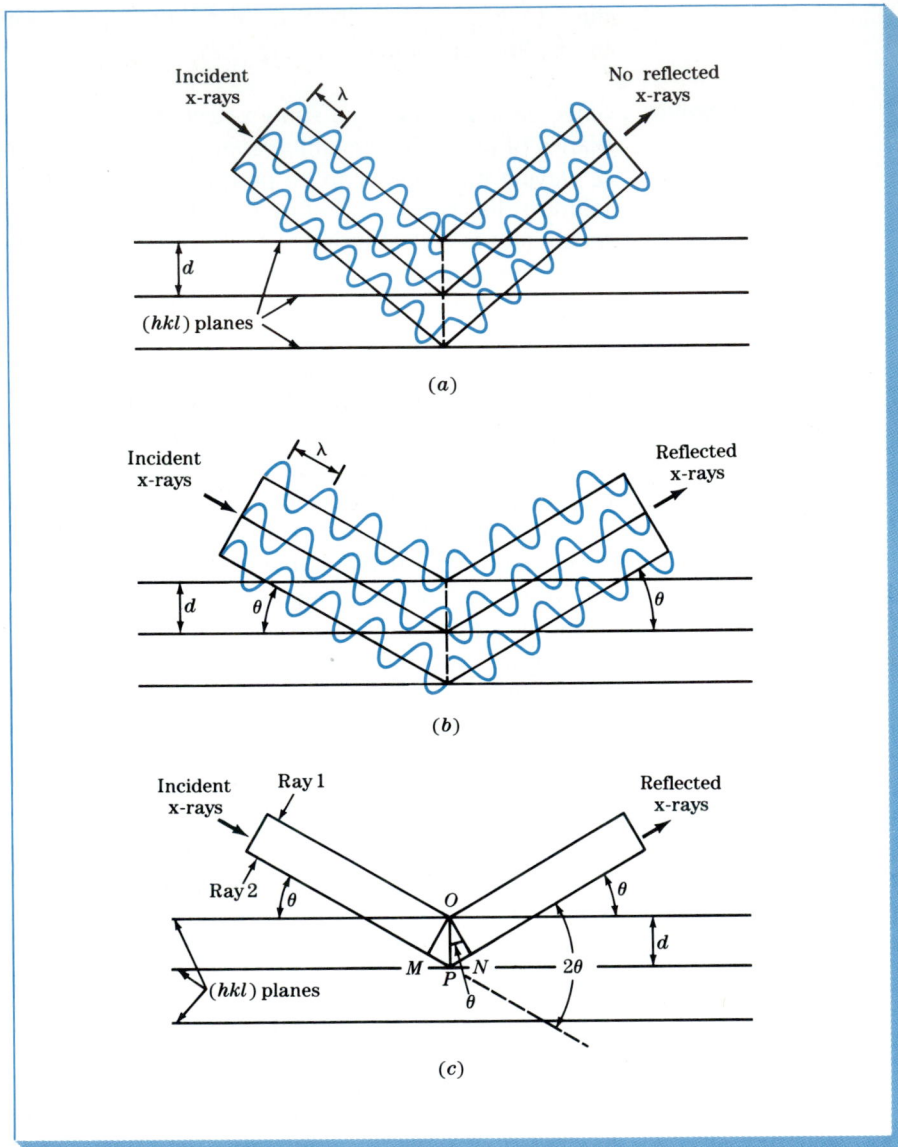

FIGURE 3.28 The reflection of an x-ray beam by the (*hkl*) planes of a crystal. (*a*) No reflected beam is produced at an arbitrary angle of incidence. (*b*) At the Bragg angle *θ*, the reflected rays are in phase and reinforce one another. (*c*) Similar to (*b*) except that the wave representation has been omitted. (*After A. G. Guy and J. J. Hren, "Elements of Physical Metallurgy," 3d ed., Addison-Wesley, 1974, p. 201.*)

Example Problem 3.15

A sample of BCC iron was placed in an x-ray diffractometer using incoming x-rays with a wavelength $\lambda = 0.1541$ nm. Diffraction from the {110} planes was obtained at $2\theta = 44.704°$. Calculate a value for the lattice constant a of BCC iron. (Assume first-order diffraction with $n = 1$.)

Solution:

$$2\theta = 44.704° \qquad \theta = 22.35°$$

$$\lambda = 2d_{hkl} \sin \theta \qquad\qquad (3.10)$$

$$d_{110} = \frac{\lambda}{2 \sin \theta} = \frac{0.1541 \text{ nm}}{2(\sin 22.35°)}$$

$$= \frac{0.1541 \text{ nm}}{2(0.3803)} = 0.2026 \text{ nm}$$

Rearranging Eq. (3.4) gives

$$a = d_{hkl}\sqrt{h^2 + k^2 + l^2}$$

Thus

$$a \text{ (Fe)} = d_{110}\sqrt{1^2 + 1^2 + 0^2}$$

$$= (0.2026 \text{ nm})(1.414) = 0.287 \text{ nm} \blacktriangleleft$$

X-Ray Diffraction Analysis of Crystal Structures

The powder method of x-ray diffraction analysis   The most commonly used x-ray diffraction technique is the *powder method.* In this technique a powdered specimen is utilized so that there will be a random orientation of many crystals to ensure that some of the particles will be oriented in the x-ray beam to satisfy the diffraction conditions of Bragg's law. Modern x-ray crystal analysis uses an x-ray diffractometer which has a radiation counter to detect the angle and intensity of the diffracted beam (Fig. 3.29). A recorder automatically plots the intensity of the diffracted beam as the counter moves on a goniometer[1] circle (Fig. 3.30) which is in synchronization with the specimen over a range of 2θ values. Figure 3.31 shows an x-ray diffraction recorder chart for the intensity of the diffracted beam vs. the diffraction angles 2θ for a powdered pure-metal specimen. In this way both the angles of the diffracted beams and their intensities can be recorded at one time. Sometimes a powder camera with an enclosed filmstrip is used instead of the diffractometer, but this method is much slower and in most cases less convenient.

---

[1]A goniometer is an instrument for measuring angles.

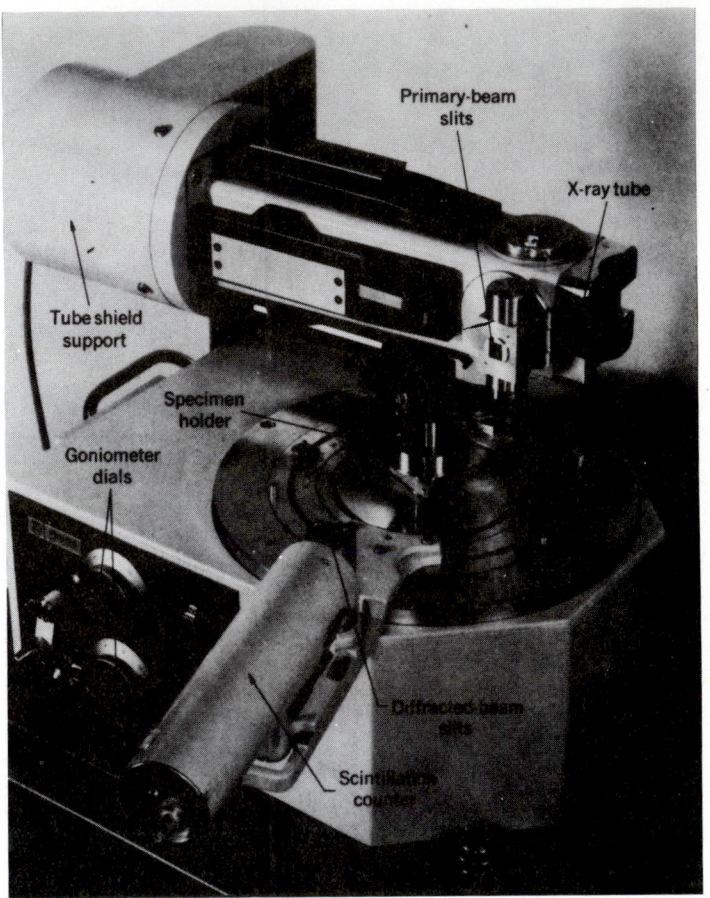

Primary-beam
slits

X-ray tube

Tube shield
support

Specimen
holder

Goniometer
dials

Diffracted beam
slits

Scintillation
counter

FIGURE 3.29 An x-ray diffractometer (with x-radiation shields removed). (*Philips Electronic Instruments, Inc.*)

Diffraction conditions for cubic unit cells   X-ray diffraction techniques enable the structures of crystalline solids to be determined. The interpretation of x-ray diffraction data for most crystalline substances is complex and beyond the scope of this book, and so only the simple case of diffraction in pure cubic metals will be considered. The analysis of x-ray diffraction data for cubic unit cells can be simplified by combining Eq. (3.4),

$$d_{hkl} = \frac{a}{\sqrt{h^2 + k^2 + l^2}}$$

with the Bragg equation $\lambda = 2d \sin \theta$, giving

$$\lambda = \frac{2a \sin \theta}{\sqrt{h^2 + k^2 + l^2}} \qquad (3.11)$$

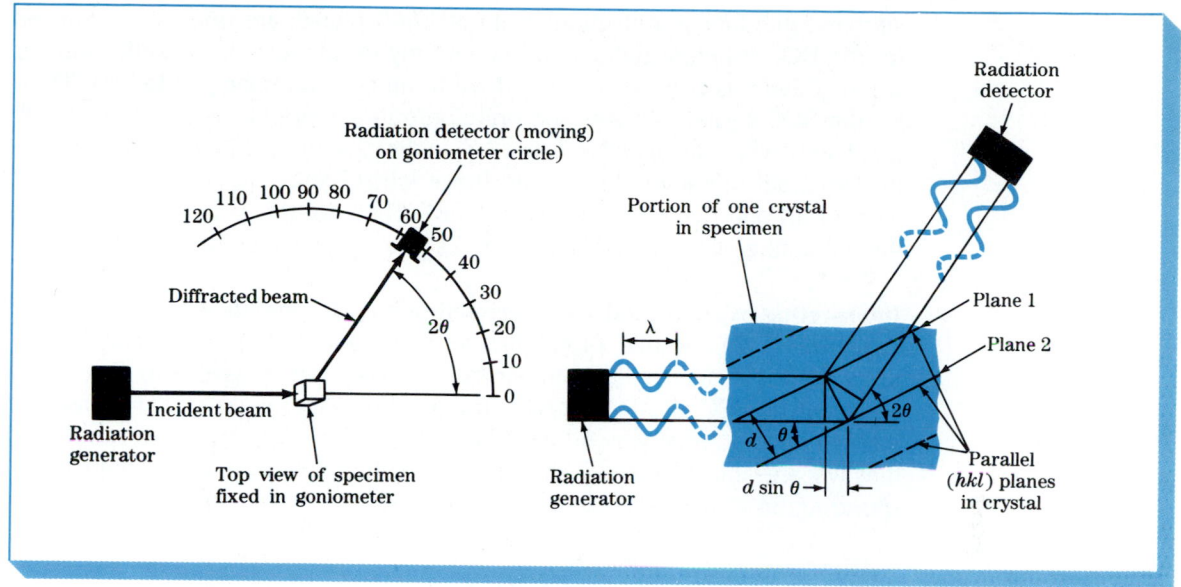

FIGURE 3.30 Schematic illustration of the diffractometer method of crystal analysis and of the conditions necessary for diffraction. (*After A. G. Guy, "Essentials of Materials Science," McGraw-Hill, 1976.*)

FIGURE 3.31 Record of the diffraction angles for a tungsten sample obtained by the use of a diffractometer with copper radiation. (*After A. G. Guy and J. J. Hren, "Elements of Physical Metallurgy," 3d ed., Addison-Wesley, 1974, p. 208.*)

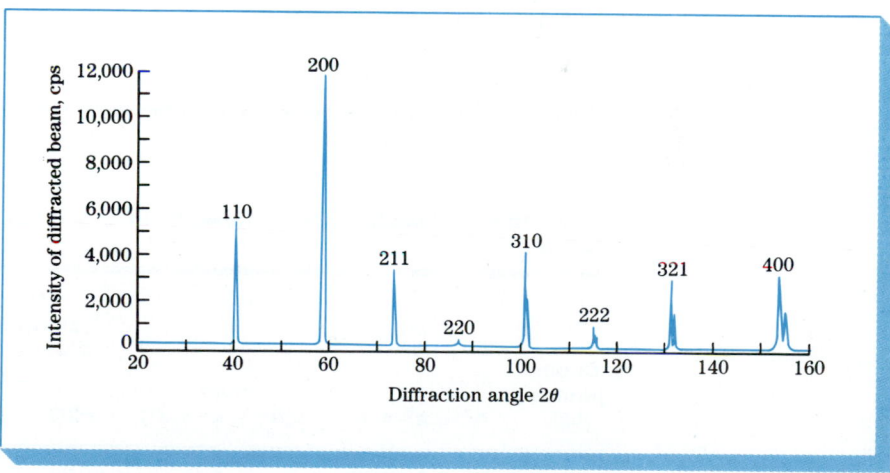

This equation can be used along with x-ray diffraction data to determine if a cubic crystal structure is body-centered or face-centered cubic. The rest of this subsection will describe how this is done.

To use Eq. (3.11) for diffraction analysis, we must know which crystal planes are the diffracting planes for each type of crystal structure. For the simple cubic lattice, reflections from all (hkl) planes are possible. However, for the BCC structure diffraction occurs only on planes whose Miller indices when added together ($h + k + l$) total to an even number (Table 3.6). Thus, for the BCC crystal structure the principal diffracting planes are {110}, {200}, {211}, etc., which are listed in Table 3.7. In the case of the FCC crystal structure, the principal diffracting planes are those whose Miller indices are either all even or all odd (zero is considered even). Thus, for the FCC crystal structure the diffracting planes are {111}, {200}, {220}, etc., which are listed in Table 3.7.

Interpreting experimental x-ray diffraction data for metals with cubic crystal structures   We can use x-ray diffractometer data to determine crystal structures. A simple case to illustrate how this analysis can be used is to distinguish between the BCC and FCC crystal structures of a cubic metal. Let us assume that we have a metal which has either a BCC or an FCC crystal structure and that we are able to identify the principal diffracting planes and their corresponding 2θ values, as indicated for the metal tungsten in Fig. 3.31.

TABLE 3.6   Rules for Determining the Diffracting {*hkl*} Planes in Cubic Crystals

Bravais lattice	Reflections present	Reflections absent
BCC	($h + k + l$) = even	($h + k + l$) = odd
FCC	(h, k, l) all odd or all even	(h, k, l) not all odd or all even

TABLE 3.7   Miller Indices of the Diffracting Planes for BCC and FCC Lattices

Cubic planes {*hkl*}	$h^2 + k^2 + l^2$	Sum $\Sigma(h^2 + k^2 + l^2)$	Cubic diffracting planes {*hkl*} FCC	Cubic diffracting planes {*hkl*} BCC
{100}	$1^2 + 0^2 + 0^2$	1		
{110}	$1^2 + 1^2 + 0^2$	2	· · ·	110
{111}	$1^2 + 1^2 + 1^2$	3	111	
{200}	$2^2 + 0^2 + 0^2$	4	200	200
{210}	$2^2 + 1^2 + 0^2$	5		
{211}	$2^2 + 1^2 + 1^2$	6	· · ·	211
· · ·		7		
{220}	$2^2 + 2^2 + 0^2$	8	220	220
{221}	$2^2 + 2^2 + 1^2$	9		
{310}	$3^2 + 1^2 + 0^2$	10	· · ·	310

By squaring both sides of Eq. (3.11) and solving for $\sin^2 \theta$, we obtain

$$\sin^2 \theta = \frac{\lambda^2(h^2 + k^2 + l^2)}{4a^2} \tag{3.12}$$

From x-ray diffraction data we can obtain experimental values of 2θ for a series of principal diffracting $\{hkl\}$ planes. Since the wavelength of the incoming radiation and the lattice constant a are both constants, we can eliminate these quantities by forming the ratio of two $\sin^2 \theta$ values as

$$\frac{\sin^2 \theta_A}{\sin^2 \theta_B} = \frac{h_A^2 + k_A^2 + l_A^2}{h_B^2 + k_B^2 + l_B^2} \tag{3.13}$$

where θ_A and θ_B are two diffracting angles associated with the principal diffracting planes $\{h_A k_A l_A\}$ and $\{h_B k_B l_B\}$, respectively.

Using Eq. (3.13) and the Miller indices of the first two sets of principal diffracting planes listed in Table 3.7 for BCC and FCC crystal structures, we can determine values for the $\sin^2 \theta$ ratios for both BCC and FCC structures.

For the BCC crystal structure the first two sets of principal diffracting planes are the $\{110\}$ and $\{200\}$ planes (Table 3.7). Substitution of the Miller $\{hkl\}$ indices of these planes into Eq. (3.13) gives

$$\frac{\sin^2 \theta_A}{\sin^2 \theta_B} = \frac{1^2 + 1^2 + 0^2}{2^2 + 0^2 + 0^2} = 0.5 \tag{3.14}$$

Thus, if the crystal structure of the unknown cubic metal is BCC, the ratio of the $\sin^2 \theta$ values which correspond to the first two principal diffracting planes will be 0.5.

For the FCC crystal structure the first two sets of principal diffracting planes are the $\{111\}$ and $\{200\}$ planes (Table 3.7). Substitution of the Miller $\{hkl\}$ indices of these planes into Eq. (3.13) gives

$$\frac{\sin^2 \theta_A}{\sin^2 \theta_B} = \frac{1^2 + 1^2 + 1^2}{2^2 + 0^2 + 0^2} = 0.75 \tag{3.15}$$

Thus, if the crystal structure of the unknown cubic metal is FCC, the ratio of the $\sin^2 \theta$ values which correspond to the first two principal diffracting planes will be 0.75.

Example Problem 3.16 uses Eq. (3.13) and experimental x-ray diffraction data for the 2θ values for the principal diffracting planes to determine whether an unknown cubic metal is BCC or FCC. X-ray diffraction analysis is usually much more complicated than Example Problem 3.16, but the principles used are the same. Both experimental and theoretical x-ray diffraction analysis has been and continues to be used for the determination of the crystal structure of materials.

Example Problem 3.16

An x-ray diffractometer recorder chart for an element which has either the BCC or the FCC crystal structure shows diffraction peaks at the following 2θ angles: 40, 58, 73, 86.8, 100.4, and 114.7. The wavelength of the incoming x-ray used was 0.154 nm.
(a) Determine the cubic structure of the element.
(b) Determine the lattice constant of the element.
(c) Identify the element.

Solution:

(a) *Determination of the crystal structure of the element.* First, the $\sin^2\theta$ values are calculated from the 2θ diffraction angles.

2θ, deg	θ, deg	$\sin\theta$	$\sin^2\theta$
40	20	0.3420	0.1170
58	29	0.4848	0.2350
73	36.5	0.5948	0.3538
86.8	43.4	0.6871	0.4721
100.4	50.2	0.7683	0.5903
114.7	57.35	0.8420	0.7090

Next the ratio of the $\sin^2\theta$ values of the first and second angles is calculated:

$$\frac{\sin^2\theta}{\sin^2\theta} = \frac{0.117}{0.235} = 0.498 \approx 0.5$$

The crystal structure is BCC since this ratio is ≈ 0.5. If the ratio had been ≈ 0.75, the structure would have been FCC.

(b) *Determination of the lattice constant.* Rearranging Eq. (3.12) and solving for a^2 gives

$$a^2 = \frac{\lambda^2}{4} \frac{h^2 + k^2 + l^2}{\sin^2\theta} \tag{3.16}$$

or

$$a = \frac{\lambda}{2} \sqrt{\frac{h^2 + k^2 + l^2}{\sin^2\theta}} \tag{3.17}$$

Substituting into Eq. (3.17) $h = 1$, $k = 1$, and $l = 0$ for the h, k, l Miller indices of the first set of principal diffracting planes for the BCC crystal structure, which are the {110} planes, the corresponding value for $\sin^2\theta$, which is 0.117, and 0.154 nm for λ, the incoming radiation, gives

$$a = \frac{0.154 \text{ nm}}{2} \sqrt{\frac{1^2 + 1^2 + 0^2}{0.117}} = 0.318 \text{ nm} \blacktriangleleft$$

(c) *Identification of the element.* The element is tungsten since this element has a lattice constant of 0.316 nm and is BCC.

SUMMARY

Atomic arrangements in crystalline solids can be described by a network of lines called a *space lattice.* Each space lattice can be described by specifying the atom positions in a repeating *unit cell.* There are seven crystal systems based on the geometry of the axial lengths and interaxial angles of the unit cells. These seven systems have a total of 14 sublattices (unit cells) based on the internal arrangements of atomic sites within the unit cells.

In metals the most common crystal structure unit cells are: *body-centered cubic* (BCC), *face-centered cubic* (FCC), and *hexagonal close-packed* (HCP) (which is a dense variation of the simple hexagonal structure).

Crystal directions in cubic crystals are the vector components of the directions resolved along each of the component axes and reduced to smallest integers. They are indicated as [*uvw*]. Families of directions are indexed by the direction indices enclosed by pointed brackets as ⟨*uvw*⟩. *Crystal planes* in cubic crystals are indexed by the reciprocals of the axial intercepts of the plane (followed by the elimination of fractions) as (*hkl*). Cubic crystal planes of a form (family) are indexed with braces as {*hkl*}. Crystal planes in hexagonal crystals are commonly indexed by four indices *h*, *k*, *i*, and *l* enclosed in parentheses as (*hkil*). These indices are the reciprocals of the intercepts of the plane on the a_1, a_2, a_3, and *c* axes of the hexagonal crystal structure unit cell. Crystal directions in hexagonal crystals are the vector components of the direction resolved along each of the four coordinate axes and reduced to smallest integers as [*uvtw*].

Using the hard-sphere model for atoms, calculations can be made for the volume, planar, and linear density of atoms in unit cells. Planes in which atoms are packed as tightly as possible are called *close-packed planes,* and directions in which atoms are in closest contact are called *close-packed directions.* Atomic packing factors for different crystal structures can also be determined by assuming the hard-sphere atomic model. Some metals have different crystal structures at different ranges of temperature and pressure, a phenomenon called *polymorphism.*

Crystal structures of crystalline solids can be determined by using x-ray diffraction analysis techniques. X-rays are diffracted in crystals when the *Bragg's law* ($n\lambda = 2d \sin \theta$) conditions are satisfied. By using the x-ray diffractometer and the *powder method,* the crystal structure of many crystalline solids can be determined.

DEFINITIONS

Sec. 3.1

Crystal: a solid composed of atoms, ions, or molecules arranged in a pattern that is repeated in three dimensions.

Crystal structure: a regular three-dimensional pattern of atoms or ions in space.

Space lattice: a three-dimensional array of points each of which has identical surroundings.

Lattice point: one point in an array in which all the points have identical surroundings.

Unit cell: a convenient repeating unit of a space lattice. The axial lengths and axial angles are the lattice constants of the unit cell.

Sec. 3.3

Body-centered cubic (BCC) unit cell: a unit cell with an atomic packing arrangement in which one atom is in contact with eight identical atoms located at the corners of an imaginary cube.

Face-centered cubic (FCC) unit cell: a unit cell with an atomic packing arrangement in which 12 atoms surround a central atom. The stacking sequence of layers of close-packed planes in the FCC crystal structure is *ABCABC....*

Hexagonal close-packed (HCP) unit cell: a unit cell with an atomic packing arrangement in which 12 atoms surround a central identical atom. The stacking sequence of layers of close-packed planes in the HCP crystal structure is *ABABAB....*

Atomic packing factor (APF): the volume of atoms in a selected unit cell divided by the volume of the unit cell.

Sec. 3.5

Indices of direction in a cubic crystal: a direction in a cubic unit cell is indicated by a vector drawn from the origin at one point in a unit cell through the surface of the unit cell; the position coordinates (x, y, and z) of the vector where it leaves the surface of the unit cell (with fractions cleared) are the indices of direction. These indices, designated u, v, and w are enclosed in brackets as [uvw]. Negative indices are indicated by a bar over the index.

Sec. 3.6

Indices for cubic crystal planes (Miller indices): the reciprocals of the intercepts (with fractions cleared) of a crystal plane with the x, y, and z axes of a unit cube are called the Miller indices of that plane. They are designated h, k, and l for the x, y, and z axes, respectively, and are enclosed in parentheses as (hkl). Note that the selected crystal plane must *not* pass through the origin of the x, y, and z axes.

Sec. 3.9

Volume density ρ_v: mass per unit volume; this quantity is usually expressed in Mg/m^3 or g/cm^3.

Planar density ρ_p: the equivalent number of atoms whose centers are intersected by a selected area divided by the selected area.

Linear density ρ_l: the number of atoms whose centers lie on a specific direction on a specific length of line in a unit cube.

Sec. 3.10

Polymorphism (as pertains to metals): the ability of a metal to exist in two or more crystal structures. For example, iron can have a BCC or an FCC crystal structure, depending on the temperature.

PROBLEMS

3.1.1 Define a crystalline solid.

3.1.2 Define a crystal structure. Give examples of materials which have crystal structures.

3.1.3 Define a space lattice.

3.1.4 Define a unit cell of a space lattice. What lattice constants define a unit cell?

3.2.1 What are the 14 Bravais unit cells?

3.3.1 What are the three most common metal crystal structures? List five metals which have each of these crystal structures.

3.3.2 How many atoms per unit cell are there in the BCC crystal structure?

3.3.3 What is the coordination number for the atoms in the BCC crystal structure?

3.3.4 What is the relationship between the length of the side a of the BCC unit cell and the radius of its atoms?

3.3.5 Potassium at 20°C is BCC and has an atomic radius of 0.238 nm. Calculate a value for its lattice constant a in nanometers.

3.3.6 Tungsten at 20°C is BCC and has an atomic radius of 0.141 nm. Calculate a value for its lattice constant a in nanometers.

3.3.7 Tantalum at 20°C is BCC and has a lattice constant of 0.33026 nm. Calculate a value for the atomic radius of a tungsten atom in nanometers.

3.3.8 Barium at 20°C is BCC and has a lattice constant of 0.5019 nm. Calculate a value for the atomic radius of a chromium atom in nanometers.

3.3.9 How many atoms per unit cell are there in the FCC crystal structure?

3.3.10 What is the coordination number for the atoms in the FCC crystal structure?

3.3.11 Copper is FCC and has a lattice constant of 0.3615 nm. Calculate a value for the atomic radius of a copper atom in nanometers.

3.3.12 Calcium is FCC and has a lattice constant of 0.5582 nm. Calculate a value for the atomic radius of an aluminum atom in nanometers.

3.3.13 Iridium is FCC and has an atomic radius of 0.135 nm. Calculate a value for its lattice constant a in nanometers.

3.3.14 Aluminum is FCC and has an atomic radius of 0.143 nm. Calculate a value for its lattice constant a in nanometers.

3.3.15 Calculate the atomic packing factor for the FCC structure.

3.3.16 How many atoms per unit cell are there in the HCP crystal structure?

3.3.17 What is the coordination number for the atoms in the HCP crystal structure?

3.3.18 What is the ideal c/a ratio for HCP metals?

3.3.19 Of the following HCP metals, which have higher or lower c/a ratios than the ideal ratio: Zr, Ti, Zn, Mg, Co, Cd, and Be?

3.3.20 Calculate the volume in cubic nanometers of the beryllium crystal structure unit cell. Beryllium is HCP at 20°C with $a = 0.22856$ nm and $c = 0.35832$ nm.

3.3.21 Magnesium at 20°C is HCP. The height c of its unit cell is 0.52105 nm and its c/a ratio is 1.623. Calculate a value for its lattice constant a in nanometers.

3.3.22 Cadmium at 20°C is HCP. Using a value of 0.148 nm for the atomic radius of cadmium atoms, calculate a value for its unit-cell volume. Assume a packing factor of 0.74.

3.4.1 How are atomic positions located in cubic unit cells?

3.4.2 List the atom positions for the eight corner and six face-centered atoms of the FCC unit cell.

3.5.1 How are the indices for a crystallographic direction in a cubic unit cell determined?

3.5.2 Draw the following directions in a BCC unit cell and list the position coordinates of the atoms whose centers are intersected by the direction vector:
(*a*) [100]   (*b*) [110]   (*c*) [111]

3.5.3 Draw direction vectors in unit cubes for the following cubic directions:
(*a*) [1$\bar{1}\bar{1}$]   (*b*) [1$\bar{1}$0]   (*c*) [$\bar{1}$2$\bar{1}$1]   (*d*) [$\bar{1}\bar{1}$3]

3.5.4 Draw direction vectors in unit cubes for the following cubic directions:
(*a*) [1$\bar{2}$1]   (*d*) [20$\bar{1}$]   (*g*) [$\bar{2}$21]   (*j*) [20$\bar{3}$]
(*b*) [21$\bar{2}$]   (*e*) [10$\bar{2}$]   (*h*) [$\bar{3}$23]   (*k*) [1$\bar{2}$2]
(*c*) [3$\bar{2}$1]   (*f*) [2$\bar{3}\bar{3}$]   (*i*) [1$\bar{2}$3]   (*l*) [$\bar{2}\bar{2}$3]

3.5.5 What are the indices of the directions shown in the unit cubes of Fig. P3.5.5?

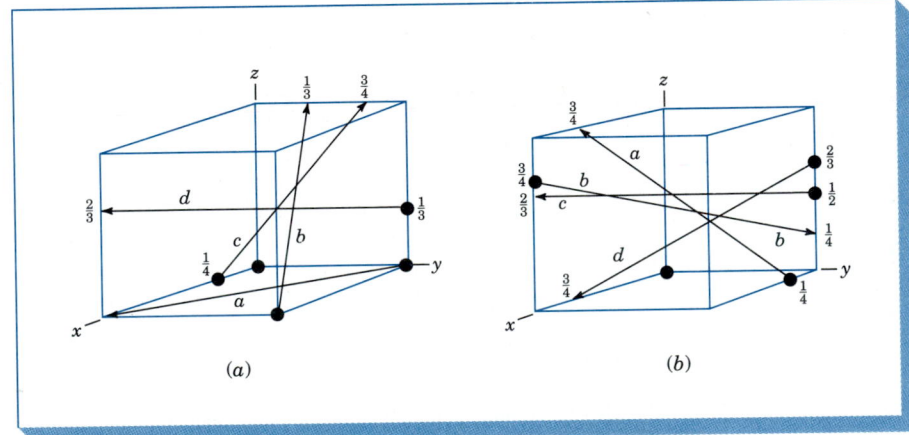

FIGURE P3.5.5

3.5.6 A direction vector passes through a unit cube from the ($\frac{1}{4}$, 0, $\frac{1}{4}$) to the ($\frac{1}{4}$, $\frac{3}{4}$, 1) positions. What are its direction indices?

3.5.7 A direction vector passes through a unit cube from the ($\frac{3}{4}$, $\frac{3}{4}$, 0) to the ($\frac{1}{2}$, $\frac{1}{2}$, 0) positions. What are its direction indices?

3.5.8 What are the crystallographic directions of a family or form? What generalized notation is used to indicate them?

3.5.9 What are the directions of the ⟨100⟩ family or form for a unit cube?

3.5.10 What are the directions of the ⟨111⟩ family or form for a unit cube?

3.5.11 What ⟨110⟩-type directions lie on the (111) plane of a cubic unit cell?

3.5.12 What ⟨111⟩-type directions lie on the (110) plane of a cubic unit cell?

3.6.1 How are the Miller indices for a crystallographic plane in a cubic unit cell determined? What generalized notation is used to indicate them?

3.6.2   Draw in unit cubes the crystal planes which have the following Miller indices:
   (a) $(21\bar{3})$ (d) $(\bar{3}\bar{3}1)$ (g) $(1\bar{2}0)$ (j) $(12\bar{2})$
   (b) $(21\bar{2})$ (e) $(31\bar{2})$ (h) $(23\bar{2})$ (k) $(1\bar{3}3)$
   (c) $(\bar{3}\bar{2}1)$ (f) $(20\bar{3})$ (i) $(\bar{3}32)$ (l) $(2\bar{2}3)$

3.6.3   What are the Miller indices of the cubic crystallographic planes shown in Fig. P3.6.3?

3.6.4   What is the notation used to indicate a family or form of cubic crystallographic planes?

3.6.5   What are the {100} family of planes of the cubic system?

3.6.6   Draw the following crystallographic planes in a BCC unit cell and list the position of the atoms whose centers are intersected by each of the planes:
   (a) (100)   (b) (110)   (c) (111)

3.6.7   Draw the following crystallographic planes in an FCC unit cell and list the position coordinates of the atoms whose centers are intersected by each of the planes:
   (a) (100)   (b) (110)   (c) (111)

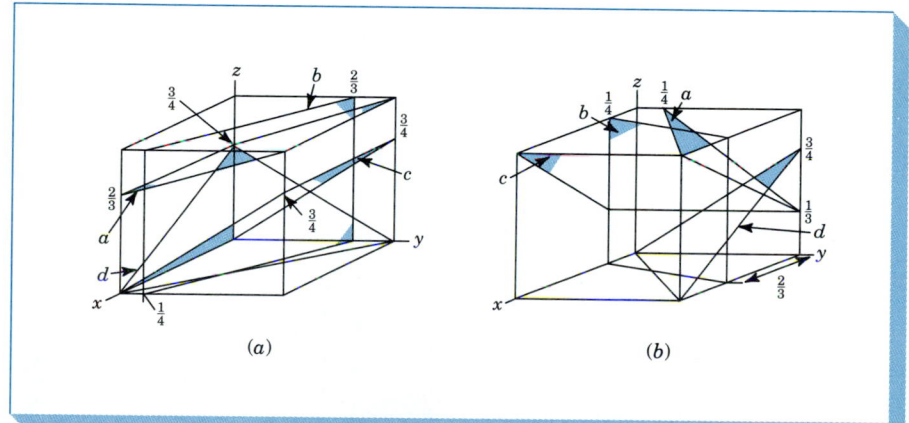

(a)   (b)

FIGURE P3.6.3

3.6.8   A cubic plane has the following axial intercepts: $a = \frac{1}{3}$, $b = -\frac{2}{3}$, $c = \frac{1}{2}$. What are the Miller indices of this plane?

3.6.9   A cubic plane has the following axial intercepts: $a = -\frac{1}{2}$, $b = \frac{1}{2}$, $c = \frac{2}{3}$. What are the Miller indices of this plane?

3.6.10 A cubic plane has the following axial intercepts: $a = 1$, $b = \frac{2}{3}$, $c = -\frac{1}{2}$. What are the Miller indices of this plane?

3.6.11 Determine the Miller indices of the cubic crystal plane which intersects the following position coordinates: $(1, 0, 0)$; $(1, \frac{1}{2}, \frac{1}{4})$; $(\frac{1}{2}, \frac{1}{2}, 0)$

3.6.12 Determine the Miller indices of the cubic crystal plane which intersects the following position coordinates: $(\frac{1}{2}, 0, \frac{1}{2})$; $(0, 0, 1)$; $(1, 1, 1)$

3.6.13 Determine the Miller indices of the cubic crystal plane which intersects the following position coordinates: $(1, \frac{1}{2}, 1)$; $(\frac{1}{2}, 0, \frac{3}{4})$; $(1, 0, \frac{1}{2})$

3.6.14 Determine the Miller indices of the cubic crystal plane which intersects the following position coordinates: $(0, 0, \frac{1}{2})$; $(1, 0, 0)$; $(\frac{1}{2}, \frac{1}{4}, 0)$

3.6.15 Nickel is FCC and has a lattice constant of 0.35236 nm. Calculate the following interplanar spacings in nm.

(a) d_{111} (b) d_{200} (c) d_{220}

3.6.16 Molybdenum is BCC and has a lattice constant of 0.31468 nm. Calculate the following interplanar spacings in nm.

(a) d_{110} (b) d_{220} (c) d_{310}

3.6.17 The d_{321} interplanar spacing in a BCC metal is 0.084165 nm. (a) What is its lattice constant "a"? (b) What is the atomic radius of the metal? (c) What could this metal be based on its lattice constant "a"?

3.6.18 The d_{222} interplanar spacing in an FCC metal is 0.11327 nm. (a) What is its lattice constant "a"? (b) What is the atomic radius of the metal? (c) What could this metal be based on its lattice constant "a"?

3.7.1   How are crystallographic planes indicated in HCP unit cells?

3.7.2   What notation is used to describe HCP crystal planes?

3.7.3   Draw the hexagonal crystal planes whose Miller-Bravais indices are:

(a) $(10\bar{1}1)$ (d) $(1\bar{2}12)$ (g) $(\bar{1}2\bar{1}2)$ (j) $(\bar{1}100)$
(b) $(01\bar{1}1)$ (e) $(21\bar{1}1)$ (h) $(2\bar{2}00)$ (k) $(\bar{2}111)$
(c) $(\bar{1}2\bar{1}0)$ (f) $(1\bar{1}01)$ (i) $(10\bar{1}2)$ (l) $(\bar{1}012)$

3.7.4   Determine the Miller-Bravais indices of the hexagonal crystal planes in Fig. P3.7.4.

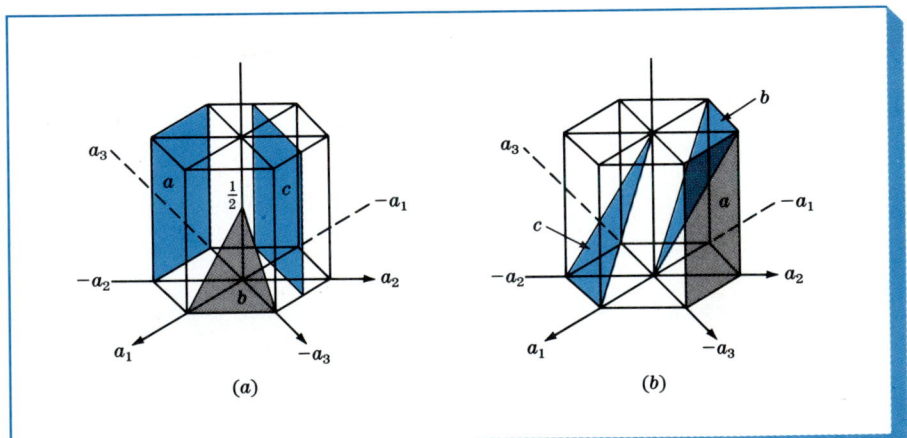

(a)   (b)

FIGURE P3.7.4

3.8.1   What is the difference in the stacking arrangement of close-packed planes in (a) the HCP crystal structure and (b) the FCC crystal structure?

3.8.2   What are the densest-packed planes in (a) the FCC structure and (b) the HCP structure?

3.8.3   What are the closest-packed directions in (a) the FCC structure and (b) the HCP structure?

3.9.1   The lattice constant for BCC niobium at 20°C is 0.33007 nm, and its density is 8.60 g/cm^3. Calculate a value for its atomic mass.

3.9.2   Calculate a value for the density of FCC silver in grams per cubic centimeters from its lattice constant a of 0.40856 nm and its atomic mass of 107.87 g/mol.

3.9.3   Calculate the planar atomic density in atoms per square millimeter for the following crystal planes in BCC tungsten, which has a lattice constant of 0.31648 nm: (a) (100), (b) (110), (c) (111).

3.9.4   Calculate the planar atomic density in atoms per square millimeter for the following crystal planes in FCC nickel, which has a lattice constant of 0.35236 nm: (a) (100), (b) (110), (c) (111).

3.9.5   Calculate the planar atomic density in atoms per square millimeter for the (0001) plane in HCP cobalt, which has an a constant of 0.2506 nm and a c constant of 0.4069 nm.

3.9.6   Calculate the linear atomic density in atoms per millimeter for the following directions in BCC tantalum, which has a lattice constant of 0.33026 nm: (a) [100], (b) [110], (c) [111].

3.9.7   Calculate the linear atomic density in atoms per millimeter for the following directions in FCC platinum, which has a lattice constant of 0.39239 nm: (a) [100], (b) [110], (c) [111].

3.10.1   What is polymorphism with respect to metals?

3.10.2   Titanium goes through a polymorphic change from BCC to HCP crystal structure upon cooling through 882°C. Calculate the percentage change in volume when the crystal structure changes from BCC to HCP. The lattice constant a of the BCC unit cell at 882°C is 0.332 nm, and the HCP unit cell has $a = 0.2950$ nm and $c = 0.4683$ nm.

3.10.3   Pure iron goes through a polymorphic change from BCC to FCC upon heating through 912°C. Calculate the volume change associated with the change in crystal structure from BCC to FCC if at 912°C the BCC unit cell has a lattice constant $a = 0.293$ nm and the FCC unit cell $a = 0.363$ nm.

3.11.1   What are x-rays, and how are they produced?

3.11.2   Draw a schematic diagram of an x-ray tube used for x-ray diffraction, and indicate on it the path of the electrons and x-rays.

3.11.3   What is the characteristic x-ray radiation? What is its origin?

3.11.4   Distinguish between destructive interference and constructive interference of reflected x-ray beams through crystals.

3.11.5   Derive Bragg's law by using the simple case of incident x-ray beams being diffracted by parallel planes in a crystal.

3.11.6   A sample of BCC metal was placed in an x-ray diffractometer using x-rays with a wavelength of $\lambda = 0.1541$ nm. Diffraction from the {310} planes was obtained at $2\theta = 101.502°$. Calculate a value for the lattice constant a for this BCC elemental metal. (Assume first-order diffraction, $n = 1$.)

3.11.7   X-rays of an unknown wavelength are diffracted by a nickel sample. The 2θ angle was 102.072° for the {220} planes. What is the wavelength of the x-rays used? (The lattice constant of nickel = 0.352236 nm; assume first-order diffraction, $n = 1$.)

*3.11.8   An x-ray diffractometer recorder chart for an element which has either the BCC or the FCC crystal structure showed diffraction peaks at the following 2θ angles: 38.184°, 44.392°, 64.576°, and 77.547°. (Wavelength of the incoming radiation was 0.154056 nm.)

(a) Determine the crystal structure of the element.

(b) Determine the lattice constant of the element.

(c) Identify the element.

*3.11.9   An x-ray diffractometer recorder chart for an element which has either the BCC or the FCC crystal structure showed diffraction peaks at the following 2θ angles: 25.062°, 35.698°, 44.116°, and 51.405°. (Wavelength λ of the incoming radiation was 0.154056 nm.)

(a) Determine the crystal structure of the element.

(b) Determine the lattice constant of the element.

(c) Identify the element.

*3.11.10 An x-ray diffractometer recorder chart for an element which has either the BCC or the FCC crystal structure showed diffraction peaks at the following 2θ angles: 38.116°, 44.277°, 64.426°, and 77.472°. (Wavelength of the incoming radiation was 0.154056 nm.)

(a) Determine the crystal structure of the element.

(b) Determine the lattice constant of the element.

(c) Identify the element.

*3.11.11 An x-ray diffractometer recorder chart for an element which has either the BCC or the FCC crystal structure showed diffraction peaks at the following 2θ angles: 40.113°, 46.659°, 68.080°, and 82.090°. (Wavelength λ of the incoming radiation was 0.154056 nm.)

(a) Determine the crystal structure of the element.

(b) Determine the lattice constant of the element.

(c) Identify the element.

*Data from International Centre for Diffraction Data

4

Solidification, Crystalline Imperfections, and Diffusion in Solids

4.1 SOLIDIFICATION OF METALS

The solidification of metals and alloys is an important industrial process since most metals are melted and then cast into a semifinished or finished shape. Figure 4.1 shows a large, semicontinuously[1] cast aluminum ingot which will be further fabricated into aluminum alloy flat products and illustrates the large scale on which the casting process (solidification) of metals is sometimes carried out.

In general the solidification of a metal or alloy can be divided into the following steps:

1. The formation of stable nuclei in the melt (nucleation) (Fig. 4.2a)
2. The growth of nuclei into crystals (Fig. 4.2b) and the formation of a grain structure (Fig. 4.2c)

---

[1]A semicontinuously cast ingot is produced by solidifying molten metal (e.g., aluminum or copper alloys) in a mold that has a movable bottom block (see Fig. 4.8) which is slowly lowered as the metal is solidified. The prefix *semi-* is used since the maximum length of the ingot produced is determined by the depth of the pit into which the bottom block is lowered.

FIGURE 4.1 Large, semicontinuously cast aluminum alloy ingot being removed from casting pit. Ingots of this type are subsequently hot- and cold-rolled into plate or sheet. (*Courtesy of Reynolds Metals Co.*)

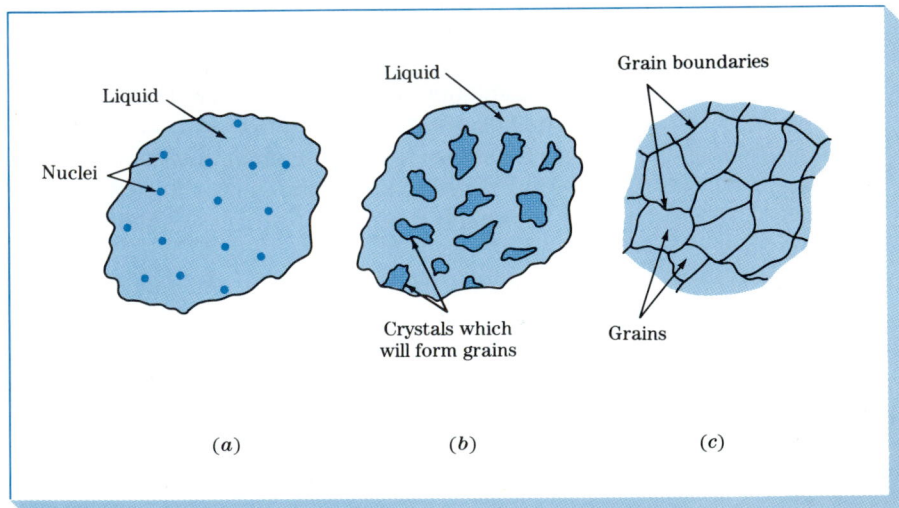

FIGURE 4.2 Schematic illustration showing the several stages in the solidification of metals: (*a*) formation of nuclei, (*b*) growth of nuclei into crystals, and (*c*) joining together of crystals to form grains and associated grain boundaries. Note that the grains are randomly oriented.

FIGURE 4.3 A grain grouping parted from an arc-cast titanium alloy ingot under the blows of a hammer. The grouping has preserved the true bonding facets of the individual grains of the original cast structure. (Magnification $\frac{1}{6}\times$.) (*After W. Rostoker and J. R. Dvorak, "Interpretation of Metallographic Structures," Academic, 1965, p. 7.*)

The shapes of some real grains formed by the solidification of a titanium alloy are shown in Fig. 4.3. The shape that each grain acquires after solidification of the metal depends on many factors, of which thermal gradients are important. The grains shown in Fig. 4.3 are *equiaxed* since their growth has been about equal in all directions.

The Formation of Stable Nuclei in Liquid Metals

The two main mechanisms by which nucleation of solid particles in liquid metal occurs are homogeneous nucleation and heterogeneous nucleation.

Homogeneous nucleation Homogeneous nucleation is considered first since it is the simplest case of nucleation. Homogeneous nucleation in a liquid melt occurs when the metal itself provides the atoms to form nuclei. Let us consider the case of a pure metal solidifying. When a pure liquid metal is cooled below its equilibrium freezing temperature to a sufficient degree, numerous homogeneous nuclei are created by slow-moving atoms bonding together. Homogeneous nucleation usually requires a considerable amount of undercooling which may be as much as several hundred degrees Celsius for some metals (see Table 4.1). For a nucleus to be stable so that it can grow into a crystal,

TABLE 4.1   Values for the Freezing Temperature, Heat of Solidification, Surface Energy, and Maximum Undercooling for Selected Metals

Metal	Freezing temp.		Heat of solidification, J/cm³	Surface energy, J/cm²	Maximum undercooling, observed, $\Delta T(°C)$
	°C	K			
Pb	327	600	−280	33.3×10^{-7}	80
Al	660	933	−1066	93×10^{-7}	130
Ag	962	1235	−1097	126×10^{-7}	227
Cu	1083	1356	−1826	177×10^{-7}	236
Ni	1453	1726	−2660	255×10^{-7}	319
Fe	1535	1808	−2098.	204×10^{-7}	295
Pt	1772	2045	−2160	240×10^{-7}	332

Source: B. Chalmers, "Solidification of Metals," Wiley, 1964.

it must reach a *critical size.* A cluster of atoms bonded together which is less than the critical size is called an *embryo,* and one which is larger than the critical size is called a *nucleus.* Because of their instability, embryos are continuously being formed and redissolved in the molten metal due to the agitation of the atoms.

Energies involved in homogeneous nucleation   In the homogeneous nucleation of a solidifying pure metal, two kinds of energy changes must be considered: (1) the *volume (or bulk) free energy* released by the liquid to solid transformation and (2) *the surface energy* required to form the new solid surfaces of the solidified particles.

When a pure liquid metal such as lead is cooled below its equilibrium freezing temperature, the driving energy for the liquid-to-solid transformation is the difference in the volume (bulk) free energy ΔG_v of the liquid and that of the solid. If ΔG_v is the change in free energy between the liquid and solid per unit volume of metal, then the free-energy change for a *spherical nucleus* of radius r is $\frac{4}{3}\pi r^3 \, \Delta G_v$ since the volume of a sphere is $\frac{4}{3}\pi r^3$. The change in volume free energy vs. radius of an embryo or nucleus is shown schematically in Fig. 4.4 as the lower curve and is a negative quantity since energy is released by the liquid to solid transformation.

However, there is an opposing energy to the formation of embryos and nuclei, the energy required to form the surface of these particles. The energy needed to create a surface for these spherical particles ΔG_s, is equal to the specific surface free energy of the particle, γ, times the area of the surface of

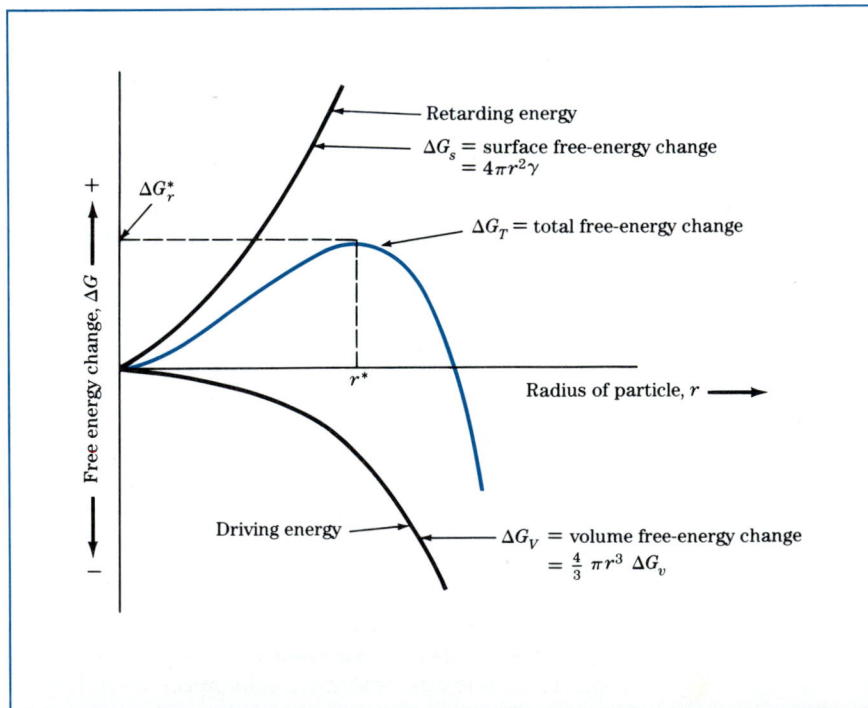

FIGURE 4.4 Free-energy change ΔG vs. radius of embryo or nucleus created by the solidifying of a pure metal. If the radius of the particle is greater than r^*, a stable nucleus will continue to grow.

the sphere, or $4\pi r^2\gamma$, where $4\pi r^2$ is the surface area of a sphere. This retarding energy ΔG_s for the formation of the solid particles is shown in Fig. 4.4 by an upward curve in the positive upper half of the figure. The total free-energy associated with the formation of an embryo or nucleus, which is the sum of the volume free-energy and surface free-energy changes, is shown in Fig. 4.4 as the middle curve. In equation form the total free-energy change for the formation of a spherical embryo or nucleus of radius r formed in a freezing pure metal is

$$\Delta G_T = \tfrac{4}{3}\pi r^3 \, \Delta G_v + 4\pi r^2\gamma \qquad (4.1)$$

where ΔG_T = total free energy change
$\quad r$ = radius of embryo or nucleus
$\quad \Delta G_v$ = volume free energy
$\quad \gamma$ = specific surface free energy

In nature a system can change spontaneously from a higher to a lower energy state. In the case of the freezing of a pure metal, if the solid particles formed upon freezing have radii less than the critical radius r^*, the energy of the system will be lowered if they redissolve. These small embryos can, therefore, redissolve in the liquid metal. However, if the solid particles have radii greater than r^*, the energy of the system will be lowered when these particles (nuclei) grow into larger particles or crystals (Fig. 4.2b). When r reaches the critical radius r^*, ΔG_T has its maximum value of ΔG^* (Fig. 4.4).

A relation among the size of the critical nucleus, surface free energy, and volume free energy for the solidification of a pure metal can be obtained by differentiating Eq. (4.1). The differential of the total free energy ΔG_T with respect to r is zero when $r = r^*$, since the total free energy vs. radius of the embryo or nucleus plot is then at a maximum and the slope $d(\Delta G_T)/dr = 0$. Thus,

$$\frac{d(\Delta G_T)}{dr} = \frac{d}{dr}\left(\frac{4}{3}\pi r^3 \, \Delta G_v + 4\pi r^2\gamma\right)$$

$$= \tfrac{12}{3}\pi r^{*2} \, \Delta G_v + 8\pi r^*\gamma = 0 \qquad \text{or} \qquad r^* = -\frac{2\gamma}{\Delta G_v}$$

Critical radius vs. undercooling The greater the degree of undercooling ΔT below the equilibrium melting temperature of the metal, the greater the change in volume free energy ΔG_V. However, the change in free energy due to the surface energy ΔG_s does not change much with temperature. Thus, the critical nucleus size is determined mainly by ΔG_V. Near the freezing temperature, the critical nucleus size must be infinite since ΔT approaches zero. As the amount of undercooling increases, the critical nucleus size decreases. Figure 4.5 shows the variation in critical nucleus size for copper as a function of undercooling. The maximum amount of undercooling for homogeneous nucleation in the pure metals listed in Table 4.1 is from 80 to 332°C. The critical-sized nucleus is related to the amount of undercooling by the relation

$$r^* = -\frac{2\gamma T_m}{\Delta H_s \, \Delta T} \qquad (4.2)$$

where r^* = critical radius of nucleus
$\quad \gamma$ = surface free energy
$\quad \Delta H_s$ = latent heat of solidification
$\quad \Delta T$ = amount of undercooling at which nucleus is formed

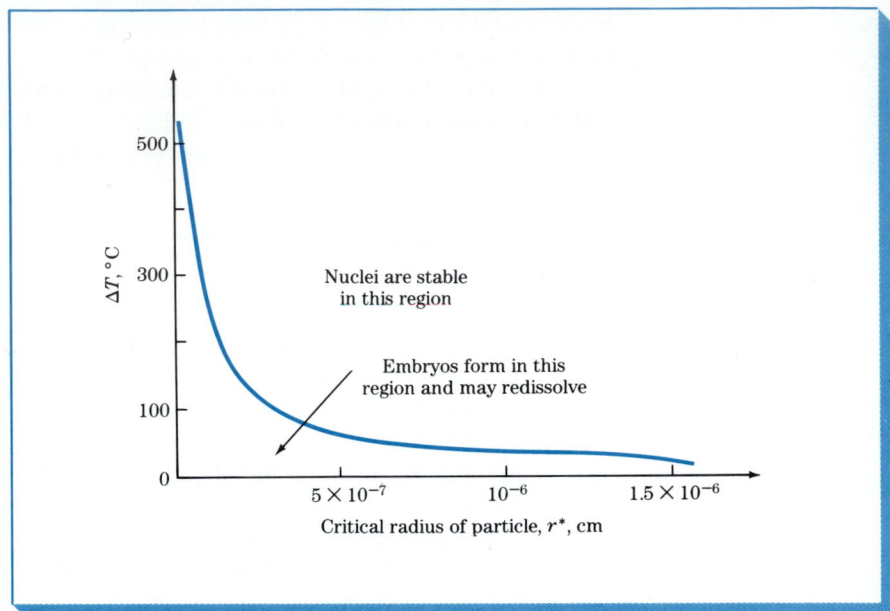

FIGURE 4.5 Critical radius of copper nuclei vs. degree of undercooling ΔT. (*After B. Chalmers, "Principles of Solidification," Wiley, 1964.*)

Example Problem 4.1 shows how a value for the number of atoms in a critical nucleus can be calculated from experimental data.

Example Problem 4.1

(*a*) Calculate the critical radius (in centimeters) of a homogeneous nucleus that forms when pure liquid copper solidifies. Assume ΔT (undercooling) = $0.2T_m$. Use data from Table 4.1.

(*b*) Calculate the number of atoms in the critical-sized nucleus at this undercooling.

Solution:

(*a*) Calculation of critical radius of nucleus:

$$r^* = -\frac{2\gamma T_m}{\Delta H_s \, \Delta T} \tag{4.2}$$

$$\Delta T = 0.2T_m = 0.2(1083°C + 273) = (0.2 \times 1356 \text{ K}) = 271 \text{ K}$$

$$\gamma = 177 \times 10^{-7} \text{ J/cm}^2 \qquad \Delta H_s = -1826 \text{ J/cm}^3 \qquad T_m = 1083°C = 1356 \text{ K}$$

$$r^* = \frac{-2(177 \times 10^{-7} \text{ J/cm}^2)(1356 \text{ K})}{(-1826 \text{ J/cm}^3)(271 \text{ K})} = 9.70 \times 10^{-8} \text{ cm} \blacktriangleleft$$

(b) Calculation of number of atoms in critical-sized nucleus:

$$\text{Vol. of critical-sized nucleus} = \tfrac{4}{3}\pi r^{*3} = \tfrac{4}{3}\pi(9.70 \times 10^{-8}\,\text{cm})^3 = 3.82 \times 10^{-21}\,\text{cm}^3$$

$$\text{Vol. of unit cell of Cu } (a = 0.361\,\text{nm}) = a^3$$
$$= (3.61 \times 10^{-8}\,\text{cm})^3$$
$$= 4.70 \times 10^{-23}\,\text{cm}^3$$

Since there are four atoms per FCC unit cell,

$$\text{Volume/atom} = \frac{4.70 \times 10^{-23}\,\text{cm}^3}{4} = 1.175 \times 10^{-23}\,\text{cm}^3$$

Thus, the number of atoms per homogeneous critical nucleus is

$$\frac{\text{Volume of nucleus}}{\text{Volume/atom}} = \frac{3.82 \times 10^{-21}\,\text{cm}^3}{1.175 \times 10^{-23}\,\text{cm}^3} = 325 \text{ atoms} \blacktriangleleft$$

Heterogeneous nucleation   Heterogeneous nucleation is nucleation that occurs in a liquid on the surfaces of its container, insoluble impurities, or other structural material which lower the critical free energy required to form a stable nucleus. Since large amounts of undercooling do not occur during industrial casting operations and usually range between 0.1 to 10°C, the nucleation must be heterogeneous and not homogeneous.

For heterogeneous nucleation to take place, the solid nucleating agent (impurity solid or container) must be wetted by the liquid metal. Also the liquid should solidify easily on the nucleating agent. Figure 4.6 shows a nucleating agent (substrate) which is wetted by the solidifying liquid creating a low contact angle θ between the solid metal and the nucleating agent. Heterogeneous

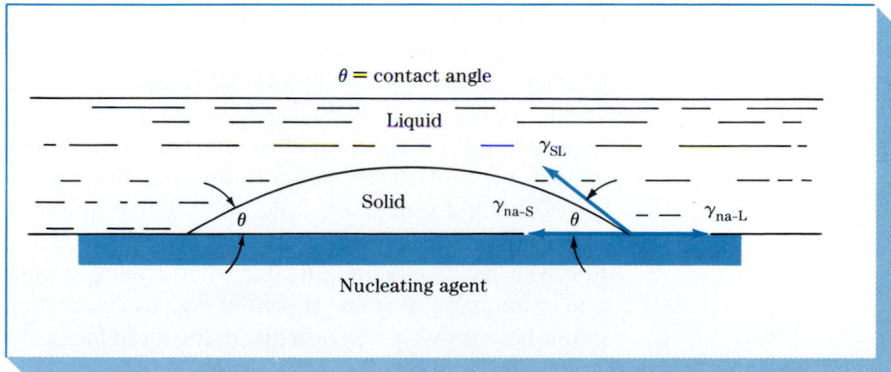

FIGURE 4.6 Heterogeneous nucleation of a solid on a nucleating agent. na = nucleating agent, SL = solid-liquid, S = solid, L = liquid; θ = contact angle. (*After J. H. Brophy, R. M. Rose, and John Wulff, "Structure and Properties of Materials," vol. II: "Thermodynamics of Structure," Wiley, 1964, p. 105.*)

nucleation takes place on the nucleating agent because the surface energy to form a stable nucleus is lower on this material than if the nucleus is formed in the pure liquid itself (homogeneous nucleation). Since the surface energy is lower for heterogeneous nucleation, the total free-energy change for the formation of a stable nucleus will be lower and the critical size of the nucleus will be smaller. Thus a much smaller amount of undercooling is required to form a stable nucleus produced by heterogeneous nucleation.

Growth of Crystals in Liquid Metal and Formation of a Grain Structure

After stable nuclei have been formed in a solidifying metal, these nuclei grow into crystals, as shown in Fig. 4.2b. In each solidifying crystal the atoms are arranged in an essentially regular pattern, but the orientation of each crystal varies (Fig. 4.2b). When solidification of the metal is finally completed, the crystals join together in different orientations and form crystal boundaries at which changes in orientation take place over a distance of a few atoms (Fig. 4.2c). Solidified metal containing many crystals is said to be *polycrystalline*. The crystals in the solidified metal are called *grains,* and the surfaces between them, *grain boundaries.*

The number of nucleation sites available to the freezing metal will affect the grain structure of the solid metal produced. If relatively few nucleation sites are available during solidification, a coarse, or large-grain, structure will be produced. If many nucleation sites are available during solidification, a fine-grain structure will result. Almost all engineering metals and alloys are cast with a fine-grain structure since this is the most desirable type for strength and uniformity of finished metal products.

When a relatively pure metal is cast into a stationary mold without the use of *grain refiners,*[1] two major types of grain structures are usually produced:

1. Equiaxed grains
2. Columnar grains

If the nucleation and growth conditions in the liquid metal during solidification are such that the crystals can grow approximately equally in all directions, *equiaxed grains* will be produced. Equiaxed grains are commonly found adjacent to a cold mold wall, as shown in Fig. 4.7. Large amounts of undercooling near the wall create a relatively high concentration of nuclei during solidification, a condition necessary to produce the equiaxed grain structure.

Columnar grains are long, thin coarse grains which are created when a metal solidifies relatively slowly in the presence of a steep temperature gradient. Relatively few nuclei are available when columnar grains are produced. Equiaxed and columnar grains are shown in Fig. 4.7. Note that in Fig. 4.7b the columnar grains have grown perpendicular to the mold faces since large thermal gradients were present in those directions.

---

[1] A grain refiner is a material added to a molten metal to attain finer grains in the final grain structure.

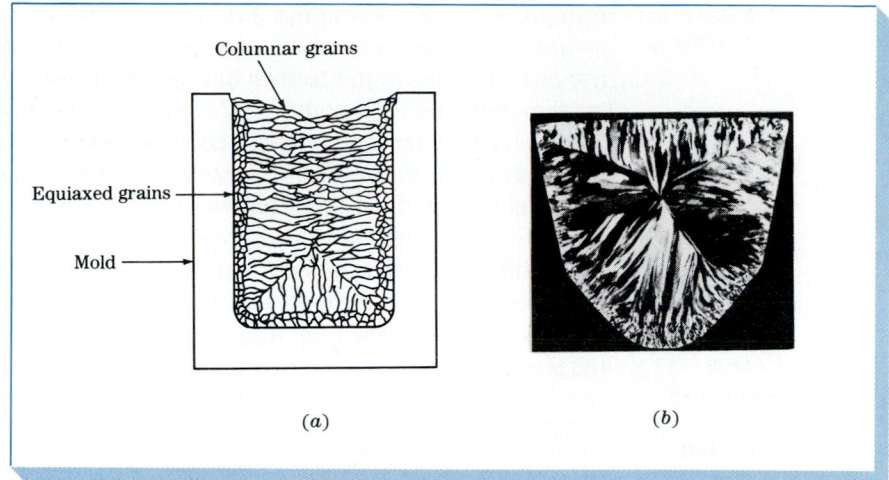

FIGURE 4.7 (*a*) Schematic drawing of a solidified metal grain structure produced by using a cold mold. (*b*) Transverse section through an ingot of aluminum alloy 1100 (99.0% Al) cast by the Properzi method (a wheel and belt method). Note the consistency with which columnar grains have grown perpendicular to each mold face. (*After "Metals Handbook," vol. 8, 8th ed., American Society for Metals, 1973, p. 164.*)

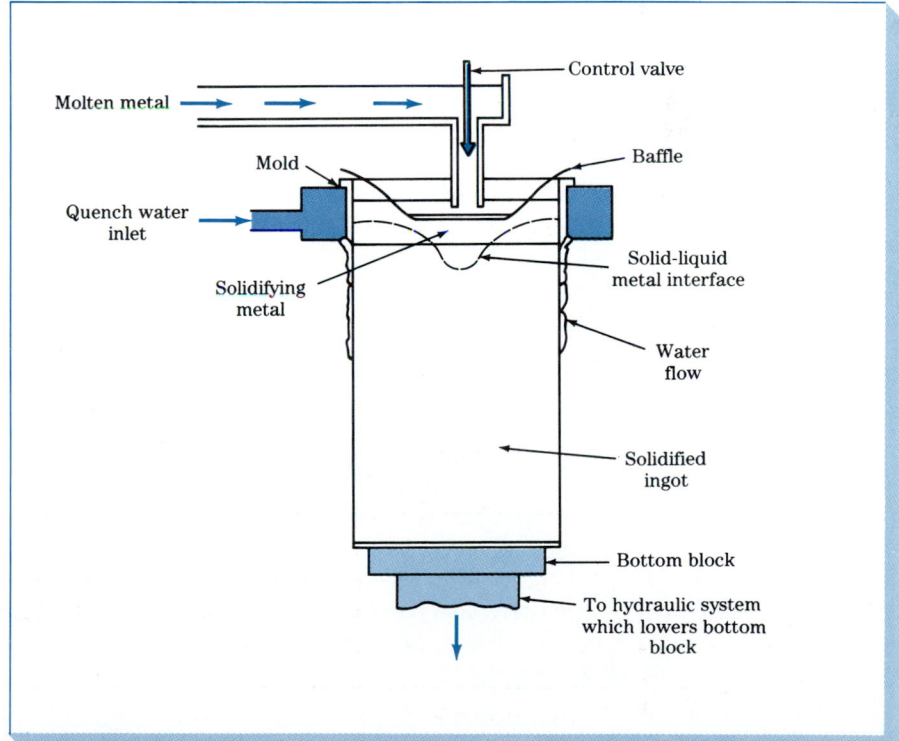

FIGURE 4.8 Schematic of an aluminum alloy ingot being cast in a direct-chill semicontinuous casting unit.

Grain Structure of Industrial Castings

In industry, metals and alloys are cast into various shapes. If the metal is to be further fabricated after casting, large castings of simple shapes are produced first and then fabricated further into semifinished products. For example, in the aluminum industry, common shapes for further fabrication are sheet ingots (Fig. 4.1), which have rectangular cross sections, and extrusion[1] ingots, which have circular cross sections. For some applications, the molten metal is cast into essentially its final shape as, for example, an automobile piston (Fig. 6.3).

The large aluminum alloy sheet ingot shown in Fig. 4.1 was cast by a direct-chill semicontinuous casting process. In this casting method the molten metal is cast into a mold with a movable bottom block which is slowly lowered after the mold is filled (Fig. 4.8). The mold is water-cooled by a water box, and water is also sprayed down the sides of the solidified surface of the ingot.

FIGURE 4.9 Continuous casting of steel ingots. (*a*) General setup; (*b*) Close-up of the mold arrangement. (*After "Making, Shaping, and Treating of Steel," 10th ed., Association of Iron and Steel Engineers, 1985.*)

---

[1] *Extrusion* is the process of converting a metal ingot into lengths of uniform cross section by forcing solid plastic metal through a die or orifice of the desired cross-sectional outline.

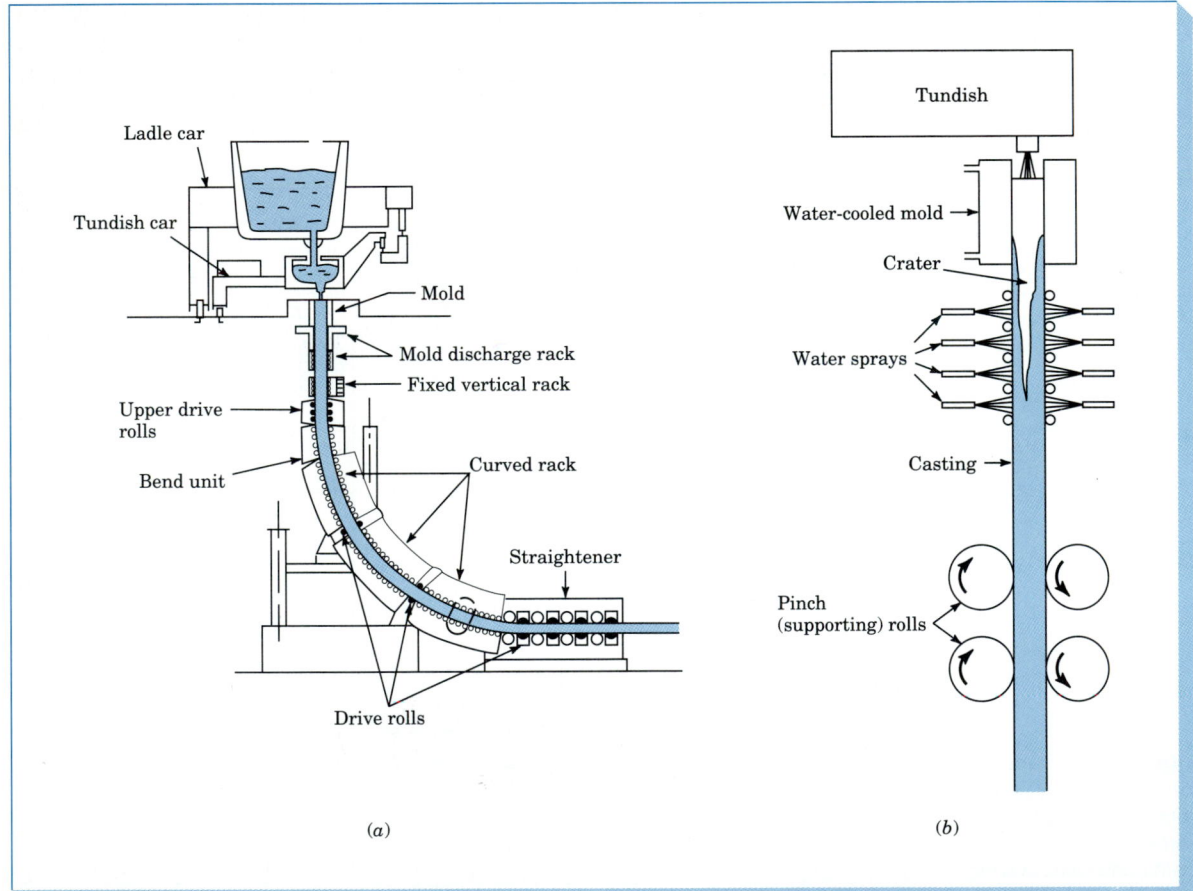

(*a*)   (*b*)

FIGURE 4.9c Continuous casting of steel. Note the steel ladle containing the molten steel in the mid-upper part of the photo. (*Courtesy of Bethlehem Steel Co.*)

FIGURE 4.9d Slabs of continuously cast steel being cut to required length by automatic flame torches. (*Courtesy of Bethlehem Steel Co.*)

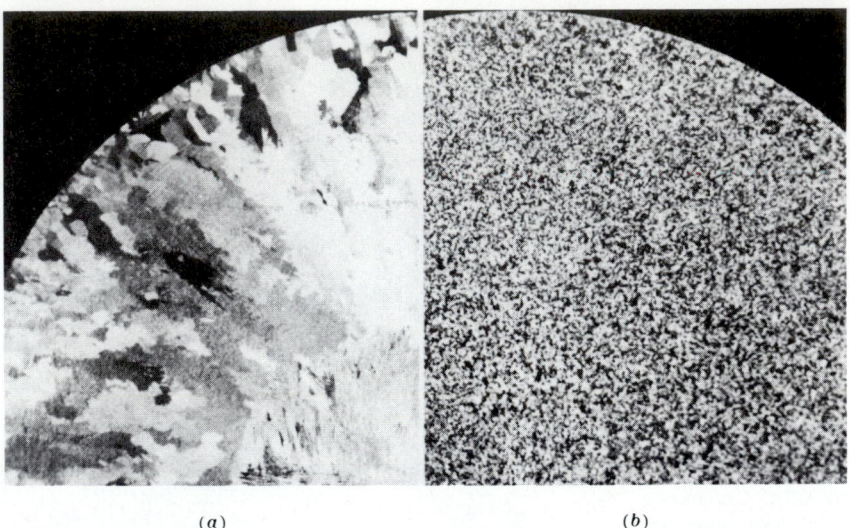

(a)   (b)

FIGURE 4.10 Parts of transverse sections through two 6-in-diameter ingots of alloy 6063 (Al–0.7% Mg–0.4% Si) that were direct-chill semicontinuous cast. Ingot section (a) was cast without the addition of a grain refiner; note columnar grains and colonies of featherlike crystals near the center of the section. Ingot section (b) was cast with the addition of a grain refiner and shows a fine, equiaxed grain structure. (Tucker's reagent; actual size.) (*After "Metals Handbook," vol. 8, 8th ed., American Society for Metals, 1973, p. 164.*)

In this way large ingots about 15 ft (4.6 m) long can be cast continuously, as shown in Fig. 4.1. In the U.S. steel industry about 85 percent of the crude steel is cast continuously.[1] Figure 4.9a shows the general setup for continuous casting of steel, Fig. 4.9b a close-up of the mold arrangement, Fig. 4.9c a photo of the actual process, and Fig. 4.9d a photo of the cast slabs exiting the casting setup.

To produce cast ingots with a fine-grain size, grain refiners are usually added to the liquid metal before casting. For aluminum alloys, small amounts of grain-refining elements such as titanium, boron, or zirconium are included in the liquid metal just before the casting operation so that a fine dispersion of heterogeneous nuclei will be available during solidification. Figure 4.10 shows the effect of using a grain refiner while casting 6-in (15-cm)-diameter aluminum extrusion ingots. The ingot section cast without the grain refiner has large co-lumnar grains (Fig. 4.10a, and the section cast with the grain refiner has a fine, equiaxed grain structure (Fig. 4.10b).

4.2 SOLIDIFICATION OF SINGLE CRYSTALS

Almost all engineering crystalline materials are composed of many crystals and are therefore *polycrystalline*. However, there are a few which consist of only one crystal and are therefore *single crystals*. For example, solid-state

---

[1]1993 data.

electronic components such as transistors and some types of diodes are made from single crystals of semiconducting elements and compounds. Single crystals are necessary for these applications since grain boundaries would disrupt the electrical properties of devices made from semiconducting materials.

In growing single crystals, solidification must take place around a single nucleus so that no other crystals are nucleated and grow. To accomplish this, the interface temperature between the solid and liquid must be slightly lower than the melting point of the solid and the liquid temperature must increase beyond the interface. To achieve this temperature gradient, the latent heat of solidification[1] must be conducted through the solidifying solid crystal. The growth rate of the crystal must be slow so that the temperature at the liquid-solid interface is slightly below the melting point of the solidifying solid.

In industry single crystals of silicon 6 to 8 in (15 to 20 cm) in diameter have been grown for semiconducting device applications. One of the commonly used techniques to produce high-quality (minimization of defects) silicon single crystals is the Czochralski method. In this process high-purity polycrystalline silicon is first melted in a nonreactive crucible and held at a temperature just above the melting point. A high-quality seed crystal of silicon of the desired orientation is lowered into the melt while it is rotated. Part of the surface of the seed crystal is melted in the liquid to remove the outer strained region and to produce a surface for the liquid to solidify on. The seed crystal continues to rotate and is slowly raised from the melt. As it is raised from the melt, silicon from the liquid in the crucible adheres and grows on the seed crystal, producing a much larger diameter single crystal of silicon (Figs. 4.11 and 4.12).

---

[1] The latent heat of solidification is the thermal energy released when a metal solidifies.

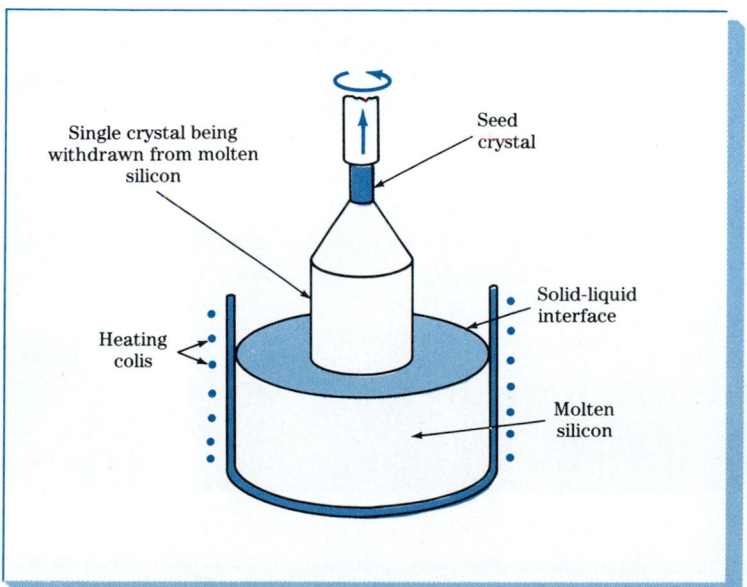

FIGURE 4.11 Formation of single crystal of silicon by the Czochralski process.

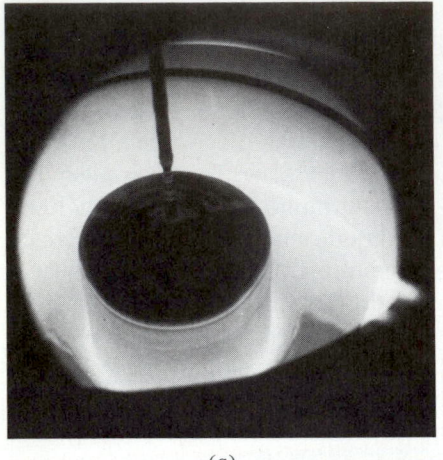 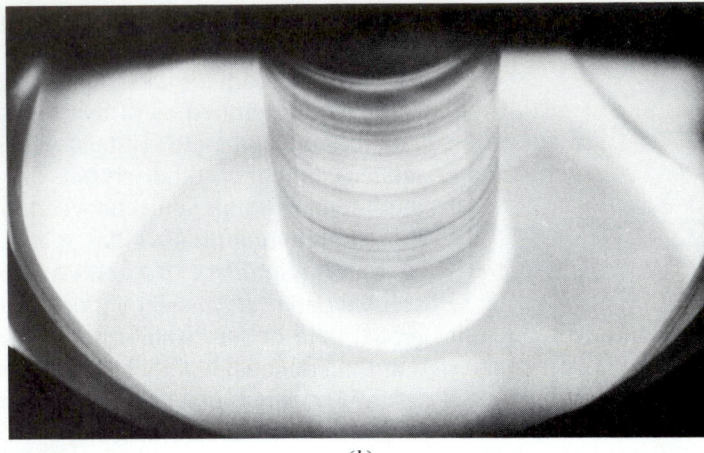

(a)   (b)

FIGURE 4.12 Formation of silicon single crystal by Czochralski method: (a) seed at top of growing single crystal; (b) new large single crystal at later stage in crystal growing process. (*Courtesy of Monsanto Co.*)

After the single crystal has been produced, it is ground to a precision diameter and then sliced into wafers about 1 mm thick (Fig. 4.13a). The wafers are then chemically etched and polished with successively finer polishing abrasives until a defect-free mirror finish is achieved (Fig. 4.13b). The wafers can then be used for semiconductor device fabrication.

There are other methods for producing single crystals, but the Czochralski method is one of the most important in the semiconductor industry. Also, there

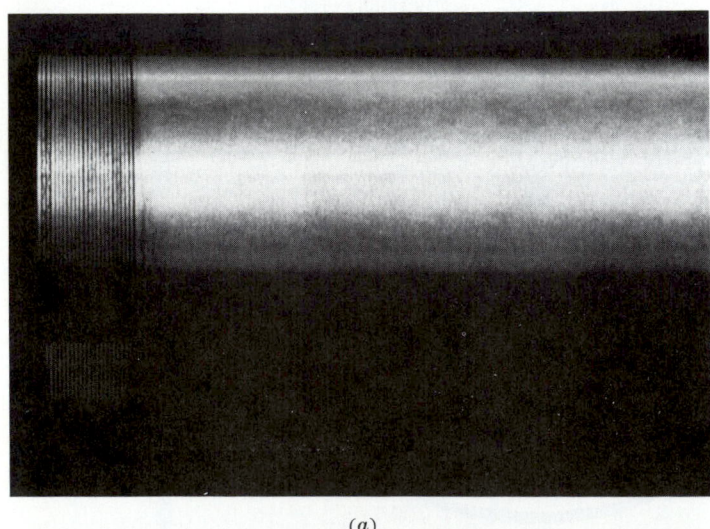

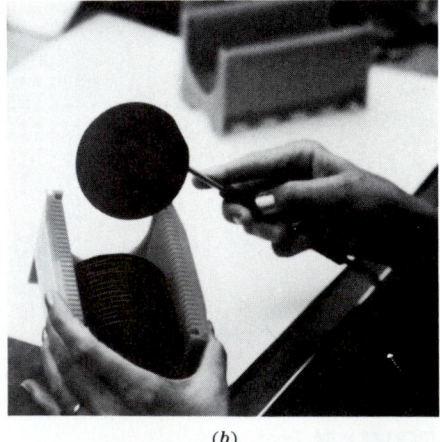

(a)   (b)

FIGURE 4.13 (a) Silicon single crystal, after being ground to precision diameter, is sliced into wafers. (b) Silicon wafers for the semiconductor device industry being inspected for defects. (*Courtesy of Monsanto Co.*)

are other uses for single crystals besides the production of silicon wafers. For example, single crystals are useful in research in the study of mechanical properties since the effects of grain boundaries and randomly oriented grains are eliminated.

4.3 METALLIC SOLID SOLUTIONS

Although very few metals are used in the pure or nearly pure state, a few are used in the nearly pure form. For example, high-purity copper of 99.99 percent purity is used for electronic wires because of its very high electrical conductivity. High-purity aluminum (99.99% Al) (called *superpure aluminum*) is used for decorative purposes because it can be finished with a very bright metallic surface. However, most engineering metals are combined with other metals or nonmetals to provide increased strength, higher corrosion resistance, or other desired properties.

A *metal alloy,* or simply an *alloy,* is a mixture of two or more metals or a metal (metals) and a nonmetal (nonmetals). Alloys can have structures that are relatively simple, such as that of cartridge brass which is essentially a binary alloy (two metals) of 70% Cu and 30% Zn. On the other hand, alloys can be extremely complex, such as the nickel-base superalloy Inconel 718 used for jet engine parts which has about 10 elements in its nominal composition.

The simplest type of alloy is that of the solid solution. A solid solution is a *solid* that consists of two or more elements atomically dispersed in a single-phase structure. In general there are two types of solid solutions: *substitutional* and *interstitial.*

Substitutional Solid Solutions In substitutional solid solutions formed by two elements, solute atoms can substitute for parent solvent atoms in a crystal lattice. Figure 4.14 shows a

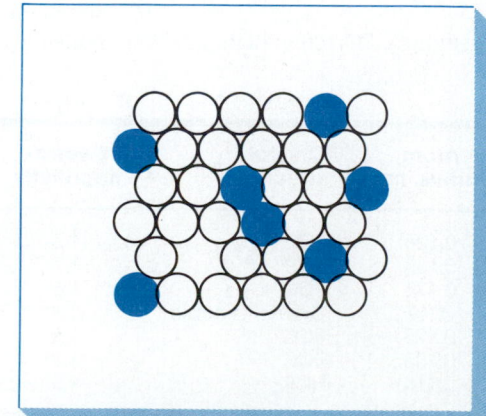

FIGURE 4.14
Substitutional solid solution. The dark circles represent one type of atom and the white another. The plane of atoms is a (111) plane in an FCC crystal lattice.

(111) plane in an FCC crystal lattice in which some solute atoms of one element have substituted for solvent atoms of the parent element. The crystal structure of the parent element or solvent is unchanged, but the lattice may be distorted by the presence of the solute atoms, particularly if there is a significant difference in atomic diameters of the solute and solvent atoms.

The fraction of atoms of one element that can dissolve in another can vary from a fraction of an atomic percent to 100 percent. The following conditions are favorable for extensive solid solubility of one element in another:

1. The diameters of the atoms of the elements must not differ by more than about 15 percent.
2. The crystal structures of the two elements must be the same.
3. There should be no appreciable difference in the electronegativities of the two elements so that compounds will not form.
4. The two elements should have the same valence.

If the atomic diameters of the two elements which form a solid solution differ, there will be a distortion of the crystal lattice. Since the atomic lattice can only sustain a limited amount of contraction or expansion, there is a limit in the difference in atomic diameters that atoms can have and still maintain a solid solution with the same kind of crystal structure. When the atomic diameters differ by more than about 15 percent, the "size factor" becomes unfavorable for extensive solid solubility.

Example Problem 4.2

Using the data in the table below, predict the relative degree of atomic solid solubility of the following elements in copper:

(a) Zinc
(b) Lead
(c) Silicon
(d) Nickel
(e) Aluminum
(f) Beryllium

Use the scale very high, 70–100%; high, 30–70%; moderate, 10–30%; low 1–10%; and very low, <1%.

Element	Atom radius, nm	Crystal structure	Electro-negativity	Valence
Copper	0.128	FCC	1.8	+2
Zinc	0.133	HCP	1.7	+2
Lead	0.175	FCC	1.6	+2, +4
Silicon	0.117	Diamond cubic	1.8	+4
Nickel	0.125	FCC	1.8	+2
Aluminum	0.143	FCC	1.5	+3
Beryllium	0.114	HCP	1.5	+2

Solution:

A sample calculation for the atomic radius difference for the Cu–Zn system is

$$\text{Atomic radius difference} = \frac{\text{final radius} - \text{initial radius}}{\text{initial radius}}(100\%)$$

$$= \frac{R_{Zn} - R_{Cu}}{R_{Cu}}(100\%) \qquad (4.3)$$

$$= \frac{0.133 - 0.128}{0.128}(100\%) = +3.9\%$$

System	Atomic radius difference, %	Electronegativity difference	Predicted relative degree of solid solubility	Observed maximum solid solubility, at %
Cu–Zn	+3.9	0.1	High	38.3
Cu–Pb	+36.7	0.2	Very low	0.1
Cu–Si	−8.6	0	Moderate	11.2
Cu–Ni	−2.3	0	Very high	100
Cu–Al	+11.7	0.3	Moderate	19.6
Cu–Be	−10.9	0.3	Moderate	16.4

The predictions can be made principally on the atomic radius difference. In the case of the Cu–Si system, the difference in the crystal structures is important. There is very little electronegativity difference for all these systems. The valences are all the same except for Al and Si. In the final analysis the experimental data must be referred to.

If the solute and solvent atoms have the same crystal structure, then extensive solid solubility is favorable. If the two elements are to show complete solid solubility in all proportions, then both elements must have the same crystal structure. Also, there cannot be too great a difference in the electronegativities of the two elements forming solid solutions, or else the highly electropositive element will lose electrons, the highly electronegative element will acquire electrons, and compound formation will result. Finally, if the two solid elements have the same valence, solid solubility will be favored. If there is a shortage of electrons between the atoms, the binding between them will be upset, resulting in conditions unfavorable for solid solubility.

Interstitial Solid Solutions

In interstitial solutions the solute atoms fit into the spaces between the solvent or parent atoms. These spaces or voids are called *interstices*. Interstitial solid solutions can form when one atom is much larger than another. Examples of atoms that can form interstitial solid solutions due to their small size are hydrogen, carbon, nitrogen, and oxygen.

An important example of an interstitial solid solution is that formed by

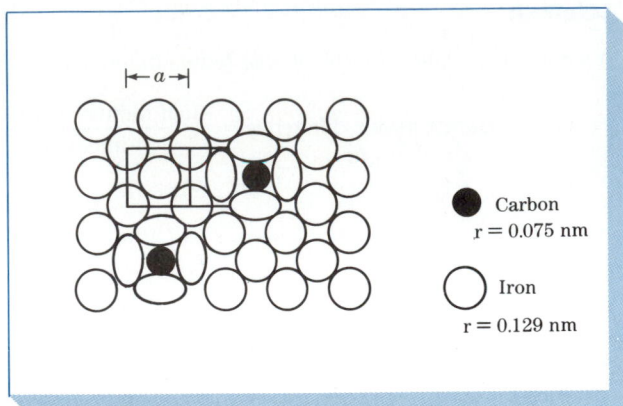

carbon in FCC γ iron which is stable between 912–1394°C. The atomic radius of γ iron is 0.129 nm and that of carbon is 0.075 nm, and so there is an atomic radius difference of 42 percent. However, in spite of this difference, a maximum of 2.08 percent of the carbon can dissolve interstitially in iron at 1148°C. Figure 4.15a illustrates this schematically by showing distortion around the carbon atoms in the γ iron lattice.

The radius of the largest interstitial hole in FCC γ iron is 0.053 nm (see Example Problem 4.3), and since the atomic radius of the carbon atom is 0.075 nm, it is not surprising that the maximum solid solubility of carbon in γ iron is only 2.08 percent. The radius of the largest interstitial void in BCC α iron is only 0.036 nm, and as a result just below 723°C only 0.025 percent of the carbon can be dissolved interstitially.

Example Problem 4.3

Calculate the radius of the largest interstitial void in the FCC γ iron lattice. The atomic radius of the iron atom is 0.129 nm in the FCC lattice, and the largest interstitial voids occur at the $(\frac{1}{2}, 0, 0)$, $(0, \frac{1}{2}, 0)$, $(0, 0, \frac{1}{2})$, etc., type positions.

Solution:

Figure 4.15b shows a (100) FCC lattice plane on the yz plane. Let the radius of an iron atom be R and that of the interstitial void at the position $(0, \frac{1}{2}, 0)$ be r. Then, from Fig. 4.15b,

$$2R + 2r = a \qquad (4.4)$$

Also from Fig. 4.15b,

$$(2R)^2 = (\tfrac{1}{2}a)^2 + (\tfrac{1}{2}a)^2 = \tfrac{1}{2}a^2 \qquad (4.5)$$

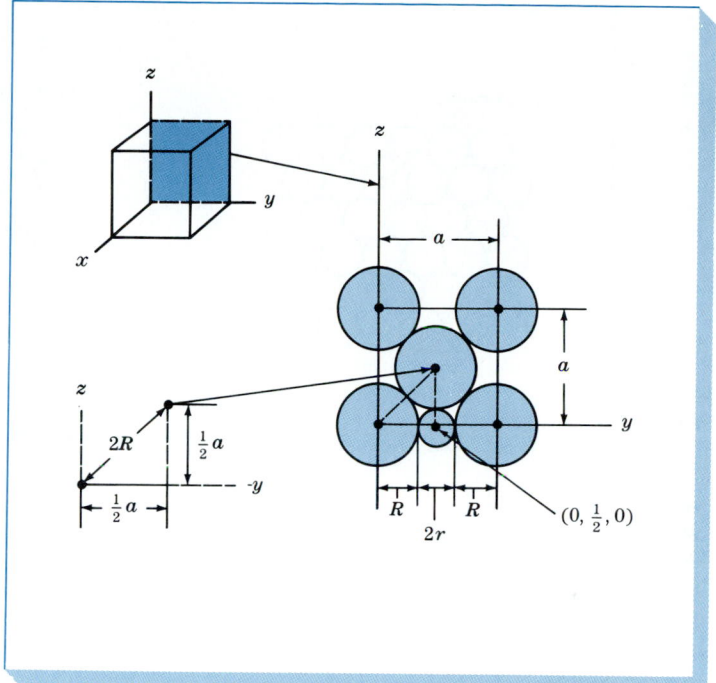

FIGURE 4.15b (100) plane of the FCC lattice containing an interstitial atom at the $(0, \frac{1}{2}, 0)$ position coordinate.

Solving for a gives

$$2R = \frac{1}{\sqrt{2}} a \quad \text{or} \quad a = 2\sqrt{2}R \tag{4.6}$$

Combining Eqs. (4.4) and (4.6) gives

$$2R + 2r = 2\sqrt{2}R$$

$$r = (\sqrt{2} - 1)R = 0.414R$$

$$= (0.414)(0.129 \text{ nm}) = 0.053 \text{ nm} \blacktriangleleft$$

4.4 CRYSTALLINE IMPERFECTIONS

In reality crystals are never perfect and contain various types of imperfections and defects which affect many of their physical and mechanical properties, which in turn affect many important engineering properties of materials such as the cold formability of alloys, the electronic conductivity of semiconductors, the rate of migration of atoms in alloys, and the corrosion of metals.

Crystal lattice imperfections are classified according to their geometry and shape. The three main divisions are (1) zero-dimension point defects, (2) one-

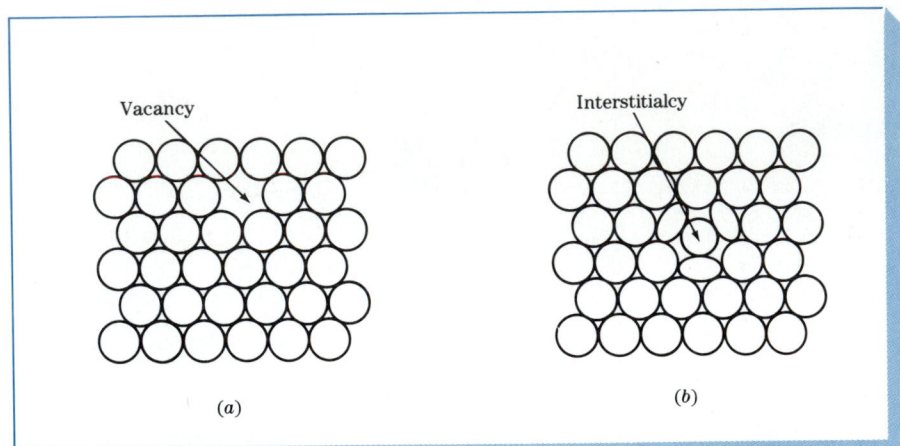

FIGURE 4.16 (*a*) Vacancy point defect. (*b*) Self-interstitial, or interstitialcy, point defect in a close-packed solid-metal lattice.

dimension or line defects (dislocations), and (3) two-dimension defects which include external surfaces and internal grain boundaries. Three-dimension macroscopic or bulk defects could also be included. Examples of these defects are pores, cracks, and foreign inclusions.

Point Defects   The simplest point defect is the vacancy, an atom site from which an atom is missing (Fig. 4.16*a*). *Vacancies* may be produced during solidification as a result of local disturbances during the growth of crystals, or they may be created by atomic rearrangements in an existing crystal due to atomic mobility. In metals the equilibrium concentration of vacancies rarely exceeds about 1 in 10,000 atoms. Vacancies are equilibrium defects in metals, and their energy of formation is about 1 eV.

Additional vacancies in metals can be introduced by plastic deformation, rapid cooling from higher temperatures to lower ones to entrap the vacancies, and by bombardment with energetic particles such as neutrons. Nonequilibrium vacancies have a tendency to cluster, causing divacancies or trivacancies to form. Vacancies can move by exchanging positions with their neighbors. This process is important in the migration or diffusion of atoms in the solid state, particularly at elevated temperatures where atomic mobility is greater.

Sometimes an atom in a crystal can occupy an interstitial site between surrounding atoms in normal atom sites (Fig. 4.16*b*). This type of point defect is called a *self-interstitial,* or *interstitialcy.* These defects do not generally occur naturally because of the structural distortion they cause, but they can be introduced into a structure by irradiation.

In ionic crystals point defects are more complex due to the necessity to maintain electrical neutrality. When two oppositely charged ions are missing from an ionic crystal, a cation-anion divacancy is created which is known as a *Schottky imperfection* (Fig. 4.17). If a positive cation moves into an interstitial site in an ionic crystal, a cation vacancy is created in the normal ion site. This

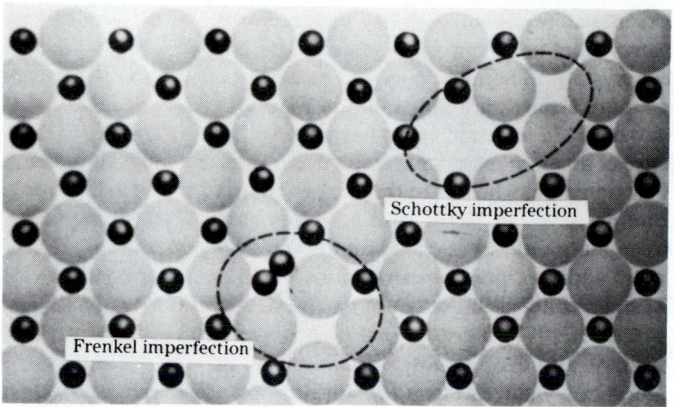

FIGURE 4.17 Two-dimensional representation of an ionic crystal illustrating a Schottky defect and a Frenkel defect. (*After Wulff et al., "Structure and Properties of Materials," vol. 1: "Structure," Wiley, 1964, p. 78.*)

Schottky imperfection

Frenkel imperfection

vacancy-interstitialcy pair is called a *Frenkel*[1] *imperfection* (Fig. 4.17). The presence of these defects in ionic crystals increases their electrical conductivity.

Impurity atoms of the substitutional or interstitial type are also point defects and may be present in metallic or covalently bonded crystals. For example, very small amounts of substitutional impurity atoms in pure silicon can greatly affect its electrical conductivity for use in electronic devices. Impurity ions are also point defects in ionic crystals.

Line Defects (Dislocations)

Line imperfections, or *dislocations,* in crystalline solids are defects that cause lattice distortion centered around a line. Dislocations are created during the solidification of crystalline solids. They are also formed by the permanent or plastic deformation of crystalline solids and by vacancy condensation and by atomic mismatch in solid solutions.

The two main types of dislocations are the *edge* and *screw types.* A combination of the two gives *mixed dislocations,* which have edge and screw components. An edge dislocation is created in a crystal by the insertion of an extra half plane of atoms, as shown in Fig. 4.18*a* just above the symbol ⊥. The inverted "tee," ⊥, indicates a positive edge dislocation, whereas the upright "tee," ⊤, indicates a negative edge dislocation.

The displacement distance of the atoms around the dislocation is called the *slip* or *Burgers vector* **b** and is *perpendicular* to the edge-dislocation line (Fig. 4.18*b*). Dislocations are nonequilibrium defects and store energy in the distorted region of the crystal lattice around the dislocation. The edge dislocation has a region of compressive strain where the extra half plane is and a region of tensile strain below the extra half plane of atoms (Fig. 4.19*a*).

The screw dislocation can be formed in a perfect crystal by applying upward and downward shear stresses to regions of a perfect crystal which have been separated by a cutting plane, as shown in Fig. 4.20*a*. These shear stresses introduce a region of distorted crystal lattice in the form of a spiral ramp of

[1]Yakov Ilyich Frenkel (1894–1954). Russian physicist who studied defects in crystals. His name is associated with the vacancy-interstitialcy defect found in some ionic crystals.

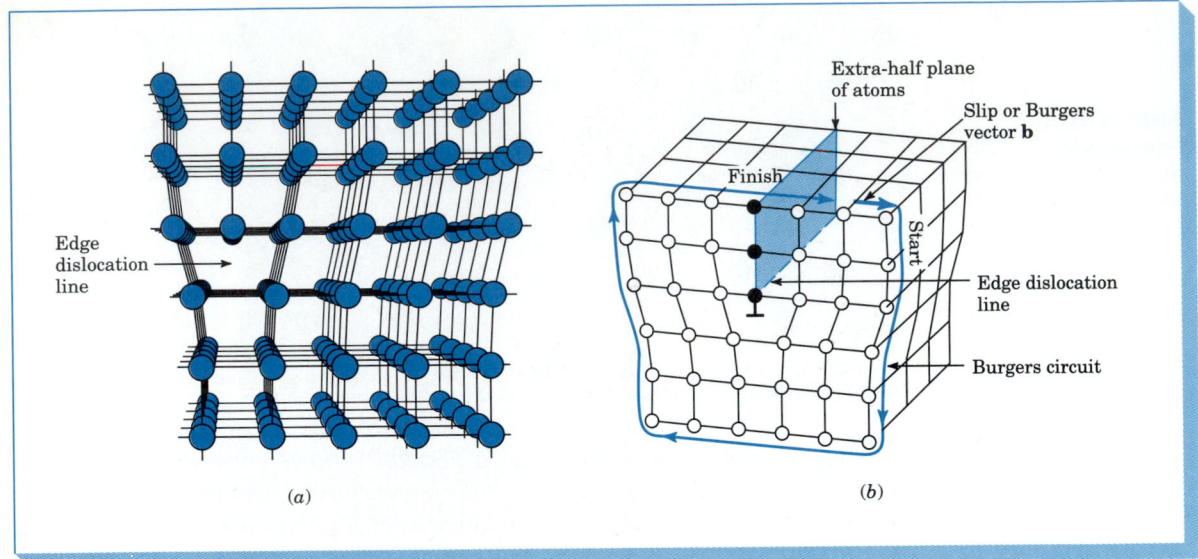

FIGURE 4.18 (*a*) Positive edge dislocation in a crystalline lattice. A linear defect occurs in the region just above the inverted "tee," ⊥, where an extra half plane of atoms has been wedged in. (*After A. G. Guy, "Essentials of Materials Science," McGraw-Hill, 1976, p. 153.*) (*b*) Edge dislocation which indicates the orientation of its Burgers or slip vector **b**. (*After M. Eisenstadt, "Introduction to Mechanical Properties of Materials," Macmillan, 1971, p. 117.*)

FIGURE 4.19 Strain fields surrounding (*a*) an edge dislocation and (*b*) a screw dislocation. (*After John Wulff et al., "The Structure and Properties of Materials," vol. 3: "Mechanical Behavior," Wiley, 1965, p. 69.*)

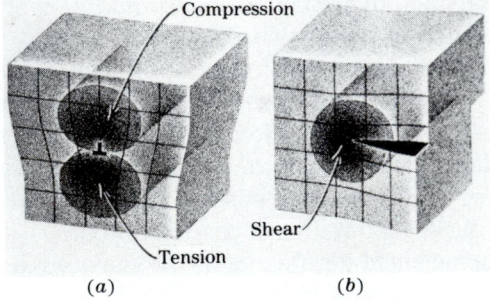

distorted atoms or screw dislocation (Fig. 4.20*b*). The region of distorted crystal is not well-defined and is at least several atoms in diameter. A region of shear strain is created around the screw dislocation in which energy is stored (Fig. 4.19*b*). The slip or Burgers vector of the screw dislocation is *parallel* to the dislocation line, as shown in Fig. 4.20*b*.

Most dislocations in crystals are of the mixed type, having edge and screw components. In the curved dislocation line *AB* in Fig. 4.21, the dislocation is of the pure screw type at the left where it enters the crystal and of the pure edge type on the right where it leaves the crystal. Within the crystal, the dislocation is of the mixed type, with edge and screw components.

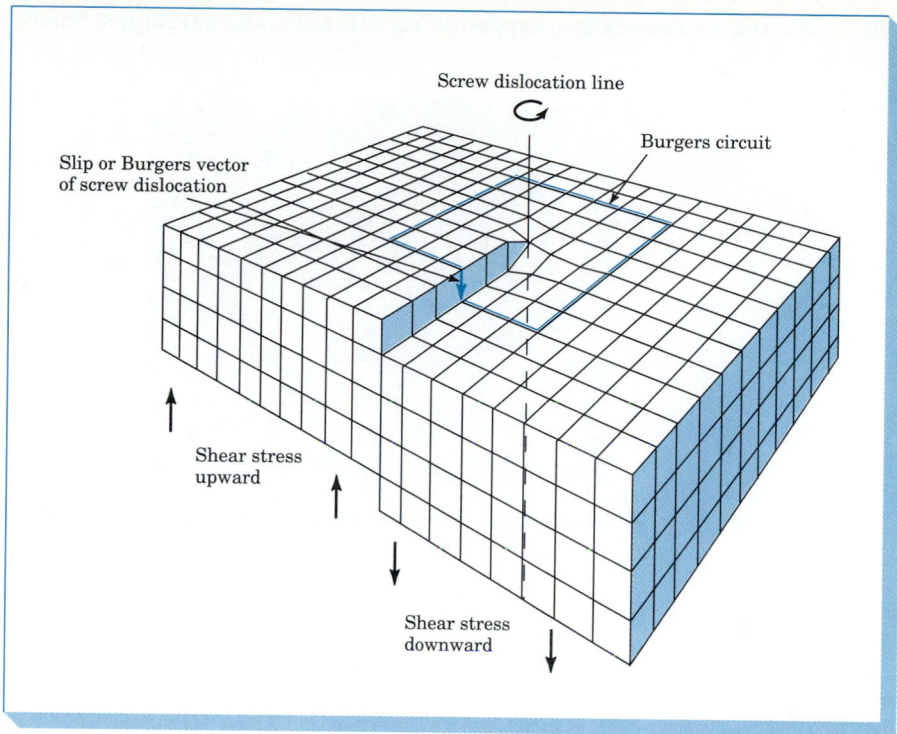

FIGURE 4.20 Screw dislocation in a cubic crystal lattice. The screw dislocation is created by a cutting plane undergoing upward and downward shearing stresses. A screw dislocation consists of a spiral ramp of distorted atoms and is represented by a line in the drawing. The extent of the distortion is not defined but is at least several atoms or more. The slip (Burgers) vector of the screw dislocation is parallel to the dislocation line. (*After W. T. Read, "Dislocations in Crystals," McGraw-Hill, 1953.*)

FIGURE 4.21 Mixed dislocation in a crystal. Dislocation line *AB* is pure screw type where it enters crystal on left and pure edge type where it leaves crystal on right. (*After John Wulff et al., "Structure and Properties of Materials," vol. 3: "Mechanical Properties," Wiley, 1965, p. 65.*)

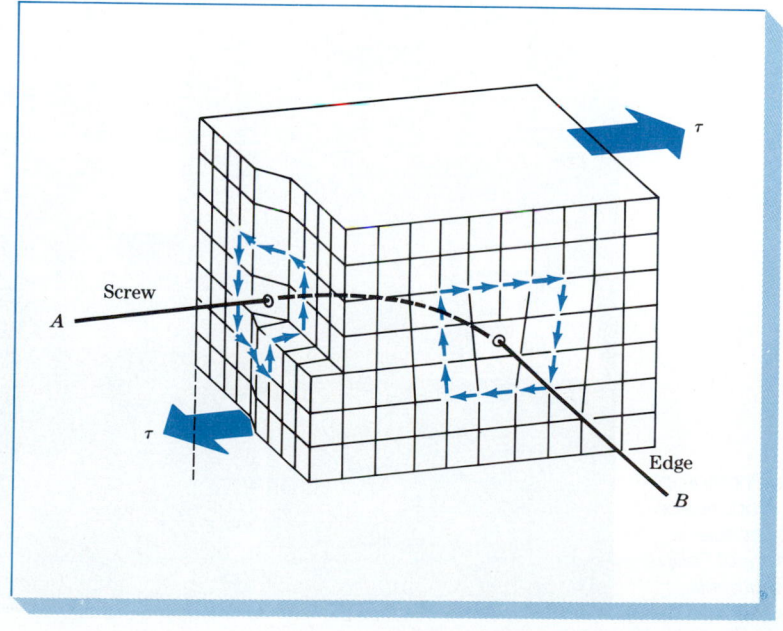

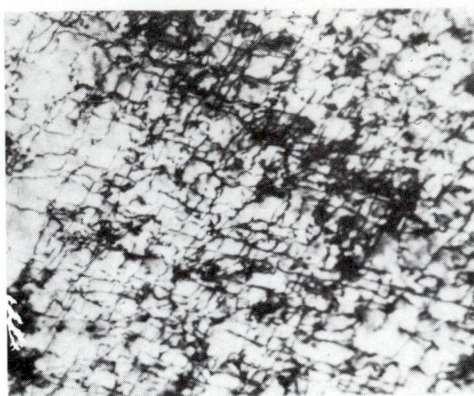

FIGURE 4.22 Dislocation structure in iron deformed 14 percent at −195°C. The dislocations appear as dark lines because electrons have been scattered along the linear irregular atomic arrangements of the dislocations. (Thin foil specimen; magnification 40,000×.) (*After "Metals Handbook," vol. 8, 8th ed., American Society for Metals, 1973, p. 219.*)

Dislocations can be observed on the image screen of a transmission electron microscope, as shown in Fig. 4.22 for a thin foil of iron deformed 14 percent at −195°C. A picture of a transmission electron microscope that can be used to produce such images is shown in Fig. 4.23. Figure 4.24 shows

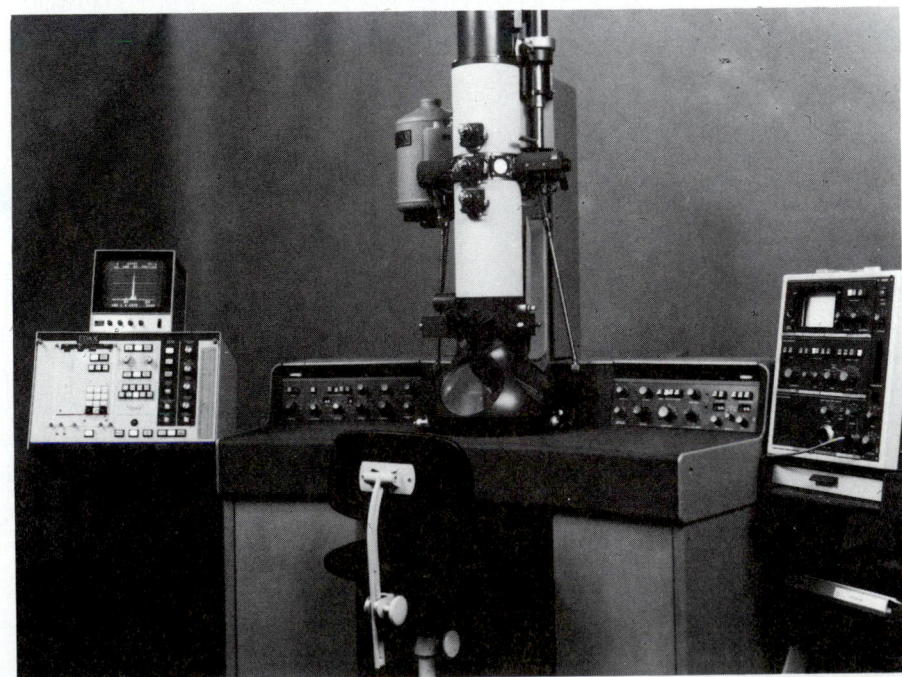

FIGURE 4.23 Photograph of a modern transmission electron microscope (TEM). (*Courtesy of Philips Electronic Instruments, Inc.*)

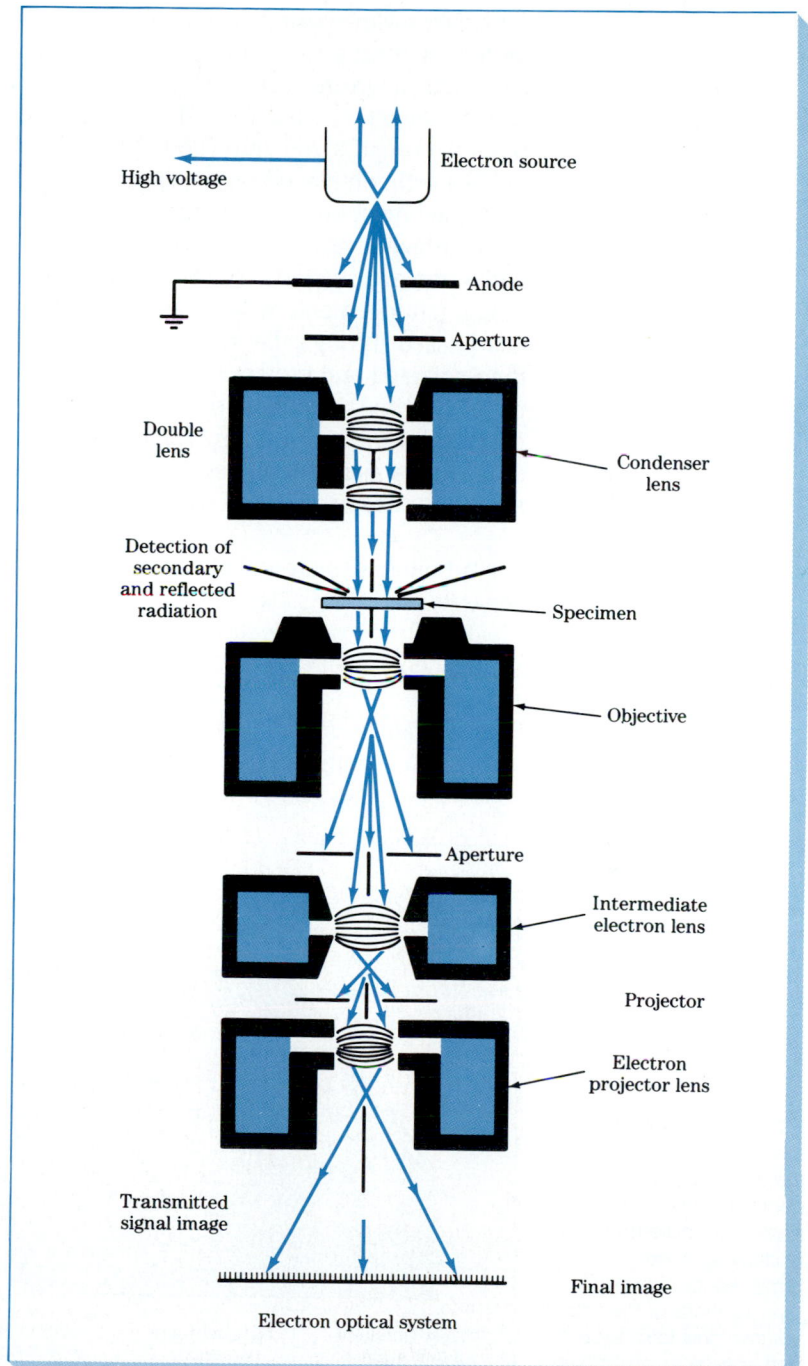

FIGURE 4.24 Schematic arrangement of electron lens system in a transmission electron microscope. All the lenses are enclosed in a column which is evacuated during operation. The path of the electron beam from the electron source to the final projected transmitted image is indicated by arrows. A specimen thin enough to allow an electron beam to be transmitted through it is placed between the condenser and objective lenses as indicated. (*After L. E. Murr, "Electron and Ion Microscopy and Microanalysis," Marcel Decker, 1982, p. 105.*)

schematically the basic principles of its operation. An electron beam is produced by a heated tungsten filament at the top of an evacuated column and is accelerated down the column by a high voltage (usually from 75 to 120 kV). Electromagnetic coils are used to condense the electron beam, and then it is passed through a very thin (i.e., about 100 nm thick or less) section of a thin metal specimen placed on the specimen stage (Fig. 4.24). The specimen area examined must be very thin so that some of the electrons entering are able to pass through it. As the electrons pass through the specimen, some are absorbed and some are scattered so that they change direction. Differences in crystal atomic arrangements will cause electron scattering. After the electron beam has passed through the specimen, it is focused with the objective coil and then enlarged and projected on a fluorescent screen (Fig. 4.24). A region in a metal specimen which tends to scatter electrons to a high degree will appear dark on the viewing screen. Thus dislocations which have an irregular linear atomic arrangement will appear as dark lines on the electron microscope screen, as observed in Fig. 4.22.

Grain Boundaries (Planar Defects)

Grain boundaries are interfacial imperfections in polycrystalline materials that separate grains (crystals) of different orientations. In metals grain boundaries are created during solidification when crystals formed from different nuclei grow simultaneously and meet each other (Fig. 4.2). The shape of the grain boundaries is determined by the restrictions imposed by the growth of neighboring grains. Grain-boundary surfaces of an approximately equiaxed grain structure are shown schematically in Fig. 4.25 and of real grains in Fig. 4.3.

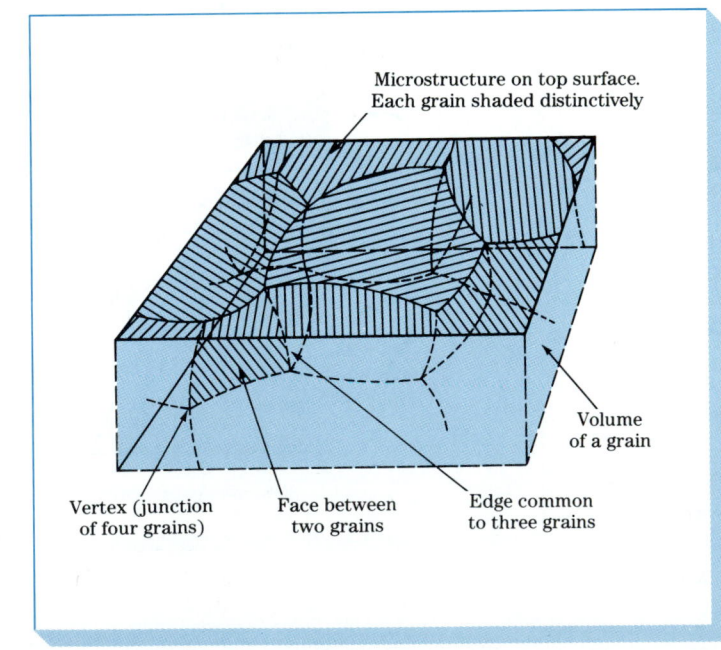

Microstructure on top surface. Each grain shaded distinctively

Volume of a grain

Vertex (junction of four grains)

Face between two grains

Edge common to three grains

FIGURE 4.25 Sketch showing the relation of the two-dimensional microstructure of a crystalline material to the underlying three-dimensional network. Only portions of the total volume and total face of any one grain are shown. (After A. G. Guy, "Essentials of Materials Science," McGraw-Hill, 1976.)

FIGURE 4.26 Grain boundaries on the surface of polished and etched samples as revealed in the optical microscope. (*a*) Low-carbon steel (magnification 100×). (*After "Metals Handbook," vol. 7, 8th ed., American Society for Metals, 1972, p. 4.*) (*b*) Magnesium oxide (magnification 225×). [*After R. E. Gardner and G. W. Robinson, J. Am. Ceram. Soc.,* **45**:46 (1962).]

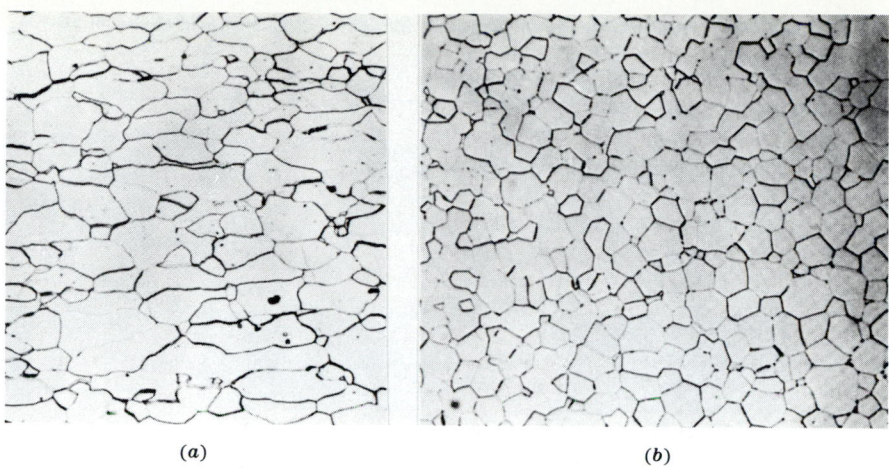

(*a*)　　　　　　　　　　　　　　(*b*)

The grain boundary itself is a narrow region between two grains of about two to five atomic diameters in width and is a region of atomic mismatch between adjacent grains. The atomic packing in grain boundaries is lower than

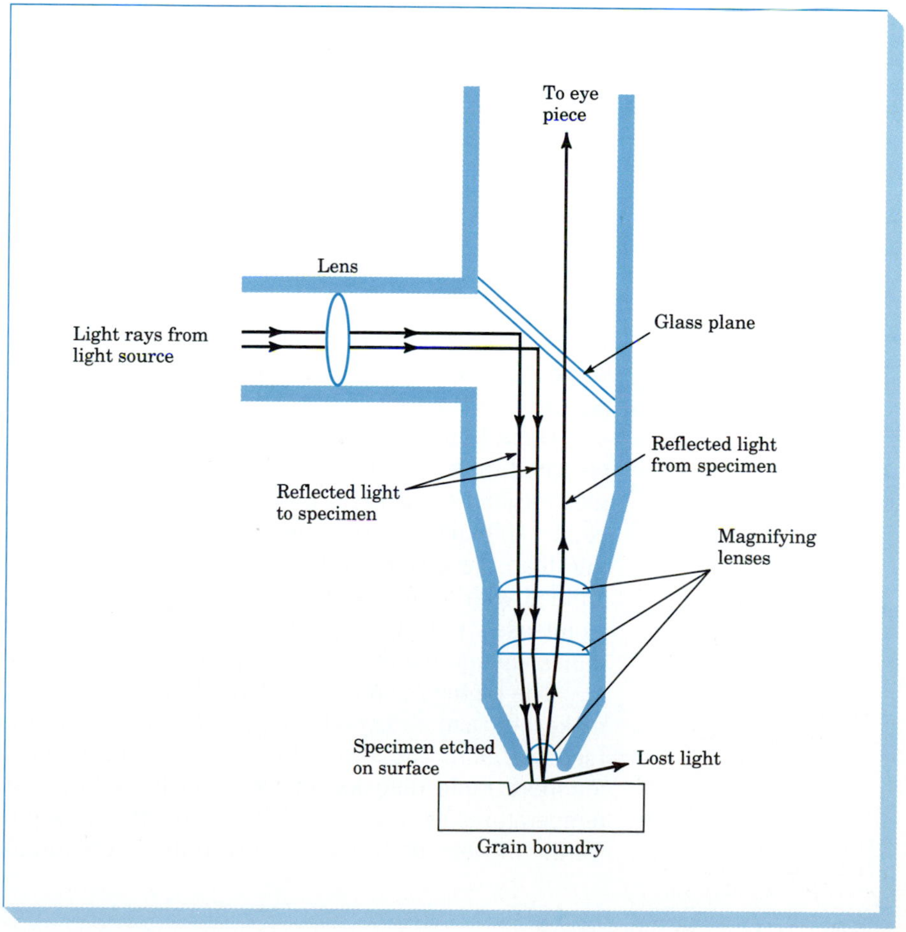

FIGURE 4.27 Schematic diagram of a side view of an optical microscope used to reflect magnified light from a polished and etched metal sample. The region that is roughened by etching does not reflect light well and appears dark.

To eye piece

Lens

Light rays from light source

Glass plane

Reflected light from specimen

Reflected light to specimen

Magnifying lenses

Specimen etched on surface

Lost light

Grain boundry

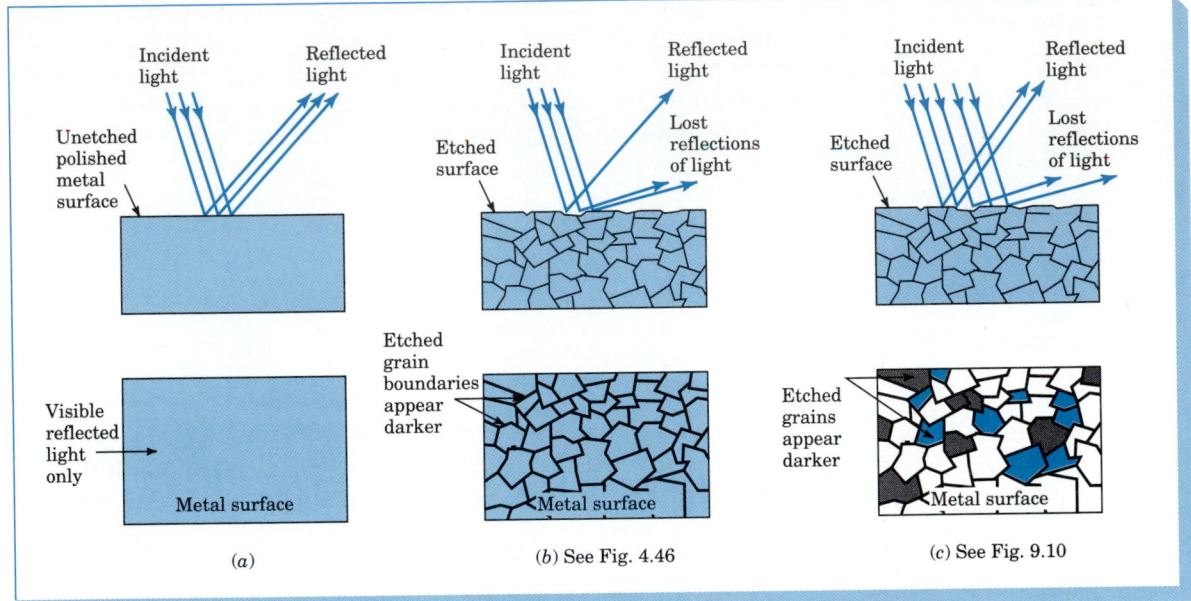

FIGURE 4.28 The effect of etching a polished surface of a steel metal sample on the microstructure observed in the optical microscope. (*a*) In the as-polished condition no microstructural features are observed. (*b*) After etching a very low-carbon steel, only grain boundaries are chemically attacked severely, and so they appear as dark lines in the optical microstructure. (*c*) After etching a medium-carbon steel polished sample, dark (pearlite) and light (ferrite) regions are observed in the microstructure. The darker pearlite regions have been more severely attacked by the etchant and thus do not reflect much light.

within the grains because of the atomic mismatch. Grain boundaries also have some atoms in strained positions which raise the energy of the grain-boundary region.

Grain boundaries in a metallic or ceramic material can be identified in a prepared material sample as dark lines (Fig. 4.26). Metallic and ceramic samples are first polished to produce a smooth surface and then are chemically etched so that the grain boundaries will be chemically attacked at a more rapid rate than the grains producing tiny grooves along the boundaries. When examined with an optical microscope (Fig. 4.27), the incident light will not be as intensely reflected at the grain boundaries, and as a result, the grain boundaries will appear as dark lines in the eyepiece of the microscope (Fig. 4.28).

The higher energy of the grain boundaries and their more open structure make them a more favorable region for the nucleation and growth of precipitates (see Sec. 9.5). The lower atomic packing of the grain boundaries also allows for more rapid diffusion of atoms in the grain-boundary region. At ordinary temperatures grain boundaries also restrict plastic flow by making it difficult for the movement of dislocations in the grain-boundary region (Fig. 6.42).

Grain Size   The grain size of polycrystalline metals is important since the amount of grain-boundary surface has a significant effect on many properties of metals, especially strength. At lower temperatures (less than about one-half of their melting temperature) grain boundaries strengthen metals by restricting dislocation movement under stress. At elevated temperatures grain-boundary sliding may occur, and grain boundaries can become regions of weakness in polycrystalline metals.

One method of measuring grain size is the ASTM[1] method, in which the *grain-size number n* is defined by

$$N = 2^{n-1} \qquad\qquad (4.7)$$

where N is the number of grains per square inch on a polished and etched material surface at a magnification of $100\times$ and n is an integer referred to as the *ASTM grain-size number*. Grain-size numbers with the nominal number of grains per square inch at $100\times$ and grains per square millimeter at $1\times$ are listed in Table 4.2. Figure 4.29 shows some examples of nominal grain sizes for low-carbon sheet steel samples.

---

[1]ASTM is an abbreviation for the American Society for Testing and Materials.

TABLE 4.2   ASTM Grain Sizes

Grain-size no.	Nominal number of grains	
	Per sq mm at 1×	Per sq in at 100×
1	15.5	1.0
2	31.0	2.0
3	62.0	4.0
4	124	8.0
5	248	16.0
6	496	32.0
7	992	64.0
8	1980	128
9	3970	256
10	7940	512

Source: "Metals Handbook," vol. 7, 8th ed., American Society for Metals, 1972, p. 4.

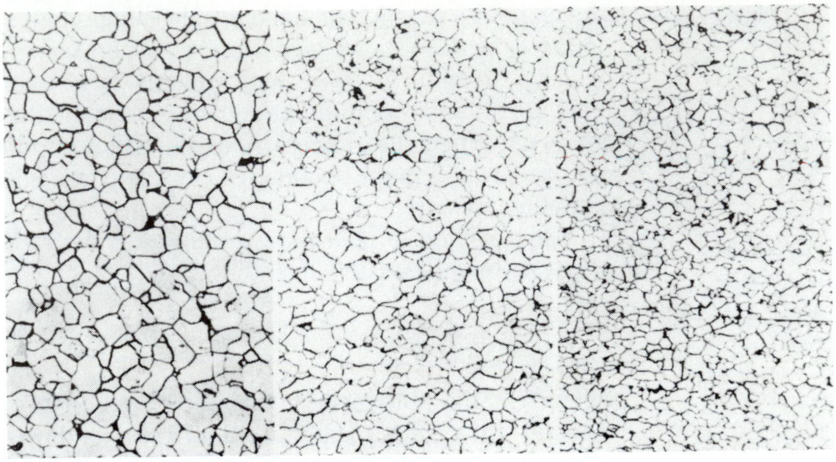

FIGURE 4.29 Several nominal ASTM grain sizes of low-carbon sheet steels: (*a*) no.7, (*b*) no.8, (*c*) no.9. (Etch: nital; magnification 100×.) (*After "Metals Handbook," vol. 7, 8th ed., American Society for Metals, 1972, p. 4.*)

(*a*)   (*b*)   (*c*)

Example Problem 4.4

An ASTM grain-size determination is being made from a photomicrograph of a metal which has a magnification of 100×. What is the ASTM grain-size number of the metal if there are 64 grains per square inch?

$$N = 2^{n-1}$$

where N = no. of grains per square inch at $100\times$
n = ASTM grain-size number

Thus,

$$64 \text{ grains/in}^2 = 2^{n-1}$$

$$\log 64 = (n - 1)(\log 2)$$

$$1.806 = (n - 1)(0.301)$$

$$n = 7 \blacktriangleleft$$

Example Problem 4.5

If there are 60 grains per square inch on a photomicrograph of a metal at $200\times$, what is the ASTM grain-size number of the metal?

Solution:

If there are 60 grains per square inch at $200\times$, then at $100\times$ we will have

$$N = \left(\frac{200}{100}\right)^2 (60 \text{ grains/in}^2) = 240 = 2^{n-1}$$

$$\log 240 = (n - 1)(\log 2)$$

$$2.380 = (n - 1)(0.301)$$

$$n = 8.91 \blacktriangleleft$$

Note that the ratio of the magnification change must be squared since we are concerned with the number of grains per square inch.

The Scanning Electron Microscope

The scanning electron microscope (SEM) is an instrument which impinges a beam of electrons in a pinpointed spot on the surface of a target specimen and collects and displays the electronic signals given off by the target material. Figure 4.30 is a photograph of a recent model of a scanning electron microscope, and Fig. 4.31 schematically illustrates its principles of operation. Basically, an electron gun produces an electron beam in an evacuated column

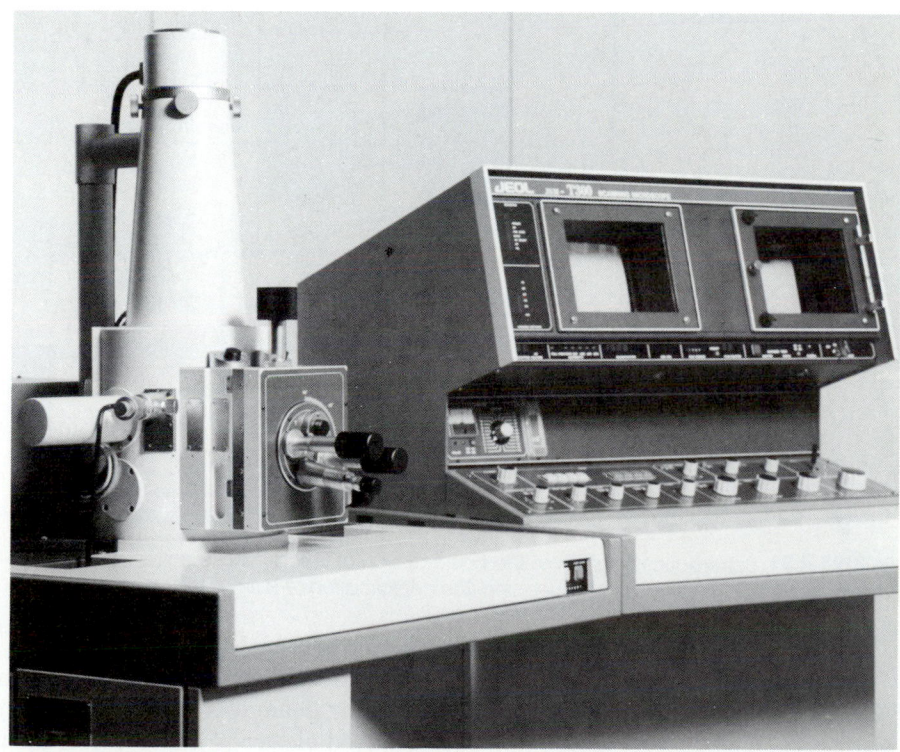

FIGURE 4.30 Photograph of a commercial scanning electron microscope (SEM). Note the column (which is evacuated during use) on the left and the viewing screen on the right. (*Courtesy of JEOL Ltd.*)

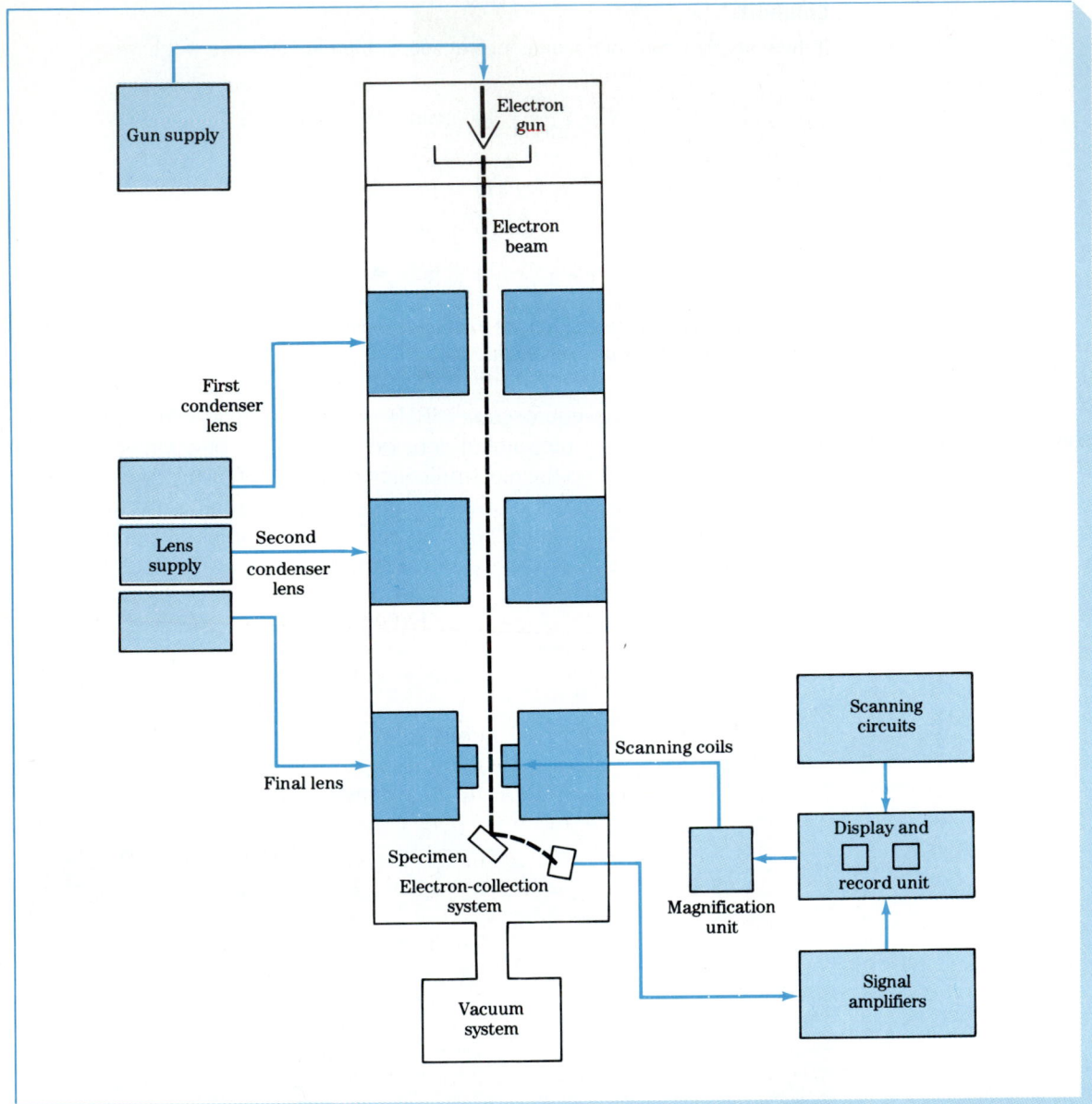

FIGURE 4.31 Schematic diagram of the basic design of a scanning electron microscope (SEM). (*After V. A. Phillips, "Modern Metallographic Techniques and Their Applications," Wiley, 1971, p. 425.*)

which is focused and directed so that it impinges on a small spot on the target. Scanning coils allow the beam to scan a small area of surface of the sample. Low-angle backscattered electrons interact with the protuberances of the sur-

FIGURE 4.32 Scanning electron fractograph of intergranular corrosion fracture near a circumferential weld in a thick-wall tube made of type 304 stainless steel. (Magnification 180×). (After *"Metals Handbook,"* vol. 9: *"Fractography and Atlas of Fractographs,"* 8th ed., American Society for Metals, 1974, p. 77.)

face and generate secondary[1] backscattered electrons to produce an electronic signal, which in turn produces an image having a depth of field of up to about 300 times that of the optical microscope (about 10 μm at 10,000 diameters magnification). The resolution of many SEM instruments is about 5 nm, with a wide range of magnification (about 15 to 100,000×).

The SEM is particularly useful in materials analysis for the examination of fractured surfaces of metals. Figure 4.32 shows an SEM fractograph of an intergranular corrosion fracture. Notice how clearly the metal grain surfaces are delineated and the depth of perception. SEM fractographs are used to determine whether a fractured surface is intergranular (Fig. 4.32), transgranular, or a mixture of both.

4.5   RATE PROCESSES IN SOLIDS

Many processes involved in the production and utilization of engineering materials are concerned with the rate at which atoms move in the solid state. In many of these processes reactions occur in the solid state which involve the spontaneous rearrangement of atoms into new and more stable atomic arrangements. In order for these reactions to proceed from the unreacted to the reacted state, the reacting atoms must have sufficient energy to overcome an

---

[1] Secondary electrons are electrons that are ejected from the target metal atoms after being struck by primary electrons from the electron beam.

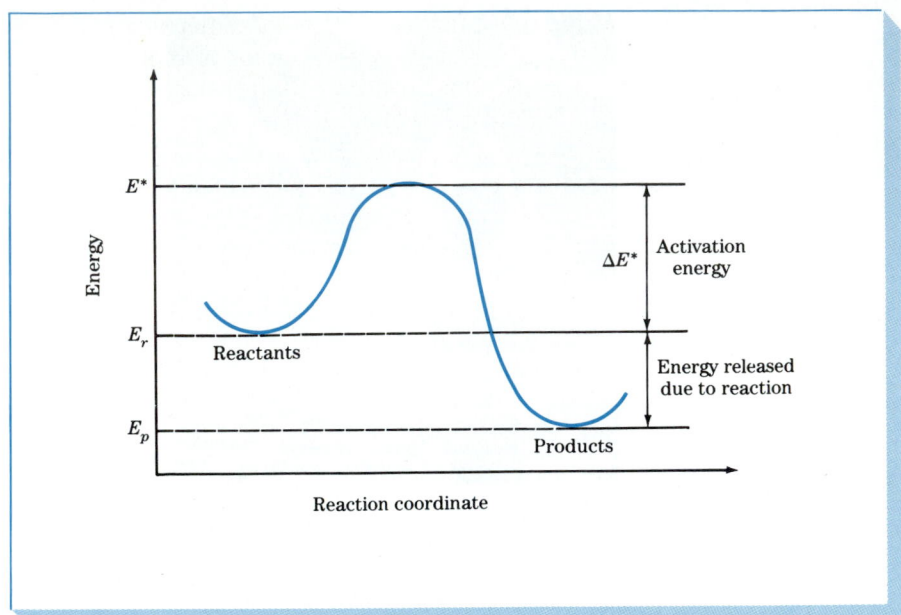

FIGURE 4.33 Energy of reacting species as it proceeds from the unreacted to the reacted state.

activation energy barrier. The additional energy required above the average energy of the atoms is called the *activation energy* ΔE^*, which is usually measured in joules per mole or calories per mole. Figure 4.33 illustrates the activation energy for a thermally activated solid-state reaction. Atoms possessing an energy level E_r (energy of the reactants) $+ \Delta E^*$ (activation energy) will have sufficient energy to react spontaneously to reach the reacted state E_p (energy of the products). The reaction shown in Fig. 4.33 is exothermic, indicating that energy is given off in the reaction.

At any temperature only a fraction of the molecules or atoms in a system will have sufficient energy to reach the activation energy level of E^*. As the temperature of the system is increased, more and more molecules or atoms will attain the activation energy level. Boltzmann studied the effect of temperature on increasing the energies of gas molecules. On the basis of statistical analysis, Boltzmann's results showed that the probability of finding a molecule or atom at an energy level E^* greater than the average energy E of all the molecules or atoms in a system at a particular temperature T in kelvins was

$$\text{Probability} \propto e^{-(E^* - E)/kT} \tag{4.8}$$

where k = Boltzmann's constant = 1.38×10^{-23} J/(atom·K).

The fraction of atoms or molecules in a system having energies greater than E^* in a system where E^* is much greater than the average energy of any atom or molecule can be written as

$$\frac{n}{N_{\text{total}}} = Ce^{-E^*/kT} \tag{4.9}$$

where n = number of atoms or molecules with an energy greater than E^*
N_{total} = total number of atoms or molecules present in system
k = Boltzmann's constant
T = temperature, K
C = a constant

The number of vacancies at equilibrium at a particular temperature in a metallic crystal lattice can be expressed by the following relationship, which is similar to Eq. (4.9):

$$\frac{n_v}{N} = Ce^{-E_v/kT} \tag{4.10}$$

where n_v = number of vacancies per cubic meter of metal
N = total number of atom sites per cubic meter of metal
E_v = activation energy to form a vacancy, eV
T = absolute temperature, K
k = Boltzmann's constant = 8.62×10^{-5} eV/K
C = constant

In Example Problem 4.6 the equilibrium concentration of vacancies present in pure copper at 500°C is calculated by using Eq. (4.10) and assuming that $C = 1$. According to this calculation, there is only about 1 vacancy for every 1 million atoms!

Example Problem 4.6

Calculate (a) the equilibrium number of vacancies per cubic meter in pure copper at 500°C and (b) the vacancy fraction at 500°C in pure copper. Assume the energy of formation of a vacancy in pure copper is 0.90 eV. Use Eq. (4.10) with $C = 1$. (Boltzmann's constant $k = 8.62 \times 10^{-5}$ eV/K)

Solution:

(a) The equilibrium number of vacancies per cubic meters in pure copper at 500°C is

$$n_v = Ne^{-E_v/kT} \qquad \text{(assume } C = 1) \tag{4.10a}$$

where n_v = no. of vacancies/m³
N = no. of atom sites/m³
E_v = energy of formation of a vacancy in pure copper at 500°C, eV
k = Boltzmann's constant
T = temperature, K

First, we determine a value for N by using the equation

$$N = \frac{N_0 \rho_{Cu}}{\text{at. mass Cu}} \tag{4.11}$$

where N_0 = Avogadro's constant and ρ_{Cu} = density of Cu = 8.96 Mg/m³. Thus

$$N = \frac{6.02 \times 10^{23} \text{ atoms}}{\text{at. mass}} \times \frac{1}{63.54 \text{ g/at. mass}} \times \frac{8.96 \times 10^6 \text{ g}}{m^3}$$

$$= 8.49 \times 10^{28} \text{ atoms/m}^3$$

Substituting the values of N, E_v, k, and T into Eq. (4.10a) gives

$$n_v = Ne^{-E_v/kT}$$

$$= (8.49 \times 10^{28}) \left\{ \exp\left[-\frac{0.90 \text{ eV}}{(8.62 \times 10^{-5} \text{ eV/K})(773 \text{ K})} \right] \right\}$$

$$= (8.49 \times 10^{28})(e^{-13.5}) = (8.49 \times 10^{28})(1.37 \times 10^{-6})$$

$$= 1.2 \times 10^{23} \text{ vacancies/m}^3 \blacktriangleleft$$

(b) The vacancy fraction in pure copper at 500°C is found from the ratio n_v/N from Eq. (4.10a):

$$\frac{n_v}{N} = \exp\left[-\frac{0.90 \text{ eV}}{(8.62 \times 10^{-5} \text{ eV/K})(773 \text{ K})} \right]$$

$$= e^{-13.5} = 1.4 \times 10^{-6} \blacktriangleleft$$

Thus, there is only 1 vacancy in every 10^6 atom sites!

A similar expression to the Boltzmann relationship for the energies of molecules in a gas was arrived at by Arrhenius[1] experimentally for the effect of temperature on chemical reaction rates. Arrhenius found that the rate of many chemical reactions as a function of temperature could be expressed by the relationship

Arrhenius rate equation: Rate of reaction = $Ce^{-Q/RT}$ (4.12)

where Q = activation energy, J/mol or cal/mol
　　　R = molar gas constant
　　　　 = 8.314 J/(mol·K) or 1.987 cal/(mol·K)
　　　T = temperature, K
　　　C = rate constant, independent of temperature

[1] Svante August Arrhenius (1859–1927). Swedish physical chemist who was one of the founders of modern physical chemistry and who experimentally studied reaction rates.

In working with liquids and solids the activation energy is usually expressed in terms of a mole, or 6.02×10^{23} atoms or molecules. The activation energy is also usually given the symbol Q and expressed in joules per mole or calories per mole.

The Boltzmann equation (4.9) and the Arrhenius equation (4.12) both imply that the reaction rate among atoms or molecules in many cases depends on the number of reacting atoms or molecules that have activation energies of E^* or greater. The rates of many solid-state reactions of particular interest to materials scientists and engineers obey the Arrhenius rate law, and so the Arrhenius equation is often used to analyze experimental solid-state rate data.

The Arrhenius equation (4.12) is commonly rewritten in the natural logarithmic form as

$$\ln \text{ rate } = \ln \text{ constant } - \frac{Q}{RT} \tag{4.13}$$

This equation is that of a straight line of the type

$$y = b + mx \tag{4.14}$$

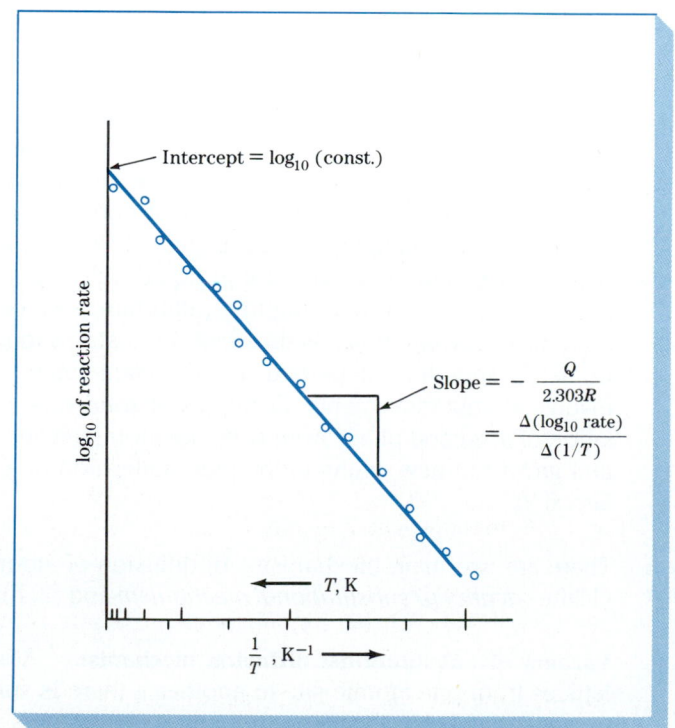

FIGURE 4.34 Typical Arrhenius plot of experimental rate data. (After J. Wulff et al., "Structure and Properties of Materials," vol. II: "Thermodynamics of Structure," Wiley, 1966, p. 64.)

Intercept $= \log_{10} (\text{const.})$

$\log_{10}$ of reaction rate

$\text{Slope} = -\dfrac{Q}{2.303R}$

$= \dfrac{\Delta(\log_{10} \text{ rate})}{\Delta(1/T)}$

T, K

$\dfrac{1}{T}, \text{K}^{-1}$

where b is the y intercept and m is the slope of the line. The ln rate term of Eq. (4.13) is equivalent to the y term of Eq. (4.14), and the ln constant term of Eq. (4.13) is equivalent to the b term of Eq. (4.14). The $-Q/R$ quantity of Eq. (4.13) is equivalent to the slope m of Eq. (4.14). Thus a plot of ln rate vs. $1/T$ K produces a straight line of slope $-Q/R$.

The Arrhenius equation (4.12) can also be rewritten in common logarithmic form as

$$\log_{10} \text{rate} = \log_{10} \text{constant} - \frac{Q}{2.303RT} \tag{4.15}$$

The 2.303 is the conversion factor from natural to common logarithms. This equation is also an equation of a straight line. A schematic plot of $\log_{10}$ rate vs. $1/T$ K is given in Fig. 4.34.

Thus, if a plot of experimental ln reaction rate vs. $1/T$ K data produces a straight line, an activation energy for the process involved can be calculated from the slope of the line. We shall use the Arrhenius equation to explore the effects of temperature on the diffusion of atoms and the electrical conductivity of pure elemental semiconductors later.

4.6 ATOMIC DIFFUSION IN SOLIDS

Diffusion in Solids in General

Diffusion can be defined as the mechanism by which matter is transported through matter. Atoms in gases, liquids, and solids are in constant motion and migrate over a period of time. In gases atomic movement is relatively rapid, as indicated by the rapid movement of cooking odors or smoke particles. Atomic movements in liquids are in general slower than in gases, as evidenced by the movement of colored dye in liquid water. In solids atomic movements are restricted due to bonding to equilibrium positions. However, thermal vibrations occurring in solids do allow some atoms to move. Diffusion of atoms in metals and alloys is particularly important since most solid-state reactions involve atomic movements. Examples of solid-state reactions are the precipitation of a second phase from solid solution (see Sec. 9.5) and the nucleation and growth of new grains in the recrystallization of a cold-worked metal (see Sec. 6.9).

Diffusion Mechanisms

There are two main mechanisms of diffusion of atoms in a crystalline lattice: (1) the *vacancy or substitutional mechanism* and (2) the *interstitial mechanism.*

Vacancy or substitutional diffusion mechanism Atoms can move in crystal lattices from one atomic site to another if there is sufficient activation energy present provided by the thermal vibration of the atoms and if there are vacancies

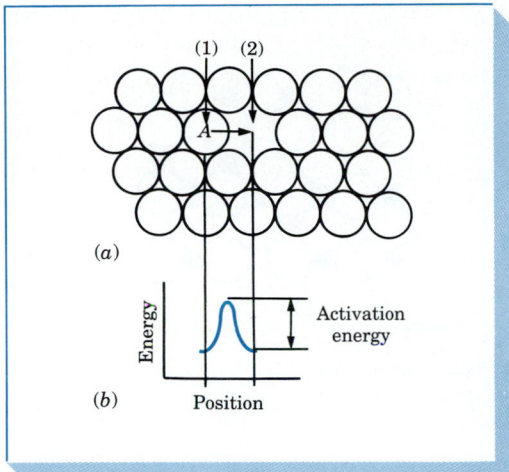

(1) (2)

A

(a)

Energy

Activation
energy

(b)   Position

FIGURE 4.35 Activation energy associated with the movement of atoms in a metal. (*a*) Diffusion of copper atom *A* at position (1) on the (111) plane of a copper crystal crystal can move to position (2) (a vacancy site), if sufficient activation energy is provided as indicated in (*b*).

or other crystal defects in the lattice for atoms to move into. Vacancies in metals and alloys are equilibrium defects, and therefore some are always present to enable substitutional diffusion of atoms to take place. As the temperature of the metal increases, more vacancies are present and more thermal energy is available, and so the diffusion rate is higher at higher temperatures.

Consider the example of vacancy diffusion shown in Fig. 4.35 on a (111) plane of copper atoms in a copper crystal lattice. If an atom which is next to the vacancy has sufficient activation energy, it can move into the vacant site and thereby contribute to the *self-diffusion* of copper atoms in the lattice. The activation energy for self-diffusion is equal to the sum of the activation energy to form a vacancy and the activation energy to move the vacancy. Table 4.3 lists some activation energies for self-diffusion in pure metals. Note that in general as the melting point of the metal is increased, the activation energy is also. This relationship exists because the higher-melting-temperature metals tend to have stronger bonding energies between their atoms.

TABLE 4.3   Self-Diffusion Activation Energies for Some Pure Metals

Metal	Melting point, °C	Crystal structure	Temperature range studied, °C	Activation energy	
				kJ/mol	kcal/mol
Zinc	419	HCP	240–418	91.6	21.9
Aluminum	660	FCC	400–610	165	39.5
Copper	1083	FCC	700–990	196	46.9
Nickel	1452	FCC	900–1200	293	70.1
α iron	1530	BCC	808–884	240	57.5
Molybdenum	2600	BCC	2155–2540	460	110

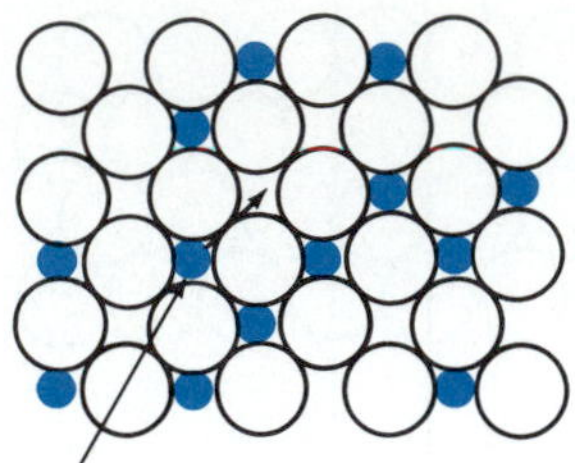

FIGURE 4.36 A schematic diagram of an interstitial solid solution. The large circles represent atoms on a (100) plane of an FCC crystal lattice. The small dark circles are interstitial atoms which occupy interstitial sites. The interstitial atoms can move into adjacent interstitial sites which are vacant. There is an activation energy associated with interstitial diffusion.

Interstitial atom
diffusing into
interstitial vacancy

Diffusion can also occur by the vacancy mechanism in solid solutions. Atomic size differences and bonding energy differences between the atoms are factors which affect the diffusion rate.

Interstitial diffusion mechanims   The interstitial diffusion of atoms in crystal lattices takes place when atoms move from one interstitial site to another neighboring interstitial site without permanently displacing any of the atoms in the matrix crystal lattice (Fig. 4.36). For the interstitial mechanism to be operative the size of the diffusing atoms must be relatively small compared to the matrix atoms. Small atoms such as hydrogen, oxygen, nitrogen, and carbon can diffuse interstitially in some metallic crystal lattices. For example, carbon can diffuse interstitially in BCC alpha iron and FCC gamma iron (see Fig. 4.15a). In the interstitial diffusion of carbon in iron, the carbon atoms must squeeze between the iron matrix atoms.

Steady-State Diffusion   Consider the diffusion of solute atoms in the x direction between two parallel atomic planes perpendicular to the paper separated by a distance x as shown in Fig. 4.37. We will assume that over a period of time the concentration of atoms at plane 1 is C_1 and that of plane 2 is C_2. That is, there is no change in the concentration of solute atoms at these planes for the system with time. Such diffusion conditions are said to be *steady-state conditions*. This type of diffusion takes place when a nonreacting gas diffuses through a metal foil. For example, steady-state diffusion conditions are attained when hydrogen gas diffuses through a foil of palladium if the hydrogen gas is at high pressure on one side and low pressure on the other.

If in the diffusion system shown in Fig. 4.37 no chemical interaction occurs between the solute and solvent atoms, because there is a concentration difference between planes 1 and 2, there will be a net flow of atoms from the

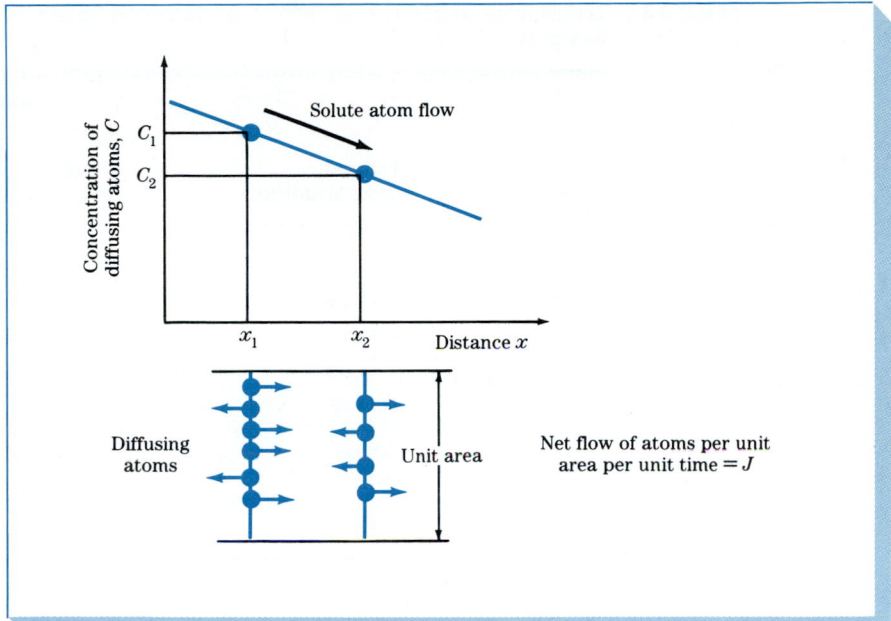

FIGURE 4.37 Steady-state diffusion of atoms in a concentration gradient. An example is hydrogen gas diffusing through a palladium metal foil.

higher concentration to the lower concentration. The *flux* or flow of atoms in this type of system can be represented by the equation

$$J = -D\frac{dC}{dx} \tag{4.16}$$

where J = flux or net flow of atoms

D = proportionality constant called the *diffusivity* (atomic conductivity) or *diffusion coefficient*

$\dfrac{dC}{dx}$ = concentration gradient

A negative sign is used because the diffusion is from a higher to a lower concentration; i.e., there is a negative diffusion gradient.

This equation is called *Fick's*[1] *first law of diffusion* and states that for steady-state diffusion conditions (i.e., no change in system with time), the net flow of atoms by atomic diffusion is equal to the diffusivity D times the diffusion gradient dC/dx. The SI units for this equation are

$$J\left(\frac{\text{atoms}}{\text{m}^2 \cdot \text{s}}\right) = D\left(\frac{\text{m}^2}{\text{s}}\right)\frac{dC}{dx}\left(\frac{\text{atoms}}{\text{m}^3} \times \frac{1}{\text{m}}\right) \tag{4.17}$$

---

[1] Adolf Eugen Fick (1829–1901). German physiologist who first put diffusion on a quantitative basis by using mathematical equations. Some of his work was published in the *Annals of Physics (Leipzig)*, **170:**59(1855).

TABLE 4.4 **Diffusivities at 500°C and 1000°C for Selected Solute-Solvent Diffusion Systems**

| Solute | Solvent (host structure) | Diffusivity, m²/s | |
		500°C (930°F)	1000°C (1830°F)
1. Carbon	FCC iron	$(5 \times 10^{-15})^*$	3×10^{-11}
2. Carbon	BCC iron	10^{-12}	(2×10^{-9})
3. Iron	FCC iron	(2×10^{-23})	2×10^{-16}
4. Iron	BCC iron	10^{-20}	(3×10^{-14})
5. Nickel	FCC iron	10^{-23}	2×10^{-16}
6. Manganese	FCC iron	(3×10^{-24})	10^{-16}
7. Zinc	Copper	4×10^{-18}	5×10^{-13}
8. Copper	Aluminum	4×10^{-14}	10^{-10} M†
9. Copper	Copper	10^{-18}	2×10^{-13}
10. Silver	Silver (crystal)	10^{-17}	10^{-12} M
11. Silver	Silver (grain boundary)	10^{-11}	
12. Carbon	HCP titanium	3×10^{-16}	(2×10^{-11})

*Parentheses indicate that the phase is metastable.
†M—Calculated, although temperature is above melting point.
Source: L. H. Van Vlack, "Elements of Materials Science and Engineering," 5th ed., Addision-Wesley, 1985.

Table 4.4 lists some values of atomic diffusivities of selected interstitial and substitutional diffusion systems. The diffusivity values depend on many variables, of which the following are important:

1. *The type of diffusion mechanism.* Whether the diffusion is interstitial or substitutional will affect the diffusivity. Small atoms can diffuse interstitially in the crystal lattice of larger solvent atoms. For example, carbon diffuses interstitially in the BCC or FCC iron lattices. Copper atoms diffuse substitutionally in an aluminum solvent lattice since both the copper and the aluminum atoms are about the same size.

2. *The temperature at which the diffusion takes place* greatly affects the value of the diffusivity. As the temperature is increased, the diffusivity also increases, as shown in Table 4.4 for all the systems by comparing the 500°C values with those for 1000°C. The effect of temperature on diffusivity in diffusion systems will be discussed further in Sec. 4.8.

3. *The type of crystal structure of the solvent lattice* is important. For example, the diffusivity of carbon in BCC iron is 10^{-12} m²/s at 500°C, which is much *greater* than 5×10^{-15} m²/s, the value for the diffusivity of carbon in FCC iron at the same temperature. The reason for this difference is that the BCC crystal structure has a lower atomic packing factor of 0.68 as compared to that of FCC crystal structure, which is 0.74. Also, the interatomic spaces between the iron atoms are wider in the BCC crystal structure than in the

FCC one, and so the carbon atoms can diffuse between the iron atoms in the BCC structure more easily than in the FCC one.

4. *The type of crystal imperfections present* in the region of solid-state diffusion is also important. More open structures allow for more rapid diffusion of atoms. For example, diffusion takes place more rapidly along grain boundaries than in the grain matrix in metals and ceramics. Excess vacancies will increase diffusion rates in metals and alloys.

5. *The concentration of the diffusing species* is important in that higher concentrations of diffusing solute atoms will affect the diffusivity. This aspect of solid-state diffusion is very complex.

Non-Steady-State Diffusion   Steady-state diffusion in which conditions do not change with time is not commonly encountered with engineering materials. In most cases non-steady-state diffusion in which the concentration of solute atoms at any point in the material changes with time takes place. For example, if carbon is being diffused into the surface of a steel camshaft to harden its surface, the concentration of carbon under the surface at any point will change with time as the diffusion process progresses. For cases of non-steady-state diffusion in which the diffusivity is independent of time, Fick's second law of diffusion applies, which is

$$\frac{dC_x}{dt} = \frac{d}{dx}\left(D\frac{dC_x}{dx}\right) \tag{4.18}$$

This law states that the rate of compositional change is equal to the diffusivity times the rate of change of the concentration gradient. The derivation and solving of this differential equation is beyond the scope of this book. However, the particular solution to this equation in which a gas is diffusing into a solid is of great importance for some engineering diffusion processes and will be used to solve some practical industrial diffusion problems.

Let us consider the case of a gas A diffusing into a solid B, as illustrated in Fig. 4.38a. As the time of diffusion increases the concentration of solute atoms at any point in the x direction will also increase, as indicated for times t_1 and t_2 in Fig. 4.38b. If the diffusivity of gas A in solid B is independent of position, then the solution to Fick's second law [Eq. (4.18)] is

$$\frac{C_s - C_x}{C_s - C_0} = \text{erf}\left(\frac{x}{2\sqrt{Dt}}\right) \tag{4.19}$$

where C_s = surface concentration of element in gas diffusing into the surface
C_0 = initial uniform concentration of element in solid
C_x = concentration of element at distance x from surface at time t
x = distance from surface
D = diffusivity of diffusing solute element
t = time

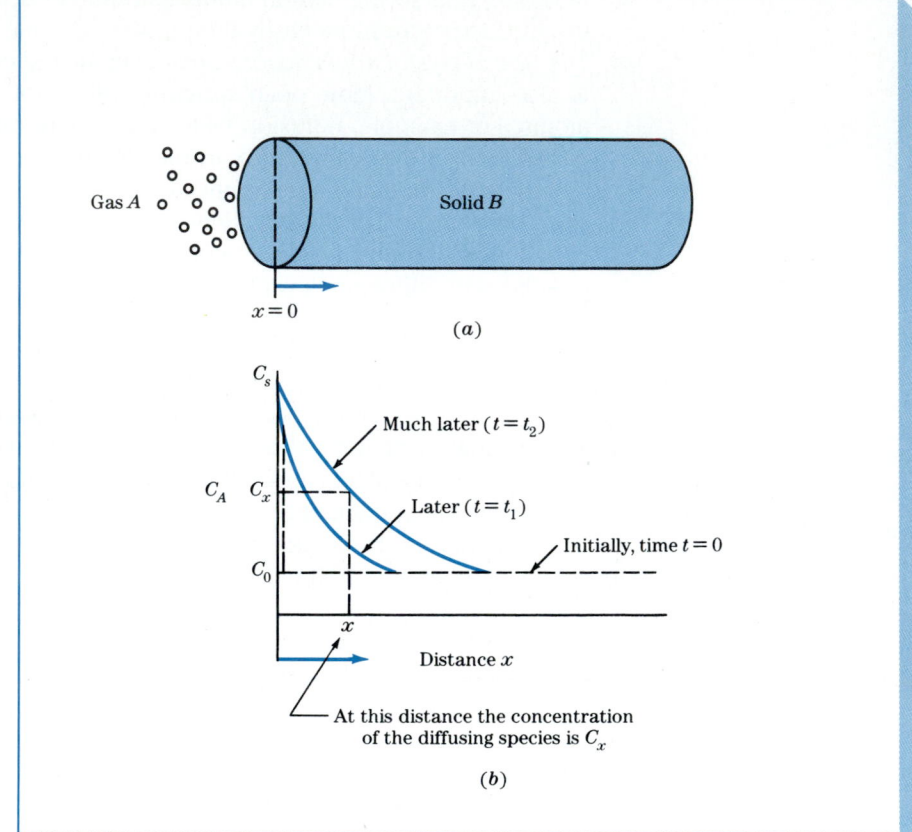

FIGURE 4.38 Diffusion of a gas into a solid. (*a*) Gas A diffuses into solid B at the surface where $x = 0$. The gas maintains a concentration of A atoms, called C_s, on this surface. (*b*) Concentration profiles of element A at various times along the solid in the x direction. The solid contained a uniform concentration of element A, called C_0, before diffusion started.

TABLE 4.5   Table of the Error Function

z	erf z	z	erf z	z	erf z	z	erf z
0	0	0.40	0.4284	0.85	0.7707	1.6	0.9763
0.025	0.0282	0.45	0.4755	0.90	0.7970	1.7	0.9838
0.05	0.0564	0.50	0.5205	0.95	0.8209	1.8	0.9891
0.10	0.1125	0.55	0.5633	1.0	0.8427	1.9	0.9928
0.15	0.1680	0.60	0.6039	1.1	0.8802	2.0	0.9953
0.20	0.2227	0.65	0.6420	1.2	0.9103	2.2	0.9981
0.25	0.2763	0.70	0.6778	1.3	0.9340	2.4	0.9993
0.30	0.3286	0.75	0.7112	1.4	0.9523	2.6	0.9998
0.35	0.3794	0.80	0.7421	1.5	0.9661	2.8	0.9999

Source: R. A. Flinn and P. K. Trojan, "Engineering Materials and Their Applications," 2d ed., Houghton Mifflin, 1981, p. 137.

The error function, erf, is a mathematical function existing by agreed definition and is used in some solutions of Fick's second law. The error function can be found in standard tables in the same way as sines and cosines. Table 4.5 is an abbreviated table of the error function.

4.7 INDUSTRIAL APPLICATIONS OF DIFFUSION PROCESSES

Many industrial manufacturing processes utilize solid-state diffusion. In this section we will consider the following two diffusion processes: (1) case hardening of steel by gas carburizing and (2) the impurity doping of silicon wafers for integrated electronic circuits.

Case Hardening of Steel by Gas Carburizing

Many rotating or sliding steel parts such as gears and shafts must have a hard outside case for wear resistance and a tough inner core for fracture resistance. In the manufacture of a carburized steel part, usually the part is machined first in the soft condition, and then, after machining, the outer layer is hardened by some case-hardening treatment such as gas carburizing. Carburized steels are low-carbon steels which have about 0.10 to 0.25% C. However, the alloy content of the carburized steels can vary considerably depending on the application for which the steel will be used. Some typical gas-carburized parts are shown in Fig. 4.39.

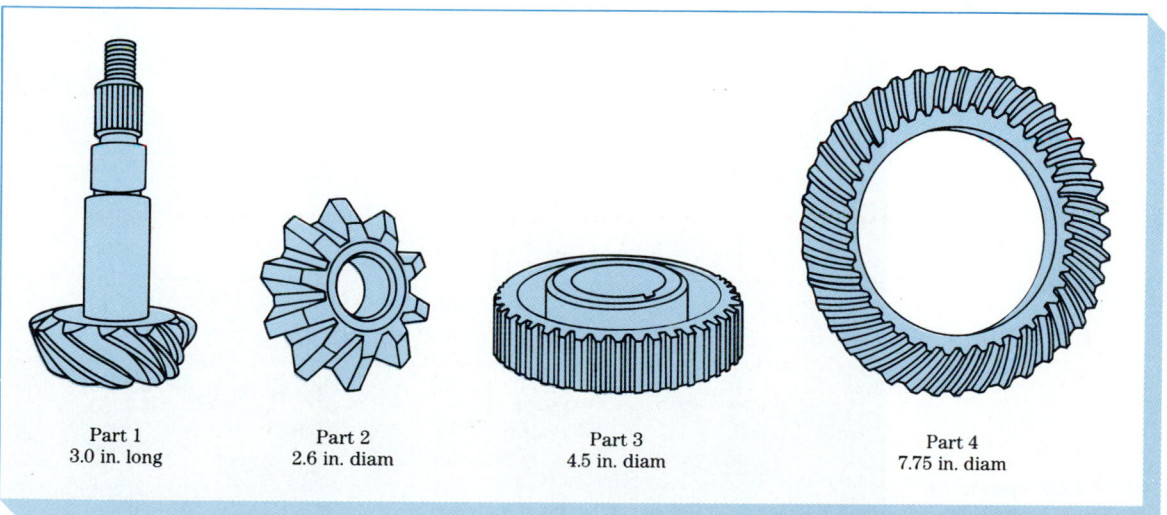

| Part 1 | Part 2 | Part 3 | Part 4 |
| 3.0 in. long | 2.6 in. diam | 4.5 in. diam | 7.75 in. diam |

FIGURE 4.39 Typical gas-carburized steel parts. (*After "Metals Handbook," vol. 2: "Heat Treating," 8th ed., American Society for Metals, 1964, p. 108.*)

In the first part of the gas-carburizing process, the steel parts are placed in a furnace in contact with gases containing methane (CH_4) or other hydrocarbon gases at about 927°C (1700°F). Figure 4.40 shows some gears about to be gas-carburized in a furnace with a nitrogen-methanol mixture for an atmosphere. The carbon from the atmosphere diffuses into the surface of the gears so that after subsequent heat treatments the gears are left with high-carbon hard cases as indicated, for example, by the darkened surface areas of the macrosection of the gear shown in Fig. 4.41.

Figure 4.42 shows some typical carbon gradients in test bars of AISI 1022 (0.22% C) plain-carbon steel carburized at 1685°F (918°C) by a carburizing atmosphere with 20% CO. Notice how the carburizing time greatly affects the carbon content vs. distance-below-the-surface profile. Example Problems 4.7 and 4.8 illustrate how the diffusion equation (4.19) can be used to determine one unknown variable, such as time of diffusion or carbon content at a particular depth below the surface of the part being carburized.

FIGURE 4.40 Parts to be carburized in a nitrogen-methanol carburizing atmosphere. (After *B. J. Sheehy, Met. Prog., September 1981, p. 120.*)

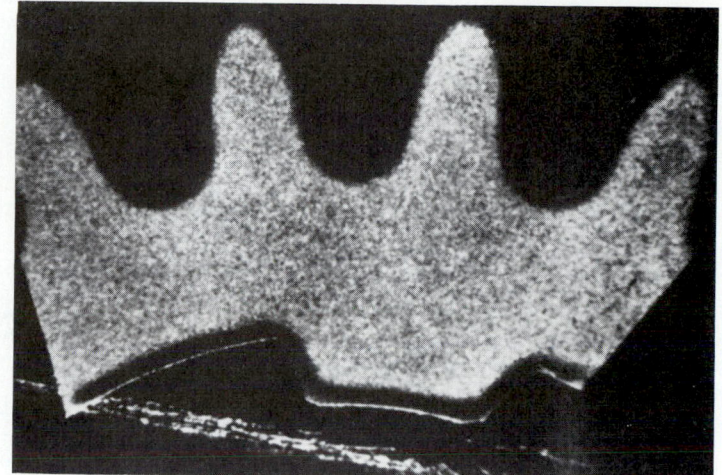

FIGURE 4.41
Macrosection of nitrogen-methanol-carburized SAE 8620 final-drive pinion gear. (*After B. J. Sheehy, Met. Prog., September 1981, p. 120.*)

Example Problem 4.7

Consider the gas carburizing of a gear of 1020 steel at 927°C (1700°F). Calculate the time in minutes necessary to increase the carbon content to 0.40% at 0.50 mm below the surface. Assume that the carbon content at the surface is 0.90% and that the steel has a nominal carbon content of 0.20%.

$$D_{927°C} = 1.28 \times 10^{-11} \text{ m}^2/\text{s}$$

Solution:

$$\frac{C_s - C_x}{C_s - C_0} = \text{erf}\left(\frac{x}{2\sqrt{Dt}}\right) \tag{4.19}$$

$C_s = 0.90\%$ $x = 0.5 \text{ mm} = 5.0 \times 10^{-4} \text{ m}$

$C_0 = 0.20\%$ $D_{927°C} = 1.28 \times 10^{-11} \text{ m}^2/\text{s}$

$C_x = 0.40\%$ $t = ? \text{ s}$

Substituting the above values in Eq. (4.19) gives

$$\frac{0.90 - 0.40}{0.90 - 0.20} = \text{erf}\left[\frac{5.0 \times 10^{-4} \text{ m}}{2\sqrt{(1.28 \times 10^{-11} \text{ m}^2/\text{s}) \, (t)}}\right]$$

$$\frac{0.50}{0.70} = \text{erf}\left(\frac{69.88}{\sqrt{t}}\right) = 0.7143$$

Let $Z = \dfrac{69.88}{\sqrt{t}}$ then $\text{erf } Z = 0.7143$

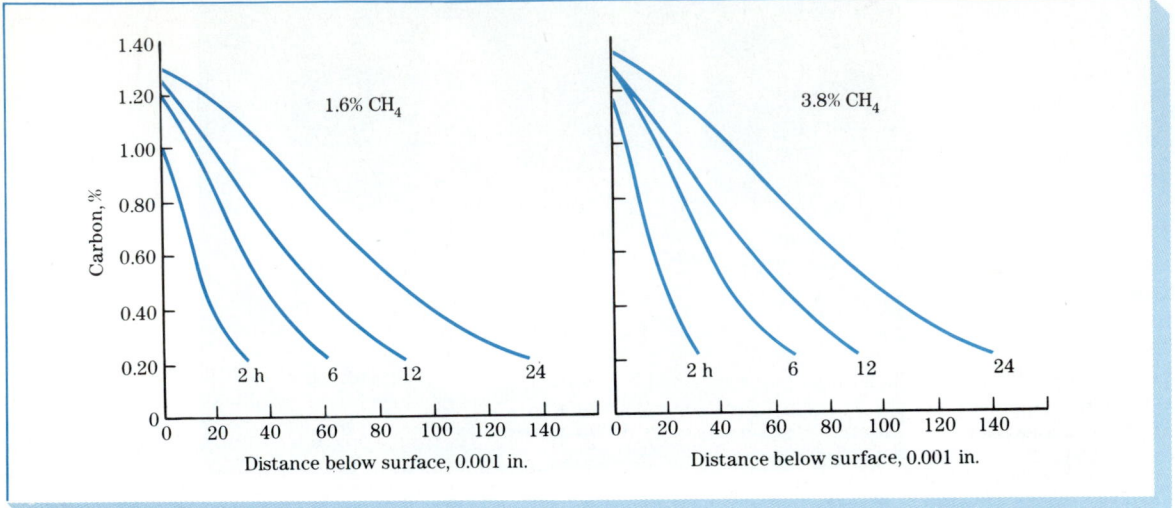

FIGURE 4.42 Carbon gradients in test bars of 1022 steel carburized at 918°C (1685°F) in a 20% CO–40% H_2 gas with 1.6 and 3.8% methane (CH_4) added. (*After "Metals Handbook," vol. 2: "Heat Treating," 8th ed., American Society for Metals, 1964, p. 100.*)

We need a number for Z whose error function (erf) is 0.7143. From Table 4.5 we find this number by interpolation (see below) to be 0.755:

erf Z	Z
0.7112	0.75
0.7143	x
0.7421	0.80

$$\frac{0.7143 - 0.7112}{0.7421 - 0.7112} = \frac{x - 0.75}{0.80 - 0.75}$$

$$x - 0.75 = (0.1003)(0.05)$$

$$x = 0.75 + 0.005 = 0.755$$

Thus

$$Z = \frac{69.88}{\sqrt{t}} = 0.755$$

$$\sqrt{t} = \frac{69.88}{0.755} = 92.6$$

$$t = 8567 \text{ s} = 143 \text{ min} \blacktriangleleft$$

Example Problem 4.8

Consider the gas carburizing of a gear of 1020 steel at 927°C (1700°F) as in Example Problem 4.7. Only in this problem calculate the *carbon content* at 0.50 mm beneath the surface of the gear after 5 h carburizing time. Assume that the carbon content of the surface of the gear is 0.90% and that the steel has a nominal carbon content of 0.20%.

Solution:

$$D_{927°C} = 1.28 \times 10^{-11} \text{ m}^2/\text{s}$$

$$\frac{C_s - C_x}{C_s - C_0} = \text{erf}\left(\frac{x}{2\sqrt{Dt}}\right) \tag{4.19}$$

$$C_s = 0.90\% \qquad x = 0.50 \text{ mm} = 5.0 \times 10^{-4} \text{ m}$$

$$C_0 = 0.20\% \qquad D_{927°C} = 1.28 \times 10^{-11} \text{ m}^2/\text{s}$$

$$C_x = ?\% \qquad t = 5 \text{ h} = 5 \text{ h} \times 3600 \text{ s/h} = 1.8 \times 10^4 \text{ s}$$

$$\frac{0.90 - C_x}{0.90 - 0.20} = \text{erf}\left[\frac{5.0 \times 10^{-4} \text{ m}}{2\sqrt{(1.28 \times 10^{-11} \text{ m/s})(1.8 \times 10^4 \text{ s})}}\right]$$

$$\frac{0.90 - C_x}{0.70} = \text{erf } 0.521$$

Let $Z = 0.521$. We need to know what the corresponding error function for the Z value of 0.521 is. To determine this number from Table 4.5, we must interpolate the data as shown in the accompanying table.

Z	erf Z
0.500	0.5205
0.521	x
0.550	0.5633

$$\frac{0.521 - 0.500}{0.550 - 0.500} = \frac{x - 0.5205}{0.5633 - 0.5205}$$

$$0.42 = \frac{x - 0.5205}{0.0428}$$

$$x - 0.5205 = (0.42)(0.0428)$$

$$x = 0.0180 + 0.5205$$

$$= 0.538$$

Therefore

$$\frac{0.90 - C_x}{0.70} = \text{erf } 0.521 = 0.538$$

$$C_x = 0.90 - (0.70)(0.538)$$

$$= 0.52\% \blacktriangleleft$$

Note that by increasing the carburizing time from about 2.4 to 5 h for the 1020 steel, the carbon content at 0.5 mm below the surface of the gear is increased from 0.4 to only 0.52%.

Impurity Diffusion into Silicon Wafers for Integrated Circuits

Impurity diffusion into silicon wafers to change their electrical conducting characteristics is an important phase in the production of modern integrated electronic circuits. In one method of impurity diffusion into silicon wafers the silicon surface is exposed to the vapor of an appropriate impurity at a temperature above about 1100°C in a quartz tube furnace, as shown schematically

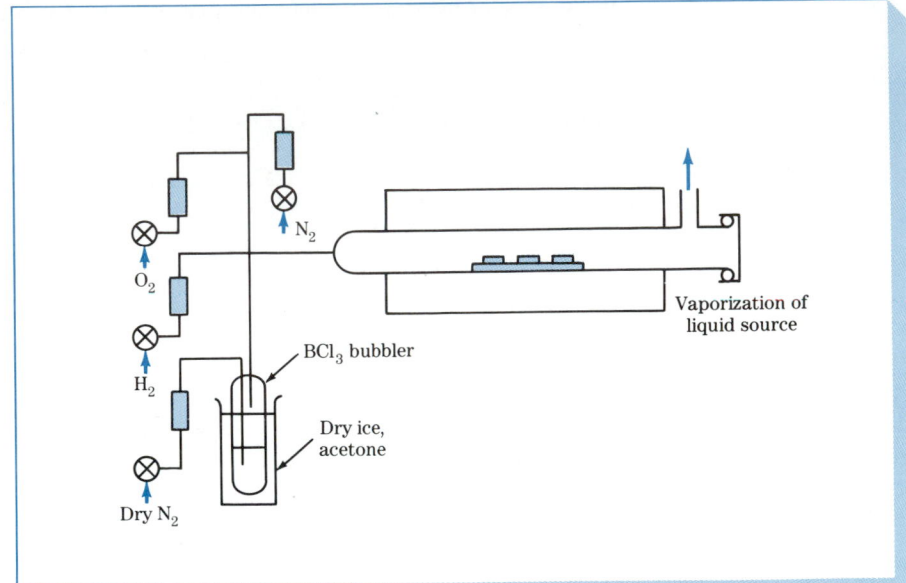

FIGURE 4.43 Diffusion method for diffusing boron into silicon wafers. (After *W. R. Runyan, "Silicon Semiconductor Technology,"* McGraw-Hill, 1965.)

in Fig. 4.43. The part of the silicon surface not to be exposed to the impurity diffusion must be masked off so that the impurities diffuse into the parts selected by the design engineer for conductivity change. Figure 4.44 shows a technician loading a rack of silicon wafers into a tube furnace for impurity diffusion.

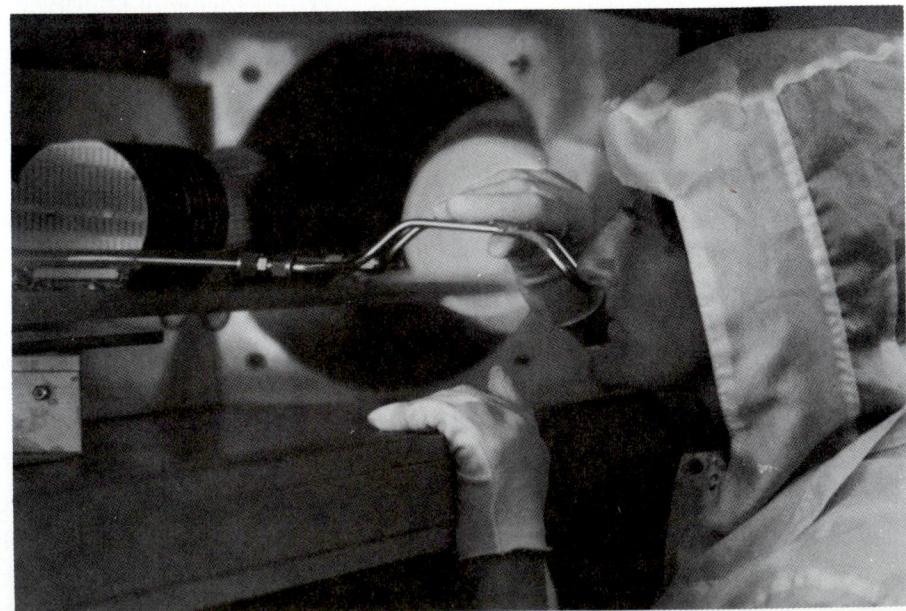

FIGURE 4.44 Loading a rack of silicon wafers into a tube furnace for impurity diffusion. (*Courtesy of Harris Corporation.*)

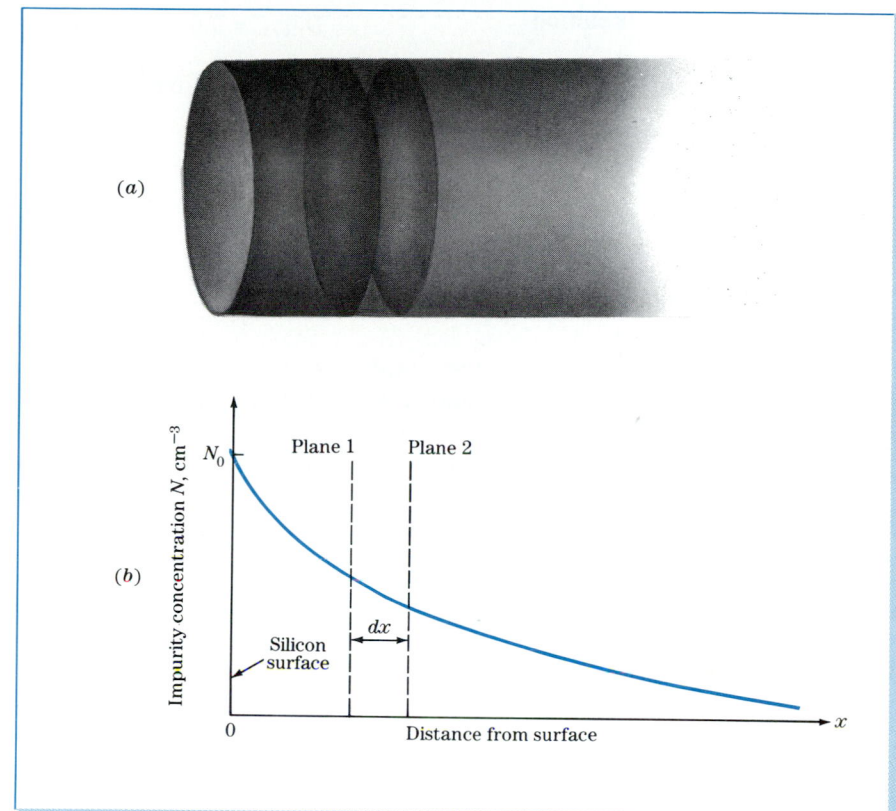

FIGURE 4.45 Impurity diffusion into a silicon wafer from one face. (*a*) A silicon wafer with thickness greatly exaggerated having an impurity concentration which diminishes from the left face toward the interior. (*b*) Graphical representation of the same impurity distribution. (*After R. M. Warner, "Integrated Circuits," McGraw-Hill, 1965, p. 70.*)

As in the case of the gas carburizing of a steel surface, the concentration of impurities diffused into the silicon surface decreases as the depth of penetration increases, as shown in Fig. 4.45. Changing the time of diffusion will also change the concentration of impurities vs. depth-of-penetration profile, as shown qualitatively in Fig. 4.38. Example Problem 4.9 illustrates how Eq. (4.19) can be used quantitatively to determine one unknown variable, such as time of diffusion or depth of penetration at a particular concentration level.

Example Problem 4.9

Consider the impurity diffusion of gallium into a silicon wafer. If gallium is diffused into a silicon wafer with no previous gallium in it at a temperature of 1100°C for 3 h, what is the depth below the surface at which the concentration is 10^{22} atoms/m³ if the surface concentration is 10^{24} atoms/m³? For gallium diffusing into silicon at 1100°C, the solution is as follows:

Solution:

$$D_{1100°C} = 7.0 \times 10^{-17} \text{ m}^2/\text{s}$$

$$\frac{C_s - C_x}{C_s - C_0} = \text{erf}\left(\frac{x}{2\sqrt{Dt}}\right) \tag{4.19}$$

$C_s = 10^{24}$ atoms/m³ $x = ?$m   (depth at which $C_x = 10^{22}$ atoms/m³)

$C_x = 10^{22}$ atoms/m³ $D_{1100°C} = 7.0 \times 10^{-17}$ m²/s

$C_0 = 0$ atoms/m³ $t = 3$ h $= 3$ h $\times 3600$ s/h $= 1.08 \times 10^4$ s

Substituting the above values into Eq. (4.19) gives

$$\frac{10^{24} - 10^{22}}{10^{24} - 0} = \text{erf}\left[\frac{x \text{ m}}{2\sqrt{(7.0 \times 10^{-17} \text{ m}^2/\text{s})(1.08 \times 10^4 \text{ s})}}\right]$$

$$1 - 0.01 = \text{erf}\left(\frac{x \text{ m}}{1.74 \times 10^{-6} \text{ m}}\right) = 0.99$$

Let

$$Z = \frac{x}{1.74 \times 10^{-6} \text{ m}}$$

Thus $\text{erf } Z = 0.99$ and $Z = 1.82$

(from Table 4.5 using interpolation). Therefore

$$x = (Z)(1.74 \times 10^{-6} \text{ m}) = (1.82)(1.74 \times 10^{-6} \text{ m})$$

$$= 3.17 \times 10^{-6} \text{ m} \blacktriangleleft$$

Note: Typical diffusion depths in silicon wafers are of the order of a few micrometers (i.e., about 10^{-6} m), while the wafer is usually several hundred micrometers thick.

4.8 EFFECT OF TEMPERATURE ON DIFFUSION IN SOLIDS

Since atomic diffusion involves atomic movements, it is to be expected that increasing the temperature of a diffusion system will increase the diffusion rate. By experiment, it has been found that the temperature dependence of the diffusion rate of many diffusion systems can be expressed by the following Arrhenius type equation:

$$D = D_0 e^{-Q/RT} \tag{4.20}$$

where D = diffusivity, m²/s

D_0 = proportionality constant, m²/s, independent of temperature in range for which equation is valid

Q = activation energy of diffusing species, J/mol or cal/mol

R = molar gas constant

= 8.314 J/(mol·K) or 1.987 cal/(mol·K)

T = temperature, K

Example Problem 4.10 applies Eq. (4.20) to determine the diffusivity of carbon diffusing in γ iron at 927°C when given values for D_0 and the activation energy Q.

.040 in 2.5 × 10^{-2} m
in

Example Problem 4.10

Calculate the value of the diffusivity D in meters squared per second for the diffusion of carbon in γ iron (FCC) at 927°C (1700°F). Use values of $D_0 = 2.0 \times 10^{-5}$ m²/s, $Q = 142$ kJ/mol, and $R = 8.314$ J/(mol·K).

Solution:

$$D = D_0 e^{-Q/RT} \tag{4.20}$$

$$= (2.0 \times 10^{-5} \text{ m}^2/\text{s}) \left\{ \exp \frac{-142{,}000 \text{ J/mol}}{[8.314 \text{ J/(mol·K)}](1200 \text{ K})} \right\}$$

$$= (2.0 \times 10^{-5} \text{ m}^2/\text{s})(e^{-14.23})$$

$$= (2.0 \times 10^{-5} \text{ m}^2/\text{s})(0.661 \times 10^{-6})$$

$$= 1.32 \times 10^{-11} \text{ m}^2/\text{s} \blacktriangleleft$$

TABLE 4.6   Diffusivity Data for Some Metallic Systems

Solute	Solvent	D_0, m²/s	Q kJ/mol	Q kcal/mol
Carbon	FCC iron	2.0×10^{-5}	142	34.0
Carbon	BCC iron	22.0×10^{-5}	122	29.3
Iron	FCC iron	2.2×10^{-5}	268	64.0
Iron	BCC iron	20.0×10^{-5}	240	57.5
Nickel	FCC iron	7.7×10^{-5}	280	67.0
Manganese	FCC iron	3.5×10^{-5}	282	67.5
Zinc	Copper	3.4×10^{-5}	191	45.6
Copper	Aluminum	1.5×10^{-5}	126	30.2
Copper	Copper	2.0×10^{-5}	197	47.1
Silver	Silver	4.0×10^{-5}	184	44.1
Carbon	HCP titanium	51.0×10^{-5}	182	43.5

Source: Data from L. H. Van Vlack, "Elements of Materials Science and Engineering," 5th ed., Addison-Wesley, 1985.

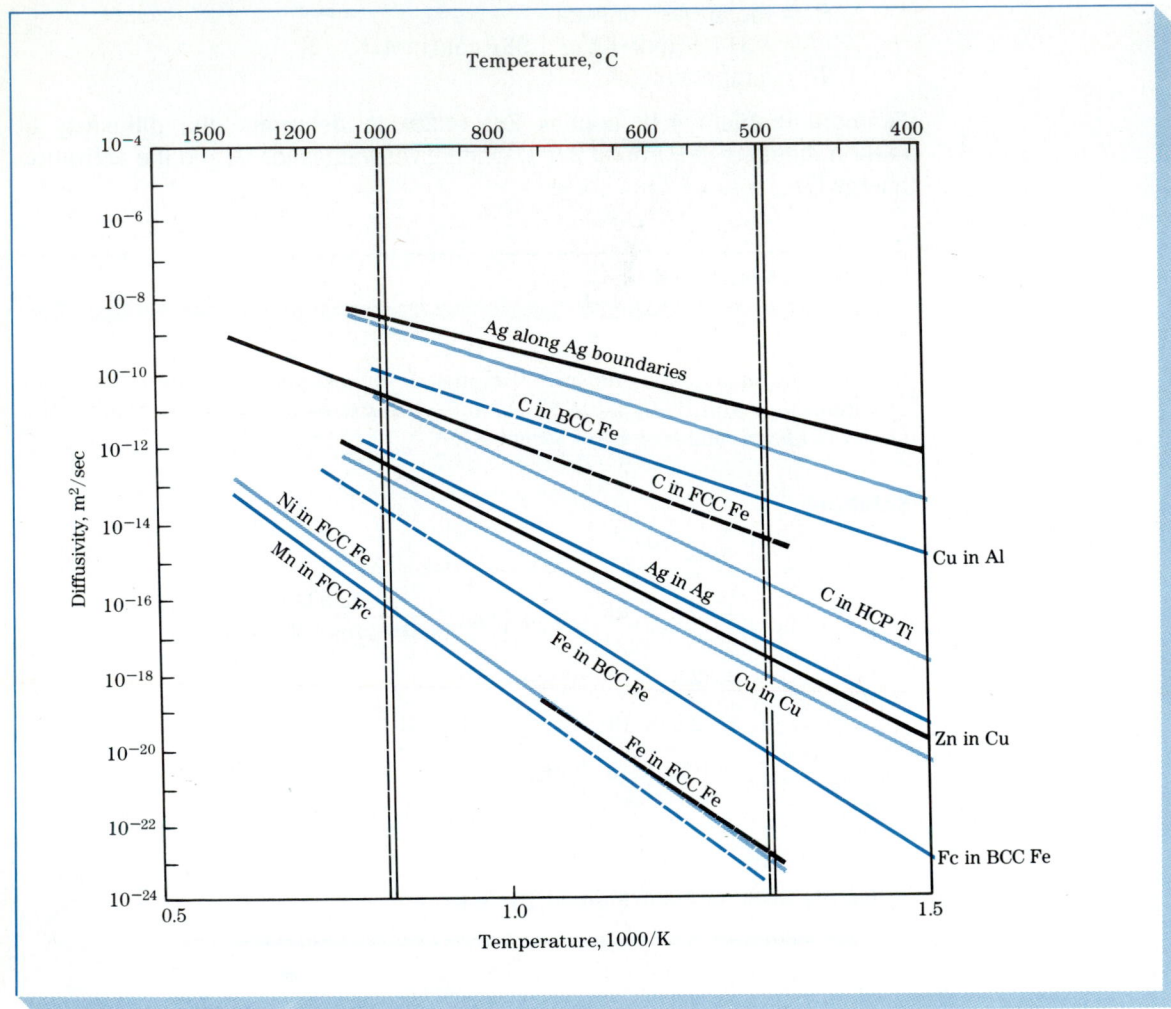

FIGURE 4.46 Arrhenius plots of diffusivity data for some metallic systems. (*After L. H. Van Vlack, "Elements of Materials Science and Engineering," 5th ed., Addison-Wesley, 1985, p. 137.*)

The diffusion equation $D = D_0 e^{-Q/RT}$ [Eq. (4.20)] can be written in logarithmic form as the equation of a straight line as was done in Eqs. (4.13) and (4.15) for the general Arrhenius rate law equation:

$$\ln D = \ln D_0 - \frac{Q}{RT} \qquad (4.21)$$

or

$$\log_{10} D = \log_{10} D_0 - \frac{Q}{2.303RT} \qquad (4.22)$$

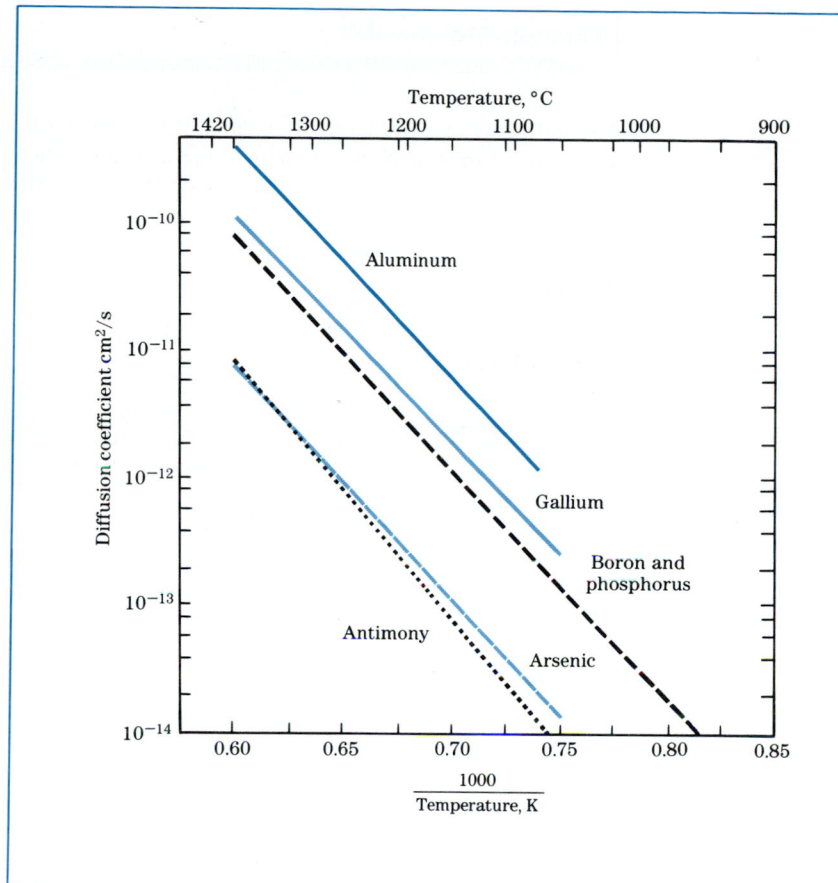

FIGURE 4.47 Diffusion coefficients as a function of temperature for some impurity elements in silicon. [*After C. S. Fuller and J. A. Ditzenberger, J. Appl. Phys., 27:544(1956).*]

If diffusivity values for a diffusion system are determined at two temperatures, values for Q and D_0 can be determined by solving two simultaneous equations of the Eq. (4.22) type. If these Q and D_0 values are substituted into Eq. (4.22), a general equation for $\log_{10} D$ vs. $1/T$ K over the temperature range investigated can be created. Example Problem 4.11 shows how the activation energy for a binary diffusion system can be calculated directly by using the relationship $D = D_0 e^{-Q/RT}$ [Eq. (4.20)] when the diffusivities are known for two temperatures.

Table 4.6 on page 172 lists D_0 and Q values for some metallic systems used to produce the Arrhenius diffusivity plots in Fig. 4.46. Figure 4.47 shows similar plots for the diffusion of impurity elements into silicon, which are useful for the fabrication of integrated circuits for the electronics industry.

Example Problem 4.11

The diffusivity of silver atoms in solid silver metal is 1.0×10^{-17} m²/s at 500°C and 7.0×10^{-13} m²/s at 1000°C. Calculate the activation energy (joules per mole) for the diffusion of Ag in Ag in the temperature range 500 to 1000°C.

Solution:

Using Eq. (4.20), $T_2 = 1000°C + 273 = 1273$ K, $T_1 = 500°C + 273 = 773$ K, and $R = 8.314$ J/(mol·K):

$$\frac{D_{1000°C}}{D_{500°C}} = \frac{\exp(-Q/RT_2)}{\exp(-Q/RT_1)} = \exp\left[-\frac{Q}{R}\left(\frac{1}{T_2} - \frac{1}{T_1}\right)\right]$$

$$\frac{7.0 \times 10^{-13}}{1.0 \times 10^{-17}} = \exp\left\{-\frac{Q}{R}\left[\left(\frac{1}{1273 \text{ K}} - \frac{1}{773 \text{ K}}\right)\right]\right\}$$

$$\ln(7.0 \times 10^4) = -\frac{Q}{R}(7.855 \times 10^{-4} - 12.94 \times 10^{-4}) = \frac{Q}{8.314}(5.08 \times 10^{-4})$$

$$11.16 = Q(6.11 \times 10^{-5})$$

$$Q = 183{,}000 \text{ J/mol} = 183 \text{ kJ/mol} \blacktriangleleft$$

SUMMARY

Most metals and alloys are melted and cast into semifinished or finished shapes. During the solidification of a metal into a casting, nuclei are formed which grow into grains, creating solidified cast metal with a polycrystalline grain structure. For most industrial applications a very small grain size is desirable. Large single crystals are rarely made in industry. However, an exception is the large single crystals of silicon produced for the semiconductor industry. For this material special solidification conditions must be used and the silicon must be of very high purity.

Crystal imperfections are present in all real crystalline materials, even at the atomic- or ionic-size level. Vacancies or empty atomic sites in metals can be explained in terms of the thermal agitation of atoms and are considered equilibrium lattice defects. Dislocations (line defects) occur in metal crystals and are created in large numbers by the solidification process. Dislocations are not equilibrium defects and increase the internal energy of the metal. Images of dislocations can be observed in the transmission electron microscope. Grain boundaries are surface imperfections in metals created by crystals of different orientations meeting each other during solidification.

Atomic diffusion occurs in metallic solids mainly by (1) a vacancy or substitution mechanism and (2) an interstitial mechanism. In the vacancy mechanism atoms of about the same size jump from one position to another, using the vacant atomic sites. In the interstitial mechanism very small atoms move through the interstitial spaces between the larger atoms of the parent matrix. Fick's first law of diffusion states that diffusion takes place because of a difference in concentration of a diffusing species from one place to another and is applicable for steady-state conditions (i.e., conditions which do not change with time). Fick's second law of diffusion is applicable for non-steady-state conditions (i.e., conditions in which the concentrations of the diffusing species change with time). In this book the use of Fick's second law has been restricted to the case of a gas diffusing into a solid. The rate of diffusion depends greatly on temperature, and this dependence is expressed by the diffusivity, a measure of the diffusion rate. Diffusivity $D = D_0 e^{-Q/RT}$. Diffusion processes are used commonly in industry. In this chapter we have examined the diffusion-gas-carburizing process for surface-hardening steel and the diffusion of controlled amounts of impurities into silicon wafers for integrated circuits.

DEFINITIONS

Sec. 4.1

Nuclei: small particles of a new phase formed by a phase change (i.e., solidification) which can grow until the phase change is complete.

Embryos: small particles of a new phase formed by a phase change (i.e., solidification) which are not of critical size and which can redissolve.

Critical radius r^* of nucleus: the minimum radius which a particle of a new phase formed by nucleation must have to become a stable nucleus.

Homogeneous nucleation (as pertains to the solidification of metals): the formation of very small regions of a new solid phase (called *nuclei*) in a pure metal which can grow until solidification is complete. The pure homogeneous metal itself provides the atoms which make up the nuclei.

Heterogeneous nucleation (as pertains to the solidification of metals): the formation of very small regions (called *nuclei*) of a new solid phase at the interfaces of solid impurities. These impurities lower the critical size at a particular temperature of stable solid nuclei.

Grain: a single crystal in a polycrystalline aggregate.

Equiaxed grains: grains which are approximately equal in all directions and which have random crystallographic orientations.

Columnar grains: long, thin grains in a solidified polycrystalline structure. These grains are formed in the interior of solidified metal ingots when heat flow is slow and uniaxial during solidification.

Polycrystalline structure: a crystalline structure which contains many grains.

Sec. 4.3

Solid solution: an alloy of two or more metals or a metal(s) and a nonmetal(s) which is a single-phase atomic mixture.

Substitutional solid solution: a solid solution in which solute atoms of one element can replace those of solvent atoms of another element. For example, in a Cu–Ni solid solution the copper atoms can replace the nickel atoms in the solid-solution crystal lattice.

Interstitial solid solution: a solid solution formed in which the solute atoms can enter the interstices or holes in the solvent-atom lattice.

Alloy: a mixture of two or more metals or a metal (metals) and a nonmetal (nonmetals).

Grain-size number: a nominal (average) number of grains per unit area at a particular magnification.

Sec. 4.4

Vacancy: a point imperfection in a crystal lattice where an atom is missing from an atomic site.

Interstitialcy (self-interstitial): a point imperfection in a crystal lattice where an atom of the same kind as those of the matrix lattice is positioned in an interstitial site between the matrix atoms.

Frenkel imperfection: a point imperfection in an ionic crystal in which a cation vacancy is associated with an interstitial cation.

Schottky imperfection: a point imperfection in an ionic crystal in which a cation vacancy is associated with an anion vacancy.

Dislocation: a crystalline imperfection in which a lattice distortion is centered around a line. The displacement distance of the atoms around the dislocation is called the *slip* or *Burgers vector* **b**. For an *edge dislocation* the slip vector is perpendicular to the dislocation line, while for a *screw dislocation* the slip vector is parallel to the dislocation line. A *mixed dislocation* has both edge and screw components.

Grain boundary: a surface imperfection which separates crystals (grains) of different orientations in a polycrystalline aggregate.

Sec. 4.5

Activation energy: the additional energy required above the average energy for a thermally activated reaction to take place.

Arrhenius rate equation: an empirical equation which describes the rate of a reaction as a function of temperature and an activation energy barrier.

Sec. 4.6

Self-diffusion: the migration of atoms in a pure material.

Substitutional diffusion: the migration of solute atoms in a solvent lattice in which the solute and solvent atoms are approximately the same size. The presence of vacancies makes the diffusion possible.

Interstitial diffusion: the migration of interstitial atoms in a matrix lattice.

Volume diffusion: atomic migration in the grain interiors of a polycrystalline aggregate.

Grain-boundary diffusion: atomic migration at the grain boundaries of a polycrystalline aggregate.

Fick's first law of diffusion in solids: the flux of a diffusing species is proportional to the concentration gradient at constant temperature.

Fick's second law of diffusion in solids: the rate of change of composition is equal to the diffusivity times the rate of change of the concentration gradient at constant temperature.

Diffusivity: a measure of the rate of diffusion in solids at a constant temperature. Diffusivity D can be expressed by the equation $D = D_0 e^{-Q/RT}$, where Q is the activation energy and T is the temperature in kelvins. D_0 and R are constants.

Steady-state conditions: for a diffusing system there is no change in the concentration of the diffusing species with time at different places in the system.

Non-steady-state conditions: for a diffusing system the concentration of the diffusing species changes with time at different places in the system.

PROBLEMS

4.1.1 Describe and illustrate the solidification process of a pure metal in terms of the nucleation and growth of crystals.

4.1.2 Define the homogeneous nucleation process for the solidification of a pure metal.

4.1.3 In the solidification of a pure metal what are the two energies involved in the transformation? Write the equation for the total free-energy change involved in the transformation of liquid to produce a strain-free solid nucleus by homogeneous nucleation. Also illustrate graphically the energy changes associated with the formation of a nucleus during solidification.

4.1.4 In the solidification of a metal what is the difference between an embryo and a nucleus? What is the critical radius of a solidifying particle?

4.1.5 During solidification, how does the degree of undercooling affect the critical nucleus size? Assume homogeneous nucleation.

4.1.6 Distinguish between homogeneous and heterogeneous nucleation for the solidification of a pure metal.

4.1.7 Calculate the size (radius) of the critically sized nucleus for pure silver when homogeneous nucleation take place.

4.1.8 Calculate the number of atoms in a critically sized nucleus for the homogeneous nucleation of pure silver.

4.1.9 Calculate the size (radius) of the critical nucleus for pure lead when nucleation takes place homogeneously.

4.1.10 Calculate the number of atoms in a critically sized nucleus for the homogeneous nucleation of pure lead.

4.1.11 Describe the grain structure of a metal ingot which was produced by slow-cooling the metal in a stationary open mold.

4.1.12 Distinguish between equiaxed and columnar grains in a solidified metal structure.

4.1.13 How can the grain size of a cast ingot be refined? How is grain refining accomplished industrially for aluminum alloy ingots?

4.2.1 What special techniques must be used to produce single crystals?

4.2.2 How are large silicon single crystals for the semiconductor industry produced?

4.3.1 What is a metal alloy? What is a solid solution?

4.3.2 Distinguish between a substitutional solid solution and an interstitial solid solution.

4.3.3 What are the conditions that are favorable for extensive solid solubility of one element in another?

4.3.4 Using the data in the table below, predict the relative degree of solid solubility of the following elements in aluminum:

(a) Copper   (d) Zinc
(b) Manganese   (e) Silicon
(c) Magnesium

Use the scale very high, 70–100%; high, 30–70%; moderate, 10–30%; low, 1–10%; and very low, <1%.

Element	Atom radius, nm	Crystal structure	Electro-negativity	Valence
Aluminum	0.143	FCC	1.5	+3
Copper	0.128	FCC	1.8	+2
Manganese	0.112	Cubic	1.6	+2, +3, +6, +7
Magnesium	0.160	HCP	1.3	+2
Zinc	0.133	HCP	1.7	+2
Silicon	0.117	Diamond cubic	1.8	+4

4.3.5 Using the data in the table below, predict the relative degree of atomic solid solubility of the following elements in iron:

(a) Nickel   (d) Titanium
(b) Chromium   (e) Manganese
(c) Molybdenum

Use the scale very high, 70–100%; high, 30–70%; moderate, 10–30%; low, 1–10%; and very low, <1%.

Element	Atom radius, nm	Crystal structure	Electro-negativity	Valence
Iron	0.124	BCC	1.7	+2, +3
Nickel	0.125	FCC	1.8	+2
Chromium	0.125	BCC	1.6	+2, +3, +6
Molybdenum	0.136	BCC	1.3	+3, +4, +6
Titanium	0.147	HCP	1.3	+2, +3, +4
Manganese	0.112	Cubic	1.6	+2, +3, +6, +7

4.3.6 Calculate the radius of the largest interstitial void in the BCC α iron lattice. The atomic radius of the iron atom in this lattice is 0.124 nm, and the largest interstitial voids occur at the $(\frac{1}{4}, \frac{1}{2}, 0)$; $(\frac{1}{2}, \frac{3}{4}, 0)$; $(\frac{3}{4}, \frac{1}{2}, 0)$; and $(\frac{1}{2}, \frac{1}{4}, 0)$, etc., type positions.

4.4.1 Describe and illustrate the following types of point imperfections which can be present in metal lattices: (a) vacancy, (b) divacancy, and (c) interstitialcy.

4.4.2 Describe and illustrate the following imperfections which can exist in crystal lattices: (a) Frenkel imperfection and (b) Schottky imperfection.

4.4.3   Describe and illustrate the edge and screw type dislocations. What type of strain fields surround both types of dislocations?

4.4.4   Describe the structure of a grain boundary. Why are grain boundaries favorable sites for the nucleation and growth of precipitates?

4.4.5   Why are grain boundaries easily observed in the optical microscope?

4.4.6   How is the grain size of polycrystalline materials measured by the ASTM method?

4.4.7   If there are 800 grains per square inch on a photomicrograph of a metal at 100×, what is its ASTM grain-size number?

4.4.8   If there are 550 grains per square inch on a photomicrograph of a ceramic material at 250×, what is the ASTM grain-size number of the material?

4.4.9   Determine, by counting, the ASTM grain-size number of the low-carbon sheet steel shown in Fig. P4.4.9. This micrograph is at 100×.

4.4.10 Determine the ASTM grain-size number of the type 430 stainless steel micrograph shown in Fig. P4.4.10. This micrograph is at 200×.

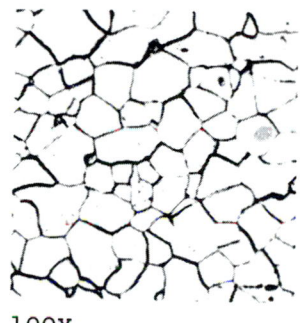

100X

FIGURE P4.4.9

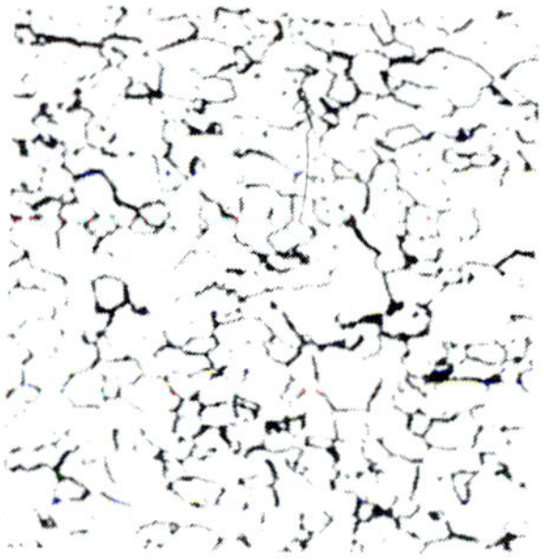

200X

FIGURE P4.4.10

4.5.1   What is a thermally activated process? What is the activation energy for such a process?

4.5.2   Write an equation for the number of vacancies present in a metal at equilibrium at a particular temperature and define each of the terms. Give the units for each term and use electron volts for the activation energy.

4.5.3   (a) Calculate the equilibrium concentration of vacancies per cubic meter in pure aluminum at 550°C. Assume that the energy of formation of a vacancy in pure aluminum is 0.76 eV. (b) What is the vacancy fraction at 600°C?

4.5.4   (a) Calculate the equilibrium concentration of vacancies per cubic meter in pure tin at 150°C. Assume the energy of formation of a vacancy in pure tin is 0.51 eV. (b) What is the vacancy fraction at 200°C?

4.6.1   Write the Arrhenius rate equation in the (a) exponential and (b) common logarithmic forms.

4.6.2   Draw a typical Arrhenius plot of $\log_{10}$ of the reaction rate vs. reciprocal absolute temperature and indicate the slope of the plot.

4.6.3   Describe the substitutional and interstitial diffusion mechanisms in solid metals.

4.6.4   Write the equation for Fick's first law of diffusion and define each of the terms in SI units.

4.6.5   What factors affect the diffusion rate in solid metal crystals?

4.6.6 Write the equation for Fick's second law of diffusion in solids and define each of the terms.

4.6.7 Write the equation for the solution to Fick's second law for the diffusion of a gas into the surface of a solid metal crystal lattice.

4.7.1 Describe the gas-carburizing process for steel parts. Why is the carburization of steel parts carried out?

4.7.2 Consider the gas carburizing of a gear of 1020 steel (0.20 wt %) at 927°C (1700°F). Calculate the time necessary to increase the carbon content to 0.40 wt % at 0.50 mm below the surface of the gear. Assume the carbon content at the surface to be 1.10 wt % and that the nominal carbon content of the steel gear before carburizing is 0.20 wt %. D (C in γ iron) at 927°C = 1.28×10^{-11} m²/s.

4.7.3 The surface of a steel gear made of 1022 steel (0.22 wt % C) is to be gas-carburized at 927°C (1700°F). Calculate the time necessary to increase the carbon content to 0.35 wt % at 0.035 in below the surface of the gear. Assume the carbon content of the surface to be 1.20 wt %. D (C in γ iron) at 927°C = 1.28×10^{-11} m²/s.

4.7.4 A gear made of 1022 steel (0.22 wt % C) is to be gas-carburized at 927°C (1700°F). Calculate the carbon content at 0.040 in below the surface of the gear after a 3-h carburizing time. Assume the carbon content at the surface of the gear is 1.00 wt %. D (C in γ iron) at 927°C = 1.28×10^{-11} m²/s.

4.7.5 A gear made of 1020 steel (0.20 wt % C) is to be gas-carburized at 927°C (1700°F). Calculate the carbon content at 0.050 in below the surface of the gear after an 8-h carburizing time. Assume the carbon content at the surface of the gear is 1.20 wt %. D (C in γ iron) at 927°C = 1.28×10^{-11} m²/s.

4.7.6 The surface of a steel gear made of 1018 steel (0.18 wt % C) is to be gas-carburized at 927°C. Calculate the time necessary to increase the carbon content to 0.40 wt % at 0.95 mm below the surface. Assume the carbon content of the surface of the gear is 1.10 wt %. D (C in γ iron) at 927°C = 1.28×10^{-11} m²/s.

4.7.7 A gear made of 1020 steel (0.20 wt % C) is to be gas-carburized at 927°C. Calculate the carbon content at 1.10 mm below the surface of the gear after a 7-h carburizing time. Assume the carbon content at the surface of the gear is 1.10 wt %. D (C in γ iron) at 927°C = 1.28×10^{-11} m²/s.

4.7.8 A gear made of 1022 steel (0.22 wt % C) is to be gas-carburized at 927°C. If the carburizing time is 6.0 h, at what depth in millimeters will the carbon content be 0.35 wt %? Assume the carbon content at the surface of the gear is 1.15 wt %. D (C in γ iron) at 927°C = 1.28×10^{-11} m²/s.

4.7.9 If boron is diffused into a thick slice of silicon with no previous boron in it at a temperature of 1100°C for 7 h, what is the depth below the surface at which the concentration is 10^{17} atoms/cm³ if the surface concentration is 10^{18} atoms/cm³? $D = 4 \times 10^{-13}$ cm²/s for boron diffusing in silicon at 1100°C.

4.7.10 If aluminum is diffused into a thick slice of silicon with no previous aluminum in it at a temperature of 1100°C for 8 h, what is the depth below the surface at which the concentration is 10^{16} atoms/cm³ if the surface concentration is 10^{18} atoms/cm³? $D = 2 \times 10^{-12}$ cm²/s for aluminum diffusing in silicon at 1100°C.

4.7.11 Phosphorus is diffused into a thick slice of silicon with no previous phosphorus in it at a temperature of 1100°C. If the surface concentration of the phosphorus is 1×10^{18} atoms/cm³ and its concentration at 1 μm is 1×10^{15} atoms/cm³, how long must the diffusion time be? $D = 3.0 \times 10^{-13}$ cm²/s for P diffusing in Si at 1100°C.

4.7.12 If the diffusivity in Prob. 4.7.11 had been 1.5×10^{-13} cm^2/s, at what depth in micrometers would the phosphorus concentration be 1×10^{15} atoms/cm^3?

4.7.13 Arsenic is diffused into a thick slice of silicon with no previous arsenic in it at 1100°C. If the surface concentration of the arsenic is 5.0×10^{18} atoms/cm^3 and its concentration at 1.2 μm below the silicon surface is 1.5×10^{16} atoms/cm^3, how long must the diffusion time be? ($D = 3.0 \times 10^{-14}$ cm^2/s for As diffusing in Si at 1100°C.)

4.8.1 Calculate the diffusivity in square meters per second of nickel in FCC iron at 1200°C. Use $D_0 = 7.7 \times 10^{-5}$ m^2/s; $Q = 280$ kJ/mol; $R = 8.314$ J/(mol · K).

4.8.2 Calculate the diffusivity in square meters per second of carbon in HCP titanium at 800°C. Use $D_0 = 5.10 \times 10^{-4}$ m^2/s; $Q = 182$ kJ/mol; $R = 8.314$ J/(mol · K).

4.8.3 Calculate the diffusivity in square meters per second for the diffusion of zinc in copper at 300°C. Use $D_0 = 3.4 \times 10^{-5}$ m^2/s; $Q = 191$ kJ/(mol · K).

4.8.4 The diffusivity of manganese atoms in the FCC iron lattice is 1.0×10^{-14} m^2/s at 1300°C and 1.0×10^{-15} m^2/s at 400°C. Calculate the activation energy in kilojoules per mole for this case in this temperature range. Data: $R = 8.314$ J/(mol · K).

4.8.5 The diffusivity of copper atoms in the aluminum lattice is 7.0×10^{-13} m^2/s at 600°C and 2.0×10^{-15} m^2/s at 400°C. Calculate the activation energy in kilojoules per mole for this case in this temperature range. $R = 8.314$ J/(mol · K).

4.8.6 The diffusivity of iron atoms in the BCC Fe Lattice is 4.2×10^{-23} m^2/s at 400°C and 5.6×10^{-16} m^2/s at 800°C. Calculate the activation energy in joules per mole for the diffusion of iron atoms in the BCC Fe lattice in this temperature range. Data: $R = 8.314$ J/(mol · K).

Electrical Properties of Materials

In this chapter electrical conduction in metals is considered first. The effects of impurities, alloy additions, and temperature on the electrical conductivity of metals are discussed. The energy-band model for electrical conduction in metals is then considered. Following this, the effects of impurities and temperature on the electrical conductivity of semiconducting materials are treated. Finally, the principles of operation of some basic semiconducting devices are examined, and some of the fabrication processes used to produce modern microelectronic circuitry are presented. An example of the complexity of recent microelectronic integrated circuitry is shown in Fig. 5.1.

5.1 ELECTRICAL CONDUCTION IN METALS

The Classical Model for Electrical Conduction in Metals

In metallic solids the atoms are arranged in a crystal structure (for example, FCC, BCC, and HCP) and are bound together by their outer valence electrons by metallic bonding (see Sec. 2.7). The metallic bonds in solid metals make the free movement of the valence electrons possible since they are shared by many atoms and are not bound to any particular atom. Sometimes the valence

Intel Pentium™ Processor

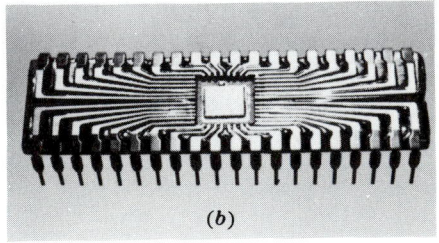

Code Cache

Code TLB

Clock Driver

Instruction Fetch

Branch Prediction Logic

Instruction Decode

Control Logic

Complex Instruction Support

Bus Interface Logic

Data TLB

Data Cache

Superscalar Integer Execution Units

Pipelined Floating Point

(a)

(b)

FIGURE 5.1 (a) The microprocessor or "computer on a chip" shown is magnified about six times on a side and incorporates about 3.1 million transistors on a single chip of silicon which is actually about 17.2 mm on each side. This microprocessor is the Intel Pentium and uses BiCMOS technology with minimum design features of 0.8 μm. (b) One-half actual size microprocessor mounted in a package with connecting wire passages shown. (*Courtesy of Intel Corporation, Santa Clara, Calif.*)

electrons are visualized as forming an electron charge cloud, as shown in Fig. 5.2a. Other times the valence electrons are considered to be individual free electrons not associated with any particular atom, as shown in Fig. 5.2b.

In the classical model for electrical conduction in metallic solids, the outer valence electrons are assumed to be completely free to move between the positive-ion cores (atoms without valence electrons) in the metal lattice. At room temperature the positive-ion cores have kinetic energy and vibrate

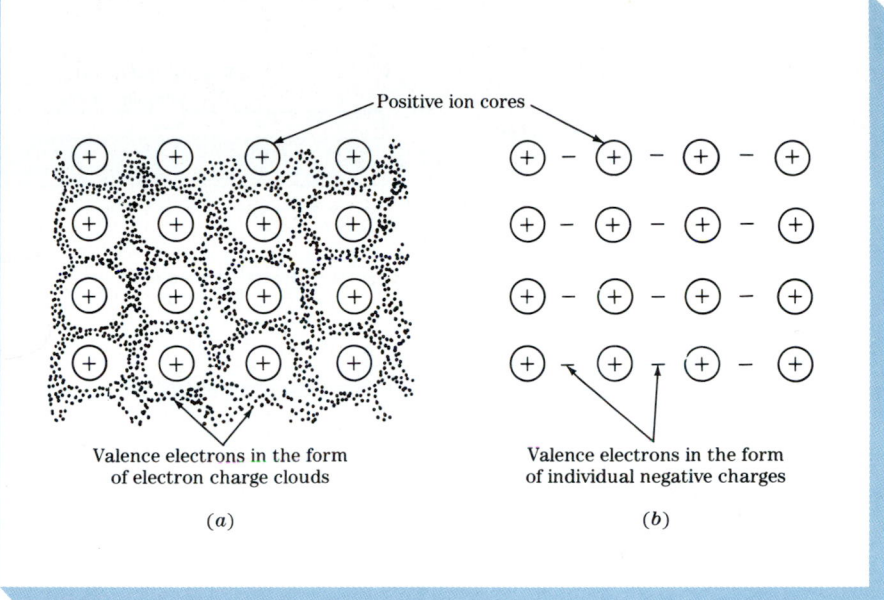

Positive ion cores

Valence electrons in the form
of electron charge clouds

(a)

Valence electrons in the form
of individual negative charges

(b)

FIGURE 5.2 Schematic arrangements of the atoms in one plane of a monovalent metal such as copper, silver, or sodium. In (a) the valence electrons are pictured as an "electron gas," and in (b) the valence electrons are visualized as free electrons of unit charge.

about their lattice positions. With increasing temperature these ions vibrate with increasing amplitudes, and there is a continual interchange of energy between the ion cores and their valence electrons. In the absence of an electric potential, the motion of the valence electrons is random and restricted, so there is no net electron flow in any direction and thus no current flow. In the presence of an applied electric potential the electrons attain a directed *drift velocity* which is proportional to the applied field but in the opposite direction.

Ohm's Law

Consider a length of copper wire whose ends are connected to a battery, as shown in Fig. 5.3. If a potential difference V is applied to the wire, a current i will flow which is proportional to the resistance R of the wire. According to *Ohm's law*, the electric current flow i is proportional to the applied voltage V and inversely proportional to the resistance of the wire, or

$$i = \frac{V}{R}$$

(5.1)

where i = electric current, A (amperes)
V = potential difference, V (volts)
R = resistance of wire, Ω (ohms)

The electrical resistance R of an electrical conductor such as the metal wire specimen of Fig. 5.3 is directly proportional to its length l and inversely proportional to its cross-sectional area A. These quantities are related by a

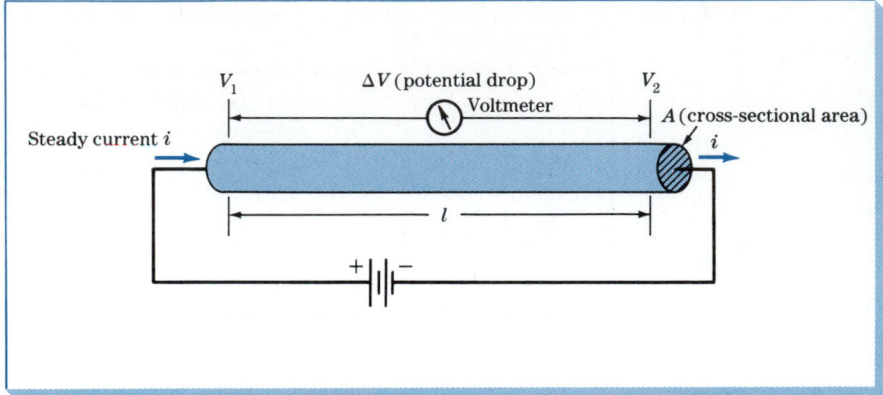

FIGURE 5.3 Potential difference ΔV applied to metal wire specimen of cross-sectional area A.

material constant called the *electrical resistivity ρ*, as

$$R = \rho \frac{l}{A} \quad \text{or} \quad \rho = R \frac{A}{l} \qquad (5.2)$$

The units for electrical resistivity, which is a constant for a material at a particular temperature, are

$$\rho = R \frac{A}{l} = \Omega \frac{m^2}{m} = \text{ohm-meter} = \Omega \cdot m$$

Frequently it is more convenient to think in terms of the passage of electric current instead of resistance, and so the quantity *electrical conductivity σ*[1] is defined as the reciprocal of electrical resistivity:

$$\sigma = \frac{1}{\rho} \qquad (5.3)$$

The units for electrical conductivity are $(\text{ohm-meter})^{-1} = (\Omega \cdot m)^{-1}$. The SI unit for the reciprocal of the ohm is the siemens (S), but this unit is rarely used and so will not be used in this book.

Table 5.1 lists the electrical conductivities of some selected metals and nonmetals. From this table it is seen that the pure metals silver, copper, and gold have the highest conductivities, about $10^7 \; (\Omega \cdot m)^{-1}$. *Electrical insulators* such as polyethylene and polystyrene, on the other hand, have very low electrical conductivities, about $10^{-14} \; (\Omega \cdot m)^{-1}$, which are about 10^{23} times less than those of the highly conductive metals. Silicon and germanium have conductivities in between those of metals and insulators and are thus classified as *semiconductors*.

---

[1]σ = Greek letter sigma

TABLE 5.1   Electrical Conductivities of Some Metals and Nonmetals at Room Temperature

Metals and alloys	$\sigma, (\Omega \cdot m)^{-1}$	Nonmetals	$\sigma, (\Omega \cdot m)^{-1}$
Silver	6.3×10^7	Graphite	10^5 (average)
Copper, commercial purity	5.8×10^7	Germanium	2.2
Gold	4.2×10^7	Silicon	4.3×10^{-4}
Aluminum, commercial purity	3.4×10^7	Polyethylene	10^{-14}
		Polystyrene	10^{-14}
		Diamond	10^{-14}

Example Problem 5.1

A wire whose diameter is 0.20 cm must carry a 20-A current. The maximum power dissipation along the wire is 4 W/m (watts per meter). Calculate the minimum allowable conductivity of the wire in (ohm-meters)$^{-1}$ for this application.

Solution:

Power $P = iV = i^2R$ where i = current, A R = resistance, Ω
                                  V = voltage, V P = power, W (watts)

$$R = \rho \frac{l}{A}$$ where ρ = resistivity, $\Omega \cdot m$
                        l = length, m
                        A = cross-sectional area of wire, m^2

Combining the two equations above gives

$$P = i^2\rho \frac{l}{A} = \frac{i^2l}{\sigma A} \quad \text{since } \rho = \frac{1}{\sigma}$$

Rearranging gives

$$\sigma = \frac{i^2l}{PA}$$

Given that $P = 4$ W (in 1 m) $i = 20$ A $l = 1$ m

and $A = \dfrac{\pi}{4} (0.0020 \text{ m})^2 = 3.14 \times 10^{-6} \text{ m}^2$

thus

$$\sigma = \frac{i^2l}{PA} = \frac{(20 \text{ A})^2(1 \text{ m})}{(4 \text{ W})(3.14 \times 10^{-6} \text{ m}^2)} = 3.18 \times 10^7 \ (\Omega \cdot m)^{-1} \blacktriangleleft$$

Therefore, for this application, the conductivity σ of the wire must be equal to or greater than $3.18 \times 10^7 \ (\Omega \cdot m)^{-1}$.

Example Problem 5.2

If a copper wire of commercial purity is to conduct 10 A of current with a maximum voltage drop of 0.4 V/m, what must be its minimum diameter? [σ (commercially pure Cu) $= 5.85 \times 10^7$ $(\Omega \cdot m)^{-1}$.]

Solution:

Ohm's law:
$$V = iR \quad \text{and} \quad R = \rho \frac{l}{A}$$

Combining the two equations gives

$$V = i\rho \frac{l}{A}$$

and rearranging gives
$$A = i\rho \frac{l}{V}$$

Substituting $(\pi/4)d^2 = A$ and $\rho = 1/\sigma$ yields

$$\frac{\pi}{4} d^2 = \frac{il}{\sigma V}$$

and solving for d gives
$$d = \sqrt{\frac{4il}{\pi \sigma V}}$$

Given that $i = 10$ A, $V = 0.4$ V for 1 m of wire, $l = 1.0$ m (chosen length of wire), and conductivity of Cu wire $\sigma = 5.85 \times 10^7$ $(\Omega \cdot m)^{-1}$, thus

$$d = \sqrt{\frac{4il}{\pi \sigma V}} = \sqrt{\frac{4(10 \text{ A})(1.0 \text{ m})}{\pi[5.85 \times 10^7 \ (\Omega \cdot m)^{-1}](0.4 \text{ V})}} = 7.37 \times 10^{-4} \text{ m} \blacktriangleleft$$

Therefore, for this application, the Cu wire must have a diameter of 7.37×10^{-4} m or greater.

Equation (5.1) is called the *macroscopic form* of Ohm's law since the values of i, V, and R are dependent on the geometrical shape of a particular electrical conductor. Ohm's law can also be expressed in a *microscopic form* which is independent of the shape of the electrical conductor as

$$\mathbf{J} = \frac{\mathbf{E}}{\rho} \quad \text{or} \quad \mathbf{J} = \sigma \mathbf{E} \tag{5.4}$$

where $\mathbf{J}$ = current density, A/m^2 ρ = electrical resistivity, $\Omega \cdot m$
$\mathbf{E}$ = electric field, V/m σ = electrical conductivity $(\Omega \cdot m)^{-1}$

The current density $\mathbf{J}$ and electric field $\mathbf{E}$ are vector quantities with magnitude and direction. Both macroscopic and microscopic forms of Ohm's law are compared in Table 5.2.

TABLE 5.2   Comparison of the Macroscopic and Microscopic Forms of Ohm's Law

Macroscopic form of Ohm's law	Microscopic form of Ohm's law
$i = \dfrac{V}{R}$	$J = \dfrac{E}{\rho}$
where i = current, A V = voltage, V R = resistance, Ω	where J = current density, A/m^2 E = electric field, V/m ρ = electrical resistivity, $\Omega \cdot$ m

Drift Velocity of Electrons in a Conducting Metal

At room temperature the positive-ion cores in a metallic conductor crystal lattice vibrate about neutral positions and therefore possess kinetic energy. The free electrons continually exchange energy with the lattice ions by elastic and inelastic collisions. Since there is no external electric field, the electron motion is random, and since there is no net electron motion in any direction, there is no net current flow.

If a uniform electric field of intensity $\mathbf{E}$ is applied to the conductor, the electrons are accelerated with a definite velocity in the direction opposite to the applied field. The electrons periodically collide with the ion cores in the lattice and lose their kinetic energy. After a collision, the electrons are free to accelerate again in the applied field, and as a result the electron velocity varies with time in a "sawtooth manner," as shown in Fig. 5.4. The average time between collisions is 2τ, where τ is the *relaxation time.*

The electrons thus acquire an average drift velocity $\mathbf{v}_d$ which is directly proportional to the applied electric field $\mathbf{E}$. The relationship between the drift velocity and the applied field is

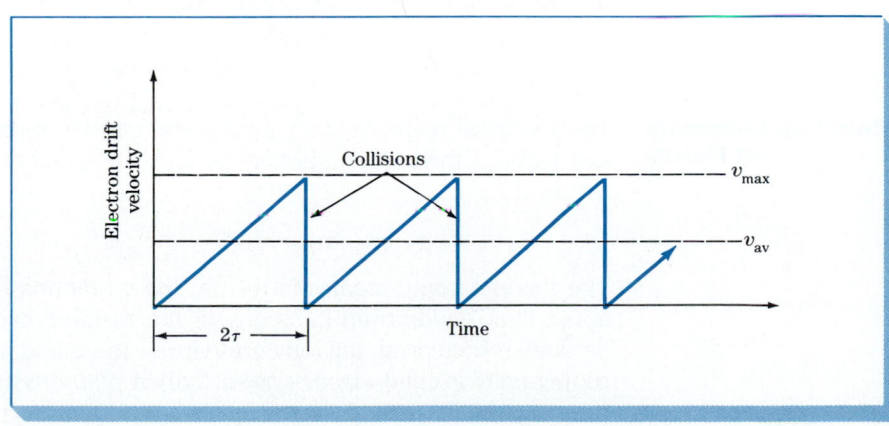

FIGURE 5.4 Electron drift velocity vs. time for classical model for electrical conductivity of a free electron in a metal.

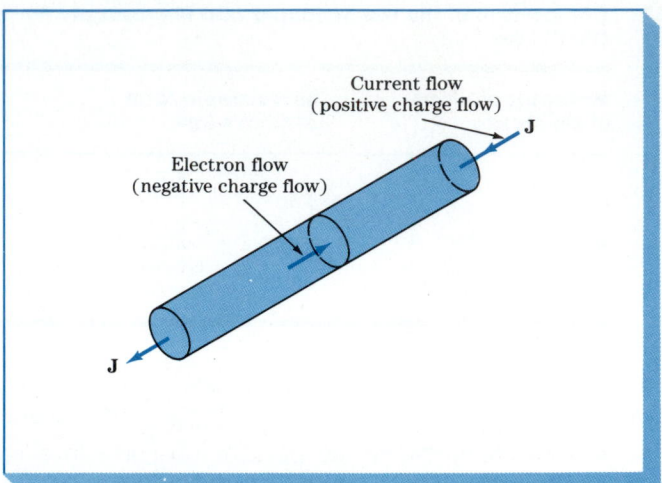

Current flow
(positive charge flow)

J

Electron flow
(negative charge flow)

J

FIGURE 5.5 A potential difference along a copper wire causes electron flow, as indicated in the drawing. Because of the negative charge on the electron, the direction of electron flow is opposite to that of conventional current flow which assumes positive charge flow.

$$\mathbf{v}_d = \mu \mathbf{E} \tag{5.5}$$

where μ(mu), the electron mobility, m²/(V·s), is the proportionality constant.

Consider the wire shown in Fig. 5.5 as having a current density $\mathbf{J}$ flowing in the direction shown. Current density by definition is equal to the rate at which charges cross any plane which is perpendicular to $\mathbf{J}$. That is, a certain number of amperes per square meter or coulombs per second per square meter flow past the plane.

The electron flow in a metal wire subjected to a potential difference depends on the number of electrons per unit volume, the electronic charge $-e$ (-1.60×10^{-19} C), and the drift velocity of the electrons, $\mathbf{v}_d$. The rate of charge flow per unit area is $-ne\mathbf{v}_d$. However, by convention, electric current is considered to be positive charge flow, and thus the current density $\mathbf{J}$ is given a positive sign. In equation form then,

$$\mathbf{J} = ne\mathbf{v}_d \tag{5.6}$$

Electrical Resistivity of Metals

The electrical resistivity of a pure metal can be approximated by the sum of two terms: a thermal component ρ_T and a residual component ρ_r:

$$\rho_{\text{total}} = \rho_T + \rho_r \tag{5.7}$$

The thermal component arises from the vibrations of the positive-ion cores about their equilibrium positions in the metallic crystal lattice. As the temperature is increased, the ion cores vibrate more and more, and a large number of thermally excited elastic waves (called *phonons*) scatter conduction electrons and decrease the mean free paths and relaxation times between collisions.

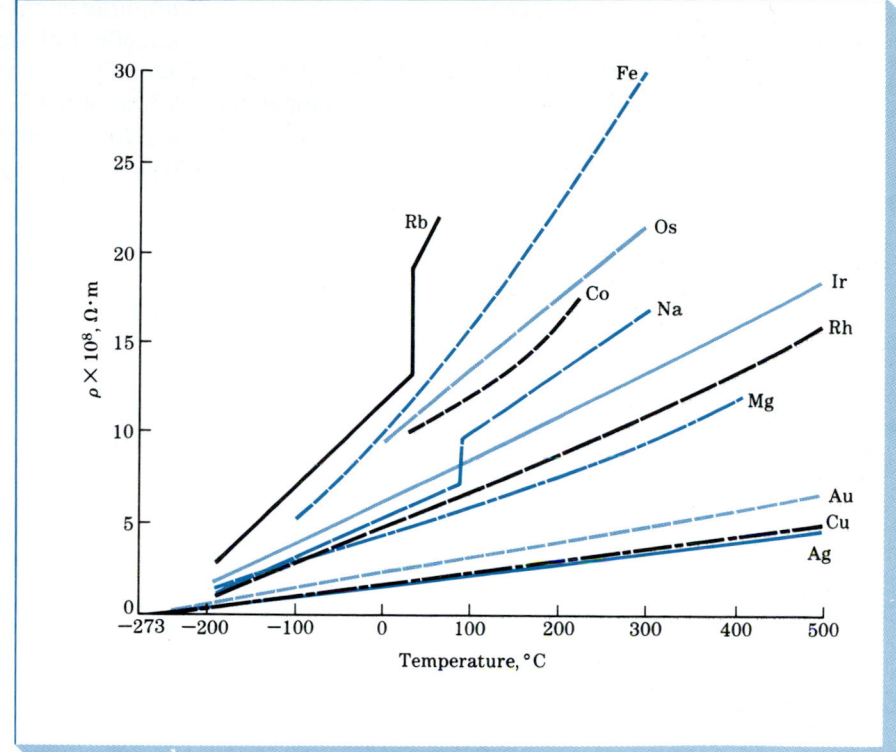

FIGURE 5.6 The effect of temperature on the electrical resistivity of selected metals. Note that there is almost a linear relationship between resistivity and temperature (°C). (*After Zwikker, "Physical Properties of Solid Materials," Pergamon, 1954, pp. 247, 249.*)

Thus as the temperature is increased, the electrical resistivities of pure metals increase, as shown in Fig. 5.6. The residual component of the electrical resistivity of pure metals is small and is caused by structural imperfections such

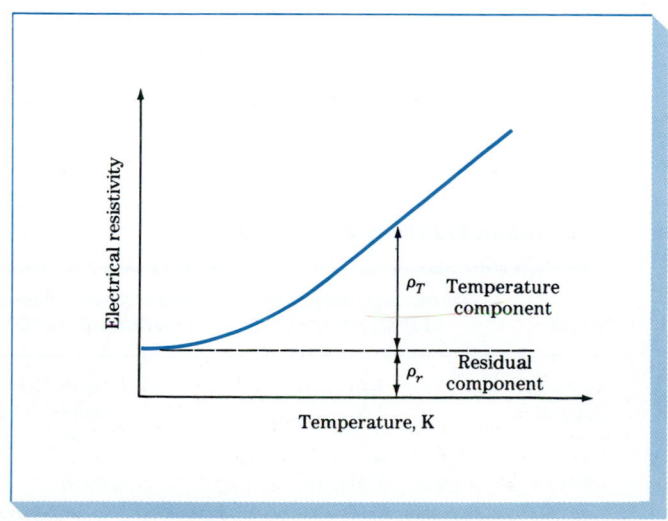

FIGURE 5.7 Schematic variation of electrical resistivity of a metal with absolute temperature. Note that at higher temperatures the electrical resistivity is the sum of a residual component ρ_r and a thermal component ρ_T.

as dislocations, grain boundaries, and impurity atoms which scatter electrons. The residual component is almost independent of temperature and becomes significant only at low temperatures (Fig. 5.7).

For most metals at temperatures above about $-200°C$, the electrical resistivity varies almost linearly with temperature, as shown in Fig. 5.6. Thus the electrical resistivities of many metals may be approximated by the equation

$$\rho_T = \rho_{0°C}(1 + \alpha_T T) \tag{5.8}$$

where $\rho_{0°C}$ = electrical resistivity at 0°C
    α_T = temperature coefficient of resistivity, $°C^{-1}$
    T = temperature of metal, °C

Table 5.3 lists the temperature resistivity coefficients for selected metals. For these metals α_T ranges from 0.0034 to 0.0045/°C.

Example Problem 5.3

Calculate the electrical resistivity of pure copper at 132°C, using the temperature resistivity coefficient for copper from Table 5.3.

Solution:

$$\rho_T = \rho_{0°C}(1 + \alpha_T T) \tag{5.8}$$

$$= 1.6 \times 10^{-6} \, \Omega \cdot cm \left(1 + \frac{0.0039}{°C} \times 132°C \right)$$

$$= 2.42 \times 10^{-6} \, \Omega \cdot cm$$

$$= 2.42 \times 10^{-8} \, \Omega \cdot m \; \blacktriangleleft$$

Alloying elements added to pure metals cause additional scattering of the conduction electrons and thus increase the electrical resistivity of pure metals. The effect of small additions of various elements on the room temperature

TABLE 5.3   Temperature Resistivity Coefficients

Metal	Electrical resistivity at 0°C, $\mu\Omega \cdot cm$	Temperature resistivity coefficient α_T, $°C^{-1}$
Aluminum	2.7	0.0039
Copper	1.6	0.0039
Gold	2.3	0.0034
Iron	9	0.0045
Silver	1.47	0.0038

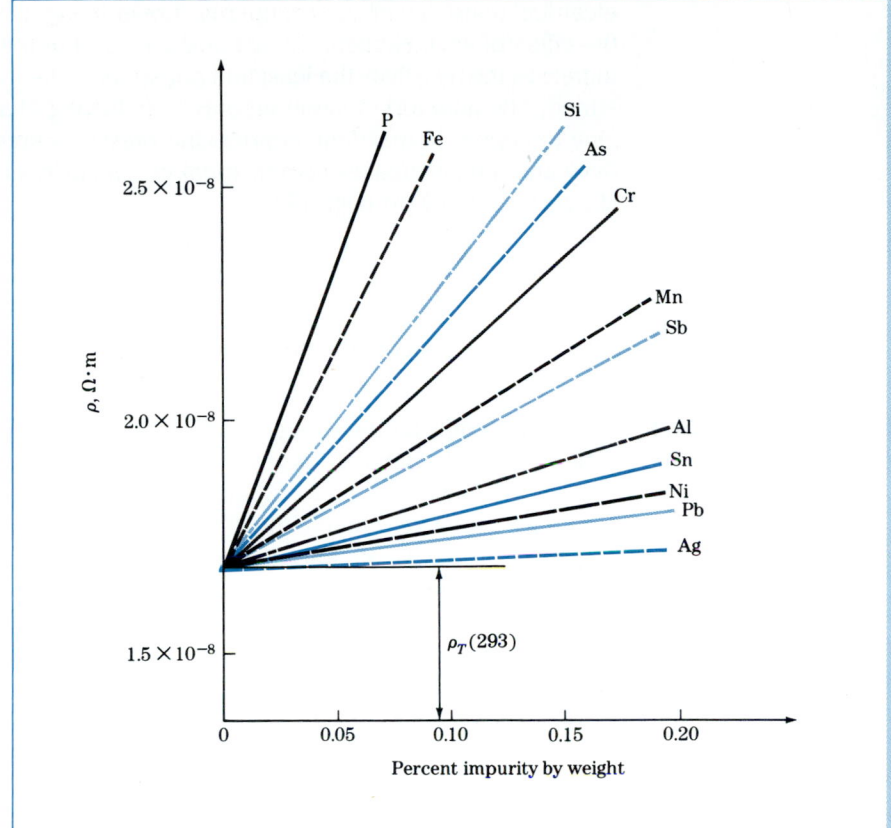

FIGURE 5.8 The effect of small additions of various elements on the room temperature electrical resistivity of copper. [*After F. Pawlek and K. Reichel, Z. Metallkd.,* **47**:347(1956).]

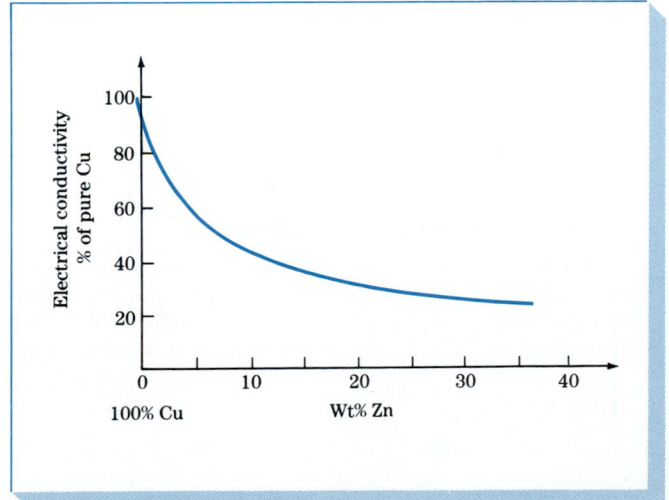

FIGURE 5.9 The effect of zinc additions to pure copper in reducing the electrical conductivity of the copper. (*After ASM data.*)

electrical resistivity of pure copper is shown in Fig. 5.8 on page 193. Note that the effect of each element varies considerably. For the elements shown, silver increases the resistivity the least and phosphorus the most for the same amount added. The addition of larger amounts of alloying elements such as 5 to 35% zinc to copper to make the copper-zinc brasses increases the electrical resistivity and thus decreases the electrical conductivity of pure copper greatly, as shown in Fig. 5.9 on page 193.

5.2 ENERGY-BAND MODEL FOR ELECTRICAL CONDUCTION

Energy-Band Model for Metals

Let us now consider the energy-band model for electrons in solid metals since it helps in understanding the mechanism of electrical conduction in metals. We use the metal sodium to explain the energy-band model since the sodium atom has a relatively simple electronic structure.

Electrons of isolated atoms are bound to their nuclei and can only have energy levels which are *sharply defined* such as the $1s^1$, $1s^2$, $2s^1$, $2s^2$, . . . states as necessitated by the Pauli principle. Otherwise, it would be possible for all electrons in an atom to descend to the lowest energy state, $1s^1$! Thus the 11 electrons in the neutral sodium atom occupy two $1s$ states, two $2s$ states, six $2p$ states, and one $3s$ state, as illustrated in Fig. 5.10a. The electrons in the lower levels ($1s^2$, $2s^2$, $2p^6$) are tightly bound and constitute the *core electrons* of the sodium atom (Fig. 5.10b). The outer $3s^1$ electron can be involved in bonding with other atoms and is called the *valence electron.*

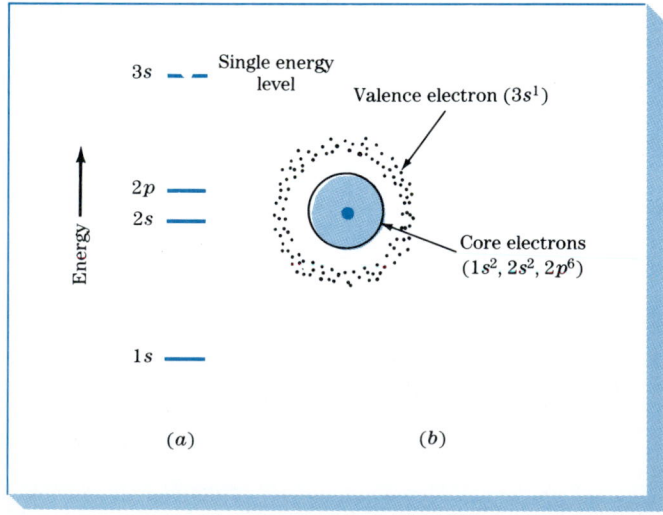

FIGURE 5.10 (*a*) Energy levels in a single sodium atom. (*b*) Arrangement of electrons in a sodium atom. The outer $3s^1$ valence electron is loosely bound and is free to be involved in metallic bonding.

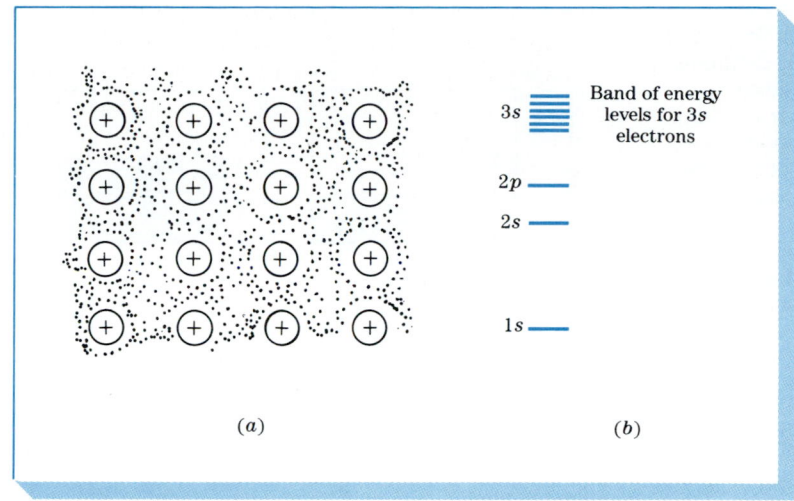

FIGURE 5.11 (*a*) Delocalized valence electrons in a block of sodium metal. (*b*) Energy levels in a block of sodium metal; note the expansion of the 3*s* level into an energy band and that the 3*s* band is shown closer to the 2*p* level since bonding has caused a lowering of the 3*s* levels of the isolated sodium atoms.

In a solid block of metal the atoms are close together and touch each other. The valence electrons which are delocalized (Fig. 5.11*a*) interact and interpenetrate each other so that their original sharp atomic energy levels are broadened into wider regions called *energy bands* (Fig. 5.11*b*). The inner electrons, since they are shielded from the valence electrons, do not form bands.

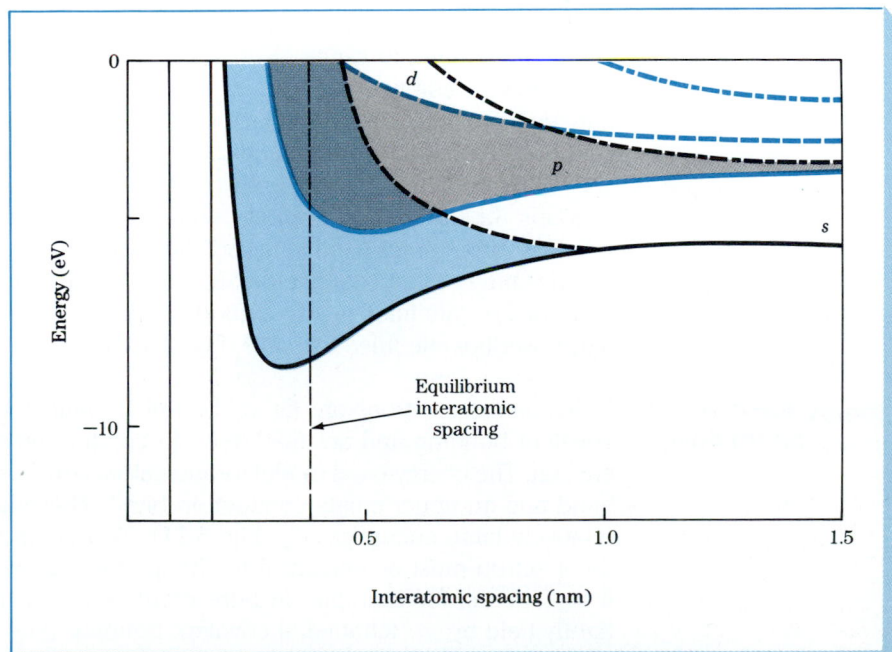

FIGURE 5.12 Valence energy bands in sodium metal. Note the splitting of the *s*, *p*, and *d* levels. [*After J. C. Slater, Phys. Rev.,* **45**:794(1934).]

FIGURE 5.13 Schematic energy-band diagrams for several metallic conductors. (*a*) Sodium, $3s^1$: the 3s band is half-filled since there is only one $3s^1$ electron. (*b*) Magnesium, $3s^2$: the 3s band is filled and overlaps the empty 3p band. (*c*) Aluminum, $3s^2 3p^1$: the 3s band is filled and overlaps the partially filled 3p band.

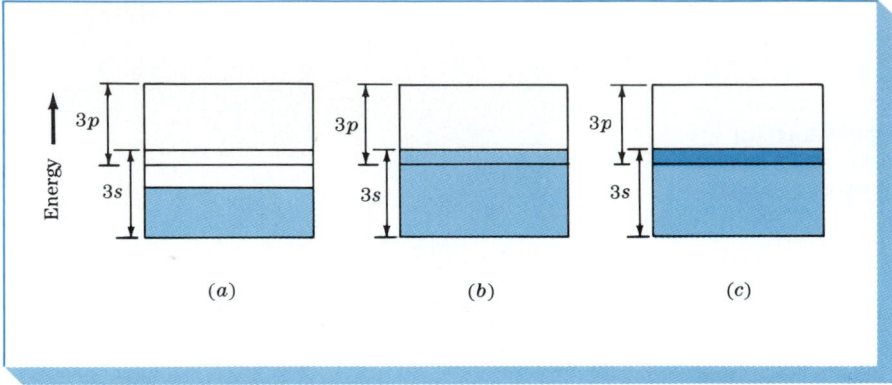

Each valence electron in a block of sodium metal, for example, must have a slightly different energy level according to the Pauli exclusion principle. Thus if there are N sodium atoms in a block of sodium, where N can be very large, there will be N distinct but only slightly different $3s^1$ energy levels in the 3s energy *band*. Each energy level is called a *state*. In the valence energy band the energy levels are so close that they form a continuous energy band.

Figure 5.12 shows part of the energy-band diagram for metallic sodium as a function of interatomic spacing. In solid metallic sodium the 3s and 3p energy bands overlap (Fig. 5.12). However, since there is only one 3s electron in the sodium atom, the 3s band is only half-filled (Fig. 5.13a). As a result very little energy is required to excite electrons in sodium from the highest filled states to the lowest empty ones. Sodium is, therefore, a good conductor since very little energy is required to produce electron flow in it. Copper, silver, and gold also have half-filled outer s bands.

In metallic magnesium both 3s states are filled. However, the 3s band overlaps the 3p band and allows some electrons into it, creating a partially filled 3sp combined band (Fig. 5.13b). Thus, in spite of the filled 3s band, magnesium is a good conductor. Similarly, aluminum which has both 3s states and one 3p state filled is also a good conductor because the partially filled 3p band overlaps the filled 3s band (Fig. 5.13c).

Energy-Band Model for Insulators   In insulators electrons are tightly bound to their bonding atoms by ionic or covalent bonding and are not "free" to conduct electricity unless highly energized. The energy-band model for insulators consists of a lower filled valence band and an upper empty conduction band. These bands are separated by a relatively large energy gap E_g (Fig. 5.14). To free an electron for conduction, the electron must be energized to "jump" the gap which may be as much as 6 to 7 eV as, for example, in pure diamond. In diamond, the electrons are tightly held by sp^3 tetrahedral covalent bonding (Fig. 5.15).

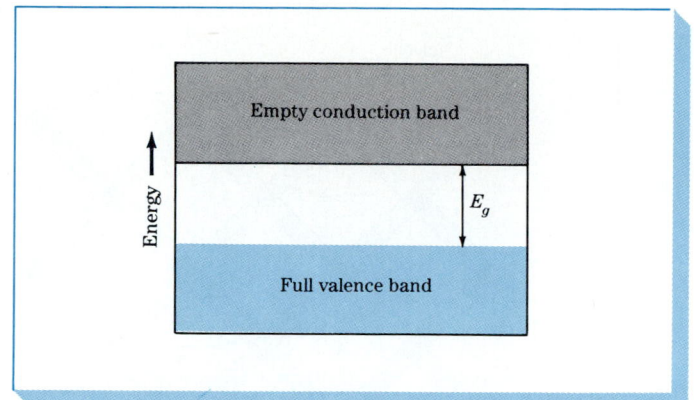

FIGURE 5.14 Energy-band diagram for an insulator. The valence band is completely filled and is separated from an empty conduction band by a large energy gap E_g.

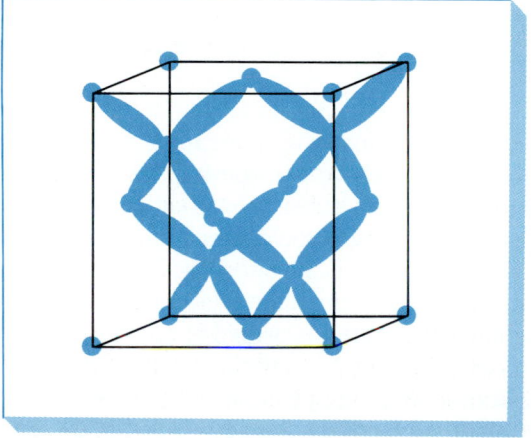

FIGURE 5.15 Diamond cubic crystal structure. The atoms in this structure are bonded together by sp^3 covalent bonds. Diamond (carbon), silicon, germanium, and gray tin (the tin polymorph stable below 13°C) all have this structure. There are 8 atoms per unit cell: $\frac{1}{8} \times 8$ at the corners, $\frac{1}{2} \times 6$ at the faces, and 4 inside the unit cube.

5.3 INTRINSIC SEMICONDUCTORS

The Mechanism of Electrical Conduction in Intrinsic Semiconductors

Semiconductors are materials whose electrical conductivities are between those of highly conducting metals and poorly conducting insulators. *Intrinsic semiconductors* are pure semiconductors whose electrical conductivity is determined by their inherent conductive properties. Pure elemental silicon and germanium are intrinsic semiconducting materials. These elements, which are in group IVA in the periodic table, have the diamond cubic structure with highly directional covalent bonds (Fig. 5.15). Tetrahedral sp^3 hybrid bonding orbitals consisting of electron pairs bond the atoms together in the crystal lattice. In this structure each silicon or germanium atom contributes four valence electrons.

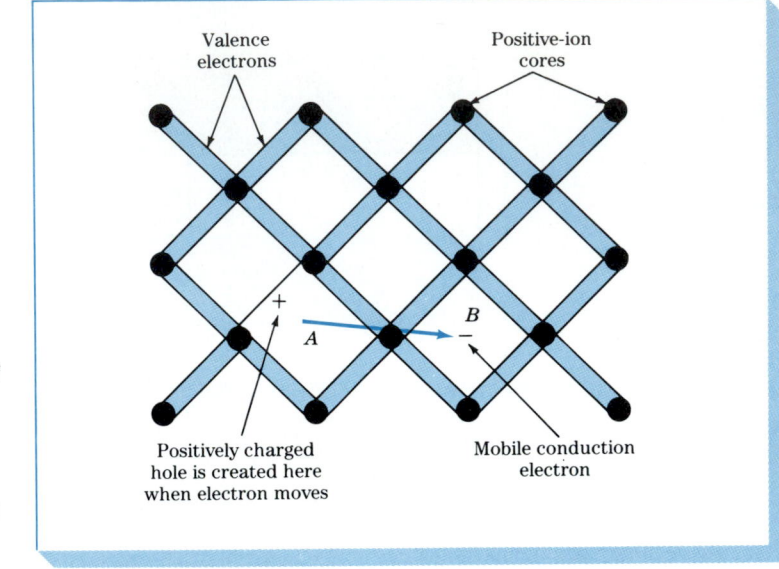

FIGURE 5.16 Two-dimensional representation of the diamond cubic lattice of silicon or germanium showing positive-ion cores and valence electrons. Electron has been excited from bond at A and has moved to point B.

Electrical conductivity in pure semiconductors such as Si and Ge can be described qualitatively by considering the two-dimensional pictorial representation of the diamond cubic crystal lattice shown in Fig. 5.16. The circles in this illustration represent the *positive-ion cores* of the Si or Ge atoms, and the joining pairs of lines indicate bonding *valence electrons.* The bonding electrons are unable to move through the crystal lattice and hence to conduct electricity unless sufficient energy is provided to excite them from their bonding positions. When a critical amount of energy is supplied to a valence electron to excite it away from its bonding position, it becomes a free conduction electron and leaves behind a positively charged "hole" in the crystal lattice (Fig. 5.16).

Electrical Charge Transport in the Crystal Lattice of Pure Silicon

In the electrical conduction process in a semiconductor such as pure silicon or germanium, both electrons and holes are charge carriers and move in an applied electric field. Conduction electrons have a negative charge and are attracted to the positive terminal of an electrical circuit (Fig. 5.17). Holes, on the other hand, behave like positive charges and are attracted to the negative terminal of an electrical circuit (Fig. 5.17). A hole has a positive charge equal in magnitude to the electron charge.

The motion of a "hole" in an electric field can be visualized by referring to Fig. 5.18. Let a hole exist at atom A where a valence electron is missing, as shown in Fig. 5.18a. When an electric field is applied in the direction shown in Fig. 5.18a, a force is exerted on the valence electrons of atom B, and one of the electrons associated with atom B will break loose from its bonding orbital and move to the vacancy in the bonding orbital of atom A. The hole will now appear at atom B and in effect will have moved from A to B in the direction of the applied field (Fig. 5.18b). By a similar mechanism the hole is

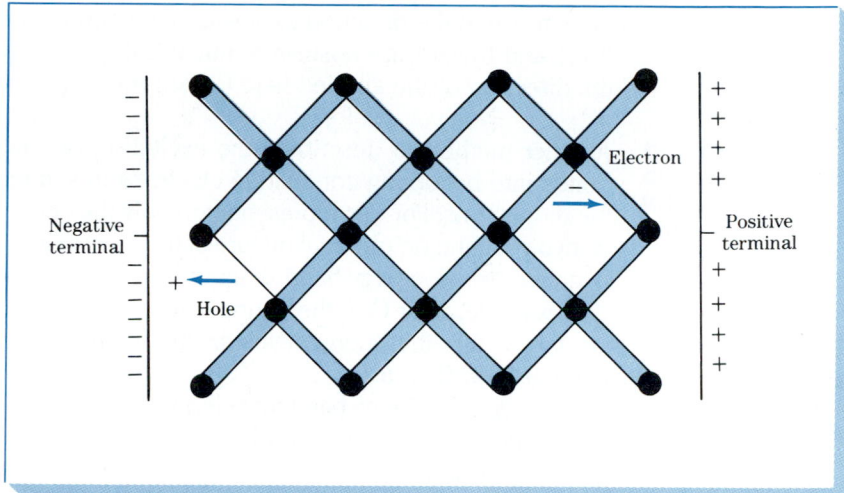

FIGURE 5.17 Electrical conduction in a semiconductor such as silicon showing the migration of electrons and holes in an applied electric field.

transported from atom B to C by an electron moving from C to B (Fig. 5.18c). The net result of this process is that an electron is transported from C to A, which is in the direction opposite to the applied field, and a hole is transported from A to C, which is in the direction of the applied field. Thus during electrical conduction in a pure semiconductor such as silicon, negatively charged elec-

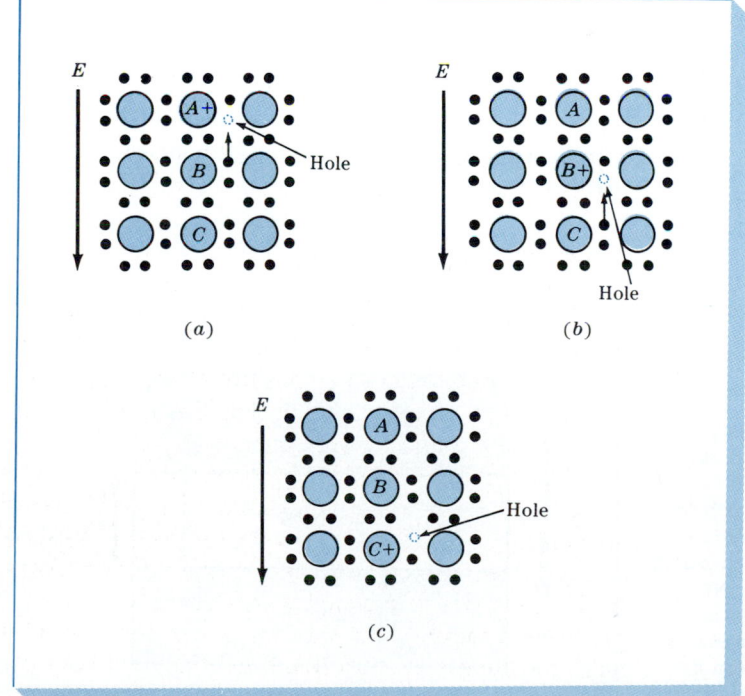

FIGURE 5.18 Schematic illustration of the movement of holes and electrons in a pure silicon semiconductor during electrical conduction caused by the action of an applied electric field. (*After S. N. Levine, "Principles of Solid State Microelectronics," Holt, 1963.*)

trons move in the direction opposite to the applied field (conventional current flow) and toward the positive terminal and positively charged holes move in the direction of the applied field toward the negative terminal.

Energy-Band Diagram for Intrinsic Elemental Semiconductors

Another method of describing the excitation of electrons from their valence bonds into becoming conduction electrons in semiconductors is with energy-band diagrams. For this representation only the energy required for the process is involved, and no physical picture of the electrons moving in the crystal lattice is given. In the energy-band diagram for intrinsic elemental semiconductors (for example, Si or Ge), the bound valence electrons of the covalently bonded crystal occupy the energy levels in the lower valence band which is almost filled at 20°C (Fig. 5.19).

Above the valence band there is a forbidden energy gap in which no energy states are allowed and which is 1.1 eV for silicon at 20°C. Above the energy gap there is an almost empty (at 20°C) conduction band. At room temperature thermal energy is sufficient to excite some electrons from the valence band to the conduction band, leaving vacant sites or holes in the valence band. Thus when an electron is excited across the energy gap into the conduction band, two charge carriers are created, a negatively charged electron and a positively charged hole. Both electrons and holes carry electric current.

Quantitative Relationships for Electrical Conduction in Elemental Intrinsic Semiconductors

During electrical conduction in intrinsic semiconductors, the current density $\mathbf{J}$ is equal to the sum of the conduction due to *both* electrons and holes. Using Eq. (5.6),

$$\mathbf{J} = nq\mathbf{v}_n^* + pq\mathbf{v}_p^* \tag{5.9}$$

---

*The subscript n (for negative) refers to electrons, and the subscript p (for positive) refers to holes.

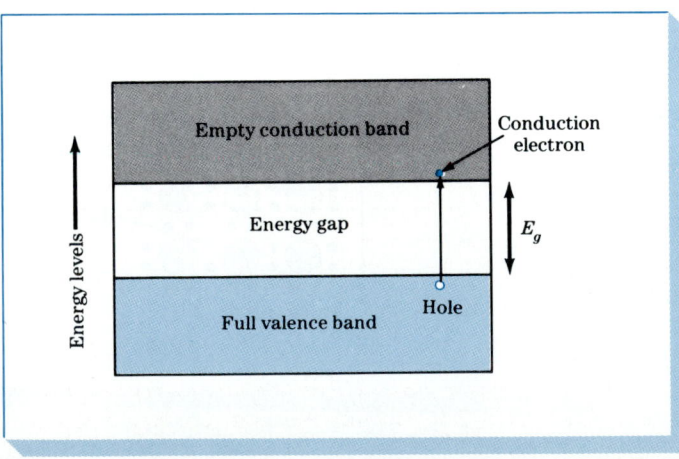

FIGURE 5.19 Energy-band diagram for an intrinsic elemental semiconductor such as pure silicon. When an electron is excited across the energy gap, an electron-hole pair is created. Thus for each electron that jumps the gap, two charge carriers, an electron and a hole, are produced.

where n = number of conduction electrons per unit volume
p = number of conduction holes per unit volume
q = absolute value of electron or hole charge, 1.60×10^{-19} C
$\boldsymbol{v}_n, \boldsymbol{v}_p$ = drift velocities of electrons and holes, respectively.

Dividing both sides of Eq. (5.9) by the electric field **E** and using Eq. (5.4), $\mathbf{J} = \sigma\mathbf{E}$,

$$\sigma = \frac{\mathbf{J}}{\mathbf{E}} = \frac{nq\boldsymbol{v}_n}{\mathbf{E}} + \frac{pq\boldsymbol{v}_p}{\mathbf{E}} \tag{5.10}$$

The quantities $\boldsymbol{v}_n/\mathbf{E}$ and $\boldsymbol{v}_p/\mathbf{E}$ are called the *electron* and *hole mobilities* since they measure how fast the electrons and holes in semiconductors drift in an applied electric field. The symbols μ_n and μ_p are used for the mobilities of electrons and holes, respectively. Substituting electron and hole mobilities for $\boldsymbol{v}_n/\mathbf{E}$ and $\boldsymbol{v}_p/\mathbf{E}$ in Eq. (5.10) enables the electrical conductivity of a semi-conductor to be expressed as

$$\sigma = nq\mu_n + pq\mu_p \tag{5.11}$$

The units for mobility μ are

$$\frac{\boldsymbol{v}}{\mathbf{E}} = \frac{\text{m/s}}{\text{V/m}} = \frac{\text{m}^2}{\text{V}\cdot\text{s}}$$

In intrinsic elemental semiconductors, electrons and holes are created in pairs and thus the number of conduction electrons equals the number of holes produced, so that

$$n = p = n_i \tag{5.12}$$

where n = intrinsic carrier concentration, carriers/unit volume.

Equation (5.11) now becomes

$$\sigma = n_i q(\mu_n + \mu_p) \tag{5.13}$$

Table 5.4 lists some of the important properties of intrinsic silicon and germanium at 300 K.

The mobilities of electrons are always greater than those of holes. For intrinsic silicon the electron mobility of 0.135 m²/(V·s) is 2.81 times greater than the hole mobility of 0.048 m²/(V·s) at 300 K (Table 5.4). The ratio of electron-to-hole mobility for intrinsic germanium is 2.05 at 300 K.

TABLE 5.4   Some Physical Properties of Silicon and Germanium at 300 K

	Silicon	Germanium
Energy gap, eV	1.1	0.67
Electron mobility μ_n, m²/(V·s)	0.135	0.39
Hole mobility μ_p, m²/(V·s)	0.048	0.19
Intrinsic carrier density n_i, carriers/m³	1.5×10^{16}	2.4×10^{19}
Intrinsic resistivity ρ, $\Omega\cdot\text{m}$	2300	0.46
Density, g/m³	2.33×10^6	5.32×10^6

Source: After E. M. Conwell, Properties of Silicon and Germanium II, *Proc. IRE,* June 1958, p. 1281.

Example Problem 5.4

Calculate the number of silicon atoms per cubic meter. The density of silicon is 2.33 Mg/m^3 (2.33 g/cm^3), and its atomic mass is 28.08 g/mol.

Solution:

$$\frac{\text{Si atoms}}{m^3} = \left(\frac{6.023 \times 10^{23} \text{ atoms}}{\text{mol}}\right)\left(\frac{1}{28.08 \text{ g/mol}}\right)\left(\frac{2.33 \times 10^6 \text{ g}}{m^3}\right)$$

$$= 5.00 \times 10^{28} \text{ atoms/m}^3 \blacktriangleleft$$

Example Problem 5.5

Calculate the electrical resistivity of intrinsic silicon at 300 K. For Si at 300 K, $n_i = 1.5 \times 10^{16}$ carriers/m^3, $q = 1.60 \times 10^{-19}$ C, $\mu_n = 0.135$ $m^2/(V \cdot s)$, and $\mu_p = 0.048$ $m^2/(V \cdot s)$.

Solution:

$$\rho = \frac{1}{\sigma} = \frac{1}{n_i q(\mu_n + \mu_p)} \qquad \text{[reciprocal of Eq. (5.13)]}$$

$$= \frac{1}{\left(\dfrac{1.5 \times 10^{16}}{m^3}\right)(1.60 \times 10^{-19} \text{ C})\left(\dfrac{0.135 \text{ m}^2}{V \cdot s} + \dfrac{0.048 \text{ m}^2}{V \cdot s}\right)}$$

$$= 2.28 \times 10^3 \ \Omega \cdot m \blacktriangleleft$$

The units for the reciprocal of Eq. (5.13) are ohm-meters as is shown by the following unit conversion:

$$\rho = \frac{1}{n_i q(\mu_n + \mu_p)} = \frac{1}{\left(\dfrac{1}{m^3}\right)(C)\left(\dfrac{1 \text{ A} \cdot \text{s}}{1 \text{ C}}\right)\left(\dfrac{m^2}{V \cdot s}\right)\left(\dfrac{1 \text{ V}}{1 \text{ A} \cdot \Omega}\right)} = \Omega \cdot m$$

Effect of Temperature on Intrinsic Semiconductivity

At 0 K the valence bands of intrinsic semiconductors such as silicon and germanium are completely filled and their conduction bands completely empty. At temperatures above 0 K, some of the valence electrons are thermally activated and excited across the energy gap into the conduction band, creating electron-hole pairs. Thus in contrast to metals whose conductivities are decreased with

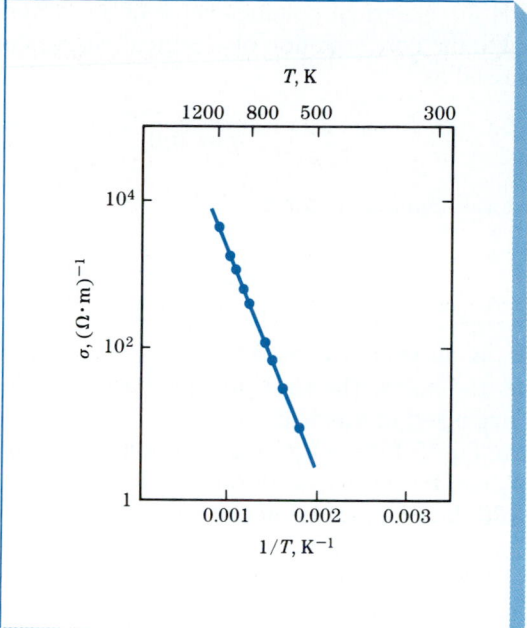

FIGURE 5.20 Electrical conductivity as a function of reciprocal absolute temperature for intrinsic silicon. (*After C. A. Wert and R. M. Thomson, "Physics of Solids," 2d ed., McGraw-Hill, 1970, p. 282.*)

increasing temperatures, the conductivities of semiconductors *increase* with increasing temperatures for the temperature range over which this process predominates.

Since electrons are thermally activated into the conduction band of semiconductors, the concentration of thermally activated electrons in semiconductors shows a temperature dependence similar to that of many other thermally activated processes. By analogy with Eq. (4.8), the concentration of electrons with sufficient thermal energy to enter the conduction band (and thus creating the same concentration of holes in the valence band), n_i, varies according to

$$n_i \propto e^{-(E_g - E_{av})/kT} \tag{5.14}$$

where E_g = band energy gap
E_{av} = average energy across band gap
k = Boltzmann's constant
T = temperature, K

For intrinsic semiconductors such as pure silicon and germanium, E_{av} is halfway across the gap, or $E_g/2$. Thus Eq. (5.14) becomes

$$n_i \propto e^{-(E_g - E_g/2)/kT} \tag{5.15a}$$

or

$$n_i \propto e^{-E_g/2kT} \tag{5.15b}$$

Since the electrical conductivity σ of an intrinsic semiconductor is proportional to the concentration of electrical charge carriers, n_i, Eq. (5.15b) can be expressed as

$$\sigma = \sigma_0 e^{-E_g/2kT} \tag{5.16a}$$

or in natural logarithmic form,

$$\ln \sigma = \ln \sigma_0 - \frac{E_g}{2kT} \tag{5.16b}$$

where σ_0 is an overall constant which depends mainly on the mobilities of electrons and holes. The slight temperature dependence of σ_0 on temperature will be neglected in this text.

Since Eq. (5.16b) is the equation of a straight line, the value of $E_g/2k$ and hence E_g can be determined from the slope of the plot of $\ln \sigma$ vs. $1/T$, K^{-1}. Figure 5.20 shows an experimental plot of $\ln \sigma$ vs. $1/T$, K^{-1} for intrinsic silicon.

Example Problem 5.6

The electrical resistivity of pure silicon is $2.3 \times 10^3 \ \Omega \cdot m$ at room temperature, 27°C (300 K). Calculate its electrical conductivity at 200°C (473 K). Assume that the E_g of silicon is 1.1 eV; $k = 8.62 \times 10^{-5}$ eV/K.

Solution:

For this problem we use Eq. (5.16a) and set up two simultaneous equations. We then eliminate σ_0 by dividing the first equation by the second.

$$\sigma = \sigma_0 \exp \frac{-E_g}{2kT} \tag{5.16a}$$

$$\sigma_{473} = \sigma_0 \exp \frac{-E_g}{2kT_{473}}$$

$$\sigma_{300} = \sigma_0 \exp \frac{-E_g}{2kT_{300}}$$

Dividing the first equation by the second to eliminate σ_0 gives

$$\frac{\sigma_{473}}{\sigma_{300}} = \exp \left(\frac{-E_g}{2kT_{473}} + \frac{E_g}{2kT_{300}} \right)$$

$$\frac{\sigma_{473}}{\sigma_{300}} = \exp \left[\frac{-1.1 \text{ eV}}{2(8.62 \times 10^{-5} \text{ eV/K})} \left(\frac{1}{473 \text{ K}} - \frac{1}{300 \text{ K}} \right) \right]$$

$$\ln \frac{\sigma_{473}}{\sigma_{300}} = 7.777$$

$$\sigma_{473} = \sigma_{300} \, (2385)$$

$$= \frac{1}{2.3 \times 10^3 \, \Omega \cdot m} \, (2385) = 1.04 \, (\Omega \cdot m)^{-1} \blacktriangleleft$$

The electrical conductivity of the silicon increased by about 2400 times when the temperature was raised from 27 to 200°C.

5.4 EXTRINSIC SEMICONDUCTORS

Extrinsic semiconductors are very dilute substitutional solid solutions in which the solute impurity atoms have different valence characteristics from the solvent atomic lattice. The concentrations of the added impurity atoms in these semi-conductors are usually in the range of 100 to 1000 parts per million (ppm).

n-Type (Negative-Type) Extrinsic Semiconductors

Consider the two-dimensional covalent bonding model for the silicon crystal lattice shown in Fig. 5.21a. If an impurity atom of a group VA element, e.g., phosphorus, replaces a silicon atom, which is a group IVA element, there will be one excess electron above the four needed for the tetrahedral covalent bonding in the silicon lattice. This extra electron is only loosely bonded to the

FIGURE 5.21 (a) The addition of a pentavalent phosphorus (P^{5+}) impurity atom to the tetravalent silicon (Si^{4+}) lattice provides a fifth electron which is weakly attached to the parent phosphorus atom. Only a small amount of energy (0.044 eV) makes this electron mobile and conductive. (b) Under an applied electric field the excess electron becomes conductive and is attracted to the positive terminal of the electrical circuit. With the loss of the extra electron, the phosphorus atom is ionized and acquires a +1 charge.

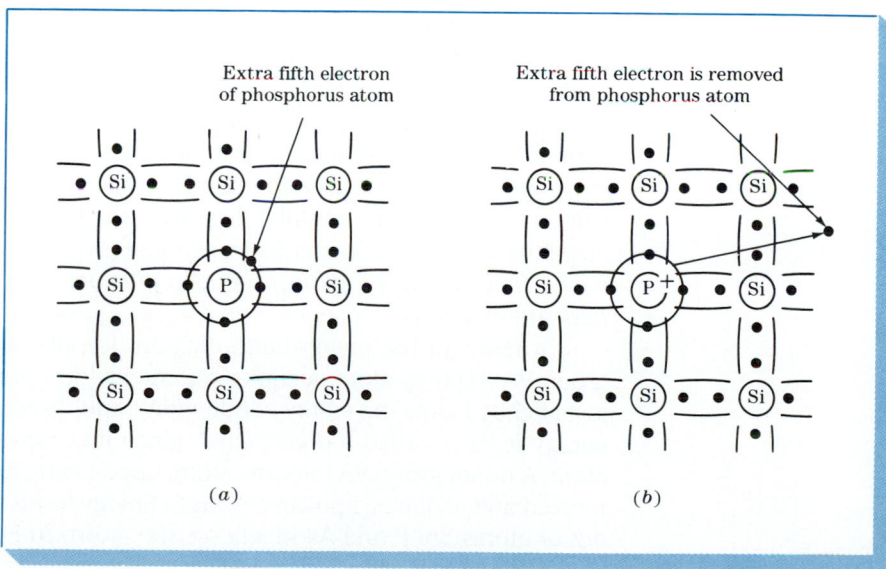

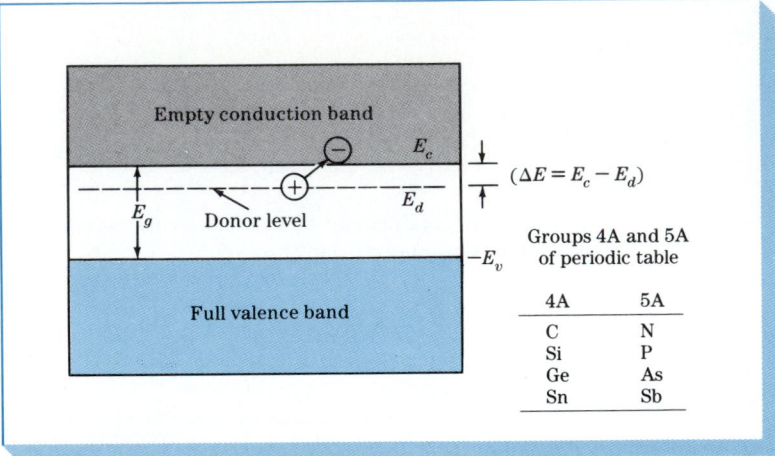

FIGURE 5.22 Energy-band diagram for an *n*-type extrinsic semiconductor showing the position of the donor level for the extra electron of a group VA element such a P, As, and Sb which is contained in the silicon crystal lattice (Fig. 5.21*a*). Electrons at the donor energy level require only a small amount of energy ($\Delta E = E_c - E_d$) to be excited into the conduction band. When the extra electron at the donor level jumps to the conduction band, a positive immobile ion is left behind.

positively charged phosphorus nucleus and has a binding energy of 0.044 eV at 27°C. This energy is about 5 percent of that required for a conduction electron to jump the energy gap of 1.1 eV of pure silicon. That is, only 0.044 eV of energy is required to remove the excess electron from its parent nucleus so that it can participate in electrical conduction. When under the action of an electrical field, the extra electron becomes a free electron available for conduction and the remaining phosphorus atom becomes ionized and acquires a positive charge (Fig. 5.21*b*).

Group VA impurity atoms such as P, As, and Sb when added to silicon or germanium provide easily ionized electrons for electrical conduction. Since these group VA impurity atoms donate conduction electrons when present in silicon or germanium crystals, they are called *donor impurity atoms.* Silicon or germanium semiconductors containing group V impurity atoms are called *n-type (negative-type) extrinsic semiconductors* since the majority charge carriers are electrons.

In terms of the energy-band diagram for silicon, the extra electron of a group VA impurity atom occupies an energy level in the forbidden energy gap just slightly below the empty conduction band, as shown in Fig. 5.22. Such an energy level is called a *donor level,* since it is provided by a donor impurity atom. A donor group VA impurity atom, upon losing its extra electron, becomes ionized and acquires a positive charge. Energy levels for the group VA impurity donor atoms Sb, P and As in silicon are shown in Fig. 5.23.

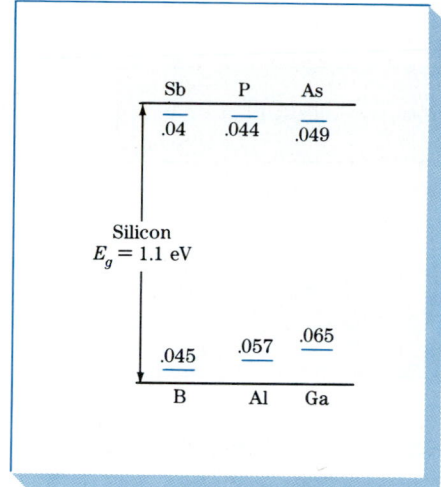

FIGURE 5.23 Ionization energies (in electron volts) for various impurities in silicon.

p-Type (Positive-Type) Extrinsic Semiconductors

When a trivalent group IIIA element such as boron (B^{3+}) is substitutionally introduced in the silicon tetrahedrally bonded lattice, one of the bonding orbitals is missing and a hole exists in the bonding structure of the silicon (Fig. 5.24a). If an external electric field is applied to the silicon crystal, one of the neighboring electrons from another tetrahedral bond can attain sufficient energy to break loose from its bond and move to the missing bond (hole) of the

FIGURE 5.24 (a) The addition of a trivalent boron (B^{3+}) impurity atom into the tetravalent (Si^{4+}) lattice creates a hole in one of the boron-silicon bonds since one electron is missing. (b) Under an applied electric field only a small amount of energy (0.045 eV) attracts an electron from a nearby silicon atom to fill this hole, thereby creating an immobile boron ion with a −1 charge. The new hole created in the silicon lattice acts as a positive charge carrier and is attracted to the negative terminal of an electrical circuit.

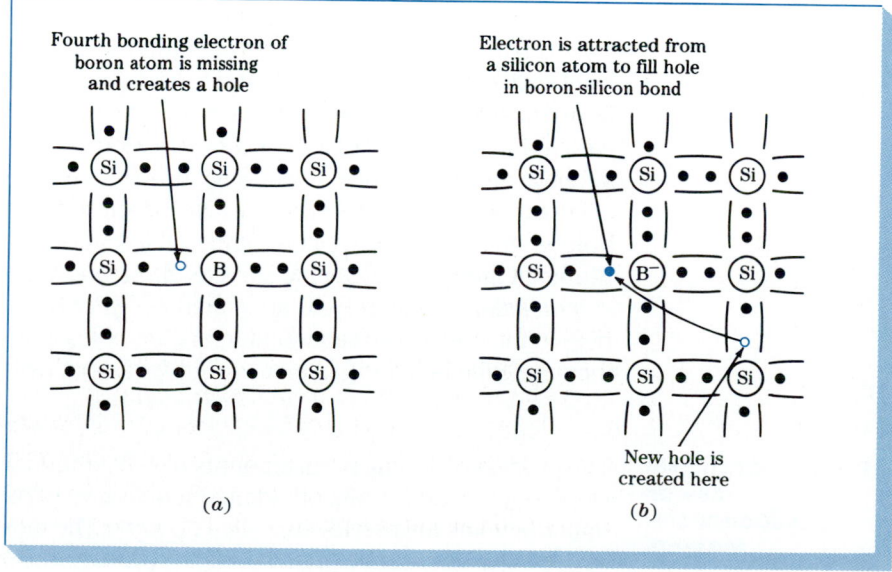

FIGURE 5.25 Energy-band diagram for a p-type extrinsic semiconductor showing the position of the acceptor level created by the addition of an atom of a group IIIA element such as Al, B, or Ga to replace a silicon atom in the silicon lattice (Fig. 5.24). Only a small amount of energy ($\Delta E = E_a - E_v$) is necessary to excite an electron from the valence band to the acceptor level, thereby creating an electron hole (charge carrier) in the valence band.

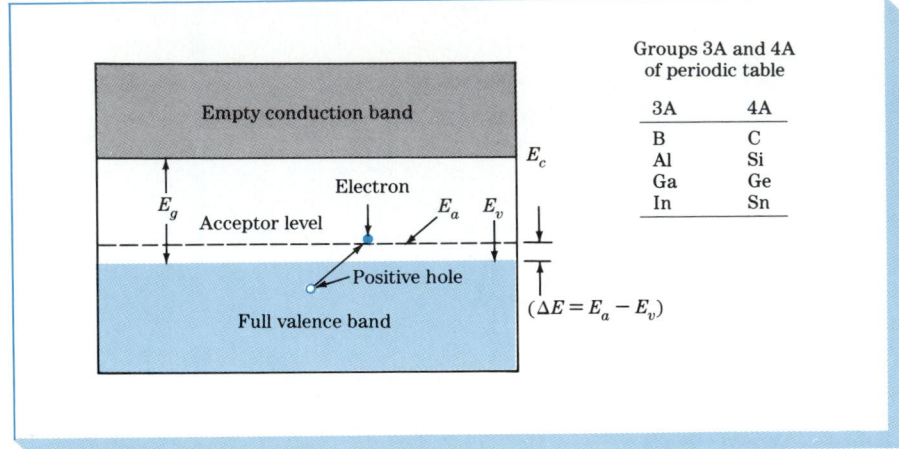

boron atom (Fig. 5.24b). When the hole associated with the boron atom is filled by an electron from a neighboring silicon atom, the boron atom becomes ionized and acquires a negative charge of -1. The binding energy associated with the removal of an electron from a silicon atom, thereby creating a hole, and the subsequent transfer of the electron to the boron atom is only 0.045 eV. This amount of energy is small compared to the 1.1 eV required to transfer an electron from the valence band to the conduction band. In the presence of an applied electric field the hole created by the ionization of a boron atom behaves as a positive charge carrier and migrates in the silicon lattice toward the negative terminal, as described in Fig. 5.17.

In terms of the energy-band diagram the boron atom provides an energy level called an *acceptor level* which is slightly higher (≈ 0.045 eV) than the uppermost level of the full valence band of silicon (Fig. 5.25). When a valence electron of a silicon atom near a boron atom fills a missing electron hole in a boron-silicon valence bond (Fig. 5.24b), this electron is elevated to the acceptor level and creates a negative boron ion. In this process an electron hole is created in the silicon lattice which acts as a positive charge carrier. Atoms of group IIIA elements such as B, Al, and Ga provide acceptor levels in silicon semiconductors and are called *acceptor atoms*. Since the majority carriers in these extrinsic semiconductors are holes in the valence bond structure, they are called *p-type (positive-carrier-type) extrinsic semiconductors*.

Doping of Extrinsic Silicon Semiconductor Material

The process of adding small amounts of substitutional impurity atoms to silicon to produce extrinsic silicon semiconducting material is called *doping*, while impurity atoms themselves are called *dopants*. The most commonly used method of doping silicon semiconductors is the *planar process*. In this process dopant atoms are introduced into selected areas of the silicon from one surface in order to form regions of *p*- or *n*-type material. The wafers are usually about 4 in (10 cm) in diameter and about a few hundred micrometers[1] thick. Figure 4.13 shows some of these silicon wafers.

In the diffusion process for doping silicon wafers, the dopant atoms are typically deposited on or near the surface of the wafer by a gaseous deposition step, followed by a drive-in diffusion which moves the dopant atoms farther into the wafer. A high temperature of about 1100°C is required for this diffusion process. More details of this process will be described in Sec. 5.6 on microelectronics.

Effect of Doping on Carrier Concentrations in Extrinsic Semiconductors

The mass action law   In semiconductors such as silicon and germanium mobile electrons and holes are constantly being generated and recombined. At constant temperature under equilibrium conditions the product of the negative free electron and positive hole concentrations is a constant. The general relation is

$$np = n_i^2 \qquad (5.17)$$

where n_i is the intrinsic concentration of carriers in a semiconductor and is a constant at a given temperature. This relation is valid for both intrinsic and extrinsic semiconductors. In an extrinsic semiconductor the increase in one type of carrier (*n* or *p*) reduces the concentration of the other through recombination so that the product of the two (*n* and *p*) is a constant at any given temperature.

The carriers whose concentration in extrinsic semiconductors is the larger are designated the *majority carriers,* and those whose concentration is the smaller are called the *minority carriers* (see Table 5.5). The concentration of electrons in an *n*-type semiconductor is denoted by n_n and that of holes in

---

[1] 1 micrometer (μm) $= 10^{-4}$ cm $= 10^4$ Å.

TABLE 5.5   **Summary of the Carrier Concentrations in Extrinsic Semiconductors**

Semiconductor	Majority-carrier concentrations	Minority-carrier concentrations
n-type	n_n (concentration of electrons in *n*-type material)	p_n (concentration of holes in *n*-type material)
p-type	p_p (concentration of holes in *p*-type material)	n_p (concentration of electrons in *p*-type material)

n-type material by p_n. Similarly, the concentration of holes in a p-type semi-conductor is given by p_p and that of electrons in p-type material by n_p.

Charge densities in extrinsic semiconductors   A second fundamental rela-tionship for extrinsic semiconductors is obtained from the fact that the total crystal must be electrically neutral. This means that the charge density in each volume element must be zero. There are two types of charged particles in extrinsic semiconductors such as Si and Ge: immobile ions and mobile charge carriers. The immobile ions originate from the ionization of donor or acceptor impurity atoms in the Si or Ge. The concentration of the positive donor ions is denoted by N_d and that of the negative acceptor ions by N_a. The mobile charge carriers originate mainly from the ionization of the impurity atoms in the Si or Ge, and their concentrations are designated by n for the negatively charged electrons and p for the positively charged holes.

Since the semiconductor must be electrically neutral, the magnitude of the total negative charge density must equal the total positive charge density. The total negative charge density is equal to the sum of the negative acceptor ions N_a and the electrons, or $N_a + n$. The total positive charge density is equal to the sum of the positive donor ions N_d and the holes, or $N_d + p$. Thus

$$N_a + n = N_d + p \tag{5.18}$$

In an n-type semiconductor created by adding donor impurity atoms to intrinsic silicon, $N_a = 0$. Since the number of electrons is much greater than the number of holes in an n-type semiconductor (that is, $n \gg p$), then Eq. (5.18) reduces to

$$n_n \approx N_d \tag{5.19}$$

Thus, in an n-type semiconductor the free-electron concentration is approxi-mately equal to the concentration of donor atoms. The concentration of holes in an n-type semiconductor is obtained from Eq. (5.17). Thus

$$p_n = \frac{n_i^2}{n_n} \approx \frac{n_i^2}{N_d} \tag{5.20}$$

The corresponding equations for p-type semiconductors of silicon and germanium are

$$p_p \approx N_a \tag{5.21}$$

and

$$n_p = \frac{n_i^2}{p_p} \approx \frac{n_i^2}{N_a} \tag{5.22}$$

Typical carrier concentrations in intrinsic and extrinsic semiconductors   For silicon at 300 K the intrinsic carrier concentration n_i is equal to 1.5×10^{16}

carriers/m^3. For extrinsic silicon doped with arsenic at a typical concentration of 10^{21} impurity atoms/m^3,

$$\text{Concentration of major carriers } n_n = 10^{21} \text{ electrons/m}^3$$

$$\text{Concentration of minority carriers } p_n = 2.25 \times 10^{11} \text{ holes/m}^3$$

Thus for extrinsic semiconductors the concentration of the majority carriers is normally much larger than that of the minority carriers. Example Problem 5.7 shows how the concentrations of majority and minority carriers can be calculated for an extrinsic silicon semiconductor.

Example Problem 5.7

A silicon wafer is doped with 10^{21} phosphorus atoms/m^3. Calculate (*a*) the majority-carrier concentration, (*b*) the minority-carrier concentration, and (*c*) the electrical resistivity of the doped silicon at room temperature (300 K). Assume complete ionization of the dopant atoms; $n_i(\text{Si}) = 1.5 \times 10^{16} \text{ m}^{-3}$, $\mu_n = 0.135 \text{ m}^2/(\text{V} \cdot \text{s})$, and $\mu_p = 0.048$ m^2/(V·s).

Solution:

Since the silicon is doped with phosphorus, a group V element, the doped silicon is *n-type*.

(*a*)
$$n_n = N_d = 10^{21} \text{ electrons/m}^3 \blacktriangleleft$$

(*b*)
$$p_n = \frac{n_i^2}{N_d} = \frac{(1.5 \times 10^{16} \text{ m}^{-3})^2}{10^{21} \text{ m}^{-3}} = 2.25 \times 10^{11} \text{ holes/m}^3 \blacktriangleleft$$

(*c*)
$$\rho = \frac{1}{q\mu_n n_n} = \frac{1}{(1.60 \times 10^{-19} \text{ C}) \left(0.135 \dfrac{\text{m}^2}{\text{V} \cdot \text{s}} \right) \left(\dfrac{10^{21}}{\text{m}^3} \right)}$$

$$= 0.0463 \ \Omega \cdot \text{m}^* \blacktriangleleft$$

Example Problem 5.8

A phosphorus-doped silicon wafer has an electrical resistivity of $8.33 \times 10^{-5} \ \Omega \cdot \text{m}$ at 27°C. Assume mobilities of charge carriers to be the constants 0.135 m^2/(V·s) for electrons and 0.048 m^2/(V·s) for holes.

---

*See Example Problem 5.5 for the conversion of units.

(a) What is its majority-carrier concentration (carriers per cubic meter) if complete ionization is assumed?

(b) What is the ratio of phosphorus to silicon atoms in this material?

Solution:

(a) Phosphorus produces an n-type silicon semiconductor. Therefore, the mobility of the charge carriers will be assumed to be that of electrons in silicon at 300 K, which is 0.1350 m²/(V·s). Thus

$$\rho = \frac{1}{n_n q \mu_n}$$

or $\quad n_n = \dfrac{1}{\rho q \mu_n} = \dfrac{1}{(8.33 \times 10^{-5}\ \Omega\cdot\text{m})(1.60 \times 10^{-19}\text{C})[0.1350\ \text{m}^2/(\text{V}\cdot\text{s})]}$

$$= 5.56 \times 10^{23}\ \text{electrons/m}^3 \blacktriangleleft$$

(b) Assuming each phosphorus atom provides one electron charge carrier, there will be 5.56×10^{23} phosphorus atoms/m³ in the material. Pure silicon contains 5.00×10^{28} atoms/m³ (Example Problem 5.4). Thus, the ratio of phosphorus to silicon atoms will be

$$\frac{5.56 \times 10^{23}\ \text{P atoms/m}^3}{5.00 \times 10^{28}\ \text{Si atoms/m}^3} = 1.11 \times 10^{-5}\ \text{P to Si atoms} \blacktriangleleft$$

Effect of Total Ionized Impurity Concentration on the Mobility of Charge Carriers in Silicon at Room Temperature

Figure 5.25a shows that the mobilities of electrons and holes in silicon at room temperature are at a maximum at low impurity concentrations and then decrease with impurity concentration, reaching a minimum value at high concentrations. Example Problem 5.9 shows how neutralizing one type of charge carrier by another leads to a lower mobility for the majority carriers.

Example Problem 5.9

A silicon semiconductor at 27°C is doped with 1.4×10^{16} boron atoms/cm³ plus 1.0×10^{16} phosphorus atoms/cm³. Calculate (a) the equilibrium electron and hole concentrations, (b) the mobilities of electrons and holes, and (c) the electrical resistivity. Assume complete ionization of the dopant atoms. n_i (Si) $= 1.50 \times 10^{10}$ cm^{-3}.

Solution:

(a) *Majority-carrier concentration:* The net concentration of immobile ions is equal to the acceptor ion concentration minus the donor ion concentration. Thus,

$$p_p \approx N_a - N_d = 1.4 \times 10^{16}\ \text{B atoms/cm}^3 - 1.0 \times 10^{16}\ \text{P atoms/cm}^3$$
$$\approx N_a \approx 4.0 \times 10^{15}\ \text{holes/cm}^3 \blacktriangleleft$$

Minority-carrier concentration: Electrons are the minority carriers. Thus,

$$n_p = \frac{n_i^2}{N_a} = \frac{(1.50 \times 10^{10}\ \text{cm}^{-3})^2}{4 \times 10^{15}\ \text{cm}^{-3}} = 5.6 \times 10^4\ \text{electrons/cm}^3 \blacktriangleleft$$

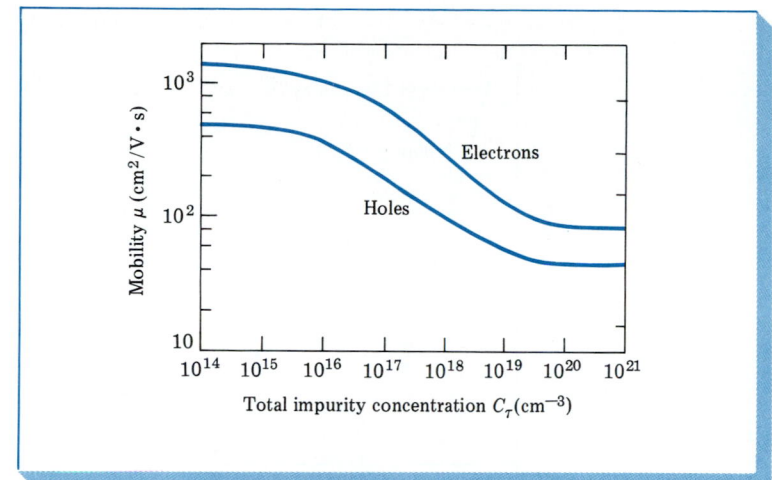

FIGURE 5.25a The effect of total ionized impurity concentration on the mobility of charge carriers in silicon at room temperature (*After A. S. Grove, "Physics and Technology of Semiconductor Devices," Wiley, 1967, p. 110.*)

(b) *Mobilities of electrons and holes:* For electrons, using the total impurity concentration $C_T = 2.4 \times 10^{16}$ ions/cm³ and Fig. 5.25a,

$$\mu_n = 900 \text{ cm}^2/(\text{V}\cdot\text{s}) \blacktriangleleft$$

For holes, using $C_T = 2.4 \times 10^{16}$ ions/cm³ and Fig. 5.25a,

$$\mu_p = 300 \text{ cm}^2/(\text{V}\cdot\text{s}) \blacktriangleleft$$

(c) *Electrical resistivity:* The doped semiconductor is *p*-type:

$$\rho = \frac{1}{q\mu_p p_p}$$

$$= \frac{1}{(1.60 \times 10^{-19} \text{ C})[300 \text{ cm}^2/(\text{V}\cdot\text{s})](4.0 \times 10^{15}/\text{cm}^3)}$$

$$= 5.2 \ \Omega\cdot\text{cm} \blacktriangleleft$$

Effect of Temperature on the Electrical Conductivity of Extrinsic Semiconductors

The electrical conductivity of an extrinsic semiconductor such as silicon which contains doped impurity atoms is affected by temperature, as shown schematically in Fig. 5.26. At lower temperatures the number of impurity atoms per unit volume activated (ionized) determines the electrical conductivity of the silicon. As the temperature is increased, more and more impurity atoms are

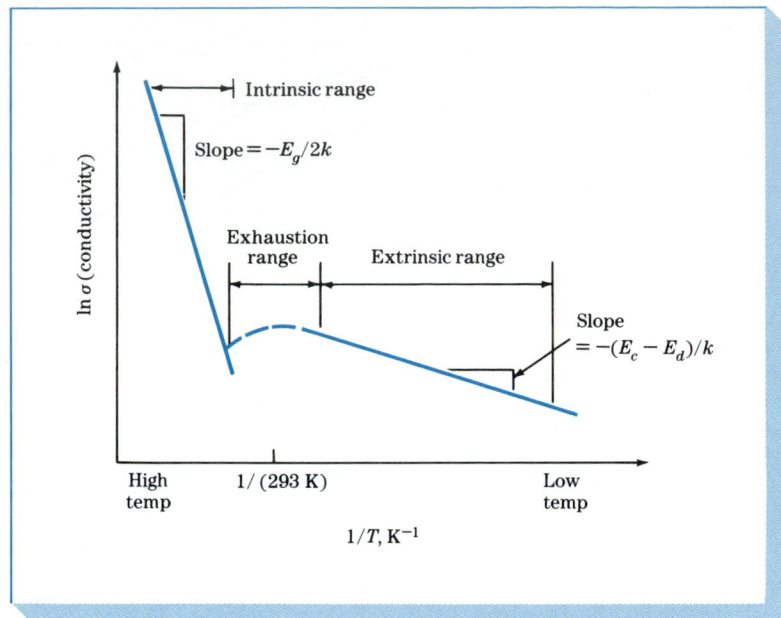

FIGURE 5.26 Schematic plot of ln σ (conductivity) vs. $1/T$, K^{-1} for an n-type extrinsic semiconductor.

ionized, and thus the electrical conductivity of extrinsic silicon increases with increasing temperature in the extrinsic range (Fig. 5.26).

In this extrinsic range only a relatively small amount of energy (≈ 0.04 eV) is required to ionize the impurity atoms. The amount of energy required to excite a donor electron into the conduction band in n-type silicon is $E_c - E_d$ (Fig. 5.22). Thus the slope of ln σ vs. $1/T$, K^{-1} for n-type silicon is $-(E_c - E_d)/k$. Correspondingly, the amount of energy required to excite an electron in p-type silicon into an acceptor level and thereby creating a hole in the valence band is $E_a - E_v$. Thus the slope of ln σ vs. $1/T$, K^{-1} for p-type silicon is $-(E_a - E_v)/k$ (Fig. 5.25).

For a certain temperature range above that where complete ionization occurs, an increase in temperature does not substantially change the electrical conductivity of an extrinsic semiconductor. For an n-type semiconductor this temperature range is referred to as the *exhaustion range* since donor atoms become completely ionized after the loss of their donor electrons (Fig. 5.26). For p-type semiconductors this range is referred to as the *saturation range* since acceptor atoms become completely ionized with acceptor electrons. To provide an exhaustion range at about room temperature (300 K), silicon doped with arsenic requires about 10^{21} carriers/m^3 (Fig. 5.27a). Donor exhaustion and acceptor saturation temperature ranges are important for semiconductor devices since they provide temperature ranges which have essentially constant electrical conductivities for operation.

As the temperature is increased beyond that of the exhaustion range, the intrinsic range is entered upon. The higher temperatures provide sufficient activation energies for electrons to jump the semiconductor gap (1.1 eV for

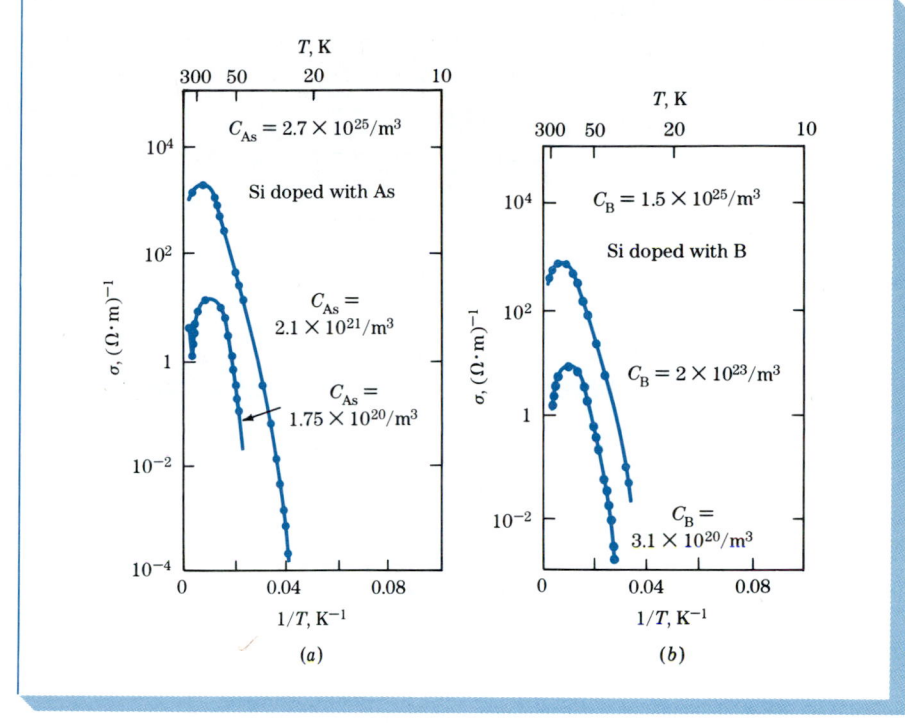

FIGURE 5.27 (*a*) Plot of ln σ vs. $1/T$, K^{-1} for As-doped Si. At the lowest level of impurity the intrinsic contribution is slightly visible at the highest temperatures; slope of the line at 40 K gives $E_i = 0.048$ eV. (*b*) Plot of ln σ vs. $1/T$, K^{-1} for B-doped Si. Slope of the line below 50 K gives $E_i = 0.043$ eV. (*After C. A. Wert and R. M. Thomson, "Physics of Solids," 2d ed., McGraw-Hill, 1970, p. 282.*)

silicon) so that intrinsic conduction becomes dominant. The slope of the ln σ vs. $1/T$, K^{-1} plot becomes much steeper and is $-E_g/2k$. For silicon-based semiconductors with an energy gap of 1.1 eV, extrinsic conduction can be used up to about 200°C. The upper limit for the use of extrinsic conduction is determined by the temperature at which intrinsic conductivity becomes important.

5.5 SEMICONDUCTOR DEVICES

The use of semiconductors in the electronics industry has become increasingly important over the past years. The ability of semiconductor manufacturers to put extremely complex electrical circuits on a single chip of silicon of about 1 cm square or less and about 200 μm thick has revolutionized the design and manufacture of countless products. An example of the complex electronic circuitry able to be put on a silicon chip in 1994 is shown in Fig. 5.1 of an advanced microprocessor or "computer on a chip." The microprocessor forms the basis for many of the latest products which utilize the progressive miniaturization of silicon-based semiconductor technology.

In this section we will first study the electron-hole interactions at a *pn*

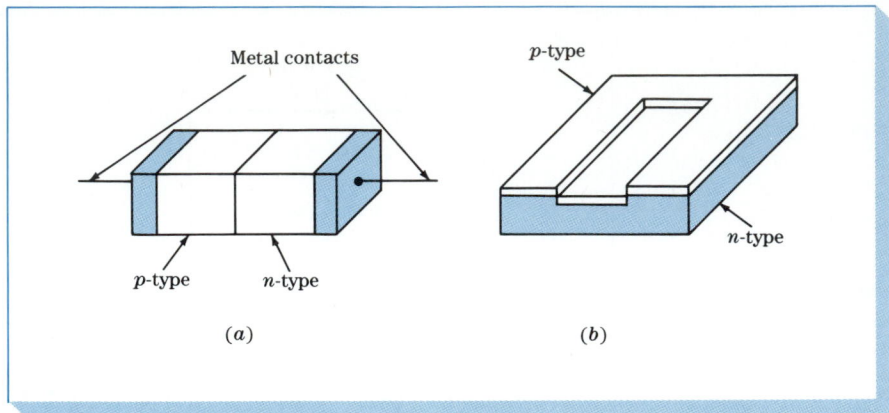

FIGURE 5.28 (a) pn junction diode grown in the form of a single crystal bar. (b) Planar pn junction formed by selectively diffusing a p-type impurity into an n-type semiconductor crystal.

junction and then examine the operation of the *pn* junction diode. We shall then look at some applications of *pn* junction diodes. Finally, we shall briefly examine the operation of the bipolar junction transistor.

The *pn* Junction

Most common semiconductor devices depend on the properties of the boundary between *p*-type and *n*-type materials, and therefore we shall examine some of the characteristics of this boundary. A *pn* junction diode can be produced by growing a single crystal of intrinsic silicon and doping it first with *n*-type material and then with *p*-type material (Fig. 5.28*a*). More commonly, however, the *pn* junction is produced by solid-state diffusion of one type of impurity (for example, *p*-type) into existing *n*-type material (Fig. 5.28*b*).

The *pn* junction diode at equilibrium   Let us consider an idealistic case in which *p*-type and *n*-type silicon semiconductors are joined together to form a junction. Before joining, both types of semiconductors are electrically neutral. In the *p*-type material holes are the majority carriers and electrons are the minority carriers. In the *n*-type material electrons are the majority carriers and holes are the minority carriers.

After joining the *p*- and *n*-type materials (i.e., after a *pn* junction is formed in actual fabrication), the majority carriers near or at the junction diffuse across the junction and recombine (Fig. 5.29*a*). Since the remaining ions near or at the junction are physically larger and heavier than the electrons and holes, they remain in their positions in the silicon lattice (Fig. 5.29*b*). After a few recombinations of majority carriers at the junction, the process stops because the electrons crossing over the junction into the *p*-type material are repelled by the large negative ions. Similarly, holes crossing over the junction are repelled by the large positive ions in the *n*-type material. The immobile ions at the junction create a zone depleted of majority carriers called a *depletion region*. Under equilibrium conditions (i.e., open-circuit conditions) there exists a potential difference or barrier to majority-carrier flow. Thus there is no net current flow under open-circuit conditions.

FIGURE 5.29 (*a*) *pn* junction diode showing majority carriers (holes in *p*-type material and electrons in *n*-type material) diffusing toward junction. (*b*) Formation of depletion region at and near *pn* junction due to loss of majority carriers in this region by recombination. Only ions remain in this region in their positions in the crystal structure.

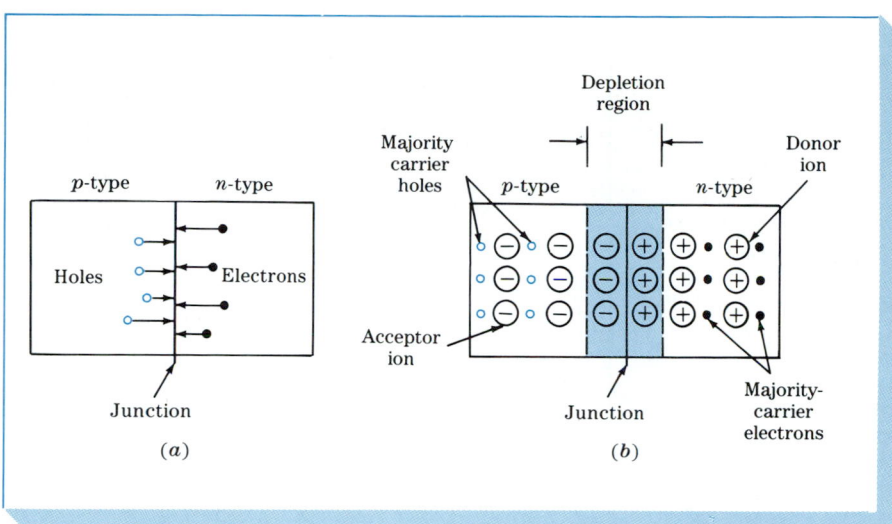

The *pn* junction diode reverse-biased   When an external voltage is applied to a *pn* junction, it is said to be *biased*. Let us consider the effect of applying an external voltage from a battery to the *pn* junction. The *pn* junction is said to be *reversed-biased* if the *n*-type material of the junction is connected to the positive terminal of the battery and if the *p*-type material is connected to the negative terminal (Fig. 5.30). With this arrangement, the electrons (majority carriers) of the *n*-type material are attracted to the positive terminal of the battery away from the junction and the holes (majority carriers) of the *p*-type material are attracted to the negative terminal of the battery away from the junction (Fig. 5.30). The movement of the majority-carrier electrons and holes away from the junction increases its barrier width, and as a result current due

FIGURE 5.30 Reverse-biased *pn* junction diode. Majority carriers are attracted away from the junction, creating a wider depletion region than when the junction is at equilibrium. Current flow due to majority carriers is reduced to near zero. However, minority carriers are biased forward, creating a small leakage current, as shown in Fig. 5.31.

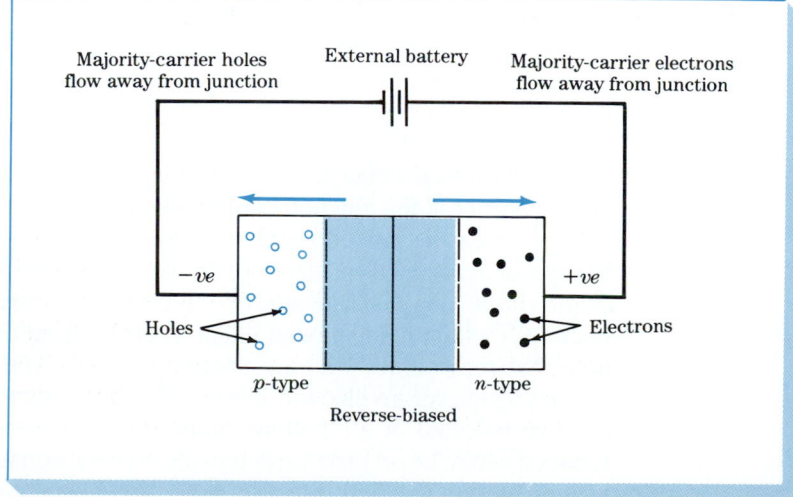

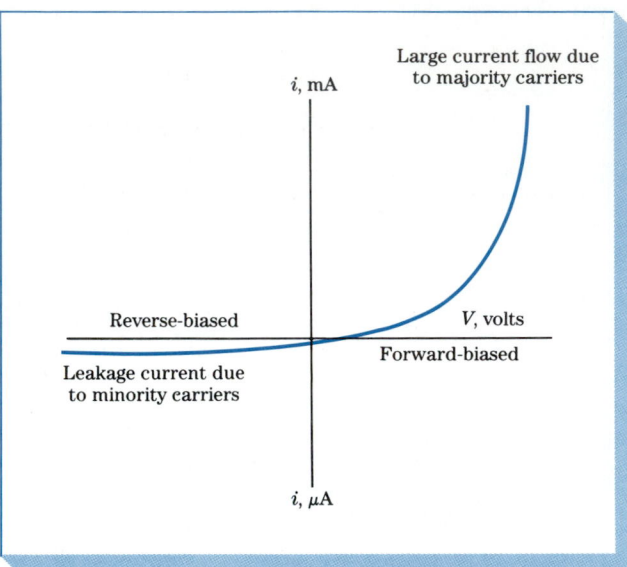

FIGURE 5.31 Schematic of current-voltage characteristics of a *pn* junction diode. When the *pn* junction diode is reversed-biased, a leakage current due to minority carriers combining exists. When the *pn* junction diode is forward-biased, a large current flows due to recombination of majority carriers.

to majority carriers will not flow. However, thermally generated minority carriers (holes in *n*-type material and electrons in *p*-type material) will be driven toward the junction so that they can combine and create a very small current flow under reverse-bias conditions. This minority or *leakage current* is usually of the order of microamperes (μA) (Fig. 5.31).

The *pn* junction diode forward-biased   The *pn* junction diode is said to be *forward-biased* if the *n*-type material of the junction is connected to the negative terminal of an external battery (or other electrical source) and if the *p*-type material is connected to the positive terminal (Fig. 5.32). In this arrangement the majority carriers are repelled toward the junction and can combine. That is, electrons are repelled away from the negative terminal of the battery toward the junction and holes are repelled away from the positive terminal toward the junction.

Under forward bias, i.e., forward bias with respect to majority carriers, the energy barrier at the junction is reduced so that some electrons and holes are able to cross the junction and subsequently recombine. During the forward biasing of a *pn* junction, electrons from the battery enter the negative material of the diode (Fig. 5.32). For every electron that crosses the junction and re-combines with a hole, another electron enters from the battery. Also, for every hole that recombines with an electron in the *n*-type material, a new hole is formed whenever an electron leaves the *p*-type material and flows toward the positive terminal of the battery. Since the energy barrier to electron flow is reduced when the *pn* junction is forward-biased, considerable current can flow, as indicated in Fig. 5.31. Electron flow (and hence current flow) can continue

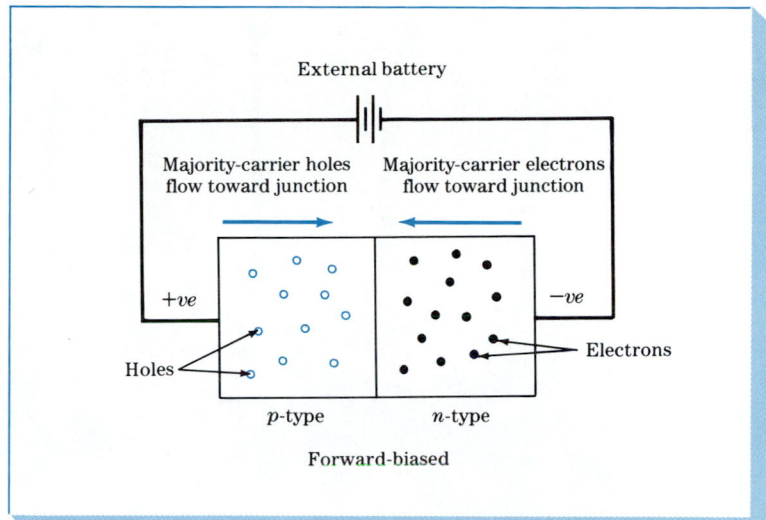

FIGURE 5.32 Forward-biased *pn* junction diode. Majority carriers are repelled toward the junction and cross over it to recombine so that a large current flows.

as long as the *pn* junction is forward-biased and the battery provides an electron source.

Some Applications for *pn* Junction Diodes

Rectifier diodes   One of the most important uses of *pn* junction diodes is to convert alternating voltage into direct voltage, a process known as *rectification.* Diodes used for this process are called *rectifier diodes.* When an ac signal is

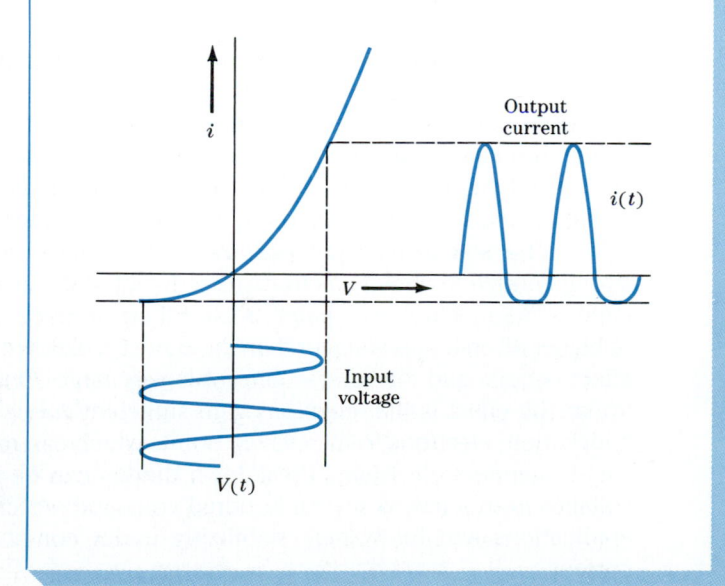

FIGURE 5.33 Voltage-current diagram illustrating the rectifying action of a *pn* junction diode to convert alternating current (ac) to direct current (dc). The output current is not completely direct current but is mostly positive. This dc signal can be smoothed out by using other electronic devices.

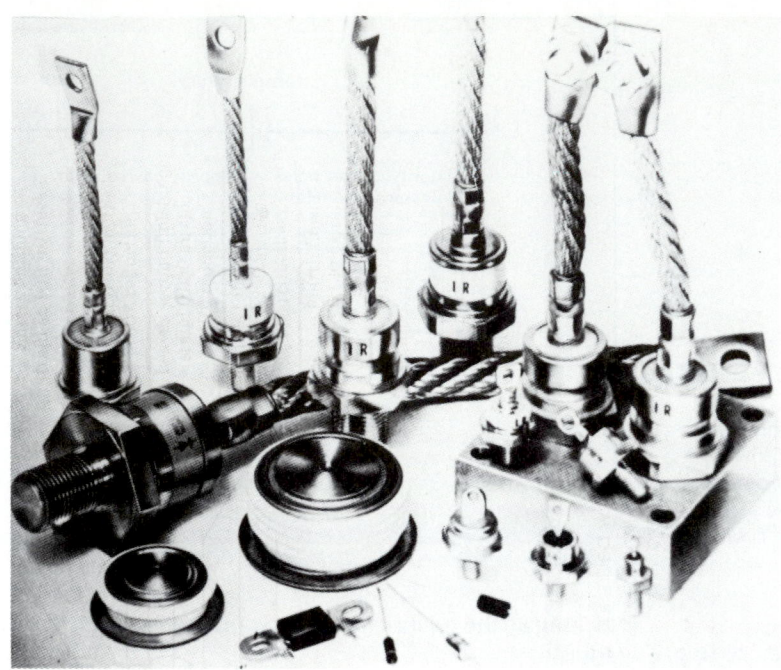

FIGURE 5.34 Photo of some low- to high-power silicon diode rectifiers. (*Courtesy of International Rectifier.*)

applied to a *pn* junction diode, the diode will conduct only when the *p* region has a positive voltage applied to it relative to the *n* region. As a result, half-wave rectification will be produced, as shown in Fig. 5.33. This output signal can be smoothed out with other electronic devices and circuits so that a steady dc signal can be produced. Solid-state silicon rectifiers are used in a wide range of current capacities which can be from tenths of an ampere to several hundred amperes or more. Voltages, too, can be as high as 1000 V or more. Figure 5.34 shows some examples of silicon diode rectifiers.

Breakdown diodes   Breakdown diodes, or *zener diodes* as they are sometimes called, are silicon rectifiers in which the reverse current (leakage current) is small, and then with only slightly more reverse-bias voltage, a breakdown voltage is reached upon which the reverse current increases very rapidly (Fig. 5.35). In the so-called zener breakdown, the electric field in the diode becomes strong enough to attract electrons directly out of the covalently bonded crystal lattice. The electron-hole pairs created then produce a high reverse current. At higher reverse-bias voltages than the zener breakdown voltages, an avalanche effect occurs, and the reverse current is very large. One theory to explain the avalanche effect is that electrons gain sufficient energy between collisions to knock more electrons from covalent bonds which can reach high-enough energies to conduct electricity. Breakdown diodes can be made with breakdown voltages from a few to several hundred volts and are used for voltage-limiting applications and for voltage stabilizing under conditions of widely varying current.

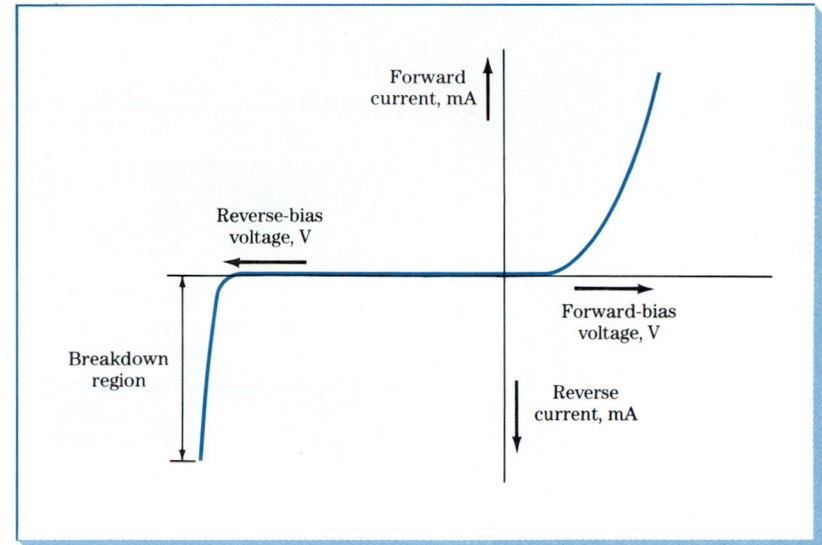

FIGURE 5.35 Zener (avalanche) diode characteristic curve. A large reverse current is produced at the breakdown-voltage region.

The Bipolar Junction Transistor

A bipolar junction transistor (BJT) is an electronic device that can serve as a current amplifier. This device consists of two *pn* junctions which occur sequentially in a single crystal of a semiconducting material such as silicon. Figure 5.36 shows schematically an *npn*-type bipolar junction transistor and

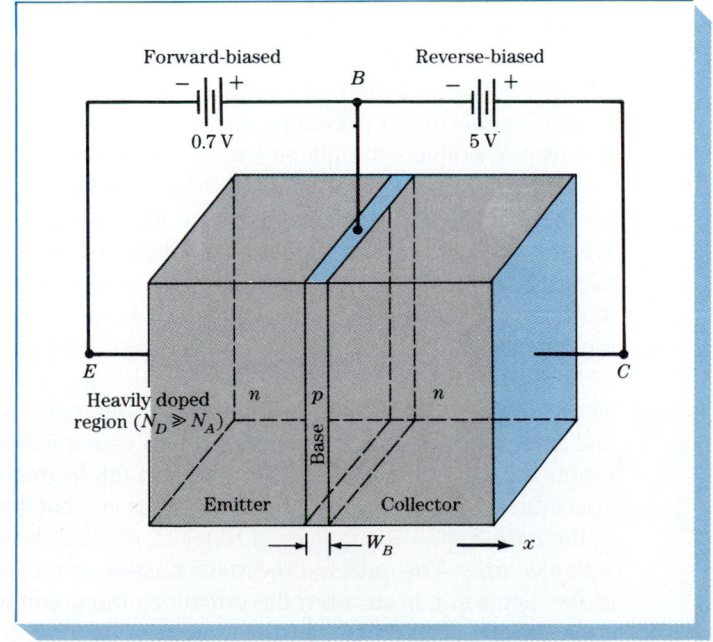

FIGURE 5.36 Schematic illustration of an *npn* bipolar junction transistor. The *n* region on the left is the emitter, the thin *p* region in the middle is the base, and the *n* region on the right is the collector. For normal operation the emitter-base junction is forward-biased and the collector-base junction is reverse-biased. (*After C. A. Holt, "Electronic Circuits," Wiley, 1978, p. 49.*)

FIGURE 5.37 Charge
carrier motion during the
normal operation of an
npn transistor. Most of the
current consists of
electrons from the emitter
which go right through
the base to the collector.
Some of the electrons,
about 1 to 5 percent,
recombine with holes
from the base current
flow. Small reverse
currents due to thermally
generated carriers are
also present, as
indicated. (*After R. J.
Smith, "Currents, Devices
and Systems," 3d ed.,
Wiley, 1976, p. 343.*)

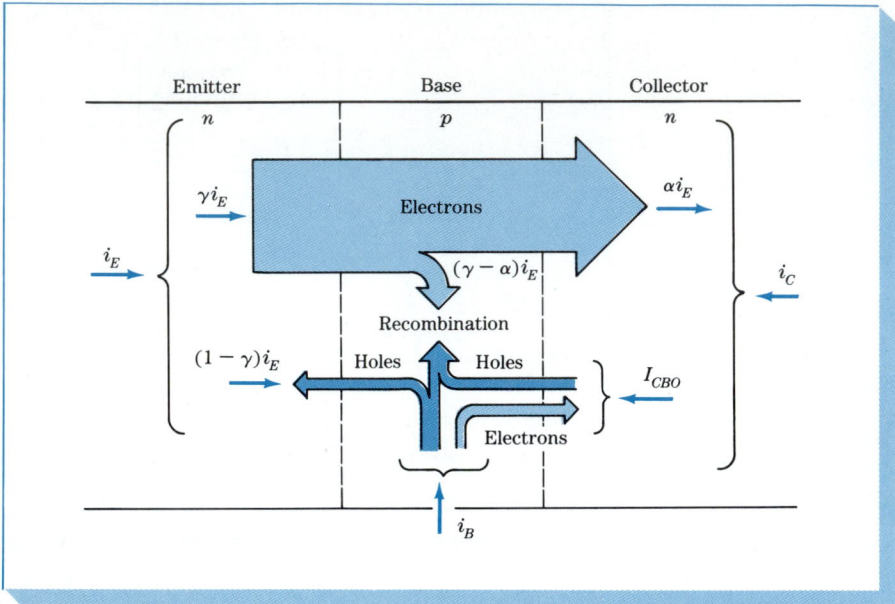

identifies the three main parts of the transistor: *emitter, base,* and *collector.*
The emitter of the transistor emits charge carriers. Since the emitter of the *npn*
transistor is *n*-type, it emits electrons. The base of the transistor controls the
flow of charge carriers and is *p*-type for the *npn* transistor. The base is made
very thin (about 10^{-3} cm thick) and is lightly doped so that only a small fraction
of the charge carriers from the emitter will recombine with the oppositely
charged majority carriers of the base. The collector of the BJT collects charge
carriers mainly from the emitter. Since the collector section of the *npn* transistor
is *n*-type, it collects mainly electrons from the emitter.

Under normal operation of the *npn* transistor, the emitter-base junction is
forward-biased and the collector-base junction is reverse-biased (Fig. 5.36).
The forward bias on the emitter-base junction causes an injection of electrons
from the emitter into the base (Fig. 5.37). Some of the electrons injected into
the base are lost by recombination with holes in the *p*-type base. However,
most of the electrons from the emitter pass right through the thin base into
the collector, where they are attracted by the positive terminal of the collector.
Heavy doping of the emitter with electrons, light doping of the base with holes,
and a very thin base are all factors which cause most of the emitter electrons
(about 95 to 99 percent) to pass right through to the collector. Very few holes
flow from the base to the emitter. Most of the current flow from the base terminal
to the base region is the flow of holes to replace those lost by recombination
with electrons. The current flow to the base is small and is about 1 to 5 percent
of the electron current from the emitter to the collector. In some respects the
current flow to the base can be thought of as a control valve since the small

base current can be used to control the much larger collector current. The bipolar transistor is so named because both types of charge carriers (electrons and holes) are involved in its operation.

5.6   MICROELECTRONICS

Modern semiconductor technology has made it possible to put thousands of transistors on a "chip" of silicon about 5 mm square and 0.2 mm thick. This ability to incorporate very large numbers of electronic elements on silicon chips has greatly increased the capability of electronic device systems (Fig. 5.1).

Large-scale integrated (LSI) microelectronic circuits are manufactured by starting with a silicon single-crystal wafer (*n*- or *p*-type) about 100 to 125 mm in diameter and 0.2 mm thick. The fabrication of the blank silicon wafers has

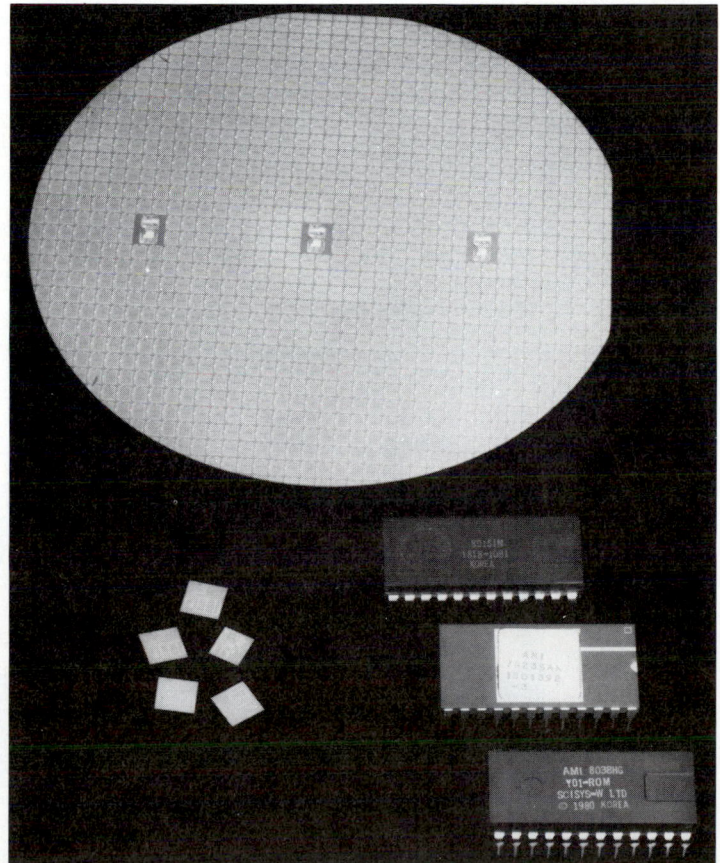

FIGURE 5.38 This photograph shows a wafer, individual integrated circuits, and three chip packages (the middle package is ceramic and the other two are plastic). The three larger devices along the middle of this wafer are process control monitors (PCMs) to monitor the technical quality of the dice on the wafer. [*Courtesy of American Microsystems, Inc. (AMI), Santa Clara, Calif.*]

already been described in Sec. 4.2. The surface of the wafer must be highly polished and free from defects on one side since the semiconductor devices are fabricated into the polished surface of the wafer. Figure 5.38 shows a silicon wafer after the microelectronic circuits have been fabricated into its surface. About 100 to 1000 chips (depending on their size) can be produced from one wafer.

First, let us examine the structure of a planar-type bipolar transistor fabricated into a silicon wafer surface. Then, we will briefly look at the structure of a more compact type of transistor called the MOSFET, or *metal oxide semiconductor field-effect transistor*, which is used in many modern semiconductor device systems. Finally, we will outline some of the basic procedures used in the manufacturing of modern microelectronic circuits.

Microelectronic Planar Bipolar Transistors

Microelectronic planar bipolar transistors are fabricated into the surface of a silicon single-crystal wafer by a series of operations that require access to only one surface of the silicon wafer. Figure 5.39 shows a schematic diagram of the cross section of an *npn* planar bipolar transistor. In its fabrication a relatively

FIGURE 5.39 Microelectronic planar bipolar *npn* transistor fabricated in a single crystal of silicon by a series of operations that require access to only one surface of the silicon chip. The entire chip is doped with *p*-type impurities, then islands of *n*-type silicon are formed. Smaller *p*- and *n*-type areas are next created within these islands in order to define the three fundamental elements of the transistor: the emitter, the base, and the collector. In this microelectronic bipolar transistor the emitter-base junction is forward-biased and the collector-base junction is reverse-biased, as in the case of the isolated *npn* transistor of Fig. 5.36. The device exhibits gain because a small signal applied to the base can control a large one at the collector. (*After J. D. Meindl, Microelectronic Circuit Elements, Sci. Am., September 1977, p. 75. Copyright © Scientific American Inc. All rights reserved.*)

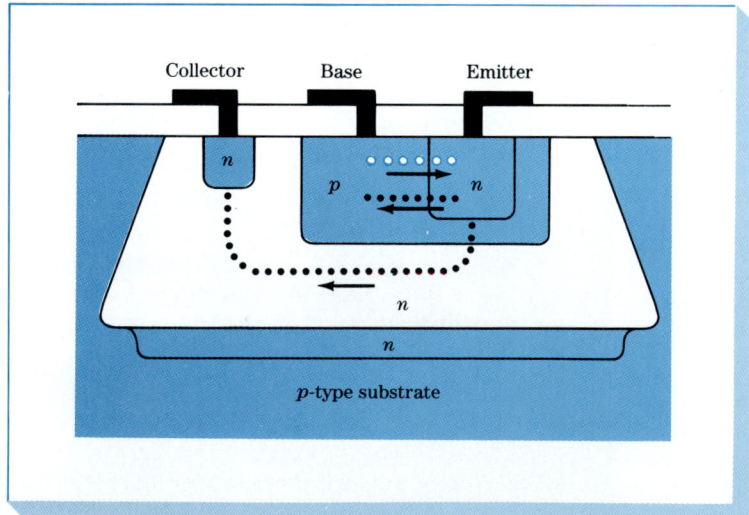

large island of n-type silicon is formed first in a p-type silicon base or substrate. Then smaller islands of p- and n-type silicon are created in the larger n-type island (Fig. 5.39). In this way the three fundamental parts of the npn bipolar transistor, the emitter, base, and collector, are formed in a planar configuration. As in the case of the individual npn bipolar transistor previously described in Sec. 5.5 (see Fig. 5.36), the emitter-base junction is forward-biased and the base-collector junction is reverse-biased. Thus, when electrons from the emitter are injected into the base, most of them go into the collector and only a small percentage ($\sim$1 to 5 percent) recombine with holes from the base terminal (see Fig. 5.37). The microelectronic planar bipolar transistor can therefore function as a current amplifier in the same way as the individual macroelectronic bipolar transistor.

Microelectronic Planar Field-Effect Transistors

In many of today's modern microelectronic systems, another type of transistor, called the *field-effect transistor,* is used because of its low cost and compactness. The most common field-effect transistor used in the United States is the n-type *m*etal *o*xide *s*emiconductor *f*ield-*e*ffect *t*ransistor, or *MOSFET*. In the n-type MOSFET, or NMOS, two islands of n-type silicon are created in a substrate of p-type silicon, as shown in Fig. 5.40. In the NMOS device the contact where the electrons enter is called the *source* and the contact where they leave is called the *drain*. Between the n-type silicon of the source and the drain, there is a p-type region on whose surface a thin layer of silicon dioxide is formed which acts as an insulator. On top of the silicon dioxide another layer of polysilicon (or metal) is deposited to form the third contact for the transistor, called the *gate*. Since silicon dioxide is an excellent insulator, the gate connection is not in direct electrical contact with the p-type material below the oxide.

For a simplified type of NMOS when no voltage is applied to the gate, the p-type material under the gate contains majority carriers which are holes and only a few electrons are attracted to the drain. However, when a positive voltage is applied to the gate, its electric field attracts electrons from the nearby n^+ source and drain regions to the thin layer beneath the surface of the silicon dioxide just under the gate so that this region becomes n-type silicon, with electrons being the majority carriers (Fig. 5.41). When electrons are present in this channel, a conducting path exists between the source and the drain. Thus electrons will flow between the source and the drain if there is a positive voltage difference between them.

The MOSFET, like the bipolar transistor, is also capable of current amplification. The gain in MOSFET devices is usually measured in terms of a voltage ratio instead of a current ratio as for the bipolar transistor. p-type MOSFETs with holes for majority carriers can be made in a similar way, using p-type islands for the source and drain in an n-type substrate. Since the current carriers are electrons in NMOS devices and holes in PMOS ones, they are known as *majority-carrier devices*.

MOSFET technology is the basis for most large-scale integrated (LSI) digital memory circuits, mainly because the individual MOSFET occupies less silicon

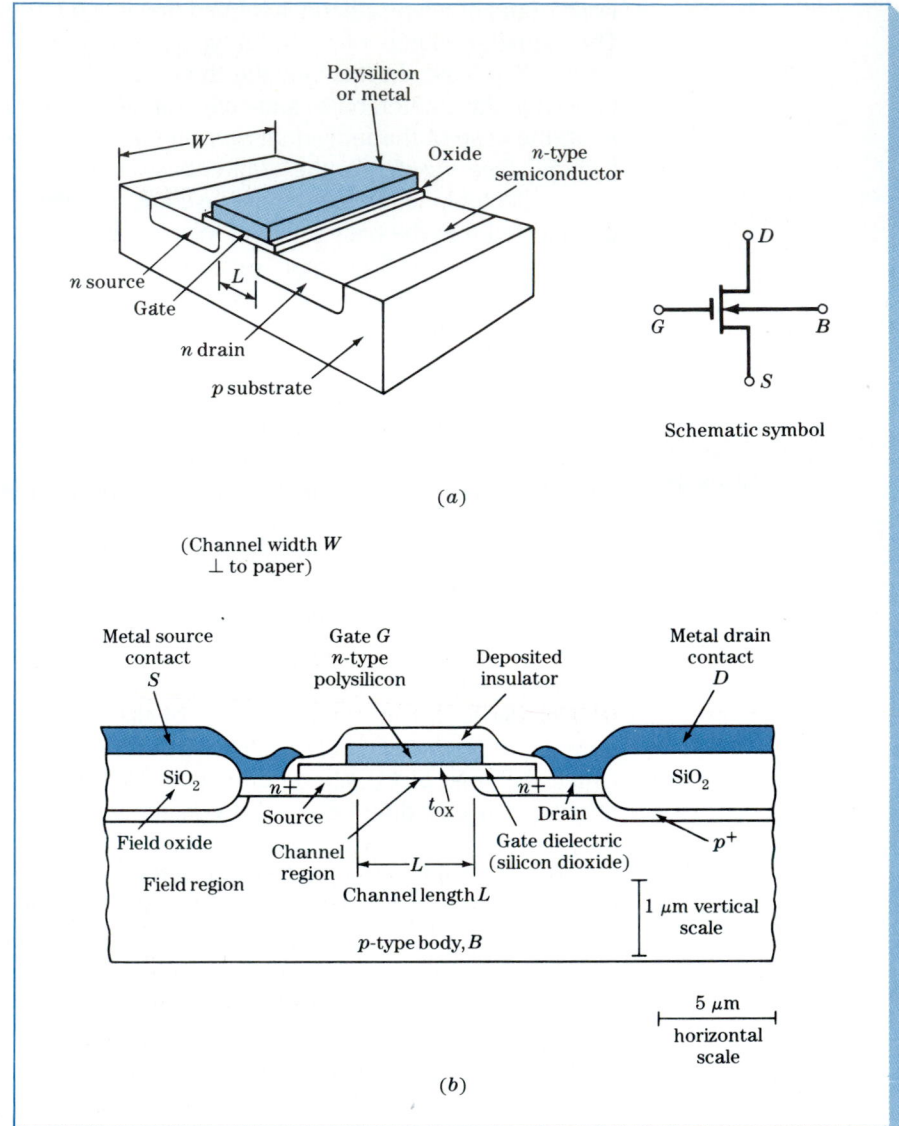

FIGURE 5.40 Schematic diagram of an NMOS field-effect transistor: (*a*) overall structure; (*b*) cross-sectional view. (*After D. A. Hodges and H. G. Jackson, "Analysis and Design of Digital Integrated Circuits," McGraw-Hill, 1983, p. 40.*)

chip area than the bipolar transistor and hence greater densities of transistors can be obtained. Also, the cost of fabrication of the MOSFET LSIs is less than that for the bipolar transistor types. However, there are some applications for which bipolar transistors are necessary.

Fabrication of Microelectronic Integrated Circuits

The design of a microelectronic integrated circuit is first laid out on a large scale, usually with computer assistance so that the most space-conserving design can be made (Fig. 5.42). In the most common fabrication process the

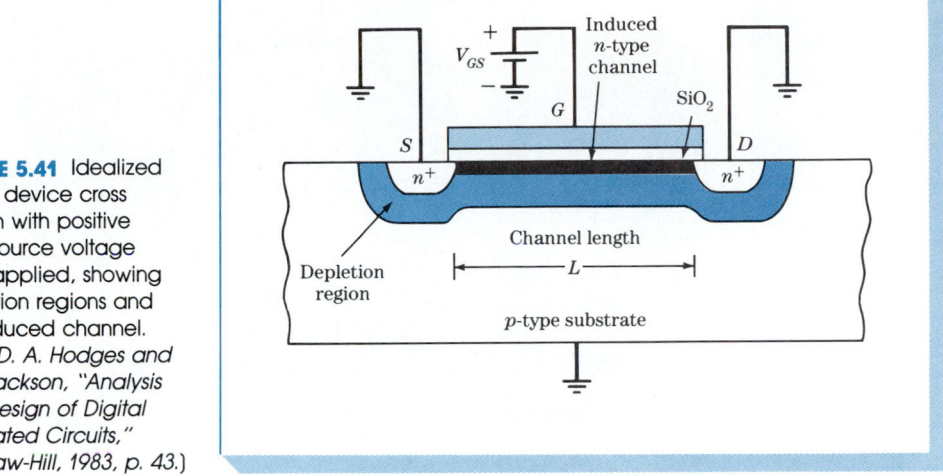

FIGURE 5.41 Idealized NMOS device cross section with positive gate-source voltage (V_{GS}) applied, showing depletion regions and the induced channel. (*After D. A. Hodges and H.G. Jackson, "Analysis and Design of Digital Integrated Circuits," McGraw-Hill, 1983, p. 43.*)

FIGURE 5.42 Engineer laying out an integrated circuit network. (*Courtesy of Harris Corporation.*)

FIGURE 5.43 This photograph depicts two types of photolithographic masks used in the fabrication of integrated circuits. At the left is the more durable chrome mask, which is used for long production runs and can be used to produce emulsion masks like the one shown on the right. Emulsion masks are less-expensive masks and tend to be used for shorter production runs, such as the fabrication of prototypes. [*Courtesy of American Microsystems, Inc. (AMI), Santa Clara, Calif.*]

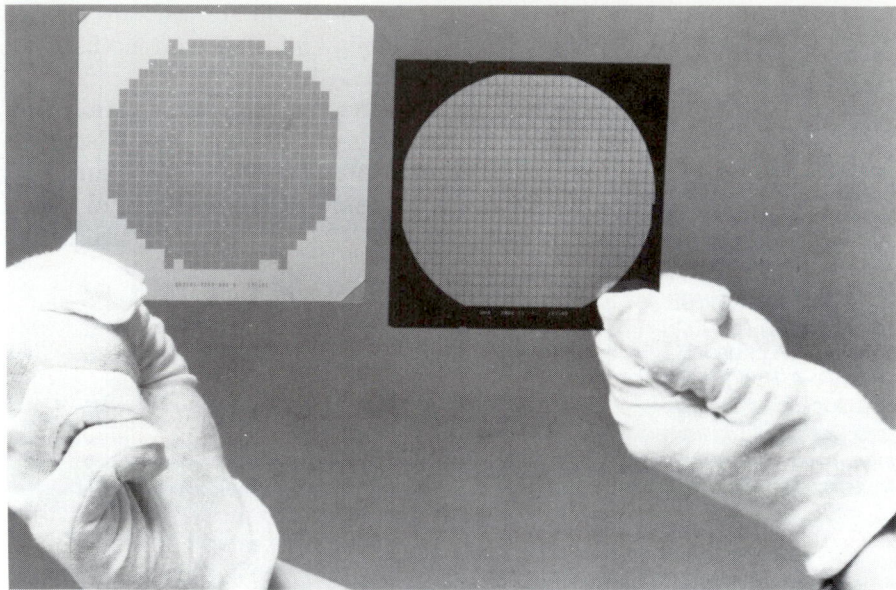

layout is used to prepare a set of photomasks, each of which contains the pattern for a single layer of the multiple-layer finished integrated circuit (Fig. 5.43).

Photolithography The process by which a microscopic pattern is transferred from a photomask to the silicon wafer surface of the integrated circuit is called *photolithography*. Figure 5.44 shows the steps necessary to form an insulating layer of silicon dioxide on the silicon surface which contains a pattern of regions of exposed silicon substrate. In one type of photolithographic process shown in step 2 of Fig. 5.44, an oxidized wafer is first coated with a layer of *photoresist*, a light-sensitive polymeric material. The important property of photoresist is that its solubility in certain solvents is greatly affected by its exposure to ultraviolet (UV) radiation. After exposure to UV radiation (step 3 of Fig. 5.44), and subsequent development, a pattern of photoresist is left wherever the mask was transparent to the UV radiation (step 4 of Fig. 5.44). The silicon wafer is then immersed in a solution of hydrofluoric acid which attacks only the exposed silicon dioxide and not the photoresist (step 5 of Fig. 5.44). In the final step of the process the photoresist pattern is removed by another chemical treatment (step 6 of Fig. 5.44). The photolithographic process has improved over the past years so that it is now possible to reproduce surface dimensions to about 2 μm apart.

Diffusion and ion implantation of dopants into the surface of silicon wafers To form active circuit elements such as bipolar and MOS transistors in integrated circuits, it is necessary to selectively introduce impurities (dopants) into

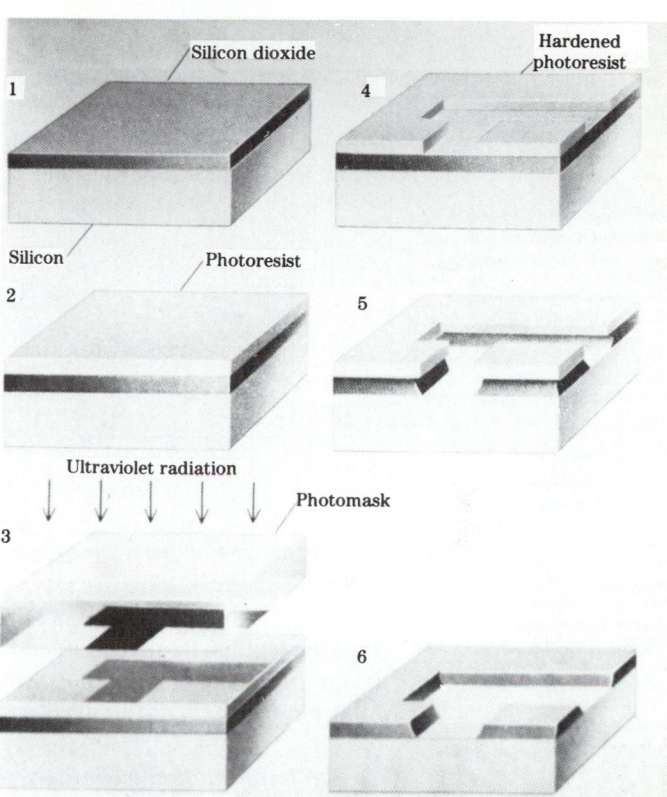

FIGURE 5.44 The steps for the photolithographic process. In this process a microscopic pattern can be transferred from a photomask to a material layer in an actual circuit. In this illustration a pattern is shown being etched into a silicon dioxide layer on the surface of a silicon wafer. The oxidized wafer (1) is first coated with a layer of a light-sensitive material called *photoresist* (2) and then exposed to ultraviolet light through the photomask (3). The exposure renders the photomask insoluble in a developer solution; hence a pattern of photoresist is left wherever the mask is transparent (4). The wafer is next immersed in a solution of hydrofluoric acid, which selectively attacks the silicon dioxide, leaving the photoresist pattern and the silicon substrate unaffected (5). In the final step the photoresist pattern is removed by another chemical treatment (6). (*After W. G. Oldham, The Fabrication of Microelectronic Circuits, Sci. Am., September 1977, p. 121. Copyright © Scientific American Inc. All rights reserved.*)

the silicon substrate to create localized *n*- and *p*-type regions. There are two main techniques for introducing dopants into the silicon wafers: (1) *diffusion* and (2) *ion implantation.*

The diffusion technique   As previously described in Sec. 4.7, the impurity atoms are diffused into the silicon wafers at high temperatures, i.e., about 1000 to 1100°C. Important dopant atoms such as boron and phosphorus move much more slowly through silicon dioxide than through the silicon crystal lattice. Thin silicon dioxide patterns can serve as masks to prevent the dopant atoms from penetrating into the silicon (Fig. 5.45a). Thus a rack of silicon wafers can be placed into a diffusion furnace at 1000 to 1100°C in an atmosphere containing phosphorus (or boron), for example. The phosphorus atoms will enter the unprotected surface of the silicon and slowly diffuse into the bulk of the wafer, as shown in Fig. 5.45a.

The important variables which control the concentration and depth of penetration are *temperature* and *time* (see Example Problem 4.10). To achieve the maximum control of concentration most diffusion operations are carried out in two steps. In the first, or *predeposit,* step, a relatively high concentration of dopant atoms is deposited near the surface of the wafer. After the predeposit

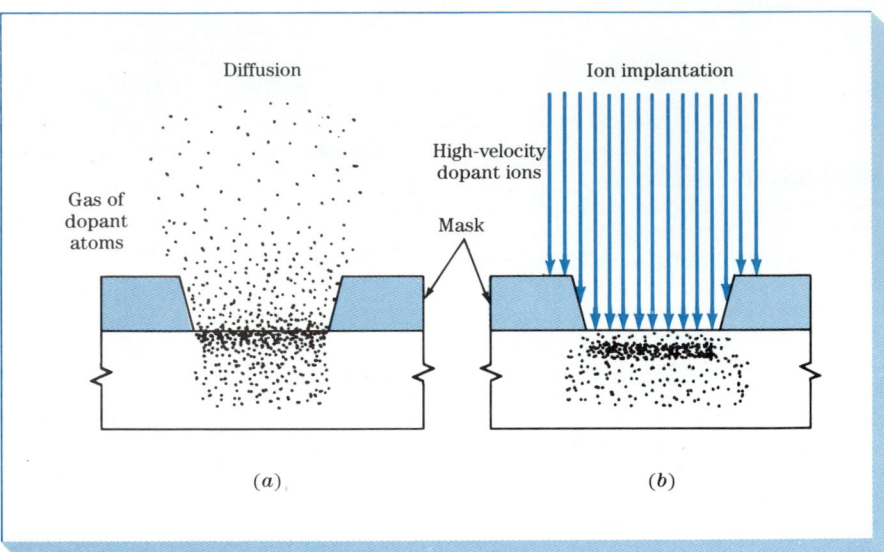

FIGURE 5.45 Selective doping processes for exposed silicon surfaces: (*a*) high-temperature diffusion of impurity atoms; (*b*) ion implantation. (*After S. Triebwasser, "Today and Tomorrow in Microelectronics," from the Proceedings of an NSF workshop held at Arlie, Va., Nov. 19–22, 1978.*)

step, the wafers are placed in another furnace, usually at a higher temperature, for the *drive-in diffusion* step which achieves the necessary concentration of dopant atoms at a particular depth below the surface of the silicon wafer.

The ion implantation technique   Another selective doping process for silicon wafers for integrated circuits is the ion implantation technique (Fig. 5.45*b*), which has the advantage that the dopant impurities can be embedded at room temperature. Figure 5.46 shows a schematic diagram of an ion implanation system. In this process the dopant atoms are ionized (electrons are removed or

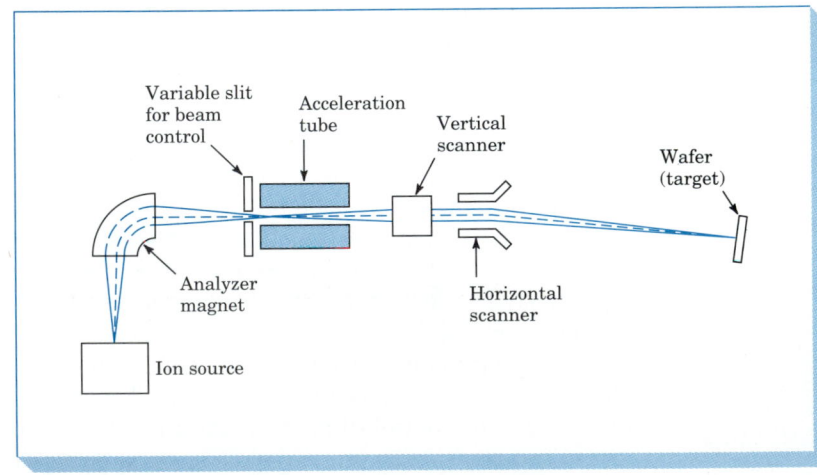

FIGURE 5.46 Ion implantation system for semiconductor wafers. (*After S. M. Sze (ed.), "VLSI Technology," McGraw-Hill, 1983.*)

added to atoms to form ions), for example, B^+ and As^+ ions, and exit from an ion source. The ions then are separated according to their masses by an analyzer magnet which separates unwanted ions. The selected ions (for example, B^+) subsequently enter an accelerator tube and are accelerated to high energies (that is, 30 to 1000 keV) by an electric field. The high-energy ions then pass through vertical and horizontal scanners and are implanted into the semiconductor substrate at various depths (Fig. 5.45*b*). A photoresist or silicon dioxide pattern can mask off regions of the surface where ion implantation is not desired.

The energized ions lose energy as they progress through the semiconductor substrate by colliding with electrons and nuclei and finally are stopped. The total distance an ion travels before stopping is called its *range R* (Fig. 5.47). The projection of this distance along the incident beam axis is called the *projected range* R_p. As expected, due to random collisions, there is a statistical deviation in each of the *x,y,* and *z* directions. If the ion beam is traveling in the *x* direction, this deviation is referred to as the *projected straggle* ΔR_p. An approximate value for the peak concentration of ions, C_p, in the ion beam direction can be calculated from the relationship

$$C_p = \frac{S}{\sqrt{2\pi}\,\Delta R_p} \tag{5.23}$$

where S is the ion dose in ions per unit area and ΔR_p is the projected straggle.

Example Problem 5.10

A silicon wafer with a series of windows in an oxide layer is undergoing ion implantation with a beam of boron ions at 100 keV. If the beam dose is 3.0×10^{15} cm^{-2} and the projected straggle range is 900 Å, what is the peak concentration of the boron ions at the projected range?

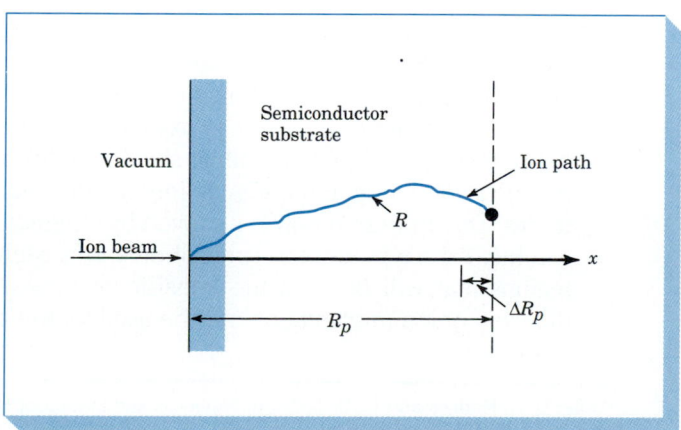

FIGURE 5.47 Schematic diagram showing range R, projected range R_p, and projected straggle ΔR_p.

Solution:

$$C_B \text{ at } R_p = \frac{S}{\sqrt{2\pi} \, \Delta R_p} = \frac{3.0 \times 10^{15} \text{ ions/cm}^2}{\sqrt{2\pi}(900 \times 10^{-8} \text{ cm})}$$

$$= 1.33 \times 10^{20} \text{ ions/cm}^3$$

Ion implanation for doping semiconductor substrates has many advantages, and as many as 10 or more ion implantations may be done on a single silicon chip, for example. These advantages include room temperature doping, doping control at a specific depth, narrow width of doping, angular doping, and overdoping, i.e., changing from one polarity to another at the same place. A disadvantage of this process is that accelerated ions cause some damage to the crystal lattice of the substrate, but most of the damage can be healed by heating at a moderate temperature.

MOS integrated circuit fabrication technology   There are many different procedures used in the fabrication of MOS integrated circuits. New innovations and discoveries for improving the equipment design and processing of ICs are constantly being made in this rapidly progressing technology. The general processing sequence for one method of producing NMOS integrated circuits is described in the following steps[1] and illustrated in Figs. 5.48 and 5.49.

1. (See Fig. 5.48*a*) A chemical vapor deposition (CVD) process deposits a thin layer of silicon nitride (Si_3N_4) on the entire wafer surface. The first photolithographic step defines areas where transistors are to be formed. The silicon nitride is removed outside the transistor areas by chemical etching. Boron (*p*-type) ions are implanted in the exposed regions to suppress unwanted conduction between the transistor sites. Next, a layer of silicon dioxide (SiO_2) about 1 μm thick is grown thermally in these inactive, or field, regions by exposing the wafer to oxygen in an electric furnace. This is known as a *selective*, or *local*, oxidation process. The Si_3N_4 is impervious to oxygen and thus inhibits growth of the thick oxide in the transistor regions.

2. (See Fig. 5.48*b*.) The Si_3N_4 is next removed by an etchant that does not attack SiO_2. A clean thermal oxide about 0.1 μm thick is grown in the transistor areas, again by exposure to oxygen in a furnace. Another CVD process deposits a layer of polycrystalline silicon (poly) over the entire wafer. The second photolithographic step defines the desired patterns for the gate electrodes. The undesired poly is removed by chemical or plasma (reactive-gas) etching. An *n*-type dopant (phosphorus or arsenic) is introduced into the regions that will become the transistor source and drain. Either thermal diffusion or ion implantation may be used for this doping process. The thick

---

[1] After D. A. Hodges and H. G. Jackson, "Analysis and Design of Digital Integrated Circuits," McGraw-Hill, 1983, pp. 16–18.

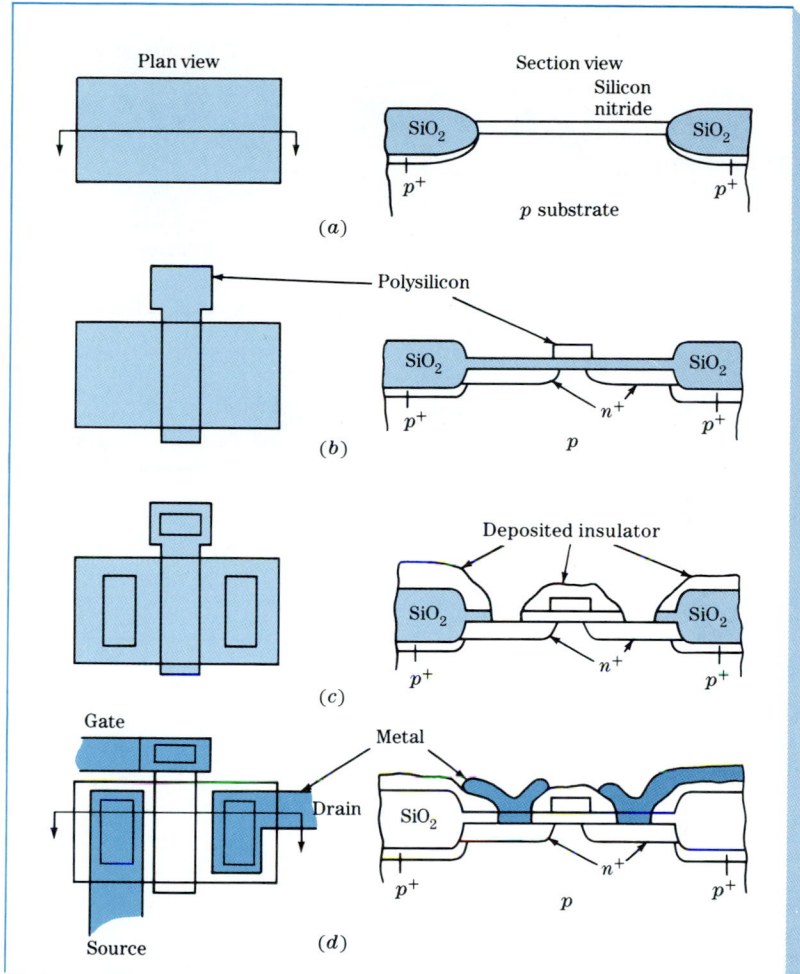

FIGURE 5.48 Steps in NMOS field-effect transistor fabrication: (a) first mask; (b) second mask: polysilicon gate; source-drain diffusion; (c) third mask: contact areas; (d) fourth mask: metal pattern. (After D. A. Hodges and H. G. Jackson, "Analysis and Design of Digital Integrated Circuits," McGraw-Hill, 1983, p. 17.)

field oxide and the poly gate are barriers to the dopant, but in the process, the poly becomes heavily doped n-type.

3. (See Fig. 5.48c.) Another CVD process deposits an insulating layer, often SiO_2, over the entire wafer. The third masking step defines the areas in which contacts to the transistors are to be made, as shown in Fig. 5.46c. Chemical or plasma etching selectively exposes bare silicon or poly in the contact areas.

4. Aluminum (Al) is deposited over the entire wafer by evaporation from a hot crucible in a vacuum evaporator. The fourth masking step patterns the Al as desired for circuit connections, as shown in Fig. 5.48d.

5. A protective passivating layer is deposited over the entire surface. A final masking step removes this insulating layer over the pads where contacts will

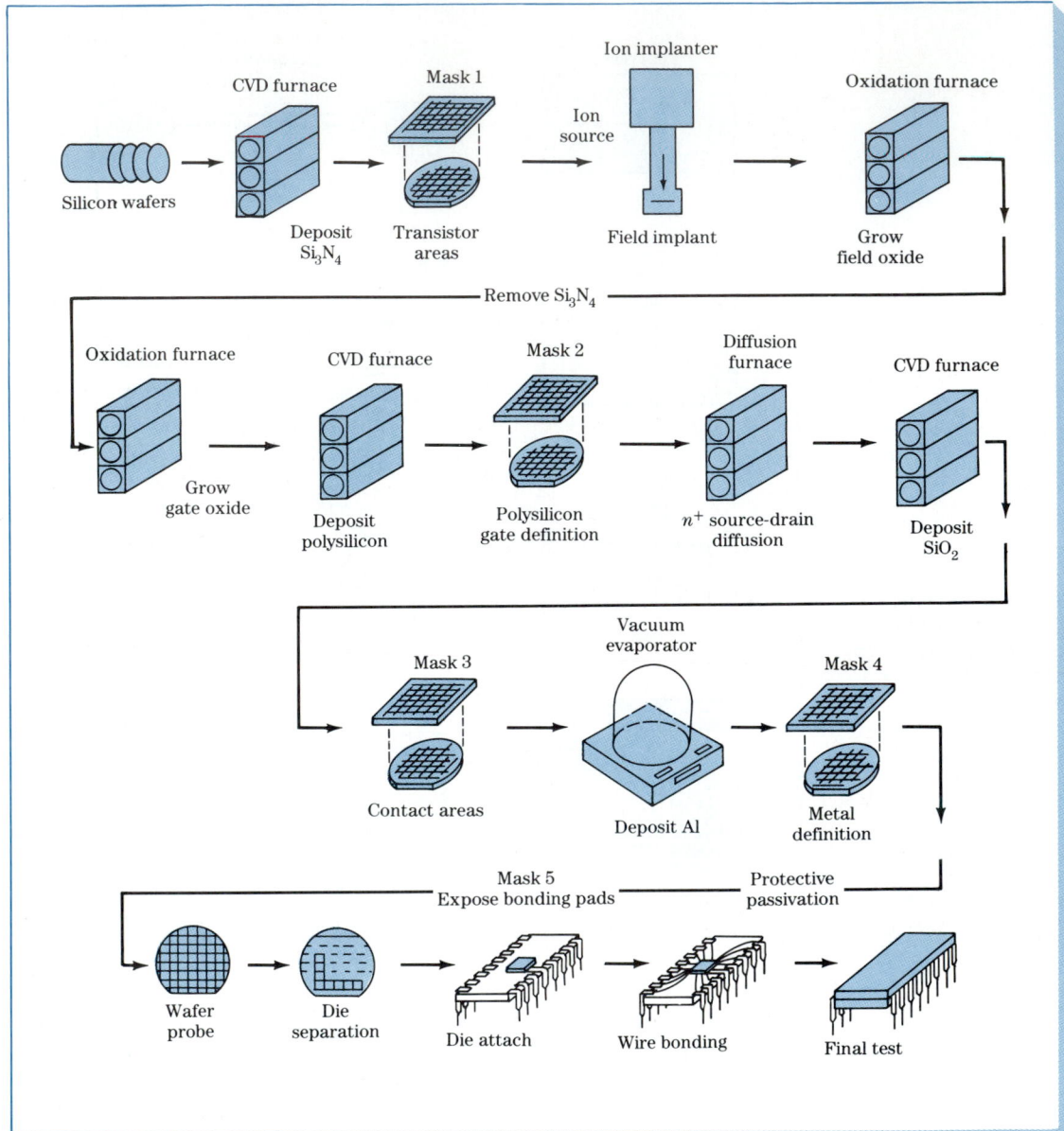

FIGURE 5.49 A manufacturing process for NMOS silicon-gate integrated circuits. (The processes used to make NMOS integrated circuits vary considerably from company to company. The above sequence is given as an outline.) (*Courtesy of Integrated Circuit Engineering Co.*)

be made. Circuits are tested by using needlelike probes on the contact pads. Defective units are marked, and the wafer is then sawed into individual chips. Good chips are packaged and given a final test.

This is the simplest process for forming NMOS circuits, and it is summarized schematically in Fig. 5.49. More-advanced NMOS circuit processes require more masking steps.

Complementary metal oxide semiconductor (CMOS) devices   It is possible to manufacture a chip containing both types of MOSFETs (NMOS and PMOS) but only by increasing the complexity of the circuitry and lowering the density of the transistors. Circuits containing both NMOS and PMOS devices are called *complementary,* or *CMOS, circuits* and can be made, for example, by isolating all NMOS devices with islands of *p*-type material (Fig. 5.50). An advantage of CMOS circuits is that the MOS devices can be arranged to achieve lower power consumption. CMOS devices are used in a variety of applications. For example, large-scale-integrated (LSI) CMOS circuits are used in almost all the modern electronic watches and calculators. Also, CMOS technology is becoming of increasing importance for use in microprocessors and computer memories.

New Trends and Developments in Microelectronic Technologies

In the 1990s (1994) the major trend for microprocessors has been to develop integrated silicon chips with higher and higher densities of transistors and circuit speeds and smaller and smaller minimum design features. Table 5.6 lists some of the design characteristics of selected microprocessors. As noted in this table the Power PC 601 chip has a minimum design feature dimension of 0.65 μm and contains 2.8 million transistors. Technologies have been developed which will enable CMOS designs to be made with minimum design features as low as 0.35 μm on a microprocessor chip.

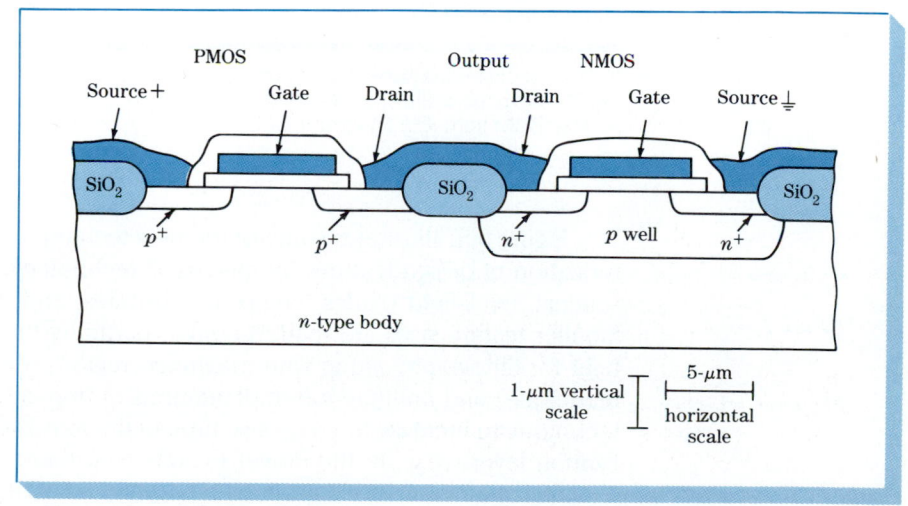

FIGURE 5.50
Complementary MOS field-effect transistors (CMOS). Both *n*- and *p*-type transistors are fabricated on the same silicon substrate. (*After D. A. Hodges and H. G. Jackson, "Analysis and Design of Digital Integrated Circuits," McGraw-Hill, 1983, p. 42.*)

TABLE 5.6   Microprocessors (1993)

| | Microprocessor | | | |
	68040	80486	Pentium	PowerPc 601
Company	Motorola Inc.	Intel Corp.	Intel Corp.	IBM Corp. and Motorola Inc.
Introduction date	1989	6/91	3/93	4/93
Architecture and organization type	CISC*	CISC	CISC	RISC[†]
Technology and performance				
Technology	0.65 μm CMOS	0.8 μm CMOS	0.8 μm BiCMOS	0.65 μm CMOS
Die size, mm	10.8 by 11.7	N.S.	17.2 by 17.2	11 by 11
Transistors, millions	1.2	1.2	3.1	2.8
Metallization layers	2	3	3	4
Operating voltage, V	5	5	5	3.6
Clock, MHz	25	50	66	80
Power, packaging, and price				
Peak power, W	6	5	16	9.1
Cooling	Ambient plus heat sink	Fan or heat sink	Fan plus heat sink	Ambient plus heat sink
Ceramic package (pins / style)	179 / PGA	168 / PGA	273 / PGA	304 / QFP
US $ price per 1000	$233	$432	$898	$545/$557

*CISI: Complex-instruction-set computer
[†]RISC: Reduced-instruction-set computer
Source: Data from *IEEE Spectrum,* December 1993, p. 21.

Figure 5.51 illustrates some of the new techniques used for the ultraminiaturization of design features by the CMOS technology. These include sidewall spacers, thick-field oxides, epitaxial substrates, and lightly doped predrains. Smaller feature sizes are made possible partly by the use of deep ultraviolet light for lithography along with magnified regions for exposure (the stepping technique) and multiple ion implantations of dopant ions. Another important technique to increase the response time of the transistors is to use many metallization layers (e.g., in the Power PC 601 chip there are four layers).

Still another technology which has appeared in recent years is the BiCMOS

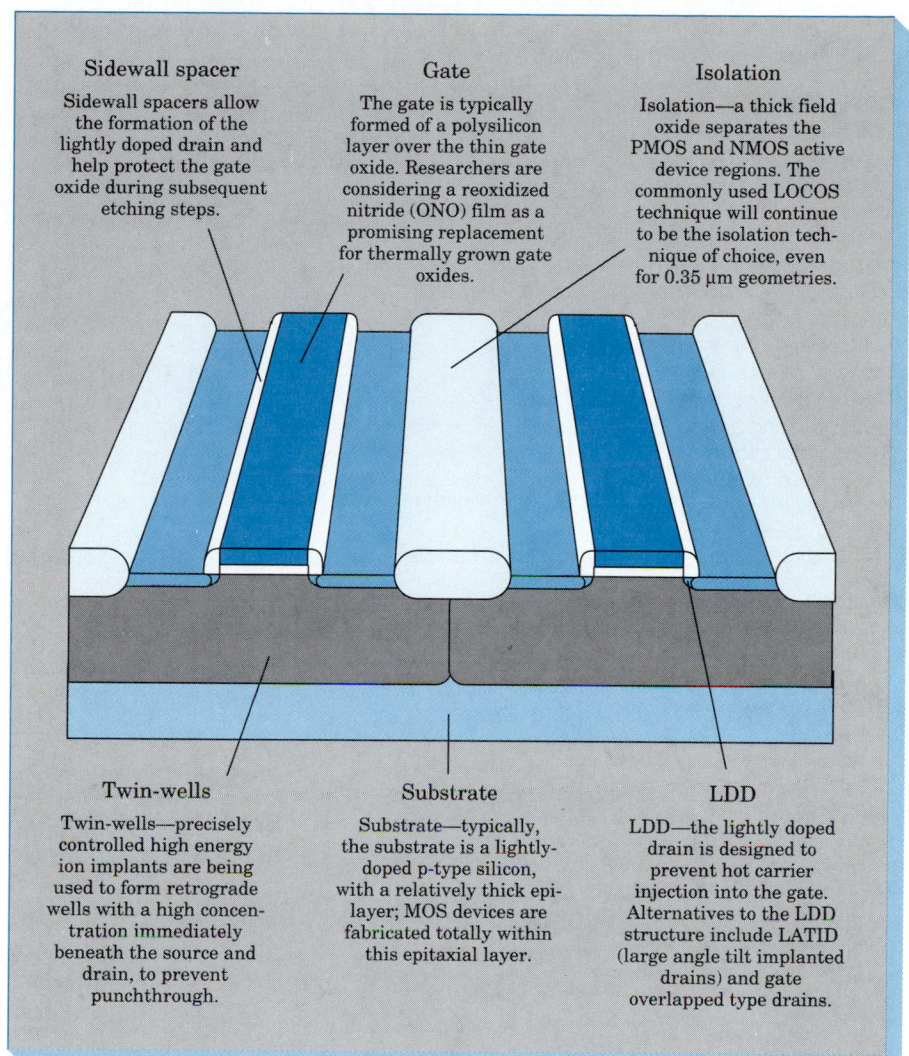

Sidewall spacer

Sidewall spacers allow the formation of a lightly doped drain and help protect the gate oxide during subsequent etching steps.

Gate

The gate is typically formed of a polysilicon layer over the thin gate oxide. Researchers are considering a reoxidized nitride (ONO) film as a promising replacement for thermally grown gate oxides.

Isolation

Isolation—a thick field oxide separates the PMOS and NMOS active device regions. The commonly used LOCOS technique will continue to be the isolation technique of choice, even for 0.35 µm geometries.

Twin-wells

Twin-wells—precisely controlled high energy ion implants are being used to form retrograde wells with a high concentration immediately beneath the source and drain, to prevent punchthrough.

Substrate

Substrate—typically, the substrate is a lightly-doped p-type silicon, with a relatively thick epi-layer; MOS devices are fabricated totally within this epitaxial layer.

LDD

LDD—the lightly doped drain is designed to prevent hot carrier injection into the gate. Alternatives to the LDD structure include LATID (large angle tilt implanted drains) and gate overlapped type drains.

FIGURE 5.51 CMOS device design in the 1990s. Shown are enhancements or changes commonly used in the newer CMOS technology. (*After Semiconductor International, April 1992, p. 56.*)

design (e.g., the Pentium microprocessor uses this technology). The BiCMOS technology combines bipolar transistors with the field-effect CMOS transistors. The bipolar transistors provide higher response speeds and the CMOS transistors provide higher density of transistors. Disadvantages of the BiCMOS designs are higher costs and longer fabrication cycle times. Figure 5.52 compares the circuit speed and circuit density of bipolar and CMOS chips for the period 1985 to 1993. Note how the bipolar chips used to have a great advantage over the CMOS technology for circuit speed, but now the new CMOS chips are approaching the bipolar chips in circuit speed.

FIGURE 5.52
Competition between bipolar and CMOS technologies. Shown here are the performance characteristics of successive generations of bipolar and CMOS chips used in Unisys products. Note how the CMOS microprocessors are approaching the speeds of bipolar chips while having higher circuit density. (*After IEEE Spectrum, January 1994, p. 48.*)

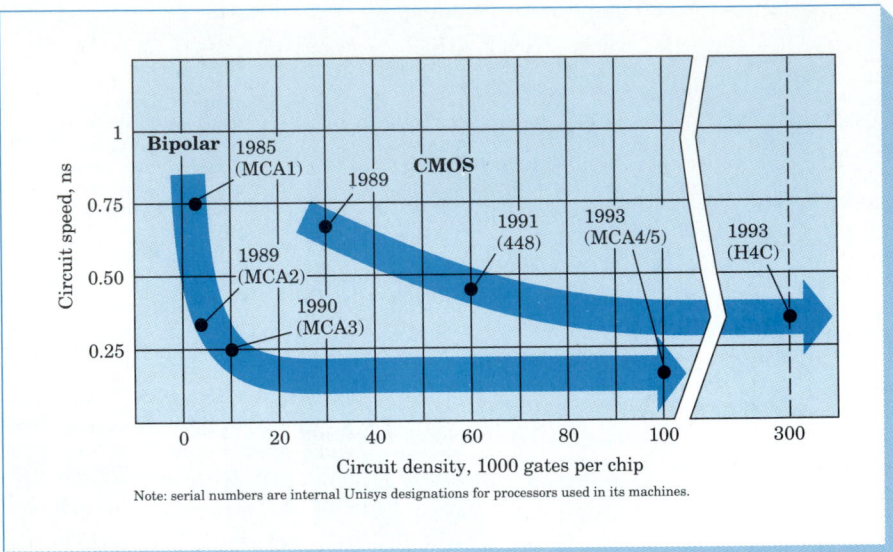

Note: serial numbers are internal Unisys designations for processors used in its machines.

5.7 COMPOUND SEMICONDUCTORS

There are many compounds of different elements which are semiconductors. One of the major types of semiconducting compounds are the MX ones, where M is a more electropositive element and X a more electronegative element. Of the MX semiconductor compounds, two important groups are the III–V (3–5) and II–VI (2–6) compounds formed by elements adjacent to the group IVA of the periodic table (Fig. 5.53). The III-V semiconductor compounds consist of M group III elements such as Al, Ga, and In combined with X group V elements such as P, As, and Sb. The II-VI compounds consist of M group II elements such as Zn, Cd, and Hg combined with X group VI elements such as S, Se, and Te.

Table 5.7 lists some electrical properties of selected compound semiconductors. From this table the following trends can be observed.

1. By increasing the molecular mass of a compound within a family by moving down in the columns of the periodic table, the energy band gap decreases, electron mobility increases (exceptions are GaAs and GaSb), and the lattice constant increases. The electrons of the larger and heavier atoms have, in general, more freedom to move and are less tightly bound to their nuclei and thus tend to have smaller band gaps and higher electron mobilities.

2. By moving across the periodic table from the group IVA elements to the III–V and II–VI materials, the increased ionic bonding character causes the

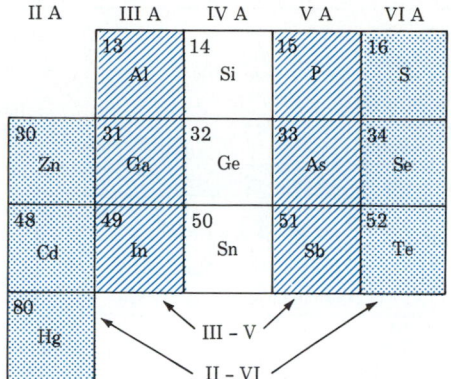

FIGURE 5.53 Part of the periodic table containing elements used in the formation of MX-type III–V and II–VI semiconductor compounds.

energy band gaps to increase and the electron mobilities to decrease. The increased ionic bonding causes a tighter binding of the electrons to their positive-ion cores, and thus II–VI compounds have larger band gaps than comparable III–V compounds.

Gallium arsenide is the most important of all the compound semiconductors and is used in many electronic devices. GaAs has been used for a long time for discrete components in microwave circuits. Today, many digital integrated circuits are made with GaAs. The GaAs metal-semiconductor field-effect transistors (MESFETs) are the most widely used GaAs transistors (Fig. 5.54).

TABLE 5.7   Electrical Properties of Intrinsic Semiconducting Compounds at Room Temperature (300 K)

Group	Material	E_g eV	μ_n m²/(V·s)	μ_p m²/(V·s)	Lattice constant	n_i carriers/m³
IVA	Si	1.10	0.135	0.048	5.4307	1.50×10^{16}
	Ge	0.67	0.390	0.190	5.257	2.4×10^{19}
IIIA–VA	GaP	2.25	0.030	0.015	5.450	
	GaAs	1.47	0.720	0.020	5.653	1.4×10^{12}
	GaSb	0.68	0.500	0.100	6.096	
	InP	1.27	0.460	0.010	5.869	
	InAs	0.36	3.300	0.045	6.058	
	InSb	0.17	8.000	0.045	6.479	1.35×10^{22}
IIA–VIA	ZnSe	2.67	0.053	0.002	5.669	
	ZnTe	2.26	0.053	0.090	6.104	
	CdSe	2.59	0.034	0.002	5.820	
	CdTe	1.50	0.070	0.007	6.481	

Source: W. R. Runyun and S. B. Watelski, in C. A. Harper (ed.), "Handbook of Materials and Processes for Electronics," McGraw-Hill, New York, 1970.

FIGURE 5.54 Cross-sectional view of a GaAs MESFET. [*After A. N. Sato et al., IEEE Electron. Devices Lett.,* **9**(5):238 (1988).]

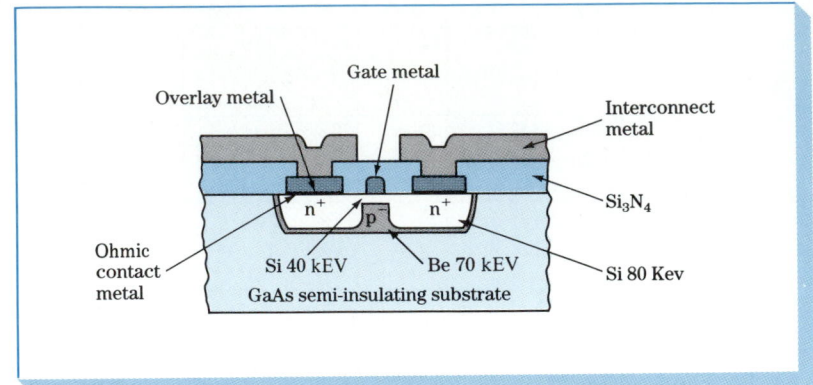

GaAs MESFETs offer some advantages over silicon as devices for use in high-speed digital integrated circuits. Some of these are:

1. Electrons travel faster in *n*-GaAs as is indicated by their higher mobility in GaAs than in Si [$\mu_n = 0.720$ m²/(V·s) for GaAs vs. 0.135 m²/(V·s) for Si].
2. Because of its larger band gap of 1.47 eV and the absence of a critical-gate oxide, GaAs devices are claimed to have better radiation resistance. This consideration is important for space and military applications.

Unfortunately, the major limitation of GaAs technology is that the yield for complex IC circuitry is much lower than for silicon due mainly to the fact that GaAs contains more defects in the base material than silicon. The cost of producing the base material is also higher for GaAs than silicon. However, the use of GaAs is expanding, and much research is being done in this area.

Example Problem 5.11

(*a*) Calculate the intrinsic electrical conductivity of GaAs at (i) room temperature (27°C) and (ii) 70°C.

(*b*) What fraction of the current is carried by the electrons in the intrinsic GaAs at 27°C?

Solution:

(*a*) (i) σ at 27°C:

$$\sigma = n_i q(\mu_n + \mu_p)$$

$$= (1.4 \times 10^{12} \text{ m}^{-3})(1.60 \times 10^{-19} \text{ C})[0.720 \text{ m}^2/(\text{V} \cdot \text{s}) + 0.020 \text{ m}^2/(\text{V} \cdot \text{s})]$$

$$= 1.66 \times 10^{-7} \ (\Omega \cdot \text{m})^{-1} \blacktriangleleft$$

(ii) σ at 70°C:

$$\sigma = \sigma_0 e^{-E_g/2kT} \qquad (5.16a)$$

$$\frac{\sigma_{343}}{\sigma_{300}} = \frac{\exp\{-1.47\text{eV}/[(2)(8.62 \times 10^{-5} \text{ eV/K})(343 \text{ K})]\}}{\exp\{-1.47\text{eV}/[(2)(8.62 \times 10^{-5} \text{ eV/K})(300 \text{ K})]\}}$$

$$\sigma_{343} = \sigma_{300}e^{3.56} = 1.66 \times 10^{-7} \text{ }(\Omega\cdot\text{m})^{-1}(35.2)$$

$$= 5.84 \times 10^{-6} \text{ }(\Omega\cdot\text{m})^{-1} \blacktriangleleft$$

(b) $\dfrac{\sigma_n}{\sigma_n + \sigma_p} = \dfrac{n_i q \mu_n}{n_i q(\mu_n + \mu_p)} = \dfrac{0.720 \text{ m}^2/(\text{V}\cdot\text{s})}{0.720 \text{ m}^2/(\text{V}\cdot\text{s}) + 0.020 \text{ m}^2(\text{V}\cdot\text{s})} = 0.973 \blacktriangleleft$

SUMMARY

In the classical model for electrical conduction in metals, the outer valence electrons of the atoms of the metal are assumed to be free to move between the positive-ion cores (atoms without their valence electrons) of the metal lattice. In the presence of an applied electric potential the free electrons attain a directed drift velocity. The movement of electrons and their associated electric charge in a metal constitute an electric current. By convention, electric current is considered to be positive charge flow, which is in the opposite direction to electron flow.

In the energy-band model for electrical conduction in metals, the valence electrons of the metal atoms interact and interpenetrate each other to form energy bands. Since the energy bands of the valence electrons of metal atoms overlap, producing partially filled composite energy bands, very little energy is required to excite the highest-energy electrons so that they become free to be conductive. In insulators the valence electrons are tightly bound to their atoms by ionic and covalent bonding and are not free to conduct electricity unless highly energized. The energy-band model for an insulator consists of a lower filled valence band and a higher empty conduction band. The valence band is separated from the conduction band by a large energy gap (about 6 to 7 eV, for example). Thus for insulators to be conductive a large amount of energy must be applied to cause the valence electrons to "jump" the gap. Intrinsic semiconductors have a relatively small energy gap (i.e., about 0.7 to 1.1 eV) between their valence and conduction bands. By doping the intrinsic semiconductors with impurity atoms to make them extrinsic, the amount of energy required to cause semiconductors to be conductive is greatly reduced.

Extrinsic semiconductors can be n-type or p-type. The n-type (negative) semiconductors have electrons for their majority carriers. The p-type (positive) semiconductors have holes (missing electrons) for their majority charge carriers. By fabricating pn junctions in a single crystal of a semiconductor such as silicon, various types of semiconducting devices can be made. For example, *pn* junction diodes and *npn* transistors can be produced by using these junctions. Modern microelectronic technology has developed to such an extent that thousands of transistors can be placed on a "chip" of semiconducting silicon less than about 0.5 cm square and about 0.2 mm thick. Complex microelectronic technology has made possible highly sophisticated microprocessors and computer memories.

DEFINITIONS

Sec. 5.1

Electric current: the time rate passage of charge through material; electric current i is the number of coulombs per second which passes a point in a material. The SI unit for electric current is the ampere (1 A = 1 C/s).

Electric current density J: the electric current per unit area. SI units: amperes/meter^2 (A/m^2).

Electrical resistance R: the measure of the difficulty of electric current to pass through a volume of material. Resistance increases with the length and increases with decreasing cross-sectional area of the material through which the current passes. SI unit: ohm (Ω).

Electrical resistivity ρ_e: a measure of the difficulty of electric current to pass through a *unit* volume of material. For a volume of material, $\rho_e = RA/l$, where R = resistance of material, Ω; l = its length, m; A = its cross-sectional area, m^2. In SI units, ρ_e = ohm-meters ($\Omega \cdot$ m).

Electrical conductivity σ_e: a measure of the ease of electric current to pass through a unit volume of material. Units: $(\Omega \cdot \text{m})^{-1}$. σ_e is the inverse of ρ_e.

Electrical conductor: a material with a high electrical conductivity. Silver is a good conductor and has a $\sigma_e = 6.3 \times 10^7 \ (\Omega \cdot \text{m})^{-1}$.

Electrical insulator: a material with a low electrical conductivity. Polyethylene is a poor conductor and has a $\sigma_e = 10^{-15}$ to $10^{-17} \ (\Omega \cdot \text{m})^{-1}$.

Semiconductor: a material whose electrical conductivity is approximately midway between the values for good conductors and insulators. For example, pure silicon is a semiconducting element and has $\sigma_e = 4.3 \times 10^{-4} \ (\Omega \cdot \text{m})^{-1}$ at 300 K.

Sec. 5.2

Energy-band model: in this model the energies of the bonding valence electrons of the atoms of a solid form a band of energies. For example, the 3s valence electrons in a piece of sodium form a 3s energy band. Since there is only one 3s electron (the 3s orbital can contain two electrons), the 3s energy band in sodium metal is half-filled.

Valence band: the energy band containing the valence electrons. In a conductor the valence band is also the conduction band. The valence band in a conducting metal is not full, and so some electrons can be energized to levels within the valence band and become conductive electrons.

Conduction band: the unfilled energy levels into which electrons can be excited to become conductive electrons. In semiconductors and insulators there is an energy gap between the filled lower valence band and the upper empty conduction band.

Sec. 5.3

Intrinsic semiconductor: a semiconducting material which is essentially pure and for which the energy gap is small enough (about 1 eV) to be surmounted by thermal excitation; current carriers are electrons in the conduction band and holes in the valence band.

Electron: a negative charge carrier with a charge of 1.60×10^{-19} C.

Hole: a positive charge carrier with a charge of 1.60×10^{-19} C.

Sec. 5.4

n-**type extrinsic semiconductor:** a semiconducting material that has been doped with an *n*-type element (e.g., silicon doped with phosphorus). The *n*-type impurities donate electrons that have energies close to the conduction band.

Donor levels: in the band theory, local energy levels near the conduction band.

p-**type extrinsic semiconductor:** a semiconducting material that has been doped with a *p*-type element (e.g., silicon doped with aluminum). The *p*-type impurities provide electron holes close to the upper energy level of the valence band.

Acceptor levels: in the band theory, local energy levels close to the valence band.

Majority carriers: the type of charge carrier most prevalent in a semiconductor; the majority carriers in an *n*-type semiconductor are conduction electrons, and in a *p*-type semiconductor they are conduction holes.

Minority carriers: the type of charge carrier in the lowest concentration in a semiconductor. The minority carriers in *n*-type semiconductors are holes, and in *p*-type semiconductors they are electrons.

Sec. 5.5

pn **junction:** an abrupt junction or boundary between *p*- and *n*-type regions within a single crystal of a semiconducting material.

Rectifier diode: a *pn* junction diode which converts alternating current to direct current (ac to dc).

Bias: voltage applied to two electrodes of an electronic device.

Forward bias: bias applied to a *pn* junction in the conducting direction; in a *pn* junction under forward bias, majority-carrier electrons and holes flow toward the junction so that a large current flows.

Reverse bias: bias applied to a *pn* junction so that little current flows; in a *pn* junction under reverse bias, majority-carrier electrons and holes flow away from the junction.

Bipolar transistor: a three-element, two-junction semiconducting device. The three basic elements of the transistor are the emitter, base, and collector. Bipolar junction transistors (BJTs) can be of the *npn* or *pnp* types. The emitter-base junction is forward-biased and the collector-base junction reverse-biased so that the transistor can act as a current amplification device.

PROBLEMS

5.1.1   Describe the classical model for electrical conduction in metals.

5.1.2   Distinguish between (*a*) positive-ion cores and (*b*) valence electrons in a metallic crystal lattice such as sodium.

5.1.3   Write equations for the (*a*) macroscopic and (*b*) microscopic forms of Ohm's law. Define the symbols in each of the equations and indicate their SI units.

5.1.4   How is electrical conductivity related numerically to electrical resistivity?

5.1.5   Give two kinds of SI units for electrical conductivity.

5.1.6   Calculate the resistance of a silver rod 6.15 mm in diameter and 1.10 m long.

5.1.7   A nichrome wire must have a resistance of 110 Ω. How long must it be (in meters) if it is 0.003 in in diameter? [σ_e(nichrome) $= 9.3 \times 10^5$ $(\Omega \cdot m)^{-1}$.]

5.1.8   A wire 0.35 cm in diameter must carry a 35-A current.
   (*a*) If the maximum power dissipation along the wire is 0.035 W/cm, what is the minimum allowable electrical conductivity of the wire (give answer in SI units)?
   (*b*) What is the current density in the wire?

5.1.9   A gold wire is to conduct a 6-A current with a maximum voltage drop of 0.003 V/cm. What must be the minimum diameter of the wire in meters?

5.1.10 Define the following quantities pertaining to the flow of electrons in a metal conductor: (*a*) drift velocity; (*b*) relaxation time; (*c*) electron mobility.

5.1.11 What causes the electrical resistivity of a metal to increase as its temperature increases? What is a phonon?

5.1.12 What structural defects contribute to the residual component of the electrical resistivity of a pure metal?

5.1.13 What effect do elements that form solid solutions have on the electrical resistivities of pure metals?

5.1.14 Calculate the electrical resistivity (in ohm-meters) of a copper wire 70 ft long and 0.025 in in diameter at 180°C.

5.1.15 At what temperature will a silver wire have the same electrical resistivity as an aluminum one at 45°C?

5.1.16 At what temperature will the electrical resistivity of a gold wire be 6.55×10^{-8} $\Omega \cdot m$?

5.2.1   Why are the valence-electron energy levels broadened into bands in a solid block of a good conducting metal such as sodium?

5.2.2   Why don't the energy levels of the inner-core electrons of a block of sodium metal also form energy bands?

5.2.3   Why is the 3*s* electron energy band in a block of sodium only half-filled?

5.2.4   What explanation is given for the good electrical conductivity of magnesium and aluminum even though these metals have filled outer 3*s* energy bands?

5.2.5   How does the energy-band model explain the poor electrical conductivity of an insulator such as pure diamond?

5.3.1   Define an intrinsic semiconductor. What are the two most important elemental semiconductors?

5.3.2   What type of bonding does the diamond cubic structure have? Make a two-dimensional sketch of the bonding in the silicon lattice, and show how electron-hole pairs are produced in the presence of an applied field.

5.3.3   Why is a hole said to be an imaginary particle? Use a sketch to show how electron holes can move in a silicon crystal lattice.

5.3.4   Define electron and electron hole mobility as pertains to charge movement in a silicon lattice. What do these quantities measure, and what are their SI units?

5.3.5   Explain, using an energy-band diagram, how electrons and electron holes are created in pairs in intrinsic silicon.

5.3.6   What is the ratio of the electron-to-hole mobility in silicon and germanium?

5.3.7   Calculate the number of germanium atoms per cubic meter.

5.3.8   Calculate the electrical resistivity of germanium at 300 K.

5.3.9   Explain why the electrical conductivity of intrinsic silicon and germanium increases with increasing temperature.

5.3.10 The electrical resistivity of pure germanium is 0.46 $\Omega \cdot$ m at 300 K. Calculate its electrical conductivity at 400°C.

5.3.11 The electrical resistivity of pure silicon is $2.3 \times 10^3 \ \Omega \cdot$ m at 300 K. Calculate its electrical conductivity at 300°C.

5.4.1   Define n-type and p-type extrinsic silicon semiconductors.

5.4.2   Draw two-dimensional lattices of silicon of the following:
(a) n-type lattice with an arsenic impurity atom present
(b) p-type lattice with a boron impurity atom present

5.4.3   Draw energy-band diagrams showing donor or acceptor levels for the following:
(a) n-type silicon with phosphorus impurity atoms
(b) p-type silicon with boron impurity atoms

5.4.4   (a) When a phosphorus atom is ionized in an n-type silicon lattice, what charge does the ionized atom acquire?
(b) When a boron atom is ionized in a p-type silicon lattice, what charge does the ionized atom acquire?

5.4.5   What are dopants as pertains to semiconductors? Explain the process of doping by diffusion.

5.4.6   What are the majority and minority carriers in an n-type silicon semiconductor? In a p-type one?

5.4.7   A silicon wafer is doped with 5.0×10^{21} phosphorus atoms/m³. Calculate (a) the electron and hole concentrations after doping and (b) the resultant electrical resistivity at 300 K. [Assume $n_i = 1.5 \times 10^{16}$/m³ and $\mu_n = 0.1350$ m²/(V · s).]

5.4.8   Phosphorus is added to make an n-type silicon semiconductor with an electrical conductivity of 200 $(\Omega \cdot$ m$)^{-1}$. Calculate the necessary number of charge carriers required.

5.4.9   A semiconductor is made by adding boron to silicon to give an electrical resistivity of 1.70 $\Omega \cdot$ m. Calculate the concentration of carriers per cubic meter in the material. [Assume $\mu_p = 0.048$ m²/(V · s).]

5.4.10 A silicon wafer is doped with 2.0×10^{16} boron atoms/cm³ plus 1.5×10^{16} phosphorus atoms/cm³ at 27°C. Calculate (a) the electron and hole concentrations (carriers per cubic centimeters), (b) the electron and hole mobilities (use Fig. 5.25a), and (c) the electrical resistivity of the material.

5.4.11 A silicon wafer is doped with 2.0×10^{15} phosphorus atoms/cm³, 4.0×10^{16} boron atoms/cm³, and 3.0×10^{16} arsenic atoms/cm³. Calculate (a) the electron and hole concentrations (carriers per cubic centimeters), (b) the electron and hole mobilities (use Fig. 5.25a), and (c) the electrical resistivity of the material.

5.4.12 An arsenic-doped silicon wafer has an electrical resistivity of $8.0 \times 10^{-4} \ \Omega \cdot$ cm at 27°C. Assume intrinsic carrier mobilities and complete ionization.
(a) What is the majority-carrier concentration (carriers per cubic centimeter)?
(b) What is the ratio of arsenic to silicon atoms in this material?

5.4.13 A boron-doped silicon wafer has an electrical resistivity of $6.3 \times 10^{-4} \ \Omega \cdot$ cm at 27°C. Assume intrinsic carrier mobilities and complete ionization.
(a) What is the majority-carrier concentration (carriers per cubic centimeter)?
(b) What is the ratio of boron-to-silicon atoms in this material?

5.4.14 Describe the origin of the three stages that appear in the plot of ln σ vs. 1/T for an extrinsic silicon semiconductor (going from low to high temperatures). Why does the conductivity decrease just before the rapid increase due to intrinsic conductivity?

5.5.1   Define the term *microprocessor*.

5.5.2   Describe the movement of majority carriers in a *pn* junction diode at equilibrium. What is the depletion region of a *pn* junction?

5.5.3   Describe the movement of the majority and minority carriers in a *pn* junction diode under reverse bias.

5.5.4   Describe the movement of the majority carriers in a *pn* junction diode under forward bias.

5.5.5   Describe how a *pn* junction diode can function as a current rectifier.

5.5.6   What is a zener diode? How does this device function? Describe a mechanism to explain its operation.

5.5.7   What are the three basic elements of a bipolar junction transistor?

5.5.8   Describe the flow of electrons and holes when an *npn* bipolar junction transistor functions as a current amplifier.

5.5.9   What fabrication techniques are used to encourage electrons from the emitter of an *npn* bipolar transistor to go right through to the collector?

5.5.10 Why is a bipolar junction transistor called bipolar?

5.6.1   Describe the structure of a planar *npn* bipolar transistor.

5.6.2   Describe how the planar bipolar transistor can function as a current amplifier.

5.6.3   Describe the structure of an *n*-type metal oxide semiconductor field-effect transistor (NMOS).

5.6.4   How do NMOSs function as a current amplifier?

5.6.5   Describe the photolithographic steps necessary to produce a pattern of an insulating layer of silicon dioxide ona silicon surface.

5.6.6   Describe the diffusion process for the introduction of dopants into the surface of a silicon wafer.

5.6.7   Describe the ion implantation process for introducing dopants into the surface of a silicon wafer.

5.6.8   A silicon wafer with a series of windows in an oxide layer is undergoing ion implantation with a beam of boron ions at 100 keV. If the beam dose is 2.5×10^{15} cm^{-2} and the projected straggle range is 700 Å, what is the peak concentration of the boron ions at the projected range?

5.6.9   Describe the general process for fabricating NMOS integrated circuits on a silicon wafer.

5.6.10 Why is silicon nitride (Si_3N_4) used in producing NOS integrated circuits on a silicon wafer?

5.6.11 What are complementary metal oxide semiconductor (CMOS) devices? What are the advantages of CMOS devices over the NMOS or PMOS devices?

5.7.1   Calculate the intrinsic electrical conductivity of GaAs at 75°C. [$E_g = 1.47$ eV; $\mu_n = 0.720$ m^2/(V · s); $\mu_p = 0.020$ m^2/(V · s); $n_i = 1.4 \times 10^{12}$ m^{-3}.]

5.7.2   Calculate the intrinsic electrical conductivity of InSb at 30 and at 50°C. [$E_g = 0.17$ eV; $\mu_n = 8.00$ m^2/(V · s); $\mu_p = 0.045$ m^2/(V · s); $n_i = 1.35 \times 10^{22}$ m^{-3}.]

5.7.3   Calculate the intrinsic electrical conductivity of (*a*) GaAs and (*b*) InSb at 90°C.

5.7.4   What fraction of the current is carried by (*a*) electrons and (*b*) holes in (i) InSb, (ii) InB, and (iii) InP at 27°C?

5.7.5   What fraction of the current is carried by (*a*) electrons and (*b*) holes in (i) GaSb and (ii) GaP at 27°C?

Mechanical Properties of Metals

This chapter first examines some of the basic methods of processing metals and alloys into useful shapes. Then stress and strain in metals are defined, and the tensile test used to determine these properties is described. This is followed by a treatment of hardness and hardness testing of metals. Next, the permanent deformation of metals and the effects of elevated temperature on the structure and properties of metals are studied. Finally, the fracture, fatigue, and creep (time-dependent deformation) of metals are considered.

6.1 THE PROCESSING OF METALS AND ALLOYS

The Casting of Metals and Alloys

Most metals are processed by first melting the metal in a furnace which functions as a reservoir for the molten metal. Alloying elements can be added to the molten metal to produce various alloy compositions. For example, solid magnesium metal may be added to molten aluminum and, after melting, can be mechanically mixed with the aluminum to produce a homogeneous melt of an aluminum-magnesium alloy. After oxide impurities and unwanted hydrogen gas are removed from the molten Al–Mg alloy, it is cast into a mold of a direct-chill semicontinuous casting unit, as shown in Fig. 4.8. Huge sheet

FIGURE 6.1 Large aluminum four-high rolling mill is shown hot rolling a slab of an aluminum alloy which will eventually be reduced in thickness to make aluminum sheet. (*Courtesy of Reynolds Metals Co.*)

ingots, such as those shown in Fig. 4.1, are produced in this way. Other types of ingots with different cross sections are cast in a similar way; for example, extrusion ingots are cast with circular cross sections.

Semifinished products are manufactured from the basic shaped ingots. Sheet[1] and plate[2] are produced by rolling sheet ingots to reduced thicknesses (Fig. 6.1). Extruded shapes such as channels and structural shapes are produced from extrusion ingots, and rod and wire are manufactured from wire bar ingots. All these products which are manufactured by hot and cold working the metal from large ingots are called *wrought alloy products.* The effects of permanent

---

[1] For this book *sheet* is defined as a rolled product rectangular in cross section and form of thickness 0.006 to 0.249 in (0.015 to 0.063 cm).
[2] For this book *plate* is defined as a rolled product rectangular in cross section and form of thickness 0.250 in (0.635 cm) or more.

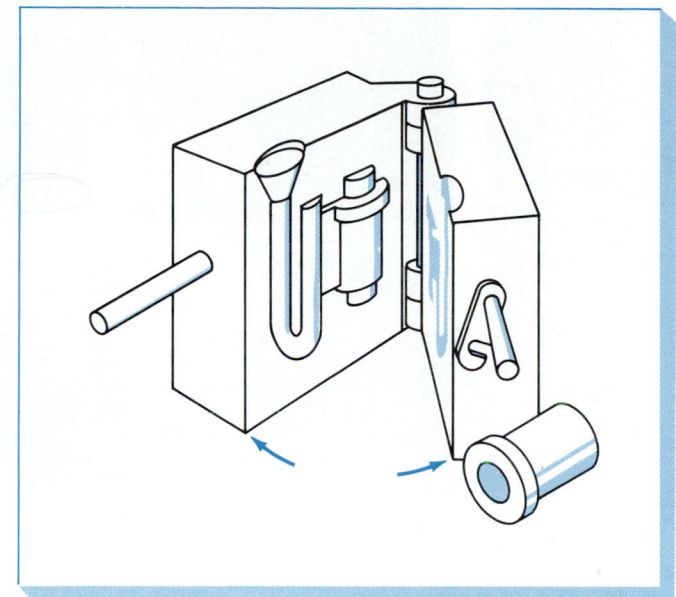

FIGURE 6.2 Permanent mold casting. Solidified casting with gate and metal core is shown in the left half of the mold. The completed casting is shown in front of the mold. (*After H. F. Taylor, M. C. Flemings, and J. Wulff, "Foundry Engineering," Wiley, 1959, p. 58.*)

deformation on the structure and properties of metals will be treated in Secs. 6.5 and 6.6.

On a smaller scale molten metal may be cast into a mold which is in the shape of the final product, and usually only a small amount of machining or other finishing operation is required to produce the final casting. Products made in this manner are called *cast products* and the alloys used to produce them, *casting alloys*. For example, pistons used in automobile engines are usually made by casting molten metal into a permanent steel mold. A schematic diagram of a simple permanent mold containing a casting is shown in Fig. 6.2. Figure 6.3*a* shows an operator pouring an aluminum alloy into a permanent mold to produce a pair of piston castings; Fig. 6.3*b* shows the castings after they have been removed from the mold. After being trimmed, heat-treated, and machined, the finished piston (Fig. 6.3*c*) is ready for installation in an automobile engine.

Hot and Cold Rolling of Metals and Alloys

Hot and cold rolling are commonly used methods for fabricating metals and alloys. Long lengths of metal sheet and plate with uniform cross sections can be produced by these processes.

Hot rolling of sheet ingots   Hot rolling of sheet ingots is carried out first since greater reductions in thickness can be taken with each rolling pass when the metal is hot. Before hot rolling, sheet and plate ingots are preheated to a high temperature and soaked in a furnace (Fig. 6.4). After removal from the preheat furnace (soaking pit) the ingots are hot-rolled in a reversing *breakdown* rolling

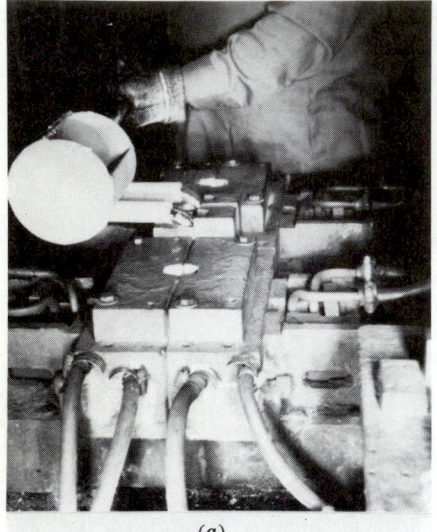

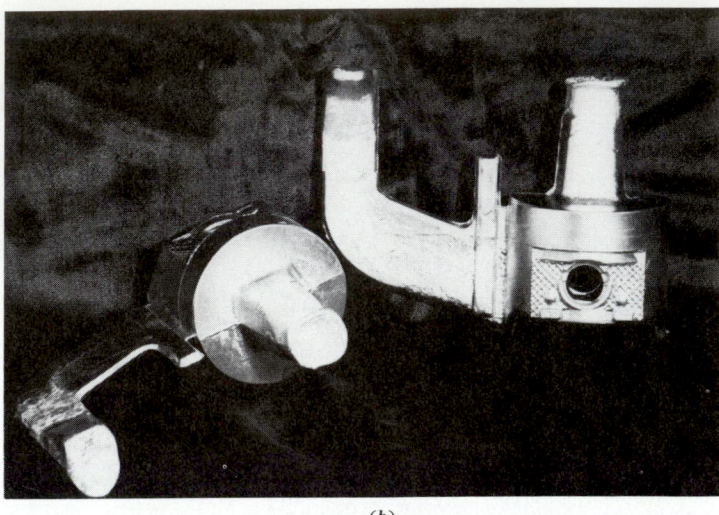

(a)

(b)

(c)

FIGURE 6.3 (*a*) Permanent mold casting of two aluminum alloy pistons simultaneously. (*b*) Aluminum alloy permanent mold piston castings after being removed from the mold shown in (*a*). (*c*) An automobile piston after heat treatment and machining is ready for installation in an engine. (*Courtesy of General Motors Corporation.*)

mill. For example, steel ingots are commonly reduced to slabs by the use of a two-high reversing mill, as illustrated in Fig. 6.5.

Hot rolling is continued until the temperature of the slab drops so low that continued rolling becomes too difficult. The slab is then reheated and hot rolling is continued, usually until the hot-rolled strip is thin enough to be wound into a coil. In most large-scale operations hot rolling of the slab is carried out by using a series of four-high rolling mills alone and in series, as shown for the hot rolling of steel strip in Fig. 6.6.

Cold rolling of metal sheet   After hot rolling, which may also include some cold rolling, the coils of metal are usually given a reheating treatment called *annealing* to soften the metal to remove any cold work introduced during the

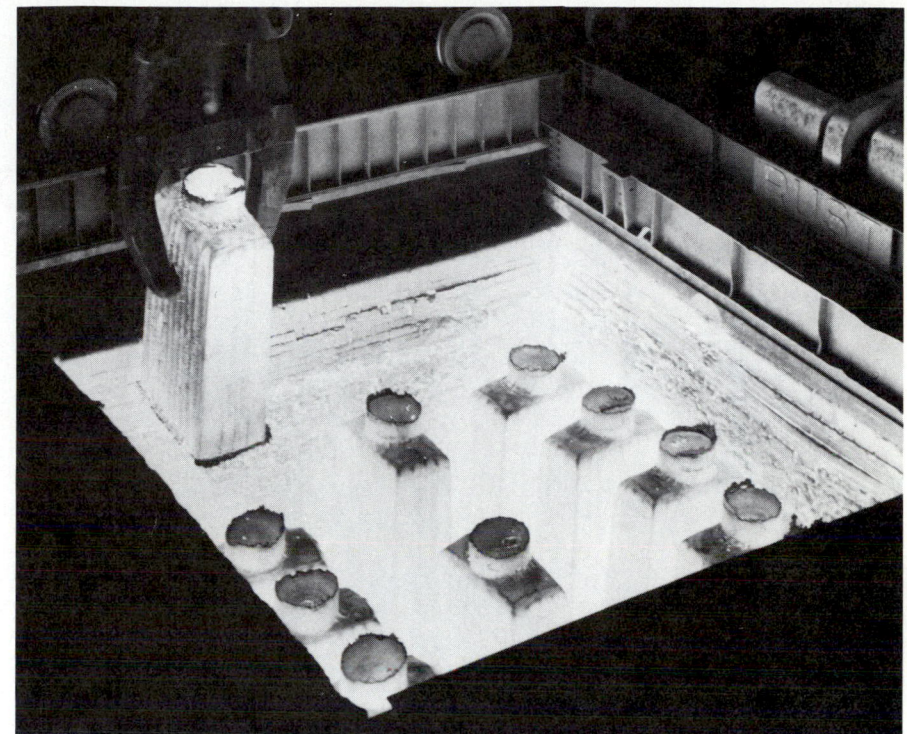

FIGURE 6.4 Before a steel ingot can be hot-rolled, it must be heated to a high temperature. The picture shows steel ingots which have been heated to about 1200°C (2200°F) in a special gas-fired furnace called a *soaking pit*. The ingots are heated for 4 to 8 h to make sure their temperature is uniform throughout. (*Courtesy of Bethlehem Steel Co.*)

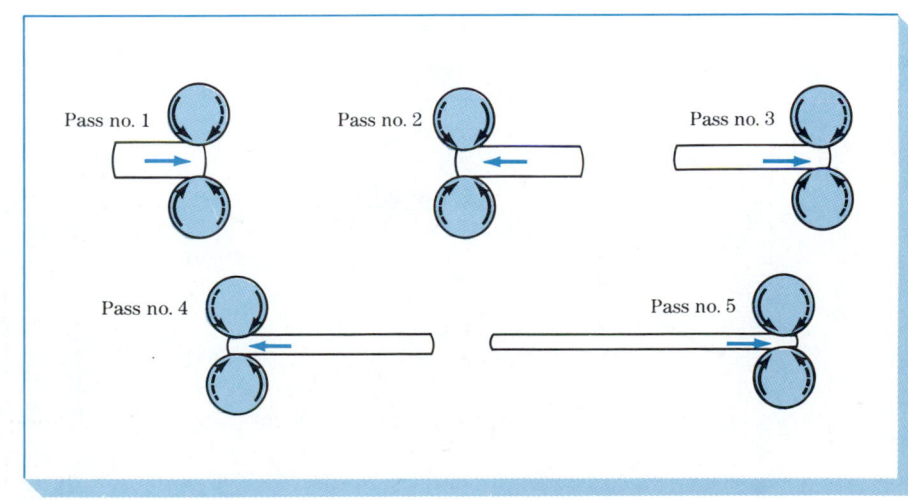

FIGURE 6.5 Diagrammatic representation of the sequence of hot-rolling operations involved in reducing an ingot to a slab on a reversing two-high mill. (*After H. E. McGannon (ed.), "The Making, Shaping, and Treating of Steel," 9th ed., United States Steel, 1971, p. 677.*)

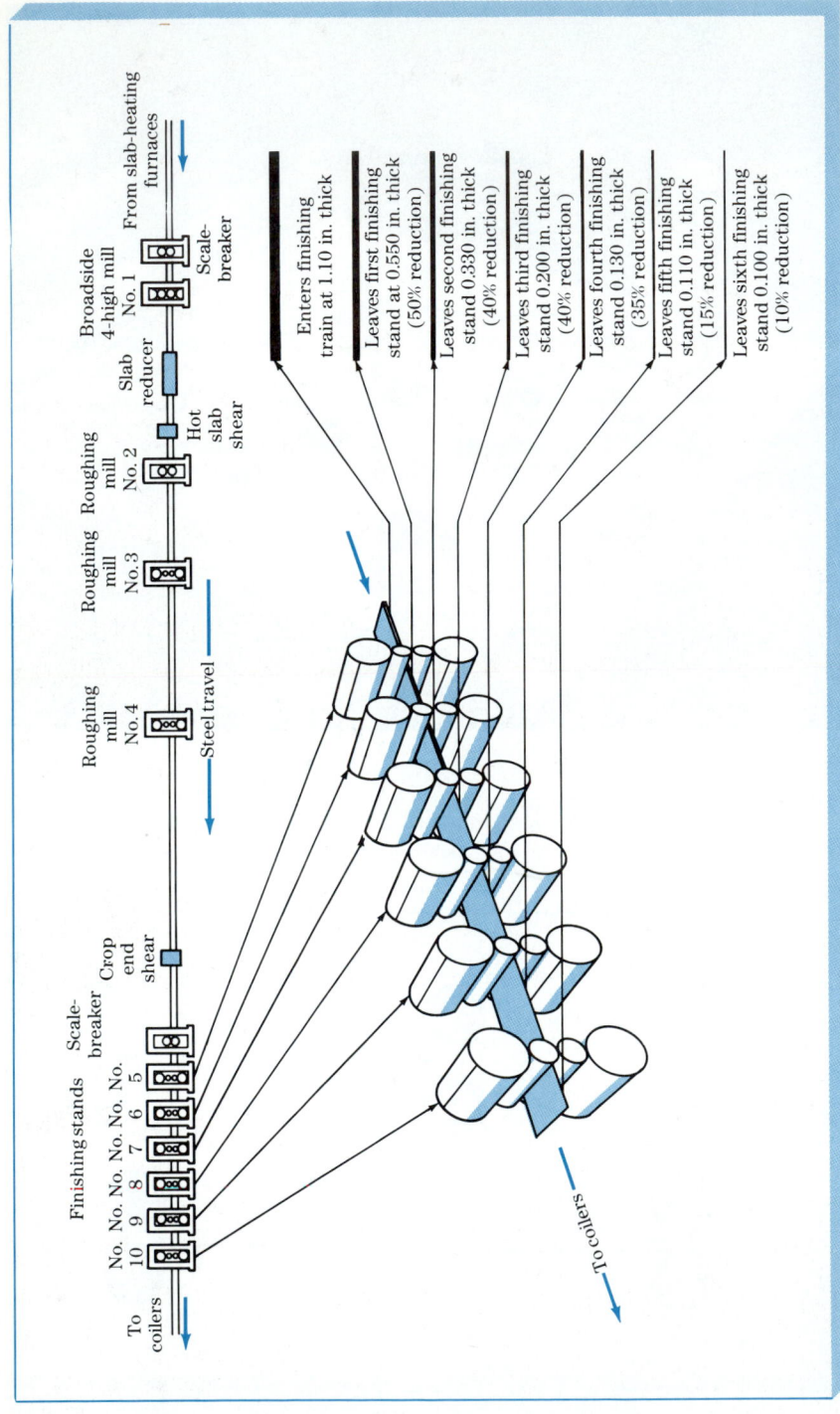

From slab-heating furnaces

Broadside 4-high mill No. 1

Scale-breaker

Slab reducer

Roughing mill No. 2

Hot slab shear

Roughing mill No. 3

Roughing mill No. 4

Steel travel

Finishing stands

Scale-breaker

Crop end shear

No. 10 No. 9 No. 8 No. 7 No. 6 No. 5

To coilers

Enters finishing train at 1.10 in. thick

Leaves first finishing stand at 0.550 in. thick (50% reduction)

Leaves second finishing stand 0.330 in. thick (40% reduction)

Leaves third finishing stand 0.200 in. thick (40% reduction)

Leaves fourth finishing stand 0.130 in. thick (35% reduction)

Leaves fifth finishing stand 0.110 in. thick (15% reduction)

Leaves sixth finishing stand 0.100 in. thick (10% reduction)

To coilers

FIGURE 6.6 Typical reductions per pass in the finishing stands of a hot-strip mill equipped with four roughing stands and six finishing stands. Drawing is not to scale. (*After H. E. McGannon (ed.), "The Making, Shaping, and Treating of Steel," 9th ed., United States Steel, 1971, p. 937.*)

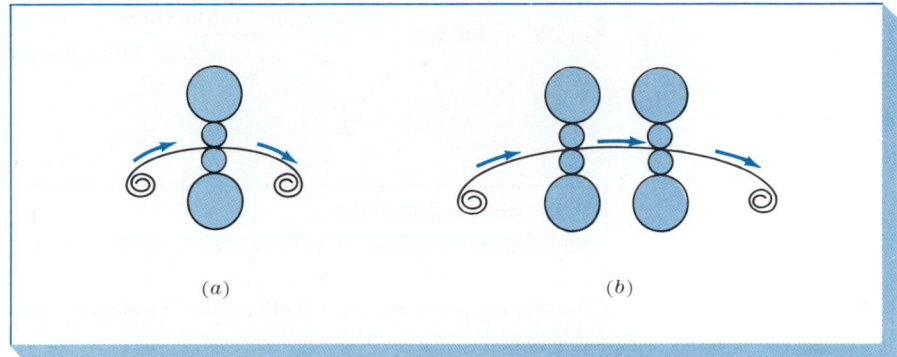

FIGURE 6.7 Schematic drawing illustrating the metal path during the cold rolling of metal sheet by four-high rolling mills: (*a*) single mill; (*b*) two mills in series.

hot-rolling operation. Cold rolling, which normally is done at room temperature, is again usually carried out with four-high rolling mills either alone or in series (Fig. 6.7). Figure 6.8 shows some sheet steel being cold-rolled on an industrial rolling mill.

The percent cold reduction of a plate or sheet of metal can be calculated as follows:

FIGURE 6.8 Cold rolling sheet steel. Mills of this type are used for cold rolling steel strip, tin plate, and nonferrous metals. (*Courtesy of Bethlehem Steel Co.*)

$$\% \text{ cold reduction} = \frac{\text{initial metal thickness} - \text{final metal thickness}}{\text{initial metal thickness}} \times 100\%$$

$$(6.1)$$

Example Problem 6.1

Calculate the percent cold reduction in cold rolling an aluminum sheet alloy from 0.120 to 0.040 in.

Solution:

$$\% \text{ cold reduction} = \frac{\text{initial thickness} - \text{final thickness}}{\text{initial thickness}} \times 100\%$$

$$= \frac{0.120 \text{ in} - 0.040 \text{ in}}{0.120 \text{ in}} \times 100\% = \frac{0.080 \text{ in}}{0.120 \text{ in}} \times 100\%$$

$$= 66.7\%$$

Example Problem 6.2

A sheet of a 70% Cu–30% Zn alloy is cold-rolled 20 percent to a thickness of 3.00 mm. The sheet is then further cold-rolled to 2.00 mm. What is the total percent cold work?

Solution:

We first determine the starting thickness of the sheet by considering the first cold reduction of 20 percent. Let x equal the starting thickness of the sheet. Then,

$$\frac{x - 3.00 \text{ mm}}{x} = 0.20$$

or

$$x - 3.00 \text{ mm} = 0.20x$$
$$x = 3.75 \text{ mm}$$

We can now determine the *total* percent cold work from the starting thickness to the finished thickness from the relationship

$$\frac{3.75 \text{ mm} - 2.00 \text{ mm}}{3.75 \text{ mm}} = \frac{1.75 \text{ mm}}{3.75 \text{ mm}} = 0.466 \text{ or } 46.6\%$$

Extrusion of Metals and Alloys

Extrusion is a plastic forming process in which a material under high pressure is reduced in cross section by forcing it through an opening in a die (Fig. 6.9). For most metals the extrusion process is used to produce cylindrical bars or hollow tubes. For the more readily extrudable metals, such as aluminum and copper and some of their alloys, shapes with irregular cross sections are also commonly produced. Most metals are extruded hot since the deformation resistance of the metal is lower than if it is extruded cold. During extrusion the metal of a billet in the container of an extrusion press is forced by a ram through a die so that the metal is continuously deformed into a long length of metal with a uniform desired cross section.

The two main types of extrusion processes are *direct extrusion* and *indirect extrusion.* In direct extrusion the metal billet is placed in a container of an extrusion press and forced directly through the die by the ram (Fig. 6.9a). In indirect extrusion a hollow ram holds the die, with the other end of the container of the extrusion press being closed by a plate (Fig. 6.9b). The frictional forces and power requirements for indirect extrusion are lower than those for direct extrusion. However, the loads that can be applied by using a hollow ram in the indirect process are more limited than those that can be used for direct extrusion.

The extrusion process is used primarily for producing bar shapes, tube, and irregular shapes of the lower-melting nonferrous metals such as aluminum and copper and their alloys. However, with the development of powerful extrusion presses and improved lubricants such as glass, some carbon and stainless steels can also be hot-extruded.

FIGURE 6.9 Two basic types of extrusion processes for metals: (*a*) direct and (*b*) indirect. (*After G. Dieter, "Mechanical Metallurgy," 2d ed., McGraw-Hill, 1976, p. 639.*)

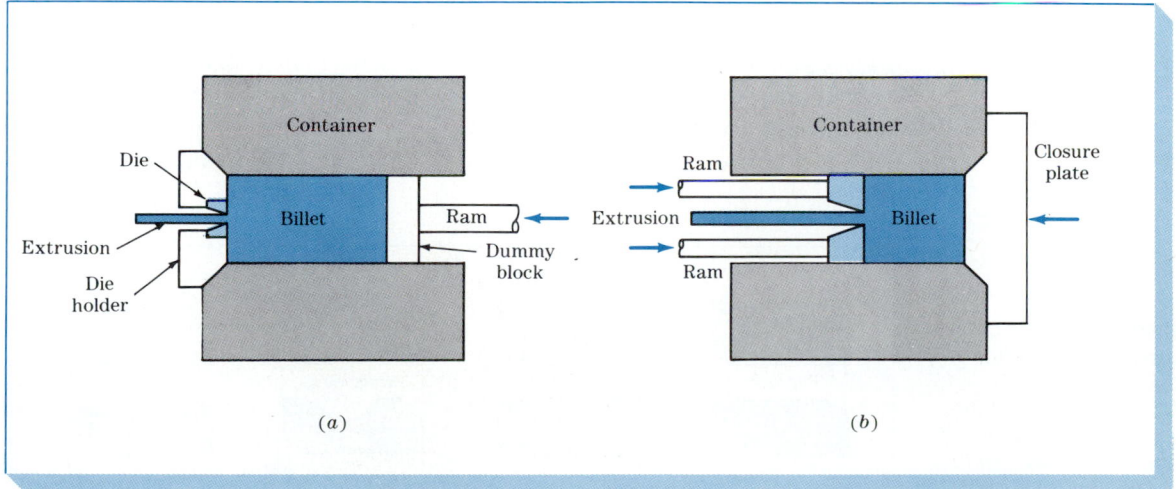

(*a*)   (*b*)

FIGURE 6.10 Heavy-duty manipulator holding an ingot in position while a 10,000-ton press squeezes the hot steel into the rough shape of the finished product. (*After H. E. McGannon (ed.), "The Making, Shaping, and Treating of Steel," 9th ed., United States Steel, 1971, p. 1044.*)

Forging

Forging is another primary method for working metals into useful shapes. In the forging process the metal is hammered or pressed into a desired shape. Most forging operations are carried out with the metal in the hot condition, although in some cases the metal may be forged cold. There are two major types of forging methods: *hammer* and *press forging.* In hammer forging a drop hammer repeatedly exerts a striking force against the surface of the metal. In press forging the metal is subjected to a slowly moving compressive force (Fig. 6.10).

Forging processes can also be classified as *open-die forging* or *closed-die forging.* Open-die forging is carried out between two flat dies or dies with very simple shapes such as vees or semicircular cavities (Fig. 6.11) and is particularly useful for producing large parts such as steel shafts for electric steam turbines and generators. In closed-die forging the metal to be forged is

FIGURE 6.11 Basic shapes for open-die forging. (*After H. E. McGannon (ed.), "The Making, Shaping, and Treating of Steel," 9th ed., United States Steel, 1971, p. 1045.*)

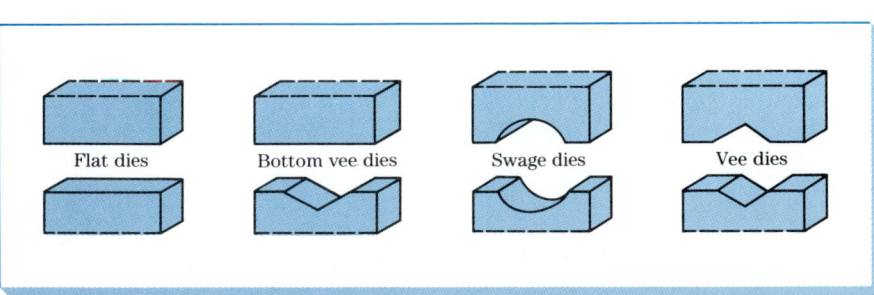

Flat dies   Bottom vee dies   Swage dies   Vee dies

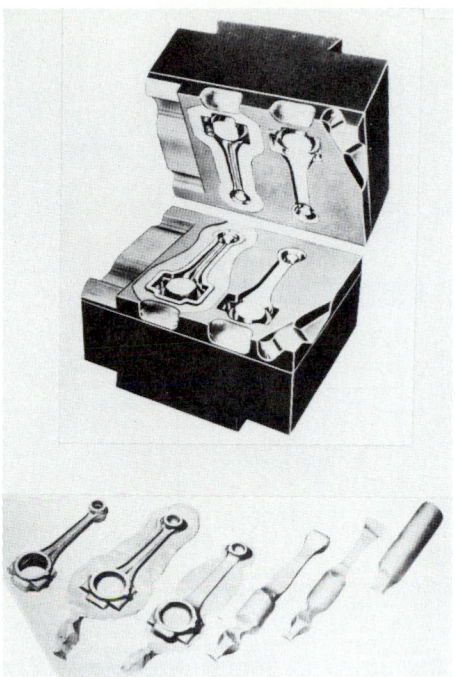

FIGURE 6.12 A set of closed forging dies used to produce an automobile connecting rod. (*Courtesy of the Forging Industry Association.*)

placed between two dies which have the upper and lower impressions of the desired shape of the forging. Closed-die forging can be carried out by using a single pair of dies or with multiple-impression dies. An example of a closed-die forging in which multiple-impression dies are used is the automobile engine connecting rod (Fig. 6.12).

In general the forging process is used for producing irregular shapes which require working to improve the structure of the metal by reducing porosity and refining the internal structure. For example, a wrench that has been forged will be tougher and less likely to break than one that is just simply cast into shape. Forging is also sometimes used to break down the as-cast ingot structure of some highly alloyed metals (e.g., some tool steels) so that the metal is made more homogeneous and less likely to crack during subsequent working.

Other Metal Forming Processes

There are many types of secondary metal forming processes whose descriptions are beyond the scope of this book. However, two of these processes, *wire drawing* and *deep drawing* of sheet metal, will be briefly described.

Wire drawing is an important metal forming process. Starting rod or wire stock is drawn through one or more tapered wire-drawing dies (Fig. 6.13). For steel wire drawing, a tungsten carbide inner "nib" is inserted inside a steel casing. The hard carbide provides a wear-resistant surface for the reduction of the steel wire. Special precautions must be taken to make sure the surface of the stock to be drawn into wire is clean and properly lubricated. Intermediate

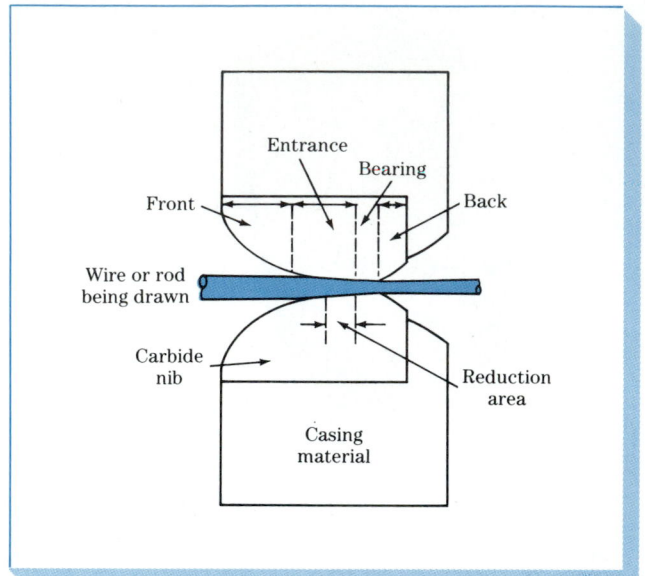

FIGURE 6.13 Section through a wire-drawing die. (After "Wire and Rods, Alloy Steel," Steel Products Manual, American Iron and Steel Institute, 1975.)

softening heat treatments are sometimes necessary when the drawn wire work hardens during processing. The procedures used vary considerably, depending on the metal or alloy being drawn and the final diameter and temper desired.

Example Problem 6.3

Calculate the percent cold reduction when an annealed copper wire is cold-drawn from a diameter of 1.27 mm (0.050 in) to a diameter of 0.813 mm (0.032 in).

Solution:

$$\% \text{ cold reduction} = \frac{\text{change in cross-sectional area}}{\text{original area}} \times 100\%$$

$$= \frac{(\pi/4)(1.27 \text{ mm})^2 - (\pi/4)(0.813 \text{ mm})^2}{(\pi/4)(1.27 \text{ mm})^2} \times 100\% \quad (6.2)$$

$$= \left[1 - \frac{(0.813)^2}{(1.27)^2} \right](100\%)$$

$$= (1 - 0.41)(100\%) = 59\% \blacktriangleleft$$

Deep drawing is another metal forming process and is used for shaping flat sheets of metal into cup-shaped articles. A metal blank is placed over a

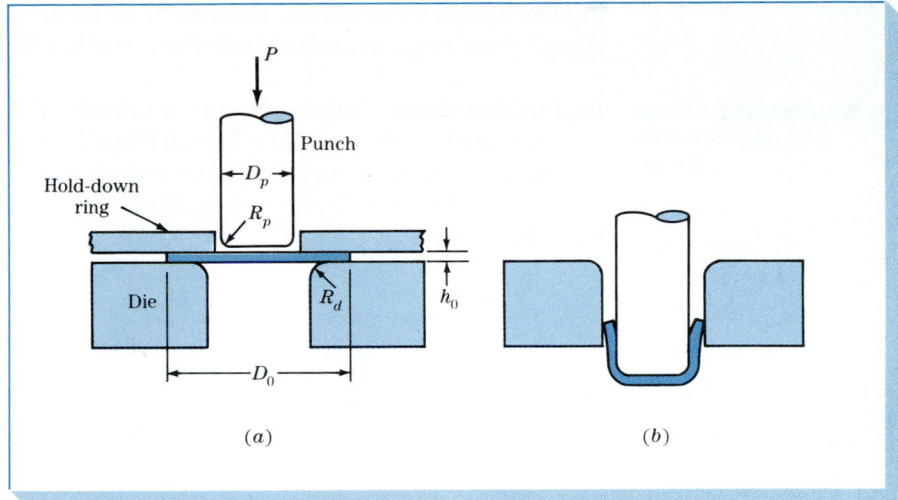

FIGURE 6.14 Deep drawing of a cylindrical cup (a) before drawing and (b) after drawing. (After G. Dieter, "Mechanical Metallurgy," 2d ed., McGraw-Hill, 1976, p. 688.)

shaped die and then is pressed into the die with a punch (Fig. 6.14). Usually a hold-down device is used to allow the metal to be pressed smoothly into the die to prevent wrinkling of the metal.

6.2 STRESS AND STRAIN IN METALS

In the first section of this chapter we briefly examined most of the principal methods by which metals are processed into semifinished wrought and cast products. Let us now investigate how the mechanical properties of strength and ductility are evaluated for engineering applications.

Elastic and Plastic Deformation

When a piece of metal is subjected to a uniaxial tensile force, deformation of the metal occurs. If the metal returns to its original dimensions when the force is removed, the metal is said to have undergone *elastic deformation*. The amount of elastic deformation a metal can undergo is small, since during elastic deformation the metal atoms are displaced from their original positions but not to the extent that they take up new positions. Thus, when the force on a metal that has been elastically deformed is removed, the metal atoms return to their original positions and the metal takes back its original shape. If the metal is deformed to such an extent that it cannot fully recover its original dimensions, it is said to have undergone *plastic deformation*. During plastic deformation, the metal atoms are *permanently* displaced from their original positions and take up new positions. The ability of some metals to be extensively plastically deformed without fracture is one of the most useful engineering properties of metals. For example, the extensive plastic deformability of steel enables automobile parts such as fenders, hoods, and doors to be stamped out mechanically without the metal fracturing.

Engineering Stress and Engineering Strain

Engineering stress   Let us consider a cylindrical rod of length l_0 and cross-sectional area A_0 subjected to a uniaxial tensile force F, as shown in Fig. 6.15. By definition, the *engineering stress* σ on the bar is equal to the average uniaxial tensile force F on the bar divided by the original cross-sectional area A_0 of the bar. Thus

$$\text{Engineering stress } \sigma = \frac{F \text{ (average uniaxial tensile force)}}{A_0 \text{ (original cross-sectional area)}} \tag{6.3}$$

The units for engineering stress are:

U.S. customary: pounds force per square inch (lb_f/in^2, or psi); lb_f = pounds force

SI: newtons per square meter (N/m^2) or pascals (Pa), where $1 \ N/m^2 = 1$ Pa

The conversion factors for psi to pascals are

$$1 \text{ psi} = 6.89 \times 10^3 \text{ Pa}$$
$$10^6 \text{ Pa} = 1 \text{ megapascal} = 1 \text{ MPa}$$
$$1000 \text{ psi} = 1 \text{ ksi} = 6.89 \text{ MPa}$$

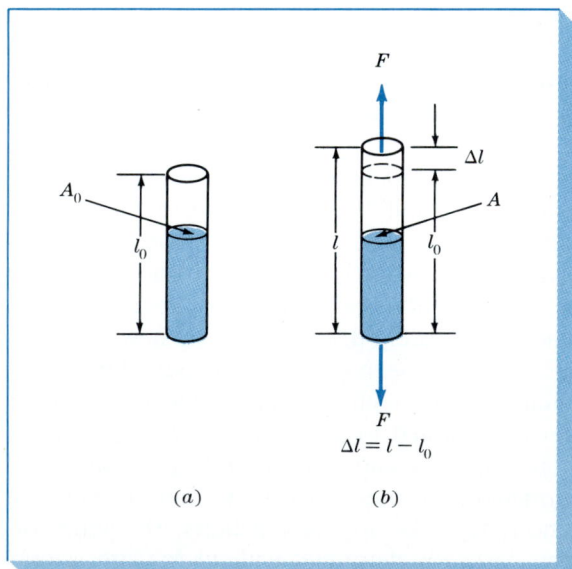

FIGURE 6.15 Elongation of a cylindrical metal rod subjected to a uniaxial tensile force F. (*a*) The rod with no force on it; (*b*) the rod subjected to a uniaxial tensile force F which elongates the rod from length l_0 to l.

Example Problem 6.4

A 0.500-in-diameter aluminum bar is subjected to a force of 2500 lb$_f$. Calculate the engineering stress in pounds per square inch (psi) on the bar.

Solution:

$$\sigma = \frac{\text{force}}{\text{original cross-sectional area}} = \frac{F}{A_0}$$

$$= \frac{2500 \text{ lb}_f}{(\pi/4)(0.500 \text{ in})^2} = 12{,}700 \text{ lb}_f/\text{in}^2 \blacktriangleleft$$

Example Problem 6.5

A 1.25-cm-diameter bar is subjected to a load of 2500 kg. Calculate the engineering stress on the bar in megapascals (MPa).

Solution:

The load on the bar has a mass of 2500 kg. In SI units the force on the bar is equal to the mass of the load times the acceleration of gravity (9.81 m/s^2), or

$$F = ma = (2500 \text{ kg})(9.81 \text{ m/s}^2) = 24{,}500 \text{ N}$$

The diameter d of the bar = 1.25 cm = 0.0125 m. Thus the engineering stress on the bar is

$$\sigma = \frac{F}{A_0} = \frac{F}{(\pi/4)(d^2)} = \frac{24{,}500 \text{ N}}{(\pi/4)(0.0125 \text{ m})^2}$$

$$= (2.00 \times 10^8 \text{ Pa})\left(\frac{1 \text{ MPa}}{10^6 \text{ Pa}}\right) = 200 \text{ MPa} \blacktriangleleft$$

Engineering strain   When a uniaxial tensile force is applied to a rod, such as shown in Fig. 6.15, it causes the rod to be elongated in the direction of the force. Such a displacement is called *strain*. By definition, *engineering strain,* which is caused by the action of a uniaxial tensile force on a metal sample, is the ratio of the change in length of the sample in the direction of the force divided by the original length of sample considered. Thus the engineering strain for the metal bar shown in Fig. 6.15 (or for a similar-type metal sample) is

$$\text{Engineering strain } \epsilon = \frac{l - l_0}{l_0} = \frac{\Delta l \text{ (change in length of sample)}}{l_0 \text{ (original length of sample)}} \quad (6.4)$$

where l_0 = original length of sample and l = new length of sample after being extended by a uniaxial tensile force. In most cases engineering strain is determined by using a small length, usually 2 in, called the *gage length*, within a much longer, for example, 8 in, sample (see Example Problem 6.6).

The *units for engineering strain* ϵ are:

U.S. customary: inches per inch (in/in)

SI: meters per meter (m/m)

Thus engineering strain has *dimensionless units*. In industrial practice it is common to convert engineering strain into *percent strain* or *percent elongation*:

% engineering strain = engineering strain × 100% = % elongation

---

Example Problem 6.6

A sample of commercially pure aluminum 0.500 in wide, 0.040 in thick, and 8 in long which has gage markings 2.00 in apart in the middle of the sample is strained so that the gage markings are 2.65 in apart (Fig. 6.16). Calculate the engineering strain and the percent engineering strain elongation which the sample undergoes.

Solution:

$$\text{Engineering strain } \epsilon = \frac{l - l_0}{l_0} = \frac{2.65 \text{ in} - 2.00 \text{ in}}{2.00 \text{ in}} = \frac{0.65 \text{ in}}{2.00 \text{ in}} = 0.325 \blacktriangleleft$$

$$\% \text{ elongation} = 0.325 \times 100\% = 32.5\% \blacktriangleleft$$

Poisson's Ratio A longitudinal elastic deformation of a metal produces an accompanying lateral dimensional change. As shown in Fig. 6.17b, a tensile stress σ_z produces an axial strain $+\epsilon_z$ and lateral contractions of $-\epsilon_x$ and $-\epsilon_y$. For isotropic[1] behavior, ϵ_x and ϵ_y are equal. The ratio

$$\nu = -\frac{\epsilon \text{ (lateral)}}{\epsilon \text{ (longitudinal)}} = -\frac{\epsilon_x}{\epsilon_z} = -\frac{\epsilon_y}{\epsilon_z} \quad (6.5)$$

---

[1] *Isotropic:* exhibiting properties with the same values when measured along axes in all directions.

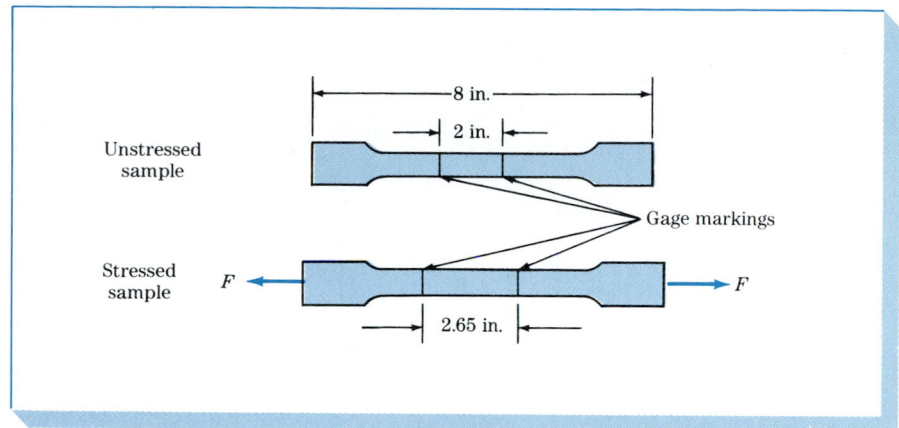

FIGURE 6.16 Flat tensile specimen before and after testing.

is called *Poisson's ratio*. For ideal materials, $\nu = 0.5$. However, for real materials, Poisson's ratio typically ranges from 0.25 to 0.4, with an average of about 0.3. Table 6.1 lists ν values for some metals and alloys.

Shear Stress and Shear Strain

Until now we have discussed the elastic and plastic deformation of metals and alloys under uniaxial tension stresses. Another important method by which a metal can be deformed is under the action of a *shear stress*. The action of a simple shear stress couple (shear stresses act in pairs) on a cubic body is

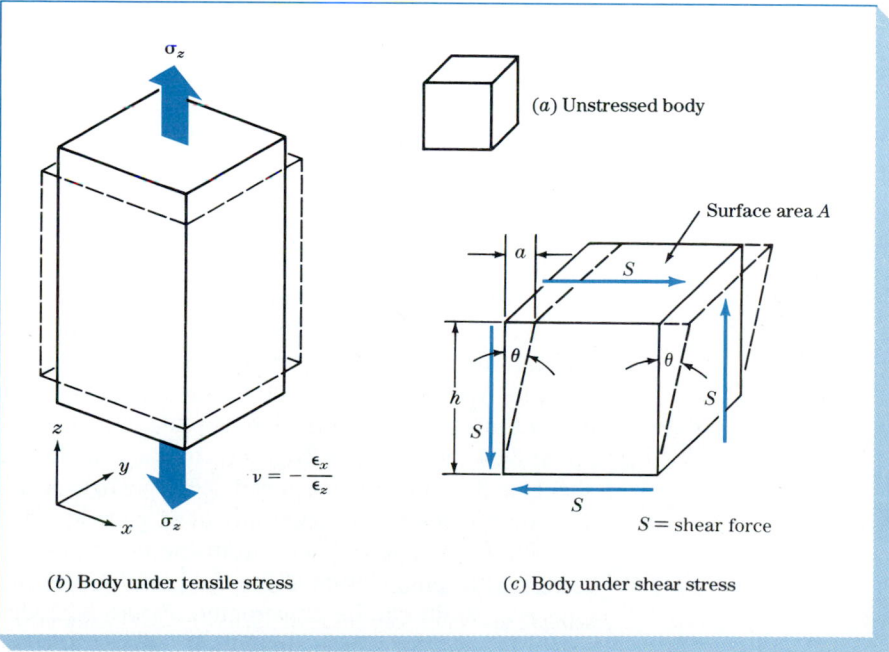

FIGURE 6.17 (*a*) Unstressed cubic body. (*b*) Cubic body subjected to tensile stress. The ratio of the elastic contraction perpendicular to the extension is designated Poisson's ratio ν. (*c*) Cubic body subjected to pure shear forces S acting over surface areas A. The shear stress τ acting on the body is equal to S/A.

(*b*) Body under tensile stress

(*c*) Body under shear stress

shown in Fig. 6.17c, where a shearing force S acts over an area A. The shear stress τ is related to the shear force S by

$$\tau \text{ (shear stress)} = \frac{S \text{ (shear force)}}{A \text{ (area over which shear force acts)}} \qquad (6.6)$$

The units for shear stress are the same as for uniaxial tensile stress:

U.S. customary: pounds force per square inch (lb_f/in^2, or psi)

SI: newtons per square meter (N/m^2) or pascals (Pa)

The shear strain γ is defined in terms of the amount of the shear displacement a in Fig. 6.17c divided by the distance h over which the shear acts, or

$$\gamma = \frac{a}{h} = \tan \theta \qquad (6.7)$$

For pure elastic shear, the proportionality between shear and stress is

$$\tau = G\gamma \qquad (6.8)$$

where G is the elastic shear modulus.

We will be concerned with shear stresses when discussing the plastic deformation of metals in Sec. 6.5.

6.3 THE TENSILE TEST AND THE ENGINEERING STRESS-STRAIN DIAGRAM

The *tensile test* is used to evaluate the strength of metals and alloys. In this test a metal sample is pulled to failure in a relatively short time at a constant rate. Figure 6.18 is a picture of a modern tensile testing machine, and Fig. 6.19 illustrates schematically how the sample is tested in tension.

The force (load) on the sample being tested is plotted by the instrument on moving chart graph paper, while the corresponding strain can be obtained from a signal from an external extensometer attached to the sample (Fig. 6.20) and also recorded on the chart paper.

The type of samples used for the tensile test vary considerably. For metals with a thick cross section such as plate, a 0.50-in-diameter round specimen is commonly used (Fig. 6.21a). For metal with thinner cross sections such as sheet, a flat specimen is used (Fig. 6.21b). A 2-in gage length within the specimen is the most commonly used gage length for tensile tests.

The force data obtained from the chart paper for the tensile test can be converted to engineering stress data, and a plot of engineering stress vs. engineering strain can be constructed. Figure 6.22 shows an engineering stress-strain diagram for a high-strength aluminum alloy.

FIGURE 6.18 Tensile testing machine. The force (load) on the sample is recorded on the chart paper in the drawer on the left. The strain which the sample undergoes is also recorded on the chart. The signal for the strain is obtained from the extensometer attached to the sample. (*Courtesy of the Instron Corporation.*)

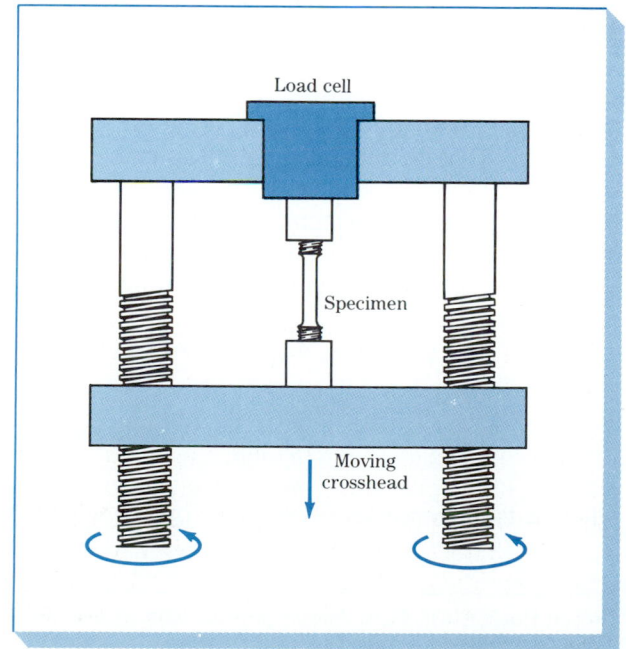

FIGURE 6.19 Schematic illustration showing how the tensile machine of Fig. 6.18 operates. Note, however, that the crosshead of the machine in Fig. 6.18 moves up. (*After H. W. Hayden, W. G. Moffatt, and John Wulff, "The Structure and Properties of Materials," vol. 3: "Mechanical Behavior," Wiley, 1965, Fig. 1.1, p. 2.*)

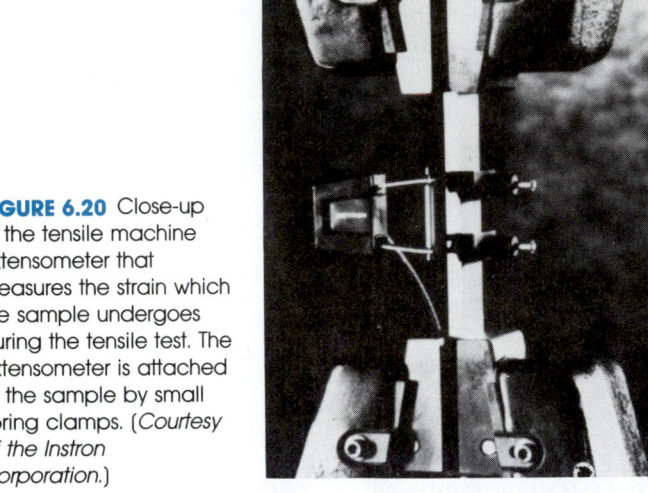

FIGURE 6.20 Close-up of the tensile machine extensometer that measures the strain which the sample undergoes during the tensile test. The extensometer is attached to the sample by small spring clamps. (*Courtesy of the Instron Corporation.*)

Mechanical Property Data Obtained from the Tensile Test and the Engineering Stress-Strain Diagram

The mechanical properties of metals and alloys which are of engineering importance for structural design and which can be obtained from the engineering tensile test are:

1. Modulus of elasticity
2. Yield strength at 0.2 percent offset
3. Ultimate tensile strength
4. Percent elongation at fracture
5. Percent reduction in area at fracture

Modulus of elasticity In the first part of the tensile test the metal is deformed elastically. That is, if the load on the specimen is released, the specimen will return to its original length. For metals the maximum elastic deformation is usually less than 0.5 percent. In general metals and alloys show a linear relationship between stress and strain in the elastic region of the engineering stress-strain diagram which is described by Hooke's law:[1]

$$\sigma \ (\text{stress}) = E\epsilon \ (\text{strain}) \tag{6.9}$$

or

$$E = \frac{\sigma \ (\text{stress})}{\epsilon \ (\text{strain})} \quad (\text{units of psi or Pa})$$

where E is the *modulus of elasticity,* or *Young's modulus.*[2]

---

[1] Robert Hooke (1635–1703). English physicist who studied the elastic behavior of solids.
[2] Thomas Young (1773–1829). English physicist.

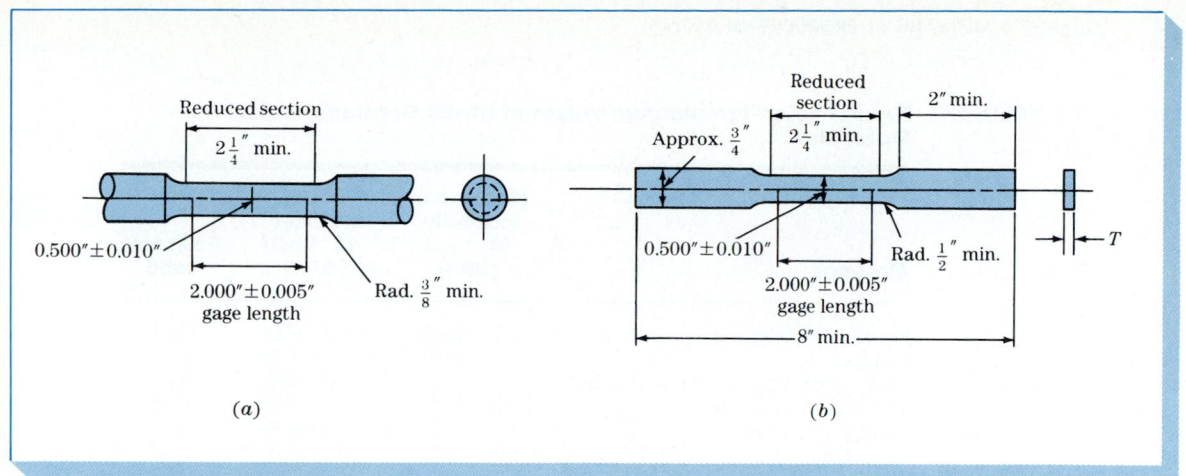

FIGURE 6.21 Examples of the geometrical shape of commonly used tension test specimens. (*a*) Standard round tension test specimen with 2-in gage length. (*b*) Standard rectangular tension test specimen with 2-in gage length. (*Based on H. E. McGannon (ed.), ASTM Standards, 1968, "The Making, Shaping, and Treating of Steel," 9th ed., United States Steel, 1971, p. 1220.*)

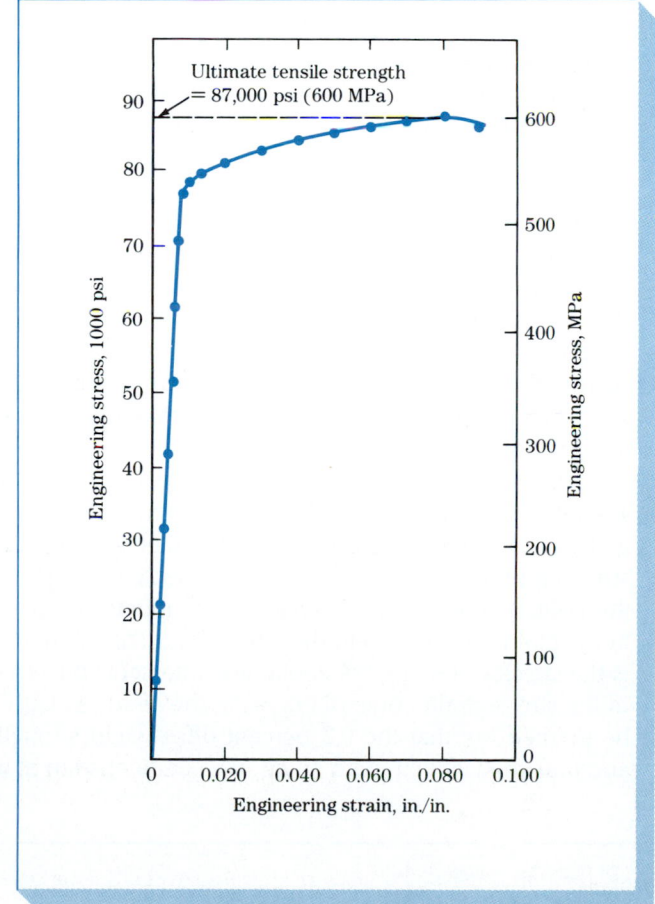

FIGURE 6.22 Engineering stress-strain diagram for a high-strength aluminum alloy (7075-T6). The specimens for the diagram were taken from ⅝-in plate and had a 0.50-in diameter with a 2-in gage length. (*Courtesy of Aluminum Company of America.*)

TABLE 6.1 Typical Room-Temperature Values of Elastic Constants for Isotropic Materials

Material	Modulus of elasticity, 10^{-6} psi (GPa)	Shear modulus, 10^{-6} psi (GPa)	Poisson's ratio
Aluminum alloys	10.5 (72.4)	4.0 (27.5)	0.31
Copper	16.0 (110)	6.0 (41.4)	0.33
Steel (plain carbon and low-alloy)	29.0 (200)	11.0 (75.8)	0.33
Stainless steel (18-8)	28.0 (193)	9.5 (65.6)	0.28
Titanium	17.0 (117)	6.5 (44.8)	0.31
Tungsten	58.0 (400)	22.8 (157)	0.27

Source: G. Dieter, "Mechanical Metallurgy," 3d ed., McGraw-Hill, 1986.

The modulus of elasticity is related to the bonding strength between the atoms in a metal or alloy. Table 6.1 lists the elastic moduli for some common metals. Metals with high elastic moduli are relatively stiff and do not deflect easily. Steels, for example, have high elastic moduli values of 30×10^6 psi (207 GPa),[1] whereas aluminum alloys have lower elastic moduli of about 10 to 11×10^6 psi (69 to 76 GPa). Note that in the elastic region of the stress-strain diagram, the modulus does not change with increasing stress.

Yield strength   The yield strength is a very important value for use in engineering structural design since it is the strength at which a metal or alloy shows significant plastic deformation. Because there is no definite point on the stress-strain curve where elastic strain ends and plastic strain begins, the yield strength is chosen to be that strength when a definite amount of plastic strain has occurred. For American engineering structural design, the yield strength is chosen when 0.2 percent plastic strain has taken place, as indicated on the engineering stress-strain diagram of Fig. 6.23.

The 0.2 percent yield strength, also called the *0.2 percent offset yield strength,* is determined from the engineering stress-strain diagram, as shown in Fig. 6.23. First, a line is drawn parallel to the elastic (linear) part of the stress-strain plot at 0.002 in/in (m/m) strain, as indicated on Fig. 6.23. Then at the point where this line intersects the upper part of the stress-strain curve, a horizontal line is drawn to the stress axis. The 0.2 percent offset yield strength is the stress where the horizontal line intersects the stress axis, and in the case of the stress-strain curve of Fig. 6.23, the yield strength is 78,000 psi. It should be pointed out that the 0.2 percent offset yield strength is arbitrarily chosen, and thus the yield strength could have been chosen at any other small amount

[1] SI prefix G = giga = 10^9.

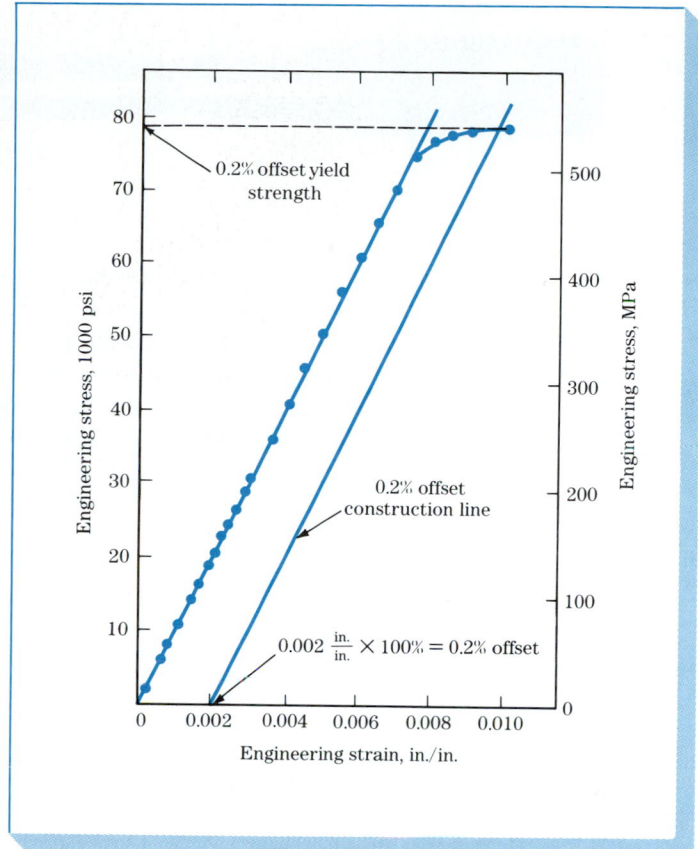

FIGURE 6.23 Linear part of engineering stress-strain diagram of Fig. 6.22 expanded on the strain axis to make a more accurate determination of the 0.2 percent offset yield stress. (*Courtesy of Aluminum Company of America.*)

of permanent deformation. For example, a 0.1 percent offset yield strength is commonly used in the United Kingdom.

Ultimate tensile strength   The ultimate tensile strength is the maximum strength reached in the engineering stress-strain curve. If the specimen develops a localized decrease in cross-sectional area (commonly called *necking*) (Fig. 6.24), the engineering stress will decrease with further strain until fracture occurs since the engineering stress is determined by using the *original* cross-sectional area of the specimen. The more ductile a metal is, the more the specimen will neck before fracture and hence the more the decrease in the stress on the stress-strain curve beyond the maximum stress. For the high-strength aluminum alloy whose stress-strain curve is shown in Fig. 6.22, there is only a small decrease in stress beyond the maximum stress because this material has relatively low ductility.

An important point to understand with respect to engineering stress-strain diagrams is that the metal or alloy continues to increase in stress up to the stress at fracture. It is only because we use the original cross-sectional area

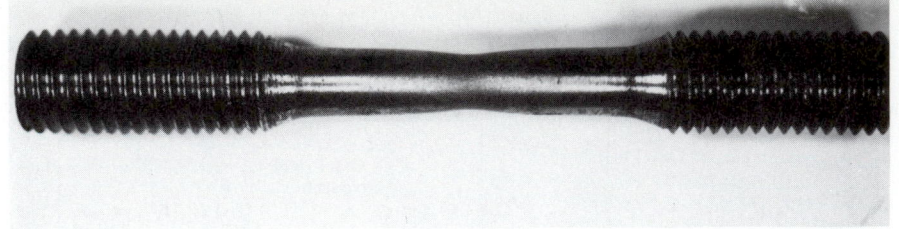

FIGURE 6.24 Necking in a mild-steel round specimen. The specimen was originally uniformly cylindrical. After being subjected to uniaxial tension forces up to almost fracture, the specimen decreased in cross section, or "necked" in the middle.

to determine engineering stress that the stress on the engineering stress-strain diagram decreases at the latter part of the test.

The ultimate tensile strength of a metal is determined by drawing a horizontal line from the maximum point on the stress-strain curve to the stress axis. The stress where this line intersects the stress axis is called the *ultimate tensile strength,* or sometimes just the *tensile strength.* For the aluminum alloy of Fig. 6.22, the ultimate tensile strength is 87,000 psi.

The ultimate tensile strength is not used much in engineering design for ductile alloys since too much plastic deformation takes place before it is reached. However, the ultimate tensile strength can give some indication of the presence of defects. If the metal contains porosity or inclusions, these defects may cause the ultimate tensile strength of the metal to be lower than normal.

Percent elongation The amount of elongation that a tensile specimen undergoes during testing provides a value for the ductility of a metal. Ductility of metals is most commonly expressed as percent elongation, starting with a gage length usually of 2 in (5.1 cm) (Fig. 6.21). In general the higher the ductility (the more deformable the metal is), the higher the percent elongation is. For example, a sheet of 0.062-in (1.6-mm) commercially pure aluminum (alloy 1100-0) in the soft condition has a high percent elongation of 35 percent, whereas the same thickness of the high-strength aluminum alloy 7075-T6 in the fully hard condition has a percent elongation of only 11 percent.

As previously mentioned, during the tensile test an extensometer can be used to continuously measure the strain of the specimen being tested. However, the percent elongation of a specimen after fracture can be measured by fitting the fractured specimen together and measuring the final elongation with calipers. The percent elongation can then be calculated from the equation

$$\% \text{ elongation} = \frac{\text{final length*} - \text{initial length*}}{\text{initial length}} \times 100\%$$

$$= \frac{l - l_0}{l_0} \times 100\% \tag{6.10}$$

---

*The initial length is the length between the gage marks on the specimen before testing. The final length is the length between these same gage marks after testing when the fractured surface of the specimen is fitted together (see Example Problem 6.6).

The percent elongation at fracture is of engineering importance not only as a measure of ductility but also as an index of the quality of the metal. If porosity or inclusions are present in the metal or if damage due to overheating the metal has occurred, the percent elongation of the specimen tested may be decreased below normal.

Percent reduction in area   The ductility of a metal or alloy can also be expressed in terms of the percent reduction in area. This quantity is usually obtained from a tensile test using a specimen 0.50 in (12.7 mm) in diameter. After the test, the diameter of the reduced cross section at the fracture is measured. Using the measurements of the initial and final diameters, the percent reduction in area can be determined from the equation

$$\% \text{ reduction in area} = \frac{\text{initial area} - \text{final area}}{\text{initial area}} \times 100\%$$

$$= \frac{A_0 - A_f}{A_0} \times 100\% \tag{6.11}$$

The percent reduction in area, like the percent elongation, is a measure of the ductility of the metal and is also an index of quality. The percent reduction in area may be decreased if defects such as inclusions and/or porosity are present in the metal specimen.

Example Problem 6.7

A 0.500-in-diameter round sample of a 1030 carbon steel is pulled to failure in a tensile testing machine. The diameter of the sample was 0.343 in at the fracture surface. Calculate the percent reduction in area of the sample.

Solution:

$$\% \text{ reduction in area} = \frac{A_0 - A_f}{A_0} \times 100\% = \left(1 - \frac{A_f}{A_0}\right)(100\%)$$

$$= \left[1 - \frac{(\pi/4)(0.343 \text{ in})^2}{(\pi/4)(0.500 \text{ in})^2}\right](100\%)$$

$$= (1 - 0.47)(100\%) = 53\% \blacktriangleleft$$

Comparison of Engineering Stress-Strain Curves for Selected Alloys   Engineering stress-strain curves for selected metals and alloys are shown in Fig. 6.25. Alloying a metal with other metals or nonmetals and heat treatment can greatly affect the tensile strength and ductility of metals. The stress-strain curves of Fig. 6.25 show a great variation in ultimate tensile strength (UTS). Elemental magnesium has a UTS of 35 ksi (1 ksi = 1000 psi), whereas SAE 1340 steel water-quenched and tempered at 700°F (370°C) has a UTS of 240 ksi.

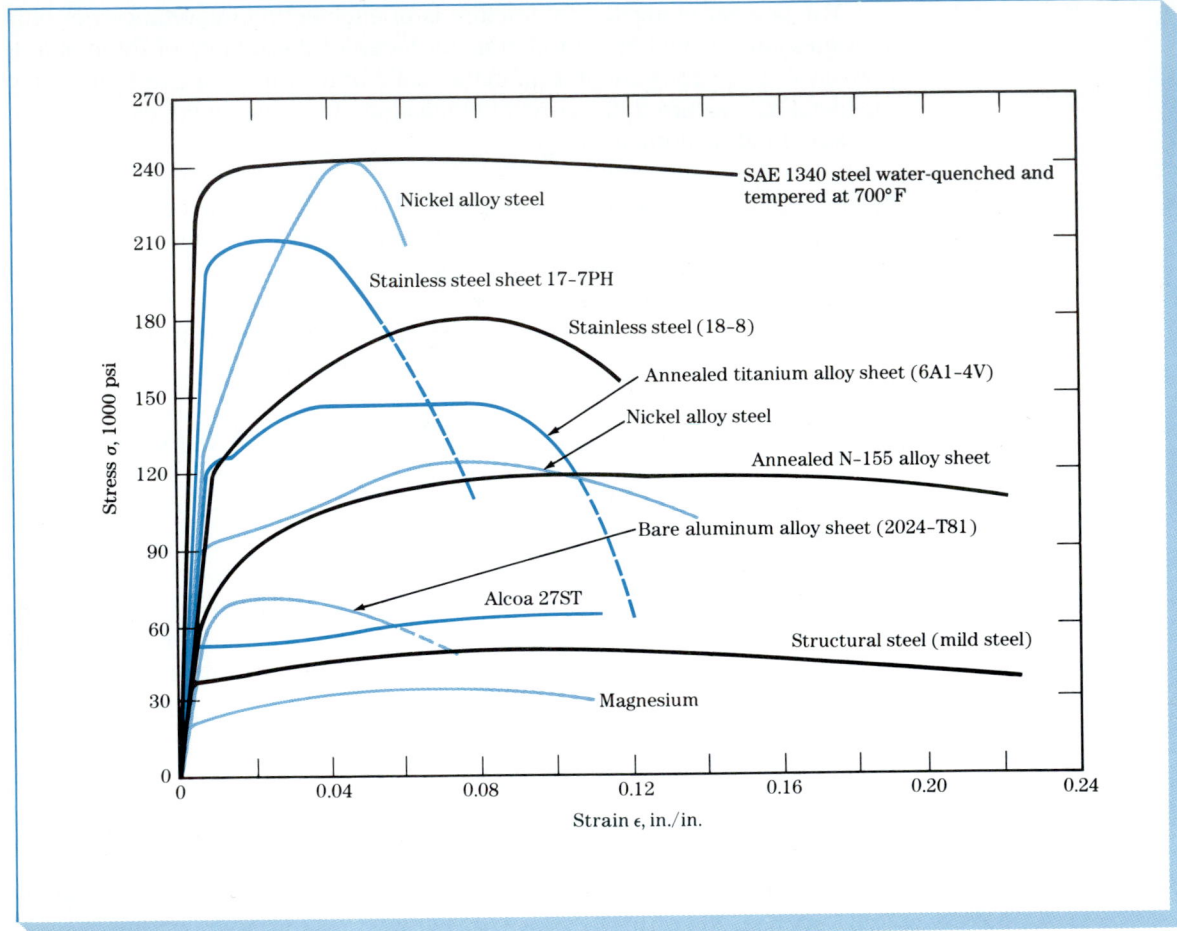

FIGURE 6.25 Engineering stress-strain curves for selected metals and alloys. (*After J. Marin, "Mechanical Behavior of Engineering Materials," Prentice-Hall, 1962, p. 24.*)

True Stress, True Strain

The engineering stress is calculated by dividing the applied force F on a tensile test specimen by its original cross-sectional area A_0 [Eq. (6.3)]. Since the cross-sectional area of the test specimen changes continuously during a tensile test, the engineering stress calculated is not precise. During the tensile test, after necking of the sample occurs (Fig. 6.24), the engineering stress decreases as the strain increases, leading to a maximum engineering stress in the engineering stress-strain curve (Fig. 6.26). Thus, once necking begins during the tensile test, the true stress is higher than the engineering stress. We define the true stress and true strain by the following:

$$\text{True stress } \sigma_t = \frac{F \text{ (average uniaxial force on the test sample)}}{A_i \text{ (instantaneous minimum cross-sectional area of sample)}} \qquad (6.12)$$

$$\text{True strain } \epsilon_t = \int_{l_0}^{l_i} \frac{dl}{l} = \ln \frac{l_i}{l_0} \qquad (6.13)$$

where l_0 is the original gage length of the sample and l_i is the instantaneous extended gage length during the test. If we assume constant volume of the gage-length section of the test specimen during the test, then $l_0 A_0 = l_i A_i$ or

$$\frac{l_i}{l_0} = \frac{A_0}{A_i} \qquad \text{and} \qquad \epsilon_t = \ln \frac{l_i}{l_0} = \ln \frac{A_0}{A_i}$$

Figure 6.26 compares engineering stress-strain and true stress-strain curves for a low-carbon steel.

Engineering designs are not based on true stress at fracture since as soon as the yield strength is exceeded, the material starts to deform. Engineers use instead the 0.2 percent offset engineering yield stress for structural designs

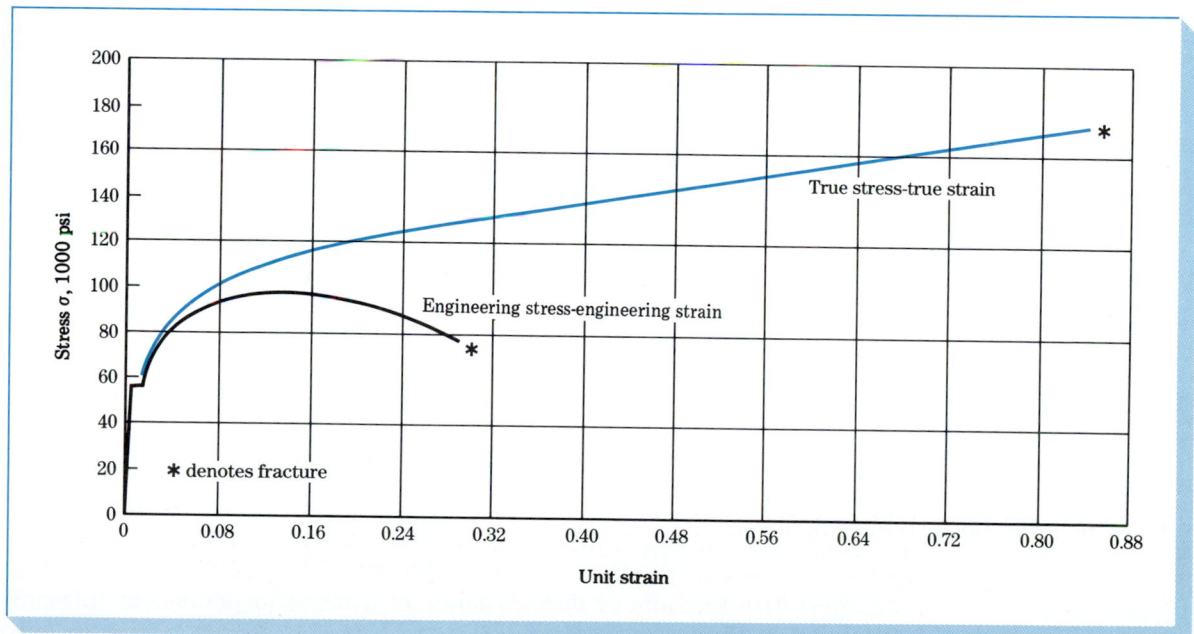

FIGURE 6.26 Comparison of the true stress–true strain curve with the engineering (nominal) stress-strain diagram for a low-carbon steel. (*From H. E. McGannon (ed.), "The Making, Shaping, and Treating of Steel," United States Steel, 1971.*)

with the proper safety factors. However, for research, sometimes the true stress-strain curves are needed.

Example Problem 6.8

Compare the engineering stress and strain with the true test and strain for the tensile test of a low-carbon steel which has the following test values.

Load applied to specimen = 17,000 lb$_f$ Initial specimen diameter = 0.500 in

Diameter of specimen under 17,000-lb$_f$ load = 0.472 in

Solution:

$$\text{Area at start } A_0 = \frac{\pi}{4}d^2 = \frac{\pi}{4}(0.500 \text{ in})^2 = 0.196 \text{ in}^2$$

$$\text{Area under load } A_i = \frac{\pi}{4}(0.472 \text{ in})^2 = 0.175 \text{ in}^2$$

Assuming no volume change during extension, $A_0 l_0 = A_i l_i$ or $l_i/l_0 = A_0/A_i$.

$$\text{Engineering stress} = \frac{F}{A_0} = \frac{17,000 \text{ lb}_f}{0.196 \text{ in}^2} = 86,700 \text{ psi} \blacktriangleleft$$

$$\text{Engineering strain} = \frac{\Delta l}{l} = \frac{l_i - l_0}{l_0} = \frac{A_0}{A_i} - 1 = \frac{0.196 \text{ in}^2}{0.175 \text{ in}^2} - 1 = 0.12$$

$$\text{True stress} = \frac{F}{A_i} = \frac{17,000 \text{ lb}_f}{0.175 \text{ in}^2} = 97,100 \text{ psi} \blacktriangleleft$$

$$\text{True strain} = \ln\frac{l_i}{l_0} = \ln\frac{A_0}{A_i} = \ln\frac{0.196 \text{ in}^2}{0.175 \text{ in}^2} = \ln 1.12 = 0.113$$

6.4 HARDNESS AND HARDNESS TESTING

Hardness is a measure of the resistance of a metal to permanent (plastic) deformation. The hardness of a metal is measured by forcing an indenter into its surface. The indenter material, which is usually a ball, pyramid, or cone, is made of a material much harder than the material being tested. For example, hardened steel, tungsten carbide, or diamond are commonly used materials for indenters. For most standard hardness tests a known load is applied slowly

TABLE 6.2   **Hardness Tests**

Test	Indenter	Shape of indentation Side view	Top view	Load	Formula for hardness number
Brinell	10-mm sphere of steel or tungsten carbide	$\leftarrow D \rightarrow$ $\quad d$	d	P	$BHN = \dfrac{2P}{\pi D(D - \sqrt{D^2 - d^2})}$
Vickers	Diamond pyramid	$136°$	$d_1 \qquad d_1$	P	$VHN = \dfrac{1.72P}{d_1^2}$
Knoop microhardness	Diamond pyramid	t $l/b = 7.11$ $b/t = 4.00$	b $\leftarrow l \rightarrow$	P	$KHN = \dfrac{14.2P}{l^2}$

Rockwell

Test	Indenter	Side view	Top view	Load		
A C D	Diamond cone	$120°$		60 kg 150 kg 100 kg	$R_A =$ $R_C =$ $R_D =$	100–500*t*
B F G	$\frac{1}{16}$-in-diameter steel sphere	t		100 kg 60 kg 150 kg	$R_B =$ $R_F =$ $R_G =$	130–500*t*
E	$\frac{1}{8}$-in-diameter steel sphere	t		100 kg	$R_E =$	

Source: After H. W. Hayden, W. G. Moffatt, and J. Wulff, "The Structure and Properties of Materials," vol. III, Wiley, 1965, p. 12.

by pressing the indenter at 90° into the metal surface being tested [Fig. 6.27*b* (2)]. After the indentation has been made, the indenter is withdrawn from the surface [Fig. 6.27*b* (3)]. An empirical hardness number is then calculated or read off a dial (or digital display), which is based on the cross-sectional area or depth of the impression.

Table 6.2 lists the types of indenters and types of impressions associated with four common hardness tests: Brinell, Vickers, Knoop, and Rockwell. The hardness number for each of these tests depends on the shape of the indentation and the applied load. Figure 6.27 shows a modern Rockwell hardness tester which has a digital readout display.

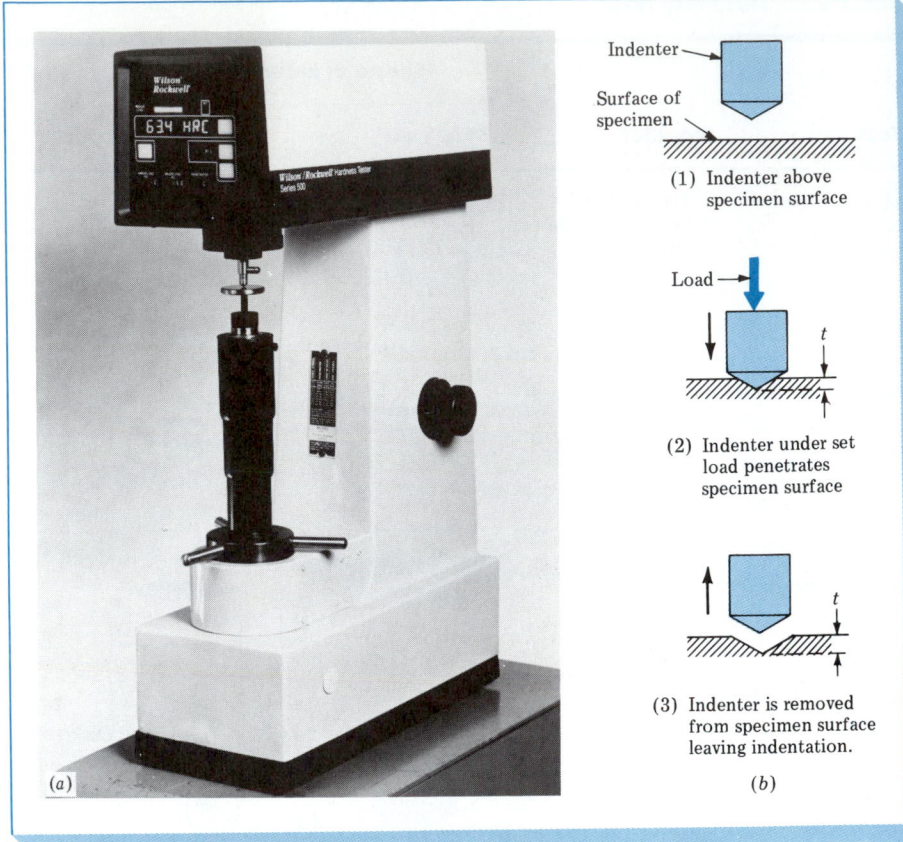

Indenter

Surface of specimen

(1) Indenter above specimen surface

Load

t

(2) Indenter under set load penetrates specimen surface

t

(3) Indenter is removed from specimen surface leaving indentation.

(a)

(b)

FIGURE 6.27 (a) A Rockwell hardness tester. (*Courtesy of the Page-Wilson Co.*) (b) Steps in the measurement of hardness with a diamond-cone indenter. The depth *t* determines the hardness of the material. The lower the value of *t*, the harder the material.

The hardness of a metal depends on the ease with which it plastically deforms. Thus a relationship between hardness and strength for a particular metal can be determined empirically. The hardness test is much simpler than the tensile test and can be nondestructive (i.e., the small indentation of the indenter may not be detrimental to the use of an object). For these reasons, the hardness test is used extensively in industry for quality control.

6.5 PLASTIC DEFORMATION OF METAL SINGLE CRYSTALS

Slipbands and Slip Lines on the Surface of Metal Crystals

Let us first consider the permanent deformation of a rod of a zinc single crystal by stressing it beyond its elastic limit. An examination of the zinc crystal after the deformation shows that step markings appear on its surface, which are called *slipbands* (Fig. 6.28a and b). The slipbands are caused by the slip or shear deformation of metal atoms on specific crystallographic planes called *slip planes*. The deformed zinc single-crystal surface illustrates the formation of slipbands very clearly since slip in these crystals is restricted primarily to slip on the HCP basal planes (Fig. 6.28c and d).

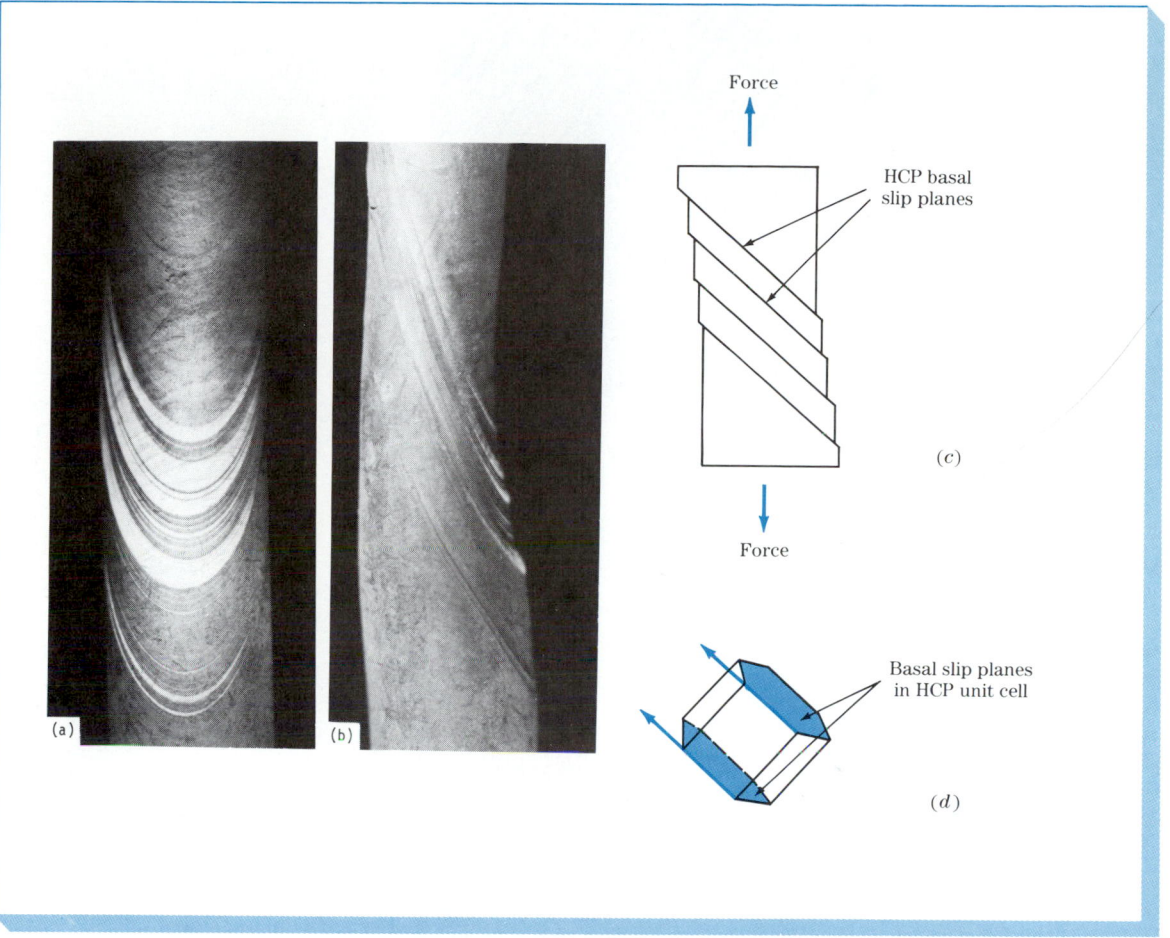

FIGURE 6.28 Plastically deformed zinc single crystal showing slipbands: (*a*) front view of real crystal, (*b*) side view of real crystal, (*c*) schematic side view indicating HCP basal slip planes in crystal, and (*d*) HCP unit cell indicating basal slip planes. (*Zinc single-crystal photos courtesy of Prof. Earl Parker of the University of California at Berkeley.*)

In single crystals of ductile FCC metals like copper and aluminum, slip occurs on multiple slip planes, and as a result the slipband pattern on the surface of these metals when they are deformed is more uniform (Fig. 6.29). A closer examination of the slipped surface of metals at high magnification shows that slip has occurred on many slip planes within the slipbands (Fig. 6.30). These fine steps are called *slip lines* and are usually about 50 to 500 atoms apart, whereas slipbands are commonly separated by about 10,000 atom diameters. Unfortunately, the terms slipband and slip line are often used interchangeably.

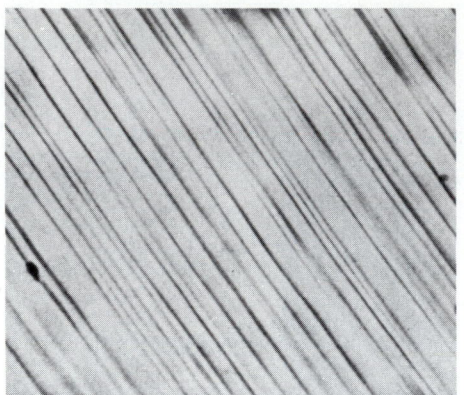

FIGURE 6.29 Slipband pattern on surface of copper single crystal after 0.9 percent deformation. (*Magnification 100×.*) [*After F. D. Rosi. Trans. AIME*, **200**:1018(1954).]

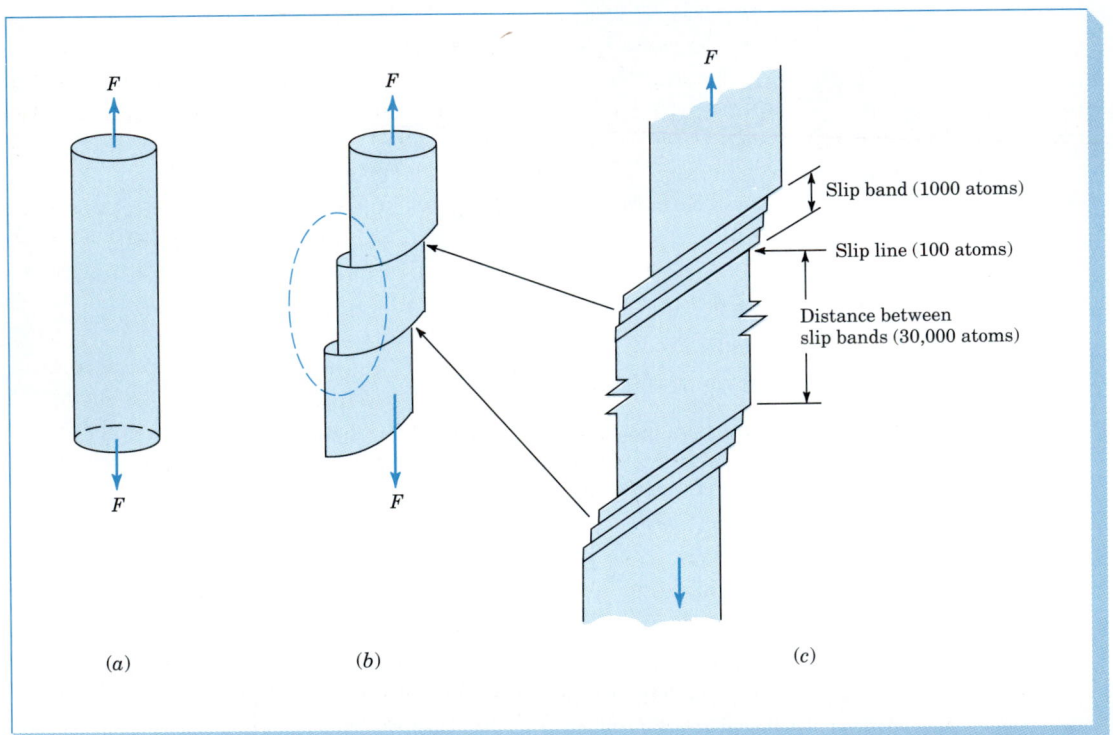

Slip band (1000 atoms)

Slip line (100 atoms)

Distance between slip bands (30,000 atoms)

(a)   (b)   (c)

FIGURE 6.30 Formation of slip lines and slipbands during the plastic deformation of a metal single crystal. (a) A cylindrical rod of a metal single crystal. (b) Slip caused by plastic deformation due to force on rod. (c) Enlarged region showing slip lines contained within slipbands (schematic); at low magnifications, the slip lines together appear as a single slipband line, as in Fig. 6.29.

FIGURE 6.31 Large groups of atoms in large metal crystals do *not* slide over each other simultaneously during plastic shear deformation, as indicated in this figure, since the process requires too much energy. A lower-energy process involving the slippage of a small group of atoms takes place instead.

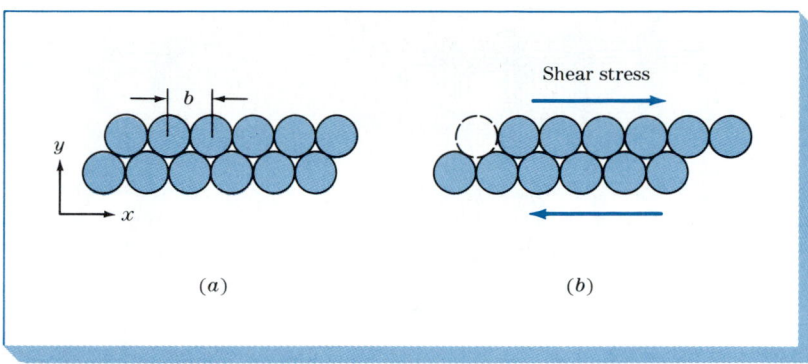

Plastic Deformation in Metal Crystals by the Slip Mechanism

Figure 6.31 shows a possible atomic model for the slippage of one block of atoms over another in a perfect metal crystal. Calculations made from this model determine that the strength of metal crystals should be about 1000 to 10,000 times greater than their observed shear strengths. Thus this mechanism for atomic slip in large real metal crystals must be incorrect.

In order for large metal crystals to deform at their observed low shear strengths, a high density of crystalline imperfections known as *dislocations* must be present. These dislocations are created in large numbers ($\sim 10^6$ cm/cm^3) as the metal solidifies, and when the metal crystal is deformed, many more are created so that a highly deformed crystal may contain as high as 10^{12} cm/cm^3 of dislocations. Figure 6.32 shows schematically how an *edge dislocation* can produce a unit of slip under a low *shear stress*. A relatively small amount of stress is required for slip by this process since only a small group of atoms slips over each other at any instant.

An analogous situation to the movement of a dislocation in a metal crystal under a shear stress can be envisaged by the movement of a carpet with a ripple in it across a very large floor. By pulling on one end of the carpet, it may be impossible to move it because of the friction between the floor and the carpet. However, by putting a ripple in the carpet (analogous to a dislocation in a metal crystal), the carpet may be moved by pushing the ripple in the carpet one step at a time across the floor (Fig. 6.32*d*).

Dislocations in real crystals can be observed in the transmission electron microscope in thin metal foils and appear as lines due to the atomic disarray at the dislocations which interfere with the transmission path of the electron beam of the microscope. Figure 6.33 shows a cellular wall pattern of dislocations created by lightly deforming an aluminum sample. The cells are relatively free from dislocations but are separated by walls of high dislocation density.

Slip Systems

Dislocations produce atomic displacements on specific crystallographic slip planes and in specific crystallographic slip directions. The slip planes are usually the most densely packed planes which are also the farthest separated.

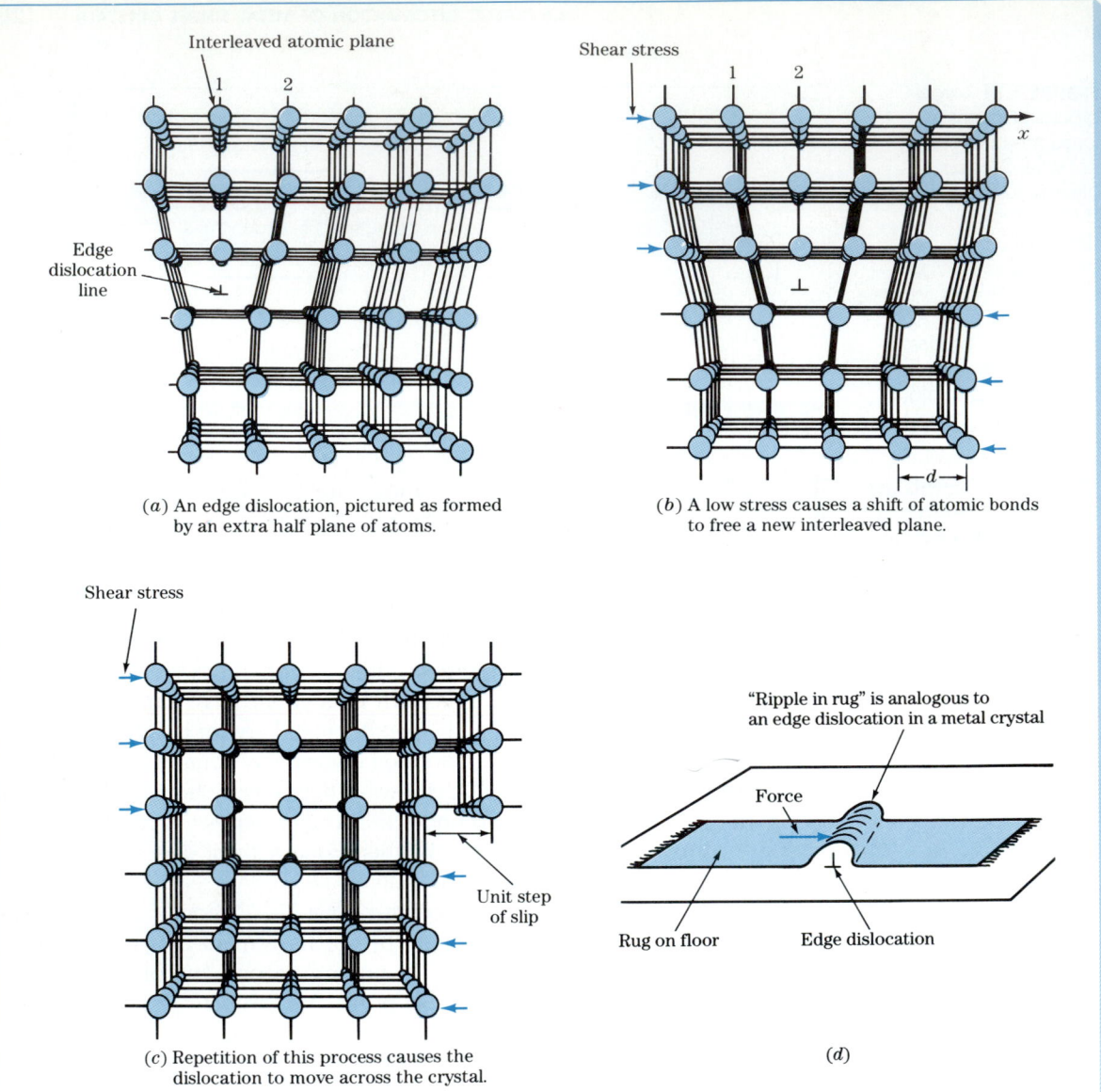

(a) An edge dislocation, pictured as formed by an extra half plane of atoms.

(b) A low stress causes a shift of atomic bonds to free a new interleaved plane.

(c) Repetition of this process causes the dislocation to move across the crystal.

(d)

FIGURE 6.32 Schematic illustration of how the motion of an edge dislocation produces a unit step of slip under a low shear stress. (a) An edge dislocation, pictured as formed by an extra half plane of atoms. (b) A low stress causes a shift of atomic bonds to free a new interleaved plane. (c) Repetition of this process causes the dislocation to move across the crystal. This process requires less energy than the one depicted in Fig. 6.30. (*After A. G. Guy, "Essentials of Materials Science," McGraw-Hill, 1976, p. 153.*) (d) The "ripple in the rug" analogy. A dislocation moves through a metal crystal during plastic deformation in a manner similar to a ripple which is pushed along a carpet lying on a floor. In both cases a small amount of relative movement is caused by the passage of the dislocation or ripple, and hence a relatively low amount of energy is expended in this process.

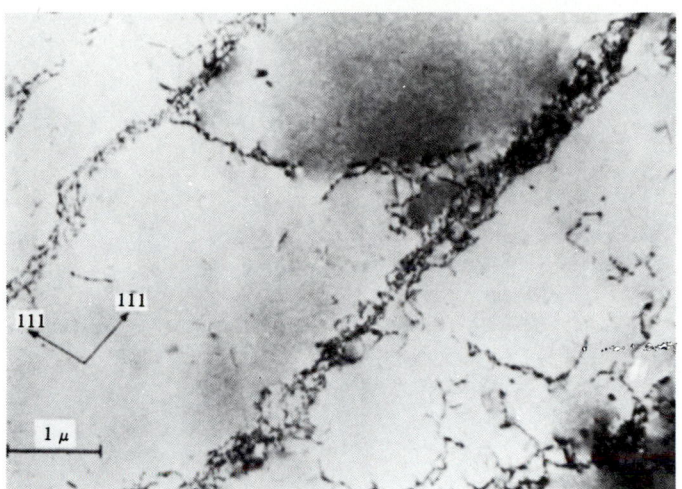

FIGURE 6.33 Dislocation cell structure in a lightly deformed aluminum sample as revealed by transmission electron microscopy. The cells are relatively free from dislocations but are separated by walls of high dislocation density. (*After P. R. Swann, in G. Thomas and J. Washburn, (eds.), "Electron Microscopy and Strength of Crystals," Wiley, 1963, p. 133.*)

Slip is favored on close-packed planes since a lower shear stress for atomic displacement is required than for less densely packed planes (Fig. 6.34). However, if slip on the close-packed planes is restricted due to local high stresses, for example, then planes of lower atomic packing can become operative. Slip in the close-packed directions is also favored since less energy is required to move the atoms from one position to another if the atoms are closer together.

A combination of a slip plane and a slip direction is called a *slip system.* Slip in metallic structures occurs on a number of slip systems which are characteristic for each crystal structure. Table 6.3 lists the predominant slip planes and slip directions for FCC, BCC, and HCP crystal structures.

FIGURE 6.34 Comparison of atomic slip on (*a*) a close-packed plane and (*b*) a non-close-packed plane. Slip is favored on the close-packed plane because less force is required to move the atoms from one position to the next closest one, as indicated by the slopes of the bars on the atoms. Note that dislocations move one atomic slip step at a time. (*After A. H. Cottrell, The Nature of Metals, "Materials," Scientific American, 1967, p. 48. Copyright © by Scientific American, Inc. All rights reserved.*)

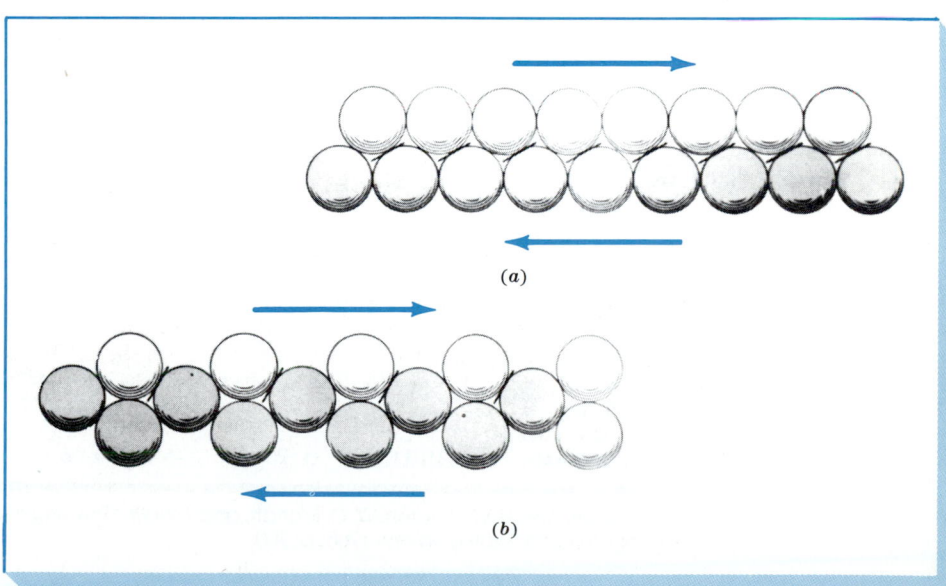

TABLE 6.3 **Slip Systems Observed in Crystal Structures**

Structure	Slip plane	Slip direction	Number of slip systems	
FCC: Cu, Al, Ni, Pb, Au, Ag, γFe, . . .	{111}	$\langle 1\bar{1}0 \rangle$	$4 \times 3 = 12$	
BCC: αFe, W, Mo, β brass	{110}	$\langle \bar{1}11 \rangle$	$6 \times 2 = 12$	
αFe, Mo, W, Na	{211}	$\langle \bar{1}11 \rangle$	$12 \times 1 = 12$	
αFe, K	{321}	$\langle \bar{1}11 \rangle$	$24 \times 1 = 24$	
HCP: Cd, Zn, Mg, Ti, Be, . . .	(0001)	$\langle 11\bar{2}0 \rangle$	$1 \times 3 = 3$	
Ti (prism planes)	{10$\bar{1}$0}	$\langle 11\bar{2}0 \rangle$	$3 \times 1 = 3$	
Ti, Mg (pyramidal planes)	{10$\bar{1}$1}	$\langle 11\bar{2}0 \rangle$	$6 \times 1 = 6$	

Source: After H. W. Hayden, W. G. Moffatt, and J. Wulff, "The Structure and Properties of Materials," vol. III, Wiley, 1965, p. 100.

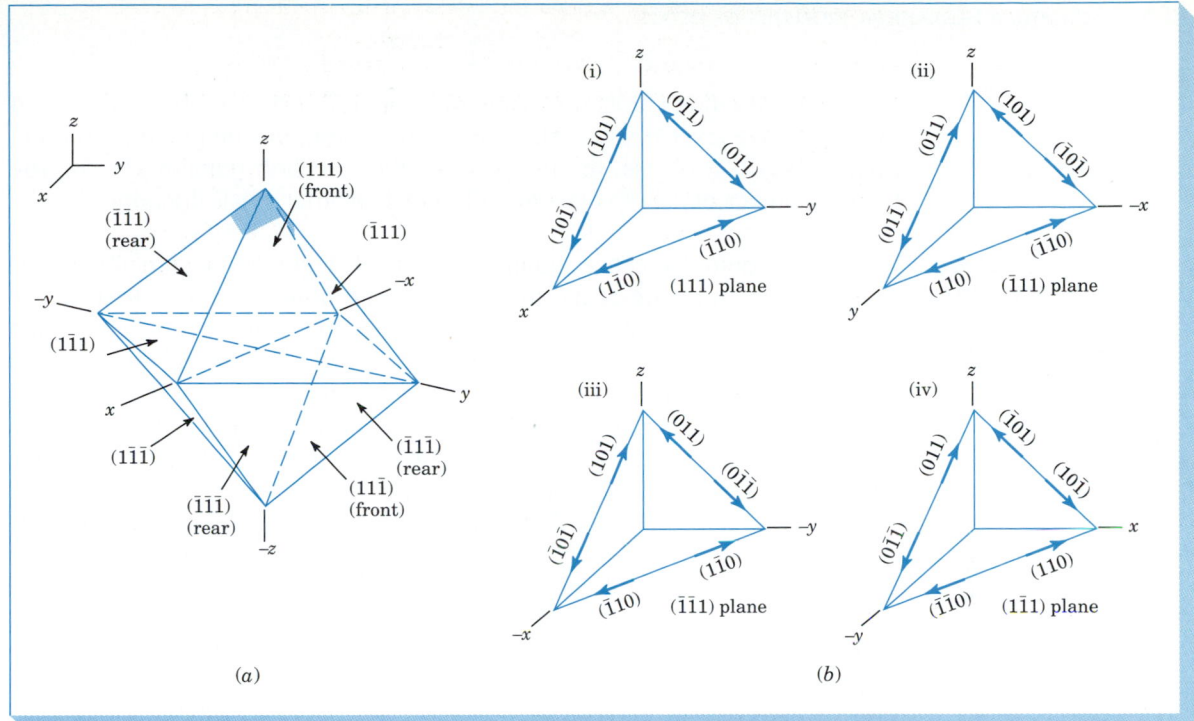

FIGURE 6.35 Slip planes and directions for the FCC crystal structure. (*a*) Only four of the eight {111} octahedral planes are considered slip planes since planes opposite each other are considered the same slip plane. (*b*) For each slip plane there are three ⟨1$\bar{1}$0⟩ slip directions since opposite directions are considered as only one slip direction. Note that slip directions are only shown for the upper four slip planes of the octahedral FCC planes. Thus there are 4 slip planes × 3 slip directions, giving a total of 12 slip systems for the FCC crystal structure.

For metals with the FCC crystal structure, slip takes place on the close-packed {111} octahedral planes and in the ⟨1$\bar{1}$0⟩ close-packed directions. There are eight {111} octahedral planes in the FCC crystal structure (Fig. 6.35). The (111) type planes at opposite faces of the octahedron which are parallel to each other are considered the same type of (111) slip plane. Thus, there are only four different types of (111) slip planes in the FCC crystal structure. Each (111)-type plane contains three [1$\bar{1}$0]-type slip directions. The reverse directions are not considered different slip directions. Thus for the FCC lattice there are 4 slip planes × 3 slip directions = 12 slip systems (Table 6.3).

The BCC structure is *not* a close-packed structure and does not have a predominant plane of highest atomic packing like the FCC structure. The {110} planes have the highest atomic density, and slip commonly takes place on these planes. However, slip in BCC metals also occurs on {112} and {123} planes. Since the slip planes in the BCC structure are not close-packed as in the case of the FCC structure, higher shear stresses are necessary for slip in BCC than in FCC metals. The slip direction in BCC metals is always of the ⟨$\bar{1}$11⟩ type. Since there are six (110)-type slip planes of which each can slip in two [$\bar{1}$11] directions, there are 6 × 2 = 12 {110}⟨$\bar{1}$11⟩ slip systems.

In the HCP structure, the basal plane (0001) is the closest-packed plane and is the common slip plane for HCP metals such as Zn, Cd, and Mg which have high *c/a* ratios (Table 6.3). However, for HCP metals such as Ti, Zr, and Be which have low *c/a* ratios, slip also occurs commonly on prism $\{10\bar{1}0\}$ and pyramidal $\{10\bar{1}1\}$ planes. In all cases the slip direction remains $\langle 11\bar{2}0 \rangle$. The limited number of slip systems in HCP metals restricts their ductilities.

Critical Resolved Shear Stress for Metal Single Crystals

The stress required to cause slip in a pure-metal single crystal depends mainly on the crystal structure of the metal, its atomic bonding characteristics, the temperature at which it is deformed, and the orientation of the active slip planes with respect to the shear stresses. Slip begins within the crystal when the shear stress on the slip plane in the slip direction reaches a required level called the *critical resolved shear stress* τ_c. Essentially, this value is the yield stress of a single crystal and is equivalent to the yield stress of a polycrystalline metal or alloy determined by a stress-strain tensile test curve.

Table 6.4 lists values for the critical resolved shear stresses of some pure-metal single crystals at room temperature. The HCP metals Zn, Cd, and Mg have low critical resolved shear stresses ranging from 0.18 to 0.77 MPa. The HCP metal titanium, on the other hand, has a very high τ_c of 13.7 MPa. It is believed that some covalent bonding mixed with metallic bonding is partly responsible for this high value of τ_c. Pure FCC metals such as Ag and Cu have low τ_c values of 0.48 and 0.65 MPa, respectively, because of their multiple slip systems.

Schmid's Law

The relationship between a uniaxial stress acting on a cylinder of a pure metal single crystal and the resulting resolved shear stress produced on a slip system

TABLE 6.4   Room-Temperature Slip Systems and Critical Resolved Shear Stress for Metal Single Crystals

Metal	Crystal structure	Purity, %	Slip plane	Slip direction	Critical shear stress, MPa
Zn	HCP	99.999	(0001)	$[11\bar{2}0]$	0.18
Mg	HCP	99.996	(0001)	[1120]	0.77
Cd	HCP	99.996	(0001)	$[11\bar{2}0]$	0.58
Ti	HCP	99.99	(1010)	$[11\bar{2}0]$	13.7
		99.9	(1010)	$[11\bar{2}0;$	90.1
Ag	FCC	99.99	(111)	$[1\bar{1}0]$	0.48
		99.97	(111)	$[1\bar{1}0]$	0.73
		99.93	(111)	$[1\bar{1}0]$	1.3
Cu	FCC	99.999	(111)	$[1\bar{1}0]$	0.65
		99.98	(111)	$[1\bar{1}0]$	0.94
Ni	FCC	99.8	(111)	$[1\bar{1}0]$	5.7
Fe	BCC	99.96	(110)	$[\bar{1}11]$	27.5
			(112)		
			(123)		
Mo	BCC	$\cdots$	(110)	$[\bar{1}11]$	49.0

Source: After G. Dieter, "Mechanical Metallurgy," 2d ed., McGraw-Hill, 1976, p. 129.

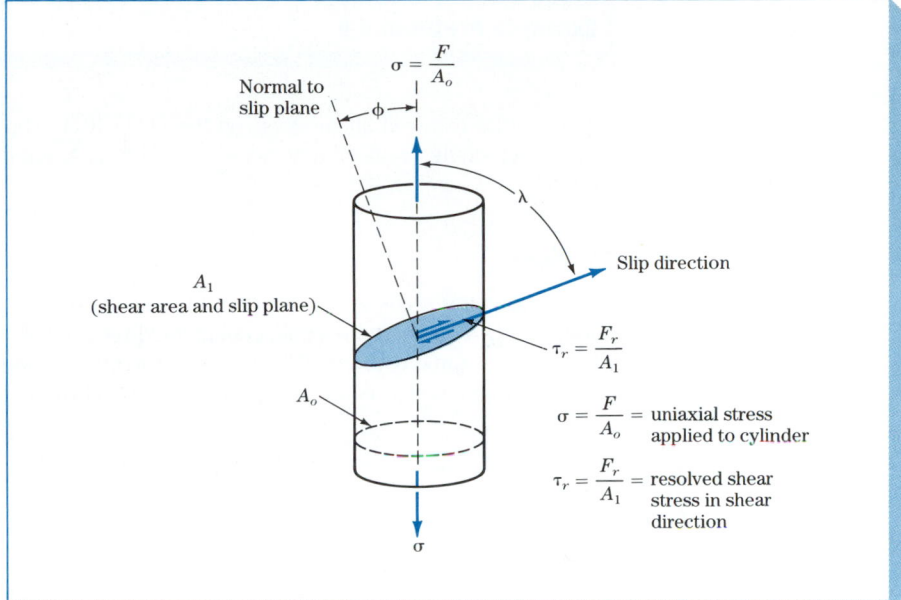

FIGURE 6.36 Axial stress σ can produce a resolved shear stress τ_r and cause dislocation motion in slip plane A_1 in the slip direction.

within the cylinder can be derived as follows. Consider a uniaxial tensile stress σ acting on a metal cylinder, as shown in Fig. 6.36. Let A_0 be the area normal to the axial force F and A_1 the area of the slip plane or shear area on which the resolved shear force F_r is acting. We can orient the slip plane and slip direction by defining the angles ϕ and λ. ϕ is the angle between the uniaxial force F and the normal to the slip plane area A_1, and λ is the angle between the axial force and the slip direction.

In order for dislocations to move in the slip system, a sufficient resolved shear stress acting in the slip direction must be produced by the applied axial force. The resolved shear stress is

$$\tau_r = \frac{\text{shear force}}{\text{shear area (slip plane area)}} = \frac{F_r}{A_1} \qquad (6.14)$$

The resolved shear force F_r is related to the axial force F by $F_r = F \cos \lambda$. The area of the slip plane (shear area) $A_1 = A_0/\cos \phi$. By dividing the shear force $F \cos \lambda$ by the shear area $A_0/\cos \phi$, we obtain

$$\tau_r = \frac{F \cos \lambda}{A_0/\cos \phi} = \frac{F}{A_0} \cos \lambda \cos \phi = \sigma \cos \lambda \cos \phi \qquad (6.15)$$

which is called *Schmid's law*. Let us now consider an example problem to calculate the resolved shear stress when a slip system is acted upon by an axial stress.

Example Problem 6.9

Calculate the resolved shear stress on the (111) [0$\bar{1}$1] slip system of a unit cell in an FCC nickel single crystal if a stress of 13.7 MPa is applied in the [001] direction of a unit cell.

Solution:

By geometry the angle λ between the applied stress and the slip direction is 45°, as shown in Fig. EP6.9a. In the cubic system the direction indices of the normal to a crystal plane are the same as the Miller indices of the crystal plane. Therefore, the normal to the (111) plane which is the slip plane is the [111] direction. From Fig. EP6.9b,

$$\cos \phi = \frac{a}{\sqrt{3}a} = \frac{1}{\sqrt{3}} \quad \text{or} \quad \phi = 54.74°$$

$$\tau_r = \sigma \cos \lambda \cos \phi = (13.7 \text{ MPa})(\cos 45°)(\cos 54.74°) = 5.6 \text{ MPa} \blacktriangleleft$$

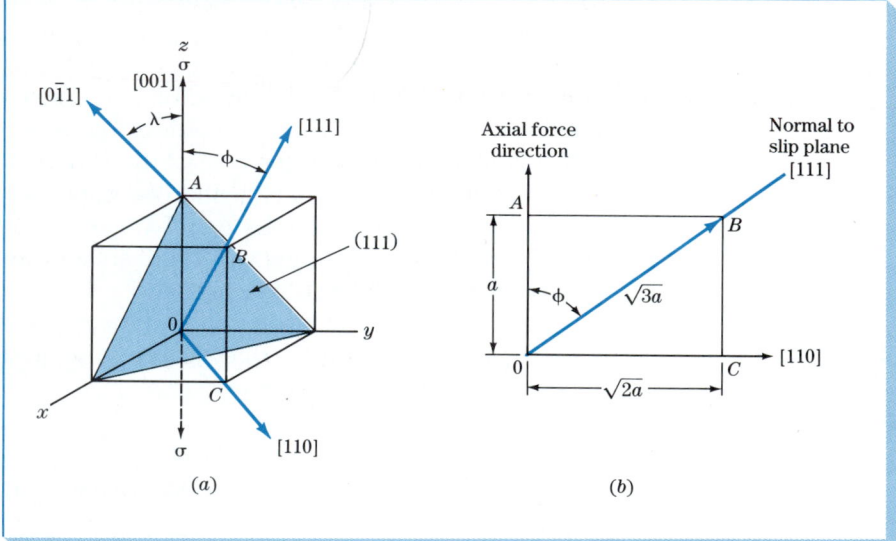

FIGURE EP6.9 An FCC unit cell is acted upon by a [001] tensile stress producing a resolved shear stress on the (111) [0$\bar{1}$1] slip system.

Twinning   A second important plastic deformation mechanism which can occur in metals is *twinning*. In this process a part of the atomic lattice is deformed so that it forms a mirror image of the undeformed lattice next to it (Fig. 6.37). The crystallographic plane of symmetry between the undeformed and deformed parts of the metal lattice is called the *twinning plane*. Twinning, like slip, occurs in a specific direction called the *twinning direction*. However, in slip the atoms on one side of the slip plane all move equal distances (Fig. 6.32),

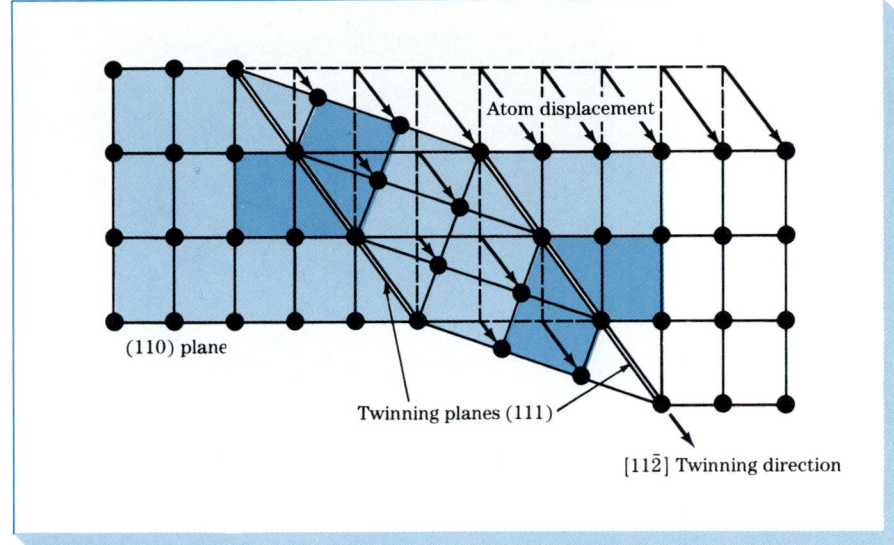

FIGURE 6.37 Schematic diagram of the twinning process in an FCC lattice. (*After H. W. Hayden, W. G. Moffatt, and J. Wulff, "The Structure and Properties of Materials," vol. III, Wiley, 1965, p. 111.*)

whereas in twinning the atoms move distances proportional to their distance from the twinning plane (Fig. 6.37). Figure 6.38 illustrates the basic difference between slip and twinning on the surface of a metal after deformation. Slip leaves a series of steps (lines) (Fig. 6.38*a*), whereas twinning leaves small but well-defined regions of the crystal deformed (Fig. 6.38*b*). Figure 6.39 shows some deformation twins on the surface of titanium metal.

Twinning only involves a small fraction of the total volume of the metal crystal, and so the amount of overall deformation that can be produced by twinning is small. However, the important role of twinning in deformation is that the lattice orientation changes that are caused by twinning may place new

FIGURE 6.38 Schematic diagram of surfaces of a deformed metal after (*a*) slip and (*b*) twinning.

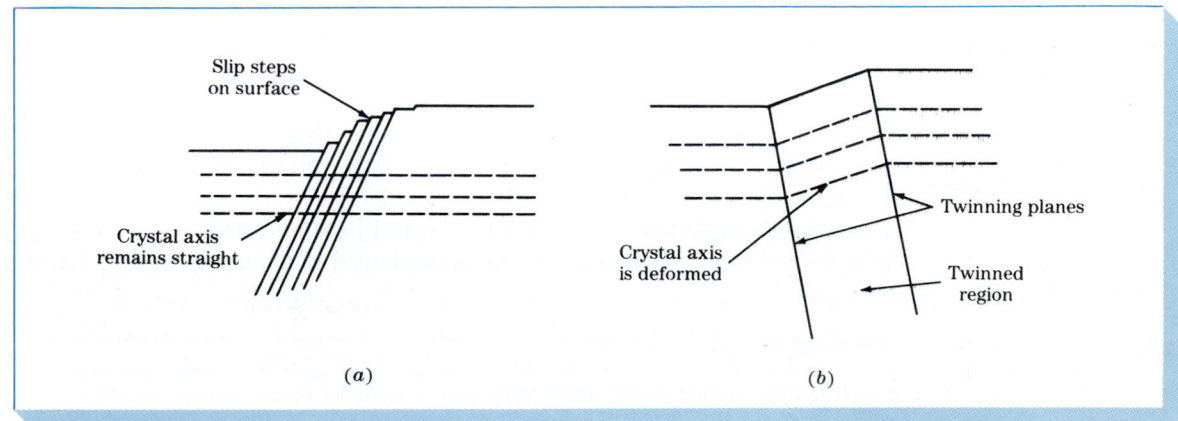

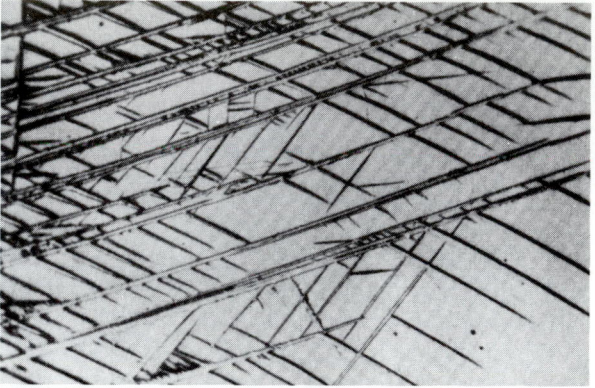

FIGURE 6.39
Deformation twins in unalloyed (99.77%) titanium. (Magnification 150×.) [After F. D. Rosi, C. A. Dube, and B. H. Alexander, Trans. AIME, **197**:259(1953).]

slip systems into favorable orientation with respect to the shear stress and thus enable additional slip to occur. Of the three major metallic unit-cell structures (BCC, FCC, and HCP), twinning is most important for the HCP structure because of its small number of slip systems. However, even with the assistance of twinning, HCP metals like zinc and magnesium are still less ductile than the BCC and FCC metals which have more slip systems.

Deformation twinning is observed at room temperature for the HCP metals. Twinning is found in the BCC metals such as Fe, Mo, W, Ta, and Cr in crystals that were deformed at very low temperatures. Twinning has also been found in some of these BCC metal crystals at room temperature when they have been subjected to very high strain rates. The FCC metals show the least tendency to form deformation twins. However, deformation twins can be produced in some FCC metals if the stress level is high enough and the temperature sufficiently low. For example, copper crystals deformed at 4 K at high stress levels can form deformation twins.

6.6 PLASTIC DEFORMATION OF POLYCRYSTALLINE METALS

Effect of Grain Boundaries on the Strength of Metals

Almost all engineering alloys are polycrystalline. Single-crystal metals and alloys are used mainly for research purposes and only in a few cases for engineering applications.[1] Grain boundaries strengthen metals and alloys by acting as barriers to dislocation movement except at high temperatures, where they become regions of weakness. For most applications where strength is impor-

[1] Single-crystal turbine blades have been developed for use in gas turbine engines to avoid grain-boundary cracking at high temperatures and stresses. See F. L. Ver Snyder and M. E. Shank, *Mater. Sci. Eng.*, **6**:213–247(1970).

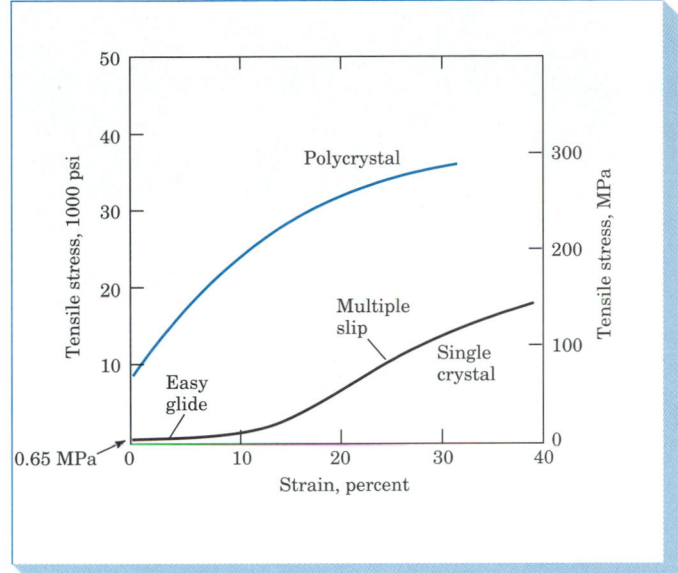

FIGURE 6.40 Stress-strain curves for single crystal and polycrystalline copper. The polycrystalline copper has higher strength due to grain boundaries which restrict slip.

tant, a fine grain size is desirable and so most metals are fabricated with a fine grain size. Figure 6.40 compares the tensile stress-strain curves for single-crystal and polycrystalline unalloyed copper at room temperature. At all strains the polycrystalline copper is stronger than the single-crystal copper. At 20 percent strain the tensile strength of the polycrystalline copper is 40 ksi (276 MPa) as compared to 8 ksi (55 MPa) for single-crystal copper.

During the plastic deformation of metals, dislocations moving along on a particular slip plane cannot go directly from one grain into another in a straight line. As shown in Fig. 6.41, slip lines change directions at grain boundaries. Thus, each grain has its own set of dislocations on its own preferred slip planes which have different orientations from those of neighboring grains. Figure 6.42 shows clearly a high-angle grain boundary which is acting as a barrier to dislocation movement and has caused dislocations to pile up at the grain boundary.

Effect of Plastic Deformation on Grain Shape and Dislocation Arrangements

Grain shape changes with plastic deformation   Let us consider the plastic deformation of annealed samples[1] of unalloyed copper which have an equiaxed grain structure. Upon cold plastic deformation the grains are sheared relative to each other by the generation, movement, and rearrangement of dislocations. Figure 6.43 shows the microstructures of samples of unalloyed copper sheet

---

[1] Samples in the annealed conditions have been plastically deformed and then reheated to such an extent that a grain structure in which the grains are approximately equal in all directions (equiaxed) is produced.

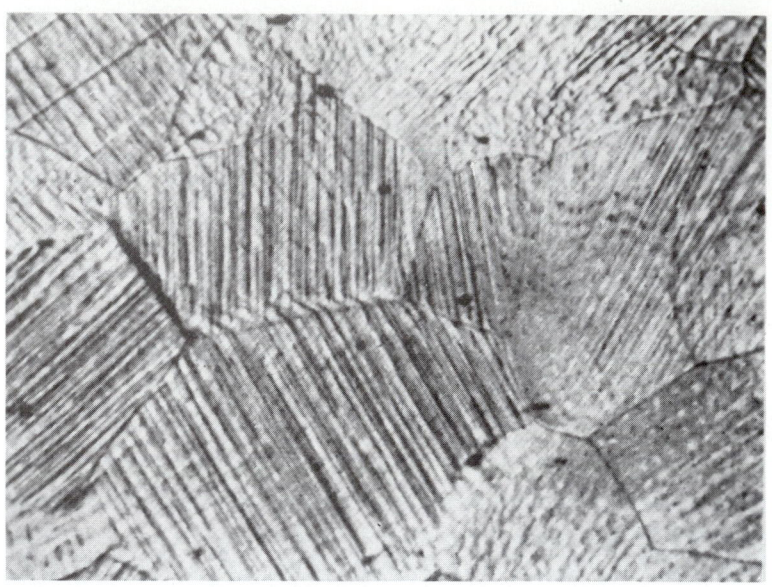

FIGURE 6.41
Polycrystalline aluminum which has been plastically deformed. Note that the slipbands are parallel within a grain but are discontinuous across the grain boundaries. (Magnification 60×.) (*After G. C. Smith, S. Charter, and S. Chiderley of Cambridge University.*)

that was cold-rolled to reductions of 30 and 50 percent, respectively. Note that with increased cold rolling the grains are more elongated in the rolling direction as a consequence of dislocation movements.

Dislocation arrangement changes with plastic deformation   The dislocations in the unalloyed copper sample after 30 percent plastic deformation form cell-

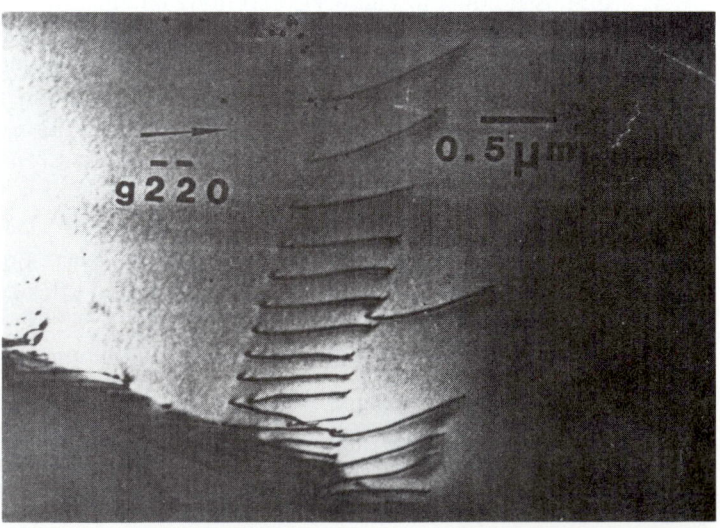

FIGURE 6.42
Dislocations piled up against a grain boundary as observed with a transmission electron microscope in a thin foil of stainless steel. (Magnification 20,000×.) [*After Z. Shen, R. H. Wagoner, and W. A. T. Clark, Scripta Met., 20:926 (1986).*]

FIGURE 6.43 Optical micrographs of deformation structures of unalloyed copper that was cold-rolled to reductions of (*a*) 30 percent and (*b*) 50 percent. (Etch: potassium dichromate; magnification 300×.) (*After J. E. Boyd in "Metals Handbook," vol. 8: "Metallography, Structures, and Phase Diagrams," 8th ed., Amercian Society for Metals, 1973, p. 221.*)

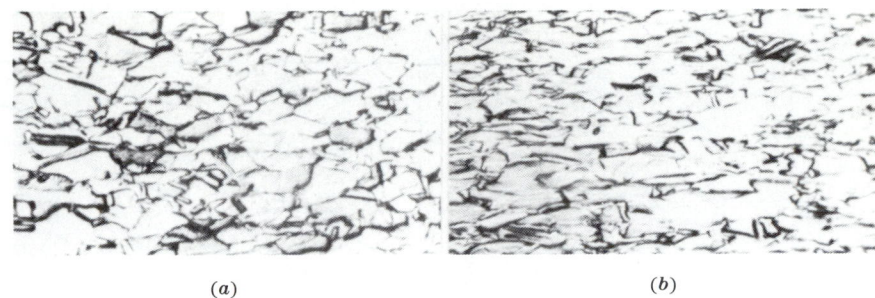

(*a*)   (*b*)

like configurations with clear areas in the centers of the cells (Fig. 6.44*a*). With increased cold plastic deformation to 50 percent reduction, the cell structure becomes denser and elongated in the direction of rolling (Fig. 6.44*b*).

Effect of Cold Plastic Deformation on Increasing the Strength of Metals

As shown by the electron micrographs of Fig. 6.44, the dislocation density increases with increased cold deformation. The exact mechanism by which the dislocation density is increased by cold working is not completely understood. New dislocations are created by the cold deformation and must interact with those already existing. As the dislocation density increases with deformation, it becomes more and more difficult for the dislocations to move through the existing "forest of dislocations," and thus the metal work or strain hardens with increased cold deformation.

When ductile metals such as copper, aluminum, and α iron which have been annealed are cold-worked at room temperature, they strain-harden because of the dislocation interaction which has been just described. Figure 6.45 shows how cold working at room temperature increases the tensile strength

FIGURE 6.44 Transmission electron micrographs of deformation structures of unalloyed copper that was cold-rolled to reductions of (*a*) 30 percent and (*b*) 50 percent. Note that these electron micrographs correspond to the optical micrographs of Fig. 6.43. (Thin-foil specimens, magnification 30,000×.) (*After J. E. Boyd in "Metals Handbook," vol. 8: "Metallography, Structures, and Phase Diagrams," 8th ed., American Society for Metals, 1973, p. 221.*)

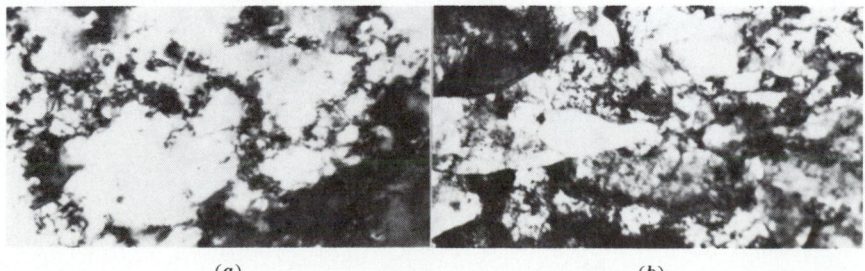

(*a*)   (*b*)

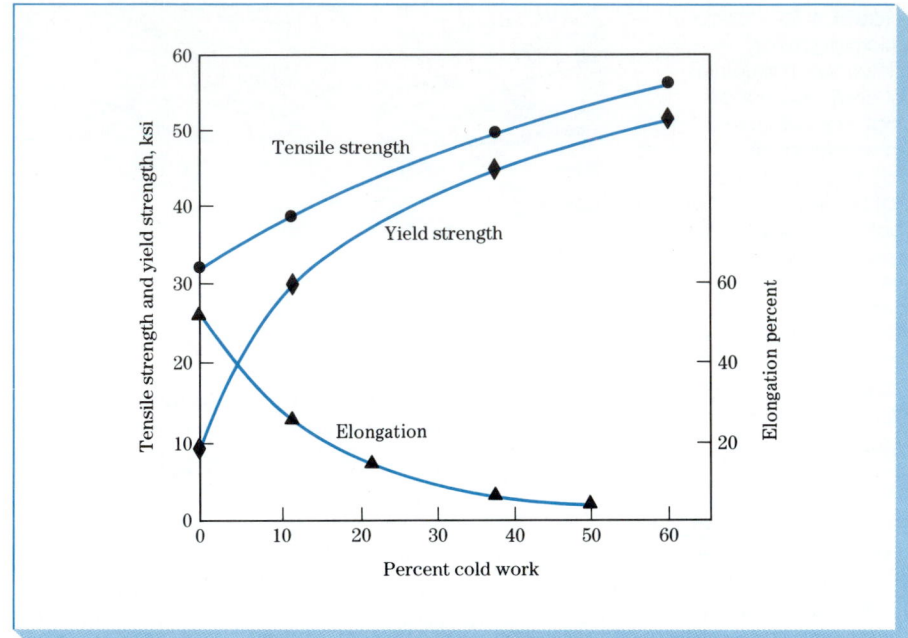

FIGURE 6.45 Percent cold work vs. tensile strength and elongation for unalloyed oxygen-free copper. Cold work is expressed as a percent reduction in cross-sectional area of the metal being reduced.

of unalloyed copper from about 30 ksi (200 MPa) to 45 ksi (320 MPa) with 30 percent cold work. Associated with the increase in tensile strength, however, is a decrease in elongation (ductility), as observed in Fig. 6.45. With 30 percent cold work, the elongation of unalloyed copper decreases from about 42 to 10 percent elongation.

Cold working or *strain hardening* is one of the most important methods for strengthening some metals. For example, pure copper and aluminum can be strengthened significantly only by this method. Thus, cold-drawn unalloyed copper wire can be produced with different strengths (within certain limitations) by varying the amount of strain hardening.

Example Problem 6.10

It is desired to produce a 0.040-in-thick sheet of oxygen-free copper with a tensile strength = 45 ksi. What percent cold work must the metal be given? What must the starting thickness of the metal be before cold rolling?

Solution:

From Fig. 6.45 the percent cold work must be 25%. Thus, the starting thickness must be

$$\frac{x - 0.040 \text{ in}}{x} = 0.25$$

$$x = 0.053 \text{ in} \blacktriangleleft$$

6.7   SOLID-SOLUTION STRENGTHENING OF METALS

Another method besides cold working by which the strength of metals can be increased is by *solid-solution strengthening*. The addition of one or more elements to a metal can strengthen it by the formation of a solid solution. The structure of *substitutional* and *interstitial solid solutions* has already been discussed in Sec. 4.3 and should be referred to for review. When substitutional (solute) atoms are mixed in the solid state with those of another metal (solvent), stress fields are created around each solute atom. These stress fields interact with dislocations and make their movement more difficult, and thus the solid solution becomes stronger than the pure metal.

Two important factors in solid-solution strengthening are:

1. *Relative-size factor.* Differences in atomic size of solute and solvent atoms affect the amount of solid-solution strengthening because of the crystal lattice distortions produced. Lattice distortions make dislocation movement more difficult and hence strengthen the metallic solid solution.

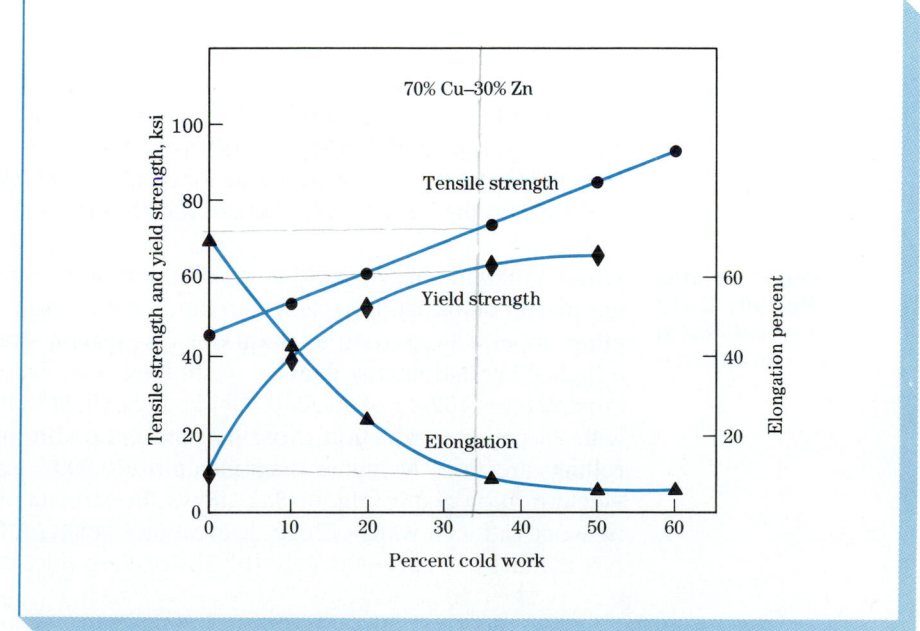

FIGURE 6.46 Percent cold work vs. tensile strength and elongation for 70 wt % Cu–30 wt % Zn alloy. Cold work is expressed as a percent reduction in cross-sectional area of the metal being reduced. [see Eq. (6.2)].

2. *Short-range order.* Solid solutions are rarely random in atomic mixing, and some kind of short-range order or clustering of like atoms takes place. As a result, dislocation movement is impeded by different bonding structures.

In addition to these factors there are others which also contribute to solid-solution strengthening but which will not be dealt with in this book.

As an example of solid-solution strengthening, let us consider a solid solution alloy of 70 wt % Cu and 30 wt % Zn (cartridge brass). The tensile strength of unalloyed copper with 30 percent cold work is about 48 ksi (330 MPa) (Fig. 6.45). However, the tensile strength of the 70 wt % Cu–30 wt % Zn alloy with 30 percent cold work is about 72 ksi (500 MPa) (Fig. 6.46). Thus, solid-solution strengthening in this case produced an increase in strength in the copper of about 24 ksi (165 MPa). On the other hand, the ductility of the copper by the 30% zinc addition after 30 percent cold work was reduced from about 65 to 10 percent (Fig. 6.46).

6.8   RECOVERY AND RECRYSTALLIZATION OF PLASTICALLY DEFORMED METALS

During the processing and fabrication of metals and alloys, sometimes it is necessary to reheat a cold-worked metal to soften it, and thereby increase its ductility. If the metal is reheated to a sufficiently high temperature for a long-enough time, the cold-worked metal structure will go through a series of changes called (1) *recovery,* (2) *recrystallization,* and (3) *grain growth.* Figure 6.47 shows these structural changes schematically as the temperature of the metal is increased along with the corresponding changes in mechanical properties. This reheating treatment which softens a cold-worked metal is called *annealing,* and the terms *partial anneal* and *full anneal* are often used to refer to degrees of softening. Let us now examine these structural changes in more detail, starting with the heavily cold worked metal structure.

Structure of a Heavily Cold Worked Metal before Reheating   When a metal is heavily cold worked, much of the strain energy expended in the plastic deformation is stored in the metal in the form of dislocations and other imperfections such as point defects. Thus a strain-hardened metal has a higher internal energy than an unstrained one. Figure 6.48a shows the microstructure (100×) of an Al–0.8% Mg alloy sheet that has been cold-worked with 85 percent reduction. Note that the grains are greatly elongated in the rolling direction. At higher magnification (20,000×) a thin-foil transmission electron micrograph (Fig. 6.49a) shows the structure to consist of a cellular network with cell walls of high dislocation density. A fully cold-worked metal has a density of approximately 10^{12} dislocation lines/cm^2.

Recovery   When a cold-worked metal is heated in the recovery temperature range which is just below the recrystallization temperature range, internal stresses in the

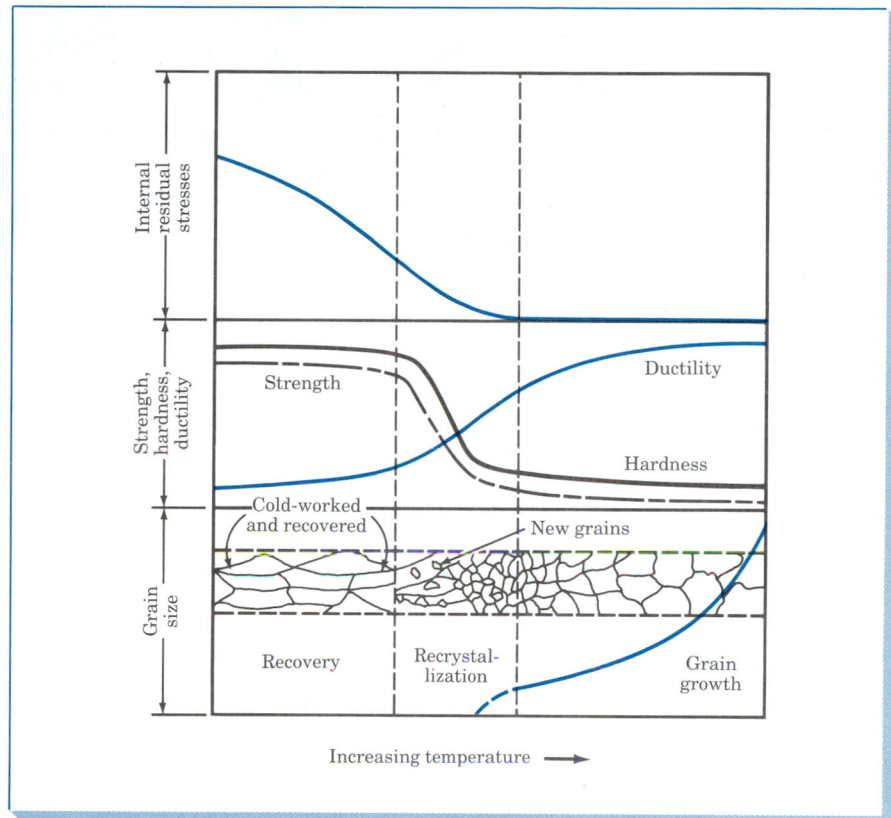

FIGURE 6.47 Effect of annealing on the structure and mechanical property changes of a cold-worked metal. (*Adapted from Z. D. Jastrzebski, "The Nature and Properties of Engineering Materials," 2d ed., Wiley, 1976, p. 228.*)

FIGURE 6.48 Aluminum alloy 5657 (0.8% Mg) sheet showing microstructures after cold rolling 85 percent and subsequent reheating (optical micrographs at 100× viewed under polarized light). (*a*) Cold-worked 85 percent; longitudinal section. Grains are greatly elongated. (*b*) Cold-worked 85 percent and stress-relieved at 302°C (575°F) for 1 h. Structure shows onset of recrystallization, which improves the formability of the sheet. (*c*) Cold-worked 85 percent and annealed at 316°C (600°F) for 1 h. Structure shows recrystallized grains and bands of unrecrystallized grains. (*After "Metals Handbook," vol. 7, 8th ed., American Society for Metals, 1972, p. 243.*)

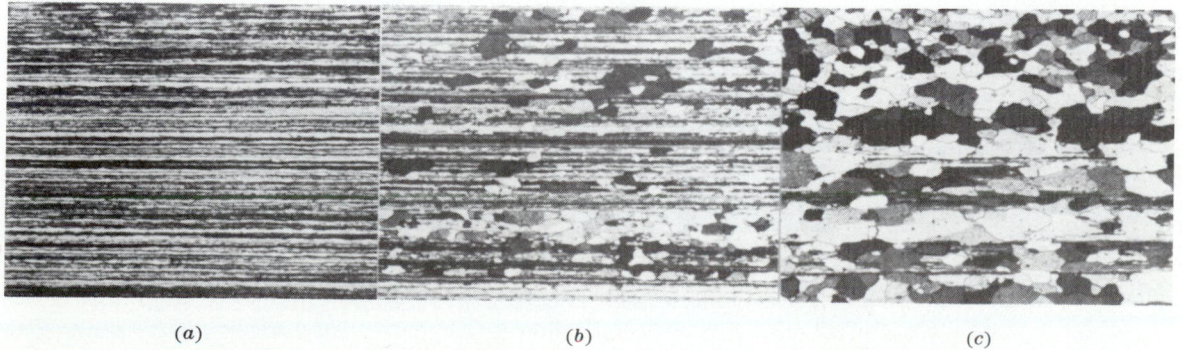

(*a*)   (*b*)   (*c*)

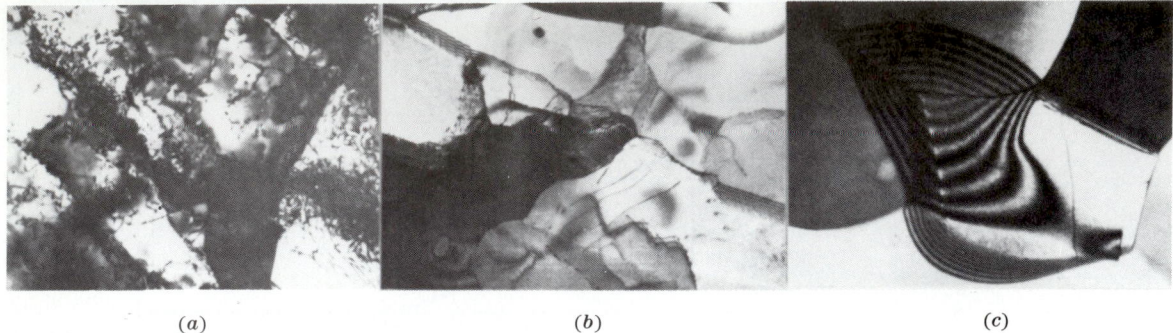

(a)   (b)   (c)

FIGURE 6.49 Aluminum alloy 5657 (0.8% Mg) sheet showing microstructures after cold rolling 85 percent and subsequent reheating. The microstructures shown in this figure correspond to those of Fig. 6.48 and were obtained by using thin-foil transmission electron microscopy. (Magnified 20,000×.) (a) Sheet was cold-worked 85 percent; micrograph shows dislocation tangles and banded cells (subgrains) caused by cold working extensively. (b) Sheet was cold-worked 85 percent and subsequently stress-relieved at 302°C (575°F) for 1 h. Micrograph shows dislocation networks and other low-angle boundaries produced by polygonization. (c) Sheet was cold-worked 85 percent and annealed at 316°C (600°F) for 1 h. Micrograph shows recrystallized structure and some subgrain growth. (*After "Metals Handbook," vol. 7, 8th ed., American Society for Metals, 1972, p. 243.*)

metal are relieved (Fig. 6.47). During recovery, sufficient thermal energy is supplied to allow the dislocations to rearrange themselves into lower energy configurations (Fig. 6.50). Recovery of many cold-worked metals (i.e., pure aluminum) produces a subgrain structure with low-angle grain boundaries, as shown in Fig. 6.49*b*. This recovery process is called *polygonization*, and often it is a structural change which precedes recrystallization. The internal energy of the recovered metal is lower than that of the cold-worked state since many dislocations are annihilated or moved into lower energy configurations by the recovery process. During recovery the strength of a cold-worked metal is reduced only slightly but its ductility is usually significantly increased (Fig. 6.47).

FIGURE 6.50 Schematic representation of polygonization in a deformed metal. (a) Deformed metal crystal showing dislocations piled up on slip planes. (b) After recovery heat treatment, dislocations move to form small-angle grain boundaries. (*After L. E. Tanner and I. S. Servi, in "Metals Handbook," vol. 8, 8th ed., American Society for Metals, 1973, p. 222.*)

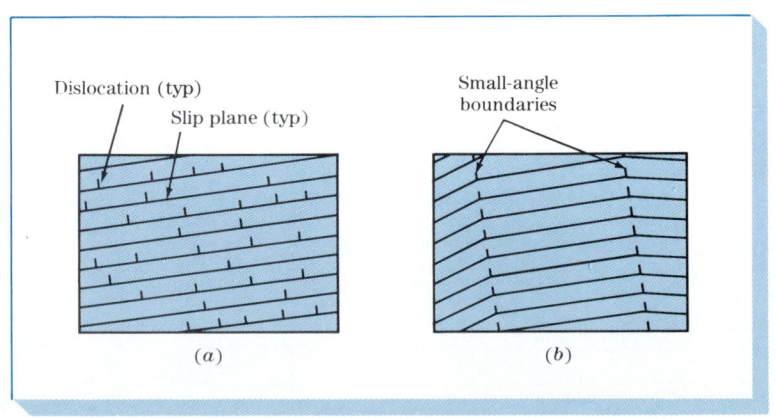

Recrystallization   Upon heating a cold-worked metal to a sufficiently high temperature, new strain-free grains are nucleated in the recovered metal structure and begin to grow (Fig. 6.48b), forming a recrystallized structure. After a long-enough time at a temperature at which recrystallization takes place, the cold-worked structure is completely replaced with a recrystallized grain structure, as shown in Fig. 6.48c.

Primary recrystallization occurs by two principal mechanisms: (1) an isolated nucleus can expand within a deformed grain (Fig. 6.51a) or (2) an original high-angle grain boundary can migrate into a more highly deformed region of the metal (Fig. 6.51b). In either case, the structure on the concave side of the moving boundary is strain-free and has a relatively low internal energy, whereas the structure on the convex side of the moving interface is highly strained with a high dislocation density and high internal energy. Grain-boundary movement is therefore away from the boundary's center of curvature. Thus the growth of an expanding new grain during primary recrystallization leads to an overall decrease in the internal energy of the metal by replacing deformed regions with strain-free regions.

The tensile strength of a cold-worked metal is greatly decreased and its ductility increased by an annealing treatment which causes the metal structure to be recrystallized. For example, the tensile strength of a 0.040-in (1-mm) sheet of 85% Cu–15% Zn brass which had been cold-rolled to 50 percent reduction was decreased from 75 to 45 ksi (520 to 310 MPa) by annealing 1 h at 400°C (Fig. 6.52a). The ductility of the sheet, on the other hand, was increased from 3 to 38 percent with the annealing treatment (Fig. 6.52b). Figure 6.53A shows a photo of the box annealing process for coils of steel. Figure 6.53B shows a schematic diagram of the continuous annealing process for sheet steel, and Fig. 6.53C shows a photo of an actual continuous annealing process line.

Important factors which affect the recrystallization process in metals and alloys are (1) amount of prior deformation of the metal, (2) temperature, (3) time, (4) initial grain size, and (5) composition of the metal or alloy. The recrystallization of a metal can take place over a range of temperatures, and the range is dependent to some extent on the above variables. Thus one cannot refer to the recrystallization temperature of a metal in the same sense as the

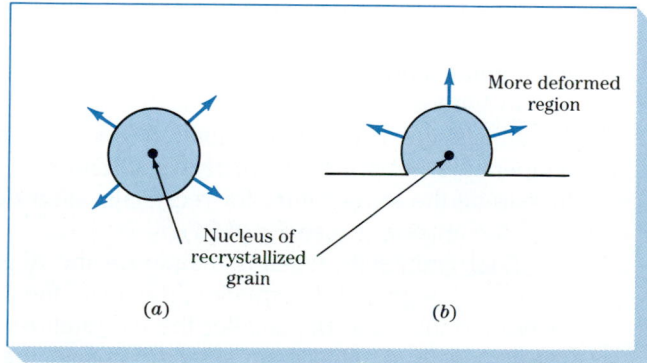

FIGURE 6.51 Schematic model of the growth of a recrystallized grain during the recrystallization of a metal. (a) Isolated nucleus expanded by growth within a deformed grain. (b) Original high-angle grain boundary migrating into a more highly deformed region of metal.

More deformed region

Nucleus of recrystallized grain

(a)   (b)

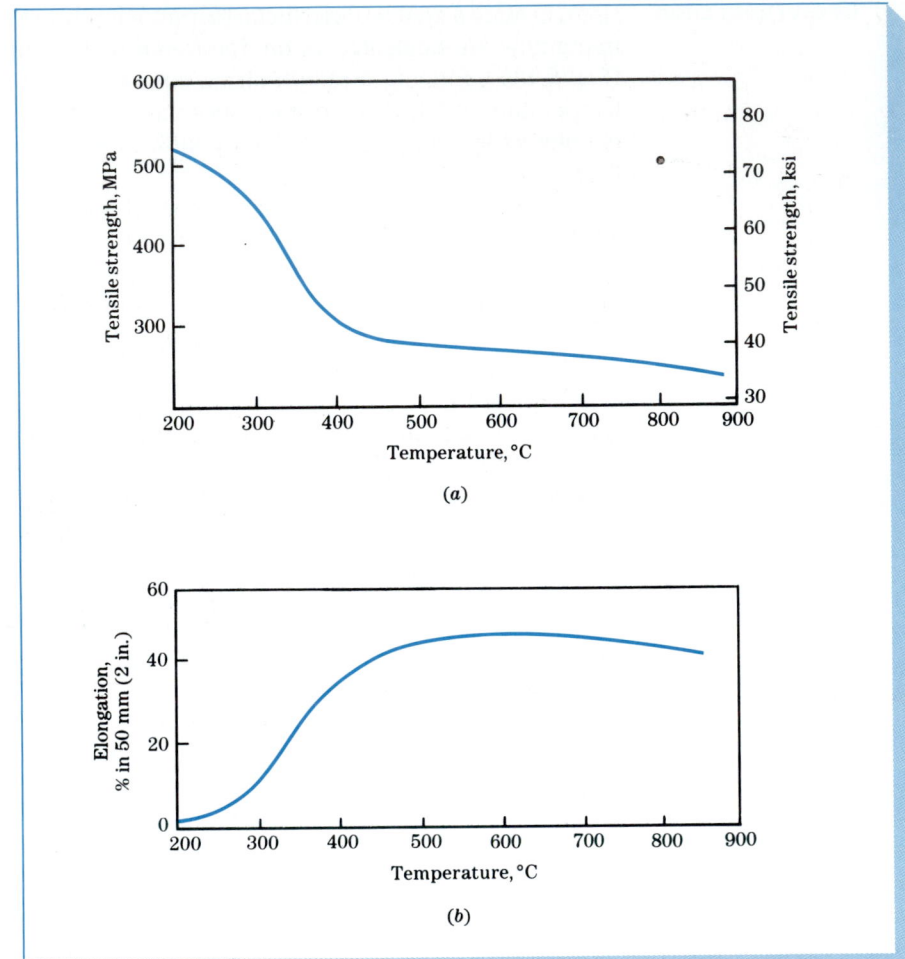

FIGURE 6.52 Effect of annealing temperature on (*a*) the tensile strength and (*b*) elongation of a 50 percent cold-rolled 85% Cu–15% Zn, 0.040-in (1-mm) thick sheet. (Annealing time was 1 h at temperature.) (*After "Metals Handbook," vol. 2, 9th ed., American Society for Metals, 1979, p. 320.*)

melting temperature of a pure metal. The following generalizations can be made about the recrystallization process:

1. A minimum amount of deformation of the metal is necessary for recrystallization to be possible.
2. The smaller the degree of deformation (above the minimum), the higher the temperature needed to cause recrystallization.
3. Increasing the temperature for recrystallization decreases the time necessary to complete it (see Fig. 6.54).
4. The final grain size depends mainly on the degree of deformation. The greater the degree of deformation, the lower the annealing temperature for recrystallization and the smaller the recrystallized grain size.
5. The larger the original grain size, the greater the amount of deformation required to produce an equivalent amount of recrystallization.

FIGURE 6.53A Furnaces for annealing coils of sheet steel. The coils are placed under cylindrical covers and then a furnace top is placed over the covered coils. In this box-annealing process the coils are held at a temperature of 1200 to 1300°F (650 to 700°C) for an average of 26 h. During the cooling period, a controlled deoxidizing atmosphere is provided to protect the surface of the steel coils. (*Courtesy of United States Steel Corporation.*)

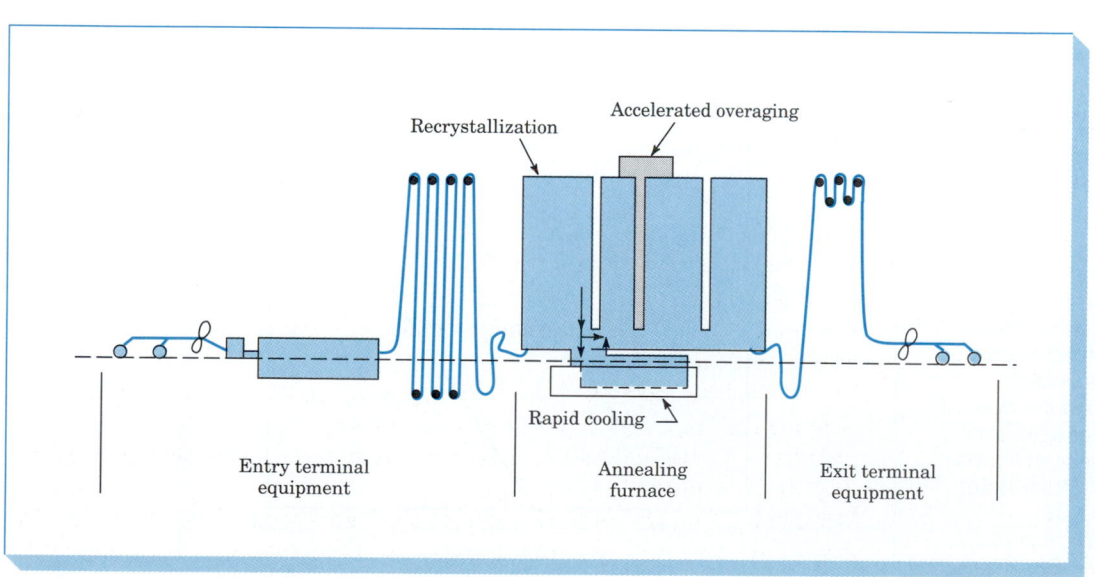

FIGURE 6.53B Continuous annealing schematic diagram. (*After W. L. Roberts, "Flat Processing of Steel," Marcel Dekker, 1988.*)

FIGURE 6.53C Photo of continuous annealing line for low-carbon sheet steel. (*Courtesy of Bethlehem Steel Co.*)

FIGURE 6.54 Time-temperature relations for the recrystallization of 99.0% Al cold-worked 75 percent. The solid line is for recrystallization finished and the dashed line for recrystallization started. Recrystallization in this alloy follows an Arrhenius-type relationship of ln t vs. $1/(T, K)$ (see Sec. 4.6). (*After "Aluminum," vol. 1, American Society for Metals, 1967, p. 98.*)

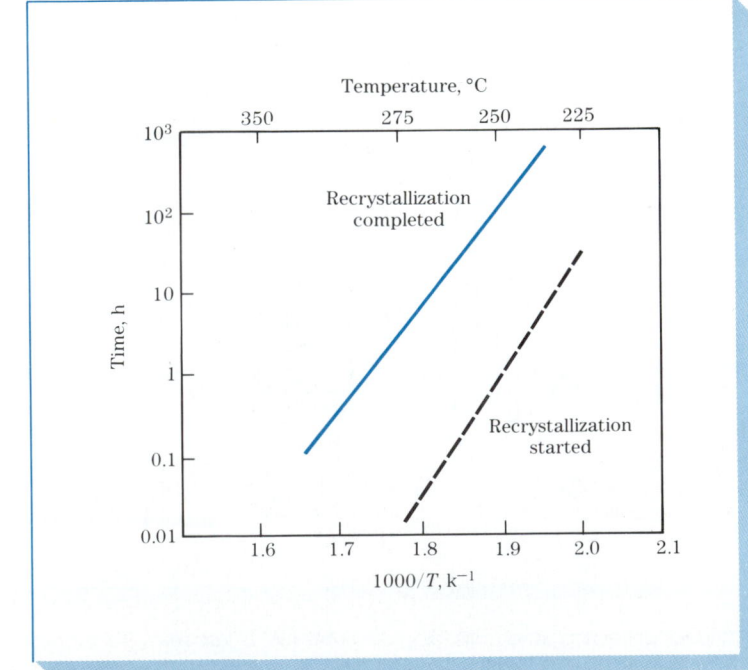

6. The recrystallization temperature decreases with increasing purity of the metal. Solid-solution alloying additions always increase the recrystallization temperature.

Example Problem 6.11

If it takes 9.0×10^3 min to recrystallize a piece of copper at 88°C and 200 min at 135°C, what is the activation energy for the process, assuming the process obeys the Arrhenius rate equation and the time to recrystallize $= Ce^{+Q/RT}$, where $R = 8.314$ J/(mol·K) and T is in kelvins?

Solution:

$$t_1 = 9.0 \times 10^3 \text{ min}; \ T_1 = 88°C + 273 = 361 \text{ K}$$
$$t_2 = 200 \text{ min}; \ T_2 = 135°C + 273 = 408 \text{ K}$$

$$t_1 = Ce^{Q/RT_1} \quad \text{or} \quad 9.0 \times 10^3 \text{ min} = Ce^{Q/R(361\,K)} \quad (6.16)$$
$$t_2 = Ce^{Q/RT_2} \quad \text{or 200 min} \quad = Ce^{Q/R(408\,K)} \quad (6.17)$$

Dividing Eq. (6.16) by (6.17) gives

$$45 = \exp\left[\frac{Q}{8.314}\left(\frac{1}{361} - \frac{1}{408}\right)\right]$$

$$\ln 45 = \frac{Q}{8.314}(0.00277 - 0.00245) = 3.80$$

$$Q = \frac{3.80 \times 8.314}{0.000319} = 99{,}038 \text{ J/mol or 99.0 kJ/mol} \blacktriangleleft$$

6.9 FRACTURE OF METALS

Fracture is the separation of a solid under stress into two or more parts. In general metal fractures can be classified as ductile or brittle but can be a mixture of the two. The *ductile fracture* of a metal occurs after extensive plastic deformation and is characterized by slow crack propagation. *Brittle fracture,* in contrast, usually proceeds along characteristic crystallographic planes called *cleavage planes* and has rapid crack propagation. Figure 6.55 shows an example of a ductile fracture in an aluminum alloy test specimen.

Ductile Fracture Ductile fracture of a metal occurs after extensive plastic deformation. For simplicity let us consider the ductile fracture of a round (0.50-in-diameter) tensile specimen. If a stress is applied to the specimen which exceeds its ultimate tensile strength and is sustained long enough, the specimen will fracture. Three

FIGURE 6.55 Ductile (cup-and-cone) fracture of an aluminum alloy.

distinct stages of ductile fracture can be recognized: (1) the specimen forms a neck, and cavities form within the necked region (Fig. 6.56a and b); (2) the cavities in the neck coalesce into a crack in the center of the specimen and propagate toward the surface of the specimen in a direction perpendicular to the applied stress (Fig. 6.56c); (3) when the crack nears the surface, the direction of the crack changes to 45° to the tensile axis and a cup-and-cone fracture results (Fig. 6.56d and e). Figure 6.57 shows internal cracks in the necked region of a deformed specimen of high-purity copper.

Brittle Fracture Many metals and alloys fracture in a brittle manner with very little plastic deformation. Brittle fracture usually proceeds along specific crystallographic planes called cleavage planes under a stress normal to the cleavage plane. Many metals with the HCP crystal structure commonly show brittle fracture because of their limited number of slip planes. A zinc single crystal, for example, under a high stress normal to the (0001) planes will fracture in a brittle manner. Many BCC metals such as α iron, molybdenum, and tungsten also fracture in a brittle manner at low temperatures and high strain rates.

Most brittle fractures in polycrystalline metals are transgranular; i.e., the cracks propagate across the matrix of the grains. However, brittle fracture can occur in an intergranular manner if the grain boundaries contain a brittle film or if the grain-boundary region has been embrittled by the segregation of detrimental elements.

Brittle fracture in metals is believed to take place in three stages:

1. Plastic deformation concentrates dislocations along slip planes at obstacles.
2. Shear stresses build up in places where dislocations are blocked, and as a result microcracks are nucleated.
3. Further stress propagates the microcracks, and stored elastic strain energy may also contribute to the propagation of the cracks.

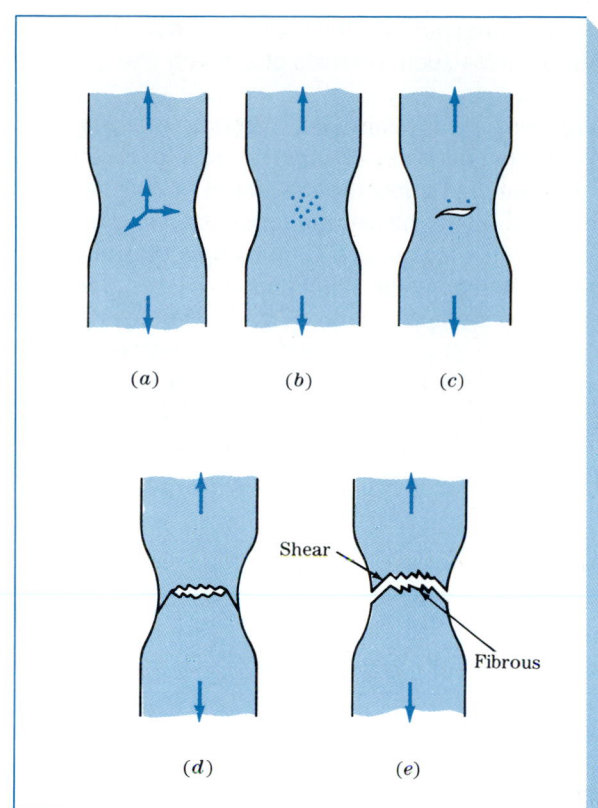

FIGURE 6.56 Stages in the formation of a cup-and-cone ductile fracture. (*After G. Dieter, "Mechanical Metallurgy," 2d ed., McGraw-Hill, 1976, p. 278.*)

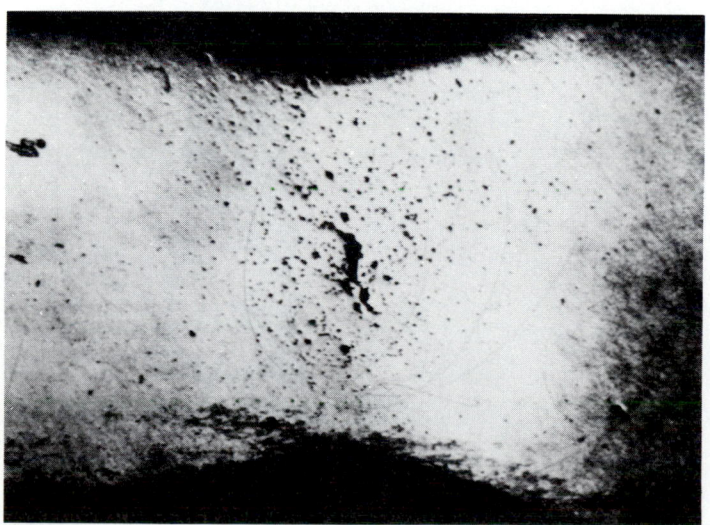

FIGURE 6.57 Internal cracking in the necked region of a polycrystalline specimen of high-purity copper. (Magnification 9×.) [*After K. E. Puttick, Philas. Mag.* **4**:964(1959).]

Low temperatures and high strain rates favor brittle fracture. Also a triaxial state of stress such as exists at a notch can contribute to brittle fracture.

Toughness and Impact Testing

Toughness is a measure of the amount of energy a material can absorb before fracturing. It becomes of engineering importance when the ability of a material to withstand an impact load without fracturing is considered. One of the simplest methods of measuring toughness is to use an impact-testing apparatus. A schematic diagram of a simple impact-testing machine is shown in Fig. 6.58. One method of using this apparatus is to place a Charpy V-notch specimen (shown in the upper part of Fig. 6.58) across parallel jaws in the machine. In the impact test a heavy pendulum released from a known height strikes the sample on its downward swing, fracturing it. By knowing the mass of the pendulum and the difference between its initial and final heights, the energy absorbed by the fracture can be measured. Figure 6.59 shows the relative effect of temperature on the impact energy of some types of materials.

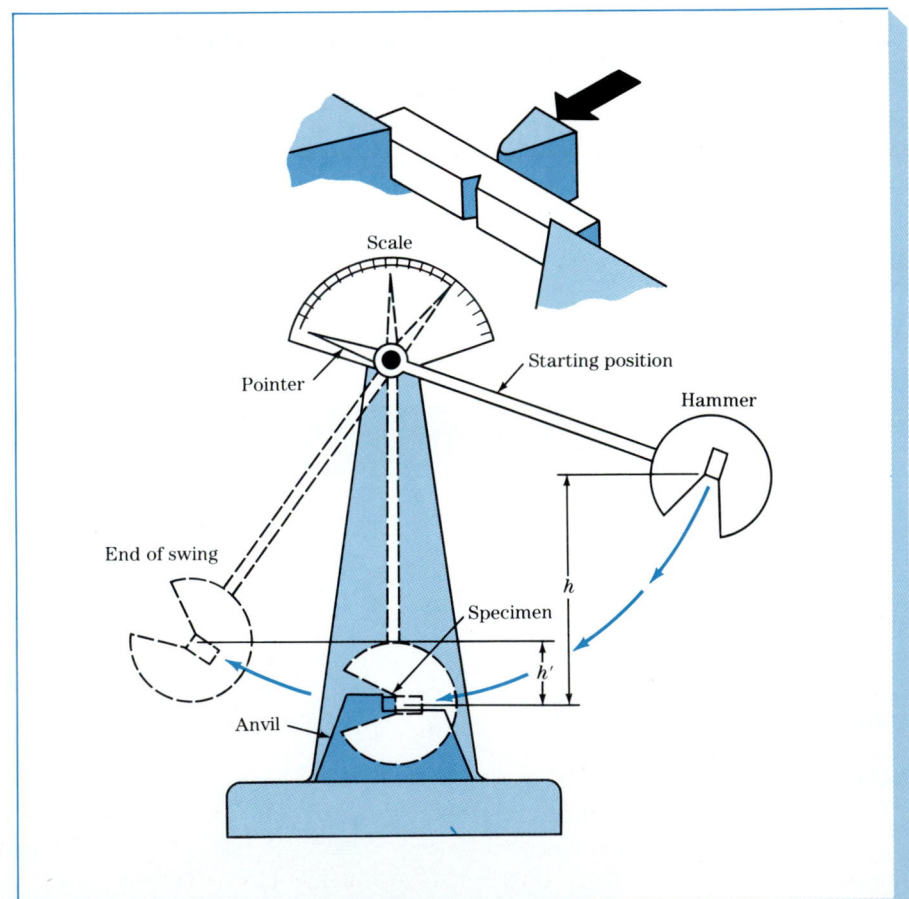

FIGURE 6.58 Schematic drawing of a standard impact-testing apparatus. (*After H. W. Hayden, W. G. Moffatt, and J. Wulff, "The Structure and Properties of Materials," vol. III, Wiley, 1965, p. 13.*)

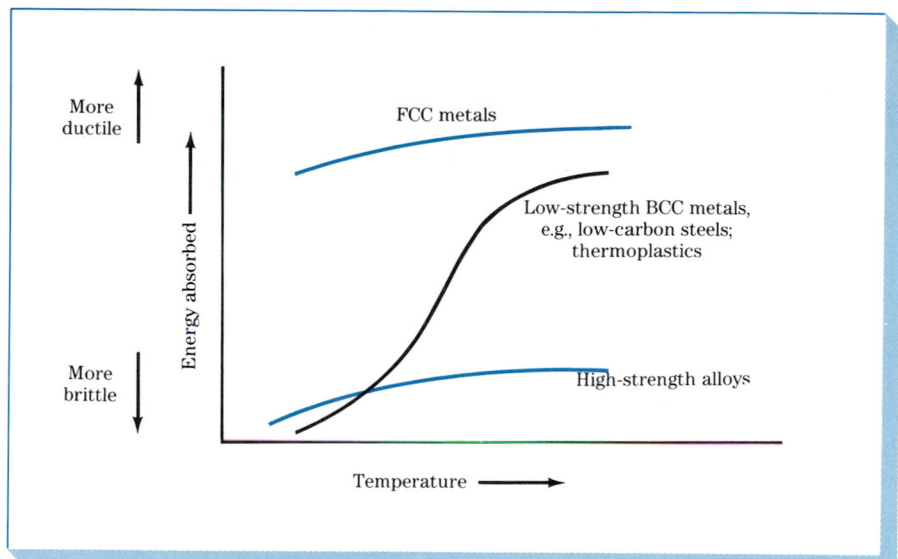

FIGURE 6.59 Effect of temperature on the energy absorbed upon impact by different types of materials.

This impact test can be used to determine the temperature range for the transition from ductile to brittle behavior for metals and alloys as the temperature is lowered. The carbon content of annealed steels affects this transition temperature range, as shown in Fig. 6.60. Low-carbon annealed steels have a lower-temperature transition range and a narrower one than high-carbon steels. Also, as the carbon content of the annealed steels is increased, the steels become more brittle and less energy is absorbed on impact during fracture.

Other more-sophisticated tests have been developed for measuring the "fracture toughness" of metals and alloys, but the discussion of these is beyond the scope of this book.

Fracture Toughness Impact tests such as the one previously described give quantitative comparative useful data with relatively simple test specimens and equipment. However, these tests do not provide property data for design purposes for material sections containing cracks or flaws. Data of this type are obtained from the discipline of fracture mechanics, in which theoretical and experimental analyses are made of the fracture of structural materials containing preexisting cracks or flaws. In this book we shall focus on the fracture toughness property of fracture mechanics and show how it can be applied for some simple design applications.

The fracture of a metal (material) starts at a place where the stress concentration is the highest, which may be at the top of a sharp crack, for example. Let us consider a plate sample under uniaxial tension which contains an edge crack (Fig. 6.61*a*) or a center-through crack (Fig. 6.61*b*). The stress at the tip of a sharp crack is highest at the tip as indicated in Fig. 6.61*c*.

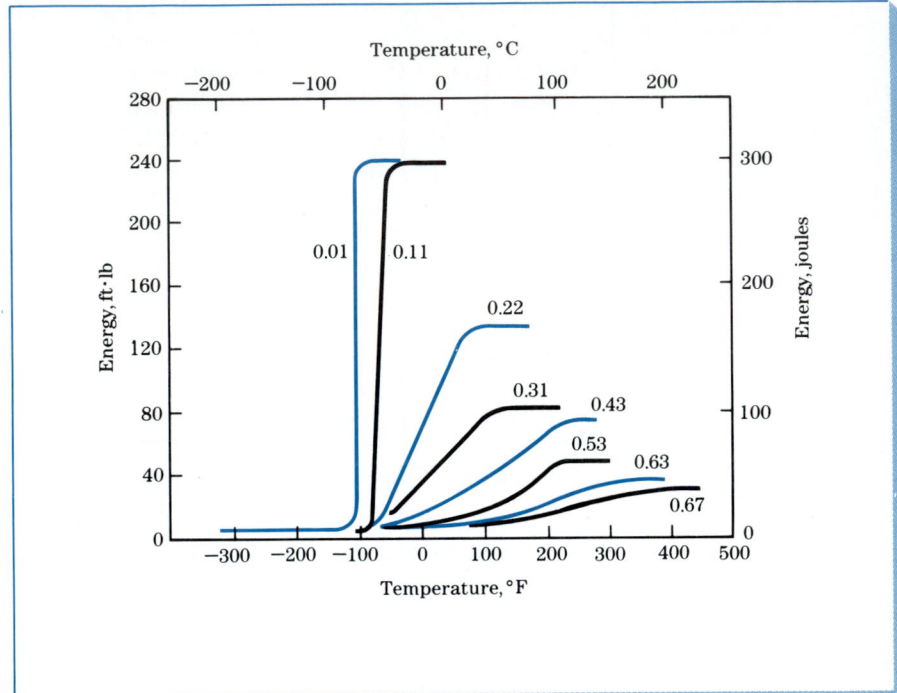

FIGURE 6.60 Effect of carbon content on the impact energy temperature plots for annealed steels. [*After J. A. Rinebolt and W. H. Harris, Trans. ASM,* **43**:*1175(1951).*]

FIGURE 6.61 Metal alloy plate under uniaxial tension (*a*) with edge crack *a*, (*b*) with center crack 2*a*. (*c*) Stress distribution vs. distance from crack tip. The stress is a maximum at the crack tip.

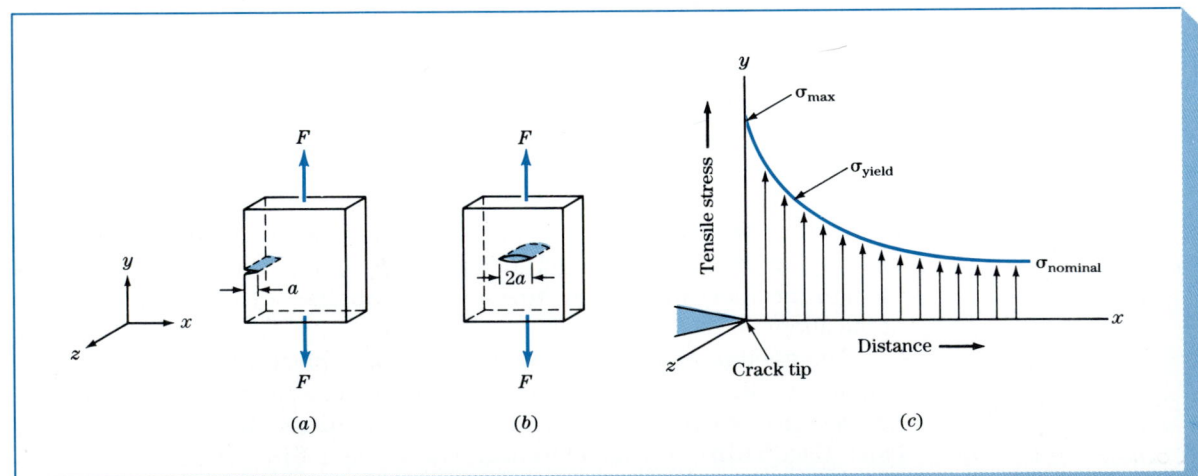

The stress intensity at the crack tip is found to be dependent on both the applied stress and the length of the crack. We use the stress-intensity factor K_1 to express the combination of the effects of the stress at the crack tip and the crack length. The subscript I (pronounced "one") indicates mode I testing in which a tensile stress causes the crack to open. By experiment, for the case of uniaxial tension on a metal plate containing an edge or internal crack (mode I testing), we find that

$$K_1 = Y\sigma\sqrt{\pi a} \tag{6.18}$$

where K_1 = stress-intensity factor
σ = applied nominal stress
a = edge-crack length or half the length of an internal through crack
Y = dimensionless geometric constant of the order of 1

The critical value of the stress-intensity factor that causes failure of the plate is called the *fracture toughness* K_{IC}, (pronounced "kay-one-see") of the material. In terms of the fracture stress σ_f and the crack length a for an edge crack (or one-half of the internal crack length),

FIGURE 6.62 Fracture-toughness test using a compact type of specimen and plain-strain conditions. (*a*) Dimensions of specimen. (*b*) Actual test at the critical stress for fracture, using a laser beam to detect this stress. (*Courtesy of White Shell Research.*)

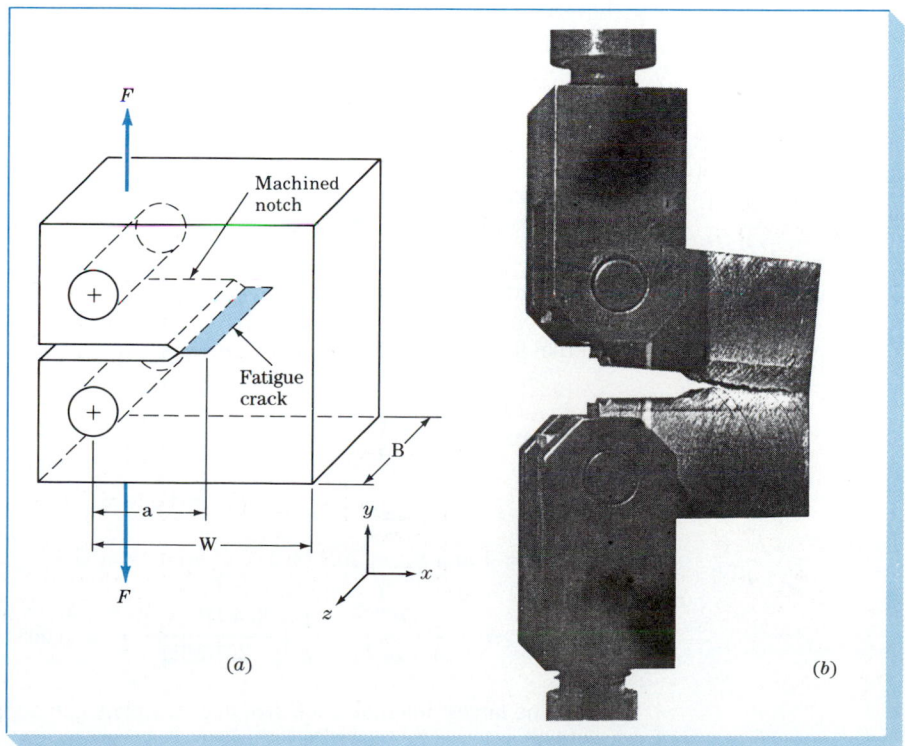

$$K_{IC} = Y\sigma_f\sqrt{\pi a} \tag{6.19}$$

Fracture-toughness (K_{IC}) values have the SI units of MPa$\sqrt{m}$ and U.S. customary units of ksi$\sqrt{in}$. Figure 6.62a is a schematic diagram of the compact type of fracture-toughness test specimen. To obtain constant values for K_{IC} the base dimension B of the specimen must be relatively large compared to the notch-depth dimension a so that so-called plain-strain conditions prevail. Plain-strain conditions require that during testing there is no strain deformation in the direction of the notch (i.e., in the z direction of Fig. 6.62a). Plain-strain conditions generally prevail when B (specimen thickness) $= 2.5(K_{IC}/\text{yield strength})^2$. Note that the fracture-toughness specimen has a machined notch and a fatigue crack at the end of the notch of about 3 mm depth to start the fracture during the test. Figure 6.62b shows a real fracture-toughness test at the time of rapid fracture.

Fracture-toughness values of materials are most useful in mechanical design when working with materials of limited toughness or ductility such as high-strength aluminum, steel, and titanium alloys. Table 6.5 lists some K_{IC} values for some of these alloys. Materials that show little plastic deformation before fracture have relatively low fracture toughness K_{IC} values and tend to be more brittle, whereas those with higher K_{IC} values are more ductile. Fracture-toughness values can be used in mechanical design to predict the allowable flaw size in alloys with limited ductility when acted upon by specific stresses. Example Problem 6.12 illustrates this design approach.

Example Problem 6.12

A structural plate component of an engineering design must support 207 MPa (30 ksi) in tension. If aluminum alloy 2024-T851 is used for this application, what is the largest internal flaw size that this material can support? (Use $Y = 1$).

Solution:

$$K_{IC} = Y\sigma_f\sqrt{\pi a} \tag{6.19}$$

Using $Y = 1$ and $K_{IC} = 26.4$ MPa $\sqrt{m}$ from Table 6.5,

$$a = \frac{1}{\pi}\left(\frac{K_{IC}}{\sigma_f}\right)^2 = \frac{1}{\pi}\left(\frac{26.4 \text{ MPa}\sqrt{m}}{207 \text{ MPa}}\right)^2 = 0.00518 \text{ m} = 5.18 \text{ mm}$$

Thus, the largest internal crack size that this plate can support is $2a$, or $(2)(5.18 \text{ mm})$ $= 10.36$ mm.

TABLE 6.5 **Typical Fracture-Toughness Values for Selected Engineering Alloys**

Material	K_{IC}		$\sigma_{yield\ strength}$	
	MPa$\sqrt{m}$	ksi$\sqrt{in}$	MPa	ksi
Aluminum alloys:				
2024-T851	26.4	24	455	66
7075-T651	24.2	22	495	72
7178-T651	23.1	21	570	83
Titanium alloy:				
Ti-6A1-4V	55	50	1035	150
Alloy steels:				
4340 (low-alloy steel)	60.4	55	1515	220
17-7 pH (precipitation hardening)	76.9	70	1435	208
350 maraging steel	55	50	1550	225

Source: R. W. Herzberg, "Deformation and Fracture Mechanics of Engineering Materials," 3d ed., Wiley, 1989.

6.10 FATIGUE OF METALS

In many types of service applications metal parts subjected to repetitive or cyclic stresses will fail at a much lower stress than that which the part can withstand under the application of a single static stress. These failures which occur under repeated or cyclic stressing are called *fatigue failures.* Examples of machine parts in which fatigue failures are common are moving parts such as shafts, connecting rods, and gears. Some estimates of failures in machines attribute about 80 percent to the direct action of fatigue failures.

A typical fatigue failure of a keyed steel shaft is shown in Fig. 6.63. A fatigue failure usually originates at a point of stress concentration such as a sharp corner or notch (Fig. 6.63), or at a metallurgical inclusion or flaw. Once nucleated, the crack propagates across the part under the cyclic or repeated stresses. During this stage of the fatigue process, clamshell or "beach" marks are created, as shown in Fig. 6.63. Finally, the remaining section becomes so small that it can no longer support the load, and complete fracture occurs. Thus there are usually two distinct types of surface areas that can be recognized: (1) a smooth surface region due to the rubbing action between the open surface region as the crack propagates across the section and (2) a rough surface area which is formed by the fracture when the load becomes too high for the remaining cross section. In Fig. 6.63 the fatigue crack had propagated almost through the entire cross section before final rupture occurred.

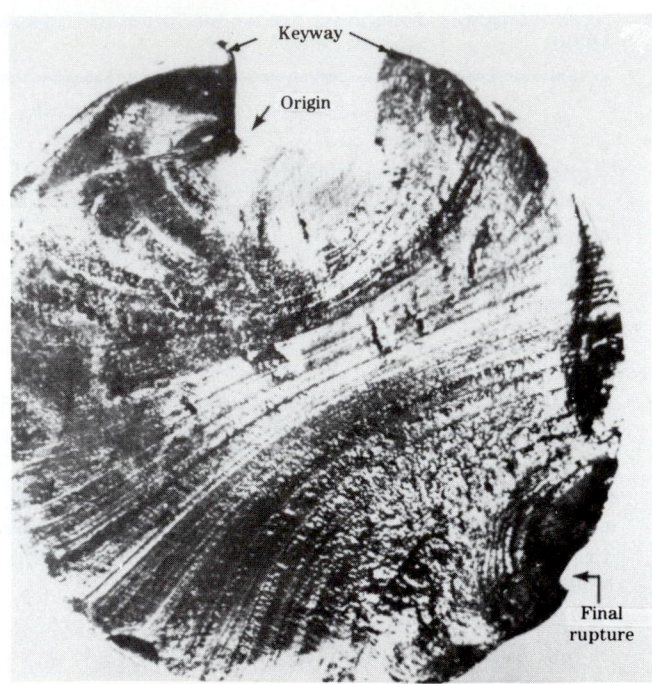

FIGURE 6.63 Light fractograph of the fatigue-fracture surface of a keyed shaft of 1040 steel (hardness ~ Rockwell C 30). The fatigue crack originated at the left bottom corner of the keyway and extended almost through the entire cross section before final rupture occurred. (Magnified $1\frac{7}{8}\times$.) (*After "Metals Handbook," vol. 9, 8th ed., American Society for Metals, 1974, p. 389.*)

Many types of tests are used to determine the fatigue life of a material. The most commonly used small-scale fatigue test is the rotating-beam test in which a specimen is subjected to alternating compression and tension stresses of equal magnitude while being rotated (Fig. 6.64). A sketch of the specimen for the R. R. Moore reversed-bending fatigue test is shown in Fig. 6.65. The

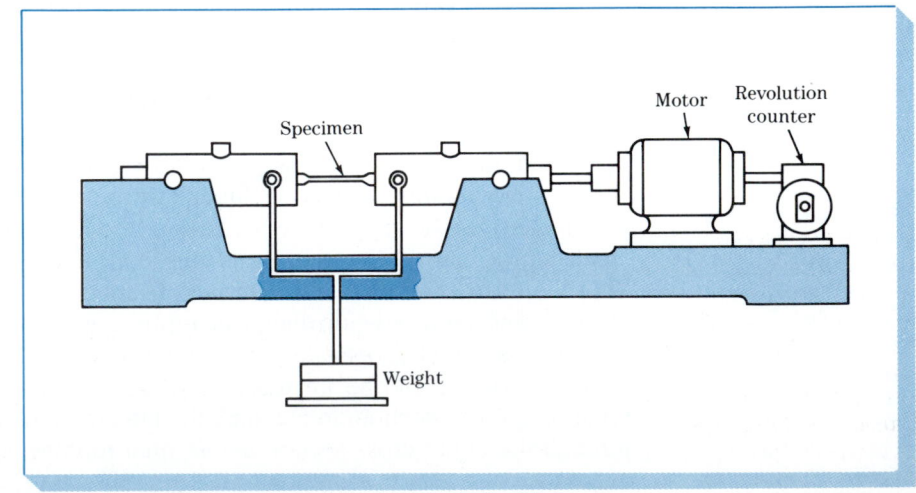

FIGURE 6.64 Schematic diagram of an R. R. Moore reversed-bending fatigue machine. (*After H. W. Hayden, W. G. Moffatt, and J. Wulff, "The Structure and Properties of Materials," vol. III, Wiley, 1965, p. 15.*)

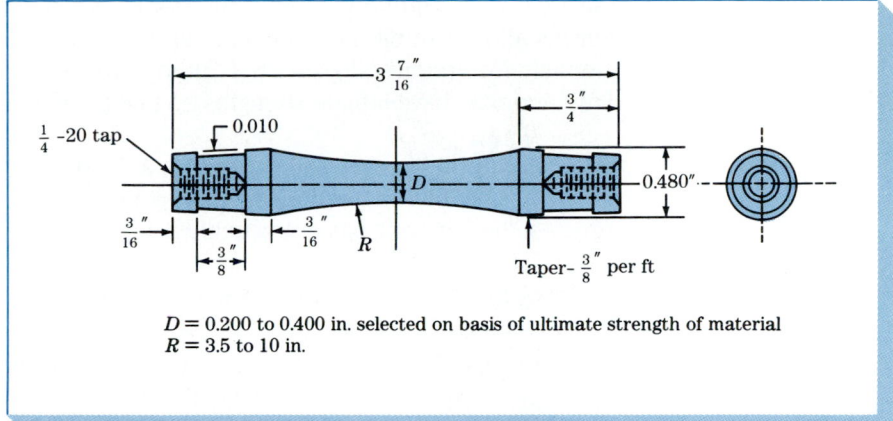

FIGURE 6.65 Sketch illustrating rotating-beam fatigue specimen (R. R. Moore type). (*From "Manual on Fatigue Testing," American Society for Testing and Materials, 1949.*)

$D = 0.200$ to 0.400 in. selected on basis of ultimate strength of material
$R = 3.5$ to 10 in.

surface of this specimen is carefully polished and tapered toward the center. Data from this test are plotted in the form of *SN* curves in which the stress *S* to cause failure is plotted against the number of cycles *N* at which failure occurs. Figure 6.66 shows typical *SN* curves for a high-carbon steel and a high-strength aluminum alloy. For the aluminum alloy, the stress to cause failure decreases as the number of cycles is increased. For the carbon steel, there is first a decrease in fatigue strength as the number of cycles is increased and then there is leveling off in the curve, with no decrease in fatigue strength as the number of cycles is increased. This horizontal part of the *SN* plot is called

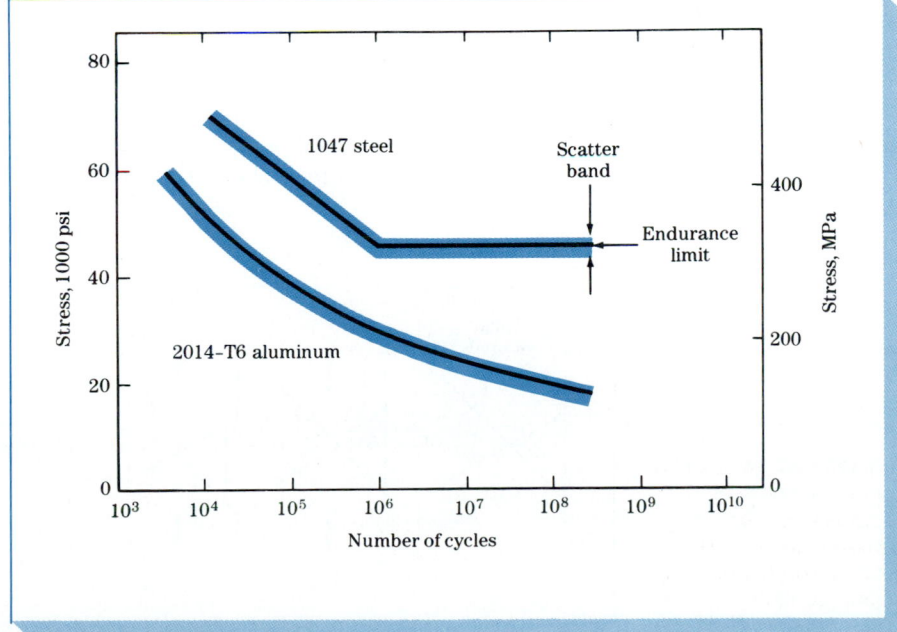

FIGURE 6.66 Stress vs. number of cycles (*SN*) curves for fatigue failure for aluminum alloy 2014-T6 and medium-carbon steel 1047. (*After H. W. Hayden, W. G. Moffatt, and J. Wulff, "The Structure and Properties of Materials," vol. III, Wiley, 1965, p. 15.*)

the *fatigue* or *endurance limit* and lies between 10^6 and 10^{10} cycles. Many ferrous alloys exhibit an endurance limit which is about one-half their tensile strength. Nonferrous alloys such as aluminum alloys do not have an endurance limit and may have fatigue strengths as low as one-third their tensile strength.

Basic structural changes which occur in a ductile metal in the fatigue process When a specimen of a ductile homogeneous metal is subjected to cyclic stresses, the following basic structural changes occur during the fatigue process:

1. *Crack initiation.* The early development of fatigue damage occurs.
2. *Slipband crack growth.* Crack initiation occurs because plastic deformation is not a completely reversible process. Plastic deformation in one direction and then alternatively in the reverse direction causes surface ridges and grooves called *slipband extrusions* and *slipband intrusions* to be created on the surface of the metal specimen (Fig. 6.67) as well as damage within the metal along *persistent slipbands.* The surface irregularities and damage along the persistent slipbands cause cracks to form at or near the surface which propagate into the specimen along the planes subjected to high shear stresses. This is called stage I of fatigue crack growth, and the rate of the crack growth is in general very low (for example, 10^{-10} m/cycle).
3. *Crack growth on planes of high tensile stress.* During stage I the crack may grow in a polycrystalline metal only a few grain diameters before it changes its direction to be perpendicular to the direction of the maximum tensile stress on the metal specimen. In this stage II of crack growth a well-defined crack propagates at a relatively rapid rate (e.g., micrometers per cycle), and fatigue striations are created as the crack advances across the cross section of the metal specimen (Fig. 6.63). These striations are useful in fatigue failure analysis in determining the origin and direction of propagation of fatigue cracks.
4. *Ultimate ductile failure.* Finally, when the crack covers a sufficient area so that the remaining metal at the cross section cannot support the applied load, the sample ruptures by ductile failure.

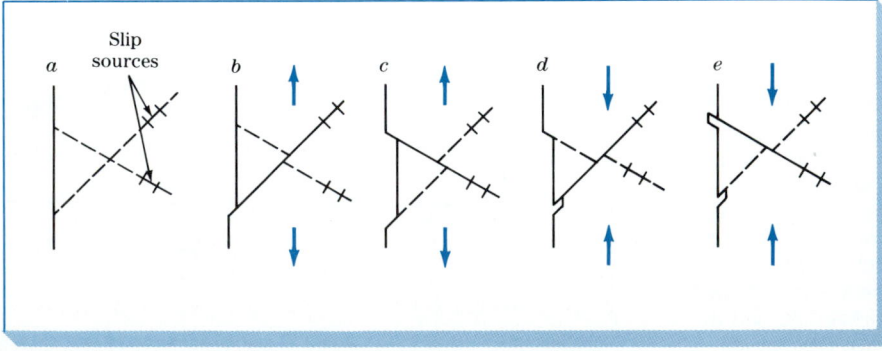

FIGURE 6.67 Mechanism for the formation of slipband extrusions and intrusions. [*After A. H. Cottrell and D. Hull, Proc. R. Soc. London,* **242A**:*211–213(1957).*]

Some major factors which affect the fatigue strength of a metal   The fatigue strength of a metal or alloy is affected by factors other than the chemical composition of the metal itself. Some of the most important of these are:

1. *Stress concentration.* Fatigue strength is greatly reduced by the presence of stress raisers such as notches, holes, keyways, or sharp changes in cross sections. For example, the fatigue failure shown in Fig. 6.63 started at the keyway in the steel shaft. Fatigue failures can be minimized by careful design to avoid stress raisers whenever possible.
2. *Surface roughness.* In general the smoother the surface finish on the metal sample, the higher the fatigue strength. Rough surfaces create stress raisers which facilitate fatigue crack formation.
3. *Surface condition.* Since most fatigue failures originate at the metal surface, any major change in the surface condition will affect the fatigue strength of the metal. For example, surface-hardening treatments for steels such as carburizing and nitriding which harden the surface increase fatigue life. Decarburizing, on the other hand, which softens a heat-treated steel surface, lowers fatigue life. The introduction of a favorable compressive residual stress pattern on the metal surface also increases fatigue life.
4. *Environment.* If a corrosive environment is present during the cyclic stress of a metal, the chemical attack greatly accelerates the rate at which fatigue cracks propagate. The combination of corrosion attack and cyclic stresses on a metal is known as *corrosion fatigue.*

6.11   CREEP AND STRESS RUPTURE OF METALS

Creep of Metals   When a metal or an alloy is under a constant load or stress, it may undergo progressive plastic deformation over a period of time. This *time-dependent strain* is called *creep.* The creep of metals and alloys is very important for some types of engineering designs, particularly those operating at elevated temperatures. For example, an engineer selecting an alloy for the turbine blades of a gas turbine engine must choose an alloy with a very low creep rate so that the blades can remain in service for a long period of time before having to be replaced due to their reaching the maximum allowable strain. For many engineering designs operating at elevated temperatures, the creep of materials is the limiting factor with respect to how high the operating temperature can be.

Let us consider creep of a pure polycrystalline metal at a temperature above one-half its absolute melting point, $\frac{1}{2}T_M$ (high-temperature creep). Let us also consider a creep experiment in which an annealed tensile specimen is subjected to a constant load of sufficient magnitude to cause extensive creep deformation. When the change of length of the specimen over a period of time is plotted against time increments, a *creep curve,* such as the one shown in Fig. 6.68, is obtained.

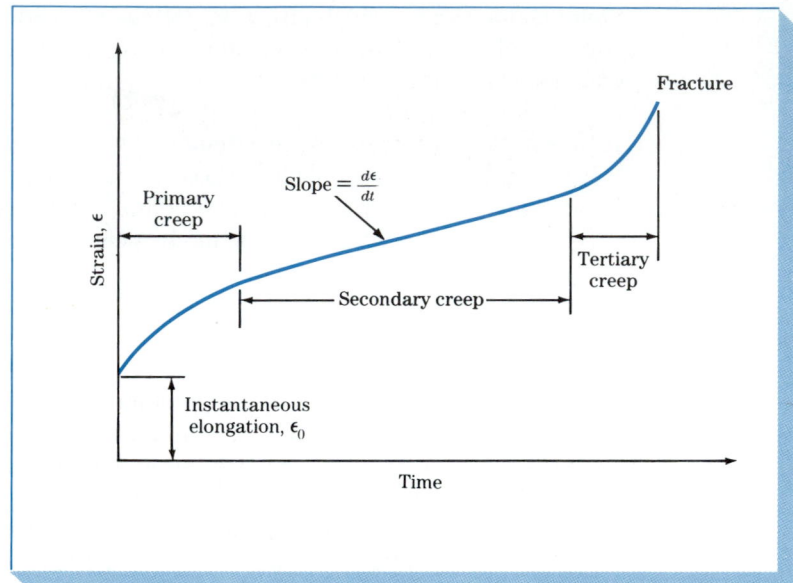

FIGURE 6.68 A typical creep curve for a metal. The curve represents the time vs. strain behavior of a metal or alloy under a constant load at constant temperature. The second stage of creep (linear creep) is of most interest to the design engineer for conditions in which extensive creep occurs.

In the idealized creep curve of Fig. 6.68, there is first an instantaneous rapid elongation of the specimen, ϵ_0. Following this, the specimen exhibits primary creep in which the strain rate decreases with time. The slope of the creep curve ($d\epsilon/dt$, or $\dot{\epsilon}$) is designated the *creep rate*. Thus during primary creep the creep rate progressively decreases with time. After primary creep, a second stage of creep occurs in which the creep rate is essentially constant and is therefore also referred to as *steady-state creep*. Finally, a third or tertiary stage of creep occurs in which the creep rate rapidly increases with time up to the strain at fracture. The shape of the creep curve depends strongly on the applied load (stress) and temperature. Higher stresses and higher temperatures increase the creep rate.

During primary creep the metal strain-hardens to support the applied load and the creep rate decreases with time as further strain hardening becomes more difficult. At higher temperatures (i.e., above about $0.5T_M$ for the metal) during secondary creep, recovery processes involving highly mobile dislocations counteract the strain hardening so that the metal continues to elongate (creep) at a steady-state rate (Fig. 6.68). The slope of the creep curve ($d\epsilon/dt = \dot{\epsilon}$) in the secondary stage of creep is referred to as the *minimum creep rate*. During secondary creep the creep resistance of the metal or alloy is the highest. Finally, for a constant-loaded specimen, the creep rate accelerates in the tertiary stage of creep due to necking of the specimen and also to the formation of voids, particularly along grain boundaries. Figure 6.69 shows intergranular cracking in a type 304L stainless steel which has undergone creep failure.

At low temperatures (i.e., below $0.4T_M$) and low stresses, metals show primary creep but negligible secondary creep since the temperature is too low

FIGURE 6.69 A jet engine turbine blade which has undergone creep deformation causing local deformation and a multiplicity of intergranular cracks. (*After J. Schijve in "Metals Handbook," vol. 10, 8th ed., American Society for Metals, 1975, p. 23.*)

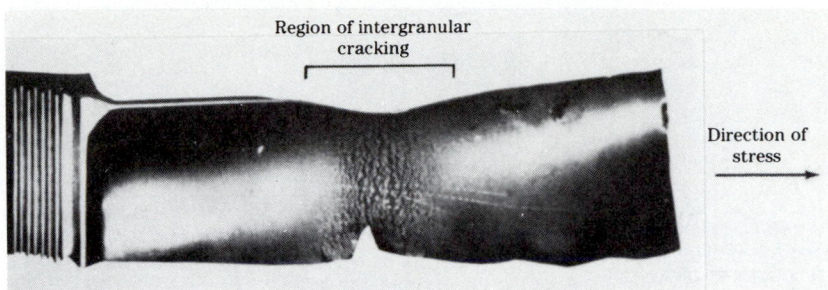

Region of intergranular cracking

Direction of stress

for diffusional recovery creep. However, if the stress on the metal is above the ultimate tensile strength, the metal will elongate as in an ordinary engineering tensile test. In general, as both the stress on the metal undergoing creep and its temperature are increased, the creep rate is also increased (Fig. 6.70).

The Creep Test The effects of temperatures and stress on the creep rate are determined by the creep test. Multiple creep tests are run using different stress levels at constant temperature or different temperatures at a constant stress, and the creep curves are plotted as shown in Fig. 6.71. The minimum creep rate or slope of the second stage of the creep curve is measured for each curve, as indicated in Fig. 6.71. The stress to produce a minimum creep rate of 10^{-5} percent/h at a

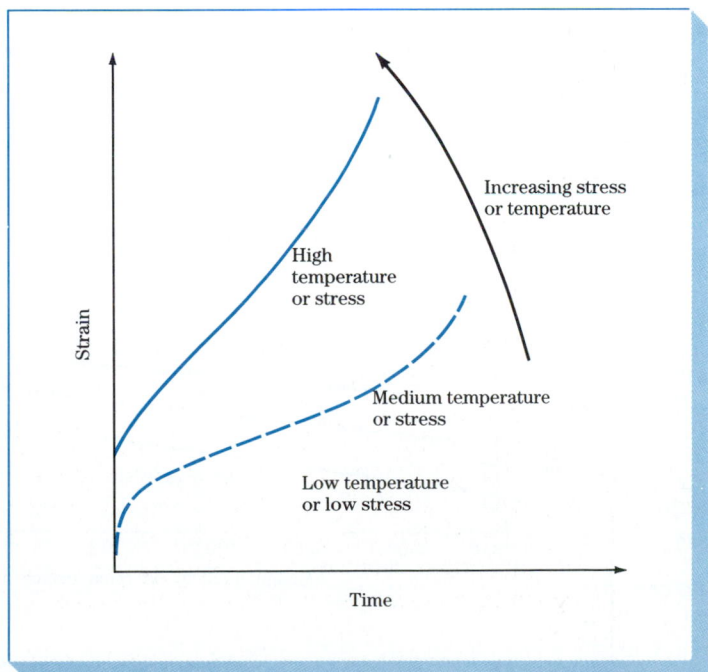

Increasing stress or temperature

High temperature or stress

Medium temperature or stress

Low temperature or low stress

Strain

Time

FIGURE 6.70 Effect of increasing stress on the shape of the creep curve of a metal (schematic). Note that as the stress increases, the strain rate increases.

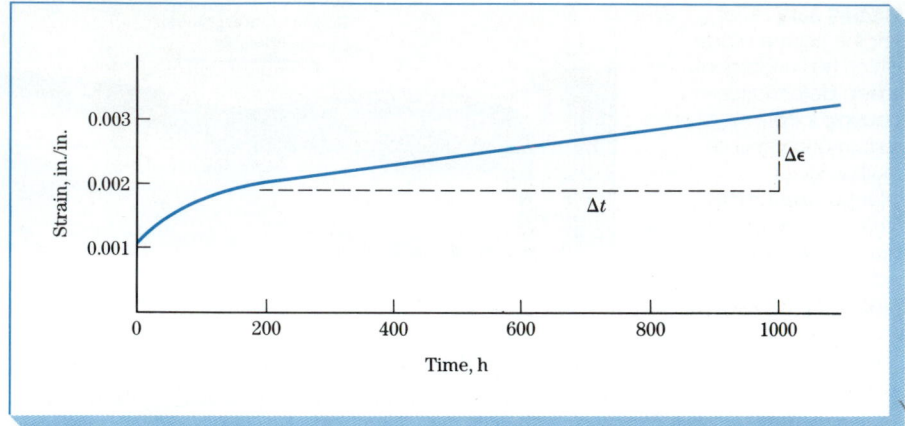

FIGURE 6.71 Creep curve for a copper alloy tested at 225°C and 230 MPa (33.4 ksi). The slope of the linear part of the curve is the steady-state creep rate. (See Example Problem 6.13.)

given temperature is a common standard for creep strength. In Fig. 6.72 the stress to produce a minimum creep rate of 10^{-5} percent/h for type 316 stainless steel can be determined by extrapolation.

Example Problem 6.13

Determine the steady-state creep rate for the copper alloy whose creep curve is shown in Fig. 6.71.

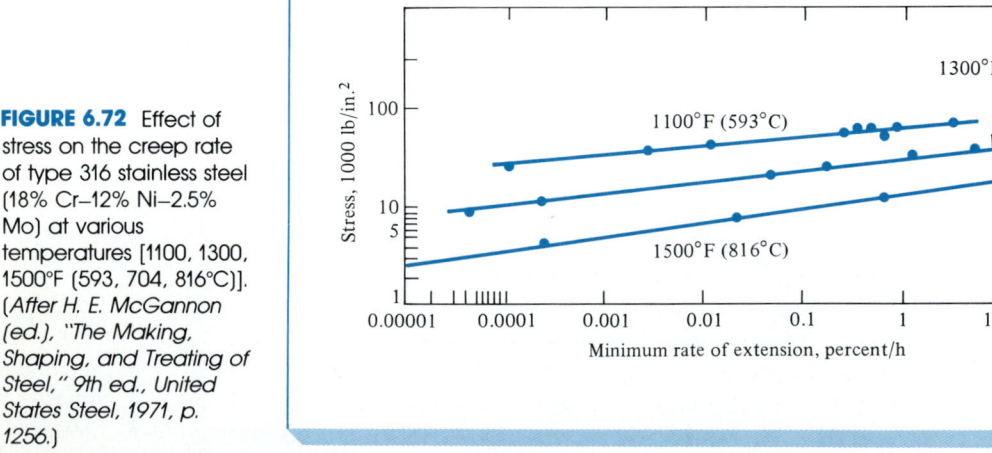

FIGURE 6.72 Effect of stress on the creep rate of type 316 stainless steel (18% Cr–12% Ni–2.5% Mo) at various temperatures [1100, 1300, 1500°F (593, 704, 816°C)]. (*After H. E. McGannon (ed.), "The Making, Shaping, and Treating of Steel," 9th ed., United States Steel, 1971, p. 1256.*)

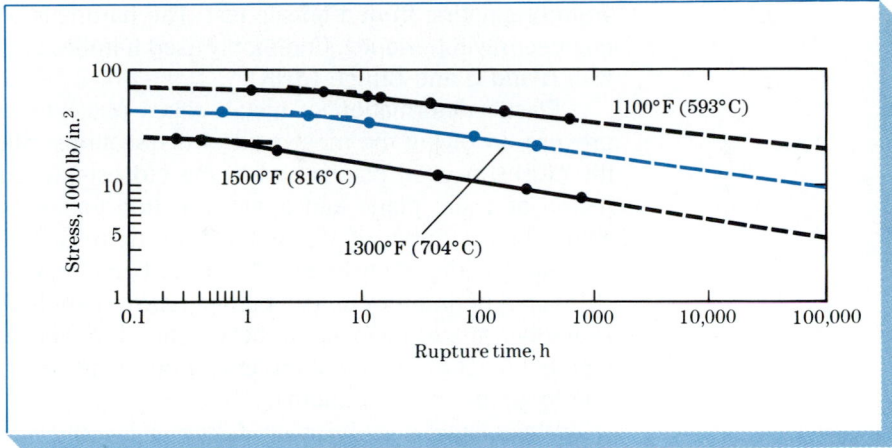

FIGURE 6.73 Effect of stress on the time to rupture of type 316 stainless steel (18% Cr–12% Ni–2.5% Mo) at various temperatures [1100, 1300, 1500°F (593, 704, 816°C)]. (*After H. E. McGannon (ed.), "The Making, Shaping, and Treating of Steel," 9th ed., United States Steel, 1971, p. 1257.*)

Solution:

The steady-state creep rate for this alloy for the creep curve shown in Fig. 6.71 is obtained by taking the slope of the linear part of the curve as indicated in the figure. Thus,

$$\text{Creep rate} = \frac{\Delta \epsilon}{\Delta t} = \frac{0.0029 - 0.0019}{1000 \text{ h} - 200 \text{ h}} = \frac{0.001 \text{ in/in}}{800 \text{ h}} = 1.2 \times 10^{-6} \text{ in/in/h} \blacktriangleleft$$

Creep-Rupture Test   The *creep-rupture* or *stress-rupture test* is essentially the same as the creep test except that the loads are higher and the test is carried out to failure of the specimen. Creep-rupture data are plotted as log stress vs. log rupture time, as shown in Fig. 6.73. In general the time for stress rupture to occur is decreased as the applied stress and temperature are increased. Slope changes as observed in Fig. 6.73 are caused by factors such as recrystallization, oxidation, corrosion, or phase changes.

SUMMARY

Metals and alloys are processed into different shapes by various manufacturing methods. Some of the most important industrial processes are casting, rolling, extruding, wire drawing, forging, and deep drawing.

When a uniaxial stress is applied to a long metal bar, the metal deforms elastically at first and then plastically, causing permanent deformation. For many engineering designs the engineer is interested in the 0.2 percent offset yield strength, ultimate tensile strength, and elongation (ductility) of a metal or alloy. These quantities are obtained from the engineering stress-strain diagram originating from a tensile test. The hardness of a metal may also be of engineering importance. Commonly used hardness scales in industry are Rockwell B and C and Brinell (BHN).

Plastic deformation of metals takes place most commonly by the slip process, involving the movement of dislocations. Slip usually takes place on the closest-packed planes and in the closest-packed directions. The combination of a slip plane and a slip direction constitutes a slip system. Metals with a high number of slip systems are more ductile than those with only a few slip systems. Many metals deform by twinning when slip becomes difficult.

Grain boundaries at lower temperatures usually strengthen metals by providing barriers to dislocation movement. However, under some conditions of high-temperature deformation, grain boundaries become regions of weakness due to grain-boundary sliding.

When a metal is plastically deformed by cold working, the metal becomes strain-hardened, resulting in an increase in its strength and a decrease in its ductility. The strain hardening can be removed by giving the metal an annealing heat treatment. When the strain-hardened metal is slowly heated to a high temperature below its melting temperature, the processes of recovery, recrystallization, and grain growth take place, and the metal is softened. By combining strain hardening and annealing, large thickness reductions of metal sections can be accomplished without fracture.

Simple tension metal failures can be classified according to ductile, brittle, or ductile-brittle types. A metal may also fracture due to fatigue failure if it is subjected to cyclic tension and compression stress of sufficient magnitude. At high temperatures and stresses a metal may undergo creep, or time-dependent deformation. The creep of metal may be so severe that fracture of the metal occurs. Extensive engineering tests are made to prevent fatigue and creep failure of manufactured products.

DEFINITIONS

Sec. 6.1

Hot working of metals: permanent deformation of metals and alloys above the temperature at which a strain-free microstructure is produced continuously (recrystallization temperature).

Cold working of metals: permanent deformation of metals and alloys below the temperature at which a strain-free microstructure is produced continuously (recrystallization temperature). Cold working causes a metal to be strain-hardened.

Percent cold reduction:

$$\% \text{ cold reduction} = \frac{\text{change in cross-sectional area}}{\text{original cross-sectional area}} \times 100\%$$

Annealing: a heat treatment given to a metal to soften it.

Extrusion: a plastic forming process in which a material under high pressure is reduced in cross section by forcing it through an opening in a die.

Forging: a primary processing method for working metals into useful shapes in which the metal is hammered or pressed into shape.

Wire drawing: a process in which wire stock is drawn through one or more tapered dies to the desired cross section.

Sec. 6.2

Elastic deformation: if a metal deformed by a force returns to its original dimensions after the force is removed, the metal is said to be elastically deformed.

Engineering stress σ: average uniaxial force divided by original cross-sectional area ($\sigma = F/A_0$).

Engineering strain ϵ: change in length of sample divided by the original length of sample ($\epsilon = \Delta l/l_0$).

Shear stress τ: shear force S divided by the area A over which the shear force acts ($\tau = S/A$).

Shear strain γ: shear displacement a divided by the distance h over which the shear acts ($\gamma = a/h$).

Sec. 6.3

Engineering stress-strain diagram: experimental plot of engineering stress vs. engineering strain; σ is normally plotted as the y axis and ϵ as the x axis.

Modulus of elasticity E: stress divided by strain (σ/ϵ) in the elastic region of an engineering stress-strain diagram for a metal ($E = \sigma/\epsilon$).

Yield strength: the stress at which a specific amount of strain occurs in the engineering tensile test. In the U.S. the yield strength is determined for 0.2 percent strain.

Ultimate tensile strength (UTS): the maximum stress in the engineering stress-strain diagram.

Sec. 6.4

Hardness: a measure of the resistance of a material to permanent deformation.

Sec. 6.5

Slip: the process of atoms moving over each other during the permanent deformation of a metal.

Slipbands: line markings on the surface of a metal due to slip caused by permanent deformation.

Slip system: a combination of a slip plane and a slip direction.

Sec. 6.6

Deformation twinning: a plastic deformation process which occurs in some metals and under certain conditions. In this process a large group of atoms are displaced together to form a region of a metal crystal lattice which is a mirror image of a similar region along a twinning plane.

Strain hardening (strengthening): the hardening of a metal or alloy by cold working. During cold working, dislocations multiply and interact, leading to an increase in strength of the metal.

Sec. 6.7

Solid-solution hardening (strengthening): strengthening a metal by alloying additions which form solid solutions. Dislocations have more difficulty moving through a metal lattice when the atoms are different in size and electrical characteristics, as is the case with solid solutions.

Sec. 6.8

Recovery: the first stage in the removal of the effects of cold working when a cold-worked metal is slowly heated with increasing temperature. During recovery, internal stresses are relieved and some dislocation rearrangement into lower-energy configurations take place. Some dislocations are annihilated.

Recrystallization: the process whereby a cold-worked metal is heated to a sufficiently high temperature for a long-enough time to form a new strain-free grain structure. During recrystallization the dislocation density of the metal is greatly reduced.

Sec. 6.9

Ductile fracture: a mode of fracture characterized by slow crack propagation. Ductile fracture surfaces of metals are usually dull with a fibrous appearance.

Brittle fracture: a mode of fracture characterized by rapid crack propagation. Brittle fracture surfaces of metals are usually shiny and have a granular appearance.

Sec. 6.11

Creep: time-dependent deformation of a material when subjected to a constant load or stress.

PROBLEMS

6.1.1   How are metal alloys made by the casting process?

6.1.2   Distinguish between wrought alloy products and cast alloy products.

6.1.3   Why are cast metal sheet ingots hot-rolled first instead of being cold-rolled?

6.1.4   What type of heat treatment is given to the rolled metal sheet after hot and "warm" rolling? What is its purpose?

6.1.5   Calculate the percent cold reduction after cold rolling 0.050-in-thick aluminum sheet to 0.035 in.

6.1.6   A 70% Cu–30% Zn brass sheet is 0.0850 cm thick and is cold-rolled with a 25 percent reduction in thickness. What must be the final thickness of the sheet?

6.1.7   A sheet of an aluminum alloy is cold-rolled 20 percent to a thickness of 0.070 in. If the sheet is then cold-rolled to a final thickness of 0.040 in, what is the total percent cold work done?

6.1.8   Describe and illustrate the following types of extrusion processes: (*a*) direct extrusion and (*b*) indirect extrusion. What is an advantage of each process?

6.1.9   Which process in Prob. 6.1.8 is used most commonly? Which metals and alloys are commonly extruded?

6.1.10   Describe the forging process. What is the difference between hammer and press forging?

6.1.11   What is the difference between open-die and closed-die forging? Illustrate. Give an example of a metal product produced by each process.

6.1.12 Describe the wire-drawing process. Why is it necessary to make sure the surface of the incoming wire is clean and lubricated?

6.1.13 Calculate the percent cold reduction when an aluminum wire is cold-drawn from a diameter of 4.24 mm to a diameter of 2.75 mm.

6.1.14 0.110-mm diameter 99.5% copper wire is to be cold-drawn with a 25 percent cold reduction. What must be the final diameter of the wire?

6.1.15 A brass wire is cold-drawn 20 percent to a diameter of 1.30 mm. It is then further cold-drawn to 1.05 mm. What is the total percent cold reduction?

6.1.16 A copper wire is cold-drawn 25 percent to a diameter of 1.10 mm. It is then further cold-drawn to 0.80 mm. What is the total percent cold reduction?

6.2.1 Distinguish between elastic and plastic deformation.

6.2.2 Define engineering stress. What are the U.S. customary and SI units for engineering stress?

6.2.3 Calculate the engineering stress in SI units on a 2.40-cm-diameter rod which is subjected to a load of 1450 kg.

6.2.4 Calculate the engineering stress in SI units on a bar 12 cm long and having a cross section of 4.00 mm × 8.50 mm which is subjected to a load of 4500 kg.

6.2.5 Calculate the engineering stress in SI units on a bar 20 cm long and having a cross section of 3.50 mm × 5.00 mm which is subjected to a load of 4200 kg.

6.2.6 Calculate the engineering stress in U.S. customary units on a 0.320-in-diameter rod which is subjected to a force of 1100 lb_f.

6.2.7 What is the relationship between engineering strain and percent elongation?

6.2.8 A tensile specimen of cartridge brass sheet has a cross section of 0.500 in × 0.015 in and a gage length of 2.00 in. Calculate the engineering strain which occurred during a test if the distance between gage markings is 2.45 in after the test.

6.3.1 A 0.505-in-diameter rod of an aluminum alloy is pulled to failure in a tension test. If the final diameter of the rod at the fractured surface is 0.435 in, what is the percent reduction in area of the sample due to the test?

6.3.2 The following engineering stress-strain data were obtained for a 0.2% C plain-carbon steel. (*a*) Plot the engineering stress-strain curve. (*b*) Determine the ultimate tensile strength of the alloy. (*c*) Determine the percent elongation at fracture.

Engineering stress, ksi	Engineering strain, in/in	Engineering stress ksi	Engineering strain, in/in
0	0	76	0.08
30	0.001	75	0.10
55	0.002	73	0.12
60	0.005	69	0.14
68	0.01	65	0.16
72	0.02	56	0.18
74	0.04	51	(Fracture) 0.19
75	0.06		

6.3.3 Plot the data of Prob. 6.3.2 as engineering stress (MPa) vs. engineering strain (mm/mm) and determine the ultimate strength of the steel.

6.3.4 The following engineering stress-train data were obtained at the beginning of a tensile test for a 0.2% C plain-carbon steel. (*a*) Plot the engineering stress-strain curve for these data. (b) Determine the 0.2 percent offset yield stress for this steel. (c) Determine the tensile elastic modulus of this steel. (Note that these data only give the beginning part of the stress-strain curve.)

Engineering stress, ksi	Engineering strain, in/in	Engineering stress ksi	Engineering strain, in/in
0	0	60	0.0035
15	0.0005	66	0.004
30	0.001	70	0.006
40	0.0015	72	0.008
50	0.0020		

6.3.5 Plot the data of Problem 6.3.4 as engineering stress (MPa) vs. engineering strain (mm/mm) and determine the 0.2 percent offset yield stress of the steel.

6.3.6 A 1.28-cm-diameter aluminum alloy test bar is subjected to a load of 110,000 N. If the diameter of the bar is 1.24 cm at this load, determine (a) the engineering stress and strain and (b) the true stress and strain.

6.3.7 A 0.500-in-diameter test bar is subjected to a load of 21,000 lb. If the diameter of the bar is 0.450 in at this load, determine the (a) the engineering stress and strain and (b) the true stress and strain.

6.3.8 A 20-cm-long rod with a diameter of 0.20 cm is loaded with a 3000-N weight. If the diameter decreases to 0.170 cm, determine (a) the engineering stress and strain at this load and (b) the true stress and strain at this load.

6.4.1 Define the hardness of a metal.

6.4.2 How is the hardness of a material determined by a hardness-testing machine?

6.4.3 What types of indenters are used in (a) the Brinell hardness test, (b) Rockwell C hardness, and (c) Rockwell B hardness?

6.5.1 What are slipbands and slip lines? What causes the formation of slipbands on a metal surface?

6.5.2 Describe the slip mechanism which enables a metal to be plastically deformed without fracture.

6.5.3 Why does slip in metals usually take place on the densest-packed planes?

6.5.4 Why does slip in metals usually take place in the closest-packed directions?

6.5.5 What are the principal slip planes and slip directions for FCC metals?

6.5.6 For the FCC octahedral crystal planes, draw each lower octahedral plane separately and indicate on each plane the six slip directions (see Fig. 6.35).

6.5.7 What are the principal slip planes and slip directions for BCC metals?

6.5.8 What are the principal slip planes and slip directions for HCP metals?

6.5.9 What other types of slip planes are important other than the basal planes for HCP metals with low c/a ratios?

6.5.10 What is the critical resolved shear stress for a pure-metal single crystal?

6.5.11 Why do pure FCC metals like Ag and Cu have low values of τ_c?

6.5.12 What is believed to be responsible for the high values of τ_c for HCP titanium?

6.5.13 A stress of 60 MPa is applied in the [001] direction on an FCC single crystal. Calculate (a) the resolved shear stress acting on the (111) [$\bar{1}$01] slip system and (b) the resolved shear stress acting on the (111) [$\bar{1}$10] slip system.

6.5.14 A stress of 50 MPa is applied in the [001] direction of a BCC single crystal. Calculate (a) the resolved shear stress acting on the (101) [$\bar{1}$11] system and (b) the resolved shear stress acting on the (110) [$\bar{1}$11] system.

6.5.15 Determine the tensile stress that must be applied to the [1$\bar{1}$0] axis of a high-purity copper single crystal to cause slip on the (1$\bar{1}$1) [0$\bar{1}$1] system. The resolved shear stress for the crystal is 0.70 MPa.

6.5.16 A stress of 2.30 MPa is applied in the [00$\bar{1}$] direction of a unit cell of an FCC copper single crystal. Calculate the resolved shear stress on the (11$\bar{1}$) plane in the following directions: (a) [$\bar{1}$0$\bar{1}$], (b) [0$\bar{1}$$\bar{1}$], and (c) [$\bar{1}$10].

6.5.17 A stress of 1.30 MPa is applied in the [001] direction of a unit cell of an FCC silver single crystal. Calculate the resolved shear stress on the (1$\bar{1}$1) plane in the following directions: (a) [$\bar{1}$01], (b) [011], and (c) [110].

6.5.18 A stress of 1.59 MPa is applied in the [001] direction of a unit cell of an FCC copper single crystal. Calculate the resolved shear stress on the ($\bar{1}$11) plane in the following directions: (a) [101], (b) [110], and (c) [0$\bar{1}$1].

6.5.19 A stress of 67 MPa is applied in the [001] direction of a unit cell of a BCC iron single crystal. Calculate the resolved shear stress for the following slip systems: (a) (011) [1$\bar{1}$1], (b) (110) [$\bar{1}$11], and (c) (0$\bar{1}$1) [111].

6.5.20 A stress of 75 MPa is applied in the [001] direction of a unit cell of a BCC iron single crystal. Calculate the resolved shear stress for the following slip systems: (a) (011) [$\bar{1}$$\bar{1}$1], (b) (1$\bar{1}$0) [111], and (c) ($\bar{1}$01) [111].

6.5.21 Describe the deformation twinning process which occurs in some metals when plastically deformed.

6.5.22 What is the difference between the slip and twinning mechanisms of plastic deformation of metals?

6.5.23 What important role does twinning play in the plastic deformation of metals with regard to deformation of metals by slip?

6.5.24 Why is deformation by twinning especially important for HCP metals?

6.6.1 By what mechanism do grain boundaries strengthen metals?

6.6.2 What experimental evidence shows that grain boundaries arrest slip in polycrystalline metals?

6.6.3 Describe the grain shape changes that occur when a sheet of alloyed copper with an original equiaxed grain structure is cold-rolled with 30 and 50 percent cold reductions.

6.6.4 What happens to the dislocation substructure in Prob. 6.6.3?

6.6.5 How is the ductility of a metal normally affected by cold working? Why?

6.6.6 An oxygen-free copper rod must have a tensile strength of 45.0 ksi and a final diameter of 0.240 in. (a) What amount of cold work must the rod undergo (see Fig. 6.45)? (b) What must the initial diameter of the rod be?

6.6.7 A 70% Cu–30% Zn brass sheet is to be cold-rolled from 0.080 to 0.065 in. (a) Calculate the percent cold work, and (b) estimate the tensile strength, yield strength, and elongation from Fig. 6.46.

6.6.8 A 70% Cu–30% Zn brass wire is cold-drawn 25 percent to a diameter of 2.40 mm. The wire is then further cold-drawn to a diameter of 2.10 mm. (a) Calculate the total percent cold work which the wire undergoes. (b) Estimate the wire's tensile and yield strengths and elongation from Fig. 6.46.

6.7.1 What is solid-solution strengthening? Describe the two main types.

6.7.2 What are two important factors which affect solid-solution hardening?

6.8.1 What are the three main metallurgical stages that a sheet of a cold-worked metal such as aluminum or copper goes through as it is heated from room temperature to an elevated temperature just below its melting point?

6.8.2 Describe the microstructure of a heavily cold worked metal of an aluminum−0.8% magnesium alloy as observed with an optical microscope at 100 × (see Fig. 6.48a). Describe the microstructure of the same material at 20,000 × (see Fig. 6.49a).

6.8.3 Describe what occurs microscopically when a cold-worked sheet of metal such as aluminum undergoes a recovery heat treatment.

6.8.4 When a cold-worked metal is heated into the temperature range where recovery takes place, how are the following affected: (a) internal residual stresses, (b) strength, (c) ductility, and hardness?

6.8.5 Describe what occurs microscopically when a cold-worked sheet of metal such as aluminum undergoes a recrystallization heat treatment.

6.8.6 When a cold-worked metal is heated into the temperature range where recrystallization takes place, how are the following affected: (a) internal residual stresses, (b) strength, (c) ductility, and (d) hardness?

6.8.7 Describe two principal mechanisms whereby primary recrystallization can occur.

6.8.8 What are five important factors which affect the recrystallization process in metals?

6.8.9 What generalizations can be made about the recrystallization temperature with respect to (a) the degree of deformation, (b) the temperature, (c) the time of heating at temperature, (d) the final grain size, and (e) the purity of the metal?

6.8.10 If it takes 150 h to completely recrystallize an 1100-H18 aluminum alloy sheet at 240°C and 12 h at 290°C, calculate the activation energy in kilojoules per mole for this process. Assume an Arrhenius-type rate behavior.

6.8.11 It takes 10 min to 50 percent recrystallize a piece of high-purity copper sheet at 140°C and 180 min at 90°C. How many minutes are required to recrystallize the sheet 50 percent at 102°C?

6.8.12 If it takes 90 h to completely recrystallize an aluminum sheet at 245°C and 5 h at 320°C, calculate the activation energy in kilojoules per mole for this process. Assume an Arrhenius-type rate behavior.

6.9.1 What are the characteristics of the surface of a ductile fracture of a metal?

6.9.2 Describe the three stages in the ductile fracture of a metal.

6.9.3 What are the characteristics of the surface of a brittle fracture of a metal?

6.9.4 Describe the three stages in the brittle fracture of a metal.

6.9.5 Describe the simple impact test which uses a Charpy V-notch sample.

6.9.6 How does the carbon content of a plain-carbon steel affect the ductile-brittle transition temperature range?

6.9.7 Determine the critical crack length for a through crack contained within a thick plate of 7178-T651 aluminum alloy which is under uniaxial tension. For this alloy $K_{IC} = 23.5$ ksi$\sqrt{in}$ and $\sigma_f = 90.0$ ksi. Assume $Y = 1$.

6.9.8 Determine the critical crack length for a through crack in a thick plate of 2024-T851 aluminum alloy which is in uniaxial tension. For this alloy $K_{IC} = 24.0$ MPa$\sqrt{m}$ and $\sigma_f = 465$ MPa. Assume $Y = 1$.

6.9.9 The critical stress intensity (K_{IC}) for a material for a component of a design is 22.5 ksi$\sqrt{in}$. What is the applied stress that will cause fracture if the component contains an internal crack 0.12 in long? Assume $Y = 1$.

6.9.10 Determine the critical crack length (mm) for a through crack in a thick 2024-T6 alloy plate which has a fracture toughness $K_{IC} = 24.2$ MPa$\sqrt{m}$ and which is under a stress of 400 MPa. Assume $Y = 1$.

6.9.11 What is the largest size (inches) internal through crack that a thick plate of aluminum alloy 7178-T651 can support at an applied stress of (*a*) $\frac{3}{4}$ of the yield strength and (*b*) $\frac{1}{2}$ of the yield strength? Assume $Y = 1$.

6.9.12 A Ti-6Al-4V alloy plate contains an internal through crack of 2.20 mm. What is the highest stress (MPa) that this material can withstand without catastrophic failure? Assume $Y = 1$.

6.9.13 Using the equation $K_{IC} = \sigma_f \sqrt{\pi a}$, plot the fracture stress (MPa) for aluminum alloy 7075-T651 vs. surface crack size a (millimeters) for a values from 0.2 to 2.0 mm. What is the minimum size surface crack which will cause catastrophic failure?

6.10.1 Describe a metal fatigue failure.

6.10.2 What two distinct types of surface areas are usually recognized on a fatigue failure surface?

6.10.3 Where do fatigue failures usually originate on a metal section?

6.10.4 What is a fatigue test *SN* curve, and how are the data for the *SN* curve obtained?

6.10.5 How does the *SN* curve of a carbon steel differ from that of a high-strength aluminum alloy?

6.10.6 Describe the four basic structural changes which take place when a homogeneous ductile metal is caused to fail by fatigue under cyclic stresses.

6.10.7 Describe four major factors which affect the fatigue strength of a metal.

6.11.1 What is metal creep?

6.11.2 For which environmental conditions is the creep of metals especially important industrially?

6.11.3 Draw a typical creep curve for a metal under constant load and at a relatively high temperature, and indicate on it all three stages of creep.

6.11.4 The following creep data were obtained for a titanium alloy at 50 ksi and 400°C. Plot the creep strain vs. time (hours) and determine the steady-state creep rate for these test conditions.

Strain, in/in	Time, h	Strain, in/in	Time, h
0.010×10^{-2}	2	0.075×10^{-2}	80
0.030×10^{-2}	18	0.090×10^{-2}	120
0.050×10^{-2}	40	0.11×10^{-2}	160

6.11.5 Describe what occurs microstructurally at each stage of creep.

6.11.6 What is the minimum creep rate with respect to the creep curve?

7

Polymeric Materials

7.1 INTRODUCTION

The word *polymer* literally means "many parts." A polymeric solid material may be considered to be one that contains many chemically bonded parts or units which themselves are bonded together to form a solid. In this chapter we shall study some aspects of the structure, properties, processing, and applications of two industrially important polymeric materials: *plastics* and *elastomers. Plastics*[1] are a large and varied group of synthetic materials which are processed by forming or molding into shape. Just as we have many types of metals such as aluminum and copper, we have many types of plastics such as polyethylene and nylon. Plastics can be divided into two classes, *thermoplastics* and *thermosetting plastics,* depending on how they are structurally chemically bonded. *Elastomers* or rubbers can be elastically deformed a large amount when a force is applied to them and can return to their original shape (or almost) when the force is released.

---

[1] The word *plastic* has many meanings. As a noun, plastic refers to a class of materials which can be molded or formed into shape. As an adjective, plastic can mean capable of being molded. Another use of plastic as an adjective is to describe the continuous permanent deformation of a metal without rupture, as in the "plastic deformation of metals."

Thermoplastics   Thermoplastics require heat to make them formable and after cooling, retain the shape they were formed into. These materials can be reheated and reformed into new shapes a number of times without significant change in their properties. Most thermoplastics consist of very long main chains of carbon atoms covalently bonded together. Sometimes nitrogen, oxygen, or sulfur atoms are also covalently bonded in the main molecular chain. Pendant atoms or groups of atoms are covalently bonded to the main-chain atoms. In thermoplastics the long molecular chains are bonded to each other by secondary bonds.

Thermosetting plastics (thermosets)   Thermosetting plastics formed into a permanent shape and cured or "set" by a chemical reaction cannot be remelted and reformed into another shape but degradate or decompose upon being heated to too high a temperature. Thus, thermosetting plastics cannot be recycled. The term *thermosetting* implies that heat (the Greek word for heat is *thermē*) is required to permanently set the plastic. There are, however, many so-called thermosetting plastics which set or cure at room temperature by a chemical reaction only. Most thermosetting plastics consist of a network of carbon atoms covalently bonded together to form a rigid solid. Sometimes nitrogen, oxygen, sulfur, or other atoms are also covalently bonded into a thermoset network structure.

Plastics are important engineering materials for many reasons. They have a wide range of properties, some of which are unattainable from any other materials, and in most cases they are relatively low in cost. The use of plastics for mechanical engineering designs offers many advantages which include elimination of parts through engineering design with plastics, elimination of many finishing operations, simplified assembly, weight savings, noise reduction, and in some cases elimination of the need for lubrication of some parts. Plastics are also very useful for many electrical engineering designs mainly because of their excellent insulative properties. Electrical-electronic applications for plastic materials include connectors, switches, relays, TV tuner components, coil forms, integrated circuit boards, and computer components. Figure 7.1 shows some examples of the use of plastic materials in engineering designs.

The amount of plastic materials used by industry has increased markedly over the past years. A good example of the increased industrial use of plastics is in the manufacture of automobiles. Engineers designing the 1959 Cadillac were amazed to find that they had put as much as 25 lb (11.4 kg) of plastics in that vehicle. In 1980 there was an average of about 200 lb (90.9 kg) of plastic used per car. The average use of plastic and plastic composites in the 1993 car was 245 lb (111.4 kg). Certainly not all industries have increased their usage of plastics like the auto industry, but there has been an overall increased usage of plastics in industry over the past decades. Let us now look into the details of the structure, properties, and applications of plastics and elastomers.

(a)

(b)

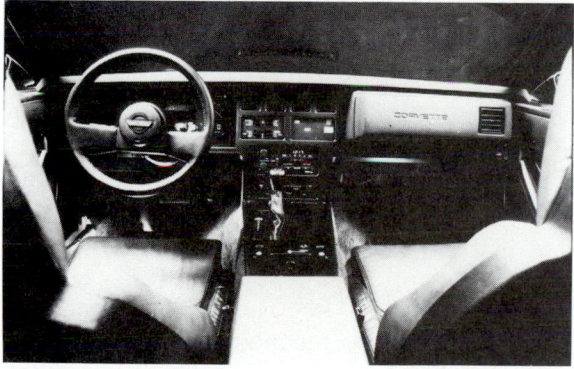

(c)

(d)

FIGURE 7.1 Some applications for engineering plastics. (*a*) New computer terminal has unpainted off-white pedestal and top and bottom covers molded of UV-resistant and heat-resistant polycarbonate. (*Courtesy of Hewlett-Packard Co.*) (*b*) The housing of this hand-held data terminal is made of toughened polybutylene terephthalate (PBT). (*Courtesy of General Electric Co.*) (*c*) and (*d*) High-strength polycarbonate is used in the cockpit interior of the 1984 Corvette for the instrument panel carrier and related framework. (*Courtesy of Mobay Chemical Co.*)

7.2 POLYMERIZATION REACTIONS

Most thermoplastics are synthesized by the process of chain-growth polymerization. In this process many (there may be thousands) small molecules are covalently bonded together to form very long molecular chains. The simple molecules that are covalently bonded into long chains are called *monomers* (from the Greek words *mono*, meaning "one," and *meros*, meaning "part"). The long-chain molecule formed from the monomer units is called a *polymer* (from the Greek words *polys*, meaning "many," and *meros*, meaning "part"). The chemical process by which the monomers are chemically combined into long-chain molecular polymers is called *chain-growth polymerization.*

Covalent Bonding Structure of an Ethylene Molecule

The ethylene molecule, C_2H_4, is chemically bonded by a double covalent bond between the carbon atoms and by four single covalent bonds between the carbon and hydrogen atoms (Fig. 7.2). A carbon-containing molecule which has one or more carbon-carbon double bonds is said to be an *unsaturated molecule.* Thus ethylene is an unsaturated carbon-containing molecule since it contains one carbon-carbon double bond.

Covalent Bonding Structure of an Activated Ethylene Molecule

When the ethylene molecule is activated so that the double bond between the two carbon atoms is "opened up," the double covalent bond is replaced by a single covalent bond, as shown in Fig. 7.3. As a result of the activation, each carbon atom of the former ethylene molecule has a free electron for covalent bonding with another free electron from another molecule. In the following discussion we shall see how the ethylene molecule can be activated and, as a result, how many ethylene monomer units can be covalently bonded together to form a long molecular chain called a *polymer.* This is the process of *chain*

FIGURE 7.2 Covalent bonding in the ethylene molecule illustrated by (*a*) electron-dot (dots represent valence electrons) and (*b*) straight-line notation. There is one double carbon-carbon covalent bond and four single carbon-hydrogen covalent bonds in the ethylene molecule. The double bond is chemically more reactive than the single bonds.

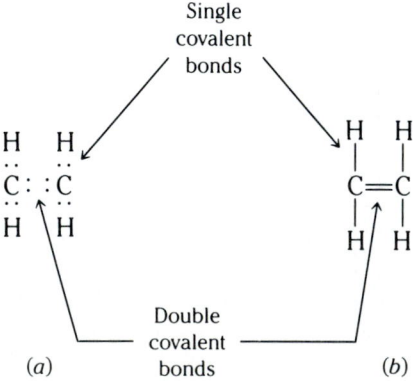

Free electron available
for covalent bonding

Half covalent bond
or free electron

H H

· C : C ·

H | H

Single
covalent
bond
(a)

H H
| |
—C—C—
| |
H | H

Single
covalent
bond
(b)

FIGURE 7.3 Covalent bonding structure of an activated ethylene molecule. (*a*) Electron-dot notation (where dots represent valence electrons). Free electrons are created at each end of the molecule which are able to be covalently bonded with free electrons from other molecules. Note that the double covalent bond between the carbon atoms has been reduced to a single bond. (*b*) Straight-line notation. The free electrons created at the ends of the molecule are indicated by half bonds which are only attached to one carbon atom.

polymerization. The polymer produced by the polymerization of ethylene is called *polyethylene.*

General Reaction for the Polymerization of Polyethylene and the Degree of Polymerization

The general reaction for the chain polymerization of ethylene monomer into polyethylene may be written as

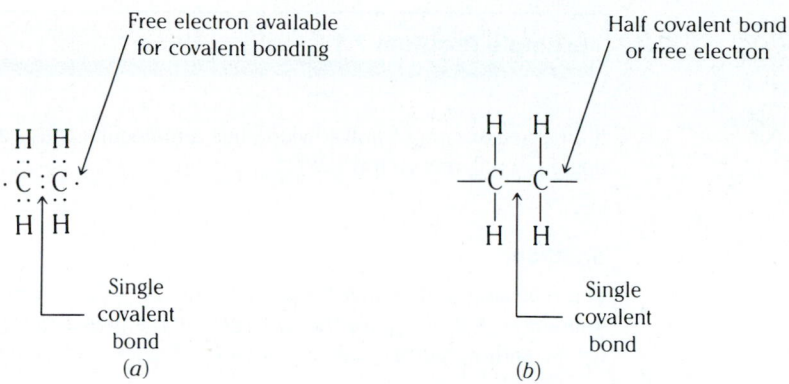

Ethylene
monomer

Polyethylene
(polymer)

The repeating subunit in the polymer chain is called a *mer*. The mer for poly-ethylene is $+CH_2-CH_2+$ and is indicated in the above equation. The *n* in the equation is known as the *degree of polymerization (DP)* of the polymer chain and is equal to the number of subunits or mers in the polymer molecular chain. The average DP for polyethylene ranges from about 3500 to 25,000, corresponding to average molecular masses ranging from about 100,000 to 700,000 g/mol.

Example Problem 7.1

If a particular type of polyethylene has a molecular mass of 150,000 g/mol, what is its degree of polymerization (DP)?

Solution:

The repeating unit or mer for polyethylene is $+CH_2-CH_2+$. This mer has a mass of 4 atoms $\times$ 1 g = 4 g for the hydrogen atoms plus a mass of 2 atoms $\times$ 12 g = 24 g for the carbon atoms, making a total of 28 g for each polyethylene mer.

$$DP = \frac{\text{molecular mass of polymer (g/mol)}}{\text{mass of a mer (g/mer)}} \quad (7.1)$$

$$= \frac{150,000 \text{ g/mol}}{28 \text{ g/mer}} = 5357 \text{ mers/mol} \blacktriangleleft$$

Chain Polymerization Steps

The reactions for the chain polymerization of monomers like ethylene into linear polymers like polyethylene can be divided into the following steps: (1) initiation, (2) propagation, and (3) termination.

Initiation For the chain polymerization of ethylene, one of many types of catalysts can be used. In this discussion we shall consider the use of organic peroxides which act as free-radical formers. A *free radical* can be defined as a group of atoms having an unpaired electron (free electron) which can covalently bond to an unpaired electron (free electron) of another molecule.

Let us first consider how a molecule of hydrogen peroxide, H_2O_2, can decompose into two free radicals, as shown by the following equations. Using electron-dot notation for the covalent bonds,

$$H\!:\!\overset{..}{\underset{..}{O}}\!:\!\overset{..}{\underset{..}{O}}\!:\!H \quad \xrightarrow{\text{heat}} \quad H\!:\!\overset{..}{\underset{..}{O}}^{\bullet} + {}^{\bullet}\overset{..}{\underset{..}{O}}\!:\!H \quad (7.2)$$

Hydrogen peroxide   Free radicals

Using straight-line notation for the covalent bonds,

Free electron

$$H-O-O-H \quad \xrightarrow{\text{heat}} \quad 2H-O^{\bullet} \quad (7.3)$$

Hydrogen peroxide   Free radicals

In the free-radical chain polymerization of ethylene, an organic peroxide can decompose in the same way as hydrogen peroxide. If R—O—O—R represents an organic peroxide, where R is a chemical group, then upon heating, this peroxide can decompose into two free radicals in a manner similar to that of hydrogen peroxide above, as

$$R—O—O—R \longrightarrow 2R—O^{\bullet} \tag{7.4}$$

Organic peroxide   Free radicals

Benzoyl peroxide is an organic peroxide which is used to initiate some chain polymerization reactions and decomposes into free radicals as illustrated below:[1]

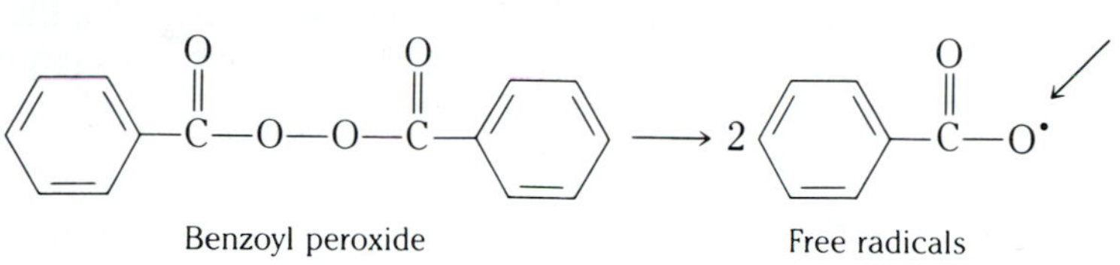

$$(7.5)$$

Benzoyl peroxide   Free radicals

One of the free radicals created by the decomposition of the organic peroxide can react with an ethylene molecule to form a new longer-chain free radical, as shown by the reaction

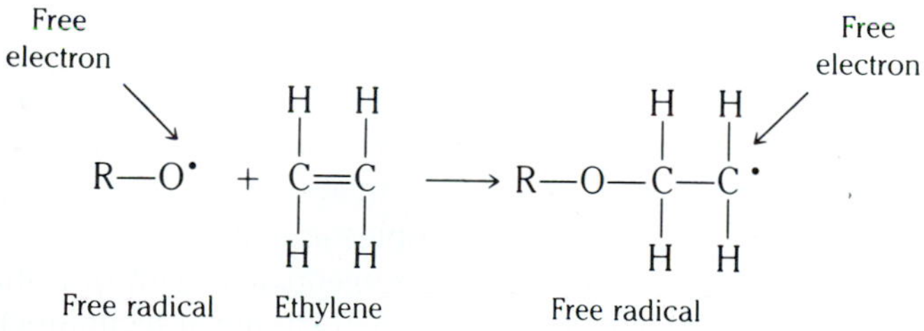

$$(7.6)$$

Free radical   Ethylene   Free radical

The organic free radical in this way acts as an initiator catalyst for the polymerization of ethylene.

---

[1] The hexagonal ring represents the benzene structure, as indicated below. Also see Sec. 2.6.

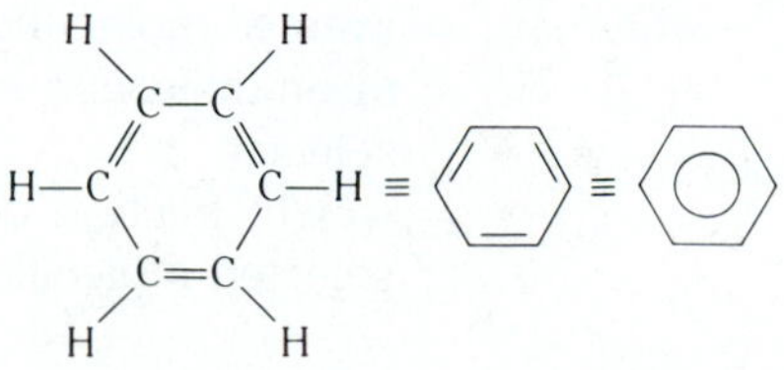

Propagation   The process of extending the polymer chain by the successive addition of monomer units is called *propagation*. The double bond at the end of an ethylene monomer unit can be "opened up" by the extended free radical and be covalently bonded to it. Thus, the polymer chain is further extended by the reaction

$$R-CH_2-CH_2^* + CH_2{=}CH_2 \longrightarrow R-CH_2-CH_2-CH_2-CH_2^* \quad (7.7)$$

The polymer chains in chain polymerization keep growing spontaneously because the energy of the chemical system is lowered by the chain polymerization process. That is, the sum of the energies of the produced polymers is lower than the sum of the energies of the monomers which produced the polymers. The degrees of polymerization (DP) of the polymers produced by chain polymerization vary within the polymeric material. Also, the average DP varies among polymeric materials. For commercial polyethylene the DP usually averages in the range from 3500 to 25,000.

Termination   *Termination* can occur by the addition of a terminator free radical or when two chains combine. Another possibility is that trace amounts of impurities may terminate the polymer chain. Termination by the coupling of two chains can be represented by the reaction

$$R(CH_2-CH_2)_m^* + R'(CH_2-CH_2)_n^* \longrightarrow R(CH_2-CH_2)_m-(CH_2-CH_2)_nR' \quad (7.8)$$

Average Molecular Weight for Thermoplastics

Thermoplastics consist of chains of polymers of many different lengths, each of which has its own molecular weight and degree of polymerization. Thus, one must speak of an average molecular weight when referring to the molecular mass of a thermoplastic material.

The average molecular weight of a thermoplastic can be determined by using special physical-chemical techniques. One method commonly used for this analysis is to determine the weight fractions of molecular weight ranges. The average molecular weight of the thermoplastic is then the sum of the weight fractions times their mean molecular weight for each particular range divided by the sum of the weight fractions. Thus

$$\overline{M}_m = \frac{\Sigma f_i M_i}{\Sigma f_i} \quad (7.9)$$

where $\overline{M}_m$ = average molecular weight for a thermoplastic
  M_i = mean molecular weight for each particular molecular weight range selected
  f_i = weight fraction of the material having molecular weights of a selected molecular weight range

Example Problem 7.2

Calculate the average molecular weight $\overline{M}_m$ for a thermoplastic material which has the mean molecular weight fractions f_i for the molecular weight ranges listed below.

Molecular weight range, g/mol	M_i	f_i	f_iM_i
5,000–10,000	7,500	0.11	825
10,000–15,000	12,500	0.17	2,125
15,000–20,000	17,500	0.26	4,550
20,000–25,000	22,500	0.22	4,950
25,000–30,000	27,500	0.14	3,850
30,000–35,000	32,500	0.10	3,250
		$\Sigma = 1.00$	$\Sigma = 19,550$

Solution:

First determine the mean values for the molecular weight ranges and then list these values, as in the column under M_i shown above. Then multiply f_i by M_i to obtain the f_iM_i values listed above. The average molecular weight for this thermoplastic is

$$\overline{M}_m = \frac{\Sigma f_iM_i}{\Sigma f_i} = \frac{19,550}{1.00} = 19,550 \text{ g/mol} \blacktriangleleft$$

Functionality of a Monomer

In order for a monomer to polymerize, it must have at least two active chemical bonds. When a monomer has two active bonds, it can react with two other monomers, and by repetition of the bonding, other monomers of the same type can form a long-chain or linear polymer. When a monomer has more than two active bonds, polymerization can take place in more than two directions, and thus three-dimensional network molecules can be built up.

The number of active bonds a monomer has is called the *functionality* of the monomer. A monomer which utilizes two active bonds for the polymerization of long chains is called *bifunctional.* Thus ethylene is an example of a bifunctional monomer. A monomer which utilizes three active bonds to form a network polymeric material is called *trifunctional.* Phenol, C_6H_5OH, is an example of a trifunctional monomer and is used in the polymerization of phenol and formaldehyde which will be discussed later.

Structure of Noncrystalline Linear Polymers

If we microscopically examine a short length of a polyethylene chain, we find that it takes on a zigzag configuration (Fig. 7.4) because the covalent bonding angle between single carbon-carbon covalent bonds is about 109°. However,

Carbon

Hydrogen

FIGURE 7.4 The molecular structure of a short length of a polyethylene chain. The carbon atoms have a zigzag arrangement because all carbon-carbon covalent bonds are directed at about 109° to each other. (*After W. G. Moffatt, G. W. Pearsall, and J. Wulff, "The Structure and Properties of Materials," vol. I: "Structure," Wiley, 1965, p. 65.*)

on a larger scale, the polymer chains in noncrystalline polyethylene are randomly entangled like spaghetti thrown into a bowl. This entanglement of a linear polymer is illustrated in Fig. 7.5. For some polymeric materials, of which polyethylene is one, there can be both crystalline and noncrystalline regions. This subject will be discussed in more detail in Sec. 7.4.

The bonding between the long molecular chains in polyethylene consists of weak, permanent dipole secondary bonds. However, the physical entanglement of the long molecular chains also adds to the strength of this type of polymeric material. Side branches also can be formed which cause loose packing of the molecular chains and favor a noncrystalline structure. Branching of linear polymers thus weakens secondary bonds between the chains and lowers the tensile strength of the bulk polymeric material.

FIGURE 7.5 A schematic representation of a polymer. The spheres represent the repeating units of the polymer chain, not specific atoms. (*After W. G. Moffatt, G. W. Pearsall, and J. Wulff, "The Structure and Properties of Materials," vol. I: "Structure," Wiley, 1965, p. 104.*)

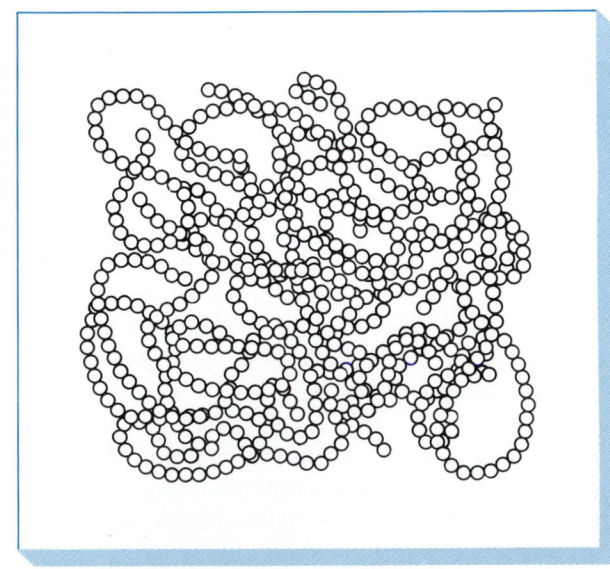

Vinyl and Vinylidene Polymers

Many useful addition (chain) polymeric materials which have carbon main-chain structures similar to polyethylene can be synthesized by replacing one or more of the hydrogen atoms of ethylene with other types of atoms or groups of atoms. If only one hydrogen atom of the ethylene monomer is replaced with another atom or group of atoms, the polymerized polymer is called a *vinyl polymer*. Examples of vinyl polymers are polyvinyl chloride, polypropylene, polystyrene, acrylonitrile, and polyvinyl acetate. The general reaction for the polymerization of the vinyl polymers is

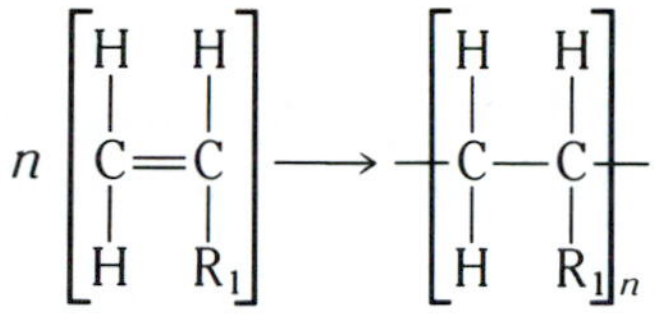

$$
n
\begin{bmatrix}
\text{H} & \text{H} \\
| & | \\
\text{C} = \text{C} \\
| & | \\
\text{H} & \text{R}_1
\end{bmatrix}
\longrightarrow
\begin{bmatrix}
\text{H} & \text{H} \\
| & | \\
\text{C} - \text{C} \\
| & | \\
\text{H} & \text{R}_1
\end{bmatrix}_n
$$

where R_1 can be another type of atom or group of atoms. Figure 7.6 shows the structural bonding of some vinyl polymers.

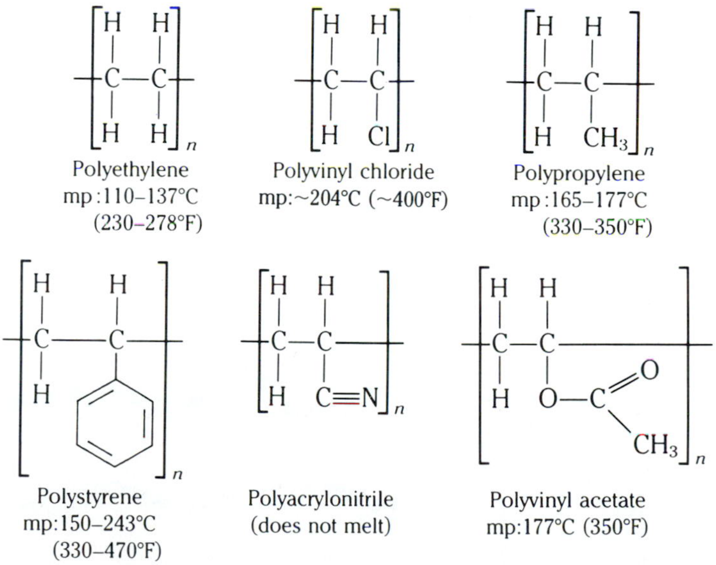

Polyethylene
mp: 110–137°C
(230–278°F)

Polyvinyl chloride
mp: ~204°C (~400°F)

Polypropylene
mp: 165–177°C
(330–350°F)

Polystyrene
mp: 150–243°C
(330–470°F)

Polyacrylonitrile
(does not melt)

Polyvinyl acetate
mp: 177°C (350°F)

FIGURE 7.6 Structural formulas for some vinyl polymers.

If both hydrogen atoms on one of the carbon atoms of the ethylene monomer are replaced by other atoms or groups of atoms, the polymerized polymer is called a *vinylidene polymer*. The general reaction for the polymerization of vinylidene polymers is

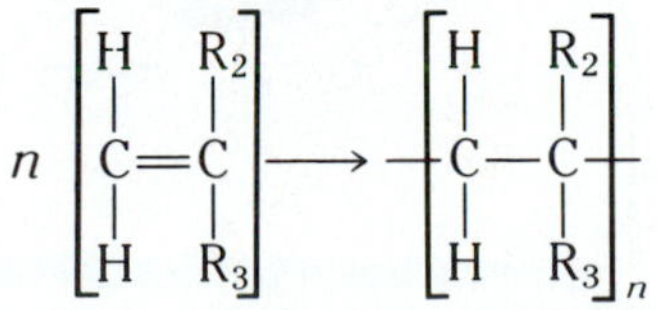

$$
n
\begin{bmatrix}
\text{H} & \text{R}_2 \\
| & | \\
\text{C} = \text{C} \\
| & | \\
\text{H} & \text{R}_3
\end{bmatrix}
\longrightarrow
\begin{bmatrix}
\text{H} & \text{R}_2 \\
| & | \\
\text{C} - \text{C} \\
| & | \\
\text{H} & \text{R}_3
\end{bmatrix}_n
$$

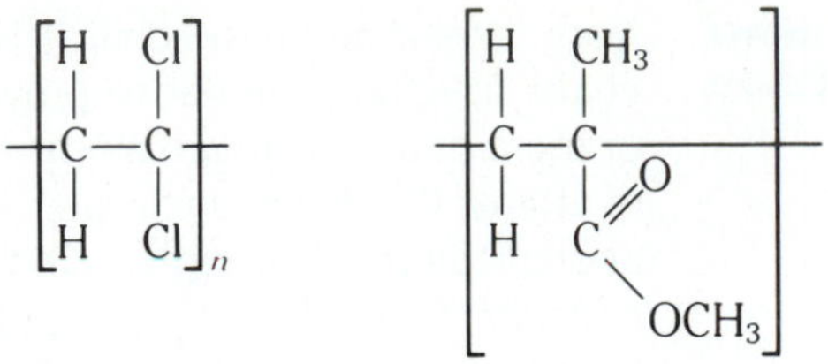

FIGURE 7.7 Structural formulas for some vinylidene polymers.

Polyvinylidene chloride
mp:177°C (350°F)

Polymethyl methacrylate
mp:160°C (320°F)

where R_2 and R_3 can be other types of atoms or atomic groups. Figure 7.7 shows the structural bonding for two vinylidene polymers.

Homopolymers and Copolymers

Homopolymers are polymeric materials which consist of polymer chains made up of single repeating units. That is, if A is a repeating unit, a homopolymer chain will have a sequence of AAAAAAA ··· in the polymer molecular chain. *Copolymers,* in contrast, consist of polymer chains made up of two or more chemically different repeating units which can be in different sequences.

Although the monomers in most copolymers are randomly arranged, four

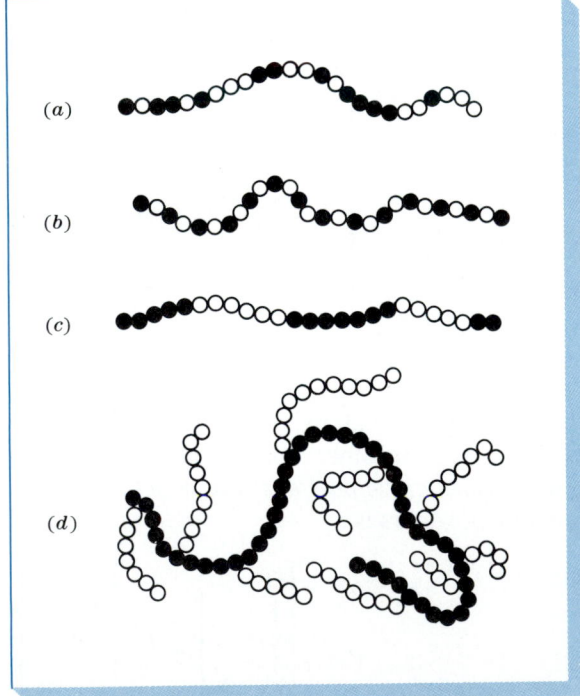

FIGURE 7.8 Copolymer arrangements. (*a*) A copolymer in which the different units are randomly distributed along the chain. (*b*) A copolymer in which the units alternate regularly. (*c*) A block copolymer. (*d*) A graft copolymer. (*After W. G. Moffatt, G. W. Pearsall, and J. Wulff, "The Structure and Properties of Materials," vol. I: "Structure," Wiley, 1965, p. 108.*)

(*a*)

(*b*)

(*c*)

(*d*)

distinct types of copolymers have been identified: random, alternating, block, and graft (Fig. 7.8).

> *Random copolymers.* Different monomers are randomly arranged within the polymer chains. If A and B are different monomers, then an arrangement might be

$$\text{AABABBBBAABABAAB} \cdots \qquad (\text{Fig. } 7.8a)$$

> *Alternating copolymers.* Different monomers show a definite ordered alternation, as

$$\text{ABABABABABAB} \cdots \qquad (\text{Fig. } 7.8b)$$

> *Block copolymers.* Different monomers in the chain are arranged in relatively long blocks of each monomer:

$$\text{AAAAA—BBBBB—} \cdots \qquad (\text{Fig. } 7.8c)$$

> *Graft copolymers.* Appendages of one type of monomer are grafted to the long chain of another:

$$
\begin{array}{ll}
\text{AAAAAAAAAAAAAAAAAAAAAA} & (\text{Fig. } 7.8d) \\
\text{B} \qquad\qquad\quad \text{B} & \\
\text{B} \qquad\qquad\quad \text{B} & \\
\text{B} \qquad\qquad\quad \text{B} &
\end{array}
$$

Chain-reaction polymerization can take place between two or more different monomers if they can enter the growing chains at relatively the same energy level and rates. An example of an industrially important copolymer is one formed with about 85% polyvinyl chloride and 15% polyvinyl acetate, which is used as a basic material for most vinyl records. A generalized polymerization reaction for the production of this copolymer is given in Fig. 7.9.

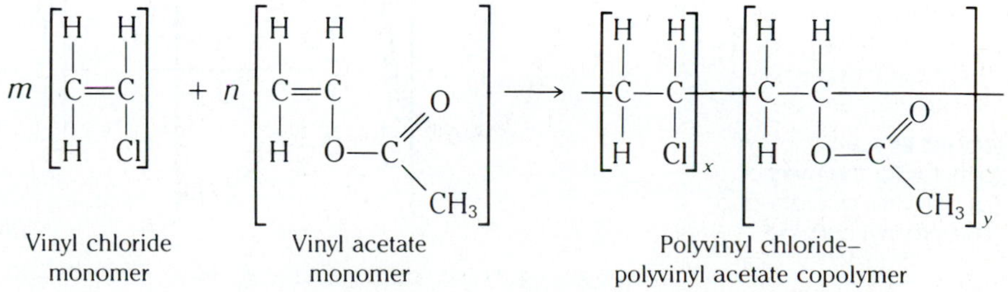

FIGURE 7.9 Generalized polymerization reaction of vinyl chloride and vinyl acetate monomers to produce a copolymer of polyvinyl chloride–polyvinyl acetate.

Example Problem 7.3

A copolymer consists of 15 wt % polyvinyl acetate (PVA) and 85 wt % polyvinyl chloride (PVC). Determine the mole fraction of each component.

Solution

Let the basis be 100 g of copolymer; therefore we have 15 g of PVA and 85 g of PVC. First we determine the number of moles of each component that we have, and then we calculate the mole fractions of each.

Moles of polyvinyl acetate. The molecular weight of the PVA mer is obtained by adding up the atomic masses of the atoms in the structural formula for the PVA mer (Fig. EP7.3a):

$$4\,C\,\text{atoms} \times 12\,\text{g/mol} + 6\,H\,\text{atoms} \times 1\,\text{g/mol} + 2\,O\,\text{atoms} \times 16\,\text{g/mol} = 86\,\text{g/mol}$$

$$\text{No. of moles of PVA in 100 g of copolymer} = \frac{15\,\text{g}}{86\,\text{g/mol}} = 0.174$$

Moles of polyvinyl chloride. The molecular weight of the PVC mer is obtained form Fig. EP7.3b.

$$2\,C\,\text{atoms} \times 12\,\text{g/mol} + 3\,H\,\text{atoms} \times 1\,\text{g/mol} + 1\,Cl\,\text{atom} \times 35.5\,\text{g/mol}$$
$$= 62.5\,\text{g/mol}$$

$$\text{No. of moles of PVC in 100 g of copolymer} = \frac{85\,\text{g}}{62.5\,\text{g/mol}} = 1.36$$

$$\text{Mole fraction of PVA} = \frac{0.174}{0.174 + 1.36} = 0.113$$

$$\text{Mole fraction of PVC} = \frac{1.36}{0.174 + 1.36} = 0.887$$

FIGURE EP7.3 Structural formulas for the mers of (a) polyvinyl acetate and (b) polyvinyl chloride.

Example Problem 7.4

Determine the mole fractions of vinyl chloride and vinyl acetate in a copolymer having a molecular weight of 10,520 g/mol and a degree of polymerization (DP) of 160.

Solution:

From Example Problem 7.3, the molecular weight of the PVC mer is 62.5 g/mol and that of the PVA mer is 86 g/mol.

Since the sum of the mole fractions of polyvinyl chloride, f_{vc}, and polyvinyl acetate, f_{va}, $= 1$, $f_{va} = 1 - f_{vc}$. Thus, the average molecular weight of the copolymer mer is

$$MW_{av}(mer) = f_{vc}MW_{vc} + f_{va}MW_{va} = f_{vc}MW_{vc} + (1 - f_{vc})MW_{va}$$

The average molecular weight of the copolymer mer is also

$$MW_{av}(mer) = \frac{MW_{av}(polymer)}{DP} = \frac{10,520 \text{ g/mol}}{160 \text{ mers}} = 65.75 \text{ g/(mol·mer)}$$

The value of f_{vc} can be obtained by equating the two equations of $MW_{av}(mer)$.

$$f_{vc}(62.5) + (1 - f_{vc})(86) = 65.75 \quad \text{or} \quad f_{vc} = 0.86$$

$$f_{va} = (1 - f_{vc}) = 1 - 0.86 = 0.14$$

Example Problem 7.5

If a vinyl chloride–vinyl acetate copolymer has a ratio of 10:1 vinyl chloride to vinyl acetate mers and a molecular weight of 16,000 g/mol, what is its degree of polymerization (DP)?

Solution:

$$MW_{av}(mer) = \tfrac{10}{11} MW_{vc} + \tfrac{1}{11} MW_{va} = \tfrac{10}{11}(62.5) + \tfrac{1}{11}(86) = 64.6 \text{ g/(mol · mer)}$$

$$DP = \frac{16,000 \text{ g/mol (polymer)}}{64.6 \text{ g/(mol·mer)}} = 248 \text{ mers}$$

Other Methods of Polymerization   **Stepwise polymerization**   In stepwise polymerization, monomers chemically react with each other to produce linear polymers. The reactivity of the functional groups at the ends of a monomer in stepwise polymerization is usually assumed to be about the same for a polymer of any size. Thus monomer units can react with each other or with produced polymers of any size. In many stepwise polymerization reactions a small molecule is produced as a by-product, so these types of reactions are sometimes called *condensation polymerization reactions.* An example of a stepwise polymerization reaction is the reaction of hexamethylene diamine with adipic acid to produce nylon 6,6 and water as a by-product, as shown in Fig. 7.10 for the reaction of one molecule of hexamethylene diamine with another of adipic acid.

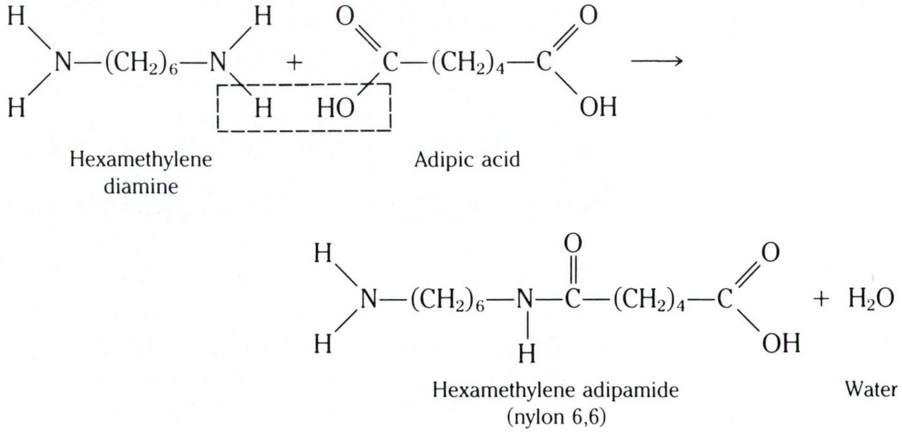

Hexamethylene diamine

Adipic acid

FIGURE 7.10
Polymerization reaction of hexamethylene diamine with adipic acid to produce a unit of nylon 6,6.

Hexamethylene adipamide
(nylon 6,6)

Water

Network polymerization   For some polymerization reactions which involve a chemical reactant with more than two reaction sites, a three-dimensional network plastic material can be produced. This type of polymerization occurs in

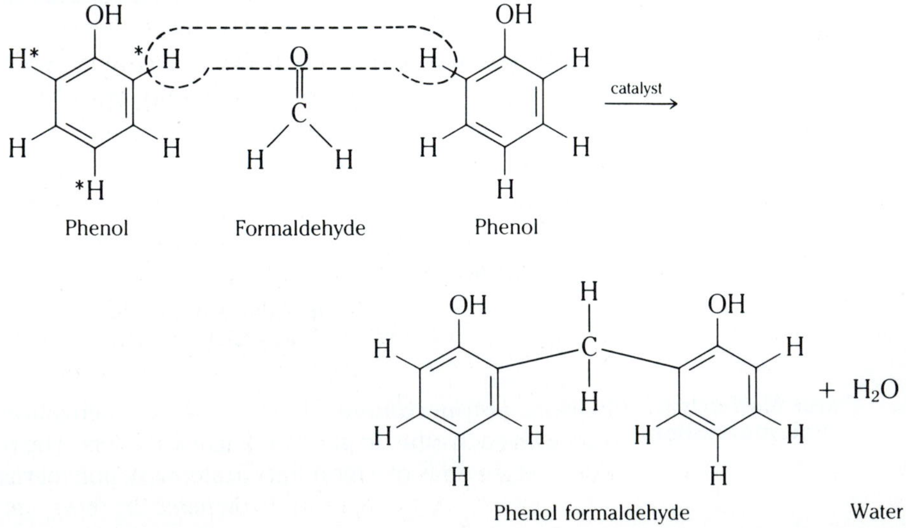

Phenol   Formaldehyde   Phenol

FIGURE 7.11
Polymerization reaction of phenol (asterisks represent reaction sites) with formaldehyde to produce a phenolic resin unit linkage.

Phenol formaldehyde   Water

the curing of thermosetting plastics such as the phenolics, epoxies, and some polyesters. The polymerization reaction of two phenol molecules and one formaldehyde molecule is shown in Fig. 7.11. Note that a molecule of water is formed as a by-product of the reaction. The phenol molecule is trifunctional, and in the presence of a suitable catalyst and sufficient heat and pressure, it can be polymerized with formaldehyde into a network thermosetting phenolic plastic material which is sometimes referred to by the trade name of *Bakelite*.

7.3   INDUSTRIAL POLYMERIZATION METHODS

At this stage one must certainly be wondering how plastic materials are produced industrially. The answer to this question is not simple since many different processes are used and new ones are constantly being developed. To start with, basic raw materials such as *natural gas, petroleum,* and *coal* are used to produce the basic chemicals for the polymerization processes. These chemicals are then polymerized by many different processes into plastic materials such as granules, pellets, powders, or liquids which are further processed into finished products. The chemical polymerization processes used to produce plastic materials are complex and diverse. The chemical engineer plays a major role in their development and industrial utilization. Some of the most important polymerization methods are outlined in the following paragraphs and illustrated in Figs. 7.12 and 7.13.

>   *Bulk polymerization* (Fig. 7.12*a*). The monomer and activator are mixed in a reactor which is heated and cooled as required. This process is used extensively for condensation polymerization where one monomer may be charged into the reactor and another added slowly. The bulk process can be used for many condensation polymerization reactions because of their low heats of reaction.
>
>   *Solution polymerization* (Fig. 7.12*b*). The monomer is dissolved in a nonreactive solvent which contains a catalyst. The heat released by the reaction is absorbed by the solvent, and so the reaction rate is reduced.
>
>   *Suspension polymerization* (Fig. 7.12*c*). The monomer is mixed with a catalyst and then dispersed as a suspension in water. In this process the heat released by the reaction is absorbed by the water. After polymerization, the polymerized product is separated and dried. This process is commonly used to produce many of the vinyl-type polymers such as polyvinyl chloride, polystyrene, polyacrylonitrile, and polymethyl methacrylate.
>
>   *Emulsion polymerization* (Fig. 7.12*d*). This polymerization process is similar to the suspension process since it is carried out in water. However, an emulsifier is added to disperse the monomer into very small particles.

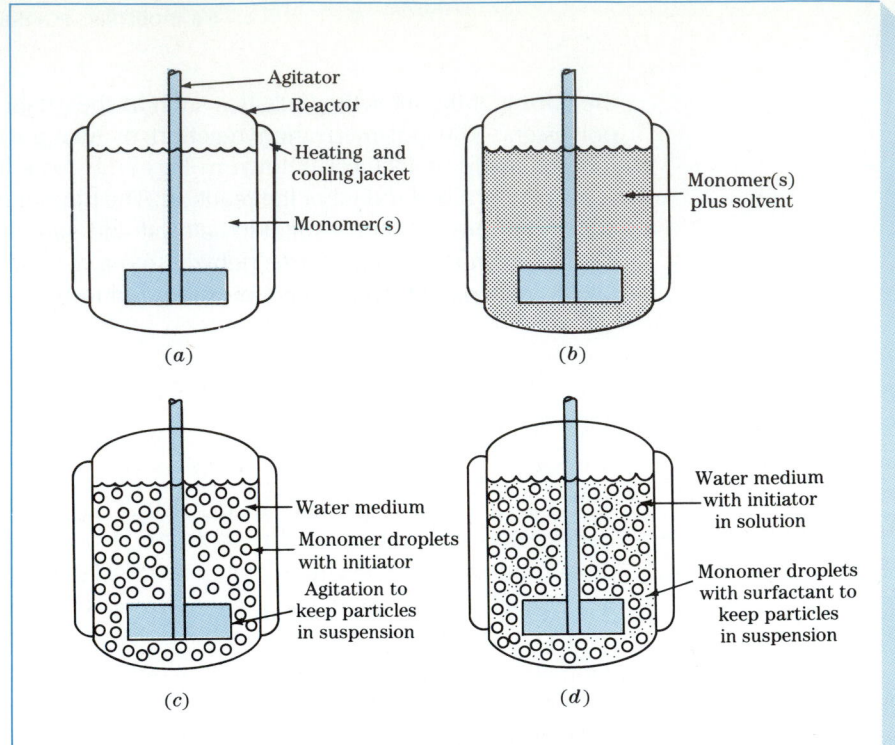

FIGURE 7.12 Schematic illustration of some commonly used industrial polymerization methods: (a) bulk, (b) solution, (c) suspension, (d) emulsion. (*After W. E. Driver, "Plastics Chemistry and Technology," Van Nostrand Reinhold, 1979, p. 19.*)

(a)

(b)

Agitator
Reactor
Heating and cooling jacket
Monomer(s)

Monomer(s) plus solvent

(c)

Water medium
Monomer droplets with initiator
Agitation to keep particles in suspension

(d)

Water medium with initiator in solution
Monomer droplets with surfactant to keep particles in suspension

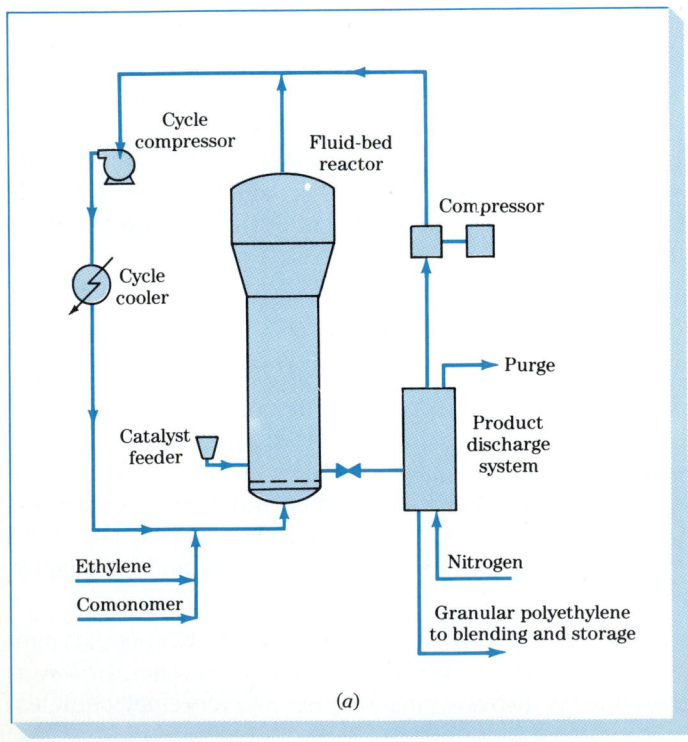

Cycle compressor

Fluid-bed reactor

Compressor

Cycle cooler

Purge

Catalyst feeder

Product discharge system

Ethylene

Comonomer

Nitrogen

Granular polyethylene to blending and storage

(a)

(b)

In addition to the batch polymerization processes just described, many types of mass continuous polymerization processes have been developed, and research and development in this area is continuously being carried out. One very important process[1] that has fairly recently been developed is Union Carbide's gas-phase Unipol process for producing low-density polyethylene. In this process gaseous ethylene monomer along with some comonomer are fed continuously into a fluidized-bed reactor into which a special catalyst is added (Fig. 7.13a). The advantages of this process are lower temperature for polymerization (100°C instead of the older process 300°C) and lower pressure (100 psi instead of 300 psi for the older process). Many industrial plants are already using the Unipol process.

7.4 CRYSTALLINITY AND STEREOISOMERISM IN SOME THERMOPLASTICS

A thermoplastic, when solidified from the liquid state, forms either a noncrystalline or a partly crystalline solid. Let us investigate some of the solidification and structural characteristics of these materials.

Solidification of Noncrystalline Thermoplastics

Let us consider the solidification and slow cooling to low temperatures of a noncrystalline thermoplastic. When noncrystalline thermoplastics solidify, there is no sudden decrease in specific volume (volume per unit mass) as the temperature is lowered (Fig. 7.14). The liquid, upon solidification, changes to a supercooled liquid which is in the solid state and shows a gradual decrease in specific volume with decreasing temperature, as indicated along the line ABC in Fig. 7.14.

Upon cooling this material to lower temperatures, a change in slope of the specific volume vs. temperature curve occurs, as indicated by C and D of curve $ABCD$ of Fig. 7.14. The average temperature between the narrow temperature range over which the slope in the curve changes is called the *glass transition temperature* T_g. Above T_g, noncrystalline thermoplastics show viscous (rubbery or flexible leathery) behavior, and below T_g, these materials show glass-brittle behavior. In some ways T_g might be considered a ductile-brittle transition temperature. Below T_g, the material is glass-brittle because molecular chain motion is very restricted. Figure 7.15 shows an experimental plot of specific volume vs. temperature for noncrystalline polypropylene which indicates a slope change for the T_g of this material at -12°C. Table 7.1 lists T_g values for some thermoplastics.

---

[1] *Chemical Engineering*, Dec. 3, 1979, p. 80.

FIGURE 7.13 Gas-phase polymerization process for low-density polyethylene. (a) Flow diagram outlining the basic steps of the process. (b) Twin reactors used in the process. (*After Chemical Engineering, Dec. 3, 1979, pp. 81, 83.*)

FIGURE 7.14 Solidification and cooling of noncrystalline and partly crystalline thermoplastics showing change in specific volume with temperature (schematic). T_g is the glass transition temperature and T_m is the melting temperature. Noncrystalline thermoplastic cools along line $ABCD$, where $A =$ liquid, $B =$ highly viscous liquid, $C =$ supercooled liquid (rubbery), and $D =$ glassy solid (hard and brittle). Partly crystalline thermoplastic cools along line $ABEF$, where $E =$ solid crystalline regions in supercooled liquid matrix and $F =$ solid crystalline regions in glassy matrix.

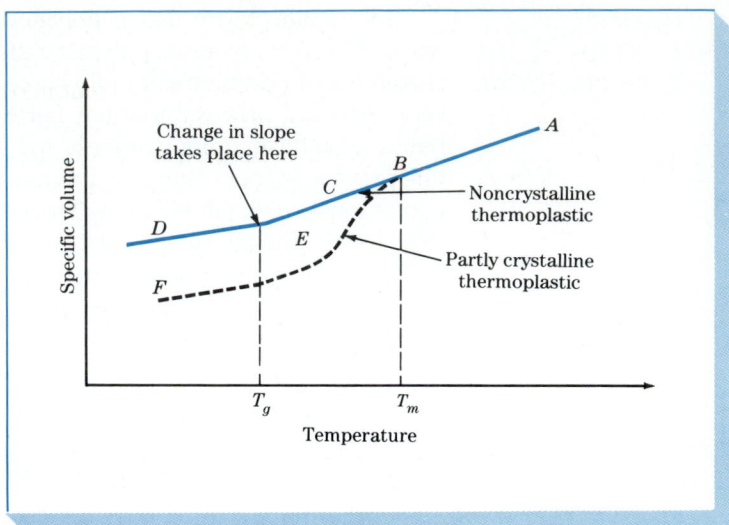

FIGURE 7.15

Experimental data of specific volume vs. temperature for the determination of the glass transition temperature of atactic polypropylene. T_g is at $-12°C$. [After D. L. Beck, A. A. Hiltz, and J. R. Knox, Soc. Plast. Eng. Trans., **3**:279(1963).].

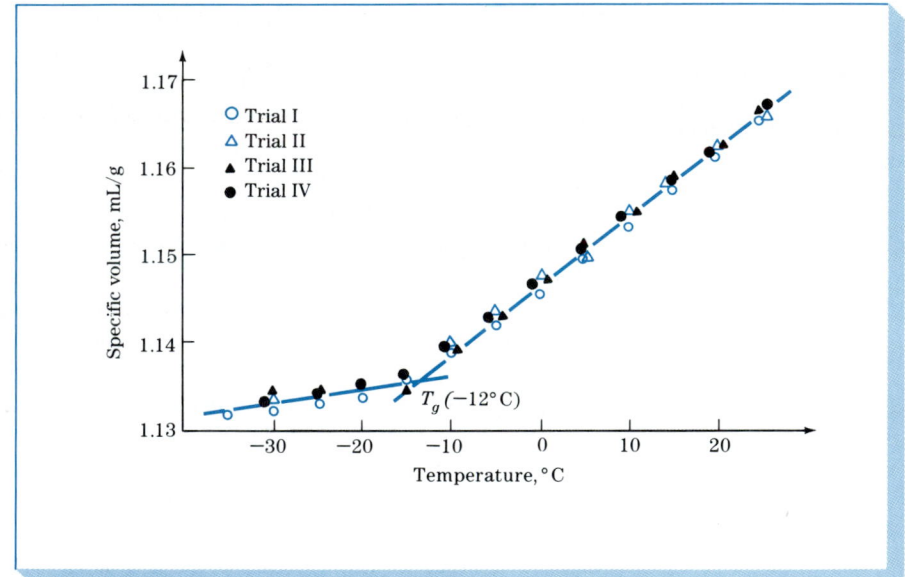

TABLE 7.1   **Glass Transition Temperature T_g*, °C, for Some Thermoplastics**

Polyethylene	−110	(nominal)
Polypropylene	−18	(nominal)
Polyvinyl acetate	29	
Polyvinyl chloride	82	
Polystyrene	75–100	
Polymethyl methacrylate	72	

* Note that the T_g of a thermoplastic is not a physical constant like the melting temperature of a crystalline solid but depends to some extent on variables such as degree of crystallinity, average molecular weight of the polymer chains, and rate of cooling of the thermoplastic.

Solidification of Partly Crystalline Thermoplastics

Let us now consider the solidification and cooling to low temperatures of a partly crystalline thermoplastic. When this material solidifies and cools, a sudden decrease in specific volume occurs, as indicated by the line *BE* in Fig. 7.14. This decrease in specific volume is caused by the more-efficient packing of the polymer chains into crystalline regions. The structure of the partly crystalline thermoplastic at *E* will thus be that of crystalline regions in a supercooled liquid (viscous solid) noncrystalline matrix. As cooling is continued, the glass transition is encountered, as indicated by the slope change of specific volume vs. temperature in Fig. 7.14 between *E* and *F*. In going through the glass transition, the supercooled liquid matrix transforms to the glassy state, and thus the structure of the thermoplastic at *F* consists of crystalline regions in a glassy noncrystalline matrix. An example of a thermoplastic which solidifies to form a partly crystalline structure is polyethylene.

Structure of Partly Crystalline Thermoplastic Materials

The exact way in which polymer molecules are arranged in a crystalline structure is still in doubt, and more research is needed in this area. The longest dimension of crystalline regions or crystallites in polycrystalline polymeric materials is usually about 5 to 50 nm, which is a small percentage of the length of a fully extended polymer molecule that may be about 5000 nm. An early model called the *fringed-micelle model* pictured long polymer chains of about 5000 nm wandering successively through a series of disordered and ordered regions along the length of the polymer molecule (Fig. 7.16a). A newer model called the *folded-chain model* pictures sections of the molecular chains folding on themselves so that a transition from crystalline to noncrystalline regions can be formed (Fig. 7.16b).

There has been an intensive study over the past years on partly crystalline thermoplastics, especially polyethylene. Polyethylene is believed to crystallize in a folded-chain structure with an orthorhombic cell, as shown in Fig. 7.17. Each length of chain between folds is about 100 carbon atoms, with each single

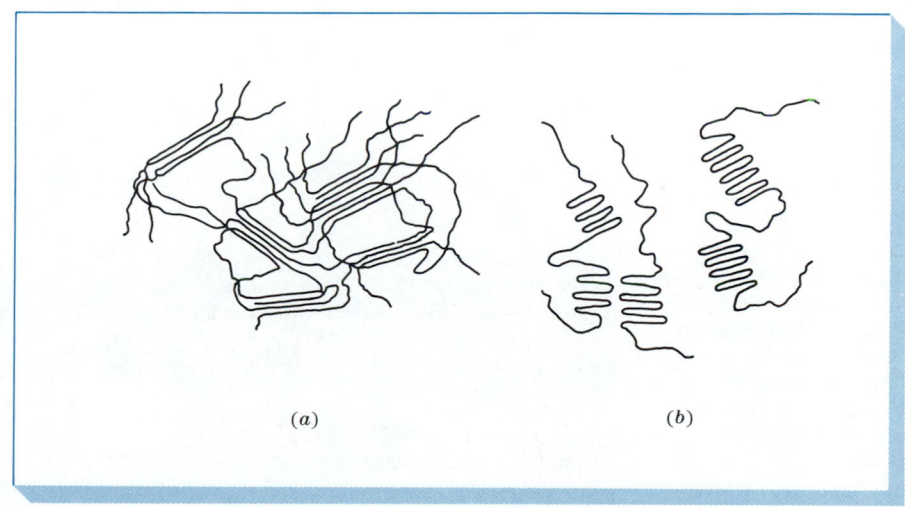

FIGURE 7.16 Two suggested crystallite arrangements for partly crystalline thermoplastic materials: (*a*) fringed-micelle model and (*b*) folded-chain model. (*After F. Rodriguez, "Principles of Polymer Systems," 2d ed., McGraw-Hill, 1982, p. 42.*)

(*a*)   (*b*)

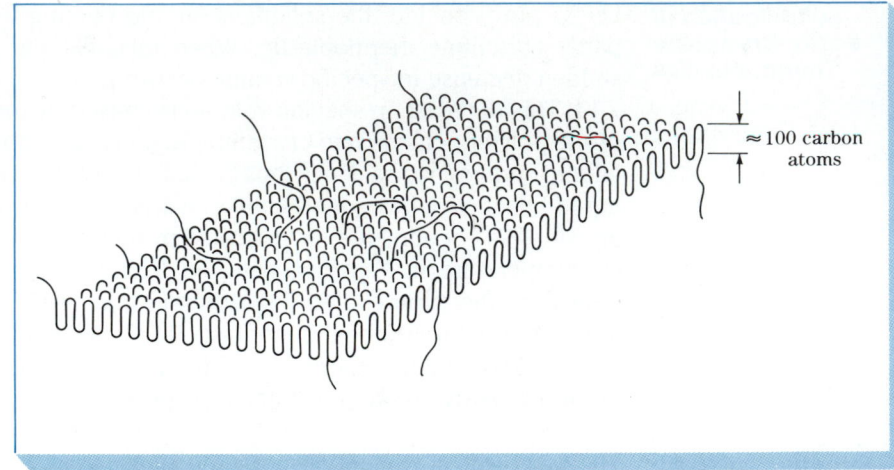

≈ 100 carbon atoms

FIGURE 7.17 Schematic folded-chain structure of a lamella of low-density polyethylene. [*After R. L. Boysen, Olefin Polymers (High-Pressure Polyethylene), in "Encyclopedia of Chemical Technology," vol. 16, Wiley, 1981, p. 405.*]

layer of the folded-chain structure being referred to as a *lamella*. Under laboratory conditions low-density polyethylene crystallizes in a spherulitic-type structure, which is shown in Fig. 7.18. The spherulitic regions which consist of crystalline lamellae are the dark areas, and the regions between the spherulitic structures are noncrystalline white areas. The spherulitic structure shown in Fig. 7.18 grows only under carefully controlled stress-free laboratory conditions.

The degree of crystallinity in partly crystalline linear polymeric materials ranges from about 5 to 95 percent of their total volume. Complete crystallization is not attainable even with polymeric materials which are highly crystallizable because of molecular entanglements and crossovers. The amount of crystalline

FIGURE 7.18 Cast-film spherulitic structure of low-density polyethylene, density 0.92 g/cm³. [*After R. L. Boysen, Olefin Polymers (High-Pressure Polyethylene), in "Encyclopedia of Chemical Technology," vol. 16, Wiley, 1981, p. 406.*]

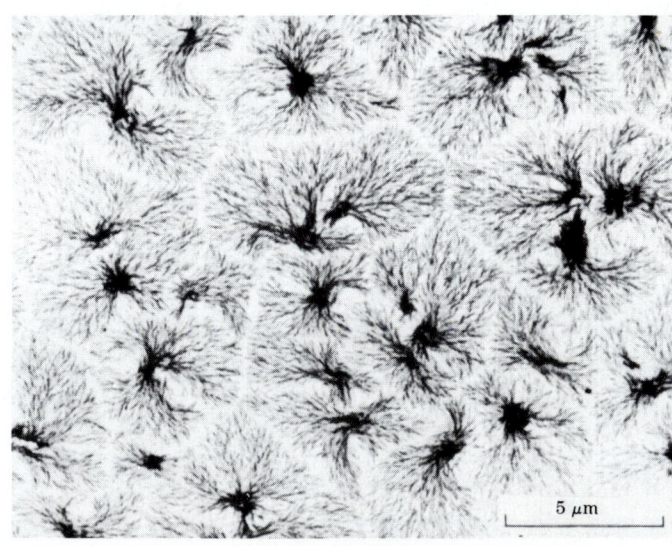

5 μm

material within a thermoplastic affects its tensile strength. In general as the degree of crystallinity increases, the strength of the material increases.

Stereoisomerism in Thermoplastics

Stereoisomers are molecular compounds which have the same chemical compositions but different structural arrangements. Some thermoplastics such as polypropylene can exist in three different stereoisomeric forms:

1. *Atactic stereoisomer.* The pendant methyl group of polypropylene is randomly arranged on either side of the main-carbon chain (Fig. 7.19*a*).
2. *Isotactic stereoisomer.* The pendant methyl group is always on the same side of the main-carbon chain (Fig. 7.19*b*).
3. *Syndiotactic stereoisomer.* The pendant group regularly alternates from one side of the main chain to the other side (Fig. 7.19*c*).

The discovery of a catalyst that made possible the industrial polymerization

FIGURE 7.19 Polypropylene stereoisomers. (*a*) Atactic isomer in which the pendant CH_3 groups are randomly arranged on either side of main-chain carbons. (*b*) Isotactic isomer in which the pendant CH_3 groups are all on the same side of the main-chain carbons. (*c*) Syndiotactic isomer in which the pendant CH_3 groups are regularly alternating from one side to the other of main-chain carbons. [*After G. Crespi and L. Luciani, Olefin Polymers (Polyethylene), in "Encyclopedia of Chemical Technology," vol. 16, Wiley, 1982, p. 454.*]

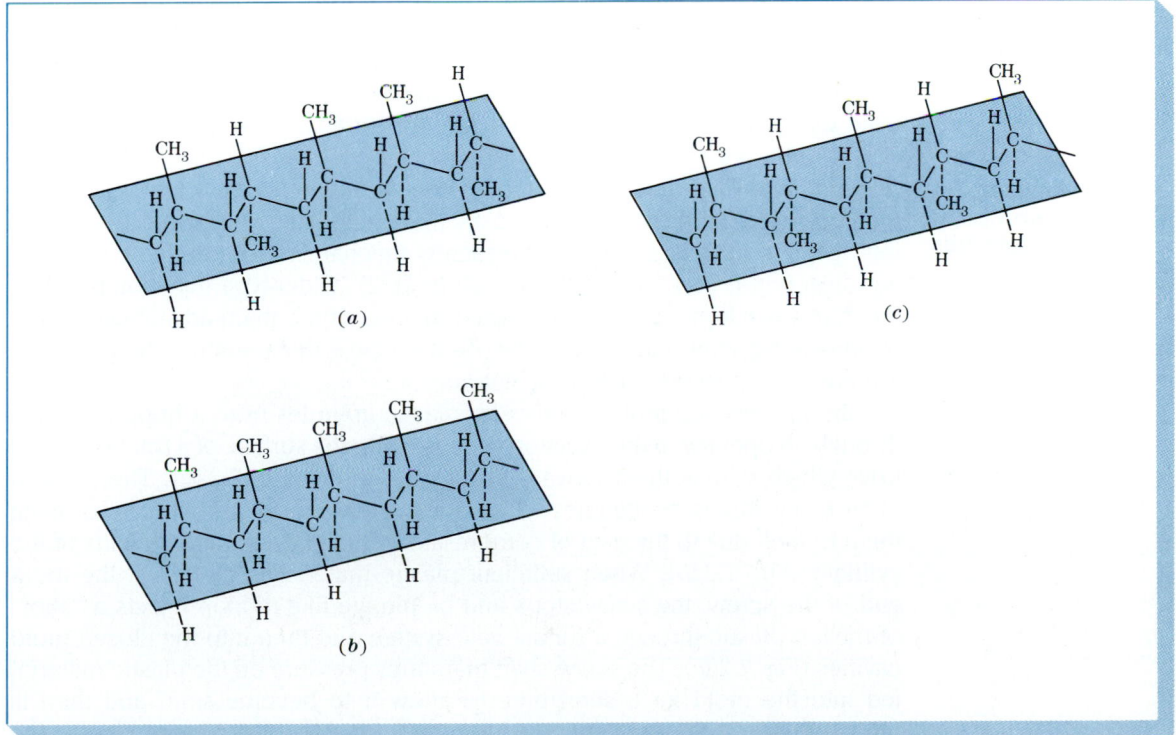

of isotactic linear-type polymers was a great breakthrough for the plastics industry. By using a stereospecific catalyst, isotactic polypropylene was able to be produced on a commercial scale. Isotactic polypropylene is a highly crystalline polymeric material with a melting point of 165 to 175°C. Because of its high crystallinity isotactic propylene has higher strengths and higher heat-deflection temperatures than atactic polypropylene. In 1983 isotactic polypropylene and copolymers accounted for about 10 percent of the total plastic sales in the United States.

7.5 PROCESSING OF PLASTIC MATERIALS

Many different processes are used to transform plastic granules and pellets into shaped products such as sheet, rods, extruded sections, pipe, or finished molded parts. The process used depends to a certain extent on whether the plastic is a thermoplastic or thermosetting one. Thermoplastics are usually heated to a soft condition and then reshaped before cooling. On the other hand, thermosetting materials, not having been completely polymerized before processing to the finished shape, use a process by which a chemical reaction occurs to cross-link polymer chains into a network polymeric material. The final polymerization can take place by the application of heat and pressure or by catalytic action at room temperature or higher temperatures.

In this section we shall discuss some of the most important processes used for thermoplastic and thermosetting materials.

Processes Used for Thermoplastic Materials

Injection molding Injection molding is one of the most important processing methods used for forming thermoplastic materials. The modern injection-molding machine utilizes a reciprocating-screw mechanism for melting the plastic and injecting it into a mold (Figs. 7.20 to 7.22). Older-type injection-molding machines use a plunger for melt injection. One of the main advantages of the reciprocating-screw method over the plunger type is that the screw drive delivers a more homogeneous melt for injection.

In the injection-molding process, plastic granules from a hopper are fed through an opening in the injection cylinder onto the surface of a rotating screw drive which carries them forward toward the mold (Fig. 7.22a). The rotation of the screw forces the granules against the heated walls of the cylinder, causing them to melt due to the heat of compression, friction, and the hot walls of the cylinder (Fig. 7.22b). When sufficient plastic material is melted at the mold end of the screw, the screw stops and by plungerlike motion injects a "shot" of melted plastic through a runner-gate system and then into the closed mold cavities (Fig. 7.22c). The screw shaft maintains pressure on the plastic material fed into the mold for a short time to allow it to become solid and then is retracted. The mold is water-cooled to rapidly cool the plastic part. Finally, the mold is opened and the part is ejected from the mold with air or by spring-

FIGURE 7.20 Front view of a 500-ton reciprocating-screw injection-molding machine for plastic materials. (*Courtesy of HPM Corporation.*)

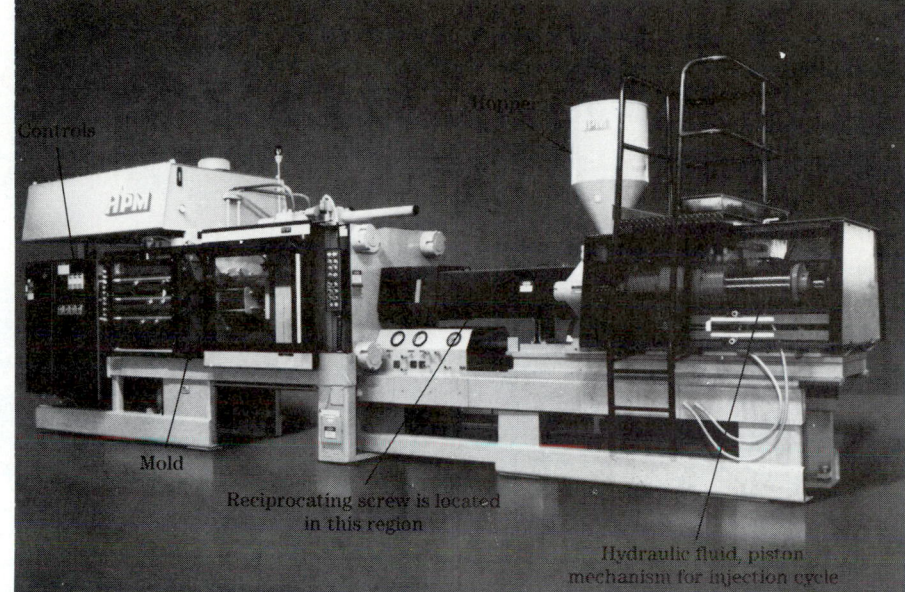

FIGURE 7.21 Cross section of reciprocating-screw injection-molding machine for plastic materials (*After J. Bown, "Injection Molding of Plastic Components,"* McGraw-Hill, 1979, p. 28.)

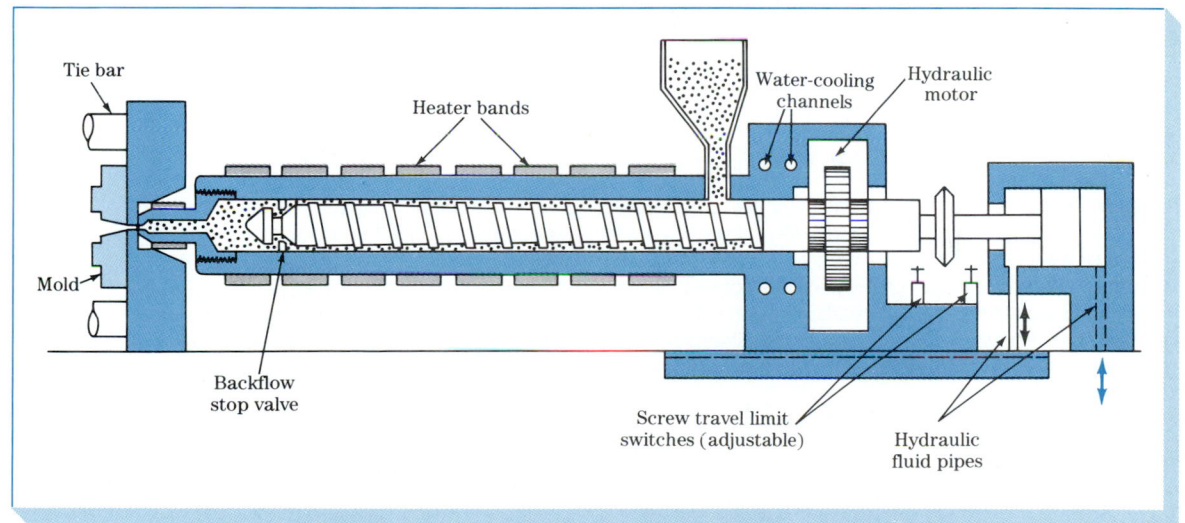

loaded ejector pins (Fig. 7.22*d*). The mold is then closed and ready for another cycle.

The main advantages of injection molding are:

1. High-quality parts can be produced at a high production rate.
2. The process has relatively low labor costs.
3. Good surface finishes can be produced on the molded part.
4. The process can be highly automated.
5. Intricate shapes can be produced.

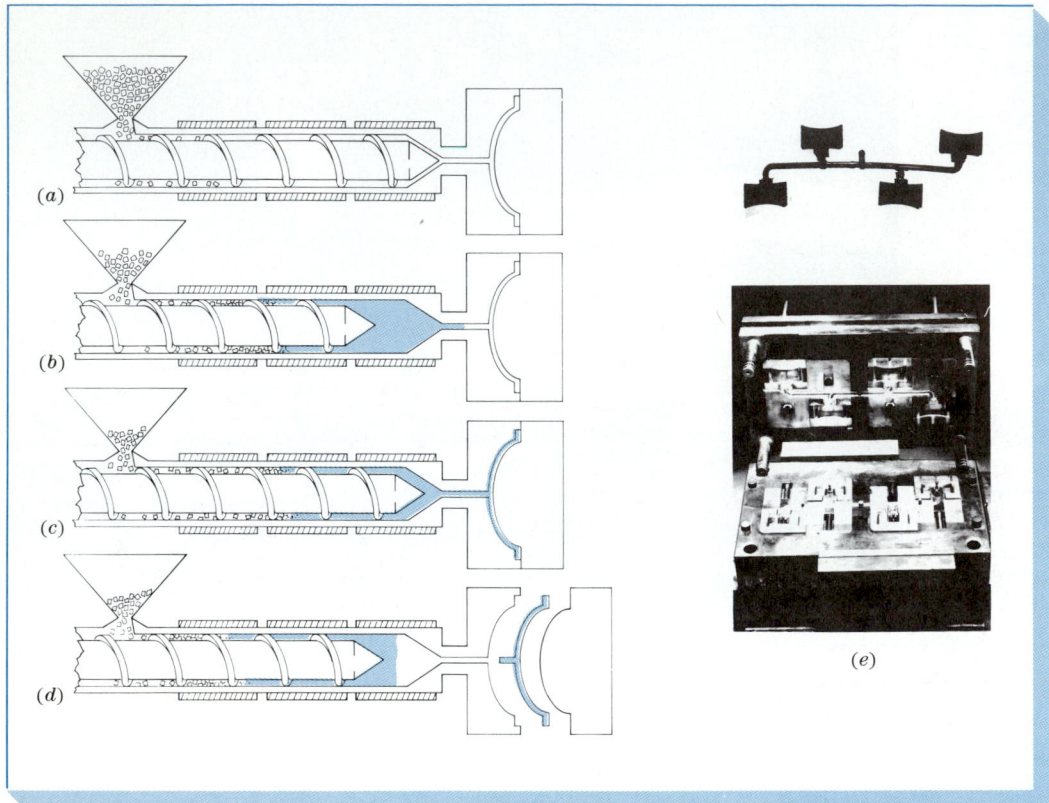

FIGURE 7.22 Sequence of operations for the reciprocating-screw injection-molding process for plastic materials. (*a*) Plastic granules are delivered by a revolving-screw barrel. (*b*) Plastic granules are melted as they travel along the revolving screw, and when sufficient material is melted at the end of the screw, the screw stops rotating. (*c*) The screw barrel is then driven forward with a plungerlike motion and injects the melted plastic through an opening into a runner and gate system and then into a closed-mold cavity. (*d*) The screw barrel is retracted and the finished plastic part ejected. (*e*) Open mold dies showing removed plastic part above. (*Courtesy of Plastics Engineering Co., Sheboygan, Wisc.*)

The main disadvantages of injection molding are:

1. High cost of the machine means that a large volume of parts must be made to pay for the machine.
2. The process must be closely controlled to produce a quality product.

Extrusion Extrusion is another of the important processing methods used for thermoplastics. Some of the products manufactured by the extrusion process are pipe, rod, film, sheet, and shapes of all kinds. The extrusion machine is also used for making compounded plastic materials for the production of raw shapes such as pellets and for the reclamation of scrap thermoplastic materials.

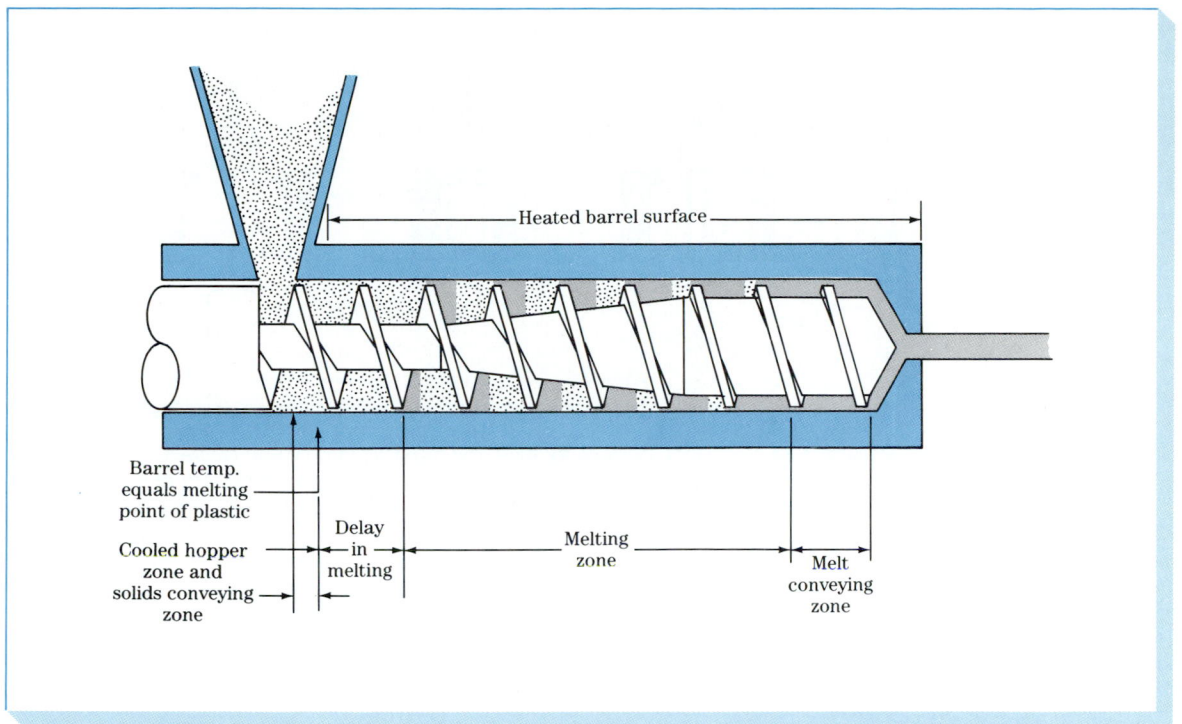

FIGURE 7.23 Schematic drawing of an extruder, showing the various functional zones: hopper, solids-conveying zone, delay in melting start, melting zone, and melt-pumping zone. [*After H. S. Kaufman and J. J. Falcetta (eds.), "Introduction to Polymer Science and Technology," Society of Plastic Engineers, Wiley, 1977, p. 462.*]

In the extrusion process thermoplastic resin is fed into a heated cylinder, and the melted plastic is forced by a rotating screw through an opening (or openings) in an accurately machined die to form continuous shapes (Fig. 7.23). After exiting from the die, the extruded part must be cooled below its glass transition temperature to ensure dimensional stability. The cooling is usually done with an air-blast or water-cooling system.

Blow molding and thermoforming   Other important processing methods for thermoplastics are blow molding and thermoforming of sheet. In blow molding, a cylinder or tube of heated plastic called a *parison* is placed between the jaws of a mold (Fig. 7.24a). The mold is closed to pinch off the ends of the cylinder (Fig. 7.24b), and compressed air is blown in, forcing the plastic against the walls of the mold (Fig. 7.24c).

In thermoforming, a heated plastic sheet is forced into the contours of a mold by pressure. Mechanical pressure may be used with mating dies, or a vacuum may be used to pull the heated sheet into an open die. Air pressure may also be used to force a heated sheet into an open die.

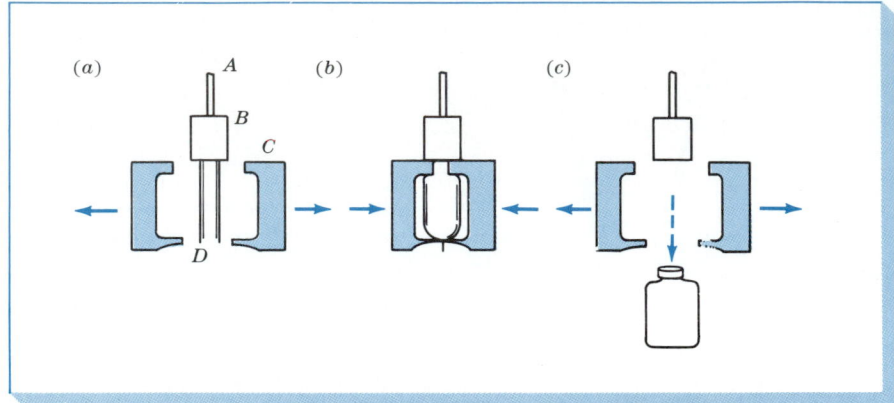

FIGURE 7.24 Sequence of steps for the blow molding of a plastic bottle. (*a*) A section of tube is introduced into the mold. (*b*) The mold is closed, and the bottom of the tube is pinched together by the mold. (*c*) Air pressure is fed through the mold into the tube, which expands to fill the mold, and the part is cooled as it is held under air pressure. *A* = air line, *B* = die, *C* = mold, *D* = tube section. (*After P. N. Richardson, Plastics Processing, in "Encyclopedia of Chemical Technology," vol. 18, Wiley, 1982, p. 198.*)

Processes Used for Thermosetting Materials

Compression molding Many thermosetting resins such as the phenol-form-aldehyde, urea-formaldehyde, and melamine-formaldehyde resins are formed into solid parts by the compression-molding process. In compression molding, the plastic resin, which may be preheated, is loaded into a hot mold containing one or more cavities (Fig. 7.25*a*). The upper part of the mold is forced down

FIGURE 7.25
Compression molding. (*a*) Cross section of open mold containing preformed powdered shape in mold cavity. (*b*) Cross section of closed mold showing molded specimen and excess flash. (*After R. B. Seymour, Plastics Technology, in "Encylopedia of Chemical Technology," vol. 15, Wiley, 1968, p. 802.*)

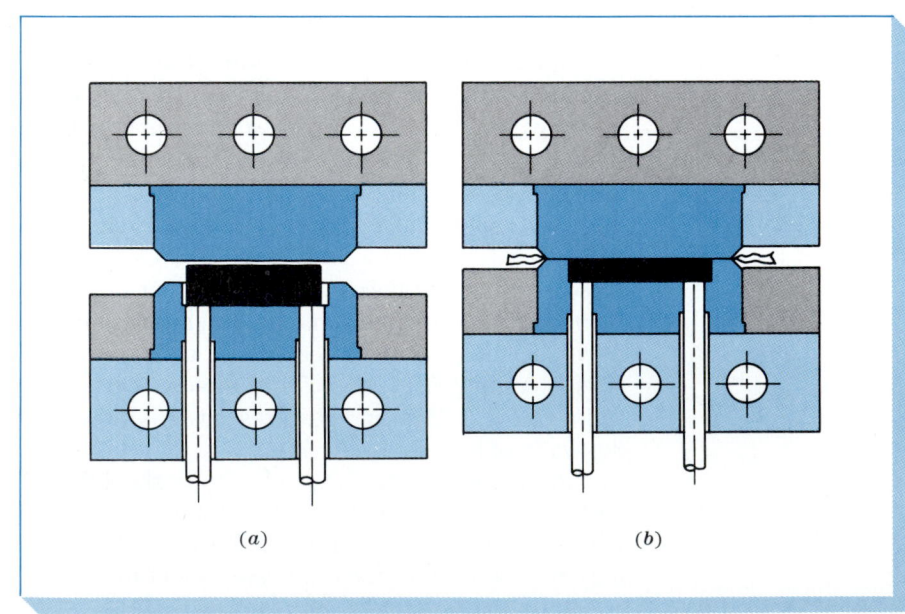

(*a*)   (*b*)

on the plastic resin, and the applied pressure and heat melts the resin and forces the liquefied plastic to fill the cavity or cavities (Fig. 7.25b). Continued heating (usually a minute or two) is required to complete the cross-linking of the thermosetting resin, and then the part is ejected from the mold. The excess flash is trimmed later from the part.

The advantages of compression molding are:

1. Because of the relative simplicity of the molds, initial mold costs are low.
2. The relatively short flow of material reduces wear and abrasion on molds.
3. Production of large parts is more feasible.
4. More-compact molds are possible because of the simplicity of the mold.
5. Expelled gases from the curing reaction are able to escape during the molding process.

The disadvantages of compression molding are:

1. Complicated part configurations are difficult to make with this process.
2. Inserts may be difficult to hold to close tolerances.
3. Flash must be trimmed from the molded parts.

Transfer molding   Transfer molding is also used for molding thermosetting plastics such as the phenolics, ureas, melamines, and alkyd resins. Transfer molding differs from compression molding in how the material is introduced into the mold cavities. In transfer molding the plastic resin is not fed directly into the mold cavity but into a chamber outside the mold cavities (Fig. 7.26a).

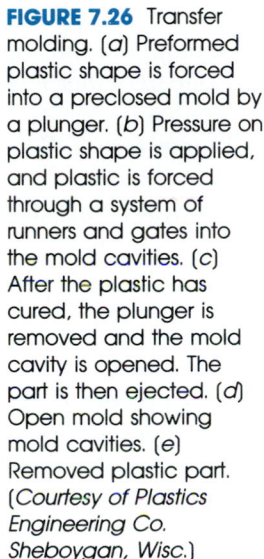

FIGURE 7.26 Transfer molding. (a) Preformed plastic shape is forced into a preclosed mold by a plunger. (b) Pressure on plastic shape is applied, and plastic is forced through a system of runners and gates into the mold cavities. (c) After the plastic has cured, the plunger is removed and the mold cavity is opened. The part is then ejected. (d) Open mold showing mold cavities. (e) Removed plastic part. (*Courtesy of Plastics Engineering Co. Sheboygan, Wisc.*)

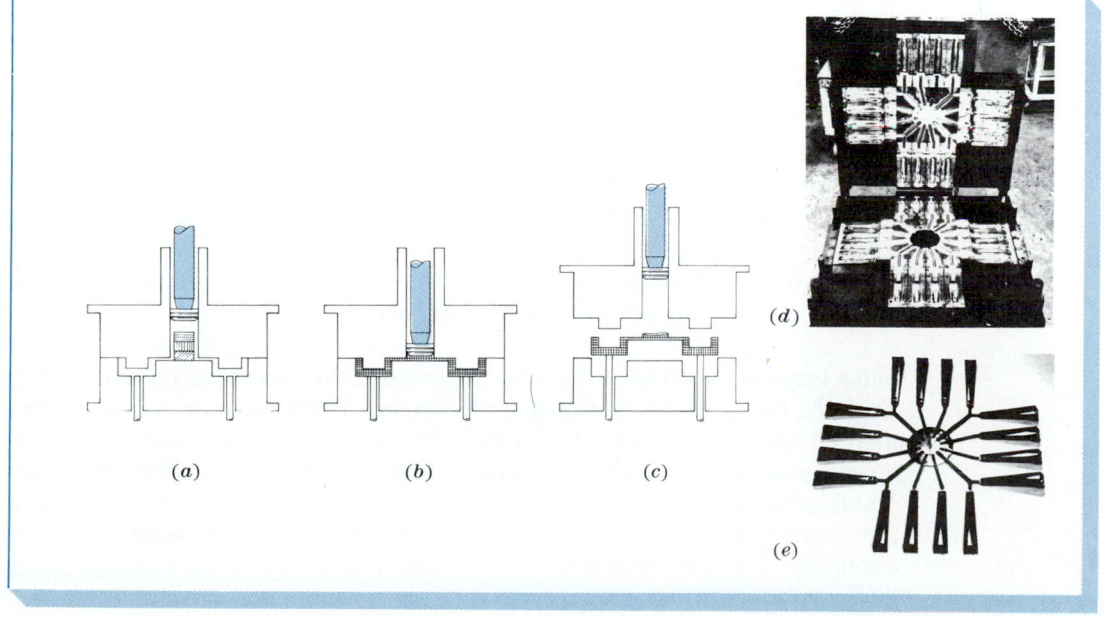

(a)   (b)   (c)

(d)

(e)

In transfer molding, when the mold is closed, a plunger forces the plastic resin (which is usually preheated) from the outside chamber through a system of runners and gates into the mold cavities (Fig. 7.26*b*). After the molded material has had time to cure so that a rigid network polymeric material is formed, the molded part is ejected from the mold (Fig. 7.26*c*). Figure 7.26*d* shows the mold cavities of an open transfer mold, and Fig. 7.26*e* shows the as-molded plastic material containing multiple thermoset plastic parts.

The advantages of transfer molding are:

1. Transfer molding has the advantage over compression molding in that no flash is formed during molding, and thus the molded part requires less finishing.
2. Many parts can be made at the same time by using a runner system (Fig. 7.26*e*).
3. Transfer molding is especially useful for making small intricate parts that would be difficult to make by compression molding.

Injection molding Using modern technology, some thermosetting compounds can be injection-molded by reciprocating-screw injection-molding machines. Special heating and cooling jackets have been added to the standard-type injection-molding machines so that the resin can be cured in the process. Good venting of the mold cavities is required for some thermosetting resins which give off reaction products during curing. In the future, injection molding will probably become more important for producing thermosetting parts because of the efficiency of this process.

7.6 GENERAL-PURPOSE THERMOPLASTICS

In this section some important aspects of the structure, chemical processing, properties, and applications of the following thermoplastics will be discussed: polyethylene, polyvinyl chloride, polypropylene, polystyrene, ABS, polymethyl methacrylate, and polytetrafluoroethylene.

Let us first, however, examine the sales tonnages, list prices, and some of the important properties of these materials.

Sales tonnages and bulk list prices for some general-purpose thermoplastics Table 7.2 lists the sales tonnages in 1988 in the United States for some selected thermoplastics, along with their 1989 bulk list prices. Four major plastic materials, polyethylene, polyvinyl chloride, polypropylene, and polystyrene, accounted for about 60 percent of the total sales tonnage. These materials have the relatively low cost of about 55¢/lb (1989 prices), which undoubtedly accounts for part of the reason for their extensive usage in industry and for many engineering applications. However, when special properties are required which

TABLE 7.2   U.S. Plastic Sales Volume (1993*) and Bulk List Prices (January 1994†) for Some General-Purpose Thermoplastics

Material	U.S. plastic sales in 1000 million pounds‡	Wt % of total U.S. plastic sales	Bulk list price, $/lb	Type of plastic
Polyethylene:				
Low-density	13,299	19.3	0.55–0.60	General-purpose molding
High-density	10,527	15.3	0.60–0.63	General-purpose molding
Polyvinyl chloride and copolymers	10,509	15.1	0.37–0.38	General-purpose homopolymer
Polypropylene and copolymers	8,961	13.0	0.55–0.60	General-purpose homopolymer
Polystyrene	5,479	8.0	0.60–0.66	General-purpose homopolymer
Styrene-acrylonitrile	112	0.16	0.97	
ABS	1,372	2.0	1.28–1.38	Medium impact
Acrylic	691	1.0	1.17	General purpose
Cellulosics	82	0.12	1.56	Acetate type
Polytetrafluoroethylene	· · ·	· · ·	6.40–7.20	

*From *Modern Plastics*, January 1994, p. 73.
†From *Plastics Technology*. Prices vary with time and are thus subject to change.
‡The total U.S. plastic sales in 1993 was 68,827 million lb.

are unobtainable with the cheaper thermoplastics, more costly plastic materials are used. For example, polytetrafluoroethylene (Teflon), which has special high-temperature and lubrication properties, costs about $6.40/lb.

Some basic properties of selected general-purpose thermoplastics   Table 7.3 lists the densities, tensile strengths, impact strengths, dielectric strengths, and maximum-use temperatures for some selected general-purpose thermoplastics. One of the most important advantages of many plastic materials for many engineering applications is their relatively low densities. Most general-purpose plastics have densities of about 1 compared to 7.8 for iron.

The tensile strengths of plastic materials are relatively low, and as a result this property can be a disadvantage for some engineering designs. Most plastic materials have a tensile strength of less than 10,000 psi (69 MPa) (Table 7.3). The tensile test for plastic materials is carried out with the same equipment used for metals (Fig. 6.18).

The impact test that is usually used for plastic materials is the notched Izod test. In this test a $\frac{1}{8} \times \frac{1}{2} \times 2\frac{1}{2}$ in sample (Fig. 7.27) is normally used and is clamped to the base of a pendulum testing machine. The amount of energy absorbed per unit length of the notch when the pendulum strikes the sample is measured and is called the *notched impact strength* of the material. This energy is usually reported in foot pounds per inch (ft · lb/in) or joules per meter (J/m). The notched impact strengths of general-purpose plastic materials of Table 7.3 range from 0.4 to 14 ft · lb/in.

Plastic materials are generally good electrical insulative materials. The electrical insulative strength of plastic materials is usually measured by their

TABLE 7.3   Some Properties of Selected General-Purpose Thermoplastics

Material	Density, g/cm³	Tensile strength, × 1000 psi*	Impact strength, Izod, ft · lb/in†	Dielectric strength, V/mil‡	Max-use temp. (no load) °F	Max-use temp. (no load) °C
Polyethylene:						
Low-density	0.92–0.93	0.9–2.5	. . .	480	180–212	82–100
High-density	0.95–0.96	2.9–5.4	0.4–14	480	175–250	80–120
Rigid, chlorinated PVC	1.49–1.58	7.5–9	1.0–5.6		230	110
Polypropylene, general-purpose	0.90–0.91	4.8–5.5	0.4–2.2	650	225–300	107–150
Styrene-acrylonitrile (SAN)	1.08	10–12	0.4–0.5	1775	140–220	60–104
ABS, general-purpose	1.05–1.07	5.9	6	385	160–200	71–93
Acrylic, general-purpose	1.11–1.19	11.0	2.3	450–500	130–230	54–110
Cellulosics, acetate	1.2–1.3	3–8	1.1–6.8	250–600	140–220	60–104
Polytetrafluoroethylene	2.1–2.3	1–4	2.5–4.0	400–500	550	288

*1000 psi = 6.9 MPa.
† Notched Izod test: 1 ft · lb/in = 53.38 J/m.
‡1 V/mil = 39.4 V/mm.
Source: Materials Engineering, May 1972.

FIGURE 7.27 (*a*) Izod impact test. (*b*) Sample used for plastic materials for the Izod impact test. (*After W. E. Driver, "Plastics Chemistry and Technology," Van Nostrand Reinhold, 1979, pp. 196–197.*)

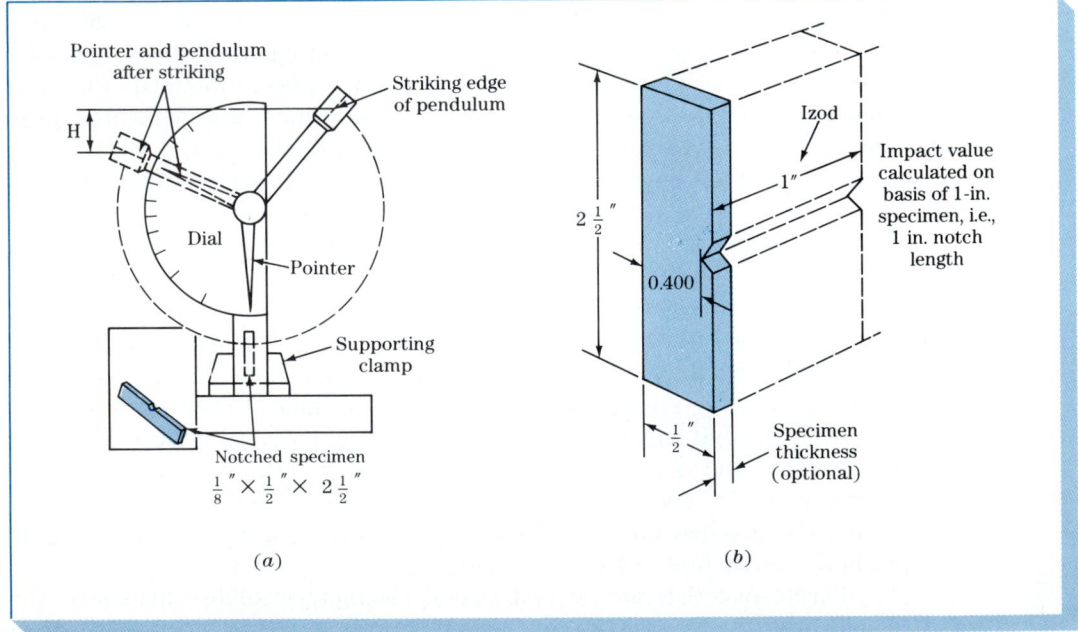

dielectric strength, which may be defined as the voltage gradient that produces electrical breakdown through the material. Dielectric strength is usually measured in volts per mil or volts per millimeter. The dielectric strengths of the plastic materials of Table 7.3 vary from 385 to 1775 V/mil.

The maximum-use temperature for most plastic materials is relatively low and varies from 130 to 300°F (54 to 149°C) for most thermoplastic materials. However, some thermoplastics have higher maximum-use temperatures as, for example, polytetrafluoroethylene which can withstand temperatures up to 550°F (288°C).

Polyethylene Polyethylene (PE) is a clear to whitish translucent thermoplastic material and is often fabricated into clear thin films. Thick sections are translucent and have a waxy appearance. With the use of colorants a wide variety of colored products is obtained.

Repeating chemical structural unit

$$\left[\begin{array}{ccc} H & H \\ | & | \\ C & C \\ | & | \\ H & H \end{array}\right]_n$$

Polyethylene
mp:110–137°C
(230–278°F)

Types of polyethylene In general there are two types of polyethylene: (1) low-density (LDPE) and (2) high-density (HDPE). Low-density has a branched-chain structure (Fig. 7.28*b*), whereas high-density polyethylene has essentially a straight-chain structure (Fig. 7.28*a*).

Low-density polyethylene was first commercially produced in the United Kingdom in 1939 by using autoclave (or tubular) reactors requiring pressures in excess of 14,500 psi (100 MPa) and a temperature of about 300°C. High-density polyethylene was first produced commercially by the Phillips and Ziegler processes by using special catalysts in 1956–1957. In these processes the pressure and temperature for the reaction to convert ethylene to polyethylene were considerably lowered. For example, the Phillips process operates at 100 to 150°C and 290 to 580 psi (2 to 4 MPa) pressure.

More recently (about 1976), a new low-pressure simplified process for producing polyethylene was developed which uses a pressure of about 100 to 300 psi (0.7 to 2 MPa) and a temperature of about 100°C. The polyethylene produced is described as linear-low-density polyethylene (LLDPE) and has a linear-chain structure with short, slanting side branches (Fig. 7.28*c*). A process for producing LLDPE has been described in Sec. 7.3 (see Fig. 7.13).

Structure and properties The chain structures of low- and high-density polyethylenes are shown in Fig. 7.28. Low-density polyethylene has a branched-chain structure which lowers its degree of crystallinity and its density (Table 7.3). The branched-chain structure also lowers the strength of low-density

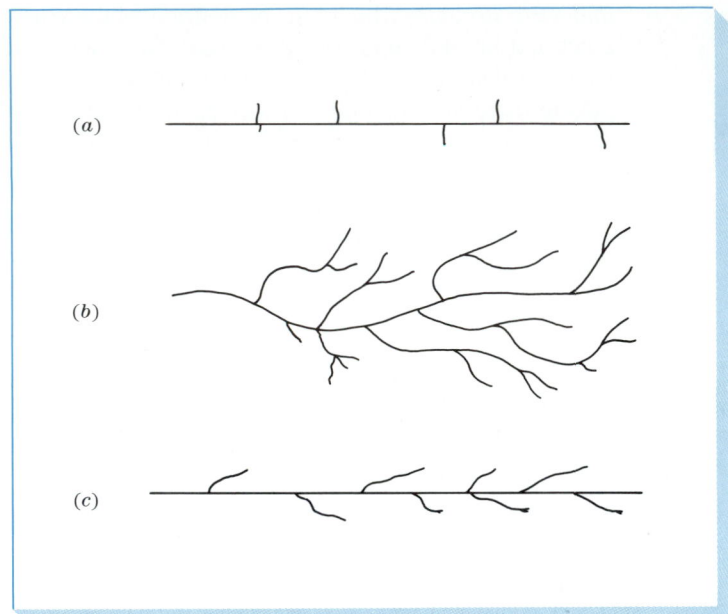

FIGURE 7.28 Chain structure of different types of polyethylene: (a) high-density, (b) low-density, (c) linear-low-density.

polyethylene because it reduces intermolecular bonding forces. High-density polyethylene, in contrast, has very little branching on the main chains, and so the chains are able to pack closer together to increase crystallinity and strength (Table 7.4).

Polyethylene is by far the most extensively used plastic material and accounted for 35 percent of the total sales in the United States in 1993. The main reason for its prime position is that it is low in cost and has many industrially important properties which include toughness at room temperature and to low temperatures with sufficient strength for many product applications, good flexibility over a wide range of temperatures even down to $-73°C$, excellent corrosion resistance, excellent insulating properties, odorless and tasteless, and low in water-vapor transmission.

TABLE 7.4   Some Properties of Low- and High-Density Polyethylenes

Property	Low-density polyethylene	Linear-low-density polyethylene	High-density polyethylene
Density, g/cm³	0.92–0.93	0.922–0.926	0.95–0.96
Tensile strength, × 1000 psi	0.9–2.5	1.8–2.9	2.9–5.4
Elongation, %	550–600	600–800	20–120
Crystallinity, %	65	· · ·	95

FIGURE 7.29 High-density polyethylene film pond liner dwarfs workers installing it. Individual sheets can be half an acre in area and weigh up to 5 tons. (*Photo, Schlegel Lining Technology, Inc.*)

Applications Applications for polyethylene include containers, electrical insulation, chemical tubing, housewares, and blow-molded bottles. Uses for polyethylene films include films for packaging and materials handling and water-pond liners (Fig. 7.29).

Polyvinyl Chloride and Copolymers Polyvinyl chloride (PVC) is a widely used synthetic plastic that has the second largest sales tonnage in the United States (Table 7.2). The widespread use of PVC is attributed mainly to its high chemical resistance and its unique ability to be mixed with additives to produce a large number of compounds with a wide range of physical and chemical properties.

Repeating chemical structural unit

$$\left[\begin{array}{cc} \overset{\displaystyle H}{\underset{\displaystyle H}{C}} & \overset{\displaystyle H}{\underset{\displaystyle Cl}{C}} \end{array}\right]_n$$

Polyvinyl chloride
mp: ~204°C (~400°F)

Structure and properties   The presence of the large chlorine atom on every other carbon atom of the main chain of polyvinyl chloride (PVC) produces a polymeric material that is essentially amorphous and does *not* recrystallize. The strong cohesive forces between the polymer chains in PVC are due mainly to the strong dipole moments caused by the chlorine atoms. The large negative chlorine atoms, however, cause some steric hindrance and electrostatic repulsion which reduces the flexibility of the polymer chains. This molecular immobility results in difficulty in the processing of the homopolymer, and only in a few applications can PVC be used without being compounded with a number of additives so that it can be processed and converted into finished products.

PVC homopolymer has a relatively high strength (7.5 to 9.0 ksi), along with brittleness. PVC has a medium heat-deflection temperature [57 to 82°C (135 to 180°F) at 66 psi], good electrical properties (425 to 1300 V/mil dielectric strength), and high solvent resistance. The high chlorine content of PVC produces flame and chemical resistance.

Polyvinyl chloride compounding   Polyvinyl chloride can only be used for a few applications without the addition of a number of compounds to the basic material so that it can be processed and converted into a finished product. Compounds added to PVC include plasticizers, heat stabilizers, lubricants, fillers, and pigments.

Plasticizers impart flexibility to polymeric materials. They are usually high-molecular-weight compounds which are selected to be completely miscible

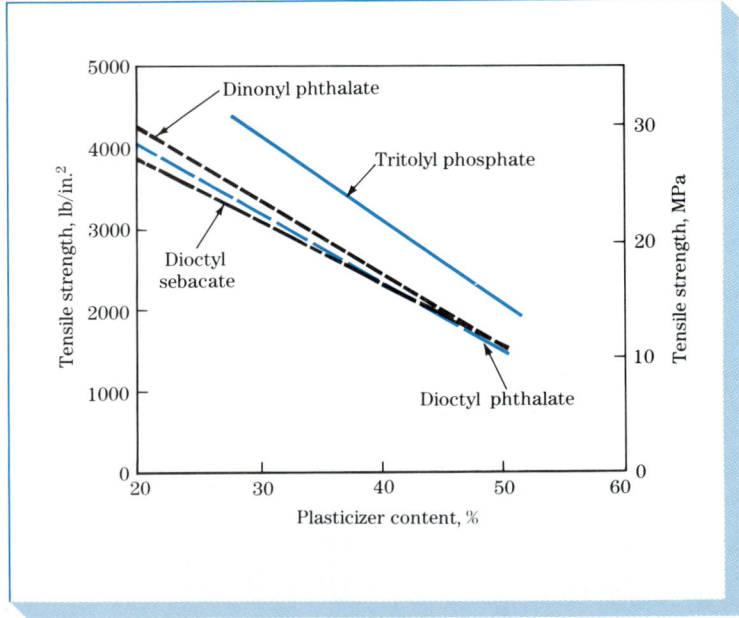

FIGURE 7.30 Effect of different plasticizers on the tensile strength of polyvinyl chloride. [*After C. A. Brighton, Vinyl Chloride Polymers (Compounding), in "Encyclopedia of Polymer Science and Technology," vol. 14, Wiley, 1971, p. 398.*]

and compatible with the basic material. For PVC, phthalate esters are commonly used as plasticizers. The effect of some plasticizers on the tensile strength of PVC is shown in Fig. 7.30.

Heat stabilizers are added to PVC to prevent thermal degradation during processing and may also help to extend the life of the finished product. Typical stabilizers used may be all organic or inorganic but are usually organometallic compounds based on tin, lead, barium-cadmium, calcium, and zinc.

Lubricants aid the melt flow of PVC compounds during processing and prevent adhesion to metal surfaces. Waxes, fatty esters, and metallic soaps are commonly used lubricants.

Fillers such as calcium carbonate are mainly added to lower the cost of PVC compounds.

Pigments, both inorganic and organic, are used to give color, opacity, and weatherability to PVC compounds.

Rigid polyvinyl chloride   Polyvinyl chloride alone can be used for some applications but is difficult to process and has low impact strength. The addition of rubbery resins can improve melt flow during processing by forming a dispersion of small, soft rubbery particles in the matrix of rigid PVC. The rubbery material serves to absorb and disperse impact energy so that the impact resistance of the material is increased. With improved properties, rigid PVC is used for many applications. In building construction rigid PVC is used for pipe, siding, window frames, gutter, and interior molding and trim. PVC is also used for electrical conduit.

Plasticized polyvinyl chloride   The addition of plasticizers to PVC produces softness, flexibility, and extensibility. These properties can be varied over a wide range by adjusting the plasticizer-polymer ratio. Plasticized polyvinyl chloride is used in many applications where it outperforms rubber, textiles, and paper. Plasticized PVC is used for furniture and auto upholstery, interior wall coverings, rainwear, shoes, luggage, and shower curtains. In transportation plasticized PVC is used for auto top coverings, electrical wire insulation, floor mats, and interior and exterior trim. Other applications include garden hoses, refrigerator gaskets, appliance components, and housewares.

Polypropylene   Polypropylene is the thirdmost-important plastic from a sales tonnage standpoint (Table 7.2) and is one of the lowest in cost since it can be synthesized from low-cost petrochemical raw materials.

Repeating chemical structural unit

$$\left[\begin{array}{c} \underset{|}{\overset{|}{C}}\!-\!\underset{|}{\overset{|}{C}} \\ \end{array}\right]_n$$

$$\begin{bmatrix} H & H \\ | & | \\ -C - C - \\ | & | \\ H & CH_3 \end{bmatrix}_n$$

Polypropylene
mp:165–177°C
(330–350°F)

Structure and properties   In going from polyethylene to polypropylene, the substitution of a methyl group on every second carbon atom of the polymer main chain restricts rotation of the chains, producing a stronger but less flexible material. The methyl groups on the chains also increase the glass transition temperature, and thus polypropylene has higher melting and heat-deflection temperatures than polyethylene. With the use of stereospecific catalysts, iso-tactic polypropylene can be synthesized which has a melting point in the 165 to 177°C (330 to 350°F) range. This material can be subjected to temperatures of about 120°C (250°F) without deformation.

Polypropylene has a good balance of attractive properties for producing many manufactured goods. These include good chemical, moisture, and heat resistance, along with low density (0.900 to 0.910 g/cm³), good surface hardness, and dimensional stability. Polypropylene also has outstanding flex life as a hinge and can be used for products with an integral hinge. Along with the low cost of its monomer, polypropylene is a very competitive thermoplastic material.

Applications   The major applications for polypropylene are housewares, appliance parts, packaging, laboratory ware, and bottles of various types. In transportation, high-impact polypropylene copolymers have replaced hard rubber for battery housings. Similar resins are used for fender liners and splash shrouds. Filled polypropylene finds application for automobile fan shrouds and heater ducts, where high resistance to heat deflection is needed. Also, polypropylene homopolymer is used extensively for primary carpet backing and as a woven material is used for shipping sacks for many industrial products. In the film market, polypropylene is used as a bag and overwrap film for soft goods because of its luster, gloss, and good stiffness. In packaging, polypropylene is used for screw closures, cases, and containers.

Polystyrene   Polystyrene is the fourth-largest tonnage thermoplastic. Homopolymer polystyrene is a clear, odorless, and tasteless plastic material which is relatively brittle unless modified. Besides crystal polystyrene, other important grades are rubber-modified impact-resistant and expandable polystyrenes. Styrene also is used to produce many important copolymers.

Repeating chemical structural unit

Polystyrene
mp:150–243°C
(330–470°F)

Structure and properties   The presence of the phenylene ring on every other carbon atom of the main chain of polystyrene produces a rigid bulky configuration with sufficient steric hindrance to make the polymer very inflexible at room temperature. The homopolymer is characterized by its rigidity, sparkling clarity, and ease of processibility but tends to be brittle. The impact properties of polystyrene can be improved by copolymerization with the polybutadiene elastomer which has the chemical structure

Polybutadiene

Copolymers of impact styrene usually have rubber levels between 3 to 12 percent. The addition of the rubber to polystyrene lowers the rigidity and heat-deflection temperature of the homopolymer.

In general polystyrenes have good dimensional stability and low-mold shrinkage and are easily processed at a low cost. However, they have poor weatherability and are chemically attacked by organic solvents and oils. Polystyrenes have good electrical insulating properties and adequate mechanical properties within operating temperature limits.

Applications   Typical applications include automobile interior parts, appliance housings, dials and knobs, and housewares.

Polyacrylonitrile   This acrylic-type polymeric material is often used in the form of fibers, and because of its strength and chemical stability, it is also used as a comonomer for some engineering thermoplastics.

Repeating chemical structural unit

Polyacrylonitrile
(does not melt)

Structure and properties   The highly electronegativity of the nitrile group on every other carbon atom of the main chain exerts mutual electrical repulsion, causing the molecular chains to be forced into extended, stiff, rodlike structures. The regularity of the rodlike structures allows them to orient themselves to produce strong fibers by hydrogen bonding between the polymer chains. As a result, the acrylonitrile fibers have high strength and good resistance to moisture and solvents.

Applications   Acrylonitrile is used in fiber form for woollike applications such as sweaters and blankets. Acrylonitrile is also used as a comonomer for producing styrene-acrylonitrile copolymers (SAN resins) and acrylonitrile-butadiene-styrene terpolymers (ABS resins).

Styrene-Acrylonitrile (SAN)   Styrene-acrylonitrile (SAN) thermoplastics are high-performance members of the styrene family.

Structure and properties   SAN resins are random, amorphous copolymers of styrene and acrylonitrile. This copolymerization creates polarity and hydrogen-bond attractive forces between the polymer chains. As a result, SAN resins have better chemical resistance, higher heat-deflection temperatures, toughness, and load-bearing characteristics than polystyrene alone. SAN thermoplastics are rigid and hard, process easily, and have the gloss and clarity of polystyrene.

Applications   Major applications for SAN resins include automotive instrument lenses, dash components, and glass-filled support panels; appliance knobs and blender and mixer bowls; medical syringes and blood aspirators; construction safety glazing; and houseware tumblers and mugs.

ABS   ABS is the name given to a family of thermoplastics. The acronym is derived from the three monomers used to produce ABS: *a*crylonitrile, *b*utadiene, and *s*tyrene. ABS materials are noted for their engineering properties such as good impact and mechanical strength combined with ease of processing.

Chemical structural units   ABS contains the three chemical structural units

A: polyacrylonitrile   B: polybutadiene   S: polystyrene

Structure and properties of ABS   The wide range of useful engineering properties exhibited by ABS is due to the contributing properties of each of its components. Acrylonitrile contributes heat and chemical resistance and toughness, butadiene provides impact strength and low-property retention, and styrene provides surface gloss, rigidity, and ease of processing. The impact strengths of ABS plastics are increased as the percent rubber content is increased but the tensile-strength properties and heat-deflection temperatures are decreased (Fig. 7.31). Table 7.5 lists some of the properties of high-, medium-, and low-impact ABS plastics.

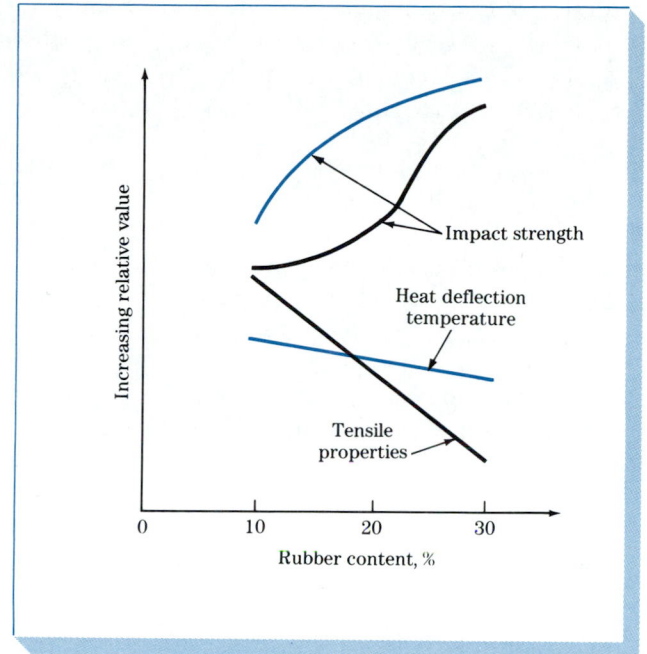

FIGURE 7.31 Percent rubber vs. some properties of ABS. (*After G. E. Teer, ABS and Related Multipolymers, in "Modern Plastics Encyclopedia," McGraw-Hill, 1981–1982.*)

The structure of ABS is *not* that of a random terpolymer. ABS can be considered a blend of a glassy copolymer (styrene-acrylonitrile) and rubbery domains (primarily a butadiene polymer or copolymer). Simply blending rubber with the glassy copolymer does not produce optimal impact properties. The best impact strength is obtained when the styrene-acrylonitrile copolymer matrix is grafted to the rubber domains to produce a two-phase structure (Fig. 7.32).

Applications   The major use for ABS is for pipe and fittings, particularly drain-waste-and-vent pipe in building. Other uses for ABS are for automotive parts, appliance parts such as refrigerator door liners and inner liners, business

TABLE 7.5   **Some Typical Properties of ABS Plastics (23°C)**

	High-impact	Medium-impact	Low-impact
Impact strength (Izod):			
ft·lb/in	7–12	4–7	2–4
J/m	375–640	215–375	105–320
Tensile strength:			
× 1000 psi	4.8–6.0	6.0–7.0	6.0–7.5
MPa	33–41	41–48	41–52
Elongation, %	15–70	10–50	5–30

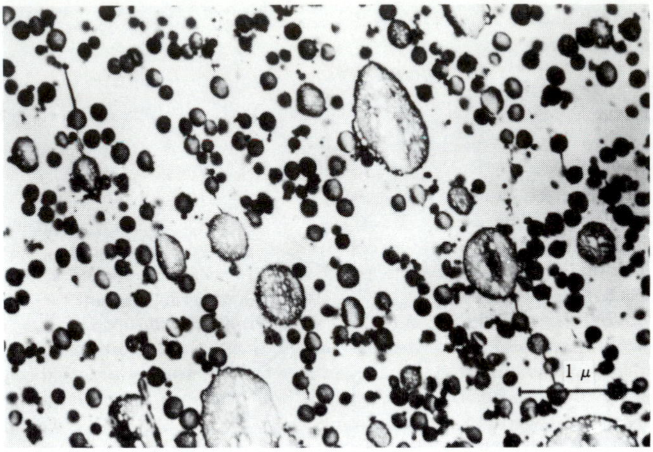

FIGURE 7.32 Electron micrograph of an ultrathin section of a type G ABS resin showing the rubbery particles in a copolymer of styrene-acrylonitrile. [*After M. Matsuo, Polym. Eng. Sci., 9:206(1969).*]

machines, computer housings and covers, telephone housings, electrical conduit, and electromagnetic interference–radio-frequency shielding applications.

Polymethyl Methacrylate (PMMA)

Polymethyl methacrylate is a hard, rigid transparent thermoplastic which has good outdoor weatherability and is more impact-resistant than glass. This material is better known by the trade names Plexiglas or Lucite and is the most important material of the group of thermoplastics known as the *acrylics*.

Repeating chemical structural unit

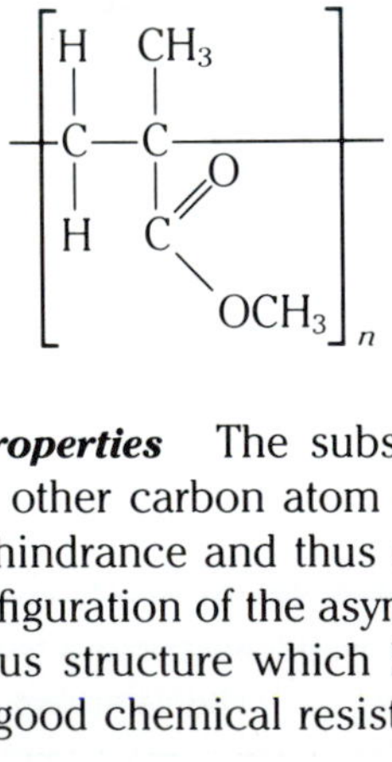

Polymethyl methacrylate
mp:160°C (320°F)

Structure and properties   The substitution of the methyl and methacrylate groups on every other carbon atom of the main carbon chain provides considerable steric hindrance and thus makes PMMA rigid and relatively strong. The random configuration of the asymmetrical carbon atoms produces a completely amorphous structure which has a high transparency to visible light. PMMA also has good chemical resistance to outdoor environments.

Applications   PMMA is used for glazing for aircraft and boats, skylights, exterior lighting, and advertising signs. Other uses include auto taillight lenses, safety shields, protective goggles, and knobs and handles.

Fluoroplastics

These materials are plastics or polymers made from monomers containing one or more atoms of fluorine. The fluoroplastics have a combination of special properties for engineering applications. As a class they have a high resistance to hostile chemical environments and outstanding electrical insulating properties. Those fluoroplastics containing a large percentage of fluorine have low coefficients of friction which give them self-lubricating and nonstick properties.

There are many fluoroplastics produced, but the two most widely used ones, polytetrafluoroethylene (PTFE) (Fig. 7.33) and polychlorotrifluoroethylene (PCTFE), will be discussed in this subsection.

Polytetrafluoroethylene (PTFE)

Repeating chemical structural unit

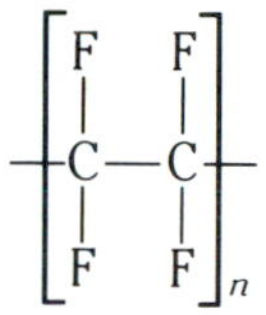

Polytetrafluoroethylene
Softens at 370°C (700°F)

FIGURE 7.33 Structure of polytetrafluorethylene.

Chemical processing   PTFE is a completely fluorinated polymer formed by the free-radical chain polymerization of tetrafluoroethylene gas to produce linear-chain polymers of —CF_2— units. The original discovery of the polymerization of tetrafluoroethylene gas into polytetrafluoroethylene (Teflon) was made by R. J. Plunkett in 1938 in a Du Pont laboratory.

Structure and properties   PTFE is a crystalline polymer with a crystalline melting point of 327°C (620°F). The small size of the fluorine atom and the regularity of the fluorinated carbon chain polymer results in a highly dense crystalline polymeric material. The density of PTFE is high for plastic materials, 2.13 to 2.19 g/cm^3.

PTFE has exceptional resistance to chemicals and is insoluble in all organics with the exception of a few fluorinated solvents. PTFE also has useful mechanical properties from cryogenic temperatures [−200°C (−330°F)] to 260°C (500°F). Its impact strength is high, but its tensile strength, wear, and creep resistance is low when compared with other engineering plastics. Fillers such as glass fibers can be used to increase strength. PTFE is slippery and waxy to the touch and has a low coefficient of friction.

Processing   Since PTFE has such a high melt viscosity, conventional extrusion and injection-molding processes cannot be used. Parts are molded by compressing granules at room temperature at 2000 to 10,000 psi (14 to 69 MPa). After compression, the preformed materials are sintered at 360 to 380°C (680 to 716°F).

Applications   PTFE is used for chemically resistant pipe and pump parts, high-temperature cable insulation, molded electrical components, tape, and non-stick coatings. Filled PTFE compounds are used for bushings, packings, gaskets, seals, O rings, and bearings.

Polychlorotrifluoroethylene (PCTFE)

Repeating chemical structural unit

$$\left[\begin{array}{cc} F & F \\ | & | \\ -C & -C- \\ | & | \\ F & Cl \end{array}\right]_n$$

Polychlorotrifluoroethylene (PCTFE)
Melting point [218 °C (420 °F)]

Structure and properties The substitution of a chlorine atom for every fourth fluorine atom produces some irregularity in the polymer chains, making the material less crystalline and more moldable. Thus, PCTFE has a lower melting point [218°C (420°F)] than PTFE and can be extruded and molded by conventional processes.

Applications Extruded, molded, and machined products of PCTFE polymeric materials are used for chemical processing equipment and electrical applications. Other applications include gaskets, O rings, seals, and electrical components.

7.7 ENGINEERING THERMOPLASTICS

In this section some important aspects of the structure, properties, and applications of engineering thermoplastics will be discussed. The definition of an engineering plastic is arbitrary since there is virtually no plastic that cannot in some form be considered an engineering plastic. A thermoplastic in this book will be considered an engineering thermoplastic if it has a balance of properties which makes it especially useful for engineering applications. In this discussion the following families of thermoplastics have been selected as engineering thermoplastics: polyamides (nylons), polycarbonates, phenylene oxide–based resins, acetals, thermoplastic polyesters, polysulfone, polyphenylene sulfide, and polyetherimide.

Let us first examine the sales tonnages and bulk list prices of these engineering thermoplastics and then some of their important properties as a group.

Sales tonnages and bulk list prices for some engineering thermoplastics Table 7.6 lists the sales tonnages in 1993 in the United States for some selected engineering thermoplastics, along with their 1994 bulk list prices. The materials listed in this table have sales tonnages of only 5 percent or less and higher prices than most general-purpose plastics.

TABLE 7.6   U.S. Plastic Sales Volume (1993*) and Bulk List Prices (January 1994†) for Some Important Engineering Thermoplastics

Material	U.S. plastic sales in 1000 metric tons‡	Wt % of total U.S. plastic sales	Bulk list price, $/lb	Type of plastic
Polyamide (nylon)	726	1.0	2.40–2.41	Nylon 6.6
Polyacetal	149	0.22	2.25–2.28	Homopolymer
Polycarbonate	617	0.90	2.42–2.48	Injection
Polyester, thermoplastic	2966	4.3	1.64–1.75	PBT, injection
Modified phenylene oxide	219	0.31	1.40–2.20	Injection
Polysulfone	⋯	⋯	4.23–4.90	
Polyphenylene sulfide	⋯	⋯	1.97	35% glass–35% filler

*From *Modern Plastics,* January 1994, p. 73.
†From *Plastics Technology.* Prices vary with time and are thus subject to change.
‡The total U.S. plastic sales in 1993 was 68,827 million lb.

Some basic properties of selected engineering thermoplastics   Table 7.7 lists the densities, tensile strengths, impact strengths, dielectric strengths, and maximum-use temperatures for some selected engineering thermoplastics. The densities of the engineering thermoplastics listed in Table 7.7 are relatively low, ranging from 1.06 to 1.42 g/cm^3. The low densities of these materials are an important property advantage for many engineering designs. As for almost all plastic materials, their tensile strengths are relatively low, with those in Table 7.7 ranging from 8000 to 12,000 psi (55 to 83 MPa). These low strengths are usually an engineering design disadvantage. As for the impact strengths of

TABLE 7.7   Some Properties of Selected Engineering Thermoplastics

Material	Density, g/cm³	Tensile strength, × 1000 psi*	Impact strength, Izod, ft·lb/in†	Dielectric strength, V/mil‡	Max-use temp. (no load) °F	Max-use temp. (no load) °C
Nylon 6.6	1.13–1.15	9–12	2.0	385	180–300	82–150
Polyacetal, homo.	1.42	10	1.4	320	195	90
Polycarbonate	1.2	9	12–16	380	250	120
Polyester:						
PET	1.37	10.4	0.8	⋯	175	80
PBT	1.31	8.0–8.2	1.2–1.3	590–700	250	120
Polyphenylene oxide	1.06–1.10	7.8–9.6	5.0	400–500	175–220	80–105
Polysulfone	1.24	10.2	1.2	425	300	150
Polyphenylene sulfide	1.34	10	0.3	595	500	260

*1000 psi = 6.9 MPa.
†Notched Izod test: 1 ft · lb/in = 53.38 J/m.
‡1 V/mil = 39.4 V/mm.

the engineering thermoplastics, polycarbonate has an outstanding impact strength, having values from 12 to 16 ft·lb/in. The low values of 1.4 and 2.0 ft·lb/in for polyacetal and nylon 6,6 are somewhat misleading since these materials are "tough" plastic materials but are notch-sensitive, as the notched Izod impact test indicates.

The electrical insulation strengths of the engineering thermoplastics listed in Table 7.7 are high as is the case for most plastic materials and range from 320 to 700 V/mil. The maximum-use temperatures for the engineering thermoplastics listed in Table 7.7 are from 180 to 500°F (82 to 260°C). Of the engineering thermoplastics of Table 7.7 polyphenylene sulfide has the highest use temperature of 500°F (260°C).

There are many other properties of engineering thermoplastics which make these materials industrially important. Engineering thermoplastics are relatively easy to process into a near-finished or finished shape, and their processing can be automated in most cases. Engineering thermoplastics have good corrosion resistance to many environments. In some cases engineering plastics have superior resistance to chemical attack. For example, polyphenylene sulfide has no known solvents below 400°F (204°C).

Polyamides (Nylons)   *Polyamides* or *nylons* are melt-processible thermoplastics whose main-chain structure incorporates a repeating amide group. Nylons are members of the engineering plastics family and offer superior load-bearing capability at elevated temperatures, good toughness, low-frictional properties, and good chemical resistance.

Chemical repeating linkage   There are many types of nylons, and the repeating unit is different for each type. They all, however, have the *amide linkage* in common:

$$\begin{array}{cc} O & H \\ \| & | \\ -C & -N- \end{array} \qquad \text{Amide linkage}$$

Chemical processing and polymerization reactions   Some types of nylons are produced by a stepwise polymerization of a dibasic organic acid with a diamine. Nylon 6,6,[1] which is the most important of the nylon family, is produced by a polymerization reaction between hexamethylene diamine and adipic acid to produce polyhexamethylene diamine (Fig. 7.10). The repeating chemical structural unit for nylon 6,6 is

---

[1] The designation 6,6 of nylon 6,6 refers to the fact that there are 6 carbon atoms in the reacting diamine (hexamethylene diamine) and also 6 carbon atoms in the reacting organic acid (adipic acid).

$$\left[\begin{array}{c} H \\ | \\ N \end{array} - (CH_2)_6 - \begin{array}{c} \\ N \\ | \\ H \end{array} - \begin{array}{c} O \\ \| \\ C \end{array} - (CH_2)_4 - \begin{array}{c} O \\ \| \\ C \end{array} \right]_n$$

Nylon 6,6
mp: 250–266°C
(482–510°F)

Other important commercial nylons made by the same type of reaction are nylons 6,9, 6,10, and 6,12 which are made with hexamethylene diamine and azelaic (9 carbons), sebacic (10 carbons), or dodecanedioic (12 carbons) acids, respectively.

Nylons can also be produced by chain polymerization of ring compounds that contain both organic acid and amine groups. For example, nylon 6 can be polymerized from ε-caprolactam (6 carbons) as shown below:

Ring opening

$$H_2C \begin{array}{c} H \\ | \\ N \\ \end{array} \begin{array}{c} O \\ \diagup \\ C \end{array} \atop H_2C \qquad CH_2 \atop H_2C - CH_2 \xrightarrow{heat} \left[\begin{array}{c} H \\ | \\ N \end{array} - CH_2 - CH_2 - CH_2 - CH_2 - CH_2 - \begin{array}{c} O \\ \| \\ C \end{array} \right]_n$$

ε-Caprolactam

Nylon 6
mp:216–225°C (420–435°F)

Structure and properties   Nylons are highly crystalline polymeric materials because of the regular symmetrical structure of their main polymer chains. The high crystallizability of the nylons is apparent by the fact that under controlled solidification conditions spherulites can be produced. Figure 7.34 shows an

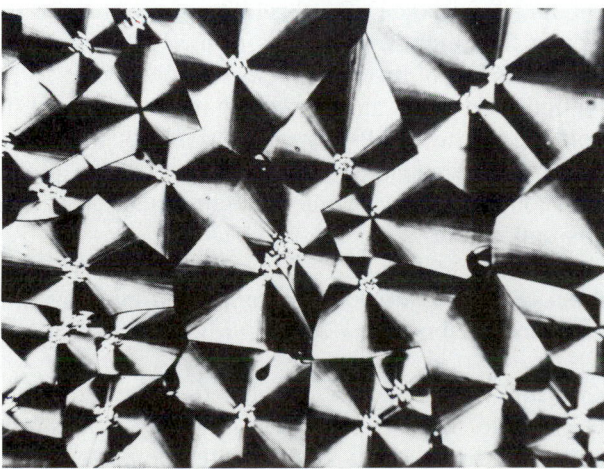

FIGURE 7.34 Complex spherulitic structure of nylon 9,6 grown at 210°C. The fact that spherulites can be grown in this nylon material emphasizes the capability of nylon materials to crystallize. (*Courtesy of J. H. Magill, University of Pittsburgh.*)

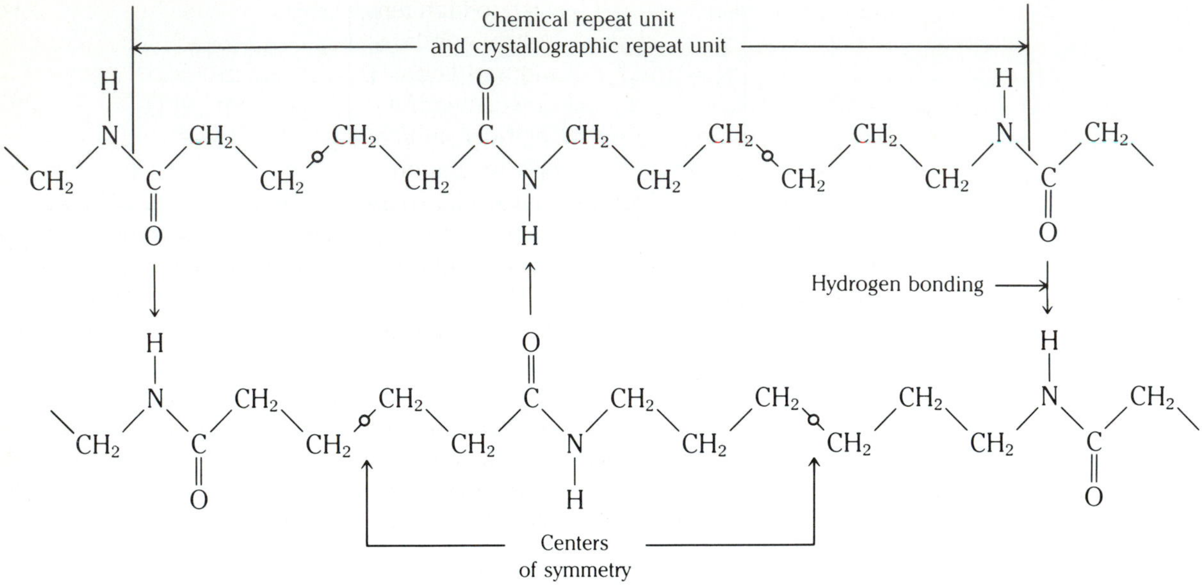

FIGURE 7.35 Schematic representation of hydrogen bonding between two molecular chains. [*After M. I. Kohan (ed.),* "*Nylon Plastics,*" *Wiley, 1973, p. 274.*]

excellent example of the formation of a complex spherulitic structure in nylon 9,6 grown at 210°C.

The high strength of the nylons is partly due to the hydrogen bonding between the molecular chains (Fig. 7.35). The amide linkage makes possible a —NHO type of hydrogen bond between the chains. As a result the nylon polyamides have high strength, high heat-deflection temperatures, and good chemical resistance. The flexibility of the main carbon chains produces molecular flexibility, which leads to low melt viscosity and easy processibility. The flexibility of the carbon chains contributes to high lubricity, low surface friction, and good abrasion resistance. However, the polarity and hydrogen bonding of the amide groups causes high water absorption which results in dimensional changes with increasing moisture content. Nylons 11 and 12 with their longer carbon chains between amide groups are less sensitive to water absorption.

Processing Most nylons are processed by conventional injection-molding or extrusion methods.

Applications Applications for nylons are found in almost all industries. Typical uses are for unlubricated gears, bearings, and antifriction parts, mechanical parts that must function at high temperatures and resist hydrocarbons and solvents, electrical parts subjected to high temperatures, and high-impact parts requiring strength and rigidity. Automobile applications include speedometer and windshield wiper gears, and trim clips. Glass-reinforced nylon is used for engine fan blades, brake and power-steering fluid reservoirs, valve covers, and steering column housings. Electrical and/or electronic applications include connectors, plugs, hook-up wire insulation, antenna mounts, and terminals. Nylon is also used in packaging and for many general-purpose applications.

Nylons 6,6 and 6 make up most of the U.S. nylon sales tonnage because they offer the most favorable combination of price, properties, and processibility. However, for some applications nylons 6,10 and 6,12, and nylons 11 and 12 as well as others are produced and sold at a premium price when their special properties are required.

Polyphthalamide (PPA)

Polypthalamide (PPA) resins are semiaromatic polyamides made by combining aromatic polyamide with some aliphatic (linear) polyamide. PPAs are produced in both semicrystalline and amorphous grades, with glass transition temperatures in the range of 124°C (255°F), PPA resins were introduced in 1991 under the trade name of Amodel resins.

Chemical structure The major chemical repeating unit of PPA resins is the aromatic polyamide unit:

PPA resins also contain some aliphatic (linear) nylon material to increase the castability of the resin. Resonating and interacting aromatic rings in PPA strengthen this material. Hydrogen bonding between the amide linkages also provide some increased strengthening.

Properties In general PPA resins are stronger, stiffer, much less sensitive to water absorption, and have much higher thermal capability than the aliphatic (linear) polyamides such as nylon 6/6. They also have significantly better creep, fatigue, and chemical resistance.

For example, the typical tensile strength of unreinforced PPA resin (Amodel AD-1000) is about 15.1 ksi (104 MPa) with an elongation of about 6.4 percent. However, the tensile strength of 45 percent glass-reinforced PPA has an increased tensile strength of about 38 ksi (262 MPa) (Amodel A-1145) with a decreased elongation of about 2.0 percent.

Applications PPA is suitable for many automotive applications such as head lamp reflectors, bearing retainers, pulleys, sensor housings, fuel line components, and various other structural and electrical components. The high heat deflection temperature of PPA enables it to be used for electrical application involving vapor phase and infrared soldering. PPA's excellent electrical characteristics combined with high heat deflection temperature, high modulus at high temperatures, and moldability make it suitable for switching devices, connectors (Fig. 7.36), and other electrical components. Mineral-filled grades have been developed for reflective surfaces and plated applications, such as decorative plumbing and hardware.

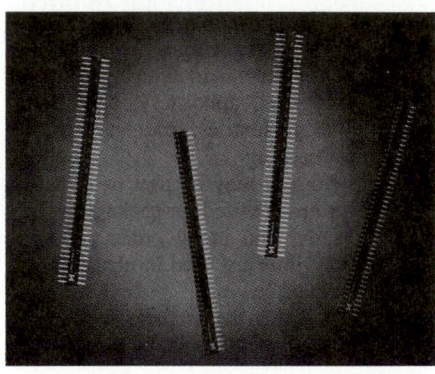

FIGURE 7.36 Subminiature electrical conductors made of PPA replacing glass-filled nylon. (*Amoco Performance Products.*)

Polycarbonate

Polycarbonates form another class of engineering thermoplastics because some of their special high-performance characteristics such as high strength, toughness, and dimensional stability are required for some engineering designs. Polycarbonate resins are manufactured in the United States, for example by General Electric under the trade name Lexan and by Miles, Inc., under the trade name Makrolon.

Basic repeating chemical structural unit

Polycarbonate
mp:270°C (520°F)

Carbonate linkage

Structure and properties   The two phenyl and two methyl groups attached to the same carbon atom in the repeating structural unit produce considerable steric hindrance and make a very stiff molecular structure. However, the carbon-oxygen single bonds in the carbonate linkage provide some molecular flexibility for the overall molecular structure, which produces high-impact energy. The tensile strengths of the polycarbonates at room temperature are relatively high at about 9 ksi (62 MPa), and their impact strengths are very high at 12 to 16 ft · lb/in (640 to 854 J/m) as measured by the Izod test. Other important properties of polycarbonates for engineering designs are their high heat-deflection temperatures, good electrical insulating properties, and transparency. The creep resistance of these materials is also good. Polycarbonates are resistant to a variety of chemicals but are attacked by solvents. Their high dimensional stability enables them to be used in precision engineering components where close tolerances are required.

Applications Typical applications for polycarbonates include safety shields, cams and gears, helmets, electrical relay covers, aircraft components, boat propellers, traffic light housings and lenses, glazing for windows and solar collectors, and housings for hand-held power tools, small appliances, and computer terminals (Fig. 7.1*a*).

Phenylene Oxide–Based Resins The phenylene oxide–based resins form a class of engineering thermoplastic materials.

Basic repeating chemical structural unit

Polyphenylene oxide

Chemical processing A patented process for the oxidative coupling of phenolic monomers is used to produce phenylene oxide-based thermoplastic resins. The PPO resins are blended with an undisclosed amount of rubberized polystyrene to provide good impact properties and are produced by General Electric under the trade name Noryl resins.

Structure and properties The repeating phenylene rings[1] create steric hindrance to rotation of the polymer molecule and electronic attraction due to the resonating electrons in the benzene rings of adjacent molecules. These factors lead to a polymeric material with high rigidity, strength, chemical resistance to many environments, dimensional stability, and heat-deflection temperature.

There are many different grades of these materials to meet the requirements of a wide range of engineering design applications. Among the principal design advantages of the polyphenylene oxide resins are excellent mechanical properties over the temperature range from -40 to $150°C$ (-40 to $300°F$), excellent dimensional stability with low creep, high modulus, and low water absorption,

---

[1] A phenylene ring is a benzene ring chemically bonded to other atoms as, for example,

Bonds with other atoms

good dielectric properties, excellent impact properties, and excellent resistance to aqueous chemical environments.

Applications Typical applications for polyphenylene oxide resins are electrical connectors, TV tuners and deflection yoke components, small appliance and business machine housings, and automobile dashboards, grills, and exterior body parts.

Acetals *Acetals* are a class of high-performance engineering thermoplastic materials. They are one of the strongest [tensile strength of 10 ksi (68.9 MPa)] and stiffest [modulus in flexure of 410 ksi (2820 MPa)] thermoplastics and have excellent fatigue life and dimensional stability. Other important characteristics include low friction coefficients, good processability, good solvent resistance, and high heat resistance to about 90°C (195°F) with no load.

Repeating chemical structural unit

$$\left[\begin{array}{c} H \\ | \\ -C-O- \\ | \\ H \end{array} \right]_n$$

Polyoxymethylene
mp:175°C (347°F)

Types of acetals At present there are two basic types of acetals: a homopolymer (Du Pont's Delrin) and a copolymer (Celanese's Celcon).

Structure and properties The regularity, symmetry, and flexibility of the acetal polymer molecules produce a polymeric material with high regularity, strength, and heat-deflection temperature. Acetals have excellent long-term load-carrying properties and dimensional stability and thus can be used for precision parts such as gears, bearings, and cams. The homopolymer is harder, more rigid, and has higher tensile strength and flexural strength than the copolymer. The copolymer is more stable for long-term high-temperature applications and has a higher elongation.

The low moisture absorption of unmodified acetal homopolymer provides it with good dimensional stability. Also, the low wear and friction characteristics of acetal make it useful for moving parts. In all moving parts, acetal's excellent fatigue resistance is an important property. However, acetals are flammable, and so their use in electrical and/or electronic applications is limited.

Applications Acetals have replaced many metal castings of zinc, brass, and aluminum and stampings of steel because of lower cost. Where the higher strength of the metals is not required, finishing and assembly operation costs can be reduced or eliminated by using acetals for many applications.

In automobiles, acetals are used for components in fuel systems, seat belts, and window handles. Machinery applications for acetals include mechanical couplings, pump impellers, gears, cams, and housings. Acetals are also used in a wide variety of consumer products such as zippers, fishing reels, and writing pens.

Thermoplastic Polyesters Two important engineering thermoplastic polyesters are polybutylene terephthalate (PBT) and polyethylene terephthalate (PET). PET is widely used for film for food packaging and as a fiber for clothing, carpeting, and tire cord. Since 1977 PET has been used as a container resin. PBT, which has a higher-molecular-weight repeating unit in its polymer chains, was introduced in 1969 as a replacement material for some applications where thermosetting plastics and metals were used. The use of PBT is continuing to expand because of its properties and relatively low cost.

Repeating chemical structural units

Polyethylene terephthalate (PET)

Polybutylene terephthalate (PBT)

Structure and properties The phenylene rings along with the carbonyl groups (—C—O—) in PBT form large, flat bulky units in the polymer chains. This regular structure crystallizes quite readily in spite of its bulkiness. The phenylene ring structure provides rigidity to this material, and the butylene units provide some molecular mobility for melt processing. PBT has good strength [7.5 ksi (52 MPa) for unreinforced grades and 19 ksi (131 MPa) for 40 percent glass-reinforced grades]. Thermoplastic polyester resins also have low moisture-absorption characteristics. The crystalline structure of PBT makes it resistant to most chemicals. Most organic compounds have little effect on PBT at moderate temperatures. PBT also has good electrical insulation properties which are nearly independent of temperature and humidity.

Applications Electrical-eiectronic applications for PBT include connectors, switches, relays, TV tuner components, high-voltage components, terminal boards, integrated circuit boards, motor brush holders, end bells, and housings (Fig. 7.1b). Industrial uses for PBT include pump impellers, housings and support brackets, irrigation valves and bodies, and water meter chambers and components. PBT is also used for appliance housings and handles. Automotive applications include large exterior-body components, high-energy ignition caps and rotors, ignition coil caps, coil bobbins, fuel injection controls, and speedometer frames and gears.

Polysulfones   Polysulfones are transparent, tough, strong, heat-resistant high-performance engineering thermoplastics.

Repeating chemical structural unit

Polysulfone
mp:315°C (600°F)

Sulfone
linkage

Structure and properties   The phenylene rings of the polysulfone repeating unit restrict rotation of the polymer chains and create strong intermolecular attraction to provide high strength and rigidity to this material. Oxygen atoms in para[1] positions of the phenylene ring with respect to the sulfone group provide the high-oxidation stability of the sulfone polymers. The oxygen atoms between the phenylene rings (ether linkage) provide chain flexibility and impact strength.

Properties of polysulfone of special significance to the design engineer are its high heat-deflection temperature of 174°C (345°F) at 245 psi (1.68 MPa) and ability to be used for long times at 150 to 174°C (300 to 345°F). Polysulfone has a high tensile strength (for thermoplastics) of 10.2 ksi (70 MPa) and a relatively low tendency to creep. Polysulfones resist hydrolysis in aqueous acid and alkaline environments because the oxygen linkages between the phenylene rings are hydrolytically stable.

Applications   Electrical-electronic applications include connectors, coil bobbins and cores, television components, capacitor film, and structural circuit boards. Polysulfone's resistance to autoclave sterilization makes it widely used for medical instruments and trays. In chemical processing and pollution control equipment, polysulfone is used for corrosion-resistant piping, pumps, tower packing, and filter modules and support plates.

Polyphenylene Sulfide   Polyphenylene sulfide (PPS) is an engineering thermoplastic which is characterized by outstanding chemical resistance along with good mechanical properties and stiffness at elevated temperatures. PPS was first produced in 1973 and is manufactured by Phillips Chemical Co. under the trade name Ryton.

---

[1] Para positions are at opposite ends of the benzene ring.

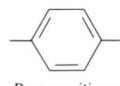

Para positions

Repeating chemical structural unit   Polyphenylene sulfide has a repeating structural unit in its main chain of para-substituted benzene rings and divalent sulfur atoms:

Polyphenylene sulfide
mp:288°C (550°F)

Structure and properties   The compact symmetrical structure of the phenylene rings separated by sulfur atoms produces a rigid and strong polymeric material. The compact molecular structure also promotes a high degree of crystallinity. Because of the presence of the sulfur atoms, PPS is highly resistant to attack by chemicals. In fact, no chemical has been found to dissolve PPS readily below 200°C (392°F). Even at high temperatures, few materials react chemically with PPS.

Unfilled PPS has a room-temperature strength of 9.5 ksi (65 MPa), whereas when 40 percent glass-filled, its strength is raised to 17 ksi (120 MPa). Because of its crystalline structure, the loss of strength with increasing temperature is gradual, and even at 200°C (392°F) considerable strength is retained.

Applications   Industrial-mechanical applications include chemical process equipment such as submersible, centrifugal, vane, and gear-type pumps. PPS compounds are specified for many under-the-hood automobile applications such as emission-control systems because they are impervious to the corrosive effects of engine exhaust gases as well as to gasoline and other automotive fluids. Electrical-electronic applications include computer components such as connectors, coil forms, and bobbins. Figure 7.37 shows some fixed internal thermostats which utilize the heat resistance and insulative properties of PPS. Corrosive-resistant and thermally stable coatings of PPS are used for oil field pipe, valves, fittings, couplings, and other equipment in the petroleum and chemical processing industries.

FIGURE 7.37 Fixed internal thermostats made partly from polyphenylene sulfide (black part of the device). The PPS provides good electrical resistance, high heat resistance, and low shrinkage during molding. (*Courtesy of Texas Instruments.*)

Polyetherimide   Polyetherimide is one of the newer amorphous high-performance engineering thermoplastics which was introduced in 1982 and is commercially available from General Electric Co. under the trademark Ultem. Polyetherimide has the following chemical structure:

Polyetherimide   Imide linkage

The stability of the imide linkage gives this material high heat resistance, creep resistance, and high rigidity. The ether linkage between the phenyl rings provides the necessary degree of chain flexibility required for good melt processibility and flow characteristics. This material has good electrical insulation properties which are stable over a wide range of temperatures and frequencies. Uses for polyetherimide include electrical-electronic, automotive, aerospace, and specialty applications. Electrical-electronic applications include high-voltage circuit breaker housings, pin connectors, high-temperature bobbins and coils, and fuse blocks. Printed wiring boards made of reinforced polyetherimide offer dimensional stability for vapor-phase soldering conditions.

Polymer Alloys   Polymer alloys consist of mixtures of structurally different homopolymers or copolymers. In thermoplastic polymer alloys different types of polymer molecular chains are bonded together by secondary intermolecular dipole forces. By contrast, in a copolymer two structurally different monomers are bonded together in a molecular chain by strong covalent bonds. The components of a polymer alloy must have some degree of compatibility or adhesion to prevent phase separation during processing. Polymer alloys are becoming more im-

TABLE 7.8   Some Commercial Polymer Alloys

Polymer alloy	Trade name of material	Supplier
ABS/polycarbonate	Bayblend MC2500	Mobay
ABS/polyvinyl chloride	Cycovin K-29	Borg-Warner Chemicals
Acetal/elastomer	Celcon C-400	Celanese
Polycarbonate/polyethylene	Lexan EM	General Electric
Polycarbonate/PBT/elastomer	Xenoy 1000	General Electric
PBT/PET	Valox 815	General Electric

Source: "Modern Plastics Encyclopedia, 1984–85," McGraw-Hill.

portant today since plastic materials with specific properties can be created and cost and performance can be optimized.

Some of the early polymer alloys were made by adding rubbery polymers such as ABS to rigid polymers such as polyvinyl chloride. The rubbery material improves the toughness of the rigid material. Today, even the newer thermoplastics are alloyed together. For example, PBT (polybutylene terephthalate) is alloyed with some PET (polyethylene terephthalate) to improve surface gloss and reduce cost. Table 7.8 lists some commercial polymer alloys.

7.8 THERMOSETTING PLASTICS (THERMOSETS)

Thermosetting plastics or thermosets are formed with a network molecular structure of primary covalent bonds. Some thermosets are cross-linked by heat or a combination of heat and pressure. Others may be cross-linked by a chemical reaction which occurs at room temperature (cold-setting thermosets). Although cured parts made from thermosets can be softened by heat, their co-valent-bonding cross-links prevent them from being restored to the flowable state that existed before the plastic resin was cured. Thermosets, therefore, cannot be reheated and remelted as can thermoplastics. This is a disadvantage for thermosets since scrap produced during the processing cannot be recycled and reused.

In general, the advantages of thermosetting plastics for engineering design applications are one or more of the following:

1. High thermal stability
2. High rigidity
3. High dimensional stability
4. Resistance to creep and deformation under load
5. Light weight
6. High electrical and thermal insulating properties

Thermosetting plastics are usually processed by using compression or transfer molding. However, in some cases thermoset injection-molding techniques have been developed so that the processing cost is lowered.

Many thermosets are used in the form of molding compounds consisting of two major ingredients: (1) a resin containing curing agents, hardeners, and plasticizers and (2) fillers and/or reinforcing materials which may be organic or inorganic materials. Wood flour, mica, glass, and cellulose are commonly used filler materials.

Let us first look at the sales tonnages and bulk list prices in the United States and at some of the important properties of some selected thermoset materials as a group for comparative purposes.

TABLE 7.9   U.S. Plastic Sales Volume (1993*) and Bulk List Prices (January 1994†) for Some Important Thermosetting Plastics

Material	U.S. plastic sales in 1000 metric tons‡	Wt % of total U.S. plastic sales	Bulk list price, $/lb	Type of plastic
Phenolic	3070	4.5	0.62–1.00	Molding compound
Polyester, unsaturated	1252	1.8	0.65–0.76	General-purpose Ortho
Melamine and urea	1997	2.9	0.95–1.03	Melamine molding compound
Alkyd	326	0.47	0.72–0.85	
Epoxy	543	0.79	1.28–1.37	General-purpose resin

*From *Modern Plastics*, January 1994, p. 73.
†From *Plastics Technology*. Prices vary with time and are thus subject to change.
‡The total U.S. plastic sales in 1993 was 68,827 million lb.

Sales tonnages and bulk list prices of some thermoset plastics   The bulk list prices of the thermosets listed in Table 7.9 are in the low to medium range for plastics, ranging from $0.62 to $1.37/lb (1994 prices). Of all the thermosets listed, the phenolics are the lowest in price and have the largest sales tonnage. Unsaturated polyesters are also relatively low in price and have a relatively large sales tonnage. Epoxy resins, which have special properties for many industrial applications, command a premium price.

Some basic properties of selected thermoset plastics   Table 7.10 lists the densities, tensile strengths, impact strengths, dielectric strengths, and maximum-use temperatures for some selected thermoset plastics. The densities of thermoset plastics tend to be slightly higher than most plastic materials, with those listed in Table 7.10 ranging from 1.34 to 2.3 g/cm³. The tensile strengths of most thermosets are relatively low, with most strengths ranging from 4000 to 15,000 psi (28 to 103 MPa). However, with a high amount of glass filling, the tensile strength of some thermosets can be increased to as high as 30,000 psi (207 MPa). Glass-filled thermosets also have much higher impact strengths, as indicated in Table 7.10. The thermosets also have good dielectric strengths, ranging from 140 to 650 V/mil. Like all plastic materials, however, their maximum-use temperature is limited. The maximum-use temperature for the thermosets listed in Table 7.10 ranges from 170 to 550°F (77 to 288°C).

Let us now examine some of the important aspects of the structure, properties, and applications of the following thermosets: phenolics, epoxy resins, unsaturated polyesters, and amino resins.

Phenolics   Phenolic thermosetting materials were the first major plastic material used by industry. The original patents for the reaction of phenol with formaldehyde to make the phenolic plastic Bakelite were issued to L. H. Baekeland in 1909.

TABLE 7.10   Some Properties of Selected Thermoset Plastics

Material	Density, g/cm³	Tensile strength, × 1000 psi*	Impact strength, Izod, ft·lb/in†	Dielectric strength, V/mil‡	Max-use temp. (no load) °F	Max-use temp. (no load) °C
Phenolic:						
Wood-flour-filled	1.34–1.45	5–9	0.2–0.6	260–400	300–350	150–177
Mica-filled	1.65–1.92	5.5–7	0.3–0.4	350–400	250–300	120–150
Glass-filled	1.69–1.95	5–18	0.3–18	140–400	350–550	177–288
Polyester:						
Glass-filled SMC	1.7–2.1	8–20	8–22	320–400	300–350	150–177
Glass-filled BMC	1.7–2.3	4–10	15–16	300–420	300–350	150–177
Melamine:						
Cellulose-filled	1.45–1.52	5–9	0.2–0.4	350–400	250	120
Flock-filled	1.50–1.55	7–9	0.4–0.5	300–330	250	120
Glass-filled	1.8–2.0	5–10	0.6–18	170–300	300–400	150–200
Urea, cellulose-filled	1.47–1.52	5.5–13	0.2–0.4	300–400	170	77
Alkyd:						
Glass-filled	2.12–2.15	4–9.5	0.6–10	350–450	450	230
Mineral-filled	1.60–2.30	3–9	0.3–0.5	350–450	300–450	150–230
Epoxy (bis A):						
No filler	1.06–1.40	4–13	0.2–10	400–650	250–500	120–260
Mineral-filled	1.6–2.0	5–15	0.3–0.4	300–400	300–500	150–260
Glass-filled	1.7–2.0	10–30	. . .	300–400	300–500	150–260

*1000 psi = 6.9 MPa.
†Notched Izod test: 1 ft · lb/in = 53.38 J/m.
‡1 V/mil = 39.4 V/mm.
Source: Materials Engineering, May 1972.

Phenolic plastics are still used today because they are low in cost and have good electrical and heat insulating properties along with good mechanical properties. They are easily molded but are limited in color (usually black or brown).

Chemistry   Phenolic resins are most commonly produced by the reaction of phenol and formaldehyde by condensation polymerization, with water as a by-product. However, almost any reactive phenol or aldehyde can be used. Two-stage (novolac) phenolic resins are commonly produced for convenience for molding. In the first stage a brittle thermoplastic resin is produced which can be melted but which will not cross-link to form a network solid. This material is prepared by reacting less than a mole of formaldehyde with a mole of phenol in the presence of an acid catalyst. The polymerization reaction is shown in Fig. 7.11.

The addition of hexamethylenetetramine (hexa) which is a basic catalyst to the first-stage phenolic resin makes it possible to create methylene cross-linkages to form a thermosetting material. When heat and pressure are applied to the hexa-containing novolac resin, the hexa decomposes, producing ammonia which provides methylene cross-linkages to form a network structure.

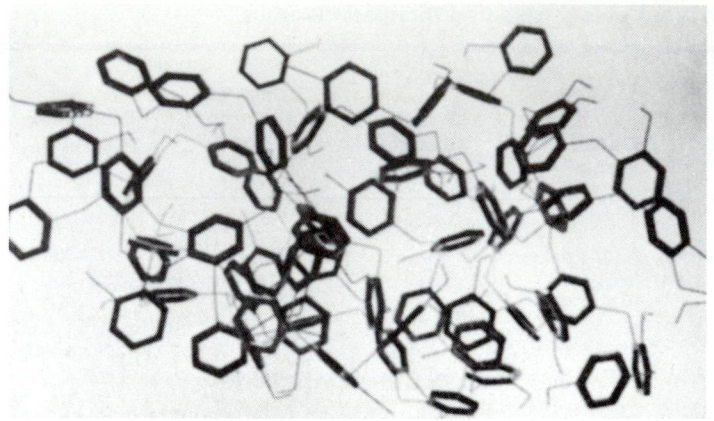

FIGURE 7.38 Three-dimensional model of polymerized phenolic resin. (*After E. G. K. Pritchett, in "Encyclopedia of Polymer Science and Technology," vol. 10, Wiley, 1969, p. 30.*)

The temperature required for the cross-linking (curing) of the novolac resin ranges from 120 to 177°C (250 to 350°F). Molding compounds are made by combining the resin with various fillers which sometimes account for up to 50 to 80 percent of the total weight of the molding compounds. The fillers reduce shrinkage during molding, lower cost, and improve strength. They also can be used to increase electrical and thermal insulating properties.

Structure and properties   The high cross-linking of the aromatic structure (Fig. 7.38) produces high hardness, rigidity, and strength combined with good heat and electrical insulating properties and chemical resistance.

Some of the various types of phenolic molding compounds manufactured are:

1. *General-purpose compounds.* These materials are usually wood flour–filled to increase impact strength and lower cost.
2. *High-impact-strength compounds.* These compounds are filled with cellulose (cotton flock and chopped fabric), mineral, and glass fibers to provide impact strengths of up to 18 ft·lb/in (961 J/m).
3. *High electrical insulating compounds.* These materials are mineral- (e.g., mica) filled to increase electrical resistance.
4. *Heat-resistant compounds.* These are mineral- (e.g., asbestos) filled and are able to withstand long-term exposure to temperatures of 150 to 180°C (300 to 350°F).

Applications   Phenolic compounds are widely used in wiring devices, electrical switchgear, connectors, and telephone relay systems. Automotive engineers use phenolic molding compounds for power-assist brake components and transmission parts (Fig. 7.39). Phenolics are widely used for handles, knobs, and end panels for small appliances. Because they are good high-temperature and moisture-resistant adhesives, phenolic resins are used in lam-

FIGURE 7.39 The three parts shown are molded of glass-reinforced phenolic material and make up the parts for the transmission reactor for a 1985 automobile. The phenolic reactor will replace one which was previously made of metal. (*Courtesy of Rogers Corp.*)

inating some types of plywood and in particleboard. Large amounts of phenolic resins are also used as a binder material for sand in the foundry and for shell molding.

Epoxy Resins   Epoxy resins are a family of thermosetting polymeric materials which do not give off reaction products when they cure (cross-link) and so have low cure shrinkage. They also have good adhesion to other materials, good chemical and environmental resistance, good mechanical properties, and good electrical insulating properties.

Chemistry   Epoxy resins are characterized by having two or more epoxy groups per molecule. The chemical structure of an epoxide group is

$$CH_2 \overset{\displaystyle O}{\underset{\displaystyle \underset{H}{|}}{\diagdown\diagup}} C -$$

Covalent half bond available for bonding

Most commercial epoxy resins have the general chemical structure where Be = benzene ring. For liquids the n in the structure is usually less than 1. For solid resins n is 2 or greater. There are also many other kinds of epoxy resins which have different structures than the one shown above.

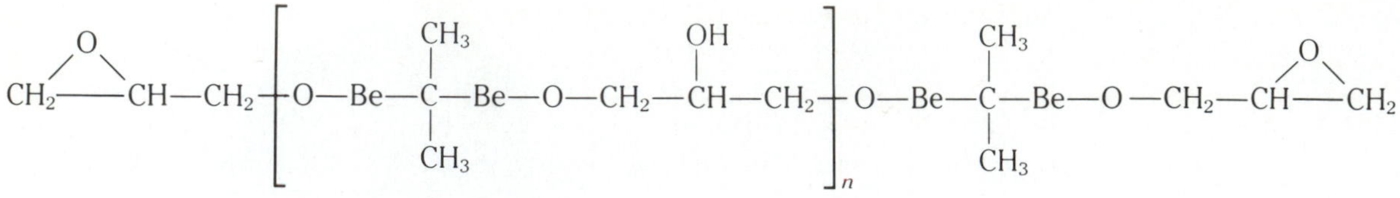

To form solid thermosetting materials, epoxy resins must be cured by using cross-linking agents and/or catalysts to develop the desired properties. The epoxy and hydroxyl groups (—OH) are the reaction sites for cross-linking. Cross-linking agents include amines, anhydrides, and aldehyde condensation products.

For curing at room temperature when the heat requirements for the epoxy solid materials are low (under about 100°C), amines such as diethylene triamine and triethylene tetramine are used as curing agents. Some epoxy resins are cross-linked by using a curing reagent, while others can react with their own reaction sites if an appropriate catalyst is present. In an epoxy reaction the epoxide ring is opened and a donor hydrogen from, for example, an amine or hydroxyl group bonds with the oxygen atom of the epoxide group. Figure 7.40 shows the reaction of epoxide groups at the ends of two linear epoxy molecules with ethylene diamine.

In the reaction of Fig. 7.40 the epoxy rings are opened up and hydrogen atoms from the diamine form —OH groups which are reaction sites for further cross-linking. An important characteristic of this reaction is that no by-product is given off. Many different kinds of amines can be used for cross-linking epoxy resins.

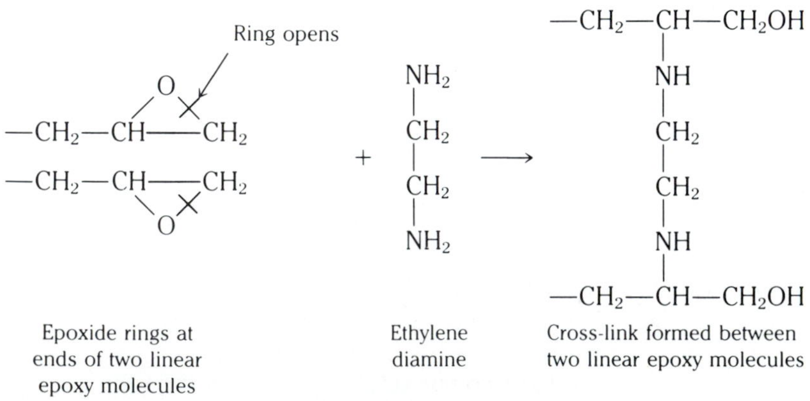

FIGURE 7.40 Reaction of epoxy rings at the ends of two linear epoxy molecules with ethylene diamine to form a cross-link. Note that no by-product is given off.

Structure and properties   The low molecular weight of uncured epoxy resins in the liquid state gives them exceptionally high molecular mobility during processing. This property allows the liquid epoxy resin to quickly and thoroughly wet surfaces. This wetting action is important for epoxies used for reinforced materials and adhesives. Also the ability to be poured into final form is important for electrical potting and encapsulating. The high reactivity of the epoxide groups with curing agents such as amines provides a high degree of cross-linking and produces good hardness, strength, and chemical resistance. Since no by-product is given off during the curing reaction, there is low shrinkage during hardening.

Applications   Epoxy resins are used for a wide variety of protective and decorative coatings because of their good adhesion and good mechanical and chemical resistance. Typical uses are can and drum linings, automotive and appliance primers, and wire coatings. In the electrical and electronics industry epoxy resins are used because of their dielectric strength, low shrinkage on curing, good adhesion, and ability to retain properties under a variety of environments such as wet and high-humidity conditions. Typical applications include high-voltage insulators, switchgear, and encapsulation of transistors. Epoxy resins are also used for laminates and for fiber-reinforced matrix materials. Epoxy resins are the predominate matrix material for most high-performance components such as those made with high-modulus fibers (e.g., graphite).

Unsaturated Polyesters   Unsaturated polyesters have reactive double carbon-carbon covalent bonds which can be cross-linked to form thermosetting materials. In combination with glass fibers, unsaturated polyesters can be cross-linked to form high-strength reinforced composite materials.

Chemistry   The ester linkage can be produced by reacting an alcohol with an organic acid, as

$$R-\overset{\overset{\displaystyle O}{\|}}{C}-O[H] + [R'OH] \xrightarrow{\text{heat}} R-[\overset{\overset{\displaystyle O}{\|}}{C}-O]-R' + H_2O$$

Organic acid   Alcohol   Ester   Water

R and R' = CH_3-, C_2H_5-, . . .

The basic unsaturated polyester resin can be formed by the reaction of a diol (an alcohol with two —OH groups) with a diacid (an acid with two —COOH groups) that contains a reactive double carbon-carbon bond. Commercial resins may have mixtures of different diols and diacids to obtain special properties; e.g., ethylene glycol can be reacted with maleic acid to form a linear polyester:

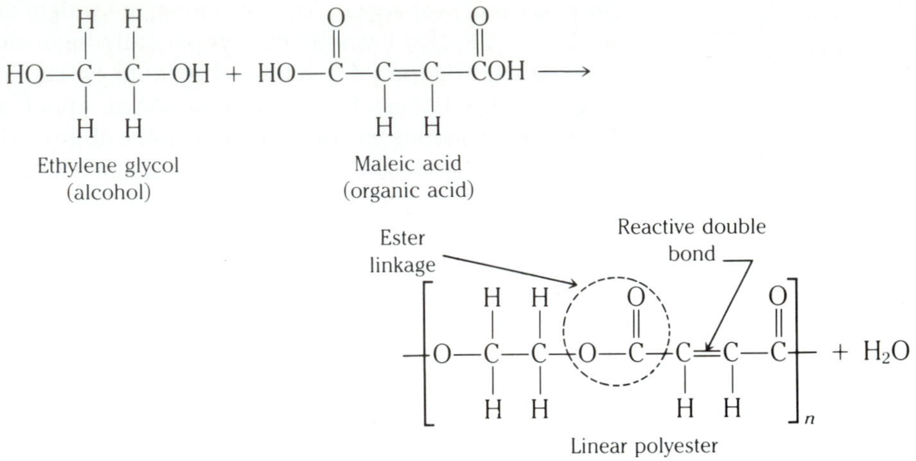

Ethylene glycol
(alcohol)

Maleic acid
(organic acid)

Ester linkage

Reactive double bond

Linear polyester

The linear unsaturated polyesters are usually cross-linked with vinyl-type molecules such as styrene in the presence of a free-radical curing agent. Peroxide curing agents are most commonly used with methyl ethyl ketone (MEK) peroxide being usually used for the room-temperature curing of polyesters. The reaction is commonly activated with a small amount of cobalt naphthanate.

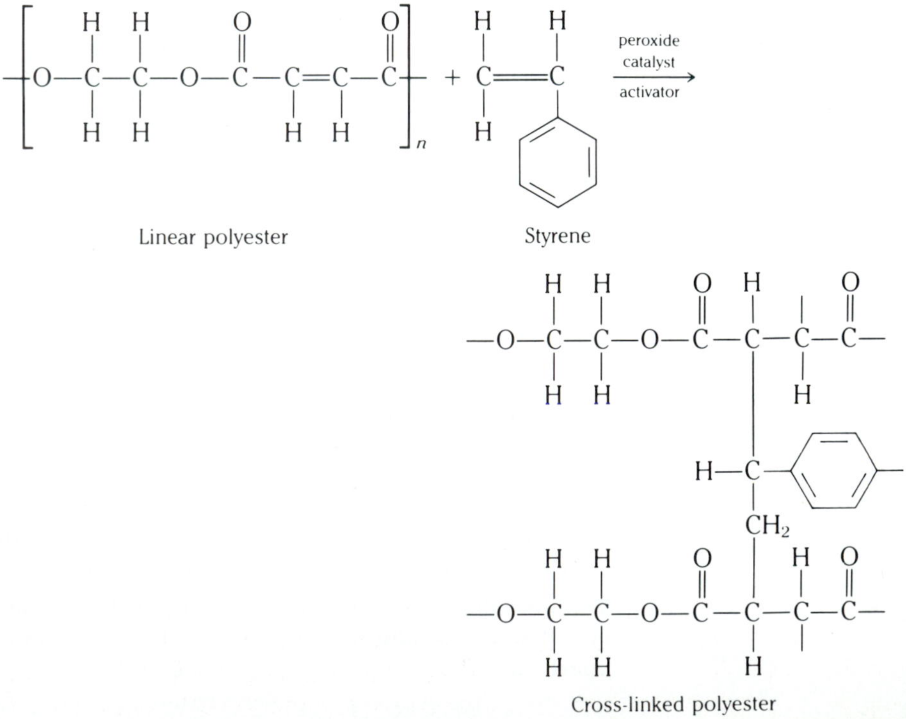

Linear polyester

Styrene

peroxide
catalyst
activator

Cross-linked polyester

Structure and properties   The unsaturated polyester resins are low-viscosity materials which are able to be mixed with high amounts of fillers and reinforcements. For example, unsaturated polyesters may contain as high as about 80 percent by weight of glass-fiber reinforcement. Glass-fiber-reinforced unsaturated polyesters when cured have outstanding strength, 25 to 50 ksi (172 to 344 MPa), and good impact and chemical resistance.

Processing   Unsaturated polyester resins can be processed by many methods, but in most cases they are molded in some way. Open-mold lay-up or spray-up techniques are used for many small-volume parts. For high-volume parts such as automobile panels compression molding is usually used. In recent years sheet molding compounds (SMC) which combine resin, reinforcement, and other additives have been produced to speed up the feeding of material into molding presses made of matched metal dies.

Applications   Glass-reinforced unsaturated polyesters are used for making automobile panels and body parts. This material is also used for small-boat hulls and in the construction industry for building panels and bathroom components. Unsaturated reinforced polyesters are also used for pipes, tanks, and ducts where good corrosion resistance is required.

Amino Resins (Ureas and Melamines)   The amino resins are thermosetting polymeric materials formed by the controlled reaction of formaldehyde with various compounds which contain the amine group $-NH_2$. The two most important types of amino resins are urea-formaldehyde and melamine-formaldehyde.

Chemistry   Both urea and melamine react with formaldehyde by condensation polymerization reactions which produce water as a by-product. The condensation reaction of urea with formaldehyde is

Urea   Formaldehyde   Urea

Urea-formaldehyde
molecule

The amine groups at the ends of the molecule shown above react with more formaldehyde molecules to produce a highly rigid network polymer structure. As in the case of the phenolic resins, urea and formaldehyde are first only partially polymerized to produce a low-molecular-weight polymer which is ground into a powder and compounded with fillers, pigments, and a catalyst. The molding compound can then be compression-molded into the final shape by applying heat [127 to 171°C (260 to 340°F)] and pressure [2 to 8 ksi (14 to 55 MPa)].

Melamine also reacts with formaldehyde by a condensation reaction, resulting in polymerized melamine-formaldehyde molecules with water being given off as a by-product:[1]

Melamine   Formaldehyde   Melamine

Melamine-formaldehyde molecule

Structure and properties   The high reactivity of the urea-formaldehyde and melamine-formaldehyde low-molecular-weight prepolymers enable highly cross-linked thermoset products to be made. When these resins are combined with cellulose (wood flour) fillers, low-cost products are obtained which have good rigidity, strength, and impact resistance. Urea-formaldehyde costs less than melamine-formaldehyde but does not have as high a heat resistance and surface hardness as melamine.

Applications   Cellulose-filled molding compounds of urea-formaldehyde are used for electrical wall plates and receptacles and for knobs and handles. Applications for cellulose-filled melamine compounds include molded dinnerware, buttons, control buttons, and knobs. Both urea and melamine water-soluble resins find application as adhesives and bonding resins for wood particleboard, plywood, boat hulls, flooring, and furniture assemblies. The amino resins are also used in binders for foundry cores and shell molds.

[1] Only one hydrogen atom is removed from each NH_2 group to form an H_2O molecule.

7.9 ELASTOMERS (RUBBERS)

Elastomers, or rubbers, are polymeric materials whose dimensions can be greatly changed when stressed and which return to their original dimensions (or almost) when the deforming stress is removed. There are many types of elastomeric materials, but only the following ones will be discussed: natural rubber, synthetic polyisoprene, styrene-butadiene rubber, nitrile rubbers, polychloroprene, and the silicones.

Natural Rubber

Production   Natural rubber is produced commercially from the latex of the *Hevea brasiliensis* tree which is cultivated in plantations mainly in the tropical regions in southeast Asia, especially in Malaysia and Indonesia. The source of natural rubber is a milky liquid known as *latex* which is a suspension containing very small particles of rubber. The liquid latex is collected from the trees and taken to a processing center where the field latex is diluted to about 15% rubber content and coagulated with formic acid (an organic acid). The coagulated material is then compressed through rollers to remove water and to produce a sheet material. The sheets are either dried with currents of hot air or by the heat of a smoke fire (rubber-smoked sheets). The rolled sheets and other types of raw rubber are usually milled between heavy rolls in which the mechanical shearing action breaks up some of the long polymer chains and reduces their average molecular weight. Natural rubber production in 1980 accounted for about 30 percent of the total world's rubber market.

Structure   Natural rubber is mainly *cis*-1,4 polyisoprene (see structural formula below) mixed with small amounts of proteins, lipids, inorganic salts, and numerous other components. *cis*-1,4 Polyisoprene is a long-chain polymer (average molecular weight of about 5×10^5 g/mol) which has the structural formula

cis-1,4 Polyisoprene
Repeating structural unit for natural rubber

The *cis*- prefix indicates that the methyl group and a hydrogen atom are on the same side of the carbon-carbon double bond, as shown by the dashed encirclement on the formula above. The 1,4 indicates that the repeating chemical units of the polymer chain covalently bond on the first and fourth carbon atoms. The polymer chains of natural rubber are long, entangled, and coiled, and at room temperature are in a state of continued agitation. The bending and coiling of the natural rubber polymer chains is attributed to the steric hindrance of the methyl group and the hydrogen atom on the same side of the carbon-carbon double bond. The arrangement of the covalent bonds in the natural rubber polymer chain is shown schematically below:

Segment of natural rubber polymer chain

There is another structural isomer[1] of polyisoprene, *trans**-1,4 polyisoprene, called *gutta-percha*, which is not an elastomer. In this structure the methyl group and hydrogen atom covalently bonded to the carbon-carbon double bond are on opposite sides of the double bond of the polyisoprene repeating unit, as shown by the encirclement below:

trans-1,4 Polyisoprene
Repeating structural unit for gutta-percha

In this structure the methyl group and hydrogen atom bonded to the double bond do not interfere with each other, and as a result the *trans*-1,4 polyisoprene molecule is more symmetrical and can crystallize into a rigid material.

Segment of gutta-percha polymer chain

Vulcanization   Vulcanization is the chemical process by which polymer molecules are joined together by cross-linking into larger molecules to restrict molecular movement. In 1839 Charles Goodyear[2] discovered a vulcanization process for rubber by using sulfur and basic lead carbonate. Goodyear found that when a mixture of natural rubber, sulfur, and lead carbonate was heated,

---

[1] Structural isomers are molecules that have the same molecular formula but different structural arrangement of their atoms.
* The prefix *trans*- is from the Latin, meaning "across."
[2] Charles Goodyear (1800–1860). American inventor who discovered the vulcanizing process for natural rubber by using sulfur and lead carbonate as chemical agents. U.S. Patent 3633 was granted to Charles Goodyear on June 15, 1844, for an "Improvement in India-Rubber Fabrics."

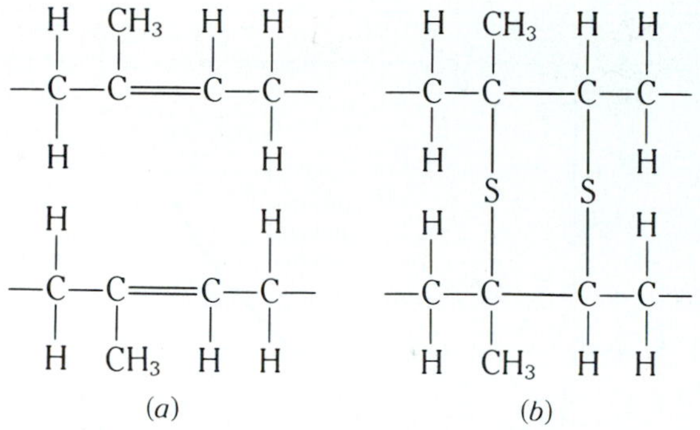

FIGURE 7.41 Schematic illustration of the vulcanization of rubber. In this process sulfur atoms form cross-links between chains in 1,4 polyisoprene. (a) cis-1,4 Polyisoprene chain before sulfur cross-linking. (b) cis-1,4 Polyisoprene chain after cross-linking with sulfur at the active double-bond sites.

(a)

(b)

the rubber changed from a thermoplastic to an elastomeric material. Although even today the reaction of sulfur with rubber is complex and not completely understood, the final result is that some of the double bonds in the polyisoprene molecules open and form sulfur atom cross-links, as shown in Fig. 7.41.

Figure 7.42 shows schematically how cross-linking the sulfur atoms gives rigidity to rubber molecules, and Fig. 7.43 shows how the tensile strength of natural rubber is increased by vulcanization. Rubber and sulfur react very slowly even at elevated temperatures, so that to shorten the cure time at elevated temperatures accelerator chemicals are usually compounded with rubber along with other additives such as fillers, plasticizers, and antioxidants.

Usually, vulcanized soft rubbers contain about 3 wt % sulfur and are heated in the 100 to 200°C range for vulcanizing or curing. If the sulfur content is increased, the cross-linking that occurs will also increase, producing harder and less-flexible material. A fully rigid structure of hard rubber can be produced with about 45% sulfur.

Oxygen or ozone will also react with the carbon double bonds of the rubber molecules in a similar way to the vulcanization sulfur reaction and cause embrittlement of the rubber. This oxidation reaction can be retarded to some extent by adding antioxidants when the rubber is compounded.

FIGURE 7.42 Model of cross-linking of cis-1,4 polyisoprene chains by sulfur atoms (black). (After W. G. Moffatt, G. W. Pearsall, and J. Wulff, "The Structure and Properties of Materials," vol. I, Wiley, 1965, p. 109.)

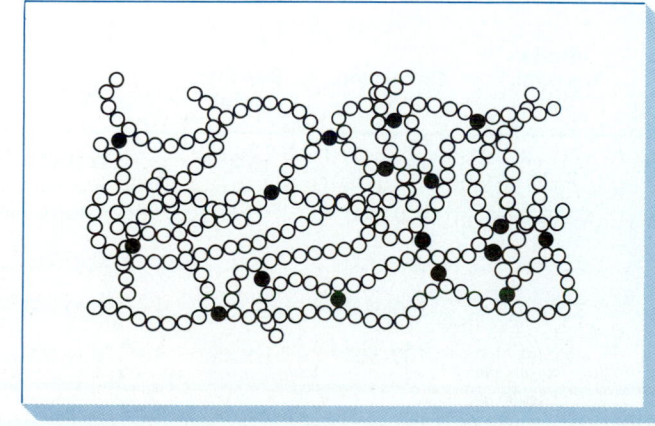

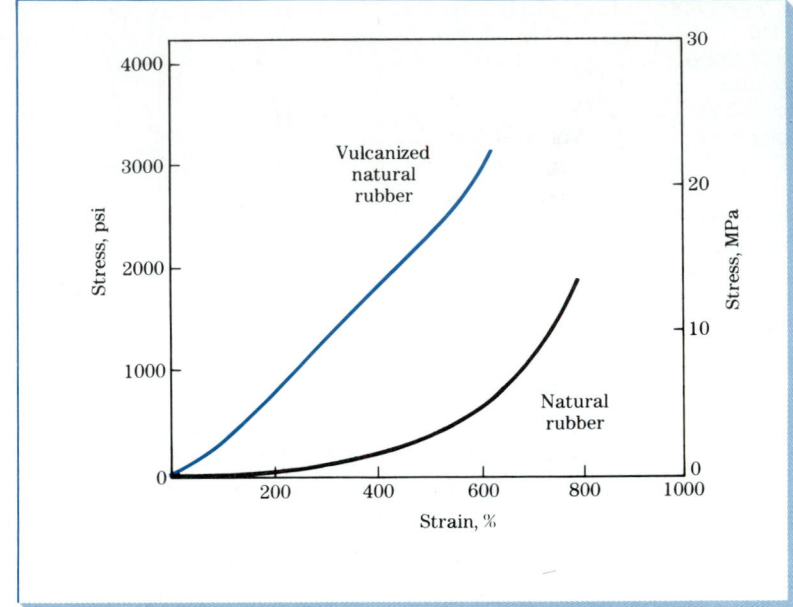

FIGURE 7.43 Stress-strain diagrams for vulcanized and unvulcanized natural rubber. The cross-linking of the sulfur atoms between the polymer chains of *cis*-1,4 polyisoprene increases the strength of the vulcanized rubber.

The use of fillers can lower the cost of the rubber product and also strengthen the material. Carbon black is commonly used as a filler for rubber, and, in general, the finer the particle size of the carbon black, the higher the tensile strength is. Carbon black also increases the tear and abrasion resistance of the rubber. Silicas (e.g., calcium silicate) and chemically altered clay are also used for fillers for reinforcing rubber.

TABLE 7.11

Some Properties of Selected Elastomers

Properties Table 7.11 compares the tensile strength, elongation, and density properties of vulcanized natural rubber with those of some other synthetic

Elastomer	Tensile strength, ksi†	Elongation, %	Density, g/cm³	Recommended operating temp. °F	Recommended operating temp. °C
Natural rubber* (*cis*-polyisoprene)	2.5–3.5	750–850	0.93	−60 to 180	−50 to 82
SBR or Buna S* (butadiene-styrene)	0.2–3.5	400–600	0.94	−60 to 180	−50 to 82
Nitrile or Buna N* (butadiene-acrylonitrile)	0.5–0.9	450–700	1.0	−60 to 250	−50 to 120
Neoprene* (polychloroprene)	3.0–4.0	800–900	1.25	−40 to 240	−40 to 115
Silicone (polysiloxane)	0.6–1.3	100–500	1.1–1.6	−178 to 600	−115 to 315

*Pure gum vulcanizate properties.   † 1000 psi = 6.89 MPa.

elastomers. Note that, as expected, the tensile strengths of these materials are relatively low and their elongations extremely high.

Synthetic Rubbers   Synthetic rubbers in 1980 accounted for about 70 percent of the total world's supply of rubber materials. Some of the important synthetic rubbers are styrene-butadiene, nitrile rubbers, and the polychloroprenes.

Styrene-butadiene rubber   The most important synthetic rubber and the most widely used is styrene-butadiene rubber (SBR), a butadiene-styrene copolymer. After polymerization, this material contains 20 to 23% styrene. The basic structure of SBR is shown in Fig. 7.44.

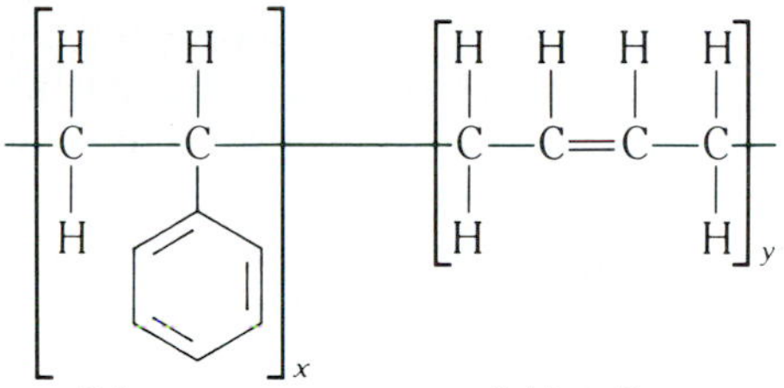

FIGURE 7.44 Chemical structure of styrene-butadiene synthetic rubber copolymer.

Polystyrene   Polybutadiene

Since the butadiene mers contain double bonds, this copolymer can be vulcanized with sulfur by cross-linking. Butadiene by itself when synthesized with a stereospecific catalyst to produce the cis isomer has even greater elasticity than natural rubber since the methyl group attached to the double bond in natural rubber is missing in the butadiene mer. The presence of styrene in the copolymer produces a tougher and stronger rubber. The phenyl side group of styrene which is scattered along the copolymer main chain reduces the tendency of the polymer to crystallize under high stresses. SBR rubber is lower in cost than natural rubber and so is used in many rubber applications. For example, for tire treads, SBR has better wear resistance but higher heat generation. A disadvantage of SBR and natural rubber is that they absorb organic solvents such as gasoline and oil and swell.

Nitrile rubbers   Nitrile rubbers are copolymers of butadiene and acrylonitrile with the proportions ranging from 55 to 82% butadiene and 45 to 18% acrylonitrile. The presence of the nitrile groups increases the degree of polarity in the main chains and the hydrogen bonding between adjacent chains. The nitrile groups provide good resistance to oils and solvents as well as improved abrasion and heat resistance. On the other hand, molecular flexibility is reduced. Nitrile rubbers are more costly than ordinary rubbers, so these copolymers are limited to special applications such as fuel hoses and gaskets where high resistance to oils and solvents is required.

Polychloroprene (neoprene)   The polychloroprene or neoprene rubbers are similar to isoprene except that the methyl group attached to the double carbon bond is replaced by a chlorine atom:

$$\left[\begin{array}{c} \overset{\displaystyle H}{\underset{\displaystyle H}{|}} \quad \overset{\displaystyle Cl}{|} \quad \overset{\displaystyle H}{|} \quad \overset{\displaystyle H}{\underset{\displaystyle H}{|}} \\ -C-C=C-C- \\ \end{array}\right]_n$$

Polychloroprene (neoprene) structural unit

The presence of the chlorine atom increases the resistance of the unsaturated double bonds to attack by oxygen, ozone, heat, light, and the weather. Neoprenes also have fair fuel and oil resistance and increased strength over that of the ordinary rubbers. However, they do have poorer low-temperature flexibility and are higher in cost. As a result, neoprenes are used in specialty applications such as wire and cable covering, industrial hoses and belts, and automotive seals and diaphragms.

Silicone rubbers   The silicon atom, like carbon, has a valence of 4 and is able to form polymeric molecules by covalent bonding. However, the silicone polymer has repeating units of silicon and oxygen, as shown below:

$$\left[\begin{array}{c} X \\ | \\ -Si-O- \\ | \\ X' \end{array}\right]_n$$

Basic repeating structural
unit for a silicone polymer

where X and X' may be hydrogen atoms or groups such as methyl (CH_3-) or phenyl (C_6H_5-). The silicone polymers based on silicon and oxygen in the main chain are called *silicones.* Of the many silicone elastomers, the most common type is the one in which the X and X' of the repeating unit are methyl groups:

$$\left[\begin{array}{c} CH_3 \\ | \\ -Si-O- \\ | \\ CH_3 \end{array}\right]_n$$

Repeating structural unit
for polydimethyl siloxane

This polymer is called *polydimethyl siloxane* and can be cross-linked at room temperature by the addition of an initiator (e.g., benzoyl peroxide) which reacts the two methyl groups together with the elimination of hydrogen gas (H_2) to form $Si-CH_2-CH_2-Si$ bridges. Other types of silicones can be cured at higher temperatures (e.g., 50 to 150°C), depending on the product and intended use.

Silicone rubbers have the major advantage of being able to be used over a wide temperature range (that is, -100 to $250°C$). Applications for the silicone rubbers include sealants, gaskets, electrical insulation, auto-ignition cable, and spark-plug boots.

Example Problem 7.6

How much sulfur must be added to 100 g of polyisoprene rubber to cross-link 5 percent of the mers? Assume all available sulfur is used and that only one sulfur atom is involved in each cross-linking bond.

Solution:

As shown in Fig. 7.41b, on the average one sulfur atom will be involved with one polyisoprene mer in the cross-linking. First we determine the molecular weight of the polyisoprene mer.

$$MW \text{ (polyisoprene)} = 5 \text{ C atoms} \times 12 \text{ g/mol} + 8 \text{ H atoms} \times 1 \text{ g/mol} = 68.0 \text{ g/mol}$$

Thus, with 100 g of polyisoprene, we have 100 g/(68.0 g/mol) = 1.47 mol of polyisoprene. For 100% cross-linking with sulfur we need 1.47 mol of S or

$$1.47 \text{ mol} \times 32 \text{ g/mol} = 47.0 \text{ g sulfur}$$

For cross-linking 5 percent of the bonds we need only

$$0.05 \times 47.0 \text{ g S} = 2.35 \text{ g S} \blacktriangleleft$$

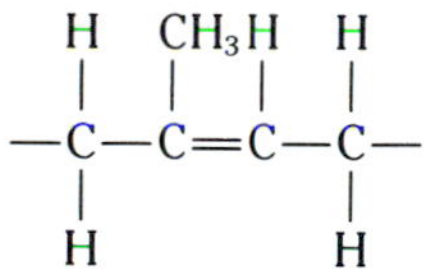

Polyisoprene mer

Thermoplastic Polyurethane Elastomers (TPUE)   Thermoplastic polyurethane elastomers are noted for their general overall toughness and abrasion resistance. They are typically used for more demanding applications since they have a high degree of flexibility (even at low temperatures) and a high resistance to abrasion. These materials are readily processable and are made extremely versatile by various formulations. They are easily compounded and are often blended with other compatible polymers.

Chemical structure   The chemical repeating unit common to all TPUEs is the urethane linkage:

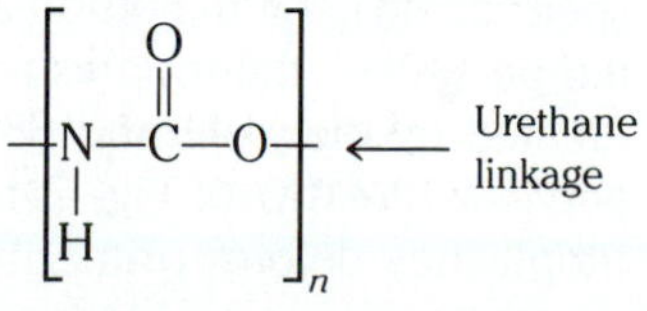

Urethane linkage

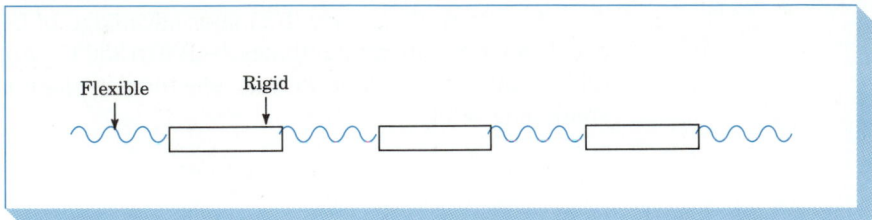

FIGURE 7.44A Thermoplastic polyurethane elastomer chain structure consisting of flexible and rigid blocks. Hydrogen bonding between the chains by the urethane linkages is of the (NH ~ O) type.

which is produced by combining isocyanate (—NCO) and hydroxyl (—OH) end groups. In general, a thermoplastic urethane elastomer can be considered a linear (segmented) block copolymer consisting of rigid and flexible blocks (Fig. 7.44A). The rigid blocks consist of repeating groups of diisocyanate and short-chain diols (sometimes called *chain extenders*). The flexible blocks consist of repeating groups of diisocyanates and long-chain diols (sometimes called *polyols*).

Properties   The most important properties of TPUEs that make them the choice material for many applications are flexibility and rigidity over a wide temperature range (especially at low temperature), abrasion and tear resistance, hydraulic stability, fuel and oil resistance, and fungus resistance. Service temperature usually varies between −51 to 121°C (−60 to 250°F)

Ultimate tensile strengths of TPUEs vary from about 3.70 to 9.0 ksi (25.5 to 62.0 MPa). Tensile modulus at 300 percent elongation ranges from about 1.10 to 4.9 ksi (76 to 34 MPa) and ultimate elongation from about 225 to 650 percent.

Applications   Parts made of TPUEs for industrial applications usually are extruded or injection-molded. Typical applications include conveyor belts for both industrial and food processing, hydraulic hose and tube for fuel lines, co-extruded cable jacketing, industrial wheels, and small gears. Typical automotive applications include fender extensions, housings, fascia, and gaskets.

7.10   DEFORMATION AND STRENGTHENING OF PLASTIC MATERIALS

Deformation Mechanisms for Thermoplastics   The deformation of thermoplastic materials can be primarily elastic, plastic (permanent), or a combination of both types. Below their glass transition temperatures, thermoplastics deform primarily by elastic deformation, as indicated by the −40 and 68°C tensile stress-strain plots for polymethyl methacrylate (PMMA) of Fig. 7.45. Above their glass transition temperatures, thermoplastics deform primarily by plastic deformation, as indicated by the 122

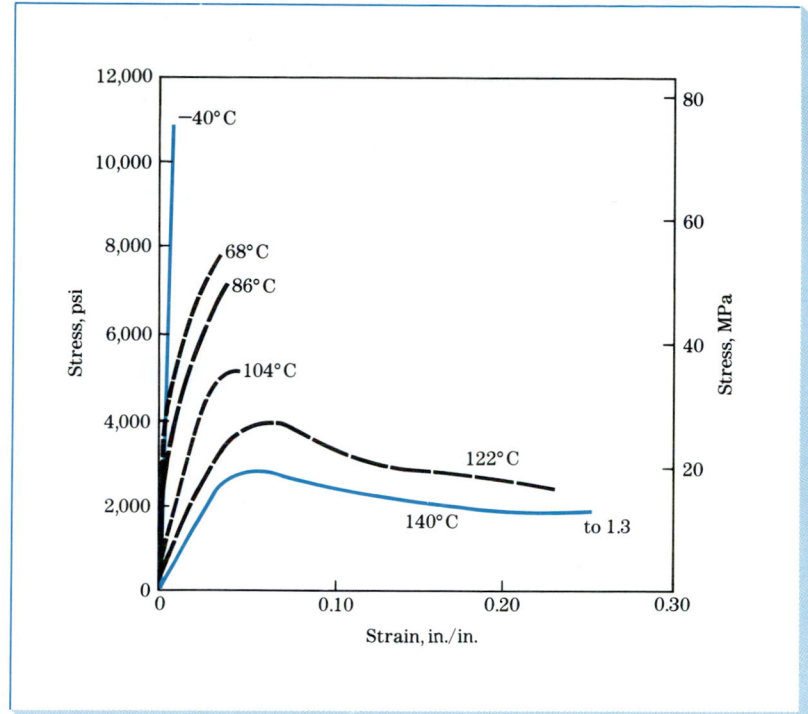

FIGURE 7.45 Tensile stress vs. strain curves for polymethyl methacrylate at various temperatures. A brittle-ductile transition occurs between 86 and 104°C. (*After T. Alfrey, "Mechanical Behavior of Polymers," Wiley-Interscience, 1967.*)

and 140°C tensile stress-strain plots for PMMA in Fig. 7.45. Thus, thermoplastics go through a brittle-ductile transition upon being heated through their glass transition temperature. PMMA goes through a ductile-brittle transition between 86 and 104°C because the T_g of PMMA is in this temperature range.

Figure 7.46 schematically illustrates the principal atomic and molecular mechanisms which occur during the deformation of long-chain polymers in a thermoplastic material. In Fig. 7.46a elastic deformation is represented as a stretching out of the covalent bonds within the molecular chains. In Fig. 7.46b elastic or plastic deformation is represented by an uncoiling of the linear polymers. Finally, in Fig. 7.46c plastic deformation is represented by the sliding of molecular chains past each other by the breaking and remaking of secondary dipole bonding forces.

Strengthening of Thermoplastics

Let us now look at the following factors, each of which in part determines the strength of a thermoplastic: (1) average molecular mass of the polymer chains, (2) the degree of crystallization, (3) the effect of bulky side groups on the main chains, (4) the effect of highly polar atoms on the main chains, (5) the effect of oxygen, nitrogen, and sulfur atoms in the main carbon chains, (6) the effect of phenyl rings in the main chains, and (7) the addition of glass-fiber reinforcement.

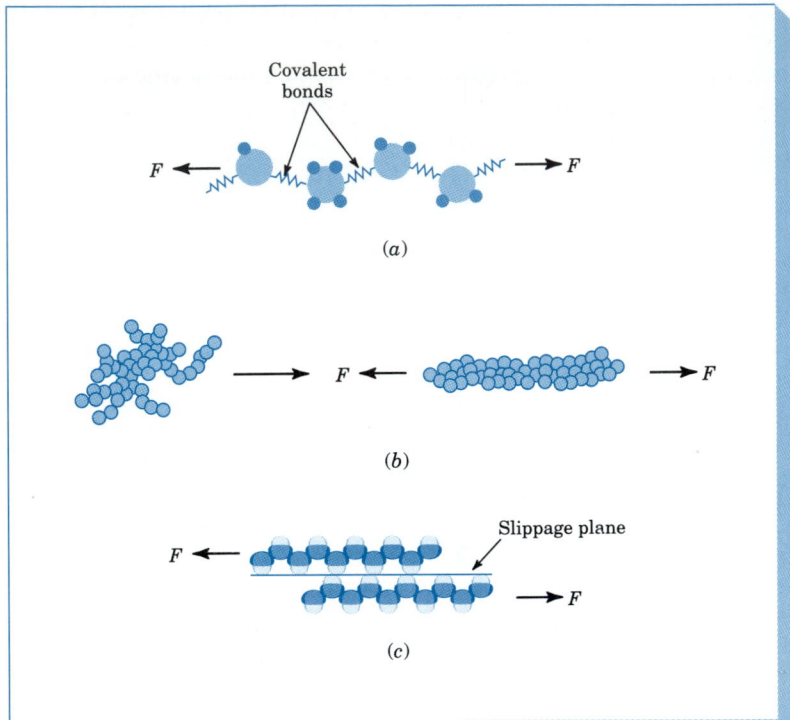

FIGURE 7.46
Deformation
mechanisms in
polymeric materials:
(a) elastic deformation
by extension of main-
chain carbon covalent
bonds, (b) elastic or
plastic deformation by
straightening out main
chains, and (c) plastic
deformation by main-
chain slippage

Strengthening due to the average molecular mass of the polymer chains The strength of a thermoplastic material is directly dependent on its average molecular mass since polymerization up to a certain molecular-mass range is necessary to produce a stable solid. However, this method is not normally used to control strength properties since in most cases after a critical molecular-mass range is reached, increasing the average molecular mass of a thermoplastic material still further does not greatly increase its strength. Table 7.12 lists the molecular mass ranges and degrees of polymerization for some thermoplastics.

TABLE 7.12   Molecular Masses and Degrees of Polymerization for Some Thermoplastics

Thermoplastic	Molecular mass, g/mol	Degree of polymerization
Polyethylene	28,000–40,000	1000–1500
Polyvinyl chloride	67,000 (average)	1080
Polystyrene	60,000–500,000	600–6000
Polyhexamethylene adipamide (nylon 6,6)	16,000–32,000	150–300

Strengthening by increasing the amount of crystallinity in a thermoplastic material   The amount of crystallinity within a thermoplastic can greatly affect its tensile strength. In general, as the degree of crystallinity of the thermoplastic increases, the tensile strength, tensile modulus of elasticity, and density of the material all increase.

Thermoplastics which are able to crystallize during solidification have simple structural symmetry about their molecular chains. Polyethylenes and nylons are examples of thermoplastics which can solidify with a considerable amount of crystallization in their structure. Figure 7.47 compares the engineering stress-strain diagrams for low-density and high-density polyethylenes. The low-density polyethylene has a lower amount of crystallinity and thus lower strength and tensile modulus than the high-density polyethylene. Since the molecular chains in the low-density polyethylene are more branched and farther apart from each other, the bonding forces between the chains are lower, and hence the lower-density polyethylene has lower strength. The yield peaks in the stress-strain curves are due to the necking down of the cross sections of the test samples during the tensile tests.

Another example of the effect of increasing crystallinity on the tensile (yield) strength of a thermoplastic material is shown in Fig. 7.48 for nylon 6,6. The increased strength of the more highly crystallized material is due to the tighter packing of the polymer chains, which leads to stronger intermolecular bonding forces between the chains.

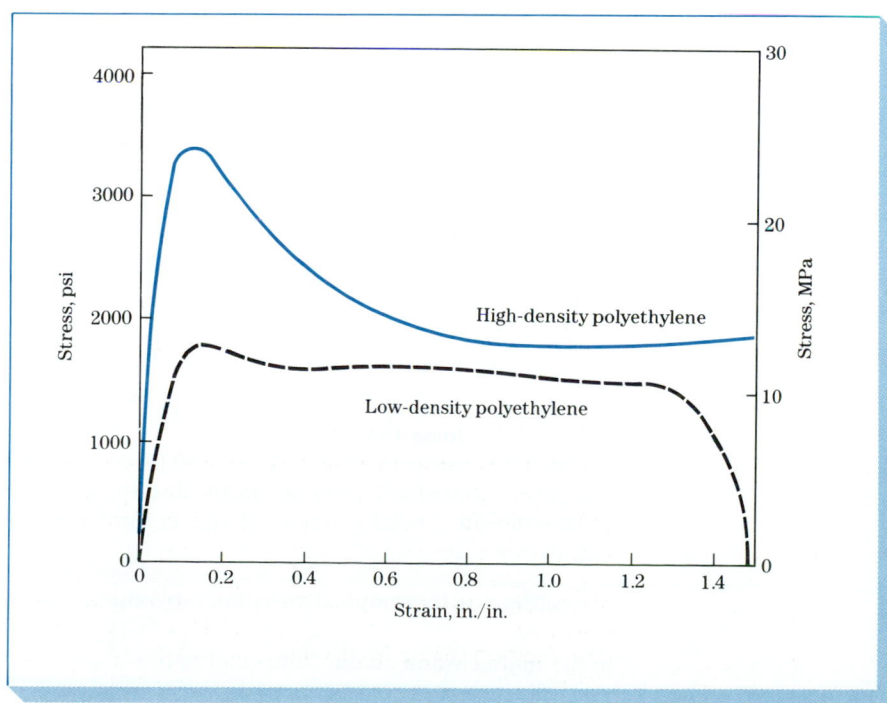

FIGURE 7.47 Tensile stress-strain curves for low-density and high-density polyethylene. The high-density polyethylene is stiffer and stronger due to a higher amount of crystallinity. [*After J. A. Sauer and K. D. Pae, Mechanical Properties of High Polymers, in H. S. Kaufman and J. J. Falcetta (eds.), "Introduction to Polymer Science and Technology," Wiley, 1977, p. 397.*]

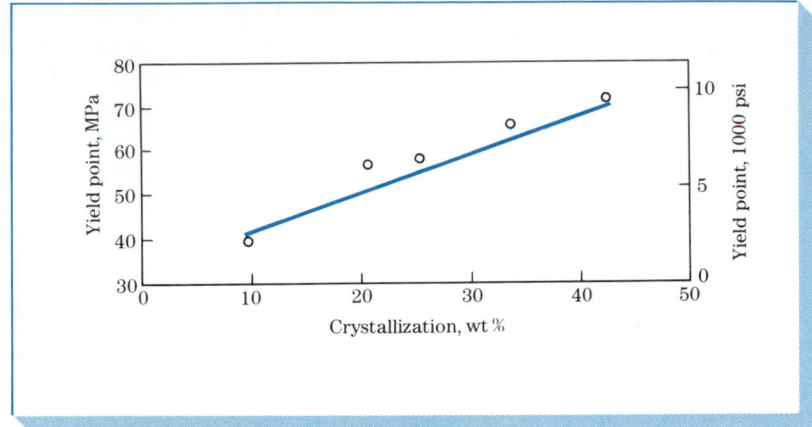

FIGURE 7.48 Yield point of dry polyamide (nylon 6,6) as a function of crystallinity. [*After R. J. Welgos, Polyamides (General), in "Encyclopedia of Chemical Technology," vol. 18, Wiley, 1982, p. 331.*]

Strengthening thermoplastics by introducing pendant atomic groups on the main carbon chains   Chain slippage during the permanent deformation of thermoplastics can be made more difficult by the introduction of bulky side groups on the main carbon chain. This method of strengthening thermoplastics is used, for example, for polypropylene and polystyrene. The tensile modulus, which is a measure of the stiffness of a material, is raised from the 0.6–1.5×10^5 psi range for high-density polyethylene to 1.5–2.2×10^5 psi for polypropylene, which has pendant methyl groups attached to the main carbon chain. The tensile elastic modulus of polyethylene is raised still further to the 4–5×10^5 psi range with the introduction of the more bulky pendant phenyl rings to the main carbon chain to make polystyrene. However, the elongation to fracture is drastically reduced from 100 to 600 percent for high-density polyethylene to 1 to 2.5 percent for polystyrene. Thus, bulky side groups on the main carbon chains of thermoplastics increase their stiffness and strength but reduce their ductility.

Strengthening thermoplastics by bonding highly polar atoms on the main carbon chain   A considerable increase in the strength of polyethylene can be attained by the introduction of a chlorine atom on every other carbon atom of the main carbon chain to make polyvinyl chloride. In this case, the large, highly polar chlorine atom greatly increases the molecular bonding forces between the polymer chains. Rigid polyvinyl chloride has a tensile strength of 6 to 11 ksi, which is considerably higher than the 2.5 to 5 ksi strength of polyethylene. Figure 7.49 shows a tensile stress-strain plot for a polyvinyl chloride test sample that has a maximum yield stress of about 8 ksi. The yield peak on the curve is due to the necking down of the central part of the test sample during extension.

Strengthening thermoplastics by the introduction of oxygen and nitrogen atoms in the main carbon chain   Introducing a $-\overset{|}{\underset{|}{C}}-O-\overset{|}{\underset{|}{C}}-$ ether linkage into the

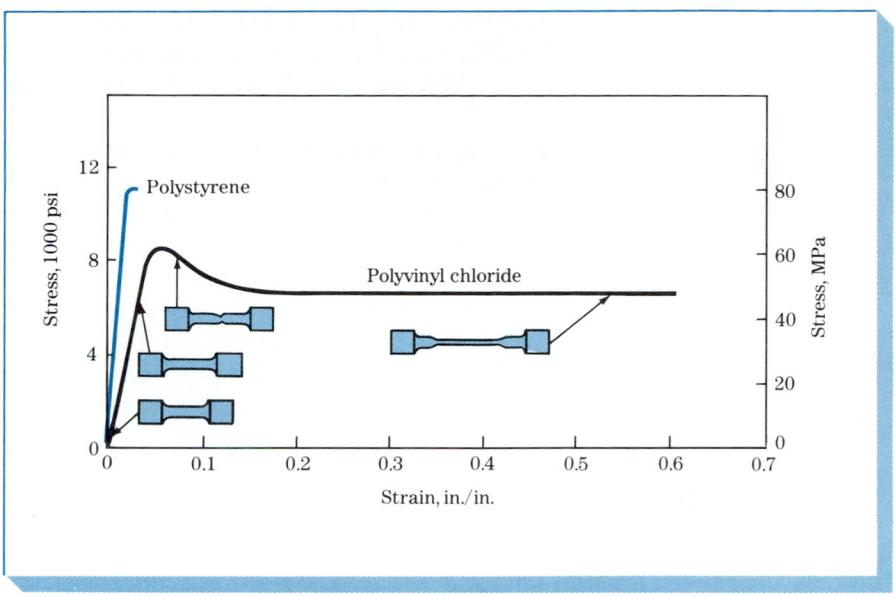

FIGURE 7.49 Tensile stress-strain data for the amorphous thermoplastic polyvinyl chloride (PVC) and polystyrene (PS). Sketches show modes of specimen deformation at various points on stress-strain curve. [*After J. A. Sauer and K. D. Pae, Mechanical Properties of High Polymers, in H. S. Kaufman and J. J. Falcetta (eds.), "Introduction to Polymer Science and Technology," Wiley, 1977, p. 331.*]

main carbon chain increases the rigidity of thermoplastics, as is the case for polyoxy methylene (acetal) which has the repeating chemical unit of

$$\left[\begin{array}{c} H \\ | \\ -C-O- \\ | \\ H \end{array}\right]_n$$

. The tensile strength of this material is in the 9 to 10 ksi range, which is considerably higher than the 2.5 to 5.5 ksi strength of high-density polyethylene. The oxygen atoms in the main carbon chains also increase the permanent dipole bonding between the polymer chains.

By introducing nitrogen in the main chains of thermoplastics, as in the case of the amide linkage $\left(\begin{array}{c} O \\ \| \\ -C-N- \end{array}\right)$, the permanent dipole forces between the polymer chains are greatly increased due to hydrogen bonding (Fig. 7.35). The relatively high tensile strength of 9 to 12 ksi of nylon 6,6 is a result of hydrogen bonding between the amide linkages of the polymer chains.

Strengthening thermoplastics by introducing phenylene rings into the main polymer chain in combination with other elements such as O, N, and S in the main chain One of the most important methods for strengthening thermoplastics is the introduction of phenylene rings in the main carbon chain. This method of strengthening is commonly used for high-strength engineering plastics. The phenylene rings cause steric hindrance to rotation within the polymer chain and electronic attraction of resonating electrons between adjacent molecules. Examples of polymeric materials which contain phenylene rings are polyphenylene oxide–based materials which have a tensile strength range of 7.8 to 9.6 ksi, thermoplastic polyesters which have tensile strengths of about 10 ksi, and polycarbonates which have strengths of about 9 ksi.

Strengthening thermoplastics by the addition of glass fibers   Some thermoplastics are reinforced with glass fibers. The glass content of most glass-filled thermoplastics ranges from 20 to 40 wt %. The optimum glass content is a trade-off between the desired strength, overall cost, and ease of processing. Thermoplastics commonly strengthened by glass fibers include the nylons, polycarbonates, polyphenylene oxides, polyphenylene sulfide, polypropylene, ABS, and polyacetal. For example, the tensile strength of nylon 6,6 can be increased from 12 to 30 ksi with a 40% fiberglass content, but its elongation is reduced from about 60 to 2.5 percent by the fiberglass addition.

Strengthening of Thermosetting Plastics

Thermosetting plastics without reinforcements are strengthened by the creation of a network of covalent bonding throughout the structure of the material. The covalent network is produced by chemical reaction within the thermosetting material after casting or during pressing under heat and pressure. The phenolics, epoxies, and polyesters (unsaturated) are examples of materials strengthened by this method. Because of their covalently bonded network, these materials have relatively high strengths, elastic moduli, and rigidities for plastic materials. For example, the tensile strength of molded phenolic resin is about 9 ksi, that of cast polyesters about 10 ksi, and that of cast epoxy resin about 12 ksi. These materials all have low ductilities because of their covalently bonded network structure.

The strength of thermosetting plastics can be considerably increased by the addition of reinforcements. For example, the tensile strength of glass-filled phenolic resins can be increased to as high as 18 ksi. Glass-filled polyester-based sheet-molding compounds can have a tensile strength as high as 20 ksi. By using unidirectional laminates of carbon-fiber-reinforced epoxy-based resin, materials with tensile strengths as high as 250 ksi in one direction can be obtained. Reinforced high-strength composite materials of this type will be discussed in Chap. 13 on Composite Materials.

Effect of Temperature on the Strength of Plastic Materials

A characteristic of thermoplastics is that they gradually soften as the temperature is increased. Figure 7.50 shows this behavior for a group of thermoplastics. As the temperature increases, the secondary bonding forces between the molecular chains become weaker and the strength of the thermoplastic decreases. When a thermoplastic material is heated through its glass transition temperature T_g, its strength decreases greatly due to a pronounced decrease in the secondary bonding forces. Figure 7.45 shows this effect for polymethyl methacrylate (PMMA), which has a T_g of about 100°C. The tensile strength of PMMA is about 7 ksi at 86°C, which is below its T_g, and decreases to about 4 ksi at 122°C, which is above its T_g. The maximum-use temperatures for some thermoplastics are listed in Tables 7.3 and 7.7.

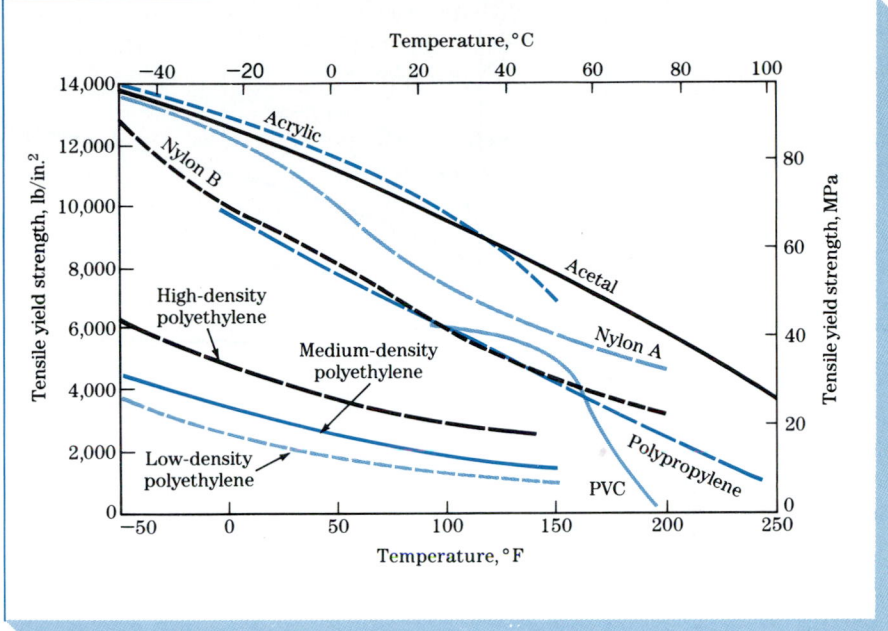

FIGURE 7.50 Effect of temperature on the tensile yield strength of some thermoplastics (*After H. E. Barker and A. E. Javitz, Plastic Molding Materials for Structural and Mechanical Applications, Electr. Manuf., May 1960.*)

Thermoset plastics also become weaker when heated, but since their atoms are bonded together primarily with strong covalent bonds in a network, they do not become viscous at elevated temperatures but degrade and char above their maximum-use temperature. In general, thermosets are more stable at higher temperatures than thermoplastics, but there are some thermoplastics which have remarkable high-temperature stability. The maximum-use temperatures of some thermosets are listed in Table 7.10.

7.11 CREEP AND FRACTURE OF POLYMERIC MATERIALS

Creep of Polymeric Materials   Polymeric materials subjected to a load may creep. That is, their deformation under a constant applied load at a constant temperature continues to increase with time. The magnitude of the strain increment increases with increased applied stress and temperature. Figure 7.51 shows how the creep strain of polystyrene increases under tensile stresses of 1760 to 4060 psi (12.1 to 30 MPa) at 77°F.

The temperature at which the creep of a polymeric material takes place is also an important factor determining the creep rate. At temperatures below the glass transition temperatures for thermoplastics, the creep rate is relatively

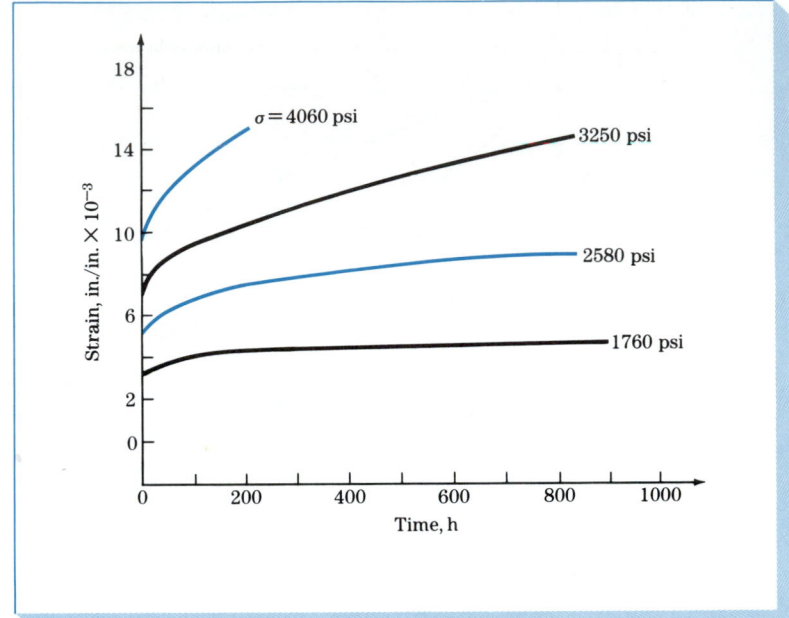

FIGURE 7.51 Creep curves for polystyrene at various tensile stresses at 77°F. [*After J. A. Sauer, J. Marin, and C. C. Hsiao, J. Appl. Phys., 20:507(1949).*]

low due to restricted molecular chain mobility. Above their glass transition temperatures, thermoplastics deform easier by a combination of elastic and plastic deformation which is referred to as *viscoelastic behavior.* Above the glass transition temperature, the molecular chains slide past each other more easily, and so this type of easier deformation is sometimes referred to as *viscous flow.*

In industry the creep of polymeric materials is measured by the creep modulus, which is simply the ratio of the initial applied stress σ_0 to the creep strain $\epsilon(t)$ after a particular time and at a constant temperature of testing. A high value for the creep modulus of a material thus implies a low creep rate. Table 7.13 lists the creep moduli of various plastics at various stress levels within the 1000 to 5000 psi range. This table shows the effect of bulky side groups and strong intermolecular forces in reducing the creep rate of polymeric materials. For example, at 73°F polyethylene has a creep modulus of 62 ksi at a stress level of 1000 psi for 10 h, whereas PMMA has a much higher creep modulus of 410 ksi at the same stress level for the same time.

Reinforcing plastics with glass fibers greatly increases their creep moduli and reduces their creep rates. For example, unreinforced nylon 6,6 has a creep modulus of 123 ksi after 10 h at 1000 psi, but when reinforced with 33% glass fiber, its creep modulus increases to 700 ksi after 10 h at 4000 psi. The addition of glass fibers to plastic materials is an important method for increasing their creep resistance as well as for strengthening them.

TABLE 7.13   Creep Modulus of Polymeric Materials at 73°F (23°C)

	Test time, h			
	10	**100**	**1000**	
	Creep modulus, ksi			Stress level, psi
Unreinforced materials:				
Polyethylene, Amoco 31-360B1	62	36		1000
Polypropylene, Profax 6323	77	58	46	1500
Polystyrene, FyRid KS1	310	290	210	Impact-modified
Polymethyl methacrylate, Plexiglas G	410	375	342	1000
Polyvinyl chloride, Bakelite CMDA 2201	· · ·	250	183	1500
Polycarbonate, Lexan 141-111	335	320	310	3000
Nylon 6,6, Zytel 101	123	101	83	1000, equil. at 50% RH
Acetal, Delrin 500	360	280	240	1500
ABS, Cycolac DFA-R	340	330	300	1000
Reinforced materials:				
Acetal, Thermocomp KF-1008, 30% glass fiber	1320	· · ·	1150	5000, 75°F (24°C)
Nylon 6,6, Zytel 70G-332, 33% glass fiber	700	640	585	4000, equil. at 50% RH
Polyester, thermosetting molding compound, Cyglas 303	1310	1100	930	2000
Polystyrene, Thermocomp CF-1007	1800	1710	1660	5000, 75°F (24°C)

Source: "Modern Plastics Encyclopedia, 1984–85," McGraw-Hill.

Stress Relaxation of Polymeric Materials

The stress relaxation of a stressed polymeric material under constant strain results in a decrease in the stress with time. The cause of the stress relaxation is that viscous flow in the polymeric material's internal structure occurs by the polymer chains slowly sliding by each other by the breaking and reforming of secondary bonds between the chains and by mechanical untangling and re-coiling of the chains. Stress relaxation allows the material to attain a lower energy state spontaneously, if there is sufficient activation energy for the process to occur. Stress relaxation of polymeric materials is thus temperature-dependent and associated with an activation energy.

The rate at which stress relaxation occurs depends on the *relaxation time* τ, which is a property of the material and which is defined as the time needed for the stress (σ) to decrease to 0.37 ($1/e$) of the initial stress σ_0. The decrease in stress with time t is given by

$$\sigma = \sigma_0 e^{-t/\tau} \tag{7.10}$$

where σ = stress after time t, σ_0 = initial stress, and τ = relaxation time.

Example Problem 7.7

A stress of 1100 psi (7.6 MPa) is applied to an elastomeric material at constant strain. After 40 days at 20°C, the stress decreased to 700 psi (4.8 MPa). (a) What is the relaxation-time constant for this material? (b) What will be the stress after 60 days at 20°C?

Solution:

(a) Since $\sigma = \sigma_0 e^{-t/\tau}$ [Eq. (7.10)] or $\ln(\sigma/\sigma_0) = -t/\tau$ and $\sigma = 700$ psi, $\sigma_0 = 1100$ psi, and $t = 40$ days,

$$\ln\left(\frac{700 \text{ psi}}{1100 \text{ psi}}\right) = -\frac{40 \text{ days}}{\tau} \qquad \tau = \frac{-40 \text{ days}}{-0.452} = 88.5 \text{ days} \blacktriangleleft$$

(b)

$$\ln\left(\frac{\sigma}{1100 \text{ psi}}\right) = -\frac{60 \text{ days}}{88.5 \text{ days}} = -0.678$$

$$\frac{\sigma}{1100 \text{ psi}} = 0.508 \quad \text{or} \quad \sigma = 559 \text{ psi} \blacktriangleleft$$

Since the relaxation time τ is the reciprocal of a rate, we can relate it to the temperature in kelvins by an Arrhenius-type rate equation [see Eq. (4.12)] as

$$\frac{1}{\tau} = Ce^{-Q/RT} \tag{7.11}$$

where C = rate constant independent of temperature, Q = activation energy for the process, T = temperature in kelvins, and R = molar gas constant = 8.314 J/(mol·K). Example Problem 7.8 shows how Eq. (7.11) can be used to determine the activation energy for an elastomeric material undergoing stress relaxation.

Example Problem 7.8

The relaxation time for an elastomer at 25°C is 40 days, while at 35°C the relaxation time is 30 days. Calculate the activation energy for this stress-relaxation process.

Solution:

Using Eq. (7.11), $1/\tau = Ce^{-Q/RT}$. For $\tau = 40$ days,

$$T_{25°C} = 25° + 273 = 298 \text{ K} \qquad T_{35°C} = 35° + 273 = 308 \text{ K}$$

$$\tfrac{1}{40} = Ce^{-Q/RT_{298}} \tag{7.12}$$

and

$$\tfrac{1}{30} = Ce^{-Q/RT_{308}} \tag{7.13}$$

Dividing Eq. (7.12) by Eq. (7.13) gives

$$\frac{30}{40} = \exp\left[-\frac{Q}{R}\left(\frac{1}{298} - \frac{1}{308}\right)\right] \qquad \text{or} \qquad \ln\left(\frac{30}{40}\right) = -\frac{Q}{R}(0.003356 - 0.003247)$$

$$-0.288 = -\frac{Q}{8.314}(0.000109) \qquad \text{or} \qquad Q = 22,000 \text{ J/mol} = 22.0 \text{ kJ/mol} \blacktriangleleft$$

Fracture of Polymeric Materials

As was the case for metals, the fracture of polymeric materials can be considered to be brittle or ductile or intermediate between the two extremes. In general, unreinforced thermosetting plastics are considered to fracture primarily in a brittle mode. Thermoplastics, on the other hand, may fracture primarily by the brittle or ductile manner. If the fracture of a thermoplastic takes place below its glass transition temperature, then its fracture mode will be primarily brittle, whereas if the fracture takes place above its glass transition temperature, its fracture mode will be ductile. Thus, temperature can greatly affect the fracture mode of thermoplastics. Thermosetting plastics heated above room temperature become weaker and fracture at a lower stress level but still fracture primarily in a brittle mode because the covalent bonding network is retained at elevated temperature. Strain rate is also an important factor in the fracture behavior of thermoplastics, with slower strain rates favoring ductile fracture because a slow strain rate allows molecular-chain realignment.

Brittle fracture of polymeric materials   The surface energy required to fracture an amorphous brittle glassy polymeric material such as polystyrene or polymethyl methacrylate (PMMA) is about 1000 times greater than that which would be required if the fracture involved just the simple breaking of carbon-carbon bonds on a fracture plane. Thus glassy polymeric materials such as PMMA are much tougher than inorganic glasses. The extra energy required to fracture glassy thermoplastics is much higher because distorted localized regions called *crazes* form before cracking occurs. A craze in a glassy thermoplastic is formed in a highly stressed region of the material and consists of an alignment of molecular chains combined with a high density of interdispersed voids.

Figure 7.52 is a diagrammatic representation of the change in molecular structure at a craze in a glassy thermoplastic such as PMMA. If the stress is

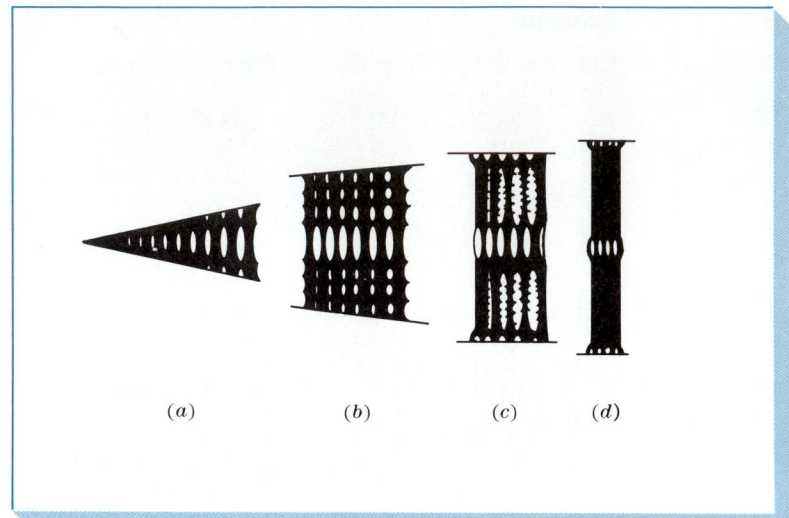

FIGURE 7.52
Diagrammatic representation of the change in the microstructure of a craze in a glassy thermoplastic as it thickens. [*After P. Beahan, M. Bevis, and D. Hull, J. Mater. Sci., 8:162(1972).*]

(a) (b) (c) (d)

intense enough, a crack forms through the craze, as shown in Fig. 7.53 and in the photo of Fig. 7.54. As the crack propagates along the craze, the stress concentration at the tip of the crack extends along the length of the craze. The work done in aligning the polymer molecules within the craze is the cause of the relatively high amount of work required for the fracture of glassy polymeric materials. This explains why the fracture energies of polystyrene and PMMA

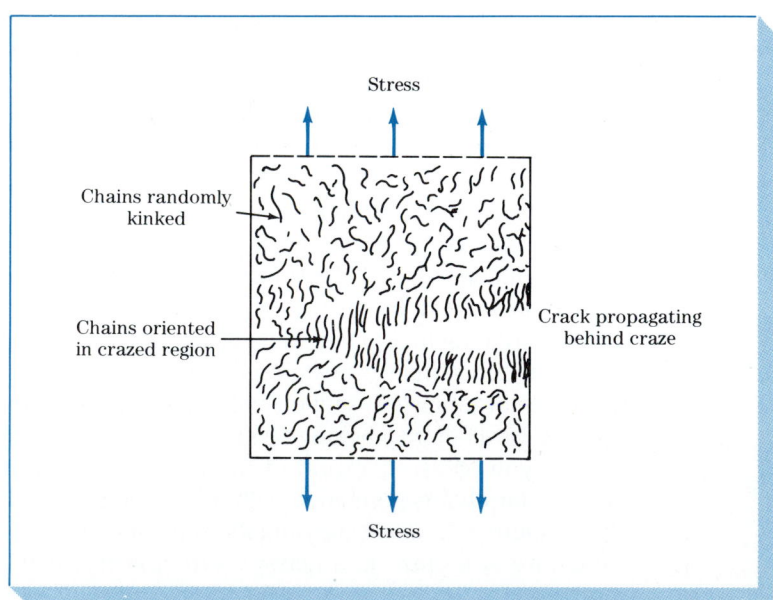

FIGURE 7.53 Schematic illustration of the structure of a craze near the end of a crack in a glassy thermoplastic.

FIGURE 7.54 Photo of a crack through the center of a craze in a glassy thermoplastic. (*After D. Hull, "Polymeric Materials," American Society of Metals, 1975, p. 511.*)

are between 300 and 1700 J/m^2 instead of about 0.1 J/m^2, which is the energy level that would be expected if only covalent bonds were broken in the fracturing process.

Ductile fracture of polymeric materials   Thermoplastics above their glass transition temperatures can exhibit plastic yielding before fracture. During plastic yielding, molecular linear chains uncoil and slip past each other and gradually align closer together in the direction of the applied stress (Fig. 7.55). Eventually,

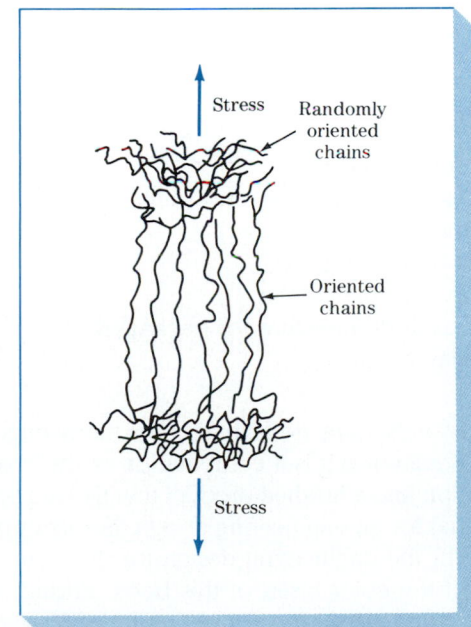

FIGURE 7.55 Plastic yielding of a thermoplastic polymeric material under stress. The molecular chains are uncoiled and slip past each other so that they align themselves in the direction of the stress. If the stress is too high, the molecular chains break, causing fracture of the material.

when the stress on the chains becomes too high, the covalent bonds of the main chains break and fracture of the material occurs. Elastomeric materials deform essentially in the same way except they undergo much more chain uncoiling (elastic deformation), but eventually if the stress on the material is too high and the extension of their molecular chains too great, the covalent bonds of the main chains will break, causing fracture of the material.

7.12   MATERIALS SELECTION FOR ENGINEERING DESIGNS USING PLASTIC MATERIALS

The increased usage of plastic materials over the past years (that is, U.S. plastics sales in 1984 was 44,513 million lb and in 1993 and 68,827 million lb) can be attributed to many reasons, the main ones being relatively low cost of material, properties of the material, and ease of processing to a finished product. The use of plastics in engineering designs has followed the general trend toward the use of more plastic materials. For use in an engineering design, the plastic part must function by itself or fit into the assembly in which it is being used. The plastic material must have the required properties for the engineering design.

There are many applications at the present time for which plastic materials are used. Table 7.14 lists some of the classes of plastics along with selected trade names, properties, applications, and prices (mid-1994). If one surveys this table, it becomes apparent that some important properties keep reoccurring:

1. Ability to reduce the number of parts in a design (see color plate 2).
2. Chemical resistance in different environments
3. Electrical insulative properties
4. Lightness
5. Ease of processability
6. Sufficient strength, stiffness, and toughness
7. Transparency
8. Low coefficient of friction
9. Colorability and plateability
10. Ability to exclude moisture for packaging
11. Dimensional stability

Plastic materials compete favorably for many engineering designs with respect to cost. Sometimes it is a combination of the cost of the material and its cost of fabrication into a finished product that tips the scale in favor of the use of a plastic material for an engineering design. For example, consider the plastics vs. metals use in the engineering design for the riding lawn mower shown in color plate 2 of the color insert of this book. Plastic materials were preferred partly because three plastic parts could replace 153 steel parts. Another consid-

TABLE 7.14   Selected Classes, Trade Names, Properties, Applications, and Prices of Some Plastic Materials for Materials Selection

Material class	Selected trade names	Selected properties	Selected applications	Grade	Prices mid-1994, $/lb U.S.
Part I: Thermoplastics					
Polyethylene (ranges from low- to high-density grades:	Dowlex Rexene Fortiflex Petrothene	Tough; low strength; low coefficient of friction; good chemical and electrical resistance	Film, pipe, sheet; blow molding; wire and cable insulation	(Railcar) LDPE, GP Mold HMW-HDPE, Blow-mold	0.55–0.60 0.60–0.63
Polypropylene	Pro-fax Rexene Fortilene	Tough; relatively low strength but better than polyethylene; good fatigue properties; low density; good chemical and electrical resistance	Containers, closures, housewares, toys; fibers and electrical insulation; In autos, copolymers are used in interior panels and trim, and batteries	(Railcar) GP homopoly. injection Random copoly. injection	0.55–0.60 0.60–0.65
Polystyrene (crystal clear)	Styron Replay Ladene	Optical clarity; fabrication ease; good thermal and dimensional stability; excellent insulator	Container bottles and lids; toys; disposable medical devices	(Railcar) GP crystal Hi Heat	0.60–0.66 0.63–0.66
Polystrene (impact-resistant)	Styron Vattra	Ease of processing; good impact strength and rigidity; good insulator; low cost	Packaging; disposables; consumer electronics; toys	HIPS: Super Hi Imp Fiber-Reinf.	0.72–0.73 0.91–0.95
Polyvinyl chloride (rigid)	Novablend	Strength, elongation; chemical resistance	Pipe and conduit; vinyl siding and windows; floor covering	(Railcar) GP Homopoly. Pipe	0.37–0.38 0.36–0.37
Polyvinyl chloride (flexible)	Unichem Ultra Kohinn	Flexibility, chemical and moisture resistance; colorability; durability; low cost	Coatings; artificial fabrics; artificial leather; electrical wire insulation; vinyl sheet; gloves	Chlorinated PVC pipe compound	1.37
Acrylics	Plexiglas Acrylite Perspex	Crystal clarity; good surface hardness; good chemical and environmental resistance; mechanical stability; weatherability	Window glazing and light transmission uses; backlighted signs; auto taillight lenses.	GP Impact	1.17 1.52
Polyethylene terephthalate (PET) (engineering-grades)	Valox Impet Petra	Reinforced or filled with glass fibers, glass flakes, minerals, or micas to improve property performance for stiffness, strength, and heat resistance; good moldability and chemical resistance	Automotive, electrical/ electronic, and appliances; motor housings, lamp sockets, sensors, switches, relays, and coil bobbins; coil forms for microwave oven transformers and small appliance housings	30% glass 55% glass PETG copoly.	1.76 2.00 0.98
Polybutylene terephthalate (PBT)	Valox Cleanex Ultradur	Easily injection-molded; used for electrical connectors for electrical properties, heat and chemical resistance, and ability to be cast into thin sections	Electrical connectors, coil bobbins, terminal blocks, and fuse holders; automotive distributor caps, door and window hardware; pump parts	Unfilled Hi-impact 30% glass, FR	1.64–1.75 1.95–2.05 1.70–1.90

TABLE 7.14   Selected Classes, Trade Names, Properties, Applications, and Prices of Some Plastic Materials for Materials Selection

Part I: Thermoplastics (Con't)

Material class	Selected trade names	Selected properties	Selected applications	Grade	Prices mid-1994, $/lb U.S.
ABS (acrylonitrile-butadiene-strene)	Magnum Cycolac Lustran	Acrylonitrile provides chemical resistance and heat stability; butadiene offers toughness and impact strength; styrene provides rigidity and processability at low cost	Automotive instrument panels and consoles; radiator grilles; headlight housings; appliance-extruded and thermoformed door liners; small appliance and computer housings; drain pipes and vents	Med impact Hi impact Pipe ABS/PC alloy	1.28–1.38 1.35–1.41 1.10–1.15 1.48–1.54
Acetal	Delrin (homopolymer) Celcon (copolymer) Tenac	Hard, strong, stiff; some notch sensitivity; good chemical resistance; dimensional stability; low coefficient of friction	Auto fuel system components, gears and window lift mechanisms; handles and cranks; industrial gears, valves, cams, and pumps; pulleys, bolts, nuts, and chain links; plumbing parts such as ballcocks, stems, valves	Homopolymer 20% glass Copolymer 25% glass	1.80–1.84 1.83–1.97 1.80–1.84 1.90–2.01
Polyphenylene oxide (modified)	Noryl Prevex	Good electrical properties over wide range of humidity and temperature conditions; good impact strength and rigidity with fiber and filler reinforcement; good processability	Automotive instrument panels and seat backs; computer housings and keyboards for computers; pump housings and impellers	Injection 20% glass 30% glass Extrusion	1.40–2.20 2.14–2.26 2.00–2.20 1.70–2.40
Nylon (linear polyamide)	Zytel Nylamid	High strength, toughness, stiffness, wear and abrasion resistance; low coefficient of friction; absorbs some moisture from environment; high-impact strength and rigidity	Gears, bearings, antifriction parts; mechanical parts subject to elevated temperatures and must resist hydrocarbons and solvents; electrical parts subject to elevated temperatures; caster wheels; housings for power tools	Type 66 Min filled 30% glass Type 6 Min filled 30% glass	2.40–2.41 1.35–1.95 2.40–2.44 2.28–3.45 1.70–1.77 1.71–2.44
Polycarbonate	Lexan Makrolon Calibre	Clarity, excellent impact resistance; high heat deflection temperature; dimensional stability; good electrical resistance	Glazing; business machine housings and instrument panels; electrical connectors and breakers; laser-read CDs; opthalmic lens	Injec. mold. 20% glass 30% glass Extrusion Fiber-rein. Compact disc	2.42–2.49 2.40–3.29 2.19–2.47 2.42–2.49 2.55–2.89 2.81

TABLE 7.14   Selected Classes, Trade Names, Properties, Applications, and Prices of Some Plastic Materials for Materials Selection

Material class	Selected trade names	Selected properties	Selected applications	Grade	Prices mid-1994, $/lb U.S.
		Part I: Thermoplastics (Con't)			
Polysulfone	Udel Ultrason	Transparent, heat-resistant, low flammability; good electrical properties to about 190°C (374°F); good thermal rigidity and stability at high temperatures; good impact strength and chemical resistance	Medical instrumentation and trays to hold instruments during sterilization; electrical connectors, coil bobbins, and cores, TV components; chemical processing equipment for corrosion-resistant piping, pumps, filter modules, and support plates, and tower packing.	GP 10% glass 30% glass	4.23–4.90 4.14–4.80 3.59–4.20
Polysulfide	Ryton Fortron Supec	Excellent combination of high-temperature resistance, chemical resistance, flowability, electrical properties, and dimensional stability.	Mainly for high-temperature electrical connectors and components; chemical resistance is used for pumps, valves, pipe fittings, and oil-field parts	40% glass 20% glass/ 35% filler 35% glass/ 35% filler	3.13–3.30 1.57 1.97
Polyetherimide	Ultem	High strength and rigidity at elevated temperatures; good electrical properties, chemical resistance, and processability	Good heat and chemical resistance for fluid and air handling parts; electrical uses which depend on high strength and dimensional stability; electrical connectors and printed circuit boards; under-the-hood sensors	Unrein. 30% glass	4.65 3.80
Polytetra-fluoroethylene	Teflon Neoflon Fluorocomp	Chemically inert in most environments; useful mechanical properties from cryogenic temperatures to 260°C (500°F); low coefficient of friction	Chemical pipes and valves; seals and rings; antistick coatings; fuel hose linings		6.40–7.20

TABLE 7.14 Selected Classes, Trade Names, Properties, Applications, and Prices of Some Plastic Materials for Materials Selection

Part II: Thermosets

Material class	Selected trade names	Selected properties	Selected applications	Grade	Prices mid-1994, $/lb U.S.
Phenolics	Bakelite Durez Plenco Valite	Heat resistant to over 150°C (300°F); good dimensional stability and creep resistance; good electrical resistance	High-temperature electrical and thermal resistance applications; electrical fixtures; adhesives	Molding compound Reinforced grades	0.62–1.00 1.15–3.00
Polyesters (unsaturated)	Rosite Premi Polyrite	Can be used unfilled, filled, or reinforced (glass); rigid; resilient; corrosion- and weather-resistant	Chair seats; auto exterior parts; boats; shower stalls; electrical components; auto patch compound; matrix for glass- and carbon-fiber-reinforced composites	GP Ortho Isophthalic Bis-A	0.68–0.79 0.73–0.81 1.20–1.50
Epoxies	Ciba-Geigy (RP-) Fiberite Farboset	Good mechanical properties; good adhesion and electrical insulative properties	Protective coatings; adhesives; electrical molding; matrix for carbon-fiber composites.	GP Resin Semiconduc. Novolac	1.28–1.37 1.93–2.28
Ureas and melamines	Perstop Cymel Fiberite Plenco	Molding compounds; good surface durability and hardness; good load bearing strength and resist organic solvents; good electrical resistance	Alpha cellulose-filled molding compounds are used in wiring devices (wall plates, receptacles, and circuit breakers), closures, housings, knobs, and handles	Molding compound Bk & Bw Wh * Iv	0.67–0.78 0.72
Silicones	Silastic Baysilon LIM (GE silicones)	Elastomeric properties; good electrical insulator; sealant; good chemical resistance	Coatings and adhesives, high-temperature insulator	Molding compound Spec. grade Silicon epoxy	5.81–6.40 8.91–31.48 3.39–3.43

eration was the improved corrosion resistance of the plastic materials. Lightness, toughness, and durability of the plastic materials were other considerations, along with the overall cost of manufacturing.

Example Problem 7.9

A new type of auto engine has been developed in which under-hood temperatures may reach 340°F (171°C). If oil vapors are also present, what type of plastic material should be used for electrical insulation for sensors or connectors which may be exposed to this environment? Use the materials listed in Table 7.14 for the selection.

Solution:

Using Table 7.14, a possible choice for this application would be 40 percent glass-reinforced polyphenylene sulfide since this material has an excellent combination of high-temperature resistance, electrical and chemical resistance, and good dimensional stability. Other plastic materials may also be satisfactory for this application.

The use of plastics in structural applications, however, has been limited; for many applications of this type, steel is still the preferred material. For example, the amount of plastics in a new automobile [about 200 to 350 lb (90.7 to 159 kg)] has remained basically unchanged for some years, and not much of it has been used in structural applications. However, plastics are serving as a niche material for automobiles, with the niche getting bigger all the time. For example, the 1994 GM Camero/Firebird sports models feature sheet-molded compound (SMC[1]) roofs and doors, and the restyled 1994 Ford Mustang has its hood and other minor parts made with SMC (Fig. 13.25). There is also an impending use of a new "toughened" (flexible) SMC for the fenders of the 1995 Lincoln Continental. Thus, the battle of plastics vs. metals for automobiles in materials selection continues.

SUMMARY

Plastics and elastomers are important engineering materials primarily because of their wide range of properties, relative ease of forming into a desired shape,

---

[1]See Sec. 13.5.

and relatively low cost. Plastic materials can be conveniently divided into two classes: *thermoplastics* and *thermosetting plastics (thermosets)*. Thermoplastics require heat to make them formable and after cooling, retain the shape they were formed into. These materials can be reheated and reused repeatedly. Thermosetting plastics are usually formed into a permanent shape by heat and pressure during which a chemical reaction takes place which bonds the atoms together to form a rigid solid. However, some thermosetting reactions take place at room temperature without the use of heat and pressure. Thermosetting plastics cannot be remelted after they are "set" or "cured" and upon heating to a high temperature, degrade or decompose.

The chemicals required for producing plastics are derived mainly from petroleum, natural gas, and coal. Plastic materials are produced by the polymerizing of many small molecules called *monomers* into very large molecules called *polymers*. Thermoplastics are composed of long-molecular-chain polymers, with the bonding forces between the chains being of the secondary permanent dipole type. Thermosetting plastics are covalently bonded throughout with strong covalent bonding between all the atoms.

The most commonly used processing methods for thermoplastics are *injection molding, extrusion,* and *blow molding,* whereas the most commonly used methods for thermosetting plastics are *compression and transfer molding* and *casting.*

There are many families of thermoplastics and thermosetting plastics. Examples of some general-purpose thermoplastics are polyethylene, polyvinyl chloride, polypropylene, and polystyrene. Examples of engineering plastics are polyamides (nylons), polyacetal, polycarbonate, saturated polyesters, polyphenylene oxide, and polysulfone. (Note that the separation of thermoplastics into general-purpose and engineering plastics is arbitrary.) Examples of thermosetting plastics are phenolics, unsaturated polyesters, melamines, and epoxies.

Elastomers or *rubbers* form a large subdivision of polymeric materials and are of great engineering importance. Natural rubber is obtained from tree plantations and is still much in demand (about 30 percent of the world's rubber supply) because of its superior elastic properties. Synthetic rubbers account for about 70 percent of the world's rubber supply, with styrene-butadiene being the most commonly used type. Other synthetic rubbers such as nitrile and polychloroprene (neoprene) are used for applications where special properties such as resistance to oils and solvents are required.

Thermoplastics have a *glass transition temperature* above which these materials behave as viscous or rubbery solids and below which they behave as brittle, glasslike solids. Above the glass transition temperature, permanent deformation occurs by molecular chains sliding past each other, breaking and remaking secondary bonds. Thermoplastics used above the glass transition temperature can be strengthened by intermolecular bonding forces by using polar pendant atoms such as chlorine in polyvinyl chloride or by hydrogen bonding as in the case of the nylons. Thermosetting plastics, because they are covalently bonded throughout, allow little deformation before fracture.

DEFINITIONS

Sec. 7.1

Thermoplastic (noun): a plastic material that requires heat to make it formable (plastic) and upon cooling, retains its shape. Thermoplastics are composed of chain polymers with the bonds between the chains being of the secondary permanent dipole type. Thermoplastics can be repeatedly softened when heated and harden when cooled. Typical thermoplastics are polyethylenes, vinyls, acrylics, cellulosics, and nylons.

Thermosetting plastic (thermoset): a plastic material that has undergone a chemical reaction by the action of heat, catalysis, etc., leading to a cross-linked network macromolecular structure. Thermoset plastics cannot be remelted and reprocessed since when they are heated they degrade and decompose. Typical thermoset plastics are phenolics, unsaturated polyesters, and epoxies.

Sec. 7.2

Monomer: a simple molecular compound that can be covalently bonded together to form long molecular chains (polymers). Example, ethylene.

Chain polymer: a high-molecular-mass compound whose structure consists of a large number of small repeating units called *mers*. Carbon atoms make up most of the main-chain atoms in most polymers.

Mer: a repeating unit in a chain polymer molecule.

Polymerization: the chemical reaction in which high-molecular-mass molecules are formed from monomers.

Copolymerization: the chemical reaction in which high-molecular-mass molecules are formed from two or more monomers.

Chain polymerization: the polymerization mechanism whereby each polymer molecule increases in size at a rapid rate once growth has started. This type of reaction occurs in three steps: (1) chain initiation, (2) chain propagation, and (3) chain termination. The name implies a chain reaction and is usually initiated by some external source. Example: the chain polymerization of ethylene into polyethylene.

Degree of polymerization: the molecular mass of a polymer chain divided by the molecular mass of its mer.

Functionality: the number of active bonding sites in a monomer. If the monomer has two bonding sites, it is said to be *bifunctional*.

Homopolymer: a polymer consisting of only one type of monomeric unit.

Copolymer: a polymer chain consisting of two or more types of monomeric units.

Cross-linking: the formation of primary valence bonds between polymer chain molecules. When extensive cross-linking occurs as in the case of thermosetting resins, cross-linking makes one supermolecule of all the atoms.

Stepwise polymerization: the polymerization mechanism whereby the growth of the polymer molecule proceeds by a stepwise intermolecular reaction. Only one type of reaction is involved. Monomer units can react with each other or with any size polymer molecule. The active group on the end of a monomer is assumed to have the same reactivity no matter what the polymer length is. Often a by-product such as water is condensed off in the polymerization process. Example: the polymerization of nylon 6,6 from adipic acid and hexamethylene diamine.

Sec. 7.3

Bulk polymerization: the direct polymerization of liquid monomer to polymer in a reaction system in which the polymer remains soluble in its own monomer.

Solution polymerization: in this process a solvent is used which dissolves the monomer, the polymer, and the polymerization initiator. Diluting the monomer with the solvent reduces the rate of polymerization, and the heat released by the polymerization reaction is absorbed by the solvent.

Suspension polymerization: in this process water is used as the reaction medium, and the monomer is dispersed rather than being dissolved in the medium. The polymer products are obtained in the form of small beads which are filtered, washed, and dried in the form of molding powders.

Sec. 7.4

Crystallinity (in polymers): the packing of molecular chains into a stereoregular arrangement with a high degree of compactness. Crystallinity in polymeric materials is never 100 percent and is favored in polymeric materials whose polymer chains are symmetrical. Example: high-density polyethylene can be 95 percent crystalline.

Glass transition temperature: the center of the temperature range where a heated thermoplastic upon cooling changes from a rubbery, leathery state to that of brittle glass.

Stereoisomers: molecules which have the same chemical composition but different structural arrangements.

Atactic stereoisomer: this isomer has pendant groups of atoms *randomly arranged* along a vinyl polymer chain. Example: atactic polypropylene.

Isotactic isomer: this isomer has pendant groups of atoms all on the *same side* of a vinyl polymer chain. Example: isotactic polypropylene.

Syndiotactic isomer: this isomer has pendant groups of atoms *regularly alternating* in positions on both sides of a vinyl polymer chain. Example: syndiotactic polypropylene.

Stereospecific catalyst: a catalyst which creates mostly a specific type of stereoisomer during polymerization. Example: the Ziegler catalyst used to polymerize propylene to mainly the isotactic polypropylene isomer.

Sec. 7.5

Injection molding: a molding process whereby a heat-softened plastic material is forced by a screw-drive cylinder into a relatively cool mold cavity which gives the plastic the desired shape.

Blow molding: a method of fabricating plastics in which a hollow tube (parison) is forced into the shape of a mold cavity by internal air pressure.

Extrusion: the forcing of softened plastic material through an orifice, producing a continuous product. Example: plastic pipe is extruded.

Compression molding: a thermoset molding process in which a molding compound (which is usually heated) is first placed in a molding cavity. Then the mold is closed and heat and pressure are applied until the material is cured.

Transfer molding: a thermoset molding process in which the molding compound is first softened by heat in a transfer chamber and then is forced under high pressure into one or more mold cavities for final curing.

Sec. 7.7

Plasticizers: chemical agents added to plastic compounds to improve flow and processibility and to reduce brittleness. Example: plasticized polyvinyl chloride.

Filler: a low-cost inert substance added to plastics to make them less costly. Fillers may also improve some physical properties such as tensile strength, impact strength, hardness, wear resistance, etc.

Sec. 7.9

Elastomer: a material that at room temperature stretches under a low stress to at least twice its length and then quickly returns to almost its original length upon removal of the stress.

***cis*-1,4 Polyisoprene:** the isomer of 1,4 polyisoprene which has the methyl group and hydrogen on the same side of the central double bond of its mer. Natural rubber consists mainly of this isomer.

***trans*-1,4 Polyisoprene:** the isomer of 1,4 polyisoprene which has the methyl group and hydrogen on opposite sides of the central double bond of its mer.

Vulcanization: a chemical reaction that causes cross-linking of polymer chains. Vulcanization usually refers to the cross-linking of rubber molecular chains with sulfur, but the word is also used for other cross-linking reactions of polymers such as those that occur in some silicone rubbers.

PROBLEMS

Calculation Problems

C7.2.1 A high-molecular-weight polyethylene has an average molecular weight of 450,000 g/mol. What is its average degree of polymerization?

C7.2.2 If a type of polyethylene has an average degree of polymerization of 7000, what is its average molecular weight?

C7.2.3 A nylon 6,6 has an average molecular weight of 10,000 g/mol. Calculate the average degree of polymerization (see Sec. 7.7 for its mer structure).

C7.2.4 An injection-molding polycarbonate material has an average molecular weight of 27,000 g/mol. Calculate its degree of polymerization (see Sec. 7.7 for the mer structure of polycarbonate).

C7.2.5 Calculate the average molecular weight M_m for a thermoplastic which has the following weight fractions f_i for the molecular weight ranges listed below:

Molecular weight range, g/mol	f_i	Molecular weight range, g/mol	f_i
0–5000	0.01	20,000–25,000	0.18
5000–10,000	0.05	25,000–30,000	0.22
10,000–15,000	0.15	30,000–35,000	0.17
15,000–20,000	0.17	35,000–40,000	0.05

C7.2.6   A copolymer consists of 75 wt % polystyrene and 25 wt % polyacrylonitrile. Calculate the mole fraction of each component in this material.

C7.2.7   An ABS copolymer consists of 25 wt % polyacrylonitrile, 35 wt % polybutadiene, and 40 wt % polystyrene. Calculate the mole fraction of each component in this material.

C7.2.8   Determine the mole fraction of polyvinyl chloride and polyvinyl acetate in a copolymer having a molecular weight of 9000 g/mol and a degree of polymerization of 125.

C7.9.1   How much sulfur must be added to 100 g of butadiene rubber to cross-link 3.0 percent of the mers? (Assume all sulfur is used to cross-link the mers and that only one sulfur atom is involved in each cross-linking bond).

C7.9.2   If 10 g of sulfur is added to 100 g of butadiene rubber, what is the maximum fraction of the cross-link sites that can be connected?

C7.9.3   How much sulfur must be added to cross-link 13 percent of the cross-link sites in 100 g of polychloroprene rubber?

C7.9.4   How many kilograms of sulfur are needed to cross-link 20 percent of the cross-link sites in 300 kg of polyisoprene rubber?

C7.9.5   If 5 kg of sulfur is added to 400 kg of butadiene rubber, what fraction of the cross-links are joined?

C7.9.6   A butadiene-styrene rubber is made by polymerizing two monomers of styrene with six monomers of butadiene. If 22 percent of the cross-link sites are to be bonded with sulfur, what weight percent sulfur is required?

C7.9.7   What weight percent sulfur must be added to polybutadiene to cross-link 15 percent of the possible cross-link sites?

C7.9.8   A butadiene-acrylonitrile rubber is made by polymerizing one acrylonitrile monomer with six butadiene monomers. How much sulfur is required to react with 100 kg of this rubber to cross-link 20 percent of the cross-link sites?

C7.9.9   If 17 percent of the cross-link sites in isoprene rubber are to be bonded, what weight percent sulfur must the rubber contain?

C7.10.1   A stress of 7.2 MPa is applied to an elastomeric material at a constant stress at 20°C. After 40 days, the stress decreases to 5.0 MPa. (a) What is the relaxation time τ for this material? (b) What will be the stress after 60 days?

C7.10.2   A polymeric material has a relaxation time of 70 days at 27°C when a stress of 7.5 MPa is applied. How many days will be required to decrease the stress to 6.2 MPa?

C7.10.3   A stress of 1100 psi is applied to an elastomer at 27°C, and after 20 days the stress is reduced to 800 psi by stress relaxation. When the temperature is raised to 50°C, the stress is reduced from 1100 to 500 psi in 30 days. Calculate the activation energy for this relaxation process using an Arrhenius-type rate equation.

C7.10.4   The stress on a sample of a rubber material at constant strain at 27°C decreases from 5.0 to 3.0 MPa in 20 days. (a) What is the relaxation time τ for this material? (b) What will be the stress on this material after (i) 10 days and (ii) 50 days?

C7.10.5   A polymeric material has a relaxation time of 110 days at 27°C when a stress of 5.0 MPa is applied. (a) How many days will be required to decrease the stress to 4.0 MPa? (b) What is the relaxation time at 45°C if the activation energy for this process is 21 kJ/mol?

Study Problems

7.1.1   Define the polymeric materials (a) plastics and (b) elastomers.

7.1.2   Define a thermoplastic plastic.

7.1.3   Describe the atomic structural arrangement of thermoplastics.

7.1.4   What types of atoms are bonded together in thermoplastic molecular chains? What are the valences of these atoms in the molecular chains?

7.1.5   What is a pendant atom or group of atoms?

7.1.6   What type of bonding exists within the molecular chains of thermoplastics?

7.1.7   What type of bonding exists between the molecular chains of thermoplastics?

7.1.8   Define thermosetting plastics.

7.1.9   Describe the atomic structural arrangement of thermosetting plastics.

7.1.10   What are some of the reasons for the great increase in the use of plastics in engineering designs over the past years?

7.1.11   What are some of the advantages of plastics for use in mechanical engineering designs?

7.1.12   What are some of the advantages of plastics for use in electrical engineering designs?

7.1.13   What are some of the advantages of plastics for use in chemical engineering designs?

7.2.1   Define the following terms: chain polymerization, monomer, and polymer.

7.2.2   Describe the bonding structure within an ethylene molecule by using (*a*) the electron-dot-cross notation and (*b*) straight-line notation for the bonding electrons.

7.2.3   What is the difference between a saturated and an unsaturated carbon-containing molecule?

7.2.4   Describe the bonding structure of an activated ethylene molecule which is ready for covalent bonding with another activated molecule by using (*a*) electron-dot-cross notation and (*b*) straight-line notation for the bonding electrons.

7.2.5   Write a general chemical reaction for the chain polymerization of ethylene monomer into the linear polymer polyethylene.

7.2.6   What is the repeating chemical unit of a polymer chain called? What is the chemical repeating unit for polyethylene?

7.2.7   Define the degree of polymerization for a polymer chain?

7.2.8   What are the three major reactions that occur during chain polymerization?

7.2.9   What is a free radical? Write a chemical equation for the formation of two free radicals from a hydrogen peroxide molecule by using (*a*) electron-dot notation and (*b*) straight-line notation for the bonding electrons.

7.2.10   Write an equation for the formation of two free radicals from a molecule of benzoyl peroxide by using straight-line notation for the bonding electrons.

7.2.11   Write an equation for the reaction of an organic free radical ($RO^\cdot$) with an ethylene molecule to form a new, longer chain-free radical.

7.2.12   What is the function of the initiator catalyst for chain polymerization?

7.2.13   Write a reaction for the free radical $R-CH_2-CH_2^\cdot$ with an ethylene molecule to extend the free radical. What type of reaction is this?

7.2.14   How is it possible for a polymer chain such as a polyethylene one to keep growing spontaneously during polymerization?

7.2.15   What are two methods by which a linear chain polymerization reaction can be terminated?

7.2.16   Why must one consider the *average* degree of polymerization and the *average* molecular weight of a thermoplastic material?

7.2.17   Define the average molecular weight of a thermoplastic?

7.2.18   What is the functionality of a polymer? Distinguish between a bifunctional and trifunctional monomer.

7.2.19   What causes a polyethylene molecular chain to have a zigzig configuration?

7.2.20   What type of chemical bonding is there between the polymer chains in polyethylene?

7.2.21   How do side branches on polyethylene main chains affect the packing of the molecular chains in a solid polymer? How does branching of the polymer chains affect the tensile strength of solid bulk polyethylene?

7.2.22   Write a general reaction for the polymerization of a vinyl-type polymer.

7.2.23   Write structural formulas for the mers of the following vinyl polymers: (a) polyethylene, (b) polyvinyl chloride, (c) polypropylene, (d) polystyrene, (e) polyacrylonitrile, and (f) polyvinyl acetate.

7.2.24   Write a general reaction for the polymerization of a vinylidene polymer.

7.2.25   Write structural formulas for the mers of the following vinylidene polymers: (a) polyvinylidene chloride and (b) polymethyl methacrylate.

7.2.26   Distinguish between a homopolymer and a copolymer.

7.2.27   Illustrate the following types of copolymers by using filled and open circles for their mers: (a) random, (b) alternating, (c) block, and (d) graft.

7.2.28   Write a general polymerization reaction for the formation of a vinyl chloride and vinyl acetate copolymer.

7.2.29   Define stepwise polymerization of linear polymers. What by-products are commonly produced by stepwise polymerization?

7.2.30   Write the equation for the reaction of a molecule of hexamethylene diamine with one of adipic acid to produce a molecule of nylon 6,6. What is the by-product of this reaction?

7.2.31   Write the reaction for the stepwise polymerization of two phenol molecules with one of formaldehyde to produce a phenol formaldehyde molecule.

7.2.32   What type of by-product is produced by this reaction?

7.3.1   What are three basic raw materials used to produce the basic chemicals needed for the polymerization of plastic materials?

7.3.2   Describe and illustrate the following polymerization processes: (a) bulk, (b) solution, (c) suspension, and (d) emulsion.

7.3.3   In which polymerization process is the heat released by the reaction absorbed by water? In which is the heat absorbed by the solvent? Which process is used when the polymerization heat of reaction is low?

7.3.4   Describe the Unipol process for producing low-density polyethylene. What are the advantages of this process?

7.4.1   During the solidification of thermoplastics, how do the specific volume vs. temperature plots differ for noncrystalline and partly crystalline thermoplastics?

7.4.2   Define the glass transition temperature T_g for a thermoplastic.

7.4.3   What are measured T_g values for (a) polyethylene, (b) polyvinyl chloride, and (c) polymethyl methacrylate? Are T_g values material constants?

7.4.4   Describe and illustrate the fringed-micelle and folded-chain models for the structure of partly crystalline thermoplastics.

7.4.5   Describe the spherulitic structure found in some partly crystalline thermoplastics.

7.4.6   Why is complete crystallinity in thermoplastics impossible?

7.4.7   How does the amount of crystallinity in a thermoplastic affect (a) its density and (b) its tensile strength? Explain.

7.4.8   What are stereoisomers with respect to chemical molecules?

7.4.9   Describe and draw structural models for the following stereoisomers of polypropylene: (a) atactic, (b) isotactic, and (c) syndiotactic.

7.4.10   What is a stereospecific catalyst? How did the development of a stereospecific catalyst for the polymerization of polypropylene affect the usefulness of commercial polypropylene?

7.5.1　In general how does the processing of thermoplastics into the desired shape differ from the processing of thermosetting plastics?

7.5.2　Describe the injection-molding process for thermoplastics.

7.5.3　Describe the operation of the reciprocal-screw injection-molding machine.

7.5.4　What are some advantages and disadvantages of the injection-molding process for molding thermoplastics?

7.5.5　What are the advantages of the reciprocating-screw injection-molding machine over the old plunger type?

7.5.6　Describe the extrusion process for processing thermoplastics.

7.5.7　Describe the blow molding and thermoforming processes for forming thermoplastics.

7.5.8　Describe the compression-molding process for thermosetting plastics.

7.5.9　What are some of the advantages and disadvantages of the compression-molding process?

7.5.10　Describe the transfer-molding process for thermosetting plastics.

7.5.11　What are some of the advantages and disadvantages of the transfer-molding process?

7.6.1　What are the four major thermoplastic materials which account for about 60 percent of the sales tonnage of plastic materials in the United States? What were their prices per pound in 1994?

7.6.2　How does the molecular-chain structure differ for the following types of polyethylene: (*a*) low-density, (*b*) high-density, and (*c*) linear-low-density?

7.6.3　How does chain branching affect the following properties of polyethylene: (*a*) amount of crystallinity, (*b*) strength, and (*c*) elongation?

7.6.4　What are some of the properties which make polyethylene such an industrially important plastic material?

7.6.5　What are some of the industrial applications for polyethylene?

7.6.6　Write the general reaction for the polymerization of polyvinyl chloride.

7.6.7　How can the higher strength of polyvinyl chloride as compared to polyethylene be explained?

7.6.8　How is the flexibility of bulk polyvinyl chloride increased?

7.6.9　What are some of the properties of polyvinyl chloride which make it an important industrial material?

7.6.10　What are plasticizers? Why are they used in some polymeric materials? How do plasticizers usually affect the strength and flexibility of polymeric materials? What types of plasticizers are commonly used for PVC?

7.6.11　How is the processibility of PVC improved to produce rigid PVC?

7.6.12　What are some of the applications for plasticized PVC?

7.6.13　Write the general reaction for the polymerization of polypropylene from propylene gas.

7.6.14　Why is the use of a stereospecific catalyst in the polymerization of polypropylene so important?

7.6.15　How does the presence of a methyl group on every other carbon atom of the main polymer chain affect the glass transition temperature of this material when compared to polyethylene?

7.6.16　What are some of the properties of polypropylene which make it an industrially important material?

7.6.17　What are some of the applications for polypropylene?

7.6.18   Write the general reaction for the polymerization of polystyrene from styrene.

7.6.19   What effect does the presence of the phenyl group on every other carbon of the main chain have on the impact properties of polystyrene?

7.6.20   How can the low-impact resistance of polystyrene be improved by copolymerization?

7.6.21   What are some of the applications for polystyrene?

7.6.22   What is the repeating chemical structural unit for polyacrylonitrile?

7.6.23   What effect does the presence of the nitrile group on every other carbon of the main polymer chains have on the molecular structure of polyacrylonitrile?

7.6.24   What special properties do polyacrylonitrile fibers have? What are some applications for polyacrylonitrile?

7.6.25   What are SAN resins? What desirable properties do SAN thermoplastics have? What are some of the applications for SAN thermoplastics?

7.6.26   What do the letters A, B, and S stand for in the ABS thermoplastic?

7.6.27   Why is ABS sometimes referred to as a terpolymer?

7.6.28   What important property advantages does each of the components in ABS contribute?

7.6.29   Describe the structure of ABS. How can the impact properties of ABS be improved?

7.6.30   What are some of the applications for ABS plastics?

7.6.31   What is the repeating chemical structural unit for polymethyl methacrylate? By what trade names is PMMA commonly known?

7.6.32   What are some of the important properties of PMMA which make it an important industrial plastic?

7.6.33   What are the fluoroplastics? What are the repeating chemical structural units for polytetrafluoroethylene and polychlorotrifluoroethylene?

7.6.34   What are some of the important properties and applications of polytetrafluoroethylene?

7.6.35   How does the presence of the chlorine atom on every other carbon atom of the main chain of polychlorotrifluoroethylene modify the crystallinity and moldability of polytetrafluoroethylene?

7.6.36   What are some of the important properties and applications of polychlorotrifluoroethylene?

7.7.1   Define an engineering thermoplastic. Why is this definition arbitrary?

7.7.2   How do the prices of engineering thermoplastics compare with those of the commodity plastics such as polyethylene, polyvinyl chloride, and polypropylene?

7.7.3   How do the densities and tensile strengths of engineering thermoplastics compare with those of polyethylene and polyvinyl chloride?

7.7.4   What is the structural formula for the amide linkage in thermoplastics? What is the general name for polyamide thermoplastics?

7.7.5   Write a chemical reaction for one molecule of a dibasic acid with a diamine to form an amide linkage. What is the by-product of this reaction?

7.7.6   In the designation nylon 6,6, what does the 6,6 stand for?

7.7.7   Write a chemical reaction for one molecule of adipic acid and one molecule of hexamethylene diamine to form an amide linkage.

7.7.8   What is the repeating structural unit for nylon 6,6?

7.7.9   How can nylons 6,9, 6,10, and 6,12 be synthesized?

7.7.10   Write the reaction for the polymerization of nylon 6 from ϵ-caprolactam.

PLATE 1   Continuous casting of steel (see also Fig. 4.9). Molten steel from ladle (above) is poured into receiving vessel (called a tundish) which controls the flow of the steel into a water-cooled mold (below) for solidification. (*Bethlehem Steel. Co.*)

(a)

(b)

(c)

PLATE 2 Plastics vs. metals in engineering designs. (*a*) A riding mower whose frame plus cover consists of three plastic parts [(*b*) and (*c*)] which replace 153 steel parts (*d*). The materials used in this design are two 9-lb (4-kg) panels that are a blend of polycarbonate and polyethelene terephthalate and a 40 percent long-glass fiber-reinforced rigid thermoplastic polyurethane 25-lb (11.4-kg) lower part. (*John Deere*)

(d)

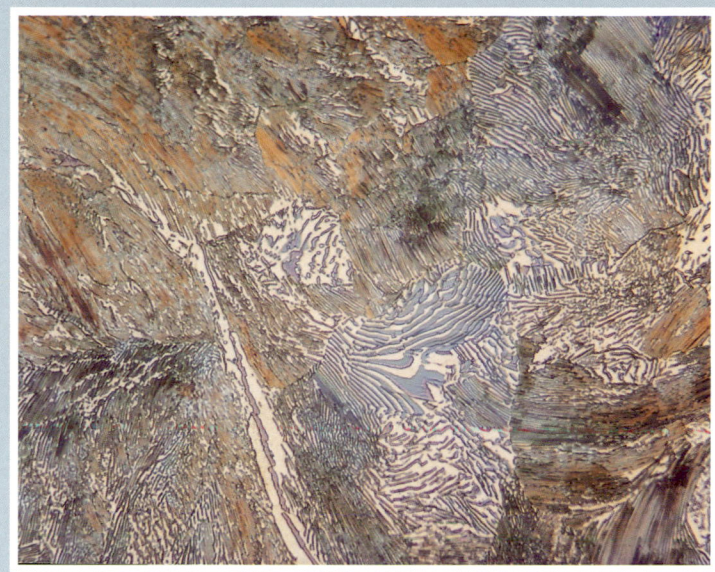

PLATE 3 Photomicrograph of an as-rolled Fe-1.00 wt %C hyper-eutectoid steel color tinted by an etchant. The eutectoid cementite in the pearlite appears blue and the proeutectoid cementite in the grain boundary appears violet. The alpha ferrite is white. (Magnification 500X.) (Also compare with microstructure of an Fe-1.2%C steel in Fig. 9.12.) (*George F. Van-der Voort, Carpenter Technology Co.*)

PLATE 4 Photomicrograph of a 70%Cu-30%Zn alpha brass cold-worked 50% and annealed 30 min at 700°C (1300°F). The etched structure reveals grains and annealing twin bands within the grains. Etchant: Klemm's; magnification: 100X. (Also see Fig. 9-50.) (*George F. Van-der Voort, Carpenter Technology Co.*)

PLATE 5 Engineering structural tests of a composite horizontal stabilizer for a Boeing 777 airplane (see lower photo). Hydraulic jacks pulled the tip of the 44-ft (13.4-m) tail structure down until it was stressed beyond the "ultimate load," which is $1\frac{1}{2}$ times greater than that expected in airline operation. (See Fig. 1.15 for the location and composite makeup of the tail section of the new Boeing 777 airliner.) (*Boeing Co.*)

7.7.11   Illustrate the bonding between polymer chains of nylon 6,6. Why is this bonding particularly strong (see Fig. 7.35)?

7.7.12   What properties do nylons have that make them useful for engineering applications? What is an important undesirable property of nylon?

7.7.13   What are some of the engineering applications for nylons?

7.7.14   What are the two main components of polyphthalamide resins? What is the principal repeating chemical unit of PPA resins? What are the main functions of each of these components?

7.7.15   What are some important properties of PPA resins for mechanical and electrical engineering designs?

7.7.16   What are some applications of polyphthalamide (PPA) resins for mechanical and electrical engineering designs?

7.7.17   What is the basic repeating chemical structural unit for polycarbonates? What is the carbonate linkage? What are the common trade names for polycarbonate?

7.7.18   What part of the polycarbonate structure makes the molecule stiff? What part of the polycarbonate molecule provides molecular flexibility?

7.7.19   What are some of the properties of polycarbonates that make them useful engineering thermoplastics?

7.7.20   What are some engineering applications for polycarbonates?

7.7.21   What is the basic repeating chemical structural unit for the polyphenylene oxide-based resins? What are the trade names for these resins?

7.7.22   What part of the structure of polyphenylene oxide provides its relatively high strength? What part of its structure provides its molecular flexibility?

7.7.23   What are some of the properties which make polyphenylene oxide resins important engineering thermoplastics?

7.7.24   What are some of the engineering applications for polyphenylene oxide resins?

7.7.25   What is the repeating chemical structural unit for the acetal high-performance engineering thermoplastics? What are the two main types of acetals and what are their trade names?

7.7.26   What part of the structure of the acetals provides high strength?

7.7.27   What are some of the properties of acetals that make them important engineering thermoplastics?

7.7.28   What outstanding property advantage do the acetals have over nylons?

7.7.29   What are some of the engineering applications for acetals?

7.7.30   What types of materials have acetals commonly replaced?

7.7.31   What are the two most-important engineering thermoplastic polyesters? What are their repeating chemical structural units?

7.7.32   What is the chemical structure of the ester linkage?

7.7.33   What part of the structure of the thermoplastic polyesters provides rigidity? What part provides molecular mobility?

7.7.34   What are some of the properties of thermoplastic polyesters which make them important engineering thermoplastics?

7.7.35   What are some engineering applications for PBT thermoplastics?

7.7.36   What is the repeating chemical structural unit for polysulfone?

7.7.37   What part of the polysulfone structure provides its high strength? What part provides chain flexibility and impact strength? What part provides high-oxidation stability?

7.7.38   What are some of the properties of polysulfone which are important for engineering designs?

7.7.39 What are some of the engineering applications for polysulfone?

7.7.40 What is the repeating chemical structural unit for polyphenylene sulfide? What engineering thermoplastic has a similar structure?

7.7.41 What is a trade name for polyphenylene sulfide?

7.7.42 What part of the structure of PPS provides its rigidity and strength? What part provides its high resistance to chemicals?

7.7.43 What properties make PPS a useful engineering thermoplastic?

7.7.44 What are some engineering applications for PPS?

7.7.45 What is the chemical structure of polyetherimide? What is its trade name?

7.7.46 What is the imide linkage?

7.7.47 What is the function of the ether linkage in polyetherimide?

7.7.48 What special properties does polyetherimide have for (*a*) electrical engineering designs and (*b*) mechanical engineering designs?

7.7.49 What are some applications for polyetherimide?

7.7.50 What are polymer alloys? How does their structure differ from copolymers?

7.7.51 Why are polymer alloys of great importance for engineering applications?

7.7.52 What type of polymer alloy is (*a*) Xenoy 1000, (*b*) Valox 815, and (*c*) Bayblend MC2500?

7.8.1 What are some of the advantages of thermosetting plastics for engineering design applications? What is the major disadvantage of thermosets which thermoplastics do not have?

7.8.2 What are the major processing methods used for thermosets?

7.8.3 What are the two major ingredients of thermosetting molding compounds?

7.8.4 What are the major advantages of phenolic plastics for industrial applications?

7.8.5 Using structural formulas, write the reaction for phenol with formaldehyde to form a phenol-formaldehyde molecule (use two phenol molecules and one formaldehyde molecule). What kind of molecule is condensed off in the reaction?

7.8.6 Why are large percentages of fillers used in phenolic molding compounds? What types of fillers are used and for what purposes?

7.8.7 What are some of the applications for phenolic compounds?

7.8.8 Write the structural formula for the epoxide group and the repeating unit of a commercial epoxy resin.

7.8.9 What are two types of reaction sites which are active in the cross-linking of commercial epoxy resins?

7.8.10 Write the reaction for the cross-linking of two epoxy molecules with ethylene diamine.

7.8.11 What are some of the advantages of epoxy thermoset resins? What are some of their applications?

7.8.12 What makes an unsaturated polyester resin "unsaturated"?

7.8.13 How are linear unsaturated polyesters cross-linked? Write a structural formula chemical reaction to illustrate the cross-linking of an unsaturated polyester.

7.8.14 How are most unsaturated polyesters reinforced?

7.8.15 What are some applications for reinforced polyesters?

7.9.1 What are elastomers? What are some elastomeric materials?

7.9.2 From what tree is most natural rubber obtained? What countries have large plantations of these trees?

7.9.3 What is natural rubber latex? Briefly describe how natural rubber is produced in the bulk form.

7.9.4 Write the formula for *cis*-1,4 polyisoprene. What does the prefix *cis*- stand for? What is the significance of the 1,4 in the name *cis*-1,4 polyisoprene?

7.9.5   What is natural rubber mainly made of? What other components are present in natural rubber?

7.9.6   To what structural arrangement is the coiling of the natural rubber polymer chains attributed? What is steric hindrance?

7.9.7   What are chemical structural isomers?

7.9.8   What is gutta-percha? What is the repeating chemical structural unit for gutta-percha?

7.9.9   What does the *trans*- prefix in the name *trans*-1,4 polyisoprene refer to?

7.9.10   Why does the trans isomer lead to a higher degree of crystallinity than the cis isomer for polyisoprene?

7.9.11   What is the vulcanization process for natural rubber? Who discovered this process and when? Illustrate the cross-linking of *cis*-1,4 polyisoprene with divalent sulfur atoms.

7.9.12   How does cross-linking with sulfur affect the tensile strength of natural rubber? Why is only about 3 wt % of sulfur used in the process?

7.9.13   What materials are used in the compounding of rubber and what is the function of each?

7.9.14   How can oxygen atoms cross-link the rubber molecules? How can the cross-linking of rubber molecules by oxygen atoms be retarded?

7.9.15   What is styrene-butadiene rubber (SBR)? What weight percent of it is styrene? What are the repeating chemical structural units for SBR?

7.9.16   Can SBR be vulcanized? Explain.

7.9.17   What are some of the advantages and disadvantages of SBR? Natural rubber?

7.9.18   What is the composition of nitrile rubbers? What effect does the nitrile group have on the main carbon chain in nitrile rubber?

7.9.19   What are some applications for nitrile rubbers?

7.9.20   Write the repeating chemical structural unit for polychloroprene. What common name is given to polychloroprene rubber? How does the presence of the chlorine atom in polychloroprene affect some of its properties?

7.9.21   What are some engineering applications for neoprene rubbers?

7.9.22   What are the silicones? What is the general repeating chemical structural unit for the silicones?

7.9.23   What is a silicone elastomer? What is the chemical structural repeating unit of the most common type of silicone rubber? What is its technical name?

7.9.24   How can a silicone rubber be cross-linked at room temperature?

7.9.25   What are some of the engineering applications for silicone rubber?

7.9.26   What is the chemical repeating unit common to all thermoplastic polyurethane elastomers (TPUEs)?

7.9.27   In general, what is the structure of a TPUE?

7.9.28   What are some significant properties of TPUEs for engineering mechanical and electrical designs?

7.9.29   What are some applications of TPUEs for mechanical and electrical engineering designs?

7.10.1   Describe the general deformation behavior of a thermoplastic plastic above and below its glass transition temperature.

7.10.2   What deformation mechanisms are involved during the elastic and plastic deformation of thermoplastics?

7.10.3   How does the average molecular mass of a thermoplastic affect its strength?

7.10.4   How does the amount of crystallinity within a thermoplastic material affect (*a*) its strength, (*b*) its tensile modulus of elasticity, and (*c*) its density?

7.10.5   Explain why low-density polyethylene is weaker than high-density polyethylene.

7.10.6   Explain why bulky side groups strengthen thermoplastics.

7.10.7   Explain how highly polar atoms bonded to the main carbon chain strengthen thermoplastics. Give examples.

7.10.8   Explain how oxygen atoms covalently bonded in the main carbon chain strengthen thermoplastics. Give an example.

7.10.9   Explain how phenylene rings covalently bonded in the main carbon chain strengthen thermoplastics. Give an example.

7.10.10   Explain why thermosetting plastics have in general high strengths and low ductilities.

7.10.11   How does increasing the temperature of thermoplastics affect their strength? What changes in bonding structure occur as thermoplastics are heated?

7.10.12   Why don't cured thermoset plastics become viscous and flow at elevated temperatures?

7.11.1   How do increases in stress and temperature affect the creep resistance of thermoplastics?

7.11.2   What is viscoelastic behavior of plastic materials?

7.11.3   Define the creep modulus of a plastic material.

7.11.4   How can the creep modulus of a thermoplastic be increased?

7.11.5   How can the extra energy required to fracture glassy thermoplastics compared to inorganic glasses be explained?

7.11.6   What is a craze in a glassy thermoplastic?

7.11.7   Describe the structure of a craze in a thermoplastic.

7.11.8   Describe the molecular structure changes that occur during the ductile fracturing of a thermoplastic.

7.12.1   Which of the following plastic materials would you select for a self-lubricating gear that would be exposed to some moisture at room temperature: (*a*) polyethylene; (*b*) polypropylene; (*c*) ABS; (*d*) nylon 6,6; (*e*) polyacetal?

7.12.2   Which of the following plastic materials would you select for a self-lubricating gear that is not exposed to moisture at room temperature: (*a*) nylon 6,6; (*b*) ABS; (*c*) polycarbonate; (*d*) polypropylene; (*e*) polyethylene?

7.12.3   Which of the following plastic materials would you select for the housing of an inexpensive computer: (*a*) polyethylene; (*b*) polypropylene; (*c*) ABS; (*d*) modified polyphenylene oxide; (*e*) polysulfone?

7.12.4   Which of the following plastic materials would you select for the housing of a more expensive computer: (*a*) polypropylene; (*b*) ABS; (*c*) modified polyphenylene oxide; (*d*) polysulfone; (*e*) polyphenylene sulfide?

7.12.5   Which of the following plastic materials would you select for electrical connectors operated at room temperature: (*a*) polyvinyl chloride; (*b*) polyethylene; (*c*) polysulfone; (*d*) polybutylene terephthalate; (*e*) ABS?

7.12.6   Which of the following plastic materials would you select for electrical connectors operated at slightly elevated temperatures: (*a*) polyethylene; (*b*) polysulfone; (*c*) polyethylene terephthalate; (*d*) polyvinyl chloride; (*e*) polyphenylene sulfide?

7.12.7   Which of the following plastic materials would you select for electrical connectors operated at elevated temperatures and in a slightly corrosive environment: (*a*) polypropylene; (*b*) polycarbonate; (*c*) polyphenylene sulfide; (*d*) polyvinyl chloride; (*e*) polyacetal?

7.12.8   Which of the following plastic materials would you select for a high-impact resistant transparent glazing application: (*a*) polyethylene; (*b*) polypropylene; (*c*) polymethyl methacrylate; (*d*) polycarbonate; (*e*) polyacetal?

Phase Diagrams

A *phase* in a material in terms of its microstructure is a region that differs in structure and/or composition from another region. *Phase diagrams* are graphical representations of what phases are present in a materials system at various temperatures, pressures, and compositions. Most phase diagrams are constructed by using equilibrium[1] conditions and are used by engineers and scientists to understand and predict many aspects of the behavior of materials. Some of the important information obtainable from phase diagrams is:

1. To show what phases are present at different compositions and temperatures under slow cooling (equilibrium) conditions
2. To indicate the equilibrium solid solubility of one element (or compound) in another
3. To indicate the temperature at which an alloy cooled under equilibrium conditions starts to solidify and the temperature range over which solidification occurs
4. To indicate the temperature at which different phases start to melt

---

[1]Equilibrium phase diagrams are determined by using slow cooling conditions. In most cases equilibrium is approached but never fully attained.

8.1 PHASE DIAGRAMS OF PURE SUBSTANCES

A pure substance such as water can exist in solid, liquid, or vapor phases, depending on the conditions of temperature and pressure. An example familiar to everyone of two phases of a pure substance in equilibrium is a glass of water containing ice cubes. In this case solid and liquid water are two separate and distinct phases that are separated by a phase boundary, the surface of the ice cubes. During the boiling of water, liquid water and water vapor are two phases in equilibrium. A graphical representation of the phases of water which exist under different conditions of temperature and pressure is shown in Fig. 8.1.

In the pressure-temperature (*PT*) phase diagram of water there exists a *triple point* at low pressure (4.579 torr) and low temperature (0.0098°C) where solid, liquid, and vapor phases of water coexist. Liquid and vapor phases exist along the vaporization line and liquid and solid phases along the freezing line, as shown in Fig. 8.1. These lines are two-phase equilibrium lines.

Pressure-temperature equilibrium phase diagrams also can be constructed for other pure substances. For example, the equilibrium *PT* phase diagram for pure iron is shown in Fig. 8.2. One major difference with this phase diagram is that there are three separate and distinct *solid phases:* alpha (α) Fe, gamma

FIGURE 8.1 Approximate pressure-temperature (*PT*) equilibrium phase diagram for pure water. (The axes of the diagram are distorted to some extent.)

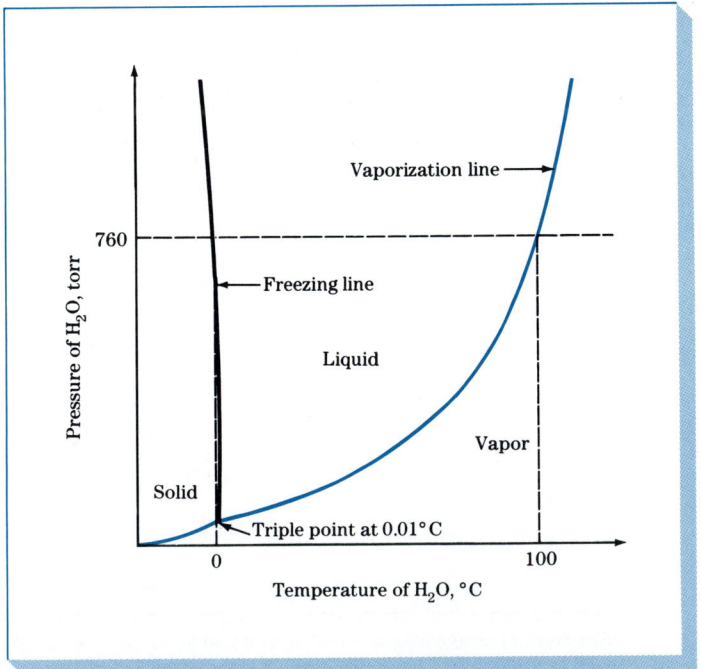

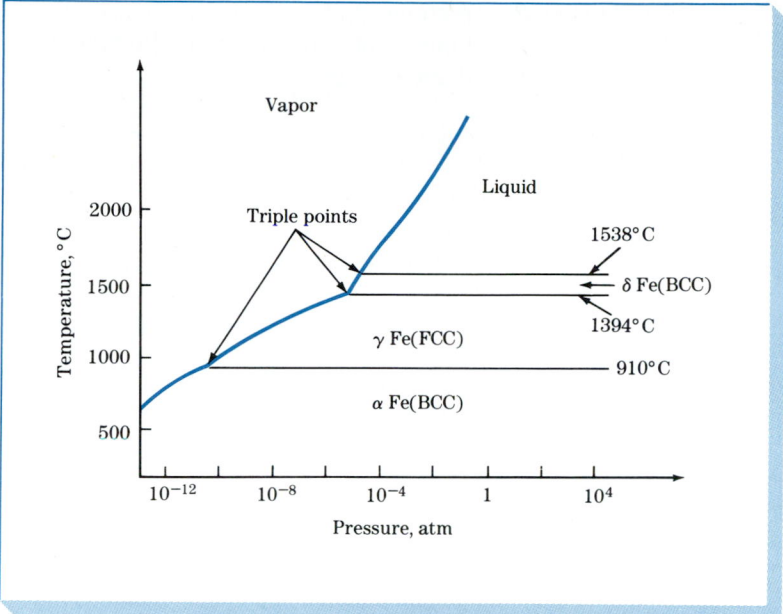

FIGURE 8.2 Approximate pressure-temperature (*PT*) equilibrium phase diagram for pure iron. (*After W. G. Moffatt et al., "Structure and Properties of Materials," vol. I, Wiley, 1964, p. 151.*)

(γ) Fe, and delta (δ) Fe. Alpha and delta iron have BCC crystal structures, whereas gamma iron has an FCC structure. The phase boundaries in the solid state have the same properties as the liquid and solid phase boundaries. For example, under equilibrium conditions, alpha and gamma iron can exist at a temperature of 910°C and 1 atm pressure. Above 910°C only single-phase gamma exists, and below 910°C only single-phase alpha exists (Fig. 8.2). There are also three triple points in the iron *PT* diagram where three different phases coexist: (1) liquid, vapor, and δFe, (2) vapor, δFe, and γFe, and (3) vapor, γFe, and αFe.

8.2 GIBBS PHASE RULE

From thermodynamic considerations, J. W. Gibbs[1] derived an equation that enables the number of phases that can coexist in equilibrium in a chosen system to be computed. This equation, called *Gibbs phase rule,* is

$$P + F = C + 2 \tag{8.1}$$

[1] Josiah Willard Gibbs (1839–1903). American physicist. He was a professor of mathematical physics at Yale University and made great contributions to the science of thermodynamics, which included the statement of the phase rule for multiphase systems.

where P = number of phases which coexist in a chosen system
C = number of components in the system
F = degrees of freedom

Usually a component C is an element, compound, or solution in the system. F, the degrees of freedom, is the number of variables (pressure, temperature, and composition) which can be changed independently without changing the state of the phase or phases in equilibrium in the chosen system.

Let us consider the application of Gibbs phase rule to the PT phase diagram of pure water (Fig. 8.1). At the triple point three phases coexist in equilibrium, and since there is one component in the system (water), the number of degrees of freedom can be calculated:

$$P + F = C + 2 \tag{8.1}$$

$$3 + F = 1 + 2$$

or $\qquad F = 0 \qquad$ (zero degrees of freedom)

Since none of the variables (temperature or pressure) can be changed and still keep the three phases of coexistence, the triple point is called an *invariant point.*

Consider next a point along the liquid-solid freezing curve of Fig. 8.1. At any point along this line there will be two phases coexisting. Thus, from the phase rule,

$$2 + F = 1 + 2$$

or $\qquad F = 1 \qquad$ (one degree of freedom)

This result tells us that there is one degree of freedom, and thus one variable (T or P) can be changed independently and still maintain a system with two coexisting phases. Thus if a particular pressure is specified, there is only one temperature at which both liquid and solid phases can coexist. For a third case, consider a point on the water PT phase diagram inside a single phase. Then there will be only one phase present ($P = 1$), and substituting into the phase-rule equation gives

$$1 + F = 1 + 2$$

or $\qquad F = 2 \qquad$ (two degrees of freedom)

This result tells us that two variables (temperature and pressure) can be varied independently and the system will still remain a single phase.

Most binary phase diagrams used in materials science are temperature composition diagrams in which pressure is kept constant, usually at 1 atm. In this case, we have the condensed phase rule, which is given by

$$P + F = C + 1 \qquad\qquad (8.1a)$$

Equation (8.1a) will apply to all subsequent binary phase diagrams discussed in this chapter.

8.3 BINARY ISOMORPHOUS ALLOY SYSTEMS

Let us now consider a mixture or alloy of two metals instead of pure substances. A mixture of two metals is called a *binary alloy* and constitutes a two-*component* system, since each metallic element in an alloy is considered a separate component. Thus pure copper is a one-component system, whereas an alloy of copper and nickel is a two-component system. Sometimes a compound in an alloy is also considered a separate component. For example, plain-carbon steels containing mainly iron and iron carbide are considered two-component systems.

In some binary metallic systems, the two elements are completely soluble in each other in both the liquid and solid states. In these systems only a single type of crystal structure exists for all compositions of the components, and therefore they are called *isomorphous systems*. In order for the two elements to have complete solid solubility in each other, they usually satisfy one or more of the following conditions formulated by Hume-Rothery[1] and known as the Hume-Rothery solid solubility rules:

1. The crystal structure of each element of the solid solution must be the same.
2. The size of the atoms of each of the two elements must not differ by more than 15 percent.
3. The elements should not form compounds with each other. That is, there should be no appreciable difference in the electronegativities of the two elements.
4. The elements should have the same valence.

All the Hume-Rothery rules are not always applicable for all pairs of elements which show complete solid solubility.

An important example of an isomorphous binary alloy system is the copper-nickel system. A phase diagram of this system with temperature as the ordinate and chemical composition in weight percent as the abscissa is shown in Fig. 8.3. This diagram has been determined for slow cooling or equilibrium conditions at atmospheric pressure and does not apply to alloys that have been

---

[1] William Hume-Rothery (1899–1968). English metallurgist who made major contributions to theoretical and experimental metallurgy and who spent years studying alloy behavior. His empirical rules for solid solubility in alloys were based on his alloy design work.

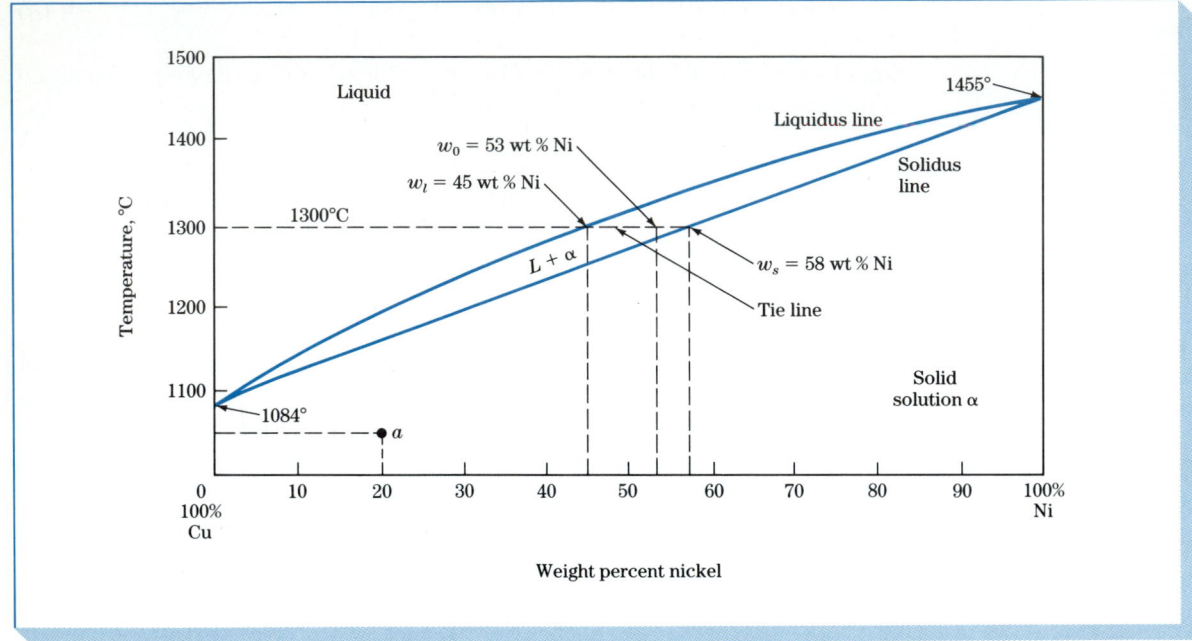

FIGURE 8.3 The copper-nickel phase diagram. Copper and nickel have complete liquid solubility and complete solid solubility. Copper-nickel solid solutions melt over a range of temperatures rather than at a fixed temperature, as is the case for pure metals. (*Adapted from "Metals Handbook," vol. 8, 8th ed., American Society for Metals, 1973, p. 294.*)

rapidly cooled through the solidification temperature range. The area above the upper line in the diagram, called the *liquidus*, corresponds to the region of stability of the liquid phase, and the area below the lower line, or *solidus*, represents the region of stability for the solid phase. The region between the liquidus and solidus represents a two-phase region where both the liquid and solid phases coexist.

In the single-phase region of solid solution α, both the temperature and the composition of the alloy must be specified in order to locate a point on the phase diagram. For example, the temperature 1050°C and 20% Ni specify the point a on the Cu–Ni phase diagram of Fig. 8.3. The microstructure of solid solution α at this temperature and composition appears the same as that of a pure metal. That is, the only observable feature in the optical microscope will be grain boundaries. However, because the alloy is a solid solution of 20% Ni in copper, the alloy will have higher strength and electrical resistivity than pure copper.

In the region between the liquidus and solidus lines, both liquid and solid phases exist. The amount of each phase present depends on the temperature and chemical composition of the alloy. Let us consider an alloy of 53 wt % Ni–47 wt % Cu at 1300°C in Fig. 8.3. Since this alloy contains both liquid and solid phases at 1300°C, neither of these phases can have the average composition of 53% Ni–47% Cu. The compositions of the liquid and solid phases at 1300°C can be determined by drawing a horizontal *tie line* at 1300°C from the liquidus line to the solidus line and then dropping vertical lines to the horizontal composition axis. The composition of the liquid phase (w_l) at 1300°C is 45 wt % Ni and that of the solid phase (w_s) is 58 wt % Ni, as indicated by the intersection of the dashed vertical lines with the composition axis.

   Binary equilibrium phase diagrams for components which are completely soluble in each other in the solid state can be constructed from a series of liquid-solid cooling curves, as shown for the Cu–Ni system in Fig. 8.4. The

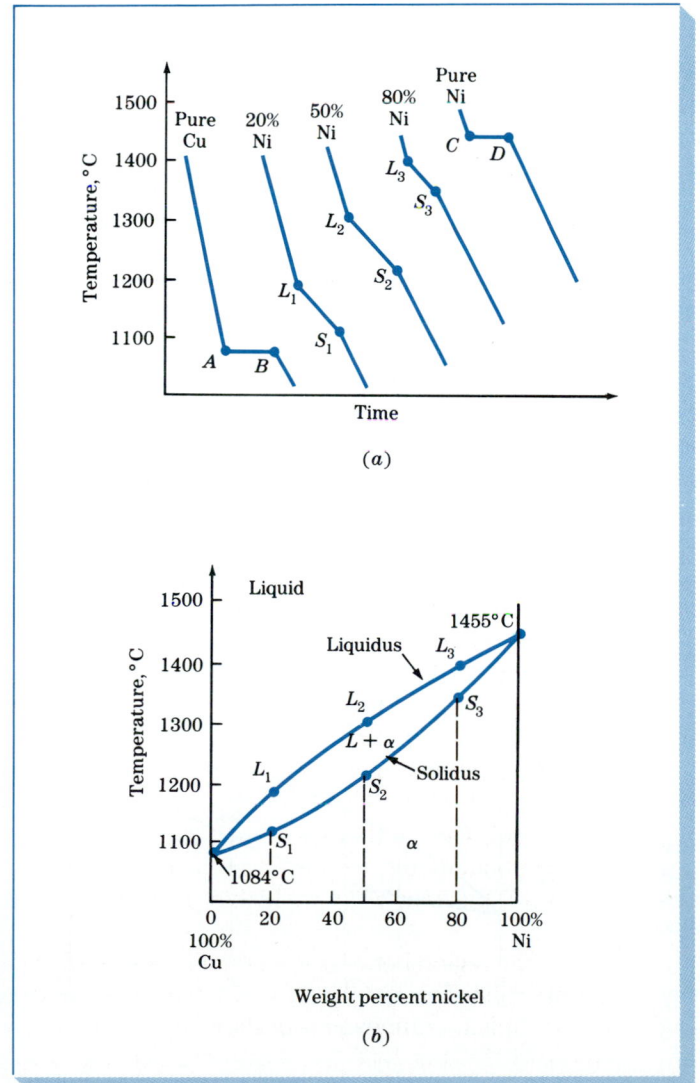

FIGURE 8.4 Construction of the Cu–Ni equilibrium phase diagram from liquid-solid cooling curves. (*a*) Cooling curves, (*b*) equilibrium phase diagram.

cooling curves for pure metals show horizontal thermal arrests at their freezing points, as shown for pure copper and nickel in Fig. 8.4a at AB and CD. Binary solid solutions exhibit slope changes in their cooling curves at the liquidus and solidus lines, as shown in Fig. 8.4a at compositions of 80% Cu–20% Ni, 50% Cu–50% Ni, and 20% Cu–80% Ni. The slope changes at L_1, L_2, and L_3 in Fig. 8.4a correspond to the liquidus points L_1, L_2, and L_3 of Fig. 8.4b. Similarly, the slope changes of S_1, S_2, and S_3 of Fig. 8.4a correspond to the points S_1, S_2, and S_3 on the solidus line of Fig. 8.4b. Further accuracy in the construction of the Cu–Ni phase diagram can be attained by determining more cooling curves at intermediate alloy compositions.

8.4 THE LEVER RULE

The weight percentages of the phases in any two-phase region of a binary equilibrium phase diagram can be calculated by using the lever rule. For example, by using the lever rule the weight percent liquid and weight percent solid for any particular temperature can be calculated for any average alloy composition in the two-phase liquid-plus-solid region of the binary copper-nickel phase diagram of Fig. 8.3.

To derive the lever-rule equations let us consider the binary equilibrium phase diagram of two elements A and B which are completely soluble in each other, as shown in Fig. 8.5. Let x be the alloy composition of interest and its weight fraction of B in A be w_0. Let T be the temperature of interest and let us construct a tie line at temperature T from the liquidus line to the solidus line

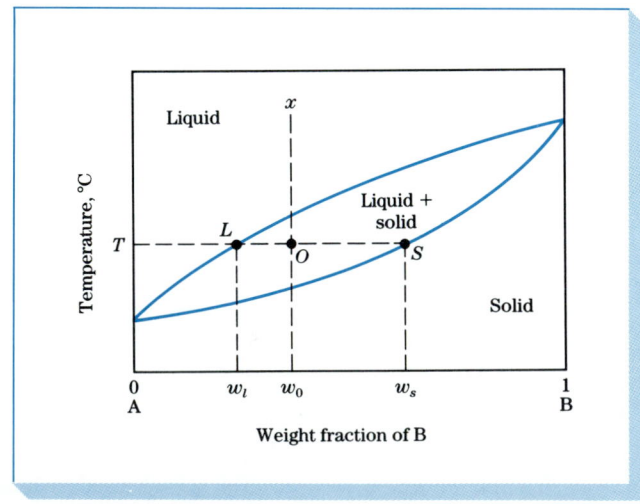

FIGURE 8.5 Binary phase diagram of two metals A and B completely soluble in each other being used to derive the lever-rule equations. At temperature T, the composition of the liquid phase is w_l and that of the solid is w_s.

(line *LS*). At temperature *T*, the alloy *x* consists of a mixture of liquid of w_l weight fraction of B and solid of w_s weight fraction of B.

The lever-rule equations can be derived by using weight balances. One equation for the derivation of the lever-rule equations is obtained from the fact that the sum of the weight fraction of the liquid phase, X_l, and the weight fraction of the solid phase, X_s, must equal 1. Thus

$$X_l + X_s = 1 \tag{8.2}$$

or
$$X_l = 1 - X_s \tag{8.2a}$$

and
$$X_s = 1 - X_l \tag{8.2b}$$

A second equation for the derivation of the lever rule can be obtained by a weight balance of B in the alloy as a whole and the sum of B in the two separate phases. Let us consider 1 g of the alloy and make this weight balance:

Grams of B in two-phase mixture	=	grams of B in liquid phase	+	grams of B in solid phase

Grams of two-phase mixture   Grams of liquid phase   Grams of solid

$$\overbrace{(1 \text{ g})(1)} \left(\frac{\%w_0}{100} \right) = \overbrace{(1 \text{ g})(X_l)} \left(\frac{\%w_l}{100} \right) + \overbrace{(1 \text{ g})(X_s)} \left(\frac{\%w_s}{100} \right) \tag{8.3}$$

Wt fraction of phase mixture   Wt fraction of liquid phase   Wt fraction of solid phase

Average wt fraction of B in phase mixture   Wt fraction of B in liquid phase   Wt fraction of B in solid phase

Thus
$$w_0 = X_l w_l + X_s w_s \tag{8.4}$$

combined with
$$X_l = 1 - X_s \tag{8.2a}$$

gives
$$w_0 = (1 - X_s)w_l + X_s w_s$$

or
$$w_0 = w_l - X_s w_l + X_s w_s$$

Rearranging,
$$X_s w_s - X_s w_l = w_0 - w_l$$

$$\boxed{\text{Wt fraction of solid phase} = X_s = \frac{w_0 - w_l}{w_s - w_l}} \tag{8.5}$$

Similarly,
$$w_0 = X_l w_l + X_s w_s \tag{8.4}$$

combined with
$$X_s = 1 - X_l \tag{8.2b}$$

gives

$$\text{Wt fraction of liquid phase} = X_l = \frac{w_s - w_0}{w_s - w_l} \qquad (8.6)$$

Equations (8.5) and (8.6) are the lever-rule equations. Effectively, the lever-rule equations state that to calculate the weight fraction of one phase of a two-phase mixture, one must use the segment of the tie line which is on the opposite side of the alloy of interest and which is farthest away from the phase for which the weight fraction is being calculated. The ratio of this line segment of the tie line to the total tie line provides the weight fraction of the phase being determined. Thus in Fig. 8.5 the weight fraction of the liquid phase is the ratio *OS/LS* and the weight fraction of the solid phase is the ratio *LO/LS*.

Weight fractions can be converted to weight percentages by multiplying by 100 percent. Example Problem 8.1 shows how the lever rule can be used to determine the weight percentage of a phase in a binary alloy at a particular temperature.

Example Problem 8.1

A copper-nickel alloy contains 47 wt % Cu and 53 wt % Ni and is at 1300°C. Use Fig. 8.3 and answer the following:

(*a*) What is the weight percent of copper in the liquid and solid phases at this temperature?
(*b*) What weight percent of this alloy is liquid and what weight percent is solid?

Solution:

(*a*) From Fig. 8.3 at 1300°C, the intersection of the 1300°C tie line with the liquidus gives 55 wt % Cu in the liquid phase and the intersection of the solidus of the 1300°C tie line gives 42 wt % Cu in the solid phase.
(*b*) From Fig. 8.3 and using the lever rule on the 1300°C tie line,

$$w_0 = 53\% \text{ Ni} \qquad w_l = 45\% \text{ Ni} \qquad w_s = 58\% \text{ Ni}$$

$$\text{Wt fraction of liquid phase} = X_l = \frac{w_s - w_0}{w_s - w_l}$$

$$= \frac{58 - 53}{58 - 45} = \frac{5}{13} = 0.38$$

$$\text{Wt \% of liquid phase} = (0.38)(100\%) = 38\% \blacktriangleleft$$

$$\text{Wt fraction of solid phase} = X_s = \frac{w_0 - w_l}{w_s - w_l}$$

$$= \frac{53 - 45}{58 - 45} = \frac{8}{13} = 0.62$$

$$\text{Wt \% of solid phase} = (0.62)(100\%) = 62\% \blacktriangleleft$$

8.5   NONEQUILIBRIUM SOLIDIFICATION OF ALLOYS

The phase diagram for the Cu–Ni system just previously referred to was constructed by using very slow cooling conditions approaching equilibrium. That is, when the Cu–Ni alloys were cooled through the two-phase liquid + solid regions, the compositions of the liquid and solid phases had to readjust continuously by solid-state diffusion as the temperature was lowered. Since atomic diffusion is very slow in the solid state, an extensive length of time is required to eliminate concentration gradients. Thus the as-cast microstructures of slowly solidified alloys usually have a *cored structure* (Fig. 8.6) caused by regions of different chemical composition.

The copper-nickel alloy system provides a good example to describe how such a cored structure originates. Consider an alloy of 70% Ni–30% Cu which is cooled from a temperature T_0 at a rapid rate (Fig. 8.7). The first solid forms at temperature T_1 and has the composition α_1 (Fig. 8.7). Upon further rapid cooling to T_2, additional layers of composition α_2 will form without much change in the composition of the solid primarily solidified. The overall composition at T_2 lies somewhere between α_1 and α_2 and will be designated α_2'. Since the tie line $\alpha_2'L_2$ is longer than α_2L_2, there will be more liquid and less solid in the rapidly cooled alloy than if it were cooled under equilibrium conditions to the same temperature. Thus solidification has been delayed at that temperature by the rapid cooling.

As the temperature is lowered to T_3 and T_4, the same processes occur and the average composition of the alloy follows the *nonequilibrium solidus* $\alpha_1\alpha_2'\alpha_3' \cdots$. At T_6 the solid freezing has less copper than the original composition of the alloy which is 30% Cu. At temperature T_7 the average compo-

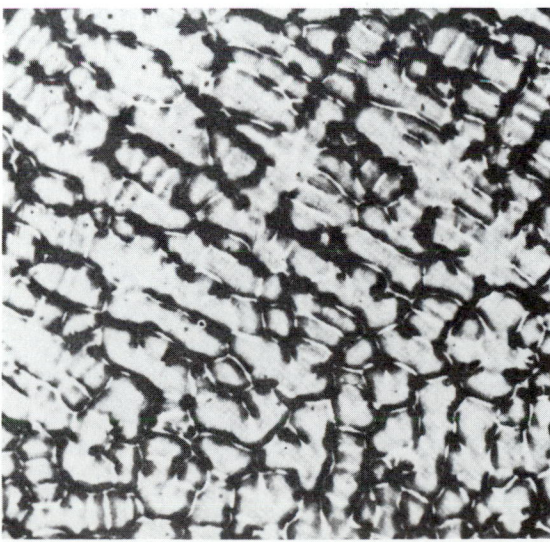

FIGURE 8.6 The microstructure of an as-cast 70% Cu–30% Ni alloy showing a cored structure. (*After W. G. Moffatt et al., "Structure and Properties of Materials," vol. I, Wiley, 1964, p. 177.*)

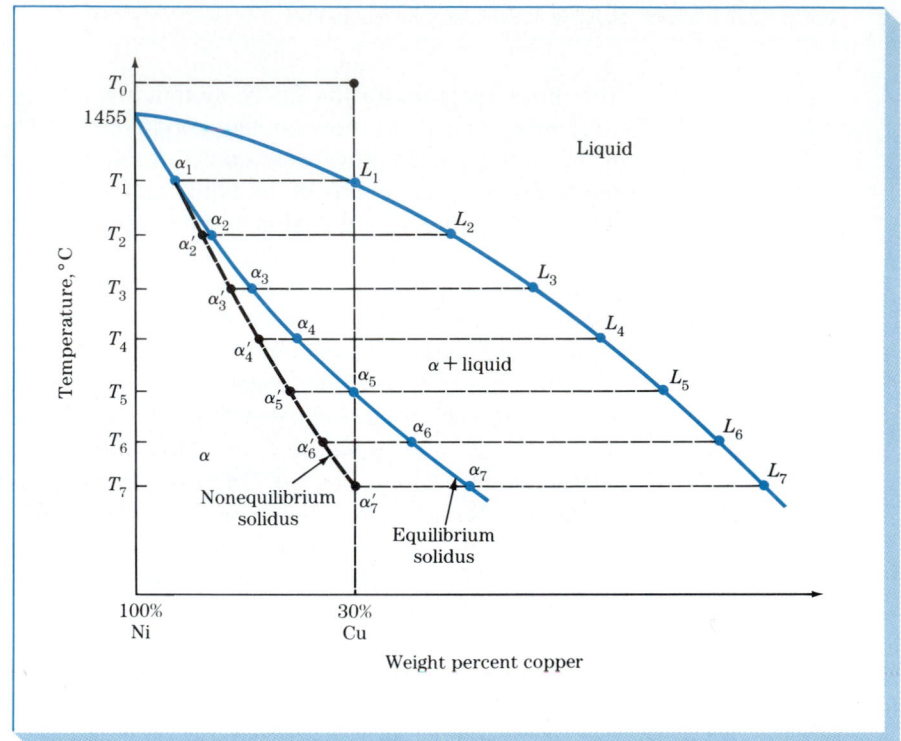

FIGURE 8.7
Nonequilibrium solidification of a 70% Ni–30% Cu alloy. This phase diagram has been distorted for illustrative purposes. Note the nonequilibrium solidus α_1 to α_7'. The alloy is not completely solidified until the nonequilibrium solidus reaches α_7' at T_7.

sition of the alloy is 30% Cu and freezing is complete. Regions in the micro-structure of the alloy will thus consist of compositions varying from α_1 to α_7' as the cored structure forms during solidification (Fig. 8.8). Figure 8.6 shows a cored microstructure of rapidly solidified 70% Cu–30% Ni alloy.

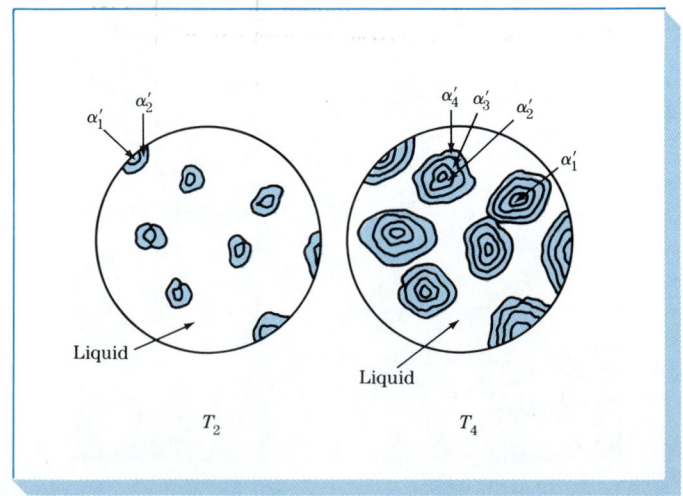

FIGURE 8.8 Schematic microstructures at temperature T_2 and T_4 of Fig. 8.7 for the nonequilibrium solidification of a 70% Ni–30% Cu alloy illustrating the development of a cored structure.

FIGURE 8.9 Large direct-chill cast aluminun alloy sheet ingots being loaded into a homogenizing furnace. By reheating the as-cast ingots at a high temperature below the melting temperature of the lowest melting phase, atomic solid-state diffusion creates a more homogeneous internal structure. (*Courtesy of Reynolds Metals Co.*)

Homogenizing heat treatment   Most as-cast microstructures are cored to some extent and thus have composition gradients. In many cases this structure is undesirable particularly if the alloy is to be subsequently worked. To eliminate the cored structure, as-cast ingots or castings are heated to elevated temperatures to accelerate solid-state diffusion. This process is called *homogenization* since it produces a homogeneous structure in the alloy. Figure 8.9 shows aluminum alloy sheet ingots being loaded into a homogenizing furnace. The homogenizing heat treatment must be carried out at a temperature which is lower than the lowest melting solid in the as-cast alloy or else melting will occur. For homogenizing the 70% Ni–30% Cu alloy just discussed, a temperature just below T_7 indicated in Fig. 8.7 should be used. If the alloy is overheated, localized melting or *liquation* may take place. If the liquid phase forms a continuous film along the grain boundaries, the alloy will lose strength and may break up during subsequent working. Figure 8.10 shows liquation in the microstructure of a 70% Ni–30% Cu alloy.

FIGURE 8.10 *Liquation* in a 70% Ni–30% Cu alloy. By heating only slightly above the solidus temperature so that melting just begins, a liquated structure such as shown in (*a*) is produced. In (*b*) the grain-boundary region was slightly melted, and then upon subsequent freezing, the melted zone became copper-rich and caused the grain boundaries to appear as broad dark lines. (*Courtesy of F. Rhines.*)

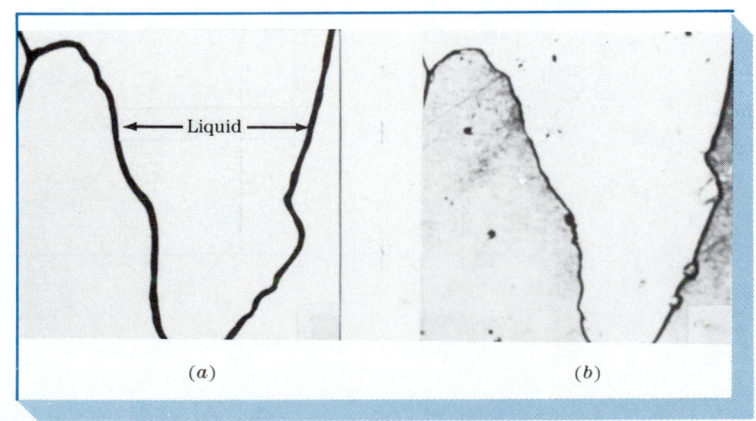

Liquid

(*a*)   (*b*)

8.6 BINARY EUTECTIC ALLOY SYSTEMS

Many binary alloy systems have components which have limited solid solubility in each other as, for example, in the lead-tin system (Fig. 8.11). The regions of restricted solid solubility at each end of the Pb–Sn diagram are designated as alpha and beta phases and are called *terminal solid solutions* since they appear at the ends of the diagram. The alpha phase is a lead-rich solid solution and can dissolve in solid solution a maximum of 19.2 wt % Sn at 183°C. The beta phase is a tin-rich solid solution and can dissolve a maximum of 2.5 wt % Pb at 183°C. As the temperature is decreased below 183°C, the maximum solid solubility of the solute elements decreases according to the *solvus* lines of the Pb–Sn phase diagram.

FIGURE 8.11 The lead-tin equilibrium phase diagram. This diagram is characterized by the limited solid solubility of each terminal phase (α and β). The eutectic invariant reaction at 61.9% Sn and 183°C is the most important feature of this system. At the eutectic point, α (19.2% Sn), β (97.5% Sn), and liquid (61.9% Sn) can coexist.

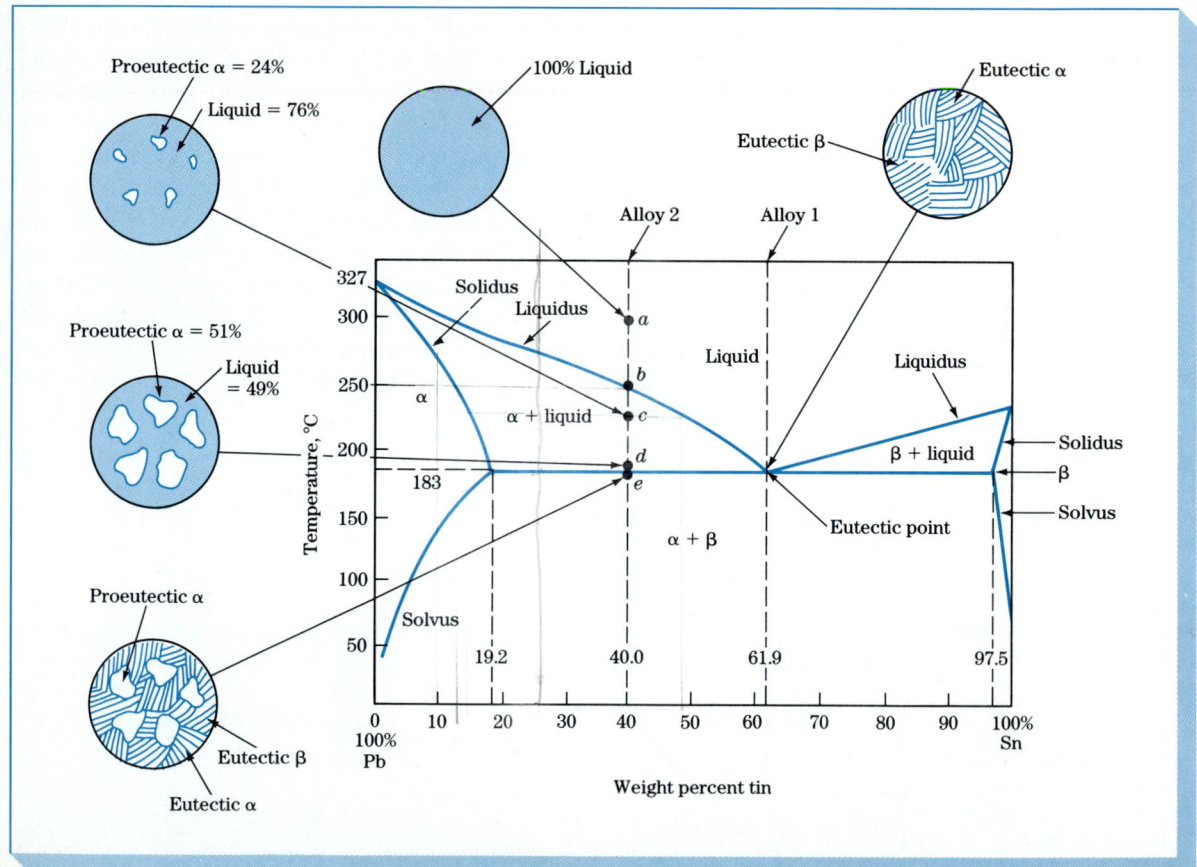

In simple binary eutectic systems like the Pb–Sn one, there is a specific alloy composition known as the *eutectic composition* which freezes at a lower temperature than all other compositions. This low temperature which corresponds to the lowest temperature at which the liquid phase can exist when cooled slowly is called the *eutectic temperature.* In the Pb–Sn system the eutectic composition (61.9% Sn and 38.1% Pb) and the eutectic temperature (183°C) determine a point on the phase diagram called the *eutectic point.* When liquid of eutectic composition is slowly cooled to the eutectic temperature, the single liquid phase transforms simultaneously into two solid forms (solid solutions α and β). This transformation is known as the *eutectic reaction* and is written as

$$\text{Liquid} \xrightarrow[\text{cooling}]{\substack{\text{eutectic} \\ \text{temperature}}} \alpha \text{ solid solution} + \beta \text{ solid solution} \qquad (8.7)$$

The eutectic reaction is called an *invariant reaction* since it occurs under equilibrium conditions at a specific temperature and alloy composition which cannot be varied. During the progress of the eutectic reaction the liquid phase is in equilibrium with the two solid solutions α and β, and thus during a eutectic reaction, three phases coexist and are in equilibrium. Since three phases in a binary phase diagram can only be in equilibrium at one temperature, a horizontal thermal arrest appears at the eutectic temperature in the cooling curve of an alloy of eutectic composition.

Slow cooling of a Pb–Sn alloy of eutectic composition   Consider the slow cooling of a Pb–Sn alloy (alloy 1 of Fig. 8.11) of eutectic composition (61.9% Sn) from 200°C to room temperature. During the cooling period from 200 to 183°C, the alloy remains liquid. At 183°C, which is the eutectic temperature, all the liquid solidifies by the eutectic reaction and forms a eutectic mixture of solid solutions α (19.2% Sn) and β (97.5% Sn) according to the reaction

$$\text{Liquid (61.9\% Sn)} \xrightarrow[\text{cooling}]{183°C} \alpha \text{ (19.2\% Sn)} + \beta \text{ (97.5\% Sn)} \qquad (8.8)$$

After the eutectic reaction has been completed, upon cooling the alloy from 183°C to room temperature, there is a decrease in solid solubility of solute in the α and β solid solutions, as indicated by the solvus lines. However, since diffusion is slow at the lower temperatures, this process does not normally reach equilibrium, and thus solid solutions α and β can still be distinguished at room temperature, as shown in the microstructure of Fig. 8.12*a*.

Slow cooling of a 60% Pb–40% Sn alloy   Next consider the slow cooling of a 40% Sn–60% Pb alloy (alloy 2 of Fig. 8.11) from the liquid state at 300°C to room temperature. As the temperature is lowered from 300°C (point *a*), the

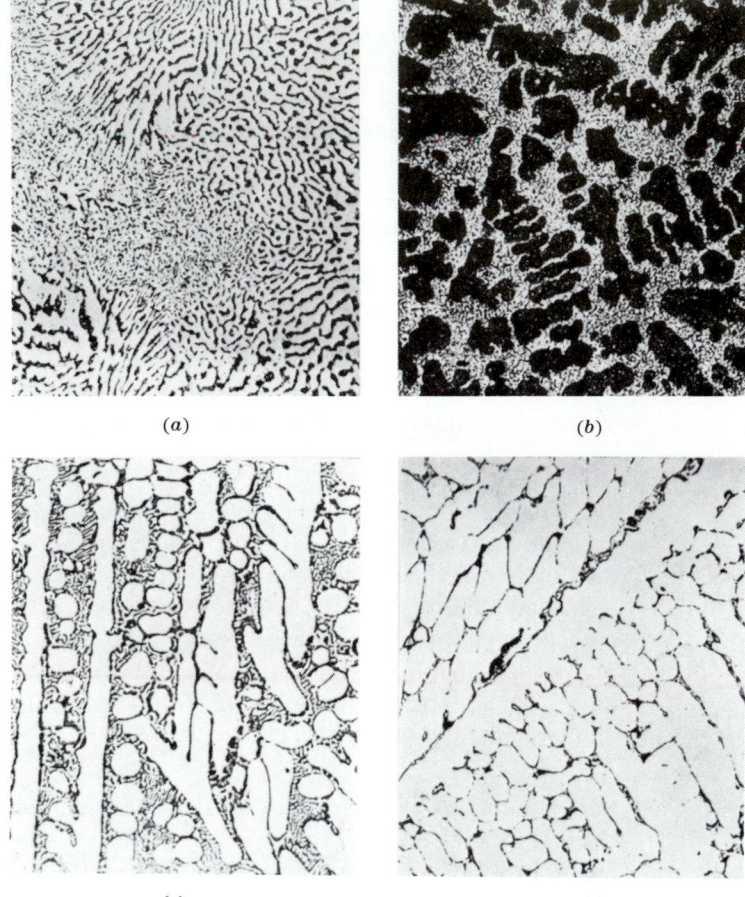

(a)

(b)

(c)

(d)

FIGURE 8.12
Microstructures of slowly cooled Pb–Sn alloys: (a) eutectic composition (63% Sn–37% Pb), (b) 40% Sn–60% Pb, (c) 70% Sn–30% Pb, (d) 90% Sn–10% Pb. (Magnification 75×.) (From J. Nutting and R. G. Baker, "Microstructure of Metals," Institute of Metals, London, 1965, p. 19.)

alloy will remain liquid until the liquidus line is intersected at point b at about 245°C. At this temperature solid solution α containing 12% Sn will begin to precipitate from the liquid. The first solid to form in this type of alloy is called *primary* or *proeutectic alpha.* The term proeutectic alpha is used to distinguish this constituent from the alpha that forms later by the eutectic reaction.

As the liquid cools from 245°C to slightly above 183°C through the two-phase liquid + alpha region of the phase diagram (points b to d), the composition of the solid phase (alpha) follows the solidus and varies from 12% Sn at 245°C to 19.2% Sn at 183°C. Likewise, the composition of the liquid phase varies from 40% Sn at 245°C to 61.9% Sn at 183°C. These composition changes are possible since the alloy is cooling very slowly and atomic diffusion occurs to equalize compositional gradients. At the eutectic temperature (183°C) all the remaining liquid solidifies by the eutectic reaction [Eq. (8.8)]. After the eutectic reaction is completed, the alloy consists of proeutectic alpha and a eutectic mixture of alpha (19.2% Sn) and beta (97.5% Sn). Further cooling

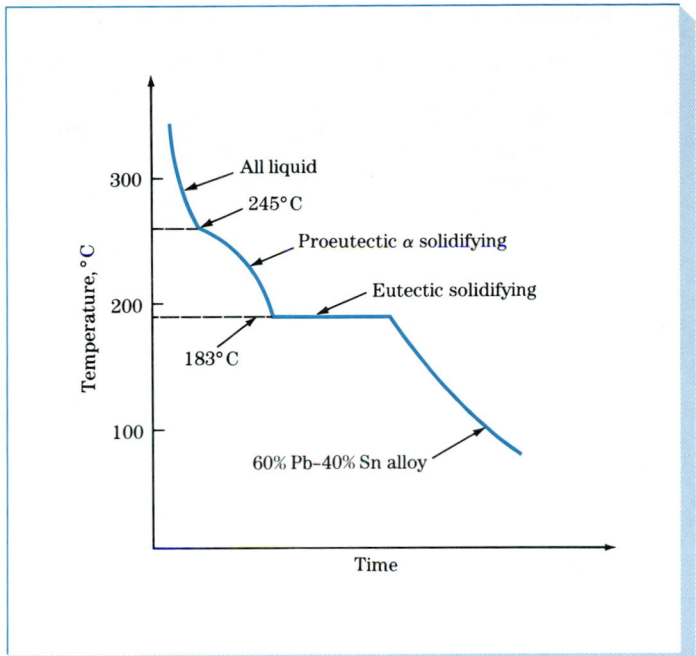

FIGURE 8.13 Schematic temperature-time cooling curve for a 60% Pb–40% Sn alloy.

below 183°C to room temperature lowers the tin content of the alpha phase and the lead content of the beta phase. However, at the lower temperatures the diffusion rate is much lower and equilibrium is not attained. Figure 8.12*b* shows the microstructure of a 40% Sn–60% Pb alloy that has been slowly cooled. Note the dark-etching dendrites of the lead-rich alpha phase surrounded by eutectic. Figure 8.13 shows a cooling curve for a 60% Pb–40% Sn alloy. Note that a slope change occurs at the liquidus at 245°C and a horizontal thermal arrest appears during the freezing of the eutectic.

Example Problem 8.2

Make phase analyses of the equilibrium (ideal) solidification of lead-tin alloys at the following points in the lead-tin phase diagram of Fig. 8.11:
(*a*) At the eutectic composition just below 183°C (eutectic temperature).
(*b*) The point *c* at 40% Sn and 230°C.
(*c*) The point *d* at 40% Sn and 183°C + ΔT.
(*d*) The point *e* at 40% Sn and 183°C − ΔT.

Solution:

(a) At the eutectic composition (61.9% Sn) just below 183°C:

	alpha	beta
Phases present:	alpha	beta
Compositions of phases:	19.2% Sn in alpha phase	97.5% Sn in beta phase
Amounts of phases:	Wt % alpha phase[1] $$= \frac{97.5 - 61.9}{97.5 - 19.2}(100\%)$$ $$= 45.5\%$$	Wt % beta phase[1] $$= \frac{61.9 - 19.2}{97.5 - 19.2}(100\%)$$ $$= 54.5\%$$

(b) The point c at 40% Sn and 230°C:

	liquid	alpha
Phases present:	liquid	alpha
Compositions of phases:	48% Sn in liquid phase	15% Sn in alpha phase
Amounts of phases:	Wt % liquid phase $$= \frac{40 - 15}{48 - 15}(100\%)$$ $$= 76\%$$	Wt % alpha phase $$= \frac{48 - 40}{48 - 15}(100\%)$$ $$= 24\%$$

(c) The point d at 40% Sn and 183°C + ΔT:

	liquid	alpha
Phases present:	liquid	alpha
Compositions of phases:	61.9% Sn in liquid phase	19.2% Sn in alpha phase
Amounts of phases:	Wt % liquid phase $$= \frac{40 - 19.2}{61.9 - 19.2}(100\%)$$ $$= 49\%$$	Wt % alpha phase $$= \frac{61.9 - 40}{61.9 - 19.2}(100\%)$$ $$= 51\%$$

(d) The point e at 40% Sn and 183°C − ΔT:

	alpha	beta
Phases present:	alpha	beta
Compositions of phases:	19.2% Sn in alpha phase	97.5% Sn in beta phase
Amounts of phases:	Wt % alpha phase $$= \frac{97.5 - 40}{97.5 - 19.2}(100\%)$$ $$= 73\%$$	Wt % beta phase $$= \frac{40 - 19.2}{97.5 - 19.2}(100\%)$$ $$= 27\%$$

---

[1] Note that in the lever-rule calculations one uses the ratio of the tie-line segment which is *farthest away* from the phase for which the weight percent is being determined to the whole tie line.

Example Problem 8.3

One kilogram of an alloy of 70% Pb and 30% Sn is slowly cooled from 300°C. Refer to the lead-tin phase diagram of Fig. 8.11 and calculate the following:

(a) The weight percent of the liquid and proeutectic alpha at 250°C.

(b) The weight percent of the liquid and proeutectic alpha just above the eutectic temperature (183°C) and the weight in kilograms of these phases.

(c) The weight in kilograms of alpha and beta formed by the eutectic reaction.

Solution:

(a) From Fig. 8.11 at 250°C,

$$\text{Wt \% liquid}^1 = \frac{30 - 12}{40 - 12}(100\%) = 64\% \blacktriangleleft$$

$$\text{Wt \% proeutectic } \alpha^1 = \frac{40 - 30}{40 - 12}(100\%) = 36\% \blacktriangleleft$$

(b) The weight percent liquid and proeutectic alpha just above the eutectic temperature, 183°C + ΔT is

$$\text{Wt \% liquid} = \frac{30 - 19.2}{61.9 - 19.2}(100\%) = 25.3\% \blacktriangleleft$$

$$\text{Wt \% proeutectic } \alpha = \frac{61.9 - 30.0}{61.9 - 19.2}(100\%) = 74.7\% \blacktriangleleft$$

$$\text{Weight of liquid phase} = 1 \text{ kg} \times 0.253 = 0.253 \text{ kg} \blacktriangleleft$$

$$\text{Weight of proeutectic } \alpha = 1 \text{ kg} \times 0.747 = 0.747 \text{ kg} \blacktriangleleft$$

(c) At 183°C − ΔT,

$$\text{Wt \% total } \alpha \text{ (proeutectic } \alpha + \text{ eutectic } \alpha) = \frac{97.5 - 30}{97.5 - 19.2}(100\%)$$

$$= 86.2\%$$

$$\text{Wt \% total } \beta \text{ (eutectic } \beta) = \frac{30 - 19.2}{97.5 - 19.2}(100\%)$$

$$= 13.8\%$$

---

[1] See note on page 436.

$$\text{Wt total } \alpha = 1 \text{ kg} \times 0.862 = 0.862 \text{ kg}$$

$$\text{Wt total } \beta = 1 \text{ kg} \times 0.138 = 0.138 \text{ kg}$$

The amount of proeutectic alpha will remain the same before and after the eutectic reaction. Thus

$$\text{Wt of } \alpha \text{ created by eutectic reaction} = \text{total } \alpha - \text{proeutectic } \alpha$$

$$= 0.862 \text{ kg} - 0.747 \text{ kg}$$

$$= 0.115 \text{ kg} \blacktriangleleft$$

$$\text{Wt of } \beta \text{ created by eutectic reaction} = \text{total } \beta$$

$$= 0.138 \text{ kg} \blacktriangleleft$$

Example Problem 8.4

A lead-tin (Pb–Sn) alloy contains 64 wt % proeutectic α and 36 wt % eutectic $\alpha + \beta$ at 183°C $-\ \Delta T$. Calculate the average composition of this alloy (see Fig. 8.11).

Solution:

Let x be the wt % Sn in the unknown alloy. Since this alloy contains 64 wt % proeutectic α, the alloy must be hypoeutectic, and x will therefore lie between 19.2 and 61.9 wt % Sn as indicated in Fig. EP8.4. At 183°C $+\ \Delta T$, using Fig. EP8.4 and the lever rule gives

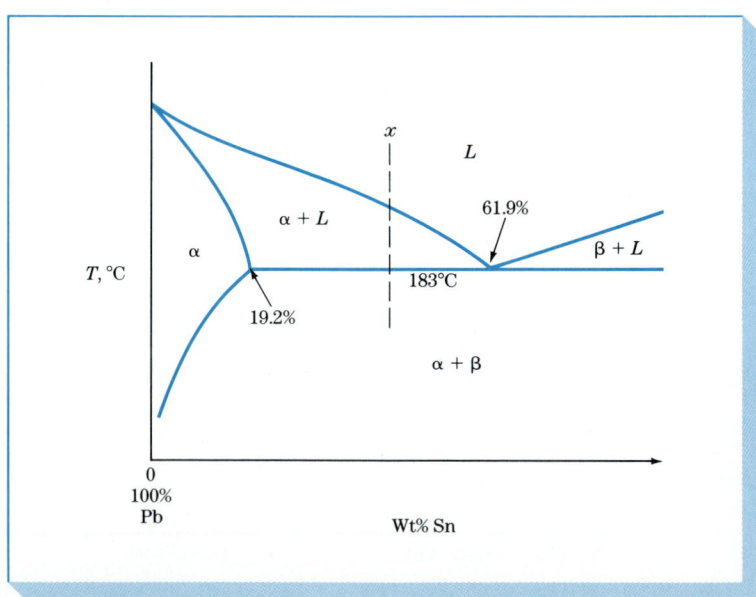

FIGURE EP8.4 Lead-rich end of the Pb–Sn phase diagram.

$$\% \text{ proeutectic } \alpha = \frac{61.9 - x}{61.9 - 19.2}(100\%) = 64\%$$

or
$$61.9 - x = 0.64(42.7) = 27.3$$

$$x = 34.6\%$$

Thus, the alloy consists of 34.6% Sn and 65.4% Pb. ◀ Note that we use the lever-rule calculation above the eutectic temperature since the percentage of the proeutectic α remains the same just above and just below the eutectic temperature.

In a binary eutectic reaction the two solid phases ($\alpha + \beta$) can have various morphologies. Figure 8.14 shows schematically some various eutectic structures. The shape that will be created depends on many factors. Of prime importance is a minimization of free energy at the α/β interfaces. An important factor that determines the eutectic shape is the manner in which the two phases (α and β) nucleate and grow. For example, rod- and plate-type eutectics form when repeated nucleation of the two phases is not required in certain direc-

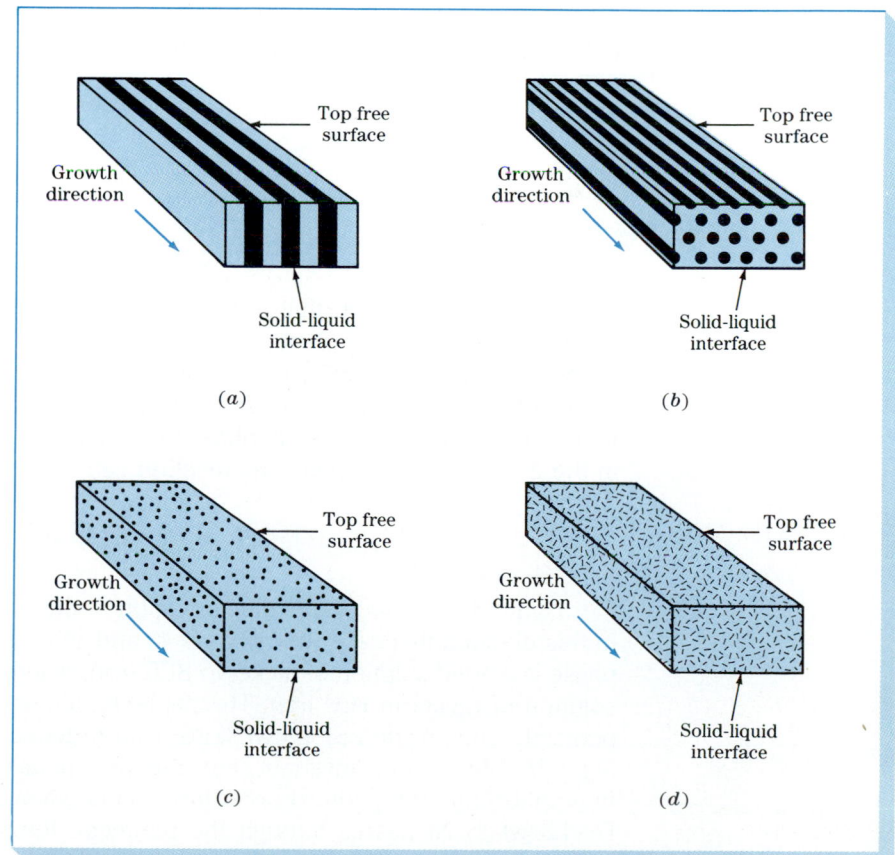

FIGURE 8.14 Schematic illustration of various eutectic structures: (a) lamellar, (b) rodlike, (c) globular, (d) acicular. (After W. C. Winegard, "An Introduction to the Solidification of Metals," Institute of Metals, London, 1964.)

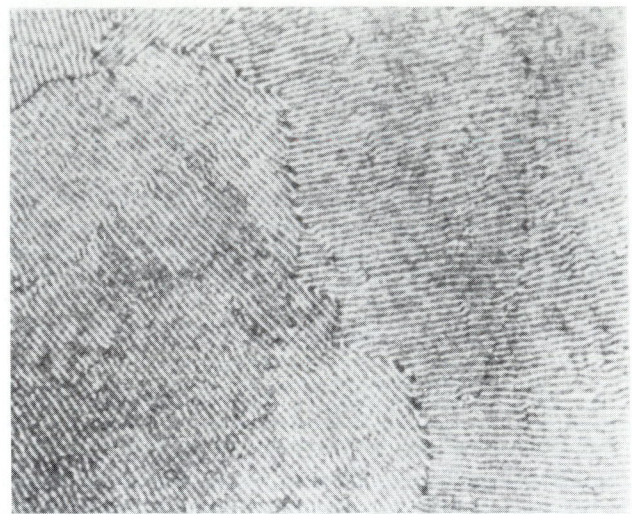

FIGURE 8.15 Lamellar eutectic structure formed by the Pb–Sn eutectic reaction. (Magnification 500×.) (*After W. G. Moffatt et al., "Structure and Properties of Materials," vol. I, Wiley, 1964.*)

tions. An example of a *lamellar eutectic structure* formed by a Pb–Sn eutectic reaction is shown in Fig. 8.15. Lamellar eutectic structures are very common. A mixed irregular eutectic structure found in the Pb–Sn system is shown in Fig. 8.12*a*.

8.7 BINARY PERITECTIC ALLOY SYSTEMS

Another type of reaction that frequently occurs in binary equilibrium phase diagrams is the *peritectic reaction.* This reaction is commonly present as part of more-complicated binary equilibrium diagrams, particularly if the melting points of the two components are quite different. In the peritectic reaction a liquid phase reacts with a solid phase to form a new and different solid phase. In the general form the peritectic reaction can be written as

$$\text{Liquid} + \alpha \xrightarrow[\text{cooling}]{} \beta \tag{8.9}$$

Figure 8.16 shows the peritectic region of the iron-nickel phase diagram. In this diagram there are solid phases (δ and γ) and one liquid phase. The δ phase is a solid solution of nickel in BCC iron, whereas the γ phase is a solid solution of nickel in FCC iron. The peritectic temperature of 1517°C and the peritectic composition of 4.3 wt % Ni in iron define the peritectic point c in Fig. 8.16. This point is invariant since the three phases δ, γ, and liquid coexist in equilibrium. The peritectic reaction occurs when a slowly cooled alloy of Fe–4.3 wt % Ni passes through the peritectic temperature of 1517°C. This reaction can be written as

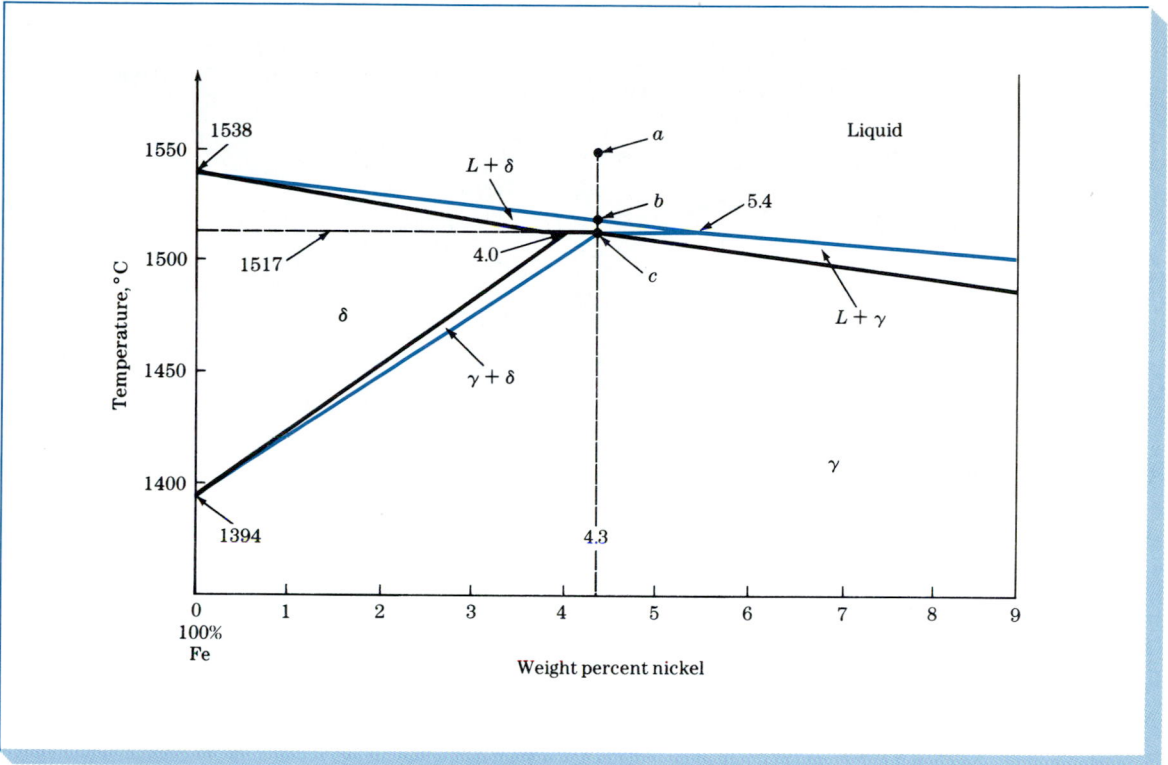

FIGURE 8.16 The peritectic region of the iron-nickel phase diagram. The peritectic point is located at 4.3% Ni and 1517°C, which is point c.

$$\text{Liquid (5.4 wt \% Ni)} + \delta \text{ (4.0 wt \% Ni)} \xrightarrow[\text{cooling}]{1517°C} \gamma \text{ (4.3 wt \% Ni)} \qquad (8.10)$$

To further understand the peritectic reaction, consider an alloy of Fe–4.3 wt % Ni (peritectic composition) which is slowly cooled from 1550°C to slightly under 1517°C (points *a* to *c* in Fig. 8.16). From 1550 to about 1525°C (points *a* to *b* in Fig. 8.16) the alloy cools as a homogeneous liquid of Fe–4.3% Ni. When the liquidus is intersected at about 1525°C (point *b*), solid δ begins to form. Further cooling to point *c* results in more and more solid δ being formed. At the peritectic temperature of 1517°C (point *c*) solid δ of 4.0% Ni and liquid of 5.4% Ni are in equilibrium, and at this temperature all the liquid reacts with all the δ solid phase to produce a new and different solid phase γ of 4.3% Ni. The alloy remains as single-phase γ solid solution until another phase change occurs at a lower temperature with which we are not concerned. The lever rule can be applied in the two-phase regions of the peritectic diagram in the same way as for the eutectic diagram.

If an alloy in the Fe–Ni system has less than 4.3% Ni and is slowly cooled from the liquid state through the liquid + δ region, there will be an excess of δ phase after the peritectic reaction is completed. Similarly, if an Fe–Ni alloy of more than 4.3% Ni but less than 5.4% Ni is slowly cooled from the liquid

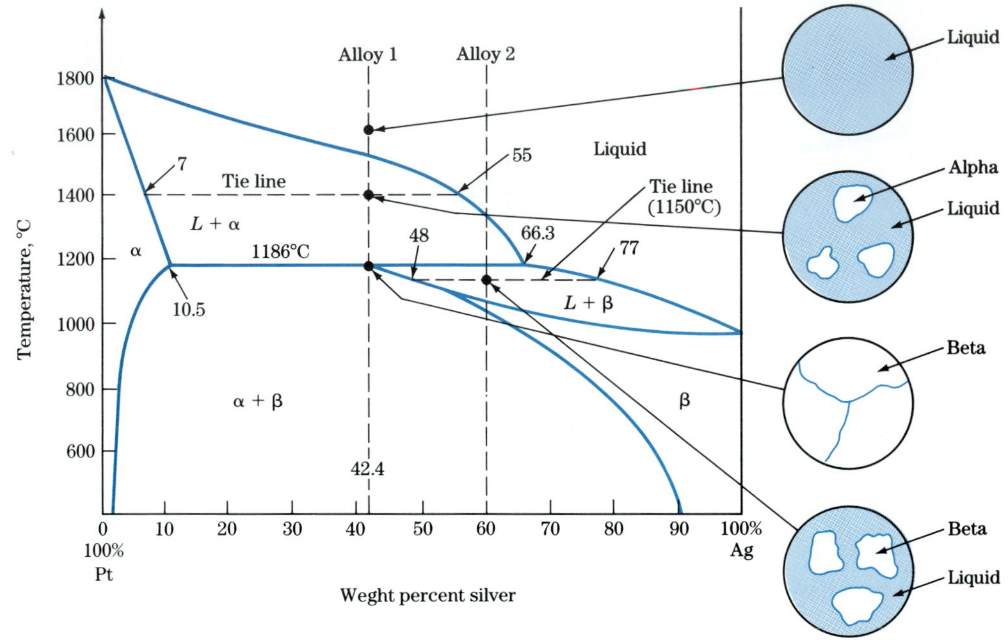

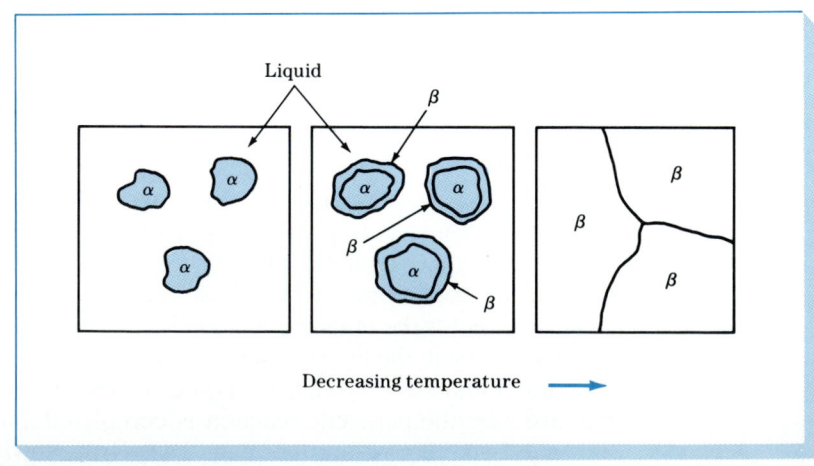

FIGURE 8.17 The platinum-silver phase diagram. The most important feature of this diagram is the peritectic invariant reaction at 42.4% Ag and 1186°C. At the peritectic point liquid (66.3% Ag), alpha (10.5% Ag), and beta (42.4% Ag) can coexist.

state through the δ + liquid region, there will be an excess of the liquid phase after the peritectic reaction is completed.

The silver-platinum binary equilibrium phase diagram is an excellent example of a system which has a single invariant peritectic reaction (Fig. 8.17). In this system the peritectic reaction $L + \alpha \rightarrow \beta$ occurs at 42.4% Ag and 1186°C. Figure 8.18 schematically illustrates how the peritectic reaction progresses

FIGURE 8.18 Schematic representation of the progressive development of the peritectic reaction liquid + $\alpha \rightarrow \beta$.

isothermally in the Pt–Ag system. In Example Problem 8.5 phase analyses are made at various points on this phase diagram. However, during the natural freezing of peritectic alloys, the departure from equilibrium is usually very large because of the relatively slow atomic diffusion rate through the solid phase created by this reaction.

Example Problem 8.5

Make phase analyses at the following points in the platinum-silver equilibrium phase diagram of Fig. 8.17.
(a) The point at 42.4% Ag and 1400°C.
(b) The point at 42.4% Ag and 1186°C + ΔT.
(c) The point at 42.4% Ag and 1186°C − ΔT.
(d) The point at 60% Ag and 1150°C.

Solution:

(a) At 42.4% Ag and 1400°C:

Phases present:	liquid	alpha
Compositions of phases:	55% Ag in liquid phase	7% Ag in alpha phase
Amounts of phases:	Wt % liquid phase	Wt % alpha phase
	$= \dfrac{42.4 - 7}{55 - 7}(100\%)$	$= \dfrac{55 - 42.4}{55 - 7}(100\%)$
	$= 74\%$	$= 26\%$

(b) At 42.4% Ag and 1186°C + ΔT:

Phases present:	liquid	alpha
Compositions of phases:	66.3% Ag in liquid phase	10.5% Ag in alpha phase
Amounts of phases:	Wt % liquid phase	Wt % alpha phase
	$= \dfrac{42.4 - 10.5}{66.3 - 10.5}(100\%)$	$= \dfrac{66.3 - 42.4}{66.3 - 10.5}(100\%)$
	$= 57\%$	$= 43\%$

(c) At 42.4% Ag and 1186°C − ΔT:

Phase present:	beta only
Composition of phase:	42.4% Ag in beta phase

Amounts of   100% beta phase
phase:

(*d*) At 60% Ag and 1150°C:

Phases present:	liquid	beta
Compositions of phases:	77% Ag in liquid phase	48% Ag in beta phase
Amounts of phases:	Wt % liquid phase	Wt % beta phase

$$= \frac{60 - 48}{77 - 48}(100\%)$$ $$= \frac{77 - 60}{77 - 48}(100\%)$$

$$= 41\%$$ $$= 59\%$$

During the equilibrium or very slow cooling of an alloy of peritectic composition through the peritectic temperature, all the solid-phase alpha reacts with all the liquid to produce a new solid-phase beta, as indicated in Fig. 8.18. However, during the rapid solidification of a cast alloy through the peritectic temperature, a nonequilibrium phenomenon called *surrounding* or *encasement* occurs. During the peritectic reaction of $L + \alpha \rightarrow \beta$, the beta phase created by the peritectic reaction surrounds or encases the primary alpha, as shown in Fig. 8.19. Since the beta phase formed is a solid phase and since solid-state diffusion is relatively slow, the beta formed around the alpha creates a diffusion barrier and the peritectic reaction proceeds at an ever-decreasing rate. Thus when a peritectic-type alloy is rapidly cast, coring occurs during the formation of the primary alpha (Fig. 8.20 along α_1 to α_4'), and encasement of the cored α by β occurs during the peritectic reaction. Figure 8.21 schematically illustrates

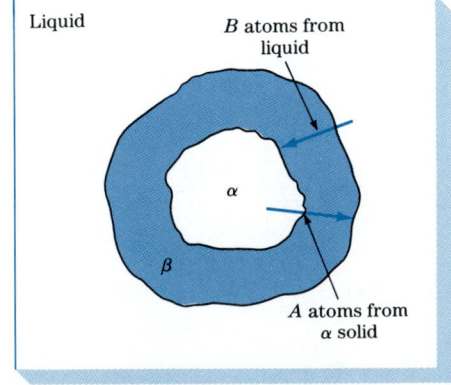

FIGURE 8.19 Surrounding during the peritectic reaction. The slow rate of atoms diffusing from the liquid to the alpha phase causes the beta phase to surround the alpha phase.

FIGURE 8.20 A hypothetical binary peritectic phase diagram to illustrate how coring occurs during natural freezing. Rapid cooling causes the nonequilibrium solidus α_1 to α_4' and the β_4 to β_7' one, which lead to cored alpha phase and cored beta phase. The surrounding phenomenon also occurs during the rapid solidification of peritectic-type alloys. (*After F. Rhines, "Phase Diagrams in Metallurgy," McGraw-Hill, 1956, p. 86.*)

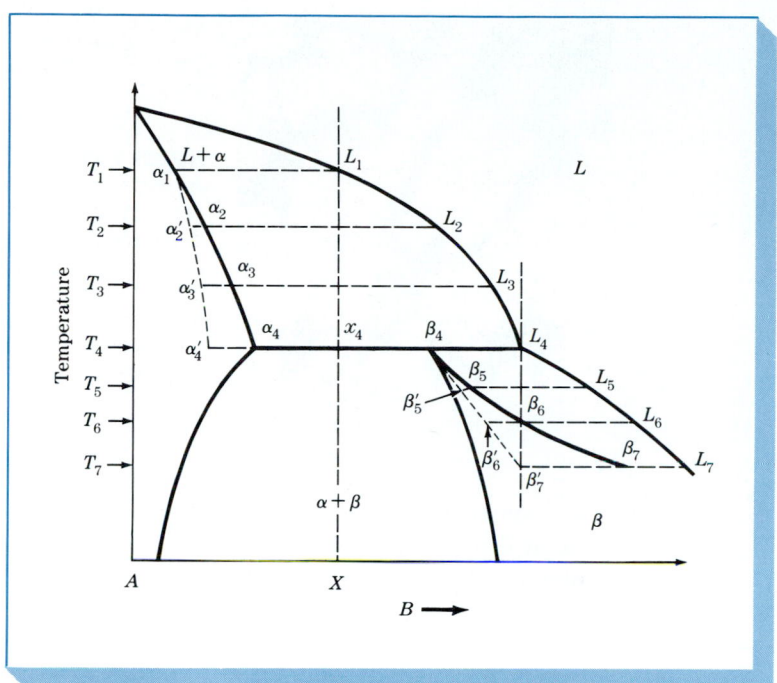

FIGURE 8.21 Schematic representation of surrounding or encasement in a cast peritectic-type alloy. A residue of cored primary α is represented by the solid circles concentric about smaller dashed circles; surrounding the cored α is a layer of β of peritectic composition. The remaining space is filled with cored β, represented by dashed curved lines. (*After F. Rhines, "Phase Diagrams in Metallurgy," McGraw-Hill, 1956, p. 87.*)

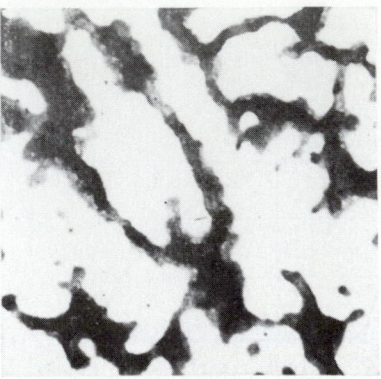

FIGURE 8.22 Cast 60% Ag–40% Pt hyperperitectic alloy. White and light gray areas are residual cored α; dark two-toned areas are β, the outer portions being of peritectic composition and the darkest central areas being the cored β that formed at temperatures below that of the peritectic reaction. (Magnification 1000×.) (*Courtesy of F. Rhines.*)

these combined nonequilibrium structures. The microstructure of a 60% Ag–40% Pt alloy that was rapidly cast is shown in Fig. 8.22. This structure shows cored alpha and its encasement by the beta phase.

8.8   BINARY MONOTECTIC SYSTEMS

Another three-phase invariant reaction that occurs in some binary phase diagrams is the *monotectic reaction* in which a liquid phase transforms into a solid phase and another liquid phase as

$$L_1 \xrightarrow{\text{cooling}} \alpha + L_2 \tag{8.11}$$

Over a certain range of compositions the two liquids are immiscible like oil in water and so constitute individual phases. A reaction of this type occurs in the copper-lead system at 955°C and 36% Pb, as shown in Fig. 8.23. The copper-lead phase diagram has an eutectic point at 326°C and 99.94% Pb, and as a result terminal solid solutions of almost pure lead (0.007% Cu) and pure copper (0.005% Pb) are formed at room temperature. Figure 8.24 shows the microstructure of a cast monotectic alloy of Cu–36% Pb. Note the distinct separation of the lead-rich phase (dark) and the copper matrix (light).

Lead is added in small amounts up to about 0.5 percent to many alloys (e.g., the Cu–Zn brasses) to make the machining of alloys easier by reducing ductility sufficiently to cause machined metal chips to break away from the workpiece. This small addition of lead reduces the strength of the alloy only

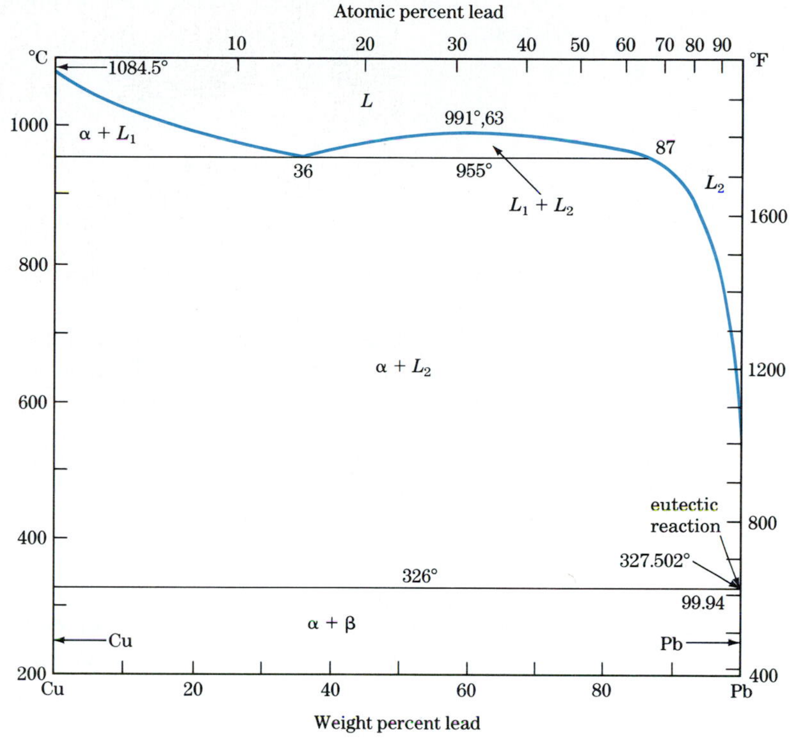

FIGURE 8.23 The copper-lead phase diagram. The most important feature of this diagram is the monotectic invariant reaction at 36% Pb and 955°C. At the monotectic point α (100% Cu), L_1 (36% Pb), and L_2 (87% Pb) can coexist. Note that copper and lead are essentially insoluble in each other. ("*Metals Handbook,*" vol. 8: "*Metallography, Structures, and Phase Diagrams,*" 8th ed., American Society for Metals, *1973, p. 296.*)

Atomic percent lead

Weight percent lead

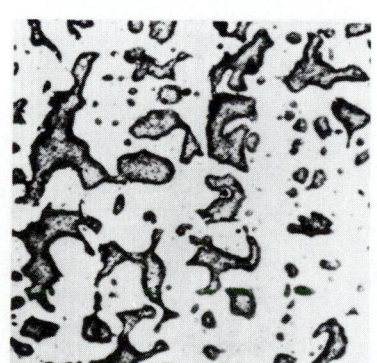

FIGURE 8.24 The microstructure of cast monotectic alloy Cu–36% Pb. Light areas are the Cu-rich matrix of the monotectic constituent; dark areas are the Pb-rich portion, which existed as L_2 at the monotectic temperature. (Magnification 100×.) (*Courtesy of F. Rhines.*)

slightly. Leaded alloys are also used for bearings where small amounts of lead smear out at wear surfaces between the bearing and shaft and thus reduce friction.

8.9 INVARIANT REACTIONS

Three invariant reactions which commonly occur in binary phase diagrams have been discussed so far: the eutectic, peritectic, and monotectic types. Table 8.1 summarizes these reactions and shows their phase-diagram characteristics at their reaction points. Two other important invariant reactions occurring in binary systems are the *eutectoid* and *peritectoid types.* Eutectic and eutectoid reactions are similar in that two solid phases are formed from one phase on cooling. However, in the eutectoid reaction the decomposing phase is solid, whereas in the eutectic reaction it is liquid. In the peritectic reaction two solid phases react to form a new solid phase, whereas in the peritectic reaction a solid phase reacts with a liquid phase to produce a new solid phase. It is interesting to note that the peritectic and peritectoid reactions are the inverse of the corresponding eutectic and eutectoid reactions. The temperatures and compositions of the reacting phases are fixed for all these

TABLE 8.1   Types of Three-Phase Invariant Reactions Occurring in Binary Phase Diagrams

Name of reaction	Equation	Phase-diagram characteristic
Eutectic	$L \xrightarrow{\text{cooling}} \alpha + \beta$	
Eutectoid	$\alpha \xrightarrow{\text{cooling}} \beta + \gamma$	
Peritectic	$\alpha + L \xrightarrow{\text{cooling}} \beta$	
Peritectoid	$\alpha + \beta \xrightarrow{\text{cooling}} \gamma$	
Monotectic	$L_1 \xrightarrow{\text{cooling}} \alpha + L_2$	

invariant reactions. That is, according to the phase rule, there are zero degrees of freedom at the reaction points.

8.10   PHASE DIAGRAMS WITH INTERMEDIATE PHASES AND COMPOUNDS

Phase Diagrams with Intermediate Phases

The phase diagrams considered so far have been relatively simple and contained only a small number of phases and have had only one invariant reaction. Many equilibrium diagrams are complex and often show intermediate phases or compounds. In phase-diagram terminology it is convenient to distinguish between two types of solid solutions: *terminal phases* and *intermediate phases.* Terminal solid-solution phases occur at the ends of phase diagrams, bordering on pure components. The α and β solid solutions of the Pb–Sn diagram (Fig. 8.11) are examples. Intermediate solid-solution phases occur in a composition range inside the phase diagram and are separated from other phases in a binary diagram by two-phase regions. The Cu–Zn phase diagram has both terminal and intermediate phases (Fig. 8.25). In this system α and η are terminal phases and β, γ, δ, and ϵ are intermediate phases. The Cu–Zn diagram contains five invariant peritectic points and one eutectoid invariant point at the lowest point of the δ intermediate-phase region.

Intermediate phases are not restricted to binary metal phase diagrams. In the ceramic phase diagram of the Al_2O_3–SiO_2 system an intermediate phase called *mullite* is formed which includes the compound $3Al_2O_3 \cdot 2SiO_2$ (Fig. 8.26). Many refractories[1] have Al_2O_3 and SiO_2 as their main components. These materials will be discussed in Chap. 10 on Ceramic Materials.

Intermediate Compounds

In some phase diagrams intermediate compounds are formed between two metals or between a metal and a nonmetal. The Mg–Ni phase diagram contains the intermediate compounds Mg_2Ni and $MgNi_2$ which are primarily metallically bonded and have fixed compositions and definite stoichiometries (Fig. 8.27). The intermetallic compound $MgNi_2$ is said to be a *congruently melting compound* since it maintains its composition right up to the melting point. On the other hand, Mg_2Ni is said to be an *incongruently melting compound* since, upon heating, it undergoes peritectic decomposition at 761°C into liquid and $MgNi_2$ phases. Other examples of intermediate compounds which occur in phase diagrams are Fe_3C and Mg_2Si. In Fe_3C the bonding is mainly metallic in character, but in Mg_2Si the bonding is mainly covalent.

---

[1] A refractory is a heat-resisting ceramic material.

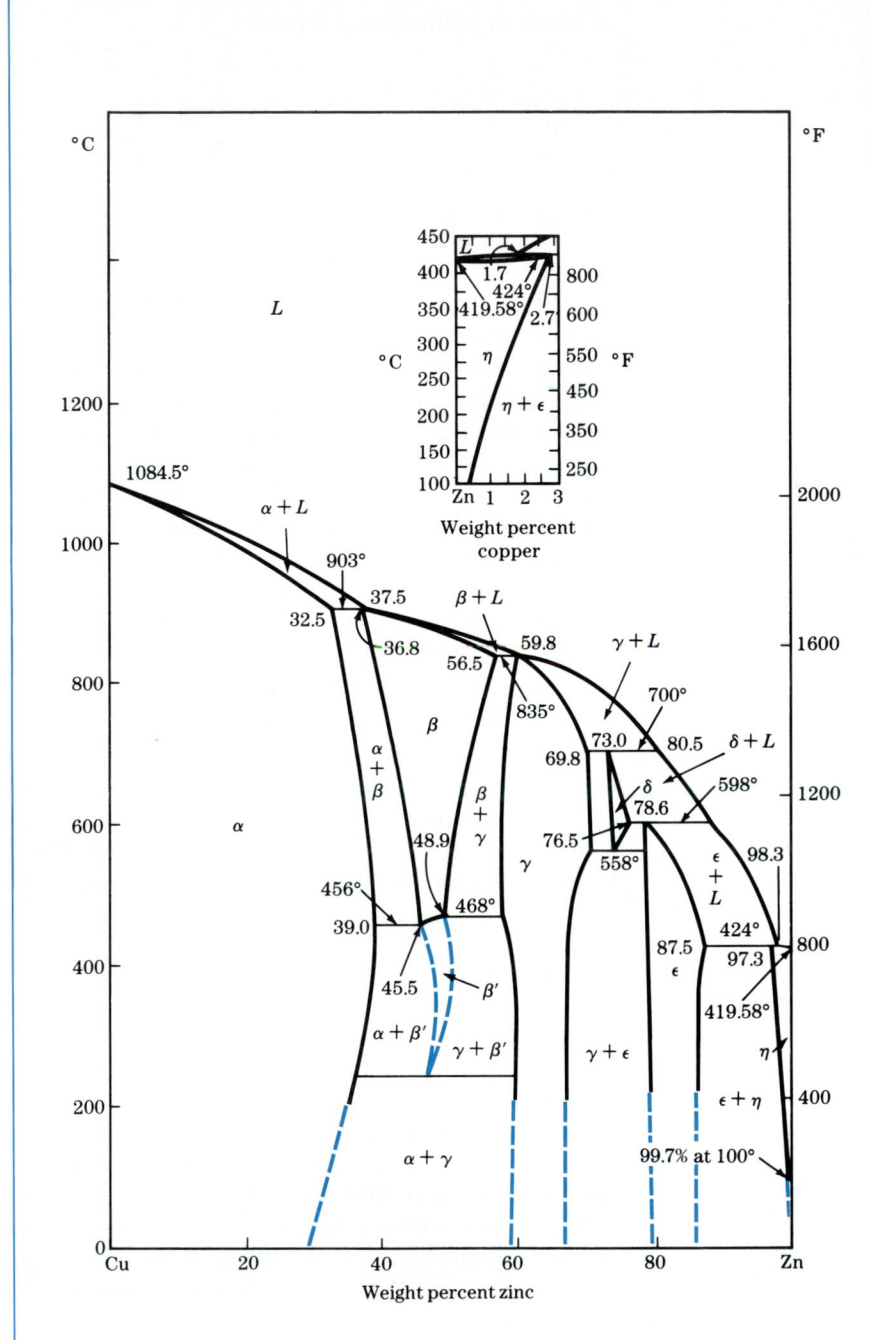

FIGURE 8.25 The copper-zinc phase diagram. This diagram has terminal phases α and η and intermediate phases β, γ, δ, and ϵ. There are five invariant peritectic points and one eutectoid point. (*After "Metals Handbook," vol. 8: "Metallography, Structures, and Phase Diagrams," 8th ed., American Society for Metals, 1973, p. 301.*)

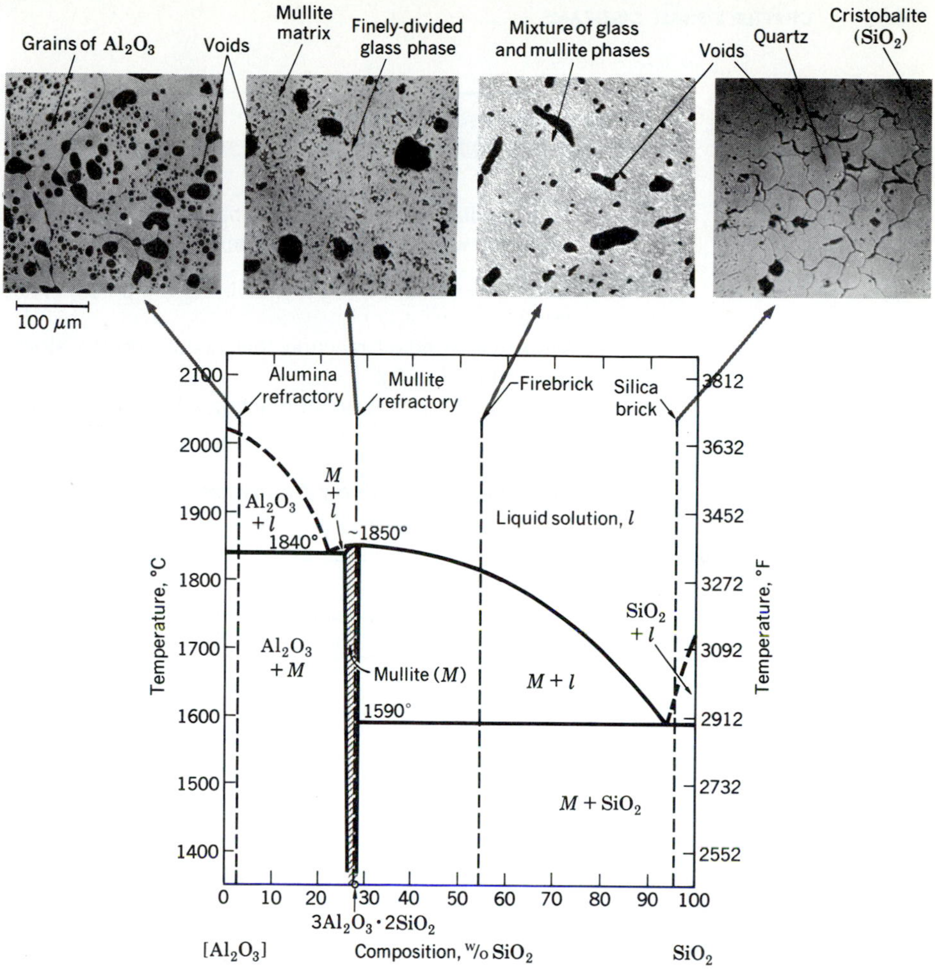

FIGURE 8.26 The phase diagram of the Al$_2$O$_3$–SiO$_2$ system which contains mullite as an intermediate phase. Typical compositions of refractories having Al$_2$O$_3$ and SiO$_2$ as their main components are shown. (*After A. G. Guy, "Essentials of Materials Science," McGraw-Hill, 1976.*)

Grains of Al$_2$O$_3$ — Voids — Mullite matrix — Finely-divided glass phase — Mixture of glass and mullite phases — Voids — Quartz — Cristobalite (SiO$_2$)

100 μm

Alumina refractory | Mullite refractory | Firebrick | Silica brick

2100 — 3812
2000 — 3632
1900 — 3452
1800 — 3272
1700 — 3092
1600 — 2912
1500 — 2732
1400 — 2552

M + l
Al$_2$O$_3$ + l
1840°
~1850°
Liquid solution, l
SiO$_2$ + l
Al$_2$O$_3$ + M
Mullite (M)
M + l
1590°
M + SiO$_2$

Temperature, °C

Temperature, °F

0 10 20 30 40 50 60 70 80 90 100
3Al$_2$O$_3$·2SiO$_2$

[Al$_2$O$_3$]   Composition, % SiO$_2$ SiO$_2$

FIGURE 8.27 The magnesium-nickel phase diagram. In this diagram there are two intermetallic compounds, Mg$_2$Ni and MgNi$_2$. (*"Metals Handbook," 8th ed., vol. 8, American Society for Metals, 1973, p. 314.*)

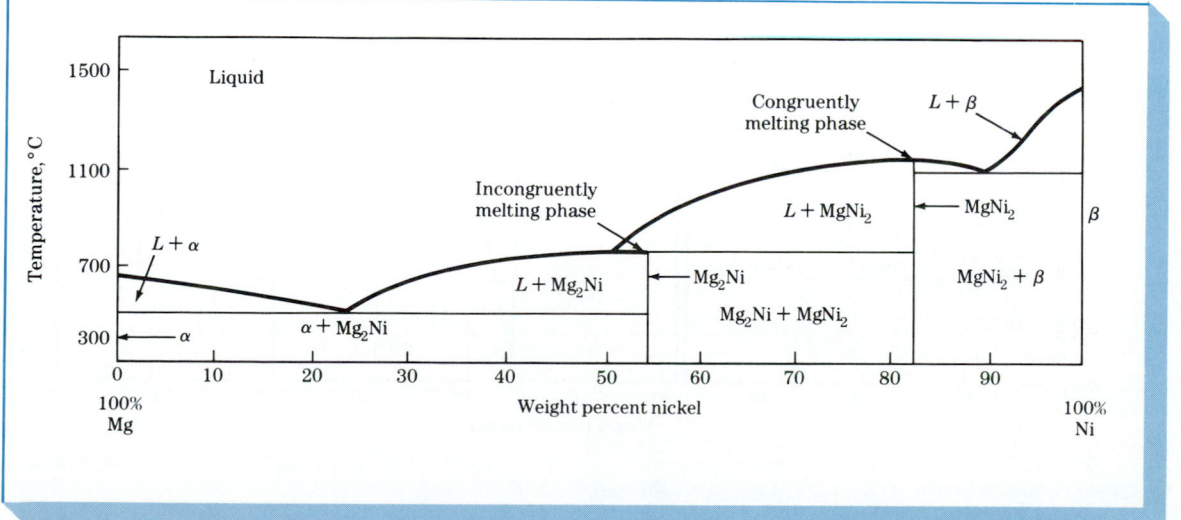

1500 — Liquid
1100 — Congruently melting phase — L + β
Incongruently melting phase — L + MgNi$_2$ — MgNi$_2$ — β
700 — L + α — L + Mg$_2$Ni — Mg$_2$Ni — MgNi$_2$ + β
300 — α — α + Mg$_2$Ni — Mg$_2$Ni + MgNi$_2$

Temperature, °C

0 10 20 30 40 50 60 70 80 90
100% Mg   Weight percent nickel   100% Ni

Example Problem 8.6

Consider the titanium-nickel (Ti–Ni) phase diagram in Fig. EP8.6. This phase diagram has six points where three phases coexist. For each of these three-phase points:

(i) List the coordinates of composition (weight percent) and temperature for each point.
(ii) Write the invariant reaction that occurs during slow cooling of the Ti–Ni alloy through each point.
(iii) Name the type of invariant reaction which takes place at each point.

FIGURE EP8.6 Titanium-nickel phase diagram. (*After Binary Alloy Phase Diagrams, ASM Int., 1986, p. 1768.*)

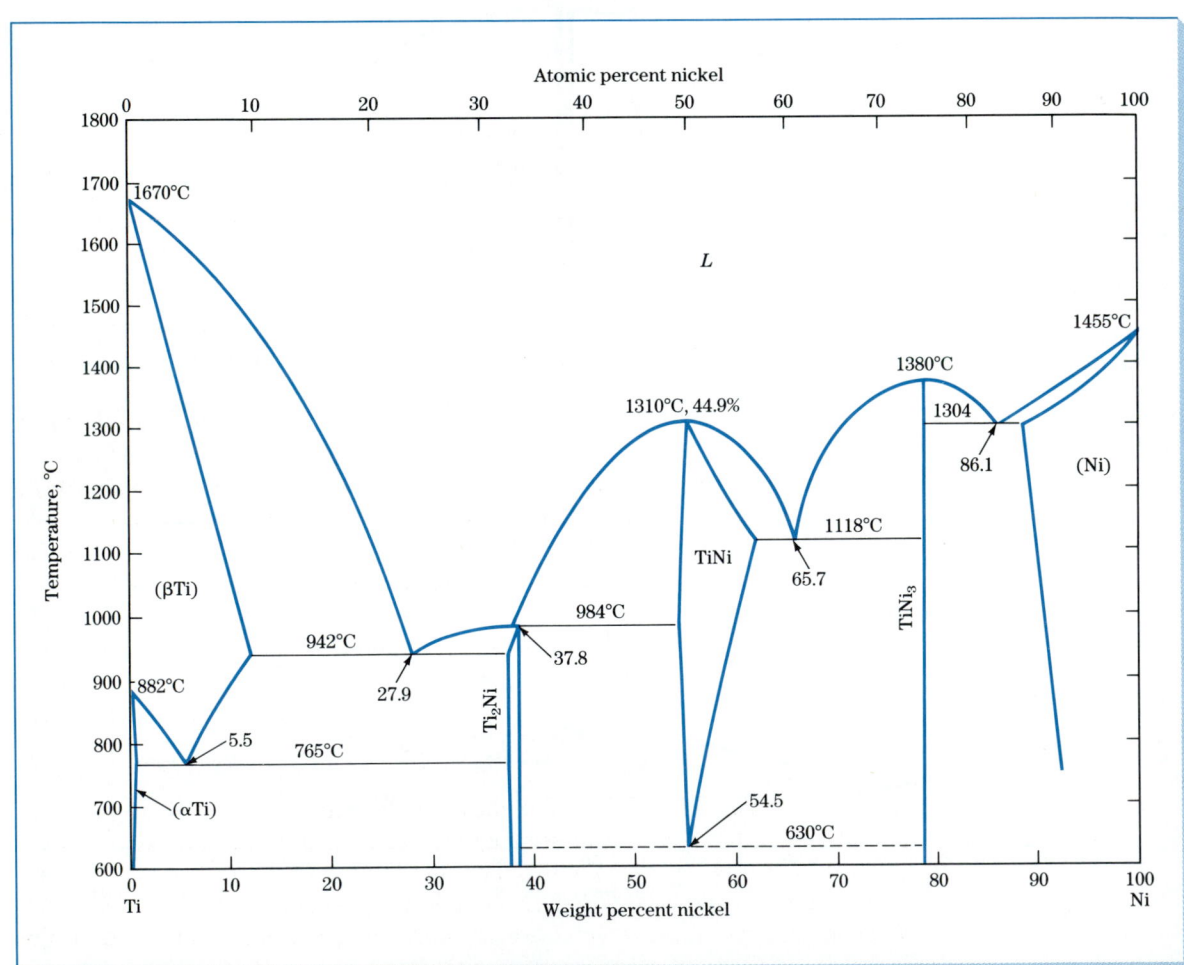

Solution:

(a) (i) 5.5 wt % Ni, 765°C
  (ii) $(\beta \text{ Ti}) \longrightarrow (\alpha \text{ Ti}) + \text{Ti}_2\text{Ni}$
  (iii) Eutectoid reaction
(b) (i) 27.9 wt % Ni, 942°C
  (ii) $L \longrightarrow (\beta \text{ Ti}) + \text{Ti}_2\text{Ni}$
  (iii) Eutectic reaction
(c) (i) 37.8 wt % Ni, 984°C
  (ii) $L + \text{TiNi} \longrightarrow \text{Ti}_2\text{Ni}$
  (iii) Peritectic reaction

(d) (i) 54.5 wt % Ni, 630°C
  (ii) $\text{TiNi} \longrightarrow \text{Ti}_2\text{Ni} + \text{TiNi}_3$
  (iii) Eutectoid reaction
(e) (i) 65.7 wt % Ni, 1118°C
  (ii) $L \longrightarrow \text{TiNi} + \text{TiNi}_3$
  (iii) Eutectic reaction
(f) (i) 86.1 wt % Ni, 1304°C
  (ii) $L \longrightarrow \text{TiNi}_3 + (\text{Ni})$
  (iii) Eutectic reaction

8.11 TERNARY PHASE DIAGRAMS

Until now we have discussed only binary phase diagrams in which there are two components. We shall now turn our attention to ternary phase diagrams which have three components. Compositions on ternary phase diagrams are usually constructed by using an equilateral triangle as a base. Compositions of ternary systems are represented on this base with the pure component at each end of the triangle. Figure 8.28 shows the composition base of a ternary phase diagram for a ternary metal alloy consisting of pure metals A, B, and C. The binary alloy compositions AB, BC, and AC are represented on the three edges of the triangle.

Ternary phase diagrams with a triangular composition base are normally constructed at a constant pressure of 1 atm. Temperature can be represented as uniform throughout the whole diagram. This type of ternary diagram is called an *isothermal section.* To show a range of temperatures at varying composi- tions, a figure with temperature on a vertical axis with a triangular composition base can be constructed. However, more commonly, temperature contour lines are drawn on a triangular composition base to indicate temperature ranges just as different elevations are shown on a flat-page map of a terrain.

Let us now consider the determination of the composition of a ternary alloy indicated by a point on a ternary diagram of the type shown in Fig. 8.28. In Fig. 8.28 the A corner of the triangle indicates 100% metal A, the B corner indicates 100% metal B, and the C corner indicates 100% metal C. The weight percent of each pure metal in the alloy is determined in the following way. A perpendicular line is drawn from a pure metal corner to the side of the triangle opposite that corner, and the distance *from* the side to the corner along the perpendicular line is measured as a fraction of 100 percent for the whole line. This percentage is the weight percent of the pure metal of that corner in the alloy. Example Problem 8.7 explains this procedure in more detail.

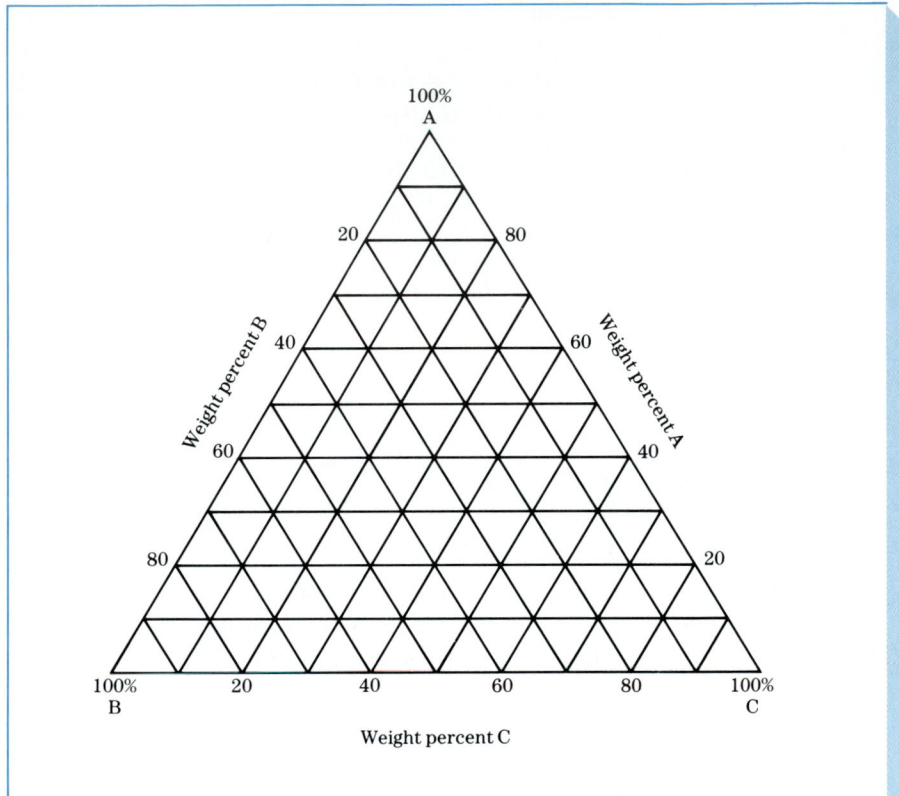

FIGURE 8.28

Composition base for a ternary phase diagram for a system with pure components A, B, and C.

Example Problem 8.7

Determine the weight percents of metals A, B, and C for a ternary alloy ABC at point x on the ternary phase diagram grid shown in Fig. 8.29.

Solution:

The composition at a point in a ternary phase diagram grid of the type shown in Fig. 8.29 is determined by separately determining the compositions of each of the pure metals from the diagram. To determine the % A at point x in Fig. 8.29, we first draw the perpendicular line AD from the corner A to point D on the side of the triangle opposite corner A. The total length of the line from D to A represents 100% A. At point D the % A in the alloy is zero. The point x is on an isocomposition line at 40% A, and

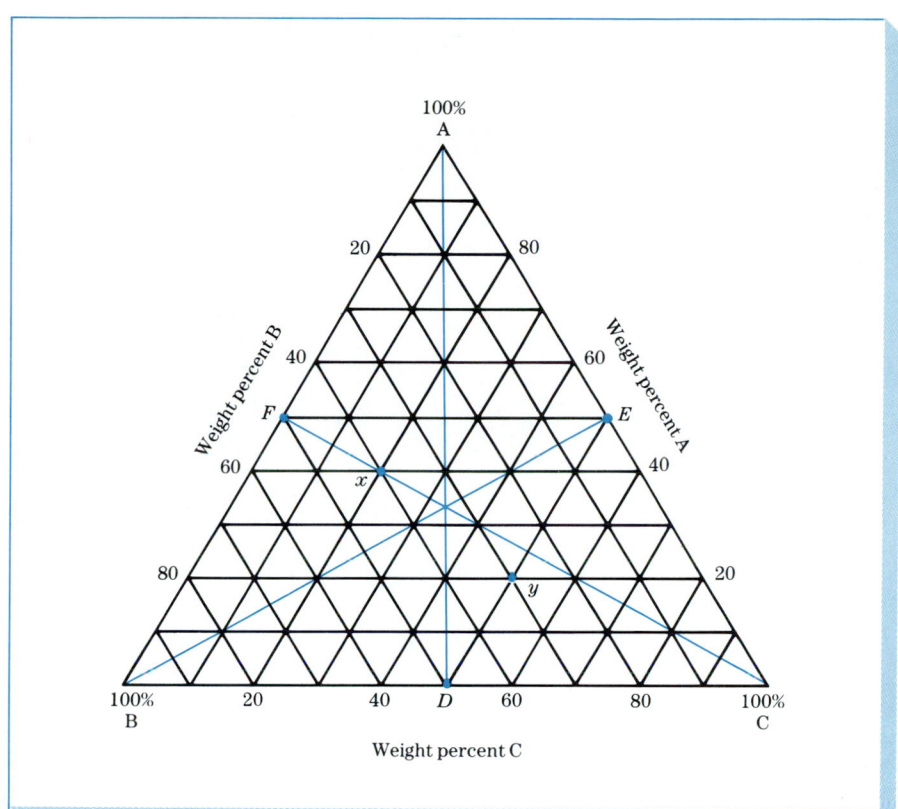

FIGURE 8.29 Ternary phase diagram composition base for an ABC alloy.

thus the percentage of A in the alloy is 40%. In a similar manner we draw line *BE* and determine that the percentage of B in the alloy is also 40%. A third line *CF* is drawn and the percentage of C in the alloy is determined to be 20%. Thus the composition of the ternary alloy at point *x* is 40% A, 40% B, and 20% C. Actually only two percentages need to be determined since the third can be obtained by subtracting the sum of two from 100%.

The ternary phase diagram of iron, chromium, and nickel is important since the commercially most important stainless steel has a composition essentially of 74% iron, 18% chromium, and 8% nickel. Figure 8.30 shows an isothermal section at 650°C (1202°F) for the iron-chromium-nickel ternary system.

Ternary phase diagrams also are important for the study of some ceramic materials. Figure 10.32 shows a ternary phase diagram of the important silica-leucite-mullite system.

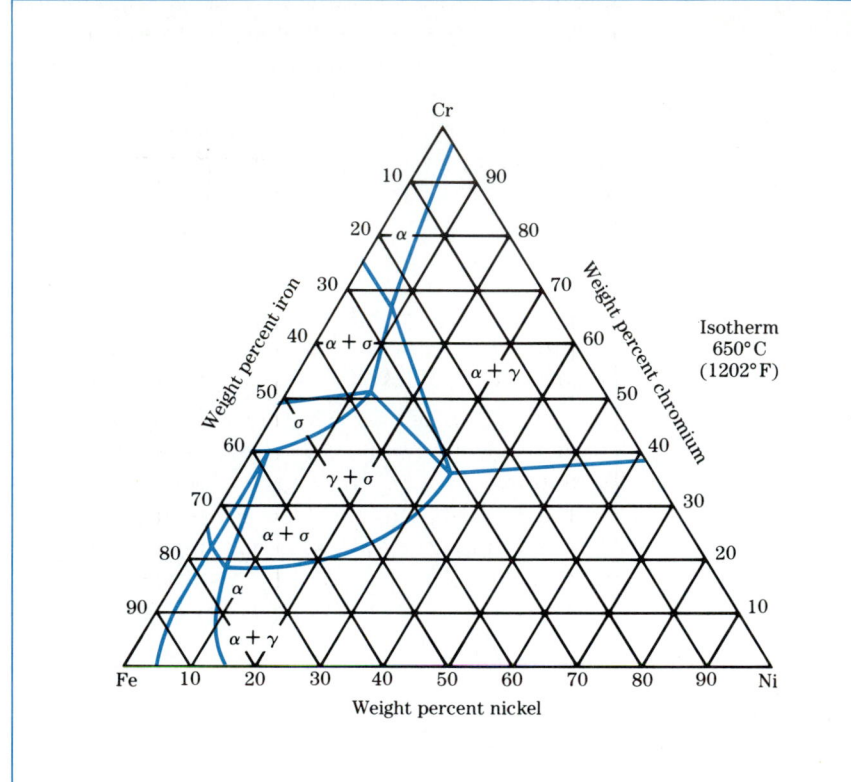

FIGURE 8.30 Ternary phase diagram of an isothermal section at 650°C (1202°F) for the iron-chromium-nickel system. (*After "Metals Handbook," vol. 8, 8th ed., American Society for Metals, 1973, p. 425.*)

SUMMARY

Phase diagrams are graphical representations of what phases are present in an alloy (or ceramic) system at various temperatures, pressures, and compositions. In this chapter emphasis has been placed on temperature-composition binary equilibrium phase diagrams. These diagrams tell us which phases are present at different compositions and temperatures for slow cooling or heating conditions that approach equilibrium. In two-phase regions of these diagrams the chemical compositions of each of the two phases is indicated by the intersection of the isotherm with the phase boundaries. The weight fraction of each phase in a two-phase region can be determined by using the lever rule along an isotherm (tie line at a particular temperature).

In binary equilibrium *isomorphous phase diagrams* the two components are completely soluble in each other in the solid state, and so there is only one solid phase. In binary equilibrium alloy (ceramic) phase diagrams, *invariant reactions* involving three phases in equilibrium often occur. The most common of these reactions are:

1. Eutectic reaction: $\qquad L \rightarrow \alpha + \beta$
2. Eutectoid reaction: $\qquad \alpha \rightarrow \beta + \gamma$
3. Peritectic reaction: $\qquad \alpha + L \rightarrow \beta$
4. Peritectoid reaction: $\qquad \alpha + \beta \rightarrow \gamma$
5. Monotectic reaction: $\qquad L_1 \rightarrow \alpha + L_2$

In many binary equilibrium phase diagrams intermediate phase(s) and/or compounds are present. The intermediate phases have a range of compositions, whereas the intermediate compounds have only one composition.

During the rapid solidification of many alloys, compositional gradients are created and *cored* structures are produced. A cored structure can be eliminated by homogenizing the cast alloy for long times at high temperatures just below the melting temperature of the lowest melting phase in the alloy. If the cast alloy is overheated slightly so that melting occurs at the grain boundaries, a *liquated* structure is produced. This type of structure is undesirable since the alloy loses strength and may break up during subsequent working.

DEFINITIONS

Sec. 8.1

System: a portion of the universe that has been isolated so that its properties can be studied.

Equilibrium: a system is said to be in equilibrium if no macroscopic changes take place with time.

Phase: a physically homogeneous and distinct portion of a material system.

Equilibrium phase diagram: a graphical representation of the pressures, temperatures, and compositions for which various phases are stable at equilibrium. In materials science the most common phase diagrams involve temperature vs. composition.

Sec. 8.2

Gibbs phase rule: the statement that at equilibrium the number of phases plus the degrees of freedom equals the number of components plus 2. $P + F = C + 2$. In the condensed form with pressure ≈ 1 atm, $P + F = C + 1$.

Degrees of freedom the number of variables (temperature, pressure, and composition) which can be changed *independently* without changing the phase or phases of the system.

Number of components of a phase diagram: the number of elements or compounds which make up the phase-diagram system. For example, the Fe–Fe_3C system is a two-component system; the Fe–Ni system is also a two-component system.

Sec. 8.3

Isomorphous system: a phase diagram in which there is only one solid phase; i.e., there is only one solid-state structure.

Liquidus: the temperature at which liquid starts to solidify under equilibrium conditions.

Solidus: the temperature during the solidification of an alloy at which the last of the liquid phase solidifies.

Sec. 8.6

Eutectic reaction (in a binary phase diagram): a phase transformation in which all the liquid phase transforms on cooling into two solid phases isothermally.

Eutectic temperature: the temperature at which a eutectic reaction takes place.

Eutectic composition: the composition of the liquid phase that reacts to form two new solid phases at the eutectic temperature.

Eutectic point: the point determined by the eutectic composition and temperature.

Invariant reactions: equilibrium phase transformations involving zero degrees of freedom.

Hypoeutectic composition: one which is to the left of the eutectic point.

Hypereutectic composition: one which is to the right of the eutectic point.

Proeutectic phase: a phase which forms at a temperature above the eutectic temperature.

Primary phase: a solid phase which forms at a temperature above that of an invariant reaction and is still present after the invariant reaction is completed.

Sec. 8.7

Peritectic reaction (in a binary phase diagram): a phase transformation in which, upon cooling, a liquid phase combines with a solid phase to produce a new solid phase.

Monotectic reaction (in a binary phase diagram): a phase transformation in which, upon cooling, a liquid phase transforms into a solid phase and a new liquid phase (of different composition than the first liquid phase).

Sec. 8.10

Terminal phase: a solid solution of one component in another for which one boundary of the phase field is a pure component.

Intermediate phase: a phase whose composition range is between those of the terminal phases.

PROBLEMS

8.1.1 Define (*a*) a phase in a material and (*b*) phase diagram.

8.1.2 In the pure water pressure-temperature equilibrium phase diagram (Fig. 8.1), what phases are in equilibrium for the following conditions?
(*a*) Along the freezing line.
(*b*) Along the vaporization line.
(*c*) At the triple point.

8.1.3   How many triple points are there in the pure iron pressure-temperature equilibrium phase diagram of Fig. 8.2? What phases are in equilibrium at each of the triple points?

8.2.1   Write the equation for Gibbs phase rule and define each of the terms.

8.2.2   Refer to the pressure-temperature equilibrium phase diagram for pure water (Fig. 8.1) and answer the following:

(a) How many degrees of freedom are there at the triple point?

(b) How many degrees of freedom are there along the freezing line?

8.3.1   What is a binary isomorphous alloy system?

8.3.2   What are the four Hume-Rothery rules for the solid solubility of one element in another?

8.3.3   A number of elements along with their crystal structures and atomic radii are listed in the table below. Which pairs might be expected to have complete solid solubility in each other?

	Crystal structure	Atomic radius, nm		Crystal structure	Atomic radius, nm
Silver	FCC	0.144	Lead	FCC	0.175
Palladium	FCC	0.137	Tungsten	BCC	0.137
Copper	FCC	0.128	Rhodium	FCC	0.134
Gold	FCC	0.144	Platinum	FCC	0.138
Nickel	FCC	0.125	Tantalum	BCC	0.143
Aluminum	FCC	0.143	Potassium	BCC	0.231
Sodium	BCC	0.185	Molybdenum	BCC	0.136

8.4.1   Derive the lever rule for the amount in weight percent of each phase in two-phase regions of a binary phase diagram. Use a phase diagram in which two elements are completely soluble in each other.

8.4.2   Consider an alloy containing 70 wt % Ni and 30 wt % Cu (see Fig. 8.3).

(a) At 1350°C make a phase analysis assuming equilibrium conditions. In the phase analysis include the following:

(i) What phases are present?

(ii) What is the chemical composition of each phase?

(iii) What amount of each phase is present?

(b) Make a similar phase analysis at 1500°C.

(c) Sketch the microstructure of the alloy at each of the above temperatures by using circular microscopic fields.

8.4.3   Describe how the liquidus and solidus of a binary isomorphous phase diagram can be determined experimentally.

8.5.1   Explain how a cored structure is produced in a 70% Cu–30% Ni alloy.

8.5.2   How can the cored structure in a 70% Cu–30% Ni alloy be eliminated by heat treatment?

8.5.3   Explain what is meant by the term *liquation*. How can a liquated structure be produced in an alloy? How can it be avoided?

8.6.1   Consider the binary eutectic copper-silver phase diagram in Fig. 8.31. Make phase

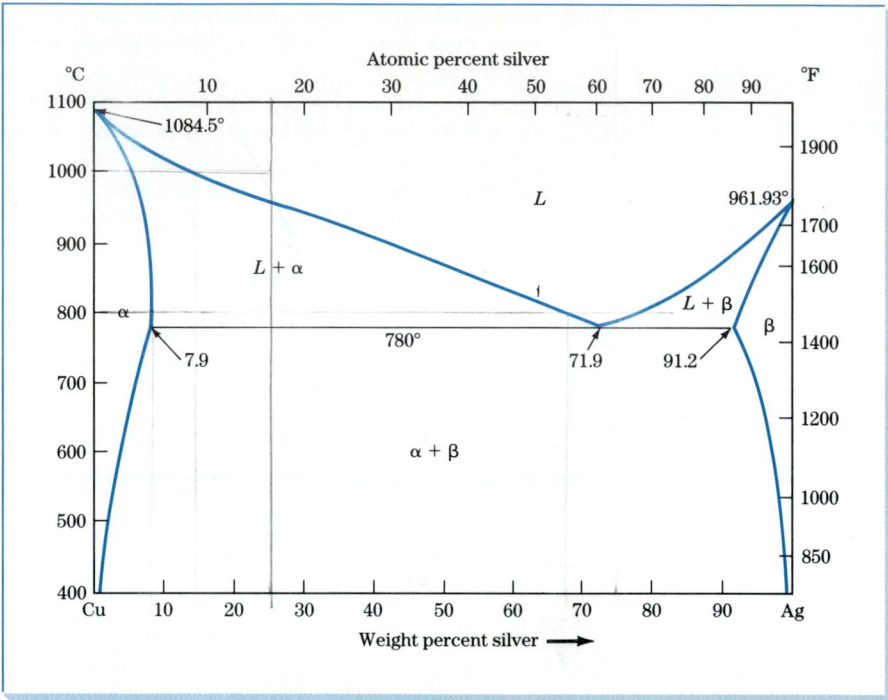

FIGURE 8.31 The copper-silver phase diagram. (After "Metals Handbook," vol. 8, 8th ed., American Society for Metals, 1973, p. 253.)

analysis of a 75 wt % Cu–25 wt % Ag alloy at the temperatures (*a*) 1000°C, (*b*) 800°C, (*c*) 780°C + ΔT, and (*d*) 780°C − ΔT. In the phase analyses include:
(i)   The phases present.
(ii) The chemical compositions of the phases.
(iii) The amounts of each phase.
(iv) Sketch the microstructure by using 2-cm-diameter circular fields.

8.6.2 If 750 g of an 80 wt % Ag–20 wt % Cu alloy is slowly cooled from 1000°C to just below 780°C (see Fig. 8.31):
(*a*) How many grams of liquid and proeutectic beta are present at 800°C?
(*b*) How many grams of liquid and proeutectic beta are present at 780°C + ΔT?
(*c*) How many grams of alpha are present in the eutectic structure at 780°C − ΔT?
(*d*) How many grams of beta are present in the eutectic structure at 780°C − ΔT?

8.6.3 A lead-tin (Pb–Sn) alloy consists of 75 wt % proeutectic β and 25 wt % eutectic $\alpha + \beta$ at 183°C − ΔT. Calculate the average composition of this alloy (see Fig. 8.11).

8.6.4 A Pb–Sn alloy (Fig. 8.11) contains 30 wt % β and 70 wt % α at 50°C. What is the average composition of Pb and Sn in this alloy?

8.6.5 An alloy of 75 wt % Pb and 25 wt % Sn is slowly cooled from 250°C to 27°C (see Fig. 8.11).
(*a*) Is this alloy hypoeutectic or hypereutectic?
(*b*) What is the composition of the first solid to form?
(*c*) What are the amounts and compositions of each phase which is present at 183°C + ΔT?

(d) What is the amount and composition of each phase which is present at 183°C − ΔT?

(e) What are the amounts of each phase present at room temperature?

8.7.1 Consider the binary peritectic iridium-rhenium phase diagram of Fig. 8.32. Make phase analyses of a 40 wt % Re–60 wt % Ir at the temperatures (a) 2900°C, (b) 2805°C + ΔT, and (c) 2805°C − ΔT. In the phase analyses include:
  (i) The phases present.
  (ii) The chemical compositions of the phases.
  (iii) The amounts of each phase.
  (iv) Sketch the microstructure by using 2-cm-diameter circular fields.

8.7.2 Consider the binary peritectic iridium-rhenium phase diagram of Fig. 8.32. Make phase analyses of a 67 wt % Ir–33 wt % Re at the temperatures (a) 3000°C, (b) 2630°C + ΔT, (c) 2805°C − ΔT, and (d) 2805°C. Include in the phase analyses the four items listed in Prob. 8.7.1.

8.7.3 Describe the mechanism that produces the phenomenon of *surrounding* in a peritectic alloy which is rapidly solidified through the peritectic reaction.

8.7.4 Can coring and surrounding occur in a peritectic-type alloy which is rapidly solidified? Explain.

8.7.5 Consider an Fe–4.1 wt % Ni alloy (Fig. 8.16) which is slowly cooled from 1550 to 1450°C. What weight percent of the alloy solidifies by the peritectic reaction?

8.7.6 Consider an Fe–5.2 wt % Ni alloy (Fig. 8.16) which is slowly cooled from 1550 to 1450°C. What weight percent of the alloy solidifies by the peritectic reaction?

8.7.7 Determine the weight percent and composition in weight percent of each phase present in Fe–4.1 wt % Ni alloy (Fig. 8.16) at 1517°C + ΔT.

8.7.8 Determine the composition in weight percent of the alloy in the Fe–Ni system (Fig. 8.16) which will produce a structure of 45 wt % δ and 55 wt % γ, just below the peritectic temperature.

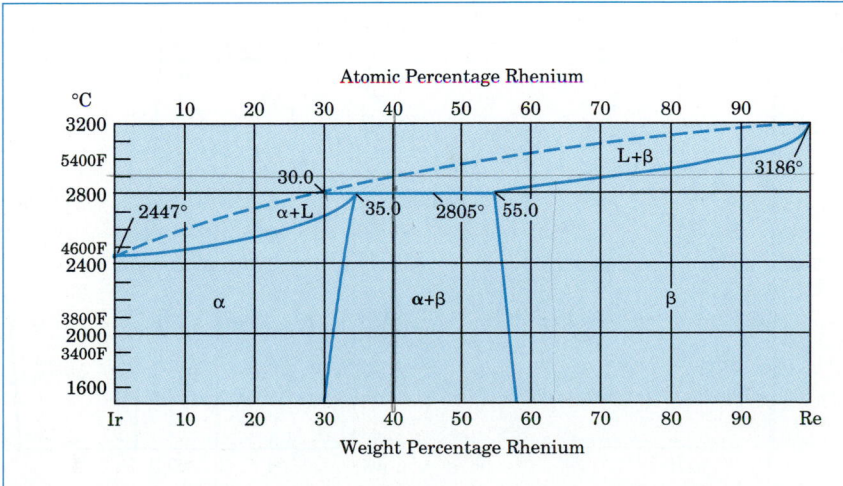

FIGURE 8.32 The iridium-Rhenium phase diagram. (After ''Metals Handbook,'' vol. 8, 8th ed., American Society for Metals, 1973, p. 332.)

8.8.1   What is a monotectic invariant reaction? How is the monotectic reaction in the copper-lead system important industrially?

8.8.2   In the copper-lead (Cu–Pb) system (Fig. 8.23) for an alloy of Cu–10 wt % Pb, determine the amounts and compositions of the phases present at (*a*) 1000°C, (*b*) 955°C + ΔT, (*c*) 955°C − ΔT, and (*d*) 200°C.

8.8.3   For an alloy of Cu–70 wt % Pb (Fig. 8.23), determine the amounts and compositions in weight percent of the phases present at (*a*) 955°C + ΔT, (*b*) 955°C − ΔT, and (*c*) 200°C.

8.8.4   What is the average composition (weight percent) of a Cu–Pb alloy which contains 30 wt % L_1 and 70 wt % α at 955°C + ΔT?

8.9.1   Write equations for the following invariant reactions: eutectic, eutectoid, peritectic, and peritectoid. How many degrees of freedom exist at invariant reaction points in binary phase diagrams?

FIGURE 8.33 Aluminum-nickel phase diagram. (*After Binary Phase Diagrams, ASM Int., 1986, p. 142*).

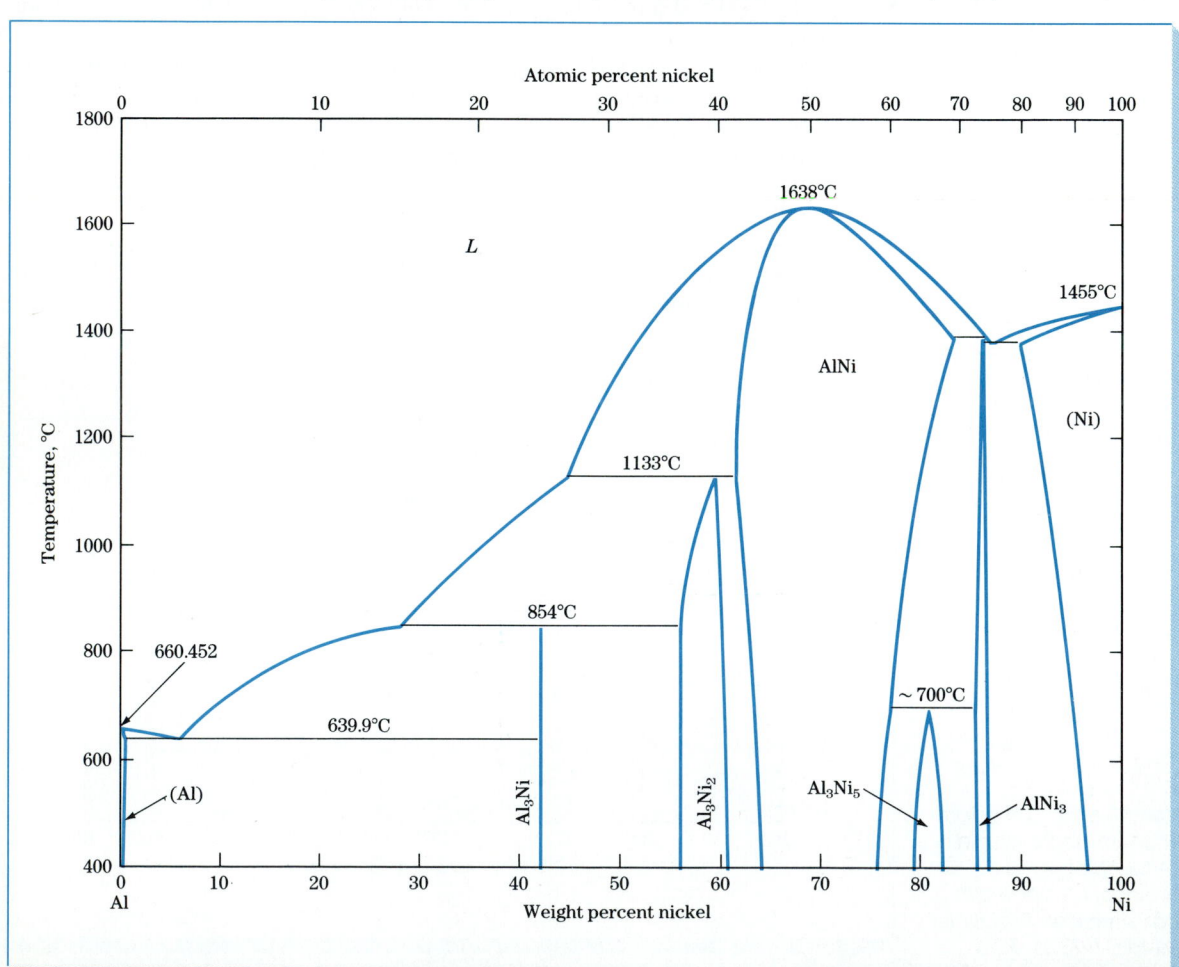

8.9.2   How are eutectic and eutectoid reactions similar? What is the significance of the *-oid* suffix?

8.10.1   Distinguish between (*a*) a terminal phase and (*b*) an intermediate phase.

8.10.2   Distinguish between (*a*) an intermediate phase and (*b*) an intermediate compound.

8.10.3   What is the difference between a congruently melting compound and an incongruently melting one?

8.10.4   Consider the Cu–Zn phase diagram of Fig. 8.25.
(*a*)   What is the maximum solid solubility in weight percent of Zn in Cu in the terminal solid solution α?
(*b*)   Identify the intermediate phases in the Cu–Zn phase diagram.
(*c*)   Identify the three-phase invariant reactions in the Cu–Zn diagram.
(i)   Determine the composition and temperature coordinates of the invariant reactions.

FIGURE 8.34   Nickel-vanadium phase diagram. (*After Binary Phase Diagrams, ASM Int., 1986, p. 1773.*)

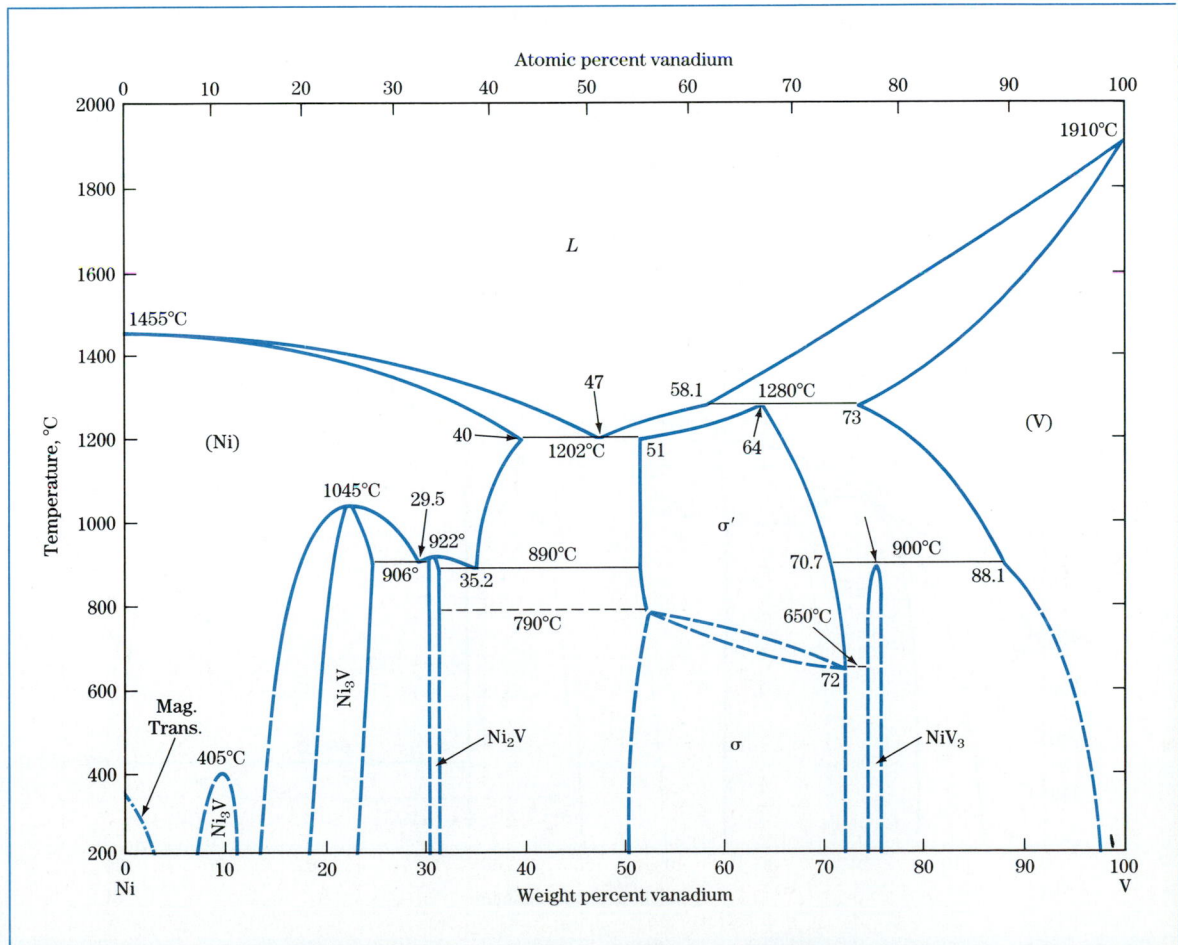

(ii) Write the equations for the invariant reactions.

(iii) Name the invariant reactions.

8.10.5 Consider the aluminum-nickel (Al–Ni) phase diagram of Fig. 8.33. For this phase diagram:

(a) Determine the coordinates of the composition and temperature of the invariant reactions

(b) Write the equations for the three-phase invariant reactions and name them.

(c) Label the two-phase regions in the phase diagram.

8.10.6 Consider the nickel-vanadium (Ni–V) phase diagram of Fig. 8.34. For this phase diagram repeat questions (a), (b), and (c) of Prob. 8.10.5.

8.10.7 Consider the titanium-aluminum (Ti–Al) phase diagram of Fig. 8.35. For this phase diagram repeat questions (a), (b), and (c) of Prob. 8.10.5.

8.11.1 What is the composition of point y in Fig. 8.29?

FIGURE 8.35 Titanium-aluminum phase diagram. (*After Binary Phase Diagrams, ASM Int., 1986, p. 1773.*)

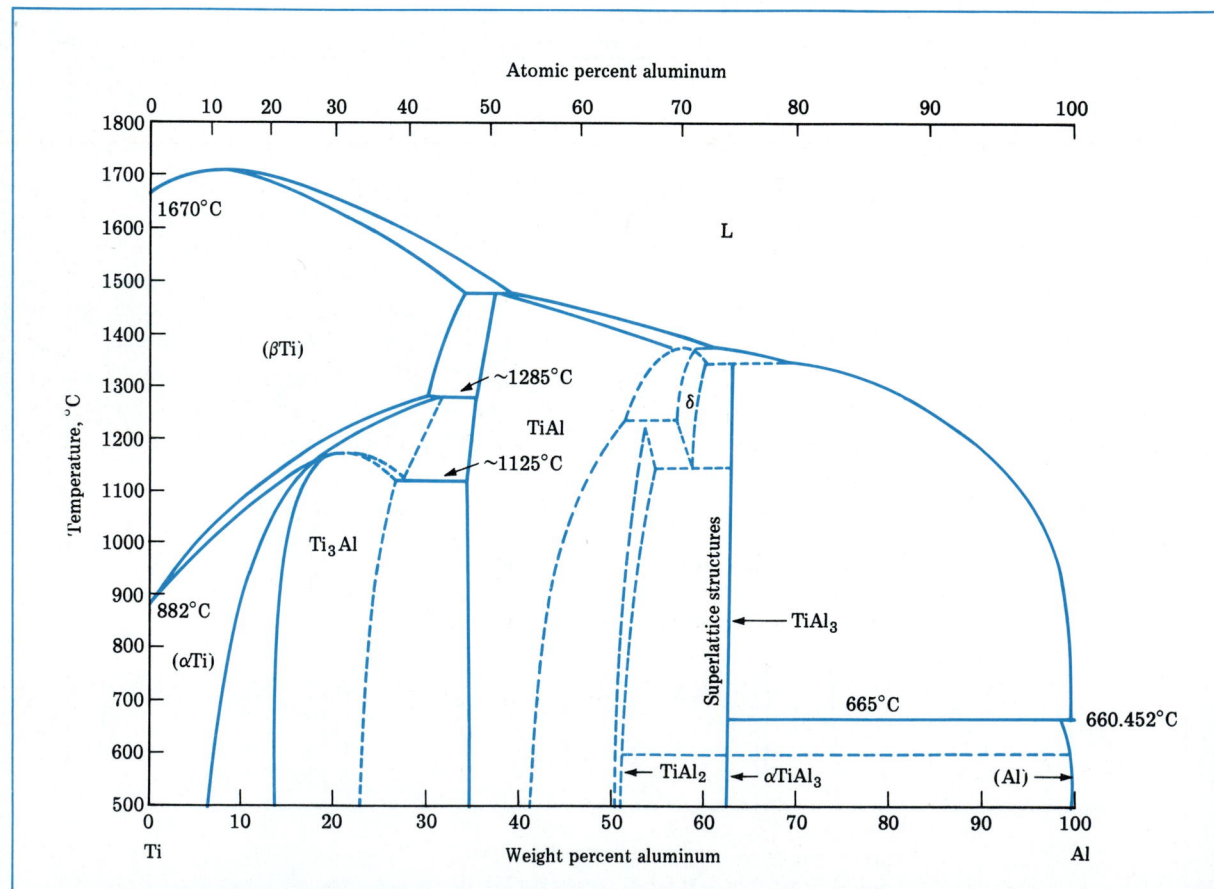

9

Engineering Alloys

Metals and alloys have many useful engineering properties and so have widespread application in engineering designs. Iron and its alloys (principally steel) account for about 90 percent of the world's production of metals mainly because of their combination of good strength, toughness, and ductility at a relatively low cost. Each metal has special properties for engineering designs and is used after a comparative cost analysis with other metals and materials (see Table 9.1).

Alloys based on iron are called *ferrous alloys,* and those based on the other metals are called *nonferrous alloys.* In this chapter we shall discuss some aspects of the processing, structure, and properties of some of the important ferrous and nonferrous alloys.

TABLE 9.1 **Approximate Prices ($/lb) of some Metals as of October 1994***

Steel†	0.23	Nickel	3.00
Aluminum	0.75	Tin	2.42
Copper	1.13	Titanium‡	3.50–4.50
Magnesium	1.53	Gold	4704
Zinc	0.47	Silver	60
Lead	0.29		

*Prices of metals vary with time.
†Hot-rolled plain-carbon steel sheet.
‡Titanium sponge. Prices for large quantity.

9.1 PRODUCTION OF IRON AND STEEL

Production of Pig Iron in a Blast Furnace

Most iron is extracted from iron ores in large blast furnaces (Fig. 9.1). In the blast furnace coke (carbon) acts as a reducing agent to reduce iron oxides (mainly Fe_2O_3) to produce raw pig iron which contains about 4% carbon along with some other impurities according to the typical reaction

$$Fe_2O_3 + 3CO \rightarrow 2Fe + 3CO_2$$

The pig iron from the blast furnace is usually transferred in the liquid state to a steelmaking furnace.

Steelmaking and Processing of Major Steel Product Forms

Plain-carbon steels are essentially alloys of iron and carbon with up to about 1.2% carbon. However, the majority of steels contain less than 0.5% carbon. Most steel is made by oxidizing the carbon and other impurities in the pig iron until the carbon content of the iron is reduced to the required level.

The most commonly used process for converting pig iron into steel is the

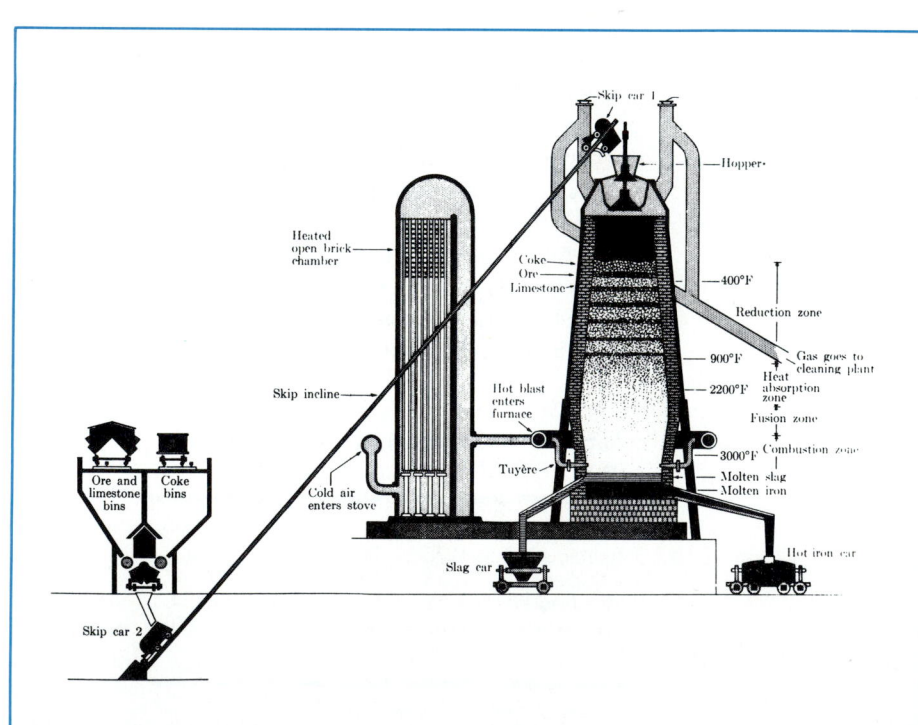

FIGURE 9.1 Cross section of the general operation of a modern blast furnace. *(After A. G. Guy, "Elements of Physical Metallurgy," 2d ed., © 1959, Addison-Wesley, Fig. 2-5, p. 21.)*

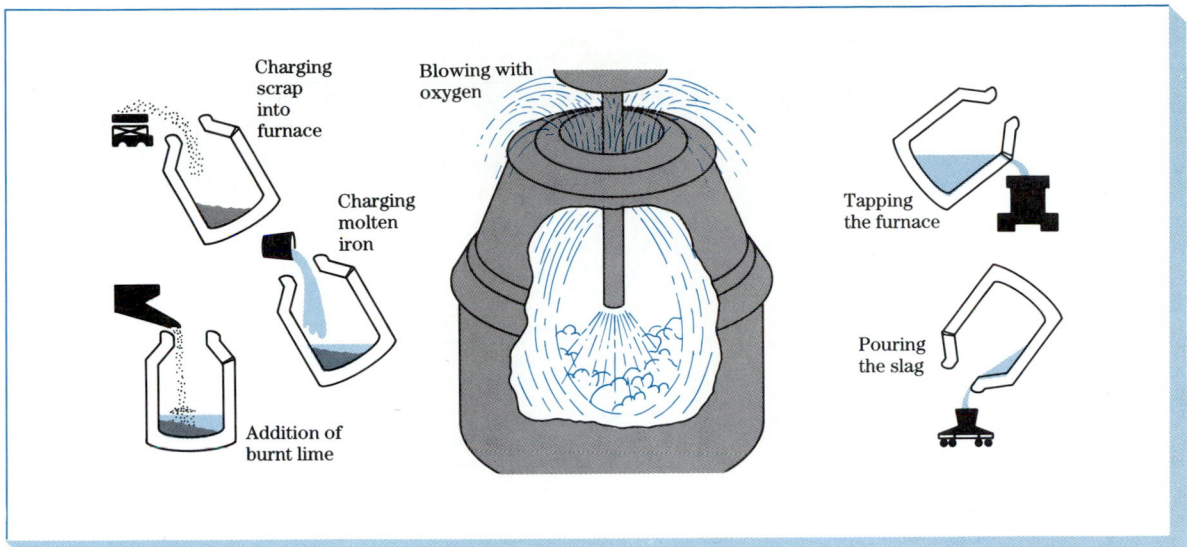

Charging
scrap
into
furnace

Charging
molten
iron

Addition of
burnt lime

Blowing with
oxygen

Tapping
the furnace

Pouring
the slag

FIGURE 9.2 Steelmaking in a basic-oxygen furnace. *(Courtesy of Inland Steel.)*

basic-oxygen process. In this process pig iron and up to about 30% steel scrap are charged into a barrel-shaped refractory-lined converter into which an oxygen lance is inserted (Fig. 9.2). Pure oxygen from the lance reacts with the liquid bath to form iron oxide. Carbon in the steel then reacts with the iron oxide to form carbon monoxide:

$$FeO + C \rightarrow Fe + CO$$

Immediately before the oxygen reaction starts, slag-forming fluxes (chiefly lime) are added in controlled amounts. In this process, the carbon content of the steel can be drastically lowered in about 22 min along with a reduction in the concentration of impurities such as sulfur and phosphorus (Fig. 9.3).

The molten steel from the converter is either cast in stationary molds or continuously cast into long slabs from which long sections are periodically cut off. Today, about 63 percent of the raw steel produced in the United States is continuously cast, and this percentage is expected to increase in future years. After being cast, the ingots are heated in a soaking pit (Fig. 6.4) and hot-rolled into slabs, billets, or blooms. The slabs are subsequently hot- and cold-rolled into steel sheet and plate (see Figs. 9.4 and 6.6 to 6.8). The billets are hot- and cold-rolled into bars, rods, and wire, while blooms are hot- and cold-rolled into shapes such as I beams and rails. Figure 9.5 is a flow diagram which summarizes the principal process steps involved in converting raw materials into major steel product forms.

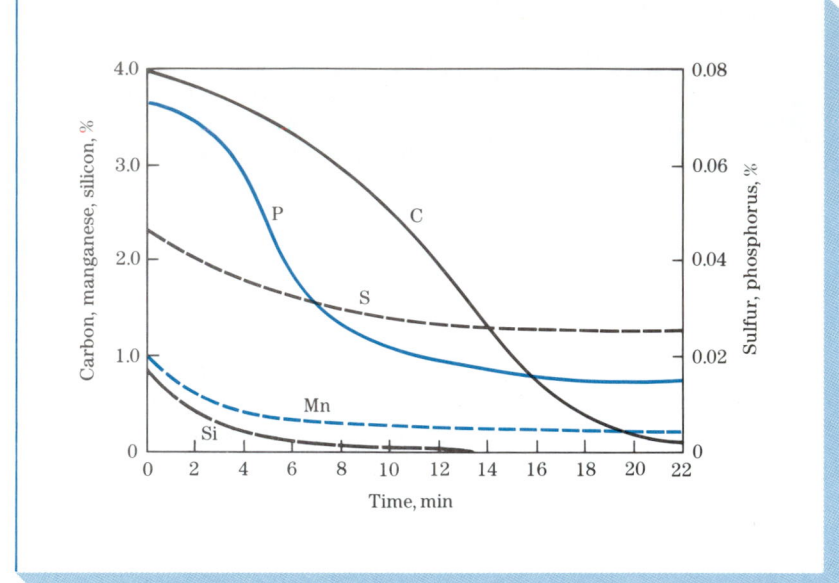

FIGURE 9.3 Schematic representation of progress of refining in a top-blown basic-lined vessel. [*After H. E. McGannon (ed.), "The Making, Shaping, and Treating of Steel," 9th ed., United States Steel Corp., 1971, p. 494.*]

FIGURE 9.4 Hot rolling of steel strip. This picture shows the roughing hot-rolling mills in the background and six finishing hot-rolling mills in the foreground. A strip of steel is exiting the last finishing stand and is being water-quenched. (*Courtesy of U.S. Steel Corp.*)

9.2 THE IRON—IRON CARBIDE PHASE DIAGRAM

Iron-carbon alloys containing from a very small amount (about 0.03%) to about 1.2% carbon, 0.25 to 1.00% manganese, and minor amounts of other elements[1] are termed *plain-carbon steels*. However, for purposes of this section of the book, plain-carbon steels will be treated as essentially iron-carbon binary alloys. The effects of other elements in steels will be dealt with in later sections.

The Iron—Iron Carbide Phase Diagram

The phases present in very slowly cooled iron-carbon alloys at various temperatures and compositions of iron with up to 6.67% carbon are shown in the $Fe-Fe_3C$ phase diagram of Fig. 9.6. This phase diagram is not a true equilibrium diagram since the compound iron carbide (Fe_3C) which is formed is not a true equilibrium phase. Under certain conditions, Fe_3C, which is called *cementite*, can decompose into the more stable phases of iron and carbon (graphite). However, for most practical conditions Fe_3C is very stable and will therefore be treated as an equilibrium phase.

Solid Phases in the Fe—Fe₃C Phase Diagram

The $Fe-Fe_3C$ diagram contains the following solid phases: α ferrite, austenite (γ), cementite (Fe_3C), and δ ferrite.

> *α ferrite.* This phase is an interstitial solid solution of carbon in the BCC iron crystal lattice. As indicated by the $Fe-Fe_3C$ phase diagram, carbon is only slightly soluble in α ferrite, reaching a maximum solid solubility of 0.02% at 723°C. The solubility of carbon in α ferrite decreases to 0.005% at 0°C.
>
> *Austenite (γ).* The interstitial solid solution of carbon in γ iron is called *austenite*. Austenite has an FCC crystal structure and a much higher solid solubility for carbon than α ferrite. The solid solubility of carbon in austenite is a maximum of 2.08% at 1148°C and decreases to 0.8% at 723°C (Fig. 9.6).
>
> *Cementite (Fe_3C).* The intermetallic compound Fe_3C is called *cementite*. Cementite has negligible solubility limits and a composition of 6.67% C and 93.3% Fe. Cementite is a hard and brittle compound.
>
> *δ ferrite.* The interstitial solid solution of carbon in δ iron is called δ *ferrite*. It has a BCC crystal structure like α ferrite but with a greater lattice constant. The maximum solid solubility of carbon in δ ferrite is 0.09% at 1465°C.

Invariant Reactions in the Fe—Fe₃C Phase Diagram

Peritectic reaction At the peritectic reaction point, liquid of 0.53% C combines with δ ferrite of 0.09% C to form γ austenite of 0.17% C. This reaction which occurs at 1495°C can be written as

---

[1] Plain-carbon steels also contain impurities of silicon, phosphorus, and sulfur as well as others.

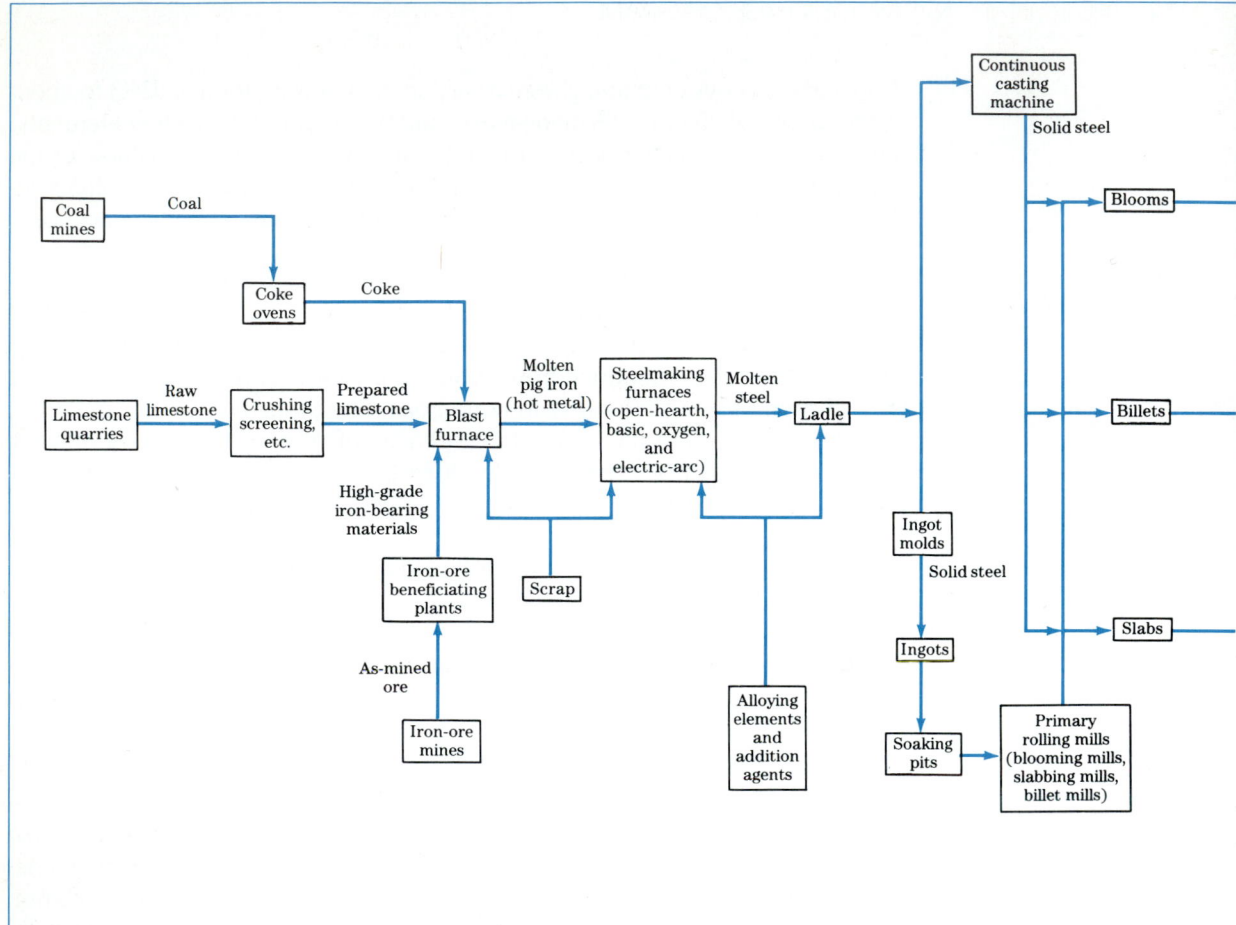

FIGURE 9.5 Flow diagram showing the principal process steps involved in converting raw materials into the major product forms, excluding coated products. [*H. E. McGannon (ed.), "The Making, Shaping, and Treating of Steel," 9th ed., United States Steel Corp., 1971, p. 2.*]

$$\text{Liquid (0.53\% C)} + \delta \text{ (0.09\% C)} \xrightarrow{1495°C} \gamma \text{ (0.17\% C)}$$

δ Ferrite is a high-temperature phase and so is not encountered in plain-carbon steels at lower temperatures.

Eutectic reaction At the eutectic reaction point, liquid of 4.3% forms γ austenite of 2.08% C and the intermetallic compound Fe_3C (cementite) which contains 6.67% C. This reaction which occurs at 1148°C can be written as

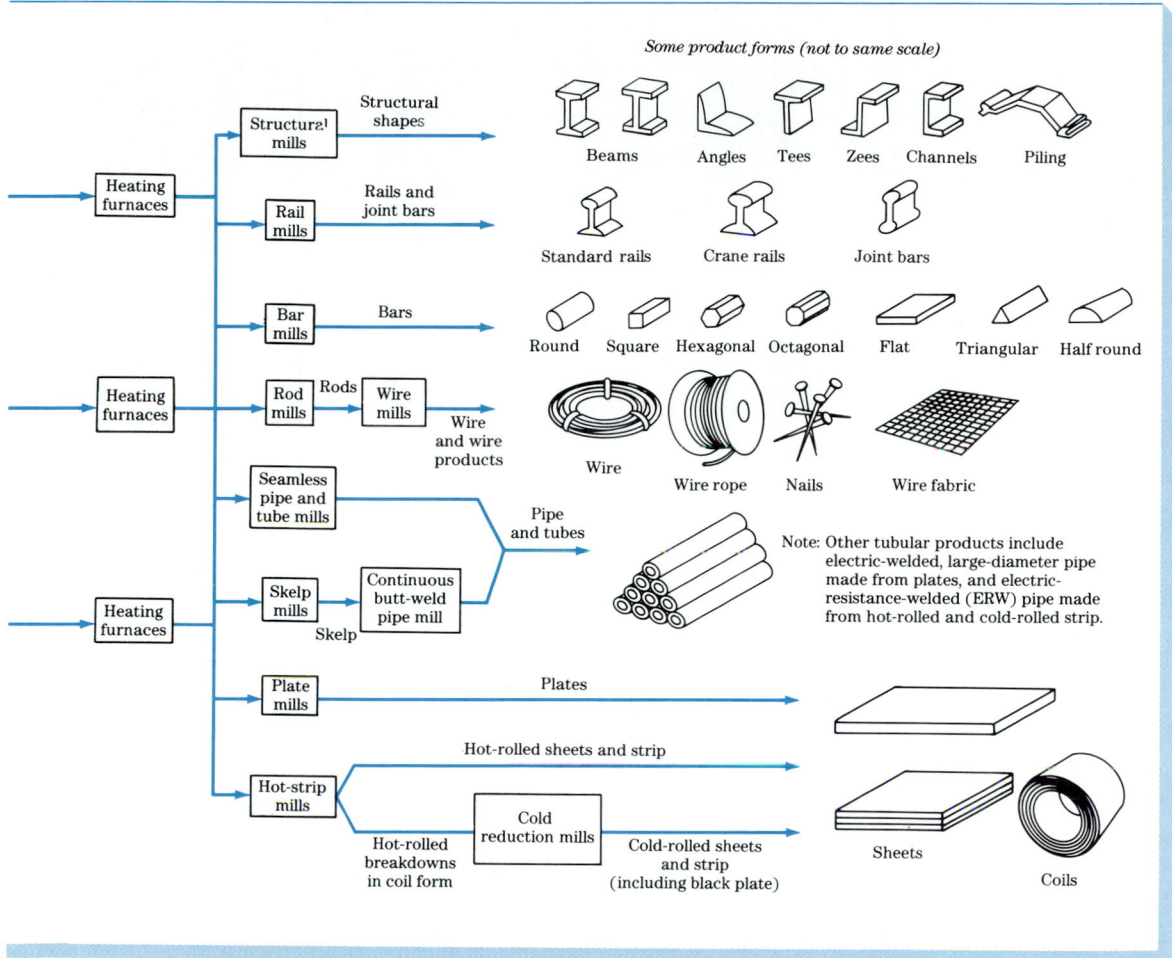

Some product forms (not to same scale)

Beams   Angles   Tees   Zees   Channels   Piling

Standard rails   Crane rails   Joint bars

Round   Square   Hexagonal   Octagonal   Flat   Triangular   Half round

Wire   Wire rope   Nails   Wire fabric

Note: Other tubular products include electric-welded, large-diameter pipe made from plates, and electric-resistance-welded (ERW) pipe made from hot-rolled and cold-rolled strip.

Plates   Sheets   Coils

$$\text{Liquid (4.3\% C)} \xrightarrow{1148°C} \gamma \text{ austenite (2.08\% C)} + Fe_3C \text{ (6.67\% C)}$$

This reaction is not encountered in plain-carbon steels because their carbon contents are too low.

Eutectoid reaction   At the eutectoid reaction point, solid austenite of 0.8% C produces α ferrite with 0.02% C and Fe_3C (cementite) which contains 6.67% C. This reaction which occurs at 723°C can be written as

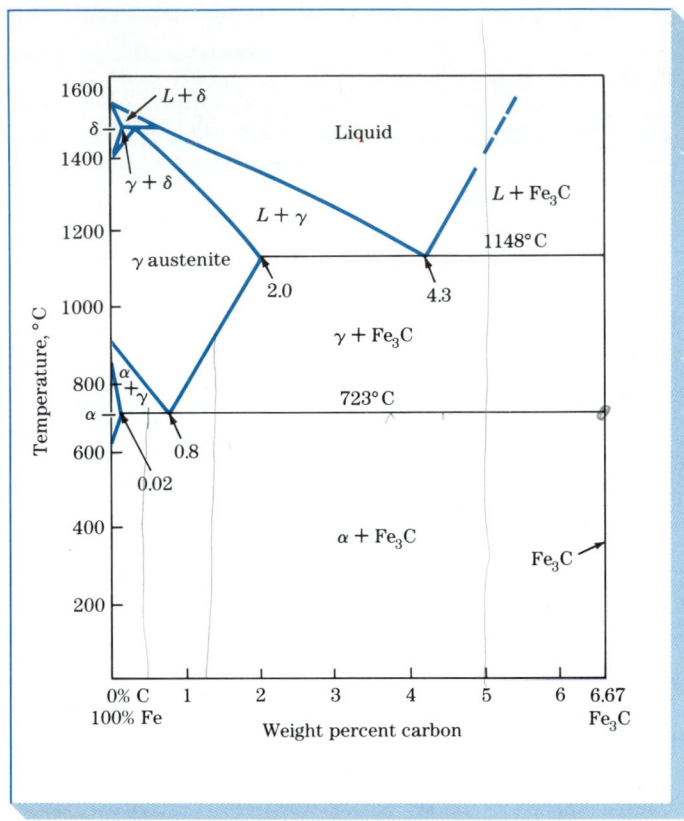

FIGURE 9.6 The iron–iron carbide phase diagram.

$$\gamma \text{ austenite } (0.8\% \text{ C}) \xrightarrow{723°C} \alpha \text{ ferrite } (0.02\% \text{ C}) + \text{Fe}_3\text{C } (6.67\% \text{ C})$$

This eutectoid reaction which takes place completely in the solid state is important for some of the heat treatments of plain-carbon steels.

A plain-carbon steel which contains 0.8% C is called a *eutectoid steel* since an all-eutectoid structure of α ferrite and Fe_3C is formed when austenite of this composition is slowly cooled below the eutectoid temperature. If a plain-carbon steel contains less than 0.8% C, it is termed a *hypoeutectoid steel,* and if the steel contains more than 0.8% C, it is designated a *hypereutectoid steel.*

Slow Cooling of Plain-Carbon Steels

Eutectoid plain-carbon steels If a sample of a 0.8% (eutectoid) plain-carbon steel is heated to about 750°C and held for a sufficient time, its structure will become homogeneous austenite. This process is called *austenitizing.* If this eutectoid steel is then cooled very slowly to just above the eutectoid temperature, its structure will remain austenitic, as indicated in Fig. 9.7 at point *a.* Further cooling to the eutectoid temperature or just below it will cause the entire structure to transform from austenite to a lamellar structure of alternate

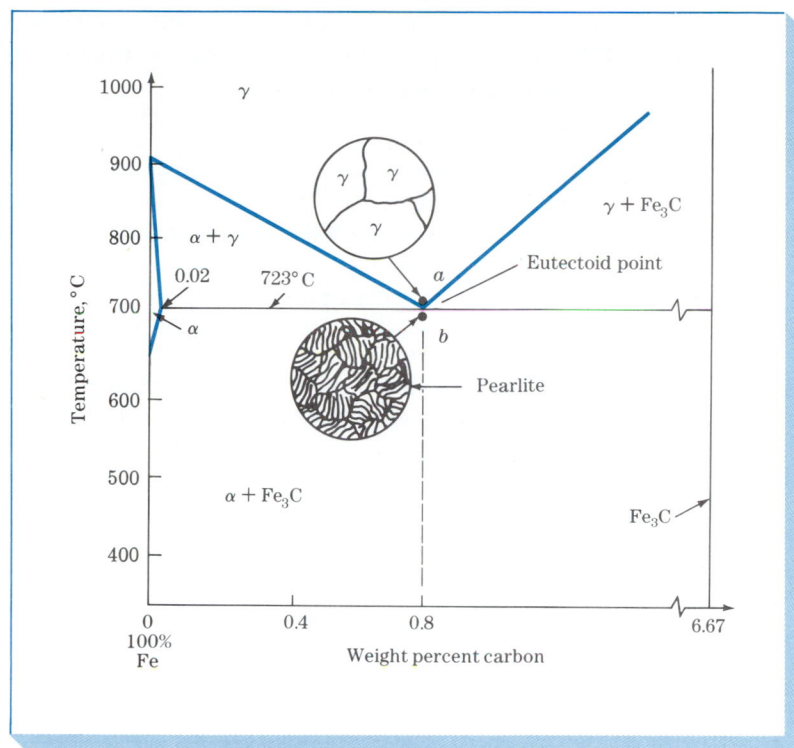

FIGURE 9.7
Transformation of a eutectoid steel (0.8% C) with slow cooling. *(After W. F. Smith, "Structure and Properties of Engineering Alloys," McGraw-Hill, 1981, p. 8.)*

plates of α ferrite and cementite (Fe_3C). Just below the eutectoid temperature, at point *b* in Fig. 9.7, the lamellar structure will appear as shown in Fig. 9.8. This eutectoid structure is called *pearlite* since it resembles mother-of-pearl.

FIGURE 9.8
Microstructure of a slowly cooled eutectoid steel. The microstructure consists of lamellar eutectoid pearlite. The dark etched phase is cementite, and the white phase is ferrite. (Etch: picral; magnification $650\times$.) *(United States Steel Corp., as presented in "Metals Handbook," vol. 8, 8th ed., American Society for Metals, 1973, p. 188.)*

Since the solubility of carbon in α ferrite and Fe_3C changes very little from 723°C to room temperature, the pearlite structure will remain essentially unchanged in this temperature interval.

Example Problem 9.1

A 0.80% C eutectoid plain-carbon steel is slowly cooled from 750°C to a temperature just slightly below 723°C. Assuming that the austenite is completely transformed to α ferrite and cementite:

(a) Calculate the weight percent eutectoid ferrite formed.
(b) Calculate the weight percent eutectoid cementite formed.

Solution:

Referring to Fig. 9.6, we first draw a tie line just below 723°C from the α ferrite phase boundary to the Fe_3C phase boundary and indicate the 0.80% C composition on the tie line as shown below.

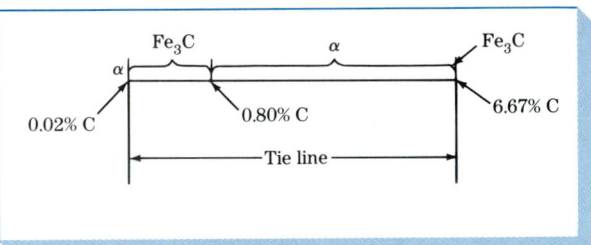

(a) The weight fraction of ferrite is calculated from the ratio of the segment of the tie line to the right of 0.80% C over the whole length of the tie line. Multiplying by 100 percent gives the weight percent ferrite:

$$\text{Wt \% ferrite} = \frac{6.67 - 0.80}{6.67 - 0.02} \times 100\% = \frac{5.87}{6.65} \times 100\% = 88.3\% \blacktriangleleft$$

(b) The weight percent cementite is calculated in a similar way by using the ratio of the segment of the tie line to the left of 0.80% C over the length of the whole tie line and multiplying by 100 percent:

$$\text{Wt \% cementite} = \frac{0.80 - 0.02}{6.67 - 0.02} \times 100\% = \frac{0.78}{6.65} \times 100\% = 11.7\% \blacktriangleleft$$

Hypoeutectoid plain-carbon steels If a sample of a 0.4% C plain-carbon steel (hypoeutectoid steel) is heated to about 900°C (point a in Fig. 9.9) for a

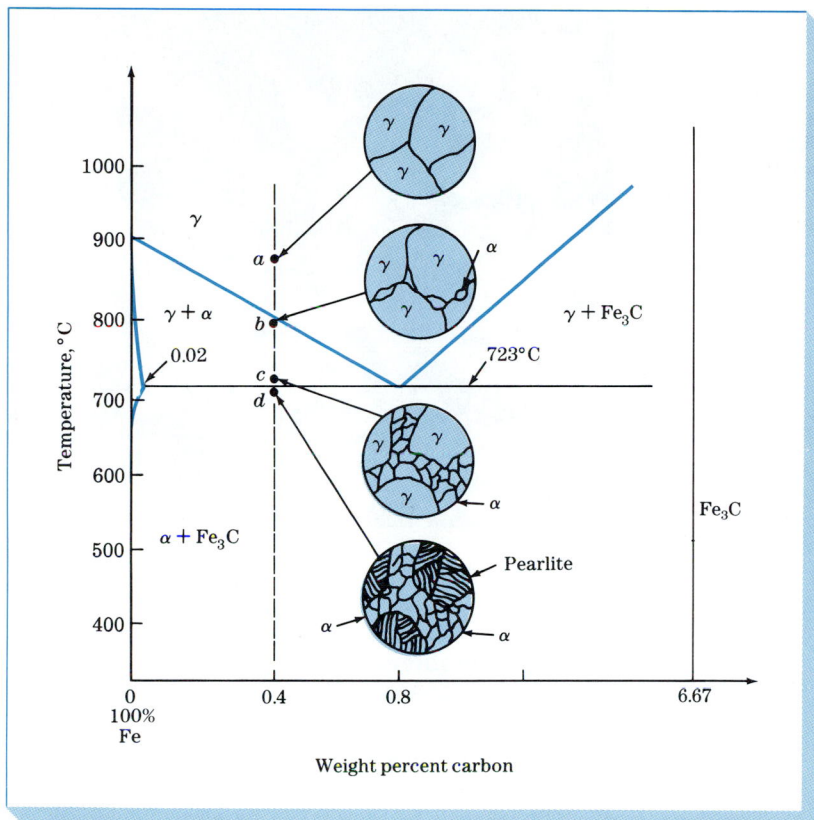

FIGURE 9.9
Transformation of a 0.4% C hypoeutectoid plain-carbon steel with slow cooling. *(After W. F. Smith, "Structure and Properties of Engineering Alloys," McGraw-Hill, 1981, p. 10.)*

sufficient time, its structure will become homogeneous austenite. Then, if this steel is slowly cooled to temperature *b* in Fig. 9.9 (about 775°C), *proeutectoid*[1] *ferrite* will nucleate and grow mostly at the austenitic grain boundaries. If this alloy is slowly cooled from temperature *b* to *c* in Fig. 9.9, the amount of proeutectoid ferrite formed will continue to increase until about 50 percent of the austenite is transformed. While the steel is cooling from *b* to *c*, the carbon content of the remaining austenite will be increased from 0.4 to 0.8%. At 723°C, if very slow cooling conditions prevail, the remaining austenite will transform isothermally into pearlite by the eutectoid reaction austenite → ferrite + cementite. The α ferrite in the pearlite is called *eutectoid ferrite* to distinguish it from the proeutectoid ferrite which forms first above 723°C. Figure 9.10 is an optical micrograph of the structure of a 0.35% C hypoeutectoid steel which was austenitized and slowly cooled to room temperature.

---

[1] The prefix *pro-* means "before," and thus the term *proeutectoid ferrite* is used to distinguish this constituent, which forms earlier, from eutectoid ferrite, which forms by the eutectoid reaction later in the cooling.

FIGURE 9.10
Microstructure of a 0.35% C hypoeutectoid plain-carbon steel slowly cooled from the austenite region. The white constituent is proeutectoid ferrite; the dark constituent is pearlite. (Etchant: 2% nital; magnification 500×.) *(After W. F. Smith, "Structure and Properties of Engineering Alloys," McGraw-Hill, 1981, p. 11.)*

Example Problem 9.2

(*a*) A 0.40% C hypoeutectoid plain-carbon steel is slowly cooled from 940°C to a temperature just slightly above 723°C.
   (i) Calculate the weight percent austenite present in the steel.
   (ii) Calculate the weight percent proeutectoid ferrite present in the steel.
(*b*) A 0.40% C hypoeutectoid plain-carbon steel is slowly cooled from 940°C to a temperature just slightly below 723°C.
   (i) Calculate the weight percent proeutectoid ferrite present in the steel.
   (ii) Calculate the weight percent eutectoid ferrite and weight percent eutectoid cementite present in the steel.

Solution:

Referring to Fig. 9.6 and using tie lines:

(*a*) (i) Wt % austenite $= \dfrac{0.40 - 0.02}{0.80 - 0.02} \times 100\% = 50\%$ ◀

   (ii) Wt % proeutectoid ferrite $= \dfrac{0.80 - 0.40}{0.80 - 0.02} \times 100\% = 50\%$ ◀

(*b*) (i) The weight percent proeutectoid ferrite present in the steel just below 723°C will be the same as that just above 723°C, which is 50%.

   (ii) The weight percent total ferrite and cementite just below 723°C are

$$\text{Wt \% total ferrite} = \frac{6.67 - 0.40}{6.67 - 0.02} \times 100\% = 94.3\%$$

$$\text{Wt \% total cementite} = \frac{0.40 - 0.02}{6.67 - 0.02} \times 100\% = 5.7\%$$

$$\text{Wt \% eutectoid ferrite} = \text{total ferrite} - \text{proeutectoid ferrite}$$
$$= 94.3 - 50 = 44.3\% \blacktriangleleft$$
$$\text{Wt \% eutectoid cementite} = \text{wt \% total cementite} = 5.7\% \blacktriangleleft$$

(No proeutectoid cementite was formed during cooling.)

Hypereutectoid plain-carbon steels   If a sample of a 1.2% C plain-carbon steel (hypereutectoid steel) is heated to about 950°C and held for a sufficient time, its structure will become essentially all austenite (point *a* in Fig. 9.11). Then, if this steel is cooled very slowly to temperature *b* in Fig. 9.11, *proeutectoid cementite* will begin to nucleate and grow primarily at the austenite grain boundaries. With further slow cooling to point *c* of Fig. 9.11, which is just above 723°C, more proeutectoid cementite will be formed at the austenite grain

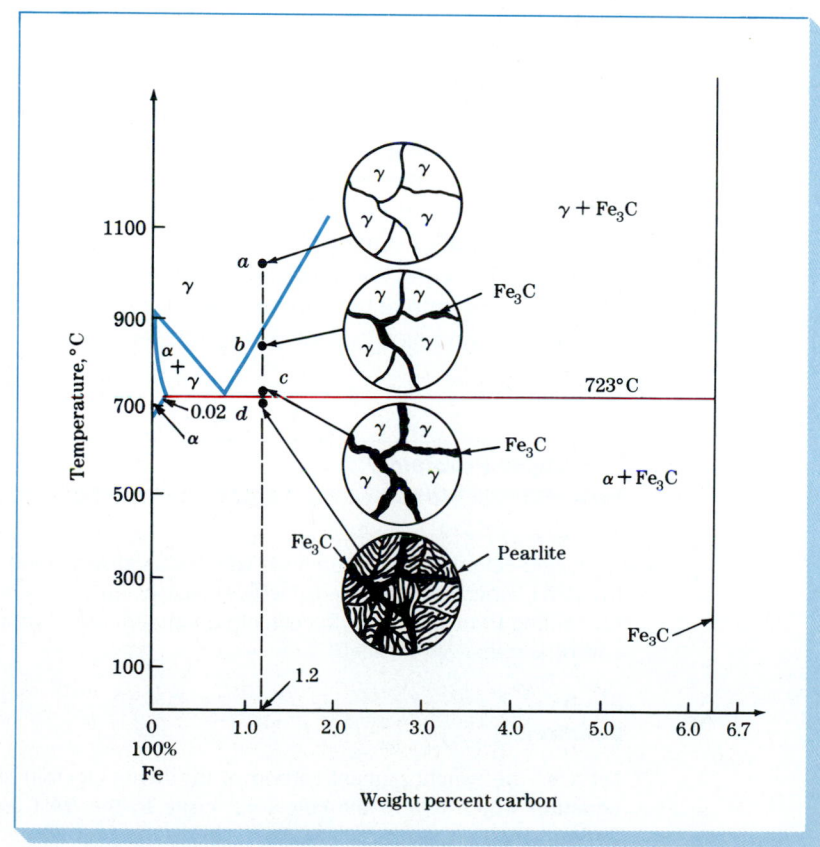

FIGURE 9.11

Transformation of a 1.2% C hypereutectoid plain-carbon steel with slow cooling. *(After W. F. Smith, "Structure and Properties of Engineering Alloys," McGraw-Hill, 1981, p. 12.)*

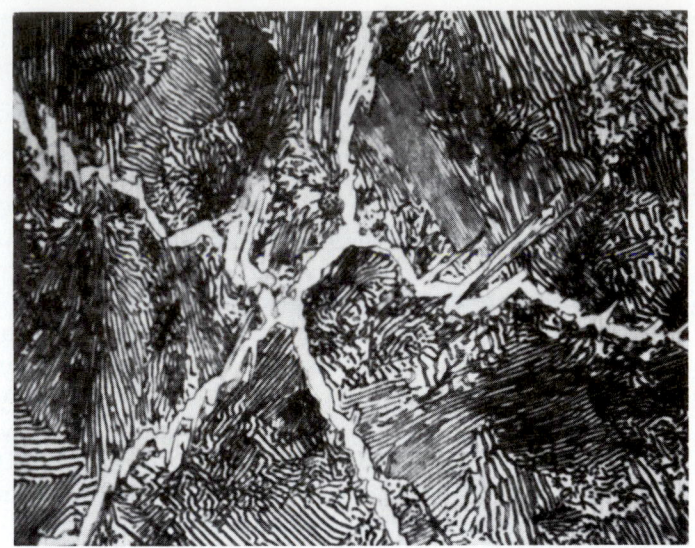

boundaries. If conditions approaching equilibrium are maintained by the slow cooling, the overall carbon content of the austenite remaining in the alloy will change from 1.2 to 0.8%.

With still further slow cooling to 723°C or just slightly below this temperature, the remaining austenite will transform to pearlite by the eutectoid reaction, as indicated at point *d* of Fig. 9.11. The cementite formed by the eutectoid reaction is called *eutectoid cementite* to distinguish it from the proeutectoid cementite formed at temperatures above 723°C. Similarly, the ferrite formed by the eutectoid reaction is termed *eutectoid ferrite*. Figure 9.12 is an optical micrograph of the structure of a 1.2% C hypereutectoid steel which was austenitized and slowly cooled to room temperature.

Example Problem 9.3

A hypoeutectoid plain-carbon steel which was slow-cooled from the austenitic region to room temperature contains 9.1 wt % eutectoid ferrite. Assuming no change in structure on cooling from just below the eutectoid temperature to room temperature, what is the carbon content of the steel?

Solution:

Let x = the weight percent carbon of the hypoeutectoid steel. Now we can use the equation which relates the eutectoid ferrite to the total ferrite and the proeutectoid ferrite, which is

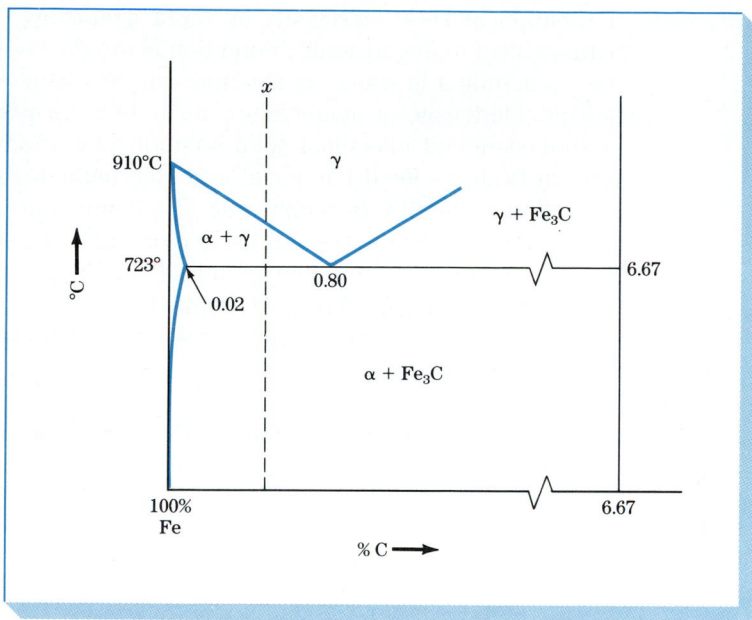

FIGURE EP9.3

Eutectoid ferrite = total ferrite − proeutectoid ferrite

Using Fig. EP9.3 and the lever rule, we can make the equation

$$
\underset{\substack{\text{Eutectoid}\\\text{ferrite}}}{0.091} = \underset{\substack{\text{Total}\\\text{ferrite}}}{\frac{6.67 - x}{6.67 - 0.02}} - \underset{\substack{\text{Proeutectoid}\\\text{ferrite}}}{\frac{0.80 - x}{0.80 - 0.02}} = \frac{6.67}{6.65} - \frac{x}{6.65} - \frac{0.80}{0.78} + \frac{x}{0.78}
$$

or $1.28x - 0.150x = 0.091 - 1.003 + 1.026 = 0.114$

$$
x = \frac{0.114}{1.13} = 0.101\% \text{ C} \blacktriangleleft
$$

9.3 HEAT TREATMENT OF PLAIN-CARBON STEELS

By varying the manner in which plain-carbon steels are heated and cooled, different combinations of mechanical properties for steels can be obtained. In this section we shall examine some of the structural and property changes that take place during some of the important heat treatments given to plain-carbon steels.

Martensite

Formation of Fe–C martensite by rapid quenching If a sample of a plain-carbon steel in the austenitic condition is rapidly cooled to room temperature by quenching it in water, its structure will be changed from austenite to *martensite*. Martensite in plain-carbon steels is a metastable phase consisting of a supersaturated interstitial solid solution of carbon in body-centered cubic iron or body-centered tetragonal iron (the tetragonality is caused by a slight distortion of the BCC iron unit cell). The temperature, upon cooling, at which the austenite-to-martensite transformation starts is called the *martensite start*, M_s, temperature, and the temperature at which the transformation finishes is called the *martensite finish*, M_f, temperature. The M_s temperature for Fe–C alloys decreases as the weight percent carbon increases in these alloys, as shown in Fig. 9.13.

Microstructure of Fe–C martensites The microstructure of martensites in plain-carbon steels depends on the carbon content of the steel. If the steel contains less than about 0.6% C, the martensite consists of *domains* of laths of different but limited orientations through a whole domain. The structure within the laths is highly distorted, consisting of regions with high densities of dislocation tangles. Figure 9.14*a* is an optical micrograph of *lath martensite* in an Fe–0.2% C alloy at 600×, while Fig. 9.15 shows the substructure of lath martensite in this alloy in an electron micrograph at 60,000×.

As the carbon content of the Fe–C martensites is increased to above about 0.6% C, a different type of martensite, called *plate martensite*, begins to form. Above about 1% C, Fe–C alloys consist entirely of plate martensite. Figure 9.14*b* is an optical micrograph of plate martensite in an Fe–1.2% C alloy at 600×.

FIGURE 9.13 Effect of carbon content on the martensite-transformation start temperature, M_s, for iron-carbon alloys. *(After A. R. Marder and G. Krauss, as presented in "Hardenability Concepts with Applications to Steel," AIME, 1978, p. 238.)*

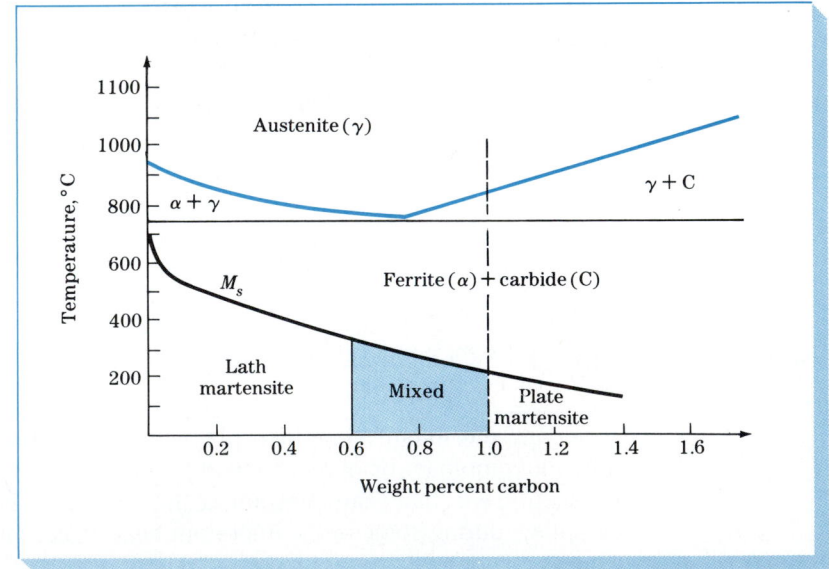

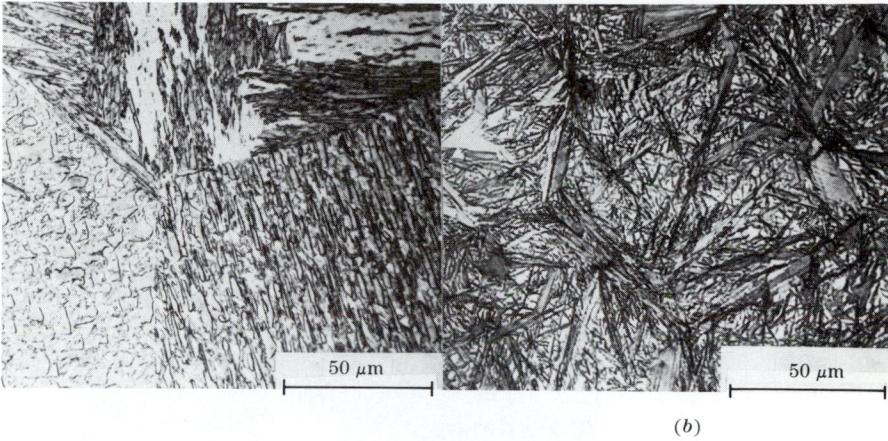

FIGURE 9.14 Effect of carbon content on the structure of martensite in plain-carbon steels: (*a*) lath type, (*b*) plate type. (Etchant: sodium bisulfite; optical micrographs.) [*After A. R. Marder and G. Krauss, Trans. ASM, 60:651(1967).*]

(*b*)

The plates in high-carbon Fe–C martensites vary in size and have a fine structure of parallel twins, as shown in Fig. 9.16. The plates are often surrounded by large amounts of untransformed (retained) austenite. Fe–C martensites with carbon contents between about 0.6 and 1.0% C have microstructures consisting of both lath- and plate-type martensites.

Structure of Fe–C martensites on an atomic scale   The transformation of austenite to martensite in Fe–C alloys (plain-carbon steels) is considered to be

FIGURE 9.15 Structure of lath martensite in an Fe–0.2% C alloy. (Note the parallel alignment of the laths.) [*After A. R. Marder and G. Krauss, ASM., 60:651(1967).*]

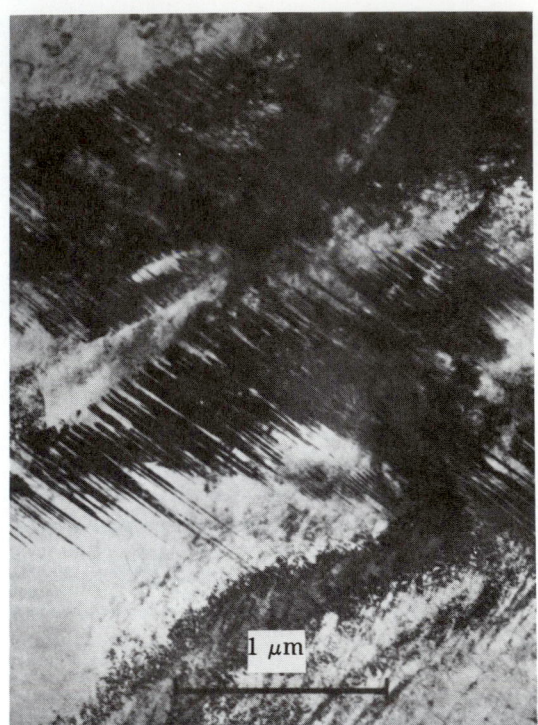

FIGURE 9.16 Plate martensite showing fine transformation twins. [*After M. Oka and C. M. Wayman, Trans. ASM, 62:370(1969).*]

diffusionless since the transformation takes place so rapidly that the atoms do not have time to intermix. There appears to be no thermal-activation energy barrier to prevent martensite from forming. Also, it is believed that no compositional change in the parent phase takes place after the reaction and that each atom tends to retain its original neighbors. The relative positions of the carbon atoms with respect to the iron atoms are the same in the martensite as they were in the austenite.

For carbon contents in Fe–C martensites of less than about 0.2% C, the austenite transforms to a BCC α ferrite crystal structure. As the carbon content of the Fe–C alloys is increased, the BCC structure is distorted into a BCT (body-centered tetragonal) crystal structure. The largest interstitial hole in the γ iron FCC crystal structure has a diameter of 0.104 nm (Fig. 9.17a), whereas the largest interstitial hole in the α iron BCC structure has a diameter of 0.072 nm (Fig. 9.17b). Since the carbon atom has a diameter of 0.154 nm, it can be accommodated in interstitial solid solution to a greater extent in the FCC γ iron lattice than in the BCC lattice. When Fe–C martensites with more than about 0.2% C are produced by rapid cooling from austenite, the reduced interstitial spacings of the BCC lattice cause the carbon atoms to distort the BCC unit cell along its c axis to accommodate the carbon atoms (Fig. 9.17c). Figure 9.18 shows how the c axis of the Fe–C martensite lattice is elongated as its carbon content increases.

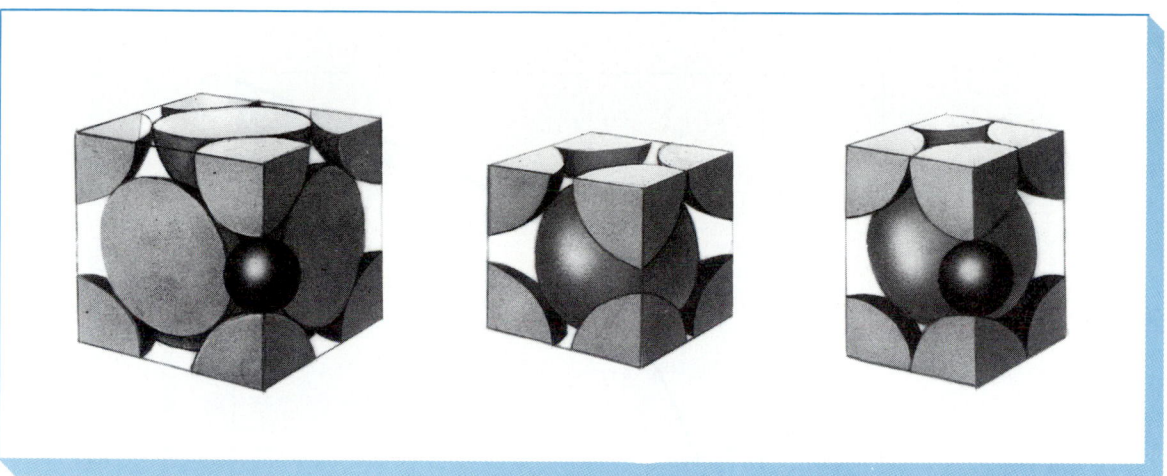

FIGURE 9.17 (*a*) FCC γ iron unit cell showing a carbon atom in a large interstitial hole along the cube edge of the cell. (*b*) BCC α iron unit cell indicating a smaller interstitial hole between cube-edge atoms of the unit cell. (*c*) BCT (body-centered tetragonal) iron unit cell produced by the distortion of the BCC unit cell by the interstitial carbon atom. *(After E. R. Parker and V. F. Zackay, Strong and Ductile Steels, Sci. Am., November 1968, p. 36; Copyright © by Scientific American, Inc; all rights reserved.)*

Hardness and strength of Fe–C martensites The hardness and strength of Fe–C martensites are directly related to their carbon content and increase as the carbon content is increased (Fig. 9.19). However, ductility and toughness also decrease with increasing carbon content, and so most martensitic plain-

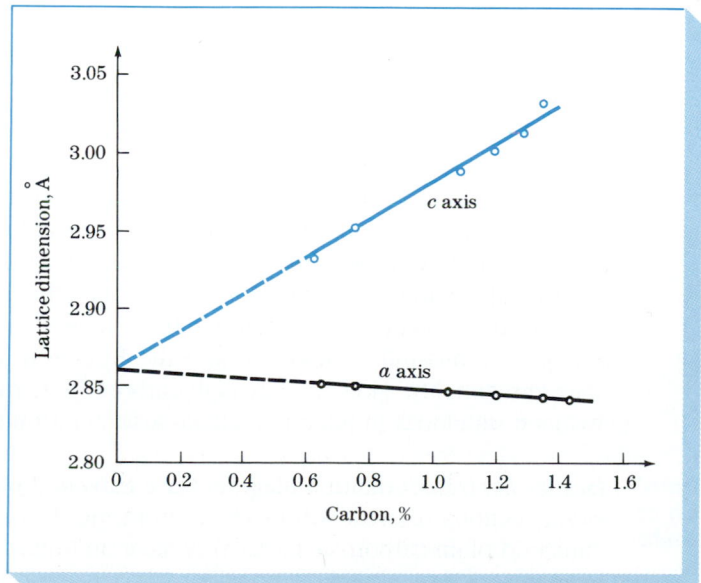

FIGURE 9.18 Variation of *a* and *c* axes of the Fe–C martensite lattice as a function of carbon content. *(After E. C. Bain and H. W. Paxton, "Alloying Elements in Steel," 2d ed., American Society for Metals, 1966, p. 36.)*

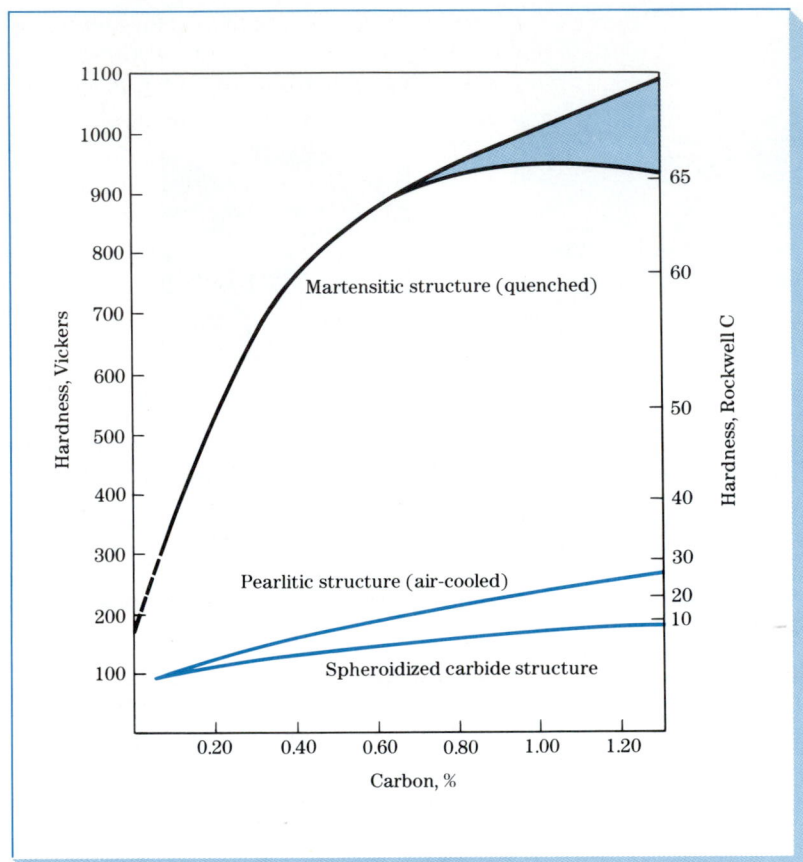

FIGURE 9.19
Approximate hardness of fully hardened martensitic plain-carbon steel as a function of carbon content. The shaded region indicates some possible loss of hardness due to the formation of retained austenite, which is softer than martensite. *(After E. C. Bain and H. W. Paxton, "Alloying Elements in Steel," 2d ed., American Society for Metals, 1966, p. 37.)*

carbon steels are tempered by reheating at a temperature below the transformation temperature of 723°C.

Low-carbon Fe–C martensites are strengthened by a high concentration of dislocations being formed (lath martensite) and by interstitial solid-solution strengthening by carbon atoms. The high concentration of dislocations in networks (lath martensite) makes it difficult for other dislocations to move. As the carbon content increases above 0.2%, interstitial solid-solution strengthening becomes more important and the BCC iron lattice becomes distorted into tetragonality. However, in high-carbon Fe–C martensites the numerous twinned interfaces in plate martensite also contribute to the hardness.

Isothermal Decomposition of Austenite

Isothermal transformation diagram for a eutectoid plain-carbon steel In previous sections the reaction products from the decomposition of austenite of eutectoid plain-carbon steels for very slow and rapid cooling conditions have been described. Let us now consider what reaction products form when austenite of eutectoid steels is rapidly cooled to temperatures below the eutectoid temperature and then *isothermally transformed.*

Isothermal transformation experiments to investigate the microstructural changes for the decomposition of eutectoid austenite can be made by using a number of small samples, each about the size of a dime. The samples are first austenitized in a furnace at a temperature above the eutectoid temperature (Fig. 9.20*a*). The samples are then rapidly cooled (quenched) in a liquid salt bath at the desired temperature below the eutectoid temperature (Fig. 9.20*b*). After various time intervals, the samples are removed from the salt bath one at a time and quenched into water at room temperature (Fig. 9.20*c*). The microstructure after each transformation time can then be examined at room temperature.

Consider the microstructural changes which take place during the isothermal transformation of a eutectoid plain-carbon steel at 705°C, as schematically shown in Fig. 9.21. After being austenitized, the samples are hot-quenched into a salt bath at 705°C. After about 6 min, coarse pearlite has formed to a small extent. After about 67 min, the austenite is completely transformed to coarse pearlite.

By repeating the same procedure for the isothermal transformation of eutectoid steels at progressively lower temperatures, an isothermal transformation (IT) diagram can be constructed, as shown schematically in Fig. 9.22 and from experimental data in Fig. 9.23. The S-shaped curve next to the temperature axis indicates the time necessary for the isothermal transformation of austenite to begin, and the second S curve indicates the time required for the transformation to be completed.

Isothermal transformations of eutectoid steels at temperatures between 723 and about 550°C produce pearlitic microstructures. As the transformation

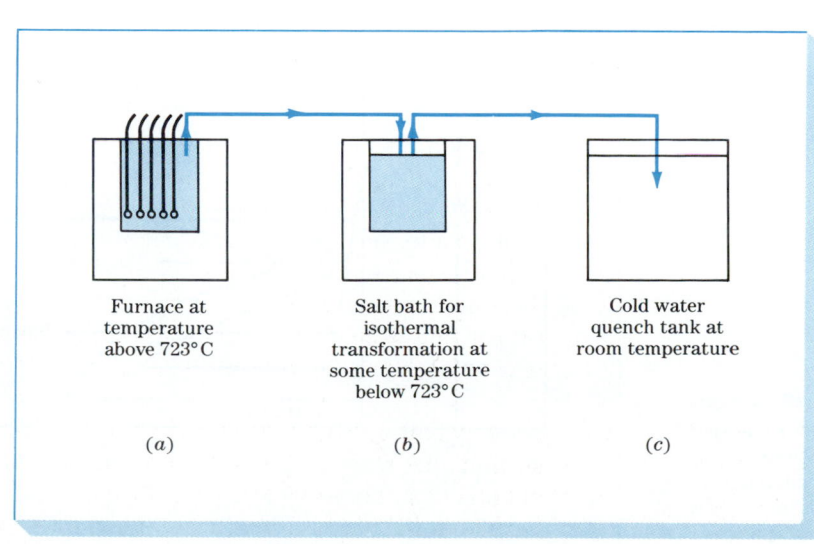

FIGURE 9.20
Experimental arrangement for determining the microscopic changes that occur during the isothermal transformation of austenite in a eutectoid plain-carbon steel. *(After W. F. Smith, "Structure and Properties of Engineering Alloys," McGraw-Hill, 1981, p. 14.)*

Furnace at temperature above 723°C

Salt bath for isothermal transformation at some temperature below 723°C

Cold water quench tank at room temperature

(*a*)

(*b*)

(*c*)

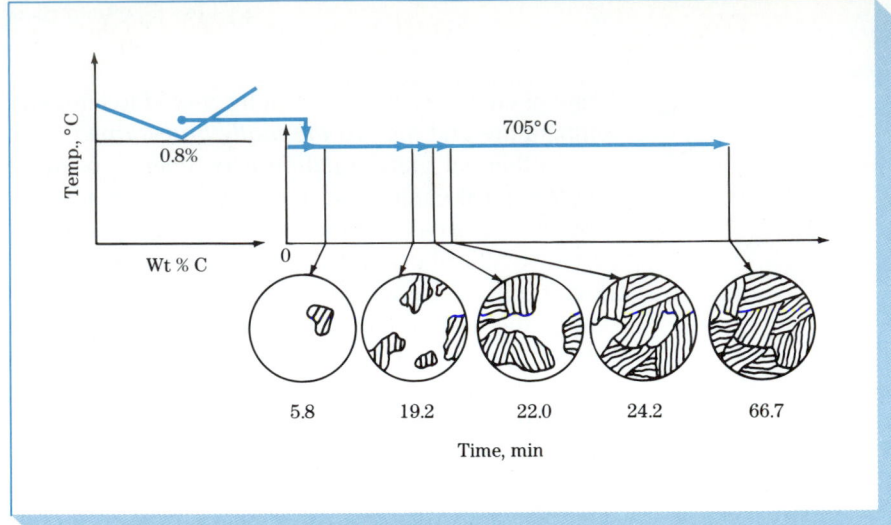

temperature is decreased in this range, the pearlite changes from a coarse to a fine structure (Fig. 9.23). Rapid quenching (cooling) of a eutectoid steel from temperatures above 723°C, where it is in the austenitic condition, transforms the austenite into martensite, as has been previously discussed.

If eutectoid steels in the austenitic condition are hot-quenched to temperatures in the 550 to 250°C range and are isothermally transformed, a structure intermediate between pearlite and martensite, called *bainite,*[1] is produced. Bainite in Fe–C alloys can be defined as an austenitic decomposition product which has a *nonlamellar eutectoid structure* of α ferrite and cementite (Fe_3C). For eutectoid plain-carbon steels, a distinction is made between *upper bainite,*

---

[1] Bainite is named after E. C. Bain, the American metallurgist who first intensively studied the isothermal transformations of steels. See E. S. Davenport and E. C. Bain, *Trans. AIME,* **90**:117 (1930).

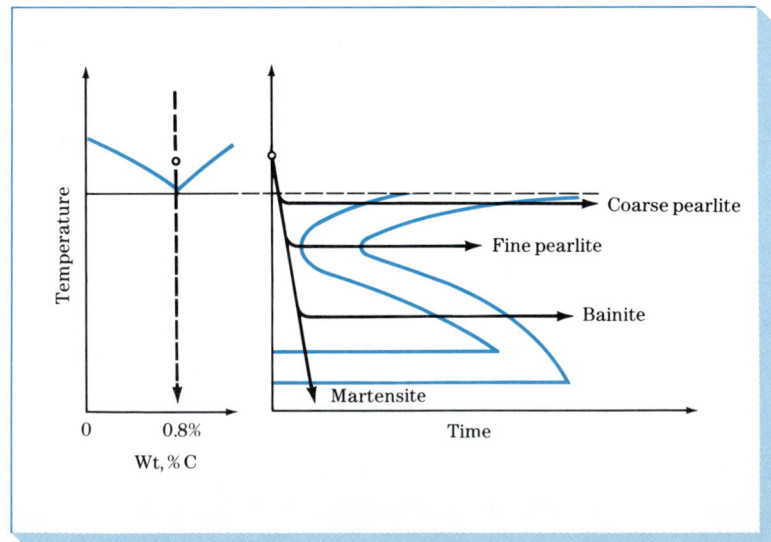

FIGURE 9.22 Isothermal transformation diagram for a eutectoid plain-carbon steel showing its relationship to the Fe–Fe$_3$C phase diagram.

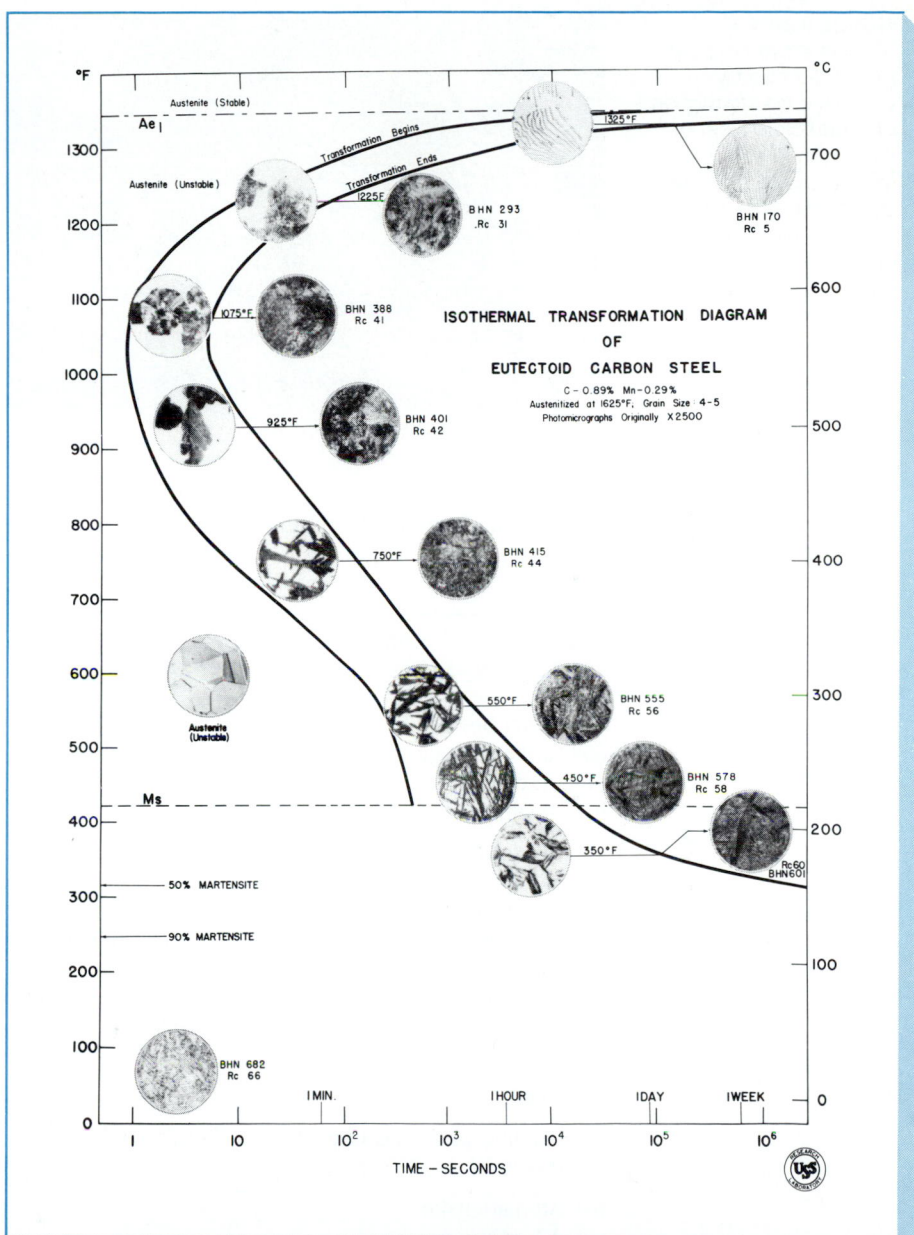

FIGURE 9.23 Isothermal transformation diagram of a eutectoid steel. *(Courtesy of United States Steel Corp., Research Laboratory.)*

which is formed by isothermal transformation at temperatures between about 550 to 350°C, and *lower bainite,* which is formed between about 350 to 250°C. Figure 9.24a shows an electron micrograph (replica-type) of the microstructure of upper bainite for a eutectoid plain-carbon steel, and Fig. 9.24b shows one for lower bainite. Upper bainite has large rodlike cementite regions, whereas lower bainite has much-finer cementite particles. As the transformation tem-

FIGURE 9.24 (*a*) Microstructure of upper bainite formed by a complete transformation of a eutectoid steel at 450°C (850°F). (*b*) Microstructure of lower bainite formed by a complete transformation of a eutectoid steel at 260°C (500°F). The white particles are Fe₃C, and the dark matrix is ferrite. (Electron micrographs, replica-type; magnification 15,000×.) [*After H. E. McGannon (ed.), "The Making, Shaping, and Treating of Steel," 9th ed., United States Steel Corp., 1971.*]

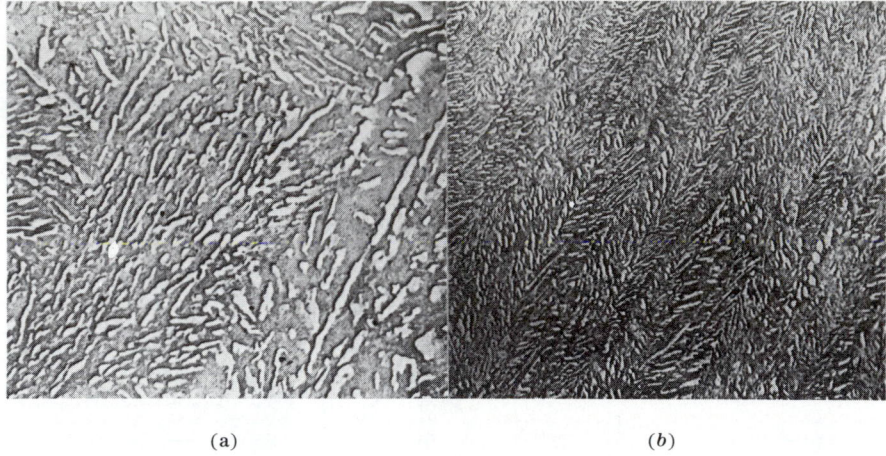

(a)   (b)

perature is decreased, the carbon atoms cannot diffuse as easily, and hence the lower bainite structure has smaller particles of cementite.

Example Problem 9.4

Small thin pieces of 0.25-mm-thick hot-rolled strips of 1080 steel are heated for 1 h at 850°C and then given the heat treatments listed below. Using the isothermal transformation diagram of Fig. 9.23, determine the microstructures of the samples after each heat treatment.

(*a*) Water-quench to room temperature.
(*b*) Hot-quench in molten salt to 690°C and hold 2 h; water-quench.
(*c*) Hot-quench to 610°C and hold 3 min; water-quench
(*d*) Hot-quench to 580°C and hold 2 s; water-quench.
(*e*) Hot-quench to 450°C and hold 1 h; water-quench
(*f*) Hot-quench to 300°C and hold 30 min; water-quench.
(*g*) Hot-quench to 300°C and hold 5 h; water quench.

Solution:

The cooling paths are indicated on Fig. EP9.4 and the microstructures obtained are listed below.

(*a*) All martensite
(*b*) All course pearlite
(*c*) All fine pearlite
(*d*) Approximately 50% fine pearlite and 50% martensite
(*e*) All upper bainite
(*f*) Approximately 50% lower bainite and 50% martensite
(*g*) All lower bainite

Isothermal transformation diagrams for noneutectoid plain-carbon steels   Isothermal transformation diagrams have been determined for noneutectoid plain-

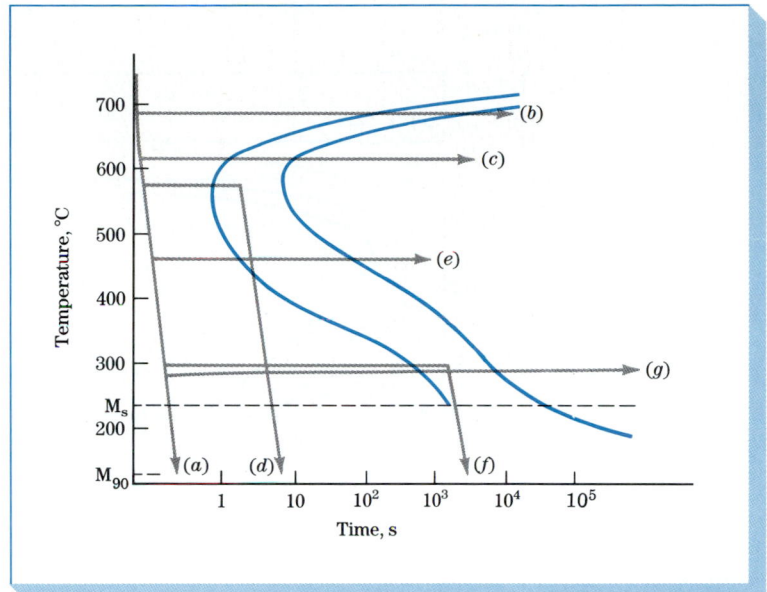

FIGURE EP9.4 Isothermal transformation diagram for an eutectoid plain-carbon steel indicating verious cooling paths.

carbon steels. Figure 9.25 shows an IT diagram for a 0.47% hypoeutectoid plain-carbon steel. Several differences between the IT diagram for a noneutectoid plain-carbon steel and for a eutectoid one (Fig. 9.23) are evident. One major difference is that the S curves of the hypoeutectoid steel have been shifted to the left, so that it is not possible to quench this steel from the austenitic region to produce an entirely martensitic structure.

A second major difference is that another transformation line has been added to the upper part of the eutectoid steel IT diagram which indicates the start of the formation of proeutectoid ferrite. Thus, at temperatures between 723 and about 765°C, only proeutectoid ferrite is produced by isothermal transformation in this temperature range.

Similar IT diagrams have been determined for hypereutectoid plain-carbon steels. However, in this case, the uppermost line of the diagram for these steels is for the start of the formation of proeutectoid cementite.

Continuous-Cooling Transformation Diagram for a Eutectoid Plain-Carbon Steel

In industrial heat-treating operations, in most cases a steel is not isothermally transformed at a temperature above the martensite start temperature but is continuously cooled from the austenitic temperature to room temperature. In continuously cooling a plain-carbon steel, the transformation from austenite to pearlite occurs over a range of temperatures rather than at a single isothermal temperature. As a result, the final microstructure after continuous cooling will be complex since the reaction kinetics change over the temperature range in which the transformation takes place. Figure 9.26 shows a continuous-cooling transformation diagram for a eutectoid plain-carbon steel superimposed over

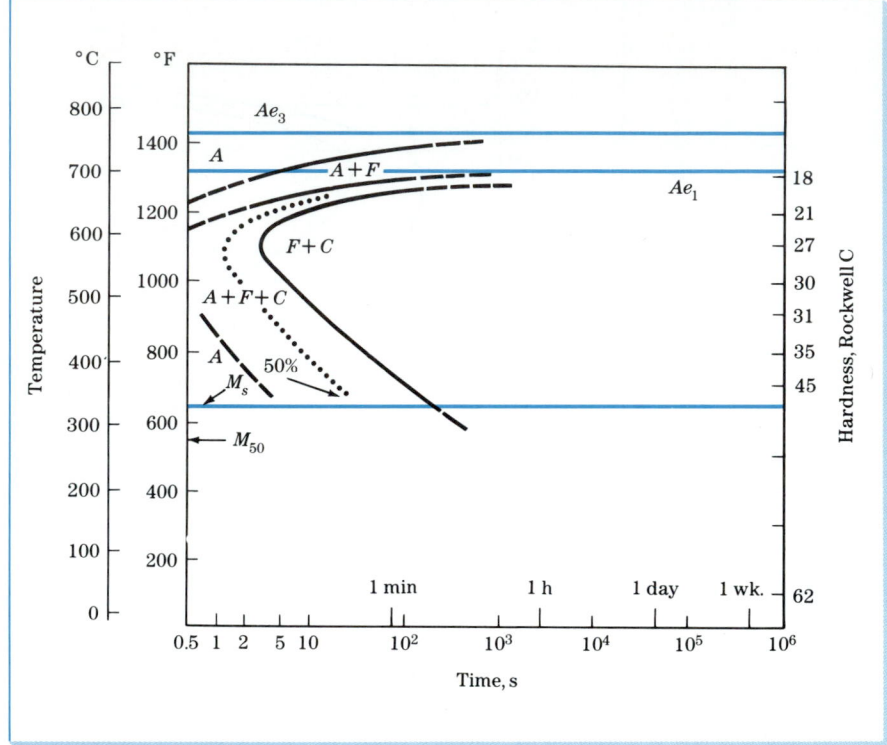

FIGURE 9.25 Isothermal transformation diagram for a hypoeutectoid steel containing 0.47% C and 0.57% Mn (austenitizing temperature: 843°C). [*After R. A. Grange, V. E. Lambert, and J. J. Harrington, Trans. ASM, 51:377(1959).*]

an IT diagram for this steel. The continuous-cooling diagram transformation start and finish lines are shifted to longer times and slightly lower temperatures in relation to the isothermal diagram. Also there are no transformation lines below about 450°C for the austenite-to-bainite transformation.

Figure 9.27 shows different rates of cooling for thin samples of eutectoid plain-carbon steels cooled continuously from the austenitic region to room temperature. Cooling curve *A* represents very slow cooling, such as would be obtained by shutting off the power of an electric furnace and allowing the steel to cool as the furnace cools. The microstructure in this case would be coarse pearlite. Cooling curve *B* represents more rapid cooling, such as would be obtained by removing an austenitized steel from a furnace and allowing the steel to cool in still air. A fine pearlite microstructure is formed in this case.

Cooling curve *C* of Fig. 9.27 starts with the formation of pearlite, but there is insufficient time to complete the austenite-to-pearlite transformation. The remaining austenite that does not transform to pearlite at the upper temperatures will transform to martensite at lower temperatures starting at about 220°C. This type of transformation, since it takes place in two steps, is called a *split transformation*. The microstructure of this steel will thus consist of a mixture

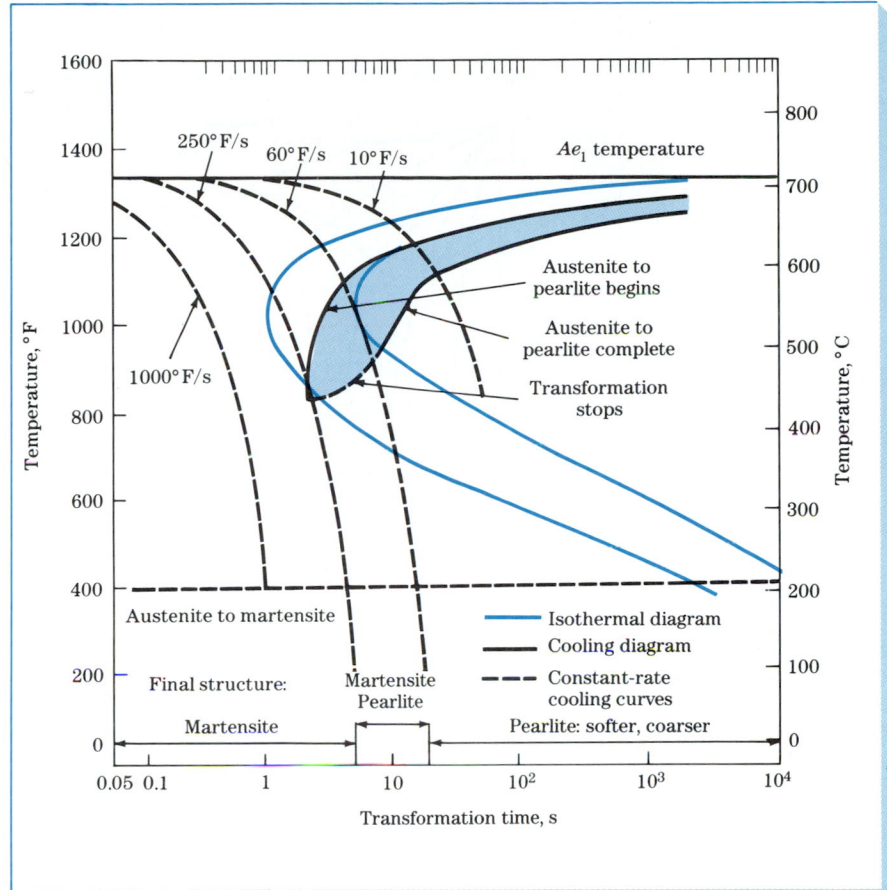

FIGURE 9.26
Continuous-cooling diagram for a plain-carbon eutectoid steel. *(After R. A. Grange and J. M. Kiefer as adapted in E. C. Bain and H. W. Paxton, "Alloying Elements in Steel," 2d ed., American Society for Metals, 1966, p. 254.)*

of pearlite and martensite. Cooling at a rate faster than curve *E* of Fig. 9.27, which is called the *critical cooling rate,* will produce a fully hardened martensitic structure.

Continuous-cooling diagrams have been determined for many hypoeutectoid plain-carbon steels and are more complex since at low temperatures some bainitic structure is also formed during continuous cooling. The discussion of these diagrams is beyond the scope of this book.

Annealing and Normalizing of Plain-Carbon Steels   In Sec. 6.8 the cold-working and annealing processes for metals were discussed and reference should be made to that section. The two most common types of annealing processes applied to commercial plain-carbon steels are *full annealing* and *process annealing.*

In full annealing, hypoeutectoid and eutectoid steels are heated in the austenite region about 40°C above the austenite-ferrite boundary (Fig. 9.28),

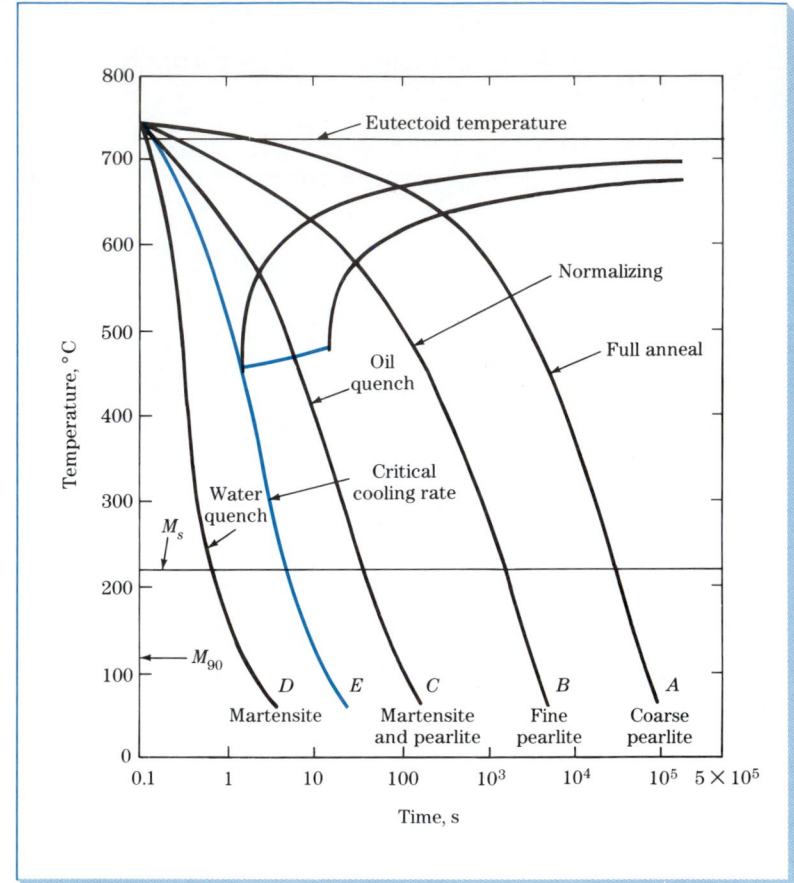

FIGURE 9.27 Variation in the microstructure of a eutectoid plain-carbon steel by continuously cooling at different rates. *(From R. E. Reed-Hill, "Physical Metallurgy Principles," 2d ed., D. Van Nostrand Co., 1973 © PWS Publishers.)*

held the necessary time at the elevated temperature, and then slowly cooled to room temperature, usually in the furnace in which they were heated. For hypereutectoid steels, it is customary to austenitize in the two-phase austenite plus cementite (Fe_3C) region, about 40°C above the eutectoid temperature. The microstructure of hypoeutectoid steels after full annealing consists of proeutectoid ferrite and pearlite (Fig. 9.10).

Process annealing, which is often referred to as a *stress relief,* partially softens cold-worked low-carbon steels by relieving internal stresses from cold working. This treatment, which is usually applied to hypoeutectoid steels with less than 0.3% C, is carried out at a temperature below the eutectoid temperature, usually between 550 to 650°C (Fig. 9.28).

Normalizing is a heat treatment in which the steel is heated in the austenitic region and then cooled in still air. The microstructure of thin sections of normalized hypoeutectoid plain-carbon steels consists of proeutectoid ferrite and fine pearlite. The purposes for normalizing vary. Some of these are:

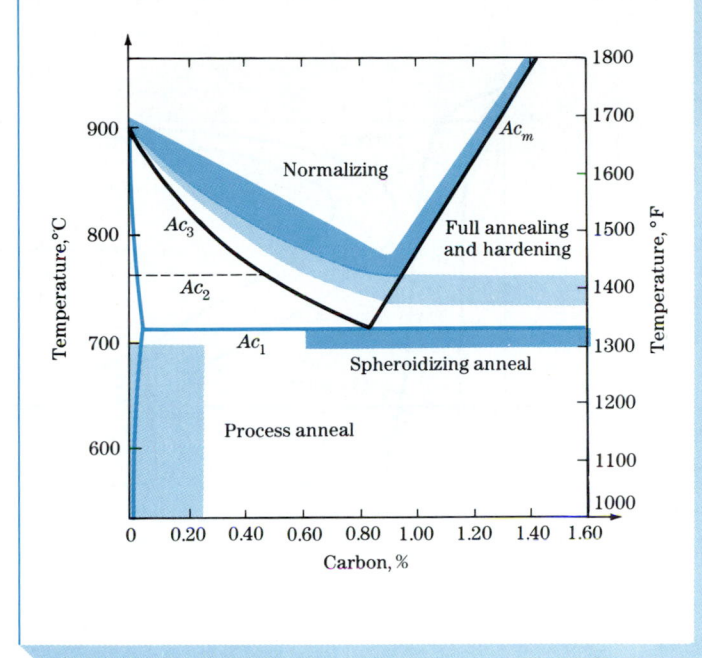

FIGURE 9.28 Commonly used temperature ranges for annealing plain-carbon steels. *(After T. G. Digges et al., "Heat Treatment and Properties of Iron and Steel," NBS Monograph 88, 1966, p. 10.)*

1. To refine the grain structure
2. To increase the strength of the steel (compared to annealed steel)
3. To reduce compositional segregation in castings or forgings and thus provide a more uniform structure

The austenitizing temperature ranges used for normalizing plain-carbon steels are shown in Fig. 9.28. Normalizing is more economical than full annealing since no furnace is required to control the cooling rate of the steel.

Tempering of Plain-Carbon Steels

The tempering process   Tempering is the process of heating a martensitic steel at a temperature below the eutectoid transformation temperature to make it softer and more ductile. Figure 9.29 schematically illustrates the customary quenching and tempering process for a plain-carbon steel. As shown in Fig. 9.29 the steel is first austenitized and then quenched at a rapid rate to produce martensite and to avoid the transformation of austenite to ferrite and cementite. The steel is then subsequently reheated at a temperature below the eutectoid temperature to soften the martensite by transforming it to a structure of iron carbide particles in a matrix of ferrite.

Microstructural changes in martensite upon tempering   Martensite is a metastable structure and decomposes upon reheating. In lath martensites of low-

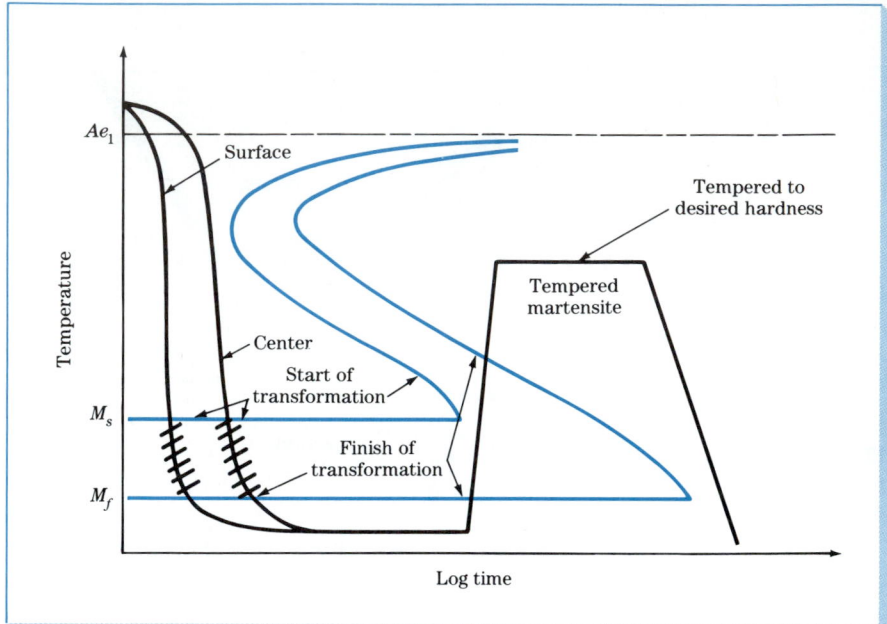

FIGURE 9.29 Schematic diagram illustrating the customary quenching and tempering process for a plain-carbon steel. *(From "Suiting the Heat Treatment to the Job," United States Steel Corp., 1968, p. 34.)*

carbon plain-carbon steels there is a high dislocation density, and these dislocations provide lower energy sites for carbon atoms than their regular interstitial positions. Thus, when low-carbon martensitic steels are first tempered in the 20 to 200°C range, the carbon atoms segregate themselves to these lower energy sites.

For martensitic plain-carbon steels with more than 0.2% carbon the main mode of carbon redistribution at tempering temperatures below 200°C is by precipitation clustering. In this temperature range a very small sized precipitate called *epsilon (ϵ) carbide* forms. The carbide which forms when martensitic steels are tempered from 200 to 700°C is *cementite, Fe₃C*. When the steels are tempered between 200 to 300°C, the shape of the precipitate is rodlike (Fig. 9.30). At higher tempering temperatures from 400 to 700°C, the rodlike carbides coalesce to form spherical-like particles. Tempered martensite which shows the coalesced cementite in the optical microscope is called *spheroidite* (Fig. 9.31).

Effect of tempering temperature on the hardness of plain-carbon steels   Figure 9.32 shows the effect of increasing tempering temperature on the hardness of several martensitic plain-carbon steels. Above about 200°C, the hardness gradually decreases as the temperature is increased up to 700°C. This gradual decrease in hardness of the martensite with increasing temperature is due mainly to the diffusion of the carbon atoms from their stressed interstitial lattice sites to form second-phase iron carbide precipitates.

0.5 μm

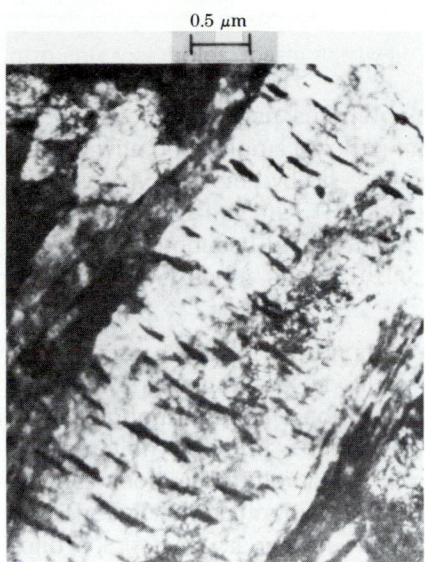

FIGURE 9.30
Precipitation of
Fe$_3$C in Fe–0.39% C
martensite tempered
1 h at 300°C. (Electron
micrograph.) [*After G. R.
Speich and W. C. Leslie,
Met. Trans.,
31:1043(1972).*]

Martempering (marquenching)   Martempering (marquenching) is a modified quenching procedure used for steels to minimize distortion and cracking that may develop during uneven cooling of the heat-treated material. The martempering process consists of (1) austenitizing the steel, (2) quenching it in hot oil or molten salt at a temperature just slightly above (or slightly below) the

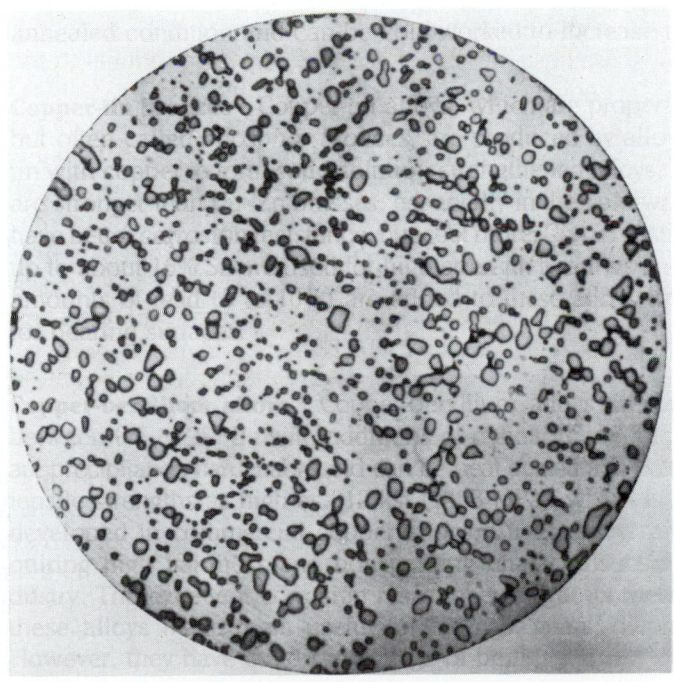

FIGURE 9.31 Spheroidite
in a 1.1% C
hypereutectoid steel.
(Magnification 1000×.)
*(After J. Vilella, E. C. Bain,
and H. W. Paxton,
"Alloying Elements in
Steel," 2d ed., American
Society for Metals, 1966,
p. 101.)*

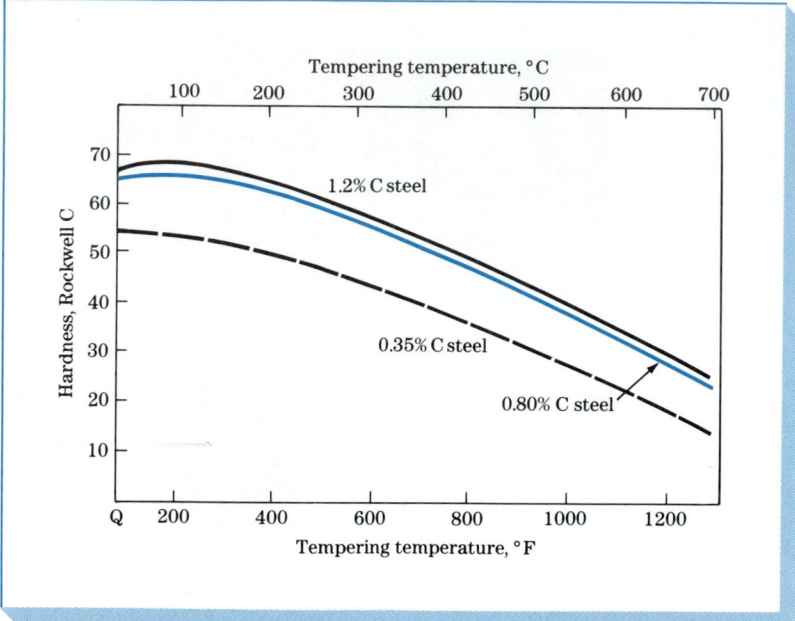

FIGURE 9.32 Hardness of iron-carbon martensites (0.35 to 1.2% C) tempered 1 h at indicated temperatures. *(After E. C. Bain and H. W. Paxton, "Alloying Elements in Steel," 2d ed., American Society for Metals, 1966, p. 38.)*

M_s temperature, (3) holding the steel in the quenching medium until the temperature is uniform throughout and stopping this isothermal treatment before the austenite-to-bainite transformation begins, and (4) cooling at a moderate rate to room temperature to prevent large temperature differences. The steel is subsequently tempered by the conventional treatment. Figure 9.33 shows a cooling path for the martempering process.

The structure of the martempered steel is *martensite* and that of the martempered (marquenched) steel which is subsequently tempered is *tempered martensite*. Table 9.2 lists some of the mechanical properties of a 0.95% C plain-carbon steel after martempering and tempering along with those obtained by conventional quenching and tempering. The major difference in these two sets of properties is that the martempered and tempered steel has higher impact energy values. It should be noted that the term martempering is misleading and a better word for the process is *marquenching*.

Austempering   Austempering is an isothermal heat treatment which produces a bainite structure in some plain-carbon steels. The process provides an alternative procedure to quenching and tempering for increasing the toughness and ductility of some steels. In the austempering process the steel is first austenitized, then quenched in a molten salt bath at a temperature just above the M_s temperature of the steel, held isothermally to allow the austenite-to-bainite transformation to take place, and then cooled to room temperature in air (Fig. 9.34). The final structure of an austempered eutectoid plain-carbon steel is *bainite*.

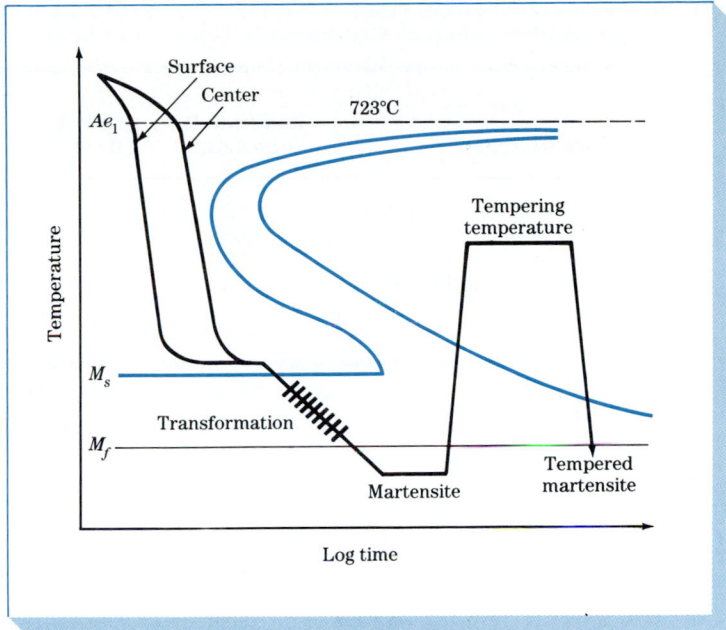

FIGURE 9.33 Cooling curve for martempering (marquenching) superimposed on a eutectoid plain-carbon steel IT diagram. The interrupted quench reduces the stresses developed in the metal during quenching. (*After "Metals Handbook," vol. 2, 8th ed., American Society for Metals, 1964, p. 37.*)

The advantages of austempering are (1) improved ductility and impact resistance of certain steels over those values obtained by conventional quenching and tempering (Table 9.2) and (2) decreased distortion of the quenched material. The disadvantages of austempering over quenching and tempering are (1) the need for a special molten salt bath and (2) the process can be used for only a limited number of steels.

FIGURE 9.34 Cooling curves for austempering a eutectoid plain-carbon steel. The structure resulting from this treatment is bainite. An advantage of this heat treatment is that tempering is unneccessary. Compare with the customary quenching and tempering process shown in Fig. 9.29. M_s and M_f are the start and finish of martensitic transformation, respectively. (*From "Suiting the Heat Treatment to the Job," United States Steel Corp., 1968, p. 34.*)

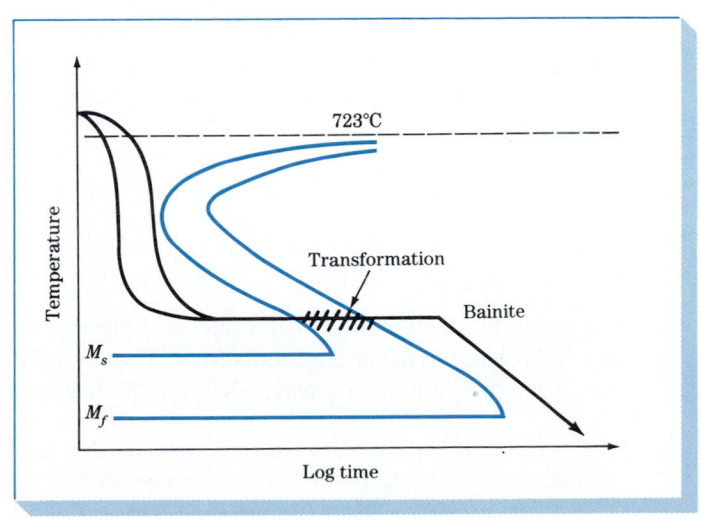

TABLE 9.2 **Some Mechanical Properties (at 20°C) of a 1095 Steel Developed by Austempering as Compared to Some Other Heat Treatments**

Heat treatment	Rockwell C hardness	Impact, ft · lb	Elongation in 1 in, %
Water-quench and temper	53.0	12	0
Water-quench and temper	52.5	14	0
Martemper and temper	53.0	28	0
Martemper and temper	52.8	24	0
Austemper	52.0	45	11
Austemper	52.5	40	8

Source: "Metals Handbook," vol. 2, 8th ed., American Society for Metals, 1964.

Classification of Plain-Carbon Steels and Typical Mechanical Properties

Plain-carbon steels are most commonly designated by a four-digit AISI-SAE[1] code. The first two digits are 10 and indicate that the steel is a plain-carbon steel. The last two digits indicate the nominal carbon content of the steel in hundredths of a percent. For example, the AISI-SAE code number 1030 for a steel indicates that the steel is a plain-carbon steel containing a nominal 0.30% carbon. All plain-carbon steels contain manganese as an alloying element to enhance strength. The manganese content of most plain-carbon steels ranges between 0.30 to 0.95%. Plain-carbon steels also contain impurities of sulfur, phosphorus, silicon, and some other elements.

Typical mechanical properties of some AISI-SAE type plain-carbon steels are listed in Table 9.3. The very low carbon plain-carbon steels have relatively low strengths but very high ductilities. These steels are used for sheet material for forming applications such as fenders and body panels for automobiles. As the carbon content of the plain-carbon steels is increased, the steels become stronger but less ductile. Medium-carbon steels (1020–1040) find application for shafts and gears. High-carbon steels (1060–1095) are used, for example, for springs, die blocks, cutters, and shear blades.

9.4 LOW-ALLOY STEELS

Plain-carbon steels can be used successfully if the strength and other engineering requirements are not too severe. These steels are relatively low in cost but have some limitations which include the following:

---

[1] AISI stands for the American Iron and Steel Institute, and SAE for the Society for Automotive Engineers.

TABLE 9.3 Typical Mechanical Properties and Applications of Plain-Carbon Steels

Alloy AISI-SAE number	Chemical composition, wt %	Condition	Tensile strength		Yield strength		Elonga-tion, %	Typical applications
			ksi	MPa	ksi	MPa		
1010	0.10 C, 0.40 Mn	Hot-rolled	40–60	276–414	26–45	179–310	28–47	Sheet and strip for drawing; wire, rod, and nails and screws; concrete reinforcement bar
		Cold-rolled	42–58	290–400	23–38	159–262	30–45	
1020	0.20 C, 0.45 Mn	As rolled	65	448	48	331	36	Steel plate and structural sections; shafts, gears
		Annealed	57	393	43	297	36	
1040	0.40 C, 0.45 Mn	As rolled	90	621	60	414	25	Shafts, studs, high-tensile tubing, gears
		Annealed	75	517	51	352	30	
		Tempered*	116	800	86	593	20	
1060	0.60 C, 0.65 Mn	As rolled	118	814	70	483	17	Spring wire, forging dies, railroad wheels
		Annealed	91	628	54	483	22	
		Tempered*	160	110	113	780	13	
1080	0.80 C, 0.80 Mn	As rolled	140	967	85	586	12	Music wire, helical springs, cold chisels, forging die blocks
		Annealed	89	614	54	373	25	
		Tempered*	189	1304	142	980	12	
1095	0.95 C, 0.40 Mn	As rolled	140	966	83	573	9	Dies, punches, taps, milling cutters, shear blades, high-tensile wire
		Annealed	95	655	55	379	13	
		Tempered*	183	1263	118	814	10	

* Quenched and tempered at 315°C (600°F).

1. Plain-carbon steels cannot be strengthened beyond about 100,000 psi (690 MPa) without a substantial loss in ductility and impact resistance.
2. Large-section thicknesses of plain-carbon steels cannot be produced with a martensitic structure throughout. That is, they are not deep-hardenable.
3. Plain-carbon steels have low corrosion and oxidation resistance.
4. Medium-carbon plain-carbon steels must be quenched rapidly to obtain a fully martensitic structure. Rapid quenching leads to possible distortion and cracking of the heat-treated part.
5. Plain-carbon steels have poor impact resistance at low temperatures.

To overcome the deficiencies of plain-carbon steels, alloy steels have been developed which contain alloying elements to improve their properties. Alloy steels in general cost more than plain-carbon steels, but for many applications they are the only materials that can be used to meet engineering requirements. The principal alloying elements added to make alloy steels are manganese, nickel, chromium, molybdenum, and tungsten. Other elements that are sometimes added include vanadium, cobalt, boron, copper, aluminum, lead, titanium, and columbium (niobium).

Classification of Alloy Steels

Alloy steels may contain up to 50 percent of alloying elements and still be considered alloy steels. In this book low-alloy steels containing from about 1 to 4 percent of alloying elements will be considered alloy steels. These steels

TABLE 9.4   Principal Types of Standard Alloy Steels

13xx	Manganese 1.75
40xx	Molybdenum 0.20 or 0.25; or molybdenum 0.25 and sulfur 0.042
41xx	Chromium 0.50, 0.80, or 0.95, molybdenum 0.12, 0.20, or 0.30
43xx	Nickel 1.83, chromium 0.50 or 0.80, molybdenum 0.25
44xx	Molybdenum 0.53
46xx	Nickel 0.85 or 1.83, molybdenum 0.20 or 0.25
47xx	Nickel 1.05, chromium 0.45, molybdenum 0.20 or 0.35
48xx	Nickel 3.50, molybdenum 0.25
50xx	Chromium 0.40
51xx	Chromium 0.80, 0.88, 0.93, 0.95, or 1.00
51xxx	Chromium 1.03
52xxx	Chromium 1.45
61xx	Chromium 0.60 or 0.95, vanadium 0.13 or min 0.15
86xx	Nickel 0.55, chromium 0.50, molybdenum 0.20
87xx	Nickel 0.55, chromium 0.50, molybdenum 0.25
88xx	Nickel 0.55, chromium 0.50, molybdenum 0.35
92xx	Silicon 2.00; or silicon 1.40 and chromium 0.70
50Bxx*	Chromium 0.28 or 0.50
51Bxx*	Chromium 0.80
81Bxx*	Nickel 0.30, chromium 0.45, molybdenum 0.12
94Bxx*	Nickel 0.45, chromium 0.40, molybdenum 0.12

*B denotes boron steel.
Source: "Alloy Steel: Semifinished; Hot-Rolled and Cold-Finished Bars," American Iron and Steel Institute, 1970.

are mainly automotive- and construction-type steels and are commonly referred to simply as *alloy steels.*

Alloy steels in the United States are usually designated by the four-digit AISI-SAE system. The first two digits indicate the principal alloying element or groups of elements in the steel, and the last two digits indicate the hundredths of percent of carbon in the steel. Table 9.4 lists the nominal compositions of the principal types of standard alloy steels.

Distribution of Alloying Elements in Alloy Steels

The way in which alloy elements distribute themselves in carbon steels depends primarily on the compound- and carbide-forming tendencies of each element. Table 9.5 summarizes the approximate distribution of most of the alloying elements present in alloy steels.

Nickel dissolves in the α ferrite of the steel since it has less tendency to form carbides than iron. Silicon combines to a limited extent with the oxygen present in the steel to form nonmetallic inclusions but otherwise dissolves in the ferrite. Most of the manganese added to carbon steels dissolves in the ferrite. Some of the manganese, however, will form carbides but will usually enter the cementite as $(Fe,Mn)_3C$.

Chromium, which has a somewhat stronger carbide-forming tendency than iron, partitions between the ferrite and carbide phases. The distribution of chromium depends on the amount of carbon present and if other stronger carbide-forming elements such as titanium and columbium are absent. Tungsten and molybdenum combine with carbon to form carbides if there is suf-

TABLE 9.5   Approximate Distribution of Alloying Elements in Alloy Steels*

Element	Dissolved in ferrite	Combined in carbide	Combined as carbide	Compound	Elemental
Nickel	Ni			Ni_3Al	
Silicon	Si			$SiO_2 \cdot M_xO_y$	
Manganese	Mn ⟷ Mn		$(Fe,Mn)_3C$	$MnS; MnO \cdot SiO_2$	
Chromium	Cr ⟷ Cr		$(Fe,Cr)_3C$		
			Cr_7C_3		
			$Cr_{23}C_6$		
Molybdenum	Mo ⟷ Mo		Mo_2C		
Tungsten	W ⟷ W		W_2C		
Vanadium	V ⟷ V		V_4C_3		
Titanium	Ti ⟷ Ti		TiC		
Columbium†	Cb ⟷ Cb		CbC		
Aluminum	Al			$Al_2O_3; AlN$	
Copper	Cu (small amount)				
Lead					Pb

*The arrows indicate the relative tendencies of the elements listed to dissolve in the ferrite or combine in carbides.
†Cb = Nb (niobium)
Source: E. C. Bain and H. W. Paxton, "Alloying Elements in Steel," 2d ed., American Society for Metals, 1966.

ficient carbon present and if other stronger carbide-forming elements such as titanium and columbium are absent. Vanadium, titanium, and columbium are strong carbide-forming elements and are found in steels mainly as carbides. Aluminum combines with oxygen and nitrogen to form the compounds Al_2O_3 and AlN, respectively.

Effects of Alloying Elements on the Eutectoid Temperature of Steels

The various alloying elements cause the eutectoid temperature of the Fe–Fe$_3$C phase diagram to be raised or lowered (Fig. 9.35). Manganese and nickel both lower the eutectoid temperature and act as *austenite-stabilizing elements* enlarging the austenitic region of the Fe–Fe$_3$C phase diagram (Fig. 9.6). In some steels with sufficient amounts of nickel or manganese, the austenitic structure may be obtained at room temperature. The carbide-forming elements such as tungsten, molybdenum, and titanium raise the eutectoid temperature of the Fe–Fe$_3$C phase diagram to higher values and reduce the austenitic phase field. These elements are called *ferrite-stabilizing elements.*

Hardenability

The *hardenability* of a steel is defined as that property which determines the depth and distribution of hardness induced by quenching from the austenitic condition. The hardenability of a steel depends primarily on (1) the composition of the steel, (2) the austenitic grain size, and (3) the structure of the steel before quenching. Hardenability should not be confused with the *hardness* of a steel, which is its resistance to plastic deformation, usually by indentation.

In industry, hardenability is most commonly measured by the *Jominy hardenability test.* In the Jominy end-quench test, the specimen consists of a

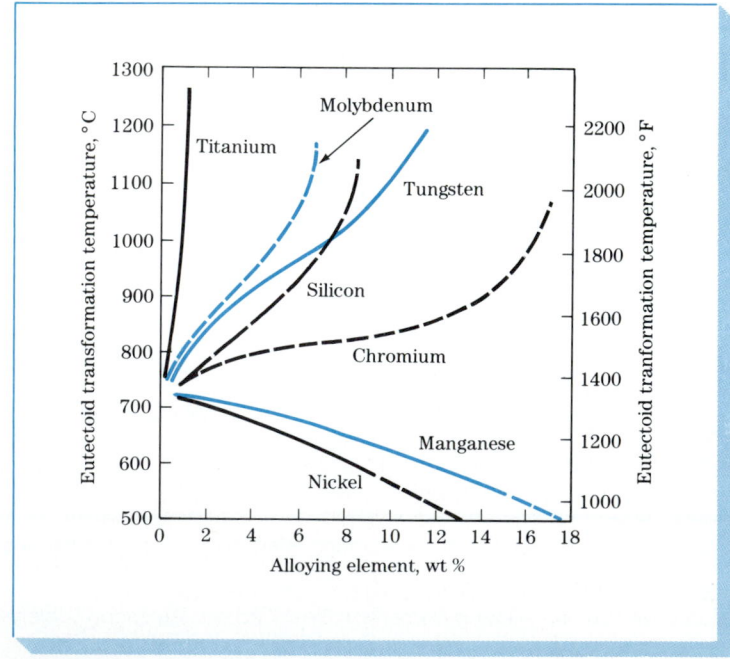

FIGURE 9.35 The effect of the percentage of alloying elements on the eutectoid temperature of the transformation of austenite to pearlite in the Fe–Fe$_3$C phase diagram. *(After "Metals Handbook," vol. 8, 9th ed., American Society for Metals, 1973, p. 191.)*

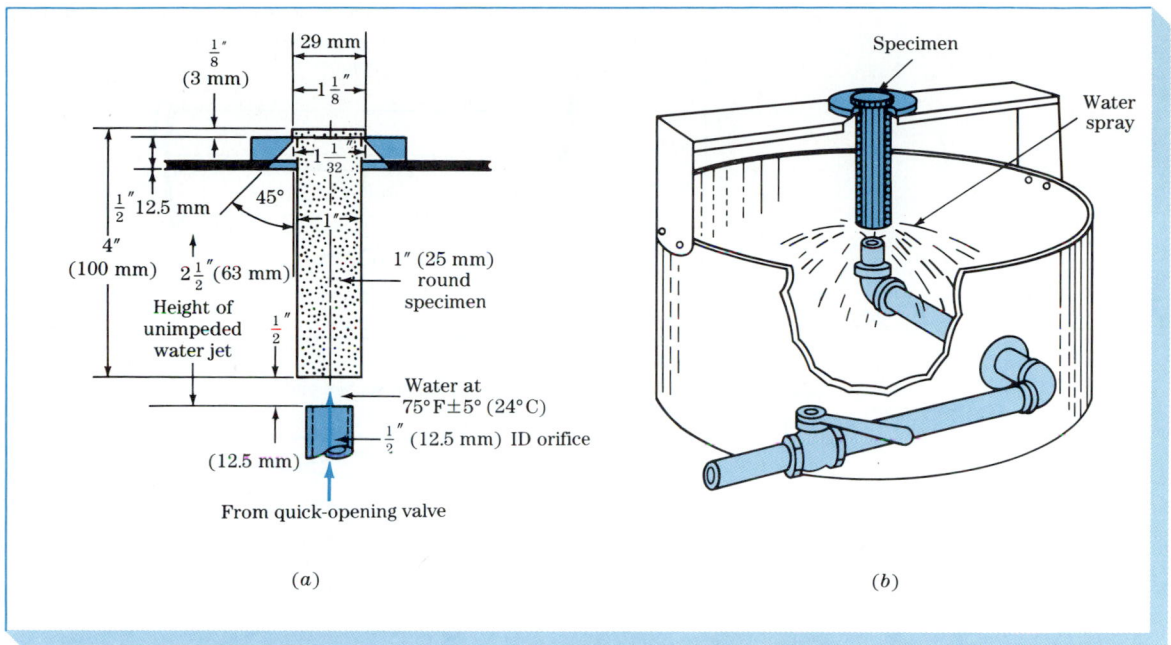

FIGURE 9.36 (*a*) Specimen and fixture for end-quench hardenability test. *(After M. A. Grossmann and E. C. Bain, "Principles of Heat Treatment," 5th ed., American Society for Metals, 1964, p. 114.)* (*b*) Schematic illustration of the end-quench test for hardenability. [*After H. E. McGannon (ed.), "The Making, Shaping, and Treating of Steel," 9th ed., United States Steel Corp., 1971, p. 1099.*]

cylindrical bar with a 1-in diameter and 4-in length and with a $\frac{1}{16}$-in flange at one end (Fig. 9.36*a*). Since prior structure has a strong effect on hardenability, the specimen is usually normalized before testing. In the Jominy test, after the sample has been austenitized, it is placed in a fixture, as shown in Fig. 9.36*b*, and a jet of water is quickly splashed at one end of the specimen. After cooling, two parallel flat surfaces are ground on the opposite sides of the test bar, and Rockwell C hardness measurements are made along these surfaces up to 2.5 in from the quenched end.

Figure 9.37 shows a hardenability plot of Rockwell C hardness vs. distance from the quenched end for a 1080 eutectoid plain-carbon steel. This steel has relatively low hardenability since its hardness decreases from a value of RC = 65 at the quenched end of the Jominy bar to RC = 50 at just $\frac{3}{16}$ in from the quenched end. Thus, thick sections of this steel cannot be made fully martensitic by quenching. Figure 9.37 correlates the end-quench hardenability data with the continuous transformation diagram for the 1080 steel and indicates the microstructural changes which take place along the bar at four distances A, B, C, and D from the quenched end.

Hardenability curves for some 0.40% C alloy steels are shown in Fig. 9.38. The 4340 alloy steel has exceptionally high hardenability and can be quenched

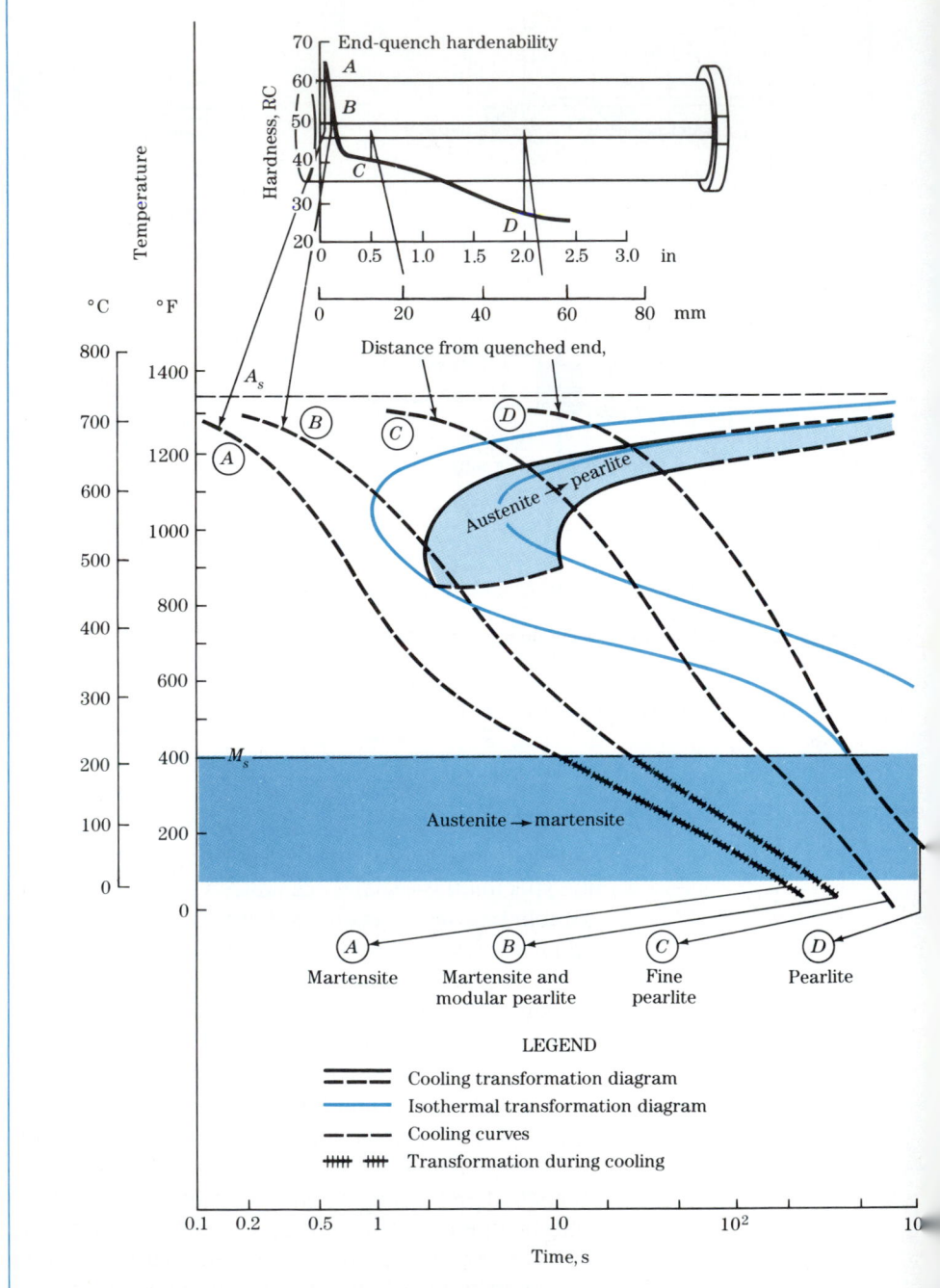

FIGURE 9.37 Correlation
of continuous-cooling
transformation diagram
and end-quench
hardenability test data
for eutectoid carbon
steel. *(After "Isothermal
Transformation
Diagrams," United States
Steel Corp., 1963, p. 181.)*

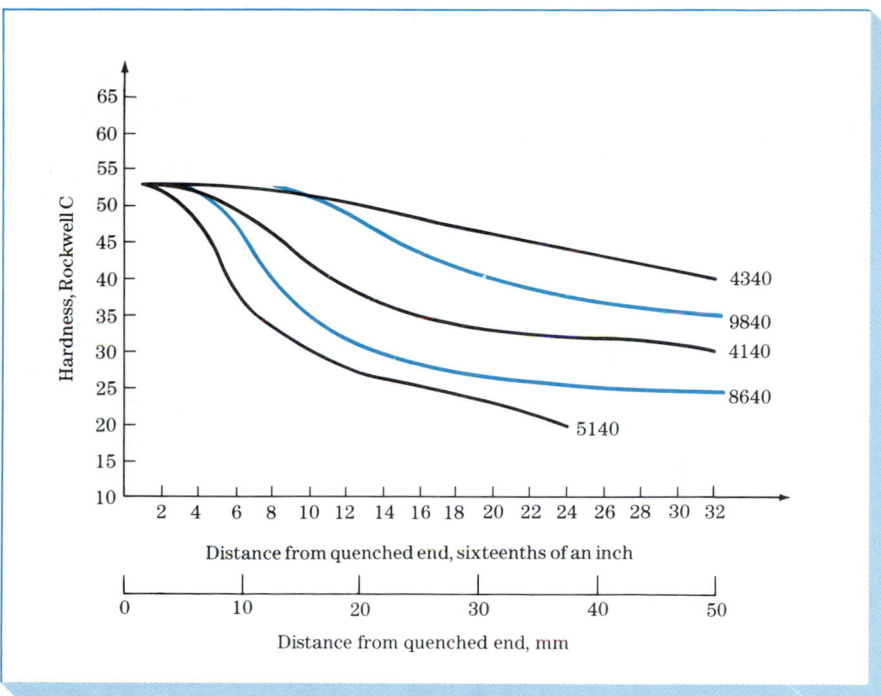

FIGURE 9.38
Comparative hardenability curves for 0.40% C alloy steels. [*After H. E. McGannon (ed.), "The Making, Shaping, and Treating of Steels," United States Steel Corp., 1971, p. 1139*].

to a hardness of RC = 40 at 2 in from the quenched end of a Jominy bar. Alloy steels thus are able to be quenched at a slower rate and still maintain relatively high hardness values.

Alloy steels such as the 4340 steel are highly hardenable because, upon cooling from the austenitic region, the decomposition of austenite to ferrite and bainite is delayed and the decomposition of austenite to martensite can be accomplished at slower rates. This delay of the austenite to ferrite plus bainite decomposition is quantitatively shown on the continuous-cooling transformation diagram of Fig. 9.39.

For most carbon and low-alloy steels a standard quench produces at the same cross-section position common cooling rates along long round steel bars of the same diameter. However, the cooling rates differ (1) for different bar diameters, (2) for different positions in the cross sections of the bars, and (3) for different quenching media. Figure 9.39a shows bar diameter vs. cooling rate curves for different cross-section locations within steel bars using quenches of (i) agitated water in (ii) agitated oil. These plots can be used to determine the cooling rate and the associated distance from the quenched end of a standard quenched Jominy bar for a selected bar diameter at a particular cross-section location in the bar using a specific quenching media. These cooling rates and their associated distances from the end of Jominy quenched bars

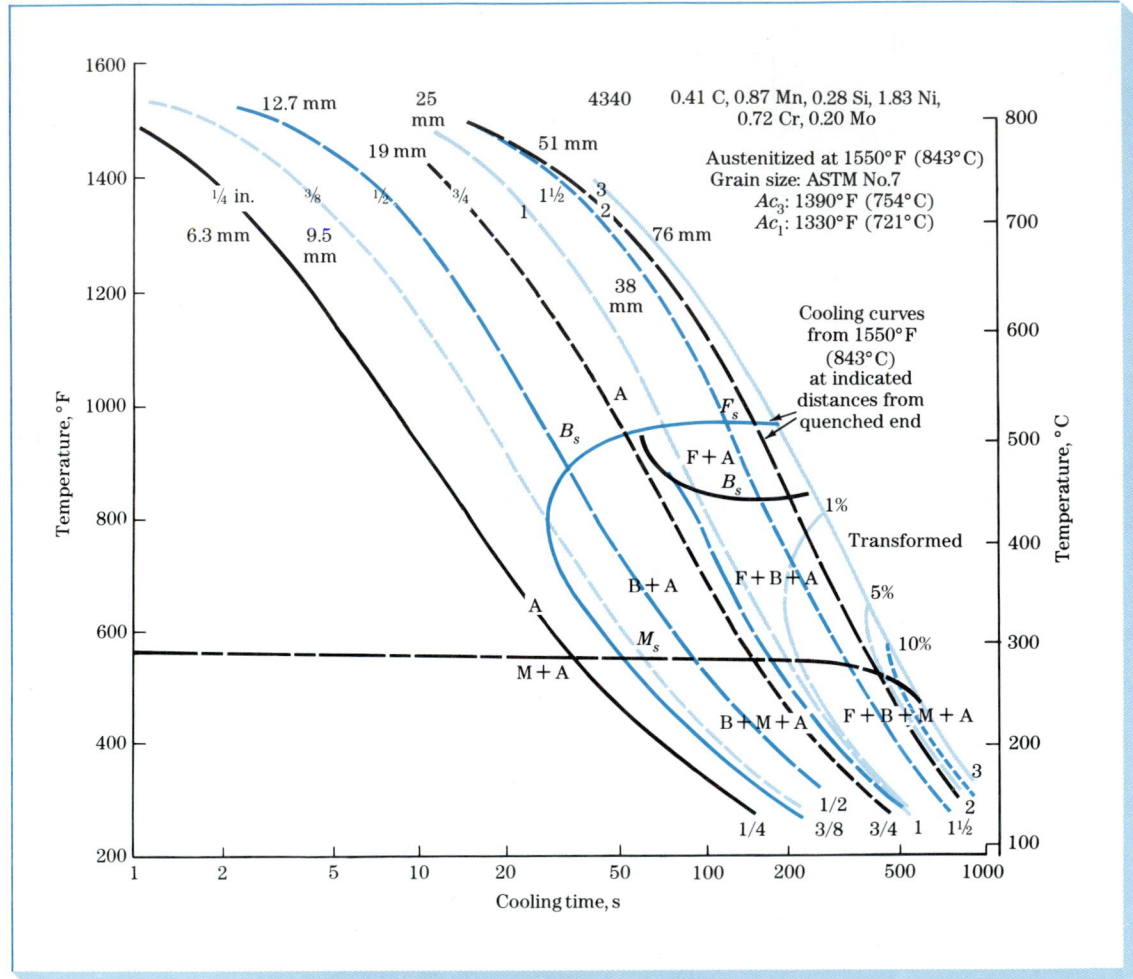

FIGURE 9.39 Continuous-cooling transformation diagram for AISI 4340 alloy steel. A = austenite, F = ferrite, B = bainite, M = martensite. (*After Metal Progress, September 1964, p. 106.*)

can be used with Jominy plots of surface hardness vs. distance from the quenched end for specific steels to determine the hardness of a particular steel at a specific location in the cross section of the steel bar in question. Example Problem 9.5 shows how the plots of Fig. 9.39a can be used to predict the hardness of a steel bar of a given diameter at a specific cross-section location quenched in a given medium. It should be pointed out that the Jominy hardness vs. distance from the quenched-end plots are usually plotted as bands of data rather than as lines so that hardnesses obtained from the line curves are actually values in the center of a range of values.

Example Problem 9.5

A austenitized 40-mm-diameter 5140 alloy steel bar is quenched in agitated oil. Predict what the Rockwell C(RC) hardness of this bar will be at (*a*) its surface and (*b*) its center.

Solution:

(*a*) Surface of bar. The cooling rate at the surface of the 40-mm steel bar quenched in agitated oil is found from part (ii) of Fig. 9.39a to be comparable to the cooling rate at 8 mm from the end of a standard quenched Jominy bar. Using Fig. 9.38 at 8 mm from the quenched end of the Jominy bar and the curve for the 5140 steel indicates that the hardness of the bar should be about 45 RC.

(*b*) Center of the bar. The cooling rate at the center of the 40-mm-diameter bar quenched in oil is found from part (ii) of Fig. 9.39a to be associated with 13 mm from the end of a quenched Jominy bar. The corresponding hardness for this distance from the end of a quenched Jominy bar for the 5140 alloy is found by using Fig. 9.38 to be about 32 RC.

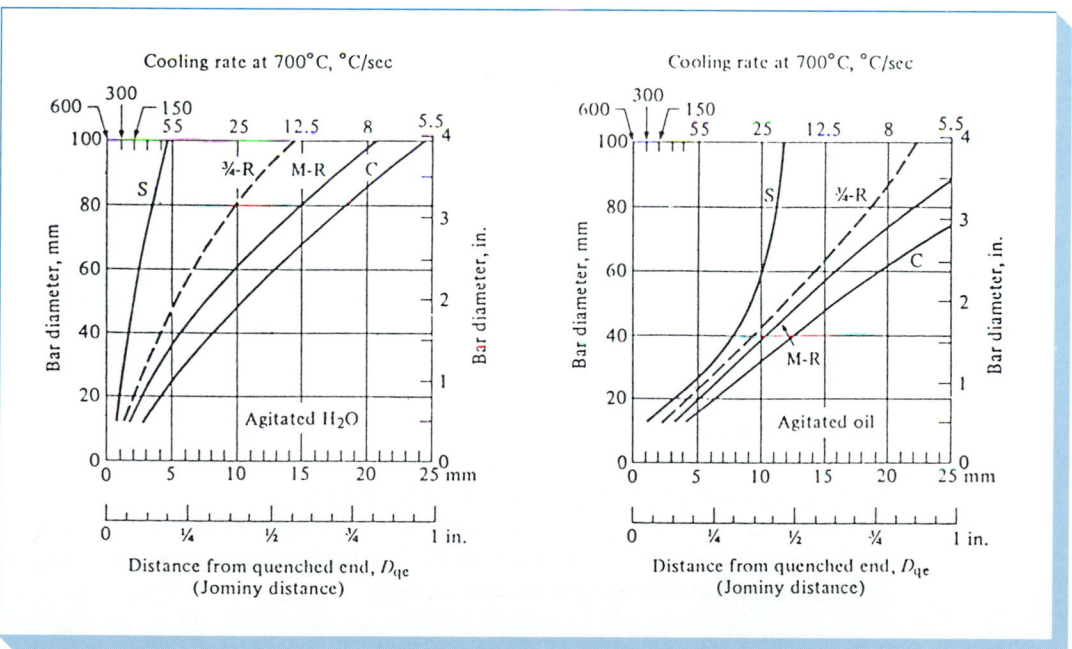

FIGURE 9.39a Cooling rates in long round steel bars quenched in (i) agitated water and (ii) agitated oil. Top abscissa, cooling rates at 700°C; bottom abscissa, equivalent positions on an end-quenched test bar. (C = center, M-R = midradius, S = surface, dashed line = approximate curve for ¾-radius positions on the cross section of bars.) *(After L. H. VanVlack, "Materials for Engineering: Concepts and Applications," Addison-Wesley, 1982, p. 155).*

Typical Mechanical Properties and Applications for Low-Alloy Steels

Table 9.6 lists some typical tensile mechanical properties and applications for some commonly used low-alloy steels. For some strength levels low alloy steels have better combinations of strength, toughness, and ductility than plain-carbon steels. However, low-alloy steels cost more and so are used only when necessary. Low-alloy steels are used to a great extent in the manufacture of automobiles and trucks for parts that require superior strength and toughness properties which cannot be obtained from plain-carbon steels. Some typical applications for low-alloy steels in automobiles are shafts, axles, gears, and springs. Low-alloy steels containing about 0.2% C are commonly carburized or surface heat-treated to produce a hard, wear-resistant surface while maintaining a tough inner core.

9.5 ALUMINUM ALLOYS

Before discussing some of the important aspects of the structure, properties, and applications of aluminum alloys, let us examine the precipitation-strengthening (hardening) process which is used to increase the strength of many aluminum and other metal alloys.

Precipitation Strengthening (Hardening)

Precipitation strengthening of a generalized binary alloy The object of precipitation strengthening is to create in a heat-treated alloy a dense and fine dispersion of precipitated particles in a matrix of deformable metal. The precipitate particles act as obstacles to dislocation movement and thereby strengthen the heat-treated alloy.

The precipitation-strengthening process can be explained in a general way by referring to the binary phase diagram of metals A and B shown in Fig. 9.40. In order for an alloy system to be able to be precipitation-strengthened for certain alloy compositions, there must be a terminal solid solution which has a decreasing solid solubility as the temperature decreases. The phase diagram of Fig. 9.40 shows this type of decrease in solid solubility in terminal solid solution α in going from point a to b along the indicated solvus.

Let us now consider the precipitation strengthening of an alloy of composition x_1 of the phase diagram of Fig. 9.40. We choose the alloy composition x_1 since there is a large decrease in the solid solubility of solid solution α in decreasing the temperature from T_2 to T_3. The precipitation-strengthening process involves the following three basic steps:

1. *Solution heat treatment* is the *first step* in the precipitation-strengthening process. Sometimes this treatment is referred to as *solutionizing*. The alloy sample which may be in the wrought or cast form is heated to a temperature between the solvus and solidus temperatures and soaked there until a uniform solid-solution structure is produced. Temperature T_1 at point c of Fig.

TABLE 9.6   Typical Mechanical Properties and Applications of Low-Alloy Steels

Alloy AISI-SAE number	Chemical composition, wt %	Condition	Tensile strength		Yield strength		Elonga-tion, %	Typical applications
			ksi	MPa	ksi	MPa		
Manganese steels								
1340	0.40 C, 1.75 Mn	Annealed	102	704	63	435	20	High-strength bolts
		Tempered*	230	1587	206	1421	12	
Chromium steels								
5140	0.40 C, 0.80 Cr, 0.80 Mn	Annealed	83	573	43	297	29	Automobile transmission gears
		Tempered*	229	1580	210	1449	10	
5160	0.60 C, 0.80 Cr, 0.90 Mn	Annealed	105	725	40	276	17	Automobile coil and leaf springs
		Tempered*	290	2000	257	1773	9	
Chromium-molybdenum steels								
4140	0.40 C, 1.0 Cr, 0.9 Mn, 0.20 Mo	Annealed	95	655	61	421	26	Gears for aircraft gas turbine engines, transmissions
		Tempered*	225	1550	208	1433	9	
Nickel-molybdenum steels								
4620	0.20 C, 1.83 Ni, 0.55 Mn, 0.25 Mo	Annealed	75	517	54	373	31	Transmission gears, chain pins, shafts, roller bearings
		Normalized	83	573	53	366	29	
4820	0.20 C, 3.50 Ni, 0.60 Mn, 0.25 Mo	Annealed	99	683	67	462	22	Gears for steel mill equipment, paper machinery, mining machinery, earth-moving equipment
		Normalized	100	690	70	483	60	
Nickel (1.83%)-chromium-molybdenum steels								
4340 (E)	0.40 C, 1.83 Ni, 0.90 Mn, 0.80 Cr, 0.20 Mo	Annealed	108	745	68	469	22	Heavy sections, landing gears, truck parts
		Tempered*	250	1725	230	1587	10	
Nickel (0.55%)-chromium-molybdenum steels								
8620	0.20 C, 0.55 Ni, 0.50 Cr, 0.80 Mn, 0.20 Mo	Annealed	77	531	59	407	31	Transmission gears
		Normalized	92	635	52	359	26	
8650	0.50 C, 0.55 Ni, 0.50 Cr, 0.80 Mn, 0.20 Mo	Annealed	103	710	56	386	22	Small machine axles, shafts
		Tempered*	250	1725	225	1552	10	

*Tempered at 600°F (315°C).

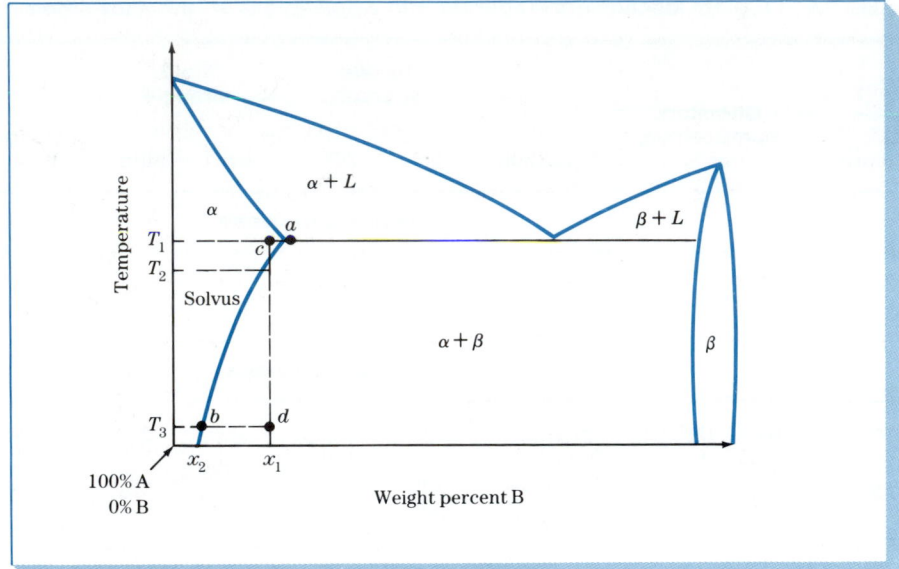

FIGURE 9.40 Binary phase diagram for two metals A and B having a terminal solid solution α which has a decreasing solid solubility of B in A with decreasing temperature.

9.40 is selected for our alloy x_1 because it lies midway between the solvus and solidus phase boundaries of solid solution α.

2. *Quenching* is the *second step* in the precipitation-strengthening process. The sample is rapidly cooled to a lower temperature, usually room temperature, and the cooling medium is usually water at room temperature. The structure of the alloy sample after water quenching consists of a supersaturated solid solution. The structure of our alloy x_1 after quenching to temperature T_3 at point d of Fig. 9.40 thus consists of a supersaturated solid solution of the α phase.

3. *Aging* is the *third basic step* in the precipication-strengthening process. Aging the solution heat-treated and quenched alloy sample is necessary so that a finely dispersed precipitate forms. The formation of a finely dispersed precipitate in the alloy is the objective of the precipitation-strengthening process. The fine precipitate in the alloy impedes dislocation movement during deformation by forcing the dislocations to either cut through the precipitated particles or go around them. By restricting dislocation movement during deformation, the alloy is strengthened.

Aging the alloy at room temperature is called *natural aging,* whereas aging at elevated temperatures is called *artificial aging.* Most alloys require artificial aging, and the aging temperature is usually between about 15 to 25 percent of the temperature difference between room temperature and the solution heat-treatment temperature.

FIGURE 9.41

Decomposition products created by the aging of a supersaturated solid solution of a precipitation-hardenable alloy. The highest energy level is for the supersaturated solid solution, and the lowest energy level is for the equilibrium precipitate. The alloy can go spontaneously from a higher energy level to a lower one if there is sufficient activation energy for the transformation and if the kinetic conditions are favorable.

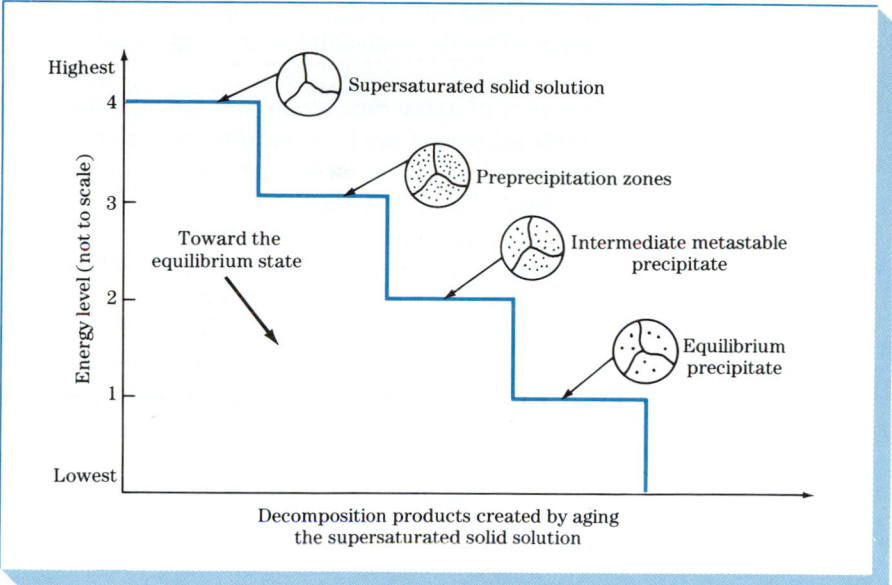

FIGURE 9.41

Decomposition products created by the aging of a supersaturated solid solution of a precipitation-hardenable alloy. The highest energy level is for the supersaturated solid solution, and the lowest energy level is for the equilibrium precipitate. The alloy can go spontaneously from a higher energy level to a lower one if there is sufficient activation energy for the transformation and if the kinetic conditions are favorable.

Decomposition products created by the aging of the supersaturated solid solution   A precipitation-hardenable alloy in the supersaturated solid-solution condition is in a high energy state, as indicated schematically by energy level 4 of Fig. 9.41. This energy state is relatively unstable, and the alloy tends to seek a lower energy state by the spontaneous decomposition of the supersaturated solid solution into metastable phases or the equilibrium phases. The driving force for the precipitation of metastable phases or the equilibrium phase is the lowering of the energy of the system when these phases form.

When the supersaturated solid solution of the precipitation-hardenable alloy is aged at a relatively low temperature where only a small amount of activation energy is available, clusters of segregated atoms called *precipitation zones,* or *GP zones,*[1] are formed. For the case of our alloy A–B of Fig. 9.40, the zones will be regions enriched with B atoms in a matrix primarily of A atoms. The formation of these zones in the supersaturated solid solution is indicated by the circular sketch at the lower energy level 3 of Fig. 9.41. Upon further aging and if sufficient activation energy is available by the aging temperature being high enough, the zones develop into or are replaced by a coarser (larger in size) intermediate metastable precipitate, indicated by the circular sketch at the still-lower energy level 2. Finally, if aging is continued (usually a higher temperature is necessary) and if there is sufficient activation energy

---

[1] Preprecipitation zones are sometimes referred to as GP zones, named after the two early scientists Guinier and Preston who first identified these structures by x-ray diffraction analyses.

available, the intermediate precipitate is replaced by the equilibrium precipitate indicated by the even-still-lower energy level 1 of Fig. 9.41.

The effect of aging time on the strength and hardness of a precipitation-hardenable alloy that has been solution heat-treated and quenched   The effect of aging on strengthening a precipitation-hardenable alloy that has been solution heat-treated and quenched is usually presented as an *aging curve*. The aging curve is a plot of strength or hardness vs. aging time (usually on a logarithmic scale) at a particular temperature. Figure 9.42 shows a schematic aging curve. At zero time, the strength of the supersaturated solid solution is indicated on the ordinate axis of the plot. As the aging time increases, preprecipitation zones form and their size increases, and the alloy becomes stronger and harder and less ductile (Fig. 9.42). A maximum strength (peak aged condition) is eventually reached if the aging temperature is sufficiently high, and is usually associated with the formation of an intermediate metastable precipitate. If aging is continued so that the intermediate precipitate coalesces and coarsens, the alloy overages and becomes weaker than in the peak aged condition (Fig. 9.42).

Precipitation strengthening (hardening) of an Al–4% Cu alloy   Let us now examine the structure and hardness changes that occur during the precipitation heat treatment of an aluminum–4% copper alloy. The heat-treatment sequence for the precipitation strengthening of this alloy is:

1. Solution heat treatment: the Al–4% Cu alloy is solutionized at about 515°C (see the Al–Cu phase diagram of Fig. 9.43).
2. Quenching: the solution heat-treated alloy is rapidly cooled in water at room temperature.
3. Aging: the alloy after solution heat treatment and quenching is artificially aged in the 130 to 190°C range.

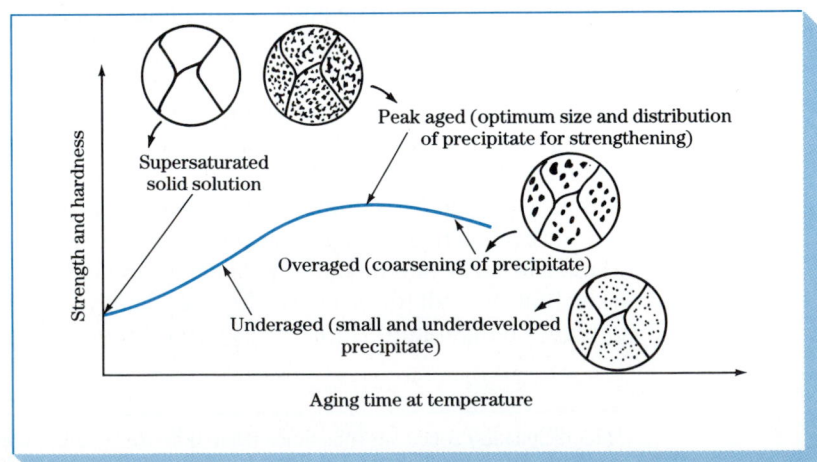

FIGURE 9.42 Schematic aging curve (strength or hardness vs. time) at a particular temperature for a precipitation-hardenable alloy.

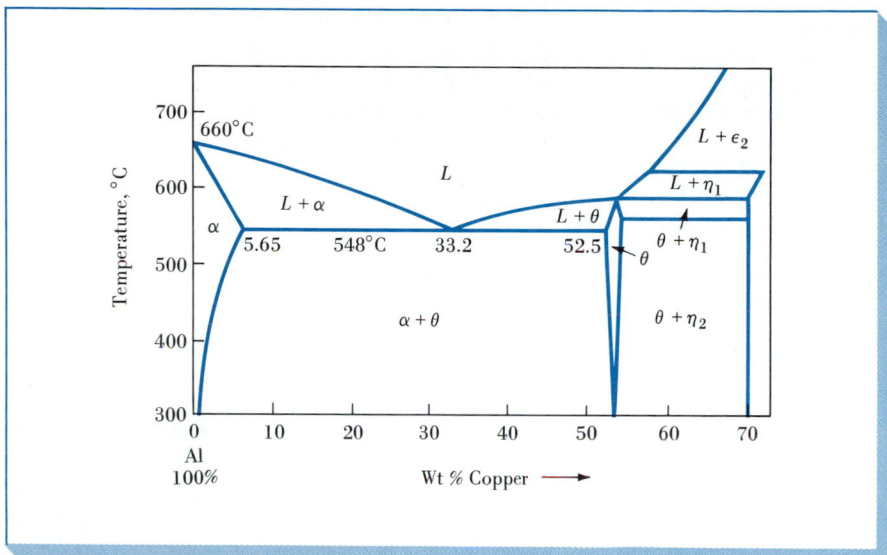

FIGURE 9.43
Aluminum-rich end of
aluminum-copper phase
diagram. [*After K. R. Van
Horn (ed.), "Aluminum,"
vol. 1, American Society
for Metals, 1967, p. 372.*]

Structures formed during the aging of the Al–4% Cu alloy   In the precipitation strengthening of Al–4% Cu alloys five sequential structures can be identified: (1) supersaturated solid-solution α, (2) GP1 zones, (3) GP2 zones (also called θ'' phase), (4) θ' phase, and (5) θ phase, $CuAl_2$. Not all these phases can be produced at all aging temperatures. GP1 and GP2 zones are produced at lower aging temperatures, and θ' and θ phases occur at higher temperatures.

> *GP1 zones.* These preprecipitation zones are formed at lower aging temperatures and are created by copper atoms segregating in the supersaturated solid-solution α. GP1 zones consist of segregated regions in the shape of disks a few atoms thick (0.4 to 0.6 nm) and about 8 to 10 nm in diameter and form on the {100} cubic planes of the matrix. Since the copper atoms have a diameter of about 11 percent less than the aluminum ones, the matrix lattice around the zones is strained tetragonally. GP1 zones are said to be *coherent* with the matrix lattice since the copper atoms just replace aluminum atoms in the lattice [Fig. 9.43*a*(i)]. GP1 zones are detected under the electron microscope by the strain fields they create (Fig. 9.44*a*).
>
> *GP2 zones (θ'' phase).* These zones also have a tetragonal structure and are coherent with the {100} of the matrix of the Al–4%Cu alloy. Their size ranges from about 1 to 4 nm thick to 10 to 100 nm in diameter as aging proceeds (Fig. 9.44*b*).
>
> *θ' phase.* This phase nucleates heterogeneously especially on dislocations and is incoherent with the matrix. {An *incoherent precipitate* is one in

FIGURE 9.43a
Schematic comparison of the nature of (i) a coherent precipitate and (ii) an incoherent precipitate. The coherent precipitate is associated with a high strain energy and low surface energy and the incoherent one is associated with a low strain energy and high surface energy.

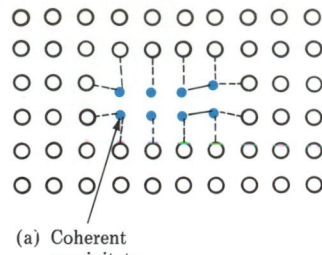

(a) Coherent precipitate

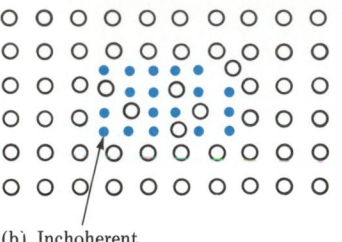

(b) Inchoherent precipitate

(This type of precipitate has its own structure)

(a)

FIGURE 9.44 Microstructures of aged Al–4% Cu alloys. (a) Al–4% Cu, heated to 540°C, water-quenched and aged 16 h at 130°C. The GP zones have been formed as disks parallel to the {100} planes of the FCC matrix and at this stage are a few atoms thick and about 100 Å in diameter. Only disks lying on one crystallographic orientation are visible. (Electron micrograph; magnification 1,000,000×.) (b) Al–4% Cu, solution-treated at 540°C, quenched in water, and aged for 1 day at 130°C. This thin-foil micrograph shows strain fields due to coherent GP2 zones. The dark regions surrounding the zones are caused by strain fields. (Electron micrograph; magnification 800,000×.) (c) Al–4% Cu alloy solution heat-treated at 540°C, quenched in water, and aged for 3 days at 200°C. This thin-foil micrograph shows the incoherent and metastable phase θ′ which forms by heterogeneous nucleation and growth. (Electron micrograph; magnification 25,000×.) (After J. Nutting and R. G. Baker, "The Microstructure of Metals," Institute of Metals, 1965, pp. 65 and 67.)

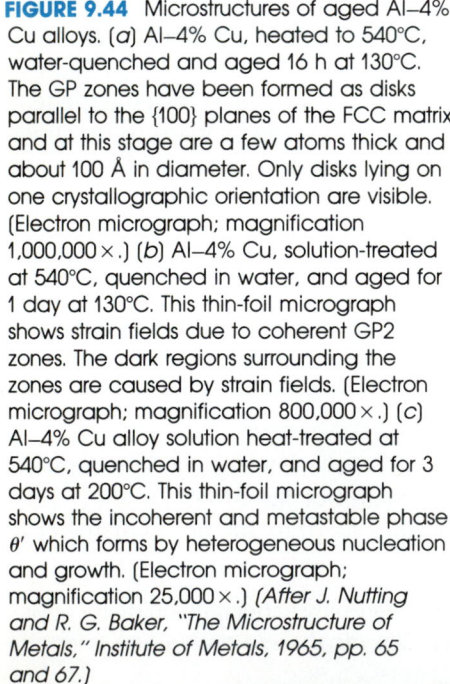

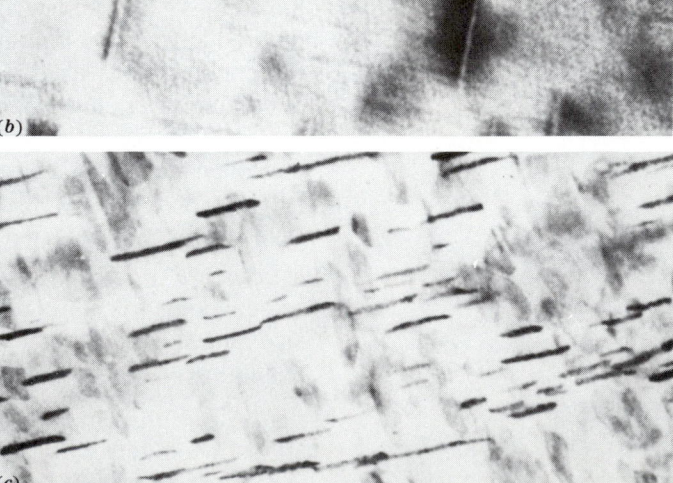

(b)

(c)

which the precipitated particle has a distinct crystal structure different from the matrix [Fig. 9.43a(ii)]}. θ' phase has a tetragonal structure with a thickness of 10 to 150 nm (Fig. 9.44c).

θ *phase.* The equilibrium phase θ is incoherent and has the composition $CuAl_2$. This phase has a BCT structure ($a = 0.607$ nm and $c = 0.487$ nm) and forms from θ' or directly from the matrix.

The general sequence of precipitation in binary aluminum-copper alloys can be represented by

Supersaturated solid solution $\rightarrow$ GP1 zones $\rightarrow$

GP2 zones (θ'' phase) $\rightarrow$ θ' $\rightarrow$ θ ($CuAl_2$)

Correlation of structures and hardness in an Al–4% Cu alloy   The hardness vs. aging time curves for an Al–4% Cu alloy aged at 130 and 190°C are shown in Fig. 9.45. At 130°C, GP1 zones are formed and increase the hardness of the alloy by impeding dislocation movement. Further aging at 130°C creates GP2 zones which increase the hardness still more by making dislocation movement still more difficult. A maximum in hardness is reached with still more aging time at 130°C at θ' forms. Aging beyond the hardness peak dissolves the GP2 zones and coarsens the θ' phase and causes the decrease in the hardness of the alloy. GP1 zones do not form during aging at 190°C in the Al–4% Cu alloy since this temperature is above the GP1 solvus. With long aging times at 190°C the equilibrium θ phase forms.

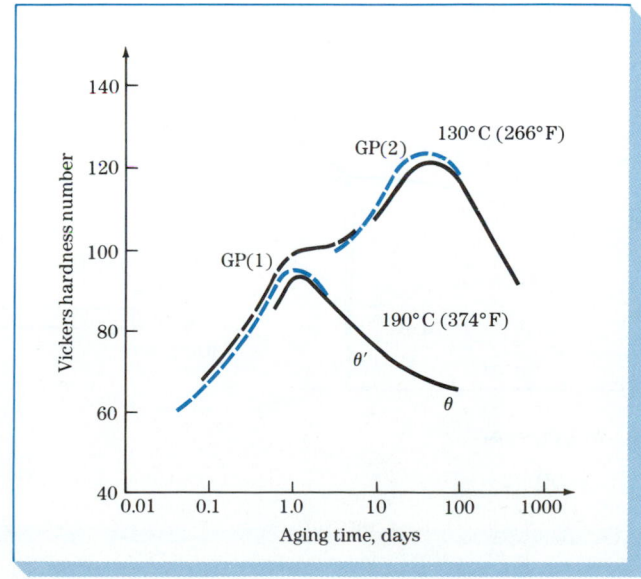

FIGURE 9.45 Correlation of structures and hardness of Al–4% Cu alloy aged at 130 and 190°C. [*After J. M. Silcock, T. J. Heal, and H. K. Hardy as presented in K. R. Van Horn (ed.), "Aluminum," vol. 1, American Society for Metals, 1967, p. 123.*]

Example Problem 9.6

Calculate the theoretical weight percent of the θ phase that could be formed at 27°C (room temperature) when a sample of Al–4.50 wt % Cu alloy is very slowly cooled from 548°C. Assume the solid solubility of Cu in Al at 27°C is 0.02 wt % and that the θ phase contains 54.0 wt % Cu.

Solution:

First, we draw a tie line xy on the Al–Cu phase diagram at 27°C between the α and θ phases, as shown in Fig. EP9.5a. Next we indicate the 4.5% Cu composition point at z. The ratio xz divided by the whole tie line xy (Fig. EP9.5b) gives the weight fraction of the θ phase. Thus,

$$\theta \text{ wt \%} = \frac{4.50 - 0.02}{54.0 - 0.02}(100\%) = \frac{4.48}{53.98}(100\%) = 8.3\% \blacktriangleleft$$

General Properties of Aluminum and Its Production

Engineering properties of aluminum Aluminum possesses a combination of properties which make it an extremely useful engineering material. Aluminum has a low density (2.70 g/cm^3), making it particularly useful for transportation manufactured products. Aluminum also has good corrosion resistance in most natural environments due to the tenacious oxide film which forms on its surface. Although pure aluminum has low strength, it can be alloyed to a strength of about 100 ksi (690 MPa). Aluminum is nontoxic and used extensively for

FIGURE EP9.5 (a) Al–Cu phase diagram with tie line xy indicated on it at 27°C and point z located at 4.5% Cu. (b) Isolated tie line xy indicating segment xz as representing the weight fraction of the θ phase.

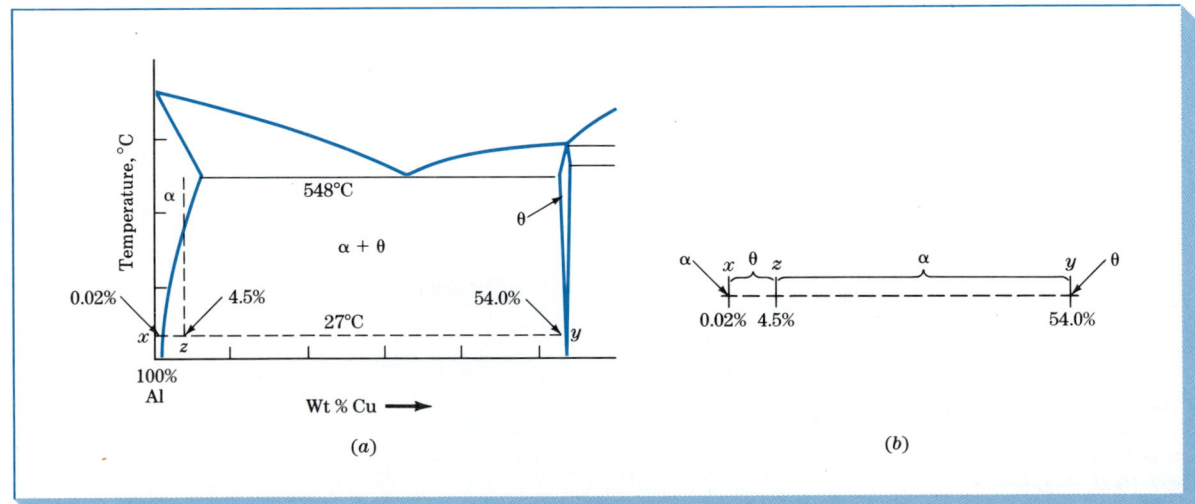

food containers and packaging. The good electrical properties of aluminum make it suitable for many applications in the electrical industry. The relatively low price of aluminum (96¢/lb in 1989) along with its many useful properties make this metal very important industrially.

Production of aluminum Aluminum is the most abundant metallic element in the earth's crust and always occurs in the combined state with other elements such as iron, oxygen, and silicon. Bauxite, which consists mainly of hydrated aluminum oxides, is the chief commercial mineral used for the production of aluminum. In the Bayer process bauxite is reacted with hot sodium hydroxide to convert the aluminum in the ore to sodium aluminate. After separation of the insoluble material, aluminum hydroxide is precipitated from the aluminate solution. The aluminum hydroxide is then thickened and calcined to aluminum oxide, Al_2O_3.

The aluminum oxide is dissolved in a molten bath of cryolite (Na_3AlF_6) and electrolyzed in an electrolytic cell (Fig. 9.46) by using carbon anodes and cathode. In the electrolysis process metallic aluminum forms in the liquid state and sinks to the bottom of the cell and is periodically tapped off. The cell-tapped aluminum usually contains from 99.5 to 99.9% aluminum with iron and silicon being the major impurities.

Aluminum from the electrolytic cells is taken to large refractory-lined furnaces where it is refined before casting. Alloying elements and alloying-element master ingots may also be melted and mixed in with the furnace charge. In the refining operation, the liquid metal is usually purged with chlorine gas to remove dissolved hydrogen gas, which is followed by a skimming of the liquid-metal surface to remove oxidized metal. After the metal has been degassed and

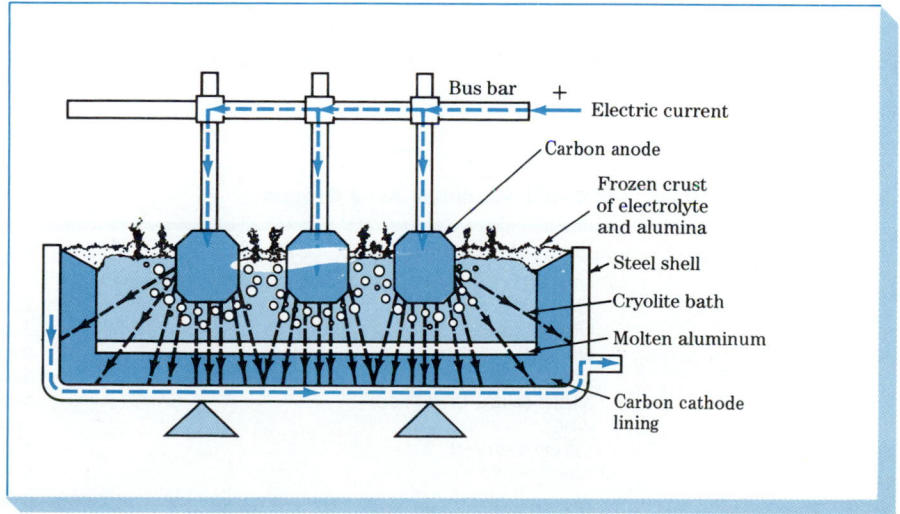

FIGURE 9.46 Electrolytic cell used to produce aluminum. (*Courtesy of Aluminum Company of America.*)

skimmed, it is screened and cast into ingot shapes for remelting or into primary ingot shapes such as sheet or extrusion ingots for further fabrication.

Wrought Aluminum Alloys

Primary fabrication Ingot shapes such as sheet and extrusion ingots are usually semicontinuously cast by the *direct-chill method.* Figure 4.8 shows schematically how an aluminum ingot is cast by this method, and Fig. 4.1 is a photograph of a large semicontinuously cast ingot being removed from the casting pit.

In the case of sheet ingots, about $\frac{1}{2}$ in of metal is removed from the ingot surfaces that will make contact with the hot-rolling-mill rolls. This operation is called *scalping* and is done to ensure a clean, smooth surface for the fabricated sheet or plate. Next the ingots are *preheated* or *homogenized* at a high temperature for about 10 to 24 h to allow atomic diffusion to make the composition of the ingot uniform. The preheating must be done at a temperature below the melting point of the constituent with the lowest melting temperature.

After reheating, the ingots are *hot-rolled* by using a four-high reversing hot-rolling mill. The ingots are usually hot-rolled to about 3 in thick and then reheated and hot-rolled down to about $\frac{3}{4}$ to 1 in with an intermediate hot-rolling mill (Fig. 6.1). Further reduction is usually carried out on a series of tandem hot-rolling mills to produce metal about 0.1 in thick. Figure 6.8 shows a typical cold-rolling operation. More than one intermediate anneal is usually required if thin sheet is to be produced.

Classification of wrought aluminum alloys Aluminum alloys produced in the wrought form (i.e., sheet, plate, extrusions, rod, and wire) are classified according to the major alloying elements they contain. A four-digit numerical designation is used to identify aluminum wrought alloys. The first digit indicates the alloy group which contains specific alloying elements. The last two digits identify the aluminum alloy or indicate the aluminum purity. The second digit indicates modification of the original alloy or impurity limits. Table 9.7 lists the wrought aluminum alloy groups.

TABLE 9.7   Wrought Aluminum Alloy Groups

Aluminum, 99.00% minimum and greater	1xxx
Aluminum alloys grouped by major alloying elements:	
Copper	2xxx
Manganese	3xxx
Silicon	4xxx
Magnesium	5xxx
Magnesium and silicon	6xxx
Zinc	7xxx
Other element	8xxx
Unused series	9xxx

Temper designations Temper designations for wrought aluminum alloys follow the alloy designation and are separated by a hyphen (for example, 1100-0). Subdivisions of a basic temper are indicated by one or more digits and follow the letter of the basic designation (for example, 1100-H14).

Basic temper designations

F—As fabricated. No control over the amount of strain hardening; no mechanical property limits.

O—Annealed and recrystallized. Temper with the lowest strength and highest ductility.

H—Strain-hardened (see subsequent subsection for subdivisions).

T—Heat-treated to produce stable tempers other than F or O (see subsequent subsection for subdivisions).

Strain-hardened subdivisions

H1—Strain-hardened only. The degree of strain hardening is indicated by the second digit and varies from quarter-hard (H12) to full-hard (H18), which is produced with approximately 75 percent reduction in area.

H2—Strain-hardened and partially annealed. Tempers ranging from quarter-hard to full-hard obtained by partial annealing of cold-worked materials with strengths initially greater than desired. Tempers are H22, H24, H26, and H28.

H3—Strain-hardened and stabilized. Tempers for age-softening aluminum-magnesium alloys that are strain-hardened and then heated at a low temperature to increase ductility and stabilize mechanical properties. Tempers are H32, H34, H36, and H38.

Heat-treated subdivisions

T1—Naturally aged. Product is cooled from an elevated-temperature shaping process and naturally aged to a substantially stable condition.

T3—Solution heat-treated, cold-worked, and naturally aged to a substantially stable condition.

T4—Solution heat-treated and naturally aged to a substantially stable condition.

T5—Cooled from an elevated-temperature shaping process and then artificially aged.

T6—Solution heat-treated and then artificially aged.

T7—Solution heat-treated and stabilized.

T8—Solution heat-treated, cold-worked, and then artificially aged.

Non-heat-treatable wrought aluminum alloys   Wrought aluminum alloys can conveniently be divided into two groups: *non-heat-treatable* and *heat-treatable alloys.* Non-heat-treatable aluminum alloys cannot be precipitation-strengthened but can only be cold-worked to increase their strength. The three main groups of non-heat-treatable wrought aluminum alloys are the 1xxx, 3xxx, and 5xxx groups. Table 9.8 lists the chemical composition, typical mechanical properties, and applications for some selected industrially important wrought aluminum alloys.

1xxx alloys. These alloys have a minimum of 99.0% aluminum with iron and silicon being the major impurities (alloying elements). An addition of 0.12% copper is added for extra strength. The 1100 alloy has a tensile strength of about 13 ksi (90 MPa) in the annealed condition and is used mainly for sheet metal work applications.

3xxx alloys. Manganese is the principal alloying element of this group and strengthens aluminum mainly by solid-solution strengthening. The most important alloy of this group is 3003, which is essentially an 1100 alloy with the addition of about 1.25% manganese. The 3003 alloy has a tensile strength of about 16 ksi (110 MPa) in the annealed condition and is used as a general-purpose alloy where good workability is required.

5xxx alloys. Magnesium is the principal alloying element of this group and is added for solid-solution strengthening in amounts up to about 5%. One of the most industrially important alloys of this group is 5052, which contains about 2.5% magnesium (Mg) and 0.2% chromium (Cr). In the annealed condition alloy 5052 has a tensile strength of about 28 ksi (193 MPa). This alloy is also used for sheet metal work, particularly for bus, truck, and marine applications.

Heat-treatable wrought aluminum alloys   Some aluminum alloys can be precipitation-strengthened by heat treatment (see pages 508 to 516). Heat-treatable wrought aluminum alloys of the 2xxx, 6xxx, and 7xxx groups are all precipitation-strengthened by similar mechanisms as described previously in Sec. 9.5 for aluminum-copper alloys. Table 9.8 lists the chemical compositions, typical mechanical properties, and applications of some of the industrially important wrought heat-treatable aluminum alloys. Figure 9.47 shows a rack of aluminum alloy extrusions about to be solution heat-treated in an industrial precipitation-strengthening heat treatment.

2xxx alloys. The principal alloying element of this group is *copper,* but magnesium is also added to most of these alloys. Small amounts of other elements are also added. One of the most important alloys of this group is 2024, which contains about 4.5% copper (Cu), 1.5% Mg, and 0.6% Mn. This alloy is strengthened mainly by solid-solution and precipitation strengthening. An intermetallic compound of the approximate

Alloy number*	Chemical composition, wt %†	Condition‡	Tensile strength		Yield strength		Elongation, %	Typical applications
			ksi	MPa	ksi	MPa		
Wrought alloys								
1100	99.0 min Al, 0.12 Cu	Annealed (-O)	13	89 (av)	3.5	24 (av)	25	Sheet metal work, fin stock
		Half-hard (-H14)	18	124 (av)	14	97 (av)	4	
3003	1.2 Mn	Annealed (-O)	17	117 (av)	5	34 (av)	23	Pressure vessels, chemical equipment, sheet metal work
		Half-hard (-H14)	23	159 (av)	23	159 (av)	17	
5052	2.5 Mg, 0.25 Cr	Annealed (-O)	28	193 (av)	9.5	65 (av)	18	Bus, truck, and marine uses, hydraulic tubes
		Half-hard (-H34)	38	262 (av)	26	179 (av)	4	
2024	4.4 Cu, 1.5 Mg, 0.6 Mn	Annealed (-O)	32	220 (max)	14	97 (max)	12	Aircraft structures
		Heat-treated (-T6)	64	442 (min)	50	345 (min)	5	
6061	1.0 Mg, 0.6 Si, 0.27 Cu, 0.2 Cr	Annealed (-O)	22	152 (max)	12	82 (max)	16	Truck and marine structures, pipelines, railings
		Heat-treated (-T6)	42	290 (min)	35	241 (min)	10	
7075	5.6 Zn, 2.5 Mg, 1.6 Cu, 0.23 Cr	Annealed (-O)	40	276 (max)	21	145 (max)	10	Aircraft and other structures
		Heat-treated (-T6)	73	504 (min)	62	428 (min)	8	
Casting alloys								
355.0	5 Si, 1.2 Cu, 0.5 Mg	Sand cast (-T6)	32	220 (min)	20	138 (min)	2.0	Pump housings, aircraft fittings, crankcases
		Permanent mold (-T6)	37	285 (min)	...	...	1.5	
356.0	7 Si, 0.3 Mg	Sand cast (-T6)	30	207 (min)	20	138 (min)	3	Transmission cases, truck-axle housings, truck wheels
		Permanent mold (-T6)	33	229 (min)	22	152 (min)	3	
332.0	9.5 Si, 3 Cu, 1.0 Mg	Permanent mold (-T5)	31	214 (min)				Automotive pistons
413.0	12 Si, 2 Fe	Die casting	43	297	21	145 (min)	2.5	Large intricate castings

* Aluminum Association number.
† Balance aluminum.
‡ O = annealed and recrystallized; H14 = strain-hardened only; H34 = strain-hardened and stabilized; T5 = cooled from elevated-temperature shaping process, then artificially aged; T6 = solution heat-treated, then artificially aged.

FIGURE 9.47 Industrial precipitation strengthening (hardening of aluminum alloy extrusions). *(Courtesy of Reynolds Metals Co.)*

composition of Al$_2$CuMg is the main strengthening precipitate. Alloy 2024 in the T6 condition has a tensile strength of about 64 ksi (442 MPa) and is used, for example, for aircraft structurals.

6xxx alloys. The principal alloying elements for the 6xxx group are *magnesium* and *silicon* which combine together to form an intermetallic compound, Mg$_2$Si, which in precipitate form strengthens this group of alloys. Alloy 6061 is one of the most important alloys of this group and has an approximate composition of 1.0% Mg, 0.6% Si, 0.3% Cu, and 0.2% Cr. This alloy in the T6 heat-treated condition has a tensile strength of about 42 ksi (290 MPa) and is used for general-purpose structurals.

7xxx alloys. The principal alloying elements for the 7xxx group of aluminum alloys are zinc, magnesium, and copper. Zinc and magnesium combine to form an intermetallic compound, MgZn$_2$, which is the basic precipitate that strengthens these alloys when they are heat-treated. The relatively high solubility of zinc and magnesium in aluminum makes it possible to create a high density of precipitates and hence to produce very great increases in strength. Alloy 7075 is one of the most important alloys of this group and has an approximate composition of 5.6% Zn, 2.5% Mg, 1.6% Cu, and 0.25% Cr. Alloy 7075 when heat-treated to the T6 temper, has a tensile strength of about 73 ksi (504 MPa) and is used mainly for aircraft structurals.

Aluminum Casting Alloys

Casting processes Aluminum alloys are normally cast by three main processes: sand casting, permanent-mold, and die casting.

Sand casting is the simplest and most versatile of the aluminum casting processes. Figure 9.48 shows how a simple sand mold for producing sand

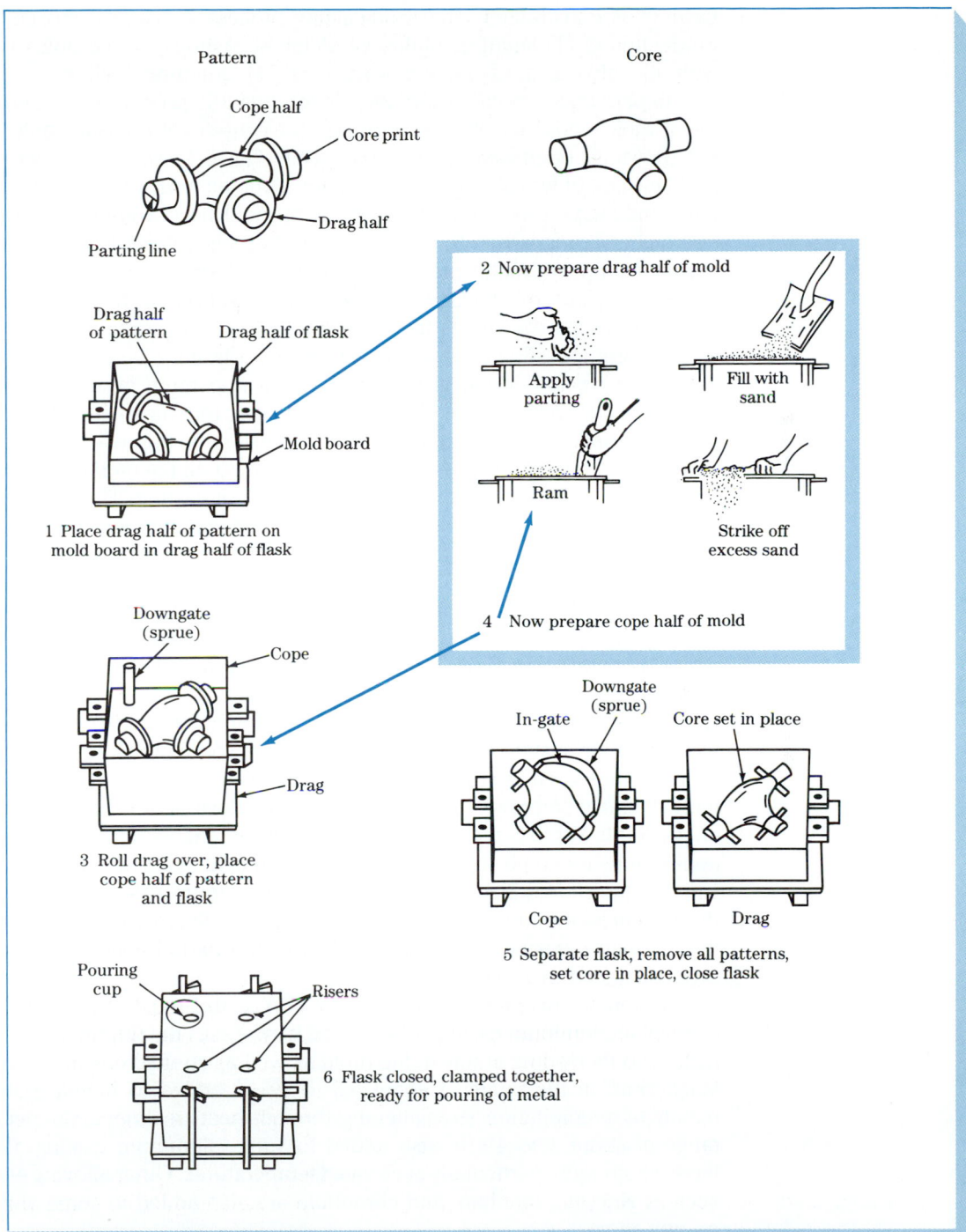

FIGURE 9.48 Steps in the construction of a simple sand mold for making a sand casting. *(After H. F. Taylor, M. C. Flemings, and J. Wulff, "Foundry Engineering," Wiley, 1959, p. 20.)*

castings is constructed. The sand-casting process is usually chosen for the production of (1) small quantities of identical castings, (2) complex castings with intricate cores, (3) large castings, and (4) structural castings.

In *permanent-mold casting* the molten metal is poured into a permanent metal mold under gravity, low pressure, or centrifugal pressure only. Figure 6.2 shows an open permanent mold, while Fig. 6.3*a* shows the permanent-mold casting of two aluminum alloy automobile pistons. Castings of the same alloy and shape produced by a permanent mold have a finer grain structure and higher strength than those cast by sand molds. The faster cooling rate of permanent-mold casting produces a finer grain structure. Also, permanent-mold castings usually have less shrinkage and gas porosity than sand castings. However, permanent molds have size limitations, and complex parts may be difficult or impossible to cast with a permanent mold.

In *die casting,* identical parts are cast at maximum production rates by forcing molten metal under considerable pressure into metal molds. Two metal die halves are securely locked together to withstand high pressure. The molten aluminum is forced into the cavities in the dies. When the metal has solidified, the dies are unlocked and opened to eject the hot casting. The die halves are locked together again and the casting cycle is repeated. Some of the advantages of die casting are (1) parts die cast are almost completely finished and can be produced at high rates, (2) dimensional tolerances of each cast part can be more closely held than with any other major casting process, (3) smooth surfaces on the casting are obtainable, (4) rapid cooling of the casting produces a fine-grain structure, and (5) the process can be automated easily.

Aluminum casting alloy compositions   Aluminum casting alloys have been developed for casting qualities such as fluidity and feeding ability as well as for properties such as strength, ductility, and corrosion resistance. As a result, their chemical compositions differ greatly from those of the wrought aluminum alloys. Table 9.8 lists the chemical compositions, mechanical properties, and applications for some selected aluminum casting alloys. These alloys are classified in the United States according to the Aluminum Association system. In this system aluminum casting alloys are grouped by the major alloying elements they contain by using a four-digit number with a period between the last two digits, as listed in Table 9.9.

Silicon in the range of about 5 to 12% is the most important alloying element in aluminum casting alloys since it increases the fluidity of the molten metal and its feeding ability in the mold as well as strengthens the aluminum. Magnesium in the range of about 0.3 to 1% is added to increase strength, mainly by precipitation strengthening through heat treatment. Copper in the range of about 1 to 4% is also added to some aluminum casting alloys to increase strength, particularly at elevated temperatures. Other alloying elements such as zinc, tin, titanium, and chromium are also added to some aluminum casting alloys.

In some cases, if the cooling rate of the solidified casting in the mold is

TABLE 9.9   Cast Aluminum Alloy Groups

Aluminum, 99.00% minimum and greater	1xx.x
Aluminum alloys grouped by major alloying elements:	
Copper	2xx.x
Silicon, with added copper and/or magnesium	3xx.x
Silicon	4xx.x
Magnesium	5xx.x
Zinc	7xx.x
Tin	8xx.x
Other element	9xx.x
Unused series	6xx.x

sufficiently rapid, a heat-treatable alloy can be produced in the supersaturated solid condition. Thus, the solution heat-treatment and quenching steps can be omitted for precipitation strengthening the casting, and only subsequent aging of the casting after it has been removed from the mold is required. A good example of the application of this type of heat treatment is in the production of precipitation-strengthened automobile pistons. The pistons shown in Fig. 6.3a, after being removed from the mold, only require an aging treatment to be precipitation-strengthened. This heat-treatment temper is called T5.

9.6   COPPER ALLOYS

General Properties of Copper

Copper is an important engineering metal and is widely used in the unalloyed condition as well as combined with other metals in the alloyed form. In the unalloyed form, copper has an extraordinary combination of properties for industrial applications. Some of these are high electrical and thermal conductivity, good corrosion resistance, ease of fabrication, medium tensile strength, controllable annealing properties, and general soldering and joining characteristics. Higher strengths are attained in a series of brass and bronze alloys which are indispensable for many engineering applications.

Production of Copper

Most copper is extracted from ores containing copper and iron sulfides. Copper sulfide concentrates obtained from lower-grade ores are smelted in a reverberatory furnace to produce a matte which is a mixture of copper and iron sulfides and is separated from a slag (waste material). The copper sulfide in the matte is then chemically converted to impure or blister copper (98% + Cu) by blowing air through the matte. The iron sulfide is oxidized first and slagged off in this operation. Subsequently most of the impurities in the blister copper are removed in a refining furnace and are removed as a slag. This fire-refined copper is called *tough-pitch copper*, and although it can be used for some applications, most tough-pitch copper is further refined electrolytically to produce 99.95% *electrolytic tough-pitch (ETP) copper.*

Classification of Copper Alloys   Copper alloys in the United States are classified according to a designation system administered by the Copper Development Association (CDA). In this system the numbers C10100 to C79900 designate wrought alloys and the numbers from C80000 to C99900 designate casting alloys. Table 9.10 lists the alloy groups of each major level, and Table 9.11 lists the chemical compositions, typical mechanical properties, and applications for some selected copper alloys.

Wrought Copper Alloys   **Unalloyed copper**   Unalloyed copper is an important engineering metal, and because of its high electrical conductivity, it is used to a large extent in the electrical industry. Electrolytic tough-pitch (ETP) copper is the least expensive of the industrial coppers and is used for the production of wire, rod, plate, and strip. ETP copper has a nominal oxygen content of 0.04%. Oxygen is almost insoluble in ETP copper and forms interdendritic Cu_2O when copper is cast. For most applications the oxygen in ETP copper is an insignificant impurity. However, if ETP copper is heated to a temperature above about 400°C in an atmosphere containing hydrogen, the hydrogen can diffuse into the solid copper and react with the internally dispersed Cu_2O to form steam according to the reaction

$$Cu_2O + H_2 \text{ (dissolved in Cu)} \rightarrow 2Cu + H_2O \text{ (steam)}$$

TABLE 9.10   Classification of Copper Alloys (Copper Development Association System)

Wrought alloys	
C1xxxx	Coppers* and high-copper alloys†
C2xxxx	Copper-zinc alloys (brasses)
C3xxxx	Copper-zinc-lead alloys (leaded brasses)
C4xxxx	Copper-zinc-tin alloys (tin brasses)
C5xxxx	Copper-tin alloys (phosphor bronzes)
C6xxxx	Copper-aluminum alloys (aluminum bronzes), copper-silicon alloys (silicon bronzes) and miscellaneous copper-zinc alloys
C7xxxx	Copper-nickel and copper-nickel-zinc alloys (nickel silvers)

Cast alloys	
C8xxxx	Cast coppers, cast high-copper alloys, cast brasses of various types, cast manganese-bronze alloys, and cast copper-zinc-silicon alloys
C9xxxx	Cast copper-tin alloys, copper-tin-lead alloys, copper-tin-nickel alloys, copper-aluminum-iron alloys, and copper-nickel-iron and copper-nickel-zinc alloys.

*"Coppers" have a minimum copper content of 99.3 percent or higher.
†High-copper alloys have less than 99.3% Cu, but more than 96%, and do not fit into the other copper alloy groups.

The large water molecules formed by the reaction do not diffuse readily and form internal holes, particularly at the grain boundaries, which makes the copper brittle (Fig. 9.49).

To avoid hydrogen embrittlement caused by Cu_2O, the oxygen can be reacted with phosphorus to form phosphorus pentoxide (P_2O_5) (alloy C12200). Another way to avoid hydrogen embrittlement is to eliminate the oxygen from the copper by casting the ETP copper under a controlled reducing atmosphere. The copper produced by this method is called *oxygen-free high-conductivity (OFHC) copper* and is alloy C10200.

Copper-zinc alloys   The copper-zinc brasses consist of a series of alloys of copper with additions of about 5 to 40% zinc. Copper forms substitutional solid solutions with zinc up to about 35% zinc, as indicated in the all-alpha phase region of the Cu–Zn phase diagram (Fig. 8.25). When the zinc content reaches about 40%, alloys with two phases, alpha and beta, form.

The microstructure of the single-phase alpha brasses consists of an alpha solid solution, as shown in Fig. 9.50 for a 70% Cu–30% Zn alloy (C26000, cartridge brass). The microstructure of the 60% Cu–40% Zn brass (C28000, Muntz metal) has two phases, alpha and beta, as shown in Fig. 9.51.

Small amounts of lead (0.5 to 3%) are added to some Cu–Zn brasses to improve machinability. Lead is almost insoluble in solid copper and is distributed in leaded brasses in small globules (Fig. 9.52).

The tensile strengths of some selected brasses are listed in Table 9.11. These alloys are of medium strength (34 to 54 ksi; 234 to 374 MPa) in the annealed condition and can be cold-worked to increase their strength.

Copper-tin bronzes   Copper-tin alloys, which are properly called *tin bronzes* but often called *phosphor bronzes,* are produced by alloying about 1 to 10% tin with copper to form solid-solution-strengthened alloys. Wrought tin bronzes are stronger than Cu–Zn brasses, especially in the cold-worked condition, and have better corrosion resistance, but cost more. Cu–Sn casting alloys containing up to about 16% Sn are used for high-strength bearings and gear blanks. Large amounts of lead (5 to 10%) are added to these alloys to provide lubrication for bearing surfaces.

Copper-beryllium alloys   Copper-beryllium alloys are produced containing between 0.6 to 2% Be with additions of cobalt from 0.2 to 2.5%. These alloys are precipitation-hardenable and can be heat-treated and cold-worked to produce tensile strengths as high as 212 ksi (1463 MPa), which is the highest strength developed in commercial copper alloys. Cu–Be alloys are used for tools requiring high hardness and nonsparking characteristics for the chemical industry. The excellent corrosion resistance, fatigue properties, and strength of these alloys make them useful for springs, gears, diaphragms, and valves. However, they have the disadvantage of being relatively costly materials.

TABLE 9.11 Typical Mechanical Properties and Applications of Copper Alloys

Alloy number	Chemical composition, wt %	Condition	Tensile strength		Yield strength		Elongation in 2 in, %	Typical applications
			ksi	MPa	ksi	MPa		
Wrought alloys								
C10100	99.99 Cu	Annealed Cold-worked	32 50	220 345	10 45	69 310	45 6	Bus conductors, waveguides, hollow conductors, lead-in wires and anodes for vacuum tubes, vacuum seals, transistor components, glass-to-metal seals, coaxial cables and tubes, klystrons, microwave tubes, rectifiers
C11000 (ETP)	99.9 Cu, 0.04 O	Annealed Cold-worked	32 50	220 345	10 45	69 310	45 6	Gutters, roofing, gaskets, auto radiators, busbars, nails, printing rolls, rivets, radio parts
C26000	70 Cu, 30 Zn	Annealed Cold-worked	47 76	325 525	15 63	105 435	62 8	Radiator cores and tanks, flashlight shells, lamp fixtures, fasteners, locks, hinges, ammunition components, plumbing accessories, pins, rivets
C28000	60 Cu, 40 Zn	Annealed Cold-worked	54 70	370 485	21 50	145 345	45 10	Architectural, large nuts and bolts, brazing rod, condenser plates, heat exchanger and condenser tubing, hot forgings
C17000	99.5 Cu, 1.7 Be, 0.20 Co	SHT* SHT, CW, PH*	60 180	410 1240	28 155	190 1070	60 4	Bellows, Bourdon tubing, diaphragms, fuse clips, fasteners, lock-washers, springs, switch parts, roll pins, valves, welding equipment
C61400	95 Cu, 7 Al, 2 Fe	Annealed Cold-worked	80 89	550 615	40 60	275 415	40 32	Nuts, bolts, stringers and threaded members, corrosion-resistant vessels and tanks, structural components, machine parts, condenser tube and piping systems, marine protective sheathing and fastening

C71500	70 Cu, 30 Ni	Annealed Cold-worked	55 84	380 580	18 79	125 545	36 3	Communication relays, condensers, condenser plates, electrical springs, evaporator and heat exchanger tubes, ferrules, resistors
Casting alloys								
C80500	99.75 Cu	As-cast	25	172	9	62	40	Electrical and thermal conductors; corrosion- and oxidation-resistant applications
C82400	96.4 Cu, 1.70 Be, 0.25 Co	As-cast Heat-treated	72 150	497 1035	37 140	255 966	20 1	Safety tools, molds for plastic parts, cams, bushings, bearings, valves, pump parts, gears
C83600	85 Cu, 5 Sn, 5 Pb, 5 Zn	As-cast	37	255	17	117	30	Valves, flanges, pipe fittings, plumbing goods, pump castings, water pump impellers and housings, ornamental fixtures, small gears
C87200	89 Cu, 4 Si	As-cast	55	379	25	172	30	Bearings, belts, impellers, pump and valve components, marine fittings, corrosion-resistant castings
C90300	93 Cu, 8 Sn, 4 Zn	As-cast	45	310	21	145	30	Bearings, bushings, pump impellers, piston rings, valve components, seal rings, steam fittings, gears
C95400	85 Cu, 4 Fe, 11 Al	As-cast Heat-treated	85 105	586 725	35 54	242 373	18 8	Bearings, gears, worms, bushings, valve seats and guides, pickling hooks
C96400	69 Cu, 30 Ni, 0.9 Fe	As-cast	68	469	37	255	28	Valves, pump bodies, flanges, elbows used for seawater corrosion resistance

*SHT = solution heat-treated; CW = cold-worked; PH = precipitation-hardened.

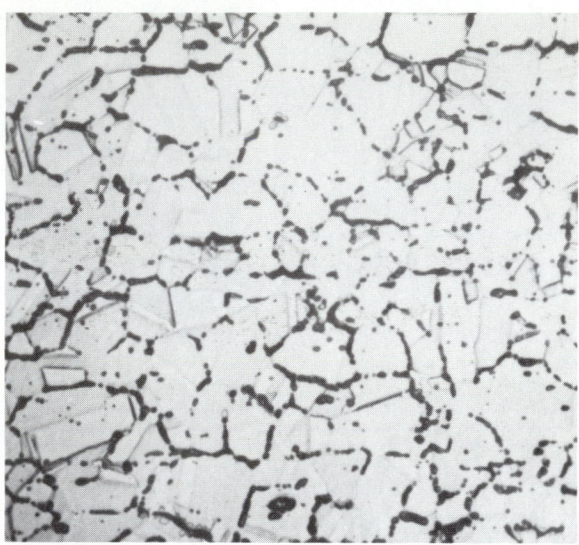

FIGURE 9.49 Electrolytic tough-pitch copper exposed to hydrogen at 850°C for $\frac{1}{2}$ h; structure shows internal holes developed by steam, which makes the copper brittle. (Etch: potassium dichromate; magnification 150×.) *(Courtesy of Amax Base Metals Research, Inc.)*

FIGURE 9.50 Microstructures of cartridge brass (70% Cu–30% Zn) in the annealed condition. (Etchant: $NH_4OH + H_2O_2$; magnification 75×.)

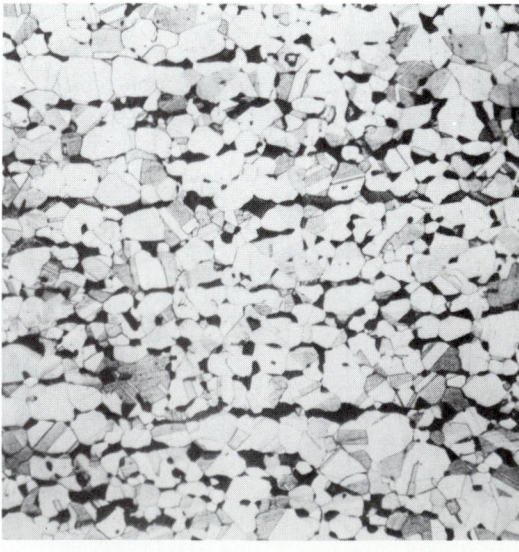

FIGURE 9.51 Hot-rolled Muntz metal sheet (60% Cu–40% Zn). Structure consists of beta phase (dark) and alpha phase (light). (Etchant: $NH_4OH + H_2O_2$; magnification 75×.) *(Courtesy of Anaconda American Brass Co.)*

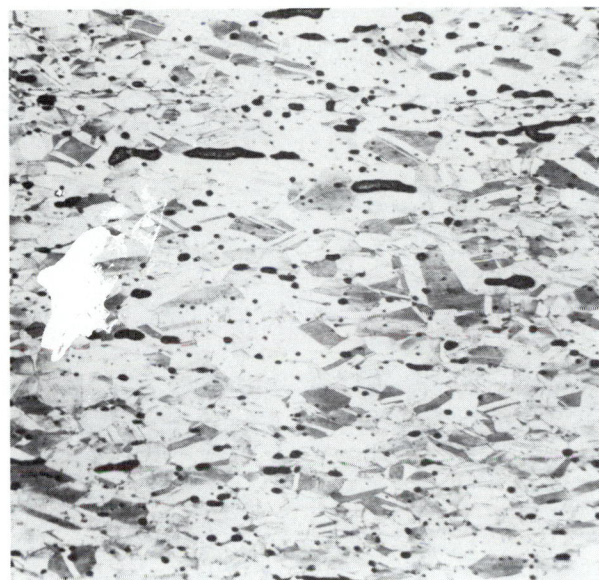

FIGURE 9.52 Free-cutting brass extruded rod showing elongated lead globules. Remainder of structure is α phase. (Etch: NH_4OH + H_2O_2; magnification 75×.) *(Courtesy of Anaconda American Brass Co.)*

9.7 STAINLESS STEELS

Stainless steels are selected as engineering materials mainly because of their excellent corrosion resistance in many environments. The corrosion resistance of stainless steels is due to their high chromium contents. In order to make a "stainless steel" stainless, there must be at least 12% chromium (Cr) in the steel. According to classical theory, chromium forms a surface oxide which protects the underlying iron-chromium alloy from corroding. To produce the protective oxide, the stainless steel must be exposed to oxidizing agents.

In general there are four main types of stainless steels: ferritic, martensitic, austenitic, and precipitation-hardening. Only the first three types will be briefly discussed in this section.

Ferritic Stainless Steels

Ferritic stainless steels are essentially iron-chromium binary alloys containing about 12 to 30% Cr. They are called ferritic since their structure remains mostly ferritic (BCC, α iron type) at normal heat-treatment conditions. Chromium, since it has the same BCC crystal structure as α ferrite, extends the α phase region and suppresses the γ phase region. As a result, the "γ loop" is formed in the Fe–Cr phase diagram and divides it into FCC and BCC regions (Fig. 9.53). Ferritic stainless steels, since they contain more than 12% Cr, do not undergo the FCC-to-BCC transformation and cool from high temperatures as solid solutions of chromium in α iron.

Table 9.12 lists the chemical compositions, typical mechanical properties, and applications of some selected stainless steels, including ferritic type 430.

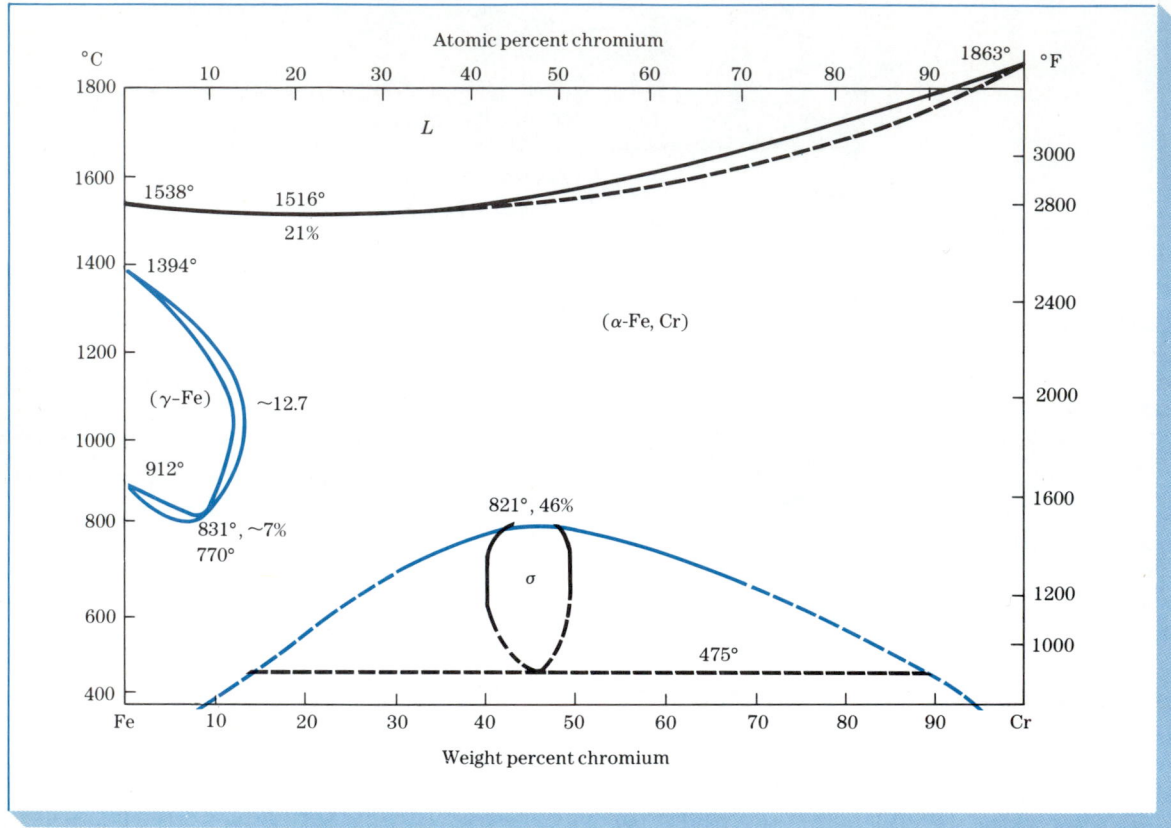

FIGURE 9.53 Iron-chromium phase diagram. *(After "Metals Handbook," vol. 8, 8th ed., American Society for Metals, 1973, p. 291.)*

The ferritic stainless steels are relatively low in cost since they do not contain nickel. They are used mainly as general construction materials in which their special corrosion and heat resistance is required. Figure 9.54 shows the microstructure of the ferritic stainless steel type 430 in the annealed condition. The presence of the carbides in this steel reduces its corrosion resistance to some extent. New ferritics have more recently been developed with very low carbon and nitrogen levels and so have improved corrosion resistance.

Martensitic Stainless Steels

Martensitic stainless steels are essentially Fe–Cr alloys containing 12 to 17% Cr with sufficient carbon (0.15 to 1.0% C) so that a martensitic structure can be produced by quenching from the austenitic phase region. These alloys are called martensitic since they are capable of developing a martensitic structure after an austenitizing and quenching heat treatment. Since the composition of martensitic stainless steels is adjusted to optimize strength and hardness, the

TABLE 9.12   Typical Mechanical Properties and Applications of Stainless Steels

Alloy number	Chemical composition, wt %*	Condition	Tensile strength		Yield strength		Elongation in 2 in, %	Typical applications
			ksi	MPa	ksi	MPa		
Ferritic stainless steels								
430	17 Cr, 0.012 C	Annealed	75	517	50	345	25	General-purpose, nonhardenable; uses: range hoods, restaurant equipment
446	25 Cr, 0.20 C	Annealed	80	552	50	345	20	High-temperature applications; heaters, combustion chambers
Martensitic stainless steels								
410	12.5 Cr, 0.15 C	Annealed Q & T†	75	517	40	276	30	General-purpose heat-treatable; machine parts, pump shafts, valves
440A	17 Cr, 0.70 C	Annealed Q & T†	105 265	724 1828	60 245	414 1690	20 5	Cutlery, bearings, surgical tools
440C	17 Cr, 1.1 C	Annealed Q & T†	110 285	759 1966	70 275	276 1897	13 2	Balls, bearings, races, valve parts
Austenitic stainless steels								
301	17 Cr, 7 Ni	Annealed	110	759	40	276	60	High work-hardening rate alloy; structural applications
304	19 Cr, 10 Ni	Annealed	84	580	42	290	55	Chemical and food processing equipment
304L	19 Cr, 10 Ni, 0.03 C	Annealed	81	559	39	269	55	Low carbon for welding: chemical tanks
321	18 Cr, 10 Ni, Ti = 5 × %C min	Annealed	90	621	35	241	45	Stabilized for welding; process equipment, pressure vessels
347	18 Cr, 10 Ni, Cb (Nb) = 10 × C min	Annealed	95	655	40	276	45	Stabilized for welding; tank cars for chemicals
Precipitation-hardening stainless steels								
17-4PH	16 Cr, 4 Ni, 4 Cu, 0.03 Cb (Nb)	Precipitation-hardened	190	1311	175	1207	14	Gears, cams, shafting, aircraft and turbine parts

* Balance Fe.   † Quenched and tempered.

FIGURE 9.54 Type 430 (ferritic) stainless steel strip annealed at 788°C (1450°F). The structure consists of a ferrite matrix with equiaxed grains and dispersed carbide particles. (Etchant: picral + HCl; magnification 100×.) *(Courtesy of United States Steel Corp., Research Laboratories.)*

corrosion resistance of these steels is relatively poor compared to the ferritic and austenitic types.

The heat treatment of martensitic stainless steels for increased strength and toughness is basically the same as that for plain-carbon and low-alloy steels. That is, the alloy is austenitized, cooled fast enough to produce a martensitic structure, and then tempered to relieve stresses and increase toughness. The high hardenability of the Fe–12 to 17% Cr alloys avoids the need for water quenching and allows a slower cooling rate to produce a martensitic structure.

Table 9.12 includes the chemical compositions, typical mechanical properties, and applications for types 410 and 440C martensitic stainless steels. The 410 stainless steel with 12% Cr is a lower-strength martensitic stainless steel and is a general-purpose heat-treatable type used for applications such as machine parts, pump shafts, bolts, and bushings.

When the carbon content of Fe–Cr alloys is increased up to about 1% C, the γ loop is enlarged. Consequently, Fe–Cr alloys with about 1% C can contain about 16% Cr and still be able to produce a martensitic structure upon austenitizing and quenching. Type 440C alloy with 16% Cr and 1% C is the martensitic stainless steel which has the highest hardness of any corrosion-resisting steel. Its high hardness is due to a hard martensitic matrix and to the presence of a large concentration of primary carbides, as shown in the 440C steel microstructure of Fig. 9.55.

Austenitic Stainless Steels

Austenitic stainless steels are essentially iron-chromium-nickel ternary alloys containing about 16 to 25% Cr and 7 to 20% Ni. These alloys are called austenitic since their structure remains austenitic (FCC, γ iron type) at all normal heat-treating temperatures. The presence of the nickel which has an FCC crystal structure enables the FCC structure to be retained at room temperature. The

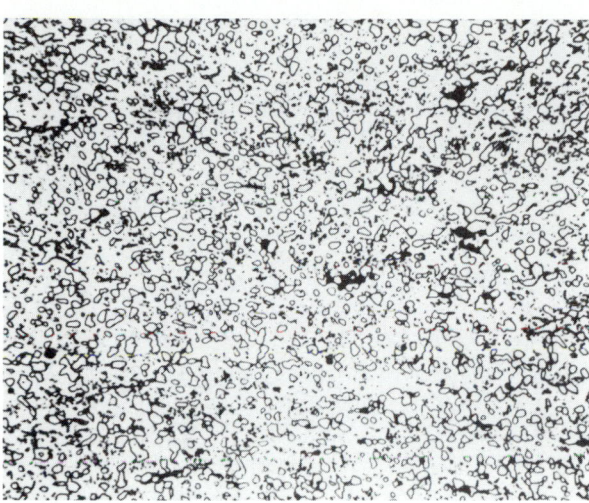

FIGURE 9.55 Type 440C (martensitic) stainless steel hardened by austenitizing at 1010°C (1850°F) and air-cooled. Structure consists of primary carbides in martensite matrix. (Etchant: HCl + picral; magnification 500×.) *(Courtesy of Allegheny Ludlum Steel Co.)*

high formability of the austenitic stainless steels is due to its FCC crystal structure. Table 9.12 includes the chemical composition, typical mechanical properties, and applications for austenitic stainless steel types 301, 304, and 347.

Austenitic stainless steels normally have better corrosion resistance than ferritic and martensitic ones because the carbides can be retained in solid solution by rapid cooling from high temperatures. However, if these alloys are to be welded or slowly cooled from high temperatures through the 870 to 600°C range, they can become susceptible to intergranular corrosion, because chromium-containing carbides precipitate at the grain boundaries. This difficulty can be circumvented to some degree either by lowering the maximum carbon content in the alloy to about 0.03% C (type 304L alloy) or by adding an alloying element such as columbium (niobium) (type 347 alloy) to combine with the carbon in the alloy (see Sec. 12.5 on intergranular corrosion). Figure 9.56 shows the microstructure of a type 304 stainless steel which has been annealed at 1065°C and air-cooled. Note that there are no carbides visible in the microstructure, as in the case of the type 430 steel (Fig. 9.54) and type 440C steel (Fig. 9.55).

9.8 CAST IRONS

General Properties Cast irons are a family of ferrous alloys with a wide range of properties, and as their name implies, they are intended to be cast into the desired shape instead of being worked in the solid state. Unlike steels, which usually contain

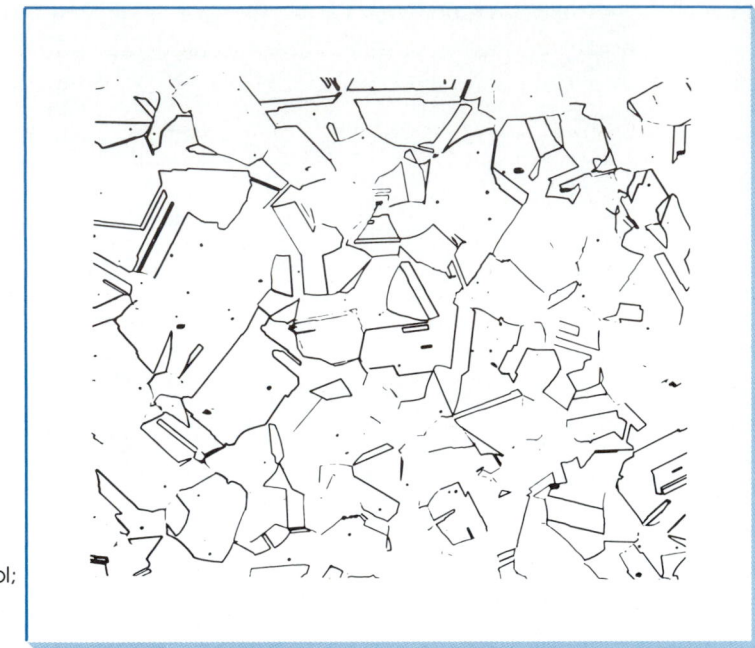

FIGURE 9.56 Type 304 (austenitic) stainless steel strip annealed 5 min at 1065°C (1950°F) and air-cooled. Structure consists of equiaxed austenite grains. Note annealing twins. (Etchant: HNO_3–acetic–HCl–glycerol; magnification 250×.) *(Courtesy of Allegheny Ludlum Steel Co.)*

less than about 1% carbon, cast irons normally contain 2 to 4% carbon and 1 to 3% silicon. Other alloying elements may also be present to control or vary certain properties.

Cast irons make excellent casting alloys since they are easily melted, are very fluid in the liquid state, and do not form undesirable surface films when poured. Cast irons solidify with slight to moderate shrinkage during casting and cooling. These alloys have a wide range of strengths and hardness and in most cases are easy to machine. They can be alloyed to produce superior wear, abrasion, and corrosion resistance. However, cast irons have relatively low impact resistance and ductility, and this limits their use for some applications. The wide industrial use of cast irons is due mainly to their comparatively low cost and versatile engineering properties.

Types of Cast Irons   Four different kinds of cast irons can be differentiated from each other by the distribution of carbon in their microstructures: *white, gray, malleable,* and *ductile iron. High-alloy cast irons* constitute a fifth type of cast iron. However, since the chemical compositions of cast irons overlap, they cannot be distinguished from each other by chemical composition analyses. Table 9.13 lists the chemical composition ranges for the four basic cast irons, and Table 9.14 presents some of their typical tensile mechanical properties and applications.

TABLE 9.13   Chemical Composition Ranges for Typical Unalloyed Cast Irons

Element	Gray Iron, %	White Iron, %	Malleable Iron (cast white), %	Ductile Iron, %
Carbon	2.5–4.0	1.8–3.6	2.00–2.60	3.0–4.0
Silicon	1.0–3.0	0.5–1.9	1.10–1.60	1.8–2.8
Manganese	0.25–1.0	0.25–0.80	0.20–1.00	0.10–1.00
Sulfur	0.02–0.25	0.06–0.20	0.04–0.18	0.03 max
Phosphorus	0.05–1.0	0.06–0.18	0.18 max	0.10 max

Source: C. F. Walton (ed), "Iron Castings Handbook," Iron Castings Society, 1981.

White Cast Iron

White cast iron is formed when much of the carbon in a molten cast iron forms iron carbide instead of graphite upon solidification. The microstructure of as-cast unalloyed white cast iron contains large amounts of iron carbides in a pearlitic matrix (Fig. 9.57). White cast irons are so-called because they fracture to produce a "white" or bright crystalline fractured surface. To retain the carbon in the form of iron carbide in white cast irons, their carbon and silicon contents must be kept relatively low (that is, 2.5–3.0% C and 0.5–1.5% Si) and the solidification rate high.

White cast irons are most often used for their excellent resistance to wear and abrasion. The large amount of iron carbides in their structure is mainly responsible for their wear resistance. White cast iron serves as the raw material for malleable cast irons.

Gray Cast Iron

Gray cast iron is formed when the carbon in the alloy exceeds the amount that can dissolve in the austenite and precipitates as graphite flakes. When a piece of solidified gray iron is fractured, the fracture surface appears gray because of the exposed graphite.

Gray cast iron is an important engineering material because of its relatively low cost and useful engineering properties, including excellent machinability at hardness levels which have good wear resistance, resistance to galling under restricted lubrication, and excellent vibrational damping capacity.

Composition and microstructure   As listed in Table 9.13 unalloyed gray cast irons usually contain 2.5 to 4% C and 1 to 3% Si. Since silicon is a graphite stabilizing element in cast irons, a relatively high silicon content is used to promote the formation of graphite. The solidification rate is also an important factor that determines the extent to which graphite forms. Moderate and slow cooling rates favor the formation of graphite. The solidification rate also affects the type of matrix formed in gray cast irons. Moderate cooling rates favor the

TABLE 9.14 Typical Mechanical Properties and Applications of Cast Irons

Alloy name and number	Chemical composition, wt %	Condition	Microstructure	Tensile strength ksi	Tensile strength MPa	Yield strength ksi	Yield strength MPa	Elongation, %	Typical applications
Gray cast irons									
Ferritic (G2500)	3.4 C, 2.2 Si, 0.7 Mn	Annealed	Ferritic matrix	26	179	…	…	…	Small cylinder blocks, cylinder heads, clutch plates
Pearlitic (G3500)	3.2 C, 2.0 Si, 0.7 Mn	As-cast	Pearlitic matrix	36	252	…	…	…	Truck and tractor cylinder blocks, heavy gear boxes
Pearlitic (G4000)	3.3 C, 2.2 Si, 0.7 Mn	As-cast	Pearlitic matrix	42	293	…	…	…	Diesel engine castings
Malleable cast irons									
Ferritic (32510)	2.2 C, 1.2 Si, 0.04 Mn	Annealed	Temper carbon and ferrite	50	345	32	224	10	General engineering service with good machinability
Pearlitic (45008)	2.4 C, 1.4 Si, 0.75 Mn	Annealed	Temper carbon and pearlite	65	440	45	310	8	General engineering service with dimensional tolerance specified
Martensitic (M7002)	2.4 C, 1.4 Si, 0.75 Mn	Quenched and tempered	Tempered martensite	90	621	70	438	2	High-strength parts: connecting rods and universal joint yokes
Ductile cast irons									
Ferritic (60-40-18)	3.5 C, 2.2 Si	Annealed	Ferritic	60	414	40	276	18	Pressure castings, such as valve and pump bodies
Pearlitic	3.5 C, 2.2 Si	As-cast	Ferritic-pearlitic	80	552	55	379	6	Crankshafts, gears, and rollers
Martensitic (120-90-02)	3.5 C, 2.2 Si	Martensitic	Quenched and tempered	120	828	90	621	2	Pinions, gears, rollers, and slides

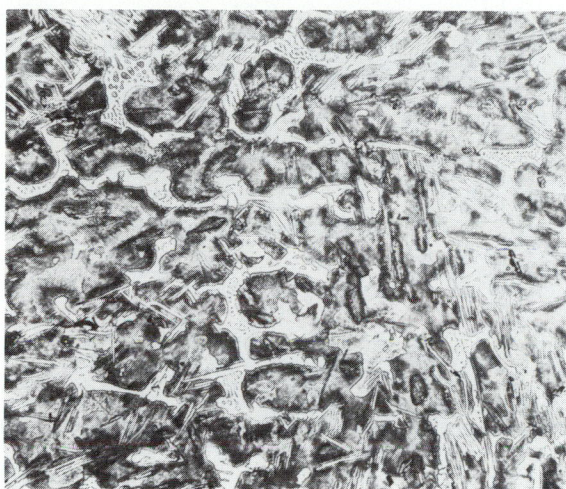

FIGURE 9.57
Microstructure of white cast iron. The white constituent is iron carbide. The gray areas are unresolved pearlite. (Etch: 2% nital; magnification 100×.) *(Courtesy of Central Foundry.)*

formation of a pearlitic matrix, whereas slow cooling rates favor a ferritic matrix. To produce a fully ferritic matrix in an unalloyed gray iron, the iron is usually annealed to allow the carbon remaining in the matrix to deposit on the graphite flakes, leaving the matrix completely ferritic.

Figure 9.58 shows the microstructure of an unalloyed as-cast gray iron with graphite flakes in a matrix of mixed ferrite and pearlite. Figure 9.59 shows a scanning electron micrograph of a hypereutectic gray iron with the matrix etched out.

Ductile Cast Irons   Ductile cast iron (sometimes called *nodular* or *spherulitic graphite cast iron*) combines the processing advantages of gray cast iron with the engineering

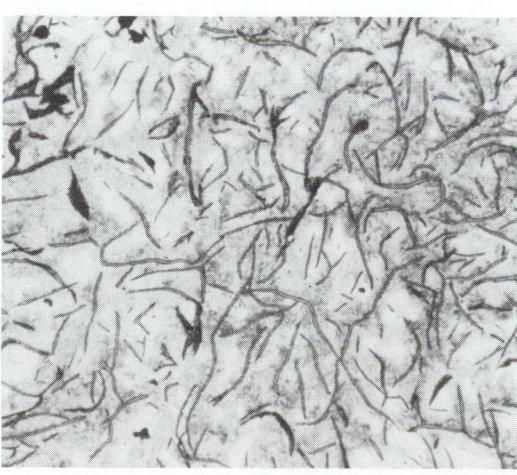

FIGURE 9.58 Class 30 as-cast gray iron cast in a sand mold. Structure is type A graphite flakes in a matrix of 20% free ferrite (light constituent) and 80% pearlite (dark constituent). (Etch: 3% nital; magnification 100×.) *(After "Metals Handbook," vol. 7, 8th ed., American Society for Metals, 1972, p. 82.)*

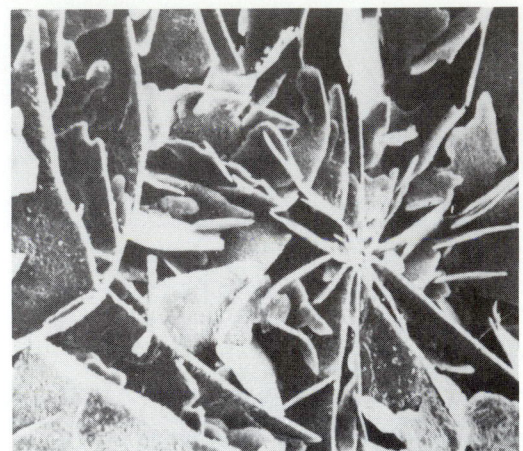

FIGURE 9.59 Scanning electron micrograph of hypereutectic gray iron with matrix etched out to show position of type B graphite in space. (Etch: 3:1 methyl acetate–liquid bromine. Magnification 130×.) (*After "Metals Handbook," vol. 7, 8th ed., American Society for Metals, 1972, p. 82.*)

advantages of steel. Ductile iron has good fluidity and castability, excellent machinability, and good wear resistance. In addition ductile cast iron has a number of properties similar to those of steel such as high strength, toughness, ductility, hot workability, and hardenability.

Composition and microstructure   The exceptional engineering properties of ductile iron are due to the spherical nodules of graphite in its internal structure, as shown in the microstructures of Figs. 9.60 and 9.61. The relatively ductile matrix regions between the nodules allow significant amounts of deformation to take place without fracture.

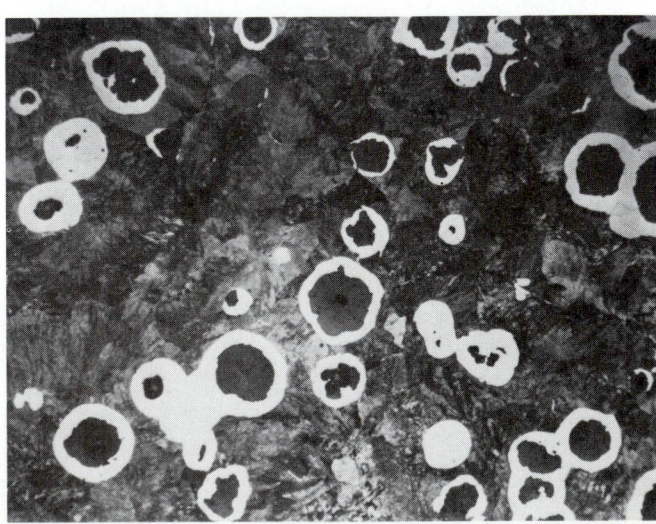

FIGURE 9.60 Grade 80-55-06 as-cast pearlitic ductile iron. Graphite nodules (spherulites) in envelopes of free ferrite (bull's-eye structure) in a matrix of pearlite. (Etch: 3% nital; magnification 100×.) (*After "Metals Handbook," vol. 7, 8th ed., American Society for Metals, 1972, pp. 88.*)

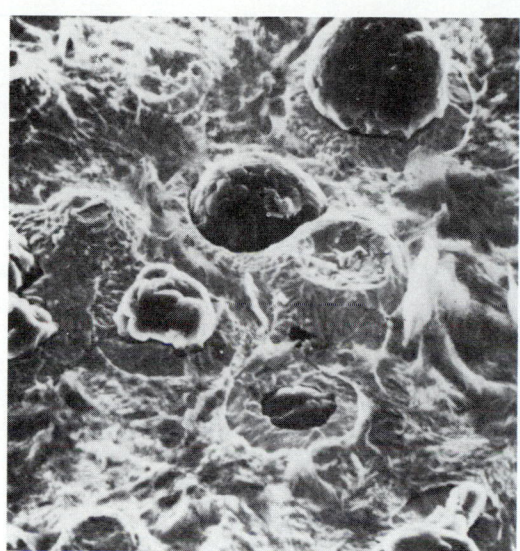

FIGURE 9.61 Scanning electron micrograph of as-cast pearlitic ductile iron with matrix etched away to show secondary graphite, and bull's-eye ferrite around primary graphite nodules. (Etch: 3:1 methyl acetate– liquid bromine; magnification 130×.) *(After "Metals Handbook," vol. 7, 8th ed., American Society for Metals, 1972, p. 88.)*

The composition of unalloyed ductile iron is similar to that of gray iron with respect to carbon and silicon contents. As listed in Table 9.13, the carbon content of unalloyed ductile iron ranges from 3.0 to 4.0% C and the silicon content from 1.8 to 2.8%. The sulfur and phosphorus levels of high-quality ductile iron, however, must be kept very low at 0.03% S maximum and 0.1% P maximum, which are about 10 times lower than the maximum levels for gray cast iron. Other impurity elements also must be kept low because they interfere with the formation of graphite nodules in ductile cast iron.

The spherical nodules in ductile cast iron are formed during the solidification of the molten iron, because the sulfur and oxygen levels in the iron are reduced to very low levels by adding magnesium to the metal just before it is cast. The magnesium reacts with sulfur and oxygen so that these elements cannot interfere with the formation of the spherical-like nodules.

The microstructure of unalloyed ductile cast iron is usually of the bull's-eye type shown in Fig. 9.60. This structure consists of "spherical" graphite nodules with envelopes of free ferrite around them in a matrix of pearlite. Other as-cast structures with all-ferrite and all-pearlite matrixes can be produced by alloying additions. Subsequent heat treatments can also be used to change the as-cast bull's-eye structure and hence the mechanical properties of as-cast ductile cast iron, as indicated in Fig. 9.62.

Malleable Cast Irons

Composition and microstructure Malleable cast irons are first cast as white cast irons which contain large amounts of iron carbides and no graphite. The chemical compositions of malleable cast irons are therefore restricted to compositions that form white cast irons. As listed in Table 9.13, the carbon and

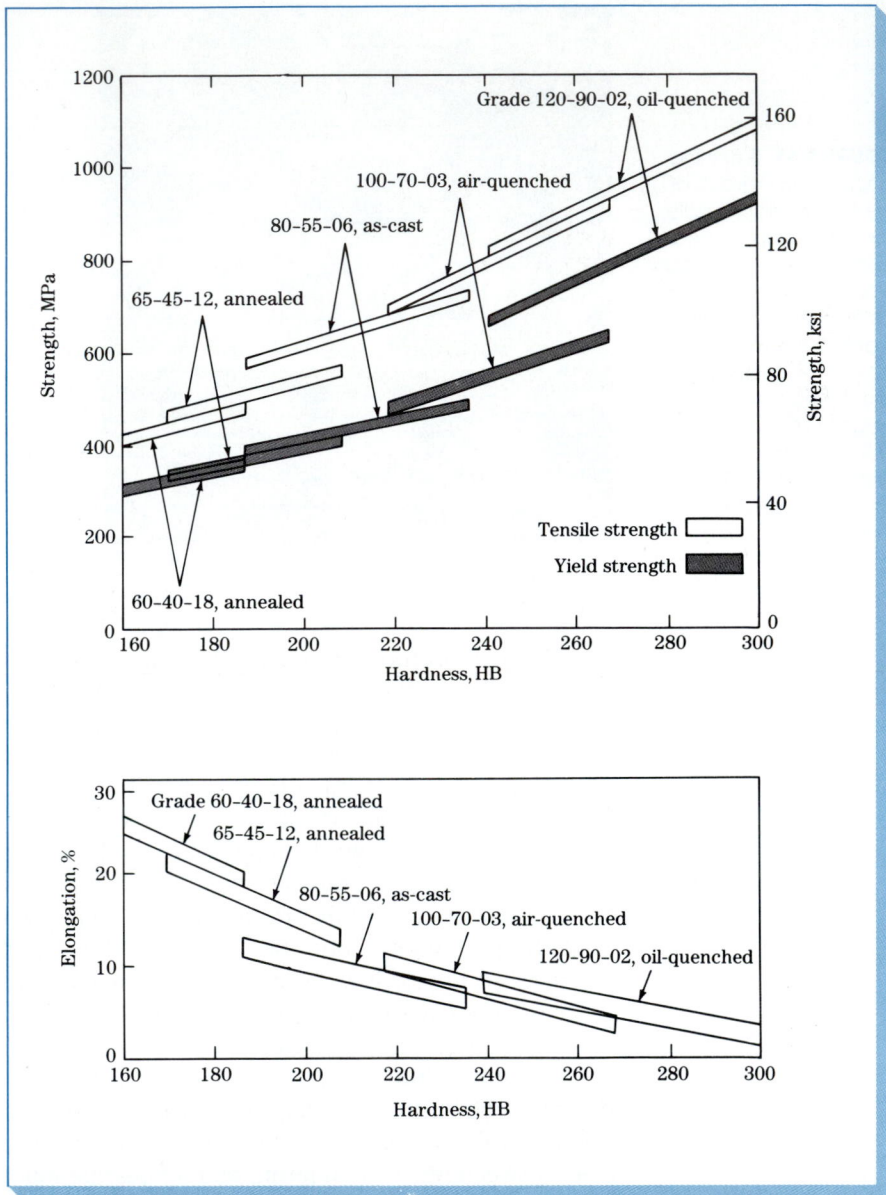

FIGURE 9.62 Tensile properties of ductile iron vs. hardness. *(After "Metals Handbook," vol. 1, 9th ed., American Society for Metals, 1978, p. 36.)*

silicon contents of malleable irons are in the 2.0 to 2.6% C and 1.1 to 1.6% Si ranges.

To produce a malleable iron structure, cold white iron castings are heated in a malleablizing furnace to dissociate the iron carbide of the white iron to graphite and iron. The graphite in the malleable cast iron is in the form of irregular nodular aggregates called *temper carbon*. Figure 9.63 is a microstruc-

FIGURE 9.63
Microstructure of ferritic malleable cast iron (grade M3210), two-stage annealed by holding 4 h at 954°C (1750°F), cooling to 704°C (1300°F) in 6 h, and air cooling. Graphite (temper carbon) nodules in a matrix of granular ferrite. (Etch: 2% nital; magnification 100×.) (After "Metals Handbook," vol. 7, 8th ed., American Society for Metals, 1972, p. 95.)

ture of a ferritic malleable cast iron which shows temper carbon in a matrix of ferrite.

Malleable cast irons are important engineering materials since they have the desirable properties of castability, machinability, moderate strength, toughness, corrosion resistance for certain applications, and uniformity since all castings are heat-treated.

Heat treatment   The heat treatment of white irons to produce malleable irons consists of two stages:

1. *Graphitization.* In this stage the white iron castings are heated above the eutectoid temperature, usually about 940°C (1720°F), and held for about 3 to 20 h depending on the composition, structure, and size of the casting. In this stage the iron carbide of the white iron is transformed to temper carbon (graphite) and austenite.
2. *Cooling.* In this stage the austenite of the iron can be transformed to three basic types of matrixes: ferrite, pearlite, or martensite.

Ferritic malleable iron. To produce a ferrite matrix, the casting, after the first-stage heating, is fast-cooled to 740 to 760°C (1360 to 1400°F) and then slowly cooled at a rate of about 3 to 11°C (5 to 20°F) per hour. During cooling the austenite is transformed to ferrite and graphite, with the graphite depositing on existing particles of temper carbon (Fig. 9.63).

Pearlitic malleable iron. To produce this iron, the castings are slowly cooled to about 870°C (1600°F) and are air-cooled. The rapid cooling in this case transforms the austenite to pearlite and as a result forms pearlitic malleable iron which consists of temper carbon nodules in a pearlite matrix.

Tempered martensitic malleable iron. This type of malleable iron is produced by cooling the castings in the furnace to a quenching temperature of

845 to 870°C (1550 to 1600°F), holding for 15 to 30 min to allow them to homogenize, and quenching in agitated oil to develop a martensitic matrix. Finally, the castings are tempered at a temperature between 590 to 725°C (1100 to 1340°F) to develop the specified mechanical properties. The final microstructure is thus temper carbon nodules in a tempered martensitic matrix.

9.9 MAGNESIUM, TITANIUM, AND NICKEL ALLOYS

Magnesium Alloys
Magnesium is a light metal (density $= 1.74$ g/cm^3) and competes with aluminum (density $= 2.70$ g/cm^3) for applications requiring a low-density metal. However, magnesium and its alloys have many disadvantages which limit their widespread usage. First of all, magnesium costs more than aluminum ($1.53/lb for Mg vs. $0.75/lb for Al in 1994). Magnesium is difficult to cast because in the molten state it burns in air and cover fluxes must be used during casting. Also, magnesium alloys have relatively low strength and poor resistance to creep, fatigue, and wear. In addition, magnesium has the HCP crystal structure which makes deformation at room temperature difficult since only three major slip systems are available. On the other hand, because of their very low density, magnesium alloys are used advantageously, for example, for aerospace applications and materials-handling equipment. Table 9.15 compares some of the physical properties and costs of magnesium with some other engineering metals.

Classification of magnesium alloys There are two major types of magnesium alloys: *wrought alloys,* mainly in the form of sheet, plate, extrusions, and forgings, and *casting alloys.* Both types have non-heat-treatable and heat-treatable grades.

TABLE 9.15 **Some Physical Properties and Costs of Some Engineering Metals***

Metal	Density at 20°C, g/cm^3	Melting point, °C	Crystal structure	Cost, $/lb (1994)
Magnesium	1.74	651	HCP	1.53
Aluminum	2.70	660	FCC	0.75
Titanium	4.54	1675	HCP $\rightleftharpoons$ BCC†	3.50–4.50‡
Nickel	8.90	1453	FCC	3
Iron	7.87	1535	BCC $\rightleftharpoons$ FCC§	0.23–0.26¶
Copper	8.96	1083	FCC	1.13

*Prices vary with time.
†Transformation occurs at 883°C.
‡Price is for a large amount.
§Transformation occurs at 910°C.
¶Prices vary with time.

Magnesium alloys are usually designated by two capital letters followed by two or three numbers. The letters stand for the two major alloying elements in the alloy, with the first letter indicating the one in highest concentration and the second letter indicating the one in second highest concentration. The first number following the letters stands for the weight percent of the first letter element (if there are only two numbers) and the second number stands for the weight percent of the second letter element. If a letter A, B, etc., follows the numbers, it indicates that there has been an A, B, etc., modification to the alloy. Some of the letter symbols used to indicate magnesium alloying elements are

A = aluminum	K = zirconium	S = silicon
Z = zinc	E = rare earths	T = tin
M = manganese	Q = silver	W = yttrium

The temper designations for magnesium alloys are the same as those for aluminum alloys and are listed in Sec. 9.5.

Example Problem 9.7

Explain the meaning of the magnesium alloy designations (*a*) AZ31B-H24 and (*b*) ZK51A-T5.

Solution:

(*a*) The designation AZ31B-H24 means that the magnesium alloy contains a nominal 3 wt % aluminum and 1 wt % zinc and that the alloy is the B modification. The H24 designation means that the alloy was cold-rolled and partially annealed back to the half-hard temper.

(*b*) The designation ZK51A-T5 means that the magnesium alloy contains a nominal 5 wt % zinc and 1 wt % zirconium and is the A modification. The T5 signifies that the alloy was artificially aged after casting.

Structure and properties   Magnesium has the HCP crystal structure, and thus the cold working of magnesium alloys can only be carried out to a limited extent. At elevated temperatures for magnesium, some slip planes other than the basal ones become active. Thus magnesium alloys are usually hot-worked or warm-worked instead of being cold-worked.

Aluminum and zinc are commonly alloyed with magnesium to form wrought magnesium alloys. Aluminum and zinc both increase the strength of magnesium by solid-solution strengthening.

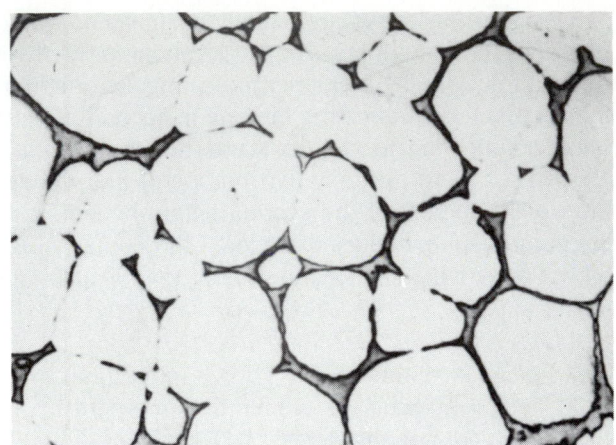

FIGURE 9.64
Microstructure of
magnesium alloy EZ33A
in the as-cast condition
showing massive Mg$_9$Re
(rare-earth) compound
grain-boundary network.
(Etch: glycol;
magnification 500×.)
(Courtesy of the Dow
Chemical Co.)

Most structural magnesium alloys are produced in the cast condition, primarily because of the difficulty in cold working them. Aluminum and zinc are alloyed with magnesium primarily for solid-solution strengthening. In the past years it has been discovered that the corrosion resistance of magnesium in normal environments can be greatly improved by restricting the iron and nickel impurity levels to 0.005 wt % and the copper level to 0.02 wt %. However, the very high electrochemical potential of −2.36 V means that electrochemical galvanic cells between more-dissimilar metals like steel must be avoided.

Rare-earth additions (primarily cerium) to magnesium produce a rigid Mg$_9$Re-type compound at the grain boundaries as shown in Fig. 9.64. Alloyed with about 3% zinc, pressure-tight sand and permanent mold castings can be made which can be used at elevated temperatures at 175 to 260°C. Table 9.16 summarizes the chemical compositions, mechanical properties, and applications of some magnesium alloys.

Titanium is a relatively light metal (density = 4.54 g/cm^3) but has high strength (96 ksi for 99.0% Ti), and so titanium and its alloys can compete favorably with aluminum alloys for some aerospace applications even though titanium costs much more ($3.50–4.50/lb for Ti* vs. $0.73/lb for Al in 1994). Titanium is also used for applications where it has superior corrosion resistance to many chemical environments such as solutions of chlorine and inorganic chloride solutions.

Titanium Alloys

Titanium metal is expensive because it is difficult to extract in the pure state from its compounds. At high temperatures titanium combines with oxygen, nitrogen, hydrogen, carbon, and iron, and so special techniques must be used to cast and work the metal.

---

*In large amounts.

Titanium has the HCP crystal structure (alpha) at room temperature which transforms to the BCC (beta) structure at 883°C. Elements such as aluminum and oxygen stabilize the α phase and increase the temperature at which the α phase transforms to the β phase. Other elements such as vanadium and molybdenum stabilize the beta phase and lower the temperature at which the β phase is stable. Still other elements such as chromium and iron reduce the transformation temperature at which the β phase is stable by causing a eutectoid reaction which produces a two-phase structure at room temperature.

Table 9.17 lists representative types of alpha, alpha-beta, and beta titanium alloys along with their nominal chemical compositions, typical mechanical properties, and applications. The extensively used titanium alloy is T–6 Al–4 V since this alloy combines high strength with workability and low density (for titanium alloys). By solution heat treating and aging, its tensile strength may reach 170 ksi (1173 MPa). This alloys is used, for example, for blades and disks in aircraft gas turbine engines as well as for chemical process equipment.

More recently (in the 1990s) beta-stabilized titanium alloys have become more prominent although still a relatively small amount of the titanium markets. These alloys provide higher strengths and workability but higher densities. For example, the beta C alloy can reach a tensile strength of 210 ksi (1448 MPa) by solution heat treatment and aging. More recently, the beta alloy Ti–10 V–2 Fe–3 Al has been used for forgings in the new 777 passenger aircraft.

Nickel Alloys   Nickel is an important engineering metal mainly because of its exceptional resistance to corrosion and high-temperature oxidation. Nickel also has the FCC crystal structure which makes it highly formable but is relatively expensive ($3/lb in 1994) and has a high density (8.9 g/cm^3) which limits its use.

Commercial nickel and Monel alloys   Commercially pure nickel because of its good strength and electrical conductivity is used for electrical and electronics parts and because of its good corrosion resistance for food-processing equipment. Nickel and copper are completely soluble in each other in the solid state at all compositions, and so many solid-solution-strengthened alloys are made with nickel and copper. Nickel is alloyed with about 32% copper to produce the Monel 400 alloy (Table 9.17) which has a relatively high strength, weldability, and excellent corrosion resistance to many environments. The 32% copper strengthens the nickel to a limited extent and lowers its cost. The addition of about 3% aluminum and 0.6% titanium increases the strength of Monel (66% Ni–30% Cu) significantly by precipitation hardening. The strengthening precipitates in this case are Ni$_3$Al and Ni$_3$Ti.

Nickel-base superalloys   A whole spectrum of nickel-base superalloys have been developed primarily for gas turbine parts which must be able to withstand high temperatures and high oxidizing conditions and be creep-resistant. Most wrought nickel-base superalloys consist of about 50 to 60% nickel, 15 to 20% chromium, and 15 to 20% cobalt. Small amounts of aluminum (0.5 to 4%) and

TABLE 9.16 Chemical Compositions, Typical Mechanical Properties, and Applications of Some Magnesium Alloys

Alloy name and number	Chemical composition, wt %					Condition*	Tensile strength		Yield strength		Elongation, %	Typical applications
	Al	Mn	Zn	Zr	Other		ksi	MPa	ksi	MPa		
Magnesium casting alloys												
Die castings:												
AM60B	6.0	0.13†					F	32	220	19	131	8	Automobile wheels
AS41A	4.2	0.35‡			1.0 Si		F	31	214	20	138	6	Automobile engines and housings; good creep resistance
AZ91D	9.0	0.15†	0.7		0.001 Ni max 0.005 Fe max		F	34	234	23	158	3	Die castings; parts for cars, lawn mowers, business machines, chain saws, hand tools, sporting goods; good corrosion resistance
Sand and permanent mold castings:												
AM100A	10.0	0.1†					T6	35	241	17	117	2	Pressure-tight sand and permanent-mold castings
AZ63A	6.0	0.15†	3.0				T6	34	234	16	110	3	Sand castings requiring good room-temperature strength and ductility
AZ81A	7.6	0.13†	0.7				T4	34	234	10	69	7	Tough leak-proof sand castings
AZ91E	8.7	0.26‡	0.7		0.001 Ni max 0.005 Fe max		T6	34	234	16	110	3	Sand and permanent-mold castings requiring room-temperature strength and ductility
ZK51A			4.6	0.7			T5	34	234	20	138	5	Sand castings; good strength at room temperature
ZK61A			6.0	0.8			T6	40	275	26	179	5	Sand castings; good strength at room temperature
EZ33A			2.6	0.7	§RE 3.2		T5	20	138	14	96	2	Pressure-tight sand and permanent-mold castings for applications at 175–260°C
ZE41A			4.2	0.7	RE 1.2		T5	29	200	19.5	134	2–5	Sand castings; good strength at room temperature; improved castability over ZK alloys
ZE63A			5.7	0.7	RE 2.5		T6	40	275	27	186	5	

*F = as fabricated; for temper designations, –H24, –T4, –T5, –T6, see aluminum alloy temper designations in Sec. 9.5.
†Minimum.
‡Nominal.
§RE = rare earth.
Source: Adapted from W. F. Smith. "Structure and Properties of engineering Alloys," 2d ed., McGraw-Hill, 1993.

TABLE 9.16 Chemical Compositions, Typical Mechanical Properties, and Applications of Some Magnesium Alloys (Con't)

Alloy name and number	Chemical composition, wt %					Condition*	Tensile strength		Yield strength		Elongation,	Typical applications
	Al	Mn	Zn	Zr	Other		ksi	MPa	ksi	MPa	%	
Wrought magnesium alloys												
Sheet and plate,												
AZ31B	3.0	0.20	1.0			O	32	220	15–18	103–124	2–9	General-purpose sheet and plate
						H24	29–39	200–287	14–29	96–200	6–8	
Extruded bars and shapes												
ZA31B	3.0	0.2	1.0			F	31–35	213–241	16–22	110–152	4–8	General-purpose extrusions
AZ61A	6.5	0.15	1.0			F	32–40	220–276	16–24	110–165	7–9	Higher properties than AZ31B
AZ80A	8.5	0.15	0.5			F	42–43	289–296	27–28	186–193	4–9	High-strength extrusions
ZK30A			2.8	0.4		F	40–44	276–303	28–33	193–227	8	High-strength extrusions
ZK60A			5.5	0.4		T5	43–46	296–317	31–38	213–262	4–6	High-strength extrusions
ZM21A		1.0	2.0			F	33–35	207–241	22–23	152–158	8–20	Excellent extrudability

*F = as fabricated; for temper designations, –H24, –T4, –T5, –T6, see aluminum alloy temper designations in Sec. 9.5.
†Minimum.
‡Nominal.
§RE = rare earth.
Source: Adapted from W. F. Smith, "Structure and Properties of engineering Alloys," 2d ed., McGraw-Hill, 1993.

TABLE 9.17 Chemical Compositions, Typical Mechanical Properties, and Applications of Some Titanium and Nickel Alloys

Alloy name and number	Chemical composition, wt %	Condition	Tensile strength		Yield strength		Elongation, %	Typical applications
			ksi	MPa	ksi	MPa		
Titanium alloys								
Commercially pure	99.0 Ti	Annealed	96	662	85	586	20	Airframe skins, heat exchangers, corrosion-resistant equipment for temperatures up to 480°C (900°F)
Alpha	Ti–5 Al–2.5 Sn	Annealed	125	862	117	807	16	Gas-turbine casings and rings; chemical process equipment
Alpha-beta	Ti–6 Al–4 V	Annealed SHT* + aging	144 170	993 1172	134 160	924 1103	14 10	Aircraft turbine disks and blades; aircraft structural components for temperatures up to 315°C (600°F); chemical process equipment
Beta	Ti–10 V–2 Fe–3 Al	SHT* + aging	185	1276	174	1200	10	Aircraft components for up to 315°C (600°F) where high strength and toughness are required; forgings
Beta C	Ti–3 Al–8 V–6 Cr–4 Zr–4 Mo	SHT* + aging	210	1448	200	1379	7	Airframe high-strength fasteners; rivets; springs; pipe for oil industry
Nickel alloys								
Nickel 200	99.5 Ni	Annealed	70	483	22	152	48	Chemical and food processing; electronic parts
Monel 400	66 Ni, 32 Cu	Annealed	80	552	38	262	45	Chemical and oil processing; marine service
Monel K500	66 Ni, 30 Cu, 2.7 Al, 0.6 Ti	Age-hardened	150	1035	110	759	25	Valves, pumps, springs, oil-well drill collars
IN 718	54 Ni, 18 Cr, 5 Nb–3 Mo–0.9 Ti–0.5 Al–18.5 Fe	Age-hardened	180	1240	150	1936	12	Gas turbines; rocket motors and spacecraft
MARM200-Hf	58.5 Ni–9 Cr–10 Co–12.5 W–1 Nb–5 Al–2 Ti–2 Hf	Age-hardened	150	1035	120	825		High-temperature gas turbine blades

*SHT = solution heat treatment.

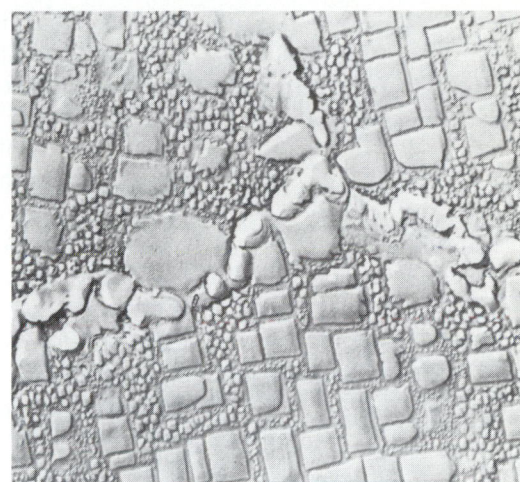

FIGURE 9.65 Astroloy forging, solution heat-treated 4 h at 1150°C, air-cooled aged at 1079°C for 4 h, oil-quenched, aged at 843°C for 4 h, air-cooled, aged at 760°C for 16 h, air-cooled. Intergranular gamma prime precipitated at 1079°C, fine gamma prime at 843 and 760°C. Carbide particles are also at grain boundaries. Matrix is gamma. (Electrolytic: H_2SO_4, H_3PO_4, HNO_3; magnification 10,000×.) *(After "Metals Handbook," vol. 7, 8th ed., American Society for Metals, 1972, p. 171.)*

titanium (1 to 4%) are added for precipitation strengthening. The nickel-base superalloys usually consist essentially of three main phases: (1) a matrix of gamma austenite, (2) a precipitate phase of Ni_3Al and Ni_3Ti called *gamma prime*, and (3) carbide particles (due to the addition of about 0.01 to 0.04% C). The gamma prime provides high-temperature strength and stability to these alloys, and the carbides stabilize the grain boundaries at high temperatures. The gamma prime increases the high-temperature strength of nickel-base superalloys by making it difficult for dislocation pairs to move under stress in the gamma prime. This type of strengthening is referred to as *antiphase boundary* strengthening. Figure 9.65 shows the microstructure of one of the nickel-base superalloys after heat treatment. In this microstructure the cuboidal and fine gamma prime and grain-boundary carbids are clearly visible.

In the past years single-crystal nickel-base superalloys have been developed that are able to operate at still higher temperatures in gas turbine engines. Table 9.19 in the problems section compares the chemical compositions of two of these alloys. Single-crystal alloys do not contain grain-boundary strengthening carbides.

9.10 MATERIALS SELECTION FOR ENGINEERING DESIGNS USING METALLIC MATERIALS*

The use of metallic materials in engineering designs is important today and will remain important in the foreseeable future mainly because of some of the following properties and attributes of metallic materials.

*This section is considered introductory only. For further information and study, the author recommends the test by "Structure and Properties of Engineering Alloys," 2d ed., W. F. Smith, McGraw-Hill, 1993.

1. Medium strength and good ductility, formability, and weldability at low cost (i.e., very low-carbon sheet steels).
2. Medium to high strengths with sufficient toughness ductility (i.e., low- to medium-carbon and low-alloy steels).
3. Low density, sufficient strength and ductility, and good corrosion resistance (i.e., aluminum and its alloys).
4. Good electrical and thermal conduction, formability, joinability, and corrosion resistance (i.e., copper and its alloys).
5. Good castability, sufficient strength, good wear resistance, and low cost (i.e., some cast irons).
6. Excellent corrosion resistance, good strength, and sufficient ductility (i.e., stainless steels).
7. High strength, relatively low density, and good corrosion resistance (titanium and its alloys).
8. Good strength, oxidation, and corrosion resistance at high temperatures (nickel alloys).
9. Very low density with sufficient strength and ductility (i.e., magnesium alloys).
10. Most of the above materials can be remelted and reused.

This list is not inclusive of all metallic materials and only highlights some of the important properties and attributes of metallic materials. There are also some negative factors, such as high cost, corrosion resistance, and pollution problems.

If the selection of the metallic materials for an engineering design has not been decided, it is recommended, as a start, that the applications of metallic materials listed in the tables of this chapter and summarized in Table 9.18 be

TABLE 9.18   **Summary of the Chemical Compositions, Mechanical Properties, and Applications of Various Metal Alloy Systems in This Chapter***

Metal alloy system	Table numbers	Prices, $/lb (October 1994)	Metal alloy system	Table Numbers	Price, $/lb (October 1994)
Plain-carbon steels	9.3	0.23–0.25[†]	Cast irons	9.13, 9.14	
Low-alloy steels	9.4, 9.6		Magnesium alloys	9.16	Mg = 1.43
Aluminum alloys	9.8	Al = 0.73	Titanium alloys	9.17	Ti-3.50–4.50[‡]
Copper alloys	9.10, 9.11	Cu = 1.13	Nickel alloys	9.17	Ni = 3.00
Stainless steels	9.12				

*Prices change with time.
†Hot-rolled strip.
‡Sponge in large lots.

reviewed. Of course, cost is a factor in many choices of materials. After a metal or alloy has been tentatively selected it is recommended that the section in this book pertinent to the selection be re-read and that further information from other sources such as the one listed below or other handbooks and company fabricator literature also be referred to. One must keep in mind, too, that times are constantly changing, and other competitive materials such as polymeric, ceramic, and composite materials are available.

SUMMARY

Engineering alloys can conveniently be subdivided into two types: ferrous and nonferrous. Ferrous alloys have iron as their principal alloying metal, whereas nonferrous alloys have a principal alloying metal other than iron. The steels, which are ferrous alloys, are by far the most important metal alloys mainly because of their relatively low cost and wide range of mechanical properties. The mechanical properties of carbon steels can be varied considerably by cold working and annealing. When the carbon content of steels is increased to above about 0.3%, they can be heat-treated by quenching and tempering to produce high strength with reasonable ductility. Alloying elements such as nickel, chromium, and molybdenum are added to plain-carbon steels to produce low-alloy steels. Low-alloy steels have good combinations of high strength and toughness and are used extensively in the automotive industry for uses such as gears, shafts, and axles.

Aluminum alloys are the most important of the nonferrous alloys mainly because of their lightness, workability, corrosion resistance, and relatively low cost. Unalloyed copper is used extensively because of its high electrical conductivity, corrosion resistance, workability, and relatively low cost. Copper is alloyed with zinc to form a series of brass alloys which have higher strength than unalloyed copper.

Stainless steels are important ferrous alloys because of their high corrosion resistance in oxidizing environments. To make a stainless steel "stainless," it must contain at least 12% Cr.

Cast irons are still another industrially important family of ferrous alloys. They are low in cost and have special properties such as good castability, wear resistance, and durability. Gray cast iron has high machinability and vibration-damping capacity due to the graphite flakes in its structure.

Other nonferrous alloys briefly discussed in this chapter are magnesium, titanium, and nickel alloys. Magnesium alloys are exceptionally light and have aerospace applications and are used for materials-handling equipment. Titanium alloys are expensive but have a combination of strength and lightness not available from any other metal alloy system and so are used extensively for aircraft structural parts. Nickel alloys have high corrosion and oxidation resistance and are therefore commonly used in the oil and chemical process industries. Nickel when alloyed with chromium and cobalt forms the basis for the nickel-base superalloys which are necessary for gas turbines for jet aircraft and some electric-power generating equipment.

In this chapter we have discussed to a limited extent the structure, properties, and applications of some of the important engineering alloys. However, it must be pointed out that many important alloys have been left out due to the limited scope of this book.

DEFINITIONS

Sec. 9.2

Austenite (γ phase in Fe–Fe$_3$C phase diagram): an interstitial solid solution of carbon in FCC iron; the maximum solid solubility of carbon in austenite is 2.0%.

Austenitizing: heating a steel into the austenite temperature range so that its structure becomes austenite. The austenitizing temperature will vary depending on the composition of the steel.

α ferrite (α phase in the Fe–Fe$_3$C phase diagram): an interstitial solid solution of carbon in BCC iron; maximum solid solubility of carbon in BCC iron is 0.02%.

Cementite: the intermetallic compound Fe$_3$C; a hard and brittle substance.

Pearlite: a mixture of α ferrite and cementite (Fe$_3$C) phases in parallel plates (lamellar structure) produced by the eutectoid decomposition of austenite.

Eutectoid α ferrite: α ferrite which forms during the eutectoid decomposition of austenite; the α ferrite in pearlite.

Eutectoid cementite (Fe$_3$C): cementite which forms during the eutectoid decomposition of austenite; the cementite in pearlite.

Eutectoid (plain-carbon steel): a steel with 0.8% C.

Hypoeutectoid (plain-carbon steel): a steel with less than 0.8% C.

Hypereutectoid (plain-carbon steel): a steel with 0.8 to 2.0% C.

Proeutectoid α ferrite: α ferrite which forms by the decomposition of austenite at temperatures above the eutectoid temperature.

Proeutectoid cementite (Fe$_3$C): Cementite which forms by the decomposition of austenite at temperatures above the eutectoid temperature.

Sec. 9.3

Martensite: a supersaturated interstitial solid solution of carbon in body-centered tetragonal iron.

Bainite: a mixture of α ferrite and very small particles of Fe$_3$C particles produced by the decomposition of austenite; a nonlamellar eutectoid decomposition product of austenite.

Spheroidite: a mixture of particles of cementite (Fe$_3$C) in an α ferrite matrix.

Isothermal transformation (IT) diagram: a time-temperature-transformation diagram which indicates the time for a phase to decompose into other phases isothermally at different temperatures.

Continuous-cooling transformation (CCT) diagram: a time-temperature-transformation diagram which indicates the time for a phase to decompose into other phases continuously at different rates of cooling.

Martempering (marquenching): a quenching process whereby a steel in the austenitic condition is hot-quenched in a liquid (salt) bath at above the M_s temperature, held for a time interval short enough to prevent the austenite from transforming, and then allowed to cool slowly to room temperature. After this treatment the steel will be in the martensitic condition, but the interrupted quench allows stresses in the steel to be relieved.

Austempering: a quenching process whereby a steel in the austenitic condition is quenched in a hot liquid (salt) bath at a temperature just above the M_s of the steel, held in the bath until the austenite of the steel is fully transformed, and then cooled to room temperature. With this process a plain-carbon eutectoid steel can be produced in the fully bainitic condition.

M_s: the temperature at which the austenite in a steel starts to transform to martensite.

M_f: the temperature at which the austenite in a steel finishes transforming to martensite.

Tempering (of a steel): the process of reheating a quenched steel to increase its toughness and ductility. In this process martensite is transformed into tempered martensite.

Plain-carbon steel: an iron-carbon alloy with 0.02 to 2% C. All commercial plain-carbon steels contain about 0.3 to 0.9% manganese along with sulfur, phosphorus, and silicon impurities.

Sec. 9.4

Hardenability: the ease of forming martensite in a steel upon quenching from the austenitic condition. A highly hardenable steel is one which will form martensite throughout in thick sections. Hardenability should not be confused with hardness. Hardness is the resistance of a material to penetration. The hardenability of a steel is mainly a function of its composition and grain size.

Jominy hardenability test: a test in which a 1-in (2.54-cm)-diameter bar by 4 in (10.2 cm) long is austenitized and then water-quenched at one end. Hardness is measured along the side of the bar up to about 2.5 in (6.35 cm) from the quenched end. A plot called the Jominy hardenability curve is made by plotting the hardness of the bar against the distance from the quenched end.

Sec. 9.8

White cast irons: iron-carbon-silicon alloys with 1.8–3.6% C and 0.5–1.9% Si. White cast irons contain large amounts of iron carbide which make them hard and brittle.

Gray cast irons: iron-carbon-silicon alloys with 2.5–4.0% C and 1.0–3.0% Si. Gray cast irons contain large amounts of carbon in the form of graphite flakes. They are easy to machine and have good wear resistance.

Ductile cast irons: iron-carbon-silicon alloys with 3.0–4.0% C and 1.8–2.8% Si. Ductile cast irons contain large amounts of carbon in the form of graphite nodules (spheres) instead of flakes as in the case of gray cast iron. The addition of magnesium (about 0.05%) before the liquid cast iron is poured enables the nodules to form. Ductile irons are in general more ductile than gray cast irons.

Malleable cast irons: iron-carbon-silicon alloys with 2.0–2.6% C and 1.1–1.6% Si. Malleable cast irons are first cast as white cast irons and then are heat-treated at about 940°C (1720°F) and held about 3 to 20 h. The iron carbide in the white iron is decomposed into irregularly shaped nodules or graphite.

PROBLEMS

9.1.1   How is raw pig iron extracted from iron oxide ores?

9.1.2   Write a typical chemical reaction for the reduction of iron oxide (Fe_2O_3) by carbon monoxide to produce iron.

9.1.3   Describe the basic oxygen process for converting pig iron into steel.

9.2.1   Why is the Fe–Fe_3C phase diagram a metastable phase diagram instead of a true equilibrium phase diagram?

9.2.2   Define the following phases which are present in the Fe–Fe_3C phase diagram: (*a*) austenite, (*b*) α ferrite, (*c*) cementite, (*d*) δ ferrite.

9.2.3   Write the reactions for the three invariant reactions which take place in the Fe–Fe_3C phase diagram.

9.2.4   What is the structure of pearlite?

9.2.5   Distinguish between the following three types of plain-carbon steels: (*a*) eutectoid, (*b*) hypoeutectoid, and (*c*) hypereutectoid.

9.2.6   Describe the structural changes which take place when a plain-carbon eutectoid steel is slowly cooled from the austenitic region just above the eutectoid temperature.

9.2.7   Describe the structural changes that take place when a 0.4% C plain-carbon steel is slowly cooled from the austenitic region just above the upper transformation temperature.

9.2.8   Distinguish between proeutectoid ferrite and eutectoid ferrite.

9.2.9   A 0.50% C hypoeutectoid plain-carbon steel is slowly cooled from about 950°C to a temperature just slightly *above* 723°C. Calculate the weight percent austenite and weight percent proeutectoid ferrite in this steel.

9.2.10   A 0.50% C hypoeutectoid plain-carbon steel is slowly cooled from 950°C to a temperature just slightly *below* 723°C.
(*a*) Calculate the weight percent proeutectoid ferrite in the steel.
(*b*) Calculate the weight percent eutectoid ferrite and weight percent eutectoid cementite in the steel.

9.2.11   A plain-carbon steel contains 92 wt % ferrite and 8 wt % Fe_3C. What is its average carbon content in weight percent?

9.2.12   A plain-carbon steel contains 48.2 wt % proeutectoid ferrite. What is its average carbon content in weight percent?

9.2.13   A plain-carbon steel contains 6.6 wt % eutectoid ferrite. What is its average carbon content?

9.2.14   A 1.05% C hypereutectoid plain-carbon steel is slowly cooled from 900°C to a temperature just slightly *above* 723°C. Calculate the weight percent proeutectoid cementite and weight percent austenite present in the steel.

9.2.15   A 1.05% C hypereutectoid plain-carbon steel is slowly cooled from 900°C to a temperature just slightly *below* 723°C.

(a) Calculate the weight percent proeutectoid cementite present in the steel.

(b) Calculate the weight percent eutectoid cementite and the weight percent eutectoid ferrite present in the steel.

9.2.16 If a hypereutectoid plain-carbon steel contains 4.7 wt % proteutectoid cementite, what is its average carbon content?

9.2.17 A hypereutectoid plain carbon steel contains 10.45 wt % eutectoid Fe_3C. What is its average carbon content in weight percent?

9.2.18 A plain-carbon steel contains 27.5 wt % proteutectoid ferrite. What is its average carbon content?

9.2.19 A 0.70% C hypoeutectoid plain-carbon steel is slowly cooled from 950°C to a temperature just slightly below 723°C.

(a) Calculate the weight percent proeutectoid ferrite in the steel.

(b) Calculate the weight percent eutectoid ferrite and eutectoid cementite in the steel.

9.2.20 A hypoeutectoid steel contains 42.0 wt % eutectoid ferrite. What is its average carbon content?

9.2.21 A hypoeutectoid steel contains 25.5 wt % eutectoid ferrite. What is its average carbon content?

9.2.22 A 0.90% C hypereutectoid plain-carbon steel is slowly cooled from 900°C to a temperature just slightly below 723°C.

(a) Calculate the weight percent proeutectoid cementite present in the steel.

(b) Calculate the weight percent eutectoid cementite and the weight percent eutectoid ferrite present in the steel.

9.3.1 Define an Fe–C martensite.

9.3.2 Describe the following types of Fe–C martensites which occur in plain-carbon steels: (a) lath martensite, (b) plate martensite.

9.3.3 Describe some of the characteristics of the Fe–C martensite transformation which occurs in plain-carbon steels.

9.3.4 What causes the tetragonality to develop in the BCC iron lattice when the carbon content of Fe–C martensites exceeds about 0.2%?

9.3.5 What causes the high hardness and strength to be developed in Fe–C martensites of plain-carbon steels when their carbon content is high?

9.3.6 What is an isothermal transformation in the solid state?

9.3.7 Draw an isothermal transformation diagram for a plain-carbon eutectoid steel and indicate the various decomposition products on it. How can such a diagram be constructed by a series of experiments?

9.3.8 If a thin sample of a eutectoid plain-carbon steel is hot-quenched from the austenitic region and held at 700°C until transformation is complete, what will be its microstructure?

9.3.9 If a thin sample of a eutectoid plain-carbon steel is water-quenched from the austenitic region to room temperature, what will be its microstructure?

9.3.10 What does the bainite microstructure consist of? What is the microstructural difference between upper and lower bainite?

9.3.11 Draw time-temperature cooling paths for a 1080 steel on an isothermal transformation diagram that will produce the following microstructures. Start with the steels in the austenitic condition at time = 0 and 850°C. (a) 100% martensite, (b) 50% martensite and 50% coarse pearlite, (c) 100% fine pearlite, (d) 50% martesite and 50% upper bainite, (e) 100% upper bainite, and (f) 100% lower bainite.

9.3.12 How does the isothermal transformation diagram for a hypoeutectoid plain-carbon steel differ from that of a eutectoid one?

9.3.13 Draw a continuous-cooling transformation diagram for a eutectoid plain-carbon steel. How does it differ from a eutectoid isothermal transformation diagram for a plain-carbon steel?

9.3.14 Draw time-temperature cooling paths for a 1080 steel on a continuous-cooling transformation diagram that will produce the following microstructures. Start with the steels in the austenitic condition at time = 0 and 850°C. (*a*) 100% martensite, (*b*) 50% fine pearlite and 50% martensite, (*c*) 100% coarse pearlite, and (*d*) 100% fine pearlite.

9.3.15 Describe the full-annealing heat treatment for a plain-carbon steel. What types of microstructures are produced by full annealing (*a*) a eutectoid steel and (*b*) a hypoeutectoid steel?

9.3.16 Describe the process-annealing heat treatment for a plain-carbon hypoeutectoid steel with less than 0.3% C.

9.3.17 What is the normalizing heat treatment for steel? What are some of its purposes?

9.3.18 Describe the tempering process for a plain-carbon steel.

9.3.19 What types of microstructures are produced by tempering a plain-carbon steel with more than 0.2% carbon in the temperature ranges (*a*) 20–250°C, (*b*) 250–350°C, and (*c*) 400–600°C?

9.3.20 What causes the decrease in hardness during the tempering of a plain-carbon steel?

9.3.21 Describe the martempering (marquenching) process for a plain-carbon steel. Draw a cooling curve for a martempered (marquenched) austenitized eutectoid plain-carbon steel by using an IT diagram. What type of microstructure is produced after martempering this steel?

9.3.22 What are the advantages of martempering? What type of microstructure is produced after tempering a martempered steel?

9.3.23 Why is the term *martempering* a misnomer? Suggest an improved term.

9.3.24 Describe the austempering process for a plain-carbon steel. Draw a cooling curve for an austempered austenitized eutectoid plain-carbon steel by using an IT diagram.

9.3.25 What is the microstructure produced after austempering a eutectoid plain-carbon steel? Does an austempered steel need to be tempered? Explain.

9.3.26 What are the advantages of the austempering process? Disadvantages?

9.3.27 Thin pieces of 0.3-mm-thick hot-rolled strips of 1080 steel are heat-treated in the following ways. Use the IT diagram of Fig. 9.23 and other knowledge to determine the microstructure of the steel samples after each heat treatment.
(*a*) Heat 1 h at 860°C; water-quench.
(*b*) Heat 1 h at 860°C; water-quench; reheat 1 h at 350°C. What is the name of this heat treatment?.
(*c*) Heat 1 h at 860°C; quench in molten salt bath at 700°C and hold 2 h; water quench.
(*d*) Heat 1 h at 860°C; quench in molten salt bath at 260°C and hold 1 min; air-cool. What is the name of this heat treatment?
(*e*) Heat 1 h at 860°C; quench in molten salt bath at 350°C; hold 1 h; air-cool. What is the name of this heat treatment?
(*f*) Heat 1 h at 860°C; water-quench; reheat 1 h at 700°C.

9.3.28 Explain the numbering system used by the AISI and SAE for plain-carbon steels?

9.4.1   What are some of the limitations of plain-carbon steels for engineering designs?

9.4.2   What are the principal alloying elements added to plain-carbon steels to make low-alloy steels?

9.4.3   What is the AISI-SAE system for designating low-alloy steels?

9.4.4   Which elements dissolve primarily in the ferrite of carbon steels?

9.4.5   List in order of increasing carbide-forming tendency the following elements: titanium, chromium, molybdenum, vanadium, and tungsten.

9.4.6   What compounds does aluminum form in steels?

9.4.7   Name two austenite-stabilizing elements in steels.

9.4.8   Name four ferrite-stabilizing elements in steels.

9.4.9   Which elements raise the eutectoid temperature of the $Fe-Fe_3C$ phase diagram? Which elements lower it?

9.4.10 Define the hardenability of a steel. Define the hardness of a steel.

9.4.11 Describe the Jominy hardenability test.

9.4.12 Explain how the data for the plotting of the Jominy hardenability curve are obtained and how the curve is constructed.

9.4.13 Of what industrial use are Jominy hardenability curves?

9.4.14 An austenitized 45-mm-diameter steel bar of a 5140 steel is quenched in agitated oil. Predict the RC hardness at $\frac{3}{4}R$ from the center of the bar and at the center of the bar.

9.4.15 An austenitized 40-mm-diameter 9840 long steel bar is quenched in agitated water. Predict the RC hardness at its surface and center.

9.4.16 An austenitized 50-mm-diameter 4140 steel bar is quenched in agitated oil. Predict what the Rockwell C hardness of the bar will be at (*a*) its surface and (*b*) midway between its surface and center (midradius).

9.4.17 An austenitized 70-mm-diameter 4340 steel bar is quenched in agitated water. Predict what the Rockwell C hardness of the bar will be at (*a*) its surface and (*b*) its center.

9.4.18 An austenitized 60-mm-diameter 8640 steel bar is quenched in agitated oil. Predict what the Rockwell C hardness of the bar will be at (*a*) $\frac{3}{4}R$ and (*b*) the center.

9.4.19 An austenitized and quenched 4140 steel bar has a Rockwell C hardness of 35 at a point on its surface. What cooling rate did the bar experience at this point?

9.4.20 An austenitized and quenched 8640 steel bar has a Rockwell C hardness of 30 at a point on its surface. What cooling rate did the bar experience at this point?

9.4.21 An austenitized and quenched 5140 steel bar has a Rockwell C hardness of 25 at a point on its surface. What cooling rate did the bar experience at this point?

9.4.22 An austenitized 60-mm-diameter 8640 steel bar is quenched in agitated water. Plot the Rockwell C hardness of the bar vs. distance from one surface of the bar to the other across the diameter of the bar at the following points: surface, $\frac{3}{4}R$, $\frac{1}{2}R$ (midradius), and center. This type of plot is called a hardness profile across the diameter of the bar. Assume the hardness profile is symmetrical about the center of the bar.

9.4.23 An austenitized 50-mm-diameter bar of 5140 steel is quenched in agitated oil. Repeat the hardness profile of Prob. 9.4.22 for this steel.

9.4.24 An austenitized 50-mm-diameter 5140 steel bar is quenched in agitated water. Repeat the hardness profile of Prob. 9.4.22 for this steel.

9.4.25 An austenitized 60-mm-diameter 8640 steel bar is quenched in agitated oil. Repeat the hardness profile of Prob. 9.4.22 for this steel.

9.4.26 An austenitized 4340 standard steel bar is cooled at a rate of 5°C/s (25 mm from the quenched end of a Jominy bar). What will be the constituents in the microstructure of the bar at 200°C? See Fig. 9.39.

9.4.27 An austenitized 4340 standard steel bar is cooled at the rate of 25°C/s (9.50 mm from the quenched end of a Jominy bar). What will be the constituents in the microstructure of the bar at 200°C? See Fig. 9.39.

9.4.28 An austenitized 4340 standard steel bar is cooled at a very slow rate (76 mm from the quenched end of a Jominy bar). What will be the constituents in the microstructure of the bar at 200°C? See Fig. 9.39.

9.5.1 Explain how a precipitation-hardenable alloy is strengthened by heat treatment.

9.5.2 What type of phase diagram is necessary for a binary alloy to be precipitation-hardenable?

9.5.3 What are the three basic heat-treatment steps to strengthen a precipitation-hardenable alloy?

9.5.4 In what temperature range must a binary precipitation-hardenable alloy be heated for the solution heat-treatment step?

9.5.5 Why is a precipitation-hardenable alloy relatively weak just after solution heat treatment and quenching?

9.5.6 Distinguish between natural aging and artificial aging for a precipitation-hardenable alloy.

9.5.7 What is the driving force for the decomposition of a supersaturated solid solution of a precipitation-hardenable alloy?

9.5.8 What is the first decomposition product of a precipitation-hardenable alloy in the supersaturated solid-solution condition after aging at a low temperature?

9.5.9 What are GP zones?

9.5.10 Why doesn't the equilibrium precipitate form directly from the supersaturated solid solution of a precipitation-hardenable alloy if the aging temperature is low? How can the equilibrium precipitate be formed from the supersaturated solid solution?

9.5.11 What is an aging curve for a precipitation-hardenable alloy?

9.5.12 What types of precipitates are developed in an alloy that is considerably underaged at low temperatures? What types are developed upon overaging?

9.5.13 What is the difference between a coherent precipitate and an incoherent one?

9.5.14 Describe the four decomposition structures that can be developed when a supersaturated solid solution of an Al–4% Cu alloy is aged.

9.5.15 Calculate the wt % θ in an Al–4.5% Cu alloy which is slowly cooled from 548 to 27°C. Assume the solid solubility of Cu in Al at 27°C is 0.02 wt % and that the θ phase contains 54.0 wt % Cu.

9.5.16 A binary Al–9.0 wt % Cu alloy is slowly cooled from 700°C to just below 548°C (the eutectic temperature).

(a) Calculate the wt % proeutectic α present just above 548°C.

(b) Calculate the wt % eutectic α present just below 548°C.

(c) Calculate the wt % θ phase present just below 548°C.

9.5.17 What are some of the properties which make aluminum an extremely useful engineering material?

9.5.18 How is aluminum oxide extracted from bauxite ores? How is aluminum extracted from pure aluminum oxide?

9.5.19 How are aluminum wrought alloys classified?

9.5.20 What are the basic temper designations for aluminum alloys?

9.5.21 Which series of aluminum wrought alloys are non-heat-treatable? Which are heat-treatable?

9.5.22 What are the basic strengthening precipitates for the wrought heat-treatable aluminum alloys?

9.5.23 Describe the three principal casting processes used for aluminum alloys.

9.5.24 How are aluminum casting alloys classified? What is the most important alloying element for aluminum casting alloys? Why?

9.6.1 What are some of the important properties of unalloyed copper which make it an important industrial metal?

9.6.2 How is copper extracted from copper sulfide ore concentrates?

9.6.3 How are copper alloys classified by the Copper Development Association system?

9.6.4 Why can't electrolytic tough-pitch copper be used for applications in which it is heated above 400°C in a hydrogen-containing atmosphere?

9.6.5 How can the hydrogen embrittlement of ETP copper be avoided? (Give two methods.)

9.6.6 Describe the microstructures of the following Cu–Zn brasses at 75×: (a) 70% Cu–30% Zn (cartridge brass) in the annealed condition and (b) 60% Cu–40% Zn (Muntz metal) in the hot-rolled condition.

9.6.7 Why are small amounts of lead added to some Cu–Zn brasses? In what state is the lead distributed in brasses?

9.6.8 What are the highest-strength commercial copper alloys? What type of heat treatment and fabrication method makes these alloys so strong?

9.7.1 What alloying element and how much of it (weight percent) is necessary to make a stainless steel "stainless"?

9.7.2 What type of surface film protects stainless steels?

9.7.3 What are the four basic types of stainless steels?

9.7.4 What is the gamma loop in the Fe–Cr phase diagram? Is chromium an austenite- or ferrite-stabilizing element for iron? Explain the reason for your answer.

9.7.5 What is the basic composition of ferritic stainless steels?

9.7.6 Why are ferritic stainless steels considered non-heat-treatable?

9.7.7 What is the basic composition of martensitic stainless steels? Why are these steels heat-treatable?

9.7.8 What are some applications for ferritic and martensitic stainless steels?

9.7.9 What makes it possible for an austenitic stainless steel to have an austenitic structure at room temperature?

9.7.10 What makes austenitic stainless steels which are cooled slowly through the 870 to 600°C range become susceptible to intergranular corrosion?

9.7.11 How can the intergranular susceptibility of slow-cooled austenitic stainless steels be prevented?

9.8.1 What are the cast irons? What is their basic range of composition?

9.8.2 What are some of the properties of cast irons which make them important engineering materials? What are some of their applications?

9.8.3 What are the four basic types of cast irons?

9.8.4 Describe the as-cast microstructure of unalloyed white cast iron at 100×.

9.8.5 Why does the fractured surface of white cast iron appear "white"?

9.8.6 Describe the microstructure of a class 30 gray cast iron in the as-cast condition at 100×. Why does the fractured surface of a gray cast iron appear gray?

9.8.7   What are the composition ranges for the carbon and silicon in gray cast iron? Why do gray cast irons have relatively high amounts of silicon?

9.8.8   What are some of the applications for gray cast irons?

9.8.9   What casting conditions favor the formation of gray cast iron?

9.8.10 How can a fully ferritic matrix be produced in an as-cast gray iron after it has been cast?

9.8.11 What are the composition ranges for the carbon and silicon in ductile cast irons?

9.8.12 Describe the microstructure of an as-cast grade 80-55-06 ductile cast iron at $100\times$. What causes the bull's-eye structure?

9.8.13 Why are ductile cast irons in general more ductile than gray cast irons?

9.8.14 What are some applications for ductile cast irons?

9.8.15 Why does the graphite form spherical nodules in ductile cast irons instead of graphite flakes as in gray cast irons?

9.8.16 What are the composition ranges of carbon and silicon in malleable cast irons?

9.8.17 Describe the microstructure of a ferritic malleable cast iron (grade M3210) at $100\times$.

9.8.18 How are malleable cast irons produced?

9.8.19 What are some of the property advantages of malleable cast irons?

9.8.20 What are some applications for malleable cast irons?

9.9.1   What advantages do magnesium alloys have as engineering materials?

9.9.2   How are magnesium alloys designated?

9.9.3   Explain what the following magnesium alloy designations indicate: (*a*) ZE63A; (*b*) ZK61A-T6; (*c*) AZ61A-F; (*d*) EZ33A-T5.

9.9.4   What alloying elements are added to magnesium for solid-solution strengthening?

9.9.5   Why is it difficult to cold-work magnesium alloys?

9.9.6   What alloying elements are added to magnesium to provide better high-temperature strengths?

9.9.7   Why are titanium and its alloys of special engineering importance for aerospace applications?

9.9.8   Why is titanium metal so expensive?

9.9.9   What crystal-structure change takes place in titanium at 883°C?

9.9.10 What are two alpha phase stabilizing elements for titanium?

9.9.11 What are two beta phase stabilizing elements for titanium?

9.9.12 What is the most important titanium alloy?

9.9.13 What are some applications for titanium and its alloys?

9.9.14 Why is nickel an important engineering metal? What are its advantages? Disadvantages?

9.9.15 What are the Monel alloys? What are some of their applications?

9.9.16 What type of precipitates are used to strengthen the precipitation-hardenable alloy Monel K500?

9.9.17 In what respect are the nickel-base superalloys "super"?

9.9.18 What is the basic composition of most nickel-base superalloys?

9.9.19 What are the three main phases present in nickel-base superalloys?

9.10.1   Which of the following plain-carbon steels would you select for a gear that will be surface-carburized (see Table 9.3 and Sec. 4.7): (*a*) 1010; (*b*) 1020; (*c*) 1040; (*d*) 1060; (*e*) 1080?

9.10.2   Which of the following plain-carbon steels would you select for a railway wheel (see Table 9.3): (*a*) 1010; (*b*) 1020; (*c*) 1040; (*d*) 1060; (*e*) 1080?

9.10.3   Which of the following low-alloy steels would you select for an automobile transmission gear (see Table 9.6): (*a*) 1340; (*b*) 5140; (*c*) 5160; (*d*) 4820; (*e*) 4340?

9.10.4   Which of the following low-alloy steels would you select for an aircraft landing gear support (see Table 9.6): (*a*) 1340; (*b*) 5140; (*c*) 5160; (*d*) 4820; (*e*) 4340?

9.10.5   Which of the following wrought aluminum alloys would you select for a high-strength aircraft structure (see Table 9.8): (*a*) 6061-T6; (*b*) 7075-T6; (*c*) 5052-H34; (*d*) 3003-H14; (*e*) 3003-H18?

9.10.6   Which of the following wrought copper alloys would you select for small copper-alloy springs (see Table 9.11): (*a*) C28000; (*b*) C10100; (*c*) 26000; (*d*) C17000; (*e*) C61400?

9.10.7   Which of the following stainless steels would you select for a high-work-hardening wrought alloy (see Table 9.12): (*a*) 430; (*b*) 301; (*c*) 410; (*d*) 304; (*e*) 440C?

9.10.8   Which of the following cast irons would you select for a truck cylinder block (see Table 9.14): (*a*) ferritic (32510); (*b*) martensitic (7002); (*c*) pearlitic (G3500); (*d*) ferritic 60-40-18; (*e*) martensitic (120-90-02)?

9.10.9   Which of the following magnesium alloy would you select for a chain-saw housing (see Table 9.16): (*a*) AZ61A; (*b*) AZ31A; (*c*) AZ91D; (*d*) AM60B; (*e*) AS41A?

9.10.10 Which of the following titanium alloys would you select for a corrosion-resistant heat exchanger (see Table 9.17): (*a*) 99.0% Ti; (*b*) 99.99% Ti; (*c*) Ti–6 Al–4 V; (*d*) Ti–10 V–2 Fe–3 Al; (*e*) Ti–5 Al–2.5 Sn?

9.10.11 Which of the following nickel alloys would you select for a pump operating under corrosion conditions (see Table 9.17): (*a*) nickel 200; (*b*) Monel 400; (*c*) Monel K500; (*d*) nickel 718; (*e*) superalloy?

Open-ended problems for Sec. 9.10:

9.10.12 Some new sheet steels for auto external panels are being made with <0.01% C and 0.5% P. After fabrication, the autos are painted, and during the baking cycle of the painting, the steel is hardened to some extent by precitation hardening by a phosphorus compound. What are some of the advantages and disadvantages of selecting this material for auto panels?

9.10.13 A new European auto is being made with an aluminum alloy space frame. The materials selected for the space frame are Al–Mg–Si alloy extrusions and Al–Si–Mg die casting which are welded together. What are some of the advantages and disadvantages of these selected materials for this application?

9.10.14 Engineers designing the airframe for the new Boeing 777 passenger airplane have chosen Ti–10 V–2 Fe–3 Al alloy large forgings for its main landing gear. Why has such a material been chosen for this application? Disadvantages?

9.10.15 The Pratt & Whitney Co. have designed a new high-temperature single-crystal nickel-based superalloy designated PWA 1484 which contains 3 wt % rhenium. What might be the function of the rhenium in the new higher-temperature PWA 1484 superalloy as compared with the older PWA 1480? What significant compositional changes have been made in this alloy design? (See Table 9.19.)

TABLE 9.19   **Comparison of the Chemical Composition (wt %) of PWA 1480 and PWA 1484 Alloys**

	Ni	Cr	Ti	Mo	W	Re	Ta	Al	Co	Hf
PWA 1480	Bal	10	1.5	...	4	...	12	5	5	
PWA 1484	Bal	5	...	2	6	3	8.7	5.6	10	0.1

10

Ceramic Materials

10.1 INTRODUCTION

Ceramic materials are inorganic, nonmetallic materials which consist of metallic and nonmetallic elements bonded together primarily by ionic and/or covalent bonds. The chemical compositions of ceramic materials vary considerably, from simple compounds to mixtures of many complex phases bonded together.

The properties of ceramic materials also vary greatly due to differences in bonding. In general, ceramic materials are typically hard and brittle with low toughness and ductility. Ceramics are usually good electrical and thermal insulators due to the absence of conduction electrons. Ceramic materials normally have relatively high melting temperatures and high chemical stability in many hostile environments due to the stability of their strong bonds. Because of these properties ceramic materials are indispensable for many engineering designs. Two examples of the strategic importance of ceramic materials in new high technology are shown in Fig. 10.1.

In general, ceramic materials used for engineering applications can be divided into two groups: traditional ceramic materials and the engineering ceramic materials. Typically, traditional ceramics are made from three basic

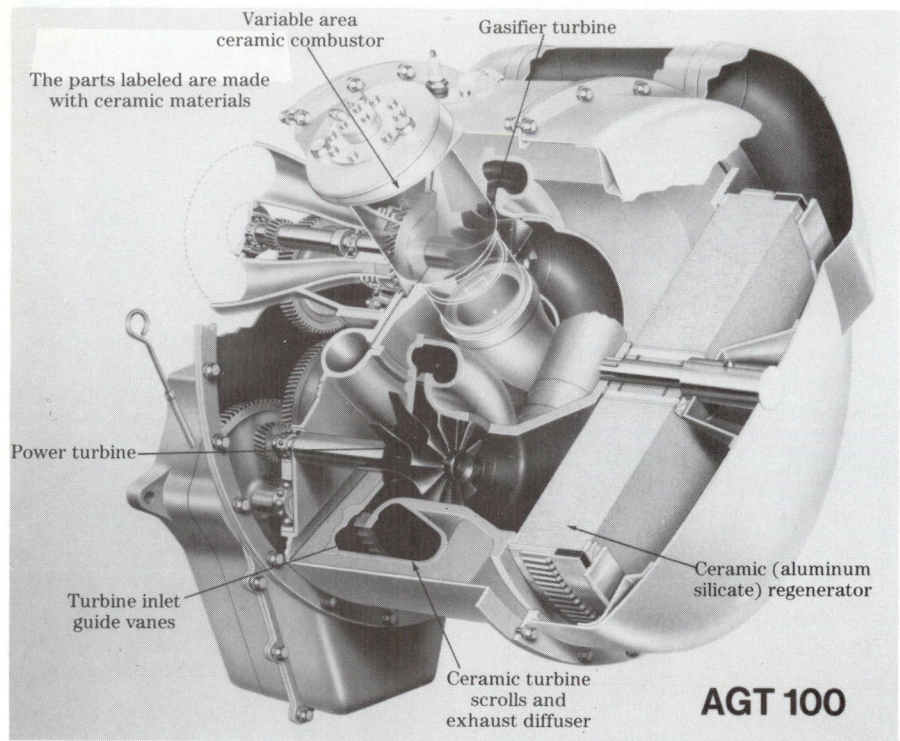

Variable area
ceramic combustor

Gasifier turbine

The parts labeled are made
with ceramic materials

Power turbine

Turbine inlet
guide vanes

Ceramic turbine
scrolls and
exhaust diffuser

Ceramic (aluminum
silicate) regenerator

AGT 100

(a)

(b)

Integrated circuit chips

Multilayer
ceramic substrate

Alumina pistons
for heat removal

FIGURE 10.1 Ceramic materials are strategically important for some new high technologies. Shown are two examples. (a) The high-temperature components of the experimental AGT-100 automotive gas turbine engine are to be made of ceramic materials. The higher operating temperatures of "ceramic" engines should increase fuel efficiencies. (*Courtesy of the Allison Gas Turbine Division of General Motors Co.*) (b) The thermal-conduction module cutaway shown has a large multilayered ceramic base which supports, interconnects, and insulates a large number of integrated circuit chips. (*Courtesy of IBM, East Fishkill, New York.*)

components: clay, silica (flint), and feldspar. Examples of traditional ceramics are bricks and tiles used in the construction industries and electrical porcelain in the electrical industry. The engineering ceramics, in contrast, typically consist of pure or nearly pure compounds such as aluminum oxide (Al_2O_3), silicon carbide (SiC), and silicon nitride (Si_3N_4). Examples of the use of the engineering ceramics in high technology are silicon carbide in the high-temperature areas of the experimental AGT-100 automotive gas turbine engine (Fig. 10.1a) and aluminum oxide in the support base for integrated circuit chips in a thermal-conduction module (Fig. 10.1b).

In this chapter we shall first examine some simple ceramic crystal structures and then look at some of the more complicated silicate ceramic structures. Then we shall explore some of the methods for processing ceramic materials, and following that, study some of the electrical, mechanical, and thermal properties of ceramic materials. Finally, we shall examine some aspects of the structure and properties of glasses.

10.2 SIMPLE CERAMIC CRYSTAL STRUCTURES

Ionic and Covalent Bonding in Simple Ceramic Compounds

Let us first consider some simple ceramic crystal structures. Some ceramic compounds with relatively simple crystal structures are listed in Table 10.1 with their melting points.

In the ceramic compounds listed, the atomic bonding is a mixture of ionic and covalent types. Approximate values for the percentages of ionic and covalent character for the bonds between the atoms in these compounds can be obtained by considering the electronegativity differences between the different types of atoms in the compounds and by using Pauling's equation for percent ionic character [Eq. (2.10)]. Table 10.2 shows that the percent ionic or covalent character varies considerably in simple ceramic compounds. The amount of ionic or covalent bonding between the atoms of these compounds is important since it determines to some extent what type of crystal structure will form in the bulk ceramic compound.

TABLE 10.1   Some Simple Ceramic Compounds with Their Melting Points

Ceramic compound	Melting point, °C	Ceramic compound	Melting point, °C
Hafnium carbide, HfC	4150	Boron carbide, B_4C	2450
Titanium carbide, TiC	3120	Aluminum oxide, Al_2O_3	2050
Tungsten carbide, WC	2850	Silicon dioxide,* SiO_2	1715
Magnesium oxide, MgO	2798	Silicon nitride, Si_3N_4	1900
Silicon carbide, SiC	2500	Titanium dioxide, TiO_2	1605

*Cristobalite.

TABLE 10.2   Percent Ionic and Covalent Bonding Character for Some Ceramic Compounds

Ceramic compound	Bonding atoms	Electronegativity difference	% Ionic character	% Covalent character
Magnesium oxide, MgO	Mg—O	2.3	73	27
Aluminum oxide, Al_2O_3	Al—O	2.0	63	37
Silicon dioxide, SiO_2	Si—O	1.7	51	49
Silicon nitride, Si_3N_4	Si—N	1.2	30	70
Silicon carbide, SiC	Si—C	0.7	11	89

Simple Ionic Arrangements Found in Ionically Bonded Solids

In ionic (ceramic) solids the packing of the ions is determined primarily by the following factors:

1. The relative size of the ions in the ionic solid (assume the ions to be hard spheres with definite radii)
2. The need to balance electrostatic charges to maintain electrical neutrality in the ionic solid

When ionic bonding between atoms takes place in the solid state, the energies of the atoms are lowered by the formation of the ions and their bonding into an ionic solid. Ionic solids tend to have their ions packed together as densely as possible to lower the overall energy of the solid as much as possible. The limitations to dense packing are the relative sizes of the ions and the necessity for maintaining charge neutrality.

Size limitations for the dense packing of ions in an ionic solid   Ionic solids consist of cations and anions. In ionic bonding some atoms lose their outer electrons to become *cations* and others gain outer electrons to become *anions*.

FIGURE 10.2 Stable and unstable coordination configurations for ionic solids. (*After W. D. Kingery, H. K. Bowen, and D. R. Uhlmann, "Introduction to Ceramics," 2d ed., Wiley, 1976.*)

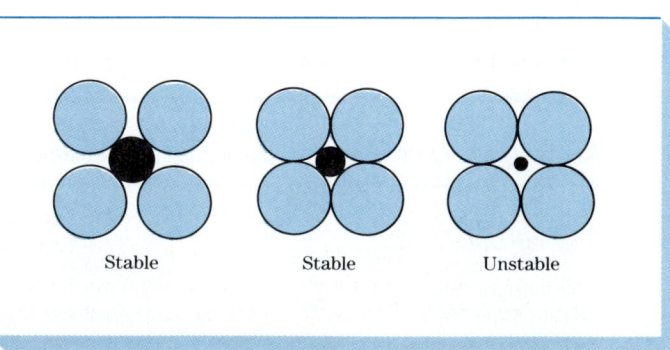

Stable   Stable   Unstable

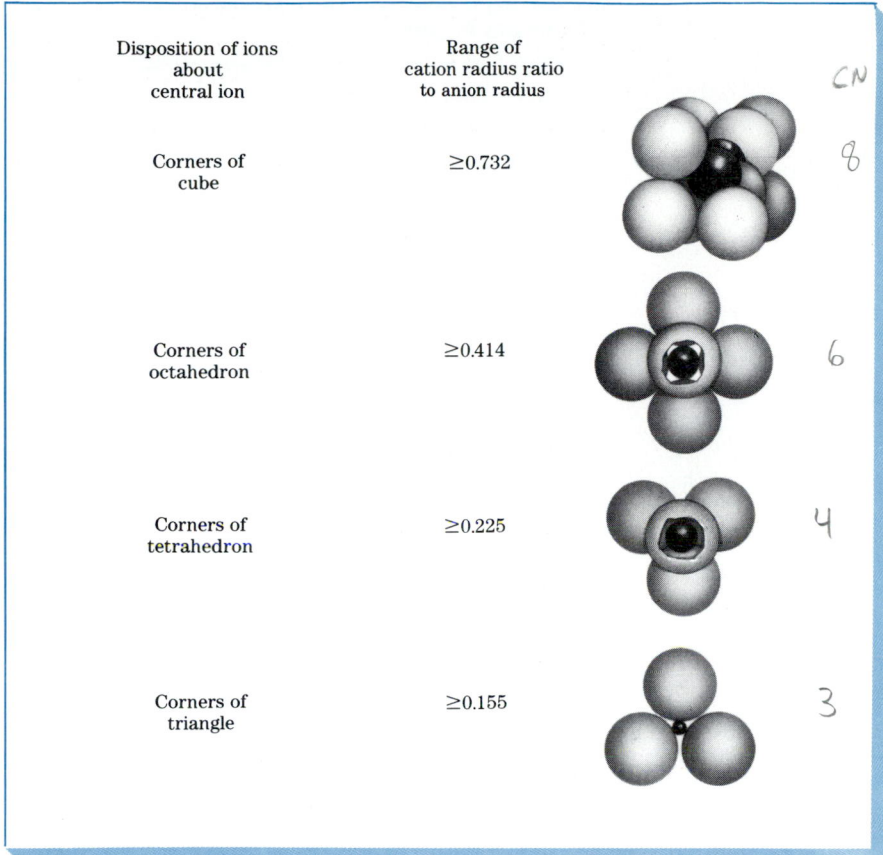

Disposition of ions about central ion	Range of cation radius ratio to anion radius	CN
Corners of cube	≥0.732	8
Corners of octahedron	≥0.414	6
Corners of tetrahedron	≥0.225	4
Corners of triangle	≥0.155	3

FIGURE 10.3 Radius ratios for coordination numbers of 8, 6, 4, and 3 anions surrounding a central cation in ionic solids. (*After W. D. Kingery, H. K. Bowen, and D. R. Uhlmann, "Introduction to Ceramics," 2d ed., Wiley, 1976.*)

Thus, cations are normally smaller than the anions they bond with. The number of anions that surround a central cation in an ionic solid is called the *coordination number* (CN) and corresponds to the number of nearest neighbors surrounding a central cation. For stability as many anions as possible surround a central cation. However, the anions must make contact with the central cation and charge neutrality must be maintained.

Figure 10.2*a* and *b* shows stable configurations for the coordination of anions around a central cation in an ionic solid. If the anions do not touch the central cation, the structure becomes unstable because the central cation can "rattle around in its cage of anions" (Fig. 10.2*c*). The ratio of the radius of the central cation to that of the surrounding anions is called the *radius ratio*, r_{cation}/r_{anion}. The radius ratio when the anions just touch each other and contact the central cation is called the *critical (minimum) radius ratio*. Allowable radius ratios for ionic solids with coordination numbers of 3, 4, 6, and 8 are listed in Fig. 10.3 along with illustrations showing the coordinations.

Example Problem 10.1

Calculate the critical (minimum) radius ratio r/R for the triangular coordination (CN = 3) of three anions of radii R surrounding a central cation of radius r in an ionic solid.

Solution:

Figure 10.4a shows three large anions of radii R surrounding and just touching a central cation of radius r. Triangle ABC is an equilateral triangle (each angle = 60°), and line AD bisects angle CAB. Thus angle DAE = 30°. To find the relationship between R and r, triangle ADE is constructed as shown in Fig. 10.4b. Thus

$$AD = R + r$$

$$\cos 30° = \frac{AE}{AD} = \frac{R}{R + r} = 0.866$$

$$R = 0.866(R + r)$$

$$= 0.866R + 0.866r$$

$$0.866r = R - 0.866R = R(0.134)$$

$$\frac{r}{R} = 0.155 \blacktriangleleft$$

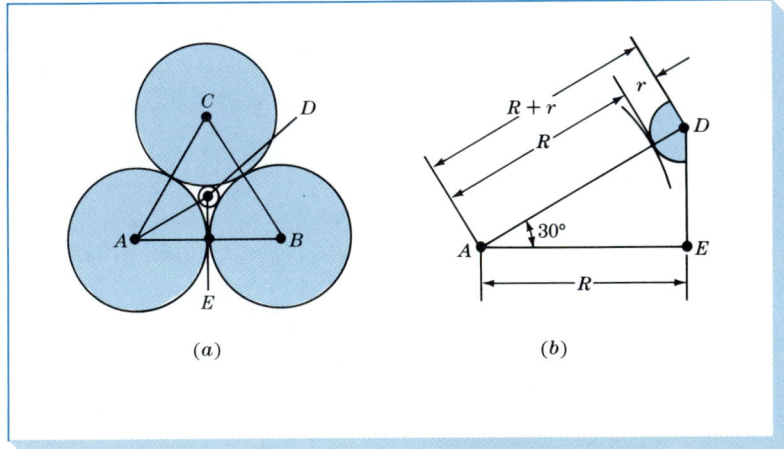

FIGURE 10.4
Diagram for triangular coordination.

Example Problem 10.2

Predict the coordination number for the ionic solids CsCl and NaCl. Use the following ionic radii for the prediction:

$$\text{Cs}^+ = 0.170 \text{ nm} \qquad \text{Na}^+ = 0.102 \text{ nm} \qquad \text{Cl}^- = 0.181 \text{ nm}$$

Solution:

The radius ratio for CsCl is

$$\frac{r\,(\text{Cs}^+)}{R\,(\text{Cl}^-)} = \frac{0.170 \text{ nm}}{0.181 \text{ nm}} = 0.94$$

Since this ratio is greater than 0.732, CsCl should show cubic coordination (CN = 8), which it does.

The radius ratio for NaCl is

$$\frac{r\,(\text{Na}^+)}{R\,(\text{Cl}^-)} = \frac{0.102 \text{ nm}}{0.181 \text{ nm}} = 0.56$$

Since this ratio is greater than 0.414 but less than 0.732, NaCl should show octahedral coordination (CN = 6), which it does.

Cesium Chloride (CsCl) Crystal Structure

The chemical formula for solid cesium chloride is CsCl, and since this structure is principally ionically bonded, there are equal numbers of Cs^+ and Cl^- ions. Because the radius ratio for CsCl is 0.94 (see Example Problem 10.2), cesium chloride has cubic coordination (CN = 8), as shown in Fig. 10.5. Thus, eight chloride ions surround a central cesium cation at the $(\tfrac{1}{2}, \tfrac{1}{2}, \tfrac{1}{2})$ position in the CsCl unit cell. Ionic compounds which also have the CsCl crystal structure are CsBr, TlCl, and TlBr. The intermetallic compounds AgMg, LiMg, AlNi, and β–Cu–Zn also have this structure. The CsCl structure is not of much importance for ceramic materials but does illustrate how higher radius ratios lead to higher coordination numbers in ionic crystal structures.

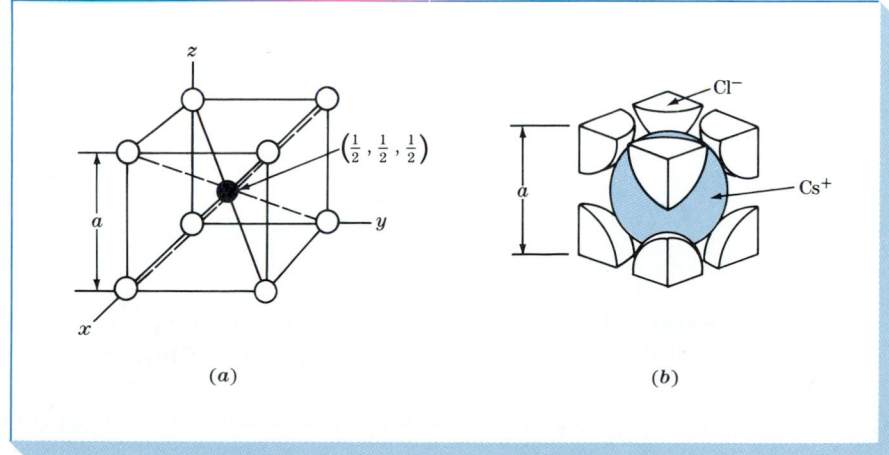

FIGURE 10.5 Cesium chloride (CsCl) crystal structure unit cell. (*a*) Ion-site unit cell. (*b*) Hard-sphere unit cell. In this crystal structure eight chloride ions surround a central cation in cubic coordination (CN = 8). In this unit cell there is one Cs^+ and one Cl^- ion.

Example Problem 10.3

Calculate the ionic packing factor for CsCl. Ionic radii are $Cs^+ = 0.170$ nm and $Cl^- = 0.181$ nm.

Solution:

The ions touch each other across the cube diagonal of the CsCl unit cell, as shown in Fig. 10.6. Let $r = Cs^+$ ion and $R = Cl^-$ ion. Thus

$$\sqrt{3}a = 2r + 2R$$
$$= 2(0.170 \text{ nm} + 0.181 \text{ nm})$$
$$a = 0.405 \text{ nm}$$

$$\text{CsCl ionic packing factor} = \frac{\frac{4}{3}\pi r^3 \text{ (1 Cs}^+ \text{ ion)} + \frac{4}{3}\pi R^3 \text{ (1 Cl}^- \text{ ion)}}{a^3}$$

$$= \frac{\frac{4}{3}\pi(0.170 \text{ nm})^3 + \frac{4}{3}\pi(0.181 \text{ nm})^3}{(0.405 \text{ nm})^3}$$

$$= 0.68 \blacktriangleleft$$

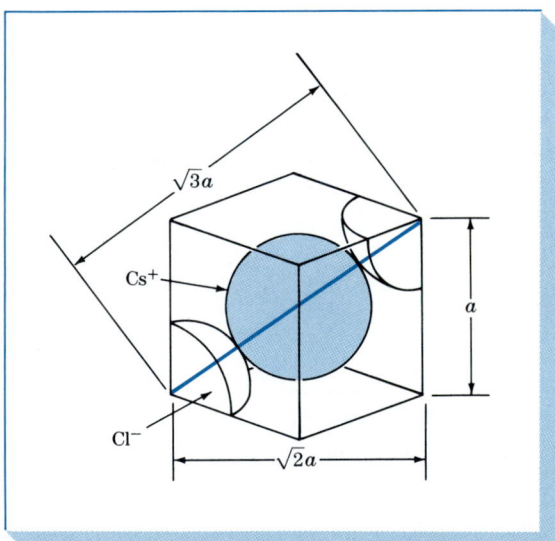

FIGURE 10.6

Sodium Chloride (NaCl) Crystal Structure
The sodium chloride or rock salt crystal structure is highly ionically bonded and has the chemical formula NaCl. Thus, there are an equal number of Na^+ and Cl^- ions to maintain charge neutrality. Figure 10.7a shows a lattice-site NaCl unit cell and Fig. 2.13b a hard-sphere model of the NaCl unit cell. Figure 10.7a has negative Cl^- anions occupying regular FCC atom lattice sites and positive Na^+ cations occupying the interstitial sites between the FCC atom

FIGURE 10.7 (*a*) NaCl lattice-point unit cell indicating positions of the Na$^+$ (radii = 0.102 nm) and Cl$^-$ (radii = 0.181 nm) ions. (*b*) Octahedron showing octahedral coordination of six Cl$^-$ anions around a central Na$^+$ cation.

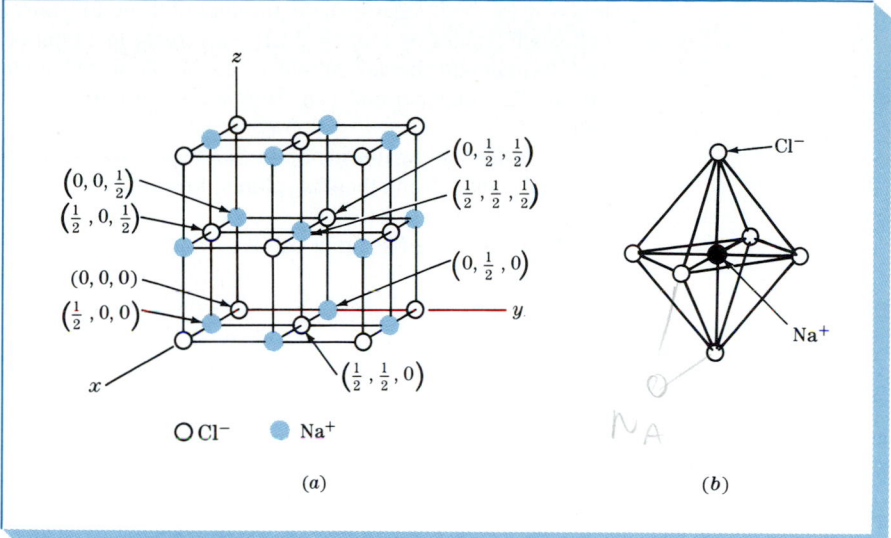

(*a*) (*b*)

sites. The centers of the Na$^+$ and Cl$^-$ ions occupy the following lattice positions, which are indicated in Fig. 10.7*a*:

Na$^+$: $(\frac{1}{2}, 0, 0)$ $(0, \frac{1}{2}, 0)$ $(0, 0, \frac{1}{2})$ $(\frac{1}{2}, \frac{1}{2}, \frac{1}{2})$

Cl$^-$: $(0, 0, 0)$ $(\frac{1}{2}, \frac{1}{2}, 0)$ $(\frac{1}{2}, 0, \frac{1}{2})$ $(0, \frac{1}{2}, \frac{1}{2})$

Since each central Na$^+$ cation is surrounded by six Cl$^-$ anions, the structure has octahedral coordination (that is, CN = 6), as shown in Fig. 10.7*b*. This type of coordination is predicted from the radius ratio calculation of r_{Na^+}/R_{Cl^-} = 0.102 nm/0.181 nm = 0.56, which is greater than 0.414 but less than 0.732. Other ceramic compounds which have the NaCl structure include MgO, CaO, NiO, and FeO.

Example Problem 10.4

Calculate the density of NaCl from a knowledge of its crystal structure (Fig. 10.7*a*), the ionic radii of Na$^+$ and Cl$^-$ ions, and the atomic masses of Na and Cl. The ionic radius of Na$^+$ = 0.102 nm and that of Cl$^-$ = 0.181 nm. The atomic mass of Na = 22.99 g/mol and that of Cl = 35.45 g/mol.

Solution:

As shown in Fig. 10.7*a*, the Cl$^-$ ions in the NaCl unit cell form an FCC-type atom lattice, and the Na$^+$ ions occupy the interstitial spaces between the Cl$^-$ ions. There is the equivalent of one Cl$^-$ ion at the corners of the NaCl unit cell since 8 corners × $\frac{1}{8}$

ion = 1 ion, and there is the equivalent of three Cl^- ions at the faces of the NaCl unit cell since 6 faces $\times \frac{1}{2}$ ion = 3 Cl^- ions, making a total of four Cl^- ions per NaCl unit cell. To maintain charge neutrality in the NaCl unit cell, there must also be the equivalent of four Na^+ ions per unit cell. Thus there are four Na^+Cl^- ion pairs in the NaCl unit cell.

To calculate the density of the NaCl unit cell, we shall first determine the mass of one NaCl unit cell and then its volume. Knowing these two quantities, we can calculate the density m/V.

The mass of a NaCl unit cell is

$$m = \frac{(4Na^+ \times 22.99 \text{ g/mol}) + (4Cl^- \times 35.45 \text{ g/mol})}{6.02 \times 10^{23} \text{ atoms (ions)/mol}}$$

$$= 3.88 \times 10^{-22} \text{ g}$$

The volume of the NaCl unit cell is equal to a^3, where a is the lattice constant of the NaCl unit cell. The Cl^- and Na^+ ions contact each other along the cube edges of the unit cell, as shown in Fig. 10.8, and thus

$$a = 2(r_{Na^+} + R_{Cl^-})$$
$$= 2(0.102 \text{ nm} + 0.181 \text{ nm}) = 0.566 \text{ nm}$$
$$= 0.566 \text{ nm} \times 10^{-7} \text{ cm/nm} = 5.66 \times 10^{-8} \text{ cm}$$
$$V = a^3 = 1.81 \times 10^{-22} \text{ cm}^3$$

The density of NaCl is

$$\rho = \frac{m}{V} = \frac{3.88 \times 10^{-22} \text{ g}}{1.81 \times 10^{-22} \text{ cm}^3} = 2.14 \frac{\text{g}}{\text{cm}^3} \blacktriangleleft$$

The handbook value for the density of NaCl is 2.16 g/cm³.

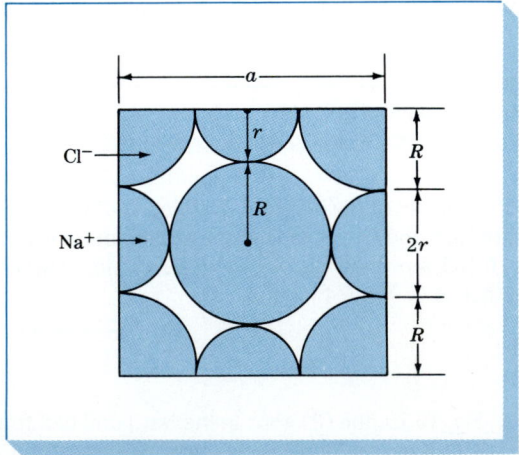

FIGURE 10.8 Cube face of NaCl unit cell. Ions contact along the cube edge, and thus $a = 2r + 2R = 2(r + R)$.

Example Problem 10.5

Calculate the linear density of Ca^{2+} and O^{2-} ions in ions per nanometer in the [110] direction of CaO which has the NaCl structure. (Ionic radii: $Ca^{2+} = 0.106$ nm and $O^{2-} = 0.132$ nm.)

Solution:

From Fig. 10.7 and Fig. EP10.5, we see that the [110] direction passes through $2O^{2-}$ ion diameters in traversing from (0, 0, 0), to (1, 1, 0) ion positions. The length of the [110] distance across the base face of a unit cube is $\sqrt{2}a$, where a is the length of a side of the cube, or lattice constant. From Fig. 10.8 of the cube face of the NaCl unit cell, we see that $a = 2r + 2R$. Thus, for CaO,

$$a = 2(r_{Ca^{2+}} + R_{O^{2-}}) = 2(0.106 \text{ nm} + 0.132 \text{ nm}) = 0.476 \text{ nm}$$

The linear density of the O^{2-} ions in the [110] direction is

$$\rho_L = \frac{2O^{2-}}{\sqrt{2}a} = \frac{2O^{2-}}{\sqrt{2}(0.476 \text{ nm})} = 2.97O^{2-}/\text{nm} \blacktriangleleft$$

The linear density of Ca^{2+} ions in the [110] direction is also $2.97Ca^{2+}$/nm if we shift the origin of the [110] direction from (0, 0, 0) to (0, $\frac{1}{2}$, 0). Thus, the solution to the problem is that there are $2.97(Ca^{2+}$ or $O^{2-})$/nm in the [110] direction.

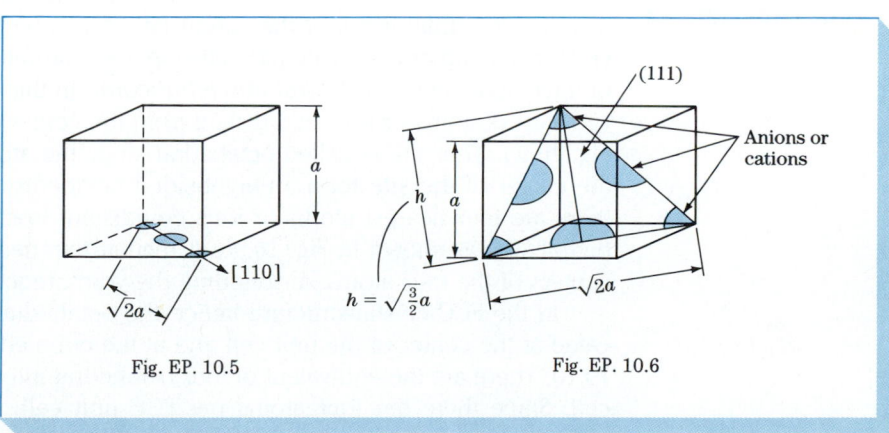

Fig. EP. 10.5   Fig. EP. 10.6

Example Problem 10.6

Calculate the planar density of Ca^{2+} and O^{2-} ions in ions per square nanometer on the (111) plane of CaO which has the NaCl structure. (Ionic radii: $Ca^{2+} = 0.106$ nm and $O^{2-} = 0.132$ nm.)

Solution:

If we consider the anions (O^{2-} ions) to be located at the FCC positions of a cubic unit cell as shown for the Cl^- ions of Fig. 10.7 and Fig. EP10.6, then the (111) plane contains the equivalent of two anions. [$3 \times 60° = 180° = \frac{1}{2}$ anion $+ (3 \times \frac{1}{2})$ anions at each midpoint of the sides of the (111) planar triangle of Fig. EP10.6 $=$ a total of 2 anions within the (111) triangle.] The lattice constant for the unit cell $a = 2(r + R) = 2(0.106$ nm $+ 0.132$ nm$) = 0.476$ nm. The planar area $A = \frac{1}{2} bh$, where $h = \frac{\sqrt{3}}{2} a^2$. Thus,

$$A = (\tfrac{1}{2}\sqrt{2}a)(\sqrt{\tfrac{3}{2}}a) = \frac{\sqrt{3}}{2}a^2 = \frac{\sqrt{3}}{2}(0.476 \text{ nm})^2 = 0.196 \text{ nm}^2$$

The planar density for the O^{2-} anions is

$$\frac{2(O^{2-} \text{ ions})}{0.196 \text{ nm}^2} = 10.2 O^{2-} \text{ ions/nm}^2 \blacktriangleleft$$

The planar density for the Ca^{2+} cations is the same if we consider the Ca^{2+} to be located at the FCC lattice points of the unit cell, and thus

$$\rho_{\text{planar}} (CaO) = 10.2(Ca^{2+} \text{ or } O^{2-})/\text{nm}^2 \blacktriangleleft$$

Interstitial Sites in FCC and HCP Crystal Lattices

There are empty spaces or voids among the atoms or ions which are packed into a crystal-structure lattice. These voids are *interstitial sites* in which atoms or ions other than those of the parent lattice can be fitted in. In the FCC and HCP crystal structures, which are close-packed structures, there are two types of interstitial sites: *octahedral* and *tetrahedral.* In the octahedral site there are six nearest atoms or ions equidistant from the center of the void, as shown in Fig. 10.9a. This site is called octahedral since the atoms or ions surrounding the center of the site form an eight-sided octahedron. In the tetrahedral site there are four nearest atoms or ions equidistant from the center of the tetrahedral site, as shown in Fig. 10.9b. A regular tetrahedron is formed when the centers of the four atoms surrounding the void are joined.

In the FCC crystal-structure lattice the octahedral interstitial sites are located at the center of the unit cell and at the cube edges, as indicated in Fig. 10.10. There are the equivalent of four octahedral interstitial sites per FCC unit cell. Since there are four atoms per FCC unit cell, there is one octahedral interstitial site per atom in the FCC lattice. Figure 10.11a indicates the lattice positions for octahedral interstitial sites in an FCC unit cell.

FIGURE 10.9 Interstitial sites in FCC and HCP crystal-structure lattices. (a) Octahedral interstitial site formed at the center where six atoms contact each other. (b) Tetrahedral interstitial site formed at the center where four atoms contact each other. (After W. D. Kingery, H. K. Bowen, and D. R. Uhlmann, "Introduction to Ceramics," 2d ed., Wiley, 1976.)

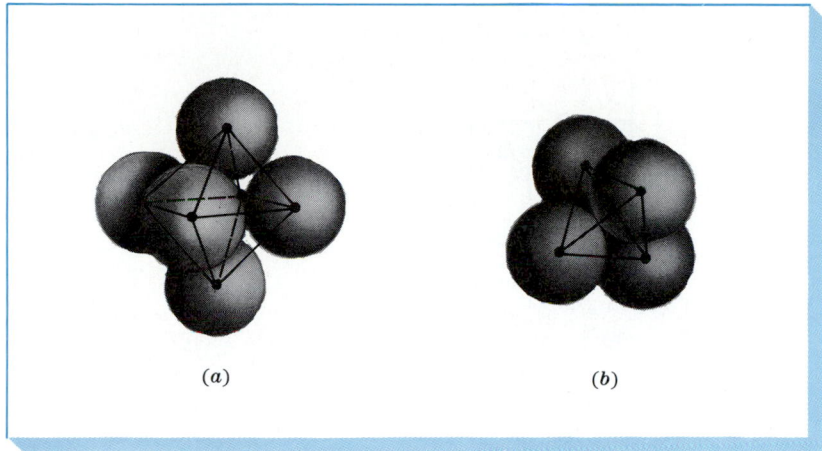

(a)   (b)

The tetrahedral sites in the FCC lattice are located at the $(\frac{1}{4}, \frac{1}{4}, \frac{1}{4})$-type positions, as indicated in Figs. 10.10 and 10.11b. In the FCC unit cell there are eight tetrahedral sites per unit cell or *two* per atom of the parent FCC unit cell. In the HCP crystal structure, because of similar close packing to the FCC structure, there is also the same number of octahedral interstitial sites as atoms in the HCP unit cell and twice as many tetrahedral sites as atoms.

Zinc Blende (ZnS) Crystal Structure

The zinc blende structure has the chemical formula ZnS and the unit cell shown in Fig. 10.12, which has the equivalent of four zinc and four sulfur atoms. One type of atom (either S or Zn) occupies the lattice points of an FCC unit cell, and the other type (either S or Zn) occupies half the tetrahedral interstitial sites of the FCC unit cell. In the ZnS crystal-structure unit cell shown in Fig. 10.12, sulfur atoms occupy the FCC unit-cell atom positions, as indicated

FIGURE 10.10 Location of octahedral and tetrahedral interstitial void sites in an FCC ionic crystal-structure unit cell. The octahedral sites are located at the center of the unit cell and at the centers of the cube edges. Since there are 12 cube edges, one-fourth of a void is located within the cube at each edge. Thus, there is the equivalent of 12 × $\frac{1}{4}$ = 3 voids within the FCC unit cell at the cube edges. Therefore, there is the equivalent of four octahedral voids per FCC unit cell (one at the center and the equivalent of three at the cube edges). The tetrahedral voids are located at the $(\frac{1}{4}, \frac{1}{4}, \frac{1}{4})$-type sites, which are indicated by points with tetrahedrally directed rays. Thus there are a total of eight tetrahedral void sites located within the FCC unit cell. (After W. D. Kingery, "Introduction to Ceramics," Wiley, 1960, p. 104.)

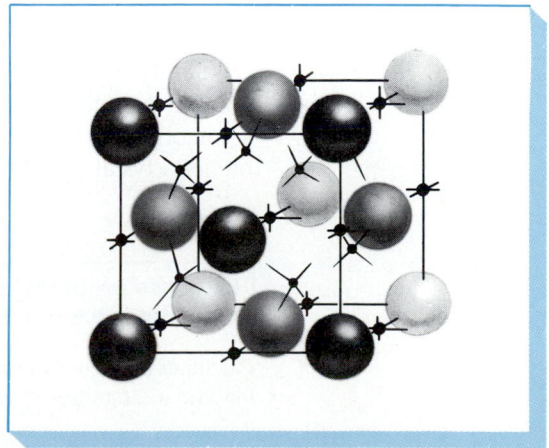

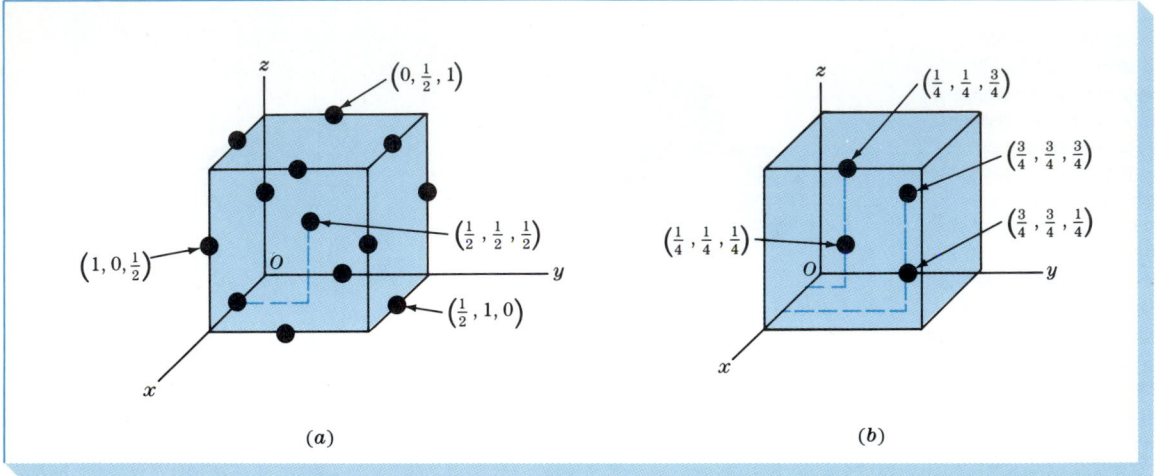

(a)   (b)

FIGURE 10.11 Location of interstitial sites in the FCC atom unit cell. (*a*) The octahedral sites in the FCC unit cell are located at the center of the unit cell and at the centers of the cube edges. (*b*) The tetrahedral sites in the FCC unit cell are located at the unit-cell positions indicated below. Only representative positions are located in the figure.

$$(\tfrac{1}{4}, \tfrac{1}{4}, \tfrac{1}{4}) \quad (\tfrac{1}{4}, \tfrac{1}{4}, \tfrac{3}{4}) \quad (\tfrac{3}{4}, \tfrac{1}{4}, \tfrac{1}{4}) \quad (\tfrac{3}{4}, \tfrac{1}{4}, \tfrac{3}{4})$$
$$(\tfrac{1}{4}, \tfrac{3}{4}, \tfrac{1}{4}) \quad (\tfrac{1}{4}, \tfrac{3}{4}, \tfrac{3}{4}) \quad (\tfrac{3}{4}, \tfrac{3}{4}, \tfrac{1}{4}) \quad (\tfrac{3}{4}, \tfrac{3}{4}, \tfrac{3}{4})$$

by the open circles, and Zn atoms occupy half the tetrahedral interstitial positions of the FCC unit cell, as indicated by the shaded circles. The position coordinates of the S and Zn atoms in the ZnS crystal structure can thus be indicated as

S atoms: $(0, 0, 0)$ $(\tfrac{1}{2}, \tfrac{1}{2}, 0)$ $(\tfrac{1}{2}, 0, \tfrac{1}{2})$ $(0, \tfrac{1}{2}, \tfrac{1}{2})$
Zn atoms: $(\tfrac{3}{4}, \tfrac{1}{4}, \tfrac{1}{4})$ $(\tfrac{1}{4}, \tfrac{1}{4}, \tfrac{3}{4})$ $(\tfrac{1}{4}, \tfrac{3}{4}, \tfrac{1}{4})$ $(\tfrac{3}{4}, \tfrac{3}{4}, \tfrac{3}{4})$

According to Pauling's equation [Eq. (2.10)], the Zn—S bond has 87 percent covalent character, and so the ZnS crystal structure must be essentially covalently bonded. As a result, the ZnS structure is tetrahedrally covalently bonded and the Zn and S atoms have a coordination number of 4. Many semiconducting compounds such as CdS, InAs, InSb, and ZnSe have the zinc blende crystal structure.

Example Problem 10.7

Calculate the density of zinc blende (ZnS). Assume the structure to consist of ions and that the ionic radius of $Zn^{2+} = 0.060$ nm and that of $S^{2-} = 0.174$ nm.

FIGURE 10.12 Zinc blende (ZnS) crystal structure. In this unit cell the sulfur atoms occupy the FCC atom unit-cell sites (equivalent of four atoms). The zinc atoms occupy half the tetrahedral interstitial sites (four atoms). Each Zn or S atom has a coordination number of 4 and is tetrahedrally covalently bonded to other atoms. (*After W. D. Kingery, H. K. Bowen, and D. R. Uhlmann, "Introduction to Ceramics," 2d ed., Wiley, 1976.*)

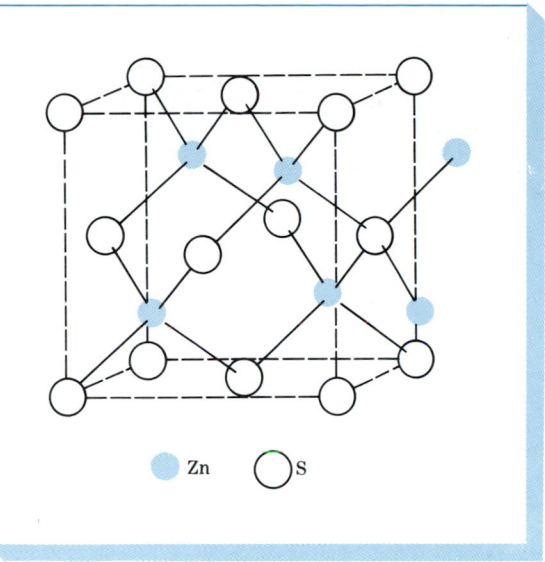

●Zn ○S

Solution:

$$\text{Density} = \frac{\text{mass of unit cell}}{\text{volume of unit cell}}$$

There are four zinc ions and four sulfur ions per unit cell. Thus

$$\text{Mass of unit cell} = \frac{(4Zn^{2+} \times 65.37 \text{ g/mol}) + (4S^{2-} \times 32.06 \text{ g/mol})}{6.02 \times 10^{23} \text{ atoms/mol}}$$

$$= 6.47 \times 10^{-22} \text{ g}$$

Volume of unit cell $= a^3$

From Fig. 10.13,

$$\frac{\sqrt{3}}{4}a = r_{Zn^{2+}} + R_{S^{2-}} = 0.060 \text{ nm} + 0.174 \text{ nm} = 0.234 \text{ nm}$$

$$a = 5.40 \times 10^{-8} \text{ cm}$$
$$a^3 = 1.57 \times 10^{-22} \text{ cm}^3$$

Thus

$$\text{Density} = \frac{\text{mass}}{\text{volume}} = \frac{6.47 \times 10^{-22} \text{ g}}{1.57 \times 10^{-22} \text{ cm}^3} = 4.12 \text{ g/cm}^3 \blacktriangleleft$$

The handbook value for the density of ZnS (cubic) is 4.10 g/cm³.

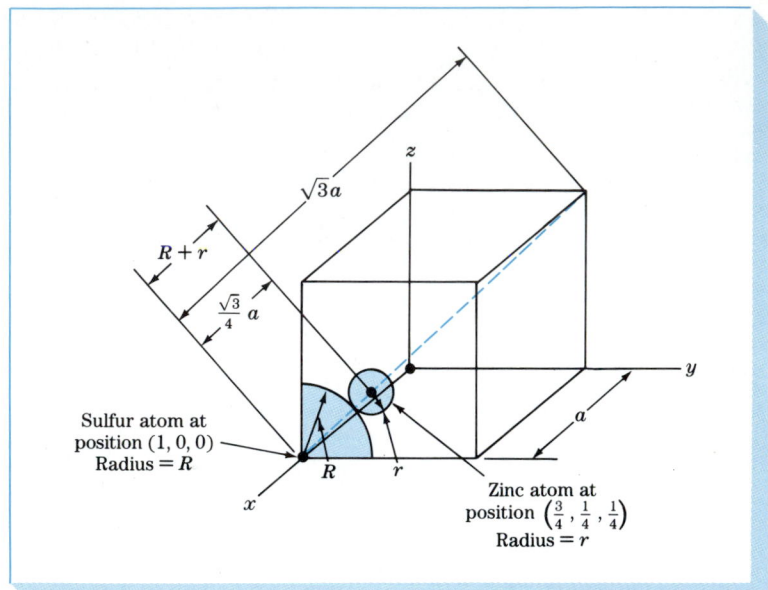

FIGURE 10.13 Zinc blende structure showing the relationship between the lattice constant a of the unit cell and the radii of the sulfur and zinc atoms (ions):

$$\frac{\sqrt{3}}{4}a = r_{Zn^{2+}} + R_{S^{2-}}$$

or

$$a = \frac{4}{\sqrt{3}}(r + R)$$

Sulfur atom at position $(1, 0, 0)$
Radius = R

Zinc atom at position $\left(\frac{3}{4}, \frac{1}{4}, \frac{1}{4}\right)$
Radius = r

Calcium Fluorite (CaF₂) Crystal Structure

The calcium fluoride structure has the chemical formula CaF_2 and the unit cell shown in Fig. 10.14. In this unit cell the Ca^{2+} ions occupy the FCC lattice sites, while the F^- ions are located at the eight tetrahedral sites. The four remaining octahedral sites in the FCC lattice remain vacant. Thus, there are four Ca^{2+} ions and eight F^- ions per unit cell. Examples of compounds which have this structure are UO_2, BaF_2, $AuAl_2$, and $PbMg_2$. The compound ZrO_2 has a distorted

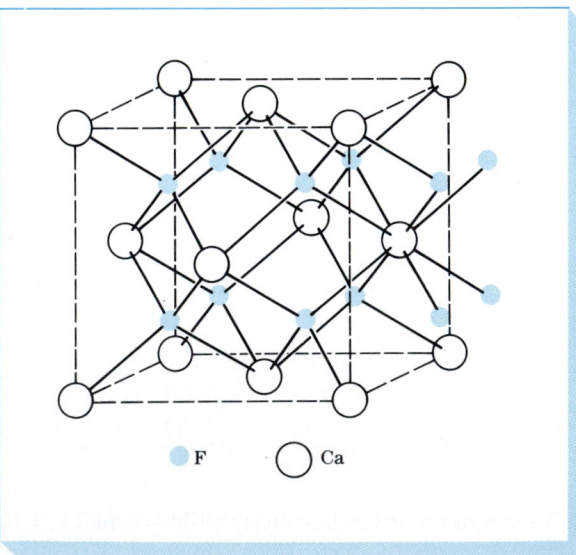

FIGURE 10.14 Calcium fluoride (CaF₂) crystal structure (also called fluorite structure). In this unit cell the Ca²⁺ ions are located at the FCC unit-cell sites (four ions). Eight fluoride ions occupy all the tetrahedral interstitial sites. (*After W. D. Kingery, H. K. Bowen, and D. R. Uhlmann, "Introduction to Ceramics," 2d ed., Wiley, 1976.*)

F   Ca

(monoclinic) CaF_2 structure. The large number of unoccupied octahedral interstitial sites in UO_2 allows this material to be used as a nuclear fuel since fission products can be accommodated in these vacant positions.

Example Problem 10.8

Calculate the density of UO_2 (uranium oxide) which has the calcium fluoride, CaF_2, structure. (Ionic radii: $U^{4+} = 0.105$ nm and $O^{2-} = 0.132$ nm.)

Solution:

$$\text{Density} = \frac{\text{mass/unit cell}}{\text{volume/unit cell}}$$

There are four uranium ions and eight oxide ions per unit cell (CaF_2 type). Thus,

$$\text{Mass of a unit cell} = \frac{(4U^{4+} \times 238 \text{ g/mol}) + (8O^{2-} \times 16 \text{ g/mol})}{6.02 \times 10^{23} \text{ ions/mol}}$$

$$= 1.794 \times 10^{-21} \text{ g}$$

$$\text{Volume of a unit cell} = a^3$$

From Fig. 10.13,

$$\frac{\sqrt{3}}{4}a = r_{U^{4+}} + R_{O^{2-}}$$

$$a = \frac{4}{\sqrt{3}}(0.105 \text{ nm} + 0.132 \text{ nm}) = 0.5473 \text{ nm} = 0.5473 \times 10^{-7} \text{ cm}$$

$$a^3 = (0.5473 \times 10^{-7} \text{ cm})^3 = 0.164 \times 10^{-21} \text{ cm}^3$$

$$\text{Density} = \frac{\text{mass}}{\text{volume}} = \frac{1.79 \times 10^{-21} \text{ g}}{0.164 \times 10^{-21} \text{ cm}^3} = 10.9 \text{ g/cm}^3 \blacktriangleleft$$

The handbook value for the density of UO_2 is 10.96 g/cm³.

Antifluorite Crystal Structure   The antifluorite structure consists of an FCC unit cell with anions (for example, O^{2-} ions) occupying the FCC lattice points. Cations (for example, Li^+) occupy the eight tetrahedral sites in the FCC lattice. Examples of compounds with this structure are Li_2O, Na_2O, K_2O, and Mg_2Si.

Corundum (Al_2O_3) Crystal Structure   In the corundum (Al_2O_3) structure the oxygen ions are located at the lattice sites of a hexagonal close-packed unit cell, as shown in Fig. 10.15. In the HCP

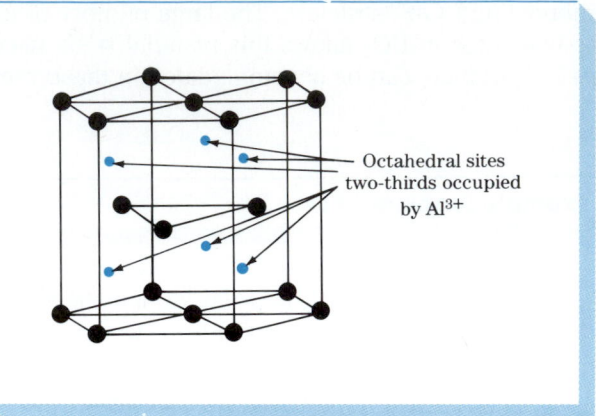

Octahedral sites
two-thirds occupied
by Al^{3+}

FIGURE 10.15 Corundum (Al_2O_3) crystal structure. Oxygen ions (O^{2-}) occupy the HCP unit-cell sites. Aluminum ions (Al^{3+}) occupy only two-thirds of the octahedral interstitial sites to maintain electrical neutrality.

crystal structure as in the FCC structure there are as many octahedral interstitial sites as there are atoms in the unit cell. However, since aluminum has a valence of $+3$ and oxygen a valence of -2, there can be only *two* Al^{3+} ions for every three O^{2-} ions to maintain electrical neutrality. Thus the aluminum ions can only occupy two-thirds of the octahedral sites of the HCP Al_2O_3 lattice, which leads to some distortion of this structure.

Perovskite ($CaTiO_3$) Crystal Structure   In the perovskite ($CaTiO_3$) structure the Ca^{2+} and O^{2-} ions form an FCC unit cell with the Ca^{2+} ions at the corners of the unit cell and the O^{2-} ions in the centers of the faces of the unit cell (Fig. 10.16). The highly charged Ti^{4+} ion is located at the octahedral interstitial site at the center of the unit cell and is coordinated to six O^{2-} ions. $BaTiO_3$ has the perovskite structure above 120°C, but below this temperature its structure is slightly changed. Other compounds having this structure are $SrTiO_3$, $CaZrO_3$, $SrZrO_3$, $LaAlO_3$, and many others. This structure is important for piezoelectric materials (see Sec. 10.6).

Spinel ($MgAl_2O_4$) Crystal Structure   A number of oxides have the $MgAl_2O_4$ or spinel structure which has the general formula AB_2O_4, where A is a metal ion with a $+2$ valence and B is a metal ion with a $+3$ valence. In the spinel structure the oxygen ions form an FCC lattice and the A and B ions occupy tetrahedral and octahedral interstitial sites, depending on the particular type of spinel. Compounds with the spinel structure are widely used for nonmetallic magnetic materials for electronic applications and will be studied in more detail in Chap. 11 on Magnetic Materials.

Graphite   Graphite is a polymorphic form of carbon and is not a compound of a metal and a nonmetal. However, since graphite is sometimes considered a ceramic

FIGURE 10.16 Perovskite (CaTiO₃) crystal structure. Calcium ions occupy FCC unit-cell corners, and oxygen ions occupy FCC unit-cell face-centered sites. The titanium ion occupies the octahedral interstitial site at the center of the cube. (*After W. D. Kingery, H. K. Bowen, and D. R. Uhlmann, "Introduction to Ceramics," 2d ed., Wiley, 1976.*)

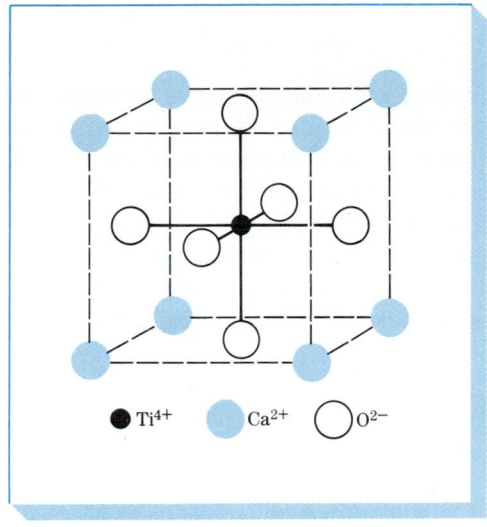

● Ti^{4+} ● Ca^{2+} ○ O^{2-}

material, its structure is included in this section. Graphite has a layered structure in which the carbon atoms in the layers are strongly covalently bonded in hexagonal arrays, as shown in Fig. 10.17. The layers are bonded together by weak secondary bonds and so can slide past each other easily. The ease of sliding of the layers gives graphite its lubricating properties.

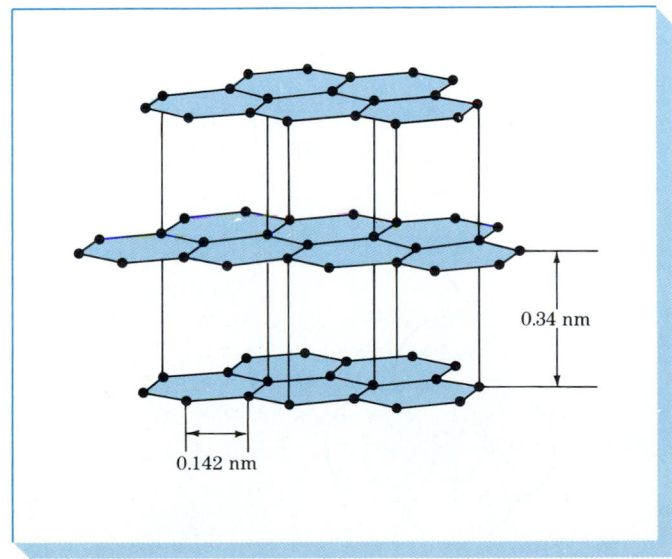

0.34 nm

0.142 nm

FIGURE 10.17 The structure of crystalline graphite. Carbon atoms form layers of strongly covalently bonded hexagonal arrays. There are weak secondary bonds between the layers.

10.3   SILICATE STRUCTURES

Many ceramic materials contain silicate structures which consist of silicon and oxygen atoms (ions) bonded together in various arrangements. Also a large number of naturally occurring minerals such as clays, feldspars, and micas are silicates since silicon and oxygen are the two most abundant elements found in the earth's crust. Many silicates are useful for engineering materials because of their low cost, availability, and special properties. Silicate structures are particularly important for the engineering construction materials of glass, portland cement, and brick. Many important electrical insulative materials also are made with silicates.

Basic Structural Unit of the Silicate Structures

The basic building block of the silicates is the silicate (SiO_4^{4-}) tetrahedron (Fig. 10.18). The Si—O bond in the SiO_4^{4-} structure is about 50 percent covalent and 50 percent ionic according to calculations from Pauling's equation [Eq. (2.10)]. The tetrahedral coordination of SiO_4^{4-} satisfies the directionality requirement of covalent bonding and the radius ratio requirement of ionic bonding. The radius ratio of the Si—O bond is 0.29, which is in the tetrahedral coordination range for stable-ion close packing. Because of the small, highly charged Si^{4+} ion, strong bonding forces are created within the SiO_4^{4-} tetrahedrons, and as a result the SiO_4^{4-} units are normally joined corner to corner and rarely edge to edge.

Island, Chain, and Ring Structures of Silicates

Since each oxygen of the silicate tetrahedron has one electron available for bonding, many different types of silicate structures can be produced. Island silicate structures are produced when positive ions bond with oxygens of the SiO_4^{4-} tetrahedra. For example, Fe^{2+} and Mg^{2+} ions combine with SiO_4^{4-} to form the olivine silicate island mineral which has the basic chemical formula $(Mg,Fe)_2SiO_4$.

If two corners of each SiO_4^{4-} tetrahedron are bonded with the corners of other tetrahedra, a chain (Fig. 10.19a) or ring structure with the unit chemical

FIGURE 10.18 The atom (ion) bonding arrangement of the SiO_4^{4-} tetrahedron. In this structure four oxygen atoms surround a central silicon atom. Each oxygen atom has one electron for bonding with another atom.

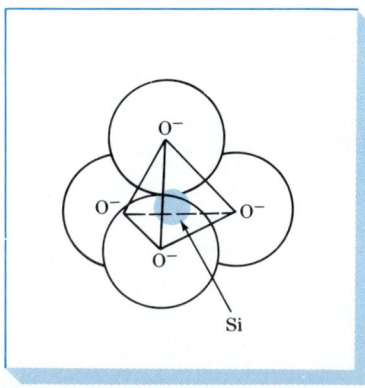

formula of SiO_3^{2-} results. The mineral enstatite ($MgSiO_3$) has a chain silicate structure, and the mineral beryl [$Be_3Al_2(SiO_3)_6$] has a ring silicate structure.

Sheet Structures of Silicates

Silicate sheet structures form when three corners in the same plane of a silicate tetrahedron are bonded to the corners of three other silicate tetrahedra, as shown in Fig. 10.19b. This structure has the unit chemical formula of $Si_2O_5^{2-}$. These silicate sheets are able to bond with other types of structural sheets because there is still one unbonded oxygen on each silicate tetrahedron (Fig. 10.19b). For example, the negatively charged silicate sheet can bond with a positively charged sheet of $Al_2(OH)_4^{2+}$ to form a composite sheet of kaolinite, as shown schematically in Fig. 10.20. The mineral kaolinite consists (in its pure form) of very small flat plates roughly hexagonal in shape, with their average size being about 0.7 μm in diameter and 0.05 μm thick (Fig. 10.21). The crystal plates are made of a series (up to about 50) of parallel sheets

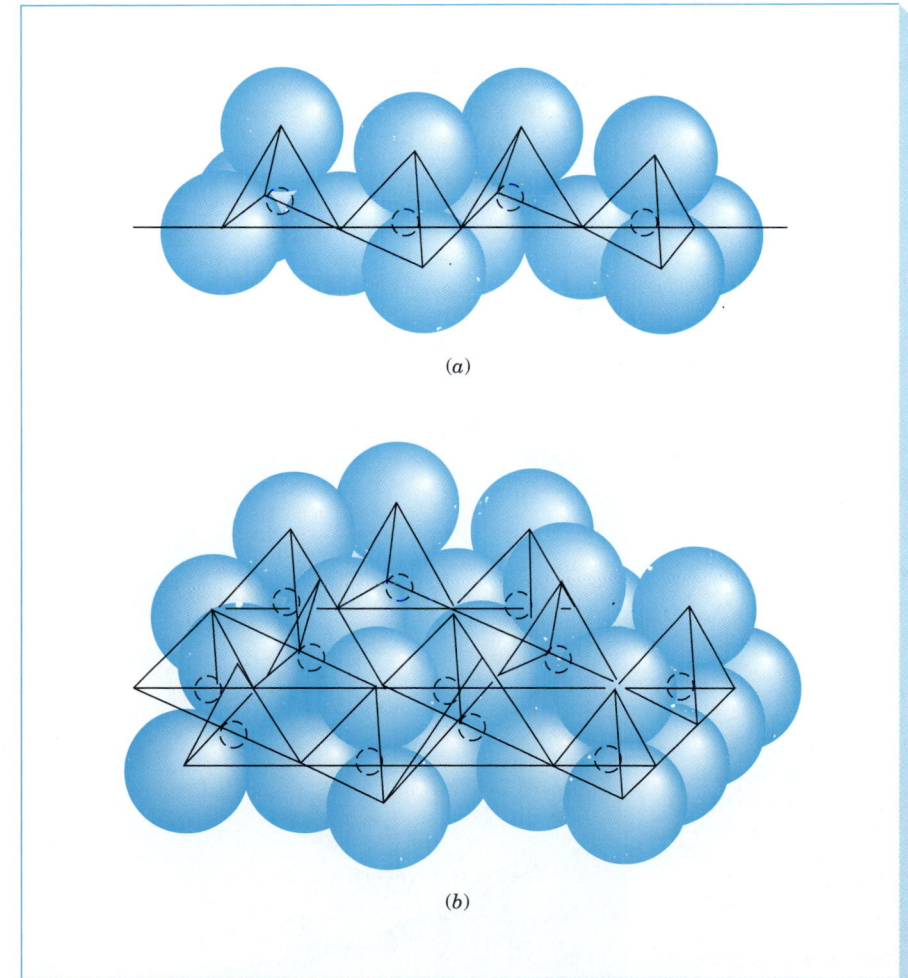

(a)

(b)

FIGURE 10.19 (a) Silicate chain structure. Two of the four oxygen atoms of the SiO_4^{4-} tetrahedra are bonded to other tetrahedra to form silicate chains. (b) Silicate sheet structure. Three of the four oxygen atoms of the SiO_4^{4-} tetrahedra are bonded to other tetrahedra to form silicate sheets. (*After L. H. Van Vlack, "Physical Ceramics for Engineers," Addison-Wesley, 1964, p. 44.*)

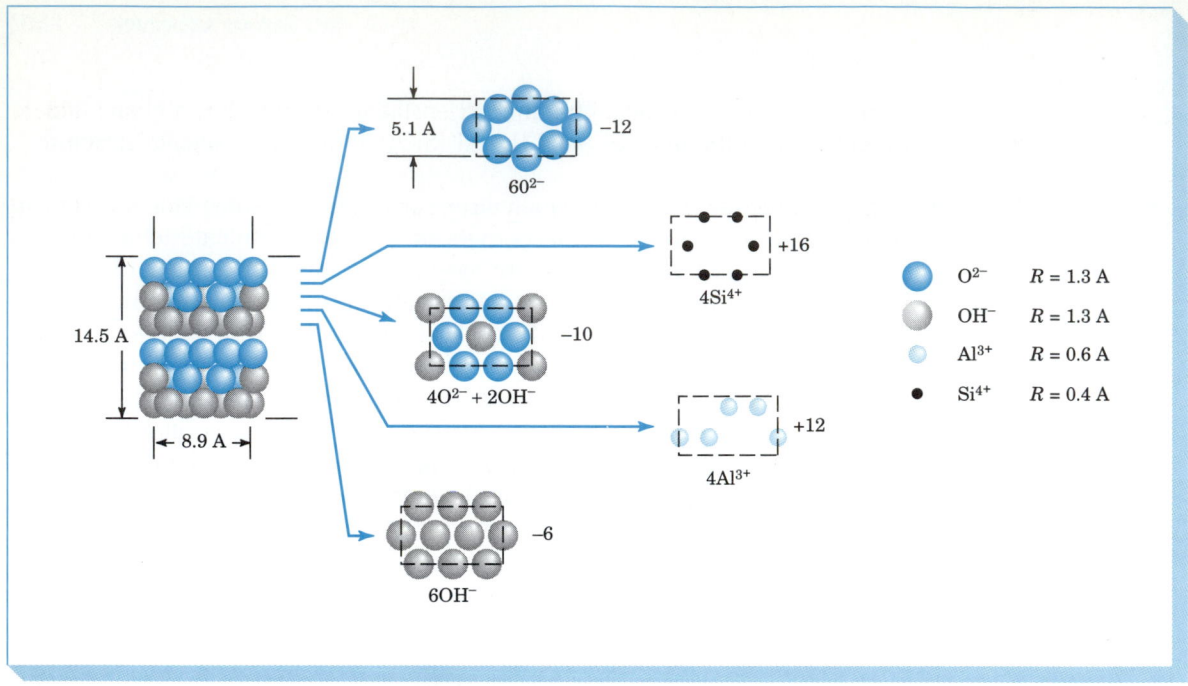

FIGURE 10.20 Ionic structure of kaolinite, $Al_2Si_2O_5(OH)_4$. This layered structure consists essentially of $Al_2(OH)_4^{2+}$ complex ions neutralized by $Si_2O_5^{2-}$ ions. All the primary ionic bonds are charge-balanced. (*After F. H. Norton, "Elements of Ceramics," Addison-Wesley, 1952, p. 9.*)

Within the figure:

5.1 A   −12 $6O^{2-}$

14.5 A   8.9 A

+16 $4Si^{4+}$

−10 $4O^{2-} + 2OH^-$

+12 $4Al^{3+}$

−6 $6OH^-$

O^{2-}	$R = 1.3$ A
OH^-	$R = 1.3$ A
Al^{3+}	$R = 0.6$ A
Si^{4+}	$R = 0.4$ A

bonded together by weak secondary bonds. Many high-grade clays consist mainly of the mineral kaolinite.

Another example of a sheet silicate is the mineral talc, in which a sheet

FIGURE 10.21 Kaolinite crystals as observed with the electron microscope (replica technique). (*After C. E. Hall as shown in F. H. Norton, "Elements of Ceramics," 2d ed., Addison-Wesley, 1974, p. 16.*)

of $Mg_3(OH)_2^{4+}$ bonds with two outer-layer $Si_2O_5^{2-}$ sheets (one on each side) to form a composite sheet with the unit chemical formula $Mg_3(OH)_2(Si_2O_5)_2$. The composite talc sheets are bonded together by weak secondary bonds, and thus this structural arrangement allows the talc sheets to slide over each other easily.

Silicate Networks   **Silica**   When all four corners of the SiO_4^{4-} tetrahedra share oxygen atoms, an SiO_2 network called *silica* is produced (Fig. 10.22). Crystalline silica exists in several polymorphic forms that correspond to different ways in which the silicate tetrahedra are arranged with all corners shared. There are three basic silica structures: *quartz, tridymite,* and *cristobalite,* and each of these has two or three modifications. The most stable forms of silica and the temperature ranges in which they exist at atmospheric pressure are low quartz below 573°C, high quartz between 573 to 867°C, high tridymite between 867 to 1470°C, and high cristobalite between 1470 and 1710°C (Fig. 10.22). Above 1710°C silica is liquid. Silica is an important component of many traditional ceramics and many different types of glasses.

Feldspars   There are many naturally occurring silicates which have infinite three-dimensional silicate networks. One of the industrially important network silicates are the feldspars which are also one of the main components of traditional ceramics. In the feldspar silicate structural network some Al^{3+} ions replace some Si^{4+} ions to form a network with a net negative charge. This negative charge is balanced with large ions of alkali and alkaline earth ions such as Na^+, K^+, Ca^{2+}, and Ba^{2+} which fit into interstitial positions. Table 10.3 summarizes the ideal compositions of some silicate minerals.

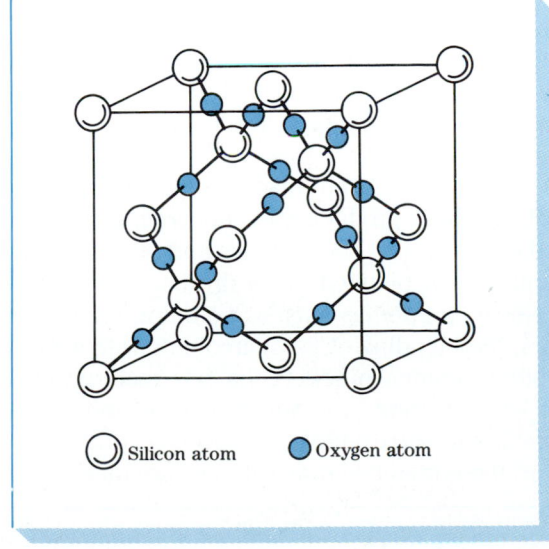

FIGURE 10.22 Structure of high cristobalite which is a form of silica (SiO_2). Note that each silicon atom is surrounded by four oxygen atoms and that each oxygen atom forms part of two SiO_4 tetrahedra. (*After W. D. Kingery, H. K. Bowen, and D. R. Uhlmann, "Introduction to Ceramics," 2d ed., Wiley, 1976.*)

◯ Silicon atom   ● Oxygen atom

TABLE 10.3   Ideal Silicate Mineral Compositions

Silica:	
Quartz	
Tridymite	Common crystalline phases of SiO_2
Cristobalite	
Alumina silicate:	
Kaolinite (china clay)	$Al_2O_3 \cdot 2SiO_2 \cdot 2H_2O$
Pyrophyllite	$Al_2O_3 \cdot 4SiO_2 \cdot H_2O$
Metakaolinite	$Al_2O_3 \cdot 2SiO_2$
Sillimanite	$Al_2O_3 \cdot SiO_2$
Mullite	$3Al_2O_3 \cdot 2SiO_2$
Alkali alumina silicate:	
Potash feldspar	$K_2O \cdot Al_2O_3 \cdot 6SiO_2$
Soda feldspar	$Na_2O \cdot Al_2O_3 \cdot 6SiO_2$
(Muscovite) mica	$K_2O \cdot 3Al_2O_3 \cdot 6SiO_2 \cdot 2H_2O$
Montmorillonite	$Na_2O \cdot 2MgO \cdot 5Al_2O_3 \cdot 24SiO_2 \cdot (6 + n)H_2O$
Leucite	$K_2O \cdot Al_2O_3 \cdot 4SiO_2$
Magnesium silicate:	
Cordierite	$2MgO \cdot 5SiO_2 \cdot 2Al_2O_3$
Steatite	$3MgO \cdot 4SiO_2$
Talc	$3MgO \cdot 4SiO_2 \cdot H_2O$
Chrysotile (asbestos)	$3MgO \cdot 2SiO_2 \cdot 2H_2O$
Forsterite	$2MgO \cdot SiO_2$

Source: O. H. Wyatt and D. Dew-Hughes, "Metals, Ceramics and Polymers," Cambridge, 1974.

10.4   PROCESSING OF CERAMICS

Most traditional and technical ceramic products are manufactured by compacting powders or particles into shapes which are subsequently heated to a high-enough temperature to bond the particles together. The basic steps in the processing of ceramics by the agglomeration of particles are (1) material preparation, (2) forming or casting, and (3) thermal treatment by drying (which is usually not required) and firing by heating the ceramic shape to a high-enough temperature to bond the particles together.

Materials Preparation

Most ceramic products are made by the agglomeration of particles.[1] The raw materials for these products vary, depending on the required properties of the finished ceramic part. The particles and other ingredients such as binders and lubricants may be blended wet or dry. For ceramic products which do not have very "critical" properties such as common bricks, sewer pipe, and other clay products, the blending of the ingredients with water is common practice. For some other ceramic products the raw materials are ground dry along with binders and other additives. Sometimes wet and dry processing of raw materials are combined. For example, to produce one type of high-alumina (Al_2O_3) insulator, the particulate raw materials are milled with water along with a wax

---

[1] The production of glass products and the casting of concrete are two major exceptions.

binder to form a slurry which is subsequently spray-dried to form small spherical pellets (Fig. 10.23).

Forming   Ceramic products made by agglomerating particles may be formed by a variety of methods in the dry, plastic, or liquid conditions. Cold-forming processes are predominant in the ceramic industry, but hot-forming processes are also used to some extent. Pressing, slip casting, and extrusion are commonly used ceramic forming methods.

Pressing   Ceramic particulate raw materials can be pressed in the dry, plastic, or wet condition into a die to form shaped products.

Dry pressing   This method is used commonly for products such as structural refractories (high-heat-resistant materials) and electronic ceramic components. Dry pressing may be defined as the simultaneous uniaxial compaction and shaping of a granular powder along with small amounts of water and/or organic binder in a die. Figure 10.24 shows a series of operations for the dry pressing of ceramic powders into a simple shape. After cold pressing, the parts are usually fired (sintered) to achieve the required strength and microstructural properties. Dry pressing is used extensively because it can form a wide variety of shapes rapidly with uniformity and close tolerances. For example, aluminas, titanates, and ferrites can be dry-pressed into sizes from a few mils to several inches in linear dimensions at a rate of up to about 5000 per minute.

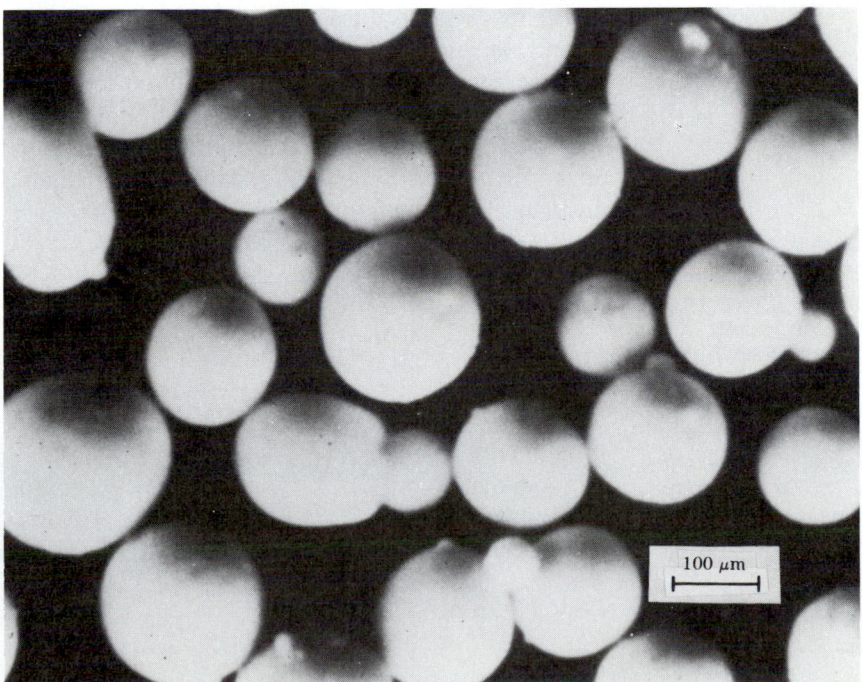

FIGURE 10.23 Spray-dried pellets of high-alumina ceramic body. [*After J. S. Owens et al., Am. Ceram. Soc. Bull.,* **56**:437(1977).]

100 μm

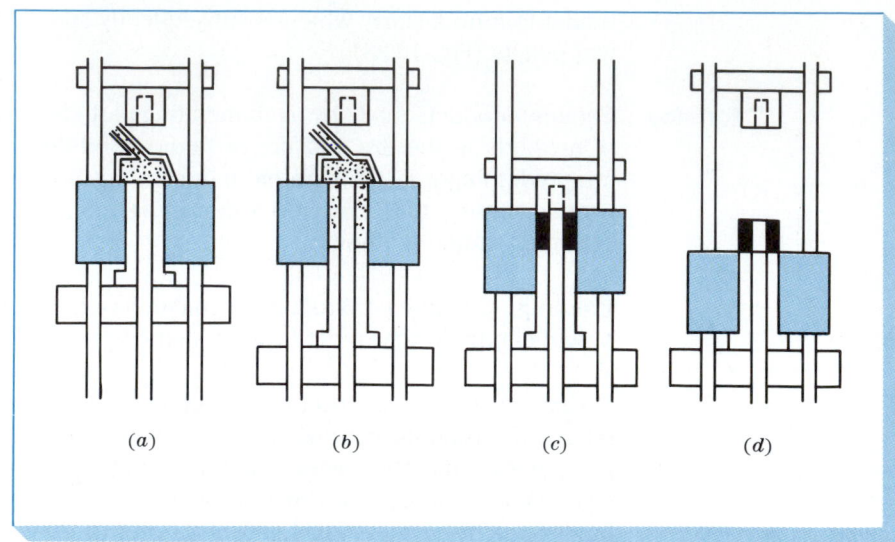

FIGURE 10.24 Dry pressing of ceramic particles: (*a*) and (*b*) filling, (*c*) pressing, and (*d*) ejection. (*After J. S. Reed and R. B. Runk, "Ceramic Fabrication Processes," vol. 9: "Treatise in Materials Science and Technology," Academic, 1976, p. 74.*)

(*a*)   (*b*)   (*c*)   (*d*)

Isostatic pressing   In this process the ceramic powder is loaded into a flexible (usually rubber) airtight container (called a *bag*) that is inside a chamber of hydraulic fluid to which pressure is applied. Figure 10.25 shows a cross section of a spark plug insulator in an isostatic pressing mold. The force of the applied pressure compacts the powder uniformly in all directions, with the final product taking the shape of the flexible container. After cold isostatic pressing the part must be fired (sintered) to achieve the required properties and microstructure. Ceramic parts manufactured by isostatic pressing include refractories, bricks and shapes, spark plug insulators, radomes, carbide tools, crucibles, and bearings. Figure 10.26 shows the stages for the manufacturing of a spark plug insulator by isostatic pressing.

FIGURE 10.25 Cross section of insulator blank in isostatic pressing mold. Spray-dried nearly spherical pellets (Fig. 10.23) are fed by gravity into top of mold and compressed by isostatic pressure, normally in the 3000 to 6000 psi range. The hydraulic fluid enters the side of the mold through the holes shown in the cross section. (*Courtesy of Champion Spark-Plug Co.*)

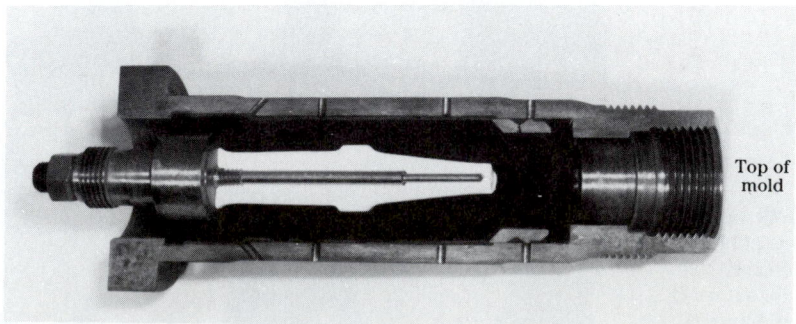

Top of mold

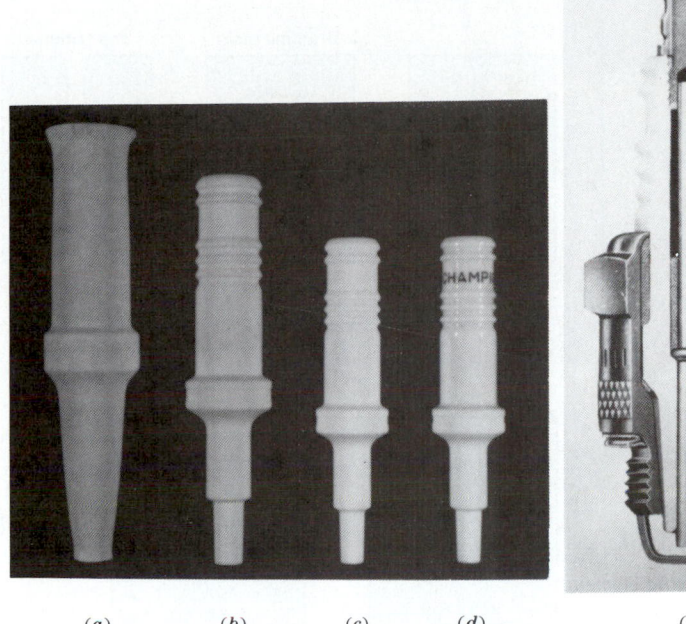
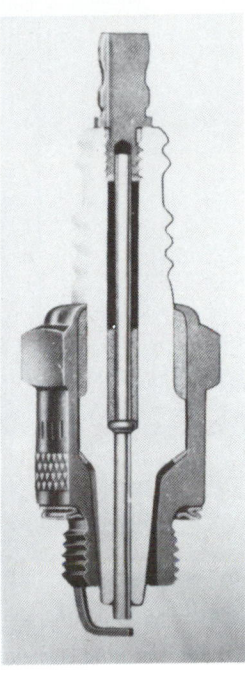

FIGURE 10.26 Stages of spark plug insulator manufacture by the isostatic processing method. (*a*) Pressed blank. (*b*) Turned (ground) insulator. (*c*) Fired insulator. (*d*) Glazed and decorated finished insulator. (*e*) Cross section of assembled automotive spark plug showing position of insulator. (*Courtesy of Champion Spark Plug Co.*)

(*a*)   (*b*)   (*c*)   (*d*)   (*e*)

Hot pressing   In this process ceramic parts of high density and improved mechanical properties are produced by combining the pressing and firing operations. Both uniaxial and isostatic methods are used.

Slip casting   Ceramic shapes can be cast by using a unique process called *slip casting*, illustrated in Fig. 10.27. The main steps in slip casting are:

1. Preparation of a powdered ceramic material and a liquid (usually clay and water) into a stable suspension called a *slip*.
2. Pouring the slip into a porous mold which is usually made of plaster of paris and allowing the liquid portion of the slip to be partially absorbed by the mold. As the liquid is removed from the slip, a layer of semihard material is formed against the mold surface.
3. When a sufficient wall thickness has been formed, the casting process is interrupted and the excess slip is poured out of the cavity (Fig. 10.27*a*). This is known as *drain casting*. Alternatively, a solid shape may be made by allowing the casting to continue until the whole mold cavity is filled, as illustrated in Fig. 10.27*b*. This type of slip casting is called *solid casting*.

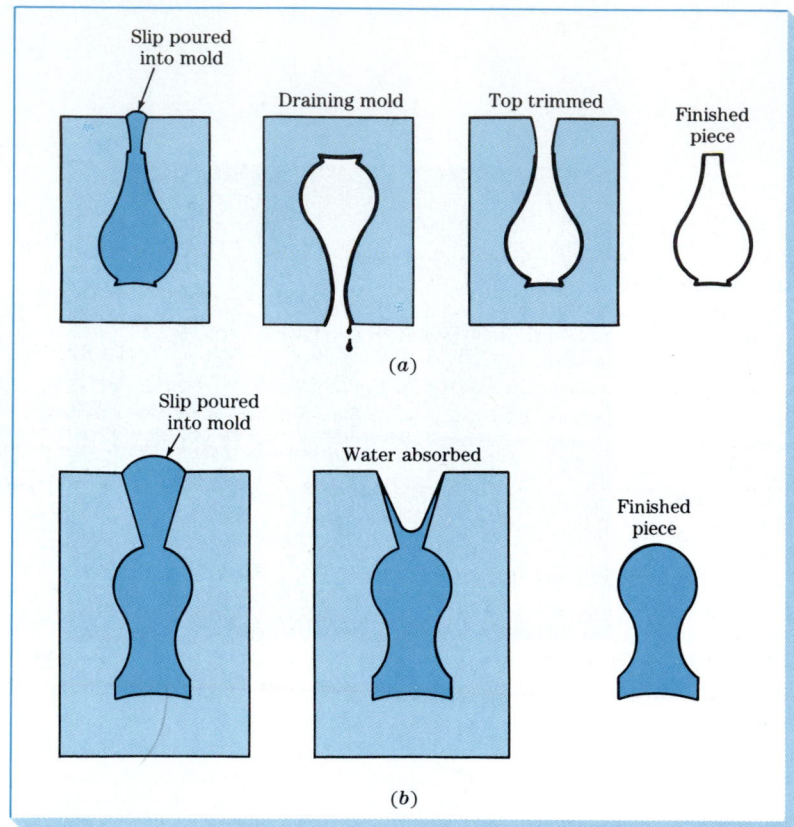

Slip poured into mold

Draining mold

Top trimmed

Finished piece

(a)

Slip poured into mold

Water absorbed

Finished piece

(b)

FIGURE 10.27 Slip casting of ceramic shapes. (*a*) Drain casting in porous plaster of paris mold. (*b*) Solid casting. (*After W. D. Kingery, "Introduction to Ceramics," Wiley, 1960, p. 52.*)

4. The material in the mold is allowed to dry to provide adequate strength for handling and the subsequent removal of the part from the mold.
5. Finally the cast part is fired to attain the required microstructure and properties.

Slip casting is advantageous for forming thin-walled and complex shapes of uniform thickness. Slip casting is especially economical for development parts and short production runs. Several new variations of the slip-casting process are pressure and vacuum casting in which the slip is shaped under pressure or vacuum.

Extrusion Single cross sections and hollow shapes of ceramic materials can be produced by extruding these materials in the plastic state through a forming die. This method is commonly used to produce, for example, refractory brick, sewer pipe, hollow tile, technical ceramics, and electrical insulators. The means most commonly used is the vacuum-auger-type extrusion machine in which the plastic ceramic material (e.g., clay and water) is forced through a hard

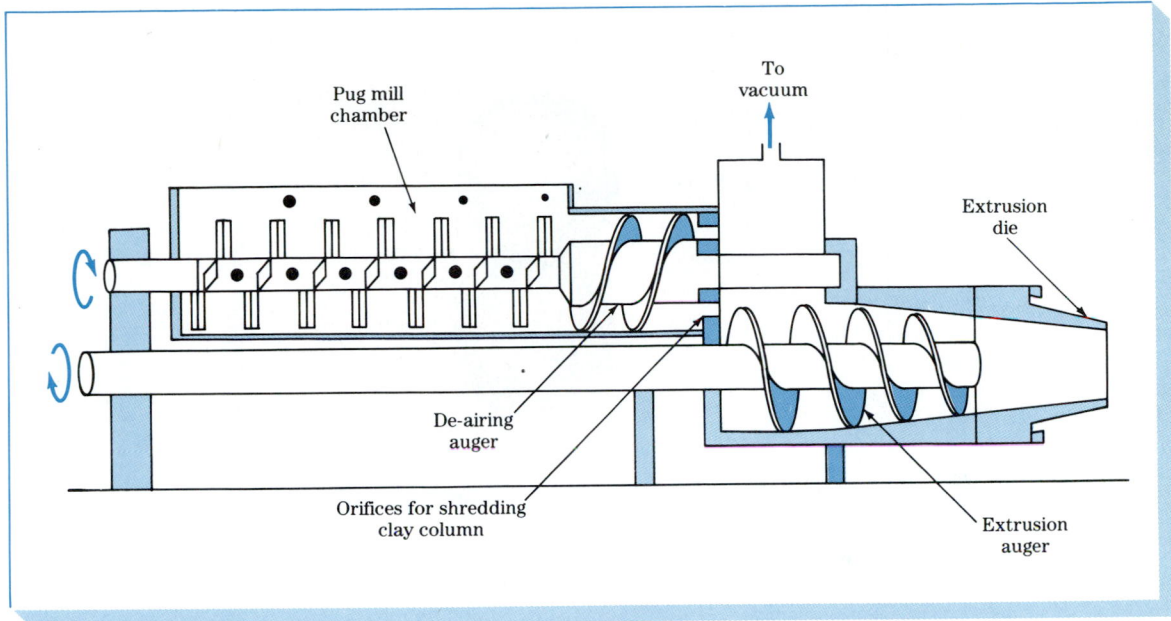

FIGURE 10.28 Cross section of combination mixing mill (pug mill) for ceramic materials and vacuum-auger extrusion machine. (*After W. D. Kingery, "Introduction to Ceramics," Wiley, 1960.*)

steel or alloy die by a motor-driven auger (Fig. 10.28). Special technical ceramics are frequently produced by using a piston extrusion under high pressure so that close tolerances can be attained.

Thermal Treatments   Thermal treatment is an essential step in the manufacturing of most ceramic products. In this subsection we shall consider the following thermal treatments: drying, sintering, and vitrification.

Drying and binder removal   The purpose of drying ceramics is to remove water from the plastic ceramic body before it is fired at higher temperatures. Generally, drying to remove water is carried out at or below 100°C and can take as long as 24 h for a large ceramic part. The bulk of organic binders can be removed from ceramic parts by heating in the range of 200 to 300°C, although some hydrocarbon residues may require heating to much higher temperatures.

Sintering   The process by which small particles of a material are bonded together by solid-state diffusion is called *sintering*. In ceramic manufacturing this thermal treatment results in the transformation of a porous compact into a dense, coherent product. Sintering is commonly used to produce ceramic shapes made of, for example, alumina, beryllia, ferrites, and titanates.

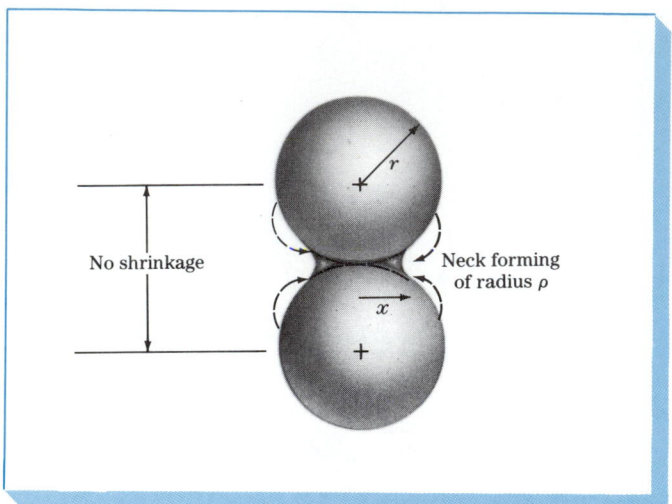

FIGURE 10.29 Formation of a neck during the sintering of two fine particles. Atomic diffusion takes place at the contacting surfaces and enlarges the contact area to form a neck. (*After J. H. Brophy, R. M. Rose, and J. Wulff, "The Structure and Properties of Materials," vol. II: "Thermodynamics of Structure," Wiley, 1964, p. 139.*)

In the sintering process particles are coalesced by solid-state diffusion at very high temperatures but below the melting point of the compound being sintered. For example, the alumina spark plug insulator shown in Fig. 10.26*a* is sintered at 1600°C (the melting point of alumina is 2050°C). In sintering, atomic diffusion takes place between the contacting surfaces of the particles so that they become chemically bonded together (Fig. 10.29). As the process proceeds, larger particles are formed at the expense of the smaller ones, as illustrated in the sintering of MgO compacts shown in Fig. 10.30*a*, *b*, and *c*.

FIGURE 10.30 Scanning electron micrographs of fractured surfaces of MgO compacts (compressed powders) sintered at 1430°C in static air for (*a*) 30 min (fractional porosity = 0.39); (*b*) 303 min (f.p. = 0.14); (*c*) 1110 min (f.p. = 0.09); as-annealed surface of (*c*) is shown in (*d*). [*After B. Wong and J. A. Pask, J. Am. Ceram. Soc., 62:141(1979).*]

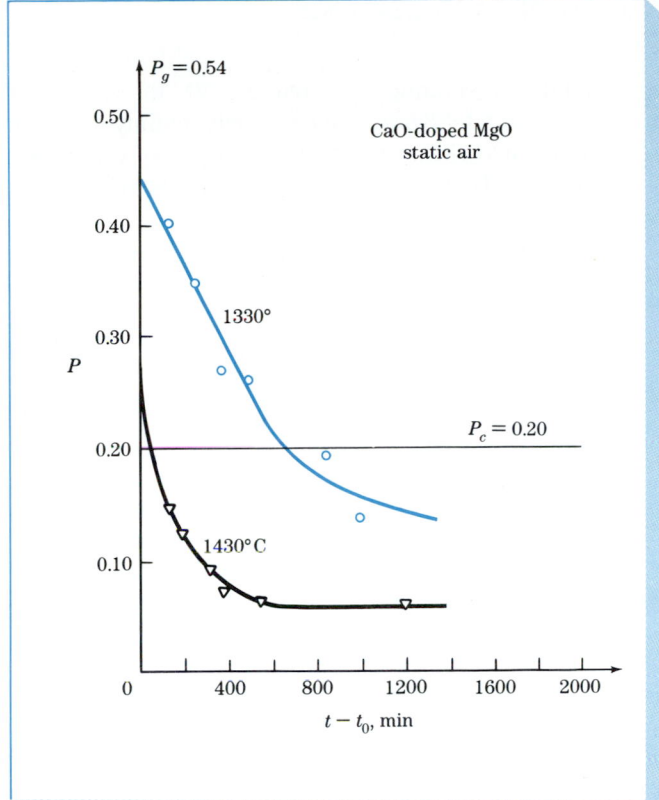

FIGURE 10.31 Porosity vs. time for MgO compacts doped with 0.2 wt % CaO and sintered in static air at 1330 and 1430°C. Note that the higher sintering temperature produces a more rapid decrease in porosity and a lower porosity level. [*After B. Wong and J. A. Pask, J. Am. Ceram. Soc., 62:141(1979).*]

As the particles get larger with the time of sintering, the porosity of the compacts decreases (Fig. 10.31). Finally, at the end of the process, an "equilibrium grain size" is attained (Fig. 10.30*d*). The driving force for the process is the lowering of the energy of the system. The high surface energy associated with the original individual small particles is replaced by the lower overall energy of the grain-boundary surfaces of the sintered product.

Vitrification   Some ceramic products such as porcelain, structural clay products, and some electronic components contain a glass phase. This glass phase serves as a reaction medium by which diffusion can take place at a lower temperature than in the rest of the ceramic solid material. During the firing of these types of ceramic materials, a process called *vitrification* takes place whereby the glass phase liquefies and fills the pore spaces in the material. This liquid glass phase may also react with some of the remaining solid refractory material. Upon cooling, the liquid phase solidifies to form a vitreous or glassy matrix that bonds the unmelted particles together.

10.5   TRADITIONAL AND ENGINEERING CERAMICS

Traditional Ceramics

Traditional ceramics are made from three basic components: *clay, silica* (flint), and *feldspar*. Clay consists mainly of hydrated aluminum silicates ($Al_2O_3 \cdot SiO_2 \cdot H_2O$) with small amounts of other oxides such as TiO_2, Fe_2O_3, MgO, CaO, Na_2O, and K_2O. Table 10.4 lists the chemical compositions of several industrial clays.

The clay in traditional ceramics provides workability of the material before firing hardens it and constitutes the major body material. The silica (SiO_2), also called *flint* or quartz, has a high melting temperature and is the refractory component of traditional ceramics. Potash (potassium) feldspar, which has the basic composition $K_2O \cdot Al_2O_3 \cdot 6SiO_2$, has a low melting temperature and makes a glass when the ceramic mix is fired and bonds the refractory components together.

TABLE 10.4   Chemical Compositions of Some Clays

Type of clay	Weight percentages of major oxides									Ignition loss
	Al_2O_3	SiO_2	Fe_2O_3	TiO_2	CaO	MgO	Na_2O	K_2O	H_2O	
Kaolin	37.4	45.5	1.68	1.30	0.004	0.03	0.011	0.005	13.9	
Tenn. ball clay	30.9	54.0	0.74	1.50	0.14	0.20	0.45	0.72	· · ·	11.4
Ky. ball clay	32.0	51.7	0.90	1.52	0.21	0.19	0.38	0.89	· · ·	12.3

Source: P. W. Lee, "Ceramics," Reinhold, 1961.

TABLE 10.5   Some Triaxial Whiteware Chemical Compositions

Type body	China clay	Ball clay	Feld-spar	Flint	Other
Hard porcelain	40	10	25	25	
Electrical insulation ware	27	14	26	33	
Vitreous sanitary ware	30	20	34	18	
Electrical insulation	23	25	34	18	
Vitreous tile	26	30	32	12	
Semivitreous whiteware	23	30	25	21	
Bone china	25	· · ·	15	22	38 bone ash
Hotel china	31	10	22	35	2 CaCO_3
Dental porcelain	5	· · ·	95		

Source: W. D. Kingery, H. K. Bowen, and D. R. Uhlmann, "Introduction to Ceramics," 2d ed., Wiley, 1976, p. 532.

Structural clay products such as building brick, sewer pipe, drain tile, roofing tile, and floor tile are made of natural clay which contains all three basic components. Whiteware products such as electrical porcelain, dinner china, and sanitary ware are made from components of clay, silica, and feldspar for which the composition is controlled. Table 10.5 lists the chemical compositions of some triaxial whitewares. The term *triaxial* is used since there are three major components in their compositions.

Typical composition ranges for different types of whitewares are illustrated in the silica-leucite-mullite ternary phase diagram of Fig. 10.32 (see Sec. 8.11 on Ternary Phase Diagrams for interpreting these diagrams). The composition ranges of some whitewares are indicated by the circled areas.

FIGURE 10.32 Areas of triaxial whiteware compositions shown on the silica-leucite-mullite phase-equilibrium diagram. (*After W. D. Kingery, H. K. Bowen, and D. R. Uhlmann, "Introduction to Ceramics," 2d ed., Wiley, 1976, p. 533.*)

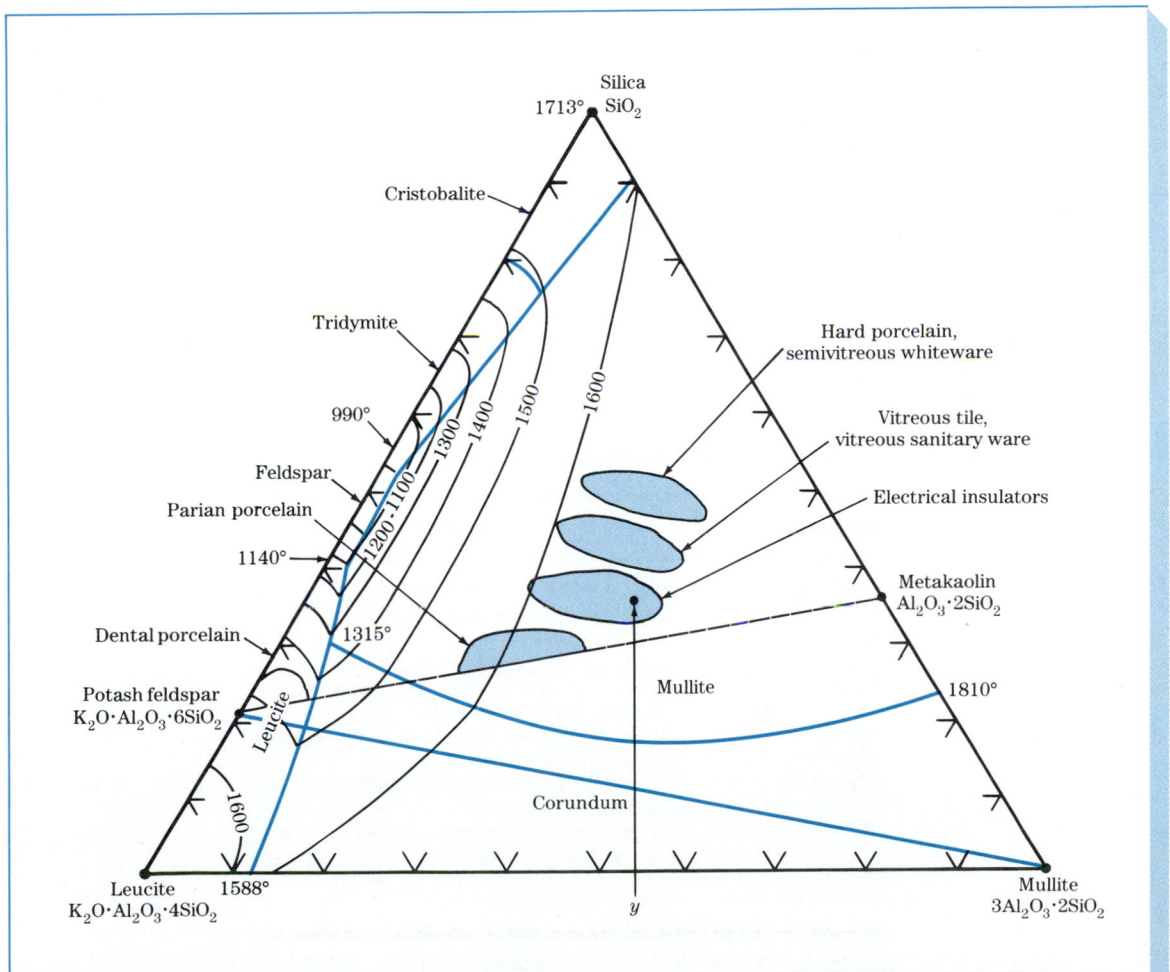

The changes that occur in the structure of triaxial bodies during firing are not completely understood due to their complexity. Table 10.6 is an approximate summary of what probably occurs during the firing of a whiteware body.

Figure 10.33 is an electron micrograph of the microstructure of an electrical insulator porcelain. As observed in this micrograph, the structure is very heterogeneous. Large quartz grains are surrounded by a solution rim of high-silica glass. Mullite needles which cross feldspar relicts and fine mullite-glass mixtures are present.

Triaxial porcelains are satisfactory as insulators for 60-cycle use, but at high frequencies dielectric losses become too high. The considerable amounts of alkalies derived from the feldspar used as a flux increase the electrical conductivity and dielectric losses of triaxial porcelains.

Engineering Ceramics

In contrast to the traditional ceramics which are mainly based on clay, engineering or technical ceramics are mainly pure compounds or nearly pure compounds of chiefly oxides, carbides, or nitrides. Some of the important engineering ceramics are alumina (Al_2O_3), silicon nitride (Si_3N_4), silicon carbide (SiC), and zirconia (ZrO_2) combined with some other refractory oxides. The melting temperatures of some of the engineering ceramics are listed in Table 10.1, and the mechanical properties of some of these materials are given in Table 10.9. A brief description of a few of the properties, processes, and applications of some of the important engineering ceramics follows.

Alumina (Al_2O_3)   Alumina was originally developed for refractory tubing and high-purity crucibles for high-temperature use and now has wide application.

TABLE 10.6   Life History of a Triaxial Body

Temperature, °C	Reactions
Up to 100	Loss of moisture
100–200	Removal of adsorbed water
450	Dehydroxylation
500	Oxidation of organic matter
573	Quartz inversion to high form. Little overall volume damage
980	Spinel forms from clay. Start of shrinkage
1000	Primary mullite forms.
1050–1100	Glass forms from feldspar, mullite grows, shrinkage continues.
1200	More glass, mullite grows, pores closing, some quartz solution
1250	60% glass, 21% mullite, 19% quartz, pores at minimum

Source: F. Norton, "Elements of Ceramics," 2d ed., Addison-Wesley, 1974, p. 140.

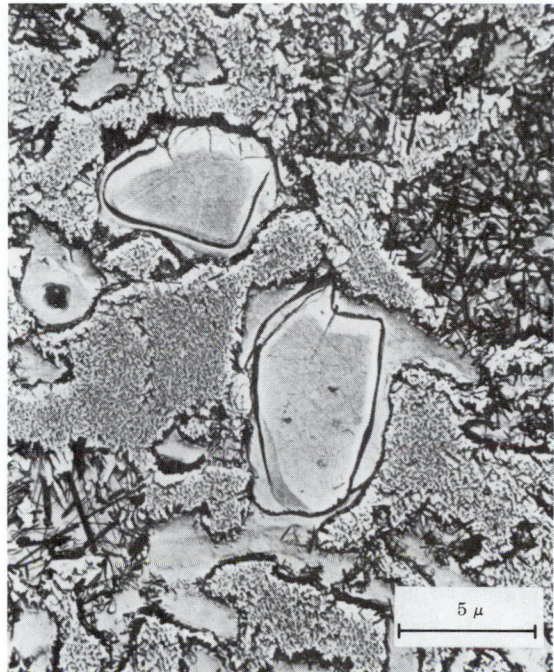

FIGURE 10.33 Electron micrograph of an electrical insulator porcelain (etched 10 s, 0°C, 40% HF, silica replica). (*After S. T. Lundin as shown in W. D. Kingery, H. K. Bowen, and D. R. Uhlmann, "Introduction to Ceramics," 2d ed., Wiley, 1976, p. 539.*)

A classic example of the application of alumina is in spark plug insulator material (Fig. 10.26). Aluminum oxide is commonly doped with magnesium oxide, cold-pressed, and sintered, producing the type of microstructure shown in Fig. 10.34. Note the uniformity of the alumina grain structure as compared to the microstructure of the electrical porcelain of Fig. 10.33. Alumina is used commonly for high-quality electrical applications where low dielectric loss and high resistivity are needed.

Silicon nitride (Si₃N₄) Silicon nitride–based ceramic materials have a useful combination of mechanical engineering properties which include reasonable strength, fracture toughness, and refractoriness. Si_3N_4 dissociates significantly at temperatures above 1800°C and so cannot be directly sintered. However, over the past years other methods of manufacturing Si_3N_4 have been developed.

Four major processing methods have been developed for silicon nitride: reaction-bonded silicon nitride (RBSN), hot-pressed silicon nitride (HPSN), sintered silicon nitride (SSN), and hot-isostatic-pressed silicon nitride (HIP-SN). In reaction-bonded silicon nitride a compact of silicon powder is nitrided in a flow of nitrogen gas. This leaves a microporous material with moderate strength (Table 10.7A).

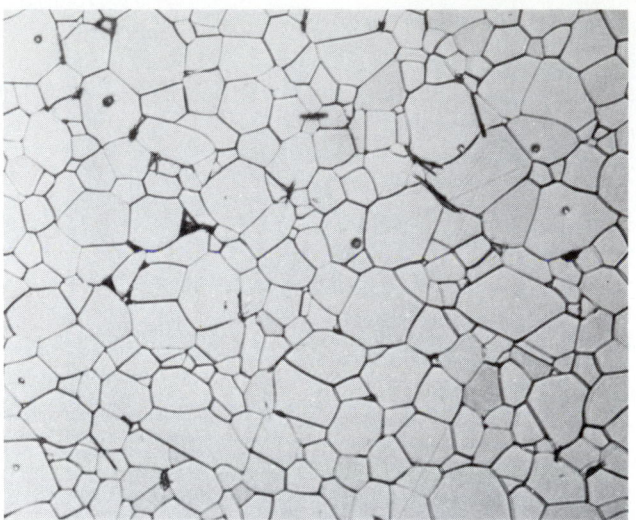

To promote sintering, varying amounts of oxide additives are used. For example, hot-pressed silicon nitride can be produced using a 1 to 5% MgO addition. Other hot-pressed or hot-isostatic-pressed silicon nitride ceramics can be made with varying MgO and/or Y_2O_3 additions. These additive react with the Si_3N_4 starting powder and its SiO_2 surface layer to create a grain boundary oxynitride whose characteristics control the bonding process at high temperatures.

Another silicon nitride–based ceramic is SiAlON, a solid-solution alloy of silicon nitride and aluminum oxide. Representative properties of some of these silicon nitride–based ceramics are listed in Table 10.7A. Applications for Si_3N_4–based ceramics include cutting tool materials, antifriction roller and ball bear-

TABLE 10.7A   **Selected Properties of Some Silicon Nitride Ceramics***

	RBSN	SSN	SiAlON	HIP-SN
Density, Mg/m^3	2.7	3.3	3.3	3.3
Youngs modulus, GPa	200	300	300	300
Strength at RT, MPa	300†	900†	800†	900†
K_{IC}, $MPa\sqrt{m}$	2	6	6.5	5.5–6.5
Wear in rolling contact fatigue	N/A	Low	Moderate	Extremely low
Dielectric constant at 10^{10} Hz	6–7	6.5	...	6.5–7.5
$/net shape part	Lowest	Medium	Medium	Highest

*RBSN = reaction-bonded silicon nitride; SSN = sintered silicon nitride; HIP-SN = hot-isostatic-pressed silicon nitride.
†Four-point bend.
‡Direct tension.

ings, and components for spark-ignited and diesel engines. For example, an Si_3N_4 ceramic turbocharger has been used for many years in one type of automobile spark-ignited engine. Si_3N_4 is also being explored for use in advanced engine applications (Fig. 1.8).

Silicon Carbide (SiC)   Silicon carbide–type high-performance ceramics have the important properties of high hardness, chemical inertness, abrasive resistance, and outstanding resistance to oxidation at high temperatures. However, SiC is relatively brittle with low fracture toughness and is difficult to produce as a fine-grained dense ceramic part.

To make reaction-sintered (bonded) silicon carbide, a powder compact of SiC and carbon (graphite) is infiltrated with molten silicon which reacts with the carbon to form more SiC, bonding the original SiC grains together. The process cannot be continued until all the pores among the original grains are filled, since the silicon must be able to be infiltrated. Thus, the final product contains about 8 to 20% free silicon.

SiC can be sintered by using alpha SiC powder of suitable quality without the application of external pressure if selected sintering aids are added. Boron and aluminum compounds are commonly used for the pressureless sintering of SiC. It has been speculated that boron decreases the grain-boundary energy and hence facilitates the high-temperature densification of SiC by solid-state diffusion.

The flexural strength of pressureless sintered SiC is about 460 MPa (67 ksi), and its fracture toughness varies from about 3 to 5 MPa$\sqrt{m}$ (2.7 to 4.5 ksi$\sqrt{in}$). Most common applications for SiC ceramics utilize their high hardness, chemical resistance, and abrasion resistance. Applications in the chemical process industries include seals and valves, lens molds, rocket nozzles, wear plates for spray drying, and wire dies. Other applications that utilize wear and erosion resistance are thrust bearings, ball bearings, pump impellers, and extrusion dies. Because of the heat- and creep-resistant characteristics of SiC, it is used for rocket nozzle throats, heat exchanger tubes, and diffusion furnace components.

Much recent research on SiC ceramics is focused on ceramic-matrix composites involving the incorporation of second phases of particulates, whiskers, and fibers. These ceramic-matrix silicon carbide composites have higher strengths and toughness and are less flaw-sensitive than silicon carbide alone.

Zirconia (ZrO_2)   Pure zirconia is polymorphic and transforms from the tetragonal to monoclinic structure at about 1170°C with an accompanying volume expansion and so is subject to cracking. However, by combining ZrO_2 with other refractory oxides such as CaO, MgO, and Y_2O_3, the cubic structure can be stabilized at room temperature and has found some applications. By combining ZrO_2 with 9% MgO and using special heat treatments, a partially stabilized zirconia (PSZ) can be produced that has especially high fracture toughness which has led to new ceramic applications. (See Sec. 10.7 on the fracture toughness of ceramics for more details.)

10.6 ELECTRICAL PROPERTIES OF CERAMICS

Ceramic materials are used for many electrical and electronic applications. Many types of ceramics are used for electrical insulators for low- and high-voltage electric currents. Ceramic materials also find application in various types of capacitors, especially where miniaturization is required. Other types of ceramics called piezoelectrics can convert weak pressure signals into electrical signals, and vice versa.

Before discussing the electrical properties of the various types of ceramic materials, let us first briefly look at some of the basic properties of insulators, or *dielectrics* as they are sometimes called.

Basic Properties of Dielectrics

There are three important properties common to all insulators or dielectrics: (1) the *dielectric constant,* (2) the *dielectric breakdown* strength, and (3) the loss factor.

Dielectric constant   Consider a simple parallel-plate capacitor[1] with metal plates of area A separated by distance d, as shown in Fig. 10.35. Consider first the case in which the space between the plates is a vacuum. If a voltage V is applied across the plates, one plate will acquire a net charge of $+q$ and the other a net charge of $-q$. The charge q is found to be directly proportional to the applied voltage V as

$$q = CV \quad \text{or} \quad C = \frac{q}{V} \tag{10.1}$$

where C is a proportionality constant called the *capacitance* of the capacitor. The SI unit of capacitance is coulombs per volt (C/V), or the *farad* (F). Thus

---

[1] A capacitor is a device which stores electric energy.

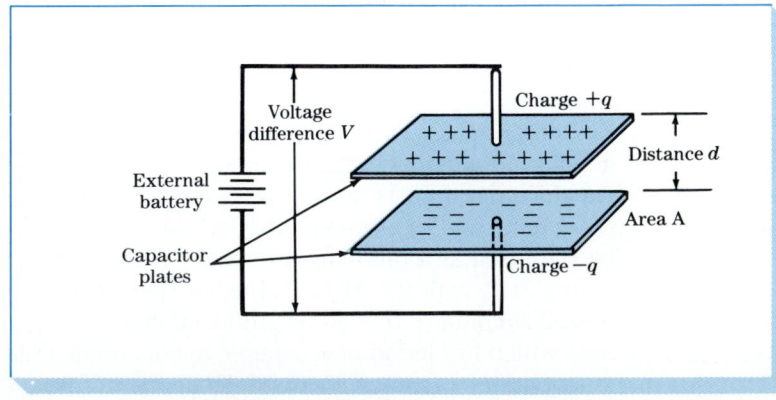

FIGURE 10.35 Simple parallel-plate capacitor.

$$1 \text{ farad} = \frac{1 \text{ coulomb}}{\text{volt}}$$

Since the farad is a much larger unit of capacitance than is normally encountered in electrical circuitry, the commonly used units are the *picofarad* ($1 \text{ pF} = 10^{-12} \text{ F}$) and the *microfarad* ($1 \text{ } \mu\text{F} = 10^{-6} \text{ F}$).

The capacitance of a capacitor is a measure of its ability to store electric charge. The more charge stored at the upper and lower plates of a capacitor, the higher is its capacitance.

The capacitance C for a parallel-plate capacitor whose area dimensions are much greater than the separation distance of the plates is given by

$$C = \epsilon_0 \frac{A}{d} \tag{10.2}$$

where ϵ_0 = permittivity of free space = 8.854×10^{-12} F/m.

When a dielectric (electrical insulator) fills the space between the plates (Fig. 10.36), the capacitance of the capacitor is increased by a factor κ, which is called the *dielectric constant* of the dielectric material. For a parallel-plate capacitor with a dielectric between the capacitor plates,

$$C = \frac{\kappa \epsilon_0 A}{d} \tag{10.3}$$

Table 10.7 lists the dielectric constants for some ceramic insulator materials.

The energy stored in a capacitor of a given volume at a given voltage is increased by the factor of the dielectric constant when the dielectric material is present. By using a material with a very high dielectric constant, very small capacitors with high capacitances can be produced.

Dielectric strength   Another property besides the dielectric constant that is important in evaluating dielectrics is the *dielectric strength*. This quantity is a

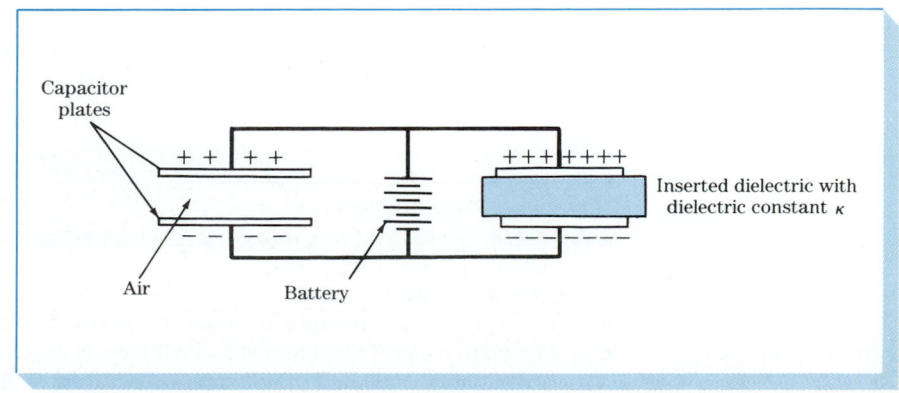

FIGURE 10.36 Two parallel-plate capacitors under the same applied voltage. The capacitor on the right has a dielectric (insulator inserted between the plates), and as a result the charge on the plates is increased by a factor of κ above that on the plates of the capacitor without the dielectric.

Capacitor plates

Air

Battery

Inserted dielectric with dielectric constant κ

TABLE 10.7   Electrical Properties of Some Ceramic Insulator Materials

Material	Volume resistivity, $\Omega \cdot m$	Dielectric strength		Dielectric constant κ		Loss factor	
		V/mil	kV/mm	60 Hz	10^6 Hz	60 Hz	10^6 Hz
Electrical porcelain insulators	10^{11}–10^{13}	55–300	2–12	6	. . .	0.06	
Steatite insulators	>10^{12}	145–280	6–11	6	6	0.008–0.090	0.007–0.025
Fosterite insulators	>10^{12}	250	9.8	. . .	6	. . .	0.001–0.002
Alumina insulators	>10^{12}	250	9.8	. . .	9	. . .	0.0008–0.009
Soda-lime glass	. . .	. . .	. . .	. . .	7.2	. . .	0.009
Fused silica	. . .	8	. . .	. . .	3.8	. . .	0.00004

Source: Materials Selector, *Mater. Eng.,* December 1982.

measure of the ability of the material to hold energy at high voltages. Dielectric strength is defined as the voltage per unit length (electric field or voltage gradient) at which failure occurs and thus is the maximum electric field that the dielectric can maintain without electrical breakdown.

Dielectric strength is most commonly measured in volts per mil (1 mil = 0.001 in) or kilovolts per millimeter. If the dielectric is subjected to a voltage gradient that is too intense, the strain of the electrons or ions in trying to pass through the dielectric may exceed its dielectric strength. If the dielectric strength is exceeded, the dielectric material begins to break down and the passage of current (electrons) occurs. Table 10.7 lists the dielectric strengths for some ceramic insulator materials.

Dielectric loss factor   If the voltage used to maintain the charge on a capacitor is sinusoidal, as is generated by an alternating current, the current leads the voltage by 90° when a loss-free dielectric is between the plates of a capacitor. However, when a real dielectric is used in the capacitor, the current leads the voltage by 90° − δ, where the angle δ is called the *dielectric loss angle*. The product of $\kappa \tan \delta$ is designated the *loss factor* and is a measure of the electric energy lost (as heat energy) by a capacitor in an ac circuit. Table 10.7 lists the loss factors for some ceramic insulator materials.

Example Problem 10.9

A simple parallel-plate capacitor is to be made to store 5.0×10^{-6} C at a potential of 8000 V. The separation distance between the plates is to be 0.30 mm. Calculate the area (in square meters) that the plates must have if the dielectric between the plates is (*a*) a vacuum ($\kappa = 1$) and (*b*) alumina ($\kappa = 9$). ($\epsilon_0 = 8.85 \times 10^{-12}$ F/m.)

Solution:

$$C = \frac{q}{V} = \frac{5.0 \times 10^{-6} \text{ C}}{8000 \text{ V}} = 6.25 \times 10^{-10} \text{ F}$$

$$A = \frac{Cd}{\epsilon_0 \kappa} = \frac{(6.25 \times 10^{-10} \text{ F})(0.30 \times 10^{-3} \text{ m})}{(8.85 \times 10^{-12} \text{ F/m})(\kappa)}$$

(a) For vacuum, $\kappa = 1$: $A = 0.021 \text{ m}^2$
(b) For alumina, $\kappa = 9$: $A = 2.35 \times 10^{-3} \text{ m}^2$

As can be seen by these calculations, the insertion of a material with a high dielectric constant can appreciably reduce the area of the plates required.

Ceramic Insulator Materials

Ceramic materials have electrical and mechanical properties which make them especially suitable for many insulator applications in the electrical and electronic industries. The ionic and covalent bonding in ceramic materials restricts electron and ion mobility and thus makes these materials good electrical insulators. These bondings make most ceramic materials strong but relatively brittle. The chemical compositions and microstructure of electrical- and electronic-grade ceramics must be more closely controlled than for structural ceramics such as bricks or tiles. Some aspects of the structure and properties of several insulator ceramic materials will now be discussed.

Electrical porcelain Typical electrical porcelain consists of approximately 50% clay ($Al_2O_3 \cdot 2SiO_2 \cdot 2H_2O$), 25% silica ($SiO_2$), and 25% feldspar ($K_2O \cdot Al_2O_3 \cdot 6SiO_2$). This composition makes a material that has good green-body plasticity and a wide firing temperature range at a relatively low cost. The major disadvantage of electrical insulator material is that it has a high power-loss factor compared to other electrical insulator materials (Table 10.7) which is due to highly mobile alkali ions. Figure 10.33 shows the microstructure of an electrical porcelain material.

Steatite Steatite porcelains are good electrical insulators because they have low power-loss factors, low moisture absorption, and good impact strength and are used extensively by the electronic and electrical appliance industries. Industrial steatite compositions are based on about 90% talc ($3MgO \cdot 4SiO_2 \cdot H_2O$) and 10% clay. The microstructure of fired steatite consists of enstatite ($MgSiO_3$) crystals bonded together by a glassy matrix. Figure 10.37 shows some examples of steatite insulator parts for electrical applications.

Fosterite Fosterite has the chemical formula Mg_2SiO_4 and thus has no alkali ions in a vitreous phase so that it has a higher resistivity and lower electrical loss with increasing temperature than the steatite insulators. Fosterite also has lower loss dielectric properties at high frequencies (Table 10.7).

Alumina Alumina ceramics have aluminum oxide (Al_2O_3) as the crystalline phase bonded with a glassy matrix. The glass phase, which is normally alkali-

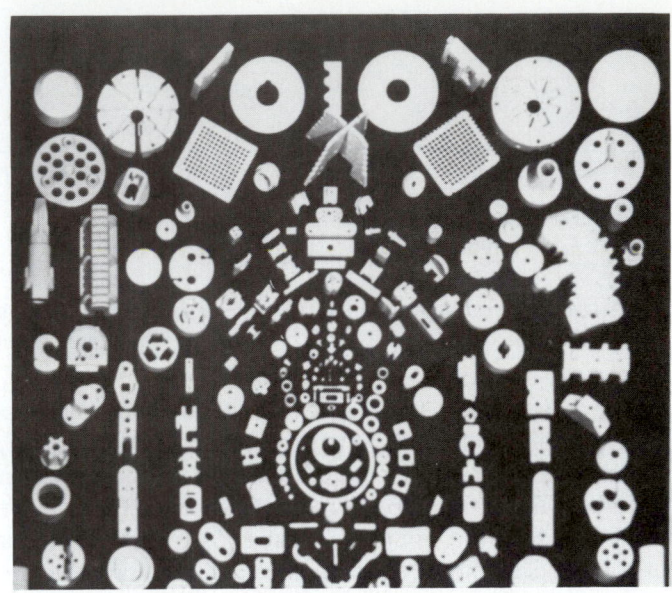

FIGURE 10.37 Some electronic and electrical insulator parts made of steatite. (*Courtesy of Wisconsin Porcelain Co.*)

free, is compounded from mixtures of clay, talc, and alkaline earth fluxes and is usually alkali-free. Alumina ceramics have relatively high dielectric strengths and low dielectric losses along with relatively high strengths. Sintered alumina (99% Al_2O_3) is widely used as a substrate for electronic-device applications (Fig. 10.1b) because of its low dielectric losses and smooth surface (Fig. 10.34). Alumina is also used for ultralow loss applications where a large energy transfer through a ceramic window is necessary as, for example, for radomes.

Ceramic Materials for Capacitors

Ceramic materials are commonly used as dielectric materials for capacitors, with disk ceramic capacitors being by far the most common type of ceramic capacitor (Fig. 10.38). These very small flat-disk ceramic capacitors consist mainly of barium titanate ($BaTiO_3$) along with other additives (Table 10.8). $BaTiO_3$ is used because of its very high dielectric constant of 1200 to 1500. With additives its dielectric constant can be raised to values of many thousands. Figure 10.38b shows the stages in the manufacture of one type of disk ceramic capacitor. In this type of capacitor a silver layer on the top and bottom of the disk provides the metal "plates" of the capacitor. For very high capacitances with a minimum-size device, small multilayered ceramic capacitors have been developed.

Ceramic chip capacitors are used in some ceramic-based thick-film hybrid electronic circuits. Chip capacitors can provide appreciably higher capacitance per unit area values and be added to the thick-film circuit by a simple soldering or bonding operation.

Ceramic Semiconductors

Some ceramic compounds have semiconducting properties that are important for the operation of some electrical devices. One of these devices is the *thermistor,* or thermally sensitive resistor, which is used for temperature measure-

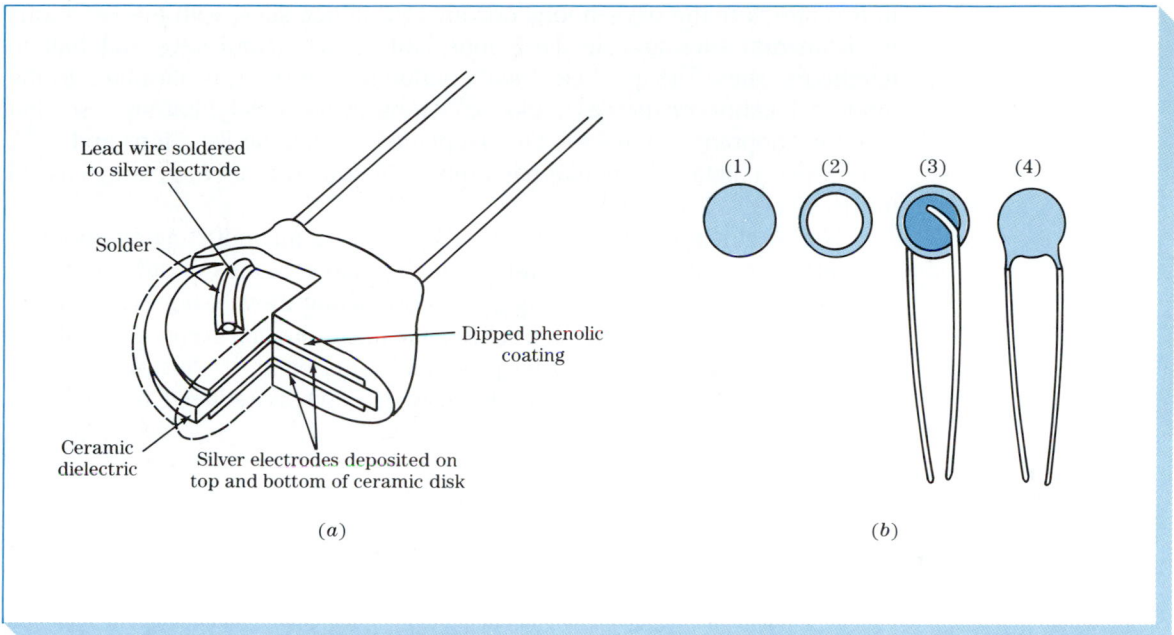

FIGURE 10.38 Ceramic capacitors. (*a*) Section showing construction. (*Courtesy of Sprague Products Co.*) (*b*) Steps in manufacture: (1) after firing ceramic disk; (2) after applying silver electrodes; (3) after soldering leads; (4) after applying dipped phenolic coating. (*Courtesy of Radio Materials Co.*)

ment and control. For our discussion we shall be concerned with the negative temperature coefficient (NTC) type of thermistor whose resistance decreases with increasing temperature. That is, as the temperature increases, the thermistor becomes more conductive, as in the case of a silicon semiconductor.

The most commonly used ceramic semiconducting materials used for NTC thermistors are sintered oxides of the elements Mn, Ni, Fe, Co, and Cu. Solid-solution combinations of the oxides of these elements are used to obtain the necessary range of electrical conductivities with temperature changes.

Let us first consider the ceramic compound magnetite, Fe_3O_4, which has a relatively low resistivity of about 10^{-5} $\Omega \cdot m$ as compared to a value of about 10^8 $\Omega \cdot m$ for most regular transition metal oxides. Fe_3O_4 has the inverse spinel structure with the composition $FeO \cdot Fe_2O_3$ that can be written as

$$Fe^{2+}(Fe^{3+}, Fe^{3+})O_4$$

TABLE 10.8   Representative Formulations for Some Ceramic Dielectric Materials for Capacitors

Dielectric constant κ	Formulation
325	$BaTiO_3 + CaTiO_3 +$ low % $Bi_2Sn_3O_9$
2100	$BaTiO_3 +$ low % $CaZrO_3$ and Nb_2O_5
6500	$BaTiO_3 +$ low % $CaZrO_3$ or $CaTiO_3 + BaZrO_3$

Source: C. A. Harper (ed.), "Handbook of Materials and Processes for Electronics," McGraw-Hill, 1970, pp. 6–61.

In this structure the oxygen ions occupy FCC lattice sites, with the Fe^{2+} ions in octahedral sites and the Fe^{3+} ions half in octahedral sites and half in tetrahedral sites. The good electrical conductivity of Fe_3O_4 is attributed to the random location of the Fe^{2+} and Fe^{3+} ions in the octahedral sites so that electron "hopping" (transfer) can take place between the Fe^{2+} ions and Fe^{3+} ions while maintaining charge neutrality. The structure of Fe_3O_4 is discussed further in Sec. 11.9.

The electrical conductivities of metal oxide semiconducting compounds for thermistors can be controlled by forming solid solutions of different metal oxide compounds. By combining a low-conducting metal oxide with a high-conducting one with a similar structure, a semiconducting compound with an intermediate conductivity can be produced. This effect is illustrated in Fig. 10.39, which shows how the conductivity of Fe_3O_4 is reduced gradually by

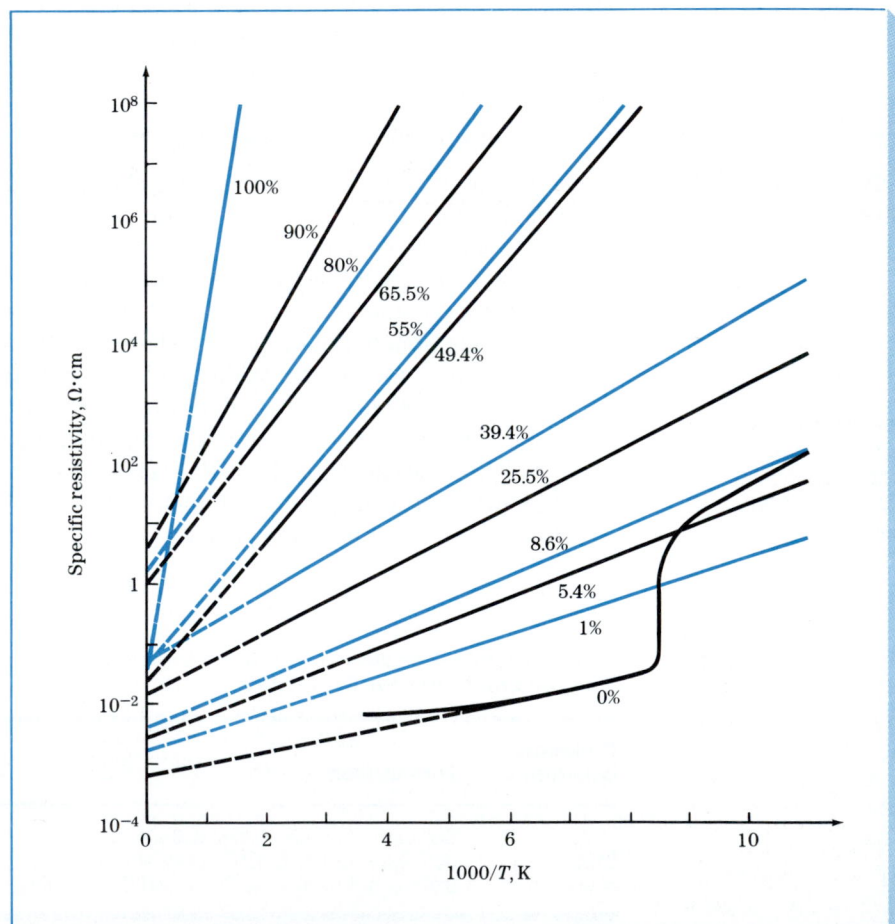

FIGURE 10.39 Specific resistivity of solid solution of Fe_3O_4 and $MgCr_2O_4$. Mole percent $MgCr_2O_4$ is indicated on the curves. [*After E. J. Verwey, P. W. Haagman, and F. C. Romeijn, J. Chem. Phys., 15:18(1947).*]

adding increasing amounts in solid solution of $MgCr_2O_4$. Most NTC thermistors with controlled temperature coefficients of resistivity are made of solid solutions of Mn, Ni, Fe, and Co oxides.

Ferroelectric Ceramics

Ferroelectric domains Some ceramic ionic crystalline materials have unit cells which do not have a center of symmetry, and as a result their unit cells contain a small electric dipole. An industrially important ceramic material in this class is barium titanate, $BaTiO_3$. Above 120°C, $BaTiO_3$ has the regular cubic symmetrical perovskite crystal structure (Fig. 10.40a). Below 120°C, the central Ti^{4+} ion and the surrounding O^{2-} ions of the $BaTiO_3$ unit cell shift slightly in opposite directions to create a small electric dipole moment (Fig. 10.40b). This shifting of the ion positions at the critical temperature of 120°C, called the *Curie temperature,* changes the crystal structure of $BaTiO_3$ from cubic to slightly tetragonal.

On a larger scale solid barium titanate ceramic material has a domain structure (Fig. 10.41) in which the small electric dipoles of the unit cells line up in one direction. The resultant dipole moment of a unit volume of this material is the sum of the small dipole moments of the unit cells. If polycrystalline barium titanate is slowly cooled through its Curie temperature in the presence of a strong electric field, the dipoles of all the domains tend to line up in the direction of the electric field to produce a strong dipole moment per unit volume of the material.

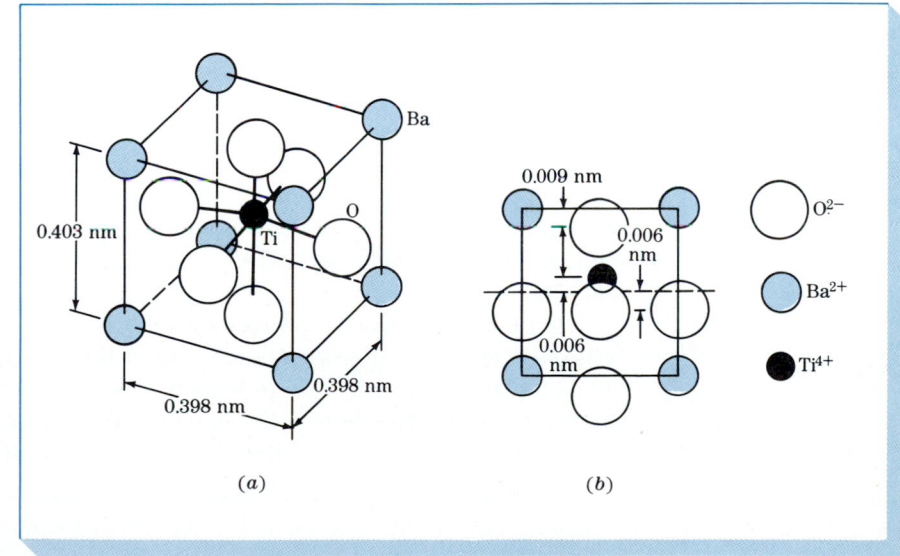

FIGURE 10.40 (a) The structure of $BaTiO_3$ above 120°C is cubic. (b) The structure of $BaTiO_3$ below 120°C (its Curie temperature) is slightly tetragonal due to a slight shift of the Ti^{4+} central ion with respect to the surrounding O^{2-} ions of the unit cell. A small electric dipole moment is present in this asymmetrical unit cell. (After K. M. Ralls, T. H. Courtney, and J. Wulff, "An Introduction to Materials Science and Engineering," Wiley, 1976, p. 610.)

FIGURE 10.41

Microstructure of barium titanate ceramic showing different ferroelectric domain orientations as revealed by etching. (Magnification 500×.) [*After R. D. DeVries and J. E. Burke, J. Am. Ceram. Soc.,* **40**:200 (1957).]

The piezoelectric[1] effect   Barium titanate and many other ceramic materials exhibit what is called the *piezoelectric effect,* illustrated schematically in Fig. 10.42. Let us consider a sample of a ferroelectric ceramic material which has a resultant dipole moment due to the alignment of many small unit dipoles, as indicated in Fig. 10.42*a*. In this material there will be an excess of positive charge at one end and negative charge at the other end in the direction of the polarization. Now let us consider the sample when compressive stresses are applied, as shown in Fig. 10.42*b*. The compressive stresses reduce the length of the sample between the applied stresses and thus reduce the distance between the unit dipoles, which in turn reduces the overall dipole moment per unit volume of the material. The change in dipole moment of the material changes the charge density at the ends of the sample and thus changes the voltage difference between the ends of the sample if they are insulated from each other.

On the other hand, if an electric field is applied across the ends of the sample, the charge density at each end of the sample will be changed (Fig. 10.42*c*). This change in charge density will cause the sample to change dimensions in the direction of the applied field. In the case of Fig. 10.42*c* the sample is slightly elongated due to an increased amount of positive charge

---

[1] The prefix *piezo-* means "pressure" and comes from the Greek word *piezein*, which means "to press."

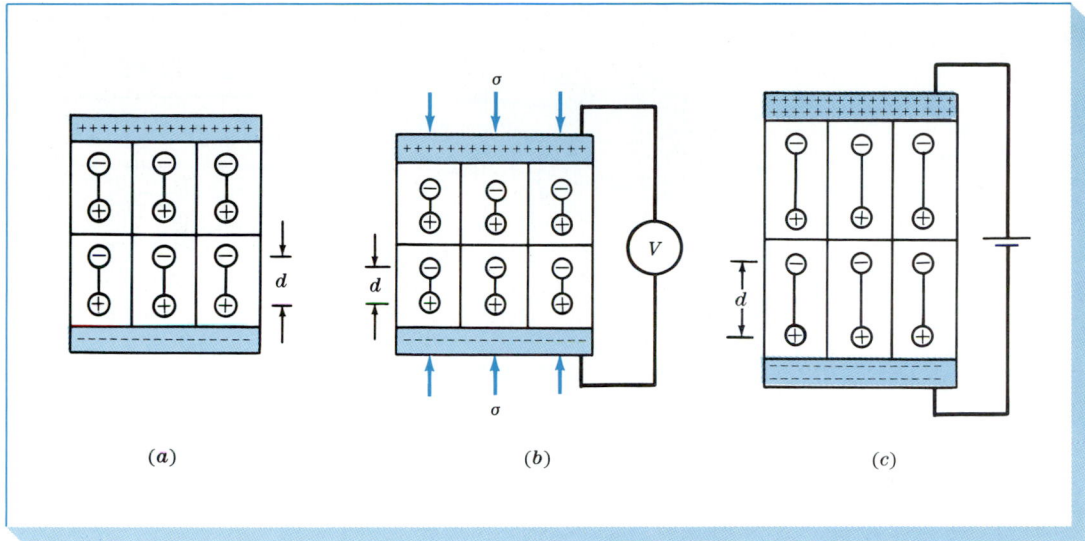

FIGURE 10.42 (*a*) Schematic illustration of electric dipoles within a piezoelectric material. (*b*) Compressive stresses on material cause a voltage difference to develop due to change in electric dipoles. (*c*) Applied voltage across ends of sample causes dimensional change and changes the electric dipole moment. (*After L. H. Van Vlack, "Elements of Materials Science and Engineering," 4th ed., Addison-Wesley, 1980, Fig. 8-6.3, p. 305.*)

atttracting the negative poles of the dipoles, and the reverse at the other end of the sample. Thus the piezoelectric effect is an electromechanical effect by which mechanical forces on a ferroelectric material can produce an electrical response, or electrical forces a mechanical response.

Piezoelectric ceramics have many industrial applications. Examples for the case of converting mechanical forces into electrical responses are the piezoelectric compression accelerometer (Fig. 10.43*a*), which can measure vibratory accelerations occurring over a wide range of frequencies, and the phonograph cartridge in which electrical responses are "picked up" from a stylus vibrating in record grooves. An example for the case of converting electrical forces into mechanical responses is the ultrasonic cleaning transducer which is caused to vibrate by ac power input so that it can induce violent agitation of the liquid in a tank (Fig. 10.43*b*). Another example of this type is the underwater sound transducer in which electric power input causes the transducer to vibrate to transmit sound waves.

Piezoelectric materials   Although $BaTiO_3$ is commonly used as a piezoelectric material, it has largely been replaced by other piezoelectric ceramic materials. Of particular importance are the ceramic materials made from solid solutions of lead zirconate ($PbZrO_3$) and lead titanate ($PbTiO_3$) to make what are called

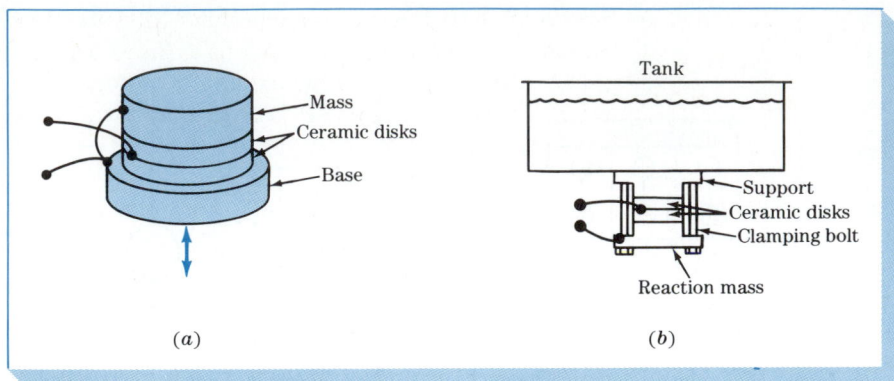

FIGURE 10.43
(a) Piezoelectric compression accelerometer. (b) Piezoelectric ceramic elements in an ultrasonic cleaning apparatus. (*Courtesy of the Vernitron Piezoelectric Division, Bedford, Ohio.*)

PZT ceramics. The PZT materials have a broader range of piezoelectric properties, including a higher Curie temperature than $BaTiO_3$.

10.7   MECHANICAL PROPERTIES OF CERAMICS

General   As a class of materials, ceramics are relatively brittle. The observed tensile strength of ceramic materials varies greatly, ranging from very low values of less than 100 psi (0.69 MPa) to about 10^6 psi (7×10^3 MPa) for whiskers of ceramics such as Al_2O_3 prepared under carefully controlled conditions. However, as a class of materials, few ceramics have tensile strengths above 25,000 psi (172 MPa). Ceramic materials also have a large difference between their tensile and compressive strengths, with the compressive strengths usually being about 5 to 10 times higher than the tensile strengths, as indicated in Table 10.9 for the 99% Al_2O_3 ceramic material. Also, many ceramic materials are

TABLE 10.9   Mechanical Properties of Selected Engineering Ceramic Materials

Material	Density, g/cm³	Compressive strength		Tensile strength		Flexural strength		Fracture toughness	
		MPa	ksi	MPa	ksi	MPa	ksi	MPa√m	ksi√m
Al₂O₃ (99%)	3.85	2585	375	207	30	345	50	4.	3.63
Si₃N₄ (hot-pressed)	3.19	3450	500	...	...	690	100	6.6	5.99
Si₃N₄ (reaction-bonded)	2.8	770	112	...	...	255	37	3.6	3.27
SiC (sintered)	3.1	3860	560	170	25	550	80	4	3.63
ZrO₂, 9% MgO (partially stabilized)	5.5	1860	270	...	...	690	100	8+	7.26+

hard and have low impact resistance due to their ionic-covalent bindings. However, there are many exceptions to the above generalizations. For example, plasticized clay is a ceramic material which is soft and easily deformable due to weak secondary bonding forces between layers of strongly ionic-covalently bonded atoms.

Mechanisms for the Deformation of Ceramic Materials

The lack of plasticity in crystalline ceramics is due to their ionic and covalent chemical bonds. In metals plastic flow takes place mainly by the movement of line faults (dislocations) in the crystal structure over special crystal slip planes (see Sec. 6.5). In metals dislocations move under relatively low stresses due to the nondirectional nature of the metallic bond and because all atoms involved in the bonding have an equally distributed negative charge at their surfaces. That is, there are no positive or negatively charged ions involved in the metallic bonding process.

In covalent crystals and covalently bonded ceramics, the bonding between atoms is specific and directional, involving the exchange of electron charge between pairs of electrons. Thus, when covalent crystals are stressed to a sufficient extent, they exhibit brittle fracture due to a separation of electron-pair bonds without their subsequent reformation. Covalently bonded ceramics, therefore, are brittle in both the single-crystal and polycrystalline states.

The deformation of primarily ionically bonded ceramics is different. Single crystals of ionically bonded solids such as magnesium oxide and sodium chloride show considerable plastic deformation under compressive stresses at room temperature. Polycrystalline ionically bonded ceramics, however, are brittle, with cracks forming at the grain boundaries.

Let us briefly examine some conditions under which an ionic crystal can be deformed, as illustrated in Fig. 10.44. The slip of one plane of ions over

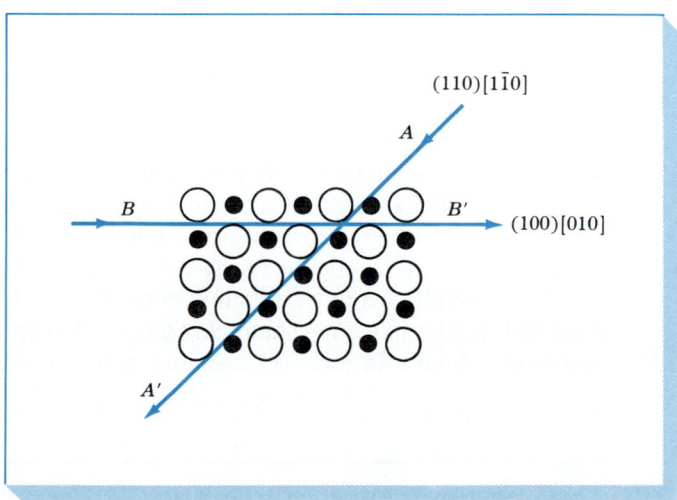

FIGURE 10.44 Top view of NaCl crystal structure indicating (*a*) slip on the (110) plane and in the [110] direction (line *AA'*) and (*b*) slip on the (100) plane in the [010] direction (line *BB'*).

another involves ions of different charge coming into contact, and thus attractive and repulsion forces may be produced. Most ionically bonded crystals having the NaCl-type structure slip on the $\{110\}<1\bar{1}0>$ systems because slip on the $\{110\}$ family of planes involves only ions of unlike charge, and hence the slip planes remain attracted to each other by coulombic forces during the slip process. Slip of the $\{110\}$ type is indicated by the line AA' of Fig. 10.44. On the other hand, slip on the $\{100\}$ family of planes is rarely observed because ions of the same charge come into contact, which will tend to separate the planes of ions slipping over each other. This $\{100\}$-type slip is indicated by the line BB' of Fig. 10.44. Many ceramic materials in the single-crystal form show considerable plasticity. However, in polycrystalline ceramics adjacent grains must change shape during deformation. Since there are limited slip systems in ionically bonded solids, cracking occurs at the grain boundaries and subsequent brittle fracture occurs. Since most industrially important ceramics are polycrystalline, most ceramic materials tend to be brittle.

Factors Affecting the Strength of Ceramics Materials

The mechanical failure of ceramic materials occurs mainly from structural defects. The principal sources of fracture in ceramic polycrystals are identified to include surface cracks produced during surface finishing, voids (porosity), inclusions, and large grains produced during processing.[1]

Pores in brittle ceramic materials are regions where stress concentrates, and when the stress at a pore reaches a critical value, a crack forms and propagates since there are no large energy-absorbing processes in these materials such as those that operate in ductile metals during deformation. Thus, once cracks start to propagate, they continue to grow until fracture occurs. Pores are also detrimental to the strength of ceramic materials because they decrease the cross-sectional area over which a load is applied and hence lower the stress a material can support. Thus, the size and volume fraction of pores in ceramic materials are important factors affecting their strength. Figure 10.45 shows how an increasing volume fraction of pores decreases the transverse tensile strength of alumina.

Flaws in processed ceramics may also be critical in determining the fracture strength of a ceramic material. A large flaw may be the major factor affecting the strength of a ceramic. In fully dense ceramic materials in which there are no large pores, the flaw size is usually related to the grain size. For porosity-free ceramics the strength of a pure ceramic material is a function of its grain size, with finer grain-size ceramics having smaller-size flaws at their grain boundaries and hence being stronger than large grain-size ones.

The strength of a polycrystalline ceramic material is thus determined by many factors which include chemical composition, microstructure, and surface condition as major factors. Temperature and environment also are important

[1] A. G. Evans, *J. Am. Ceram. Soc.*, **65**:127(1982).

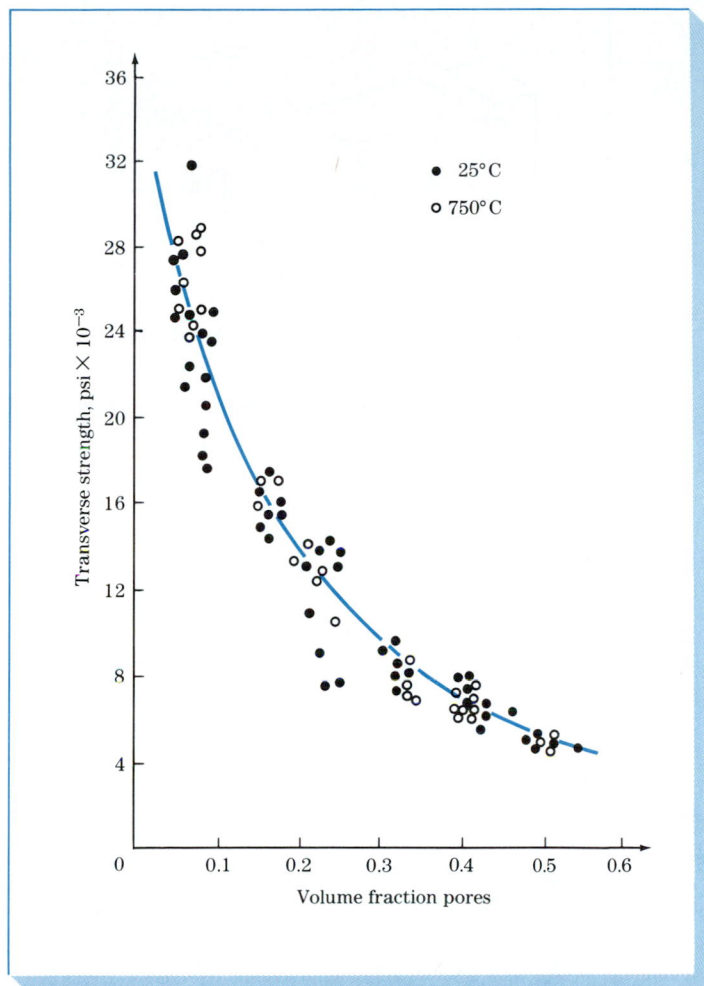

FIGURE 10.45 The effect of porosity on the transverse strength of pure alumina. [*After R. L. Coble and W. D. Kingery, J Am. Ceram. Soc.*, **39***:377(1956).*]

as well as the type of stress and how it is applied. However, the failure of most ceramic materials at room temperature usually originates at the largest flaw.

Toughness of Ceramic Materials

Ceramic materials, because of their combination of covalent-ionic bonding, have inherently low toughness. A great amount of research has been carried out in the past years to improve the toughness of ceramic materials. By using processes such as hot pressing ceramics with additives and reaction bonding, engineering ceramics with improved toughness have been produced (Table 10.9).

Fracture-toughness tests can be made on ceramic specimens to determine K_{IC} values in a manner similar to the fracture-toughness testing of metals (see Sec. 6.9). K_{IC} values for ceramic materials are usually obtained by using a four-

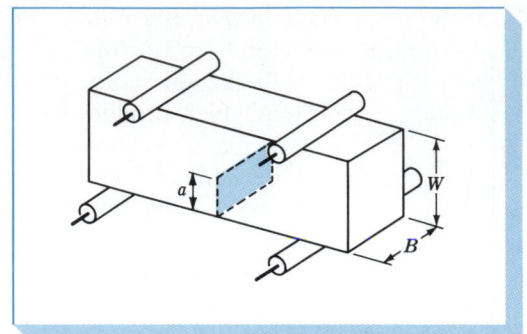

FIGURE 10.46 Setup for four-point beam fracture-toughness test of a ceramic material using a single-edge notch.

point bend test with a single-edge or chevron-notched beam specimen (Fig. 10.46). The fracture-toughness equation,

$$K_{IC} = Y\sigma_f\sqrt{\pi a} \tag{10.4}$$

which relates fracture-toughness K_{IC} values to the fracture stress, and the largest flaw size can also be used for ceramic materials. In Eq. (10.4), K_{IC} is measured in MPa$\sqrt{m}$ (ksi$\sqrt{in}$), the fracture stress σ_f in MPa(ksi), and a (half the size of the largest internal flaw) in meters (inches). Y is a dimensionless constant equal to about 1. Example Problem 10.10 shows how this equation can be used to determine the largest-size flaw that a particular engineering ceramic of a known fracture toughness and strength can tolerate without fracture.

Example Problem 10.10

A reaction-bonded silicon nitride ceramic has a strength of 300 MPa and a fracture toughness of 3.6 MPa$\sqrt{m}$. What is the largest-size internal crack that this material can support without fracturing? Use $Y = 1$ in the fracture-toughness equation.

Solution:

$$\sigma_f = 300 \text{ MPa} \qquad K_{IC} = 3.6 \text{ MPa}\sqrt{m} \qquad a = ? \qquad Y = 1$$

$$K_{IC} = Y\sigma_f\sqrt{\pi a}$$

or

$$a = \frac{K_{IC}^2}{\pi\sigma_f^2} = \frac{(3.6 \text{ MPa}\sqrt{m})^2}{\pi(300 \text{ MPa})^2} = 4.58 \times 10^{-5} \text{ m} = 45.8 \ \mu\text{m}$$

Thus, the largest internal crack $= 2a = 2(45.8 \ \mu\text{m}) = 91.6 \ \mu\text{m}$. ◀

Transformation Toughening of Partially Stabilized Zirconia (PSZ)

Recently it has been discovered that phase transformations in zirconia combined with some other refractory oxides (that is, CaO, MgO, and Y_2O_3) can produce ceramic materials with exceptionally high fracture toughness. Let us now look into the mechanisms which produce transformation toughening in a ZrO_2–9 mol % MgO ceramic material. Pure zirconia, ZrO_2, exists in three different crystal structures: *monoclinic* from room temperature to 1170°C, *tetragonal* from 1170 to 2370°C, and *cubic* (the fluorite structure of Fig. 10.14) above 2370°C.

The transformation of pure ZrO_2 from the tetragonal to monoclinic structure is martensitic and cannot be suppressed by rapid cooling. Also, this transformation is accompanied by a volume increase of about 9 percent, and so it is impossible to fabricate articles from pure zirconia. However, by the addition of about 10 mol % of other refractory oxides such as CaO, MgO, and Y_2O_3, the cubic form of zirconia is stabilized so that it can exist at room temperature in the metastable state, and articles can be fabricated from this material. Cubic ZrO_2 combined with stabilizing oxides so that it retains the cubic structure at room temperature is referred to as *fully stabilized zirconia.*

Recent developments have produced zirconia-refractory oxide ceramic materials with enhanced toughness and strength by taking advantage of their phase transformations. One of the more important zirconia compound ceramics is partially stabilized zirconia (PSZ) which contains 9 mol % MgO. If a mixture of ZrO_2–9 mol % MgO is sintered at about 1800°C, as indicated in the ZrO_2–MgO phase diagram of Fig. 10.47*a*, and then rapidly cooled to room temperature, it will be in the all-metastable cubic structure. However, if this material is reheated to 1400°C and held for a sufficient time, a fine metastable submicroscopic precipitate with the tetragonal structure is precipitated, as shown in Fig. 10.47*b*. This material is known as *partially stabilized zirconia (PSZ).* Under the action of stresses which cause small cracks in the ceramic material, the tetragonal phase transforms to the monoclinic phase, causing a volume expansion of the precipitate which retards the crack propagation by a kind of crack-closing mechanism. By impeding the advances of cracks, the ceramic is "toughened" (Fig. 10.47*c*). Partially stabilized zirconia has a fracture toughness of 8+ $MPa\sqrt{m}$, which is higher than the fracture toughness of all the other engineering ceramic materials listed in Table 10.9.

Fatigue Failure of Ceramics

Fatigue failure in metals occurs under repeated cyclic stresses due to the nucleation and growth of cracks within a work-hardened area of a specimen. Because of the ionic-covalent bonding of the atoms in a ceramic material, there is an absence of plasticity in ceramics during cyclic stressing. As a result, fatigue fracture in ceramics is rare. Recently, results of stable fatigue-crack growth at room temperature under compression–compression stress cycling in notched plates of polycrystalline alumina have been reported. A straight fatigue crack was produced after 79,000 compression cycles (Fig. 10.48*a*). Microcrack propagation along grain boundaries led to final intergranular fatigue failure (Fig. 10.48*b*). Much research is being carried out to make tougher

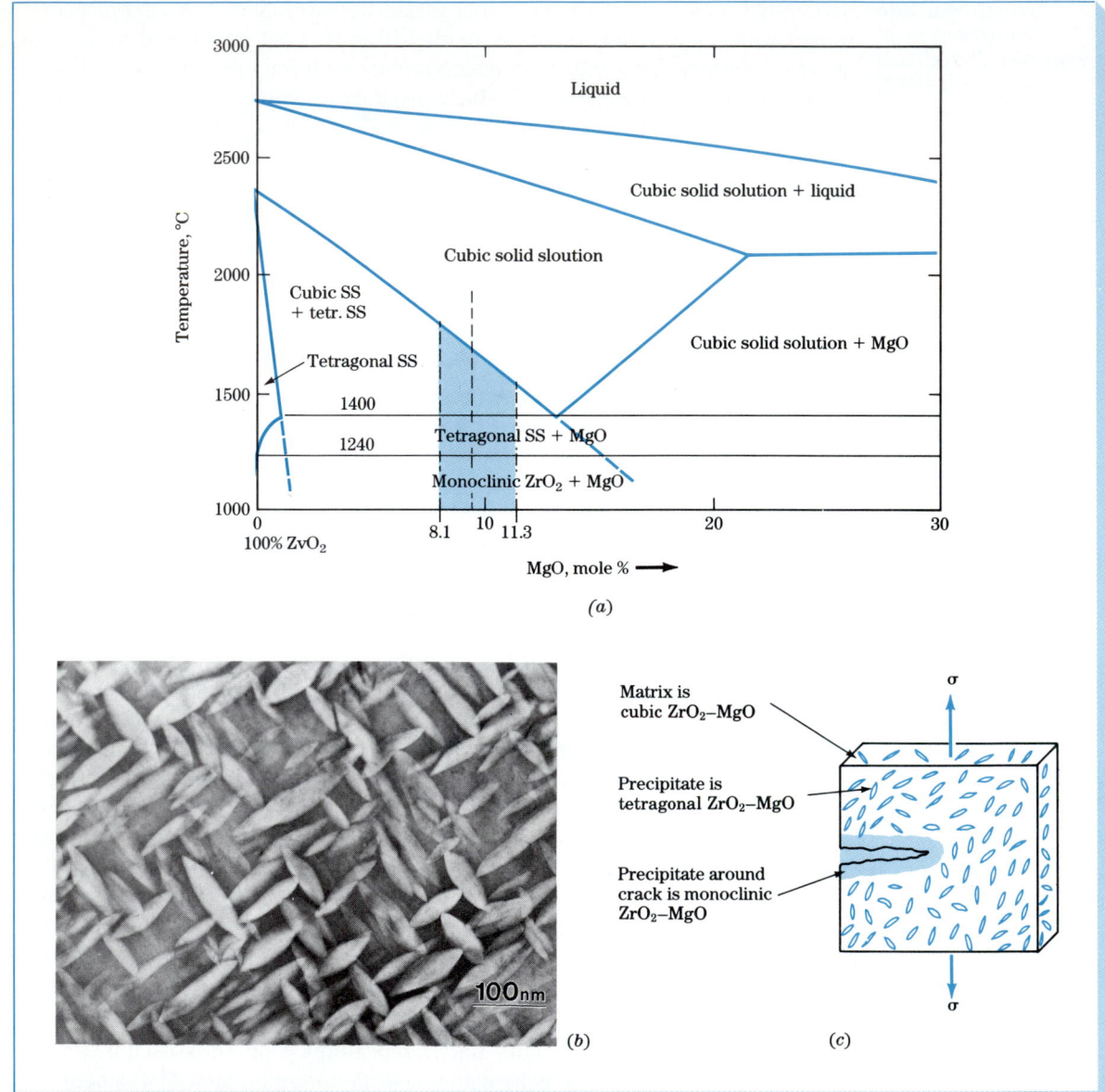

FIGURE 10.47 (*a*) Phase diagram of high-ZrO$_2$ part of the ZrO$_2$–MgO binary phase diagram. The shaded area represents the region used for combining MgO with ZrO$_2$ to produce partially stabilized zirconia. (*After A. H. Heuer, "Advances in Ceramics," vol. 3, "Science and Technology of Zirconia," American Ceramic Society, 1981.*) (*b*) Transmission electron micrograph of optimally aged MgO–partially stabilized ZrO$_2$ showing the tetragonal oblate spheroid precipitate. Upon the application of sufficient stress, these particles transform to the monoclinic phase with a volume expansion. (*Courtesy of A. H. Heuer.*) (*c*) Schematic diagram illustrating the transformation of the tetragonal precipitate to the monoclinic phase around a crack in a partially stabilized ZrO$_2$–9 mol % MgO ceramic specimen.

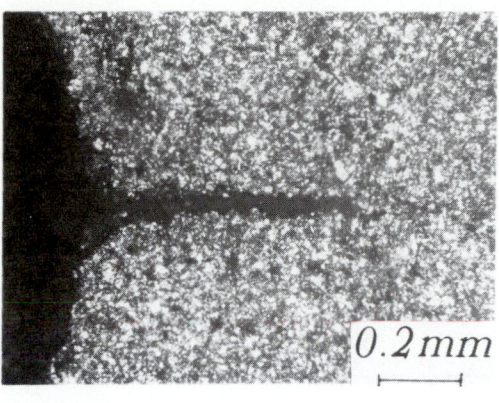

FIGURE 10.48 Fatigue cracking of polycrystalline alumina under cyclic compression. (*a*) Optical micrograph showing fatigue crack (the compression axis is vertical). (*b*) Scanning electron fractograph of the fatigue area of the same specimen where the intergranular mode of failure is evident. [*After S. Suresh and J. R. Brockenbrough, Acta Metall.* **36:** *1455 (1988).*]

ceramics which can support cyclic stresses for applications such as turbine rotors.

Ceramic Abrasive Materials   The high hardness of some ceramic materials makes them useful as abrasive materials for cutting, grinding, and polishing other materials of lower hardness. Fused alumina (aluminum oxide) and silicon carbide are two of the most commonly used manufactured ceramic abrasives. Abrasive products such as sheets and wheels are made by bonding individual ceramic particles together. Bonding materials include fired ceramics, organic resins, and rubbers. The ceramic particles must be hard with sharp cutting edges. Also, the abrasive product must have a certain amount of porosity to provide channels for air or liquid to flow through in the structure. Aluminum oxide grains are tougher than silicon carbide ones but are not as hard, and so silicon carbide is normally used for the harder materials.

By combining zirconium oxide with aluminum oxide, improved abrasives were developed[1] which have higher strength, hardness, and sharpness than aluminium oxide alone. One of these ceramic alloys contains 25% ZrO_2 and 75% Al_2O_3 and another 40% ZrO_2 and 60% Al_2O_3. Another important ceramic abrasive is cubic boron nitride which has the trade name Borazon.[2] This material is almost as hard as diamond but has better heat stability than diamond.

---

[1] ZrO_2–Al_2O_3 ceramic abrasive alloys were developed by the Norton Co. in the 1960s.

[2] Borazon, a product of the General Electric Co., was developed in the 1950s.

10.8 THERMAL PROPERTIES OF CERAMICS

In general most ceramic materials have low thermal conductivities due to their strong ionic-covalent bonding and are good thermal insulators. Figure 10.49 compares the thermal conductivities of many ceramic materials as a function of temperature. Because of their high heat resistance, ceramic materials are used as *refractories,* which are materials that resist the action of hot environments, both liquid and gaseous. Refractories are used extensively by the metals, chemical, ceramic, and glass industries.

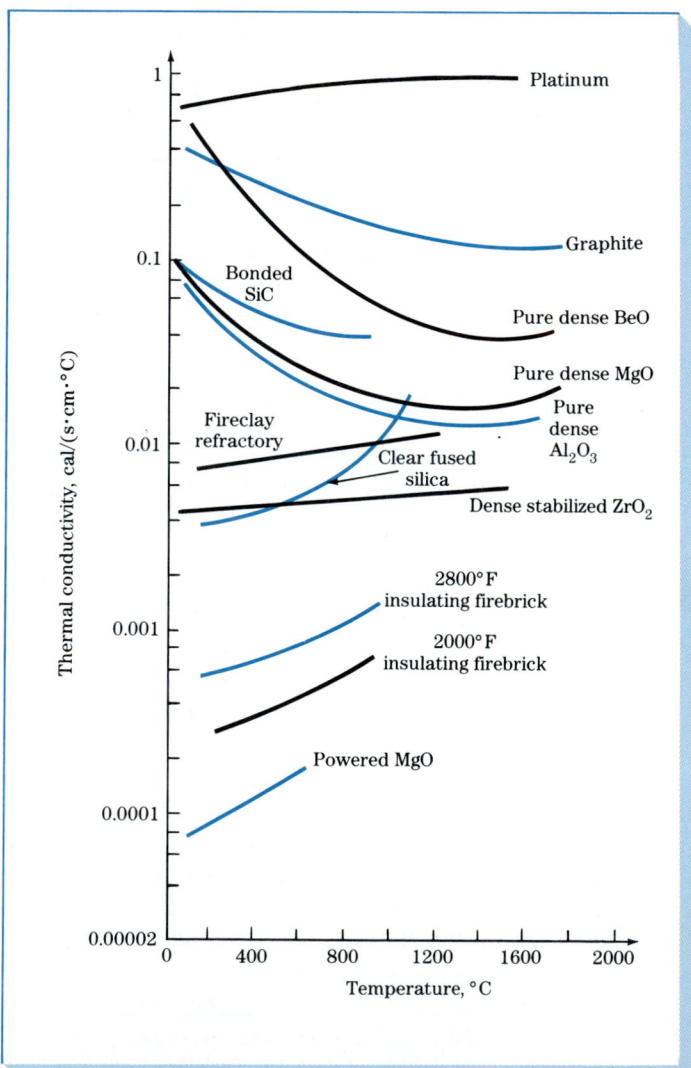

FIGURE 10.49 Thermal conductivity (logarithmic scale) of ceramic materials over a wide temperature range. (*After W. D. Kingery, H. K. Bowen, and D. R. Uhlmann, "Introduction to Ceramics," 2d ed., Wiley, 1976, p. 643.*)

TABLE 10.10   Compositions and Applications for Some Refractory Brick Materials

	Composition, wt %			
	SiO$_2$	**Al$_2$O$_3$**	**MgO**	**Other**
Acidic types:				
Silica brick	95–99			
Superduty fireclay brick	53	42		
High-duty fireclay brick	51–54	37–41		
High-alumina brick	0–50	45–99+		
Basic types:				
Magnesite	0.5–5		91–98	0.6–4 CaO
Magnesite-chrome	2–7	6–13	50–82	18–24 Cr$_2$O$_3$
Dolomite (burned)			38–50	38–58 CaO
Special types:				
Zircon	32			66 ZrO$_2$
Silicon carbide	6	2		91 SiC

Applications for some refractories:

Superduty fireclay bricks: linings for aluminum-melting furnaces, rotary kilns, blast furnaces, and hot-metal transfer ladles

High-duty fireclay brick: linings for cement and lime kilns, blast furnaces, and incinerators

High-alumina brick: boiler furnaces, spent-acid regenerating furnaces, phosphate furnaces, glass-tank refiner walls, carbon black furnaces, continuous-casting tundish linings, coal gasification reactor linings, and petroleum coke kilns

Silica brick: chemical reactor linings, glass tank parts, ceramic kilns, and coke ovens

Magnesite brick: basic-oxygen-process furnace linings for steelmaking

Zircon brick: glass-tank bottom paving and continuous-casting nozzles

Source: "Harbison-Walker Handbook of Refractory Practice," Harbison-Walker Refractories, Pittsburgh, 1980.

Ceramic Refractory Materials

Many pure ceramic compounds with high melting points such as aluminum oxide and magnesium oxide could be used as industrial refractory materials, but they are expensive and difficult to form into shapes. Therefore, most industrial refractories are made of mixtures of ceramic compounds. Table 10.10 lists the compositions of some refractory brick compositions and gives some of their applications.

Important properties of ceramic refractory materials are low- and high-temperature strengths, bulk density, and porosity. Most ceramic refractories have bulk densities which range from 2.1 to 3.3 g/cm^3 (132 to 206 lb/ft^3). Dense refractories with low porosity have higher resistance to corrosion and erosion and to penetration by liquids and gases. However, for insulating refractories a high amount of porosity is desirable. Insulating refractories are mostly used as backing for brick or refractory material of higher density and refractoriness.

Industrial ceramic refractory materials are commonly divided into acidic and basic types. Acidic refractories are based mainly on SiO$_2$ and Al$_2$O$_3$ and basic ones on MgO, CaO, and Cr_2O_3. Table 10.10 lists the compositions of many types of industrial refractories and gives some of their applications.

Acidic Refractories

Silica refractories have high refractoriness, high mechanical strength, and rigidity at temperatures almost to their melting points.

Fireclays are based on a mixture of plastic fireclay, flint clay, and clay (coarse-particle) grog. In the unfired (green) condition these refractories consist of a mixture of particles varying from coarse to extremely fine. After firing, the fine particles form a ceramic bond between the larger particles.

High-alumina refractories contain from 50 to 99% alumina and have higher fusion temperatures than fireclay bricks. They can be used for more severe furnace conditions and at higher temperatures than fireclay bricks but are more expensive.

Basic Refractories

Basic refractories consist mainly of magnesia (MgO), lime (CaO), chrome ore, or mixtures of two or more of these materials. As a group, basic refractories have high bulk densities, high melting temperatures, and good resistance to chemical attack by basic slags and oxides but are more expensive. Basic refractories containing a high percentage (92 to 95%) of magnesia are used extensively for linings in the basic-oxygen steelmaking process.

Ceramic Tile Insulation for the Space Shuttle Orbiter

The development of the thermal protection system for the space shuttle orbiter is an excellent example of modern materials technology applied to engineering design. So that the space shuttle orbiter could be used for at least 100 missions, new insulative ceramic tile materials were developed.

About 70 percent of the orbiter's external surface is protected from heat by approximately 24,000 individual ceramic tiles made from a silica-fiber compound. Figure 10.50 shows the microstructure of the high-temperature reusable-

FIGURE 10.50
Microstructure of LI900 high-temperature reusable-surface insulation (ceramic tile material used for space shuttle); structure consists of 99.7% pure silica fibers. (Magnification 1200×.) (*Courtesy of Lockheed Missiles and Space Co.*)

10 μm   10 KV   00   025   S

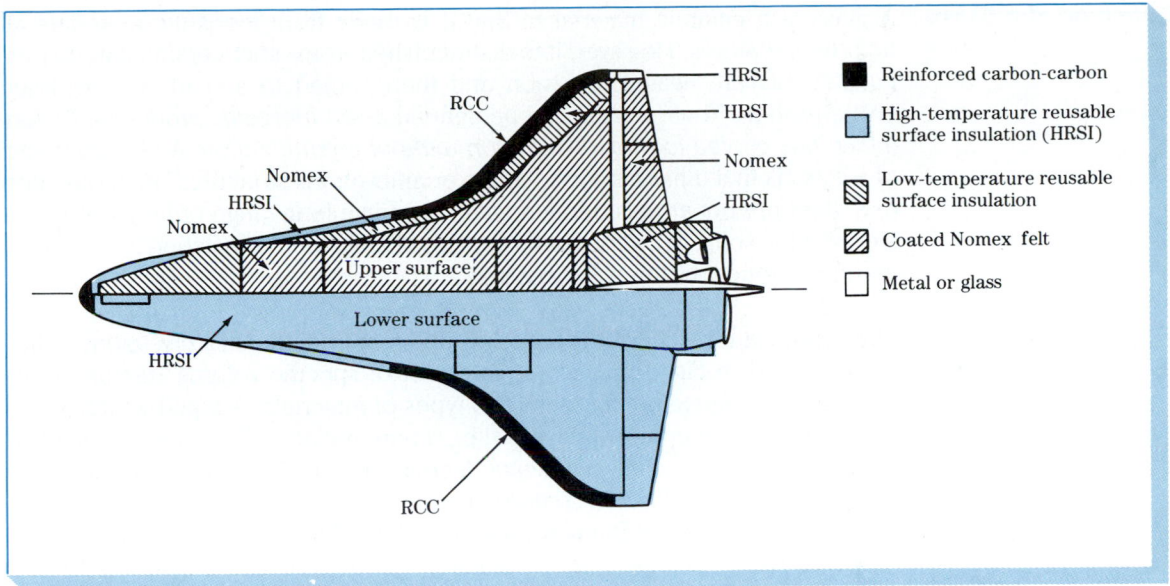

FIGURE 10.51 Space shuttle thermal protection systems. (*Courtesy of NASA.*)

surface insulation (HRSI) tile material, and Fig. 10.51 indicates the surface area where it is attached to the body of the orbiter. This material has a density of only 4 kg/ft^3 (9 lb/ft^3) and is able to withstand temperatures as high as 1260°C (2300°F). The effectiveness of this insulative material is indicated by the ability of a technician to hold a piece of the ceramic tile only about 10 s after it has been removed from a furnace at 1260°C (2300°F).

10.9 GLASSES

Glasses have special properties not found in other engineering materials. The combination of transparency and hardness at room temperature along with sufficient strength and excellent corrosion resistance to most normal environments make glasses indispensable for many engineering applications such as construction and vehicle glazing. In the electrical industry glass is essential for various types of lamps because of its insulative properties and ability to provide a vacuumtight enclosure. In the electronics industry electron tubes also require the vacuumtight enclosure provided by glass along with its insulative properties for lead-in connectors. The high chemical resistance of glass makes it useful for laboratory apparatus and for corrosion-resistant liners for pipes and reaction vessels in the chemical industry.

Definition of a Glass A glass is a ceramic material in that it is made from inorganic materials at high temperatures. However, it is distinguished from other ceramics in that its constituents are heated to fusion and then cooled to a rigid state without crystallization. Thus, a *glass* can be defined as *an inorganic product of fusion which has cooled to a rigid condition without crystallization.* A characteristic of a glass is that it has a noncrystalline or amorphous structure. The molecules in a glass are not arranged in a regular repetitive long-range order as exists in a crystalline solid. In a glass the molecules change their orientation in a random manner throughout the solid material.

Glass Transition Temperature The solidification behavior of a glass is different from that of a crystalline solid, as illustrated in Fig. 10.52, which is a plot of specific volume (reciprocal of density) vs. temperature for these two types of materials. A liquid which forms a crystalline solid upon solidifying (i.e., a pure metal) will normally crystallize at its melting point with a significant decrease in specific volume, as indicated by path *ABC* in Fig. 10.52. In contrast, a liquid which forms a glass upon cooling does not crystallize but follows a path like *AD* in Fig. 10.52. Liquid of this type becomes more viscous as its temperature is lowered and transforms from a rubbery, soft plastic state to a rigid, brittle glassy state in a narrow temperature range where the slope of the specific volume vs. temperature curve is markedly decreased. The point of intersection of the two slopes of this curve define a transformation point called the *glass transition temperature* T_g. This point is structure-sensitive, with faster cooling rates producing higher values of T_g.

Structure of Glasses **Glass-forming oxides** Most inorganic glasses are based on the glass-forming oxide silica, SiO_2. The fundamental subunit in silica-based glasses is the SiO_4^{4-}

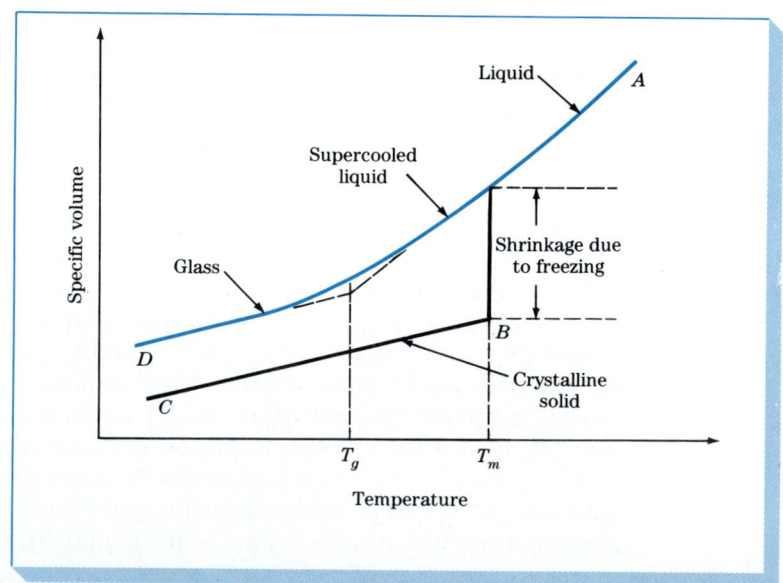

FIGURE 10.52
Solidification of crystalline and glassy (amorphous) materials showing changes in specific volume. T_g is the glass transition temperature of the glassy material. T_m is the melting temperature of the crystalline material.

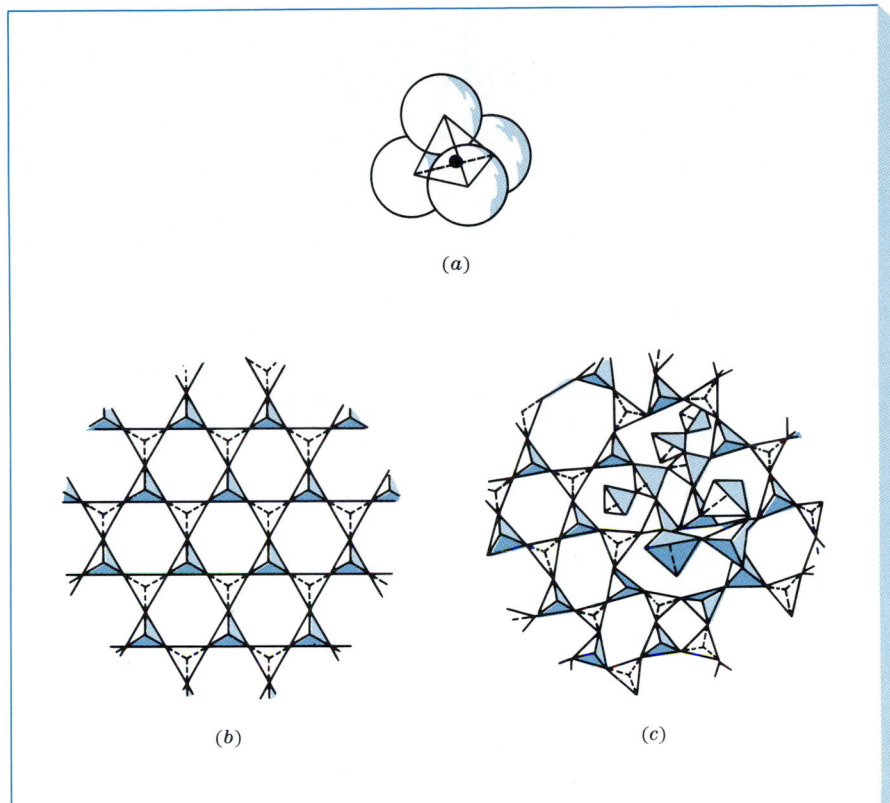

(a)

(b)   (c)

FIGURE 10.53 Schematic representation of (a) a silicon-oxygen tetrahedron, (b) ideal crystalline silica (cristobalite) in which the tetrahedra have long-range order, and (c) a simple silica glass in which the tetrahedra have no long-range order. (*Courtesy of Corning Glass Works.*)

tetrahedron in which a silicon (Si^{4+}) atom (ion) in the tetrahedron is covalently ionically bonded to four oxygen atoms (ions), as shown in Fig. 10.53a. In crystalline silica, for example, cristobalite, the Si—O tetrahedra are joined corner to corner in a regular arrangement, producing long-range order as idealized in Fig. 10.53b. In a simple silica glass the tetrahedra are joined corner to corner to form a *loose network* with no long-range order (Fig. 10.53c).

Boron oxide, B_2O_3, is also a glass-forming oxide and by itself forms subunits which are flat triangles with the boron atom slightly out of the plane of the oxygen atoms. However, in borosilicate glasses which have additions of alkali and alkaline earth oxides, BO_3^{3-} triangles can be converted to BO_4^{4-} tetrahedra, with the alkali or alkaline earth cations providing the necessary electroneutrality. Boron oxide is an important addition to many types of commercial glasses such as borosilicate and aluminoborosilicate glasses.

Glass-modifying oxides Oxides which break up the glass network are known as *network modifiers*. Alkali oxides such as Na_2O and K_2O and alkaline earth oxides such as CaO and MgO are added to silica glass to lower its viscosity

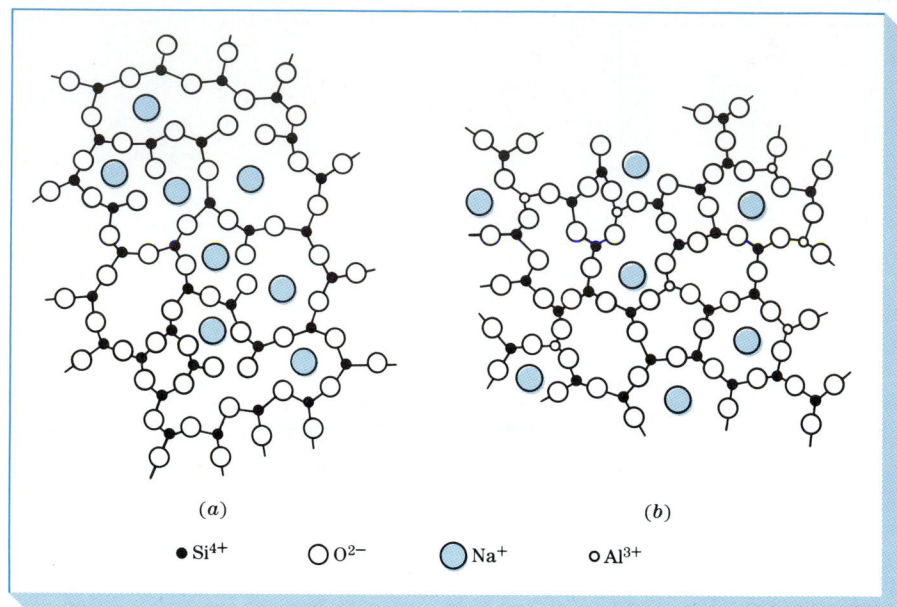

FIGURE 10.54 (a) Network-modified glass (soda-lime glass); note that the metallic (Na$^+$) ions do not form part of the network. (b) Intermediate-oxide glass (alumina-silica) glass; note that the small metallic (Al^{3+}) ions form part of the network. (After O. H. Wyatt and D. Dew-Hughes, "Metals, Ceramics, and Polymers," Cambridge, 1974, p. 263.)

(a)   (b)

● Si^{4+} ○ O^{2-} ◉ Na$^+$ ○ Al^{3+}

so that it can be worked and formed easier. The oxygen atoms from these oxides enter the silica network at points joining the tetrahedra and break up the network, producing oxygen atoms with an unshared electron (Fig. 10.54a). The Na$^+$ and K$^+$ ions from the Na$_2$O and K$_2$O do not enter the network but remain as metal ions ionically bonded in the interstices of the network. By filling some of the interstices, these ions promote crystallization of the glass.

Intermediate oxides in glasses   Some oxides cannot form a glass network by themselves but can join into an existing network. These oxides are known as *intermediate oxides.* For example, aluminum oxide, Al$_2$O$_3$, can enter the silica network as AlO$_4^{4-}$ tetrahedra, replacing some of the SiO$_4^{4-}$ groups (Fig. 10.54b). However, since the valence of Al is +3 instead of the necessary +4 for the tetrahedra, alkali cations must supply the necessary other electrons to produce electrical neutrality. Intermediate oxides are added to silica glass to obtain special properties. For example, aluminosilicate glasses are able to withstand higher temperatures than common glass. Lead oxide is another intermediate oxide that is added to some silica glasses. Depending on the composition of the glass, intermediate oxides may sometimes act as network modifiers as well as taking part in the network of the glass.

Composition of Glasses   The compositions of some important types of glasses are listed in Table 10.11 along with some remarks about their special properties and applications. Fused silica glass, which is the most important single-component glass, has a high spectral transmission and is not subject to radiation damage which causes

TABLE 10.11   Compositions of Some Glasses

Glass	SiO$_2$	Na$_2$O	K$_2$O	CaO	B$_2$O$_3$	Al$_2$O$_3$	Other	Remarks
1. (Fused) silica	99.5+							Difficult to melt and fabricate but usable to 1000°C. Very low expansion and high thermal shock resistance
2. 96% silica	96.3	<0.2	<0.2		2.9	0.4		Fabricate from relatively soft borosilicate glass; heat to separate SiO$_2$ and B$_2$O$_3$ phases; acid leach B$_2$O$_3$ phase; heat to consolidate pores.
3. Soda-lime: plate glass	71–73	12–14		10–12		0.5–1.5	MgO, 1–4	Easily fabricated. Widely used in slightly varying grades, for windows, containers, and electric bulbs.
4. Lead silicate: Electrical	63	7.6	6	0.3	0.2	0.6	PbO, 21 MgO, 0.2	Readily melted and fabricated with good electrical properties. High lead absorbs x-rays; high refractive used in achromatic lenses. Decorative crystal glass.
5. High-lead	35		7.2				PbO, 58	
6. Borosilicate: Low expansion	80.5	3.8	0.4		12.9	2.2		Low expansion, good thermal shock resistance, and chemical stability. Widely used in chemical industry.
7. Low electrical loss	70.0		0.5		28.0	1.1	PbO, 1.2	Low dielectric loss.
8. Aluminoborosilicate: Standard apparatus	74.7	6.4	0.5	0.9	9.6	5.6	B$_2$O, 2.2	Increased alumina, lower boric oxide improves chemical durability.
9. Low alkali (E-glass)	54.5	0.5		22	8.5	14.5		Widely used for fibers in glass resin composites.
10. Aluminosilicate	57	1.0		5.5	4	20.5	MgO, 12	High-temperature strength, low expansion.
11. Glass-ceramic	40–70					10–35	MgO, 10–30 TiO$_2$, 7–15	Crystalline ceramic made by devitrifying glass. Easy fabrication (as glass), good properties. Various glasses and catalysts.

Source: O. H. Wyatt and D. Dew-Hughes, "Metals, Ceramics, and Polymers," Cambridge, 1974, p. 261.

browning of other glasses. It is therefore the ideal glass for space vehicle windows, wind tunnel windows, and for optical systems in spectrophotometric devices. However, silica glass is difficult to process and expensive.

Soda-lime glass   The most commonly produced glass is soda-lime glass which accounts for about 90 percent of all the glass produced. In this glass the basic composition is 71 to 73% SiO_2, 12 to 14% Na_2O, and 10 to 12% CaO. The Na_2O and CaO decrease the softening point of this glass from 1600 to about 730°C so that the soda-lime glass is easier to form. An addition of 1 to 4% MgO is added to the soda-lime glass to prevent devitrification, and an addition of 0.5 to 1.5% Al_2O_3 is used to increase durability. Soda-lime glass is used for flat glass, containers, pressed and blown ware, and lighting products where high chemical durability and heat resistance are not needed.

Borosilicate glasses   The replacement of alkali oxides by boric oxide in the silica glassy network produces a lower expansion glass. When B_2O_3 enters the silica network, it weakens its structure and considerably lowers the softening point of the silica glass. The weakening effect is attributed to the presence of planar three-coordinate borons. Borosilicate glass (Pyrex glass) is used for laboratory equipment, piping, ovenware, and sealed-beam headlights.

Lead glasses   Lead oxide is usually a modifier in the silica network but can also act as a network former. Lead glasses with high lead oxide contents are low melting and are useful for solder sealing glasses. High-lead glasses are also used for shielding from high-energy radiation and find application for radiation windows, fluorescent lamp envelopes, and television bulbs. Because of their high refractive indexes, lead glasses are used for some optical glasses and for decorative-purpose glasses.

Viscous Deformation of Glasses   A glass behaves as a viscous (supercooled) liquid above its glass transition temperature. Under stress, groups of silicate atoms (ions) can slide past each other, allowing permanent deformation of the glass. Interatomic bonding forces resist deformation above the glass transition temperature but are unable to prevent viscous flow of the glass if the applied stress is sufficiently high. As the temperature of the glass is progressively increased above its glass transition temperature, the viscosity of the glass decreases and viscous flow becomes easier. The effect of temperature on the viscosity of a glass follows an Arrhenius-type equation except that the sign of the exponential term is positive instead of negative as is usually the case (i.e., for diffusivity, the Arrhenius-type equation is $D = D_0 e^{(-Q/RT)}$). The equation relating viscosity to temperature for viscous flow in a glass is

$$\eta^* = \eta_0 e^{+Q/RT} \tag{10.5}$$

$^*\eta$ = Greek letter eta, pronounced "eight-ah."

where η = viscosity of the glass, P* or Pa·s; η_0 = preexponential constant, P or Pa·s; Q = molar activation energy for viscous flow; R = universal molar gas constant; and T = absolute temperature. Example Problem 10.11 shows how a value for the activation energy for viscous flow in a glass can be determined from this equation using viscosity-temperature data.

The effect of temperature on the viscosity of some commercial types of glasses is shown in Fig. 10.55. For the comparison of glasses, several viscosity reference points are used, indicated by horizontal lines in Fig. 10.55. These are working, softening, annealing, and strain points. Their definitions are:

1. *Working point:* viscosity = 10^4 poises (10^3 Pa·s). At this temperature glass fabrication operations can be carried out.
2. *Softening point:* viscosity = 10^8 poises (10^7 Pa·s). At this temperature the glass will flow at an appreciable rate under its own weight. However, this point cannot be defined by a precise viscosity because it depends on the density and surface tension of the glass.
3. *Annealing point:* viscosity = 10^{13} poises (10^{12} Pa·s). Internal stresses can be relieved at this temperature.

*IP (poise) = 1 dyne · s/cm²; 1 Pa · s (pascal-second) = 1 N · s/m²; 1 P = 0.1 Pa · s.

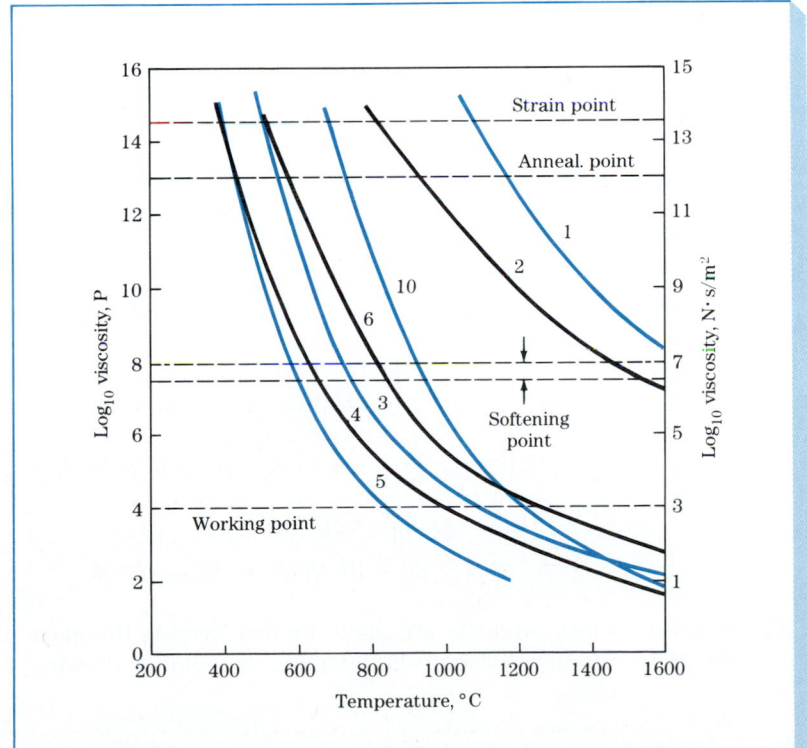

FIGURE 10.55 Effect of temperature on the viscosities of various types of glasses. Numbers on curves refer to compositions in Table 10.11. (*After O. H. Wyatt and D. Dew-Hughes, "Metals, Ceramics, and Polymers," Cambridge, 1974, p. 259.*)

4. *Strain point:* viscosity $= 10^{14.5}$ poises ($10^{13.5}$ Pa·s). Below this temperature the glass is rigid and stress relaxation occurs only at a slow rate. The interval between the annealing and strain points is commonly considered the annealing range of a glass.

Glasses are usually melted at a temperature which corresponds to a viscosity of about 10^2 poises (10 Pa·s). During forming, the viscosities of glasses are compared qualitatively. A *hard glass* has a high softening point, whereas a *soft glass* has a lower softening point. A *long glass* has a large temperature difference between its softening and strain points. That is, the long glass solidifies slower than a *short glass* as the temperature is decreased.

Example Problem 10.11

A 96% silica glass has a viscosity of 10^{13} P at its annealing point of 940°C and a viscosity of 10^8 P at its softening point of 1470°C. Calculate the activation energy in kilojoules per mole for the viscous flow of this glass in this temperature range.

Solution:

$$\text{Annealing point of glass} = T_{ap} = 940°C + 273 = 1213 \text{ K} \qquad \eta_{ap} = 10^{13} \text{ P}$$
$$\text{Softening point of glass} = T_{sp} = 1470°C + 273 = 1743 \text{ K} \qquad \eta_{sp} = 10^{8} \text{ P}$$

$$R = \text{gas constant} = 8.314 \text{ J/(mol·K)} \qquad Q = ? \text{ J/mol}$$

Using Eq. (10.5), $\eta = \eta_0 e^{Q/RT}$,

$$\begin{aligned}\eta_{ap} &= \eta_0 e^{Q/RT_{ap}} \\ \eta_{sp} &= \eta_0 e^{Q/RT_{sp}}\end{aligned} \quad \text{or} \quad \frac{\eta_{ap}}{\eta_{sp}} = \exp\left[\frac{Q}{R}\left(\frac{1}{T_{ap}} - \frac{1}{T_{sp}}\right)\right] = \frac{10^{13} \text{ P}}{10^{8} \text{ P}} = 10^{5}$$

$$10^{5} = \exp\left[\frac{Q}{8.314}\left(\frac{1}{1213 \text{ K}} - \frac{1}{1743 \text{ K}}\right)\right]$$

$$\ln 10^{5} = \frac{Q}{8.314}(8.244 \times 10^{-4} - 5.737 \times 10^{-4}) = \frac{Q}{8.314}(2.507 \times 10^{-4})$$

$$11.51 = Q(3.01 \times 10^{-5})$$

$$Q = 3.82 \times 10^{5} \text{ J/mol} = 382 \text{ kJ/mol} \blacktriangleleft$$

Forming Methods for Glasses   Glass products are made by first heating the glass to a high temperature to produce a viscous liquid and then molding, drawing, or rolling it into a desired shape.

Forming sheet and plate glass   About 85 percent of the flat glass produced in

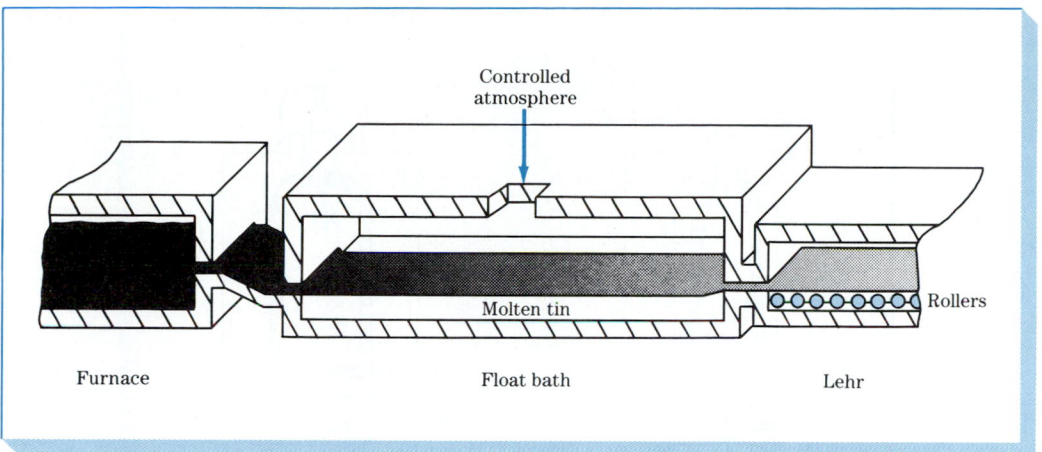

FIGURE 10.56 Diagram of the float-glass process. (*After D. C. Boyd and D. A. Thompson, "Glass,"*
vol. II: "Kirk-Othmer Encyclopedia of Chemical Technology," 3d ed., Wiley, 1980, p. 862.)

the United States is made by the float process in which a ribbon of glass moves
out of the melting furnace and floats on the surface of a bath of molten tin
(Fig. 10.56). The glass ribbon is cooled while moving across the molten tin
and while under a chemically controlled atmosphere (Fig. 10.56). When its
surfaces are sufficiently hard, the glass sheet is removed from the furnace
without being marked by rollers and passed through a long annealing furnace
called a *lehr,* where residual stresses are removed.

Blowing, pressing, and casting of glass   Deep items such as bottles, jars, and
light bulb envelopes are usually formed by blowing air to force molten glass
into molds (Fig. 10.57). Flat items such as optical and sealed-beam lenses are
made by pressing a plunger into a mold containing molten glass.

Many articles can be made by casting the glass into an open mold. A large
borosilicate glass telescope mirror 6 m in diameter was made by casting the
glass. Funnel-shaped items such as television tubes are formed by centrifugal
casting. Gobs of molten glass from feeder are dropped on to a spinning mold
which causes the glass to flow upward to form a glass wall of approximately
uniform thickness.

Tempered Glass   This type of glass is strengthened by rapid air cooling of the surface of the
glass after it has been heated to near its softening point. The surface of the
glass cools first and contracts, while the interior is warm and readjusts to the
dimensional change with little stress (Fig. 10.58*a*). When the interior cools
and contracts, the surfaces are rigid, and so tensile stresses are created in the
interior of the glass and compressive stresses on the surfaces (Figs. 10.58*b*
and 10.59). This "tempering" treatment increases the strength of the glass
because applied tensile stresses must surpass the compressive stresses on the

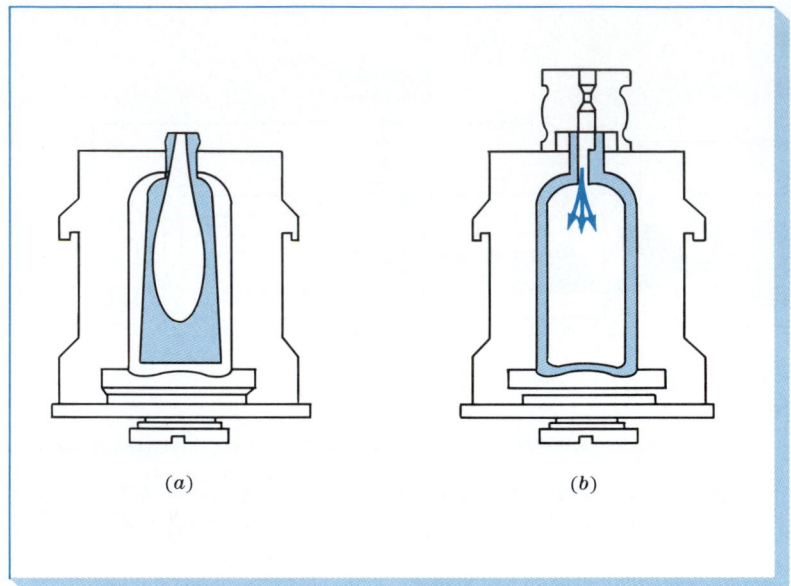

FIGURE 10.57 (*a*) Reheat and (*b*) final blow stages of a glass blowing machine process. (*After W. Giegerich and W. Trier, "Glass Machines Construction and Operation of Machines for the Forming of Hot Glass," Springer-Verlag, 1969.*)

(*a*)   (*b*)

surface before fracture occurs. Tempered glass has a higher resistance to impact than annealed glass and is about four times stronger than annealed glass. Auto side windows and safety glass for doors are items that are thermally tempered.

Chemically Strengthened Glass   The strength of glass can be increased by special chemical treatments. For example, if a sodium aluminosilicate glass is immersed in a bath of potassium nitrate at a temperature about 50°C below its stress point (~500°C) for 6 to 10 h, the smaller sodium ions near the surface of the glass are replaced by larger potassium ions. The introduction of the larger potassium ions into the surface

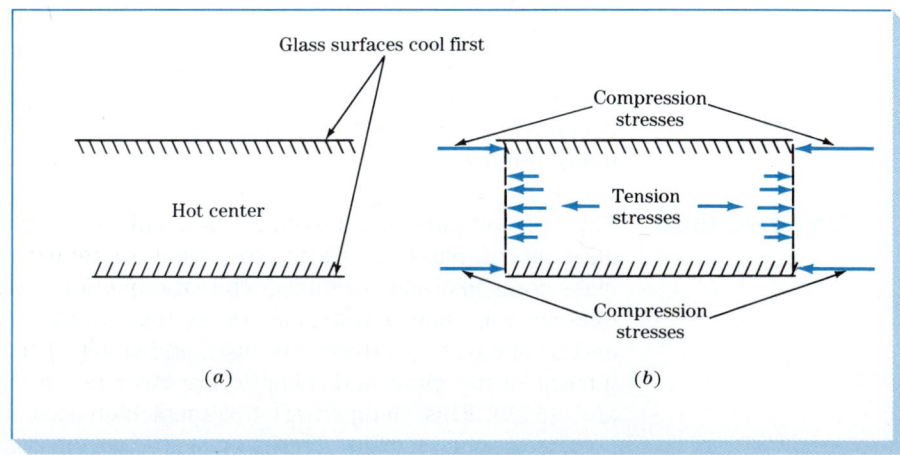

Glass surfaces cool first

Hot center

Compression stresses

Tension stresses

Compression stresses

(*a*)   (*b*)

FIGURE 10.58 Cross section of tempered glass (*a*) after surface has cooled from high temperature near glass-softening temperature and (*b*) after center has cooled.

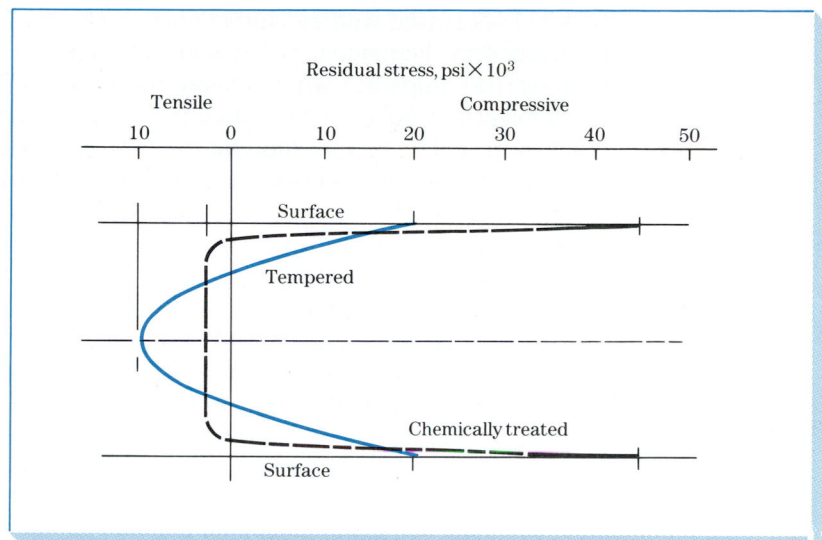

FIGURE 10.59
Distribution of residual stresses across the sections of glass thermally tempered and chemically strengthened. (*After E. B. Shand, "Engineering Glass," vol. 6: "Modern Materials," Academic, 1968, p. 270*).

of the glass produces compressive stresses at the surface and corresponding tensile stresses at its center. This chemical tempering process can be used on thinner cross sections than can thermal tempering since the compressive layer is much thinner, as shown in Fig. 10.59. Chemically strengthened glass is used for supersonic aircraft glazing and ophthalmic lenses.

SUMMARY

Ceramic materials are inorganic, nonmetallic materials consisting of metallic and nonmetallic elements bonded together primarily by ionic and/or covalent bonds. As a result, the chemical compositions and structures of ceramic materials vary considerably. They may consist of a single compound such as, for example, pure aluminum oxide, or they may be composed of a mixture of many complex phases such as the mixture of clay, silica, and feldspar which is contained in electrical porcelain.

The properties of ceramic materials also vary greatly due to differences in bonding. In general, most ceramic materials are typically hard and brittle with low impact resistance and ductility. Consequently, in most engineering designs, high stresses in ceramic materials are usually avoided, especially if they are tensile stresses. Ceramic materials are usually good electrical and thermal insulators due to the absence of conduction electrons, and thus many ceramics are used for electrical insulation and refractories. Some ceramic materials can be highly polarized with electric charge and are used for dielectric materials for capacitors. Permanent polarization of some ceramic materials produces piezoelectric properties which enable these materials to be used as electromechanical transducers. Other ceramic materials, for example, Fe_3O_4, are semiconductors and find application for thermistors for temperature measurement.

The processing of ceramic materials usually involves the agglomeration of small particles by a variety of methods in the dry, plastic, or liquid states. Cold-forming processes predominate in the ceramics industry, but hot-forming processes are also used. Pressing, slip casting, and extrusion are commonly used ceramic-forming processes. After forming, ceramic materials are usually given a thermal treatment such as sintering or vitrification. During sintering, the small particles of a formed article are bonded together by solid-state diffusion at high temperatures. In vitrification a glass phase serves as a reaction medium to bond the unmelted particles together.

Glasses are inorganic ceramic products of fusion which are cooled to a rigid solid without crystallization. Most inorganic glasses are based on a network of ionically covalently bonded silica (SiO_2) tetrahedra. Additions of other oxides such as Na_2O and CaO modify the silica network to provide a more workable glass. Other additions to glasses create a spectrum of properties. Glasses have special properties such as transparency, hardness at room temperature, and excellent resistance to most environments which make them important for many engineering designs.

DEFINITIONS

Sec. 10.1

Ceramic materials: inorganic, nonmetallic materials which consist of metallic and nonmetallic elements bonded together primarily by ionic and/or covalent bonds.

Sec. 10.2

Coordination number (CN): the number of equidistant nearest neighbors to an atom or ion in a unit cell of a crystal structure. For example, in NaCl, CN = 6 since six equidistant Cl^- anions surround a central Na^+ cation.

Radius ratio (for an ionic solid): the ratio of the radius of the central cation to that of the surrounding anions.

Critical (minimum) radius ratio: the ratio of the central cation to that of the surrounding anions when all the surrounding anions just touch each other and the central cation.

Octahedral interstitial site in the FCC crystal structure: the space enclosed when the nuclei of six surrounding atoms (ions) form an octahedron.

Tetrahedral interstitial site in the FCC crystal structure: the space enclosed when the nuclei of four surrounding atoms (ions) form a tetrahedron.

Sec. 10.4

Dry pressing: the simultaneous uniaxial compaction and shaping of ceramic granular particles (and binder) in a die.

Isostatic pressing: the simultaneous compaction and shaping of a ceramic powder (and binder) by pressure applied uniformly in all directions.

Slip casting: a ceramic shape-forming process in which a suspension of ceramic particles and water are poured into a porous mold and then some of the water from the cast material diffuses into the mold, leaving a solid shape in the mold. Sometimes excess liquid within the cast solid is poured from the mold, leaving a cast shell.

Sintering (of a ceramic material): the process in which fine particles of a ceramic material become chemically bonded together at a temperature high enough for atomic diffusion to occur between the particles.

Firing (of a ceramic material): heating a ceramic material to a high-enough temperature to cause a chemical bond to form between the particles.

Vitrification: melting or formation of a glass; the vitrification process is used to produce a viscous liquid glass in a ceramic mixture upon firing. Upon cooling the liquid phase solidifies and forms a vitreous or glassy matrix which bonds the unmelted particles of the ceramic material together.

Sec. 10.6

Dielectric: an electrical insulator material.

Capacitor: an electric device consisting of conducting plates or foils separated by layers of dielectric material and which is capable of storing electric charge.

Capacitance: a measure of the ability of a capacitor to store electric charge. Capacitance is measured in farads; the units commonly used in electrical circuitry are the picofarad ($1 \text{ pF} = 10^{-12} \text{ F}$) and the microfarad ($1 \ \mu\text{F} = 10^{-6} \text{ F}$).

Dielectric constant: the ratio of the capacitance of a capacitor using a material between the plates of a capacitor compared to that of the capacitor when there is a vacuum between the plates.

Dielectric strength: the voltage per unit length (electric field) at which a dielectric material allows conduction, that is, the maximum electric field that a dielectric can withstand without electrical breakdown.

Thermistor: a ceramic semiconductor device which changes in resistivity as the temperature changes and is used to measure and control temperature.

Ferroelectric material: a material which can be polarized by applying an electric field.

Polarization: the alignment of small electric dipoles in a dielectric material to produce a net dipole moment in the material.

Curie temperature (of a ferroelectric material): the temperature at which a ferroelectric material on cooling undergoes a crystal structure change which produces spontaneous polarization in the material. For example, the Curie temperature of $BaTiO_3$ is 120°C.

Piezoelectric effect: an electromechanical effect by which mechanical forces on a ferroelectric material can produce an electrical response and electrical forces produce a mechanical response.

Transducer: a device which is actuated by power from one source and transmits power in another form to a second system. For example, a transducer can convert input sound energy into an output electrical response.

Sec. 10.8

Refractory (ceramic) material: a material that can withstand the action of a hot environment.

Sec. 10.9

Glass: a ceramic material that is made from inorganic materials at high temperatures and is distinguished from other ceramics in that its constituents are heated to fusion and then cooled to the rigid condition without crystallization.

Glass transition temperature: the center of the temperature range in which a non-crystalline solid changes from being glass-brittle to being viscous.

Glass-forming oxide: an oxide which forms a glass easily; also an oxide which contributes to the network of silica glass when added to it, such as B_2O_3.

Glass-modifying oxide: an oxide which breaks up the silica network when added to silica glass; modifiers lower the viscosity of silica glass and promote crystallization. Examples are Na_2O, K_2O, CaO, and MgO.

Glass intermediate oxides: an oxide which may act either as a glass former or as a glass modifier depending on the composition of the glass. Example, Al_2O_3.

Glass reference points (temperatures).

   Working point: at this temperature the glass can easily be worked.

   Softening point: at this temperature the glass flows at an appreciable rate.

   Annealing point: at this temperature stresses in the glass can be relieved.

   Strain point: at this temperature the glass is rigid.

Float glass: flat glass which is produced by having a ribbon of molten glass cool to the glass-brittle state while floating on the top of a flat bath of molten tin and under a reducing atmosphere.

Thermally tempered glass: glass which has been reheated to near its softening temperature and then rapidly cooled in air to introduce compressive stresses near its surface.

Chemically tempered glass: glass which has been given a chemical treatment to introduce large ions into its surface to cause compressive stresses at its surface.

PROBLEMS

10.1.1   Define a ceramic material.

10.1.2   What are some properties common to most ceramic materials?

10.1.3   Distinguish between traditional and engineering ceramic materials and give examples of each.

10.2.1   Using Pauling's equation [Eq. (2.10)], compare the percent covalent character of the following compounds: hafnium carbide, titanium carbide, tantalum carbide, boron carbide, and silicon carbide.

10.2.2   What two main factors affect the packing of ions in ionic solids?

10.2.3   Define (*a*) coordination number and (*b*) critical radius ratio for the packing of ions in ionic solids.

10.2.4   Using Fig. 10.60, calculate the critical radius ratio for octahedral coordination.

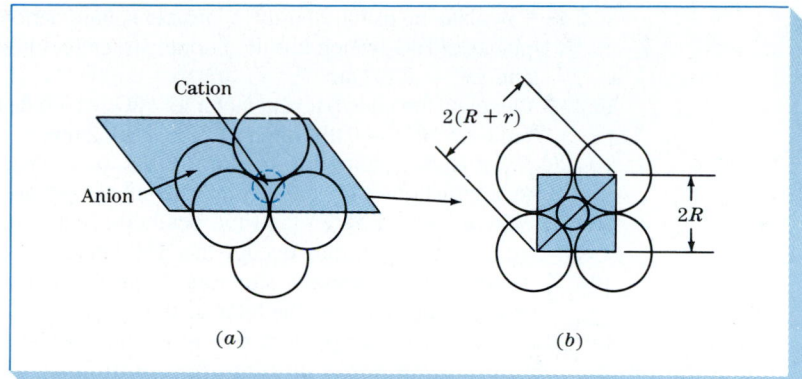

Figure 10.60 (*a*) Octahedral coordination of six anions (radii = *R*) around a central cation of radius *r*. (*b*) Horizontal section through center of (*a*).

10.2.5 Predict the coordination number for (*a*) SrO and (*b*) RbCl. Ionic radii are Sr^{2+} = 0.127 nm, O^{2-} = 0.132 nm, Rb^+ = 0.149 nm, Cl^- = 0.181 nm.

10.2.6 Calculate the density in grams per cubic centimeter of CsI which has the CsCl structure. Ionic radii are Cs^+ = 0.165 nm and I^- = 0.220 nm.

10.2.7 Calculate the density in grams per cubic centimeter of CsBr which has the CsCl structure. Ionic radii are Cs^+ = 0.165 nm and Br^- = 0.196 nm.

10.2.8 Calculate the linear densities in ions per nanometer in the [110] and [111] directions for (*a*) BaO and (*b*) CoO. Ionic radii are Ba^{2+} = 0.143 nm, Co^{2+} = 0.082 nm, and O^{2-} = 0.132 nm. Both BaO and CoO have the NaCl structure.

10.2.9 Calculate the planar densities in ions per square nanometer on the (111) and (110) planes for (*a*) FeO and (*b*) KBr. Ionic radii are Fe^{2+} = 0.087 nm, O^{2-} = 0.132 nm, K^+ = 0.133 nm, and Br^- = 0.196 nm. Both FeO and KBr have the NaCl structure.

10.2.10 Calculate the density in grams per cubic centimeter of (*a*) NiO and (*b*) CdO. Ionic radii are Ni^{2+} = 0.078 nm, Cd^{2+} = 0.103 nm, and O^{2-} = 0.132 nm. Both NiO and CdO have the NaCl structure.

10.2.11 Calculate the ionic packing factor for (*a*) CaO and (*b*) BaO. Ionic radii are Ca^{2+} = 0.106 nm, Ba^{2+} = 0.143 nm, and O^{2-} = 0.132 nm. Both CaO and BaO have the NaCl structure.

10.2.12 CdSe has the zinc blende crystal structure. Calculate the density of CdSe. Ionic radii are Cd^{2+} = 0.103 nm and Se^{2-} = 0.191 nm.

10.2.13 CdTe has the zinc blende crystal structure. Calculate the density of CdTe. Ionic radii are Cd^{2+} = 0.103 nm and Te^{2-} = 0.211 nm.

10.2.14 Draw the unit cell for BaF_2 which has the fluorite, CaF_2, crystal structure. If the Ba^{2+} ions occupy the FCC lattice sites, which sites do the F^- ion occupy?

10.2.15 Calculate the density in grams per cubic centimeter of ZrO_2 which has the CaF_2 crystal structure. Ionic radii are Zr^{4+} = 0.087 nm and O^{2-} = 0.132 nm.

10.2.16 What fraction of the octahedral interstitial sites are occupied in the CaF_2 structure?

10.2.17 Calculate the linear density in ions per nanometer in the [111] and [110] directions for ZrO_2 which has the fluorite structure. Ionic radii are Zr^{4+} = 0.087 nm and O^{2-} = 0.132 nm.

10.2.18 Calculate the planar density in ions per square nanometer in the (111) and (110) planes of HfO_2 which has the fluorite structure. Ionic radii are $Hf^{4+} = 0.084$ nm and $O^{2-} = 0.132$ nm.

10.2.19 Calculate the ionic packing factor for HfO_2 which has the fluorite structure. Ionic radii are $Hf^{4+} = 0.084$ nm and $O^{2-} = 0.132$ nm.

10.2.20 What is the antifluorite structure? What ionic compounds have this structure? What fraction of the tetrahedral interstitial sites are occupied by cations?

10.2.21 Why are only two-thirds of the octahedral interstitial sites filled by Al^{3+} ions when the oxygen ions occupy the HCP lattice sites in Al_2O_3?

10.2.22 Describe the perovskite structure. What fraction of the octahedral interstitial sites are occupied by the tetravalent cation?

10.2.23 Calculate the ionic packing factor for $SrSnO_3$ which has the perovskite structure. Ionic radii are $Sr^{2+} = 0.127$ nm, $Sn^{4+} = 0.074$ nm, and $O^{2-} = 0.132$ nm. Assume the lattice constant $a = 2(r_{Sn^{4+}} + r_{O^{2-}})$.

10.2.24 Calculate the density in grams per cubic centimeter of $SrHfO_3$ which has the perovskite structure. Ionic radii are $Sr^{2+} = 0.127$ nm, $Hf^{4+} = 0.084$ nm, and $O^{2-} = 0.132$ nm. Assume $a = 2(r_{Hf^{4+}} + r_{O^{2-}})$.

10.2.25 What is the spinel crystal structure?

10.2.26 Draw a section of the graphite structure. Why are the layers of graphite able to slide past each other easily?

10.3.1 Describe and illustrate the following silicate structures: (a) island, (b) chain, and (c) sheet.

10.3.2 Describe the structure of a sheet of kaolinite.

10.3.3 Describe the bonding arrangement in the cristobalite (silica) network structure.

10.3.4 Describe the feldspar network structure.

10.4.1 What are the basic steps in the processing of ceramic products by the agglomeration of particles?

10.4.2 What types of ingredients are added to ceramic particles in preparing ceramic raw materials for processing?

10.4.3 Describe two methods for preparing ceramic raw materials for processing.

10.4.4 Describe the dry-pressing method for producing ceramic products such as technical ceramic compounds and structural refractories. What are the advantages of dry pressing ceramic materials?

10.4.5 Describe the isostatic-pressing method for producing ceramic products.

10.4.6 Describe the four stages in the manufacture of a spark plug insulator.

10.4.7 What are the advantages of hot pressing ceramics materials?

10.4.8 Describe the steps in the slip-casting process for ceramic products?

10.4.9 What is the difference between (a) drain and (b) solid slip casting?

10.4.10 What are the advantages of slip casting?

10.4.11 What types of ceramic products are produced by extrusion? What are the advantages of this process? Limitations?

10.4.12 What are the purposes of drying ceramic products before firing?

10.4.13 What is the sintering process? What occurs to the ceramic particles during sintering?

10.4.14 What is the vitrification process? In what type of ceramic materials does vitrification take place?

10.5.1 What are the three basic components of traditional ceramics?

10.5.2 What is the approximate composition of kaolin clay?

10.5.3 What is the role of clay in traditional ceramics?

10.5.4   What is flint? What role does it have in traditional ceramics?

10.5.5   What is feldspar? What role does it have in traditional ceramics?

10.5.6   List some examples of whiteware ceramic products.

10.5.7   Why is the term *triaxial* used to describe some whitewares?

10.5.8   Determine the composition of the ternary compound at point *y* in Fig. 10.32.

10.5.9   Why are triaxial porcelains not satisfactory for use at high frequencies?

10.5.10 What kinds of ions cause an increase in the conductivity of electrical porcelain?

10.5.11 What is the composition of most technical ceramics?

10.5.12 How are pure single-compound technical ceramic particles processed to produce a solid product? Give an example.

10.6.1   What are three major applications for ceramic materials in the electrical-electronics industries?

10.6.2   Define the terms *dielectric, capacitor,* and *capacitance.* What is the SI unit for capacitance? What units are commonly used for capacitance in the electronics industry?

10.6.3   What is the dielectric constant of a dielectric material? What is the relationship among capacitance, dielectric constant, and the area and distance of separation between the plates of a capacitor?

10.6.4   What is the dielectric strength of a dielectric material? What units are used for dielectric strength? What is dielectric breakdown?

10.6.5   What is the dielectric loss angle and dielectric loss factor for a dielectric material? Why is a high dielectric loss factor undesirable?

10.6.6   A simple plate capacitor can store 7.0×10^{-5} C at a potential of 12,000 V. If a barium titanate dielectric material with $\kappa = 2100$ is used between the plates which have an area of 5.0×10^{-5} m^2, what must be the separation distance between the plates?

10.6.7   A simple plate capacitor stores 6.5×10^{-5} C at a potential of 12,000 V. If the area of the plates is 3.0×10^{-5} m^2 and the distance between the plates is 0.18 mm, what must be the dielectric constant of the material between the plates?

10.6.8   What is the approximate composition of electrical porcelain? What is a major disadvantage of electrical porcelain as electrical insulative material?

10.6.9   What is the approximate composition of steatite? What desirable electrical properties does steatite have as an insulative material?

10.6.10 What is the composition of fosterite? Why is fosterite an excellent insulator material?

10.6.11 Why is sintered alumina widely used as a substance for electronic device applications?

10.6.12 Why is $BaTiO_3$ used for high-value, small, flat-disk capacitors? How is the capacitance of $BaTiO_3$ capacitors varied? What are the four major stages in the manufacture of a flat-disk ceramic capacitor?

10.6.13 What is a thermistor? What is an NTC thermistor?

10.6.14 What materials are used to make NTC thermistors?

10.6.15 What is believed to be the mechanism for electrical conduction in Fe_3O_4?

10.6.16 How is the electrical conductivity of metal oxide semiconductors for thermistors changed?

10.6.17 What change occurs in the unit cell of $BaTiO_3$ when it is cooled below 120°C? What is this transformation temperature called?

10.6.18 What are ferroelectric domains? How can they be lined toward one direction?

10.6.19 Describe the piezoelectric effect for producing an electrical response with the application of pressure on a ferroelectric material. Do the same for producing a mechanical response by the application of an electrical force.

10.6.20 Describe several devices that utilize the piezoelectric effect.

10.6.21 What are the PZT piezoelectric materials? In what ways are they superior to $BaTiO_3$ piezoelectric materials?

10.7.1 What causes the lack of plasticity in crystalline ceramics?

10.7.2 Explain the plastic deformation mechanism for some single-crystal ionic solids such as NaCl and MgO. What is the preferred slip system?

10.7.3 What structural defects are the main cause of failure of polycrystalline ceramic materials?

10.7.4 How do (a) porosity and (b) grain size affect the tensile strength of ceramic materials?

10.7.5 A reaction-bonded silicon nitride ceramic has a strength of 320 MPa and fracture toughness of 3.8 MPa$\sqrt{m}$. What is the largest-sized internal flaw that this material can support without fracturing? (Use $Y = 1$ in the fracture-toughness equation.)

10.7.6 The maximum-sized internal flaw in a hot-pressed silicon carbide ceramic is 25.0 μm. If this material has a fracture toughness of 4.5 MPa$\sqrt{m}$, what is the maximum stress that this material can support? (Use $Y = 1$.)

10.7.7 A partially stabilized zirconia advanced ceramic has a strength of 380 MPa and a fracture toughness 9.8 MPa$\sqrt{m}$. What is the largest-sized internal flaw (expressed in micrometers) that this material can support? (Use $Y = 1$.)

10.7.8 A fully stabilized, cubic polycrystalline ZrO_2 sample has a fracture toughness of $K_{IC} = 4.8$ MPa$\sqrt{m}$ when tested on a four-point bend test.

   (a) If the sample fails at a stress of 400 MPa, what is the size of the largest surface flaw? Assume $Y = \sqrt{\pi}$.

   (b) The same test is performed with a partially stabilized ZrO_2 specimen. This material is transformation-toughened and has a $K_{IC} = 13.0$ MPa$\sqrt{m}$. If this material has the same flaw distribution as the fully stabilized sample, what stress must be applied to cause failure?

10.7.9 What are the two most important industrial abrasives?

10.7.10 What are important properties for industrial abrasives?

10.8.1 Why do most ceramic materials have low thermal conductivities?

10.8.2 What are refractories? What are some of their applications?

10.8.3 What are the two main types of ceramic refractory materials?

10.8.4 Give the composition and several applications for the following refractories: (a) silica, (b) fireclay, and (c) high-alumina.

10.8.5 What do most basic refractories consist of? What are some important properties of basic refractories? What is a main application for these materials?

10.8.6 What is the high-temperature reusable-surface insulation which can withstand temperatures as high as 1260°C made of?

10.9.1 Define a glass.

10.9.2 What are some of the properties of glasses which make them indispensable for many engineering designs?

10.9.3 How is a glass distinguished from other ceramic materials?

10.9.4 How does the specific volume vs. temperature plot for a glass differ from that for a crystalline material when these materials are cooled from the liquid state?

10.9.5 Define the glass transition temperature.

10.9.6 Name two glass-forming oxides. What are their fundamental subunits and their shape?

10.9.7 How does the silica network of a simple silica glass differ from crystalline (cristobalite) silica?

10.9.8 How is it possible for BO_3^{3-} triangles to be converted to BO_4^{4-} tetrahedra and still maintain neutrality in some borosilicate glasses?

10.9.9 What are glass network modifiers? How do they affect the silica-glass network? Why are they added to silica glass?

10.9.10 What are glass intermediate oxides? How do they affect the silica-glass network? Why are they added to silica glass?

10.9.11 What is fused silica glass? What are some of its advantages and disadvantages?

10.9.12 What is the basic composition of soda-lime glass? What are some of its advantages and disadvantages? What are some applications for soda-lime glass?

10.9.13 What is the purpose of (a) MgO and (b) Al_2O_3 additions to soda-lime glass?

10.9.14 Define the following viscosity reference points for glasses: working, softening, annealing, and strain.

10.9.15 Distinguish between hard and soft glasses and long and short glasses.

10.9.16 A silica glass between 810°C (strain point) and 1420°C (softening point) has viscosities of $10^{14.5}$ and 10^8 P, respectively. Calculate a value for the activation energy in this temperature region.

10.9.17 An electrical lead silicate glass between 400°C (strain point) and 650°C (softening point) has viscosities of $10^{14.6}$ and $10^{8.0}$ P, respectively.
(a) Calculate a value for the activation energy for viscous flow in this region.
(b) At 500°C, what is its viscosity?

10.9.18 An aluminosilicate glass between 750°C (strain point) and 950°C (softening point) has viscosities of $10^{14.6}$ and $10^{7.6}$ P respectively.
(a) Calculate a value for the activation energy for viscous flow in this temperature region.
(b) At what temperature in kelvins will its viscosity be $10^{11.0}$ P?

10.9.19 A silica glass between 810 an 1420°C has viscosities of $10^{14.0}$ and $10^{8.2}$ P, respectively. Calculate a value for the activation energy in this temperature region.

10.9.20 Describe the float-glass process for the production of flat-glass products. What is its major advantage?

10.9.21 What is tempered glass? How is it produced? Why is tempered glass considerably stronger in tension than annealed glass? What are some applications for tempered glass?

10.9.22 What is chemically strengthened glass? Why is chemically strengthened glass stronger in tension than annealed glass?

Magnetic Materials

Magnetic materials are important industrial materials necessary for many engineering designs, particularly in the area of electrical engineering. In general there are two main types: *soft* and *hard magnetic materials.* Soft magnetic materials are used for applications in which the material must be easily magnetized and demagnetized such as cores for distribution power transformers (Fig. 11.1*a*), small electronic transformers, and stator and rotor materials for motors and generators. On the other hand, hard magnetic materials are used for applications requiring permanent magnets which do not demagnetize easily such as the permanent magnets in loudspeakers, telephone receivers, synchronous and brushless motors, and automotive starting motors.

In this chapter we shall first look into the origin of magnetism in ferromagnetic materials and then briefly examine some of the basic units and relationships associated with magnetism and magnetic materials. Next we shall investigate some of the important properties of magnetic fields and look at the formation and movement of domains in ferromagnetic materials. Following that, we shall discuss some aspects of the structure and properties of some industrial soft and hard ferromagnetic materials. Finally, we shall briefly look at ferrimagnetism and the structure and properties of ferrites which are ceramic magnetic materials.

(a)   (b)

FIGURE 11.1 (a) A new magnetic material for engineering designs: metallic glass material is used for the magnetic cores of distribution electric power transformers. The use of highly magnetically soft amorphous metallic glass alloys for transformer cores reduces core energy losses by about 70 percent compared with those made with conventional iron-silicon alloys. (*Courtesy of General Electric Co.*) (b) Strip of metallic glass ribbon. (*Courtesy of Allied Metglas Products.*)

11.1 MAGNETIC FIELDS AND QUANTITIES

Magnetic Fields   Let us begin our study of magnetic materials by first reviewing some of the fundamental properties of magnetism and magnetic fields. The metals *iron, cobalt,* and *nickel* are the only three elemental metals which when magnetized at room temperature can produce a strong magnetic field around themselves, and are said to be *ferromagnetic.* The presence of a magnetic field surrounding a magnetized iron bar can be revealed by scattering small iron particles on a sheet of paper which is placed just above the iron bar (Fig. 11.2). As shown in Fig. 11.2, the bar magnet has two magnetic poles, and magnetic field lines appear to leave one pole and enter the other.

In general, magnetism is dipolar in nature, and no magnetic monopole has ever been discovered. There are always two magnetic poles or centers of a magnetic field which are separated by a definite distance, and this dipole behavior extends down to the small magnetic dipoles found in some atoms.

Magnetic fields are also produced by current-carrying conductors. Figure 11.3 illustrates the formation of a magnetic field around a long coil of copper wire, called a *solenoid,* whose length is long with respect to its radius. For a solenoid of *n* turns and length *l,* the magnetic field strength *H* is

FIGURE 11.2 The magnetic field surrounding a bar magnet is revealed by the arrangement of iron filings lying on a sheet of paper above the magnet. Note that the bar magnet is dipolar and that magnetic lines of force appear to leave one of the magnet and return to the other. (*Courtesy of the Physical Science Study Committee, as appearing in D. Halliday and R. Resnick, "Fundamentals of Physics," Wiley, 1974, p. 612.*)

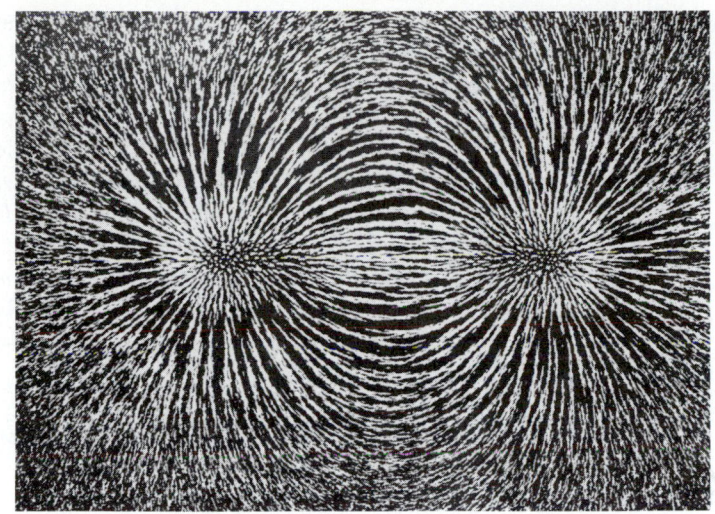

$$H = \frac{0.4\pi n i}{l} \qquad (11.1)$$

where i is the current. The magnetic field strength H has SI units of amperes per meter (A/m) and cgs units of oersteds (Oe). The conversion equality between SI and cgs units for H is 1 A/m = $4\pi \times 10^{-3}$ Oe.

FIGURE 11.3 (*a*) Schematic illustration of a magnetic field created around a coil of copper wire, called a solenoid, by the passage of current through the wire. (*b*) Schematic illustration of the increase in magnetic field around the solenoid when an iron bar is placed inside the solenoid and current is passed through the wire. (*After C. S. Barrett, W. D. Nix, and A. S. Tetelman, "Principles of Engineering Materials," Prentice-Hall, 1973, p. 459.*)

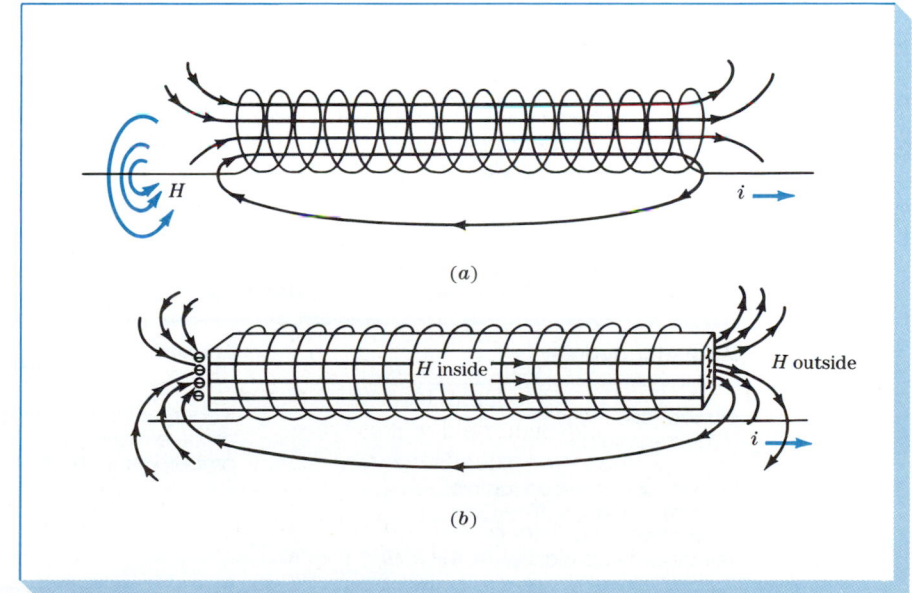

Magnetic Induction   Now let us place a demagnetized iron bar inside the solenoid and apply a magnetizing current to the solenoid, as shown in Fig. 11.3*b*. The magnetic field outside the solenoid is now found to be stronger with the magnetized bar inside the solenoid. The enhanced magnetic field outside the solenoid is due to the sum of the solenoid field itself and the external magnetic field of the magnetized bar. The new additive magnetic field is called the *magnetic induction,* or *flux density,* or simply *induction* and is given the symbol *B*.

The magnetic induction *B* is the sum of the applied field *H* and the external field that arises from the magnetization of the bar inside the solenoid. The induced magnetic moment per unit volume due to the bar is called the *intensity of magnetization,* or simply *magnetization,* and is given the symbol *M*. In the SI system of units,

$$B = \mu_0 H + \mu_0 M = \mu_0(H + M) \tag{11.2}$$

where μ_0 = *permeability of free space* = $4\pi \times 10^{-7}$ tesla-meters per ampere (T·m/A).† μ_0 has no physical meaning and is only needed in Eq. (11.2) because SI units were chosen. The SI units for *B* are webers‡ per square meter (Wb/m²), or teslas (T), and the SI units for *H* and *M* are amperes per meter (A/m). The cgs unit for *B* is the gauss (G) and for *H*, the oersted (Oe). Table 11.1 summarizes these magnetic units.

An important point to note is that for ferromagnetic materials, in many cases the magnetization $\mu_0 M$ is often much greater than the applied field $\mu_0 H$, and so we can often use the relation $B \approx \mu_0 M$. Thus, for ferromagnetic materials, sometimes the quantities *B* (magnetic induction) and *M* (magnetization) are used interchangeably.

---

† Nikola Tesla (1856–1943). American inventor of Yugoslavian birth who in part developed the polyphase induction motor and invented the Tesla coil (an air transformer). 1 T = 1 Wb/m² = 1 V·s/m².

‡ 1 Wb = 1 V·s.

TABLE 11.1   Summary of the Units for the Magnetic Quantities

Magnetic quantity	SI units	cgs units
B (magnetic induction)	weber/meter² (Wb/m²) or tesla (T)	gauss (G)
H (applied field)	ampere/meter (A/m)	oersted (Oe)
M (magnetization)	ampere/meter (A/m)	
Numerical conversion factors:		
1 A/m = $4\pi \times 10^{-3}$ Oe		
1 Wb/m² = 1.0×10^4 G		
Permeability constant $\mu_0 = 4\pi \times 10^{-7}$ T·m/A		

Magnetic Permeability   As previously pointed out, when a ferromagnetic material is placed in an applied magnetic field, the intensity of the magnetic field is increased. This increase in magnetization is measured by a quantity called *magnetic permeability* μ, which is defined as the ratio of the magnetic induction B to the applied field H, or

$$\mu = \frac{B}{H} \tag{11.3}$$

If there is only a vacuum in the applied magnetic field, then

$$\mu_0 = \frac{B}{H} \tag{11.4}$$

where $\mu_0 = 4\pi \times 10^{-7}$ T·m/A = permeability of free space, as previously stated.

An alternative method used for defining magnetic permeability is the quantity *relative permeability* μ_r, which is the ratio of μ/μ_0. Thus

$$\mu_r = \frac{\mu}{\mu_0} \tag{11.5}$$

and

$$B = \mu_0 \mu_r H \tag{11.6}$$

The relative permeability μ_r is a dimensionless quantity.

The relative permeability is a measure of the intensity of the induced magnetic field. In some ways the magnetic permeability of magnetic materials is analogous to the dielectric constant of dielectric materials. However, the magnetic permeability of a ferromagnetic material is not a constant but changes as the material is magnetized, as indicated in Fig. 11.4. The magnetic permeability of a magnetic material is usually measured by either its initial permeability μ_i or its maximum permeability μ_{max}. Figure 11.4 indicates how μ_i and μ_{max} are measured from slopes of the *B-H* initial magnetizing curve for a magnetic material. Magnetic materials which are easily magnetized have high magnetic permeabilities.

Magnetic Susceptibility   Since the magnetization of a magnetic material is proportional to the applied field, a proportionality factor called the *magnetic susceptibility* χ_m is defined as

$$\chi_m = \frac{M}{H} \tag{11.7}$$

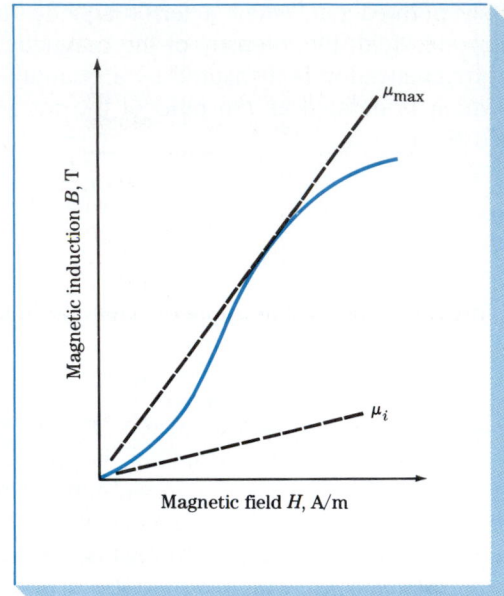

FIGURE 11.4 *B-H* initial magnetization curve for a ferromagnetic material. The slope μ_i is the initial magnetic permeability, and the slope μ_{max} is the maximum magnetic permeability.

which is a dimensionless quantity. Weak magnetic responses of materials are often measured in terms of magnetic susceptibility.

11.2 TYPES OF MAGNETISM

Magnetic fields and forces originate from the movement of the basic electric charge, the electron. When electrons move in a conducting wire, a magnetic field is produced around the wire, as shown for the solenoid of Fig. 11.3. Magnetism in materials is also due to the motion of electrons, but in this case

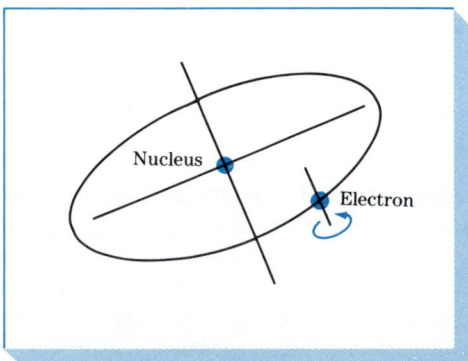

FIGURE 11.5 A schematic drawing of the Bohr atom indicating an electron spinning on its own axis and revolving about its nucleus. The spin of the electron on its axis and its orbital motion around its nucleus are the origins of magnetism in materials.

TABLE 11.2   Magnetic Susceptibilities of Some Diamagnetic and Paramagnetic Elements

Diamagnetic element	Magnetic susceptibility $\chi_m \times 10^{-6}$	Paramagnetic element	Magnetic susceptibility $\chi_m \times 10^{-6}$
Cadmium	−0.18	Aluminum	+0.65
Copper	−0.086	Calcium	+1.10
Silver	−0.20	Oxygen	+106.2
Tin	−0.25	Platinum	+1.10
Zinc	−0.157	Titanium	+1.25

the magnetic fields and forces are caused by the intrinsic spin of electrons and their orbital motion about their nuclei (Fig. 11.5).

Diamagnetism An external magnetic field acting on the atoms of a material slightly unbalances their orbiting electrons and creates small magnetic dipoles within the atoms which oppose the applied field. This action produces a negative magnetic effect known as *diamagnetism.* The diamagnetic effect produces a very small negative magnetic susceptibility of the order of $\chi_m \approx -10^{-6}$ (Table 11.2). Diamagnetism occurs in all materials, but in many its negative magnetic effect is canceled by positive magnetic effects. Diamagnetic behavior has no significant engineering importance.

Paramagnetism Materials which exhibit a small positive magnetic susceptibility in the presence of a magnetic field are called *paramagnetic,* and the magnetic effect is termed *paramagnetism.* The paramagnetic effect in materials disappears when the applied magnetic field is removed. Paramagnetism produces magnetic susceptibilities in materials ranging from about 10^{-6} to 10^{-2} and is produced in many materials. Table 11.2 lists the magnetic susceptibilities of paramagnetic materials at 20°C. Paramagnetism is produced by the alignment of individual magnetic dipole moments of atoms or molecules in an applied magnetic field. Since thermal agitation randomizes the directions of the magnetic dipoles, an increase in temperature decreases the paramagnetic effect.

The atoms of some transition and rare-earth elements possess incompletely filled inner shells with unpaired electrons. These unpaired inner electrons in atoms, since they are not counterbalanced by other bonding electrons in solids, cause strong paramagnetic effects and in some cases produce very much stronger ferromagnetic and ferrimagnetic effects, which will be discussed subsequently.

Ferromagnetism Diamagnetism and paramagnetism are induced by an applied magnetic field, and the magnetization remains only as long as the field is maintained. A third type of magnetism, called *ferromagnetism,* is of great engineering importance.

Unpaired $3d$ electrons	Atom	Number of electrons	Electronic configuration $3d$ orbitals	$4s$ electrons
3	V	23	↑ / ↑ / ↑ / — / —	2
5	Cr	24	↑ / ↑ / ↑ / ↑ / ↑	1
5	Mn	25	↑ / ↑ / ↑ / ↑ / ↑	2
4	Fe	26	↑↓ / ↑ / ↑ / ↑ / ↑	2
3	Co	27	↑↓ / ↑↓ / ↑ / ↑ / ↑	2
2	Ni	28	↑↓ / ↑↓ / ↑↓ / ↑ / ↑	2
0	Cu	29	↑↓ / ↑↓ / ↑↓ / ↑↓ / ↑↓	1

FIGURE 11.6 Magnetic moments of neutral atoms of $3d$ transition elements.

Large magnetic fields that can be retained or eliminated as desired can be produced in ferromagnetic materials. The most important ferromagnetic elements from an industrial standpoint are iron (Fe), cobalt (Co), and nickel (Ni). Gadolinium (Gd), a rare-earth element, is also ferromagnetic below 16°C but has little industrial application.

The ferromagnetic properties of the transition elements Fe, Co, and Ni are due to the way the spins of the inner unpaired electrons are aligned in their crystal lattices. The inner shells of individual atoms are filled with pairs of electrons with opposed spins, and so there are no resultant magnetic dipole moments due to them. In solids the outer valence electrons of atoms are combined with each other to form chemical bonds, and so there is no significant magnetic moment due to these electrons. In Fe, Co, and Ni the unpaired inner $3d$ electrons are responsible for the ferromagnetism which these elements exhibit. The iron atom has four unpaired $3d$ electrons, the cobalt atom three, and the nickel atom two (Fig. 11.6).

In a solid sample of Fe, Co, or Ni at room temperature, the spins of the $3d$ electrons of adjacent atoms align in a parallel direction by a phenomenon called *spontaneous magnetization*. This parallel alignment of atomic magnetic dipoles occurs only in microscopic regions called *magnetic domains*. If the

FIGURE 11.7 Magnetic exchange interaction energy as a function of the ratio of atomic spacing to the diameter of the 3d orbit for some 3d transition elements. Those elements which have positive exchange energies are ferromagnetic; those with negative exchange energies are antiferromagnetic.

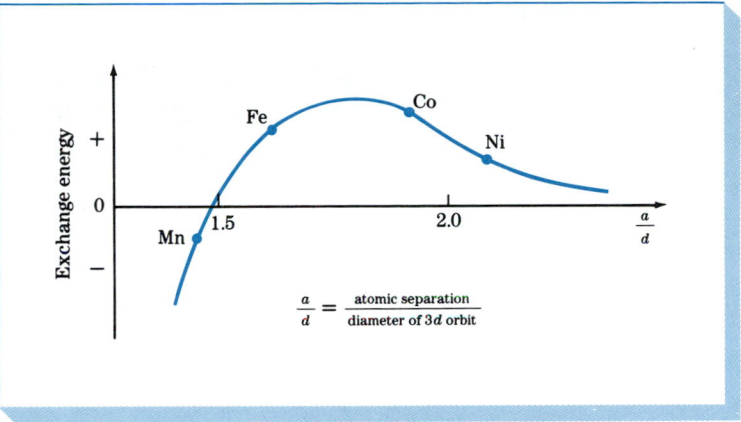

domains are randomly oriented, then there will be no net magnetization in a bulk sample. The parallel alignment of the magnetic dipoles of atoms of Fe, Co, and Ni is due to the creation of a positive exchange energy between them. For this parallel alignment to occur the ratio of the atomic spacing to the diameter of the 3d orbit must be in the range from about 1.4 to 2.7 (Fig. 11.7). Thus Fe, Co, and Ni are ferromagnetic, but manganese (Mn) and chromium (Cr) are not.

Magnetic Moment of a Single Unpaired Atomic Electron

Each electron spinning on its own axis (Fig. 11.5) behaves as a magnetic dipole and has a dipole moment called the *Bohr magneton* μ_B. This dipole moment has the value of

$$\mu_B = \frac{eh}{4\pi m} \tag{11.8}$$

where e = electronic charge, h = Planck's constant, and m = electron mass. In SI units $\mu_B = 9.27 \times 10^{-24}$ A·m². In most cases electrons in atoms are paired, and so the positive and negative magnetic moments cancel. However, unpaired electrons in inner electron shells can have small positive dipole moments, as is the case for the 3d electrons of Fe, Co, and Ni.

Example Problem 11.1

Using the relationship $\mu_B = eh/4\pi m$, show that the numerical value for a Bohr magneton is 9.27×10^{-24} A·m².

Solution:

$$\mu_B = \frac{eh}{4\pi m} = \frac{(1.60 \times 10^{-19}\ \text{C})(6.63 \times 10^{-34}\ \text{J·s})}{4\pi(9.11 \times 10^{-31}\ \text{kg})}$$

$$= 9.27 \times 10^{-24}\ \text{C·J·s/kg}$$

$$= 9.27 \times 10^{-24}\ \text{A·m}^2\ \blacktriangleleft$$

The units are consistent, as indicated below:

$$\frac{\text{C·J·s}}{\text{kg}} = \frac{(\text{A·s})\ (\text{N·m})\ (\text{s})}{\text{kg}} = \frac{\text{A·}\cancel{s}}{\cancel{\text{kg}}}\left(\frac{\cancel{\text{kg}}\cdot\text{m·m}}{\cancel{s}^2}\right)(\cancel{s}) = \text{A·m}^2$$

Example Problem 11.2

Calculate a theoretical value for the saturation magnetization M_s in amperes per meter and saturation induction B_s in teslas for pure iron, assuming all magnetic moments due to the four unpaired $3d$ Fe electrons are aligned in a magnetic field. Use equation $B_s \approx \mu_0 M_s$ and assume that $\mu_0 H$ can be neglected. Pure iron has a BCC unit cell with a lattice constant $a = 0.287$ nm.

Solution:

The magnetic moment of an iron atom is assumed to be 4 Bohr magnetons. Thus

$$M_s = \left[\frac{\dfrac{2\ \text{atoms}}{\text{unit cell}}}{\dfrac{(2.87 \times 10^{-10}\ \text{m})^3}{\text{unit cell}}}\right]\left(\frac{4\ \text{Bohr magnetons}}{\text{atom}}\right)\left(\frac{9.27 \times 10^{-24}\ \text{A·m}^2}{\text{Bohr magneton}}\right)$$

$$= \left(\frac{0.085 \times 10^{30}}{\text{m}^3}\right)(4)(9.27 \times 10^{-24}\ \text{A·m}^2) = 3.15 \times 10^6\ \text{A/m}\ \blacktriangleleft$$

$$B_s \approx \mu_0 M_s \approx \left(\frac{4\pi \times 10^{-7}\ \text{T·m}}{\text{A}}\right)\left(\frac{3.15 \times 10^6\ \text{A}}{\text{m}}\right) \approx 3.96\ \text{T}\ \blacktriangleleft$$

Example Problem 11.3

Iron has a saturation magnetization of 1.71×10^6 A/m. What is the average number of Bohr magnetons per atom that contribute to this magnetization? Iron has the BCC crystal structure with $a = 0.287$ nm.

Solution:

The saturation magnetization M_s in amperes per meter can be calculated from Eq. (11.9) below as

$$M_s = \left(\frac{\text{atoms}}{\text{m}^3}\right)\left(\frac{N\mu_B \text{ of Bohr magnetons}}{\text{atom}}\right)\left(\frac{9.27 \times 10^{-24} \text{ A}\cdot\text{m}^2}{\text{Bohr magneton}}\right)$$

$$= \text{ans. in A/m} \tag{11.9}$$

$$\text{Atomic density (no. of atoms/m}^3) = \frac{2 \text{ atoms/BCC unit cell}}{(2.87 \times 10^{-10} \text{ m})^3/\text{unit cell}}$$

$$= 8.46 \times 10^{28} \text{ atoms/m}^3$$

We rearrange Eq. (11.9) and solve for $N\mu_B$. After substituting values for M_s, atomic density, and μ_B, we can calculate the value of $N\mu_B$.

$$N\mu_B = \frac{M_s}{(\text{atoms/m}^3)(\mu_B)}$$

$$= \frac{1.71 \times 10^6 \text{ A/m}}{(8.46 \times 10^{28} \text{ atoms/m}^3)(9.27 \times 10^{-24} \text{ A}\cdot\text{m}^2)} = 2.18\mu_B/\text{atom} \blacktriangleleft$$

Antiferromagnetism   Another type of magnetism which occurs in some materials is antiferromagnetism. In the presence of a magnetic field, magnetic dipoles of atoms of antiferromagnetic materials align themselves in opposite directions (Fig. 11.8b). The elements manganese and chromium in the solid state at room temperature exhibit antiferromagnetism and have a negative exchange energy because the ratio of their atomic spacing to diameter of the 3d orbit is less than about 1.4 (Fig. 11.7).

Ferrimagnetism   In some ceramic materials, different ions have different magnitudes for their magnetic moments, and when these magnetic moments are aligned in an antiparallel manner, there is a net magnetic moment in one direction (Fig. 11.8c). As a group ferrimagnetic materials are called *ferrites*. There are many types of ferrites. One group is based on magnetite, Fe_3O_4, which is the magnetic lodestone of the ancients. Ferrites have low conductivities which make them useful for many electronics applications.

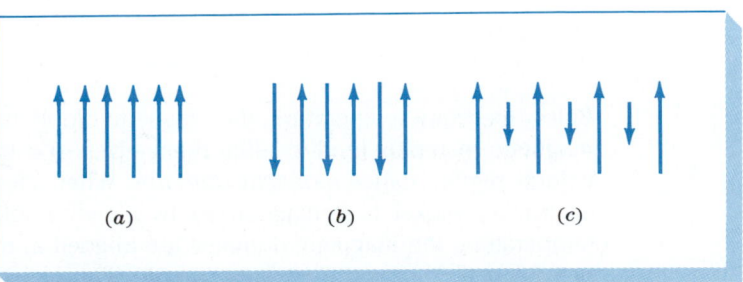

FIGURE 11.8 Alignment of magnetic dipoles for different types of magnetism: (a) ferromagnetism, (b) antiferromagnetism, and (c) ferrimagnetism.

(a)   (b)   (c)

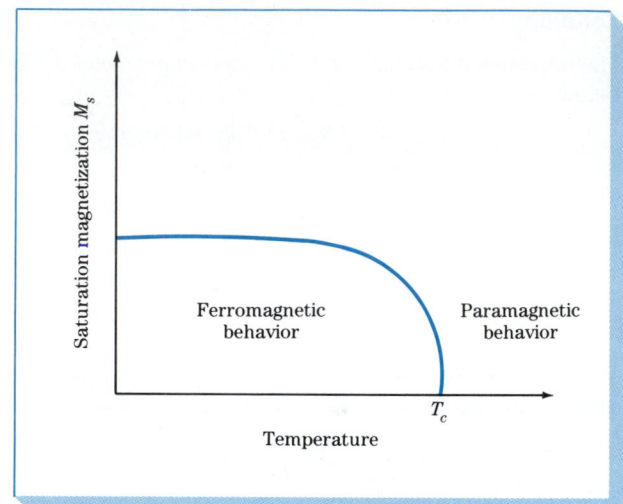

FIGURE 11.9 Effect of temperature on the saturation magnetization M_s of a ferromagnetic material below its Curie temperature T_c. Increasing the temperature randomizes the magnetic moments.

11.3   EFFECT OF TEMPERATURE ON FERROMAGNETISM

At any finite temperature above 0 K, thermal energy causes the magnetic dipoles of a ferromagnetic material to deviate from perfect parallel alignment. Thus, the exchange energy that causes the parallel alignment of the magnetic dipoles in ferromagnetic materials is counterbalanced by the randomizing effects of thermal energy (Fig. 11.9). Finally, as the temperature increases, some temperature is reached where the ferromagnetism in a ferromagnetic material completely disappears, and the material becomes paramagnetic. This temperature is called the *Curie temperature*. When a sample of a ferromagnetic material is cooled from a temperature above its Curie temperature, ferromagnetic domains reform and the material becomes ferromagnetic again. The Curie temperatures of Fe, Co, and Ni are 770, 1123, and 358°C, respectively.

11.4   FERROMAGNETIC DOMAINS

Below the Curie temperature, the magnetic dipole moments of atoms of ferromagnetic materials tend to align themselves in a parallel direction in small-volume regions called *magnetic domains*. When a ferromagnetic material such as iron or nickel is demagnetized by slowly cooling from above its Curie temperature, the magnetic domains are aligned at random so that there is no net magnetic moment for a bulk sample (Fig. 11.10).

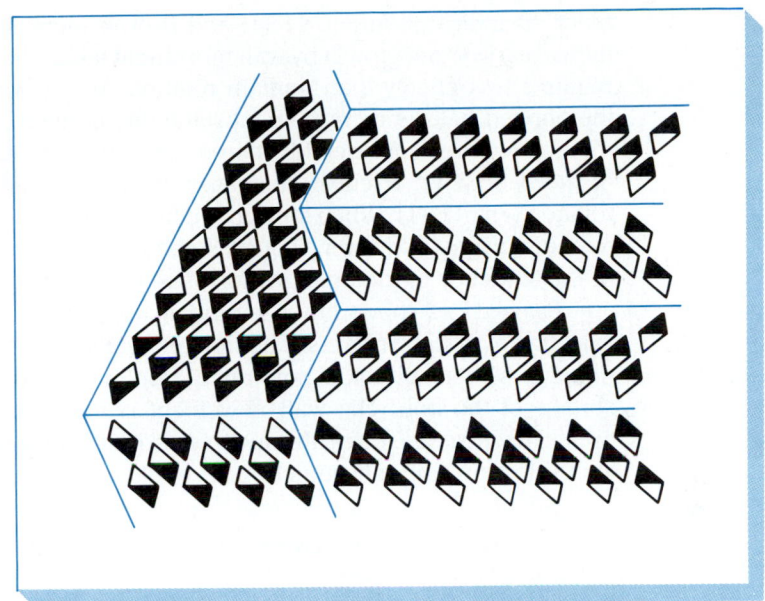

FIGURE 11.10 Schematic illustration of magnetic domains in a ferromagnetic metal. All the magnetic dipoles in each domain are aligned, but the domains themselves are aligned at random so that there is no net magnetization. (*After R. M. Rose, L. A. Shepard, and J. Wulff, "Structure and Properties of Materials," vol. IV: "Electronic Properties," Wiley, 1966, p. 193.*)

When an external magnetic field is applied to a demagnetized ferromagnetic material, magnetic domains whose moments are initially parallel to the applied magnetic field grow at the expense of the less favorably oriented domains (Fig. 11.11). The domain growth takes place by domain wall move-

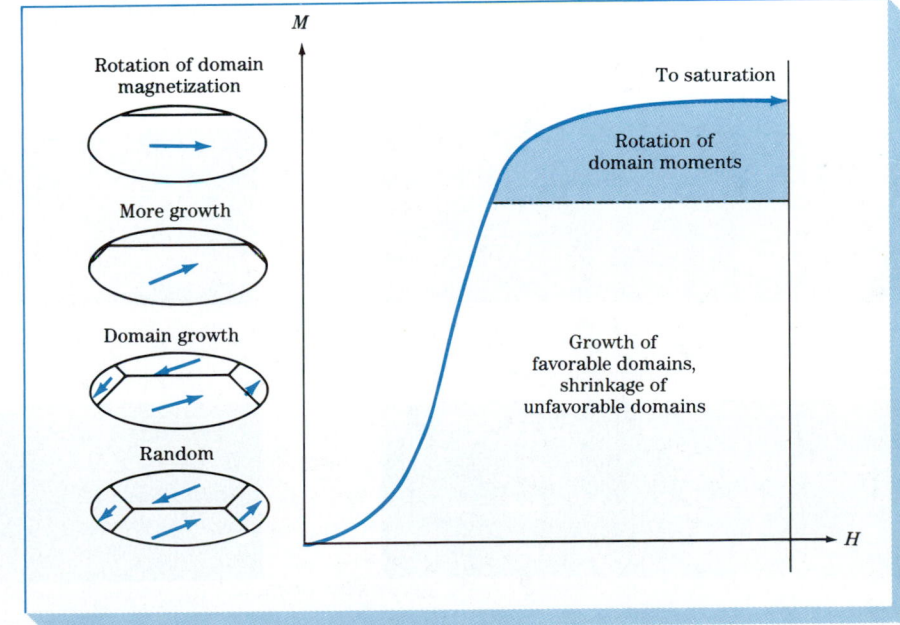

FIGURE 11.11 Magnetic domain growth and rotation as a demagnetized ferromagnetic material is magnetized to saturation by an applied magnetic field. (*After R. M. Rose, L. A. Shepard, and J. Wulff, "Structure and Properties of Materials," vol. IV: "Electronic Properties," Wiley, 1966, p. 193.*)

ment, as indicated in Fig. 11.11, and B or M increases rapidly as the H field increases. Domain growth by wall movement takes place first since this process requires less energy than domain rotation. When domain growth finishes, if the applied field is increased substantially, domain rotation occurs. Domain rotation requires considerably more energy than domain growth, and the slope of the B or M vs. H curve decreases at the high fields required for domain rotation (Fig. 11.11). When the applied field is removed, the magnetized sample remains magnetized, even though some of the magnetization is lost because of the tendency of the domains to rotate back to their original alignment.

Figure 11.12 shows how the domain walls move under an applied field in iron single-crystal whiskers. The domain walls are revealed by the Bitter technique in which a colloidal solution of iron oxide is deposited on the polished surface of the iron. The wall movement is followed by observation with an optical microscope. Using this technique, much information has been obtained about domain wall movement under applied magnetic fields.

FIGURE 11.12 Movement of domain boundaries in an iron crystal produced by the application of an applied magnetic field. Note that as the applied field is increased, the domains with their dipoles aligned in the direction of the field enlarge and those with their dipoles opposed get smaller. (*Courtesy of R. W. DeBlois, The General Electric Co., and C. D. Graham, the University of Pennsylvania.*)

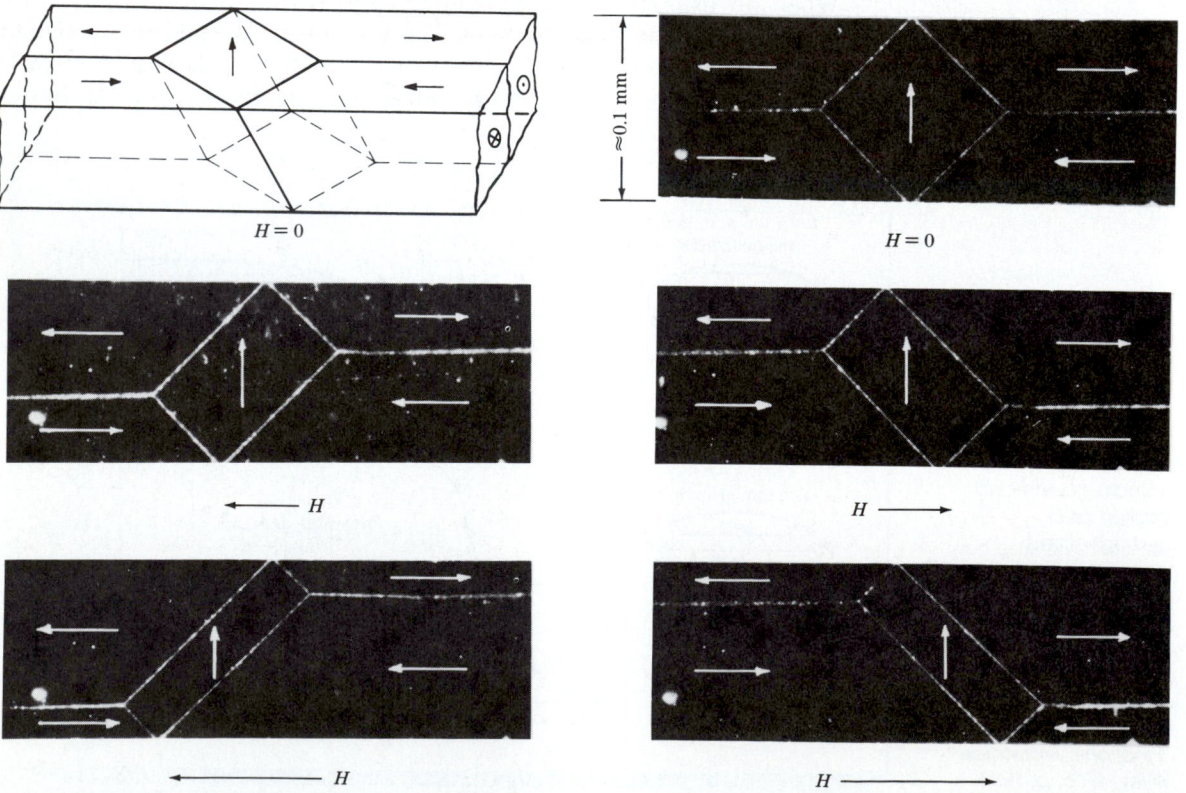

11.5   TYPES OF ENERGIES THAT DETERMINE THE STRUCTURE OF FERROMAGNETIC DOMAINS

The domain structure of a ferromagnetic material is determined by many types of energies, with the most stable structure being attained when the overall potential energy of the material is a minimum. The total magnetic energy of a ferromagnetic material is the sum of the contributions of the following energies: (1) exchange energy, (2) magnetostatic energy, (3) magnetocrystalline anisotropy energy, (4) domain wall energy, and (5) magnetostrictive energy. Let us now briefly discuss each of these energies.

Exchange Energy   The potential energy *within* a domain of a ferromagnetic solid is minimized when all its atomic dipoles are aligned in one direction. This alignment is associated with a positive exchange energy. However, even though the potential energy within a domain is minimized, its external potential energy is increased by the formation of an external magnetic field (Fig. 11.13*a*).

Magnetostatic Energy   Magnetostatic energy is the potential magnetic energy of a ferromagnetic material produced by its external field (Fig. 11.13*a*). This potential energy can be minimized in a ferromagnetic material by domain formation, as illustrated in Fig. 11.13. For a unit volume of a ferromagnetic material, a single-domain structure has the highest potential energy, as indicated by Fig. 11.13*a*. By dividing the single domain of Fig. 11.13*a* into two domains (Fig. 11.13*b*), the intensity and extent of the external magnetic field is reduced. By further sub-

FIGURE 11.13 Schematic illustration showing how reducing the domain size in a magnetic material decreases the magnetostatic energy by reducing the external magnetic field. (*a*) One domain, (*b*) two domains, and (*c*) four domains.

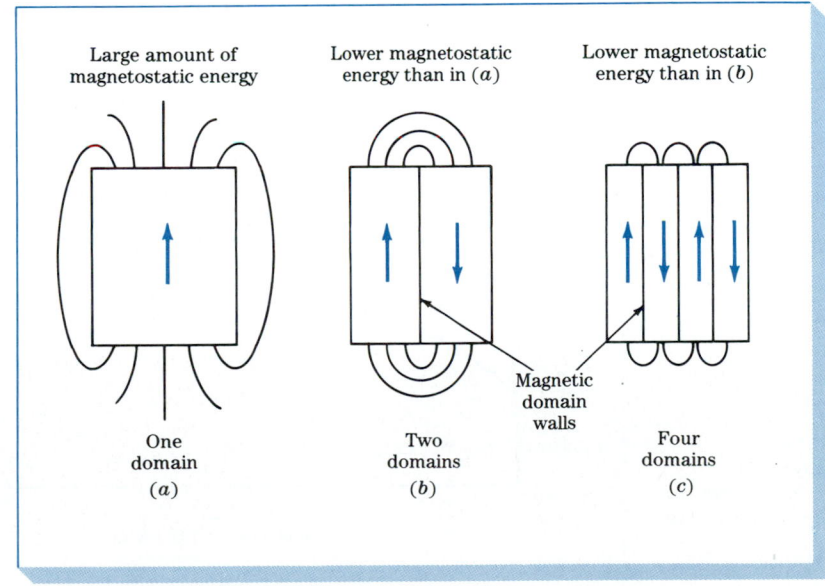

Large amount of magnetostatic energy

Lower magnetostatic energy than in (*a*)

Lower magnetostatic energy than in (*b*)

Magnetic domain walls

One domain
(*a*)

Two domains
(*b*)

Four domains
(*c*)

dividing the single domain into four domains, the external magnetic field is reduced still more (Fig. 11.13c). Since the intensity of the external magnetic field of a ferromagnetic material is directly related to its magnetostatic energy, the formation of multiple domains reduces the magnetostatic energy of a unit volume of material.

Magnetocrystalline Anisotropy Energy

Before considering domain boundary (wall) energy, let us look at the effects of crystal orientation on the magnetization of ferromagnetic materials. Magnetization vs. applied field curves for a single crystal of a ferromagnetic material vary, depending on the crystal orientation relative to the applied field. Figure 11.14 shows magnetic induction B vs. applied field H curves for magnetizations in the $\langle 100 \rangle$ and $\langle 111 \rangle$ directions for single crystals of BCC iron. As indicated in Fig. 11.14, saturation magnetization occurs easiest (or with the lowest applied field) for the $\langle 100 \rangle$ directions and with the highest applied field in the $\langle 111 \rangle$ directions. The $\langle 111 \rangle$ directions are said to be the hard directions for magnetization in BCC iron. For FCC nickel the easy directions of magnetization are the $\langle 111 \rangle$ directions and the $\langle 100 \rangle$ the hard directions; the hard directions for FCC nickel are just the opposite as those for BCC iron.

For polycrystalline ferromagnetic materials such as iron and nickel, grains at different orientations will reach saturation magnetization at different field strengths. Grains whose orientations are in the easy direction of magnetization will saturate at low applied fields, while grains oriented in the hard directions must rotate their resultant moment in the direction of the applied field and thus will reach saturation under much higher fields. The work done to rotate all the domains because of this anisotropy is called the *magnetocrystalline anisotropy energy.*

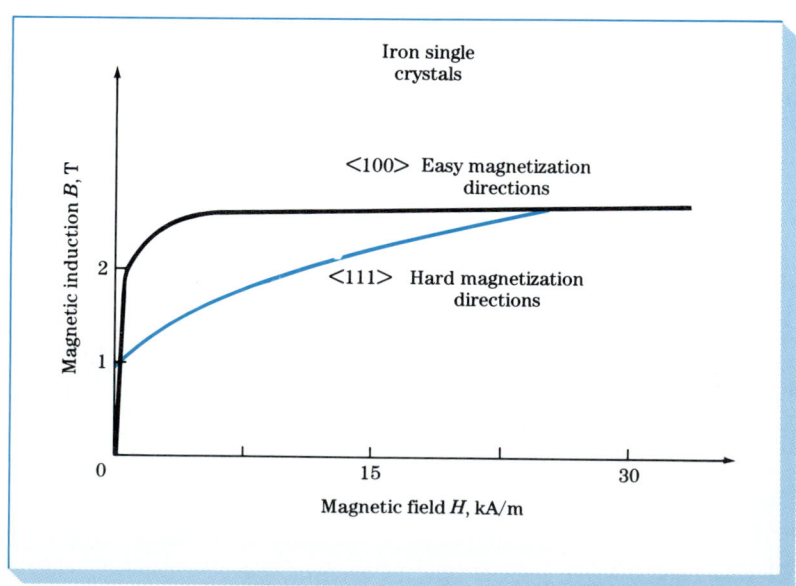

FIGURE 11.14
Magnetocrystalline anisotropy in BCC iron. Iron is magnetized easier in the $\langle 100 \rangle$ directions than in the $\langle 111 \rangle$ directions.

Domain Wall Energy   A *domain wall* is the boundary between two domains whose overall magnetic moments are at different orientations and is analogous to a grain boundary where crystal orientation changes from one grain to another. In contrast to a grain boundary at which grains change orientation abruptly and which is about 3 atoms wide, a domain changes orientation gradually with a domain boundary being about 300 atoms wide. Figure 11.15a shows a schematic drawing of a domain boundary of 180° change in magnetic moment direction which takes place gradually across the boundary.

The reason for the large width of a domain wall is due to a balance between two forces: exchange and magnetocrystalline anisotropy. When there is only a small difference in orientation between the dipoles (Fig. 11.15a), the exchange forces between the dipoles are minimized and the exchange energy is reduced (Fig. 11.15b). Thus, the exchange forces will tend to widen the domain wall. However, the wider the wall is, the greater will be the number of dipoles forced to lie in directions different from those of easy magnetization, and the magnetocrystalline anisotropy energy will be increased (Fig. 11.15b). Thus, the equilibrium wall width will be reached at the width where the sum of the exchange and magnetocrystalline anisotropy energies is a minimum (Fig. 11.15b).

FIGURE 11.15 Schematic illustration of (*a*) magnetic dipole arrangements at domain (Bloch) wall and (*b*) relationship among magnetic exchange energy, magnetocrystalline anisotropy energy, and wall width. The equilibrium wall width is about 100 nm. (*Adapted from C. S. Barrett, W. D. Nix, and A. S. Tetelman, "The Principles of Engineering Materials," Prentice-Hall, 1973, p. 485.*)

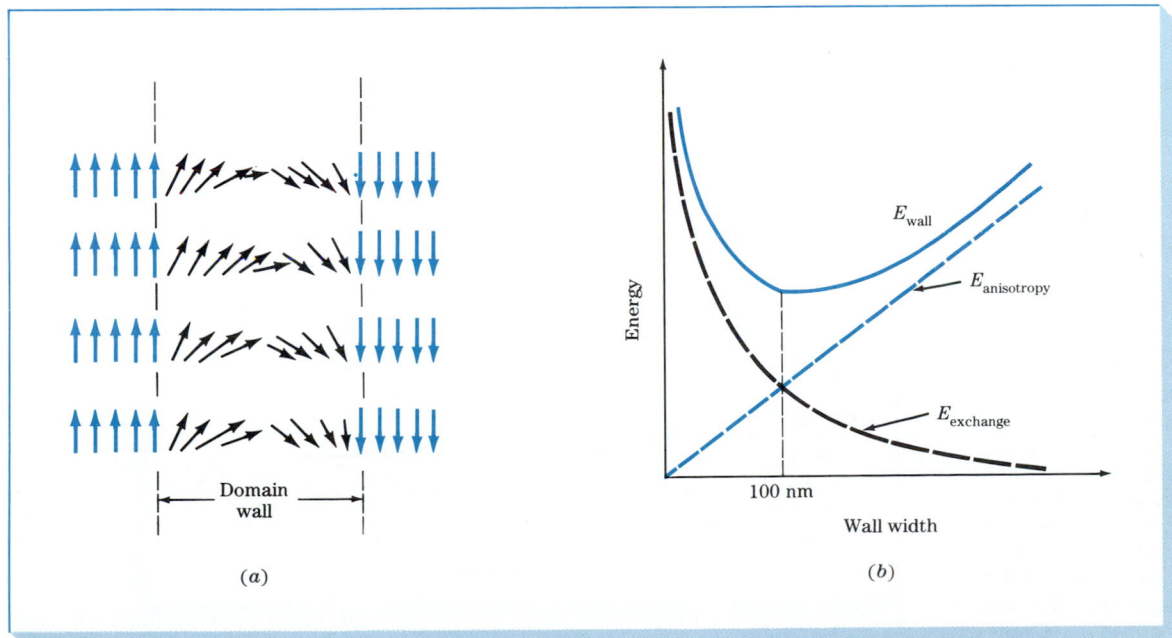

**Magnetostrictive
Energy**

When a ferromagnetic material is magnetized, its dimensions change slightly, and the sample being magnetized either expands or contracts in the direction of magnetization (Fig. 11.16). This magnetically induced reversible elastic strain ($\Delta l/l$) is called *magnetostriction* and is of the order of 10^{-6}. The energy due to the mechanical stresses created by magnetostriction is called *magnetostrictive energy*. For iron, the magnetostriction is positive at low fields and negative at high fields (Fig. 11.16).

The cause of magnetostriction is attributed to the change in the bond length between the atoms in a ferromagnetic metal when their electron-spin dipole moments are rotated into alignment during magnetization. The fields of the dipoles may attract or repel each other, leading to the contraction or expansion of the metal during magnetization.

Let us now consider the effect of magnetostriction on the equilibrium configuration of the domain structure of cubic crystalline materials, such as shown in Fig. 11.17a and b. Because of the cubic symmetry of the crystals, the formation of triangular-shaped domains, called *domains of closure,* at the ends of the crystal eliminates the magnetostatic energy associated with an external magnetic field and hence lowers the energy of the material. It might appear that very large domains such as those shown in Fig. 11.17a and b would be the lowest energy and most stable configuration since there is minimum wall energy. However, this is not the case since magnetostrictive stresses introduced during magnetization tend to be larger for larger domains. Smaller magnetic domains, such as those shown in Fig. 11.17c, reduce magnetostrictive stresses but increase domain wall area and energy. Thus, the equilibrium domain

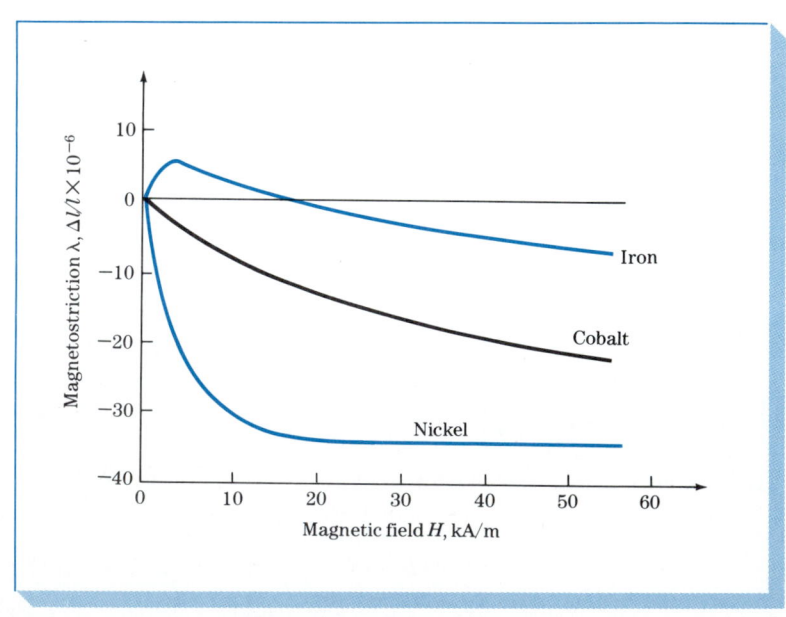

FIGURE 11.16
Magnetostrictive behavior of Fe, Co, and Ni ferromagnetic elements. Magnetostriction is a fractional elongation (or contraction) and in this illustration is in units of micrometers per meter.

FIGURE 11.17

Magnetostriction in cubic magnetic materials. Exaggeration of (*a*) negative and (*b*) positive magnetostriction pulling apart the domain boundaries of a magnetic material. (*c*) Lowering of magnetostrictive stresses by the creation of a smaller-domain-size structure.

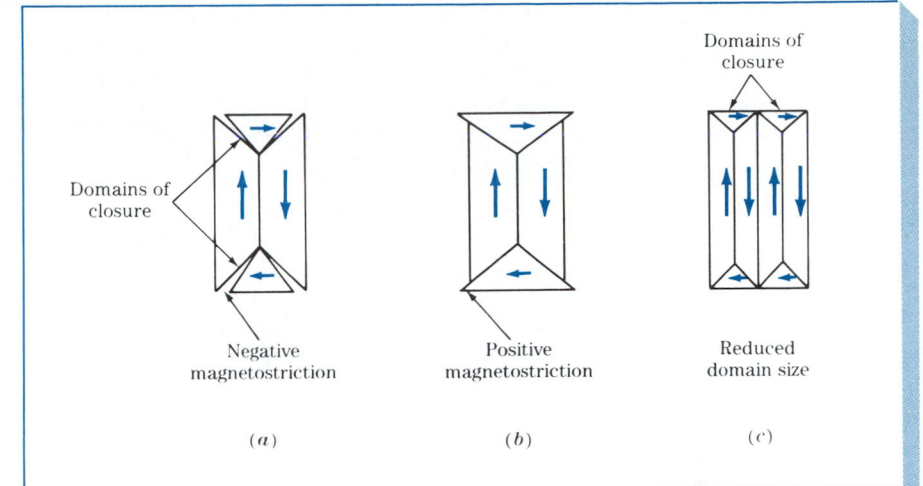

Domains of closure

Domains of closure

Negative magnetostriction

Positive magnetostriction

Reduced domain size

(*a*)   (*b*)   (*c*)

configuration is reached when the sum of the magnetostrictive and domain wall energies is a minimum.

In summary, the domain structure formed in ferromagnetic materials is determined by the various contributions of exchange, magnetostatic, magnetocrystalline anisotropic, domain wall, and magnetostrictive energies to its total magnetic energy. The equilibrium or most stable configuration is that for which the total magnetic energy is the lowest.

11.6 THE MAGNETIZATION AND DEMAGNETIZATION OF A FERROMAGNETIC METAL

Ferromagnetic metals such as Fe, Co, and Ni acquire large magnetizations when placed in a magnetizing field, and remain in the magnetized condition to a lesser extent after the magnetizing field is removed. Let us consider the effect of an applied field H on the magnetic induction B of a ferromagnetic metal during magnetizing and demagnetizing, as shown in the B vs. H graph of Fig. 11.18. First, let us demagnetize a ferromagnetic metal such as iron by slowly cooling it from above its Curie temperature. Then, let us apply a magnetizing field to the sample and follow the effect of the applied field on the magnetic induction of the sample.

As the applied field increases from zero, B increases from zero along curve *OA* of Fig. 11.18 until *saturation induction* is reached at point *A*. Upon decreasing the applied field to zero, the original magnetization curve is not retraced, and there remains a magnetic flux density called the *remanent induction* B_r (point *C* in Fig. 11.18). To decrease the magnetic induction to zero,

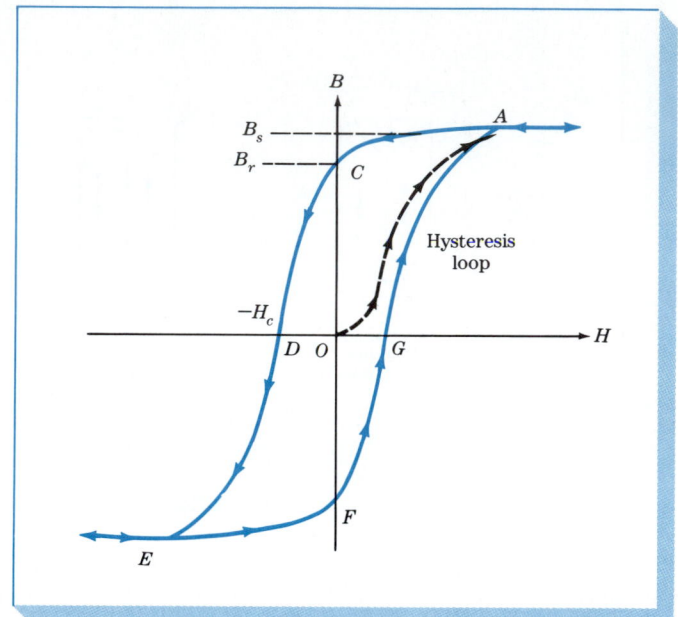

FIGURE 11.18 Magnetic induction B vs. applied field H hysteresis loop for a ferromagnetic material. The curve OA traces out the initial B vs. H relationship for the magnetization of a demagnetized sample. Cyclic magnetization and demagnetization to saturation induction traces out hysteresis loop $ACDEFGA$.

a reverse (negative) applied field of the amount H_c, called the *coercive force*, must be applied (point D in Fig. 11.18). If the negative applied field is increased still more, eventually the material will reach saturation induction in the reverse field at point E of Fig. 11.18. Upon removing the reverse field, the magnetic induction will return to the remanent induction at point F in Fig. 11.18, and upon application of a positive applied field, the B-H curve will follow FGA to complete a loop. Further application of reverse and forward applied fields to saturation induction will produce the repetitive loop of $ACDEFGA$ to be traced out. This magnetization loop is referred to as a *hysteresis loop*, and its internal area is a measure of energy lost or the work done by the magnetizing and demagnetizing cycle.

11.7   SOFT MAGNETIC MATERIALS

A *soft magnetic material* is easily magnetized and demagnetized, whereas a *hard magnetic material* is difficult to magnetize and demagnetize. In early days soft and hard magnetic materials were physically soft and hard, respectively. Today, however, the physical hardness of a magnetic material does not necessarily indicate that it is magnetically soft or hard.

Soft materials such as the iron–3 to 4% silicon alloys used in cores for transformers, motors, and generators have narrow hysteresis loops with low

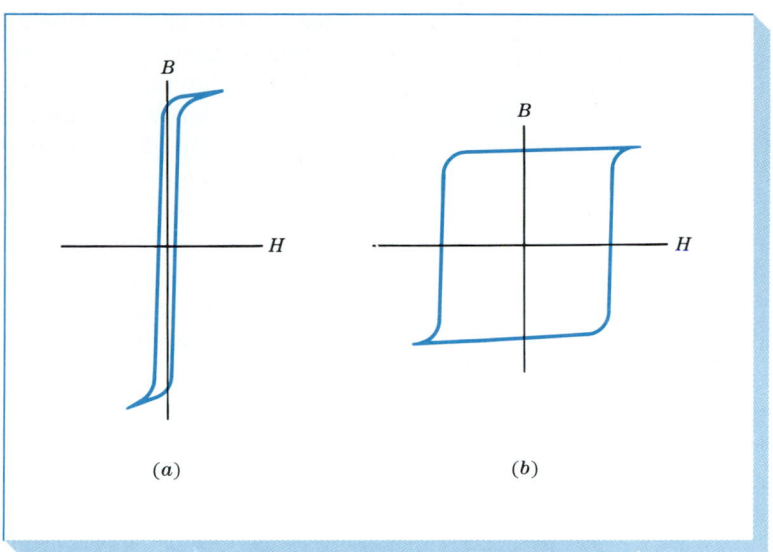

FIGURE 11.19 Hysteresis loops for (*a*) a soft magnetic material and (*b*) a hard magnetic material. The soft magnetic material has a narrow hysteresis loop which makes it easy to magnetize and demagnetize, whereas the hard magnetic material has a wide hysteresis loop which makes it difficult to magnetize and demagnetize.

coercive forces (Fig. 11.19). On the other hand, hard magnetic materials used for permanent magnets have wide hysteresis loops with high coercive forces (Fig. 11.19*b*).

Desirable Properties for Soft Magnetic Materials

For a ferromagnetic material to be "soft," its hysteresis loop should have as low a coercive force as possible. That is, its hysteresis loop should be as thin as possible so that the material magnetizes easily and has a high magnetic permeability. For most applications, a high saturation induction is also an important property of soft magnetic materials. Thus, a very thin and high hysteresis loop is desirable for most soft magnetic materials (Fig. 11.19*a*).

Energy Losses for Soft Magnetic Materials

Hysteresis energy losses Hysteresis losses are due to dissipated energy required to push the domain walls back and forth during the magnetization and demagnetization of the magnetic material. The presence of impurities, crystalline imperfections, and precipitates in soft magnetic materials all act as barriers to impede domain wall movement during the magnetization cycle and so increase hysteresis energy losses. Plastic strain, by increasing the dislocation density of a magnetic material, also increases hysteresis losses. In general, the internal area of a hysteresis loop is a measure of the energy lost due to magnetic hysteresis.

In the magnetic core of an ac electrical power transformer using 60 cycles/s electric current goes through the entire hysteresis loop 60 times per second, and in each cycle there is some energy lost due to movement of the domain walls of the magnetic material in the transformer core. Thus, increasing the ac electrical input frequency of electromagnetic devices increases the hysteresis energy losses.

Eddy-current energy losses   A fluctuating magnetic field caused by ac electrical input into a conducting magnetic core produces transient voltage gradients which create stray electric currents. These induced electric currents are called *eddy currents* and are a source of energy loss due to electrical resistance heating. Eddy-current energy losses in electrical transformers can be reduced by using a laminated or sheet structure in the magnetic core. An insulation layer between the conducting magnetic material prevents the eddy currents from going from one sheet to the next. Another approach to reducing eddy-current losses, particularly at higher frequencies, is to use a soft magnetic material which is an insulator. Ferrimagnetic oxides and other similar type magnetic materials are used for some high-frequency electromagnetic applications and will be discussed in Sec. 11.9.

Iron-Silicon Alloys   The most extensively used soft magnetic materials are the iron–3 to 4% silicon alloys. Before about 1900, low-carbon plain-carbon steels were used for low-frequency (60-cycle) power application devices such as transformers, motors, and generators. However, with these magnetic materials, core losses were relatively high.

The addition of about 3 to 4% silicon to iron to make iron-silicon alloys has several beneficial effects for reducing core losses in magnetic materials:

1. Silicon increases the electrical resistivity of low-carbon steel and thus reduces the eddy-current losses.
2. Silicon decreases the magnetoanisotropy energy of iron and increases magnetic permeability and thus decreases hysteresis core losses.
3. Silicon additions (3 to 4%) also decrease magnetostriction and lower hysteresis energy losses and transformer noise ("hum").

However, on the detrimental side, silicon decreases the ductility of iron so that only up to about 4% silicon can be alloyed with iron. Silicon also decreases the saturation induction and Curie temperature of iron.

A further decrease in eddy-current energy losses in transformer cores was achieved by using a laminated (stacked-sheet) structure. For a modern power transformer core, a multitude of thin iron-silicon sheets about 0.010 to 0.014 in (0.025 to 0.035 cm) thick are stacked on top of each other with a thin layer of insulation between them. The insulation material is coated on both sides of the iron-silicon sheets and prevents stray eddy currents from flowing perpendicular to the sheets.

Still another decrease in transformer-core energy loss was achieved in the 1940s by the production of grain-oriented iron-silicon sheet. By using a combination of cold work and recrystallization treatments, a cube-on-edge (COE) $\{110\}$ $\langle 001 \rangle$ grain-oriented material was produced on an industrial scale for Fe–3% Si sheet (Fig. 11.20). Since the [001] direction is an easy axis for magnetization for Fe–3% Si alloys, the magnetic domains in COE materials are oriented for easy magnetization upon the application of an applied field in the

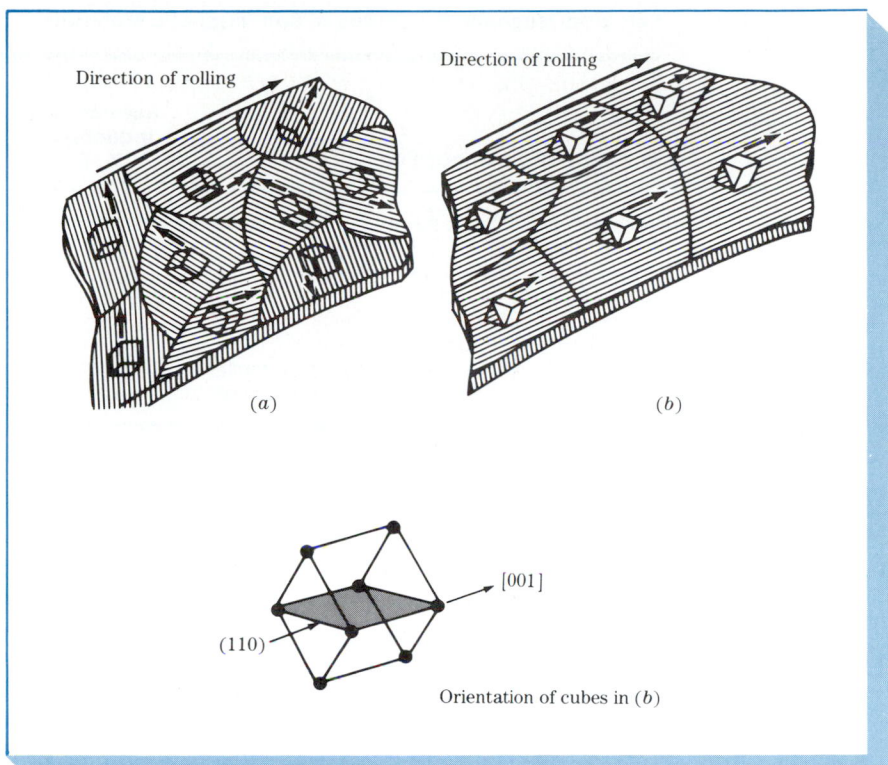

FIGURE 11.20 (*a*) Random and (*b*) preferred orientation (110) [001] texture in polycrystalline iron—3 to 4% silicon sheet. The small cubes indicate the orientation of each grain. (*After R. M. Rose, L. A. Shepard, and J. Wulff, "Structure and Properties of Materials," vol. IV: "Electronic Properties," Wiley, 1966, p. 211.*)

direction parallel to the rolling direction of the sheet. Thus, the COE material has a higher permeability and lower hysteresis losses than Fe–Si sheet with a random texture (Table 11.3).

Metallic Glasses

Metallic glasses are a relatively new class of metallic-type materials whose dominant characteristic is a noncrystalline structure, unlike normal metal alloys which have a crystalline structure. The atoms in normal metals and alloys when cooled from the liquid state arrange themselves into an orderly crystal lattice. Table 11.4 lists the atomic compositions of eight metallic glasses of engineering importance. These materials have important soft magnetic properties and consist essentially of various combinations of ferromagnetic Fe, Co, and Ni with the metalloids B and Si. Applications for these exceptionally soft magnetic materials include low-energy core-loss power transformers, magnetic sensors, and recording heads.

Metallic glasses are produced by a rapid solidification process in which molten metallic glass is cooled extremely rapidly (about $10^{6\circ}$C/s) as a thin film on a rotating copper-surfaced mold (Fig. 11.21*a*). This process produces a continuous ribbon of metallic glass of about 0.001 in (0.0025 cm) thick and 6 in (15 cm) wide.

Metallic glasses have some remarkable properties. They are very strong

TABLE 11.3   Selected Magnetic Properties of Soft Magnetic Materials

Material and composition	Saturation induction B_s, T	Coercive force H_c, A/cm	Initial relative permeability μ_i
Magnetic iron, 0.2-cm sheet	2.15	0.88	250
M36 cold-rolled Si–Fe (random)	2.04	0.36	500
M6 (110) [001], 3.2% Si–Fe (oriented)	2.03	0.06	1,500
45 Ni–55 Fe (45 Permalloy)	1.6	0.024	2,700
75 Ni–5 Cu–2 Cr–18 Fe (Mumetal)	0.8	0.012	30,000
79 Ni–5 Mo–15 Fe–0.5 Mn (Supermalloy)	0.78	0.004	100,000
48% $MnO–Fe_2O_3$, 52% $ZnO–Fe_2O_3$ (soft ferrite)	0.36		1,000
36% $NiO–Fe_2O_3$, 64% $ZnO–Fe_2O_3$ (soft ferrite)	0.29		650

Source: G. Y. Chin and J. H. Wernick, "Magnetic Materials, Bulk," vol. 14: "Kirk-Othmer Encyclopedia of Chemical Technology," 3d ed., Wiley, 1981, p. 686.

[up to 650 ksi (4500 MPa)], very hard with some flexibility, and very corrosion-resistant. The metallic glasses listed in Table 11.4 are magnetically very soft as indicated by their maximum permeabilities and thus are able to be magnetized and demagnetized very easily. Domain walls in these materials are able to move with exceptional ease, mainly because there are no grain boundaries and no long-range crystal anisotropy. Figure 11.21*b* shows some magnetic domains in a metallic glass that were produced by bending the metallic glass ribbon. Magnetically soft metallic glasses have very narrow hysteresis loops, as is indicated in Fig. 11.21*c*, and thus have very low hysteresis energy losses. This property has enabled the development of multilayered metallic glass power transformer cores which have 70 percent of the core losses of conventional iron-silicon cores (Fig. 11.1). Much research and development work in the application of metallic glasses for low-loss power transformers is in progress.

TABLE 11.4   Metallic Glasses: Compositions, Properties, and Applications

Alloy, atomic %	Saturation induction B_s, T	Maximum permeability	Applications
$Fe_{78}B_{13}Si_9$	1.56	600,000	Power transformers, low core losses
$Fe_{81}B_{13.5}Si_{3.5}C_2$	1.61	300,000	Pulse transformers, magnetic switches
$Fe_{67}Co_{18}B_{14}Si_1$	1.80	4,000,000	Pulse transformers, magnetic switches
$Fe_{77}Cr_2B_{16}Si_5$	1.41	35,000	Current transformers, sensing cores
$Fe_{74}Ni_4Mo_3B_{17}Si_2$	1.28	100,000	Low core losses at high frequencies
$Co_{69}Fe_4Ni_1Mo_2B_{12}Si_{12}$	0.70	600,000	Magnetic sensors, recording heads
$Co_{66}Fe_4Ni_1B_{14}Si_{15}$	0.55	1,000,000	Magnetic sensors, recording heads
$Fe_{40}Ni_{38}Mo_4B_{18}$	0.88	800,000	Magnetic sensors, recording heads

Source: Metglas Magnetic Alloys, Allied Metglas Products.

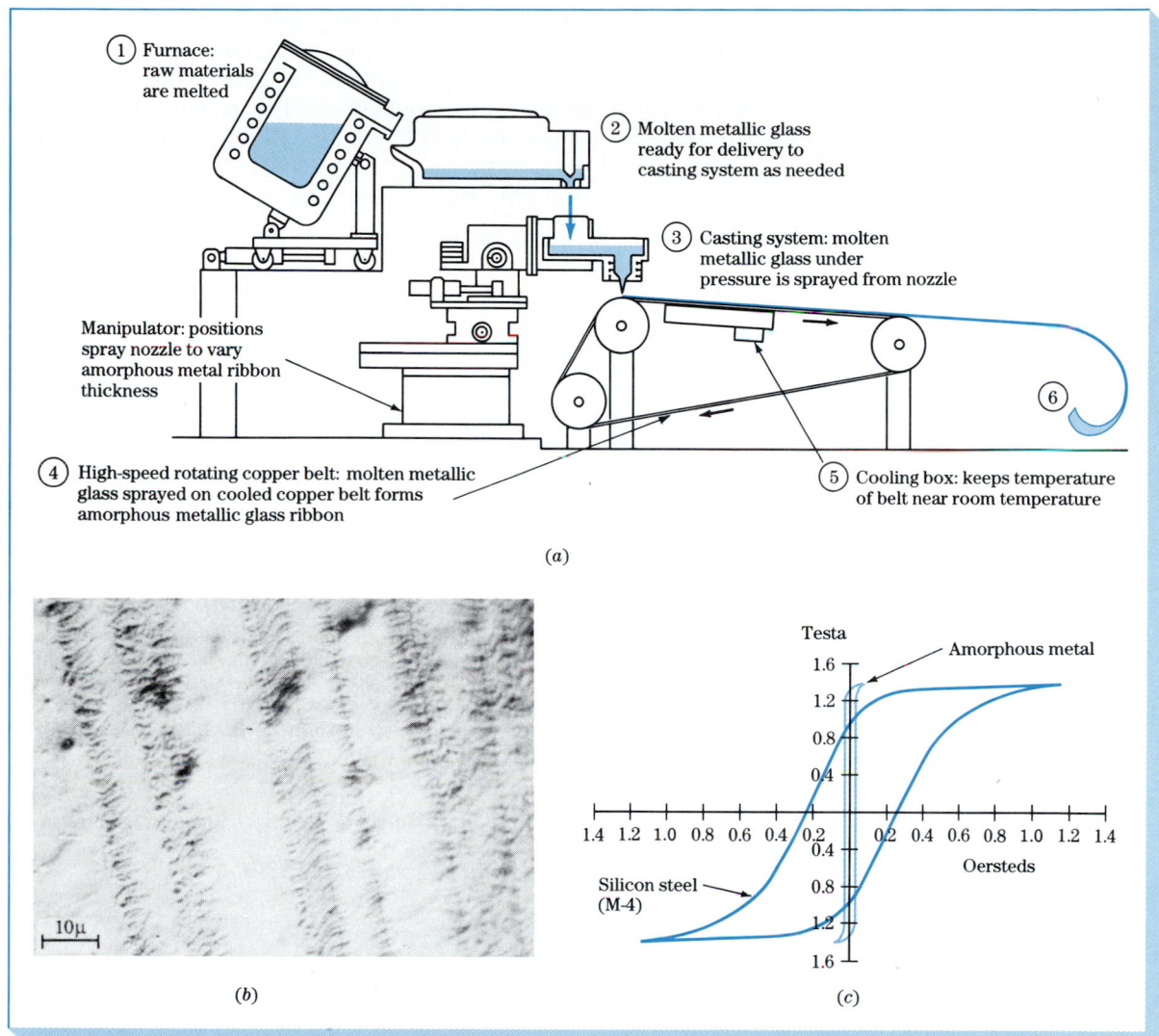

(a)

(b)

(c)

FIGURE 11.21 (a) Schematic drawing of the rapid solidification process for the production of metallic glass ribbon. (*After New York Times, Jan. 11, 1989, p. D7.*) (b) Induced magnetic domains in a metallic glass. (*After V. Lakshmanan and J. C. M. Li, Mater. Sci. Eng., 1988, p. 483.*) (c) Comparison of the hysteresis loops of a ferromagnetic metallic glass and M-4 iron-silicon ferromagnetic sheet. (*Electric World, September 1985.*)

Nickel-Iron Alloys   The magnetic permeabilities of commercially pure iron and iron-silicon alloys are relatively low at low applied fields. Low initial permeability is not as important for power applications such as transformer cores since this equipment is operated at high magnetizations. However, for high-sensitivity communications equipment used to detect or transmit small signals, nickel-iron alloys which have much higher permeabilities at low fields are commonly used.

In general two broad classes of Ni–Fe alloys are commercially produced, one with about 50% Ni and another with about 79% Ni. The magnetic properties of some of these alloys are listed in Table 11.3. The 50% Ni alloy is characterized by moderate permeability ($\mu_i = 2500$; $\mu_{max} = 25,000$) and high saturation induction [$B_s = 1.6$ T (16,000 G)]. The 79% Ni alloy has high permeability ($\mu_i = 100,000$; $\mu_{max} = 1,000,000$) but lower saturation induction [$B_s = 0.8$ T (8000 G)]. These alloys are used in audio and instrument transformers, instrument relays, and for rotor and stator laminations. Tape-wound cores, such as the cutaway section shown in Fig. 11.22, are commonly used for electronic transformers.

The Ni–Fe alloys have such high permeabilities because their magnetoanisotropy and magnetostrictive energies are low at the compositions used. The highest initial permeability in the Ni–Fe system occurs at 78.5% Ni–21.5% Fe, but rapid cooling below 600°C is necessary to suppress the formation of an ordered structure. The equilibrium ordered structure in the Ni–Fe system has an FCC unit cell with Ni atoms at the faces and Fe atoms at the face corners. The addition of about 5% Mo to the 78.5% Ni (balance Fe) alloy also suppresses the ordering reaction so that moderate cooling of the alloy from above 600°C is sufficient to prevent ordering.

The initial permeability of the Ni–Fe alloys containing about 56 to 58% Ni can be increased three to four times by annealing the alloy in the presence of a magnetic field after the usual high-temperature anneal. The magnetic anneal causes directional ordering of the atoms of the Ni–Fe lattice and thereby increases the initial permeability of these alloys. Figure 11.23 shows the effect of magnetic annealing on the hysteresis loop of a 65% Ni–35% Fe alloy.

FIGURE 11.22 Tape-wound magnetic cores. (*a*) Encapsulated core. (*b*) Cross section of tape-wound core with phenolic encapsulation. Note that there is a silicone rubber cushion between the magnetic alloy tape and the phenolic encapsulation case. The magnetic properties of annealed high Ni–Fe tape-wound alloys are sensitive to strain damage. (*Courtesy of Magnetics, Inc.*)

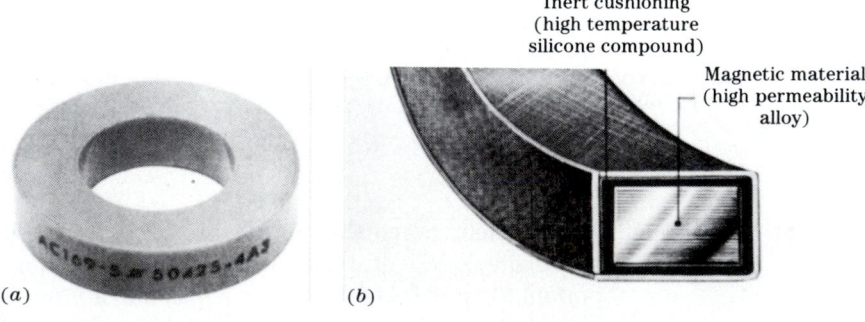

Inert cushioning
(high temperature
silicone compound)

Magnetic material
(high permeability
alloy)

(*a*)   (*b*)

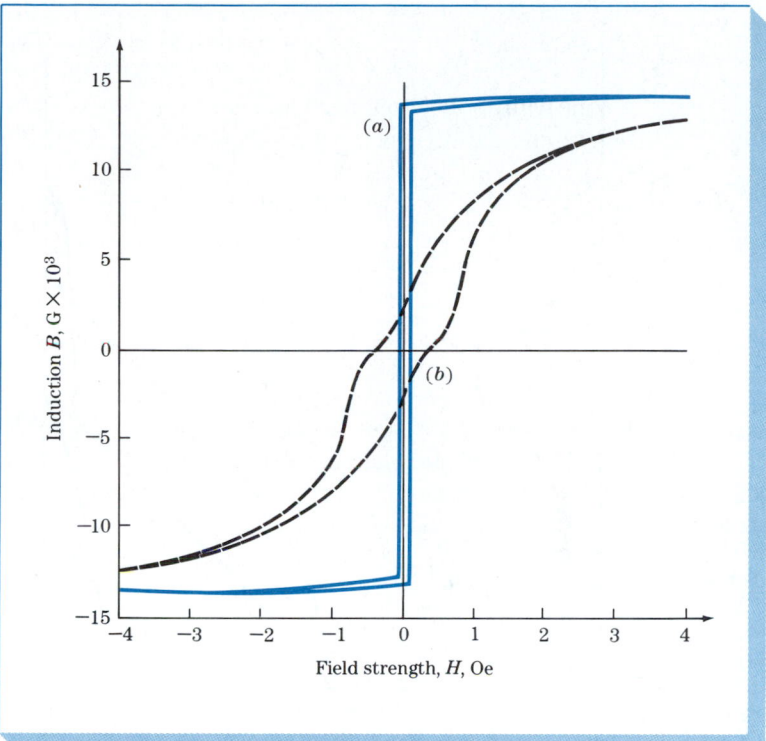

FIGURE 11.23 The effect of magnetic annealing on the hysteresis loop of a 65% Ni–35% Fe alloy. (*a*) 65 Permalloy annealed with field present; (*b*) 65 Permalloy annealed with field absent. (*After K. M. Bozorth, "Ferromagnetism," Van Nostrand, 1951, p. 121.*)

11.8   HARD MAGNETIC MATERIALS

Properties of Hard Magnetic Materials

Hard or permanent magnetic materials are characterized by a high coercive force H_c and a high remanent magnetic induction B_r, as indicated schematically in Fig. 11.19*b*. Thus the hysteresis loops of hard magnetic materials are wide and high. These materials are magnetized in a magnetic field strong enough to orient their magnetic domains in the direction of the applied field. Some of the applied energy of the field is converted into potential energy which is stored in the permanent magnet produced. A permanent magnet in the fully magnetized condition is thus in a relatively high energy state as compared to a demagnetized magnet.

Hard magnetic materials are difficult to demagnetize once magnetized. The demagnetizing curve for a hard magnetic material is chosen as the second quadrant of its hysteresis loop, and can be used for comparing the strengths of permanent magnets. Figure 11.24 compares the demagnetizing curves of various hard magnetic materials.

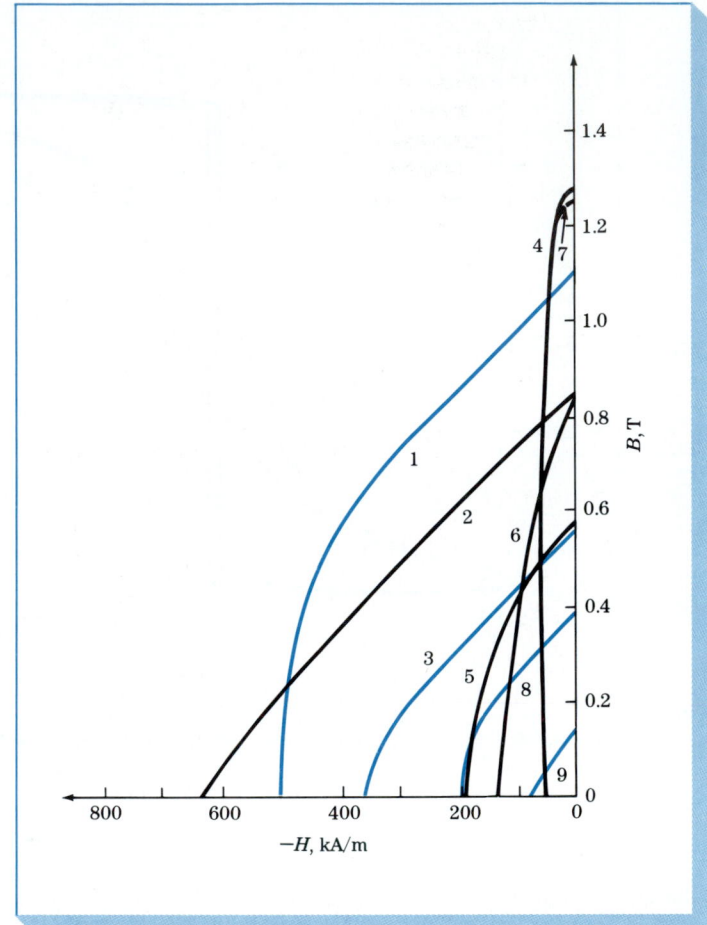

FIGURE 11.24

Demagnetization curves of various hard magnetic materials. 1: Sm(Co,Cu)$_{7.4}$; 2: SmCo$_5$; 3: bonded SmCo$_5$; 4: alnico 5; 5: Mn–Al–C; 6: alnico 8; 7: Cr–Co–Fe; 8: ferrite; and 9: bonded ferrite. (*After G. Y. Chin and J. H. Wernick, "Magnetic Materials, Bulk," vol. 14: "Kirk-Othmer Encyclopedia of Chemical Technology," 3d ed., Wiley, 1981, p. 673.*)

The power or external energy of a permanent (hard) magnetic material is directly related to the size of its hysteresis loop. The magnetic potential energy of a hard magnetic material is measured by its *maximum energy product*, which is the maximum value of the product of B (magnetic induction) and H (the demagnetizing field) determined from the demagnetizing curve of the material. Figure 11.25 shows the external energy (BH) curve for a hypothetical hard magnetic material and its maximum energy product, (BH)$_{max}$. Basically, the maximum energy product of a hard magnetic material is the area occupied by the largest rectangle that can be inscribed in the second quadrant of the hysteresis loop of the material. The SI units for the energy product BH are kJ/m^3, and for the cgs system, G·Oe. The SI units for the energy product (BH)$_{max}$

FIGURE 11.25 A schematic diagram of the energy product (B vs. BH) curve of a hard magnetic material such as an alnico alloy is shown by the circular dashed line to the right of the B (induction) axis. The maximum energy product $(BH)_{max}$ is indicated at the intersection of the vertical dashed line and the BH axis.

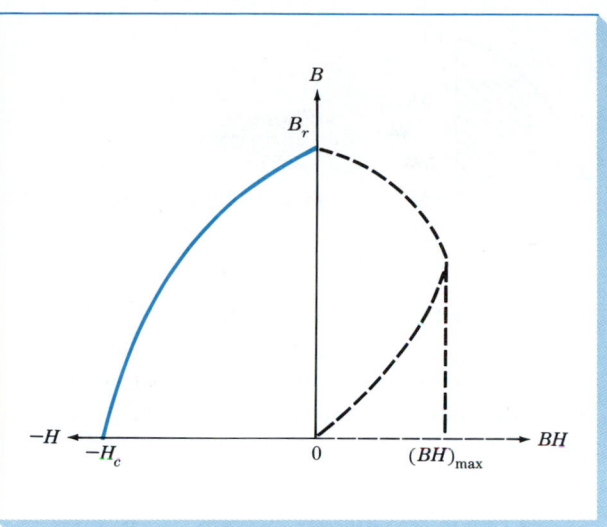

of joules per cubic meter are equivalent units to the product of the units of B in teslas and H in amperes per meter as is shown below.

$$\left[B \left(T \cdot \frac{Wb}{m^2} \cdot \frac{1}{T} \cdot \frac{V \cdot s}{Wb} \right) \right] \left[H \left(\frac{A}{m} \cdot \frac{J}{V \cdot A \cdot s} \right) \right] = BH \left(\frac{J}{m^3} \right)$$

Example Problem 11.4

Estimate the maximum energy product $(BH)_{max}$ for the Sm $(Co,Cu)_{7.4}$ alloy of Fig. 11.24.

Solution:

We need to find the area of the largest rectangle that can be located within the second-quadrant demagnetization curve of the alloy shown in Fig. 11.24. Four trial areas are listed below:

$$\begin{aligned}
\text{Trial 1} &\sim (0.8 \text{ T} \times 250 \text{ kA/m}) = 200 \text{ kJ/m}^3 \quad \text{(see Fig. EP11.4)} \\
\text{Trial 2} &\sim (0.6 \text{ T} \times 380 \text{ kA/m}) = 228 \text{ kJ/m}^3 \\
\text{Trial 3} &\sim (0.55 \text{ T} \times 420 \text{ kA/m}) = 231 \text{ kJ/m}^3 \\
\text{Trial 4} &\sim (0.50 \text{ T} \times 440 \text{ kA/m}) = 220 \text{ kJ/m}^3
\end{aligned}$$

The highest value is about 231 kJ/m³, which compares to 240 kJ/m³ listed for the Sm (Cu,Co) alloy in Table 11.5.

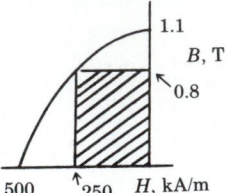

FIGURE EP11.4 Trial 1. $500 \quad 250 \quad H, kA/m \qquad (BH)_{max} \simeq (0.8 \text{ T} \times 250 \text{ kA/m}) = 200 \text{ kJ/m}^3$

Alnico Alloys

Properties and compositions   The alnico (*al*uminum-*ni*ckel-*co*balt) alloys are the most important commercial hard magnetic materials in use today and account for about 35 percent of the hard-magnet market in the United States. These alloys are characterized by a high energy product [$(BH)_{max}$ = 40 to 70 kJ/m³ (5 to 9 MG·Oe)], a high remanent induction [B_r = 0.7 to 1.35 T (7 to 13.5 kG)], and a moderate coercivity [H_c = 40 to 160 kA/m (500 to 2010 Oe)]. Table 11.5 lists some magnetic properties of several alnico and other permanent magnetic alloys.

The alnico family of alloys are iron-based alloys with additions of Al, Ni, and Co plus about 3% Cu. A few %Ti is added to the high-coercivity alloys, alnicos 6 to 9. Figure 11.26 shows bar graphs of the compositions of some of the alnico alloys. Alnicos 1 to 4 are isotropic, whereas alnicos 5 to 9 are anisotropic due to being heat-treated in a magnetic field while the precipitates form. The alnico alloys are brittle and so are produced by casting or by powder metallurgy processes. Alnico powders are used primarily to produce large amounts of small articles of complex shapes.

Structure   Above their solution heat-treatment temperature of about 1250°C, the alnico alloys are single-phase with a BCC crystal structure. During cooling

TABLE 11.5   Selected Magnetic Properties of Hard Magnetic Materials

Material and composition	Remanent induction B_r, T	Coercive force H_c, kA/m	Maximum energy product $(BH)_{max}$, kJ/m³
Alnico 1, 12 Al, 21 Ni, 5 Co, 2 Cu, bal Fe	0.72	37	11.0
Alnico 5, 8 Al, 14 Ni, 25 Co, 3 Cu, bal Fe	1.28	51	44.0
Alnico 8, 7 Al, 15 Ni, 24 Co, 3 Cu, bal Fe	0.72	150	40.0
Rare-earth–Co, 35 Sm, 65 Co	0.90	675–1200	160
Rare-earth–Co, 25.5 Sm, 8 Cu, 15 Fe, 1.5 Zr, 50 Co	1.10	510–520	240
Fe–Cr–Co, 30 Cr, 10 Co, 1 Si, 59 Fe	1.17	46	34.0
MO · Fe₂O₃ (M = Ba, Sr) (hard ferrite)	0.38	235–240	28.0

Source: G. Y. Chin and J. H. Wernick, "Magnetic Materials, Bulk," vol. 14: "Kirk-Othmer Encyclopedia of Chemical Technology," 3d ed., Wiley, 1981, p. 686.

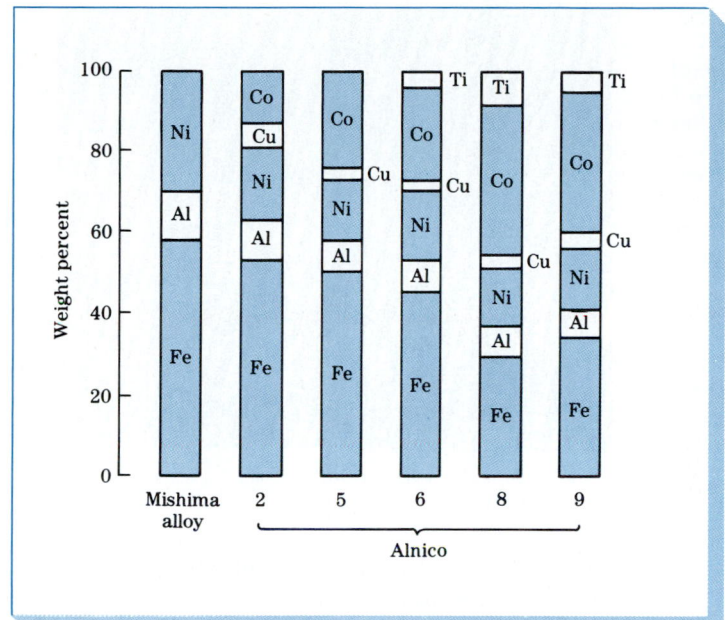

FIGURE 11.26
Development of the chemical compositions of the alnico alloys. The original alloy was discovered by Mishima in Japan in 1931. (*After B. D. Cullity, "Introduction to Magnetic Materials," Addison-Wesley, 1972, p. 566.*)

to about 750 to 850°C, these alloys decompose into two other BCC phases, α and α'. The matrix α phase is rich in Ni and Al and is weakly magnetic. The α' precipitate is rich in Fe and Co and thus has a higher magnetization than the Ni–Al–rich α phase. The α' phase tends to be rodlike, aligned in the $\langle 100 \rangle$ directions and about 10 nm in diameter and about 100 nm long.

If the 800°C heat treatment is carried out in a magnetic field, the α' precipitate forms fine elongated particles in the direction of the magnetic field (Fig. 11.27) in a matrix of the α phase. The high coercivity of the alnicos is attributed to the difficulty of rotating single-domain particles of the α' phase based on shape anisotropy. The larger the aspect (length-to-width) ratio of the rods and the smoother their surface, the greater the coercivity of the alloy. Thus, the forming of the precipitate in a magnetic field makes the precipitate longer and thinner and so increases the coercivity of the alnico magnetic material. It is believed that the addition of titanium to some of the highest-strength alnicos increases their coercivities by increasing the aspect ratio of the α' rods.

Rare-Earth Alloys Rare-earth alloy magnets are produced on a large scale in the United States and have magnetic strengths superior to those of any commercial magnetic material. They have maximum energy products $(BH)_{max}$ to 240 kJ/m³ (30 MG·Oe) and coercivities to 3200 kA/m (40 kOe). The origin of magnetism in the rare-earth transition elements is due almost entirely to their unpaired 4f electrons in the same way as the magnetism in Fe, Co, and Ni is due to their unpaired

FIGURE 11.27 Replica electron micrograph showing the structure of alnico 8 (Al–Ni–Co–Fe–Ti) alloy after an 800°C heat treatment for 9 min in an applied magnetic field. The α phase (Ni–Al–rich) is light, and the α' phase (Fe–Co–rich) is dark. The α' phase is highly ferromagnetic, which is elongated in the direction of the applied field, creating anisotropy of the coercive force. (*Courtesy of K. J. deVos, 1966.*)

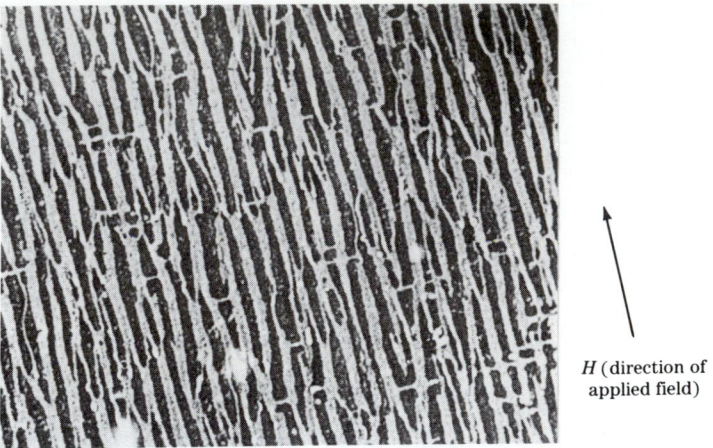

H (direction of applied field)

$3d$ electrons. There are two main groups of commercial rare-earth magnetic materials: one based on single-phase $SmCo_5$ and the other on precipitation-hardened alloys of the approximate composition $Sm(Co,Cu)_{7.5}$.

$SmCo_5$ single-phase magnets are the most widely used type. The mechanism of coercivity in these materials is based on nucleation and/or pinning of domain walls at surfaces and grain boundaries. These materials are fabricated by powder metallurgy techniques using fine particles (1 to 10 μm). During pressing the particles are aligned in a magnetic field. The pressed articles are then carefully sintered to prevent particle growth. The magnetic strengths of these materials are high, with $(BH)_{max}$ values in the range of 130 to 160 kJ/m^3 (16 to 20 MG·Oe).

In the precipitation-hardened $Sm(Co,Cu)_{7.5}$ alloy, part of the Co is substituted by Cu in $SmCo_5$ so that a fine precipitate (about 10 nm) is produced at a low aging temperature (400 to 500°C). The precipitate formed is coherent with the $SmCo_5$ structure. The coherency mechanism here is mainly based on the homogeneous pinning of domain walls at the precipitated particles. These materials are also made commercially by powder metallurgy processes using magnetic alignment of the particles. The addition of small amounts of iron and zirconium promote the development of higher coercivities. Typical values for an $Sm(Co_{0.68}Cu_{0.10}Fe_{0.21}Zr_{0.01})_{7.4}$ commercial alloy are $(BH)_{max} = 240$ kJ/m^3 (30 MG·Oe) and $B_r = 1.1$ T (11,000 G). Figures 11.24 and 11.28 show the outstanding improvement in magnetic strengths achieved with rare-earth magnetic alloys.

Sm–Co magnets are used in medical devices such as thin motors in implantable pumps and valves and in aiding eyelid motion. Rare-earth magnets are also used for electronic wristwatches and traveling-wave tubes. Direct-current and synchronous motors and generators are produced by using rare-earth magnets, resulting in a size reduction.

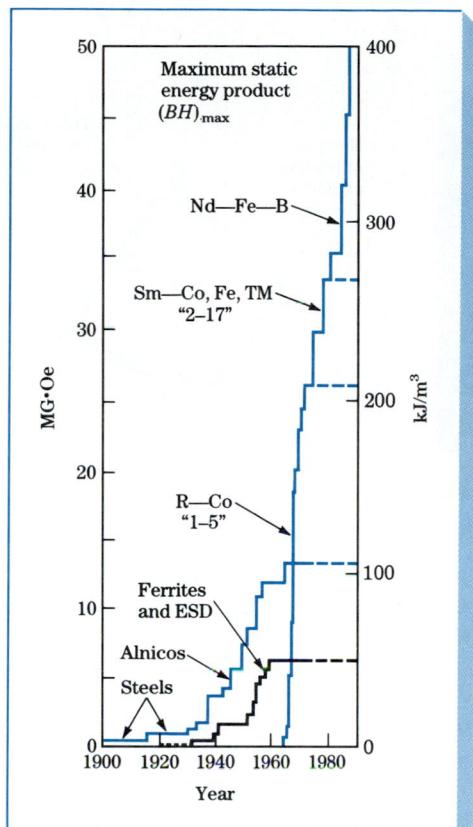

FIGURE 11.28 Progress in permanent-magnet quality in the twentieth century as measured by the maximum energy product $(BH)_{max}$. *(After K. J. Stnat, Soft and Hard Magnetic Materials with Applications, ASM Inter., 1986, p. 64.)*

Neodymium-Iron-Boron Magnetic Alloys

Hard Nd–Fe–B magnetic materials with $(BH)_{max}$ products as high as 300 kJ/m³ (45 MG·Oe) were discovered in about 1984, and today these materials are produced by both powder metallurgy and rapid-solidification melt-spun ribbon processes. Figure 11.29a shows the microstructure of a $Nd_2Fe_{14}B$-type rapidly solidified ribbon. In this structure highly ferromagnetic $Nd_2Fe_{14}B$ matrix grains are surrounded by a nonferromagnetic Nd-rich thin intergranular phase. The high coercivity and associated $(BH)_{max}$ energy product of this material result from the difficulty of nucleating reverse magnetic domains which usually nucleate at the grain boundaries of the matrix grains (Fig. 11.29b). The intergranular nonferromagnetic Nd-rich phase forces the $Nd_2Fe_{14}B$ matrix grains to nucleate their reverse domains in order to reverse the magnetization of the material. This process maximizes H_c and $(BH)_{max}$ for the whole bulk aggregate of the material. Applications for Nd–Fe–B permanent magnets include all types of electric motors, especially those like automotive starting motors where reduction in weight and compactness are desirable.

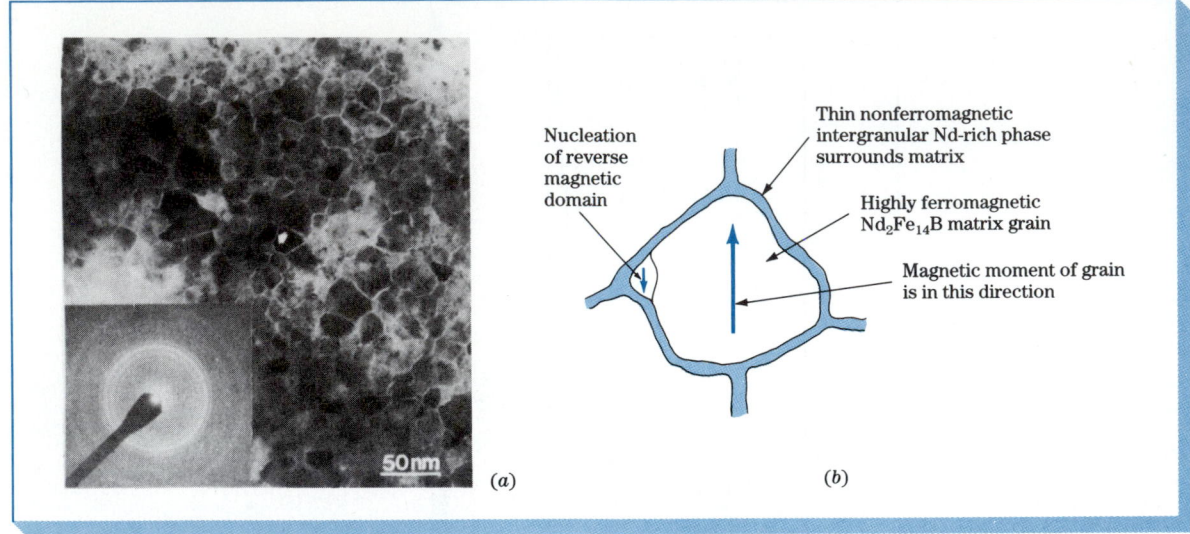

FIGURE 11.29 (a) Bright-field electron transmission micrograph of an optimally quenched Nd–Fe–B ribbon showing random oriented grains surrounded by a thin intergranular phase marked by the arrow. (After J. J. Croat and J. F. Herbst, MRS Bull., June 1988, p. 37.) (b) Single $Nd_2Fe_{14}B$ grain showing nucleation of reverse magnetic domain.

Iron-Chromium-Cobalt Magnetic Alloys

A family of magnetic Fe–Cr–Co alloys were developed in 1971 and are analogous to the alnico alloys in metallurgical structure and permanent magnetic properties, but the former are cold-formable at room temperature. A typical composition for an alloy of this type is 61% Fe–28% Cr–11% Co. Typical magnetic properties of Fe–Cr–Co alloys are B_r = 1.0 to 1.3 T (10 to 13 kG), H_c = 150 to 600 A/cm (190 to 753 Oe), and $(BH)_{max}$ = 10 to 45 kJ/m³ (1.3 to 1.5 MG·Oe). Table 11.5 lists some typical properties of an Fe–Cr–Co magnetic alloy.

The Fe–Cr–Co alloys have a BCC structure at elevated temperatures above about 1200°C. Upon slow cooling (about 15°C/h) from above 650°C, precipitates of a Cr-rich α_2 phase (Fig. 11.30a) with particles of about 30 nm (300 Å) form in a matrix of an Fe-rich α_1 phase. The mechanism for coercivity in the Fe–Cr–Co alloys is the pinning of domain walls by the precipitated particles since the magnetic domains extend through both phases. Particle shape (Fig. 11.30b) is important since elongation of the particles by deformation before a final aging

FIGURE 11.30

Transmission electron micrographs of an Fe–34% Cr–12% Co alloy showing (a) spherical precipitates produced before deformation and (b) elongated and aligned particles after deformation and alignment by final heat treatment. [After S. Jin et al., J. Appl. Phys., 53:4300(1982).]

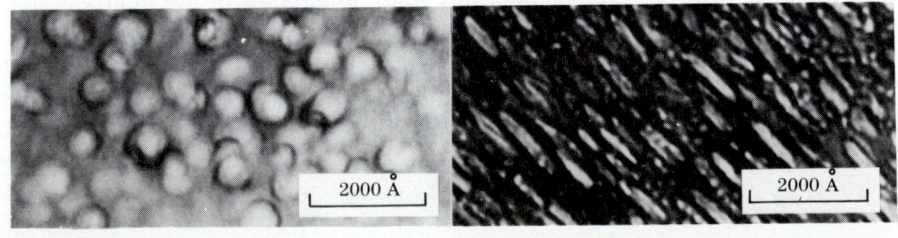

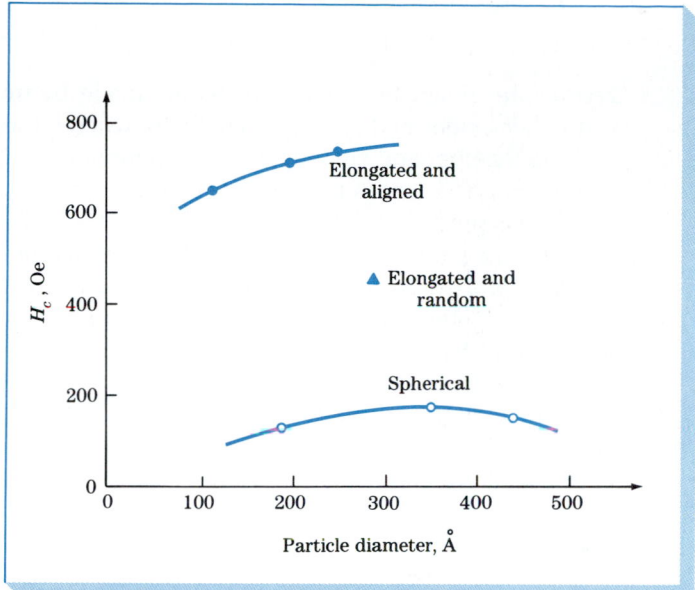

FIGURE 11.31 Coercivity vs. particle diameter for different-shape particles in an Fe–34% Cr–12% Co alloy. Note the great increase in coercivity by changing from a spherical shape to an elongated shape. [*After S. Jin et al., J. Appl. Phys., 53:4300(1982).*]

treatment greatly increases the coercivity of these alloys, as clearly indicated in Fig. 11.31.

Fe–Cr–Co alloys are especially important for engineering applications where their cold ductility allows high-speed room-temperature forming. The permanent magnet for many modern telephone receivers is an example of this cold-deformable permanent magnetic alloy (Fig. 11.32).

FIGURE 11.32 Use of ductile permanent Fe–Cr–Co alloy in a telephone receiver. (*a*) Disassembled receiver showing permanent magnet ring and (*b*) cross-sectional view of the U-type telephone receiver indicating position of permanent magnet. [*After S. Jin et al., IEEE Trans. Magn., 17:2935(1981).*]

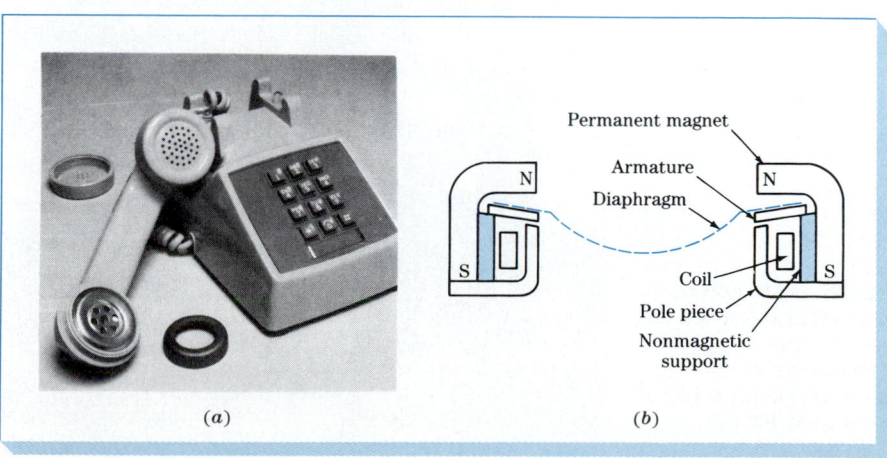

11.9 FERRITES

Ferrites are magnetic ceramic materials made by mixing iron oxide (Fe_2O_3) with other oxides and carbonates in the powdered form. The powders are then pressed together and sintered at a high temperature. Sometimes finishing machining is necessary to produce the desired part shape (Fig. 11.33). The magnetizations produced in the ferrites is large enough to be of commercial value, but their magnetic saturations are not as high as those for ferromagnetic materials. Ferrites have domain structures and hysteresis loops similar to those of ferromagnetic materials. As in the case of the ferromagnetic materials, there are *soft* and *hard ferrites.*

Magnetically Soft Ferrites Soft ferrite materials exhibit ferrimagnetic behavior. In soft ferrites there is a net magnetic moment due to two sets of unpaired inner-electron spin moments in opposite directions which do not cancel each other (Fig. 11.8c).

FIGURE 11.33 Various soft ferrite parts which are used for many electrical and electronic applications. (*Courtesy of Magnetics, Inc.*)

FIGURE 11.34 (*a*) Unit cell of soft ferrite of the type MO·Fe$_2$O$_3$. This unit cell consists of eight subcells. (*b*) The subcell for the FeO·Fe$_2$O$_3$ ferrite. The magnetic moments of the ions in the octahedral sites are aligned in one direction by the applied magnetic field, and those in the tetrahedral sites are aligned in the opposite direction. As a result, there is a net magnetic moment for the subcell and hence the material.

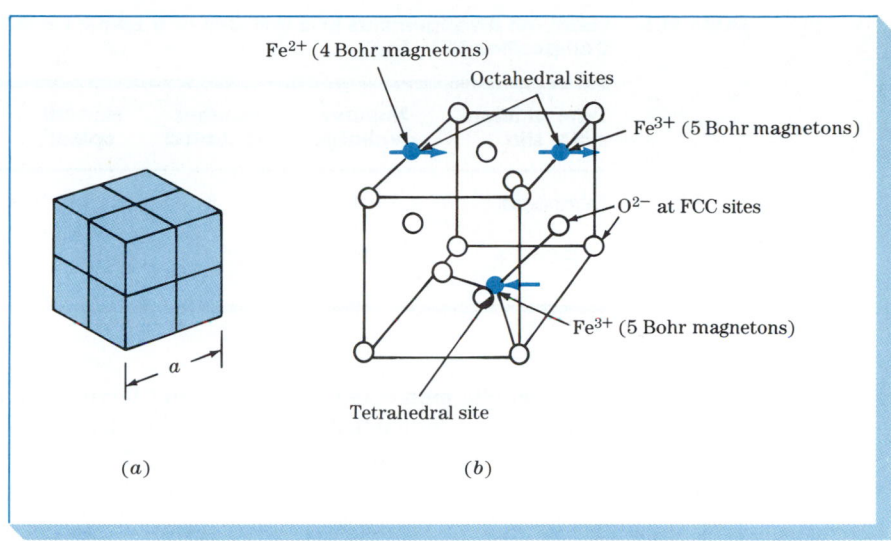

Fe^{2+} (4 Bohr magnetons)

Octahedral sites

Fe^{3+} (5 Bohr magnetons)

O^{2-} at FCC sites

Fe^{3+} (5 Bohr magnetons)

Tetrahedral site

(*a*)   (*b*)

Composition and structure of cubic soft ferrites   Most cubic soft ferrites have the composition MO·Fe$_2$O$_3$, where M is a divalent metal ion such as Fe^{2+}, Mn^{2+}, Ni^{2+}, or Zn^{2+}. The structure of the soft ferrites is based on the inverse spinel structure which is a modification of the spinel structure of the mineral spinel (MgO·Al$_2$O$_3$). Both the spinel and inverse spinel structures have cubic unit cells consisting of eight subcells, as shown in Fig. 11.34*a*. Each of the subcells consists of one molecule of MO·Fe$_2$O$_3$. Since each subunit contains one MO·Fe$_2$O$_3$ molecule and since there are seven ions in this molecule, each unit cell contains a total of 7 ions × 8 subcells = 56 ions per unit cell. Each subunit cell has an FCC crystal structure made up of the four ions of the MO·Fe$_2$O$_3$ molecule (Fig. 11.34*b*). The much smaller metal ions (M^{2+} and Fe^{3+}) which have ionic radii of about 0.07 to 0.08 nm occupy interstitial spaces between the larger oxygen ions (ionic radius ≈ 0.14 nm).

As previously pointed out in Sec. 10.2, in an FCC unit cell there are the equivalent of four octahedral and eight tetrahedral interstitial sites. In the normal spinel structure only half the octahedral sites are occupied, and so only $\frac{1}{2}$(8 subcells × 4 sites/subcell) = *16 octahedral sites are occupied* (Table 11.6). In the normal spinel structure there are 8 × 8 (tetrahedral sites per subcell) = 64 sites/unit cell. However, in the normal spinel structure only one-eighth of the 64 sites are occupied so that only *eight of the tetrahedral sites are occupied* (Table 11.6).

In the normal spinel-structure unit cell there are eight MO·Fe$_2$O$_3$ molecules. In this structure the 8 M^{2+} ions occupy 8 tetrahedral sites and the 16 Fe^{3+} ions occupy 16 octahedral sites. However, in the inverse spinel structure there is a different arrangement of the ions: the 8 M^{2+} ions occupy 8 octahedral sites, and the 16 Fe^{3+} ions are divided so that 8 occupy octahedral sites and 8 tetrahedral sites (Table 11.6).

TABLE 11.6 **Metal Ion Arrangements in a Unit Cell of a Spinel Ferrite of Composition MO · Fe₂O₃**

Type of inter-stitial site	Number available	Number occupied	Normal spinel	Inverse spinel
Tetrahedral	64	8	8 M^{2+}	8 Fe^{3+}
Octahedral	32	16	16 Fe^{3+}	← 8 Fe^{3+}, 8 M^{2+} → →

Net magnetic moments in inverse spinel ferrites To determine the net magnetic moment for each $MO \cdot Fe_2O_3$ ferrite molecule, we must know the $3d$ inner-electron configuration of the ferrite ions. Figure 11.35 gives this information. When the Fe atom is ionized to form the Fe^{2+} ion, there are *four* unpaired $3d$ electrons left after the loss of two $4s$ electrons. When the Fe atom is ionized to form the Fe^{3+} ion, there are *five* unpaired electrons left after the loss of two $4s$ and one $3d$ electrons.

Since each unpaired $3d$ electron has a magnetic moment of 1 Bohr magneton, the Fe^{2+} ion has a moment of 4 Bohr magnetons and the Fe^{3+} ion a moment of 5 Bohr magnetons. In an applied magnetic field, the magnetic moments of the octahedral and tetrahedral ions oppose each other (Fig. 11.34b). Thus, in the case of the ferrite $FeO \cdot Fe_2O_3$, the magnetic moments of the eight Fe^{3+} ions in octahedral sites will cancel the magnetic moments of the eight

FIGURE 11.35 Electronic configurations and ionic magnetic moments for some $3d$ transition-element ions.

Ion	Number of electrons	Electron configuration 3d orbitals					Ionic magnetic moment (Bohr magnetons)
Fe^{3+}	23	↑	↑	↑	↑	↑	5
Mn^{2+}	23	↑	↑	↑	↑	↑	5
Fe^{2+}	24	↑↓	↑	↑	↑	↑	4
Co^{2+}	25	↑↓	↑↓	↑	↑	↑	3
Ni^{2+}	26	↑↓	↑↓	↑↓	↑	↑	2
Cu^{2+}	27	↑↓	↑↓	↑↓	↑↓	↑	1
Zn^{2+}	28	↑↓	↑↓	↑↓	↑↓	↑↓	0

TABLE 11.7   Ion Arrangements and Net Magnetic Moments per Molecule in Normal and Inverse Spinel Ferrites

Ferrite	Structure	Tetrahedral sites occupied	Octahedral sites occupied		Net magnetic moment (μ_B/molecule)
$FeO \cdot Fe_2O_3$	Inverse spinel	Fe^{3+} 5 ←	Fe^{2+} 4 →	Fe^{3+} 5 →	4
$ZnO \cdot Fe_2O_3$	Normal spinel	Zn^{2+} 0	Fe^{3+} 5 ←	Fe^{3+} 5 →	0

Fe^{3+} ions in the tetrahedral sites. Thus, the resultant magnetic moment of this ferrite will be due to the eight Fe^{2+} ions at eight octahedral sites, which have moments of 4 Bohr magnetons each (Table 11.7). A theoretical value for the magnetic saturation of the $FeO \cdot Fe_2O_3$ ferrite is calculated in Example Problem 11.5 on the basis of the Bohr magneton strength of the Fe^{2+} ions.

Example Problem 11.5

Calculate the theoretical saturation magnetization M in amperes per meter and the saturation induction B_s in teslas for the ferrite $FeO \cdot Fe_2O_3$. Neglect the $\mu_0 H$ term for the B_s calculation. The lattice constant of the $FeO \cdot Fe_2O_3$ unit cell is 0.839 nm.

Solution:

The magnetic moment for a molecule of $FeO \cdot Fe_2O_3$ is due entirely to the 4 Bohr magnetons of the Fe^{2+} ion since the unpaired electrons of the Fe^{3+} ions cancel each other. Since there are eight molecules of $FeO \cdot Fe_2O_3$ in the unit cell, the total magnetic moment per unit cell is

$$(4 \text{ Bohr magnetons/subcell})(8 \text{ subcells/unit cell}) = 32 \text{ Bohr magnetons/unit cell}$$

Thus
$$M = \left[\frac{32 \text{ Bohr magnetons/unit cell}}{(8.39 \times 10^{-10} \text{ m})^3/\text{unit cell}} \right] \left(\frac{9.27 \times 10^{-24} \text{ A} \cdot \text{m}^2}{\text{Bohr magneton}} \right)$$

$$= 5.0 \times 10^5 \text{ A/m} \blacktriangleleft$$

B_s at saturation, assuming all magnetic moments are aligned and neglecting the H term, is given by the equation $B_s \approx \mu_0 M$. Thus

$$B_s \approx \mu_0 M \approx \left(\frac{4\pi \times 10^{-7} \text{ T} \cdot \text{m}}{\text{A}} \right) \left(\frac{5.0 \times 10^5 \text{ A}}{\text{m}} \right)$$

$$= 0.63 \text{ T} \blacktriangleleft$$

Iron, cobalt, and nickel ferrites all have the inverse spinel structure, and all are ferrimagnetic due to a net magnetic moment of their ionic structures. Industrial soft ferrites usually consist of a mixture of ferrites since increased saturation magnetizations can be obtained from a mixture of ferrites. The two most common industrial ferrites are the nickel-zinc-ferrite ($Ni_{1-x}Zn_xFe_{2-y}O_4$) and the manganese-zinc-ferrite ($Mn_{1-x}Zn_xFe_{2+y}O_4$).

Properties and applications of soft ferrites

Eddy-current losses in magnetic materials   Soft ferrites are important magnetic materials because in addition to having useful magnetic properties they are insulators and have high electrical resistivities. A high electrical resistivity is important in magnetic applications that require high frequencies since if the magnetic material is conductive eddy-current energy losses would be great at the high frequencies. Eddy currents are caused by induced voltage gradients, and thus the higher the frequency, the greater the increase in eddy currents. Since soft ferrites are insulators, they can be used for magnetic applications such as transformer cores which operate at high frequencies.

Applications for soft ferrites   Some of the most important uses for soft ferrites are for low-signal, memory-core, audiovisual, and recording-head applications. At low signal levels, soft ferrite cores are used for transformers and low-energy inductors. A large tonnage usage of soft ferrites is for deflection-yoke cores, flyback transformers, and convergence coils for television receivers.

Mn–Zn and Ni–Zn spinel ferrites are used in magnetic recording heads for various types of magnetic tapes. Recording heads are made from polycrystalline Ni–Zn ferrite since the operating frequencies required (100 kHz to 2.5 GHz) are too high for metallic alloy heads because of high eddy-current losses.

Magnetic-core memories based on the 0 and 1 binary logic are used for some types of computers. The magnetic core is useful where loss of power does not cause loss of information. Since magnetic-core memories have no moving parts, they are used when high shock resistance is needed, as in some military uses.

Magnetically Hard Ferrites   A group of hard ferrites which are used for permanent magnets have the general formula $MO \cdot 6Fe_2O_3$ and are hexagonal in crystal structure. The most important ferrite of this group is *barium ferrite* ($BaO \cdot 6Fe_2O_3$), which was introduced in the Netherlands by the Philips Company in 1952 under the trade name Ferroxdure. In recent years the barium ferrites have been replaced to some extent by the strontium ferrites, which have the general formula ($SrO \cdot 6Fe_2O_3$) and which have superior magnetic properties compared with the barium ferrites. These ferrites are produced by almost the same method used for the soft ferrites, with most being wet-pressed in a magnetic field to align the easy magnetizing axis of the particles with the applied field.

The hexagonal ferrites are low in cost, low in density, and have a high coerce force, as shown in Fig. 11.24. The high magnetic strengths of these materials is due mainly to their high magnetocrystalline anisotropy. The magnetization of these materials is believed to take place by domain wall nucleation and motion because their grain size is too large for single-domain behavior. Their $(BH)_{max}$ energy products vary from 14 to 28 kJ/m^3.

These hard-ferrite ceramic permanent magnets find widespread use in generators, relays, and motors. Electronic applications include magnets for loudspeakers, telephone ringers, and receivers. They are also used for holding devices for door closers, seals, and latches and in many toy designs.

SUMMARY

Magnetic materials are important industrial materials used for many engineering designs. Most industrial magnetic materials are *ferro-* or *ferrimagnetic* and show large magnetizations. The most important ferromagnetic materials are based on alloys of Fe, Co, and Ni. More recently some ferromagnetic alloys have been made with some rare-earth elements such as Sm. In ferromagnetic materials such as Fe, there exist regions called *magnetic domains* in which atomic magnetic dipole moments are aligned parallel to each other. The magnetic domain structure in a ferromagnetic material is determined by the following energies which are minimized: exchange, magnetostatic, magnetocrystalline anisotropy, domain wall, and magnetostrictive energies. When the ferromagnetic domains in a sample are at random orientations, the sample is in a demagnetized state. When a magnetic field is applied to a ferromagnetic material sample, the domains in the sample are aligned and the material becomes magnetized and remains magnetized to some extent when the field is removed. The magnetization behavior of a ferromagnetic material is recorded by the magnetic induction vs. applied field graph called a *hysteresis loop*. When a demagnetized ferromagnetic material is magnetized by an applied field H, its magnetic induction B eventually reaches a saturation level called the *saturation induction B_s*. When the applied field is removed, the magnetic induction decreases to a value called the *remanent induction B_r*. The demagnetizing field required to reduce the magnetic induction of a magnetized ferromagnetic sample to zero is called the *coercive force H_c*.

A *soft magnetic material* is one which is easily magnetized and demagnetized. Important magnetic properties of a soft magnetic material are high permeability, high saturation induction, and low coercive force. When a soft ferromagnetic material is repeatedly magnetized and demagnetized, *hysteresis* and *eddy-current* energy losses occur. Examples of soft ferromagnetic materials include Fe–3 to 4% Si alloys used in motors and power transformers and generators and Ni–20 to 50% Fe alloys used primarily for high-sensitivity communications equipment.

A *hard magnetic material* is one which is difficult to magnetize and which remains magnetized to a great extent after the magnetizing field is removed. Important properties of a hard magnetic material are high coercive force and high saturation induction. The power of a hard magnetic material is measured by its maximum energy product, which is the maximum value of the product of B and H in the demagnetizing quadrant of its B-H hysteresis loop. Examples of hard magnetic materials are the alnicos which are used as permanent magnets for many electrical applications and some rare-earth alloys which are based on $SmCo_5$ and $Sm(Co,Cu)_{7.5}$ compositions. The rare-earth alloys are used for small motors and other applications requiring an extremely high energy-product magnetic material.

The *ferrites* which are ceramic compounds are another type of industrially important magnetic material. These materials are ferrimagnetic due to a net magnetic moment produced by their ionic structure. Most magnetically soft ferrites have the basic composition $MO \cdot Fe_2O_3$, where M is a divalent ion such as Fe^{2+}, Mn^{2+} and Ni^{2+}. These materials have the *inverse spinel structure* and are used for low-signal, memory-core, audiovisual, and recording-head applications, as examples. Since these materials are insulators, they can be used for high-frequency application where eddy currents are a problem with alternating fields. Magnetically hard ferrites with the general formula $MO \cdot 6Fe_2O_3$, where M is usually a Ba or Sr ion, are used for applications requiring low-cost, low-density permanent magnetic materials. These materials are used for loudspeakers, telephone ringers and receivers and for holding devices for doors, seals, and latches, as examples.

DEFINITIONS

Sec. 11.1

Ferromagnetic material: one which is capable of being highly magnetized. Elemental iron, cobalt, and nickel are ferromagnetic materials.

Magnetic field H: the magnetic field due to an external applied magnetic field or the magnetic field produced by a current passing through a conducting wire or coil of wire (solenoid).

Magnetization M: a measure of the increase in magnetic flux due to the insertion of a given material into a magnetic field of strength H. In SI units the magnetization is equal to the permeability of a vacuum (μ_0) times the magnetization, or $\mu_0 M$. ($\mu_0 = 4\pi \times 10^{-4}$ T·m/A.)

Magnetic induction B: the sum of the applied field H and the magnetization M due to the insertion of a given material into the applied field. In SI units, $B = \mu_0(H + M)$.

Magnetic permeability μ: the ratio of the magnetic induction B to the applied magnetic field H for a material; $\mu = B/H$.

Relative permeability μ_r: the ratio of the permeability of a material to the permeability of a vacuum; $\mu_r = \mu/\mu_0$.

Magnetic susceptibility χ_m**:** the ratio of M (magnetization) to H (applied magnetic field); $\chi_m = M/H$.

Sec. 11.2

Diamagnetism: a weak, negative, repulsive reaction of a material to an applied magnetic field; a diamagnetic material has a small negative magnetic susceptibility.

Paramagnetism: a weak, positive, attractive reaction of a material to an applied magnetic field; a paramagnetic material has a small positive magnetic susceptibility.

Ferromagnetism: the creation of a very large magnetization in a material when subjected to an applied magnetic field. After the applied field is removed, the ferromagnetic material retains much of the magnetization.

Antiferromagnetism: a type of magnetism in which magnetic dipoles of atoms are aligned in opposite directions by an applied magnetic field so that there is no net magnetization.

Ferrimagnetism: a type of magnetism in which the magnetic dipole moments of different ions of an ionically bonded solid are aligned by a magnetic field in an antiparallel manner so that there is a net magnetic moment.

Bohr magneton: the magnetic moment produced in a ferro- or ferrimagnetic material by one unpaired electron without interaction from any others; the Bohr magneton is a fundamental unit. 1 Bohr magneton $= 9.27 \times 10^{-24}$ A·m^2.

Sec. 11.3

Curie temperature: the temperature at which a ferromagnetic material when heated completely loses its ferromagnetism and becomes paramagnetic.

Sec. 11.4

Magnetic domain: a region in a ferro- or ferrimagnetic material in which all magnetic dipole moments are aligned.

Sec. 11.5

Exchange energy: the energy associated with the coupling of individual magnetic dipoles into a single magnetic domain. The exchange energy can be positive or negative.

Magnetostatic energy: the magnetic potential energy due to the external magnetic field surrounding a sample of a ferromagnetic material.

Magnetocrystalline anisotropy energy: the energy required during the magnetization of a ferromagnetic material to rotate the magnetic domains because of crystalline anisotropy. For example, the difference in magnetizing energy between the hard [111] direction of magnetization and the [100] easy direction in Fe is about 1.4×10^4 J/m^3.

Domain wall energy: the potential energy associated with the disorder of dipole moments in the wall volume between magnetic domains.

Magnetostriction: the change in length of a ferromagnetic material in the direction of magnetization due to an applied magnetic field.

Magnetostrictive energy: the energy due to the mechanical stress caused by magnetostriction in a ferromagnetic material.

Sec. 11.6

Hysteresis loop: the B vs. H or M vs. H graph traced out by the magnetization and demagnetization of a ferro- or ferrimagnetic material.

Saturation induction B_s or saturation magnetization M_s: the maximum value of induction B_s or magnetization M_s for a ferromagnetic material.

Remanent induction B_r or remanent magnetization M_r: the value of B or M in a ferromagnetic material after H is decreased to zero.

Coercive force H_c: the applied magnetic field required to decrease the magnetic induction of a magnetized ferro- or ferrimagnetic material to zero.

Sec. 11.7

Soft magnetic material: a magnetic material with a high permeability and low coercive force.

Hysteresis energy loss: the work or energy lost in tracing out a *B-H* hysteresis loop. Most of the energy lost is expended in moving the domain boundaries during magnetization.

Eddy-current energy loss: energy losses in magnetic materials while using alternating fields; the losses are due to induced currents in the material.

Iron-silicon magnetic alloys: Fe–3 to 4% Si alloys which are soft magnetic materials with high saturation inductions. These alloys are used in motors and low-frequency power transformers and generators.

Nickel-iron magnetic alloys: high-permeability soft magnetic alloys used for electrical applications where a high sensitivity is required such as for audio and instrument transformers. Two commonly used basic compositions are 50% Ni–50% Fe and 79% Ni–21% Fe.

Sec. 11.8

Hard magnetic material: a magnetic material with a high coercive force and a high saturation induction.

Energy product $(BH)_{max}$: the maximum value of B times H in the demagnetization curve of a hard magnetic material. The $(BH)_{max}$ value has SI units of J/m^3.

Alnico magnetic alloys: a family of permanent magnetic alloys having the basic composition of Al, Ni, and Co, and about 25 to 50% Fe. A small amount of Cu and Ti is added to some of these alloys.

Magnetic anneal: the heat treatment of a magnetic material in a magnetic field which aligns part of the alloy in the direction of the applied field. For example, the α' precipitate in alnico 5 alloy is elongated and aligned by this type of heat treatment.

Rare-earth magnetic alloys: a family of permanent magnetic alloys with extremely high energy products. $SmCo_5$ and $Sm(Co,Cu)_{7.4}$ are the two most important commercial compositions of these alloys.

Iron-chromium-cobalt magnetic alloys: a family of permanent magnetic alloys containing about 30% Cr–10 to 23% Co and balance iron. These alloys have the advantage of being cold-formable at room temperature.

Sec. 11.9

Soft ferrites: Ceramic compounds with the general formula $MO \cdot Fe_2O_3$, where M is a divalent ion such as Fe^{2+}, Mn^{2+}, Zn^{2+}, or Ni^{2+}. These materials are ferrimagnetic and are insulators and so can be used for high-frequency transformer cores.

Normal spinel structure: a ceramic compound having the general formula $MO \cdot M_2O_3$. The oxygen ions in this compound form an FCC lattice, with the M^{2+} ions occupying tetrahedral interstitial sites and the M^{3+} ions occupying octahedral sites.

Inverse spinal structure: a ceramic compound having the general formula $MO \cdot M_2O_3$. The oxygen ions in this compound form an FCC lattice, with the M^{2+} ions occupying octahedral sites and the M^{3+} ions occupying both octahedral and tetrahedral sites.

Hard ferrites: ceramic permanent magnetic materials. The most important family of these materials has the basic composition $MO \cdot Fe_2O_3$, where M is a barium (Ba) ion or a strontium (Sr) ion. These materials have a hexagonal structure and are low in cost and density.

PROBLEMS

11.1.1 What elements are strongly ferromagnetic at room temperature?

11.1.2 How can a magnetic field around a magnetized iron bar be revealed?

11.1.3 What are the SI and cgs units for the magnetic field strength H?

11.1.4 Define magnetic induction B and magnetization M.

11.1.5 What is the relationship between B and H?

11.1.6 What is the permeability of free-space constant μ_0?

11.1.7 What are the SI units for B and M?

11.1.8 Write an equation that relates B, H, and M, using SI units.

11.1.9 Why is the relation $B \approx \mu_0 M$ often used in magnetic property calculations?

11.1.10 Define magnetic permeability and relative magnetic permeability.

11.1.11 Is the magnetic permeability of a ferromagnetic material a constant? Explain.

11.1.12 What magnetic permeability quantities are frequently specified?

11.1.13 Define magnetic susceptibility. For what situation is this quantity often used?

11.2.1 Describe two mechanisms involving electrons by which magnetic fields are created.

11.2.2 Define diamagnetism. What is the order of magnitude for the magnetic susceptibility of diamagnetic materials at 20°C?

11.2.3 Define paramagnetism. What is the order of magnitude for the magnetic susceptibility of paramagnetic materials at 20°C?

11.2.4 Define ferromagnetism. Which elements are ferromagnetic?

11.2.5 What causes ferromagnetism in Fe, Co, and Ni?

11.2.6 How many unpaired $3d$ electrons are there per atom in Cr, Mn, Fe, Co, Ni, and Cu?

11.2.7 What are magnetic domains?

11.2.8 How does a positive exchange energy affect the alignment of magnetic dipoles in ferromagnetic materials?

11.2.9 What is the explanation given for the fact that Fe, Co, and Ni are ferromagnetic and Cr and Mn are not even though all these elements have unpaired $3d$ electrons?

11.2.10 Calculate a theoretical value for the saturation magnetization and saturation induction for nickel, assuming all unpaired $3d$ electrons contribute to the magnetization. (Ni is FCC and $a = 0.352$ nm.)

11.2.11 Calculate a theoretical value for the saturation magnetization of pure cobalt metal assuming all unpaired $3d$ electrons contribute to the magnetization. (Co is HCP with $a = 0.25071$ nm and $c = 0.40686$ nm.)

11.2.12 Calculate a theoretical value for the saturation magnetization of pure gadolinium below 16°C assuming all seven unpaired $4f$ electrons contribute to the magnetization. (Gd is HCP with $a = 0.364$ nm and $c = 0.578$ nm.)

11.2.13 Cobalt has a saturation magnetization of 1.42×10^6 A/m. What is its average magnetic moment in Bohr magnetons per atom?

11.2.14 Nickel has an average of 0.604 Bohr magnetons/atom. What is its saturation induction?

11.2.15 Gadolinium at very low temperatures has an average of 7.1 Bohr magnetons/atom. What is its saturation magnetization?

11.2.16 Define antiferromagnetism. Which elements show this type of behavior?

11.2.17 Define ferrimagnetism. What are ferrites? Give an example of a ferrimagnetic compound.

11.3.1 What effect does increasing temperature above 0 K have on the alignment of magnetic dipoles in ferromagnetic materials?

11.3.2 What is the Curie temperature?

11.4.1 How can a ferromagnetic material be demagnetized? What is the arrangement of the magnetic domains in a demagnetized ferromagnetic material?

11.4.2 When a demagnetized ferromagnetic material is slowly magnetized by an applied magnetic field, what changes occur first in domain structure?

11.4.3 After domain growth due to magnetization of a ferromagnetic material by an applied field has finished, what change in domain structure occurs with a further substantial increase in the applied field?

11.4.4 What changes in domain structure occur in a ferromagnetic material when the applied field which magnetized a sample to saturation is removed?

11.4.5 How may the domain structure of a ferromagnetic material be revealed for observation in the optical microscope?

11.5.1 What are the five types of energies which determine the domain structure of a ferromagnetic material?

11.5.2 Define magnetic exchange energy. How can the exchange energy of a ferromagnetic material be minimized with respect to magnetic dipole alignment?

11.5.3 Define magnetostatic energy. How can the magnetostatic energy of a ferromagnetic material sample be minimized?

11.5.4 Define magnetocrystalline anisotropy energy. What are the easy directions of magnetization for (a) Fe and (b) Ni?

11.5.5 Define magnetic domain wall energy. What is the average width in terms of number of atoms for a ferromagnetic domain wall?

11.5.6 What two energies determine the domain wall width? What energy is minimized when the wall is widened? What energy is minimized when the wall is made narrower?

11.5.7 Define magnetostriction and magnetostrictive energy. What is the cause of magnetostriction in ferromagnetic materials?

11.5.8 What are domains of closure? How are magnetostrictive stresses created by domains of closure?

11.5.9 How does the domain size affect the magnetostrictive energy of a magnetized ferromagnetic material sample?

11.5.10 How does domain size affect the amount of domain wall energy in a sample?

11.6.1 Draw a hysteresis B-H loop for a ferromagnetic material and indicate on it (a) the saturation induction B_s, (b) the remanent induction B_r, and (c) the coercive force H_c.

11.6.2   Describe what happens to the magnetic induction when a ferromagnetic material is magnetized, demagnetized, and remagnetized by an applied magnetic field.

11.6.3   What happens to the magnetic domains of a ferromagnetic material sample during magnetization and demagnetization?

11.7.1   Define a soft magnetic material and a hard magnetic material.

11.7.2   What type of a hysteresis loop does a soft ferromagnetic material have?

11.7.3   What are desirable magnetic properties for a soft magnetic material?

11.7.4   What are hysteresis energy losses? What factors affect hysteresis losses?

11.7.5   How does the ac frequency affect the hysteresis losses of soft ferromagnetic materials? Explain.

11.7.6   What are eddy currents? How are they created in a ferromagnetic material?

11.7.7   How can eddy currents be reduced in metallic magnetic transformer cores?

11.7.8   Why does the addition of 3 to 4% silicon to iron reduce transformer-core energy losses?

11.7.9   What disadvantages are there to the addition of silicon to iron for transformer-core materials?

11.7.10 Why does a laminated structure increase the electrical efficiency of a power transformer?

11.7.11 Why does grain-oriented iron-silicon transformer sheet steel increase the efficiency of a transformer core?

11.7.12 What is the structure of a metallic glass? How are magnetic glass ribbons produced?

11.7.13 What are some special properties of metallic glasses?

11.7.14 Why are magnetic metallic glasses so easily magnetized and demagnetized?

11.7.15 What are the advantages of metallic glasses for power transformers? Disadvantages?

11.7.16 Calculate the weight percent of the elements in the metallic glass with the atomic percent composition $Fe_{78}B_{13}Si_9$.

11.7.17 What are some engineering advantages of using nickel-iron alloys for electrical applications?

11.7.18 What compositions of Ni–Fe alloys are especially important for electrical applications?

11.7.19 How does ordering in an Ni–Fe alloy affect the magnetic properties of a 78.5% Ni–21.5% Fe alloy? How can ordering be prevented?

11.7.20 How does magnetic annealing increase the magnetic properties of a 65% Ni–35% Fe alloy?

11.8.1   What are important magnetic properties for a hard magnetic material?

11.8.2   What is the maximum energy product for a hard magnetic material? How is it calculated? What are the SI and cgs units for the energy product?

11.8.3   Estimate the maximum energy product for the $SmCo_5$ rare-earth hard magnetic alloy (curve 2) of Fig. 11.24.

11.8.4   Estimate the maximum energy product for the alnico 5 alloy (curve 4) of Fig. 11.24.

11.8.5   Approximately how much energy in kilojoules per cubic meter would be required to demagnetize a fully magnetized 2-cm^3 block of alnico 8 alloy?

11.8.6   What elements are included in the alnico magnetic materials?

11.8.7   What two processes are used to produce alnico permanent magnets?

11.8.8   What is the basic structure of an alnico 8 magnetic material?

11.8.9   How does precipitation in a magnetic field affect the shape of the precipitates in an alnico 8 alloy? How does the shape of the precipitates affect the coercivity of this material?

11.8.10 What is the origin of ferromagnetism in rare-earth magnetic alloys?

11.8.11 How does the maximum energy products of the alnicos compare with those of the rare-earth alloy magnetic alloys?

11.8.12 What are the two main groups of rare-earth alloys?

11.8.13 What is believed to be the basic mechanism of coercivity for the $SmCo_5$ magnetic alloys?

11.8.14 What are some applications for rare-earth magnetic alloys?

11.8.15 What fabrication advantage do the Fe–Cr–Co magnetic alloys have for making permanent magnetic alloy parts?

11.8.16 What is a typical chemical composition of an Fe–Cr–Co magnetic alloy?

11.8.17 What is the basic structure of an Fe–Cr–Co magnetic alloy? What is the mechanism for coercivity of the Fe–Cr–Co type magnetic alloys?

11.8.18 How does plastic deformation before the final aging treatment affect the shape of the precipitated particles and the coercivity of the Fe–Cr–Co magnetic alloys?

11.8.19 For what type of applications are the Fe–Cr–Co alloys particularly suited?

11.9.1   What are the ferrites? How are they produced?

11.9.2   What is the basic composition of the cubic soft ferrites?

11.9.3   Describe the unit cell of the spinel structure $MgO \cdot Al_2O_3$, including which ions occupy tetrahedral and octahedral interstitial sites.

11.9.4   Describe the unit cell of the inverse spinel structure, including which ions occupy tetrahedral and octahedral interstitial sites.

11.9.5   What is the net magnetic moment per molecule for each of the following ferrites: (a) $FeO \cdot Fe_2O_3$, (b) $NiO \cdot Fe_2O_3$, (c) $MnO \cdot Fe_2O_3$?

11.9.6   Calculate the theoretical saturation magnetization in amperes per meter and saturation induction in teslas for the ferrite $NiO \cdot Fe_2O_3$ ($a = 0.834$ nm for $NiO \cdot Fe_2O_3$).

11.9.7   What are the compositions of the two most commonly used ferrites? Why are mixtures of ferrites used instead of a single pure ferrite?

11.9.8   Why is a high electrical resistivity necessary for a magnetic material which is to be used for a transformer core operating at a high frequency?

11.9.9   What are some industrial applications for soft ferrites?

11.9.10 Why are magnetic core memories particularly useful for high-shock-resistance applications?

11.9.11 What is the basic composition of the hexagonal hard ferrites?

11.9.12 What are the advantages of the hard ferrites for industrial use?

11.9.13 What are some applications for hard ferrite magnetic materials?

12

Corrosion

12.1 GENERAL

Corrosion may be defined as the deterioration of a material resulting from chemical attack by its environment. Since corrosion is caused by chemical reaction, the rate at which the corrosion takes place will depend to some extent on the temperature and the concentration of the reactants and products. Other factors such as mechanical stress and erosion may also contribute to corrosion.

Most corrosion of materials refers to the chemical attack of metals that occurs most commonly by electrochemical attack, since metals have free electrons which are able to set up electrochemical cells within the metals themselves. Most metals are corroded to some extent by water and the atmosphere. Metals can also be corroded by direct chemical attack from chemical solutions and even liquid metals.

The corrosion of metals can be regarded in some ways as reverse extractive metallurgy. Most metals exist in nature in the combined state, for example, as oxides, sulfides, carbonates, or silicates. In these combined states the energies of the metals are lower. In the metallic state the energies of metals are higher, and thus there is a spontaneous tendency for metals to react chemically to form compounds. For example, iron oxides exist commonly in nature and are reduced by thermal energy to iron, which is in a higher energy state. There is, therefore, a tendency for the metallic iron to spontaneously return to iron oxide by corroding (rusting) so that it can exist in a lower energy state (Fig. 12.1).

(a)   (b)

FIGURE 12.1 (*a*) Iron ore (iron oxide). (*b*) Corrosion products in the form of rust (iron oxide) on a steel (iron) sample that has been exposed to the atmosphere. By rusting, metallic iron in the form of steel has returned to its original lower energy state. (*Courtesy of the LaQue Center for Corrosion Technology, Inc.*)

Nonmetallic materials such as ceramics and polymers do not suffer electrochemical attack but can be deteriorated by direct chemical attack. For example, ceramic refractory materials can be chemically attacked at high temperatures by molten salts. Organic polymers can be deteriorated by the chemical attack of organic solvents. Water is absorbed by some organic polymers, which causes changes in dimensions or property changes. The combined action of oxygen and ultraviolet radiation will deteriorate some polymers even at room temperature.

Corrosion, therefore, is a destructive process as far as the engineer is concerned and represents an enormous economic loss. Thus it is not surprising that the engineer working in industry must be concerned about corrosion control and prevention. The purpose of this chapter is to serve as an introduction to this important subject.

12.2 ELECTROCHEMICAL CORROSION OF METALS

Oxidation-Reduction Reactions   Since most corrosion reactions are electrochemical in nature, it is important to understand the basic principles of electrochemical reactions. Consider a piece of zinc metal placed in a beaker of dilute hydrochloric acid, as shown in Fig. 12.2. The zinc dissolves or corrodes in the acid, and zinc chloride and hydrogen gas are produced as indicated by the chemical reaction

$$Zn + 2HCl \rightarrow ZnCl_2 + H_2 \qquad (12.1)$$

This reaction can be written in a simplified ionic form, omitting the chloride ions, as

$$Zn + 2H^+ \rightarrow Zn^{2+} + H_2 \qquad (12.2)$$

FIGURE 12.2 Reaction of hydrochloric acid with zinc to produce hydrogen gas. (*Courtesy of the LaQue Center for Corrosion Technology, Inc.*)

This equation consists of two half-reactions: one for the oxidation of the zinc and the other for the reduction of the hydrogen ions to form hydrogen gas. These half-cell reactions can be written as

$$Zn \rightarrow Zn^{2+} + 2e^- \qquad \text{(oxidation half-cell reaction)} \qquad (12.3a)$$
$$2H^+ + 2e^- \rightarrow H_2 \qquad \text{(reduction half-cell reaction)} \qquad (12.3b)$$

Some important points about the oxidation-reduction half-cell reactions are:

1. *Oxidation reaction.* The oxidation reaction by which metals form ions that go into aqueous solution is called the *anodic reaction,* and the local regions on the metal surface where the oxidation reaction takes place are called *local anodes.* In the anodic reaction electrons are produced which remain in the metal, and the metal atoms form cations (for example, $Zn \rightarrow Zn^{2+} + 2e^-$).
2. *Reduction reaction.* The reduction reaction in which a metal or nonmetal is reduced in valence charge is called the *cathodic* reaction. The local regions on the metal surface where metal ions or nonmetal ions are reduced in valence charge are called *local cathodes.* In the cathode reaction there is a *consumption of electrons.*
3. Electrochemical corrosion reactions involve oxidation reactions that produce electrons and reduction reactions that consume them. Both oxidation and reduction reactions must occur at the same time and same overall rate to prevent a buildup of electric charge in the metal.

FIGURE 12.3

Experimental setup for the determination of the standard emf of zinc. In one beaker a Zn electrode is placed in a solution of 1 M Zn^{2+} ions. In the other there is a standard hydrogen reference electrode consisting of a platinum electrode immersed in a solution of 1 M H^+ ions which contains H_2 gas at 1 atm. The overall reaction which occurs when the two electrodes are connected by an external wire is

$Zn(s) + 2H^+(aq) \rightarrow$
$\quad Zn^{2+}(aq) + H_2(g)$

(After R. E. Davis, K. D. Gailey, and K. W. Whitten, "Principles of Chemistry," CBS College Publishing, 1984, p. 635.)

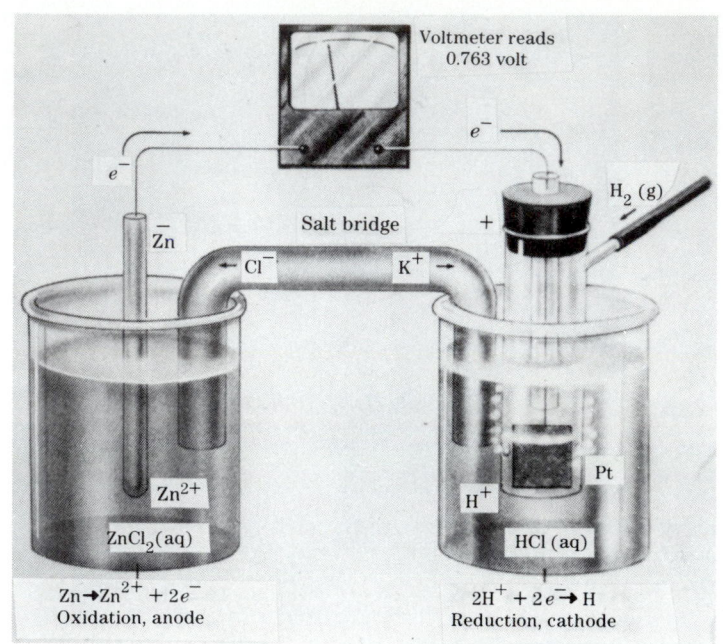

Voltmeter reads 0.763 volt

e^-

Salt bridge

H_2 (g)

Zn

$\leftarrow Cl^-$ $K^+ \rightarrow$

Zn^{2+}

$ZnCl_2(aq)$

$Zn \rightarrow Zn^{2+} + 2e^-$
Oxidation, anode

H^+

Pt

HCl (aq)

$2H^+ + 2e^- \rightarrow H$
Reduction, cathode

Standard Electrode Half-Cell Potentials for Metals

Every metal has a different tendency to corrode in a particular environment. For example, zinc is chemically attacked or corroded by dilute hydrochloric acid, whereas gold is not. One method for comparing the tendency for metals to form ions in aqueous solution is to compare their half-cell oxidation or reduction potentials (voltages) to that of a standard hydrogen–hydrogen-ion half-cell potential. Figure 12.3 shows an experimental setup for the determination of half-cell standard electrode potentials.

For this determination two beakers of aqueous solutions are used which are separated by a salt bridge so that mechanical mixing of the solutions is prevented (Fig. 12.3). In one beaker an electrode of the metal whose standard potential is to be determined is immersed in a 1 M solution of its ions at 25°C. In Fig. 12.3 an electrode of Zn is immersed in a 1 M solution of Zn^{2+} ions. In the other beaker a platinum electrode is immersed in a 1 M solution of H^+ ions in which hydrogen gas is bubbled in. A wire in series with a switch and a voltmeter connects the two electrodes. When the switch is just closed, the voltage between the half-cells is measured. The potential due to the hydrogen half-cell reaction $H_2 \rightarrow 2H^+ + 2e^-$ is arbitrarily assigned zero voltage. Thus, the voltage of the metal (zinc) half-cell reaction $Zn \rightarrow Zn^{2+} + 2e^-$ is measured directly against the hydrogen standard half-cell electrode. As indicated in Fig. 12.3, the standard half-cell electrode potential for the $Zn \rightarrow Zn^{2+} + 2e^-$ reaction is -0.763 V.

TABLE 12.1 Standard Electrode Potentials at 25°C*

Oxidation (corrosion) reaction	Electrode potential, $E°$ (volts vs. standard hydrogen electrode)
$Au \rightarrow Au^{3+} + 3e^-$	$+1.498$
$2H_2O \rightarrow O_2 + 4H^+ + 4e^-$	$+1.229$
$Pt \rightarrow Pt^{2+} + 2e^-$	$+1.200$
$Ag \rightarrow Ag^+ + e^-$	$+0.799$
$2Hg \rightarrow Hg_2^{2+} + 2e^-$	$+0.788$
$Fe^{2+} \rightarrow Fe^{3+} + e^-$	$+0.771$
$4(OH)^- \rightarrow O_2 + 2H_2O + 4e^-$	$+0.401$
$Cu \rightarrow Cu^{2+} + 2e^-$	$+0.337$
$Sn^{2+} \rightarrow Sn^{4+} + 2e^-$	$+0.150$
$H_2 \rightarrow 2H^+ + 2e^-$	0.000
$Pb \rightarrow Pb^{2+} + 2e^-$	-0.126
$Sn \rightarrow Sn^{2+} + 2e^-$	-0.136
$Ni \rightarrow Ni^{2+} + 2e^-$	-0.250
$Co \rightarrow Co^{2+} + 2e^-$	-0.277
$Cd \rightarrow Cd^{2+} + 2e^-$	-0.403
$Fe \rightarrow Fe^{2+} + 2e^-$	-0.440
$Cr \rightarrow Cr^{3+} + 3e^-$	-0.744
$Zn \rightarrow Zn^{2+} + 2e^-$	-0.763
$Al \rightarrow Al^{3+} + 3e^-$	-1.662
$Mg \rightarrow Mg^{2+} + 2e^-$	-2.363
$Na \rightarrow Na^+ + e^-$	-2.714

More cathodic (less tendency to corrode) — applies to the upper portion of the table.

More anodic (greater tendency to corrode) — applies to the lower portion of the table.

* Reactions are written as anodic half-cells. The more negative the half-cell reaction, the more anodic the reaction is and the greater the tendency for corrosion or oxidation to occur.

Table 12.1 lists the standard half-cell potentials of some selected metals. Those metals which are more reactive than hydrogen are assigned negative potentials and are said to be *anodic to hydrogen*. In the standard experiment shown in Fig. 12.3, these metals are oxidized to form ions, and hydrogen ions are reduced to form hydrogen gas. The equations for the reactions involved are

$$M \rightarrow M^{n+} + ne^- \quad \text{(metal oxidized to ions)} \tag{12.4a}$$

and

$$2H^+ + 2e^- \rightarrow H_2 \quad \text{(hydrogen ions reduced to hydrogen gas)} \tag{12.4b}$$

Those metals which are less reactive than hydrogen are assigned positive potentials and are said to be *cathodic to hydrogen*. In the standard experiment of Fig. 12.3, the ions of the metal whose potential is being determined are reduced to the atomic state (and may plate out on the metal electrode), and hydrogen gas is oxidized to hydrogen ions. The equations for the reactions involved are

$$M^{n+} + ne^- \rightarrow M \qquad \text{(metal ions reduced to atoms)} \qquad (12.5a)$$

$$\text{and} \qquad H_2 \rightarrow 2H^+ + 2e^- \qquad \begin{array}{l}\text{(hydrogen gas oxidized} \\ \text{to hydrogen ions)}\end{array} \qquad (12.5b)$$

12.3 GALVANIC CELLS

Macroscopic Galvanic Cells with Electrolytes That Are One Molar

Since most metallic corrosion involves electrochemical reactions, it is important to understand the principles of the operation of an electrochemical galvanic couple (cell). A macroscopic galvanic cell can be constructed with two dissimilar metal electrodes each immersed in a solution of their own ions. A galvanic cell of this type is shown in Fig. 12.4, which has a zinc electrode immersed in a 1 M solution of Zn^{2+} ions and another of copper immersed in a 1 M solution of Cu^{2+} ions with the solutions at 25°C. The two solutions are separated by a porous wall to prevent their mechanical mixing, and an external wire in series with a switch and a voltmeter connects the two electrodes. When the switch is just closed, electrons flow from the zinc electrode through the

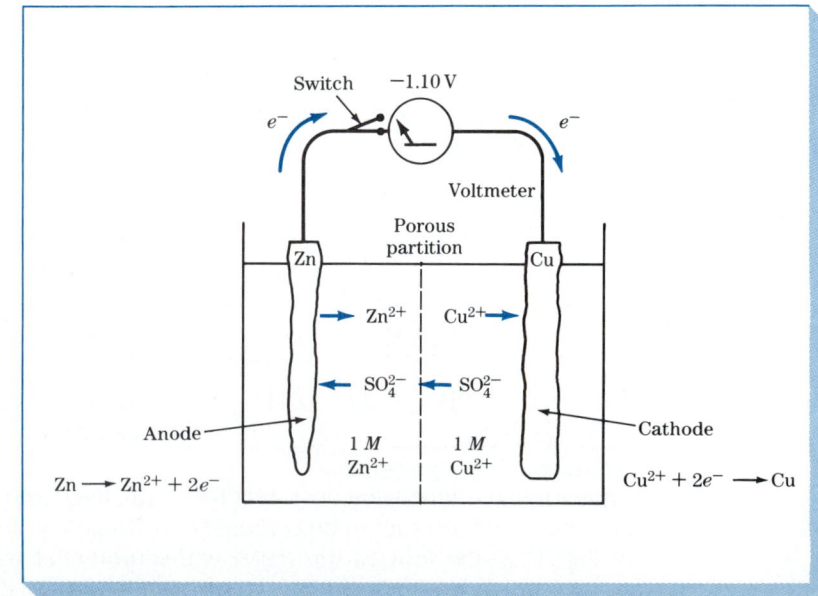

FIGURE 12.4 A macroscopic galvanic cell with zinc and copper electrodes. When the switch is closed and the electrons flow, the voltage difference between the zinc and copper electrodes is −1.10 V. The zinc electrode is the anode of the cell and corrodes.

external wire to the copper electrode, and a voltage of -1.10 V shows on the voltmeter.

In an electrochemical galvanic-couple reaction for two metals each immersed in a 1 M solution of its own ions, the electrode which has the more negative oxidation potential will be the electrode that is oxidized. A reduction reaction will take place at the electrode which has the more positive potential. Thus, for the Zn–Cu galvanic cell illustrated in Fig. 12.4, the Zn electrode will be oxidized to Zn^{2+} ions and Cu^{2+} ions will be reduced to Cu at the Cu electrode.

Let us now calculate the electrochemical potential of the Zn–Cu galvanic cell when the switch connecting the two electrodes is just closed. First, we write the oxidation half-cell reactions for zinc and copper, using Table 12.1:

$$Zn \rightarrow Zn^{2+} + 2e^{-} \qquad E° = -0.763 \text{ V}$$
$$Cu \rightarrow Cu^{2+} + 2e^{-} \qquad E° = +0.337 \text{ V}$$

We see that the Zn half-cell reaction has the more negative potential (-0.763 V for Zn vs. $+0.337$ V for Cu). Thus, the Zn electrode will be oxidized to Zn^{2+} ions, and Cu^{2+} ions will be reduced to Cu at the Cu electrode. The overall electrochemical potential of the cell is obtained by adding the oxidation half-cell potential of the Zn to the reduction half-cell potential of the Cu. Note that the sign of oxidation half-cell potential must be changed to opposite polarity when the half-cell reaction is written as a reduction reaction.

Note change of sign

Oxidation: $Zn \rightarrow Zn^{2+} + 2e^{-}$ $E° = -0.763$ V

Reduction: $Cu^{2+} + 2e^{-} \rightarrow Cu$ $E° = 0.337$ V

Overall reaction: $Zn + Cu^{2+} \rightarrow Zn^{2+} + Cu$ $E°_{cell} = -1.100$ V
(by adding)

For a galvanic couple, the electrode which is oxidized is called the *anode* and the electrode where the reduction takes place is called the *cathode*. At the anode *metal ions and electrons are produced,* and since the electrons remain in the metal electrode, the *anode is assigned negative polarity.* At the cathode *electrons are consumed,* and it is assigned positive polarity. In the case of the Zn–Cu cell described above, copper ions are plated out on the copper cathode.

Example Problem 12.1

A galvanic cell consists of an electrode of zinc in a 1 M $ZnSO_4$ solution and another of nickel in a 1 M $NiSO_4$ solution. The two electrodes are separated by a porous wall so that mixing of the solutions is prevented. An external wire with a switch connects the two electrodes. When the switch is just closed:

(a) At which electrode does oxidation occur?
(b) Which electrode is the anode of the cell?
(c) Which electrode corrodes?
(d) What is the emf of this galvanic cell when the switch is just closed?

Solutions:

The half-cell reactions for this cell are

$$Zn \rightarrow Zn^{2+} + 2e^- \qquad E° = -0.763 \text{ V}$$

$$Ni \rightarrow Ni^{2+} + 2e^- \qquad E° = -0.250 \text{ V}$$

(a) Oxidation occurs at the zinc electrode since the zinc half-cell reaction has a more negative $E°$ potential of -0.763 V as compared to -0.250 V for the nickel half-cell reaction.
(b) The zinc electrode is the anode since oxidation occurs at the anode.
(c) The zinc electrode corrodes since the anode in a galvanic cell corrodes.
(d) The emf of the cell is obtained by adding the half-cell reactions together:

Anode reaction:	$Zn \rightarrow Zn^{2+} + 2e$	$E° = -0.763$ V
Cathode reaction:	$Ni^{2+} + 2e^- \rightarrow Ni$	$E° = +0.250$ V
Overall reaction:	$Zn + Ni^{2+} \rightarrow Zn^{2+} + Ni$	$E°_{cell} = -0.513$ V ◄

Galvanic Cells with Electrolytes That Are Not One Molar   Most electrolytes for real corrosion galvanic cells are not 1 M but are usually dilute solutions which are much lower than 1 M. If the concentration of the ions in an electrolyte surrounding an anodic electrode is less than 1 M, the driving force for the reaction to dissolve or corrode the anode is greater since there is a lower concentration of ions to cause the reverse reaction. Thus there will be a more negative emf half-cell anodic reaction:

$$M \rightarrow M^{n+} + ne^- \tag{12.6}$$

The effect of metal ion concentration C_{ion} on the standard emf $E°$ at 25°C is given by the *Nernst*[1] *equation.* For a half-cell anodic reaction in which only one kind of ion is produced, the Nernst equation can be written in the form

$$E = E° + \frac{0.0592}{n} \log C_{ion} \tag{12.7}$$

---

[1] Walter Hermann Nernst (1864–1941). German chemist and physicist who did fundamental work on electrolyte solutions and thermodynamics.

where E = new emf of half-cell
$E°$ = standard emf of half-cell
n = number of electrons transferred (for example, $M \rightarrow M^{n+} + ne^-$)
C_{ion} = molar concentration of ions

For the cathode reaction the sign of the final emf is reversed. Example Problem 12.2 shows how the emf of a macroscopic galvanic cell in which the electrolytes are not $1\,M$ can be calculated by using the Nernst equation.

Example Problem 12.2

A galvanic cell at 25°C consists of an electrode of zinc in a $0.10\,M$ $ZnSO_4$ solution and another of nickel in a $0.05\,M$ $NiSO_4$ solution. The two electrodes are separated by a porous wall and connected by an external wire. What is the emf of the cell when a switch between the two electrodes is just closed?

Solution:

First, assume the dilutions from $1\,M$ solutions will not affect the order of the potentials of Zn and Ni in the standard electrode potential series. Thus, zinc with a more negative electrode potential of -0.763 V will be the anode of the Zn–Ni electrochemical cell and nickel will be the cathode. Next, use the Nernst equation to modify the standard equilibrium potentials.

$$E_{cell} = E° + \frac{0.0592}{n} \log C_{ion} \tag{12.7}$$

Anode reaction: $E_A = -0.763\ \text{V} + \dfrac{0.0592}{2} \log 0.10$

$$= -0.763\ \text{V} - 0.0296\ \text{V} = -0.793\ \text{V}$$

Cathode reaction: $E_C = -\left(-0.250\ \text{V} + \dfrac{0.0592}{2} \log 0.05\right)$

$$= +0.250\ \text{V} + 0.0385\ \text{V} = +0.288\ \text{V}$$

Emf of cell $= E_A + E_C = -0.793\ \text{V} + 0.288\ \text{V} = -0.505\ \text{V}$ ◀

Galvanic Cells with Acid or Alkaline Electrolytes with No Metal Ions Present   Let us consider a galvanic cell in which iron and copper electrodes are immersed in an aqueous acidic electrolyte in which there are no metal ions initially present. The iron and copper electrodes are connected by an external wire, as shown in Fig. 12.5. The standard electrode potential for iron to oxidize

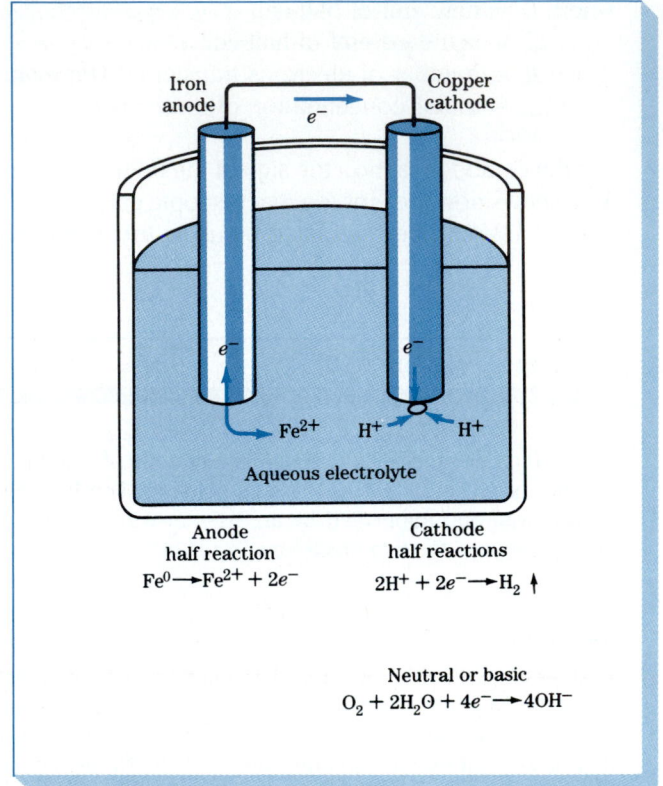

FIGURE 12.5 Electrode reactions in an iron-copper galvanic cell in which there are no metal ions initially present in the electrolyte. (*After J. Wulff et al., "The Structure and Properties of Materials," vol. II, Wiley, 1964, p. 164.*)

is -0.440 V and that for copper is $+0.337$ V. Therefore, in this couple iron will be the anode and will oxidize since it has the more negative half-cell oxidation potential. The half-cell reaction at the iron anode will therefore be

$$\text{Fe} \rightarrow \text{Fe}^{2+} + 2e^- \qquad \text{(anodic half-cell reaction)} \qquad (12.8a)$$

Since there are no copper ions in the electrolyte to be reduced to copper atoms for a cathode reaction, hydrogen ions in the acid solution will be reduced to hydrogen atoms that will subsequently combine to form diatomic hydrogen (H_2) gas. The overall reaction at the cathode will thus be

$$2\text{H}^+ + 2e^- \rightarrow \text{H}_2 \qquad \text{(cathodic half-cell reaction)} \qquad (12.8b)$$

However, if the electrolyte also contains an oxidizing agent, the cathode reaction becomes

$$\text{O}_2 + 4\text{H}^+ + 4e^- \rightarrow 2\text{H}_2\text{O} \qquad (12.8c)$$

If the electrolyte is neutral or basic and oxygen is present, oxygen and water

TABLE 12.2   Some Common Cathode Reactions for Aqueous Galvanic Cells

Cathode reaction	Example
1. Metal deposition: $\quad M^{n+} + ne^- \rightarrow M$	Fe–Cu galvanic couple in aqueous solution with Cu^{2+} ions; $Cu^{2+} + 2e^- \rightarrow$ Cu
2. Hydrogen evolution: $\quad 2H^+ + 2e^- \rightarrow H_2$	Fe–Cu galvanic couple in acid solution with no copper ions present
3. Oxygen reduction (acid solutions): $\quad O_2 + 4H^+ + 4e^- \rightarrow$ $\quad 2H_2O$	Fe–Cu galvanic couple in oxidizing acidic solution with no copper ions present
4. Oxygen reduction (neutral or basic solutions): $\quad O_2 + 2H_2O + 4e^- \rightarrow$ $\quad 4OH^-$	Fe–Cu galvanic couple in neutral or alkaline solution with no copper ions present

molecules will react to form hydroxyl ions, with the cathode reaction becoming

$$O_2 + 2H_2O + 4e^- \rightarrow 4OH^- \qquad (12.8d)$$

Table 12.2 lists four common reactions which occur in aqueous galvanic cells.

Microscopic Galvanic Cell Corrosion of Single Electrodes

If a single electrode of zinc is placed in a dilute solution of air-free hydrochloric acid, it will be corroded electrochemically since microscopic *local anodes and cathodes* will develop on its surface due to inhomogeneities in structure and composition (Fig. 12.6a). The oxidation reaction which will occur at the local anodes is

$$Zn \rightarrow Zn^{2+} + 2e^- \qquad \text{(anodic reaction)} \qquad (12.9a)$$

and the reduction reaction which will occur at the local cathodes is

$$2H^+ + 2e^- \rightarrow H_2 \qquad \text{(cathodic reaction)} \qquad (12.9b)$$

Both reactions will occur simultaneously and at the same rate on the metal surface.

Another example of single-electrode corrosion is the *rusting of iron*. If a piece of iron is immersed in oxygenated water, ferric hydroxide [$Fe(OH)_3$] will form on its surface as indicated, in Fig. 12.6b. The oxidation reaction which occurs at microscopic local anodes is

$$Fe \rightarrow Fe^{2+} + 2e^- \qquad \text{(anodic reaction)} \qquad (12.10a)$$

Since the iron is immersed in oxygenated neutral water, the reduction reaction occurring at the local cathodes is

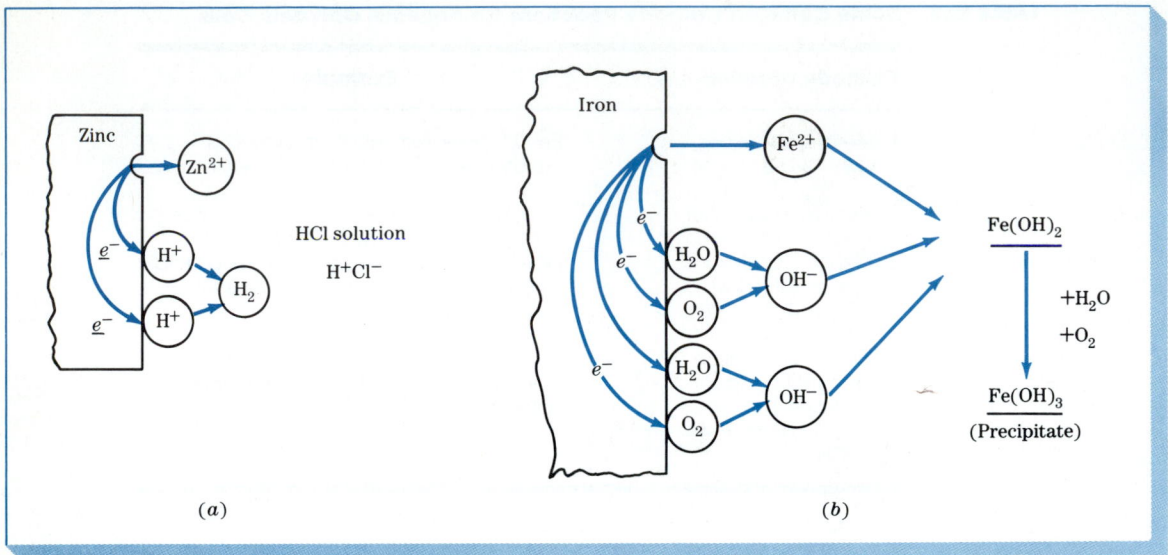

FIGURE 12.6 Electrochemical reactions for (a) zinc immersed in dilute hydrochloric acid and (b) iron immersed in oxygenated neutral water solution.

$$O_2 + 2H_2O + 4e^- \rightarrow 4\,OH^- \quad \text{(cathodic reaction)} \quad (12.10b)$$

The overall reaction is obtained by adding the two reactions (12.10a) and (12.10b) to give

$$2Fe + 2H_2O + O_2 \rightarrow 2Fe^{2+} + 4OH^- \rightarrow 2Fe(OH)_2 \downarrow \quad (12.10c)$$
$$\text{Precipitate}$$

The ferrous hydroxide, $Fe(OH)_2$, precipitates from solution since this compound is insoluble in oxygenated aqueous solutions. It is further oxidized to ferric hydroxide, $Fe(OH)_3$, which has the red-brown rust color. The reaction for the oxidation of ferrous to ferric hydroxide is

$$2Fe(OH)_2 + H_2O + \tfrac{1}{2}O_2 \rightarrow 2Fe(OH)_3 \downarrow \quad (12.10d)$$
$$\text{Precipitate} \quad \text{(rust)}$$

Example Problem 12.3

Write the anodic and cathodic half-cell reactions for the following electrode-electrolyte conditions. Use $E°$ values in Table 12.1 as basis for answers.

(a) Copper and zinc electrodes immersed in a dilute cupric sulfate ($CuSO_4$) solution.
(b) A copper electrode immersed in an oxygenated-water solution.
(c) An iron electrode immersed in an oxygenated-water solution.
(d) Magnesium and iron electrodes connected by an external wire immersed in an oxygenated 1% NaCl solution.

Solution:

(a) Anode reaction: $Zn \rightarrow Zn^{2+} + 2e^-$ $E° = -0.763$ V (oxidation)
Cathode reaction: $Cu^{2+} + 2e^- \rightarrow Cu$ $E° = -0.337$ V (reduction)
Comment: Zinc has a more negative potential and is thus the anode. It is therefore oxidized.
(b) Little or no corrosion takes place since the potential difference between that for the oxidation of copper (0.337 V) and that for the formation of water from hydroxyl ions (0.401 V) is so small.
(c) Anode reaction: $Fe \rightarrow Fe^{2+} + 2e^-$ $E° = -0.440$ V (oxidation)
Cathode reaction: $O_2 + 2H_2O + 4e^- \rightarrow 4OH^-$ $E° = -0.401$ V
Comment: Fe has a more negative potential and thus is the anode. Fe is therefore oxidized.
(d) Anode reaction: $Mg \rightarrow Mg^{2+} + 2e^-$ $E° = -2.36$ V
Cathode reaction: $O_2 + 2H_2O + 4e^- \rightarrow 4OH^-$ $E° = -0.401$ V
Comment: Magnesium has a more negative oxidation potential and is thus the anode. Mg is therefore oxidized.

Concentration Galvanic Cells

Ion-concentration cells   Consider an ion-concentration cell consisting of two iron electrodes, one immersed in a dilute Fe^{2+} electrolyte and the other in a concentrated Fe^{2+} electrolyte, as shown in Fig. 12.7. In this galvanic cell the electrode in the dilute electrolyte will be the anode since according to the Nernst equation this electrode will have a more-negative potential with respect to the other.

Let us compare the half-cell potential for an iron electrode immersed in a 0.001 M dilute Fe^{2+} electrolyte with the half-cell potential of another iron electrode immersed in a more concentrated 0.01 M dilute Fe^{2+} electrolyte. The two electrodes are connected by an external wire, as shown in Fig. 12.7. The general Nernst equation for a half-cell oxidation reaction for $Fe \rightarrow Fe^{2+} + 2e^-$, since $n = 2$, is

$$E_{Fe^{2+}} = E° + 0.0296 \log C_{ion} \tag{12.11}$$

For 0.001 M solution: $E_{Fe^{2+}} = -0.440$ V $+ 0.0296 \log 0.001 = -0.529$ V

For 0.01 M solution: $E_{Fe^{2+}} = -0.440$ V $+ 0.0296 \log 0.01 = -0.499$ V

Since -0.529 V is more negative than -0.499 V, the iron electrode in the more dilute solution will be the anode of the electrochemical cell and hence will be oxidized and corroded. Thus, the ion-concentration cell produces corrosion in the region of the more dilute electrolyte.

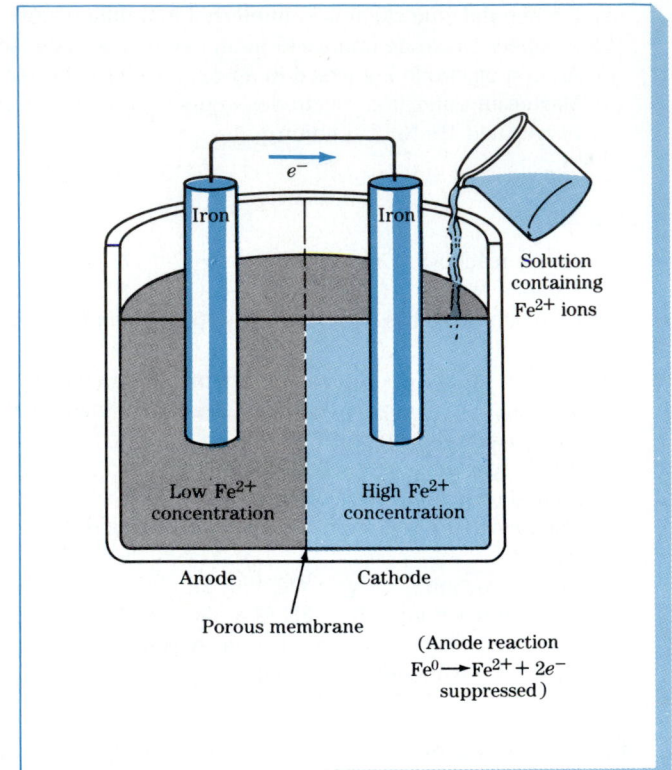

FIGURE 12.7 An ion-concentration galvanic cell composed of two iron electrodes. When the electrolyte is of different concentrations at each electrode, the electrode in the more dilute electrolyte becomes the anode. (*After J. Wulff et al., "The Structure and Properties of Materials," vol. II, Wiley, 1964, p. 163.*)

Example Problem 12.4

One end of an iron wire is immersed in an electrolyte of 0.02 M Fe^{2+} ions and the other in an electrolyte of 0.005 M Fe^{2+} ions. The two electrolytes are separated by a porous wall.

(*a*) Which end of the wire will corrode?
(*b*) What will be the potential difference between the two ends of the wire when it is just immersed in the electrolytes?

Solution

(*a*) The end of the wire that will corrode will be the one immersed in the more dilute electrolyte, which is the 0.005 M one. Thus the wire end in the 0.005 M solution will be the anode.

(b) Using the Nernst equation with $n = 2$ [Eq. (12.11)] gives

$$E_{Fe^{2+}} = E° + 0.0296 \log C_{ion} \qquad (12.11)$$

For 0.005 M solution: $E_A = -0.440 \text{ V} + 0.0296 \log 0.005$
$$= -0.508 \text{ V}$$

For 0.02 M solution: $E_C = -(-0.440 \text{ V} + 0.0296 \log 0.02)$
$$= +0.490 \text{ V}$$

$$E_{cell} = E_A + E_C = -0.508 \text{ V} + 0.490 \text{ V} = -0.018 \text{ V} \blacktriangleleft$$

Oxygen-concentration cells Oxygen-concentration cells can develop when there is a difference in oxygen concentration on a moist surface of a metal that can be oxidized. Oxygen-concentration cells are particularly important in the corrosion of easily oxidized metals such as iron which do not form protective oxide films.

   Consider an oxygen-concentration cell consisting of two iron electrodes, one in a water electrolyte with a low oxygen concentration and another in an electrolyte with a high oxygen concentration, as shown in Fig. 12.8. The anode and cathode reactions for this cell are

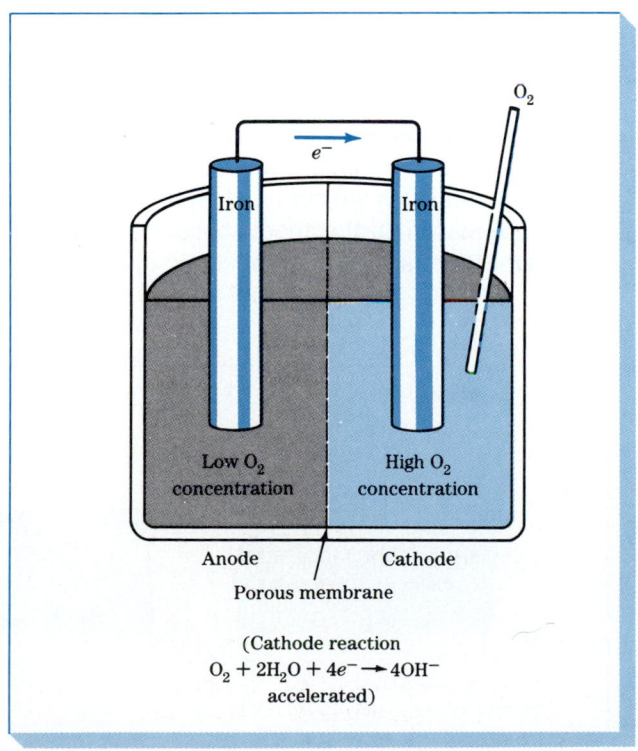

FIGURE 12.8 An oxygen-concentration cell. The anode in this cell is the electrode which has the low oxygen concentration surrounding it. (*After J. Wulff et al., "The Structure and Properties of Materials," vol. II, Wiley, 1964, p. 165.*)

| Anode reaction: | $Fe \rightarrow Fe^{2+} + 2e^-$ | (12.12a) |
| Cathode reaction: | $O_2 + 2H_2O + 4e^- \rightarrow 4OH^-$ | (12.12b) |

Which electrode is the anode in this cell? Since the cathode reaction requires oxygen and electrons, the high concentration of oxygen must be at the cathode. Also since electrons are required at the cathode, they must be produced by the anode which will have the low oxygen concentration.

In general, therefore, for an oxygen-concentration cell, the regions that are low in oxygen will be anodic to the cathode regions that are high in oxygen. Thus corrosion will be accelerated in the regions of a metal surface where the oxygen content is relatively low such as in cracks and crevices and under accumulations of surface deposits. The effects of oxygen-concentration cells will be discussed further in Sec. 12.5 dealing with different types of corrosion.

Galvanic Cells Created by Differences in Composition, Structure, and Stress

Microscopic galvanic cells can exist in metals or alloys because of differences in composition, structure, and stress concentrations. These metallurgical factors can seriously affect the corrosion resistance of a metal or alloy. They create anodic and cathodic regions of varying dimensions which can cause galvanic-cell corrosion. Some of the important metallurgical factors affecting corrosion resistance are:

1. Grain–grain-boundary galvanic cells
2. Multiple-phase galvanic cells
3. Impurity galvanic cells

Grain–grain-boundary electrochemical cells In most metals and alloys grain boundaries are more chemically active (anodic) than the grain matrix. Thus the grain boundaries are corroded or chemically attacked, as illustrated in Fig. 12.9a. The reason for the anodic behavior of the grain boundaries is that they have higher energies due to the atomic disarray in that area and also because

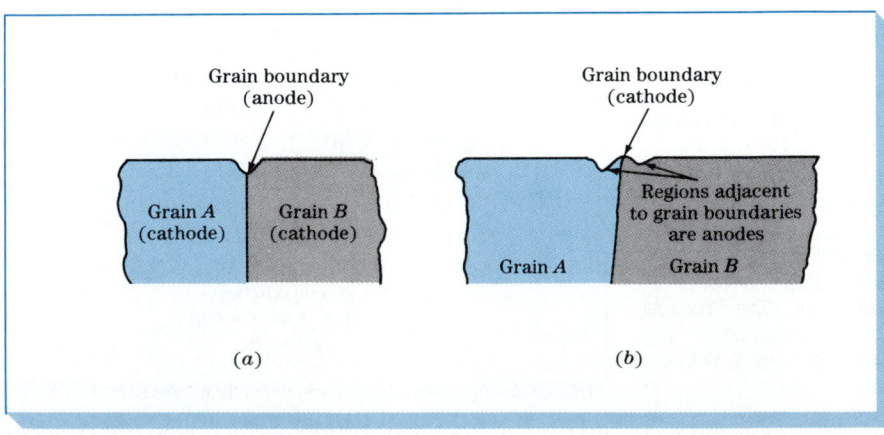

FIGURE 12.9 Corrosion at or near grain boundaries. (a) Grain boundary is the anode of galvanic cell and corrodes. (b) Grain boundary is the cathode, and regions adjacent to the grain boundary serve as anodes.

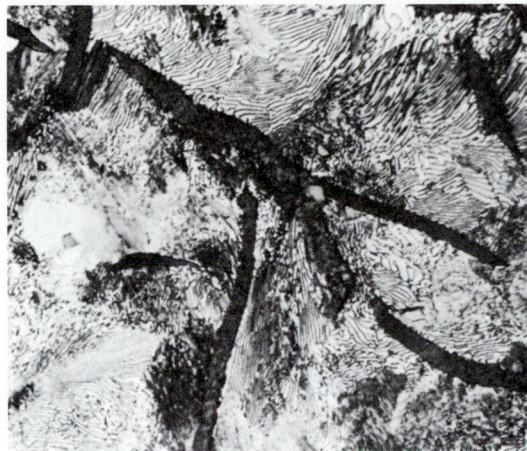

FIGURE 12.10 Class 30 gray cast iron. Structure consists of graphite flakes in a matrix of pearlite (alternating lamellae of light-etching ferrite and darker cementite). (*After "Metals Handbook," vol. 7, 8th ed., American Society for Metals, 1972, p. 83.*)

solute segregation and impurities migrate to the grain boundaries. For some alloys the situation is reversed, and chemical segregation causes the grain boundaries to become more noble or cathodic than the regions adjacent to the grain boundaries. This condition causes the regions adjacent to the grain boundaries to corrode preferentially, as illustrated in Fig. 12.9*b*.

Multiple-phase electrochemical cells   In most cases a single-phase alloy has higher corrosion resistance than a multiple-phase alloy since electrochemical cells are created in the multiphase alloy due to one phase being anodic to another that acts as the cathode. Hence corrosion rates are higher for the multiphase alloy. A classical example of multiphase galvanic corrosion can occur in pearlitic gray cast iron. The microstructure of pearlitic gray cast iron consists of graphite flakes in a matrix of pearlite (Fig. 12.10). Since graphite is much more cathodic (more noble) than the surrounding pearlite matrix, highly active galvanic cells are created between the graphite flakes and the anodic pearlite matrix. In an extreme case of galvanic corrosion of pearlitic gray cast iron, the matrix can corrode to such an extent that the cast iron is left as a network of interconnected graphite flakes (Fig. 12.11).

Another example of the effect of second phases in reducing the corrosion resistance of an alloy is the effect of tempering on the corrosion resistance of a 0.95% carbon steel. When this steel is in the martensitic condition after quenching from the austenitic condition, its corrosion rate is relatively low (Fig. 12.12), because the martensite is a single-phase supersaturated solid solution of carbon in interstitial positions of a body-centered tetragonal lattice of iron. After tempering in the 200 to 500°C range, a fine precipitate of ϵ carbide and cementite (Fe_3C) is formed. This two-phase structure sets up galvanic cells which accelerate the corrosion rate of the steel, as observed in Fig. 12.12. At higher tempering temperatures above about 500°C, the cementite coalesces into larger particles and the corrosion rate decreases.

FIGURE 12.11 Graphite residue remaining as a result of corrosion of cast-iron elbow. (*Courtesy of the LaQue Center for Corrosion Technology, Inc.*)

Impurities   Metallic impurities in a metal or alloy can lead to the precipitation of intermetallic phases which have different oxidation potentials than the matrix of the metal. Thus very small anodic or cathodic regions are created which can lead to galvanic corrosion when coupled with the matrix metal. Higher corrosion resistance is obtained with purer metals. However, most engineering metals and alloys contain a certain level of impurity elements since it costs too much to remove them.

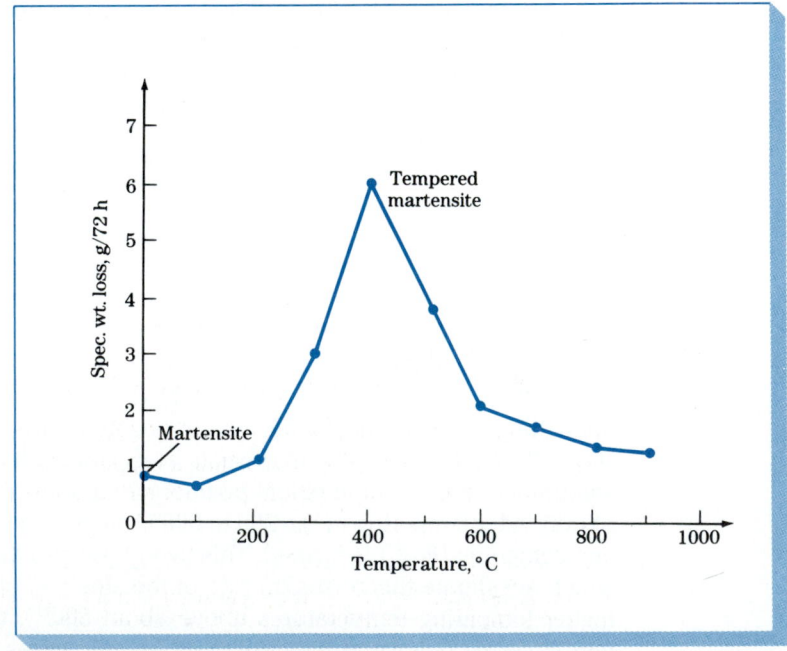

FIGURE 12.12 Effect of heat treatment on corrosion of 0.95% C steel in 1% H_2SO_4. Polished specimens 2.5 × 2.5 × 0.6 cm, tempering time probably 2 h. (*Heyn and Bauer.*)

12.4   CORROSION RATES (KINETICS)

Until now our study of the corrosion of metals has been centered on equilibrium conditions and the *tendency* of metals to corrode, which has been related to the standard electrode potentials of metals. However, corroding systems are *not at equilibrium,* and thus thermodynamic potentials do not tell us about the rates of corrosion reactions. The kinetics of corroding systems are very complex and not completely understood. In this section we will examine some of the basic aspects of corrosion kinetics.

Rate of Uniform Corrosion or Electroplating of a Metal in an Aqueous Solution

The amount of metal uniformly corroded from an anode or electroplated on a cathode in an aqueous solution in a time period can be determined by using Faraday's[1] equation of general chemistry, which states

$$w = \frac{ItM}{nF} \tag{12.13}$$

where w = weight of metal, g, corroded or electroplated in an aqueous solution in time t, s
I = current flow, A
M = atomic mass of the metal, g/mol
n = number of electrons/atom produced or consumed in the process
F = Faraday's constant = 96,500 C/mol or 96,500 A·s/mol

Sometimes the uniform aqueous corrosion of a metal is expressed in terms of a current density i, which is often expressed in amperes per square centimeter. Replacing I by iA converts Eq. (12.13) to

$$w = \frac{iAtM}{nF} \tag{12.14}$$

where i = current density, A/cm^2, and A = area, cm^2, if the centimeter is used for length. The other quantities are the same as in Eq. (12.13).

Example Problem 12.5

A copper electroplating process utilizes 15 A of current by chemically dissolving (corroding) a copper anode and electroplating a copper cathode. If it is assumed that there are no side reactions, how long will it take to corrode 8.50 g of copper from the anode?

---

[1] Michael Faraday (1791–1867). English scientist who made basic experiments in electricity and magnetism. He made experiments to show how ions of a compound migrated under the influence of an applied electric current to electrodes of opposite polarity.

Solution:

The time to corrode the copper from the anode can be determined from Eq. (12.13):

$$w = \frac{ItM}{nF} \quad \text{or} \quad t = \frac{wnF}{IM}$$

In this case,

$$w = 8.5 \text{ g} \quad n = 2 \text{ for Cu} \to Cu^{2+} + 2e^- \quad F = 96{,}500 \text{ A} \cdot \text{s/mol}$$
$$M = 63.5 \text{ g/mol for Cu} \quad I = 15 \text{ A} \quad t = ? \text{ s}$$

or
$$t = \frac{(8.5 \text{ g})(2)(96{,}500 \text{ A} \cdot \text{s/mol})}{(15 \text{ A})(63.5 \text{ g/mol})} = 1722 \text{ s or } 28.7 \text{ min} \blacktriangleleft$$

Example Problem 12.6

A mild steel cylindrical tank 1 m high and 50 cm in diameter contains aerated water to the 60-cm level and shows a loss in weight due to corrosion of 304 g after 6 weeks. Calculate (*a*) the corrosion current and (*b*) the current density involved in the corrosion of the tank. Assume uniform corrosion on the tank's inner surface and that the steel corrodes in the same manner as pure iron.

Solution:

(*a*) We will use Eq. (12.13) to solve for the corrosion current:

$$I = \frac{wnF}{tM}$$

$$w = 304 \text{ g} \quad n = 2 \text{ for Fe} \to Fe^{2+} + 2e^- \quad F = 96{,}500 \text{ A} \cdot \text{s/mol}$$
$$M = 55.85 \text{ g/mol for Fe} \quad t = 6 \text{ wk} \quad I = ? \text{ A}$$

We must convert the time, 6 weeks, into seconds and then we can substitute all the values into Eq. (12.13):

$$t = 6 \text{ wk} \left(\frac{7 \text{ days}}{\text{wk}}\right)\left(\frac{24 \text{ h}}{\text{day}}\right)\left(\frac{3600 \text{ s}}{\text{h}}\right) = 3.63 \times 10^6 \text{ s}$$

$$I = \frac{(304 \text{ g})(2)(96{,}500 \text{ A} \cdot \text{s/mol})}{(3.63 \times 10^6 \text{ s})(55.85 \text{ g/mol})} = 0.289 \text{ A} \blacktriangleleft$$

(*b*) The current density is

$$i \text{ (A/cm}^2) = \frac{I \text{ (A)}}{\text{area (cm}^2)}$$

Area of corroding surface of tank = area of sides + area of bottom

$$= \pi D h + \pi r^2$$
$$= \pi (50 \text{ cm})(60 \text{ cm}) + \pi (25 \text{ cm})^2$$
$$= 9420 \text{ cm}^2 + 1962 \text{ cm}^2 = 11{,}380 \text{ cm}^2$$

$$i = \frac{0.289 \text{ A}}{11{,}380 \text{ cm}^2} = 2.53 \times 10^{-5} \text{ A/cm}^2 \blacktriangleleft$$

In experimental corrosion work the uniform corrosion of a metal surface exposed to a corrosive environment is measured in a variety of ways. One common method is to measure the weight loss of a sample exposed to a particular environment and then after a period of time express the corrosion rate as a weight loss per unit area of exposed surface per unit time. For example, uniform surface corrosion is often expressed as milligram weight loss per square decimeter per day (mdd). Another commonly used method is to express corrosion rate in terms of loss in depth of material per unit time. Examples of corrosion rate in this system are millimeters per year (mm/yr) and mils per year (mils*/yr). For uniform electrochemical corrosion in aqueous environments, the corrosion rate may be expressed as a current density (see Example Problem 12.8).

Example Problem 12.7

The wall of a steel tank containing aerated water is corroding at a rate of 54.7 mdd. How long will it take for the wall thickness to decrease by 0.50 mm?

Solution:

The corrosion rate is 54.7 mdd, or 54.7 mg of metal is corroded on each square decimeter of surface per day.

$$\text{Corrosion rate in g/(cm}^2 \cdot \text{day)} = \frac{54.7 \times 10^{-3} \text{ g}}{100 \text{ (cm}^2 \cdot \text{day)}} = 5.47 \times 10^{-4} \text{ g/(cm}^2 \cdot \text{day)}$$

The density of Fe = 7.87 g/cm³. Dividing the corrosion rate in g/(cm² · day) by the density gives the depth of corrosion per day as

$$\frac{5.47 \times 10^{-4} \text{ g/(cm}^2 \cdot \text{day)}}{7.87 \text{ g/cm}^3} = 0.695 \times 10^{-4} \text{ cm/day}$$

---

*1 mil = 0.001 in.

The number of days required for a decrease in 0.50 mm can be obtained by ratio as

$$\frac{x \text{ days}}{0.50 \text{ mm}} = \frac{1 \text{ day}}{0.695 \times 10^{-3} \text{ mm}}$$

$$x = 719 \text{ days} \blacktriangleleft$$

Example Problem 12.8

A sample of zinc corrodes uniformly with a current density of 4.27×10^{-7} A/cm^2 in an aqueous solution. What is the corrosion rate of the zinc in milligrams per decimeter per day (mdd)? The reaction for the oxidation of zinc is $Zn \rightarrow Zn^{2+} + 2e^-$.

Solution:

To make the conversion from current density to mdd, we will use Faraday's equation [Eq. (12.14)] to calculate the milligrams of zinc corroding on an area of 1 dm^2/day (mdd).

$$w = \frac{iAtM}{nF} \tag{12.14}$$

w (mg)

$$= \left[\frac{(4.27 \times 10^{-7} \text{ A/cm}^2)(100 \text{ cm}^2)(24 \text{ h} \times 3600 \text{ s/h})(65.38 \text{ g/mol})}{(2)(96,500 \text{ A} \cdot \text{s/mol})} \right] \left(\frac{1000 \text{ mg}}{\text{g}} \right)$$

$$= 1.25 \text{ mg of zinc which corrodes on an area of 1 dm}^2 \text{ in 1 day}$$

or the corrosion rate is 1.25 mdd. ◀

Corrosion Reactions and Polarization

Let us now consider the electrode kinetics of the corrosion reaction of zinc being dissolved by hydrochloric acid, as indicated by Fig. 12.13. The anodic half-cell reaction for this electrochemical reaction is

$$Zn \rightarrow Zn^{2+} + 2e^- \qquad \text{(anodic reaction)} \qquad (12.15a)$$

The electrode kinetics for this reaction can be represented by an electrochemical potential E (volts) vs. log current density plot, as shown in Fig. 12.14. The zinc electrode in equilibrium with its ions can be represented by a point that represents its equilibrium potential $E° = -0.763$ V and a corresponding exchange current density $i_0 = 10^{-7}$ A/cm^2 (point A in Fig. 12.14). The exchange current density i_0 is the rate of oxidation and reduction reactions at an equilibrium electrode expressed in terms of current density. Exchange current

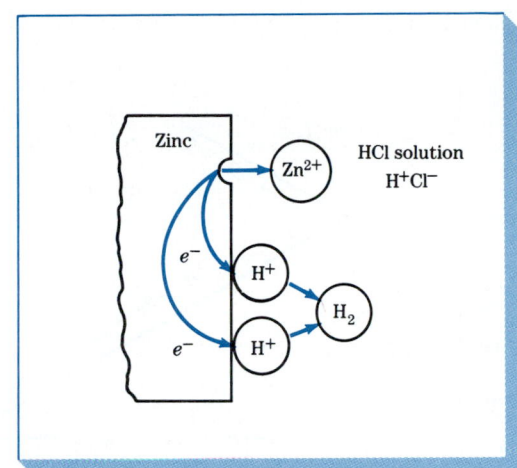

FIGURE 12.13
Electrochemical
dissolution of zinc in
hydrochloric acid.

$Zn \rightarrow Zn^{2+} + 2e^-$
     (anodic reaction)

$2H^+ + 2e^- \rightarrow H_2$
     (cathodic reaction)

densities must be determined experimentally when there is no net current. Each electrode with its specific electrolyte will have its own i_0 value.

The cathodic half-cell reaction for the corrosion reaction of zinc being dissolved in hydrochloric acid is

$$2H^+ + 2e^- \rightarrow H_2 \qquad \text{(cathodic reaction)} \qquad (12.15b)$$

The hydrogen-electrode reaction occurring on the zinc surface under equilibrium conditions can also be represented by the reversible hydrogen electrode potential $E° = 0.00$ V, and its corresponding exchange current density for this reaction on a zinc surface is 10^{-10} A/cm^2 (point B in Fig. 12.14).

When the zinc begins to react with the hydrochloric acid (corrosion starts), since the zinc is a good electrical conductor, the zinc surface must be at a constant potential. This potential is E_{corr} (Fig. 12.14, point C). Thus when the zinc starts to corrode, the potential of the cathodic areas must become more negative to reach about -0.5 V (E_{corr}) and that of the anodic areas more positive to reach -0.5 V (E_{corr}). At point C in Fig. 12.14 the rate of zinc dissolution is equal to the rate of hydrogen evolution. The current density corresponding to this rate of reaction is called i_{corr} and therefore is equal to the rate of zinc dissolution or corrosion. Example Problem 12.8 shows how a current density for a uniformly corroding surface can be expressed in terms of a certain weight loss per unit area per unit time (for example, mdd units).

Thus, when a metal corrodes by the short circuiting of microscopic galvanic-cell action, net oxidation and reduction reactions occur on the metal surface. The potentials of the local anodic and cathodic regions are no longer at equilibrium but change potential to reach a constant intermediate value of E_{corr}. The displacement of the electrode potentials from their equilibrium values

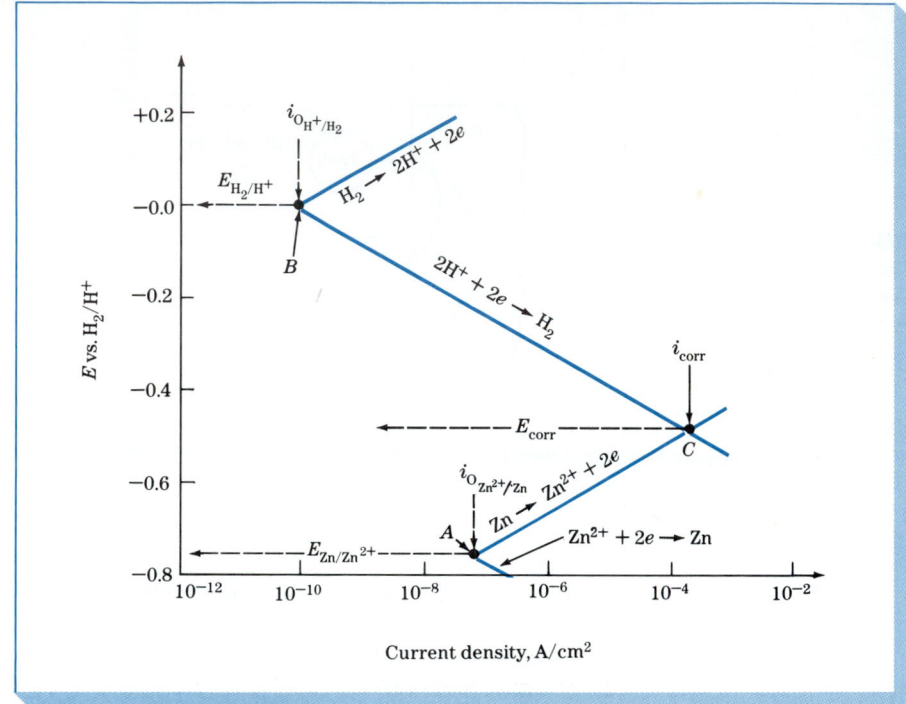

FIGURE 12.14 Electrode kinetic behavior of pure zinc in acid solution (schematic). (After M. G. Fontana and N. D. Greene, "Corrosion Engineering," 2d ed., McGraw-Hill, 1978, p. 314.)

to a constant potential of some intermediate value and the creation of a net current flow is called *polarization*.

Polarization of electrochemical reactions can conveniently be divided into two types: *activation polarization* and *concentration polarization*.

Activation polarization Activation polarization refers to electrochemical reactions which are controlled by a slow step in a reaction sequence of steps at the metal-electrolyte interface. That is, there is a critical activation energy needed to surmount the energy barrier associated with the slowest step. This type of activation energy is illustrated by considering the cathodic hydrogen reduction on a metal surface, $2H^+ + 2e^- \rightarrow H_2$. Figure 12.15 shows schematically some of the intermediate steps possible in the hydrogen reduction at a zinc surface. In this process hydrogen ions must migrate to the zinc surface, and then electrons must combine with the hydrogen ions to produce hydrogen atoms. The hydrogen atoms must combine to form diatomic hydrogen molecules which in turn must combine to form bubbles of hydrogen gas. The slowest of these steps will control the cathodic half-cell reaction. There is also an activation-polarization barrier for the anodic half-cell reaction which is the barrier for zinc atoms to leave the metal surface to form zinc ions and go into the electrolyte.

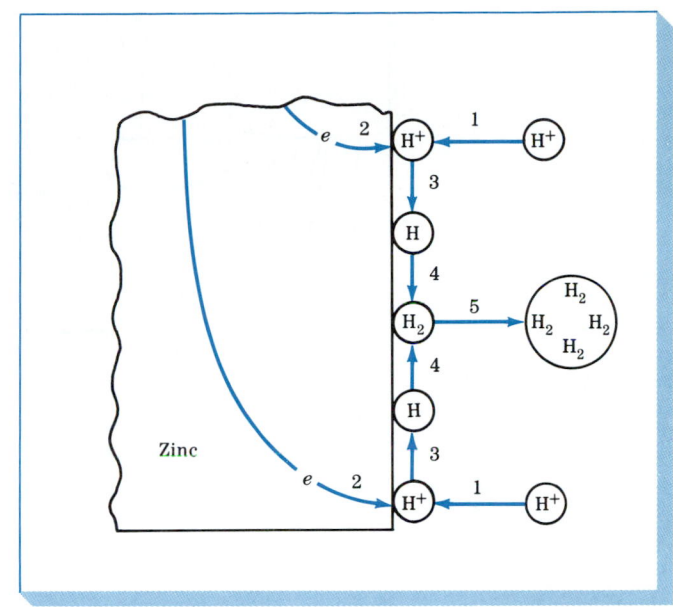

FIGURE 12.15 Hydrogen-reduction reaction at zinc cathode under activation polarization. The steps in the formation of hydrogen gas at the cathode are: (1) migration of hydrogen ions to zinc surface, (2) flow of electrons to hydrogen ions, (3) formation of atomic hydrogen, (4) formation of diatomic hydrogen molecules, and, (5) formation of hydrogen gas bubble which breaks away from zinc surface. The slowest of these steps will be the rate-limiting step for this activation-polarization process (*After M. G. Fontana and N. D. Greene, "Corrosion Engineering," 2d ed., McGraw-Hill, 1978, p. 15.*)

Concentration polarization   Concentration polarization is associated with electrochemical reactions which are controlled by the diffusion of ions in the electrolyte. This type of polarization is illustrated by considering the diffusion of hydrogen ions to a metal surface to form hydrogen gas by the cathodic reaction $2H^+ + 2e^- \rightarrow H_2$, as shown in Fig. 12.16. In this case the concentration of hydrogen ions is low, and thus the reduction rate of the hydrogen ions at the metal surface is controlled by the diffusion of these ions to the metal surface.

For concentration polarization any changes in the system which increase the diffusion rate of the ions in the electrolyte will decrease the concentration-polarization effects and increase the reaction rate. Thus stirring the electrolyte will decrease the concentration gradient of positive ions and increase the reaction rate. Increasing the temperature will also increase the diffusion rate of ions and hence increase the reaction rate.

The total polarization at the electrode in an electrochemical reaction is equal to the sum of the effects of activation polarization and concentration polarization. Activation polarization is usually the controlling factor at low reaction rates and concentration polarization at higher reaction rates. When polarization occurs mostly at the anode, the corrosion rate is said to be *anodically controlled,* and when polarization occurs mostly at the cathode, the corrosion rate is said to be *cathodically controlled.*

Passivation   The *passivation* of a metal as pertains to corrosion refers to the formation of a protective surface layer of reaction product which inhibits further reaction.

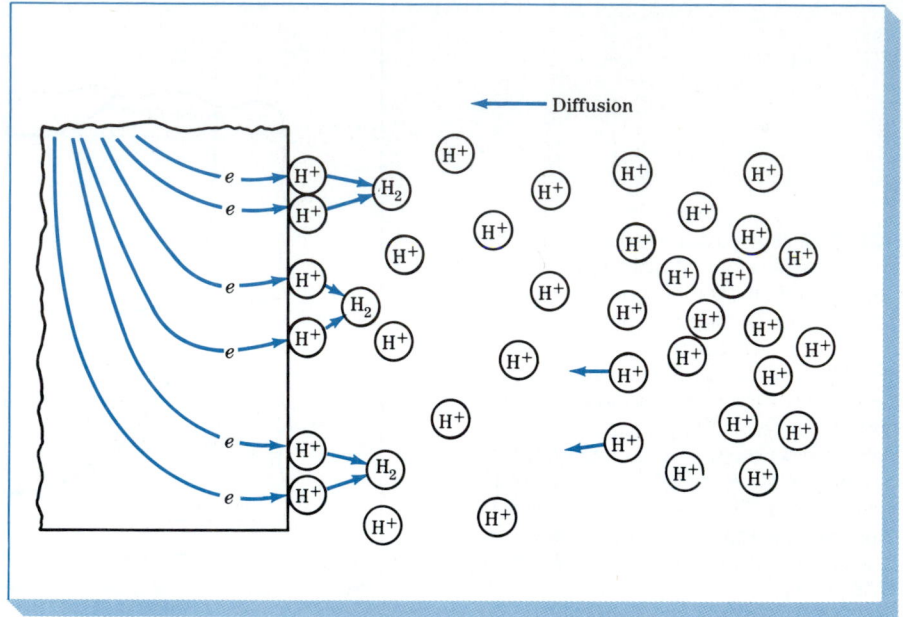

FIGURE 12.16
Concentration polarization during the cathodic hydrogen-ion reduction reaction $2H^+ + 2e^- \rightarrow H_2$. The reaction at the metal surface is controlled by the rate of diffusion of hydrogen ions to the metal surface. (*After M. G. Fontana and N. D. Greene, "Corrosion Engineering," 2d ed., McGraw-Hill, 1978, p. 15.*)

In other words, the passivation of metals refers to their loss of chemical reactivity in the presence of a particular environmental condition. Many important engineering metals and alloys become passive and hence very corrosion-resistant in moderate to strong oxidizing environments. Examples of metals and alloys which show passivity are stainless steels, nickel and many nickel alloys, and titanium and aluminum and many of their alloys.

There are two main theories regarding the nature of the passive film: (1) the oxide-film theory and (2) the adsorption theory. For the oxide-film theory it is believed that the passive film is always a diffusion-barrier layer of reaction products (e.g., metal oxides or other compounds) which separate the metal from its environment and which slow down the reaction rate. For the adsorption theory it is believed that passive metals are covered by chemisorbed films of oxygen. Such a layer is supposed to displace the normally adsorbed H_2O molecules and slow down the rate of anodic dissolution involving the hydration of metal ions. The two theories have in common that there is a protective film formed on the metal surface to create the passive state which results in increased corrosion resistance.

The passivation of metals in terms of corrosion rate can be illustrated by a polarization curve which shows how the potential of a metal varies with current density, as shown in Fig. 12.17. Let us consider the passivation behavior of a metal M as the current density is increased. At point A in Fig. 12.17 the metal is at its equilibrium potential E and its exchange current density i_0. As the electrode potential is made more positive, the metal behaves as an active metal, and its current density and hence its dissolution rate increase exponentially.

When the potential becomes more positive and reaches the potential E_{pp},

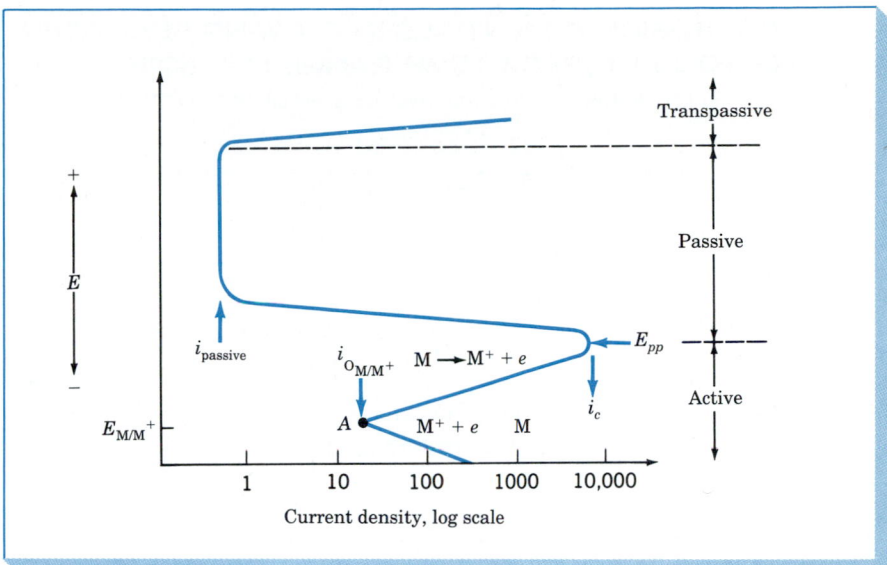

FIGURE 12.17
Polarization curve of a passive metal. (*After M. G. Fontana and N. D. Greene, "Corrosion Engineering," 2d ed., McGraw-Hill, 1978, p. 321.*)

primary passive potential, the current density and hence the corrosion rate decrease to a low value indicated as $i_{passive}$. At the potential E_{pp}, the metal forms a protective film on its surface which is responsible for the decreased reactivity. As the potential is made still more positive, the current density remains at $i_{passive}$ over the passive region. A still further increase in potential beyond the passive region makes the metal active again, and the current density increases in the transpassive region.

The Galvanic Series   Since many important engineering metals form passive films, they do not behave in galvanic cells as the standard electrode potentials would indicate. Thus for practical applications where corrosion is an important factor, a new type of series called the *galvanic series* has been developed for anodic-cathodic relationships. Thus, a galvanic series should be determined experimentally for every corrosive environment. A galvanic series for metals and alloys exposed to flowing seawater is listed in Table 12.3. The different potentials for active and passive conditions of some stainless steels are shown. In this table zinc is shown to be more active than the aluminum alloys, which is the reverse of the behavior shown in the standard electrode potentials of Table 12.1.

12.5 TYPES OF CORROSION

The types of corrosion can be conveniently classified according to the appearance of the corroded metal. Many forms can be identified, but all of them are interrelated to varying extents. These include:

TABLE 12.3   Galvanic Series in Flowing Seawater

CORROSION POTENTIALS IN FLOWING SEAWATER (8 TO 13 FT./SEC.) TEMP RANGE 50° - 80°F

VOLTS: SATURATED CALOMEL HALF-CELL REFERENCE ELECTRODE

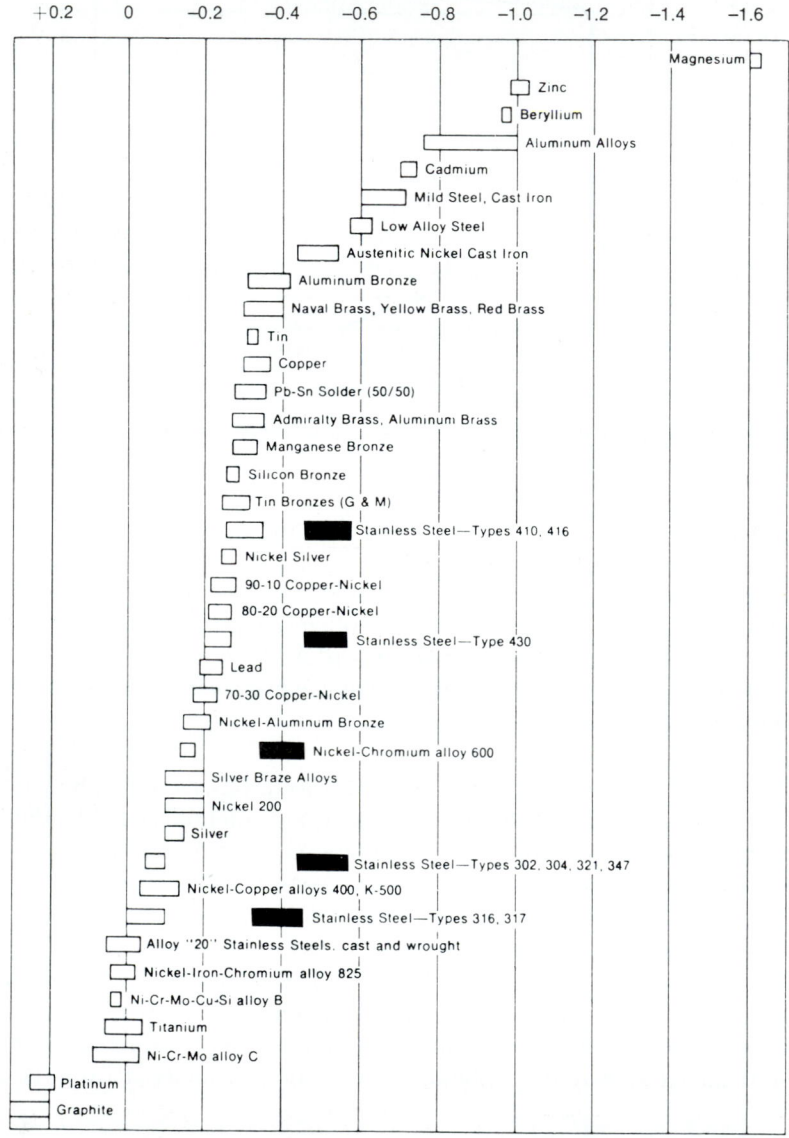

Alloys are listed in the order of the potential they exhibit in flowing seawater. Certain alloys indicated by the symbol: ■ in low-velocity or poorly aerated water, and at shielded areas, may become active and exhibit a potential near –0.5 volts.

Source: Courtesy of the LaQue Center for Corrosion Technology, Inc.

Uniform or general attack corrosion   Stress corrosion

Galvanic or two-metal corrosion   Erosion corrosion

Pitting corrosion   Cavitation damage

Crevice corrosion   Fretting corrosion

Intergranular corrosion   Selective leaching or dealloying

Uniform or General Attack Corrosion

Uniform corrosion attack is characterized by an electrochemical or chemical reaction that proceeds uniformly on the entire metal surface exposed to the corrosion environment. On a weight basis, uniform attack represents the greatest destruction of metals, particularly steels. However, it is relatively easy to control by (1) protective coatings, (2) inhibitors, and (3) cathodic protection. These methods will be discussed in Sec. 12.7 on Corrosion Control.

Galvanic or Two-Metal Corrosion

Galvanic corrosion between dissimilar metals has been discussed in Secs. 12.2 and 12.3. Care must be taken in attaching dissimilar metals together because the difference in their electrochemical potential can lead to corrosion.

Galvanized steel, which is steel coated with zinc, is an example where one metal (zinc) is sacrificed to protect the other (steel). The zinc which is hot-dipped or electroplated on the steel is anodic to the steel and hence corrodes and protects the steel which is the cathode in this galvanic cell (Fig. 12.18a). Table 12.4 shows typical weight losses for uncoupled and coupled zinc and steel in aqueous environments. When the zinc and steel are uncoupled, they both corrode at about the same rate. However, when they are coupled, the zinc corrodes at the anode of a galvanic cell and hence protects the steel.

Another case of the use of two dissimilar metals in an industrial product is in the tin plate used for the "tin can." Most tin plate is produced by electrodepositing a thin layer of tin on steel sheet. The nontoxic nature of tin salts makes tin plate useful for food-container material. Tin (standard emf of -0.136 V) and iron (standard emf of -0.441 V) are close in electrochemical behavior.

FIGURE 12.18 Anodic-cathodic behavior of steel with zinc and tin outside layers exposed to the atmosphere. (a) Zinc is anodic to steel and corrodes (standard emf of Zn $= -0.763$ V and Fe $= -0.440$ V). (b) Steel is anodic to tin and corrodes (the tin layer was perforated before the corrosion began) (standard emf of Fe $= -0.440$ V and Sn $= -0.136$ V). (After M. G. Fontana and N. D. Greene, "Corrosion Engineering," 2d ed., McGraw-Hill, 1978.)

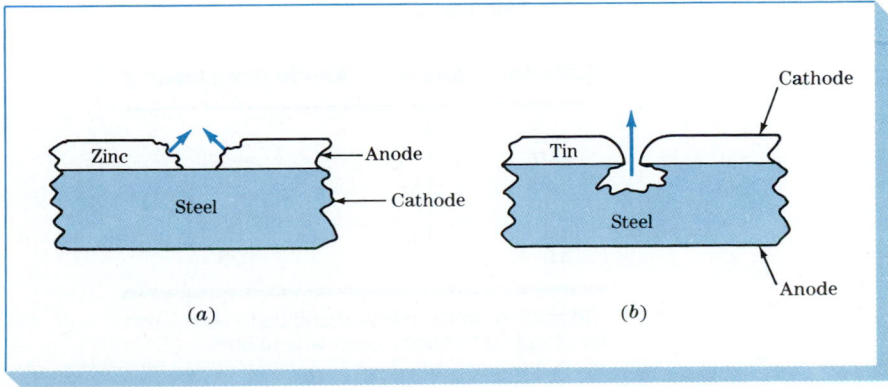

TABLE 12.4 **Change in Weight (in Grams) of Coupled and Uncoupled Steel and Zinc**

Environment	Uncoupled		Coupled	
	Zinc	Steel	Zinc	Steel
0.05 M MgSO$_4$	0.00	−0.04	−0.05	+0.02
0.05 M Na$_2$SO$_4$	−0.17	−0.15	−0.48	+0.01
0.05 M NaCl	−0.15	−0.15	−0.44	+0.01
0.005 M NaCl	−0.06	−0.10	−0.13	+0.02

Source: M. G. Fontana and N. D. Greene, "Corrosion Engineering," 2d ed., McGraw-Hill, 1978.

Slight changes in oxygen availability and ion concentrations that build up on the surface will change their relative polarity. Under conditions of atmospheric exposure, tin is normally cathodic to steel. Thus if the outside of a piece of perforated tin plate is exposed to the atmosphere, the steel will corrode and not the tin (Fig. 12.18*b*). However, in the absence of the oxygen of the air, tin is anodic to steel, which makes tin a useful container material for food and beverages. As can be seen in this example oxygen availability is an important factor in galvanic corrosion.

Another important consideration in galvanic two-metal corrosion is the ratio of the cathodic to anodic areas. This is called the *area effect*. An unfavorable cathodic–anodic area ratio consists of a large cathodic area and a small anodic area. With a certain amount of current flow to a metal couple, e.g., copper and iron electrodes of different sizes, the current density is much greater for the smaller electrode than for the larger one. Thus the smaller anodic electrode will corrode much faster. Table 12.5 shows that as the cathode-to-

TABLE 12.5 **Effect of Area on Galvanic Corrosion of Iron Coupled to Copper in 3% Sodium Chloride**

Relative areas		Anode (iron) loss,* g
Cathode	Anode	
1.01	1	0.23
2.97	1	0.57
5.16	1	0.79
8.35	1	0.94
11.6	1	1.09
18.5	1	1.25

* Tests run at 86°F in aerated and agitated solution for about 20 h. Anode area was 14 cm².

anode ratio for an iron-copper couple was increased from 1 to 18.5, the weight loss of iron at the anode rose from 0.23 to 1.25 g. This area effect is also illustrated in Fig. 12.19 for copper-steel couples immersed in seawater. Copper rivets (cathodes) with a small area caused only a slight increase in the corrosion of steel plates (Fig. 12.19*a*). On the other hand, copper plates (cathodes) caused severe corrosion of steel rivets (anodes), as shown in Fig. 12.19*b*. Thus a large cathode area–to–a small anode area ratio should be avoided.

Pitting Corrosion   *Pitting* is a form of localized corrosive attack that produces holes or pits in a metal. This form of corrosion is very destructive for engineering structures if it causes perforation of the metal. However, if perforation does not occur, minimum pitting is sometimes acceptable in engineering equipment. Pitting is often difficult to detect because small pits may be covered by corrosion products. Also the number and depth of pits can vary greatly, and so the extent of pitting damage may be difficult to evaluate. As a result, pitting, beause of its localized nature, can often result in sudden, unexpected failures.

Figure 12.20 shows an example of pitting in a stainless steel exposed to an aggressive corrosive environment. The pitting for this example was accelerated, but in most service conditions, pitting may require months or years to perforate a metal section. Pitting usually requires an initiation period, but once started, the pits grow at an ever-increasing rate. Most pits develop and grow in the direction of gravity and on the lower surfaces of engineering equipment.

Pits are initiated at places where local increases in corrosion rates occur. Inclusions, other structural heterogeneities, and compositional heterogeneities

FIGURE 12.19 Effect of area relationships between cathode and anode for copper-steel couples immersed in seawater. (*a*) Small cathode (copper rivets) and large anode (steel plates) cause only slight damage to steel. (*b*) Small anode (steel rivets) and large cathode (copper plates) cause severe corrosion of steel rivets. (*Courtesy of the LaQue Center for Corrosion Technology, Inc.*)

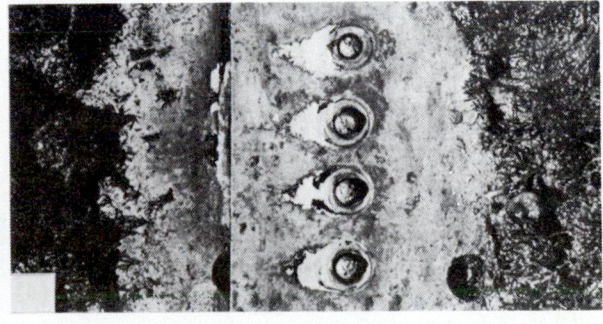

(*a*)   (*b*)

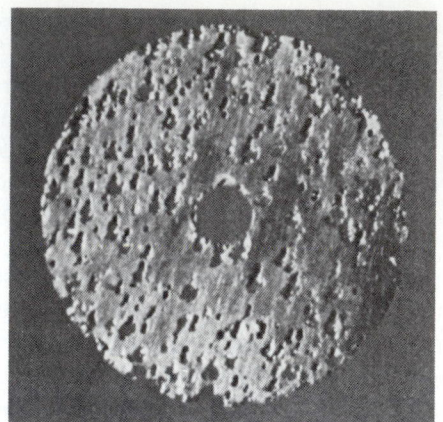

FIGURE 12.20 Pitting of a stainless steel in an aggressive corrosive environment. (*Courtesy of the LaQue Center for Corrosion Technology, Inc.*)

on the metal surface are common places where pits initiate. Differences in ion and oxygen concentrations create concentration cells which can also initiate pits. The propagation of a pit is believed to involve the dissolution of the metal in the pit while maintaining a high degree of acidity at the bottom of the pit. The propagation process for a pit in an aerated saltwater environment is illustrated in Fig. 12.21 for a ferrous metal. The anodic reaction of the metal at the bottom of the pit is $M \rightarrow M^{n+} + e^-$. The cathodic reaction takes place at the metal surface surrounding the pit and is the reaction of oxygen with water and the electrons from the anodic reaction: $O_2 + 2H_2O + 4e^- \rightarrow 4OH^-$. Thus the metal surrounding the pit is cathodically protected. The increased concentration of metal ions in the pit brings chloride ions in to maintain charge neutrality. The metal chloride then reacts with water to produce the metal hydroxide and free acid as

$$M^+Cl^- + H_2O \rightarrow MOH + H^+Cl^- \tag{12.16}$$

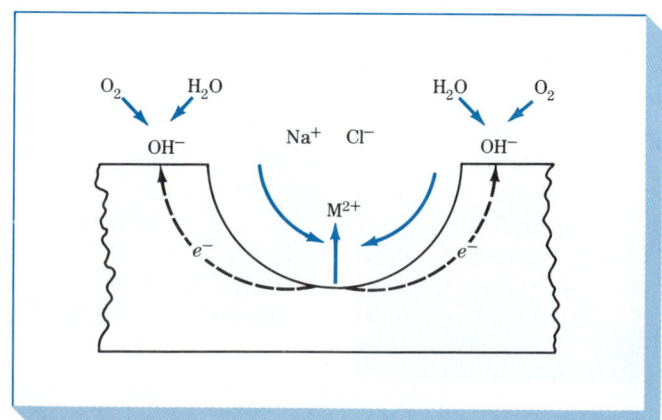

FIGURE 12.21 A schematic diagram of the growth of a pit in a stainless steel in an aerated salt solution. (*After M. G. Fontana and N. D. Greene, "Corrosion Engineering," 2d ed., McGraw-Hill, 1978.*)

TABLE 12.6   Relative Pitting Resistance of Some Corrosion-Resistant Alloys

Type 304 stainless steel	
Type 316 stainless steel	Increasing
Hastelloy F, Nionel, or Durimet 20	pitting resistance
Hastelloy C or Chlorimet 3	↓
Titanium	

Source: M. G. Fontana and N. D. Greene, "Corrosion Engineering," 2d ed., McGraw-Hill, 1978.

In this way a high acid concentration builds up at the bottom of the pit which makes the anodic reaction rate increase, and the whole process becomes *autocatalytic.*

To avoid pitting corrosion in the design of engineering equipment, materials which do not have pitting-corrosion tendencies should be used. However, if this is not possible for some designs, then materials with the best corrosion resistance must be used. For example, if stainless steels must be used in the presence of some chloride ions, type 316 alloy with 2% molybdenum in addition to the 18% Cr and 8% Ni has better pitting resistance than type 304 alloy which just contains the 18% Cr and 8% Ni as the main alloying elements. A qualitative guide for the pitting-resistance order of some corrosion-resistant materials is listed in Table 12.6. However, it is recommended that corrosion tests be made of various alloys before final selection of the corrosion-resistant alloy.

Crevice Corrosion   Crevice corrosion is a form of localized electrochemical corrosion which can occur in crevices and under shielded surfaces where stagnant solutions can exist. Crevice corrosion is of engineering importance when it occurs under gaskets, rivets, and bolts, between valve disks and seats, under porous deposits, and in many other similar situations. Crevice corrosion occurs in many alloy systems such as stainless steels and titanium, aluminum and copper alloys. Figure 12.22 shows an example of crevice-corrosion attack of a mooring pennant.

For crevice corrosion to occur a crevice must be wide enough to allow liquid to enter but sufficiently narrow enough to keep the liquid stagnant. Therefore, crevice corrosion usually occurs with an opening of a few micrometers (mils) or less in width. Fibrous gaskets which can act as wicks to absorb an electrolytic solution and keep it in contact with the metal surface make ideal locations for crevice corrosion.

Fontana and Greene[1] have proposed a mechanism for crevice corrosion which is similar to that which they proposed for pitting corrosion. Figure 12.23

---

[1] "Corrosion Engineering," 2d ed., McGraw-Hill, 1978.

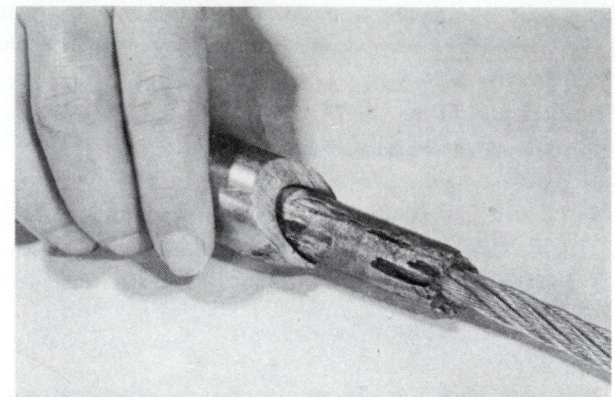

FIGURE 12.22 Crevice corrosion of a mooring pennant. (*Courtesy of the LaQue Center for Corrosion Technology, Inc.*)

illustrates this mechanism for the crevice corrosion of a stainless steel in an aerated sodium chloride solution. This mechanism assumes that initially the anodic and cathodic reactions on the surface of the crevice are

Anodic reaction: $\qquad$ $M \rightarrow M^+ + e^-$ $\qquad\qquad$ (12.17a)

Cathodic reaction: $\qquad$ $O_2 + 2H_2O + 4e^- \rightarrow 4OH^-$ $\qquad$ (12.17b)

Since the solution in the crevice is stagnant, the oxygen needed for the cathodic reaction is used up and not replaced. However, the anodic reaction $M \rightarrow M^+ + e^-$ continues to operate, creating a high concentration of positively charged ions. To balance the positive charge, negatively charged ions, mainly chloride ions, migrate into the crevice, forming M^+Cl^-. This chloride is hydrolyzed by water to form the metal hydroxide and free acid, as

$$M^+Cl^- + H_2O \rightarrow MOH + H^+Cl^- \qquad (12.18)$$

FIGURE 12.23 A schematic diagram of crevice-corrosion mechanism. (*After M. G. Fontana and N. D. Greene, "Corrosion Engineering," 2d ed., McGraw-Hill, 1978.*)

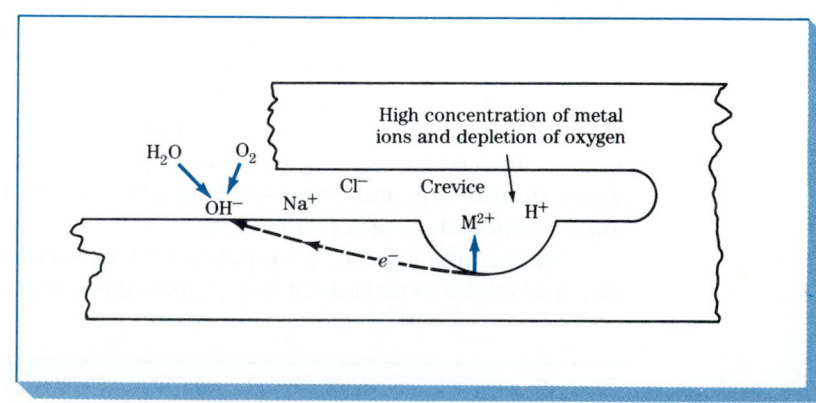

This buildup of acid breaks down the passive film and causes corrosion attack which is autocatalytic as in the case just discussed for pitting corrosion.

For type 304 (18% Cr–8% Ni) stainless steel, Peterson et al.[1] have concluded from their tests that the acidification within the crevice is probably mainly due to the hydrolysis of chromic ions, as

$$Cr^{3+} + 3H_2O \rightarrow Cr(OH)_3 + 3H^+ \tag{12.19}$$

since they found only traces of Fe^{3+} in the crevice.

To prevent or minimize crevice corrosion in engineering designs, the following methods and procedures can be used:

1. Use soundly welded butt joints instead of riveted or bolted ones in engineering structures.
2. Design vessels for complete drainage where stagnant solutions may accumulate.
3. Use nonabsorbent gaskets, such as Teflon, if possible.

Intergranular Corrosion

Intergranular corrosion is localized corrosion attack at and/or adjacent to the grain boundaries of an alloy. Under ordinary conditions if a metal corrodes uniformly, the grain boundaries will only be slightly more reactive than the matrix. However, under other conditions, the grain-boundary regions can be very reactive, resulting in intergranular corrosion which causes loss of strength of the alloy and even disintegration at the grain boundaries.

For example, many high-strength aluminum alloys and some copper alloys which have precipitated phases for strengthening are susceptible to intergranular corrosion under certain conditions. However, one of the most important examples of intergranular corrosion occurs in some austenitic (18% Cr–8% Ni) stainless steels which are heated into or slowly cooled through the 500 to 800°C (950 to 1450°F) temperature range. In this so-called sensitizing temperature range, chromium carbides ($Cr_{23}C_6$) can precipitate at the grain-boundary interfaces, as shown in Fig. 12.24a. When chromium carbides have precipitated along the grain boundaries in austenitic stainless steels, these alloys are said to be in the *sensitized condition*.

If an 18% Cr–8% Ni austenitic stainless steel contains more than about 0.02 wt % carbon, chromium carbides ($Cr_{23}C_6$) can precipitate at the grain boundaries of the alloy if heated in the 500 to 800°C range for a long-enough time. Type 304 is an austenitic stainless steel with 18% Cr–8% Ni and between about 0.06 to 0.08 wt % carbon. Hence this alloy if heated in the 500 to 800°C range for a sufficient time will be put into the sensitized condition and be susceptible to intergranular corrosion. When the chromium carbides form at the grain boundaries, they deplete the regions adjacent to the boundaries of

[1] M. H. Peterson, T. J. Lennox, and R. E. Groover, *Mater. Prot.*, January 1970, p. 23.

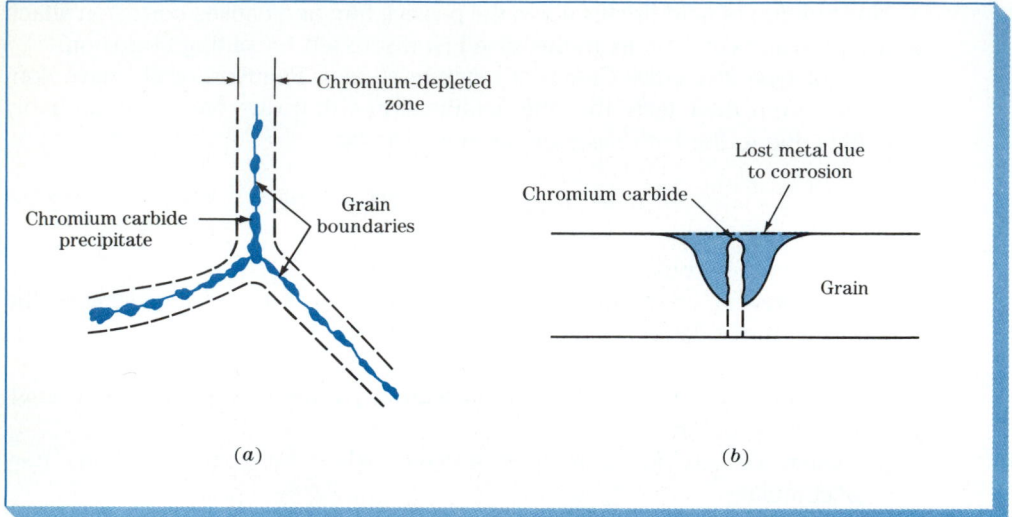

FIGURE 12.24 (*a*) Schematic representation of chromium carbide precipitation at grain boundaries in a sensitized type 304 stainless steel. (*b*) Cross section at grain boundary showing intergranular corrosion attack adjacent to the grain boundaries.

chromium so that the chromium level in these areas decreases below the 12% chromium level necessary for passive or "stainless" behavior. Thus when, for example, type 304 stainless steel in the sensitized condition is exposed to a corrosive environment, the regions next to the grain boundaries will be severely attacked. These areas become anodic to the rest of the grain bodies which are cathodic, thereby creating galvanic couples. Figure 12.24*b* shows this schematically.

Failure of welds made with type 304 stainless steel or similar alloys can occur by the same chromium carbide precipitation mechanism as previously described. This type of weld failure is called *weld decay* and is characterized by a weld decay zone somewhat removed from the center line of the weld, as shown in Fig. 12.25. The metal in the weld decay zone was held in the temperature range for sensitizing (500 to 800°C) for too long a time so that chromium carbides precipitated in the grain boundaries of the heat-affected zones of the weld. If a welded joint in the sensitized condition is not subsequently reheated to redissolve the chromium carbides, it will be subject to intergranular corrosion when exposed to a corrosive environment, and the weld could fail.

Intergranular corrosion of austenitic stainless steels can be controlled by the following methods:

1. Use a high-temperature solution heat treatment after welding. By heating the welded joint at 500 to 800°C followed by water quenching, the chromium carbides can be redissolved and returned to solid solution.

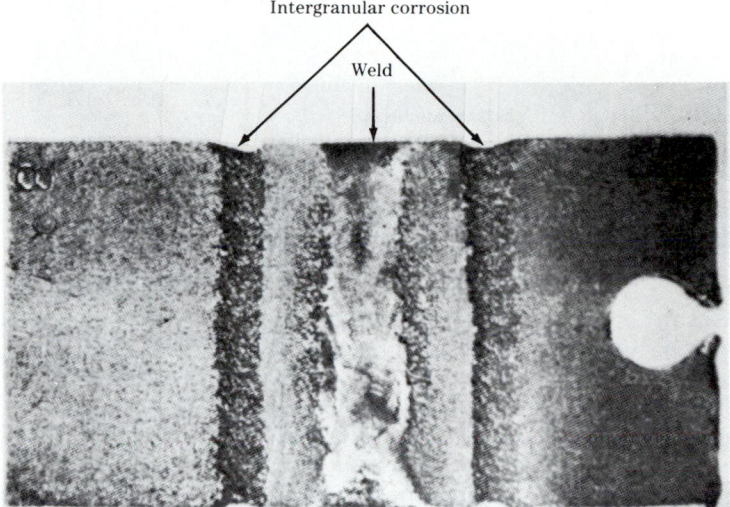

Intergranular corrosion

Weld

FIGURE 12.25
Intergranular corrosion of a stainless steel weld. The weld decay zones had been held at the critical temperature range needed for precipitation of chromium carbides during cooling. (*After H. H. Uhlig, "Corrosion and Corrosion Control," Wiley, 1963, p. 267.*)

2. Add an element which will combine with the carbon in the steel so that chromium carbides cannot form. Columbium and titanium additions in alloy types 347 and 321, respectively, are used. These elements have a greater affinity for carbon than chromium. Alloys with Ti or Cb additions are said to be in the *stabilized condition.*
3. Lower the carbon content to about 0.03 wt % or less so that significant amounts of chromium carbides cannot precipitate. Type 304L stainless steel, for example, has its carbon at such a low level.

Stress Corrosion   Stress-corrosion cracking (SCC) of metals refers to cracking caused by the combined effects of tensile stress and a specific corrosion environment acting on the metal. During SCC the metal's surface is usually attacked very little while highly localized cracks propagate through the metal section, as shown, for example, in Fig. 12.26. The stresses which cause SCC can be residual or applied. High residual stresses which cause SCC may result, for example, from thermal stresses introduced by unequal cooling rates, poor mechanical design for stresses, phase transformations during heat treatment, cold working, and welding.

Only certain combinations of alloys and environments will cause SCC. Table 12.7 lists some of the alloy-environment systems in which SCC occurs. There appears to be no general pattern to the environments which produce SCC in alloys. For example, stainless steels crack in chloride environments but not in ammonia-containing ones. In contrast, brasses (Cu–Zn alloys) crack in ammonia-containing environments but not in chloride ones. New combinations of alloys and environments which cause SCC are continuously being discovered.

Mechanism of stress-corrosion cracking   The mechanisms involved in SCC are not completely understood since there are so many different alloy-envi-

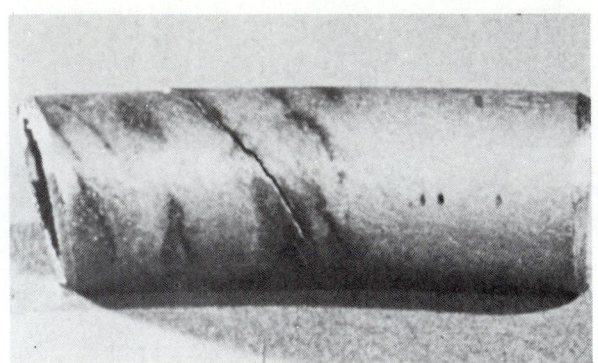

FIGURE 12.26 Stress-corrosion cracks in a pipe. (*Courtesy of the LaQue Center for Corrosion Technology, Inc.*)

ronment systems which involve many different mechanisms. Most SCC mechanisms involve crack initiation and propagation stages. In many cases the crack initiates at a pit or other discontinuity on the metal surface. After the crack has been started, the tip can advance, as shown in Fig. 12.27. A high stress builds up at the tip of the crack due to tensile stresses acting on the metal. Anodic dissolution of the metal takes place by localized electrochemical corrosion at

TABLE 12.7   Environments That May Cause Stress Corrosion of Metals and Alloys

Material	Environment	Material	Environment
Aluminum alloys	$NaCl–H_2O_2$ solutions NaCl solutions Seawater Air, water vapor	Ordinary steels	NaOH solutions $NaOH–Na_2SiO_2$ solutions Calcium, ammonium, and sodium nitrate solutions
Copper alloys	Ammonia vapors and solutions Amines Water, water vapor		Mixed acids ($H_2SO_4–HNO_2$) HCN solutions
Gold alloys	$FeCl_2$ solutions Acetic acid–salt solutions		Acidic H_2S solutions Seawater
Inconel	Caustic soda solutions		Molten Na–Pb alloys
Lead	Lead acetate solutions	Stainless steels	Acid-chloride solutions
Magnesium alloys	$NaCl–K_2CrO_4$ solutions Rural and coastal atmospheres Distilled water		such as $MgCl_2$ and $BaCl_2$ $NaCl–H_2O_2$ solutions Seawater
Monel	Fused caustic soda Hydrofluoric acid Hydrofluosilicic acid		H_2S $NaOH–H_2S$ solutions Condensing steam from chloride waters
Nickel	Fused caustic soda	Titanium alloys	Red fuming nitric acid, seawater, N_2O_4, methanol-HCl

Source: M. G. Fontana and N. D. Greene, "Corrosion Engineering," 2d ed., McGraw-Hill, 1978, p. 100.

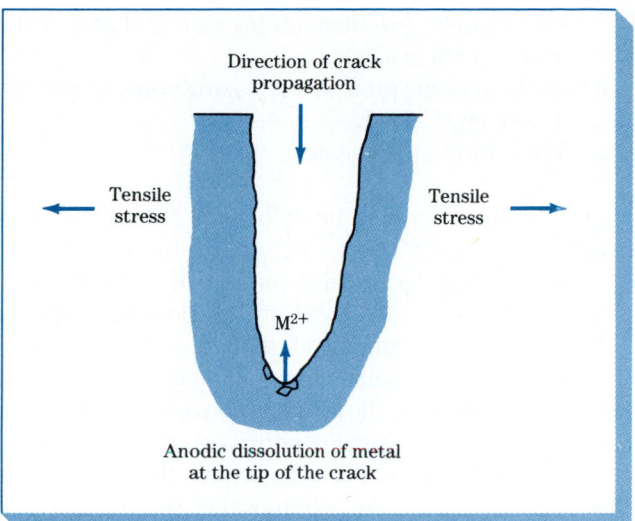

FIGURE 12.27
Development of a stress-corrosion crack in a metal by anodic dissolution. (*After R. W. Staehle.*)

the tip of the crack as it advances. The crack grows in a plane perpendicular to the tensile stress until the metal fractures. If either the stress or the corrosion is stopped, the crack stops growing. A classical experiment was made by Priest et al.[1] who showed that an advancing crack could be stopped by cathodic protection. When cathodic protection was removed, the crack started to grow again.

Tensile stress is necessary for both the initiation and propagation of cracks and is important in the rupturing of surface films. Decreasing the stress level increases the time necessary for cracking to occur. Temperature and environment are also important factors for stress-corrosion cracking.

Prevention of stress-corrosion cracking   Since the mechanisms of SCC are not completely understood, the methods of preventing it are general and empirical ones. One or more of the following methods will prevent or reduce SCC in metals.

1. Lower the stress of the alloy below that which causes cracking. This may be done by lowering the stress on the alloy or by giving the material a stress-relief anneal. Plain-carbon steels can be stress-relieved at 600 to 650°C (1100 to 1200°F), and austenitic stainless steels can be stress-relieved in the range 815 to 925°C (1500 to 1700°F).
2. Eliminate the detrimental environment.
3. Change the alloy if neither environment nor stress level can be changed.

---

[1] D. K. Priest, F. H. Beck, and M. G. Fontana, *Trans. ASM,* **47**:473(1955).

For example, use titanium instead of stainless steel for heat exchangers in contact with seawater.

4. Apply cathodic protection by using consumable anodes or an external power supply (see Sec. 12.7).
5. Add inhibitors if possible.

Erosion Corrosion

Erosion corrosion can be defined as the acceleration in the rate of corrosion attack in a metal due to the relative motion of a corrosive fluid and a metal surface. When the relative motion of corrosive fluid is rapid, the effects of mechanical wear and abrasion can be severe. Erosion corrosion is characterized by the appearance on the metal surface of grooves, valleys, pits, rounded holes, and other metal surface damage configurations which usually occur in the direction of the flow of the corrosive fluid.

Studies of the erosion-corrosion action of silica sand slurries in mild-steel pipe have led researchers to believe that the increased corrosion rate of the slurry action is due to the removal of surface rust and salt films by the abrasive action of the silica particles of the slurry, thus permitting much easier access of dissolved oxygen to the corroding surface. Figure 12.28 shows severe wear patterns caused by the erosion corrosion of an experimental section of mild-steel pipe.

Cavitation Damage

This type of erosion corrosion is caused by the formation and collapse of air bubbles or vapor-filled cavities in a liquid near a metal surface. Cavitation damage occurs at metal surfaces where high-velocity liquid flow and pressure changes exist, such as are encountered with pump impellers and ship propellers. Calculations indicate that rapidly collapsing vapor bubbles can produce localized pressures as high as 60,000 psi. With repeated collapsing vapor bubbles, considerable damage to a metal surface can be done. By removing surface films and tearing metal particles away from the metal surface, cavitation damage can increase corrosion rates and cause surface wear.

Fretting Corrosion

Fretting corrosion occurs at interfaces between materials under load subjected to vibration and slip. Fretting corrosion appears as grooves or pits surrounded

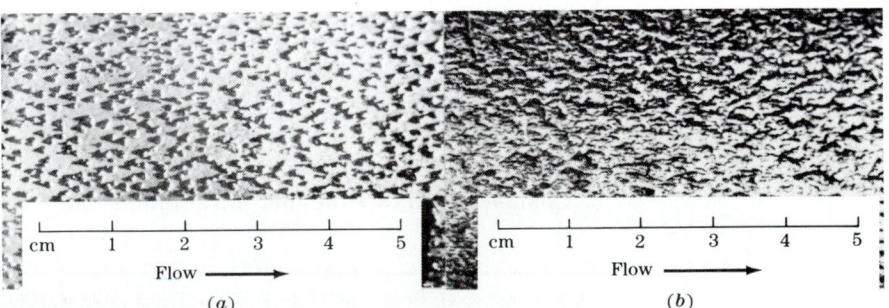

FIGURE 12.28 Erosion-corrosion wear pattern of silica slurry in mild-steel pipe, showing (*a*) pitting after 21 days and (*b*) irregular wavy pattern after 42 days. Slurry velocity is 3.5 m/s. [*After J. Postlethwaite et al., Corrosion,* **34**:245(1978).]

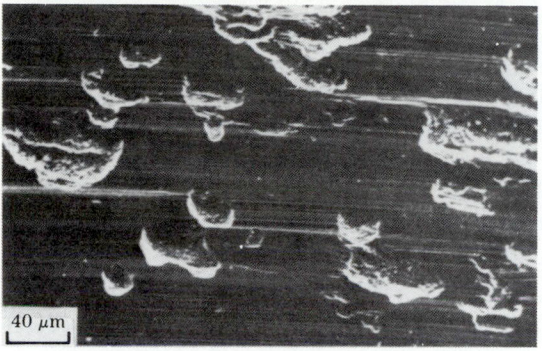

FIGURE 12.29 Scanning electron micrograph showing fretting corrosion on the surface of a Ti–6 Al–4 V alloy developed at 600°C using a sphere-on-flat-configuration with 40 μm amplitude of slip and after 3.5×10^6 cycles. [*After M. M. Hamdy and R. B. Waterhouse, Wear,* **71**:237(1981).]

40 μm

by corrosion products. In the case of the fretting corrosion of metals, metal fragments between the rubbing surfaces are oxidized and some oxide films are torn loose by the wearing action. As a result, there is an accumulation of oxide particles which act as an abrasive between the rubbing surfaces. Fretting corrosion commonly occurs between tight-fitting surfaces such as are found between shafts and bearing or sleeves. Figure 12.29 shows the effects of fretting corrosion on the surface of a Ti–6 Al–4 V alloy.

Selective Leaching

Selective leaching is the preferential removal of one element of a solid alloy by corrosion processes. The most common example of this type of corrosion is dezincification in which the selective leaching of zinc from copper in brasses occurs. Similar processes also occur in other alloy systems such as the loss of nickel, tin, and chromium from copper alloys, iron from cast iron, nickel from alloy steels, and cobalt from stellite.

In the dezincification of a 70% Cu–30% Zn brass, for example, the zinc is preferentially removed from the brass, leaving a spongy, weak matrix of copper. The mechanism of dezincification involves the following three steps:[1]

1. Dissolution of the brass
2. Remaining of the zinc ions in solution
3. Plating of the copper back on the brass

Since the copper remaining does not have the strength of the brass, the strength of the alloy is considerably lowered.

Dezincification can be minimized or prevented by changing to a brass with a lower zinc content (that is, 85% Cu–15% Zn brass) or to a cupronickel (70 to 90% Cu–10 to 30% Ni). Other possibilities are to change the corrosive environment or use cathodic protection.

[1] After M. G. Fontana and N. D. Greene, "Corrosion Engineering," 2d ed., McGraw-Hill, 1978.

12.6 OXIDATION OF METALS

Until now we have been concerned with corrosion conditions in which a liquid electrolyte was an integral part of the corrosion mechanism. Metals and alloys, however, also react with air to form external oxides. The high-temperature oxidation of metals is particularly important in the engineering design of gas turbines, rocket engines, and high-temperature petrochemical equipment.

Protective Oxide Films

The degree to which an oxide film is protective to a metal depends upon many factors, of which the following are important:

1. The volume ratio of oxide to metal after oxidation should be close to 1.
2. The film should have good adherence.
3. The melting point of the oxide should be high.
4. The oxide film should have a low vapor pressure.
5. The oxide film should have a coefficient of expansion nearly equal to that of the metal.
6. The film should have high-temperature plasticity to prevent fracture.
7. The film should have low conductivity and low diffusion coefficients for metal ions and oxygen.

The calculation of the volume ratio of oxide to metal after oxidation is a first step that can be taken to find out if an oxide of a metal might be protective. This ratio is called the *Pilling-Bedworth*[1] *(P.B.) ratio* and can be expressed in equation form as

$$\text{P.B. ratio} = \frac{\text{volume of oxide produced by oxidation}}{\text{volume of metal consumed by oxidation}} \tag{12.20}$$

If a metal has a P.B. ratio of less than 1, as is the case for the alkali metals (for example, Na has a P.B. ratio of 0.576), the metal oxide will be porous and unprotective. If the P.B. ratio is more than 1, as is the case for Fe (Fe_2O_3 = 2.15), compressive stresses will be present and the oxide will tend to crack and spall off. If the P.B. ratio is close to 1, the oxide may be protective but some of the other factors listed above have to be satisfied. Thus, the P.B. ratio alone does not determine if an oxide is to be protective. Example Problem 12.9 shows how the P.B. ratio for aluminum can be calculated.

---

[1] N. B. Pilling and R. E. Bedworth, *J. Inst. Met.*, **29**:529(1923).

Example Problem 12.9

Calculate the ratio of the oxide volume to metal volume (Pilling-Bedworth ratio) for the oxidation of aluminum to aluminum oxide, Al_2O_3. The density of aluminum = 2.70 g/cm^3 and that of aluminum oxide = 3.70 g/cm^3.

Solution:

$$\text{P.B. ratio} = \frac{\text{volume of oxide produced by oxidation}}{\text{volume of metal consumed by oxidation}} \qquad (12.20)$$

Assuming that 100 g of aluminum is oxidized,

$$\text{Volume of aluminum} = \frac{\text{mass}}{\text{density}} = \frac{100 \text{ g}}{2.70 \text{ g/cm}^3} = 37.0 \text{ cm}^3$$

To find the volume of Al_2O_3 associated with the oxidation of 100 g of Al, we first find the mass of Al_2O_3 produced by the oxidation of 100 g of Al, using the following equation:

$$
\begin{array}{cc}
100 \text{ g} & X \text{ g} \\
4Al + 3O_2 \rightarrow & 2Al_2O_3 \\
4 \times \dfrac{26.98 \text{ g}}{\text{mol}} & 2 \times \dfrac{102.0 \text{ g}}{\text{mol}}
\end{array}
$$

or

$$\frac{100 \text{ g}}{4 \times 26.98} = \frac{X \text{ g}}{2 \times 102}$$

$$X = 189.0 \text{ g Al}_2O_3$$

Then we find the volume associated with the 189.0 g of Al_2O_3 using the relationship volume = mass/density. Therefore,

$$\text{Volume of Al}_2O_3 = \frac{\text{mass of Al}_2O_3}{\text{density of Al}_2O_3} = \frac{189.0 \text{ g}}{3.70 \text{ g/cm}^3} = 51.1 \text{ cm}^3$$

Thus

$$\text{P.B. ratio} = \frac{\text{volume of Al}_2O_3}{\text{volume of Al}} = \frac{51.1 \text{ cm}^3}{37.0 \text{ cm}^3} = 1.38 \blacktriangleleft$$

Comment: The ratio 1.38 is close to 1 so that Al_2O_3 has a favorable P.B. ratio to be a protective oxide. Al_2O_3 is a protective oxide since it forms a tight coherent film on aluminum. Some of the Al_2O_3 molecules at the oxide-metal interface penetrate into the aluminum metal, and vice versa.

Mechanisms of Oxidation

When an oxide film forms on a metal by the oxidation of a metal by gaseous oxygen, it is formed by an electrochemical process and not simply by the chemical combination of a metal and oxygen as $M + \frac{1}{2}O_2 \rightarrow MO$. The oxidation and reduction partial reactions for the formation of divalent ions are

Oxidation partial reaction: $M \rightarrow M^{2+} + 2e^-$ (12.21a)

Reduction partial reaction: $\frac{1}{2}O_2 + 2e^- \rightarrow O^{2-}$ (12.21b)

In the very early stages of oxidation, the oxide layer is discontinuous and begins by the lateral extension of discrete oxide nuclei. After the nuclei interlace, the mass transport of the ions occurs in a direction normal to the surface (Fig. 12.30). In most cases the metal diffuses as cations and electrons across the oxide film, as indicated in Fig. 12.30a. In this mechanism the oxygen is reduced to oxygen ions at the oxide-gas interface, and the zone of oxide formation is at this surface (cation diffusion). In some other cases, for example, for some heavy metal oxides, the oxygen may diffuse as O^{2-} ions to the metal-oxide interface and electrons diffuse to the oxide-gas interface, as shown in Fig. 12.30b. In this case the oxide forms at the metal-oxide interface (Fig. 12.30b). This is anion diffusion. The oxide movements indicated in Fig. 12.30 are determined principally by the movements of inert markers at the oxide-gas interface. In the case of cation diffusion the markers are buried in the oxide, whereas in the case of anion diffusion the markers remain at the surface of the oxide.

FIGURE 12.30 Oxidation of flat surfaces of metals. (a) When cations diffuse, the initially formed oxide drifts toward the metal. (b) When anions diffuse, the oxide drifts in the opposite direction. [After L. L. Shreir (ed.) "Corrosion," vol. 1, 2d ed., Newnes-Butterworth, 1976, p. 1:242.]

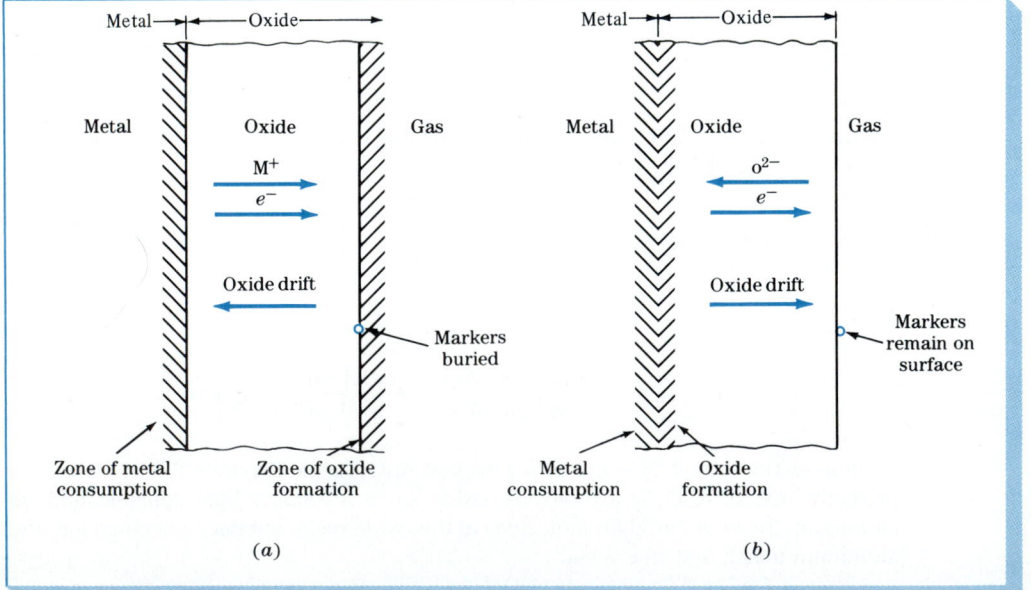

The detailed mechanisms of the oxidation of metals and alloys can be very complex, particularly when layers of different composition and defect structures are produced. Iron, for example, when oxidized at high temperatures, forms a series of iron oxides: FeO, Fe_3O_4, and Fe_2O_3. The oxidation of alloys is further complicated by the interaction of the alloying elements.

Oxidation Rates (Kinetics) From an engineering standpoint the rate at which metals and alloys oxidize is very important since the oxidation rate of many metals and alloys determines the useful life of equipment. The rate of oxidation of metals and alloys is usually measured and expressed as the weight gained per unit area. During the oxidation of different metals, various empirical rate laws have been observed; some of the common ones are shown in Fig. 12.31.

The simplest oxidation rate follows the linear law

$$w = k_L t \qquad (12.22)$$

where w = weight gain per unit area
t = time
k_L = linear rate constant

Linear oxidation behavior is shown by metals which have porous or cracked oxide films, and thus the transport of reactant ions occurs at faster rates than the chemical reaction. Examples of metals which oxidize linearly are potassium, which has an oxide-to-metal volume ratio of 0.45, and tantalum, which has one of 2.50.

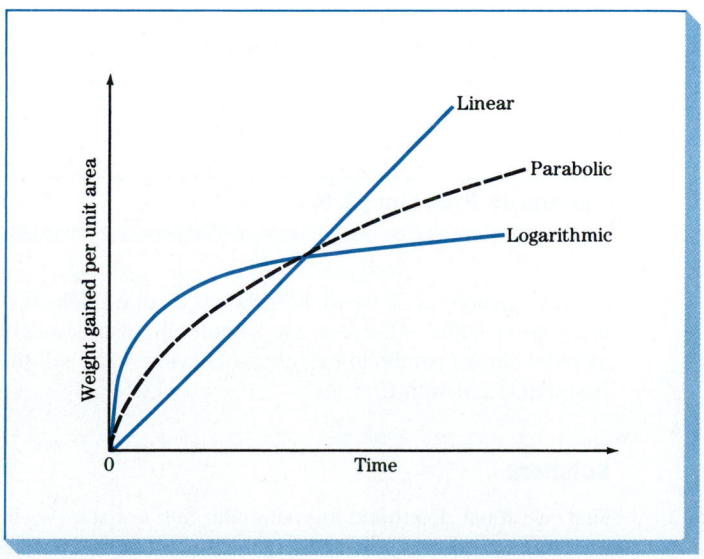

FIGURE 12.31 Oxidation rate laws.

When ion diffusion is the controlling step in the oxidation of metals, pure metals should follow the parabolic relation

$$w^2 = k_p t + C \qquad (12.23)$$

where w = weight gain per unit area
$\quad t$ = time
$\quad k_p$ = parabolic rate constant
$\quad C$ = a constant

Many metals oxidize according to the parabolic rate law, and these are usually associated with thick coherent oxides. Iron, copper, and cobalt are examples of metals which show parabolic oxidation behavior.

Some metals such as Al, Cu, and Fe oxidize at ambient or slightly elevated temperatures to form thin films which follow the logarithmic rate law

$$w = k_e \log (Ct + A) \qquad (12.24)$$

where C and A are constants and k_e is the logarithmic rate constant. These metals when exposed to oxygen at room temperature oxidize very rapidly at the start, but after a few days exposure, the rate decreases to a very low value.

Some metals which exhibit linear rate behavior tend to oxidize catastrophically at high temperatures due to rapid exothermic reactions on their surfaces. As a result, a chain reaction occurs at their surface, causing the temperature and rate of reaction to increase. Metals such as molybdenum, tungsten, and vanadium which have volatile oxides may oxidize catastrophically. Also, alloys containing molybdenum and vanadium even in small amounts often show catastrophic oxidation which limits their use in high-temperature oxidizing atmospheres. The addition of large amounts of chromium and nickel to iron alloys improves their oxidation resistance and retards the effects of catastrophic oxidation due to some other elements.

Example Problem 12.10

A 1-cm^2 sample of 99.94 wt % nickel, 0.75 mm thick, is oxidized in oxygen at 1 atm pressure at 600°C. After 2 h, the sample showed a weight gain of 70 μg/cm^2. If this material shows parabolic oxidation behavior, what will the weight gain be after 10 h? [Use Eq(12.23) with $C = 0$.]

Solution:

First, we must determine the parabolic rate constant k_p from the parabolic oxidation rate equation $y^2 = k_p t$, where y is the thickness of the oxide produced in time t. Since the weight gain of the sample during oxidation is proportional to the growth in thickness of the oxide and can be more precisely measured, we shall replace y, the oxide thickness, by x, the weight gain per unit area of the sample during oxidation. Thus, $x^2 = k'_p t$ and

$$k'_p = \frac{x^2}{t} = \frac{(70 \ \mu g/cm^2)^2}{2 \ h} = 2.45 \times 10^3 \ \mu g^2/(cm^4 \cdot h)$$

For time $t = 10 \ h$, the weight gain in micrograms per square centimeter is

$$x = \sqrt{k'_p t} = \sqrt{[2.45 \times 10^3 \ (\mu g^2/(cm^4 \cdot h))](10 \ h)}$$

$$= 156 \ \mu g/cm^2 \ \blacktriangleleft$$

12.7 CORROSION CONTROL

Corrosion can be controlled or prevented by many different methods. From an industrial standpoint, the economics of the situation usually dictate the methods used. For example, an engineer may have to determine whether it is more economical to periodically replace certain equipment or to fabricate it with materials that are highly corrosion-resistant but more expensive so that it will last longer. Some of the common methods of corrosion control or prevention are shown in Fig. 12.32.

Materials Selection   **Metallic materials**   One of the most common methods of corrosion control is to use materials that are corrosion-resistant for a particular environment. When selecting materials for an engineering design for which the corrosion resistance of the materials is important, corrosion handbooks and data should be con-

FIGURE 12.32 Common methods of corrosion control.

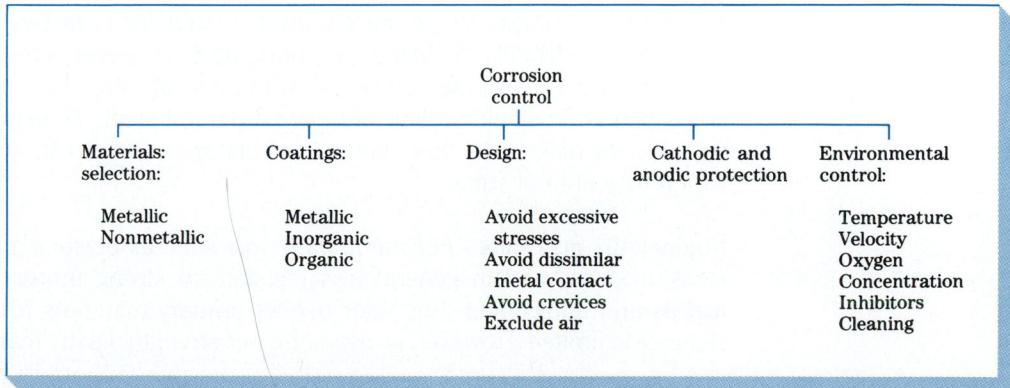

TABLE 12.8   Combinations of Metals and Environments that Give Good Corrosion Resistance for the Cost

1. Stainless steels–nitric acid
2. Nickel and nickel alloys–caustic
3. Monel–hydrofluoric acid
4. Hastelloys (Chlorimets)–hot hydrochloric acid
5. Lead–dilute sulfuric acid
6. Aluminum–nonstaining atmospheric exposure
7. Tin–distilled water
8. Titanium–hot strong oxidizing solutions
9. Tantalum–ultimate resistance
10. Steel–concentrated sulfuric acid

Source: M. G. Fontana and N. D. Greene, "Corrosion Engineering," 2d ed., McGraw-Hill, 1978.

sulted to make sure the proper material is used. Further consultation with corrosion experts of companies which produce the materials would also be helpful to ensure that the best choices are made.

There are, however, some general rules which are reasonably accurate and which can be applied when selecting corrosion-resistant metals and alloys for engineering applications. These are:[1]

1. For reducing or nonoxidizing conditions such as air-free acids and aqueous solutions, nickel and copper alloys are often used.
2. For oxidizing conditions, chromium-containing alloys are used.
3. For extremely powerful oxidizing conditions, titanium and its alloys are commonly used.

Some of the "natural" metal-corrosive environment combinations that give good corrosion resistance for the low cost are listed in Table 12.8.

One material that is often misused by fabricators not familiar with corrosion properties of metals is stainless steel. Stainless steel is not a specific alloy but is a generic term used for a large class of steels with chromium contents above about 12%. Stainless steels are commonly used for corrosive environments which are moderately oxidizing, i.e., nitric acid. However, stainless steels are less resistant to chloride-containing solutions and are more susceptible to stress-corrosion cracking than ordinary structural steel. Thus great care must be taken to make sure that stainless steels are not used in applications for which they are not suited.

Nonmetallic materials   *Polymeric materials* such as plastics and rubbers are weaker, softer, and in general less resistant to strong inorganic acids than metals and alloys, and thus their use as primary materials for corrosion resistance is limited. However, as newer higher-strength plastic materials become

[1]After M. G. Fontana and N. D. Greene, "Corrosion Engineering," 2d ed., McGraw-Hill, 1978.

available, they will become more important. *Ceramic materials* have excellent corrosion and high-temperature resistance but have the disadvantage of being brittle with low tensile strengths. Nonmetallic materials are therefore mainly used in corrosion control in the form of liners, gaskets, and coatings.

Coatings   Metallic, inorganic, and organic coatings are applied to metals to prevent or reduce corrosion.

Metallic coatings   Metallic coatings which differ from the metal to be protected are applied as thin coatings to separate the corrosive environment from the metal. Metal coatings are sometimes applied so that they can serve as sacrificial anodes which can corrode instead of the underlying metal. For example, the zinc coating on steel to make galvanized steel is anodic to the steel and corrodes sacrificially.

Many metal parts are protected by electroplating to produce a thin protective layer of metal. In this process the part to be plated is made the cathode of an electrolytic cell. The electrolyte is a solution of a salt of the metal to be plated, and direct current is applied to the part to be plated and another electrode. The plating of a thin layer of tin on steel sheet to produce tin plate for tin cans is an example of the application of this method. The plating can also have several layers, as is the case for the chrome plate used on automobiles. This plating consists of three layers: (1) an inner flashing of copper for adhesion of the plating to the steel, (2) an intermediate layer of nickel for good corrosion resistance, and (3) a thin layer of chromium primarily for appearance.

Sometimes a thin layer of metal is roll-bonded to the surfaces of the metal to be protected. The outer thin layer of metal provides corrosion resistance to the inner core metal. For example, some steels are "clad" with a thin layer of stainless steel. This cladding process is also used to provide some high-strength aluminum alloys with a corrosion-resistant outer layer. For these *Alclad* alloys as they are called, a thin layer of relatively pure aluminum is roll-bonded to the outer surface of the high-strength core alloy.

Inorganic coatings (ceramics and glass)   For some applications it is desirable to coat steel with a ceramic coating to attain a smooth durable finish. Steel is commonly coated with a porcelain coating which consists of a thin layer of glass fused to the steel surface so that it adheres well and has a coefficient of expansion adjusted to the base metal. Glass-lined steel vessels are used in some chemical industries because of their ease of cleaning and corrosion resistance.

Organic coatings   Paints, varnishes, lacquers, and many other organic polymeric materials are commonly used to protect metals from corrosive environments. These materials provide thin, tough, and durable barriers to protect the substrate metal from corrosive environments. On a weight basis, the use of organic coatings protects more metal from corrosion than any other method.

However, suitable coatings must be selected and be applied properly on well-prepared surfaces. In many cases poor performance of paints, for example, can be attributed to poor application and preparation of surfaces. Care must also be taken *not* to apply organic coatings for applications where the substrate metal could be rapidly attacked if the coating film cracks.

Design   The proper engineering design of equipment can be as important for corrosion prevention as the selection of the proper materials. The engineering designer must consider the materials along with the necessary mechanical, electrical, and thermal property requirements. All these considerations must be balanced with economic limitations. In designing a system, specific corrosion problems may require the advice of corrosion experts. However, some important general design rules are as follows:[1]

1. Make allowance for the penetration action of corrosion along with the mechanical strength requirements when considering the thickness of the metal used. This is especially important for pipes and tanks which contain liquids.
2. Weld rather than rivet containers to reduce crevice corrosion. If rivets are used, choose rivets which are cathodic to the materials being joined.
3. If possible, use galvanically similar metals for the whole structure. Avoid dissimilar metals which can cause galvanic corrosion. If galvanically dissimilar metals are bolted together, use nonmetallic gaskets and washers to prevent electrical contact between the metals.
4. Avoid excessive stress and stress concentrations in corrosive environments to prevent stress-corrosion cracking. This is especially important when using stainless steels, brasses, and other materials susceptible to stress-corrosion cracking in certain corrosive environments.
5. Avoid sharp bends in piping systems where flow occurs. Areas in which the fluid direction changes sharply promote erosion corrosion.
6. Design tanks and other containers for easy draining and cleaning. Stagnant pools of corrosive liquids set up concentration cells which promote corrosion.
7. Design systems for easy removal and replacement of parts that are expected to fail rapidly in service. For example, pumps in chemical plants should be easily removed.
8. Design heating systems so that hot spots do not occur. Heat exchangers, for example, should be designed for uniform temperature gradients.

In summary, design systems with conditions as uniform as possible and avoid heterogeneity.

Alteration of Environment   Environmental conditions can be very important in determining the severity of corrosion. Important methods for reducing corrosion by environmental changes

---

[1] After M. G. Fontana and N. D. Greene, "Corrosion Engineering," 2d ed., McGraw-Hill, 1978.

are (1) lowering the temperature, (2) decreasing the velocity of liquids, (3) removing oxygen from liquids, (4) reducing ion concentrations, and (5) adding inhibitors to electrolytes.

1. Lowering the temperature of a system usually reduces corrosion because of the lower reaction rates at lower temperatures. However, there are some exceptions in which the reverse is the case. For example, boiling seawater is less corrosive than hot seawater because of the decrease in oxygen solubility with increasing temperature.
2. Decreasing the velocity of a corrosive fluid reduces erosion corrosion. However, for metals and alloys that passivate, stagnant solutions should be avoided.
3. Removing oxygen from water solutions is sometimes helpful in reducing corrosion. For example, boiler-water feed is deaerated to reduce corrosion. However, for systems which depend on oxygen for passivation, deaeration is undesirable.
4. Reducing the concentration of corrosive ions in a solution which is corroding a metal can decrease the corrosion rate of the metal. For example, reducing the chloride ion concentration in a water solution will reduce its corrosive attack on stainless steels.
5. Adding *inhibitors* to a system can decrease corrosion. Inhibitors are essentially retarding catalysts. Most inhibitors have been developed by empirical experiment, and many are proprietary in nature. Their actions also vary considerably. For example, the *absorption-type* inhibitors are absorbed on a surface and form a protective film. The *scavenger-type* inhibitors react to remove corrosion agents such as oxygen from solution.

Cathodic and Anodic Protection

Cathodic protection   Corrosion control can be achieved by a method called *cathodic protection*[1] in which electrons are supplied to the metal structure to be protected. For example, the corrosion of a steel structure in an acid environment involves the following electrochemical equations:

$$Fe \rightarrow Fe^{2+} + 2e^-$$
$$2H^+ + 2e^- \rightarrow H_2$$

If electrons are supplied to the steel structure, metal dissolution (corrosion) will be suppressed and the rate of hydrogen evolution increased. Thus if electrons are continually supplied to the steel structure, corrosion will be suppressed. Electrons for cathodic protection can be supplied by (1) an external dc power supply, as shown in Fig. 12.33, or by (2) galvanic coupling with a more anodic metal than the one being protected. Cathodic protection of a steel

---

[1] For an interesting article on the application of cathodic protection for the preservation of submerged steel in the Arabian Gulf, see R. N. Duncan and G. A. Haines, Forty Years of Successful Cathodic Protection in the Arabian Gulf, *Mater. Perform.*, **21**:9(1982).

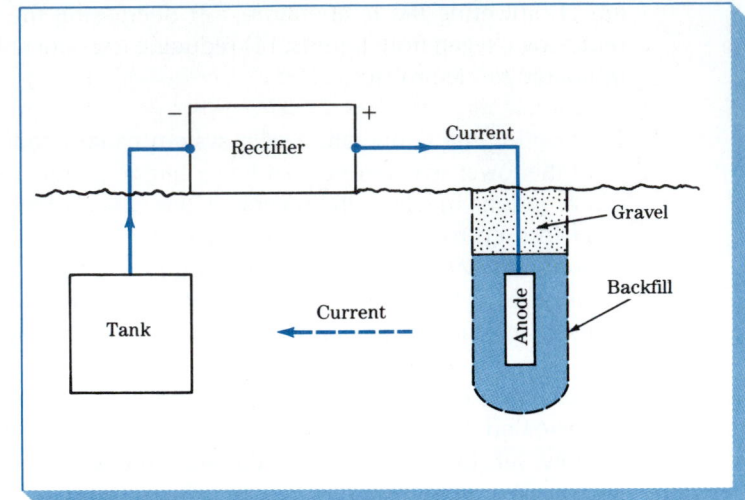

FIGURE 12.33 Cathodic protection of an underground tank by using impressed currents. (*After M. G. Fontana and N. D. Green, "Corrosion Engineering," 2d ed., McGraw-Hill, 1978, p. 207.*)

pipe by galvanic coupling to a magnesium anode is illustrated in Fig. 12.34. Magnesium anodes which corrode instead of the metal to be protected are most commonly used for cathodic protection because of their high negative potential and current density.

Anodic protection   Anodic protection is relatively new and is based on the formation of protective passive films on metal and alloy surfaces by externally impressed anodic currents. Carefully controlled anodic currents by a device called a *potentiostat* can be applied to protect metals that passivate, such as austenitic stainless steels, to make them passive and hence lower their cor-

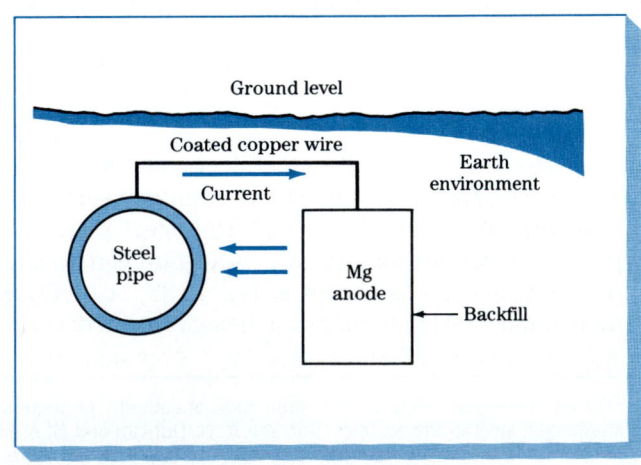

FIGURE 12.34 Protection of an underground pipeline with a magnesium anode. (*After M. G. Fontana and N. D. Greene, "Corrosion Engineering," 2d ed., McGraw-Hill, 1978, p. 207.*)

rosion rate in a corrosive environment.[1] Advantages of anodic protection are that it can be applied in weak to very corrosive conditions and uses very small applied currents. A disadvantage of anodic protection is that it requires complex instrumentation and has a high installation cost.

Example Problem 12.11

A 2.2-kg sacrificial magnesium anode is attached to the steel hull of a ship. If the anode completely corrodes in 100 days, what is the average current produced by the anode in this period?

Solution:

Magnesium corrodes according to the reaction $Mg \rightarrow Mg^{2+} + 2e^-$. We will use Eq. (12.13) and solve for I, the average corrosion current in amperes:

$$w = \frac{ItM}{nF} \quad \text{or} \quad I = \frac{wnF}{tM}$$

$$w = 2.2 \text{ kg}\left(\frac{1000 \text{ g}}{\text{kg}}\right) = 2200 \text{ g} \quad n = 2 \quad F = 96{,}500 \text{ A·s/mol}$$

$$t = 100 \text{ days}\left(\frac{24 \text{ h}}{\text{day}}\right)\left(\frac{3600 \text{ s}}{\text{h}}\right) = 8.64 \times 10^6 \text{ s} \quad M = 24.31 \text{ g/mol} \quad I = ? \text{ A}$$

$$I = \frac{(2200 \text{ g})(2)(96{,}500 \text{ A·s/mol})}{(8.64 \times 10^6 \text{ s})(24.31 \text{ g/mol})} = 2.02 \text{ A} \blacktriangleleft$$

SUMMARY

Corrosion may be defined as the deterioration of a material resulting from chemical attack by its environment. Most corrosion of materials involves the chemical attack of metals by electrochemical cells. By studying equilibrium conditions, the tendencies of pure metal to corrode in a standard aqueous environment can be related to the standard electrode potentials of the metals. However, since corroding systems are not at equilibrium, the kinetics of cor-

---

[1] S. J. Acello and N. D. Greene, *Corrosion*, **18**:286(1962).

rosion reactions must also be studied. Some examples of kinetic factors affecting corrosion reaction rates are the polarization of the corrosion reactions and the formation of passive films on the metals.

There are many types of corrosion. Some of the important types discussed are uniform or general attack corrosion, galvanic or two-metal corrosion, pitting corrosion, crevice corrosion, intergranular corrosion, stress corrosion, erosion corrosion, cavitation damage, fretting corrosion, and selective leaching or dealloying.

The oxidation of metals and alloys is also important for some engineering designs such as gas turbines, rocket engines, and high-temperature petrochemical installations. The study of the rates of oxidation of metals for some applications is very important. At high temperatures care must be taken to avoid catastrophic oxidation.

Corrosion can be controlled or prevented by many different methods. To avoid corrosion, materials that are corrosion-resistant for a particular environment should be used where feasible. For many cases corrosion can be prevented by the use of metallic, inorganic, or organic coatings. The proper engineering design of equipment can also be very important for many situations. For some special cases, corrosion can be controlled by using cathodic or anodic protection systems.

DEFINITIONS

Sec. 12.1

Corrosion: the deterioration of a material resulting from chemical attack by its environment.

Sec. 12.2

Anode: the metal electrode in an electrolytic cell which dissolves as ions and supplies electrons to the external circuit.

Cathode: the metal electrode in an electrolytic cell which accepts electrons.

Electromotive force series: an arrangement of metallic elements according to their standard electrochemical potentials.

Sec. 12.3

Galvanic cell: two dissimilar metals in electrical contact with an electrolyte.

Ion-concentration cell: galvanic cell formed when two pieces of the same metal are electrically connected by an electrolyte but are in solutions of different ion concentrations.

Oxygen-concentration cell: galvanic cell formed when two pieces of the same metal are electrically connected by an electrolyte but are in solutions of different oxygen concentration.

Sec. 12.4

Cathodic polarization: the slowing down or the stopping of cathodic reactions at a cathode of an electrochemical cell due to (1) a slow step in the reaction sequence at the metal-electrolyte interface (*activation polarization*) or (2) a shortage of reactant or accumulation of reaction products at the metal-electrolyte interface (*concentration polarization*).

Passivation: the formation of a film of atoms or molecules on the surface of an anode so that corrosion is slowed down or stopped.

Galvanic (seawater) series: an arrangement of metallic elements according to their electrochemical potentials in seawater with reference to a standard electrode.

Sec. 12.5

Pitting corrosion: local corrosion attack resulting from the formation of small anodes on a metal surface.

Intergranular corrosion: preferential corrosion occurring at grain boundaries or at regions adjacent to the grain boundaries.

Stress corrosion: preferential corrosive attack of a metal under stress in a corrosive environment.

Weld decay: corrosion attack at or adjacent to a weld as the result of galvanic action resulting from structural differences in the weld.

Selective leaching: the preferential removal of one element of a solid alloy by corrosion processes.

Sec. 12.6

Pilling-Bedworth (P.B.) ratio: the ratio of the volume of oxide formed to the volume of metal consumed by oxidation.

Sec. 12.7

Cathodic protection: the protection of a metal by connecting it to a sacrificial anode or by impressing a dc voltage to make it a cathode.

Anodic protection: the protection of a metal which forms a passive film by the application of an externally impressed anodic current.

PROBLEMS

12.1.1 Define corrosion as it pertains to materials.

12.1.2 What are some of the factors that affect the corrosion of metals?

12.1.3 Which is in a lower energy state: (*a*) elemental iron or (*b*) Fe_2O_3 (iron oxide)?

12.1.4 Give several examples of environmental deterioration of (*a*) ceramic materials and (*b*) polymeric materials.

12.2.1 What is the oxidation reaction called in which metals form ions that go into aqueous solution in an electrochemical corrosion reaction? What types of ions are produced by this reaction? Write the oxidation half-cell reaction for the oxidation of pure zinc metal in aqueous solution.

12.2.2   What is the reduction reaction called in which a metal or nonmetal is reduced in valence charge in an electrochemical corrosion reaction? Are electrons produced or consumed by this reaction?

12.2.3   What is a standard half-cell oxidation-reduction potential?

12.2.4   Describe a method used to determine the standard half-cell oxidation-reduction potential of a metal by using a hydrogen half-cell.

12.2.5   List five metals which are cathodic to hydrogen and give their standard oxidation potentials. List five metals which are anodic to hydrogen and give their standard oxidation potentials.

12.3.1   Consider a magnesium-iron galvanic cell consisting of a magnesium electrode in a solution of 1 M $MgSO_4$ and an iron electrode in a solution of 1 M $FeSO_4$. Each electrode and its electrolyte are separated by a porous wall, and the whole cell is at 25°C. Both electrodes are connected with a copper wire.
   (a) Which electrode is the anode?
   (b) Which electrode corrodes?
   (c) In which direction will the electrons flow?
   (d) In which direction will the anions in the solutions move?
   (e) In which direction will the cations in the solutions move?
   (f) Write the equation for the half-cell reaction at the anode.
   (g) Write the equation for the half-cell reaction at the cathode.

.627

12.3.2   A standard galvanic cell has electrodes of zinc and tin. Which electrode is the anode? Which electrode corrodes? What is the emf of the cell?

12.3.3   A standard galvanic cell has electrodes of iron and lead. Which electrode is the anode? Which electrode corrodes? What is the emf of the cell?

.403

12.3.4   The emf of a standard Ni–Cd galvanic cell is -0.153 V. If the standard half-cell emf for the oxidation of Ni is -0.250 V, what is the standard half-cell emf of cadmium if cadmium is the anode?

12.3.5   What is the emf with respect to the standard hydrogen electrode of a cadmium electrode which is immersed in an electrolyte of 0.04 M $CdCl_2$? Assume the cadmium half-cell reaction to be $Cd = Cd^{2+} + 2e^-$.

12.3.6   A galvanic cell consists of an electrode of nickel in a 0.04 M solution of $NiSO_4$ and an electrode of copper in a 0.08 M solution of $CuSO_4$ at 25°C. The two electrodes are separated by a porous wall. What is the emf of the cell?

12.3.7   A galvanic cell consists of an electrode of zinc in a 0.03 M solution of $ZnSO_4$ and an electrode of copper in a solution of 0.06 M $CuSO_4$ at 25°C. What is the emf of the cell?

12.3.8   A nickel electrode is immersed in a solution of $NiSO_4$ at 25°C. What must the molarity of the solution be if the electrode shows a potential of -0.2842 V with respect to a standard hydrogen electrode?

12.3.9   A copper electrode is immersed in a solution of $CuSO_4$ at 25°C. What must the molarity of the solution be if the electrode shows a potential of $+0.2985$ V with respect to a standard hydrogen electrode?

12.3.10 One end of a zinc wire is immersed in an electrolyte of 0.07 M Zn^{2+} ions and the other in one of 0.002 M Zn^{2+} ions, with the two electrolytes being separated by a porous wall.
   (a) Which end of the wire will corrode?
   (b) What will be the potential difference between the two ends of the wire when it is just immersed in the electrolytes?

12.3.11 Magnesium (Mg^{2+}) concentrations of 0.04 M and 0.007 M occur in an electrolyte at opposite ends of a magnesium wire at 25°C.
(a) Which end of the wire will corrode?
(b) What will the potential difference between the ends of the wire be?

12.3.12 Consider an oxygen-concentration cell consisting of two zinc electrodes. One is immersed in a water solution with low oxygen concentration and the other in a water solution with a high oxygen concentration. The zinc electrodes are connected by an external copper wire.
(a) Which electrode will corrode?
(b) Write half-cell reactions for the anodic reaction and the cathodic reaction.

12.3.13 In metals, which region is more chemically reactive (anodic), the grain matrix or the grain-boundary regions? why?

12.3.14 Consider a 0.95% carbon steel. In which condition is the steel more corrosion-resistant: (a) martensitic or (b) tempered martensitic with ϵ carbide and Fe_3C formed in the 200 to 500°C range? Explain.

12.3.15 Why are pure metals in general more corrosion-resistant than impure ones?

12.4.1 An electroplating process uses 15 A of current by chemically corroding (dissolving) a copper anode. What is the corrosion rate of the anode in grams per hour?

12.4.2 A cadmium electroplating process uses 10 A of current and chemically corrodes a cadmium anode. How long will it take to corrode 8.2 g of cadmium from the anode?

12.4.3 A mild steel tank 60 cm high with a 30 cm × 30 cm square bottom is filled with aerated water up to the 45-cm level and shows a corrosion loss of 350 g over a 4-week period. Calculate (a) the corrosion current and (b) the corrosion density associated with the corrosion current. Assume the corrosion is uniform over all surfaces and that the mild steel corrodes in the same way as pure iron.

12.4.4 A cylindrical steel tank is coated with a thick layer of zinc on the inside. The tank is 50 cm in diameter, 70 cm high, and filled to the 45-cm level with aerated water. If the corrosion current is 5.8×10^{-5} A/cm², how much zinc in grams per minute is being corroded?

12.4.5 A heated mild steel tank containing water is corroding at a rate of 90 mdd. If the corrosion is uniform, how long will it take to corrode the wall of the tank by 0.40 mm?

12.4.6 A mild steel tank contains a solution of ammonium nitrate and is corroding at the rate of 6000 mdd. If the corrosion on the inside surface is uniform, how long will it take to corrode the wall of the tank by 1.05 mm?

12.4.7 A tin surface is corroding uniformly at a rate of 2.40 mdd. What is the associated current density for this corrosion rate?

12.4.8 A copper surface is corroding in seawater at a current density of 2.30×10^{-6} A/cm². What is the corrosion rate in mdd?

12.4.9 If a zinc surface is corroding at a current density of 3.45×10^{-7} A/cm², what thickness of metal will be corroded in 210 days?

12.4.10 A galvanized (zinc-coated) steel sheet is found to uniformly corrode at the rate of 12.5×10^{-3} mm/year. What is the average current density associated with the corrosion of this material?

12.4.11 A galvanized (zinc-coated) steel sheet is found to uniformly corrode with an average current density of 1.32×10^{-7} A/cm². How many years will it take to uniformly corrode a thickness of 0.030 mm of the zinc coating?

12.4.12 A new aluminum container develops pits right through its walls in 350 days by pitting corrosion. If the average pit is 0.170 mm in diameter and the container wall is 1.00 mm thick, what is the average current associated with the formation of a single pit? What is the current density for this corrosion using the surface area of the pit for this calculation? Assume the pits have a cylindrical shape.

12.4.13 A new aluminum container develops pits right through its walls with an average current density of 1.30×10^{-4} A/cm^2. If the average pit is 0.70 mm in diameter and the aluminum wall is 0.90 mm thick, how many days will it take for a pit to corrode through the wall? Assume the pit has a cylindrical shape and that the corrosion current acts uniformly over the surface area of the pit.

12.4.14 What is an exchange current density? What is the corrosion current i_{corr}?

12.4.15 Define and give an example of (a) activation polarization and (b) concentration polarization.

12.4.16 Define the passivation of a metal or alloy. Give examples of some metals and alloys which show passivity.

12.4.17 Briefly describe the following theories of metal passivity: (a) the oxide theory and (b) the adsorption theory.

12.4.18 Draw a polarization curve for a passive metal and indicate on it (a) the primary passive voltage E_{pp} and (b) the passive current i_p.

12.4.19 Describe the corrosion behavior of a passive metal in (a) the active region, (b) the passive region, and (c) the transpassive region of a polarization curve. Explain the reasons for the different behavior in each region.

12.5.1 Explain the electrochemical behavior on the exterior and interior of tin plate used as a food container.

12.5.2 Explain the difference in corrosion behavior of (a) a large cathode and a small anode, and (b) a large anode and a small cathode. Which of the two conditions is more favorable from a corrosion-prevention standpoint and why?

12.5.3 What is pitting corrosion? Where are pits usually initiated? Describe an electrochemical mechanism for the growth of a pit in a stainless steel immersed in an aerated sodium chloride solution.

12.5.4 From an engineering standpoint what metals would be used where pitting resistance is important?

12.5.5 What is crevice corrosion? Describe an electrochemical mechanism for the crevice corrosion of a stainless steel in an aerated sodium chloride solution.

12.5.6 From an engineering design standpoint what should be done to prevent or minimize crevice corrosion?

12.5.7 What is intergranular corrosion? Describe the metallurgical condition that can lead to intergranular corrosion in an austenitic stainless steel.

12.5.8 For an austenitic stainless steel, distinguish between (a) the sensitized condition and (b) the stabilized condition.

12.5.9 Describe three methods of avoiding intergranular corrosion in austenitic stainless steels.

12.5.10 What is stress-corrosion cracking (SCC)? Describe a mechanism of SCC.

12.5.11 From an engineering design standpoint what can be done to avoid or minimize stress-corrosion cracking?

12.5.12 What is erosion corrosion? What is cavitation damage?

12.5.13 Describe fretting corrosion.

12.5.14 What is selective leaching of an alloy? Which types of alloys are especially susceptible to this kind of corrosion?

12.5.15 Describe a mechanism for the dezincification of a 70–30 brass.

12.6.1 What factors are important if a metal is to form a protective oxide?

12.6.2 Calculate the oxide-to-metal volume (Pilling-Bedworth) ratios for the oxidation of the metals listed in the following table and comment on whether their oxides might be protective or not.

Metal	Oxide	Density of metal, g/cm³	Density of oxide, g/cm³
Tungsten, W	WO_3	19.35	12.11
Sodium, Na	Na_2O	0.967	2.27
Hafnium, Hf	HfO_2	13.31	9.68
Copper, Cu	CuO	8.92	6.43
Manganese, Mn	MnO	7.20	5.46
Tin, Sn	SnO	6.56	6.45

(handwritten annotations: 183.85 Tungsten, 22.99 Sodium, 178.49 Hafnium, 63.54 Copper, 54.94 Manganese, 118.69 Tin)

12.6.3 Describe the anion and cation diffusion mechanisms of oxide formation on metals.

12.6.4 Using equations, describe the following oxidation of metals behavior: (a) linear, (b) parabolic, (c) logarithmic. Give examples.

12.6.5 A 1-cm², 0.75-mm-thick sample of 99.9 wt % Ni is oxidized in oxygen at 1 atm pressure at 500°C. After 7 h the sample shows a 60 μg/cm² weight gain. If the oxidation process follows parabolic behavior, what will the weight gain be after 20 h of oxidation?

2.6.6 A sample of pure iron oxidizes according to the linear oxidation rate law. After 3 h at 720°C, a 1-cm² sample shows a weight gain of 7 μg/cm². How long an oxidation time will it take for the sample to show a weight gain of 55 μg/cm²?

12.6.7 What is catastrophic oxidation? What metals are subject to this behavior? What metals when added to iron alloys retard this behavior?

12.7.1 What types of alloys are used for moderate oxidizing conditions for corrosion resistance?

12.7.2 What types of alloys are used for corrosion resistance for highly oxidizing conditions?

12.7.3 List six combinations of metals and environments which have good corrosion resistance.

12.7.4 List some of the applications of nonmetallic materials for corrosion control.

12.7.5 What is the function of each of the three layers of "chrome plate"?

12.7.6 What are Alclad alloys?

12.7.7 Describe eight engineering design rules which may be important to reduce or prevent corrosion.

12.7.8 Describe four methods of altering the environment to prevent or reduce corrosion.

12.7.9 Describe two methods by which cathodic protection can be used to protect a steel pipe from corroding.

12.7.10 If a sacrificial zinc anode shows a 1.05-kg corrosion loss in 55 days, what is the average current produced by the corrosion process in this period?

12.7.11 If a sacrificial magnesium anode corrodes with an average current of 0.80 A for 100 days, what must the loss of metal by the anode in this time period be?

12.7.12 What is anodic protection? For what metals and alloys can it be used? What are some of its advantages and disadvantages?

Composite Materials

13.1 INTRODUCTION

What is a composite material? Unfortunately there is no widely accepted definition of what a composite material is. A dictionary defines a composite as something made up of distinct parts (or constituents). At the atomic level materials such as some metal alloys and polymeric materials could be called composite materials since they consist of different and distinct atomic groupings. At the microstructural level (about 10^{-4} to 10^{-2} cm) a metal alloy such as a plain-carbon steel containing ferrite and pearlite could be called a composite material since the ferrite and pearlite are distinctly visible constituents as observed in the optical microscope. At the macrostructural level (about 10^{-2} cm or greater) a glass-fiber-reinforced plastic, in which the glass fibers can be distinctly recognized by the naked eye, could be considered a composite material. Now we see that the difficulty in defining a composite material is the size limitations we impose on the constituents which make up the material. In engineering design a composite material usually refers to a material consisting of constituents in the micro-macrosize range, and even favors the macrosize range. For the purposes of this book, the following is a definition for a composite material:

A *composite material* is a materials system composed of a mixture or combination of two or more micro- or macroconstituents that differ in form and chemical composition and which are essentially insoluble in each other.

The engineering importance of a composite material is that two or more distinctly different materials combine together to form a composite material which possesses properties that are superior, or important in some other manner, to the properties of the individual components. A multitude of materials fit into this category, and thus a discussion of all these composite materials is far beyond the scope of this book. In this chapter only some of the most important composite materials used in engineering will be discussed. These are fiber-reinforced plastics, concrete, asphalt, wood, and several miscellaneous types of composite materials. Examples of the usage of composite materials in engineering designs are shown in Fig. 13.1.

13.2   FIBERS FOR REINFORCED-PLASTIC COMPOSITE MATERIALS

Three main types of synthetic fibers are used in the United States to reinforce plastic materials: glass, aramid,[1] and carbon. Glass is by far the most widely used reinforcement fiber and is the lowest in cost. Aramid and carbon fibers have high strengths and low densities and so are used in many applications, particularly aerospace, in spite of their higher cost.

Glass Fibers for Reinforcing Plastic Resins

Glass fibers are used to reinforce plastic matrices to form structural composites and molding compounds. Glass-fiber plastic composite materials have the following favorable characteristics: high strength-to-weight ratio; good dimensional stability; good resistance to heat, cold, moisture, and corrosion; good electrical insulation properties; ease of fabrication; and relatively low cost.

The two most important types of glass used to produce glass fibers for composites are *E* (*electrical*) and *S* (*high-strength*) glasses.

E glass is the most commonly used glass for continuous fibers. Basically, E glass is a lime-aluminum-borosilicate glass with zero or low sodium and potassium levels. The basic composition of E glass ranges from 52–56% SiO_2, 12–16% Al_2O_3, 16–25% CaO, and 8–13% B_2O_3. E glass has a tensile strength of about 500 ksi (3.44 GPa) in the virgin condition and a modulus of elasticity of 10.5 Msi (72.3 GPa).

S glass has a higher strength-to-weight ratio and is more expensive than E glass and is used primarily for military and aerospace applications. The

---

[1]Aramid fiber is an aromatic polyamide polymer fiber with a very rigid molecular structure.

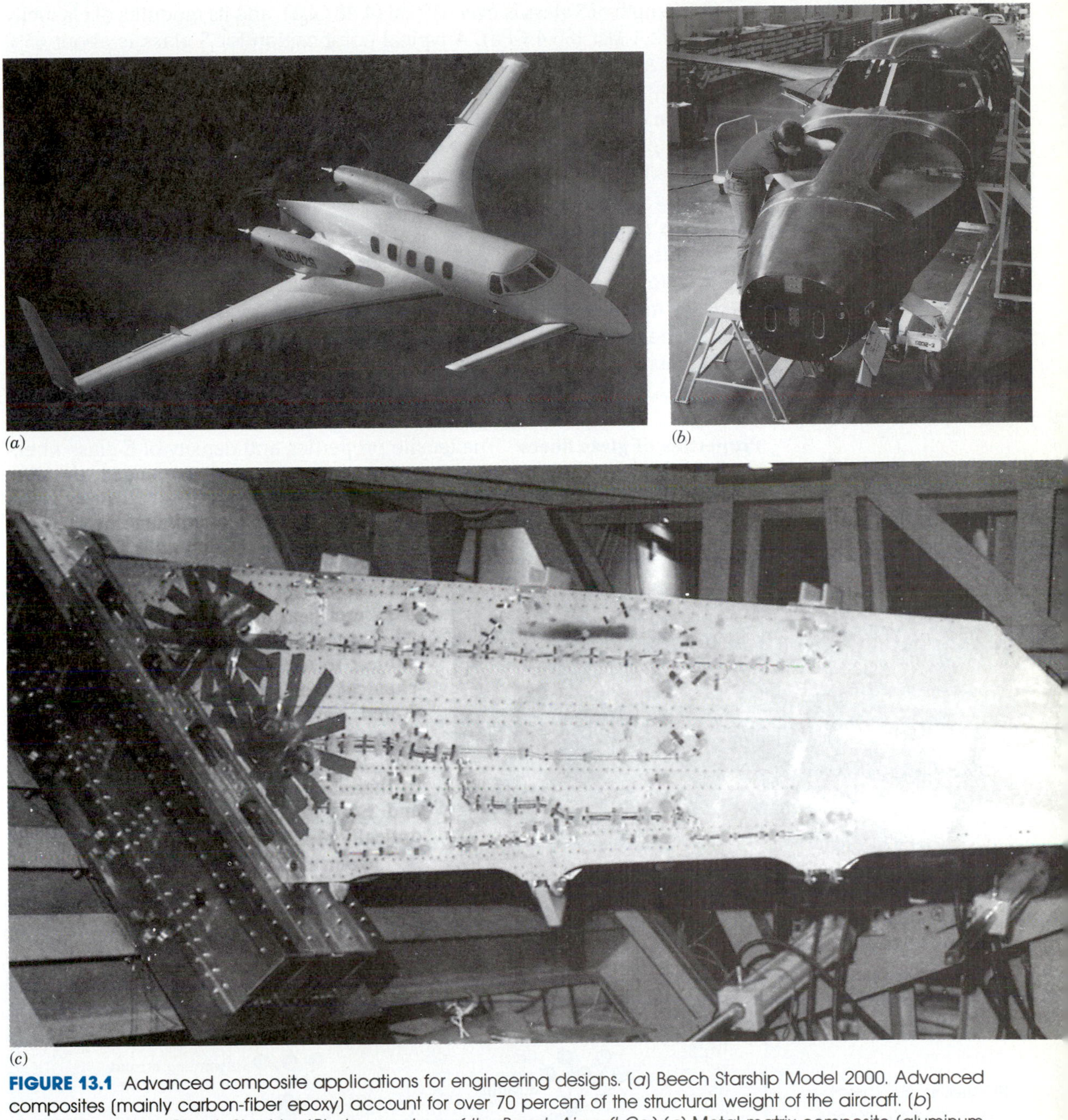

(a)

(b)

(c)

FIGURE 13.1 Advanced composite applications for engineering designs. (*a*) Beech Starship Model 2000. Advanced composites (mainly carbon-fiber epoxy) account for over 70 percent of the structural weight of the aircraft. (*b*) Construction of the Beech Starship. (*Photos courtesy of the Beech Aircraft Co.*) (*c*) Metal-matrix composite (aluminum alloy with silicon carbide reinforcement) being tested by Lockheed for the vertical tail section of an advanced tactical aircraft. (*After Aerospace America, March 1989, p. 27.*)

tensile strength of S glass is over 650 ksi (4.48 GPa), and its modulus of elasticity is about 12.4 Msi (85.4 GPa). A typical composition for S glass is about 65% SiO_2, 25% Al_2O_3, and 10% MgO.

Production of glass fibers and types of fiberglass reinforcing materials   Glass fibers are produced by drawing monofilaments of glass from a furnace containing molten glass and gathering a large number of these filaments to form a strand of glass fibers (Fig. 13.2). The strands then are used to make glass-fiber yarns (Fig. 13.3A) or rovings which consist of a collection of bundles of continuous filaments (Fig. 13.3C). The rovings may be in continuous strands or woven to make woven roving (Fig. 13.3D). Glass-fiber reinforcing mats (Fig. 13.4) are made of continuous strands (Fig. 13.4A) or chopped strands (Fig. 13.4C). The strands are usually held together with a resinous binder. Combination mats are made with woven roving chemically bonded to chopped-strand mat (Fig. 13.4D).

Properties of glass fibers   The tensile properties and density of E-glass fibers are compared with those properties of carbon and aramid fibers in Table 13.1.

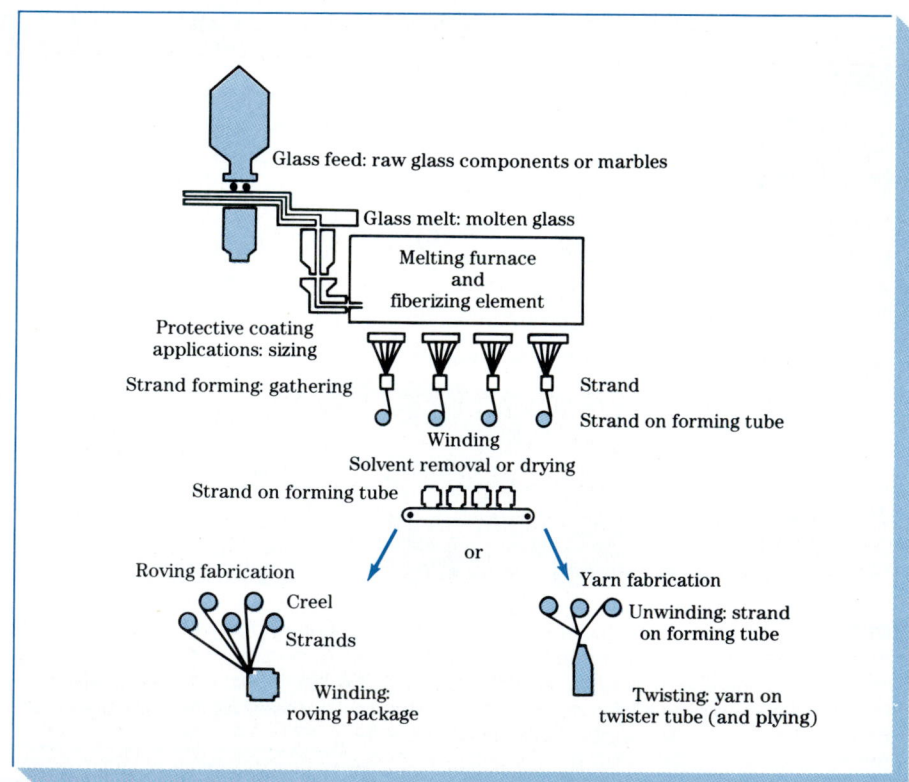

FIGURE 13.2 Glass-fiber manufacturing process. (After M. M. Schwartz, "Composite Materials Handbook," McGraw-Hill, 1984, p. 2-24.)

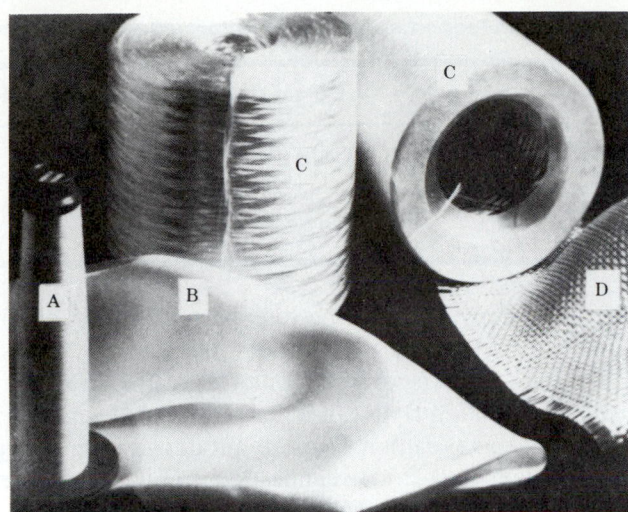

FIGURE 13.3 Fiberglass reinforcements for plastics: (*A*) fiberglass yarn, (*B*) woven fabric of fiberglass yarn, (*C*) continuous-strand roving, and (*D*) woven roving. (*Courtesy of Owens/Corning Fiberglas Co.*)

It is noted that glass fibers have a lower tensile strength and modulus than carbon and aramid fibers but a higher elongation. The density of glass fibers is also higher than carbon and aramid fibers. However, because of their low cost and versatility, glass fibers are by far the most commonly used reinforcing fibers for plastics (Table 13.1).

Carbon Fibers for Reinforced Plastics

Composite materials made by using carbon fibers for reinforcing plastic resin matrices such as epoxy are characterized by having a combination of light weight, very high strength, and high stiffness (modulus of elasticity). These properties make the use of carbon-fiber-plastic composite materials especially attractive for aerospace applications, such as the aircraft shown in Fig. 13.1.

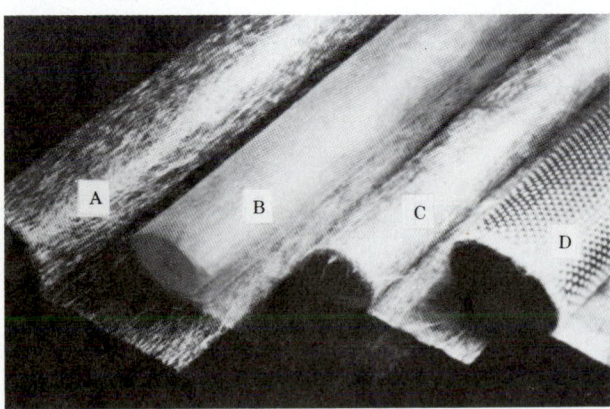

FIGURE 13.4 Fiberglass reinforcing mats: (*A*) continuous-strand mat, (*B*) surfacing mat, (*C*) chopped-strand mat, and (*D*) combination of woven roving and chopped-strand mat. (*Courtesy of Owens/Corning Fiberglas Co.*)

TABLE 13.1   Comparative Yarn Properties for Fiber Reinforcements for Plastics

Property	E glass (HTS)	Carbon (type HT)	Aramid (Kevlar 49)
Tensile strength, ksi (MPa)	350 (2410)	450 (3100)	525 (3617)
Tensile modulus, Msi (GPa)	10 (69)	32 (220)	18 (124)
Elongation to break, %	3.5	1.40	2.5
Density, g/cm³	2.54	1.75	1.48

Unfortunately, the relatively high cost of the carbon fibers restricts their use in many industries such as the auto industry.

Carbon fibers for these composites are produced mainly from two sources, polyacrylonitrile (PAN) (see Sec. 7.8) and pitch, which are called *precursors*.

In general carbon fibers are produced from PAN precursor fibers by three processing stages: (1) stabilization, (2) carbonization, and (3) graphitization (Fig. 13.5). In the *stabilization stage* the PAN fibers are first stretched to align the fibrillar networks within each fiber parallel to the fiber axis, and then they are oxidized in air at about 200 to 220°C (392 to 428°F) while held in tension.

The *second stage* in the production of high-strength carbon fibers is *carbonization*. In this process the stabilized PAN-based fibers are pyrolyzed (heated) until they become transformed into carbon fibers by the elimination of O, H, and N from the precursor fiber. The carbonization heat treatment is usually carried out in an inert atmosphere in the 1000 to 1500°C (1832–2732°F) range. During the carbonization process turbostratic graphitelike fibrils or ribbons are formed within each fiber which greatly increase the tensile strength of the material.

A *third stage*, or *graphitization treatment*, is used if an increase in the modulus of elasticity is desired at the expense of high tensile strength. During graphitization, which is carried out above about 1800°C (3272°F), the preferred orientation of the graphitelike crystallites within each fiber is increased.

The carbon fibers produced from PAN precursor material have a tensile strength ranging from about 340 to 460 ksi (2.34 to 3.17 GPa) and a tensile modulus of elasticity ranging from about 28 to 60 Msi (193 to 413 GPa). As can be seen

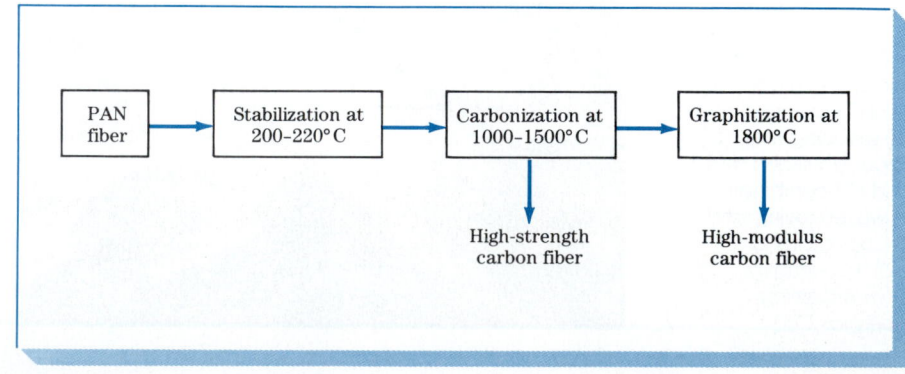

FIGURE 13.5 Processing steps for producing high-strength, high-modulus carbon fibers from polyacrylonitrile (PAN) precursor material.

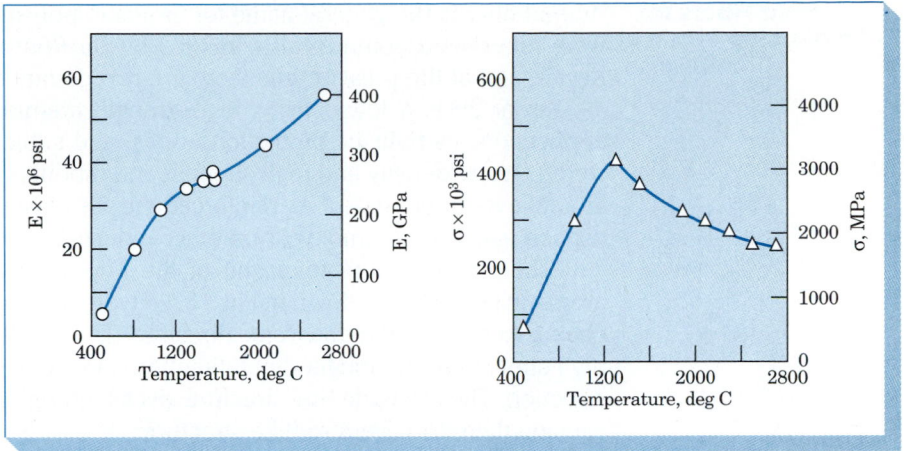

FIGURE 13.5A Effect of heat treatment temperature on (i) tensile modulus of elasticity and (ii) tensile strength of polyacrylonitrile- (PAN-) based carbon fibers. (*After J. W. Johnson, Applied Polymer Symposia No. 9, Interscience Publishers, 1969, p. 229.*)

in Fig. 13.5A(i), the tensile modulus of elasticity of PAN carbon fibers continuously increases as the pyrolysis temperature increases up to about 60 Msi (413 GPa). To achieve high tensile moduli of elasticity, the carbon fibers must possess a high degree of axial preferred orientation of the graphite basal planes (see Fig. 10.17) parallel to the fiber axis. The tensile strength of PAN carbon fibers, on the other hand, increases with increasing pyrolysis temperature up to about 1200°C and then decreases less rapidly as the temperature is further increased (Fig. 13.5A(ii)]. The decrease in tensile strength at the higher pyrolysis temperatures is believed to be caused by a greater presence of discrete flaws both in the volume of the material and on the surface.

The density of the carbonized and graphitized PAN fibers is usually about 1.7 to 2.1 g/cm^3, while their final diameter is about 7 to 10 μm. Figure 13.6 shows a photograph of a group of about 6000 carbon fibers called a *tow*.

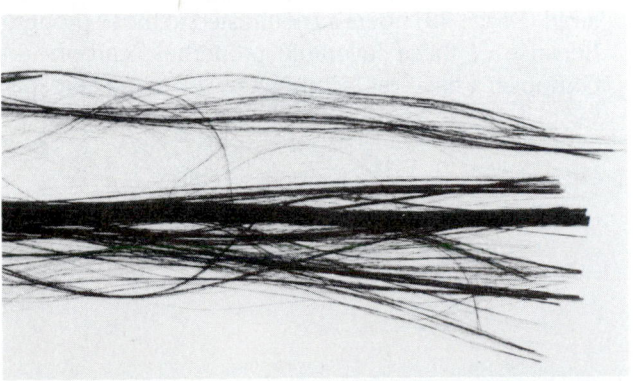

FIGURE 13.6 Photograph of a tow of about 6000 carbon fibers. (*Courtesy of the Fiberite Co., Winona, Minn.*)

Aramid Fibers for Reinforcing Plastic Resins

Aramid fiber is the generic name for aromatic polyamide fibers. Aramid fibers were introduced commercially in 1972 by Du Pont under the trade name of Kevlar, and at the present time there are two commercial types: Kevlar 29 and 49. Kevlar 29 is a low-density, high-strength aramid fiber designed for such applications as ballistic protection, ropes, and cables. Kevlar 49 is characterized by a low density and high strength and modulus. The properties of Kevlar 49 make its fibers useful as reinforcement for plastics in composites for aerospace, marine, automotive, and other industrial applications.

The chemical repeating unit of the Kevlar polymer chain is that of an aromatic polyamide, shown in Fig. 13.7. Hydrogen bonding bonds the polymer chains together in the transverse direction. Thus, collectively these fibers have high strength in the longitudinal direction and weak strength in the transverse direction. The aromatic ring structure gives high rigidity to the polymer chains, causing them to have a rodlike structure.

Kevlar aramid is used for high-performance composite applications where light weight, high strength and stiffness, damage resistance, and resistance to fatigue and stress rupture are important. Of special interest is that Kevlar-epoxy material is used for various parts of the space shuttle.

Comparison of Mechanical Properties of Carbon, Aramid, and Glass Fibers for Reinforced-Plastic Composite Materials

Figure 13.8 compares typical stress-strain diagrams for carbon, aramid, and glass fibers, and one can see that the fiber strength varies from about 250 to 500 ksi (1720 to 3440 MPa), while the fracture strain ranges from 0.4 to 4.0 percent. The tensile modulus of elasticity of these fibers ranges from 10×10^6 to 60×10^6 psi (68.9 to 413 GPa). The carbon fibers provide the best combination of high strength, high stiffness (high modulus), and low density, but have lower elongations. The aramid fiber Kevlar 49 has a combination of high strength, high modulus (but not as high as the carbon fibers), low density, and high elongation (impact resistance). The glass fibers have lower strengths and moduli and higher densities (Table 13.1). Of the glass fibers, the S-glass fibers have higher strengths and elongations than the E-glass fibers. Because the glass fibers are much less costly, they are more widely used.

Figure 13.9 compares the strength to density and the stiffness (tensile modulus) to density of various reinforcing fibers. This comparison shows the outstanding strength-to-weight and stiffness-to-weight ratios of carbon and aramid (Kevlar 49) fibers as contrasted to those properties of steel and aluminum. Because of these favorable properties, carbon- and aramid-reinforced-fiber composites have replaced metals for many aerospace applications.

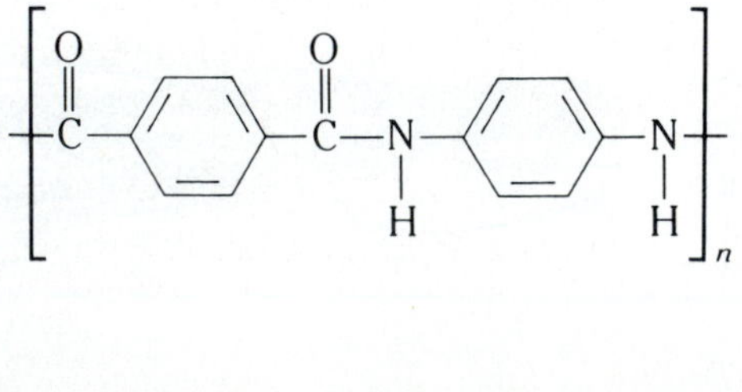

FIGURE 13.7 Repeating chemical structural unit for Kevlar fibers.

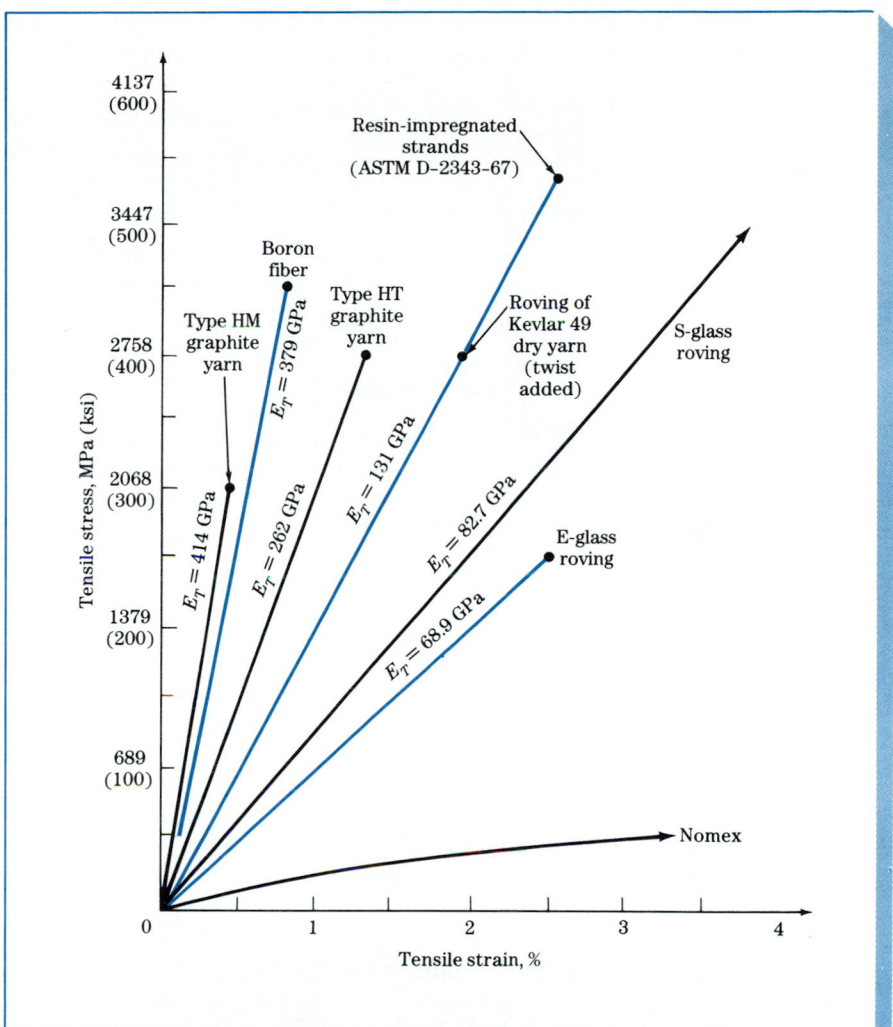

FIGURE 13.8 Stress-strain behavior of various types of reinforcing fibers. (*After "Kevlar 49 Data Manual,"* E. I. du Pont de Nemours & Co., 1974.)

13.3   FIBER-REINFORCED-PLASTIC COMPOSITE MATERIALS

Matrix Materials for Fiber-Reinforced-Plastic Composite Materials

Two of the most important matrix plastic resins for fiber-reinforced plastics are unsaturated polyester and epoxy resins. The chemical reactions responsible for the cross-linking of these thermosetting resins have already been described in Sec. 7.9.

Some of the properties of unfilled cast rigid polyester and epoxy resins are listed in Table 13.2. The polyester resins are lower in cost but are usually

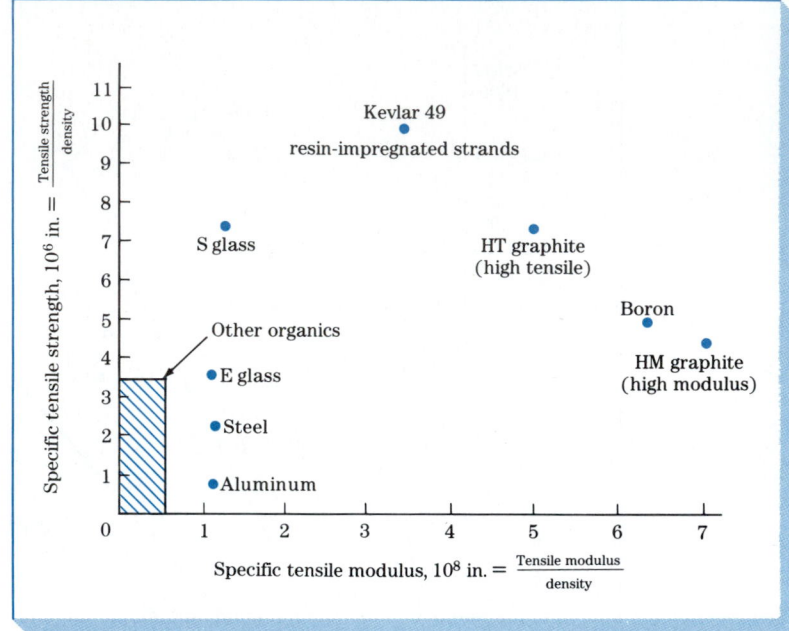

FIGURE 13.9 Specific tensile strength (tensile strength to density) and specific tensile modulus (tensile modulus to density) for various types of reinforcing fibers. (*Courtesy of E. I. du Pont de Nemours & Co., Wilmington, Del.*)

not as strong as the epoxy resins. Unsaturated polyesters are widely used for matrices of fiber-reinforced plastics. Applications for these materials include boat hulls, building panels, and structural panels for automobiles, aircraft, and appliances. Epoxy resins cost more but have special advantages such as good strength properties and lower shrinkage after curing than polyester resins. Epoxy resins are commonly used as matrix materials for carbon- and aramid-fiber composites.

Fiber-Reinforced-Plastic Composite Materials

Fiberglass-reinforced polyester resins The strength of fiberglass-reinforced plastics is mainly related to the glass content of the material and the arrangement of the glass fibers. In general, the higher the weight percent glass in the composite, the stronger the reinforced plastic is. When there are parallel strands of glass, as may be the case for filament winding, the fiberglass content may

TABLE 13.2   **Some Properties of Unfilled Cast Polyester and Epoxy Resins**

	Polyester	Epoxy
Tensile strength, ksi (MPa)	6–13 (40–90)	8–19 (55–130)
Tensile modulus of elasticity, Msi (GPa)	0.30–0.64 (2.0–4.4)	0.41–0.61 (2.8–4.2)
Flexural yield strength, ksi (MPa)	8.5–23 (60–160)	18.1 (125)
Impact strength (notched-bar Izod test), ft·lb/in (J/m) of notch	0.2–0.4 (10.6–21.2)	0.1–1.0 (5.3–53)
Density (g/cm³)	1.10–1.46	1.2–1.3

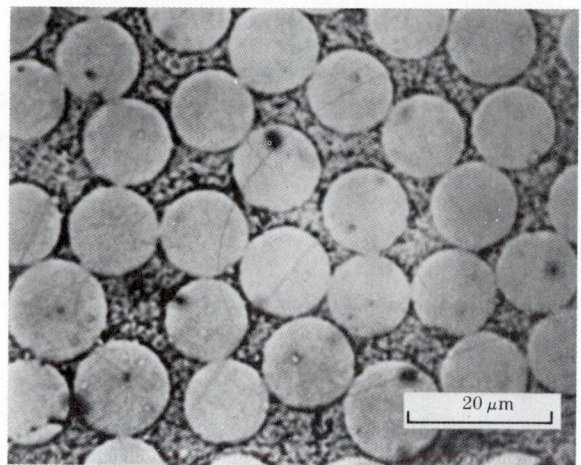

FIGURE 13.10
Photomicrograph of a cross section of a unidirectional fiberglass-polyester composite material. (*After D. Hull, "An Introduction to Composite Materials," Cambridge, 1981, p. 63.*)

be as high as 80 wt %, which leads to very high strengths for the composite material. Figure 13.10 shows a photomicrograph of a cross section of a fiberglass-polyester-resin composite material with unidirectional fibers.

Any deviation from the parallel alignment of the glass strands reduces the mechanical strength of the fiberglass composite. For example, composites made with woven fiberglass fabrics because of their interlacing have lower strengths than if all the glass strands were parallel (Table 13.3). If the roving is chopped, producing a random arrangement of glass fibers, the strength is lower for a specific direction but equal in all directions (Table 13.3).

Carbon-fiber-reinforced-epoxy resins   In carbon-fiber composite materials, the fibers contribute the high tensile properties for rigidity and strength, while the matrix is the carrier for the alignment of the fibers and contributes some impact strength. Epoxy resins are by far the most commonly used matrixes for carbon fibers, but other resins such as polyimides, polyphenylene sulfides, or polysulfones may be used for certain applications.

TABLE 13.3   Some Mechanical Properties of Fiberglass-Polyester Composites

	Woven cloth	Chopped roving	Sheet-molding compound
Tensile strength, ksi (MPa)	30–50 (206–344)	15–30 (103–206)	8–20 (55–138)
Tensile modulus of elasticity, Msi (GPa)	1.5–4.5 (103–310)	0.80–2.0 (55–138)	
Impact strength notched bar, Izod ft·lb/in (J/m) of notch	5.0–30 (267–1600)	2.0–20.0 (107–1070)	7.0–22.0 (374–1175)
Density (g/cm³)	1.5–2.1	1.35–2.30	1.65–2.0

TABLE 13.4 **Some Typical Mechanical Properties of a Commercial Unidirectional Composite Laminate of Carbon Fibers (62% by Volume) and Epoxy Resin**

Properties	Longitudinal, 0°	Transverse, 90°
Tensile strength, ksi (MPa)	270 (1860)	9.4 (65)
Tensile modulus of elasticity, Msi (GPa)	21 (145)	1.36 (9.4)
Ultimate tensile strain, %	1.2	0.70

Source: Hercules, Inc.

The major advantage of carbon fibers is that they have very high strength and moduli of elasticity (Table 13.1) combined with low density. For this reason, carbon-fiber composites are replacing metals in some aerospace applications where weight saving is important (Fig. 13.1). Table 13.4 lists some typical mechanical properties of one type of carbon-fiber-epoxy composite material which contains 62% by volume of carbon fibers. Figure 13.11 shows the exceptional fatigue properties of unidirectional carbon(graphite)-epoxy composite material as compared to those of aluminum alloy 2024-T3.

In engineering-designed structures the carbon-fiber-epoxy material is laminated so that different tailor-made strength requirements are met (Fig. 13.12). Figure 13.13 shows a photomicrograph of a five-layer bidirectional carbon-fiber-epoxy composite material after curing.

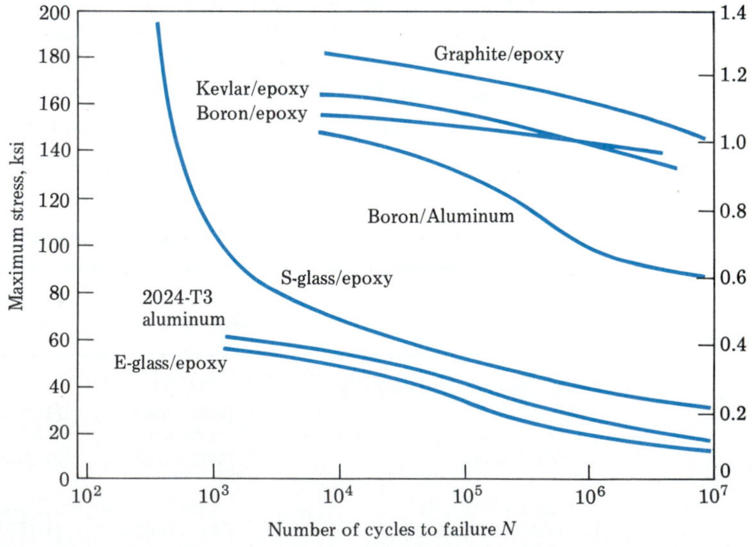

FIGURE 13.11 Fatigue properties (maximum stress vs. number of cycles to failure) for carbon (graphite)-epoxy unidirectional composite material as compared to the fatigue properties of some other composite materials and aluminum alloy 2024-T3. R (the minimum stress–maximum stress for tension-tension cyclic test) = 0.1 at room temperature. (*Courtesy of Hercules, Inc.*)

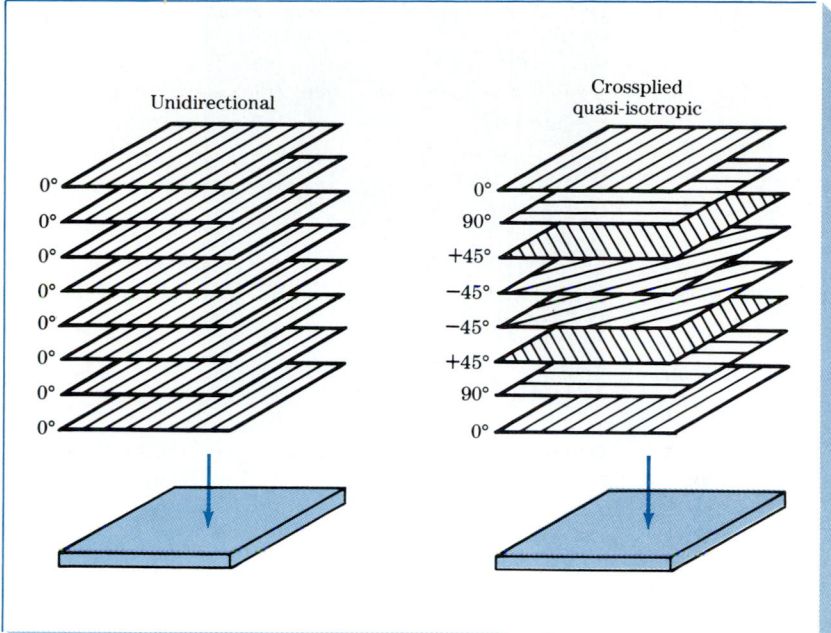

FIGURE 13.12
Unidirectional and multidirectional laminate plies for a composite laminate. (*Courtesy of Hercules, Inc.*)

Example Problem 13.1

A unidirectional Kevlar 49 fiber-epoxy composite contains 60% by volume of Kevlar 49 fibers and 40% epoxy resin. The density of the Kevlar 49 fibers is 1.48 Mg/m³ and that of the epoxy resin is 1.20 Mg/m³. (*a*) What are the weight percentages of Kevlar 49 and epoxy resin in the composite material, and (*b*) what is the average density of the composite?

Solution:

Basis is 1 m³ of composite material. Therefore we have 0.60 m³ of Kevlar 49 and 0.40 m³ of epoxy resin. Density = mass/volume, or

$$\rho = \frac{m}{V} \quad \text{and} \quad m = \rho V$$

(*a*)

$$\text{Mass of Kevlar 49} = \rho V = (1.48 \text{ Mg/m}^3)(0.60 \text{ m}^3) = 0.888 \text{ Mg}$$
$$\text{Mass of epoxy resin} = \rho V = (1.20 \text{ Mg/m}^3)(0.40 \text{ m}^3) = \underline{0.480 \text{ Mg}}$$

$$\text{Total mass} = 1.368 \text{ Mg}$$

$$\text{Wt \% Kevlar 49} = \frac{0.888 \text{ Mg}}{1.368 \text{ Mg}} \times 100\% = 64.9\%$$

$$\text{Wt \% epoxy resin} = \frac{0.480 \text{ Mg}}{1.368 \text{ Mg}} \times 100\% = 35.1\%$$

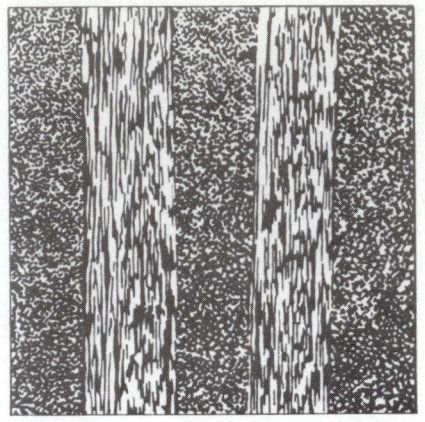

FIGURE 13.13
Photomicrograph of five-layer bidirectional composite of carbon-fiber-epoxy composite material (*After J. J. Dwyer, Composites, Am. Mach., July 13, 1979, pp. 87–96.*)

(*b*) Average density of composite is

$$\rho_c = \frac{m}{V} = \frac{1.368 \text{ Mg}}{1 \text{ m}^3} = 1.37 \text{ Mg/m}^3 \blacktriangleleft$$

Equations for Elastic Modulus of a Lamellar Continuous-Fiber-Plastic Matrix Composite for Isostrain and Isostress Conditions

Isostrain conditions   Let us consider an idealized lamellar composite test sample with alternate layers of continuous fibers and matrix material, as shown in Fig. 13.14. In this case the stress on the material causes uniform strain on all the composite layers. We shall assume that the bonding between the layers remains intact during the stressing. This type of loading on the composite sample is called the *isostrain condition*.

Let us now derive an equation relating the elastic modulus of the composite in terms of the elastic moduli of the fiber and matrix and their volume percentages. First, the load on the composite structure is equal to the sum of the

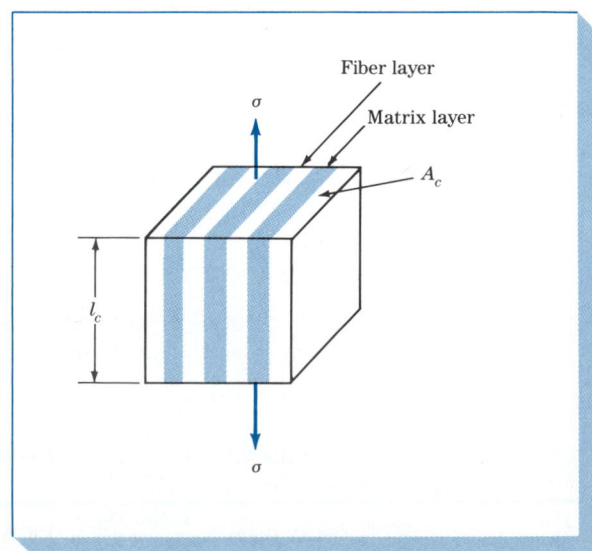

FIGURE 13.14 Composite structure consisting of layers of fiber and matrix under isostrain conditions of loading. (Volume of composite V_c = area A_c × length l_c.)

load on the fiber layers plus the load on the matrix layers, or

$$P_c = P_f + P_m \tag{13.1}$$

Since $\sigma = P/A$, or $P = \sigma A$,

$$\sigma_c A_c = \sigma_f A_f + \sigma_m A_m \tag{13.2}$$

where σ_c, σ_f, and σ_m are the stresses and A_c, A_f, and A_m are the fractional areas of the composite, fiber, and matrix, respectively. Since the lengths of the layers of matrix and fiber are equal, the areas A_c, A_f, and A_m in Eq. (13.2) can be replaced by the volume fractions V_c, V_f, and V_m:

$$\sigma_c V_c = \sigma_f V_f + \sigma_m V_m \tag{13.3}$$

Since the volume fraction of the total composite is 1, then $V_c = 1$, and Eq. (13.3) becomes

$$\sigma_c = \sigma_f V_f + \sigma_m V_m \tag{13.4}$$

For isostrain conditions and assuming a good bond between the composite layers,

$$\epsilon_c = \epsilon_f = \epsilon_m \tag{13.5}$$

Dividing Eq. (13.4) by Eq. (13.5), since all the strains are equal, gives

$$\frac{\sigma_c}{\epsilon_c} = \frac{\sigma_f V_f}{\epsilon_f} + \frac{\sigma_m V_m}{\epsilon_m} \tag{13.6}$$

Now we can substitute the modulus of elasticity E_c for σ_c/ϵ_c, E_f for σ_f/ϵ_f, and E_m for σ_m/ϵ_m, giving

$$\boxed{E_c = E_f V_f + E_m V_m} \tag{13.7}$$

This equation is known as the *rule of mixtures for binary composites* and enables a value for the elastic modulus of a composite to be calculated knowing the elastic moduli of the fiber and matrix and their volume percentages.

Equations for the loads on the fiber and matrix regions of a lamellar composite structure loaded under isostrain conditions The ratio of the loads on the fiber and matrix regions of a binary composite material stressed under isostrain conditions can be obtained from their $P = \sigma A$ ratios. Thus, since $\sigma = E\epsilon$ and $\epsilon_f = \epsilon_m$,

$$\frac{P_f}{P_m} = \frac{\sigma_f A_f}{\sigma_m A_m} = \frac{E_f \epsilon_f A_f}{E_m \epsilon_m A_m} = \frac{E_f A_f}{E_m A_m} = \frac{E_f V_f}{E_m V_m} \tag{13.8}$$

If the total load on a specimen stressed under isostrain conditions is known, then the following equation applies:

$$P_c = P_f + P_m \tag{13.9}$$

where P_c, P_f, and P_m are the loads on the total composite, fiber region, and matrix region, respectively. By combining Eq. (13.9) with Eq. (13.8), the load on each of the fiber and matrix regions can be determined if values for E_f, E_m, V_f, V_m, and P_c are known.

Example Problem 13.2

Calculate (a) the modulus of elasticity, (b) the tensile strength, and (c) the fraction of the load carried by the fiber for the following composite material stressed under isostrain conditions. The composite consists of a continuous glass-fiber-reinforced-epoxy resin produced by using 60% by volume of E-glass fibers having a modulus of elasticity of $E_f = 10.5 \times 10^6$ psi and a tensile strength of 350,000 psi and a hardened epoxy resin with a modulus of $E_m = 0.45 \times 10^6$ psi and a tensile strength of 9000 psi.

Solution:

(a) Modulus of elasticity of the composite is

$$E_c = E_f V_f + E_m V_m \tag{13.7}$$
$$= (10.5 \times 10^6 \text{ psi})(0.60) + (0.45 \times 10^6 \text{ psi})(0.40)$$
$$= 6.30 \times 10^6 \text{ psi} + 0.18 \times 10^6 \text{ psi}$$
$$= 6.48 \times 10^6 \text{ psi} \ (44.6 \text{ GPa}) \blacktriangleleft$$

(b) Tensile strength of the composite is

$$\sigma_c = \sigma_f V_f + \sigma_m V_m \tag{13.4}$$
$$= (350,000 \text{ psi})(0.60) + (9000 \text{ psi})(0.40)$$
$$= 210,000 + 3600 \text{ psi}$$
$$= 214,000 \text{ psi, or 214 ksi (1.47 GPa)} \blacktriangleleft$$

(c) Fraction of load carried by fiber is

$$\frac{P_f}{P_c} = \frac{E_f V_f}{E_f V_f + E_m V_m}$$

$$= \frac{(10.5 \times 10^6 \text{ psi})(0.60)}{(10.5 \times 10^6 \text{ psi})(0.60) + (0.45 \times 10^6 \text{ psi})(0.40)}$$

$$= \frac{6.30}{6.30 + 0.18} = 0.97 \blacktriangleleft$$

Isostress conditions   Let us now consider the case of an idealized lamellar composite structure consisting of layers of fiber and matrix in which the layers are perpendicular to the applied stress, as shown in Fig. 13.15. In this case the stress on the composite structure produces an equal stress condition on all the layers and so is called the *isostress condition*.

To derive an equation for the elastic modulus for the layered composite for this type of loading, we shall begin with an equation that states that the stress on the total composite structure is equal to the stress on the fiber layers

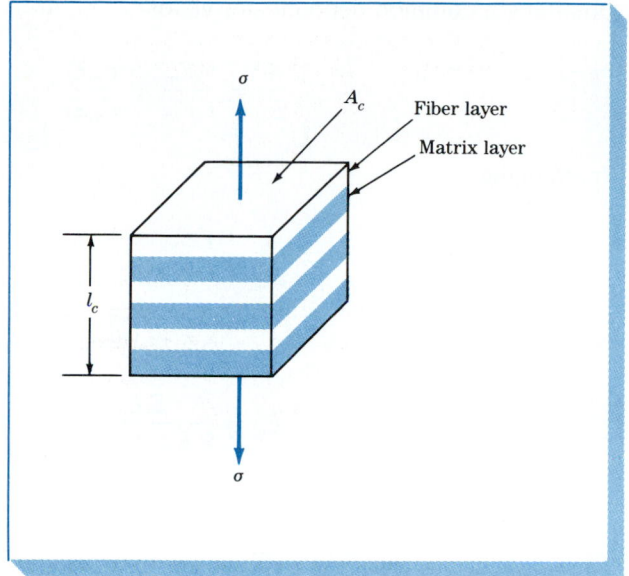

FIGURE 13.15 Composite structure consisting of layers of fiber and matrix under isostress conditions of loading. (Volume of composite V_c = area A_c × length l_c.)

and the stress on the matrix layers. Thus

$$\sigma_c = \sigma_f = \sigma_m \tag{13.10}$$

The total strain for the composite in the directions of the stresses is thus equal to the sum of the strains in the fiber and matrix layers,

$$\epsilon_c = \epsilon_f + \epsilon_m \tag{13.11}$$

Assuming that the area perpendicular to the stress does not change after the stress is applied and assuming unit length for the composite after being stressed, then

$$\epsilon_c = \epsilon_f V_f + \epsilon_m V_m \tag{13.12}$$

where V_f and V_m are the volume fractions of the fiber and matrix laminates, respectively.

Assuming Hooke's law is valid under loading, then

$$\epsilon_c = \frac{\sigma}{E_c} \qquad \epsilon_f = \frac{\sigma}{E_f} \qquad \epsilon_m = \frac{\sigma}{E_m} \tag{13.13}$$

Substituting equations of Eq. (13.13) into Eq. (13.12) gives

$$\frac{\sigma}{E_c} = \frac{\sigma V_f}{E_f} + \frac{\sigma V_m}{E_m} \tag{13.14}$$

Dividing each term of Eq. (13.14) by σ gives

$$\frac{1}{E_c} = \frac{V_f}{E_f} + \frac{V_m}{E_m} \tag{13.15}$$

Obtaining a common denominator yields

$$\frac{1}{E_c} = \frac{V_f E_m}{E_f E_m} + \frac{V_m E_f}{E_m E_f} \tag{13.16}$$

Rearranging,

$$\frac{1}{E_c} = \frac{V_f E_m + V_m E_f}{E_f E_m}$$

or

$$E_c = \frac{E_f E_m}{V_f E_m + V_m E_f} \tag{13.17}$$

Example Problem 13.3

Calculate the modulus of elasticity for a composite material consisting of 60% by volume of continuous E-glass fiber and 40% epoxy resin for the matrix when stressed under *isostress conditions* (i.e., the material is stressed perpendicular to the continuous fiber). The modulus of elasticity of the E glass is 10.5×10^6 psi and that of the epoxy resin is 0.45×10^6 psi.

Solution

$$E_c = \frac{E_f E_m}{V_f E_m + V_m E_f} \tag{13.17}$$

$$= \frac{(10.5 \times 10^6 \text{ psi})(0.45 \times 10^6 \text{ psi})}{(0.60)(0.45 \times 10^6) + (0.40)(10.5 \times 10^6)}$$

$$= \frac{4.72 \times 10^{12} \text{ psi}^2}{0.27 \times 10^6 \text{ psi} + 4.20 \times 10^6 \text{ psi}}$$

$$= 1.06 \times 10^6 \text{ psi (7.30 GPa)} \blacktriangleleft$$

Note that the isostress condition of stressing the composite material results in a modulus of elasticity for the 60% E-glass fiber–40% epoxy composite material which is about six times lower than that obtained by stressing under the isostrain condition.

A schematic representation comparing isostrain loading to isostress loading of a composite layered structure (Fig. 13.16) shows that higher-modulus values are obtained with isostrain loading for equal volumes of fibers.

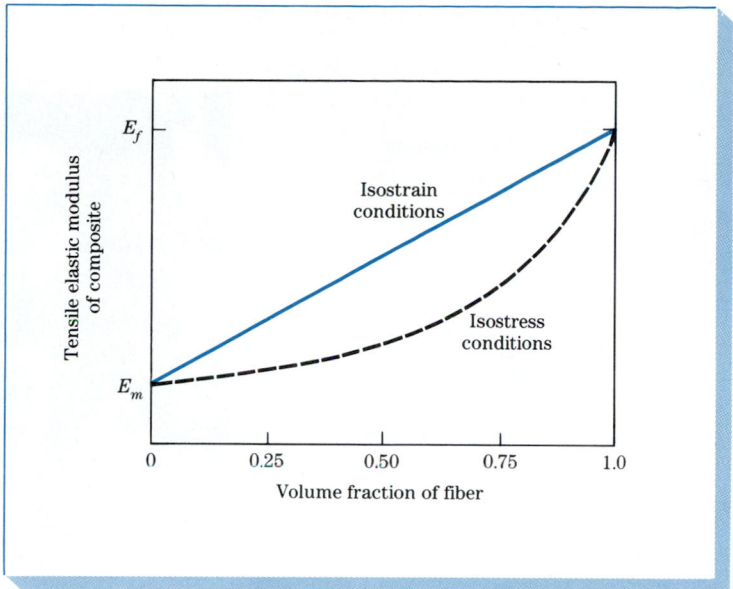

FIGURE 13.16 Schematic representation of the tensile elastic modulus as a function of the volume fraction of fiber in a reinforced-fiber-plastic matrix unidirectional composite laminate loaded under isostrain and isostress conditions. For a given volume fraction of fiber in the composite, the material loaded under isostrain conditions has a higher modulus.

13.4 OPEN-MOLD PROCESSES FOR FIBER-REINFORCED-PLASTIC COMPOSITE MATERIALS

There are many open-mold methods used for producing fiber-reinforced plastics. Some of the most important of these will now be discussed briefly.

Hand Lay-Up Process

This is the simplest method of producing a fiber-reinforced part. To produce a part with this process by using fiberglass and polyester, a gel coat is first applied to the open mold (Fig. 13.17a). Fiberglass reinforcement which is normally in the form of a cloth or mat is manually placed in the mold. The base resin mixed with catalysts and accelerators is then applied by pouring, brushing, or spraying. Rollers (Fig. 13.17b) or squeegees are used to thoroughly wet the reinforcement with the resin and to remove entrapped air. To increase the wall thickness of the part being produced, layers of fiberglass mat or woven roving and resin are added. Applications for this method include boat hulls, tanks, housings, and building panels.

Spray-Up Process

The spray-up method of producing fiber-reinforced-plastic shells is similar to the hand lay-up method and can be used to make boat hulls, tub-shower units, and other medium- to large-size shapes. In this process, if fiberglass is used, continuous-strand roving is fed through a combination chopper and spray gun (Fig. 13.18) that simultaneously deposits chopped roving and catalyzed resin into the mold. The deposited laminate is then densified with a roller or squeegee

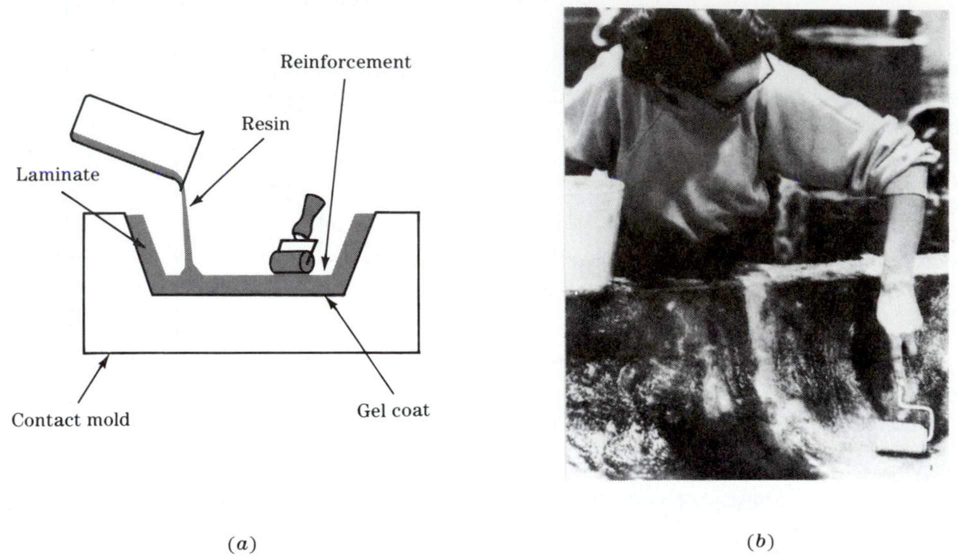

Reinforcement

Resin

Laminate

Contact mold

Gel coat

(a)

(b)

FIGURE 13.17 Hand lay-up method for molding fiber-reinforced-plastic composite materials. (*a*) Pouring the resin over the reinforcement in the mold. (*b*) Use of a roller to densify the laminate to remove entrapped air. (*Courtesy of Owens/Corning Fiberglas Co.*)

FIGURE 13.18 Spray-up method for molding fiber-reinforced-plastic composite materials; advantages of this method include greater shape complexity for the molded part and the process can be automated. (*Courtesy of Owens/Corning Fiberglas Co.*)

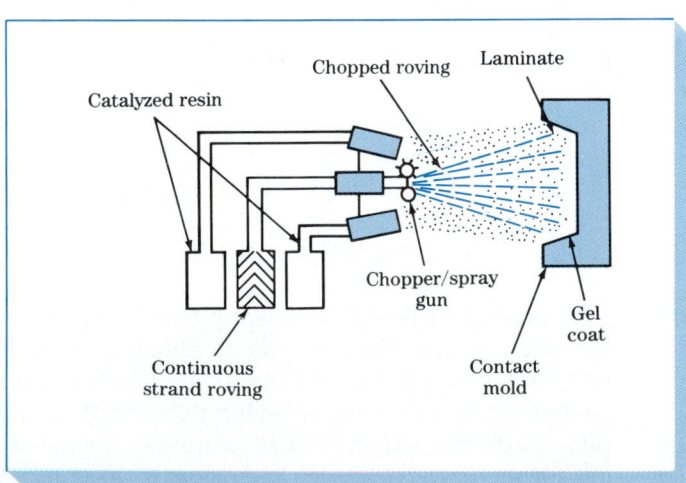

Catalyzed resin

Chopped roving   Laminate

Chopper/spray gun

Continuous strand roving

Gel coat

Contact mold

to remove air and to make sure the resin impregnates the reinforcing fibers. Multiple layers may be added to produce the desired thickness. Curing is usually at room temperature, or it may be accelerated by the application of a moderate amount of heat.

Vacuum Bag–Autoclave Process

This process is used to produce high-performance laminates usually of fiber-reinforced-epoxy systems. Composite materials produced by this method are particularly important for aircraft and aerospace applications (Fig. 13.1).

We shall now look at the various steps of this process required to produce a finished part. First, a long, thin sheet, which may be up to about 60 in (152 cm) wide, or prepreg carbon-fiber-epoxy material is laid out on a large table (Fig. 13.19). The prepreg material consists of unidirectional long carbon fibers in a partially cured epoxy matrix. Next, pieces of the prepreg sheet are cut out and placed on top of each other on a shaped tool to form a laminate (Fig. 13.20). The layers, or *plies* as they are called, may be placed in different directions to produce the desired strength pattern since the highest strength of each ply is in the direction parallel to the fibers (Fig. 13.12).

After the laminate is constructed, the tooling and the attached laminate are vacuum-bagged, with a vacuum being applied to remove entrapped air from the laminated part. Finally, the vacuum bag enclosing the laminate and the tooling is put into an autoclave for the final curing of the epoxy resin (Fig. 13.21). The conditions for curing vary depending on the material, but the carbon-fiber-epoxy composite material is usually heated at about 190°C (375°F) at a pressure of about 100 psi. After being removed from the autoclave, the composite part is stripped from its tooling and is ready for further finishing operations.

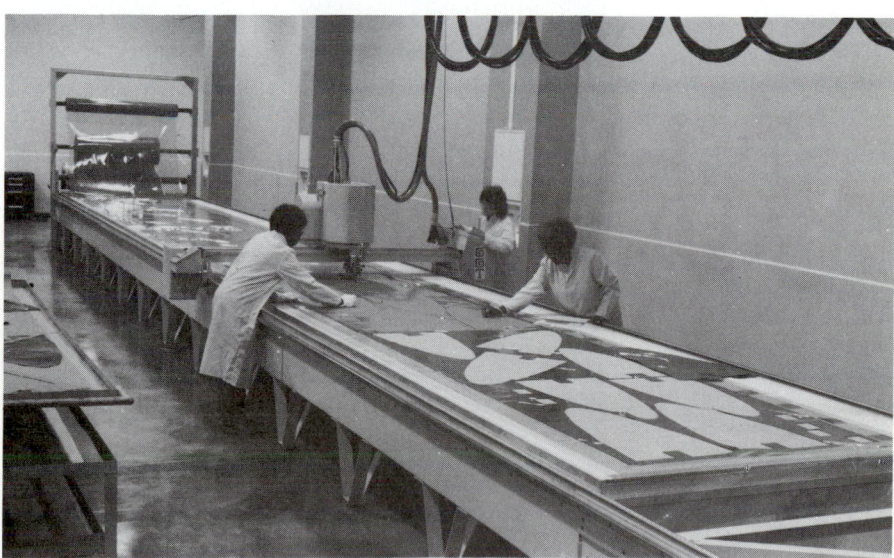

FIGURE 13.19 Carbon-fiber-epoxy prepreg sheet being cut with computerized cutter at McDonnel Douglas composite facility. (*Courtesy of McDonnell Douglas Corp.*)

FIGURE 13.20 Smooth-contour tooling mold for forming a laminate from multiple layers of prepreg fiber-reinforced-plastic composite material. (*Courtesy of Northrop Co.*)

Carbon-fiber-epoxy composite materials are used mainly in the aerospace industry where the high strength, stiffness, and lightness of the material can be fully utilized. For example, this material is used for airplane wings, elevator and rudder parts, and the cargo bay doors of the space shuttle. Cost considerations have prevented the widespread use of this material in the auto industry.

Filament-Winding Process

Another important open-mold process to produce high-strength hollow cylinders is the *filament-winding process*. In this process the fiber reinforcement

FIGURE 13.21 Carbon-fiber-epoxy laminate of AV-8B wing section and tooling being put into autoclave for curing at McDonnell Aircraft Co. plant. (*Courtesy of McDonnel Douglas Corp.*)

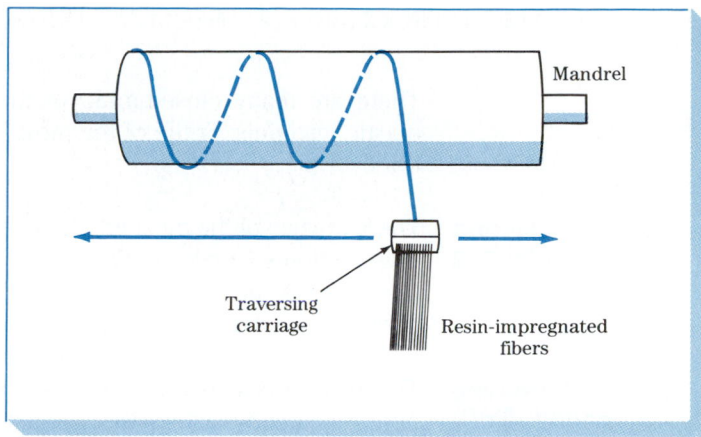

FIGURE 13.22 Filament-winding process for producing fiber-reinforced-plastic composite materials. The fibers are first impregnated with plastic resin and then wound around a rotating mandrel (drum). The carriage containing the resin-impregnated fibers traverses during the winding, laying down the impregnated fibers. (*After H. G. DeYoung, Plastic Composites Fight for Status, High Technol., October 1983, p. 63. © High Technology Publishing Co. Used with permission.*)

is fed through a resin bath and then wound on a suitable mandrel (Fig. 13.22). When sufficient layers have been applied, the wound mandrel is cured either at room temperature or at an elevated temperature in an oven. The molded part is then stripped from the mandrel. See cover photo of a polymer-matrix composite tube being filament-braided (wound) with aramid fibers and an epoxy matrix.

The high degree of fiber orientation and high fiber loading with this method produce extremely high tensile strengths in hollow cylinders. Applications for this process include chemical and fuel storage tanks, pressure vessels, and rocket motor cases (Fig. 13.23).

FIGURE 13.23 Filament winding of a rocket motor case using Kevlar 49 fiber and epoxy resin. (*Courtesy of the Du Pont Co.*)

13.5   CLOSED-MOLD PROCESSES FOR FIBER-REINFORCED-PLASTIC COMPOSITE MATERIALS

There are many closed-mold methods used for producing fiber-reinforced-plastic materials. Some of the most important of these will now be discussed briefly.

Compression and Injection Molding

These are two of the most important high-volume processes used for producing fiber-reinforced plastics with closed molds. These processes are essentially the same as those discussed in Sec. 7.5 for plastic materials except that the fiber reinforcement is mixed with the resin before processing.

The Sheet-Molding Compound (SMC) Process

The SMC process is one of the newer closed-mold processes used to produce fiber-reinforced-plastic parts, particularly in the automotive industry. This process allows excellent resin control and good mechanical strength properties (Table 13.3) to be obtained while producing high-volume, large-size, highly uniform products.

The sheet-molding compound is usually manufactured by a highly automated continuous-flow process. Continuous-strand fiberglass roving is chopped in lengths of about 2 in (5.0 cm) and deposited on a layer of resin-filler paste which is traveling on a polyethylene film (Fig. 13.24). Another layer of resin-filler paste is deposited later over the first layer to form a continuous sandwich of fiberglass and resin filler. The sandwich with top and bottom covers of polyethylene is compacted and rolled into package-sized rolls (Fig. 13.24).

The rolled-up SMC is next stored in a maturation room for about 1 to 4 days so that the sheet can carry the glass. The SMC rolls are then moved to near the press and cut into the proper charge pattern for the specific part and placed in a matched metal mold which is hot [300°F (149°C)]. The hydraulic press then is closed, and the SMC flows uniformly under pressure (1000 psi) throughout the mold to form the final product. Sometimes an in-mold coating may be injected in the middle of the pressing operation to improve the surface quality of the SMC part.

FIGURE 13.24
Manufacturing process for sheet-molding compound. The machine shown produces a sandwich of fiberglass and resin-filler paste between two thin-film sheets of polyethylene. The sheet-molding compound produced must be aged before being pressed into a finished product. (*Courtesy of Owens/Corning Fiberglas Co.*)

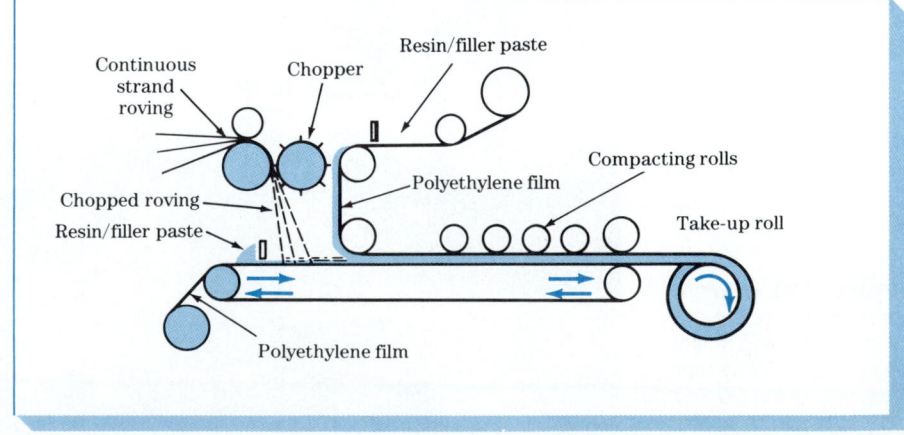

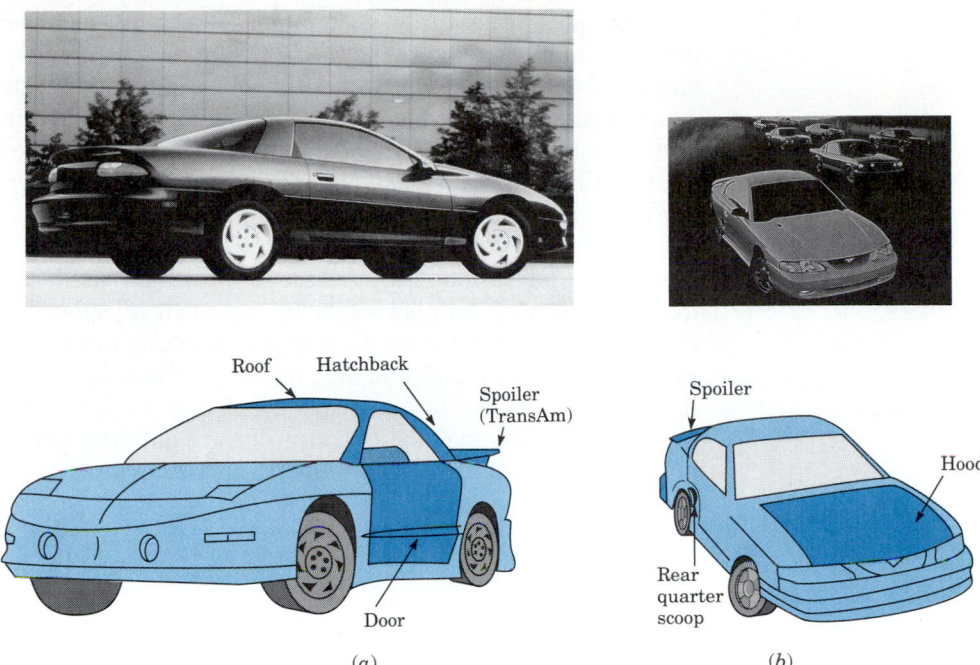

FIGURE 13.25 The use of sheet-molding compound in 1994 automobiles: (*a*) 1994 Chevrolet Camaro and (*b*) 1994 Ford Mustang. Sheet-molding compound is made by pressing the SMC at about 300°F (149°C) and 1000 psi (6.89 MPa) for 60 to 90 s. (*Photos courtesy of SMC Automotive Alliance.*)

FIGURE 13.26 The pultrusion process for producing fiber-reinforced-plastic composite materials. Fibers impregnated with resin are fed into a heated die and then are slowly drawn out as a cured composite material with a constant cross-sectional shape. (*After H. G. De Young, Plastic Composites Fight for Status, High Technol., October 1983, p. 63. © High Technology Publishing Co. Used with permission.*)

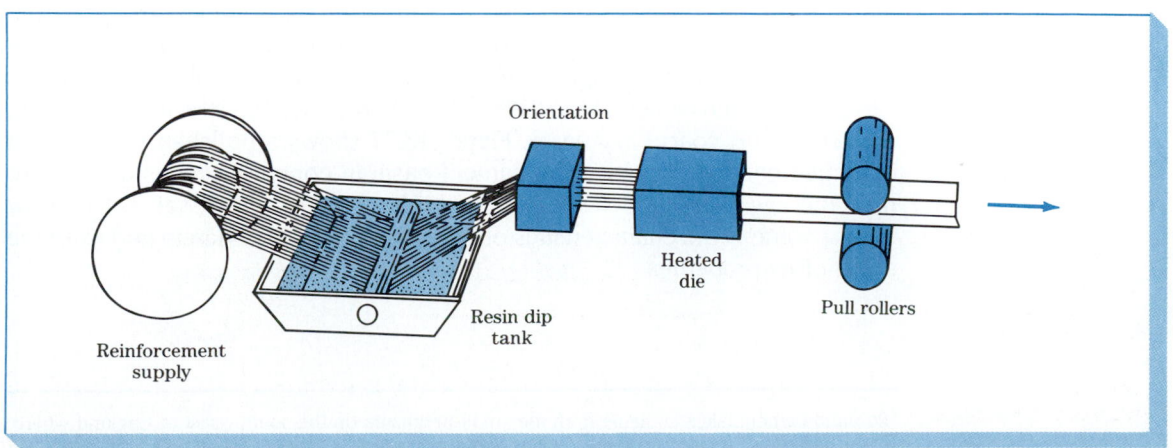

The advantages of the SMC process over the hand lay-up or spray-up processes are more-efficient high-volume production, improved surface quality, and uniformity of product. The use of SMC is particularly advantageous in the automotive industry for the production of front-end and grille-opening panels, body panels, and hoods. For example, the roof, doors, hatchback and spoiler of the 1994 Chevrolet Camaro and hood, spoiler, and rear quarter scoop of the 1994 Ford Mustang area made of sheet-molding compound (Fig. 13.25).

Continuous-Pultrusion Process   Continuous pultrusion is a process used for the manufacturing of fiber-reinforced plastics of constant cross section such as structural shapes, beams, channels, pipe, and tubing. In this process continuous-strand fibers are impregnated in a resin bath and then are drawn through a heated steel die which determines the shape of the finished stock (Fig. 13.26). Very high strengths are possible with this material because of the high fiber concentration and orientation parallel to the length of the stock being drawn.

13.6   CONCRETE

Concrete is a major engineering material used for structural construction. Civil engineers use concrete, for example, in the design and construction of bridges, buildings, dams, retainer and barrier walls, and road pavement. In 1993 about 60×10^7 metric tons of concrete was produced in the United States, which is considerably greater than the 8.7×10^7 metric tons of steel that was produced in the same year. As a construction material concrete offers many advantages, including flexibility in design since it is able to be cast, economy, durability, fire resistance, ability to be fabricated on site, and aesthetic appearance. Disadvantages of concrete from an engineering standpoint include low tensile strength, low ductility, and some shrinkage.

Concrete is a ceramic composite material composed of coarse granular material (the aggregate) embedded in a hard matrix of a cement paste (the binder) which is usually made from portland[1] cement and water. Concrete varies considerably in composition but usually contains (by absolute volume) 7 to 15% portland cement, 14 to 21% water, $\frac{1}{2}$ to 8% air, 24 to 30% fine aggregate, and 31 to 51% coarse aggregate. Figure 13.27 shows a polished section of a hardened concrete sample. The cement paste in concrete acts as a "glue" to hold the aggregate particles together in this composite material. Let us now look at some of the characteristics of the components of concrete and examine some of its properties.

---

[1]Portland cement takes its name from the small peninsula on the south coast of England where the limestone is similar to some extent to portland cement.

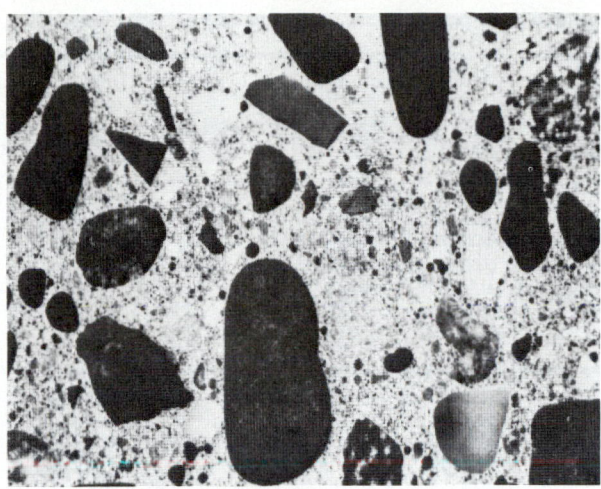

FIGURE 13.27 Cross section of hardened concrete. A cement and water paste completely coats each aggregate particle and fills the spaces between the particles to make a ceramic composite material. (*After "Design and Control of Concrete Mixtures," 12th ed., Portland Cement Association, 1979.*)

Portland Cement

Production of portland cement   The basic raw materials for portland cement are lime (CaO), silica (SiO_2), alumina (Al_2O_3), and iron oxide (Fe_2O_3). These components are appropriately proportioned to produce various types of portland cement. The selected raw materials are crushed, ground, and proportioned for the desired composition and then blended. The mixture is then fed into a rotary kiln where it is heated to temperatures from up to 1400 to 1650°C (2600 to 3000°F). In this process the mixture is chemically converted to cement clinker which is subsequently cooled and pulverized. A small amount of gypsum ($CaSO_4 \cdot 2H_2O$) is added to the cement to control the setting time of the concrete.

Chemical composition of portland cement   From a practical standpoint, portland cement can be considered to consist of four principal compounds, which are:

Compound	Chemical formula	Abbreviation
Tricalcium silicate	$3CaO \cdot SiO_2$	C_3S
Dicalcium silicate	$2CaO \cdot SiO_2$	C_2S
Tricalcium aluminate	$3CaO \cdot Al_2O_3$	C_3A
Tetracalcium aluminoferrite	$4CaO \cdot Al_2O_3 \cdot Fe_2O_3$	C_4AF

Types of portland cement   Various types of portland cement are produced by varying the amounts of the above chemical compositions. In general there are five main types whose basic chemical compositions are listed in Table 13.5.

Type I is the normal general-purpose portland cement. It is used where the concrete is not exposed to a high sulfate attack from soil or water, or where there is not an objectionable temperature increase due to heat generated by hydration of the cement. Typical applications for type I concrete are sidewalks, reinforced concrete buildings, bridges, culverts and tanks, and reservoirs.

TABLE 13.5   **Typical Compound Compositions of Portland Cement**

Cement type	ASTM C150 designation	Compositions, wt %*			
		C_3S	C_2S	C_3A	C_4AF
Ordinary	I†	55	20	12	9
Moderate heat of hydration, moderate sulfate resistance	II	45	30	7	12
Rapid hardening	III	65	10	12	8
Low heat of hydration	IV	25	50	5	13
Sulfate-resistant	V	40	35	3	14

*Missing percentages consist of gypsum, and minor components such as MgO, alkali sulfate, etc.
†This is the most common of all cement types.
Source: J. F. Young, *J. Educ. Module Mater. Sci.,* **3**:410 (1981). Used by permission, the *Journal of Materials Education,* University Park, Pa.

Type II portland cement is used where moderate sulfate attack is important such as in drainage structures where sulfate concentrations in groundwaters are higher than normal. Type II cement is normally used in hot weather for large structures such as large piers and heavy retaining walls since this cement has a moderate heat of hydration.

Type III portland cement is an early-strength type that develops high strength in an early period. It is used when concrete forms are to be removed early from a structure that must be put into use soon.

Type IV is a low-heat-of-hydration portland cement for use when the rate and amount of heat generated must be minimized. Type IV cement is used for massive concrete structures such as large gravity dams where the heat generated by the setting cement is a critical factor.

Type V is a sulfate-resisting cement used when the concrete is exposed to severe sulfate attack such as in concrete exposed to soils and groundwaters containing a high sulfate content.

Hardening of portland cement   Portland cement hardens by reactions with water which are called *hydration reactions*. These reactions are complex and not completely understood. Tricalcium silicate (C_3S) and dicalcium silicate (C_2S) constitute about 75% of portland cement by weight, and when these compounds react with water during the hardening of the cement, the principal hydration product is *tricalcium silicate hydrate*. This material is formed in extremely small particles (less than 1 μm) and is a colloidal *gel*. Calcium hydroxide is also produced by the hydration of C_3S and C_2S and is a crystalline material. These reactions are

$$2C_3S + 6H_2O \rightarrow C_3S_2 \cdot 3H_2O + 3Ca(OH)_2$$
$$2C_2S + 4H_2O \rightarrow C_3S_2 \cdot 3H_2O + Ca(OH)_2$$

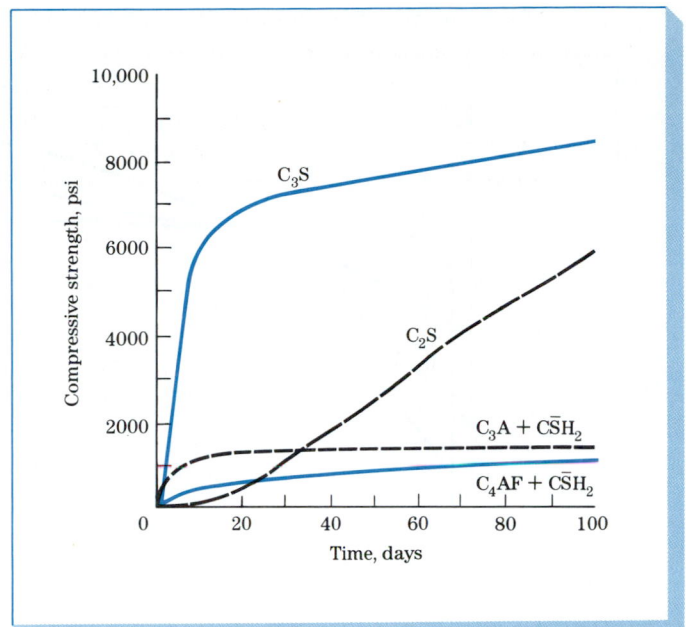

FIGURE 13.28
Compressive strength of pure-cement compound pastes as a function of curing time. Note that $C\bar{S}H_2$ is the abbreviated formula for $CaSO_4 \cdot 2H_2O$. [After J. F. Young, J. Educ. Module Mater. Sci., 3:420(1981). Used by permission, the Journal of Materials Education, University Park, Pa.]

Tricalcium silicate (C_3S) hardens rapidly and is mostly responsible for the early strength of portland cement (Fig. 13.28). Most of the hydration of C_3S takes place in about 2 days, and thus early-strength portland cements contain higher amounts of C_3S.

Dicalcium silicate (C_2S) has a slow hydration reaction with water and is mainly responsible for strength increases beyond 1 week (Fig. 13.28). Tricalcium aluminate (C_3A) hydrates rapidly with a high rate of heat liberation. C_3A contributes slightly to early-strength development and is kept to a low level in sulfate-resisting (type V) cements (Fig. 13.28). Tetracalcium aluminoferrite (C_4AF) is added to the cement to reduce the clinkering temperature during the manufacturing of the cement.

The extent to which the hydration reactions are completed determine the strength and durability of the concrete. Hydration is relatively rapid during the first few days after fresh concrete is put in place. It is important that water is retained by the concrete during the early curing period and that evaporation is prevented or reduced.

Figure 13.29 shows how the compressive strength of concretes made with different ASTM-type cements increases with curing time. Most of the compressive strength of the concretes is developed in about 28 days, but strengthening may continue for years.

Mixing Water for Concrete   Most natural water that is drinkable can be used as mixing water for making concrete. Some water unsuitable for drinking can also be used for making cement, but if the impurity content reaches certain specified levels, the water should be tested for its effect on the strength of the concrete.

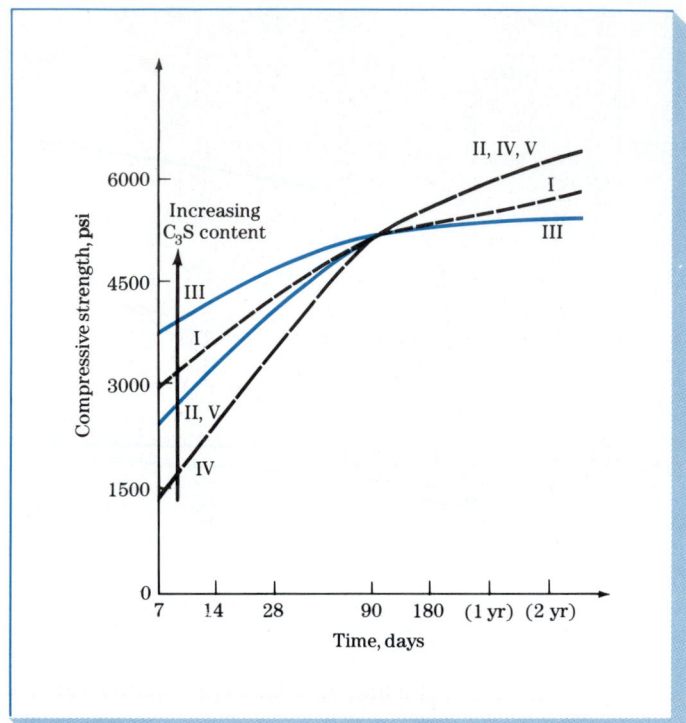

FIGURE 13.29
Compressive strengths of concretes made with different ASTM types of portland cements as a function of curing time. [After J. F. Young, J. Educ. Module Mater. Sci., 3:420(1981). Used by permission, the Journal of Materials Education, University Park, Pa.]

Aggregates for Concrete

Aggregates normally make up to about 60 to 80 percent of the concrete volume and greatly affect its properties. Concrete aggregate is usually classified as either fine or coarse. Fine aggregates consist of sand particles up to $\frac{1}{4}$ in (6 mm), while coarse aggregates are those particles retained on a no. 16 sieve (1.18-mm opening). Thus, there is some overlapping in particle-size ranges for fine and coarse aggregates. Rocks compose most of the coarse aggregate, while minerals (sand) compose most of the fine aggregates.

Air Entrainment

Air-entrained concretes are produced to improve the resistance of the concrete to freezing and thawing and also to improve the workability of some concretes. Air-entraining agents are added to some types of portland cements, and these are classified with the letter A following the type number as, for example, type IA and type IIA. Air-entraining agents contain surface-active agents which lower the surface tension at the air-water interface so that extremely small air bubbles are formed (90 percent of them are less than 100 μm) (Fig. 13.30). For satisfactory frost protection air-entrained concretes must have between 4 to 8% air by volume.

Compressive Strength of Concrete

Concrete, which is basically a ceramic composite material, has a much higher compressive strength than tensile strength. In engineering designs, therefore, concrete is primarily loaded in compression. The tensile loading capability of

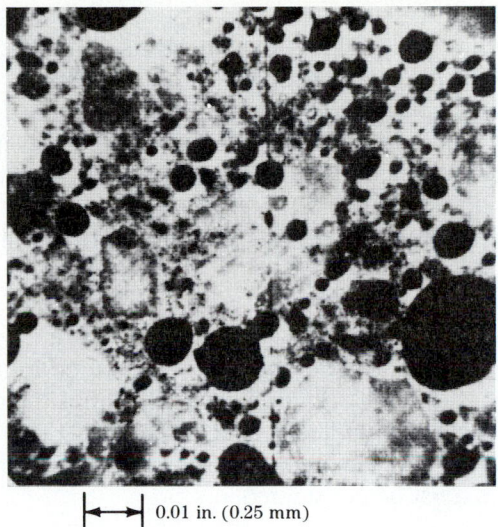

FIGURE 13.30 Polished section of air-entrained concrete as observed in the optical microscope. Most of the air bubbles in this sample appear to be about 0.1 mm in diameter. (*After "Design and Control of Concrete Mixtures," 12th ed., Portland Cement Association, 1979.*)

⊢——⊣ 0.01 in. (0.25 mm)

concrete can be increased by reinforcing it with steel rods. This subject will be discussed later.

As shown in Fig. 13.29, the strength of concrete is time-dependent since its strength develops by hydration reactions which take time to complete. The compressive strength of concrete is also greatly dependent on the water-to-cement ratio, with high ratios of water to cement producing lower-strength concrete (Fig. 13.31). However, there is a limit to how low the water-to-cement

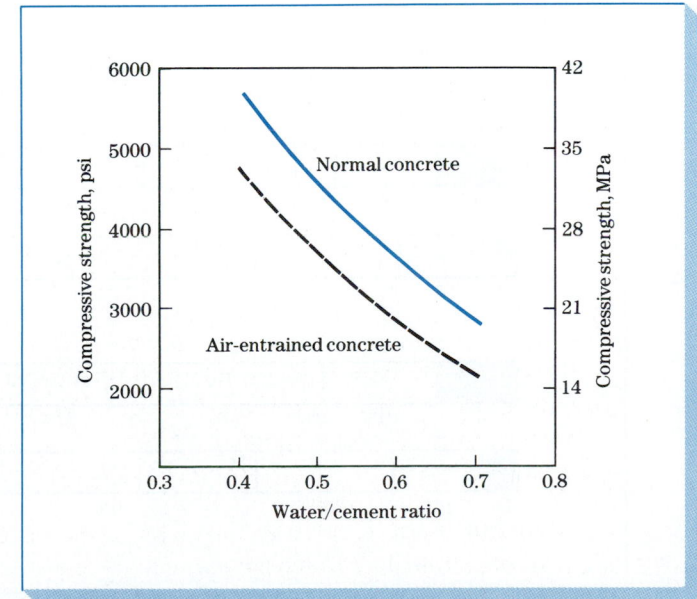

FIGURE 13.31 Effect of water-cement ratio by weight on the compressive strength of normal and air-entrained (A-E) concrete. (*After "Design and Control of Concrete Mixtures," 12th ed., Portland Cement Association, 1979, p.61.*)

ratio can be since less water makes it more difficult to work the concrete and to have it completely fill the concrete forms. With air entrainment, the concrete is more workable, and thus a lower water-to-cement ratio can be used.

Concrete compressive-strength test specimens are usually cylinders 6 in (15 cm) in diameter and 12 in (30 cm) high. However, cores or other types of specimens can be cut out of existing concrete structures.

Proportioning of Concrete Mixtures

The design of concrete mixtures should include consideration of the following factors:

1. Workability of the concrete. The concrete must be able to flow or be compacted into the shape of the form it is poured into.
2. Strength and durability. For most applications the concrete must reach certain strength and durability specifications.
3. Economy of production. For most applications cost is an important factor and thus must be considered.

Modern concrete-mixture design methods have evolved from the early 1900s arbitrary volumetric method of 1:2:4 for the ratios of cement to fine aggregate to coarse aggregate. Today, weight and absolute-volume methods of proportioning concrete mixtures are set forth by the American Concrete Institute. Example Problem 13.4 outlines a method of determining the required amounts of cement, fine and coarse aggregate, and water for a particular volume of concrete, given the weight ratios for the components, their specific gravities, and the required volume of water per unit weight of cement.

Figure 13.32 shows the ranges for the proportions of materials used in concrete by the absolute-volume method for normal and air-entrained concrete

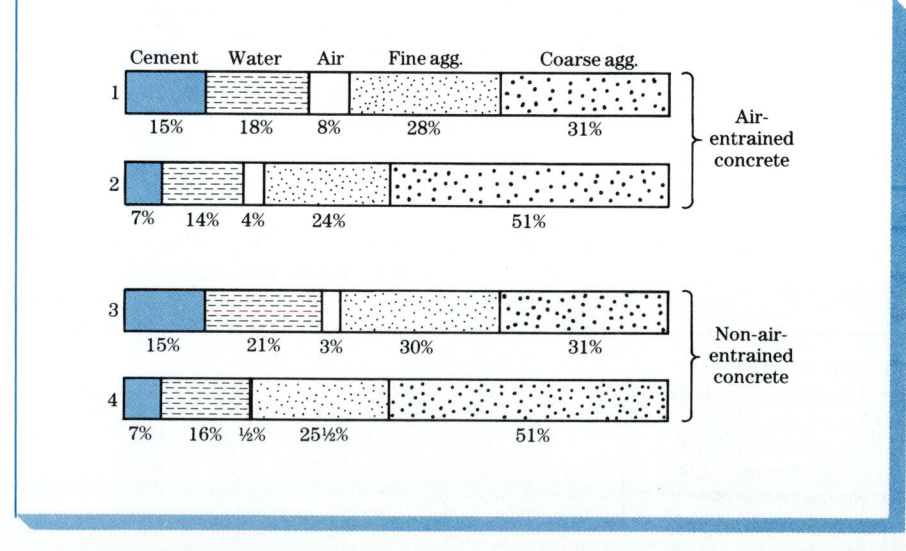

FIGURE 13.32 Ranges of proportions of components of concrete by absolute volume. The mixes for bars 1 and 3 have higher water and fine-aggregate contents, while the mixes for bars 2 and 4 have lower water and higher coarse-aggregate contents. (*After "Design and Control of Concrete Mixtures," 12th ed., Portland Cement Association, 1979, p.7.*)

mixtures. Normal concrete has a range by volume of 7 to 15% cement, 25 to 30% fine aggregate, 31 to 51% coarse aggregate, and 16 to 21% water. The air content of normal concrete ranges from $\frac{1}{2}$ to 3% but from 4 to 8% for air-entrained concrete. As previously stated the water-to-cement ratio is a determining factor for the compressive strength of concrete. A water-to-cement ratio above about 0.4 decreases the compressive strength of concrete significantly (Fig. 13.31).

Example Problem 13.4

Seventy-five cubic feet of concrete with a ratio of $1:1.8:2.8$ (by weight) of cement, sand (fine aggregate), and gravel (coarse aggregate) is to be produced. What are the required amounts of the constituents if 5.5 gal of water per sack of cement is to be used? Assume the free moisture contents of the sand and gravel are 5 and 0.5 percent, respectively. Give answers in the following units: the amount of cement in sacks, the sand and gravel in pounds, and the water in gallons. Data are given below:

Constituent	Specific gravity	Saturated surface-dry density, lb/ft³*
Cement	3.15	3.15×62.4 lb/ft³ = 197
Sand	2.65	2.65×62.4 lb/ft³ = 165
Gravel	2.65	2.65×62.4 lb/ft³ = 165
Water	1.00	1.00×62.4 lb/ft³ = 62.4

*Saturated surface-dry density (SSDD) in lb/ft³ = specific gravity × weight of 1 ft³ water = specific gravity × 62.4 lb/ft³.
1 sack of cement weighs 94 lb.
7.48 gal = 1 ft³ of water.

Solution:

First we shall calculate the absolute volumes of the constituents of the concrete per sack of cement since the problem asks for a certain number of cubic feet of concrete. Also, we shall first calculate the weight of sand and gravel required on a dry basis and then later make a correction for the moisture content of the sand and gravel.

Constituent	Ratio by weight	Weight	SSDD,* lb/ft³	Absolute vol. per sack of cement
Cement	1	1×94 lb = 94 lb	197	94 lb/197 lb/ft³ = 0.477 ft³
Sand	1.8	1.8×94 lb = 169 lb	165	169 lb/165 lb/ft³ = 1.024 ft³
Gravel	2.8	2.8×94 lb = 263 lb	165	263 lb/165 lb/ft³ = 1.594 ft³
Water	(5.5 gal)			5.5 gal/7.48 gal/ft³ = 0.735 ft³
				Tot. abs. vol. per sack of cement = 3.830 ft³

*SSDD = saturated surface-dry density

Thus, 3.830 ft^3 of concrete is produced per sack of cement on a dry basis. On this basis, the following amounts of cement, sand, and gravel are required for 75 ft^3 of concrete:

1. Required amount of cement = 75 ft^3/(3.83 ft^3/ sack of cement)
   = 19.58 sacks ◀
2. Required amount of sand = (19.58 sacks)(94 lb/sack)(1.8) = 3313 lb
3. Required amount of gravel = (19.58 sacks)(94 lb/sack)(2.8) = 5153 lb
4. Required amount of water = (19.58 sacks)(5.5 gal/sack) = 107.7 gal

On a wet basis with correction for moisture in the sand and gravel,

$$\text{Water in sand} = (3313 \text{ lb})(1.05)$$
$$= 3479 \text{ lb}; 3479 - 3313 \text{ lb} = 166 \text{ lb water}$$
$$\text{Water in gravel} = (5153 \text{ lb})(1.005)$$
$$= 5179 \text{ lb}; 5179 - 5153 \text{ lb} = 26 \text{ lb water}$$

1. Required weight of wet sand = 3313 lb + 166 lb = 3479 lb ◀
2. Required weight of wet gravel = 5153 lb + 26 lb = 5179 lb ◀

The required amount of water equals the calculated amount on the dry basis less the amount of water in the sand and gravel. Thus

$$\text{Total gal water in sand and gravel} = (166 \text{ lb} + 26 \text{ lb})\left(\frac{7.48 \text{ gal}}{\text{ft}^3}\right)\left(\frac{1 \text{ ft}^3}{62.4 \text{ lb}}\right)$$

$$= 23.0 \text{ gal}$$

Thus the water required on the dry basis minus the water in the sand and gravel equals the amount of water required on the wet basis:

$$107.7 \text{ gal} - 23.0 \text{ gal} = 84.7 \text{ gal} ◀$$

Reinforced and Prestressed Concrete

Since the tensile strength of concrete is about 10 to 15 times lower than its compressive strength, concrete is mainly used in engineering designs so that it is primarily in compression. However, when a concrete member is subjected to tensile forces such as occur in a beam, the concrete is usually cast containing steel reinforcing bars, as shown in Fig. 13.33. In this reinforced concrete, the tension forces are transferred from the concrete to the steel reinforcement through bonding. Concrete containing steel reinforcements in the form of rods, wires, wire mesh, etc., is referred to as *reinforced concrete*.

Prestressed Concrete

The tensile strength of reinforced concrete can be further improved by introducing compressive stresses into the concrete by *pretensioning* or *posttensioning* using steel reinforcements called *tendons*. The tendon may be a tensioned steel rod or cable, for example. The advantage of prestressed concrete is that the compressive stresses introduced by the steel tendons have to be counteracted before the concrete is subjected to tensile stresses.

FIGURE 13.33
Exaggerated effect of a heavy load on a reinforced-concrete beam. Note that the reinforcing steel rods are placed in the tension zone to absorb tensile stresses. (*After G. B. Wynne, "Reinforced Concrete Structures," Reston Publishing, 1981, p. 2.*)

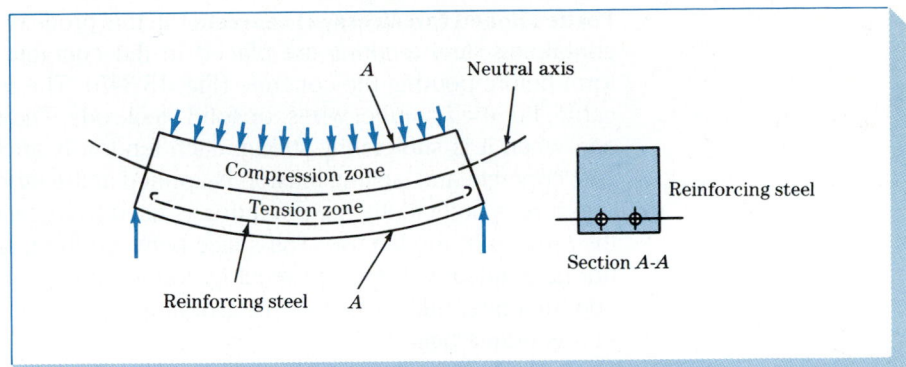

Pretensioned (prestressed) concrete   In the United States most prestressed concrete is pretensioned. In this method the tendons which are usually in the form of multiple-wire stranded cables are stretched between an external tendon anchorage and an adjustable jack for applying the tension (Fig. 13.34*a*). The concrete is then poured over the tendons which are in the tensioned state. When the concrete reaches the required strength, the jacking pressure is released. The steel strands would like to shorten elastically but are not able to because they are bonded to the concrete. In this way compressive stresses are introduced into the concrete.

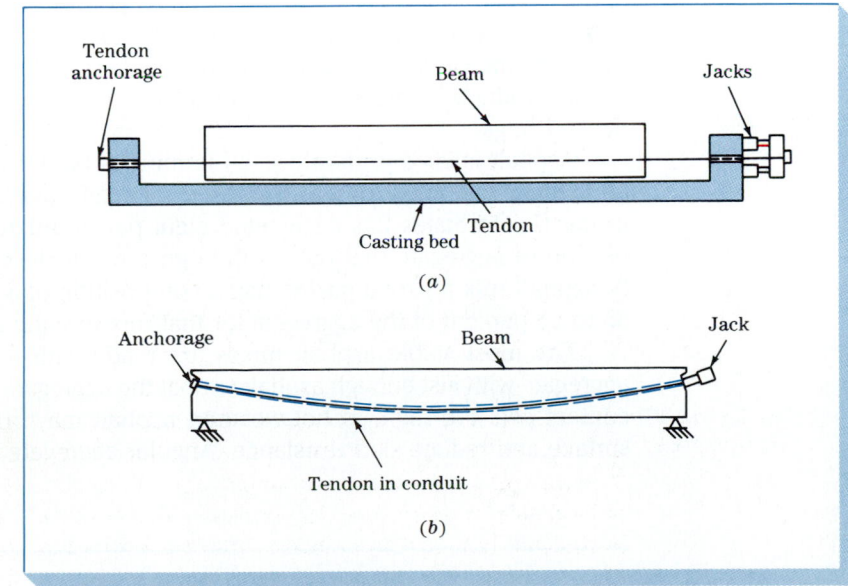

FIGURE 13.34 Schematic drawing showing the arrangement for producing (*a*) a pretensioned concrete beam and (*b*) a posttensioned concrete beam. (*After A. H. Nilson, "Design of Prestressed Concrete," Wiley, 1978, pp. 14 and 17.*)

Posttensioned (prestressed) concrete In this process, usually hollow conduits containing steel tendons are placed in the concrete (a beam, for example) form before pouring the concrete (Fig. 13.34b). The tendons may be stranded cable, bundled parallel wires, or solid steel rods. The concrete is then poured, and when it is sufficiently strong, each tendon is anchored at one end of the cast concrete and jacking tension is applied at the other end. When the jacking pressure is sufficiently high, a fitting is used to replace the jack which retains the tension in the tendon. The space between the tendons and the conduit is normally filled with cement grout by forcing it in as a paste at one end of the conduit under high pressure. The grouting improves the flexural stress capacity of a concrete beam.

13.7 ASPHALT AND ASPHALT MIXES

Asphalt is a *bitumen,* which is basically a hydrocarbon with some oxygen, sulfur, and other impurities, and has the mechanical characteristics of a thermoplastic polymeric material. Most asphalt is derived from petroleum refining, but some is processed directly from bitumen-bearing rock (rock asphalt) and from surface deposits (lake asphalt). The asphalt content of crude oils usually ranges from about 10 to 60%. In the United States about 75 percent of the asphalt consumed is used for paving roads, while the remainder is used mainly for roofing and construction.

Asphalts consist chemically of 80 to 85% carbon, 9 to 10% hydrogen, 2 to 8% oxygen, 0.5 to 7% sulfur, and small amounts of nitrogen and other trace metals. The constituents of asphalt greatly vary and are complex. They range from low-molecular-weight to high-molecular-weight polymers and condensation products consisting of chain hydrocarbons, ring structures, and condensed rings.

Asphalt is used primarily as a bituminous binder with aggregates to form an *asphalt mix,* most of which is used for road paving. The Asphalt Institute in the United States has designated eight paving mixtures based on the proportion of aggregate that passes through a no. 8 sieve.[1] For example, a type IV asphalt mix for road paving has a composition of 3.0 to 7.0% asphalt with 35 to 50 percent of the aggregate for that mix passing through a no. 8 sieve.

The most stable asphalt mixes are made with a dense-packed angular aggregate with just enough asphalt to coat the aggregate particles. If the asphalt content gets too high, in hot weather, asphalt may concentrate on the road surface and reduce skid resistance. Angular aggregate which does not polish

---

[1] A no. 8 sieve has nominal openings of 0.0937 in (2.36 mm).

easily and which interlocks produces better skid resistance than soft, easily polished aggregate. The aggregate also should bond well with the asphalt to avoid separation.

13.8 WOOD

Wood (timber) is the most widely used engineering construction material in the United States, with its annually produced tonnage exceeding all other engineering materials including concrete and steel (Fig. 1.12). In addition to the use of wood for timber and lumber for the construction of houses, buildings, and bridges, etc., wood is also used to make composite materials such as plywood, particleboard, and paper.

Wood is a naturally occurring composite material which consists mainly of a complex array of cellulose cells reinforced by a polymeric substance called *lignin* and other organic compounds. The discussion of wood in this section will first look at the macrostructure of wood, then a brief examination will be made of the microstructure of softwoods and hardwoods. Finally, some of the properties of wood will be correlated with its structure.

Macrostructure of Wood

Wood is a natural product with a complex structure, and thus we cannot expect a homogeneous product for engineering designs such as with an alloy steel bar or an injection-molded thermoplastic part. As we all know, the strength of wood is highly anisotropic, with its tensile strength being much greater in the direction parallel to the tree stem.

Layers in the cross section of a tree Let us first examine the cross section of a typical tree, as shown in Fig. 13.35. Important layered regions in this figure are indicated by the letters A to F. The name and function of each of these layers is listed below.

A. *Outer bark* layer is composed of dry, dead tissue and provides external protection for the tree.
B. *Inner bark* layer is moist and soft and carries food from the leaves to all the growing parts of the tree.
C. *Cambium layer* is the tissue layer between the bark and wood that forms the wood and bark cells.
D. *Sapwood* is the light-colored wood which forms the outer part of the tree stem. The sapwood contains some living cells which function for food storage and carry sap from the roots to the leaves of the tree.
E. *Heartwood* is the older inner region of the tree stem which is no longer

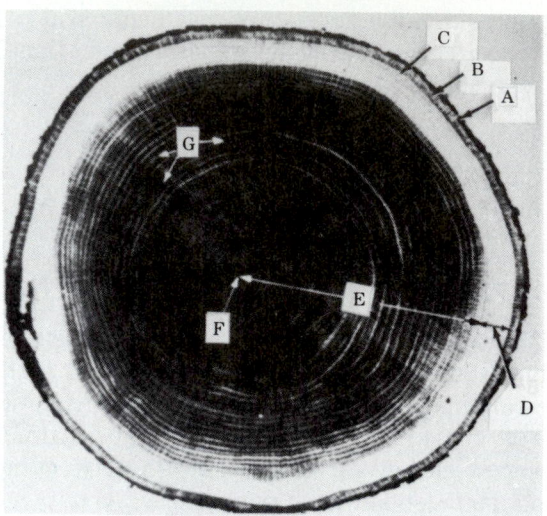

FIGURE 13.35 Cross section of a typical tree. A: outer bark; B: inner bark; C: cambium layer; D: sapwood; E: heartwood; F: pith; G: wood rays. (*After U.S. Department of Agriculture Handbook No. 72, revised 1974, p. 2-2.*)

living. The heartwood is usually darker than the sapwood and provides strength for the tree.

F. *Pith* is the soft tissue at the center of the tree around which the first growth of the tree takes place.

Also indicated in Fig. 13.35 are the wood rays which connect the tree layers from the pith to the bark and which are used for food storage and transfer of food.

Softwoods and hardwoods   Trees are classified into two major groups called *softwoods* (gymnosperms) and *hardwoods* (angiosperms). The botanical basis for their classification is that if the tree seed is exposed, the tree is a softwood type and if the seed is covered, the tree is a hardwood type. With a few exceptions, a softwood tree is one which retains its leaves and a hardwood tree is one which sheds its leaves annually. Softwood trees are often referred to as *evergreen* trees and hardwood trees as *deciduous* trees. Most softwood trees are physically soft and most hardwood trees are physically hard, but there are exceptions. Examples of softwood trees native to the United States are fir, spruce, pine, and cedar, while examples of hardwood trees are oak, elm, maple, birch, and cherry.

Annual growth rings   During each growth season in temperate climates such as in the United States, a new layer of wood is formed annually around the tree stem. These layers are called *annual growth rings* and are particularly evident in softwood tree sections (Fig. 13.36). Each ring has two subrings: *earlywood* (spring) and *latewood* (summer). In softwoods the earlywood has a lighter color and the cell size is larger.

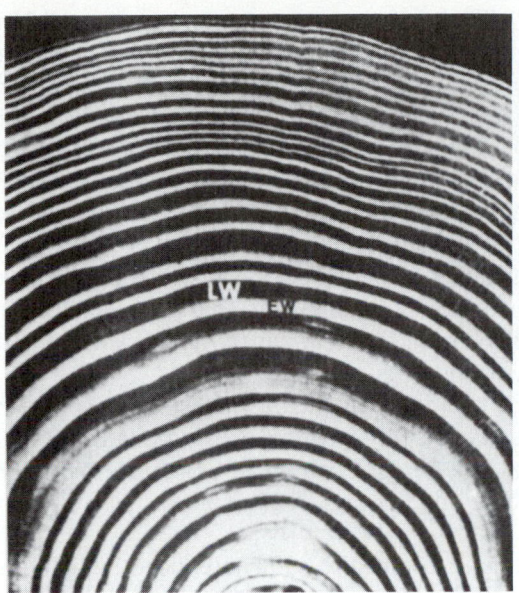

FIGURE 13.36 Annual growth rings in a softwood tree. The earlywood (EW) part of the annual ring is lighter in color than the latewood (LW) part. [*After R. J. Thomas, J. Educ. Module Mater. Sci., 2:56(1980). Used by permission, the Journal of Materials Education, University Park, Pa.*]

Axes of symmetry for wood   It is important to be able to correlate the direction in a tree with its microstructure. To do this a set of axes has been chosen, as indicated in Fig. 13.37. The axis parallel to the tree stem is called the *longitudinal axis* (*L*), while the axis perpendicular to the annual growth ring of the tree is called the *radial axis* (*R*). The third axis, the *tangential axis* (*T*), is parallel to the annual ring and is perpendicular to both the radial and longitudinal axes.

FIGURE 13.37 Axes in wood. The longitudinal axis is parallel to the grain, the tangential axis is parallel to annual growth ring, and the radial axis is perpendicular to annual growth ring. (*After U.S. Department of Agriculture Handbook No. 72, revised 1974, p. 4-2.*)

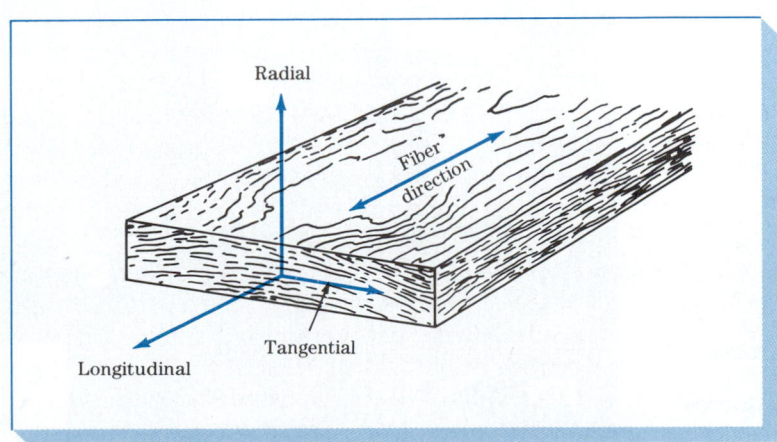

Microstructure of Softwoods

Figure 13.38 shows the microstructure of a small block of a softwood tree at 75× in which three complete growths rings can be seen. The larger cell size of the earlywood is clearly visible in this micrograph. Softwood consists mainly of long, thin-walled tubular cells called *tracheids* which can be observed in Fig. 13.38. The large open space in the center of the cells is called the *lumen* and is used for water conduction. The length of a longitudinal tracheid is about 3 to 5 mm and its diameter is about 20 to 80 μm. Holes or pits at the end of the cells allow liquid to flow from one cell to another. The longitudinal tracheids constitute about 90 percent of the volume of the softwood. The earlywood cells have a relatively large diameter, thin walls, and a large-size lumen. The latewood cells have a smaller diameter and thick walls with a smaller lumen than the earlywoods cells.

Wood rays which run in the transverse direction from the bark to the center of the tree consist of an aggregate of small *parenchyma cells* that are bricklike in shape. The parenchyma cells, which are used for food storage, are interconnected along the rays by pit pairs.

Microstructure of Hardwoods

Hardwoods in contrast to softwoods have large-diameter *vessels* for the conduction of fluids. The vessels are thin-walled structures consisting of individual elements called *vessel elements* and are formed in the longitudinal direction of the tree stem.

The wood of hardwood trees is classified as either *ring-porous* or *diffuse-porous*, depending on how the vessels are arranged in the growth rings. In a ring-porous hardwood the vessels formed in the earlywood are larger than

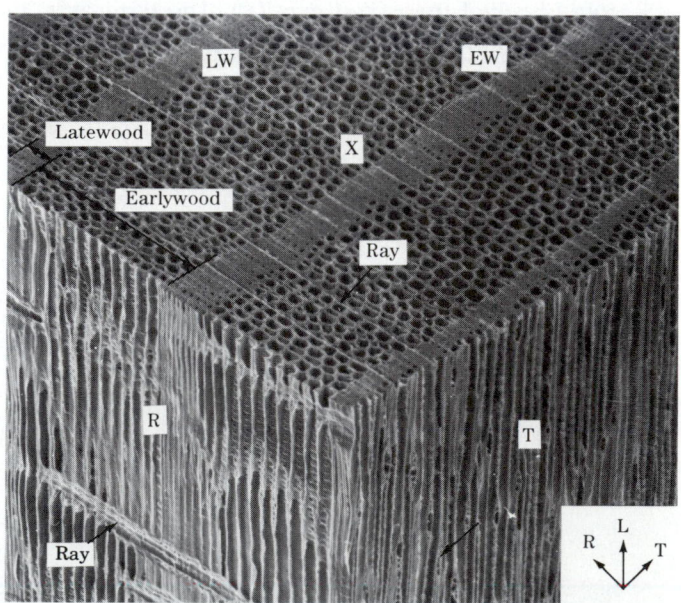

FIGURE 13.38 Scanning electron micrograph of softwood (longleaf pine) block showing three complete growth rings in cross-section surface. Note that the individual cells are larger in the earlywood (EW) than in the latewood (LW). Rays which consist of food-storing cells run perpendicular to the longitudinal direction. (Magnification 75×.) (*Courtesy of the N. C. Brown Center for Ultrastructure Studies, SUNY College of Environmental Science and Forestry.*)

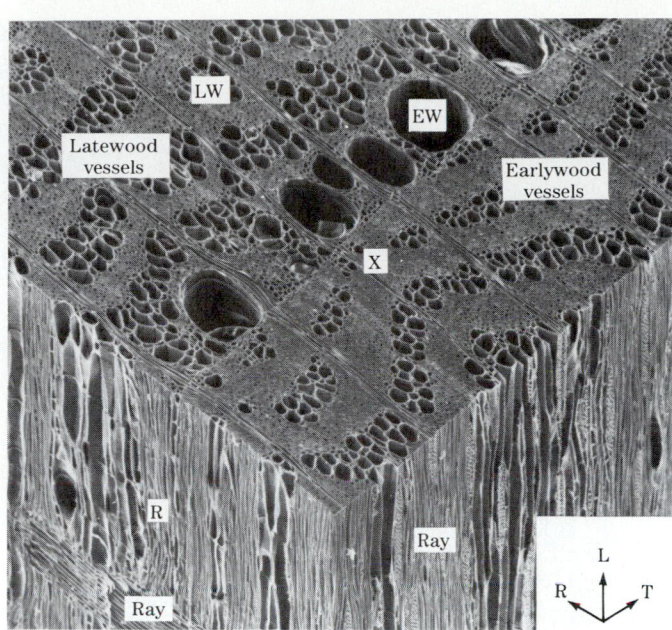

FIGURE 13.39 Scanning electron micrograph of ring-porous hardwood (American elm) block showing the abrupt change in the diameter of the earlywood (EW) and latewood (LW) vessels as observed in the cross-section surface. (Magnification 54×.) (*Courtesy of the N. C. Brown Center for Ultrastructure Studies, SUNY College of Environmental Science and Forestry.*)

those formed in the latewood (Fig. 13.39). In a diffuse-porous hardwood, the vessel diameters are essentially the same throughout all the growth ring (Fig. 13.40).

The longitudinal cells responsible for the support of the hardwood tree stem are fibers. Fibers in hardwood trees are elongated cells with close-pointed ends and are usually thick-walled. Fibers range in length from about 0.7 to 3 mm and average less than about 20 μm in diameter. The wood volume of hardwoods made up of fibers varies considerably. For example, the volume of fibers in a sweetgum hardwood is 26 percent, whereas the volume for hickory is 67 percent.

The food-storage cells of the hardwoods are the ray (transverse) and longitudinal parenchyma which are brick- or box-shaped. The rays for hardwoods are usually much larger than for softwoods, having many cells across their width.

Cell-Wall Ultrastructure

Let us now examine the structure of a wood cell at high magnification, such as the one shown telescoped in Fig. 13.41. The initial cell wall that is formed during cell division during the growth period is called the *primary wall*. During its growth the primary wall enlarges in the transverse and longitudinal directions, and after it reaches full size, the *secondary wall* forms in concentric layers growing into the center of the cell (Fig. 13.41).

The principal constituents of the wood cell are *cellulose, hemicellulose*, and *lignin*. Cellulose crystalline molecules make up between 45 to 50 percent of the solid material of wood. Cellulose is a linear polymer consisting of glucose

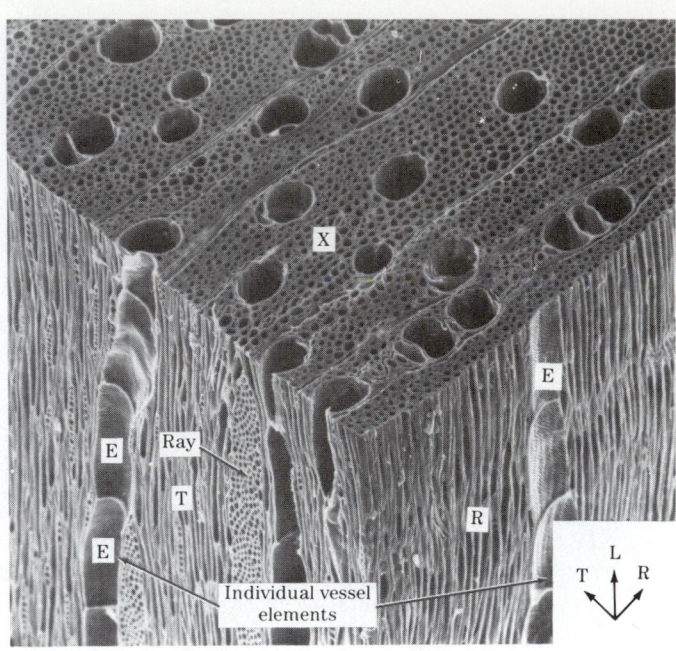

FIGURE 13.40 Scanning electron micrograph of diffuse-porous hardwood (sugar maple) block showing the rather uniform diameter of the vessels throughout the growth ring. The formation of the vessel from individual vessel elements is clearly visible. (Magnification 100×.) (*Courtesy of the N. C. Brown Center for Ultrastructure Studies, SUNY College of Environmental Science and Forestry.*)

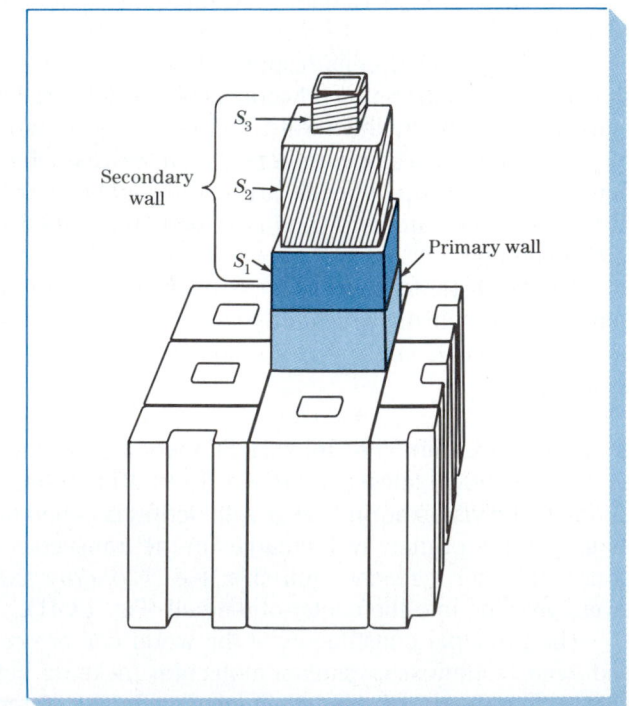

FIGURE 13.41 Schematic drawing of a telescoped wood cell in a multiwood cell structure showing the relative thicknesses of the primary and secondary walls of a wood cell. The lines on the primary and secondary walls indicate the microfibril orientations. [*After R. J. Thomas, J. Educ. Module Mater. Sci., 2:85(1980). Used by permission, the Journal of Materials Education, University Park, Pa.*]

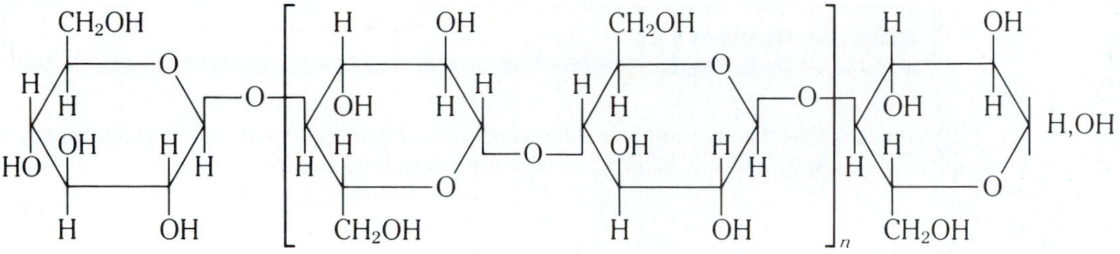

FIGURE 13.42 The structure of a cellulose molecule. (*After J. D. Wellons, "Adhesive Bonding of Woods and Other Structural Materials," University Park, Pa., Materials Education Council, 1983.*)

units (Fig. 13.42) with a degree of polymerization ranging from 5000 to 10,000. The covalent bonding within and between the glucose units creates a straight and stiff molecule with high tensile strength. Lateral bonding between the cellulose molecules is by hydrogen and permanent dipolar bonding. Hemicellulose makes up 20 to 25 percent by weight of the solid material of wood cells and is a branched amorphous molecule containing several types of sugar units. Hemicellulose molecules have a degree of polymerization between 150 to 200. The third main constituent of wood cells is lignin which constitutes about 20 to 30 percent by weight of the solid material. Lignins are very complex, cross-linked, three-dimensional polymeric materials formed from phenolic units.

The cell wall consists mainly of *microfibrils* bonded together by a lignin cement. The microfibrils themselves are believed to consist of a crystalline core of cellulose surrounded by an amorphous region of hemicellulose and lignin. The arrangement and orientations of the microfibrils vary in different layers of the cell wall, as indicated in Fig. 13.41. The lignins provide rigidity to the cell wall and allow it to resist compressive forces. In addition to solid materials, the wood cells may adsorb up to about 30 percent of their weight in water.

Properties of Wood

Moisture content   Wood, unless oven-dried to a constant weight, contains some moisture. Water occurs in wood either adsorbed in the fiber walls of the cells or as unbound water in the cell-fiber lumen. By convention, the percent of water in wood is defined by the equation

$$\text{Wood moisture content (wt \%)} = \frac{\text{wt of water in sample}}{\text{wt of dry wood sample}} \times 100\% \quad (13.18)$$

Because the percentage of water is on a dry basis, the moisture content of wood may exceed 200 percent.

Example Problem 13.5

A piece of wood containing moisture weighs 165.3 g and after oven drying to a constant weight, weighs 147.5 g. What is its percent moisture content?

Solution:

The weight of water in the wood sample is equal to the weight of the wet wood sample minus its weight after oven drying to a constant weight. Thus,

$$\% \text{ Moisture content} = \frac{\text{weight of wet wood} - \text{weight of dry wood}}{\text{weight of dry wood}} \times 100\% \quad (13.18)$$

$$= \frac{165.3 \text{ g} - 147.5 \text{ g}}{147.5 \text{ g}} \times 100\% = 12.1\% \blacktriangleleft$$

The moisture condition of wood in the living tree is referred to as the *green condition*. The average moisture content of the sapwood of softwoods in the green condition is about 150 percent, while that of the heartwood of the same species is about 60 percent. In hardwood the difference in green-condition moisture content between the sapwood and heartwood is usually much less, with both averaging about 80 percent.

Mechanical strength   Some typical mechanical properties of some types of woods grown in the United States are listed in Table 13.6. In general, woods which are botanically classified as softwoods are physically soft, and those classified as hardwoods are physically hard, although there are some exceptions. For example, balsa wood, which is physically very soft, is classified botanically as a hardwood.

The compressive strength of wood parallel to the grain is considerably higher than of wood perpendicular to the grain by a factor of about 10. For example, the compressive strength of kiln-dried (12 percent moisture content) eastern white pine is 4800 psi (33 MPa) parallel to the grain while only 440 psi (3.0 MPa) perpendicular to the grain. The reason for this difference is that in the longitudinal direction the strength of the wood is due primarily to the strong covalent bonds of the cellulose microfibrils which are oriented mainly longitudinally. The strength of the wood perpendicular to the grain is much lower because it depends on the strength of the weaker hydrogen bonds which bond the cellulose molecules laterally.

As observed in Table 13.6, wood in the green condition is weaker than kiln-dried wood. The reason for this difference is that the removal of water from the less-ordered regions of the cellulose of the microfibril allows the cell molecular structure to compact and form internal bridges by hydrogen bonding. Thus, upon losing moisture, the wood shrinks and becomes denser and stronger.

TABLE 13.6 Typical Mechanical Properties of Some Commercially Important Woods Grown in the United States

Species	Condition	Specific gravity	Static bending		Compression parallel to grain; Maximum crushing strength, psi*	Compression perpendicular to grain; fiber stress at prop. limit, psi*	Shear parallel to grain; maximum shearing strength, psi*
			Modulus of rupture, psi*	Modulus of elasticity, 10^6 psi*			
Hardwoods:							
Elm, American	Green	0.46	7,200	1.11	2910	360	1000
	Kiln dry†	0.50	11,800	1.34	5520	690	1510
Hickory, pecan	Green	0.60	9,800	1.37	3990	780	1480
	Kiln dry†	0.66	13,700	1.73	7850	1720	2080
Maple, red	Green	0.49	7,700	1.39	3280	400	1150
	Kiln dry†	0.54	13,400	1.64	6540	1000	1850
Oak, white	Green	0.60	8,300	1.25	3560	670	1250
	Kiln dry†	0.68	15,200	1.78	7440	1070	2000
Softwoods:							
Douglas fir, coast	Green	0.45	7,700	1.56	3780	380	900
	Kiln dry†	0.48	12,400	1.95	7240	800	1130
Cedar, western red	Green	0.31	5,200	0.94	2770	240	770
	Kiln dry†	0.32	7,500	1.11	4560	460	990
Pine, eastern white	Green	0.34	4,900	0.99	2440	220	680
	Kiln dry†	0.35	8,600	1.24	4800	440	900
Redwood, young growth	Green	0.34	5,900	0.96	3110	270	890
	Kiln dry†	0.35	7,900	1.10	5220	520	1110

* To obtain MPa, multiply psi by 6.89×10^{-3}.
† Kiln-dried to 12% moisture.
Source: "The Encyclopedia of Wood," Sterling Publishing Co., 1980, pp. 68–75.

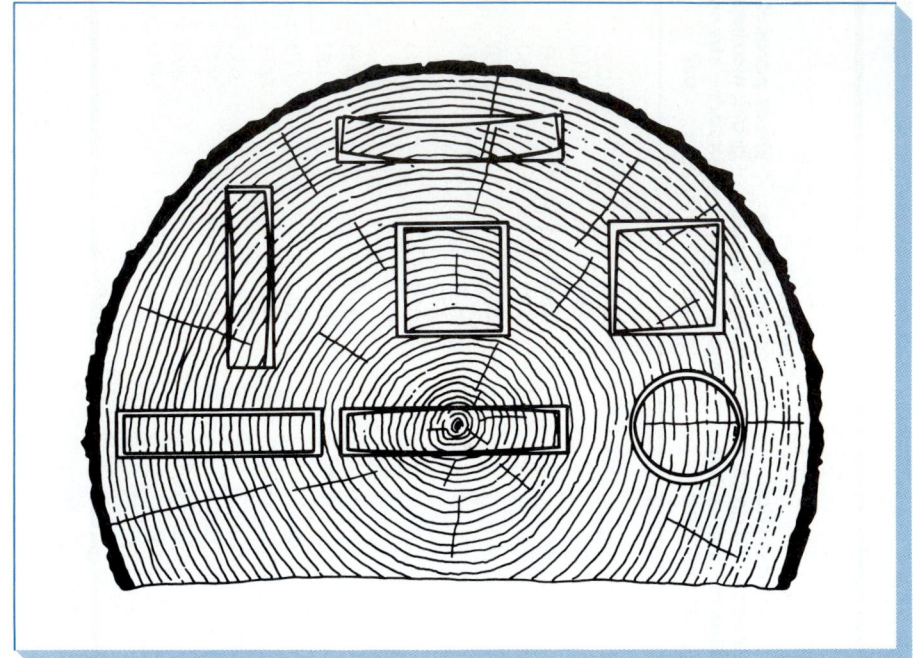

FIGURE 13.43 Shrinkage and distortion of tree section in the tangential, radial, and annual ring directions. [*After R. T. Hoyle, J. of Educ. Module Mater. Sci., 4:88(1982). Used by permission, the Journal of Materials Education, University Park, Pa.*]

Shrinkage   Greenwood shrinks as moisture is eliminated from it and causes distortion of the wood, as shown in Fig. 13.43 for the radial and tangential directions of a cross section of a tree. Wood shrinks considerably more in the transverse direction than in the longitudinal direction, with the transverse shrinkage usually ranging from 10 to 15 percent as compared to only about 0.1 percent in the longitudinal direction.

When water is eliminated from the amorphous regions on the exterior part of the microfibrils, they become closer together and the wood becomes denser. Because the long dimension of the microfibrils is mainly oriented in the longitudinal direction of the tree stem, wood shrinks mainly transversely.

13.9   SANDWICH STRUCTURES

Composite materials made by sandwiching a core material between two thin outer layers are commonly used in engineering designs. Two types of these materials are (1) honeycomb sandwich and (2) cladded sandwich structures.

Honeycomb Sandwich Structure   Bonded honeycomb sandwich construction has been used as a basic construction material in the aerospace industry for over 30 years. Most aircraft flying

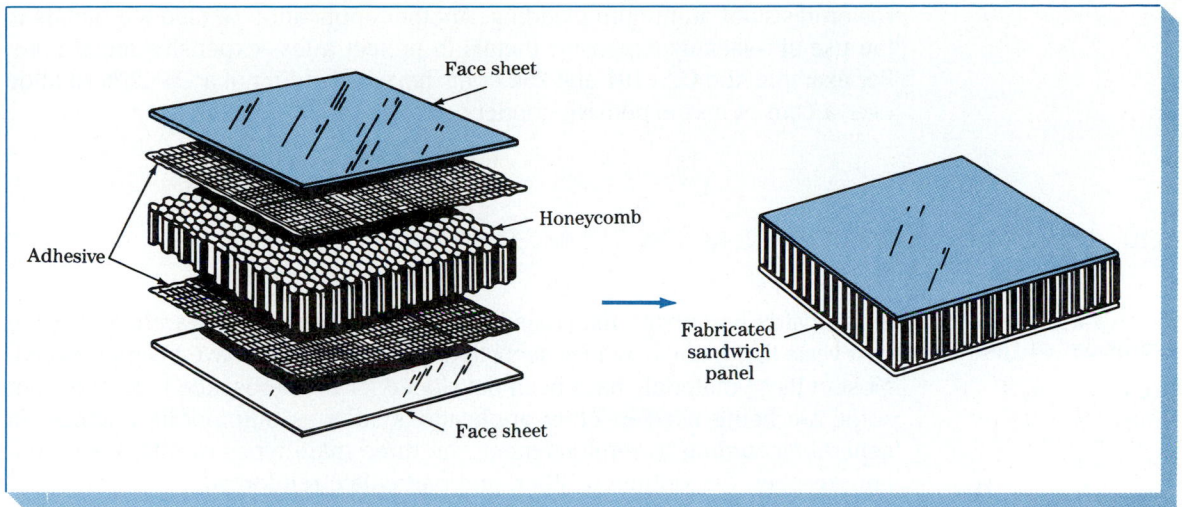

FIGURE 13.44 Sandwich panel fabricated by bonding aluminum faces to aluminum alloy honeycomb core. (*Courtesy of the Hexcel Co.*)

today depend on this construction material. Most honeycomb in use today is made of aluminum alloys such as 5052 and 2024 or glass-reinforced phenolic, glass-reinforced polyester, and aramid-fiber-reinforced materials.

Aluminum honeycomb panels are fabricated by adhesively bonding aluminum alloy face sheets to aluminum alloy honeycomb core sections, as shown in Fig. 13.44. This type of construction provides a stiff, rigid, strong, and lightweight sandwich panel.

Cladded Metal Structures

The cladded structure is used to produce composites of a metal core with thin outer layers of another metal or metals (Fig. 13.45). Usually, thin outer metal layers are hot-roll-bonded to the inner core metal to form metallurgical (atomic diffusion) bonds between the outside layers and the inner core metal. This type of composite material has many applications in industry. For example, high-strength aluminum alloys such as 2024 and 7075 have relatively poor corrosion resistance and can be protected by a thin layer of soft, highly cor-

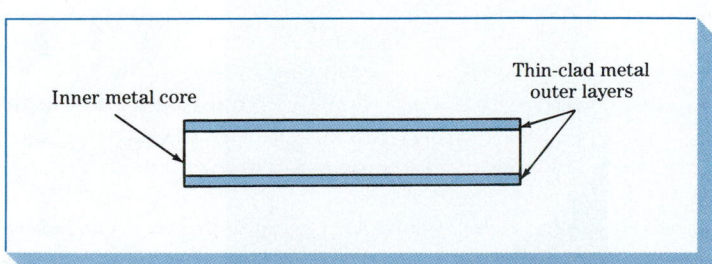

FIGURE 13.45 Cross section of a cladded metal structure.

rosion resistant aluminum cladding. Another application of cladded metals is the use of relatively expensive metals to protect a less-expensive metal core. For example, the U.S. 10¢ and 25¢ coins have a cladding of a Cu–25% Ni alloy over a core of less-expensive copper.

13.10 METAL-MATRIX AND CERAMIC-MATRIX COMPOSITES

Metal-Matrix Composites (MMCs)

Metal-matrix composite materials have been so intensely researched over the past years that many new high-strength-to-weight materials have been produced. Most of these materials have been developed for the aerospace industries, but some are being used in other applications such as automobile engines. In general, according to reinforcement, the three main types of MMCs are continuous-fiber, discontinuous-fiber, and particulate reinforced.

Continuous-fiber reinforced MMCs Continuous filaments provide the greatest improvement in stiffness (tensile modulus) and strength for MMCs. One of the first developed continuous-fiber MMCs was the aluminum alloy matrix–boron fiber reinforced system. The boron fiber for this composite is made by chemically vapor depositing boron on a tungsten-wire substrate (Fig. 13.46a). The Al–B composite is made by hot pressing layers of B fibers between aluminum foils so that the foils deform around the fibers and bond to each other. Figure 13.46b shows a cross section of a continuous-B-fiber–aluminum alloy matrix composite. Table 13.7 lists some mechanical properties for some B fiber reinforced–aluminum alloy composites. With the addition of 51 vol % B, the axial

FIGURE 13.46 (a) 100-μm-diameter boron filament surrounding a 12.5-μm-diameter tungsten-wire core. (b) Micrograph of a cross section of an aluminum alloy–boron composite (magnification 40×). (After "Engineered Materials Handbook," vol. 1, ASM International, 1987, p. 852.)

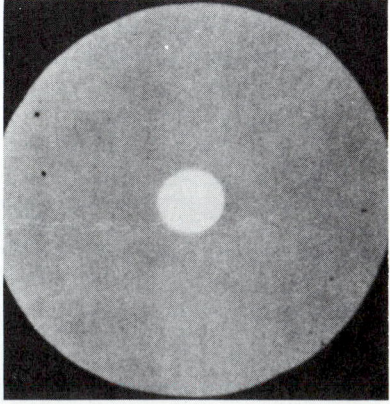
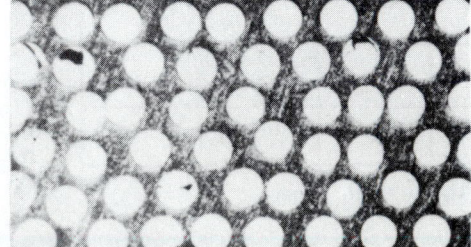

TABLE 13.7   Mechanical Properties of Metal-Matrix Composite Materials

	Tensile strength		Elastic modulus		
	MPa	ksi	GPa	Msi	Strain to failure, %
Continuous-fiber MMCs:					
Al 2024-T6 (45% B) (axial)	1458	211	220	32	0.810
Al 6061-T6 (51% B) (axial)	1417	205	231	33.6	0.735
Al 6061-T6 (47% SiC) (axial)	1462	212	204	29.6	0.89
Discontinuous-fiber MMCs:					
Al 2124-T6 (20% SiC)	650	94	127	18.4	2.4
Al 6061-T6 (20% SiC)	480	70	115	17.7	5
Particulate MMCs:					
Al 2124 (20% SiC)	552	80	103	15	7.0
Al 6061 (20% SiC)	496	72	103	15	5.5
No reinforcement:					
Al 2124-F	455	66	71	10.3	9
Al 6061-F	310	45	68.9	10	12

tensile strength of aluminum alloy 6061 was increased from 310 to 1417 MPa, while its tensile modulus was increased from 69 to 231 GPa. Applications for Al–B composites include some of the structural members in the midfuselage of the space shuttle orbiter.

Other continuous-fiber reinforcements that have been used in MMCs are silicon carbide, graphite, alumina, and tungsten fibers. A composite of Al 6061 reinforced with SiC continuous fibers is being evaluated for the vertical tail section for an advanced fighter aircraft (Fig. 13.1c). Of special interest is the projected use of SiC continuous-fiber reinforcements in a titanium aluminide matrix for hypersonic aircraft such as the National Aerospace plane (Fig. 1.1).

Discontinuous-fiber and particulate reinforced MMCs   Many different kinds of discontinuous and particulate reinforced MMCs have been produced. These materials have the engineering advantage of higher strength, greater stiffness, and better dimensional stability than the unreinforced metal alloys. In this brief treatment of MMCs we will focus on aluminum alloy MMCs.

Particulate reinforced MMCs are low-cost aluminum alloy MMCs made by using irregular-shaped particles of alumina and silicon carbide in the range of about 3 to 200 μm in diameter. The particulate, which is sometimes given a proprietary coating, can be mixed with the molten aluminum alloy and cast into remelt ingots or extrusion billets for further fabrication. Table 13.7 indicates that the ultimate tensile strength of Al alloy 6061 can be increased from 310 to 496 MPa with a 20% SiC addition, while the tensile modulus can be increased from 69 to 103 GPa. Applications for this material include sporting equipment and automobile engine parts.

(200X)

(2000X)

FIGURE 13.47 Micrographs of single-crystal silicon carbide whiskers used to reinforce metal-matrix composites. The whiskers are 1 to 3 μm in diameter and 50 to 200 μm long. (*Courtesy of American Matrix Inc.*)

Discontinuous-fiber reinforced MMCs are produced mainly by powder metallurgy and melt infiltration processes. In the powder metallurgy process, needlelike silicon carbide whiskers about 1 to 3 μm in diameter and 50 to 200 μm long (Fig. 13.47) are mixed with metal powders, consolidated by hot pressing, and then extruded or forged into the desired shape. Table 13.7 shows that the ultimate tensile strength of Al alloy 6061 can be increased from 310 to 480 MPa with a 20% SiC whisker addition, while the tensile modulus can be raised from 69 to 115 GPa. Although greater increases in strength and stiffness can be achieved with the whisker additions than with the particulate material, the powder metallurgy and melt infiltration processes are more costly. Applications for discontinuous-fiber reinforced aluminum alloy MMCs include missile guidance parts and high-performance automobile pistons.

Example Problem 13.6

A metal-matrix composite is made from a boron (B) fiber reinforced aluminum alloy (Fig. EP13.6). To form the boron fiber, a tungsten (W) wire ($r = 10$ μm) is coated with boron, giving a final radius of 75 μm. The aluminum alloy is then bonded around the boron fibers, giving a volume fraction of 0.65 for the aluminum alloy. Assuming that the rule of binary mixtures [(Eq. (13.7)] applies also to ternary mixtures, calculate the effective tensile elastic modulus of the composite material under isostrain conditions. Data: $E_W = 410$ GPa; $E_B = 379$ GPa; $E_{Al} = 68.9$ GPa.

Solution:

$$E_{comp} = f_W E_W + f_B E_B + f_{Al} E_{Al} \qquad f_{W+B} = 0.35$$

$$f_W = \frac{\text{area of W wire}}{\text{area of B fiber}} \times f_{W+B}$$

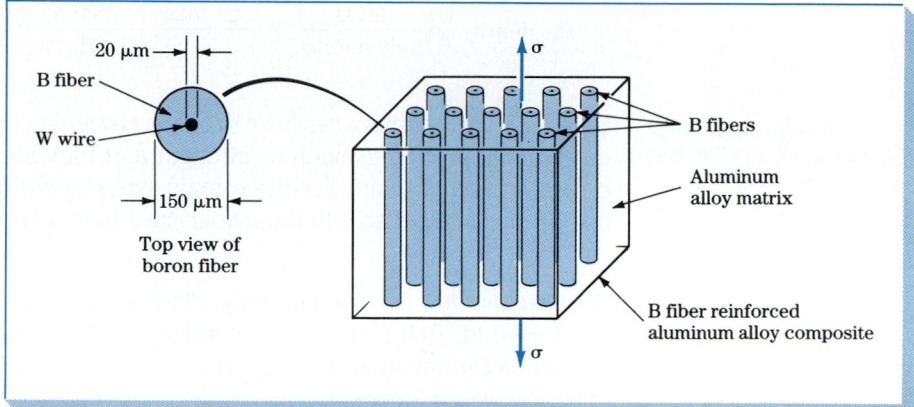

$$f_W = \frac{\pi(10\ \mu m)^2}{\pi(75\ \mu m)^2} \times 0.35 = 6.22 \times 10^{-3} \qquad f_{Al} = 0.65$$

$$f_B = \frac{\text{area of B fiber} - \text{area of W wire}}{\text{area of B fiber}} \times f_{W+B}$$

$$= \frac{\pi(75\ \mu m)^2 - \pi(10\ \mu m)^2}{\pi(75\ \mu m)^2} \times 0.35 = 0.344$$

$$E_{comp} = f_W E_W + f_B E_B + f_{Al} E_{Al}$$

$$= (6.22 \times 10^{-3})(410\ \text{GPa}) + (0.344)(379\ \text{GPa}) + (0.65)(68.9\ \text{GPa})$$

$$= 178\ \text{GPa} \blacktriangleleft$$

Note that the tensile modulus (stiffness) of the composite is about 2.5 times that of the unreinforced aluminum alloy.

Example Problem 13.7

A metal-matrix composite is made with 80% by volume of aluminum alloy 2124-T6 and 20% by volume of SiC whiskers. The density of the 2124-T6 alloy is 2.77 g/cm³ and that of the whiskers is 3.10 g/cm³. Calculate the average density of the composite material.

Solution:

Basis is 1 m³ of material; thus we have 0.80 m³ of 2124 alloy and 0.20 m³ of SiC fiber in 1 m³ of material.

$$\text{Mass of 2124 alloy in 1 m}^3 = (0.80\ \text{m}^3)(2.77\ \text{Mg/m}^3) = 2.22\ \text{Mg}$$
$$\underline{\text{Mass of SiC whiskers in 1 m}^3 = (0.20\ \text{m}^3)(3.10\ \text{Mg/m}^3) = 0.62\ \text{Mg}}$$

$$\text{Total mass in 1 m}^3 \text{ of composite} = 2.84\ \text{Mg}$$

$$\text{Av density} = \frac{\text{mass}}{\text{unit volume}} = \frac{\text{total mass of material in 1 m}^3}{1 \text{ m}^3} = 2.84 \text{ Mg/m}^3 \blacktriangleleft$$

Ceramic-Matrix Composites (CMCs)

Ceramic-matrix composites have been developed recently which have improved mechanical properties such as strength and toughness over the unreinforced ceramic matrix. Again, the three main types according to reinforcement are continuous-fiber, discontinuous-fiber, and particulate reinforced.

Continuous-fiber reinforced CMCs Two kinds of continuous fibers that have been used for CMCs are silicon carbide and aluminum oxide. In one process to make a ceramic-matrix composite, SiC fibers are woven into a mat and then chemical vapor deposition is used to impregnate SiC into the fibrous mat. In another process SiC fibers are encapsulated by a glass-ceramic material (see Example Problem 13.8). Applications for these materials include heat-exchanger tubes, thermal protection systems, and materials for corrosion-erosion environments.

Discontinuous (whisker) and particulate reinforced CMCs Ceramic whiskers (Fig. 13.47) can significantly increase the fracture toughness of monolithic ceramics (Table 13.8). A 20 vol % SiC whisker addition to alumina can increase the fracture toughness of the alumina ceramic from about 4.5 to 8.5 MPa$\sqrt{\text{m}}$. Short-fiber and particulate reinforced ceramic-matrix materials have the advantage of being able to be fabricated by common ceramic processes such as hot isostatic pressing (*HIP*ing).

Ceramic-matrix composites are believed to be toughened by three main

TABLE 13.8 **Mechanical Properties of SiC Whisker Reinforced Ceramic-Matrix Composites at Room Temperature**

Matrix	SiC whisker content, vol %	Flexural strength		Fracture toughness	
		MPa	ksi	MPa$\sqrt{\text{m}}$	ksi$\sqrt{\text{in}}$
Si$_3$N$_4$	0	400–650	60–95	5–7	4.6–6.4
	10	400–500	60–75	6.5–9.5	5.9–8.6
	30	350–450	50–65	7.5–10	6.8–9.1
Al$_2$O$_3$	0	...	...	4.5	4.1
	10	400–510	57–73	7.1	6.5
	20	520–790	75–115	7.5–9.0	6.8–8.2

Source: "Engineered Materials Handbook," vol. 1, Composites, ASM International, 1987, p. 942.

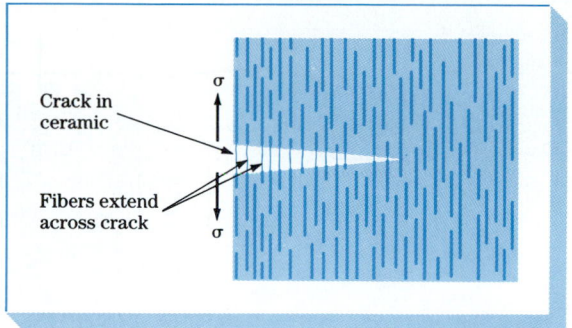

FIGURE 13.48 Schematic diagram showing how reinforcing fibers can inhibit crack propagation in ceramic-matrix materials by crack bridging and fiber pullout energy absorption.

mechanisms, all of which result from the reinforcing fibers interfering with crack propagation in the ceramic. These mechanisms are:

1. *Crack deflection.* Upon encountering the reinforcement, the crack is deflected, making its propagating path more meandering. Thus, higher stresses are required to propagate the crack.
2. *Crack bridging.* Fibers or whiskers can bridge the crack and help keep the material together, thus increasing the stress level needed to cause further cracking (Fig. 13.48).
3. *Fiber pullout.* The friction caused by fibers or whiskers being pulled out of the cracking matrix absorbs energy, and thus higher stresses must be applied to produce further cracking. Therefore, a good interfacial bond is required between the fibers and the matrix for higher strengths. There also should be a good match of coefficient of expansion between the matrix and fibers if the material is to be used at high temperatures.

Example Problem 13.8

A ceramic-matrix composite is made with continuous SiC fibers embedded in a glass-ceramic matrix (Fig. EP13.8). (*a*) Calculate the tensile elastic modulus of the composite under isostrain conditions, and (*b*) calculate the stress σ at which the cracks start to grow. Data are as follows:

Glass-ceramic matrix:
$E = 94$ GPa
$K_{IC} = 2.4$ MPa$\sqrt{\text{m}}$
Largest preexisting flaw is 10 μm in diameter

SiC fibers:
$E = 350$ GPa
$K_{IC} = 4.8$ MPa$\sqrt{\text{m}}$
Largest surface notches are 5 μm deep

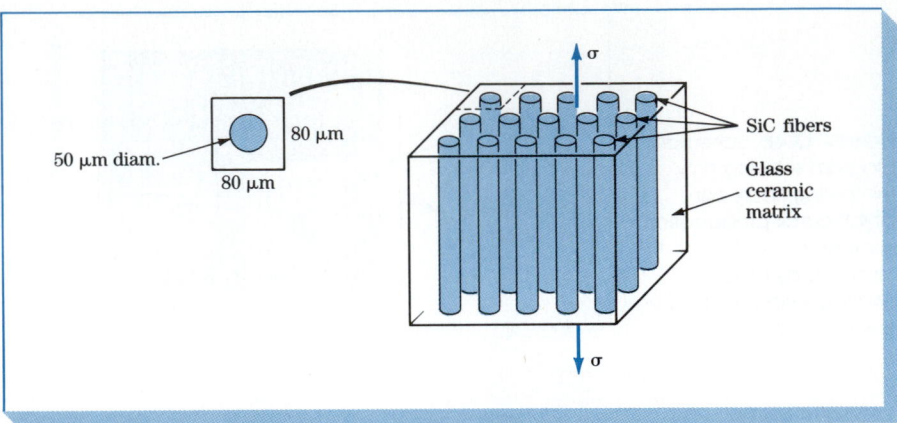

Solution:

(*a*) Calculation of E for the composite. Assuming Eq. (13.7) for isostrain conditions is valid,

$$E_{comp} = f_{GC}E_{GC} + f_{SiC}E_{SiC}$$

Since the length of the fibers is the same, we can calculate the volume fraction of the SiC fibers by calculating the area fraction occupied by the fibers on the surface at the end of the fibers. From Fig. EP13.8, each 50-μm-diameter fiber is enclosed by an 80 μm × 80 μm area. Thus,

$$f_{SiC} = \frac{\text{area of fiber}}{\text{selected tot. area}} = \frac{\pi(25 \ \mu m)^2}{(80 \ \mu m)(80 \ \mu m)} = 0.307$$

$$f_{GC} = 1 - 0.307 = 0.693$$

$$E_{comp} = (0.693)(94 \text{ GPa}) + (0.307)(350 \text{ GPa}) = 172 \text{ GPa} \blacktriangleleft$$

(*b*) Stress at which cracks first start to form in the composite. For isostrain conditions, $\epsilon_{comp} = \epsilon_{GC} = \epsilon_{SiC}$. Since $\sigma = E\epsilon$ and $\epsilon = \sigma/E$,

$$\frac{\sigma_{comp}}{E_{comp}} = \frac{\sigma_{GC}}{E_{GC}} = \frac{\sigma_{SiC}}{E_{SiC}}$$

Fracture begins in a given component when $\sigma = K_{IC}/\sqrt{\pi a}$ [Eq. (10.4)], assuming $Y = 1$. We shall calculate the minimum stress to cause crack formation in both materials and then compare our results. The component which cracks with the lower stress will determine at what stress the composite will begin to crack.

   (i) Glass ceramic. For this material the largest preexisting flaw is 10 μm in diameter. This value is $2a$ for Eq. (10.4), and thus $a = 10 \ \mu m/2$ or $a = 5 \ \mu m$.

$$\frac{\sigma_{comp}}{E_{comp}} = \frac{\sigma_{GC}}{E_{GC}} = \left(\frac{K_{IC,GC}}{\sqrt{\pi a}}\right)\left(\frac{1}{E_{GC}}\right)$$

$$\sigma_{comp} = \left(\frac{E_{comp}}{E_{GC}}\right)\left(\frac{K_{IC,GC}}{\sqrt{\pi a}}\right) = \left(\frac{172 \text{ GPa}}{94 \text{ GPa}}\right)\left[\frac{2.4 \text{ MPa}\sqrt{m}}{\sqrt{\pi(5 \times 10^{-6} \text{ m})}}\right] = 1109 \text{ MPa}$$

(ii) SiC fibers. For this material, $a = 5$ μm for surface cracks.

$$\sigma_{comp} = \left(\frac{E_{comp}}{E_{SiC}}\right)\left(\frac{K_{IC,SiC}}{\sqrt{\pi a}}\right) = \left(\frac{172 \text{ GPa}}{350 \text{ GPa}}\right)\left[\frac{4.8 \text{ MPa}\sqrt{m}}{\sqrt{\pi(5 \times 10^{-6} \text{ m})}}\right] = 596 \text{ MPa}$$

Thus, the SiC fibers in the ceramic composite will begin to crack first at an applied stress of 596 MPa. ◀

SUMMARY

A composite material with respect to materials science and engineering can be defined as a materials system composed of a mixture or combination of two or more micro- or macroconstituents that differ in form and chemical composition and which are essentially insoluble in each other.

Some fiber-reinforced-plastic composite materials are made with synthetic fibers of which glass, carbon, and aramid are important types. Of these three fibers, glass fibers are the lowest in cost and have intermediate strength and highest density as compared to the others. Carbon fibers have high strength, high modulus, and low density but are expensive and so are used only for applications requiring their especially high strength-to-weight ratio. Aramid fibers have high strength and low density but are not as stiff as carbon fibers. Aramid fibers are also relatively expensive and so are used for applications where a high strength-to-weight ratio is required along with better flexibility than carbon fibers. The most commonly used matrices for glass fibers for fiber-reinforced-plastic composites are the polyesters, whereas the most commonly used matrices for carbon-fiber-reinforced plastics are the epoxies. Carbon-fiber-reinforced-epoxy composite materials are used extensively for aircraft and aerospace applications. Glass-fiber-reinforced-polyester composite materials have a much wider usage and find application in the building, transportation, marine, and aircraft industries, for example.

Concrete is a ceramic composite material consisting of aggregate particles (i.e., sand and gravel) in a hardened cement paste matrix usually made with portland cement. Concrete as a construction material has advantages which include usable compressive strength, economy, castability on the job, durability, fire resistance, and aesthetic appearance. The low tensile strength of concrete can be increased significantly by reinforcement with steel rods. A further improvement in the tensile strength of concrete is attainable by introducing residual compressive stresses in the concrete at positions of high tensile loading by prestressing with steel reinforcements.

Wood is a natural composite material consisting essentially of cellulose fibers bonded together by a matrix of polymeric material made up mainly of lignin. The macrostructure of wood consists of sapwood, which is made up primarily of living cells and which carries the nutrients, and heartwood, which is composed of dead cells. The two main types of woods are softwoods and hardwoods. Softwoods have exposed seeds and narrow (needlelike) leaves, whereas hardwoods have covered seeds and broad leaves. The microstructure of wood consists of arrays of cells mainly in the longitudinal direction of the tree stem. Softwoods have long thin-walled tubular cells called tracheids, whereas hardwoods have a dense cell structure which contains large vessels for the conduction of fluids. Wood as a construction material has advantages which include usable strength, economy, ease of workability, and durability if properly protected.

DEFINITIONS

Sec. 13.1

Composite material: a materials system composed of a mixture or combination of two or more micro- or macroconstituents that differ in form and chemical composition and which are essentially insoluble in each other.

Sec. 13.2

Fiber-reinforced plastics: composite materials consisting of a mixture of a matrix of a plastic material such as a polyester or epoxy strengthened by fibers of high strength such as glass, carbon, or aramid. The fibers provide the high strength and stiffness, and the plastic matrix bonds the fibers together and supports them.

E-glass fibers: fibers made from E (electrical) glass which is a borosilicate glass and which is the most commonly used glass for fibers for fiberglass-reinforced plastics.

S-glass fibers: fibers made from S glass which is a magnesia-alumina-silicate glass and which is used for fibers for fiberglass-reinforced plastics when extra-high-strength fibers are required.

Roving: a collection of bundles of continuous fibers twisted or untwisted.

Carbon fibers (for a composite material): carbon fibers produced mainly from polyacrylonitrile (PAN) or pitch which are stretched to align the fibrillar network structure within each carbon fiber and which are heated to remove oxygen, nitrogen, and hydrogen from the starting or precursor fibers.

Tow (of fibers): a collection of numerous fibers in a straight-laid bundle and which is specified by a certain number of fibers, e.g., 6000 fibers/tow.

Aramid fibers: fibers produced by chemical synthesis and used for fiber-reinforced plastics. Aramid fibers have an aromatic (benzene ring type) polyamide linear structure and are produced commercially by the Du Pont Co. under the trade name of Kevlar.

Specific tensile strength: the tensile strength of a material divided by its density.

Specific tensile modulus: the tensile modulus of a material divided by its density.

Sec. 13.3

Laminate: a product made by bonding sheets of a material together, usually with heat and pressure.

Unidirectional laminate: a fiber-reinforced-plastic laminate produced by bonding together layers of fiber-reinforced sheets which all have continuous fibers in the same direction in the laminate.

Multidirectional laminate: a fiber-reinforced-plastic laminate produced by bonding together layers of fiber-reinforced sheets with some of the directions of the continuous fibers of the sheets being at different angles.

Laminate ply (lamina): one layer of a multilayer laminate.

Prepreg: a ready-to-mold plastic resin-impregnated cloth or mat which may contain reinforcing fibers. The resin is partially cured to a "B" stage and is supplied to a fabricator who uses the material as the layers for a laminated product. After the layers are laid up to produce a final shape, the layers are bonded together, usually with heat and pressure, by the curing of the laminate.

Sec. 13.4

Hand lay-up: the process of placing (and working) successive layers of reinforcing material in a mold by hand to produce a fiber-reinforced composite material.

Spray lay-up: a process in which a spray gun is used to produce a fiber-reinforced product. In one type of spray-up process chopped fibers are mixed with plastic resin and sprayed into a mold to form a composite material part.

Vacuum bag molding: a process of molding a fiber-reinforced-plastic part in which sheets of transparent flexible material are placed over a laminated part that has not been cured and which are sealed. A vacuum is then applied between the cover sheets and the laminated part so that entrapped air is mechanically worked out of the laminate and then the vacuum-bagged part is cured.

Filament winding: a process for producing fiber-reinforced plastics by winding continuous reinforcement previously impregnated with a plastic resin on a rotating mandrel. When a sufficient number of layers have been applied, the wound form is cured and the mandrel removed.

Sec. 13.5

Sheet-molding compound (SMC): a compound of plastic resin, filler, and reinforcing fiber used to make fiber-reinforced-plastic composite materials. SMC is usually made with about 25 to 30 percent fibers about 1 in (2.54 cm) long, of which fiberglass is the most commonly used fiber. SMC material is usually pre-aged to a state so that it can support itself and then cut to size and placed in a compression mold. Upon hot pressing, the SMC cures to produce a rigid part.

Pultrusion: a process for producing a fiber-reinforced-plastic part of constant cross section continuously. The pultruded part is made by drawing a collection of resin-dipped fibers through a heated die.

Sec. 13.6

Concrete (portland cement type): a mixture of portland cement, fine aggregate, coarse aggregate, and water.

Portland cement: a cement consisting predominantly of calcium silicates which react with water to form a hard mass.

Aggregate: inert material mixed with portland cement and water to produce concrete. Larger particles are called coarse aggregate (e.g., gravel), and smaller particles are called fine aggregate (e.g., sand).

Hydration reaction: reaction of water with another compound. The reaction of water with portland cement is a hydration reaction.

Air-entrained concrete: concrete in which there exists a uniform dispersion of small air bubbles. About 90 percent of the air bubbles are 100 μm or less.

Reinforced concrete: concrete containing steel wires or bars to resist tensile forces.

Prestressed concrete: reinforced concrete in which internal compressive stresses have been introduced to counteract tensile stresses resulting from severe loads.

Pretensioned (prestressed) concrete: prestressed concrete in which the concrete is poured over pretensioned steel wires or rods.

Sec. 13.7

Asphalt: a bitumen consisting mainly of hydrocarbons having a wide range of molecular weights. Most asphalt is obtained from petroleum refining.

Asphalt mixes: mixtures of asphalt and aggregate which are used mainly for road paving.

Sec. 13.8

Wood: a natural composite material consisting mainly of a complex array of cellulose fibers in a polymeric material matrix made up primarily of lignin.

Lignin: a very complex cross-linked three-dimensional polymeric material formed from phenolic units.

Sapwood: the outer part of the tree stem of a living tree that contains some living cells which store food for the tree.

Heartwood: the innermost part of the tree stem which in the living tree contains only dead cells.

Cambium: the tissue located between the wood and bark and which is capable of repeated cell division.

Softwood trees: trees which have exposed seeds and narrow leaves (needles). Examples are pine, fir, and spruce.

Hardwood trees: trees which have covered seeds and broad leaves. Examples are oak, maple, and ash.

Parenchyma: food-storing cells of trees which are short with relatively thin walls.

Wood ray: a ribbonlike aggregate of cells extending radially in the tree stem; the tissue of the ray is primarily composed of food-storing parenchyma cells.

Tracheids (longitudinal): the predominating cell found in softwoods; tracheids have the function of conduction and support.

Wood vessel: a tubular structure formed by the union of smaller cell elements in a longitudinal row.

Microfibrils: elementary cellulose-containing structures that form the wood cell walls.

Lumen: the cavity in the center of a wood cell.

PROBLEMS

13.1.1 Define a composite material with respect to a materials system.

13.2.1 What are the three main types of synthetic fibers used to produce fiber-reinforced-plastic composite materials?

13.2.2 What are some of the advantages of glass-fiber-reinforced plastics?

13.2.3 What are the differences in the compositions of E and S glasses? Which is the strongest and the most costly?

13.2.4 How are glass fibers produced? What is a glass-fiber roving?

13.2.5 What properties make carbon fibers important for reinforced plastics?

13.2.6 What are two materials used as precursors for carbon fibers?

13.2.7 What are the processing steps for the production of carbon fibers from polyacrylonitrile (PAN)? What reactions take place at each step?

13.2.8 What is a tow of carbon fibers?

13.2.9 What processing steps are carried out if a very high strength type of carbon fiber is desired? If a very high modulus type of carbon fiber is desired what processing steps are carried out?

13.2.10 What is an aramid fiber? What are two types of commercially available aramid fibers?

13.2.11 What type of chemical bonding takes place within the aramid fibers? What type of chemical bonding takes place between the aramid fibers?

13.2.12 How does the chemical bonding within and between the aramid fibers affect their mechanical strength properties?

13.2.13 Compare the tensile strength, tensile modulus of elasticity, elongation, and density properties of glass, carbon, and aramid fibers (Table 13.1 and Fig. 13.8).

13.2.14 Define specific tensile strength and specific tensile modulus. What type of reinforcing fibers of those shown in Fig. 13.9 has the highest specific modulus and what type has the highest specific tensile strength?

13.3.1 What are two of the most important matrix plastics for fiber-reinforced plastics? What are some advantages of each type?

13.3.2 How does the amount and arrangement of the glass fibers in fiberglass-reinforced plastics affect their strength?

13.3.3 What are the main property contributions of the carbon fibers in carbon-fiber-reinforced plastics? What are the main property contributions of the matrix plastic?

13.3.4 Why are some carbon-fiber-epoxy composite laminates designed with the carbon fibers of different layers oriented at different angles to each other?

13.3.5 A unidirectional carbon-fiber-epoxy-resin composite contains 64% by volume of carbon fiber and 36% epoxy resin. The density of the carbon fiber is 1.80 g/cm^3 and that of the epoxy resin is 1.22 g/cm^3. (a) What are the weight percentages of carbon fibers and epoxy resin in the composite? (b) What is the average density of the composite?

13.3.6 The average density of a carbon-fiber-epoxy composite is 1.620 g/cm^3. The density of the epoxy resin is 1.23 g/cm^3 and that of the carbon fibers is 1.75 g/cm^3. (a) What is the volume percentages of carbon fibers in the composite? (b) What are the weight percentages of epoxy resin and carbon fibers in the composite?

13.3.7 Derive an equation relating the elastic modulus of a layered composite of unidirectional fibers and a plastic matrix which is loaded under isostrain conditions.

13.3.8   Calculate the tensile modulus of elasticity of a unidirectional carbon-fiber-reinforced-plastic composite material which contains 63% by volume of carbon fibers and which is stressed under isostrain conditions. The carbon fibers have a tensile modulus of elasticity of 56.0×10^6 psi and the epoxy matrix a tensile modulus of elasticity of 0.550×10^6 psi.

13.3.9   If the tensile strength of the carbon fibers of the 67% carbon-fiber-epoxy composite material of Prob. 13.3.8 is 0.32×10^6 psi and that of the epoxy resin is 9.50×10^3 psi, calculate the strength of the composite material in psi. What fraction of the load is carried by the carbon fibers?

13.3.10 Calculate the tensile modulus of elasticity of a unidirectional Kevlar 149 fiber-epoxy composite material which contains 64% by volume of Kevlar 149 fibers and which is stressed under isostrain conditions. The Kevlar 149 fibers have a tensile modulus of elasticity of 27.5×10^6 psi and the epoxy matrix a tensile modulus of elasticity of 0.550×10^6 psi.

13.3.11 If the tensile strength of the Kevlar 149 fibers is 0.600×10^6 psi and that of the epoxy resin is 10.5×10^3 psi, calculate the strength of the composite material of Prob. 13.3.10. What fraction of the load is carried by the Kevlar 149 fibers?

13.3.12 Derive an equation relating the elastic modulus of a layered composite of unidirectional fibers and a plastic matrix which is stressed under isostress conditions.

13.3.13 Calculate the tensile modulus of elasticity for a laminated composite consisting of 64% by volume of unidirectional carbon fibers and an epoxy matrix under isostress conditions. The tensile modulus of elasticity of the carbon fibers is 350 GPa and that of the epoxy is 4.60×10^3 MPa.

13.3.14 Calculate the tensile modulus of elasticity of a laminate composite consisting of 61% by volume of unidirectional Kevlar 149 fibers and an epoxy matrix stressed under isostress conditions. The tensile modulus of elasticity of the Kevlar 149 fibers is 175 GPa and that of the epoxy is 3.80×10^3 MPa.

13.4.1   Describe the hand lay-up process for producing a fiberglass-reinforced part. What are some advantages and disadvantages of this method?

13.4.2   Describe the spray-up process for producing a fiberglass-reinforced part. What are some advantages and disadvantages of this method?

13.4.3   Describe the vacuum bag–autoclave process for producing a carbon-fiber-reinforced-epoxy part for an aircraft.

13.4.4   Describe the filament-winding process. What is a distinct advantage for this process from an engineering design standpoint?

13.5.1   Describe the sheet-molding compound manufacturing process. What are some of the advantages and disadvantages of this process?

13.5.2   Describe the pultrusion process for the manufacture of fiber-reinforced plastics. What are some advantages of this process?

13.6.1   What advantages and disadvantages does concrete offer as a composite material?

13.6.2   What are the principal components of most concretes?

13.6.3   What are the basic raw materials for portland cement? Why is portland cement so called?

13.6.4   How is portland cement made? Why is a small amount of gypsum added to portland cement?

13.6.5   What are the names, chemical formulas, and abbreviations for the four principal compounds of portland cement?

13.6.6   List the five main ASTM types of portland cement and give the general conditions for which each is used and their applications.

13.6.7   What type of chemical reactions occur during the hardening of portland cement?

13.6.8   Write the chemical reactions for C_3S and C_2S with water.

13.6.9   Which component of portland cement hardens rapidly and is mostly responsible for early strength?

13.6.10 Which component of portland cement reacts slowly and is mainly responsible for the strengthening after about 1 week.

13.6.11 Which compound is kept to a low level for sulfate-resisting portland cements?

13.6.12 Why is C_4AF added to portland cement?

13.6.13 Why is it important that during the first few days of the curing of concrete the evaporation of water from its surface be prevented or reduced?

13.6.14 What is air-entrained concrete? What is an advantage of air-entrained concrete?

13.6.15 What method is used to make air-entrained concrete? What volume percent of air is used in the concrete for frost protection?

13.6.16 How does the water-cement ratio (by weight) affect the compressive strength of concrete? What ratio gives a compressive strength of about 5500 psi to normal concrete? What is the disadvantage of too high a water-cement ratio? Of too low a water-cement ratio?

13.6.17 What major factors should be taken into account in the design of concrete mixtures?

13.6.18 What are the absolute volume percent ranges for the major components of normal concrete?

13.6.19 It is desired to produce 100 ft³ of concrete with a ratio of 1:1.9:3.8 (by weight) of cement, sand, and gravel. What are the required amounts of the components if 5.5 gal of water per sack of cement is to be used? Assume the free moisture contents of the sand and gravel are 3 and 0 percent, respectively. The specific gravities of the cement, sand, and gravel are 3.15, 2.65, and 2.65, respectively. (1 sack of cement weighs 94 lb and 1 ft³ water = 7.48 gal.) Give answers for the cement in sacks, the sand and gravel in pounds, and the water in gallons.

13.6.20 It is desired to produce 50 ft³ of concrete with a ratio of 1:1.9:3.2 (by weight) of cement, sand, and gravel. What are the required amounts of the components if 5.5 gal of water per sack of cement is to be used? Assume the free moisture contents of the sand and gravel are 4 and 0.5 percent, respectively. The specific gravities of the cement, sand, and gravel are 3.15, 2.65, and 2.65, respectively. (1 sack of cement weighs 94 lb and 1 ft³ water = 7.48 gal.) Give answers for the cement in sacks, the sand and gravel in pounds, and the water in gallons.

13.6.21 Why is concrete mainly used in compression in engineering designs?

13.6.22 What is reinforced concrete? How is it made?

13.6.23 What is the main advantage of prestressed concrete?

13.6.24 Describe how compressive stresses are introduced in pretensioned prestressed concrete.

13.6.25 Describe how compressive stresses are introduced in posttensioned prestressed concrete?

13.7.1   What is asphalt? Where is asphalt obtained?

13.7.2   What are the chemical composition ranges for asphalts?

13.7.3   What does an asphalt mix consist of? What is the asphalt content of a type IV road paving asphalt?

13.7.4   What characteristics are desirable for the aggregate for a road-paving asphalt?

13.8.1   Describe the different layers in the cross section of a tree stem. Also, give the functions of each layer.

13.8.2   What is the difference between softwoods and hardwoods? Give several examples of both. Are all hardwoods physically hard?

13.8.3   What are the subrings of the annual growth rings of trees?

13.8.4   What axis is parallel to the annual rings? What axis is parallel to the annual ring?

13.8.5   Describe the microstructure of a softwood tree.

13.8.6   What are the functions of the wood rays of a tree?

13.8.7   Describe the microstructure of a hardwood tree? What is the difference between ring-porous and diffuse-porous tree microstructures?

13.8.8   Describe the cell-wall ultrastructure of a wood cell.

13.8.9   Describe the constituents of a wood cell.

13.8.10 A piece of wood containing moisture weighs 200 g, and after oven drying to a constant weight, it weighs 130 g. What is its percent moisture content?

13.8.11 A piece of wood contains 20% moisture. What must its weight have been before oven drying if it has a constant weight of 130 g after drying?

13.8.12 A piece of wood contains 49% moisture. What must its final weight be after oven drying if it weighed 160 g before drying?

13.8.13 What is the reason for the relatively high strength of wood in the longitudinal direction of the tree stem as compared to the transverse direction?

13.8.14 What is the green condition for wood? Why is wood much weaker in the green condition than in the kiln-dried condition?

13.8.15 Why does wood shrink much more in the transverse direction than in the longitudinal direction?

13.10.1 A metal-matrix composite (MMC) is made of a 2024-Al alloy matrix and continuous boron fibers. The boron fibers are produced with a 11.5-μm-diameter tungsten-wire core which is coated with boron to make a final 105-μm-diameter fiber. A unidirectional composite is made with 51 vol % of the boron fibers in the Al 2024 matrix. Assuming the law of mixtures applies to isostrain conditions, calculate the tensile modulus of the composite in the direction of the fibers. Data are $E_B = 379$ GPa, $E_W = 410$ GPa, and $E_{Al} = 72.4$ GPa.

13.10.2 A new developmental metal-matrix composite is made for the National Aerospace plane with a matrix of the intermetallic compound titanium aluminide (Ti_3Al) and continuous silicon carbide fibers. A unidirectional composite is made with the SiC continuous fibers all in one direction. If the modulus of the composite is 210 GPa and assuming isostrain conditions, what must the volume percent of SiC fibers in the composite be if $E(SiC) = 390$ GPa and $E(Ti_3Al) = 145$ GPa?

13.10.3 A metal-matrix composite is made with a matrix of 6061-Al alloy and 45 vol % SiC continuous fibers all in one direction. If isostrain conditions prevail, what is the tensile modulus of the composite in the direction of the fibers? Data are $E(SiC) = 390$ GPa and $E(6061$ Al$) = 68.9$ GPa.

13.10.4 An MMC is made with a 2024-Al alloy with 15 vol % SiC whiskers. If the density of the composite is 2.95 g/cm^3 and that of the SiC fibers is 3.10 g/cm^3, what must the density of the 2024-Al alloy be?

13.10.5 A ceramic-matrix composite (CMC) is made with continuous SiC fibers embedded in a reaction-bonded silicon nitride (RBSN) matrix with all the SiC fibers aligned in one direction. Assuming isostrain conditions, what is the volume fraction of the SiC fibers in the composite if the composite has a tensile modulus of 260 GPa? Data are $E(SiC) = 400$ GPa and $E(RBSN) = 160$ GPa.

13.10.6 The largest preexisting flaw in the reaction-bonded silicon nitride matrix of Prob. 13.10.5 is 9.0 μm in diameter, and the largest surface notch on the SiC fibers is 6.0 μm deep. Calculate the stress at which cracks first form in the composite when stress is slowly applied under isostrain conditions and in the direction of the fibers. Data are $K_{IC}(RBSN) = 3.6$ MPa$\sqrt{m}$ and $K_{IC}(SiC) = 4.8$ MPa$\sqrt{m}$.

13.10.7 A ceramic-matrix composite is made with an aluminum oxide (Al$_2$O$_3$) matrix and continuous silicon carbide fiber reinforcement with all the SiC fibers in one direction. The composite consists of 35 vol % SiC fibers. If isostrain conditions exist, calculate the tensile modulus of the composite in the direction of the fibers. If a load of 8 MN is applied to the composite in the direction of the fibers, what is the elastic strain in the composite if the surface area over which the load is applied is 60 cm^2? Data are $E(Al_2O_3) = 360$ GPa and $E(SiC) = 350$ GPa.

13.10.8 For Prob. 13.10.7, the Al$_2$O$_3$ matrix has flaws up to 11.0 μm in diameter and the largest surface notch of the SiC fibers is 6.0 μm.
(a) Will the matrix or the fibers crack first?
(b) What stress on the composite in the direction of the fibers will cause the first crack to form? Data are $K_{IC}(Al_2O_3) = 4.0$ MPa$\sqrt{m}$ and $K_{IC}(SiC) = 4.8$ MPa$\sqrt{m}$.

Optical Properties and Superconducting Materials

Optical properties of materials play an important role in much of today's high technology (Fig. 14.1). In this chapter we shall first examine some of the basics of the refraction, reflection, and absorption of light with some classes of materials. Next we shall proceed to investigate how some materials interact with light radiation to produce luminescence. Then we shall look into the stimulated emission of radiation by lasers. In the optical fibers part of this chapter, we shall see how the development of low-light-loss optical fibers has led to the new optical-fiber communications systems.

Finally, we shall look into superconducting materials which have zero electrical resistivity below their critical temperatures, magnetic fields, and current densities. Until around 1987 the highest critical temperature for a superconducting material was about 25 K. In 1987 a spectacular discovery was made that some ceramic materials could be made superconductive up to about 100 K. This discovery set off a tremendous worldwide research effort which has high expectations for future engineering developments. In this chapter we shall examine some aspects of the structure and properties of types I and II metallic superconductors as well as the new ceramic ones.

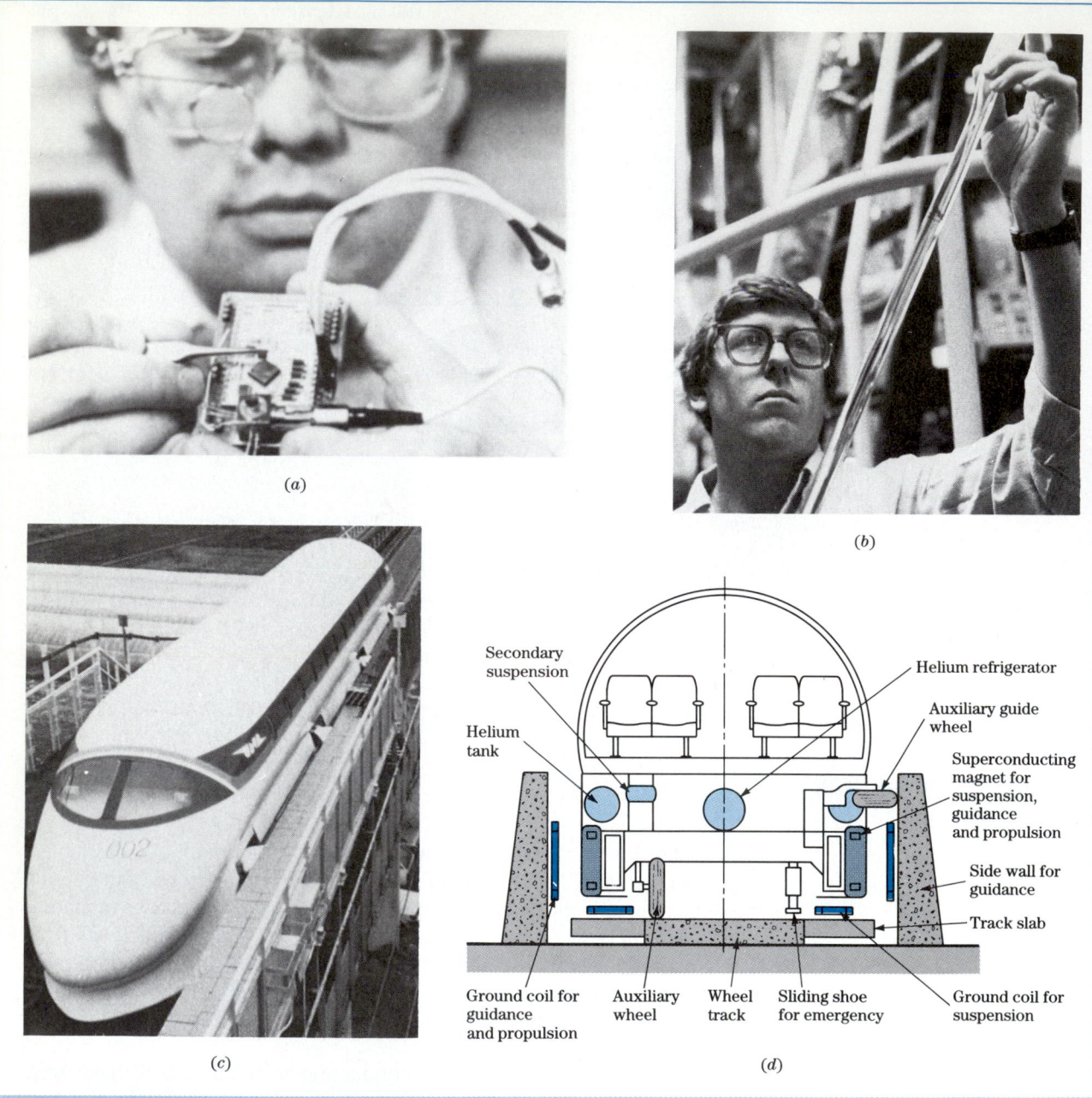

(a)

(b)

(c)

(d)

Secondary suspension

Helium tank

Ground coil for guidance and propulsion

Auxiliary wheel

Wheel track

Sliding shoe for emergency

Helium refrigerator

Auxiliary guide wheel

Superconducting magnet for suspension, guidance and propulsion

Side wall for guidance

Track slab

Ground coil for suspension

FIGURE 14.1 New technologies. (*a*) An engineer examines a laser transmitter to be used in the second transatlantic lightguide cable, which went into service in 1991. (*Courtesy of AT&T.*) (*b*) A fiber-optic glass preform is examined before being drawn into hair-thin optical fiber. (*Courtesy of AT&T.*) (*c*) A Japanese National Railways train that uses Nb–Ti superconducting magnets in liquid-helium cryostats. The train is designed to travel at 400 to 500 km/h on a test track. (*After Mechanical Engineering, June 1988, p. 61.*) (*d*) Cross section of an advanced levitated train design (Japanese National Railway). (*After the "Encyclopedia of Materials Science and Engineering," MIT Press, 1986, p. 4766.*)

14.2   LIGHT AND THE ELECTROMAGNETIC SPECTRUM

Visible light is one form of electromagnetic radiation with wavelengths extending from about 0.40 to 0.75 μm (Fig. 14.2). Visible light contains color bands ranging from violet through red, as shown in the enlarged scale of Fig. 14.2. The ultraviolet region covers the range from about 0.01 to about 0.40 μm, and the infrared region extends from about 0.75 to 1000 μm.

The true nature of light will probably never be known. However, light can be considered as having the form of waves and consisting of particles called *photons*. The energy ΔE, wavelength λ, and frequency ν of the photons are related by the fundamental equation

$$\Delta E = h\nu = \frac{hc}{\lambda} \tag{14.1}$$

where h is Planck's constant (6.62×10^{-34} J·s) and c is the speed of light in vacuum (3.00×10^{8} m/s). These equations allow us to consider the photon as a particle of energy E or as a wave with a specific wavelength and frequency.

Example Problem 14.1

A photon in a ZnS semiconductor drops from an impurity energy level at 1.38 eV below its conduction band to its valence band. What is the wavelength of the radiation given off by the photon in the transition? If visible, what is the color of the radiation? ZnS has an energy band gap of 3.54 eV.

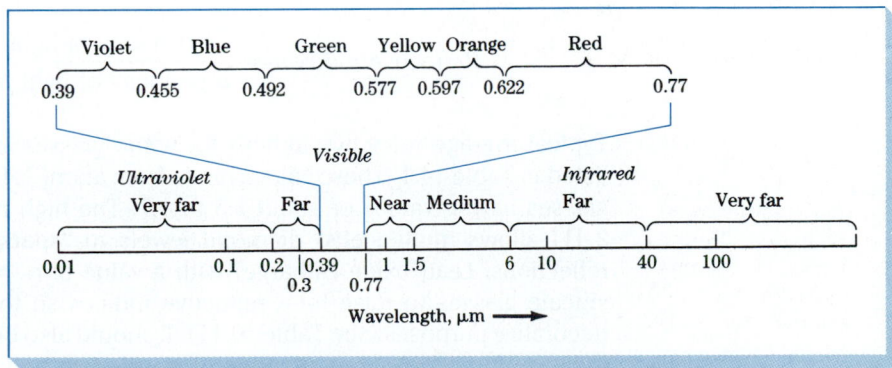

FIGURE 14.2 The electromagnetic spectrum from the ultraviolet to the infrared regions.

Solution:

The energy difference for the photon dropping from the 1.38-eV level below the conduction band to the valence band is 3.54 eV − 1.38 eV = 2.16 eV.

$$\lambda = \frac{hc}{\Delta E} \tag{14.1}$$

where $h = 6.62 \times 10^{-34}$ J·s
$c = 3.00 \times 10^{8}$ m/s
1 eV $= 1.60 \times 10^{-19}$ J

Thus,

$$\lambda = \frac{(6.62 \times 10^{-34} \text{ J} \cdot \text{s})(3.00 \times 10^{8} \text{ m/s})}{(2.16 \text{ eV})(1.60 \times 10^{-19} \text{ J}\text{eV})(10^{-9} \text{ m/nm})} = 574.7 \text{ nm} \blacktriangleleft$$

The wavelength of this photon at 574.7 nm is the visible yellow region of the electromagnetic spectrum.

14.3   REFRACTION OF LIGHT

Index of Refraction   When light photons are transmitted through a transparent material, they lose some of their energy, and as a result, the speed of light is reduced and the beam of light changes direction. Figure 14.3 shows schematically how a beam of light entering from the air is slowed down when entering a denser medium such as common window glass. Thus, the incident angle for the light beam is greater than the refracted angle for this case.

   The relative velocity of light passing through a medium is expressed by the optical property called the *index of refraction n*. The *n* value of a medium is defined as the ratio of the velocity of light in vacuum, *c*, to the velocity of light in the medium considered, *v*:

$$\text{Refractive index } n = \frac{c \text{ (velocity of light in vacuum)}}{v \text{ (velocity of light in a medium)}} \tag{14.2}$$

Typical average refractive indices for some glasses and crystalline solids are listed in Table 14.1. These values range from about 1.4 to 2.6, with most silicate glasses having values of about 1.5 to 1.7. The high refractive diamond ($n = 2.41$) allows multifaceted diamond jewels to "sparkle" by multiple internal reflections. Lead oxide (litharge) with a value of $n = 2.61$ is added to some silicate glasses to raise their refractive indices so that they can be used for decorative purposes (see Table 10.11). It should also be noted that the refractive indices of materials are a function of wavelength and frequency. For example,

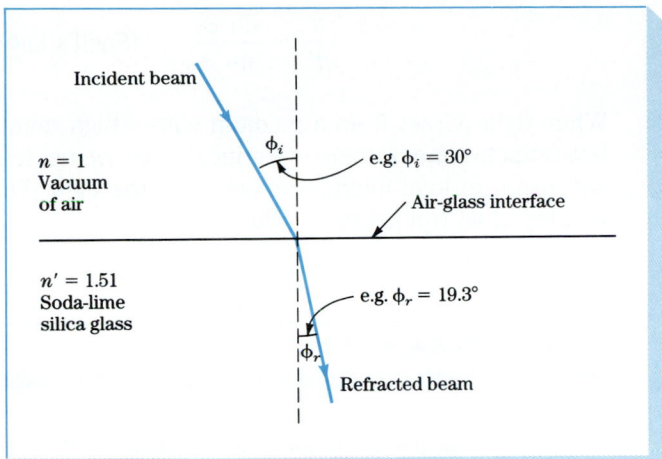

FIGURE 14.3 Refraction of light beam as it is transmitted from a vacuum (air) through soda-lime-silica glass.

the refractive index of light flint glass varies from about 1.60 at 0.40 μm to 1.57 at 1.0 μm.

Snell's Law of Light Refraction

The refractive indices for light passing from one medium of refractive index n through another of refractive index n' are related to the incident angle ϕ and the refractive angle ϕ' by the relation

TABLE 14.1   Refractive Indices for Selected Materials

Material	Average refractive index
Glass compositions:	
Silica glass	1.458
Soda-lime-silica glass	1.51–1.52
Borosilicate (Pyrex) glass	1.47
Dense flint glass	1.6–1.7
Crystalline compositions:	
Corundum, Al_2O_3	1.76
Quartz, SiO_2	1.555
Litharge, PbO	2.61
Diamond, C	2.41
Optical plastics:	
Polyethylene	1.50–1.54
Polystyrene	1.59–1.60
Polymethyl methacrylate	1.48–1.50
Polytetrafluoroethylene	1.30–1.40

$$\frac{n}{n'} = \frac{\sin \phi'}{\sin \phi} \quad \text{(Snell's law)} \quad (14.3)$$

When light passes from a medium with a high refractive index to one with a low refractive index, there is a critical angle of incidence ϕ_c, which if increased will result in total internal reflection of the light (Fig. 14.4). This ϕ_c angle is defined at ϕ' (refraction) = 90°.

Example Problem 14.2

What is the critical angle ϕ_c for light to be totally reflected when leaving a flat plate of soda-lime-silica glass ($n = 1.51$) and entering the air ($n = 1$)?

Solution:

Using Snell's law [Eq. (14.3)],

$$\frac{n}{n'} = \frac{\sin \phi'}{\sin \phi_c}$$

$$\frac{1.51}{1} = \frac{\sin 90°}{\sin \phi_c}$$

where n = refractive index of the glass
n' = refractive index of air
ϕ' = 90° for total reflection
ϕ_c = critical angle for total reflection (unknown)

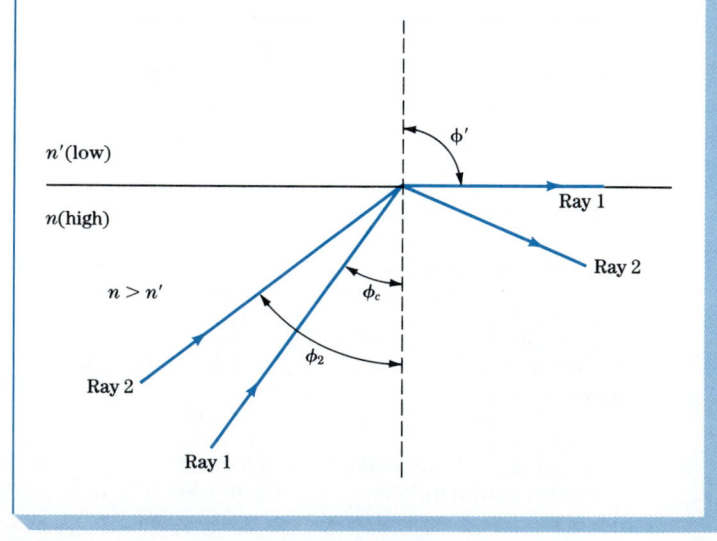

FIGURE 14.4 Diagram indicating the critical angle ϕ_c for total internal reflection of light passing from a high refractive index medium n to another of low refractive index n'. Note that ray 2, which has an incidence angle ϕ_2 greater than ϕ_c, is totally reflected back into the medium of high refractive index.

$$\sin \phi_c = \frac{1}{1.51} (\sin 90°) = 0.662$$

$$\phi_c = 41.5° \blacktriangleleft$$

Note: We shall see in Sec. 14.7 on Optical Fibers that by using a cladding of a low-refractive-index glass surrounding a core of high refractive index, an optical fiber can transmit light for long distances because the light is continually reflected internally.

14.4 ABSORPTION, TRANSMISSION, AND REFLECTION OF LIGHT

Every material absorbs light to some degree because of the interaction of light photons with the electronic and bonding structure of the atoms, ions, or molecules which make up the material. The fraction of light transmitted by a particular material thus depends on the amount of light reflected and absorbed by the material. For a particular wavelength λ, the sum of the fractions of the incoming incident light reflected, absorbed, and transmitted is equal to 1:

$$(\text{Reflected fraction})_\lambda + (\text{absorbed fraction})_\lambda + (\text{transmitted fraction})_\lambda = 1 \tag{14.4}$$

Let us now consider how these fractions vary for some selected types of materials.

Metals   Except for very thin sections, metals strongly reflect and/or absorb incident radiation for long wavelengths (radio waves) to the middle of the ultraviolet range. Since in metals the conduction band overlaps the valence band, incident radiation easily elevates electrons to higher energy levels. Upon dropping to lower energy levels, the photon energies are low and their wavelengths long. This type of action results in strongly reflected beams of light from a smooth surface, as is observed for many metals such as gold and silver. The amount of energy absorbed by metals depends on the electronic structure of each particular metal. For example, with copper and gold there is a greater absorption of the shorter wavelengths of blue and green and a greater reflection of the yellow, orange, and red wavelengths, and thus smooth surfaces of these metals show the reflected colors. Other metals such as silver and aluminum strongly reflect all parts of the visible spectrum and show a white "silvery" color.

Silicate Glasses   **Reflection of light from a single surface of a glass plate**   The proportion of incident light reflected by a single surface of a polished glass plate is very small. This amount depends mainly on the refractive index of the glass n and the angle of incidence of the light striking the glass. For normal light incidence (that is, $\phi_i = 90°$), the fraction of light reflected R (called the *reflectivity*) by a single surface can be determined from the relationship

$$R = \left(\frac{n-1}{n+1}\right)^2 \tag{14.5}$$

where n is the refractive index of the reflecting optical medium. This formula may also be used with good approximation for incident light angles up to about 20°. Using Eq. (14.5), a silicate glass with $n = 1.46$ has a calculated R value of 0.035, or a percent reflectivity of 3.5% (see Example Problem 14.3).

Example Problem 14.3

Calculate the reflectivity of ordinary incident light from the polished flat surface of a silicate glass with a refractive index of 1.46.

Solution:

Using Eq. (14.5) and $n = 1.46$ for the glass,

$$\text{Reflectivity} = \left(\frac{n-1}{n+1}\right)^2 = \left(\frac{1.46 - 1.00}{1.46 + 1.00}\right)^2 = 0.035$$

$$\% \text{ Reflectivity} = R(100\%) = 0.035 \times 100\% = 3.5\% \blacktriangleleft$$

Absorption of light by a glass plate   Glass absorbs energy from the light that it transmits so that the light intensity decreases as the light path increases. The relationship among the fraction of light entering, I_0, the fraction of light exiting, I, from a glass sheet or plate of thickness t and which is free of scattering centers, is

$$\frac{I}{I_0} = e^{-\alpha t} \tag{14.6}$$

The constant α in this relation is called the *linear absorption coefficient* and has the units cm^{-1} if the thickness is measured in centimeters. As shown in Example Problem 14.4, there is a relatively small loss of energy by absorption through a clear silicate glass plate.

Example Problem 14.4

Ordinary incident light strikes a polished glass plate 0.50 cm thick which has a refractive index of 1.50. What fraction of light is absorbed by the glass as the light passes between the surfaces of the plate? ($\alpha = 0.03$ cm^{-1}.)

Solution:

$$\frac{I}{I_0} = e^{-\alpha t} \qquad I_0 = 1.00 \qquad \alpha = 0.03 \text{ cm}^{-1}$$
$$I = ? \qquad t = 0.50 \text{ cm}$$

$$\frac{I}{1.00} = e^{-(0.03 \text{ cm}^{-1})(0.50 \text{ cm})}$$

$$I = (1.00)e^{-0.015} = 0.985$$

Thus, the fraction of light lost by absorption by the glass is: $1 - 0.985 = 0.015$, or 1.5%. ◄

Reflectance, absorption, and transmittance of light by a glass plate   The amount of incident light transmitted through a glass plate is determined by the amount of light reflected from both upper and lower surfaces as well as the amount absorbed within the plate. Let us consider the transmittance of light through a glass plate, as shown in Fig. 14.5. The fraction of incident light reaching the lower surface of the glass is $(1 - R)(I_0e^{-\alpha t})$. The fraction of incident light

FIGURE 14.5
Transmittance of light through a glass plate in which reflectance takes place at upper and lower surfaces and absorption within the plate.

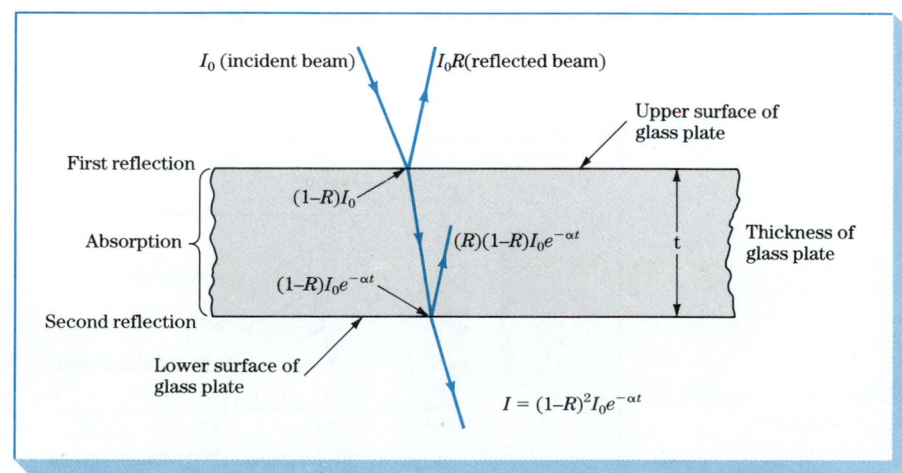

reflected from the lower surface will therefore be $(R)(1 - R)(I_0 e^{-\alpha t})$. Thus, the difference between the light reaching the lower surface of the glass plate and that which is reflected from the lower surface is the fraction of light transmitted I, which is:

$$I = [(1 - R)(I_0 e^{-\alpha t})] - [(R)(1 - R)(I_0 e^{-\alpha t})]$$
$$= (1 - R)(I_0 e^{-\alpha t})(1 - R) = (1 - R)^2 (I_0 e^{-\alpha t}) \qquad (14.7)$$

Figure 14.6 shows that about 90 percent of incident light is transmitted by silica glass if the wavelength of the incoming light is greater than about 300 nm. For shorter-wavelength ultraviolet light, much more absorption takes place and the transmittance is lowered considerably.

Plastics   Many noncrystalline plastics such as polystyrene, polymethyl methacrylate, and polycarbonate have excellent transparency. However, in some plastic materials there are crystalline regions having a higher refractive index than their non-crystalline matrix. If these regions are greater in size than the wavelength of the incoming light, the light waves will be scattered by reflection and refraction and hence the transparency of the material decreases (Fig. 14.7). For example, thin-sheet polyethylene which has a branched-chain structure and hence a lower degree of crystallinity has higher transparency than the higher-density, more crystalline linear-chain polyethylene. The transparencies of other partly crystalline plastics can range from cloudy to opaque depending mainly on their degree of crystallinity, impurity content, and filler content.

Semiconductors   In semiconductors light photons can be absorbed in several ways (Fig. 14.8). In intrinsic (pure) semiconductors such as Si, Ge, and GaAs, photons may be absorbed to create electron-hole pairs by causing electrons to jump across the energy band gap from the valence band to the conduction band (Fig. 14.8a). For this event to occur the incoming light photon must have an energy value equal to or greater than the energy gap E_g. If the energy of the photon is greater

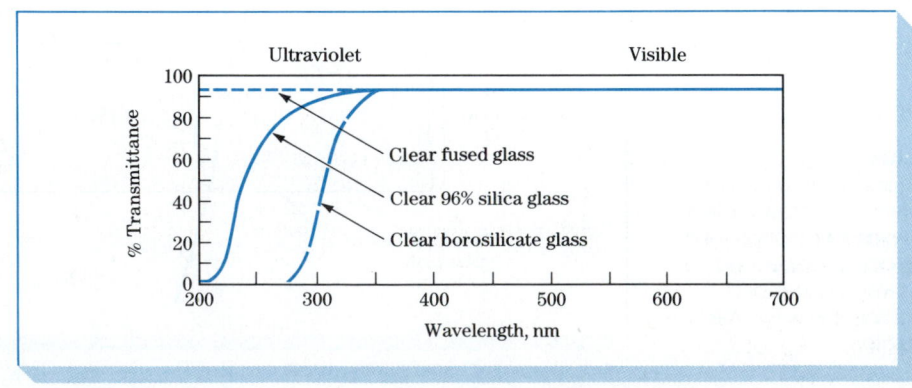

FIGURE 14.6 Percent transmittance vs. wavelength (nm) for several types of clear glasses.

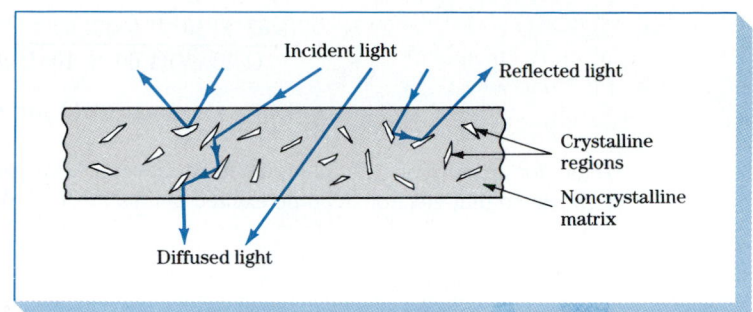

FIGURE 14.7 Multiple internal reflections at the crystalline-region interfaces reduce the transparency of partly crystalline thermoplastics.

than E_g, the excess energy is dissipated as heat. For semiconductors containing donor and acceptor impurities, much lower energy photons (and hence much longer in wavelength) are absorbed in causing electrons to jump from the valence band into acceptor levels (Fig. 14.8b) or from donor levels into the conduction band (Fig. 14.8c). Semiconductors are therefore opaque to high- and intermediate-energy (short- and intermediate-wavelength) light photons and transparent to low-energy, very long wavelength photons.

Example Problem 14.5

Calculate the minimum wavelength for photons to be absorbed by intrinsic silicon at room temperature ($E_g = 1.10$ eV).

Solution:

For absorption in this semiconductor, the minimum wavelength is given by Eq. (14.1):

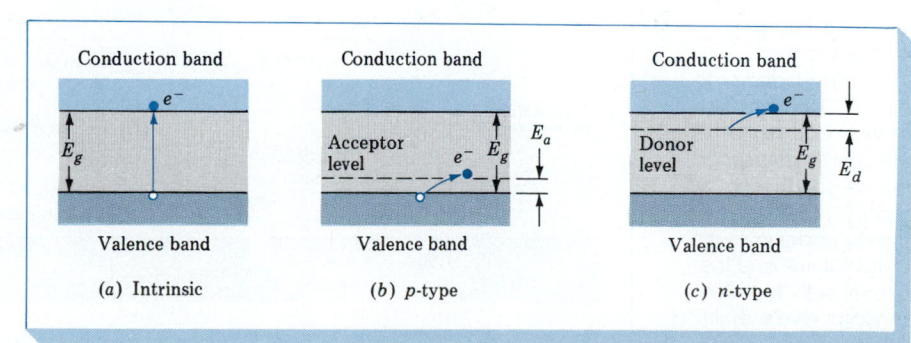

FIGURE 14.8 Optical absorption of photons in semiconductors. Absorption takes place in (a) if $h\nu > E_g$, (b) if $h\nu > E_a$, and (c) if $h\nu > E_d$.

$$\lambda_c = \frac{hc}{E_g} = \frac{(6.62 \times 10^{-34} \text{ J·s})(3.00 \times 10^8 \text{ m/s})}{(1.10 \text{ eV})(1.60 \times 10^{-19} \text{ J/eV})}$$

$$= 1.13 \times 10^{-6} \text{ m or } 1.13 \text{ } \mu\text{m} \blacktriangleleft$$

Thus, for absorption the photons must have a wavelength at least as short as 1.13 μm so that electrons can be excited across the 1.10-eV band gap.

14.5 LUMINESCENCE

Luminescence may be defined as the process by which a substance absorbs energy and then spontaneously emits visible or near-visible radiation. In this process the input energy excites electrons of a luminescent material from the valence band into the conduction band. The source of the input energy may be, for example, high-energy electrons or light photons. The excited electrons during luminescence drop to lower energy levels. In some cases the electrons may recombine with holes. If the emission takes place within 10^{-8} s after excitation, the luminescence is called *fluorescence,* and if the emission takes longer than 10^{-8} s, it is referred to as *phosphorescence.*

Luminescence is produced by materials called *phosphors* which have the capability of absorbing high-energy, short-wave radiation and spontaneously emitting lower-energy, longer-wavelength light radiation. The emission spectra of luminescent materials are controlled industrially by added impurities called *activators.* The activators provide discrete energy levels in the energy gap between the conduction and valence bands of the host material (Fig. 14.9). One mechanism postulated for the phosphorescent process is that excited

FIGURE 14.9 Energy changes during luminescence. (1) Electron-hole pairs are created by exciting electrons to the conduction band or to traps. (2) Electrons can be thermally excited from one trap to another or into the conduction band. (3) Electrons can drop to upper activator (donor) levels and then subsequently to lower acceptor levels, emitting visible light.

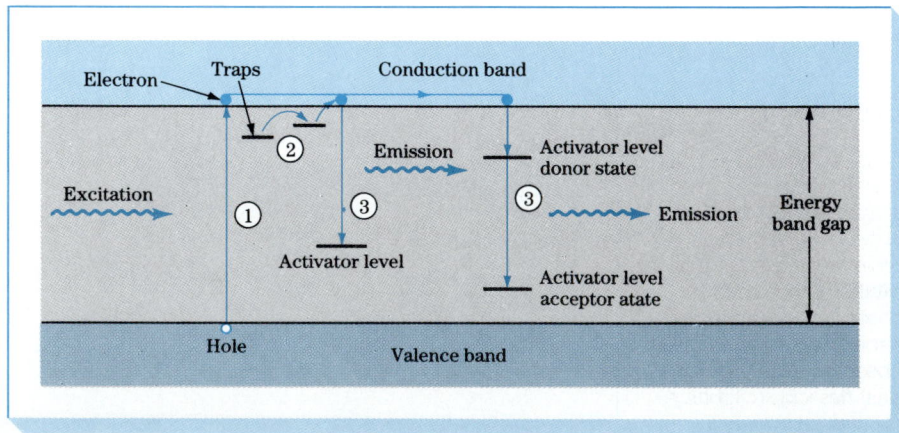

electrons are trapped in various ways at high energy levels and must get out of the traps before they can drop to lower energy levels and emit light of a characteristic spectral band. The trapping process is used to explain the delay in light emission by excited phosphors.

Luminescence processes are classified according to the energy source for electronic excitation. Two industrially important types are *photoluminescence* and *cathodoluminescence.*

Photoluminescence

In the common fluorescent lamp, photoluminescence converts ultraviolet radiation from a low-pressure mercury arc into visible light by using a halophosphate phosphor. Calcium halophosphate of the approximate composition $Ca_{10}F_2P_6O_{24}$ with about 20 percent of the F^- ions replaced with Cl^- ions is used as the host phosphor material for most lamps. Antimony, Sb^{3+}, ions provide a blue emission and manganese, Mn^{2+}, ions provide an orange-red emission band. By varying the Mn^{2+}, various shades of blue, orange, and white light may be obtained. The high-energy ultraviolet light from the excited mercury atoms causes the phosphor-coated inner wall of the fluorescent lamp tube to give off lower-energy, longer-wavelength visible light (Fig. 14.10).

Cathodoluminescence

This type of luminescence is produced by an energized cathode which generates a beam of high-energy bombarding electrons. Applications for this process include electron microscope, cathode-ray oscilloscope, and color television screen luminescences. The color television screen phosphorescence is especially interesting. The modern television set has very narrow (about 0.25 mm wide) vertical stripes of red-, green-, and blue-emitting phosphors deposited on the inner surface of the face plate of the television picture tube (Fig. 14.11). By using a steel shadow mask with small elongated holes (about 0.15 mm wide), the incoming television signal is scanned over the entire screen at 30 times per second. The small size and large number of phosphor areas consecutively exposed in the rapid scan of 15,750 horizontal lines per second and the persistence of vision of the human eye make possible a clear visible picture with good resolution. Commonly used phosphors for the colors are

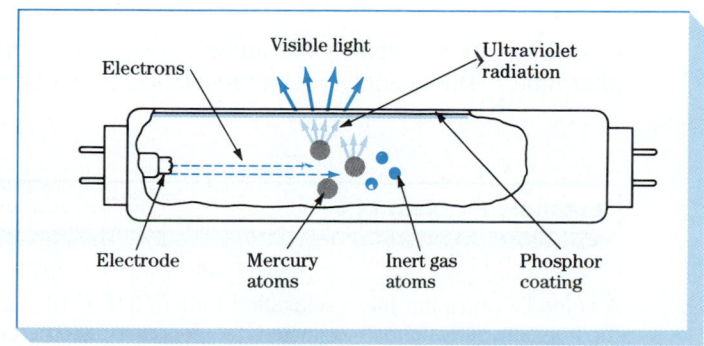

FIGURE 14.10 Cutaway diagram of a fluorescent lamp showing electron generation at an electrode and excitation of mercury atoms to provide the UV light to excite the phosphor coating on the inside of a lamp tube. The excited phosphor coating then provides visible light by luminescence.

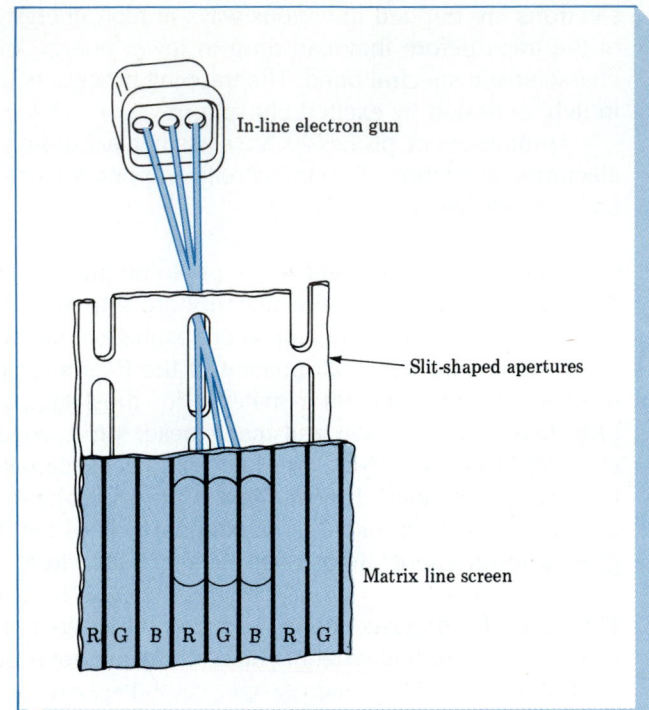

FIGURE 14.11 Diagram showing the arrangement of the R(red), G(green), and B(blue) vertical stripes of phosphors of a color television screen. Also shown are several of the elongated-shaped aperatures of the steel shadow mask. (*RCA.*)

zinc sulfide (ZnS) with an Ag^+ acceptor and Cl^- donor for the blue color, (Zn,Cd)S with a Cu^+ acceptor and Al^{3+} donor for the green color, and yttrium oxysulfide (Y_2O_2S) with 3% europium (Eu) for the red color. The phosphor materials must retain some image glow until the next scan but must not retain too much to blur the picture.

The intensity of luminescence, I, is given by

$$\ln \frac{I}{I_0} = -\frac{t}{\tau} \tag{14.8}$$

where I_0 = initial intensity of luminescence and I = fraction of luminescence after time t. The quantity τ is the relaxation time constant for the material.

Example Problem 14.6

A color TV phosphor has a relaxation time of 3.9×10^{-3} s. How long will it take for the intensity of this phosphor material to decrease to 10 percent of its original intensity?

Solution:

Using Eq. (14.8), $\ln (I/I_0) = -t/\tau$, or

$$\ln \frac{1}{10} = -\frac{t}{3.9 \times 10^{-3} \text{ s}}$$

$$t = (-2.3)(-3.9 \times 10^{-3} \text{ s}) = 9.0 \times 10^{-3} \text{ s} \blacktriangleleft$$

14.6 STIMULATED EMISSION OF RADIATION AND LASERS

Light emitted from conventional light sources such as fluorescent lamps results from the transitions of excited electrons to lower energy levels. Atoms of the same elements in these light sources give off photons of similar wavelengths independently and randomly. Consequently, the radiation emitted is in random directions and the wave trains are out of phase with each other. This type of radiation is said to be *incoherent.* In contrast, a light source called a *laser* produces a beam of radiation whose photon emissions are in phase, or *coherent,* and which are parallel, directional, and monochromatic (or nearly so). The word *laser* is an acronym whose letters stand for "*l*ight *a*mplification by *s*timulated *e*mission of *r*adiation." In lasers some "active" emitted photons stimulate many others of the same frequency and wavelength to be emitted in phase as a coherent, intense light beam (Fig. 14.12).

To understand the mechanisms involved in laser action, let us consider the operation of a solid-state ruby laser. The ruby laser shown schematically in Fig. 14.13 is a single crystal of aluminum oxide (Al_2O_3) containing $\approx 0.05\%$ chromium^{3+} ions. The Cr^{3+} ions occupy substitutional lattice sites in the Al_2O_3 crystal structure and are responsible for the pink color of the laser rod. These ions act as fluorescent centers which when excited subsequently drop to lower energy levels, causing photon emissions at specific wavelengths. The ends of

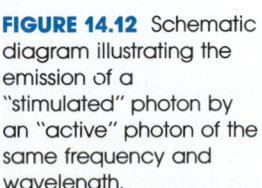

FIGURE 14.12 Schematic diagram illustrating the emission of a "stimulated" photon by an "active" photon of the same frequency and wavelength.

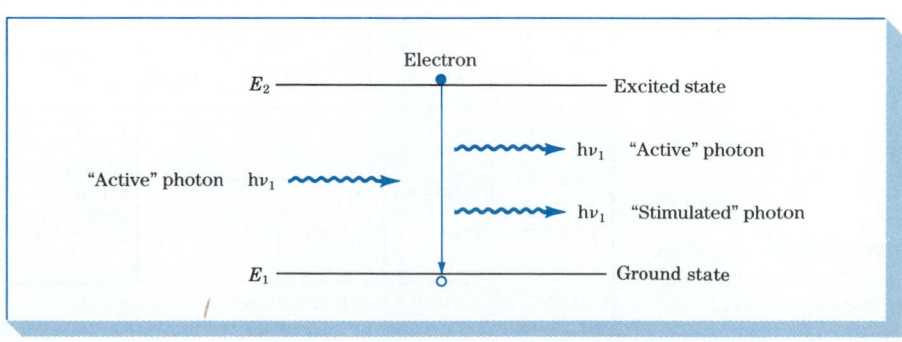

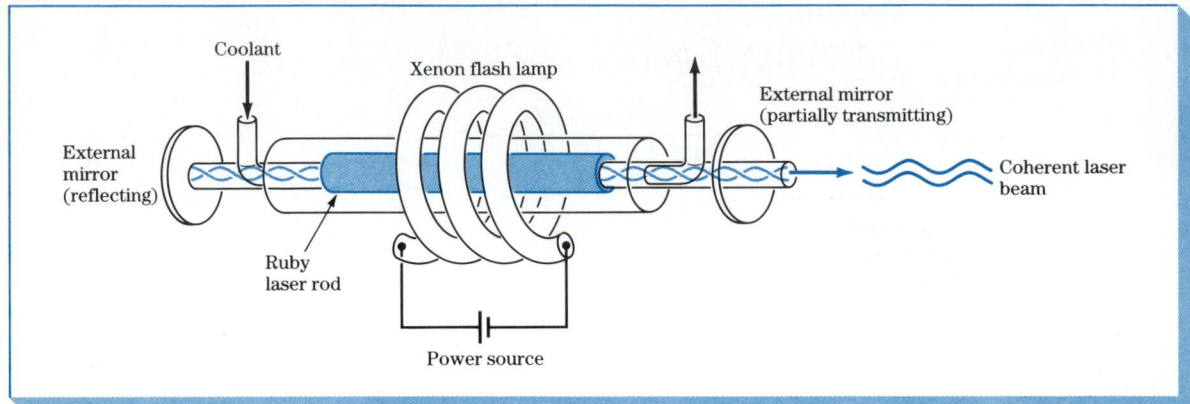

FIGURE 14.13 Schematic diagram of a pulsed ruby laser.

the ruby rod crystal are ground parallel for optical emission. A totally reflective mirror is placed parallel and near the back end of the crystal rod and another partially transmitting one at the front end of the laser which allows the coherent laser beam to pass through.

High-intensity input from a xenon flash lamp can provide the necessary energy to excite the Cr^{3+} ion electrons from the ground state to high energy levels, as indicated by the E_3 band level of Fig. 14.14. This action in laser terminology is referred to as *pumping* the laser. The excited electrons of the Cr^{3+} ions may then drop back down to the ground state or to the metastable energy level E_2 of Fig. 14.14. However, before the stimulated emission of photons can occur in the laser, there must be more electrons pumped into the high nonequilibrium metastable energy level E_2 than in the ground state (E_1). This condition of the laser is referred to as the *population inversion* of electron energy states and is schematically indicated in Fig. 14.15b as compared to the equilibrium energy-level condition of Fig. 14.15a.

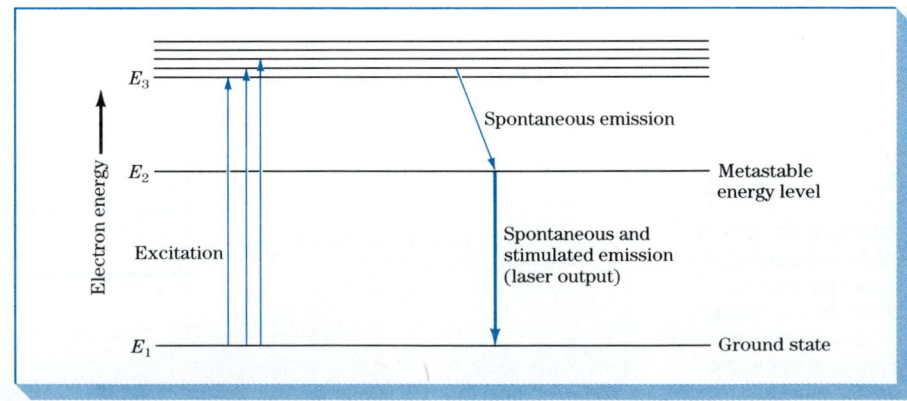

FIGURE 14.14 Simplified energy-level diagram for a three-level lasing system.

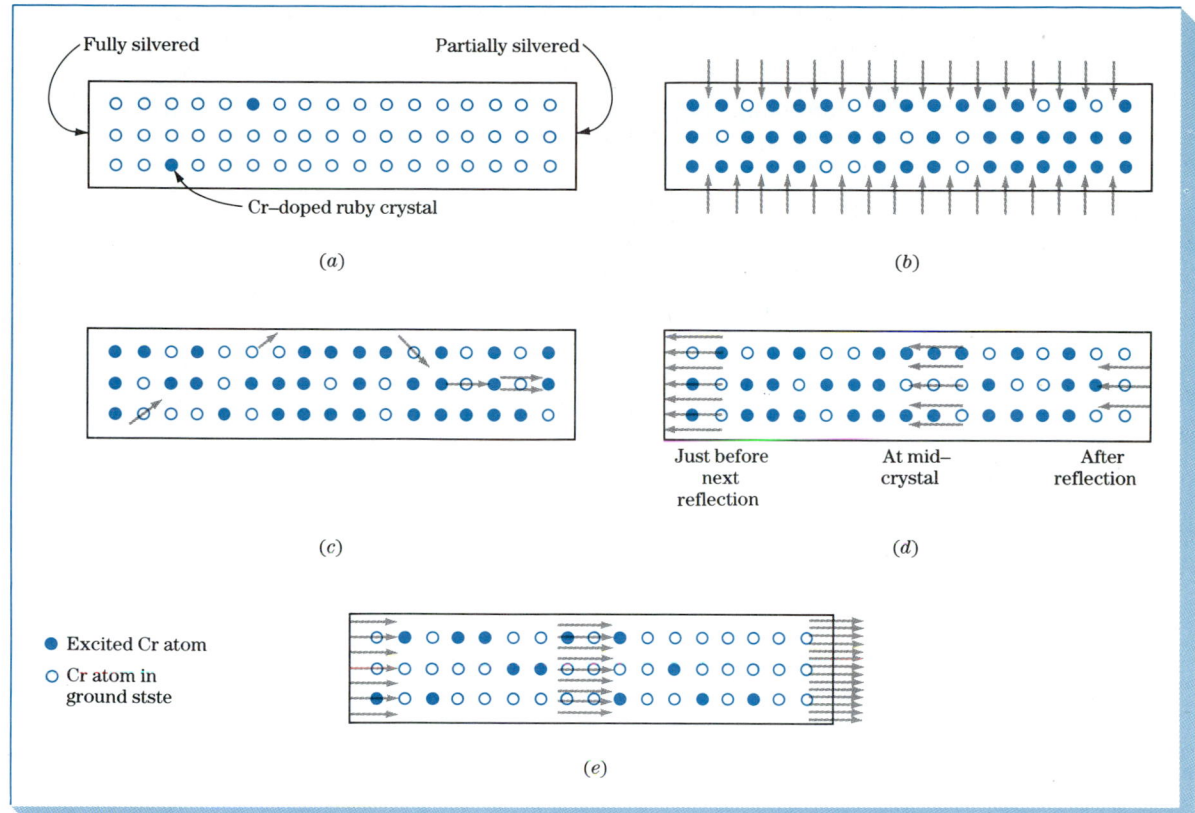

Fully silvered   Partially silvered

Cr–doped ruby crystal

(a)

(b)

(c)

Just before   At mid–   After
next   crystal   reflection
reflection

(d)

● Excited Cr atom

○ Cr atom in
  ground stste

(e)

FIGURE 14.15 Schematic steps in the functioning of a pulsed ruby laser. (*a*) At equilibrium. (*b*) Excitation by xenon flash lamp. (*c*) A few spontaneously emitted photons start the stimulated emission of photons. (*d*) Reflected back, the photons continue to stimulate the emission of more photons. (*e*) The laser beam is finally emitted. (*After R. M. Rose et al., Vol. IV, "Structure and Properties of Materials," vol. IV, Wiley, 1965.*)

The excited Cr^{3+} ions can remain in the metastable state for several milliseconds before spontaneous emission takes place by electrons dropping back to the ground state. The first few photons produced by electrons dropping from the metastable E_2 level of Fig. 14.14 to the ground level E_1 set off a stimulated-emission chain reaction causing many of the electrons to make the same jump from E_2 to E_1. This action produces a large number of photons which are in phase and moving in a parallel direction (Fig. 14.15c). Some of the photons jumping from E_2 to E_1 are lost to the outside of the rod, but many are reflected back and forth along the ruby rod by the end mirrors and stimulate more and more electrons to jump from E_2 and E_1, helping to build up a stronger coherent radiation beam (Fig. 14.15d). Finally, when a sufficiently intense coherent beam is built up inside the rod, the beam is transmitted as a high-intense-energy pulse (≈ 0.6 ms) through the partially transmitting mirror at the front end of the laser (Figs. 14.15e and Fig. 14.13). The laser beam produced by the Cr^{3+} doped aluminum oxide (ruby) crystal rod has a wavelength of 694.3 nm, which is a visible red line. This type of laser, which can only be operated intermittantly in bursts, is said to be a *pulsed* type. In contrast, most lasers are operated with a continuous beam and are called *continuous-wave (CW)* lasers.

Types of Lasers There are many types of gas, liquid, and solid lasers used in modern technology. We shall briefly describe some important aspects of several of these.

Ruby laser The structure and functioning of the ruby laser has already been described. This laser is not used much today because of the difficulties in growing the crystal rods compared to the ease of making neodymium lasers.

Neodymium-YAG lasers The neodymium-yttrium-aluminum-garnet (Nd:YAG) laser is made by combining one part per hundred of Nd atoms in a host of a YAG crystal. This laser emits in the near-infrared at 1.06-μm wavelength with continuous power up to about 250 W and with pulsed power as high as several megawatts. The YAG host material has the advantage of high thermal conductivity to remove excess heat. In materials processing, the Nd:YAG laser is used for welding, drilling, scribing, and cutting (Table 14.2).

Carbon dioxide (CO_2) lasers Carbon dioxide lasers are some of the most powerful lasers made and operate mainly in the middle-infrared at 10.6 μm. They vary from a few milliwatts of continuous power to large pulses with as high as 10,000 J of energy. They operate by electron collisions exciting nitrogen

TABLE 14.2 Selected Applications for Lasers in Materials Processing

Applications	Type of laser	Comments
1. Welding	YAG*	High average-power lasers for deep penetration and high throughput welding
2. Drilling	YAG CWCO$_2$†	High peak-power densities for drilling precision holes with a minimum heat-affected zone, low taper, and maximum depths
3. Cutting	YAG CWCO$_2$	Precision cutting of complex two- and three-dimensional shapes at high rates in metals, plastics, and ceramics
4. Surface treatment	CWCO$_2$	Transformation hardening of steel surfaces by hardening them above austenitic temperatures with a scanning, defocused beam and allowing the metal to self-quench
5. Scribing	YAG CWCO$_2$	Scribing large areas of fully fired ceramics and silicon wafers to provide individual circuit substrates
6. Photolithography	Excimer	Line-narrowed and spectrally stabilized excimer photolithographic processing in the fabrication of semiconductors

*YAG = yttrium aluminum garnet is a crystalline host used in solid-state neodymnium lasers.
†CWCO$_2$ = continuous-wave (as opposed to pulsed) carbon dioxide laser.

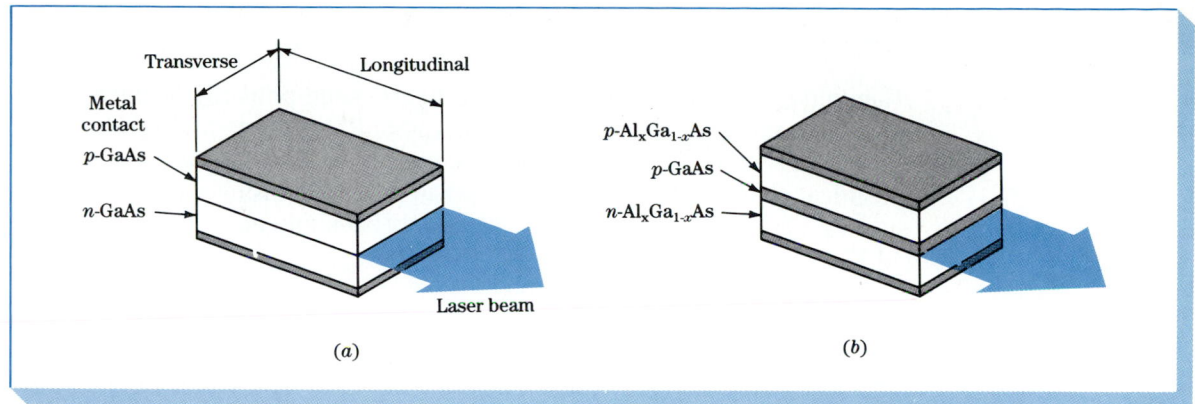

FIGURE 14.16 (*a*) Simple homojunction GaAs laser. (*b*) Double heterojunction (DH) GaAs laser. The *p*- and *n*-Al$_x$Ga$_{1-x}$As layers have wider band gaps and lower refractive indices and confine the electrons and holes within the active *p*-GaAs layer.

molecules to metastable energy levels that subsequently transfer their energy to excite CO_2 molecules which in turn give off laser radiation upon dropping to lower energy levels. Carbon dioxide lasers are used for metal processing applications such as cutting, welding, and localized heat treatment of steels (Table 14.2).

Semiconductor lasers   Semiconductor, or diode, lasers, commonly about the size of a grain of salt, are the smallest lasers produced. They consist of a *pn* junction made with a semiconducting compound such as GaAs which has a large-enough band gap for laser action (Fig. 14.16). Originally, the GaAs diode laser was made as a homojunction laser with a single *pn* junction (Fig. 14.16*a*). The resonant cavity of the laser is achieved by cleaving the crystal to make two end facets. The crystal-air interfaces cause the necessary reflections for laser action due to the difference in refractive indices of air and GaAs. The diode laser achieves population inversion by a strong forward bias of a heavily doped *pn* junction. A great number of electron-hole pairs are generated, and many of these in turn recombine to emit photons of light.

An improvement in efficiency was achieved with the double heterojunction (DH) laser (Fig. 14.16*b*). In a GaAs DH laser a thin layer of *p*-GaAs is sandwiched between *p*- and *n*-Al$_x$Ga$_{1-x}$As layers which confine the electrons and holes within the active *p*-GaAs layer. The AlGaAs layers have wider band gaps and lower refractive indices and so constrain the laser light to move in a miniature waveguide. The most widespread application of GaAs diode lasers currently is for compact disks.

14.7 OPTICAL FIBERS

Hair-thin (≈ 1.25-μm diameter) optical fibers made primarily of silica (SiO_2) glass are used for modern optical-fiber communication systems. These systems consist essentially of a transmitter (i.e., a semiconductor laser) to encode electrical signals into light signals, optical fiber to transmit the light signals, and a photodiode to convert the light signals back into electrical signals (Fig. 14.17).

Light Loss in Optical Fibers

The optical fibers used for communications systems must have extremely low light loss (attenuation) so that an entering encoded light signal can be transmitted a long distance [that is, 40 km (25 mi)] and still be detected satisfactorily. For extremely low light loss glass for optical fibers, the impurities (particularly Fe^{2+} ions) in the SiO_2 glass must be very low. The light loss (attenuation) of an optical glass fiber is usually measured in decibels per kilometer (dB/km). The light loss in a light-transmitting material in dB/km for light transmission over a length l is related to the entering light intensity I_0 and the exiting light intensity I by

$$-\text{Loss (dB/km)} = \frac{10}{l \text{ (km)}} \log \frac{I}{I_0} \qquad (14.9)$$

FIGURE 14.17 Basic elements of a fiber-optics communications system. (a) InGaAsP laser transmitter. (b) Optical fiber for transmitting light photons. (c) PIN diode photodetector.

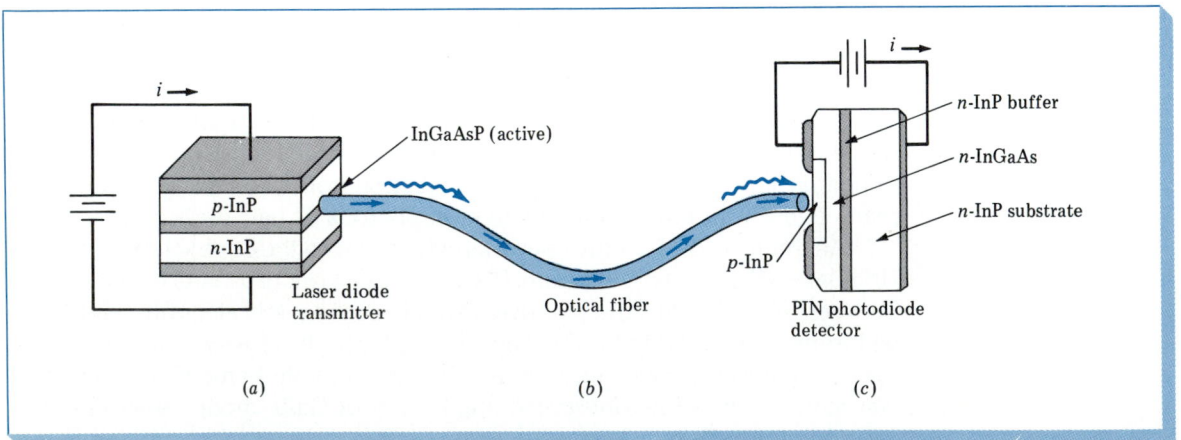

Example Problem 14.7

A low-loss silica glass fiber for optical transmission has a 0.20 dB/km light attenuation. (*a*) What is the fraction of light remaining after it has passed through 1 km of this glass fiber? (*b*) What is the fraction of light remaining after 40-km transmission?

Solution:

$$\text{Attenuation (dB/km)} = \frac{10}{l(\text{km})} \log \frac{I}{I_0} \qquad (14.9)$$

where I_0 = light intensity at source
I = light intensity at detector

(*a*) $\quad -0.20 \text{ dB/km} = \dfrac{10}{1 \text{ km}} \log \dfrac{I}{I_0} \quad$ or $\quad \log \dfrac{I}{I_0} = -0.02 \quad$ or $\quad \dfrac{I}{I_0} = 0.95 \blacktriangleleft$

(*b*) $\quad -0.20 \text{ dB/km} = \dfrac{10}{40 \text{ km}} \log \dfrac{I}{I_0} \quad$ or $\quad \log \dfrac{I}{I_0} = -0.80 \quad$ or $\quad \dfrac{I}{I_0} = 0.16 \blacktriangleleft$

Note: Single-mode optical fibers today are able to transmit communication light data about 40 km without having to repeat the signal.

Single-Mode and Multimode Optical Fibers

Optical fibers for light transmission serve as waveguides for the light signals in optical communications. The retention of light within the optical fiber is made possible by having the light pass through the central core glass which has a higher refractive index than the outer clad glass (Fig. 14.18). For the single-mode type which has a core diameter of about 8 μm and an outer clad diameter of about 125 μm, there is only one acceptable guided light ray path (Fig. 14.18*a*). In the multimode-type optical glass fiber which has a graded refractive index core, many wave modes pass through the fiber simultaneously, causing a more dispersed exiting signal than that produced by the single-mode fiber (Fig. 14.18*b*). Most new optical-fiber communication systems use single-mode fibers because they have lower light losses and are cheaper and easier to fabricate.

Fabrication of Optical Fibers

One of the most important methods for producing optical glass fibers for communication systems is the modified chemical vapor deposition (MCVD) process (Fig. 14.19). In this process high-purity dry vapor of·SiCl$_4$ with various contents of vapor of GeCl$_4$ and fluorinated hydrocarbons are passed through a rotating pure silica tube along with pure oxygen. An external oxyhydrogen torch is moved along the outer diameter of the rotating tube, allowing the contents to react to form silca glass particles doped with the desired combinations of germanium and fluorine. GeO$_2$ increases the refractive index of SiO$_2$,

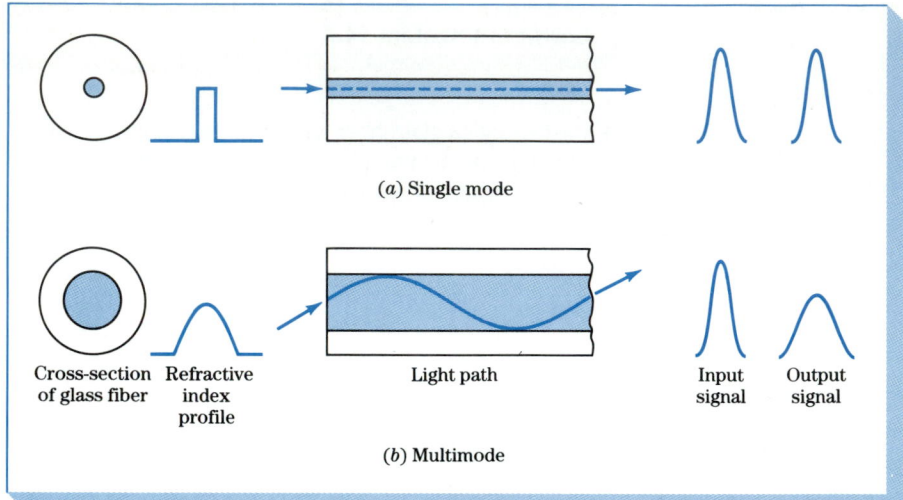

FIGURE 14.18 Comparison of (a) single-mode and (b) multimode optical fibers by cross section vs. refractive index, light path, and signal input and output. The sharper output signal of the single-mode fiber is preferred for long-distance optical communications systems.

and fluorine lowers it. Downstream from the reaction region, the glass particles migrate to the tube wall where they deposit. The moving torch which caused the reaction to form the glass particles then passes over and sinters them into a thin layer of doped glass. The thickness of the doped layer depends on the number of layers that are deposited by the repeated passes of the torch. At each pass the composition of the vapors is adjusted to produce the desired

FIGURE 14.19 Schematic of the modified chemical vapor deposition process for the production of the glass preforms for making optical glass fibers. [*After AT&T Tech. J. 66:33(1987).*]

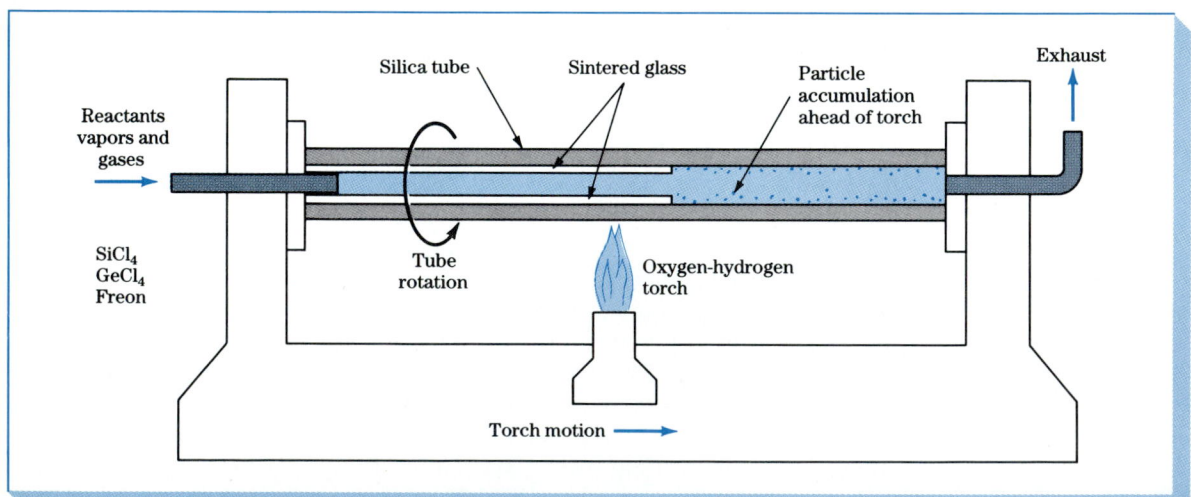

composition profile so that the glass fiber subsequently produced will have the desired refractive index profile.

In the next step the silica tube is heated to a high-enough temperature so that the glass approaches its softening point. The surface tension of the glass then causes the tube with the deposited glass layers to collapse uniformly into a solid rod called a *preform* (Fig. 14.1*b*). The glass preform from the MCVD process is then inserted into a high-temperature furnace, and glass fiber about 125 μm in diameter is drawn from it (Fig. 14.20). An in-line process applies a 60-μm-thick polymer coating to protect the glass fiber surface from damage. Spools of finished glass fibers are shown in Fig. 14.21. Very close tolerances for the core and outer diameter of the fiber are essential so that the fiber can be spliced (joined) without high light losses.

FIGURE 14.20 Setup for drawing optical glass fiber from glass preform. (*a*) Actual fabrication. (*Courtesy of AT&T.*) (*b*) Schematic setup. (*After "Encyclopedia of Materials Science and Technology," MIT Press, 1986, p. 1992.*)

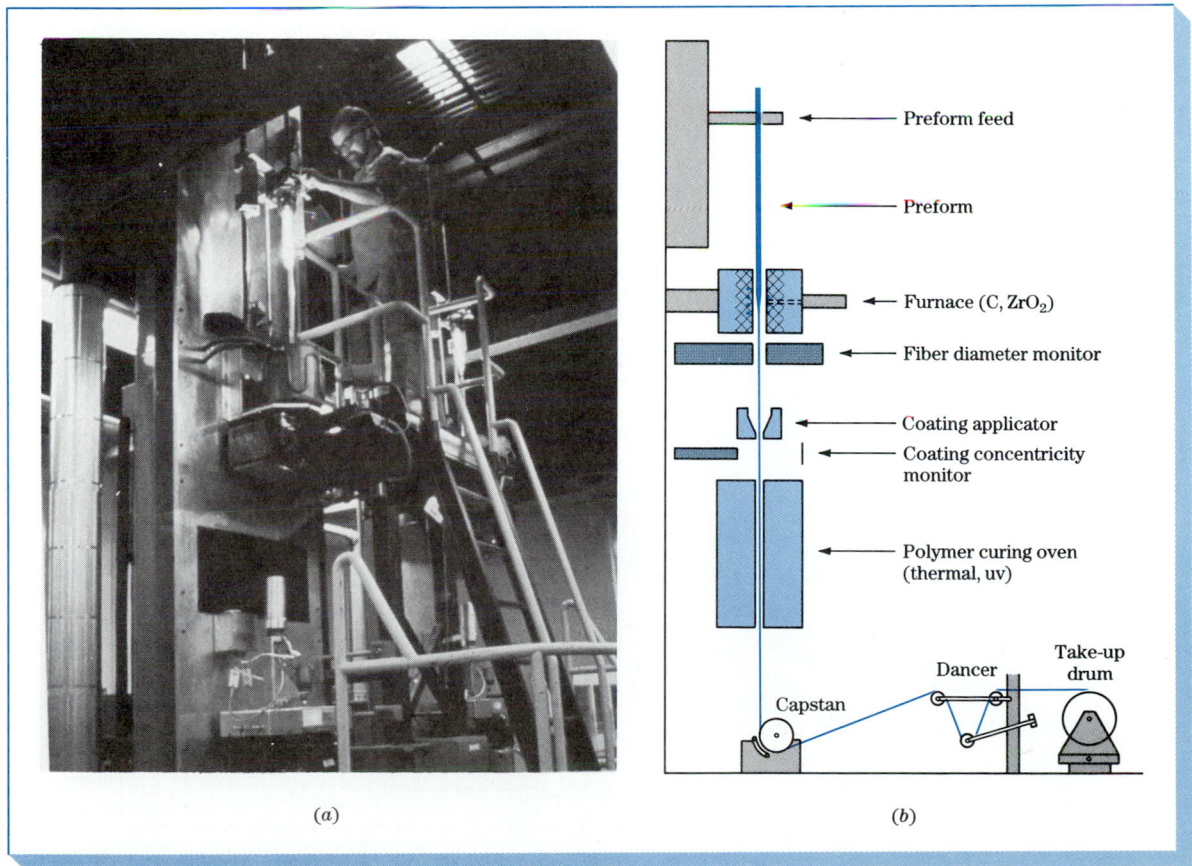

(*a*)   (*b*)

FIGURE 14.21 Spool of optical fiber. (*Courtesy of AT&T.*)

Modern Optical-Fiber Communication Systems

Most modern optical-fiber communication systems use single-mode fiber with an InGaAsP double heterojunction laser diode transmitter (Fig. 14.22a) operated at the infrared wavelength of 1.3 μm, where light losses are at a minimum. An InGaAs/InP PIN photodiode is usually used for the detector (Fig. 14.22b). With this system optical signals can be sent about 40 km (25 mi) before the signal has to be repeated. In December 1988 the first transatlantic fiber-optic communications system began operation with a capacity of 40,000 simultaneous phone calls. By 1993 there was a total of 289 undersea optical-fiber cable links.

Another advance in optical-fiber communications systems has occurred with the introduction of *erbium-doped optical-fiber amplifiers* (EDFAs). An EDFA is a length [typically about 20 to 30 m (64 to 96 ft)] of optical silica fiber

FIGURE 14.22 (*a*) Chemical substrate buried heterostructure InGaAsP laser diode used for long-distance fiber-optical communications systems. Note the focusing of the laser beam by the V channel. (*b*) PIN photodetector for optical communications systems. [*After AT&T Tech. J. 66(1987).*]

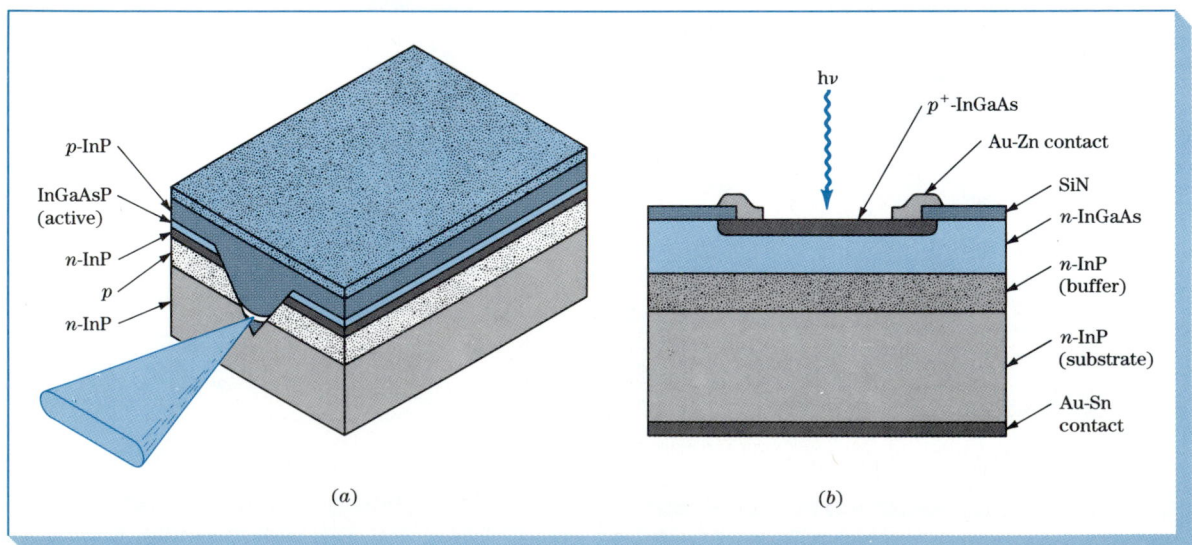

doped with the rare-earth element erbium to give fiber gain. When optically pumped with light from an outside semiconductor laser, the erbium-doped fiber boosts the power of all light signals passing through it with wavelengths centered on 1.55 μm. Thus, the erbium-doped optical fiber serves as both a lasing medium and a light guide. EDFAs can be used in optical transmission systems to boost the light signal power at the source (power amplifier), at the receiver (preamplifier), and along the fiber communication link (in-line repeater). The first EDFAs were used in 1993 in an AT&T network in a link between San Francisco and Point Arena, Calif.

14.8 SUPERCONDUCTING MATERIALS

The Superconducting State

The electrical resistivity of a normal metal such as copper decreases steadily as the temperature is decreased and reaches a low residual value near 0 K (Fig. 14.23). In contrast, the electrical resistivity of pure mercury as the temperature decreases drops suddenly at 4.2 K to an immeasurably small value. This phenomenon is called *superconductivity,* and the material which shows this behavior is called a *superconductive material.* About 26 metals are superconductive as well as hundreds of alloys and compounds.

The temperature below which a material's electrical resistivity approaches absolute zero is called the *critical temperature* T_c. Above this temperature the material is called *normal,* and below T_c it is said to be *superconducting* or *superconductive.* Besides temperature, the superconducting state also depends on many other variables, the most important of which are the magnetic field B and current density J. Thus for a material to be superconducting, the material's critical temperature, magnetic field, and current density must not be exceeded, and for each superconducting material there exists a critical surface in T, B, J space.

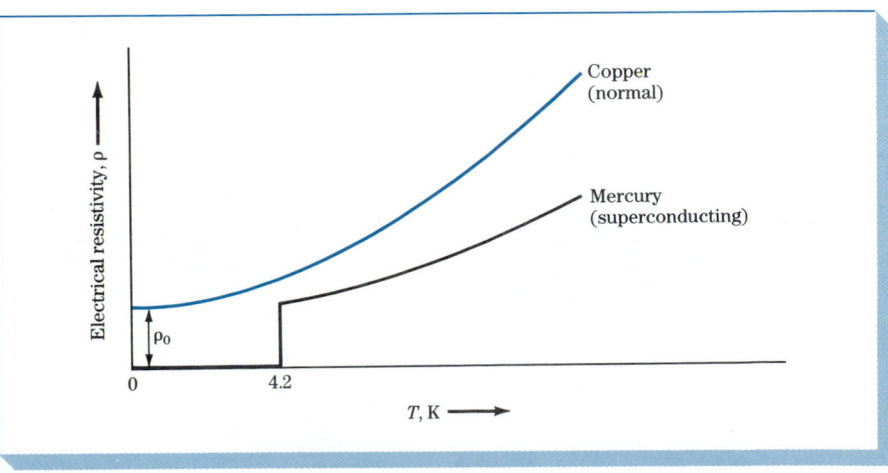

FIGURE 14.23 Electrical resistivity of a normal metal (Cu) compared to that of a superconductive metal (Hg) as a function of temperature near 0 K. The resistivity of the superconductive metal suddenly drops to an immeasurably small value.

TABLE 14.3   Critical Superconducting Temperatures T_c for Selected Metals, Intermetallic and Ceramic Compound Superconductors

Metals	T_c, K	H_0^*, T	Intermetallic compounds	T_c, K	Ceramic compounds	T_c, K
Niobium, Nb	9.15	0.1960	Nb_3Ge	23.2	$Tl_2Ba_2Ca_2Cu_3O_x$	122
Vanadium, V	5.30	0.1020	Nb_3Sn	21	$YBa_2Cu_3O_{7-x}$	90
Tantalum, Ta	4.48	0.0830	Nb_3Al	17.5	$Ba_{1-x}K_xBiO_{3-y}$	30
Titanium, Ti	0.39	0.0100	NbTi	9.5		
Tin	3.72	0.0306				

$^*H_0$ = critical field in teslas (T) at 0 K.

The critical superconducting temperatures of some selected metals, intermetallic compounds, and new ceramic compounds are listed in Table 14.3. The extremely high T_c values (90–122 K) of the newly discovered (1987) ceramic compounds are outstanding and were a surprise to the scientific community. Some aspects of their structure and properties will be discussed later in this section.

Magnetic Properties of Superconductors

If a sufficiently strong magnetic field is applied to a superconductor at any temperature below its critical temperature T_c, the superconductor will return to the normal state. The applied magnetic field necessary to restore normal electrical conductivity in the superconductor is called the *critical field H_c*. Figure 14.24a shows schematically the relationship between the critical magnetic field

FIGURE 14.24   Critical field vs. temperature (K). (a) General case. (b) Curves for several superconductors.

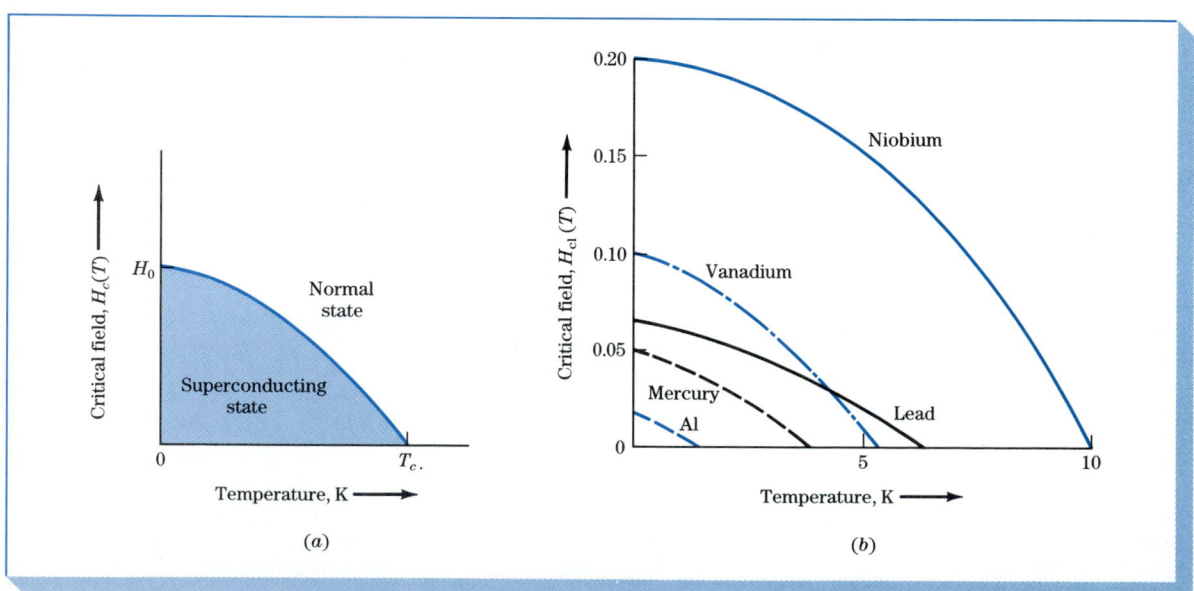

H_c and temperature (K) at zero current. It should be pointed out that a sufficiently high electrical current density, J_c will also destroy superconductivity in materials. The curve of H_c vs. T (K) can be approximated by

$$H_c = H_0 \left[1 - \left(\frac{T}{T_c}\right)^2 \right]$$
(14.10)

where H_0 is the critical field at $T = 0$ K. Equation (14.10) represents the boundary between the superconducting and normal states of the superconductor. Figure 14.24b shows the critical field vs. temperature plots for several superconducting metals.

Example Problem 14.8

Calculate the approximate value of the critical field necessary to cause the superconductivity of pure niobium metal to disappear at 6 K.

Solution:

From Table 14.3 at 0 K, the T_c for Nb is 9.15 K and its $H_0 = 0.1960$ T. From Eq. (14.10),

$$H_c = H_0 \left[1 - \left(\frac{T}{T_c}\right)^2 \right] = 0.1960 \left[1 - \left(\frac{6}{9.15}\right)^2 \right] = 0.112 \text{ T} \blacktriangleleft$$

According to their behavior in an applied magnetic field, metallic and intermetallic superconductors are classified into type I and type II superconductors. If a long cylinder of a type I superconductor such as Pb or Sn is placed in a magnetic field at room temperature, the magnetic field will penetrate normally throughout the metal (Fig. 14.25a). However, if the temperature of

FIGURE 14.25 The Meissner effect. When the temperature of a type I superconductor is lowered below T_c and the magnetic field is below H_c, the magnetic field is completely expelled from a sample except for a thin surface layer.

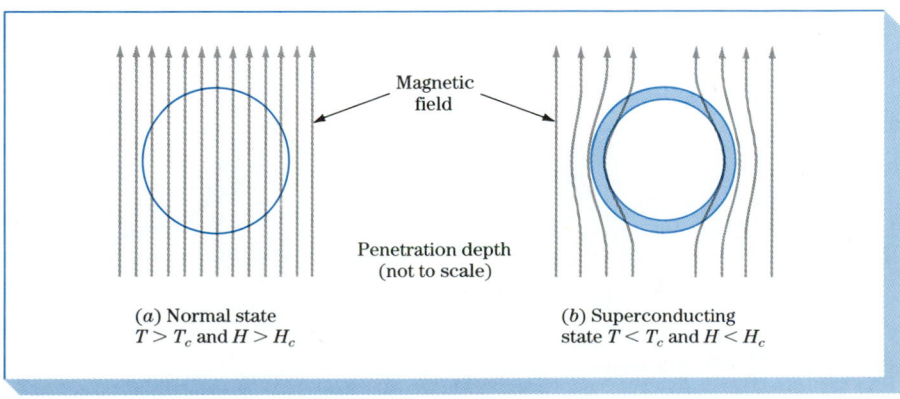

Magnetic field

Penetration depth (not to scale)

(a) Normal state
$T > T_c$ and $H > H_c$

(b) Superconducting state $T < T_c$ and $H < H_c$

the type I superconductor is lowered below its T_c (7.19 K for Pb) and if the magnetic field is below H_c, the magnetic field will be expelled from the specimen except for a very thin penetration layer of about 10^{-5} cm at the surface (Fig. 14.25b). This property of a magnetic-field exclusion in the superconducting state is called the *Meissner effect*.

Type II superconductors behave differently in a magnetic field at temperatures below T_c. They are highly diamagnetic like type I superconductors up to a critical applied magnetic field designated the lower critical field H_{c1} (Fig. 14.26), and thus the magnetic flux is excluded from the material. Above H_{c1} the field starts to penetrate the type II superconductor and continues to do so until the upper critical field H_{c2} is reached. In between H_{c1} and H_{c2} the superconductor is in the mixed state, and above H_{c2} it returns to the normal state. In the region H_{c1} and H_{c2} the superconductor can conduct electrical current within bulk material, and thus this magnetic-field region can be used for high-current, high-field superconductors such as NiTi and Ni_3Sb, which are type II superconductors.

Current Flow and Magnetic Fields in Superconductors

Type I superconductors are poor carriers of electrical current since current can only flow in the outer surface layer of a conducting specimen (Fig. 14.27a). The reason for this behavior is that the magnetic field can only penetrate the surface layer and current can only flow in this layer. In type II superconductors below H_{c1} magnetic fields behave in the same way. However, if the magnetic field is between H_{c1} and H_{c2} (mixed state), the current can be carried inside the superconductor by filaments, as indicated in Fig. 14.27b. In type II superconductors when a magnetic field between H_{c1} and H_{c2} is applied, the field penetrates the bulk of the superconductor in the form of individual quantized flux bundles called *fluxoids* (Fig. 14.28). A cylindrical supercurrent vortex surrounds each fluxoid. With increasing magnetic-field strength, more and more fluxoids enter the superconductor and form a periodic array. At H_{c2} the supercurrent vortex structure collapses and the material returns to the normal conducting state.

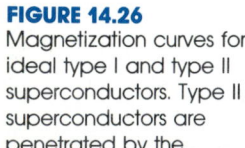

FIGURE 14.26
Magnetization curves for ideal type I and type II superconductors. Type II superconductors are penetrated by the magnetic field between H_{c1} and H_{c2}.

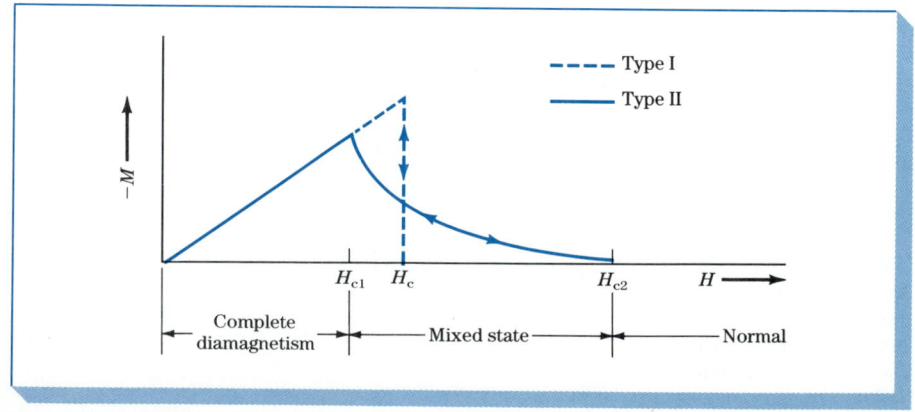

FIGURE 14.27 Cross section of a superconducting wire carrying an electrical current. (a) Type I superconductor or type II under low field ($H < H_{c1}$). (b) Type II superconductor under higher fields where current is carried by a filament network ($H_{c1} < H < H_{c2}$).

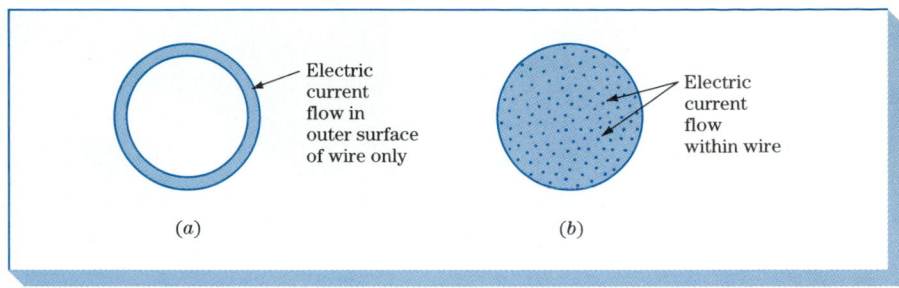

Electric current flow in outer surface of wire only

Electric current flow within wire

(a)   (b)

High-Current, High-Field Superconductors

Although ideal type II superconductors can be penetrated by an applied magnetic field in the H_{c1} to H_{c2} range, they have a small current-carrying capacity below T_c since the fluxoids are weakly tied to the crystal lattice and are relatively mobile. The mobility of the fluxoids can be greatly impeded by dislocations, grain boundaries, and fine precipitates, and thus J_c can be raised by cold working and heat treatments. Heat treatment of the Nb–45 wt % Ti alloy is used to precipitate a hexagonal α phase in the BCC matrix of the alloy to help pin down the fluxoids.

The alloy Nb–45 wt % Ti and the compound Nb$_3$Sn have become the basic materials for modern high-current, high-field superconductor technology. Commercial Ni–45 wt % Ti has been produced with a $T_c \approx 9$ K and $H_{c2} \approx 6$ T and Nb$_3$Sn with a $T_c \approx 18$ K and $H_{c2} \approx 11$ T. In today's superconductor technology these superconductors are used at liquid helium temperature (4.2 K). The Nb–45 wt % Ti alloy is more ductile and easier to fabricate than the Nb$_3$Sn compound and so is preferred for many applications even though it has lower T_c's and H_{c2}'s. Commercial wires are made of many NbTi filaments typically about 25 μm in diameter embedded in a copper matrix (Fig. 14.29). The purpose of the copper matrix is to stabilize the superconductor wire during operation so that hot spots will not develop that could cause the superconducting material to return to the normal state.

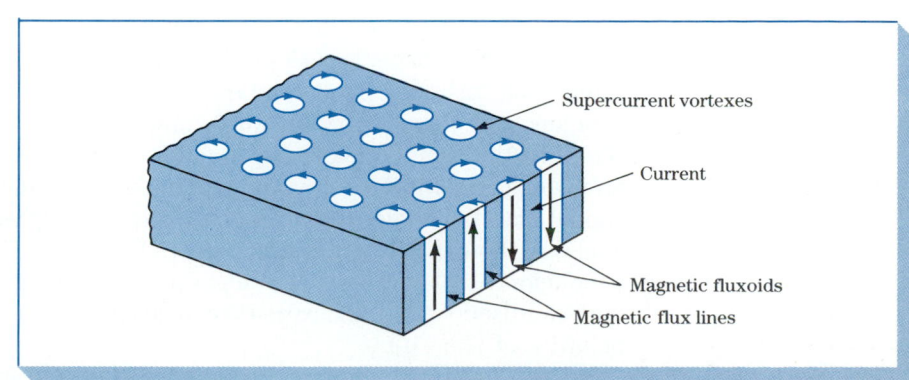

Supercurrent vortexes

Current

Magnetic fluxoids

Magnetic flux lines

FIGURE 14.28 Schematic illustration showing magnetic fluxoids in a type II superconductor with the magnetic field between H_{c1} and H_{c2}.

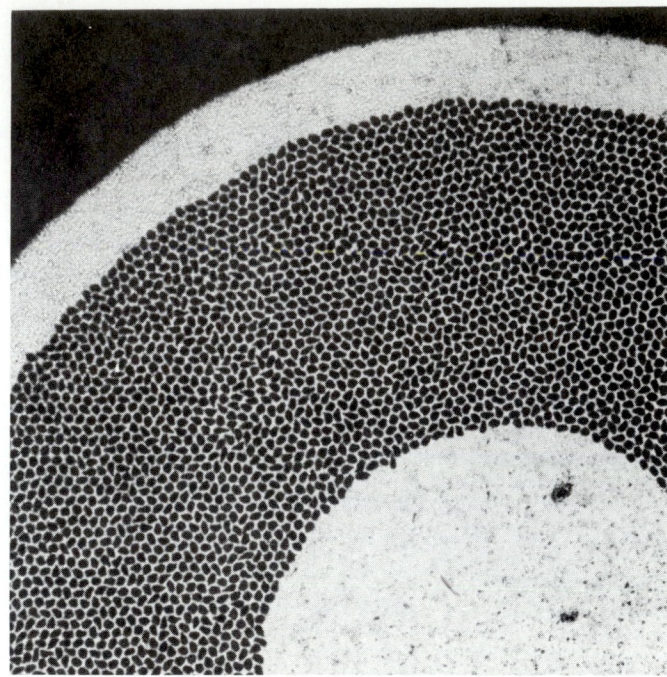

FIGURE 14.29 Cross section of Nb–46.5 wt % Ti–Cu composite wire made for the superconductor supercollider. The wire has a diameter of 0.0808 cm (0.0318 in), a Cu·NbTi volumetric ratio of 1.5, 7250 filaments of 6-μm diameters, and a J_c = 2990 A/mm^2 at 5 T and a J_c = 1256 A/mm^2 at 8 T (magnification 200 ×). (*Courtesy of IGC Advanced Superconductors Inc.*)

Applications for NbTi and Nb$_3$Sn superconductors include nuclear magnetic imaging systems for medical diagnosis and magnetic levitation of vehicles such as high-speed trains (Fig. 14.1c and d). High-field superconducting magnets are used in particle accelerators in the high-energy-physics field. The prototype for the newly proposed superconductor supercollider uses NbTi wires for its operation. On a worldwide basis the next generation of accelerators is expected to use about 2000 tons of superconducting wire.

High Critical Temperature (T_c) Superconducting Oxides

In 1987 superconductors with critical temperatures of about 90 K were discovered, surprising the scientific community since up to that time the highest T_c for a superconductor was about 23 K. The most intensely studied high T_c material has been the YBa$_2$Cu$_3$O$_y$ compound, and so our attention will be focused on some aspects of its structure and properties. From a crystal structure standpoint, this compound can be considered to have a defective perovskite structure with three perovskite cubic unit cells stacked on top of each other (Fig. 14.30). (The perovskite structure for CaTiO$_3$ is shown in Fig. 10.16). For an ideal stack of three perovskite cubic unit cells, the YBa$_2$Cu$_3$O$_y$ compound should have the composition YBa$_2$Cu$_3$O$_9$, in which y would equal 9. However, analyses show that y ranges from 6.65 to 6.90 for this material to be a superconductor. At y = 6.90, its T_c is highest (~90 K), and at y = 6.65, superconductivity disappears. Thus oxygen vacancies play a role in the superconductivity behavior of YBa$_2$Cu$_3$O$_y$.

The YBa$_2$Cu$_3$O$_y$ compound when slowly cooled from above 750°C in the

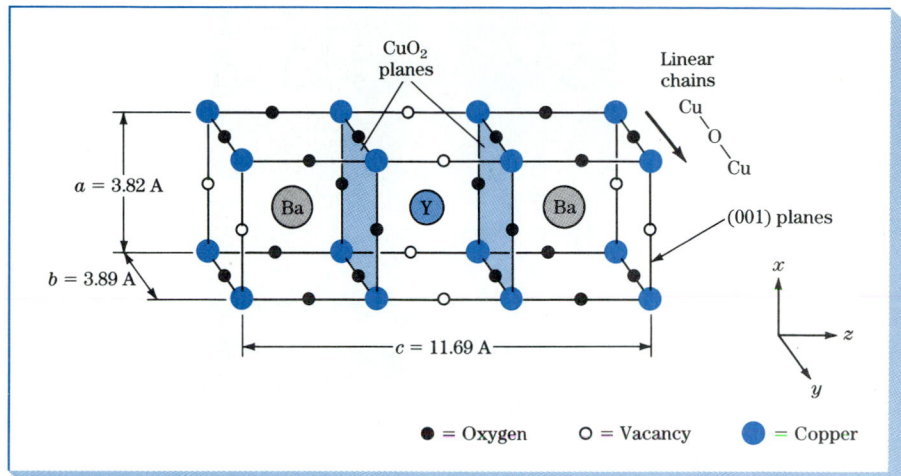

FIGURE 14.30 Idealized $YBa_2Cu_3O_7$ orthorhombic crystal structure. Note the location of the CuO_2 planes.

presence of oxygen, undergoes a tetragonal to orthorhombic crystal structure change (Fig. 14.31a). If the oxygen content is close to $y = 7$, its T_c is about 90 K (Fig. 14.31b) and its unit cell has the constants $a = 3.82$ Å, $b = 3.88$ Å, and $c = 11.6$ Å (Fig. 14.30). To have high T_c values, oxygen atoms on the (001) planes must be ordered so that oxygen vacancies are in the a direction. Su-

FIGURE 14.31 (a) Oxygen content vs. unit cell constants for $YBa_2Cu_3O_y$. (b) Oxygen content vs. T_c for $YBa_2Cu_3O_y$. (*After J. M. Tarascon and B. G. Bagley, MRS Bull., January 1989, p. 55.*)

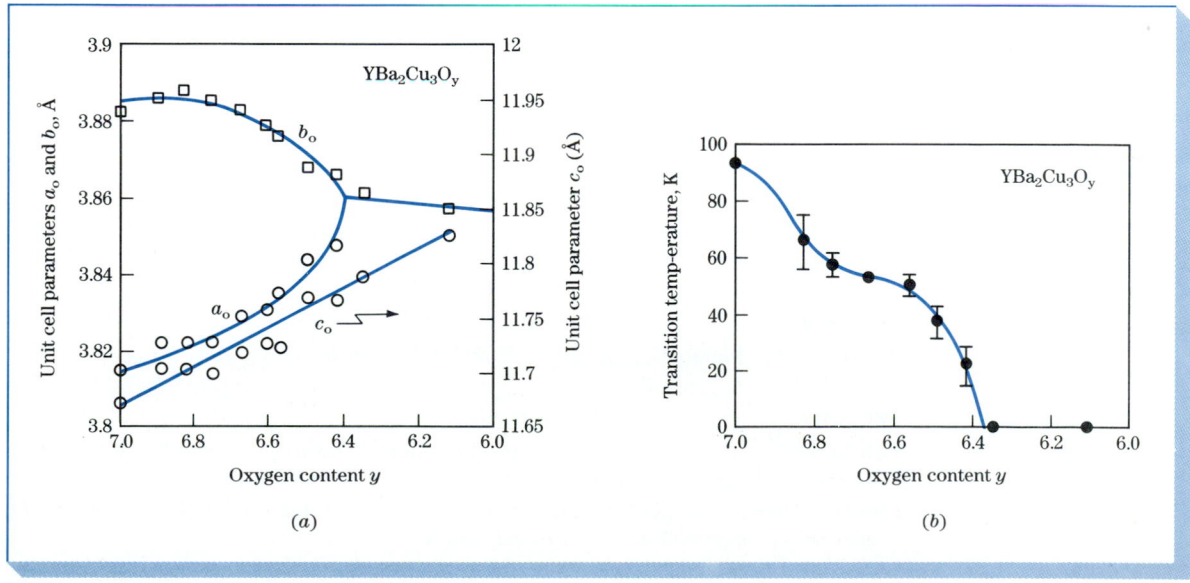

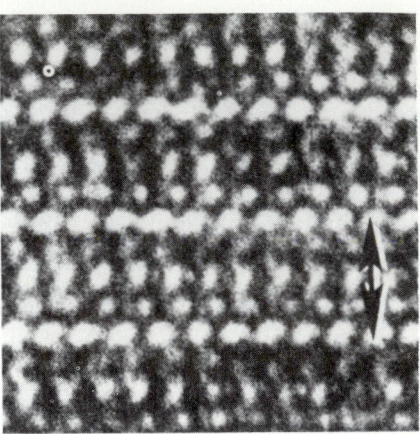

FIGURE 14.32 High-resolution TEM micrograph in the [100] direction down copper-oxygen chains and rows of Ba and Y atoms in the unit cell of $YBa_2Cu_3O_y$, as indicated by the arrow. (*After J. Narayan, JOM, January 1989, p. 18.*)

perconductivity is believed to be confined to the CuO_2 planes (Fig. 14.30), with the oxygen vacancies providing an electron coupling between the CuO_2 planes. A transmission electron micrograph (Fig. 14.32) shows the stacking of the Ba and Y atoms of the $YBa_2Cu_3O_y$ structure (see also book cover).

From an engineering viewpoint, the new high T_c superconductors hold much promise for technical advances. With T_c's at 90 K, liquid nitrogen can be used as a refrigerant to replace the much more expensive liquid helium. Unfortunately, the high-temperature superconductors are essentially ceramics which are brittle materials and also in the bulk form have low current-density capability. The first applications for these materials will probably be in thin-film technology for electronic applications such as high-speed computers.

DEFINITIONS

Sec. 14.3

Index of refraction: the ratio of the velocity of light in vacuum to that through another medium of interest.

Sec. 14.4

Absorptivity: the fraction of the incident light which is absorbed by a material.

Sec. 14.5

Luminescence: absorption of light or other energy by a material and the subsequent emission of light of longer wavelength.

Fluorescence: absorption of light or other energy by a material and the subsequent emission of light within 10^{-8} s of excitation.

Phosphorescence: absorption of light by a phosphor and its subsequent emission at times longer than 10^{-8} s.

Sec. 14.6

Laser: acronym for *l*ight *a*mplification by *s*timulated *e*mission of *r*adiation.
Laser beam: a beam of monochromatic coherent optical radiation generated by the stimulated emission of photons.
Population inversion: condition in which more atoms exist in a higher-energy state than a lower one. This condition is necessary for laser action.

Sec. 14.7

Optical communication: a method of transmitting information by the use of light.
Light attenuation: decrease in intensity of the light.
Optical wave guide: a thin-clad fiber along which light can propagate by total internal reflection and refraction.

Sec. 14.8

Superconducting state: a solid in the superconducting state that shows no electrical resistance.
Critical temperature T_c: the temperature below which a solid shows no electrical resistance.
Critical current density J_c: the current density above which superconductivity disappears.
Critical field H_c: the magnetic field above which superconductivity disappears.
Meissner effect: the expulsion of the magnetic field by a superconductor.
Type I superconductor: one which exhibits complete magnetic-flux repulsion between the normal and superconducting states.
Type II superconductor: one in which the magnetic flux gradually penetrates between the normal and superconducting states.
Lower critical field H_{c1}: the field at which magnetic flux first penetrates a type II superconductor.
Upper critical field H_{c2}: the field at which superconductivity disappears for a type II superconductor.
Fluxoid: a microscopic region surrounded by circulating supercurrents in a type II superconductor at fields between H_{c2} and H_{c1}.

PROBLEMS

14.2.1   Write the equation relating the energy of radiation to its wavelength and frequency and give the SI units for each quantity.

14.2.2   What are the approximate wavelength and frequency ranges for (*a*) visible light, (*b*) ultraviolet light, and (*c*) infrared radiation?

14.2.3   A photon in a ZnO semiconductor drops from an impurity level at 2.30 eV to its valence band. What is the wavelength of the radiation given off by the transition? If the radiation is visible, what is its color?

14.2.4 A semiconductor emits green visible radiation at a wavelength of 0.520 μm. What is the energy level from which photons drop to the valence band in order to give off this type of radiation?

14.3.1 If ordinary light is transmitted from air into a 1-cm-thick sheet of polymethacrylate, is the light sped up or slowed down upon entering the plastic? Explain.

14.3.2 Explain why cut diamonds sparkle. Why is PbO sometimes added to make decorative glasses?

14.3.3 What is Snell's law of light refraction? Use a diagram to explain.

14.3.4 What is the critical angle for light to be totally reflected when leaving a flat plate of polystyrene and entering the air?

14.4.1 Explain why metals absorb and/or reflect incident radiation up to the middle of the ultraviolet range.

14.4.2 Explain why gold is yellow in color and silver "silvery."

14.4.3 Calculate the reflectivity of ordinary light from a smooth, flat upper surface of (*a*) borosilicate glass ($n = 1.47$) and (*b*) polyethylene ($n = 1.53$).

14.4.4 Ordinary incident light strikes the flat surface of a transparent material with a linear absorption coefficient of 0.04 cm^{-1}. If the plate of the material is 0.80 cm thick, calculate the fraction of light absorbed by the plate.

14.4.5 Ordinary light strikes a flat surface of a plate of a transparent material. If the plate is 0.75 thick and absorbs 5.0 percent of the entering light, what is its linear absorption coefficient?

14.4.6 Calculate the transmittance for a flat glass plate 6.0 mm thick with an index of refraction of 1.51 and a linear absorption coefficient of 0.03 cm^{-1}.

14.4.7 Why does a sheet of 2.0-mm-thick polyethylene have lower clarity than a sheet of the same thickness of polycarbonate plastic?

14.4.8 Calculate the minimum wavelength of the radiation that can be absorbed by the following materials: (*a*) GaP, (*b*) GaSb, (*c*) InP (see Table 5.6).

14.4.9 Will visible light of a wavelength of 500 nm be absorbed or transmitted by the following materials: (*a*) CdSb, (*b*) ZnSe (*c*) diamond ($E_g = 5.40$ eV)?

14.5.1 Explain the process of luminescence.

14.5.2 Distinguish between fluorescence and phosphorescence.

14.5.3 Explain the luminescence effect operating in a fluorescent lamp.

14.5.4 Explain how the color picture is produced on a color television screen.

14.5.5 The intensity of an Al_2O_3 phosphor activated with chromium decreases to 15 percent of its original intensity in 5.6×10^{-3} s. Determine (*a*) its relaxation time and (*b*) its retained percent intensity after 5.0×10^{-2} s.

14.5.6 A Zn_2SiO_4 phosphor activated with manganese has a relaxation time of 0.015 s. Calculate the time required for the intensity of this material to decrease to 8 percent of its original value.

14.6.1 Distinguish between incoherent and coherent radiation.

14.6.2 What do the letters in the acronymn laser stand for?

14.6.3 Explain the operation of the ruby laser.

14.6.4 What does the term *population inversion* refer to in laser terminology?

14.6.5 Describe the operation and application of the following types of lasers: (*a*) neodymnium-YAG, (*b*) carbon dioxide, (*c*) double heterojunction GaAs.

14.7.1 What are optical fibers?

14.7.2 What are the basic elements of an optical-fiber communications system?

14.7.3 What type of impurities are particularly detrimental to light loss in optical fibers?

14.7.4 If the original light intensity is reduced 6.5 percent after being transmitted 300 m through an optical fiber, what is the attenuation of the light in decibels per kilometer (dB/km) for this type of optical fiber?

14.7.5 Light is attenuated in an optical fiber operating at 1.55-μm wavelength at -0.25 dB/km. If 4.2 percent of the light is to be retained at a repeater station, what must the distance between the repeaters be?

14.7.6 The attenuation of a 1.3-μm optical-fiber undersea transatlantic cable is -0.31 dB/km, and the distance between repeaters in the system is 40.2 km (25 mi). What is the percent of the light retained at a repeater if we assume 100 percent at the start of a repeater?

14.7.7 Explain how optical fibers act as waveguides.

14.7.8 Distinguish between single-mode and multimode types of optical fibers. Which type is used for modern long-distance communications systems and why?

14.7.9 How are optical fibers for communications systems fabricated? How do (a) GeO_2 and (b) F affect the refractive index of the silica glass?

14.7.10 What type of lasers are used in modern long-distance optical-fiber systems and why?

14.7.11 A single-mode optical fiber for a communications system has a core of SiO_2–GeO_2 glass with a refractive index of 1.4597 and a cladding of pure SiO_2 glass with a refractive index of 1.4580. What is the critical angle for light leaving the core to be totally reflected within the core?

14.8.1 What is the superconducting state for a material?

14.8.2 What is the significance of T_c, H_c, and J_c for a superconductor?

14.8.3 Describe the difference between type I and type II superconductors.

14.8.4 What is the Meissner effect?

14.8.5 Why are type I superconductors poor current-carrying conductors?

14.8.6 What are fluxoids? What role do they play in the superconductivity of type II superconductors in the mixed state?

14.8.7 How can fluxoids be pinned in type II superconductors? What is the consequence of pinning the fluxoids in a type II superconductor?

14.8.8 Calculate the critical magnetic field H_{c1} in teslas for niobium at 8 K. Use Eq. (14.10) and data from Table 14.3.

14.8.9 If vanadium has an H_c value of 0.06 T and is superconductive, what must its temperature be?

14.8.10 Describe the crystal structure of $YBa_2Cu_3O_7$. Use a drawing.

14.8.11 Why must the $YBa_2Cu_3O_y$ compound be cooled slowly from about 750°C in order for this compound to be highly superconductive?

14.8.12 What are some advantages and disadvantages of the new high-temperature oxide superconductors?

REFERENCES FOR FURTHER STUDY BY CHAPTER

1 "Materials Engineering 2000 and Beyond: Strategies for Competitiveness," *Advanced Materials and Processes* **145:**(1) (1994).

T. Y. Canby, "Advanced Materials—Reshaping Our Lives," *Nat. Geog.,* **176:**(6):746(1989)

"Materials Issue," *Sci. Am.,* **255:**(4)(1986).

M. B. Bever (ed.), *Encyclopedia of Materials Science and Engineering,* MIT Press-Pergamon, Cambridge, 1986.

Annual Review of Materials Science, Annual Reviews, Inc. Palo Alto, CA.

2 R. Chang, *General Chemistry,* 4th ed., McGraw-Hill, 1990.

R. McWeeny, *Coulson's Valence,* 3d ed., Oxford University Press, 1979.

L. Pauling, *The Nature of the Chemical Bond,* 3d ed., Cornell University Press, 1960.

3 C. S. Barrett and T. B. Massalski, *Structure of Metals,* 3d ed., Pergamon, 1980.

B. D. Cullity, *Elements of X-Ray Diffraction,* 2d ed., Addison-Wesley, 1978.

A. J. C. Wilson, *Elements of X-Ray Cystallography,* Addison-Wesley, 1970.

4 M. Flemings, *Solidification Processing,* McGraw-Hill, 1974.

I. Minkoff, *Solidification and Cast Structures,* Wiley, 1986.

P. G. Shewmon, *Diffusion in Solids,* 2d ed., Minerals, Mining and Materials Society, 1989.

J. P. Hirth and J. Lothe, *Theory of Dislocations,* 2d ed., Wiley, 1982.

G. Krauss (ed.), *Carburizing: Processing and Performance,* ASM International, 1989.

5 S. M. Sze (ed.), *VLSI Technology,* 2d ed., McGraw-Hill, 1988.

S. M. Sze, *Semiconductor Devices,* Wiley, 1985.

D. A. Hodges and H. G. Jackson, *Analysis and Design of Digital Integrated Circuits,* 2d ed., McGraw-Hill, 1988.

6 G. E. Dieter, *Mechanical Metallurgy,* McGraw-Hill, 1986.

R. W. Herzberg, *Deformation and Fracture Mechanics of Engineering Materials,* 3d ed., Wiley, 1989.

T. H. Courtney, *Mechanical Behavior of Materials,* McGraw-Hill, 1989.

7 "Engineering Plastics," vol. 2, *Engineered Materials Handbook,* ASM International, 1988.

G. R. Moore and D. E. Kline, *Properties and Processing of Polymers for Engineers,* Prentice-Hall, 1984.

H. S. Kaufman and J. J. Falcetta (eds.), *Introduction to Polymer Science and Technology,* Wiley, 1977.

8 T. B. Massalski, *Binary Alloy Phase Diagrams,* ASM International, 1986.

F. Rhines, *Phase Diagrams in Metallurgy,* McGraw-Hill, 1956.

9 W. F. Smith, *Structure and Properties of Engineering Alloys,* 2d ed., McGraw-Hill, 1993.

J. L. Walker et al. (ed.), *Alloying,* ASM International, 1988.

G. Krauss, *Steels: Heat Treatment and Processing Principles,* ASM International, 1990.

10 W. D. Kingery, H. K. Bowen, and D. R. Uhlmann, *Introduction to Ceramics,* 2d ed., Wiley, 1976.

"Ceramics and Glasses," vol. 4, *Engineered Materials Handbook,* ASM International, 1991.

J. B. Wachtman (ed.) *Structural Ceramics,* Academic, 1989.

11 G. Y. Chin and J. H. Wernick, "Magnetic Materials, Bulk," vol. 14, *Kirk-Othmer Encyclopedia of Chemical Technology,* 3d ed., Wiley, 1981, p. 686.

B. D. Cullity, *Introduction to Magnetic Materials,* Addison-Wesley, 1972.

J. A. Salsgiver et al. (ed.), *Hard and Soft Magnetic Materials,* ASM International, 1987.

12 "Corrosion," vol. 13, *Metals Handbook,* 9th ed., ASM International, 1987.

M. G. Fontana, *Corrosion Engineering,* 3d ed., McGraw-Hill, 1986.

H. H. Uhlig, *Corrosion and Corrosion Control,* 3d ed., 1985.

13 "Composites," vol. 1, *Engineered Materials Handbook,* ASM International, 1987.
Engineers' Guide to Composite Materials, ASM International, 1987.
K. K. Chawla, *Composite Materials,* Springer-Verlag, 1987.
B. Harris, *Engineering Composite Materials,* Institute of Metals (London), 1986.
14 C. D. Chafee, *The Rewiring of America,* Academic, 1988.
S. E. Miller and I. P. Kaminow, *Optical Fiber Communications II,* Academic, 1988.
W. H. Hatfield and J. H. Miller, *High Temperature Superconducting Materials,* Marcel Dekker, 1988.

Some Properties of Selected Elements

Element	Symbol	Melting point, °C	Density,* g/cm³	Atomic radius, nm	Crystal structure† (20°C)	Lattice constants 20°C, nm	
						a	c
Aluminum	Al	660	2.70	0.143	FCC	0.40496	
Antimony	Sb	630	6.70	0.138	Rhombohedral	0.45067	
Arsenic	As	817	5.72	0.125	Rhombohedral‡	0.4131	
Barium	Ba	714	3.5	0.217	BCC‡	0.5019	
Beryllium	Be	1278	1.85	0.113	HCP‡	0.22856	0.35832
Boron	B	2030	2.34	0.097	Orthorhombic		
Bromine	Br	−7.2	3.12	0.119	Orthorhombic		
Cadmium	Cd	321	8.65	0.148	HCP‡	0.29788	0.561667
Calcium	Ca	846	1.55	0.197	FCC‡	0.5582	
Carbon (graphite)	C	3550	2.25	0.077	Hexagonal	0.24612	0.67078
Cesium	Cs	28.7	1.87	0.190	BCC		
Chlorine	Cl	−101	1.9	0.099	Tetragonal		
Chromium	Cr	1875	7.19	0.128	BCC‡	0.28846	
Cobalt	Co	1498	8.85	0.125	HCP‡	0.2506	0.4069
Copper	Cu	1083	8.96	0.128	FCC	0.36147	
Fluorine	F	−220	1.3	0.071			
Gallium	Ga	29.8	5.91	0.135	Orthorhombic		

* Density of solid at 20°C.
‡ Other crystal structures exist at other temperatures.

Element	Symbol	Melting point, °C	Density,* g/cm³	Atomic radius, nm	Crystal structure† (20°C)	Lattice constants 20°C, nm	
						a	c
Germanium	Ge	937	5.32	0.139	Diamond cubic	0.56576	
Gold	Au	1063	19.3	0.144	FCC	0.40788	
Helium	He	−270	⋯	⋯	HCP		
Hydrogen	H	−259	⋯	0.046	Hexagonal		
Indium	In	157	7.31	0.162	FC tetragonal	0.45979	0.49467
Iodine	I	114	4.94	0.136	Orthorhombic		
Iridium	Ir	2454	22.4	0.135	FCC	0.38389	
Iron	Fe	1536	7.87	0.124	BCC‡	0.28664	
Lead	Pb	327	11.34	0.175	FCC	0.49502	
Lithium	Li	180	0.53	0.157	BCC	0.35092	
Magnesium	Mg	650	1.74	0.160	HCP	0.32094	0.52105
Manganese	Mn	1245	7.43	0.118	Cubic‡	0.89139	
Mercury	Hg	−38.4	14.19	0.155	Rhombohedral		
Molybdenum	Mo	2610	10.2	0.140	BCC	0.31468	
Neon	Ne	−248.7	1.45	0.160	FCC		
Nickel	Ni	1453	8.9	0.125	FCC	0.35236	
Niobium	Nb	2415	8.6	0.143	BCC	0.33007	
Nitrogen	N	−240	1.03	0.071	Hexagonal‡		
Osmium	Os	2700	22.57	0.135	HCP	0.27353	0.43191
Oxygen	O	−218	1.43	0.060	Cubic‡		
Palladium	Pd	1552	12.0	0.137	FCC	0.38907	
Phosphorus (white)	P	44.2	1.83	0.110	Cubic‡		
Platinum	Pt	1769	21.4	0.139	FCC	0.39239	
Potassium	K	63.9	0.86	0.238	BCC	0.5344	
Rhenium	Re	3180	21.0	0.138	HCP	0.27609	0.44583
Rhodium	Rh	1966	12.4	0.134	FCC	0.38044	
Ruthenium	Ru	2500	12.2	0.125	HCP	0.27038	0.42816
Scandium	Sc	1539	2.99	0.160	FCC	0.4541	
Silicon	Si	1410	2.34	0.117	Diamond cubic	0.54282	
Silver	Ag	961	10.5	0.144	FCC	0.40856	
Sodium	Na	97.8	0.97	0.192	BCC	0.42906	
Strontium	Sr	76.8	2.60	0.215	FCC‡	0.6087	
Sulfur (yellow)	S	119	2.07	0.104	Orthorhombic		
Tantalum	Ta	2996	16.6	0.143	BCC	0.33026	
Tin	Sn	232	7.30	0.158	Tetragonal‡	0.58311	0.31817
Titanium	Ti	1668	4.51	0.147	HCP‡	0.29504	0.46833
Tungsten	W	3410	19.3	0.141	BCC	0.31648	
Uranium	U	1132	19.0	0.138	Orthorhombic‡††	0.2858	0.4955
Vanadium	V	1900	6.1	0.136	BCC	0.3039	
Zinc	Zn	419.5	7.13	0.137	HCP	0.26649	0.49470
Zirconium	Zr	1852	6.49	0.160	HCP‡	0.32312	0.51477

* Density of solid at 20°C.
† b = 0.5877 nm.
‡ Other crystal structures exist at other temperatures.

Ionic Radii[1] of the Elements

Atomic number	Element (symbol)	Ion	Ionic radius, nm	Atomic number	Element (symbol)	Ion	Ionic radius, nm
1	H	H^-	0.154	22	Ti	Ti^{2+}	0.076
2	He					Ti^{3+}	0.069
3	Li	Li^+	0.078			Ti^{4+}	0.064
4	Be	Be^{2+}	0.034	23	V	V^{3+}	0.065
5	B	B^{3+}	0.02			V^{4+}	0.061
6	C	C^{4+}	<0.02			V^{5+}	~0.04
7	N	N^{5+}	0.01–0.02	24	Cr	Cr^{3+}	0.064
8	O	O^{2-}	0.132			Cr^{6+}	0.03–0.04
9	F	F^-	0.133	25	Mn	Mn^{2+}	0.091
10	Ne					Mn^{3+}	0.070
11	Na	Na^+	0.098			Mn^{4+}	0.052
12	Mg	Mg^{2+}	0.078	26	Fe	Fe^{2+}	0.087
13	Al	Al^{3+}	0.057			Fe^{3+}	0.067
14	Si	Si^{4-}	0.198	27	Co	Co^{2+}	0.082
		Si^{4+}	0.039			Co^{3+}	0.065
15	P	P^{5+}	0.03–0.04	28	Ni	Ni^{2+}	0.078
16	S	S^{2-}	0.174	29	Cu	Cu^+	0.096
		S^{6+}	0.034	30	Zn	Zn^{2+}	0.083
17	Cl	Cl^-	0.181	31	Ga	Ga^{3+}	0.062
18	Ar			32	Ge	Ge^{4+}	0.044
19	K	K^+	0.133	33	As	As^{3+}	0.069
20	Ca	Ca^{2+}	0.106			As^{5+}	~0.04
21	Sc	Sc^{2+}	0.083				

[1]Ionic radii can vary in different crystals due to many factors.

Atomic number	Element (symbol)	Ion	Ionic radius, nm	Atomic number	Element (symbol)	Ion	Ionic radius, nm
34	Se	Se^{2-}	0.191	63	Eu	Eu^{3+}	0.113
		Se^{6+}	0.03–0.04	64	Gd	Gd^{3+}	0.111
35	Br	Br^-	0.196	65	Tb	Tb^{3+}	0.109
36	Kr					Tb^{4+}	0.089
37	Rb	Rb^+	0.149	66	Dy	Dy^{3+}	0.107
38	Sr	Sr^{2+}	0.127	67	Ho	Ho^{3+}	0.105
39	Y	Y^{3+}	0.106	68	Er	Er^{3+}	0.104
40	Zr	Zr^{4+}	0.087	69	Tm	Tm^{3+}	0.104
41	Nb	Nb^{4+}	0.069	70	Yb	Yb^{3+}	0.100
		Nb^{5+}	0.069	71	Lu	Lu^{3+}	0.099
42	Mo	Mo^{4+}	0.068	72	Hf	Hf^{4+}	0.084
		Mo^{6+}	0.065	73	Ta	Ta^{5+}	0.068
44	Ru	Ru^{4+}	0.065	74	W	W^{4+}	0.068
45	Rh	Rh^{3+}	0.068			W^{6+}	0.065
		Rh^{4+}	0.065	75	Re	Re^{4+}	0.072
46	Pd	Pd^{2+}	0.050	76	Os	Os^{4+}	0.067
47	Ag	Ag^+	0.113	77	Ir	Ir^{4+}	0.066
48	Cd	Cd^{2+}	0.103	78	Pt	Pt^{2+}	0.052
49	In	In^{3+}	0.092			Pt^{4+}	0.055
50	Sn	Sn^{4-}	0.215	79	Au	Au^+	0.137
		Sn^{4+}	0.074	80	Hg	Hg^{2+}	0.112
51	Sb	Sb^{3+}	0.090	81	Tl	Tl^+	0.149
52	Te	Te^{2-}	0.211			Tl^{3+}	0.106
		Te^{4+}	0.089	82	Pb	Pb^{4-}	0.215
53	I	I^-	0.220			Pb^{2+}	0.132
		I^{5+}	0.094			Pb^{4+}	0.084
54	Xe			83	Bi	Bi^{3+}	0.120
55	Cs	Cs^+	0.165	84	Po		
56	Ba	Ba^{2+}	0.143	85	At		
57	La	La^{3+}	0.122	86	Rn		
58	Ce	Ce^{3+}	0.118	87	Fr		
		Ce^{4+}	0.102	88	Ra	Ra^+	0.152
59	Pr	Pr^{3+}	0.116	89	Ac		
		Pr^{4+}	0.100	90	Th	Th^{4+}	0.110
60	Nd	Nd^{3+}	0.115	91	Pa		
61	Pm	Pm^{3+}	0.106	92	U	U^{4+}	0.105
62	Sm	Sm^{3+}	0.113				

*Ionic radii can vary in different crystals due to many factors.

Source: C. J. Smithells (ed.), "Metals Reference Book," 5th ed., Butterworth, 1976.

Selected Physical Quantities and Their Units

Quantity	Symbol	Unit	Abbreviation	
Length	l	inch	in	
		meter	m	
Wavelength	λ	meter	m	
Mass	m	kilogram	kg	
Time	t	second	s	
Temperature	T	degree Celsius	°C	
		degree Fahrenheit	°F	
		kelvin	K	
Frequency	ν	hertz	Hz	$[s^{-1}]$
Force	F	newton	N	$[kg \cdot m \cdot s^{-2}]$
Stress:				
Tensile	σ	pascal	Pa	$[N \cdot m^{-2}]$
Shear	τ	pounds per square inch	lb/in² or psi	
Energy, work, quantity of heat		joule	J	$[N \cdot m]$
Power		watt	W	$[J \cdot s^{-1}]$
Current flow	i	ampere	A	
Electric charge	q	coulomb	C	$[A \cdot s]$
Potential difference, electromotive force	V, E	volt	V	
Electric resistance	R	ohm	Ω	$[V \cdot A^{-1}]$
Magnetic induction	B	tesla	T	$[V \cdot s \cdot m^{-2}]$

Greek Alphabet

Name	Lowercase	Capital	Name	Lowercase	Capital
Alpha	α	A	Nu	ν	N
Beta	β	B	Xi	ξ	Ξ
Gamma	γ	Γ	Omicron	o	O
Delta	δ	Δ	Pi	π	Π
Epsilon	ϵ	E	Rho	ρ	P
Zeta	ζ	Z	Sigma	σ	Σ
Eta	η	H	Tau	τ	T
Theta	θ	Θ	Upsilon	υ	Υ
Iota	ι	I	Phi	ϕ	Φ
Kappa	κ	K	Chi	χ	X
Lambda	λ	Λ	Psi	ψ	Ψ
Mu	μ	M	Omega	ω	Ω

SI Unit Prefixes

Multiple	Prefix	Symbol
10^{-12}	pico	p
10^{-9}	nano	n
10^{-6}	micro	μ
10^{-3}	milli	m
10^{-2}	centi	c
10^{-1}	deci	d
10^{1}	deca	da
10^{2}	hecto	h
10^{3}	kilo	k
10^{6}	mega	M
10^{9}	giga	G
10^{12}	tera	T

Example: 1 kilometer = 1 km = 10^3 meters.

Answers to Selected Problems

Chapter 2

2.2.2	1.79×10^{-22} g
2.2.3	5.88×10^{21} atoms
2.2.4	5.81×10^{21} atoms
2.2.5	1.97×10^{-22} g
2.2.6	5.07×10^{21} atoms
2.2.7	68.1% Sn, 31.9% Pb
2.2.8	66.8% Ni, 33.2% Cu
2.2.9	80.5% Cu, 19.5% Zn
2.2.10	$CuAl_2$
2.2.11	Mg_2Ni
2.3.2	6.95×10^{-19} J, 4.34 eV
2.3.3	4.68×10^{-19} J, 2.92 eV
2.3.4	(a) 1.55×10^{-19} J, (b) 2.34×10^{14} Hz, (c) 1280 nm
2.3.5	(a) 7.56×10^{-20} J, (b) 1.14×10^{14} Hz, (c) 2630 nm
2.3.6	(a) 6,406 eV, (b) 1.55×10^{18} Hz
2.3.7	(a) 5911 eV: Mn, (b) 6417 eV: Fe, (c) 6944 eV: Co
2.5.4	2.33×10^{-9} N
2.5.5	1.09×10^{-8} N
2.5.6	-6.52×10^{-19} J
2.5.7	-2.82×10^{-18} J
2.5.8	0.133 nm
2.5.9	0.127 nm
2.9.2	ZnS = 11.5%; GaP = 2.2%
2.9.3	CdS = 18.3%; InAs = 11.5%

Chapter 3

3.3.5	0.550 nm
3.3.6	0.326 nm
3.3.7	0.143 nm
3.3.8	0.217 nm
3.3.11	0.128 nm
3.3.12	0.197 nm
3.3.13	0.382 nm
3.3.14	0.404 nm
3.3.20	0.0486 nm^3
3.3.21	0.3210 nm
3.3.22	0.110 nm^3
3.5.5	(a) a. $[1\bar{1}0]$, b. $[\bar{3}23]$ c. $[\bar{1}34]$ d. $[3\bar{3}1]$ (b) a. $[1\bar{2}2]$ b. $[\bar{2}2\bar{1}]$ c. $[6\bar{6}1]$ d. $[9\overline{128}]$

3.5.6   [033]

3.5.7   [1$\bar{1}$0]

3.5.9   [100], [010], [001], [$\bar{1}$00], [0$\bar{1}$0], [00$\bar{1}$]

3.5.10   [111], [$\bar{1}\bar{1}\bar{1}$], [1$\bar{1}$1], [$\bar{1}$1$\bar{1}$], [$\bar{1}$11], [1$\bar{1}\bar{1}$], [11$\bar{1}$], [$\bar{1}\bar{1}$1]

3.5.11   [0$\bar{1}$1], [0$\bar{1}\bar{1}$], [$\bar{1}$10], [1$\bar{1}$0], [$\bar{1}$01], [10$\bar{1}$]

3.5.12   [1$\bar{1}\bar{1}$], [$\bar{1}$11], [$\bar{1}\bar{1}$1], [$\bar{1}\bar{1}\bar{1}$]

3.6.3   (*a*) a. (013) b. (5**120**) c. (0$\bar{3}$4); (*b*) a. ($\bar{6}$89) b. (**12**$\bar{5}$0) c. (20$\bar{3}$)

3.6.5   (100), (010), (001), ($\bar{1}$00), (0$\bar{1}$0), (001)

3.6.8   (6$\bar{3}$4)

3.6.9   ($\bar{4}$43)

3.6.10   (23$\bar{4}$)

3.6.11   ($\bar{1}\bar{1}$2)

3.6.12   ($\bar{1}$1$\bar{1}$)

3.6.13   (1$\bar{2}\bar{2}$)

3.6.14   (122)

3.6.15   (*a*) 0.203 nm, (*b*) 0.176 nm, (*c*) 0.125 nm

3.6.16   (*a*) 0.222 nm, (*b*) 0.111 nm, (*c*) 0.099 nm

3.6.17   (*a*) 0.315 nm, (*b*) 0.136 nm, (*c*) Mo

3.6.18   (*a*) 0.392 nm, (*b*) 0.139 nm, (*c*) Pt

3.7.4   (*a*) a. (0$\bar{1}$10), b. (10$\bar{1}$2), c. ($\bar{2}$200); (*b*) a. (01$\bar{1}$0), b. (1$\bar{1}$01), c. (1$\bar{1}$01)

3.9.1   93.1 g/mol

3.9.2   10.5 g/cm^3

3.9.3   (*a*) 9.98×10^{12} atoms/mm^2, (*b*) 1.41×10^{13} atoms/mm^2, (*c*) 5.76×10^{12} atoms/mm^2

3.9.4   (*a*) 1.61×10^{13} atoms/mm^2, (*b*) 1.14×10^{13} atoms/mm^2, (*c*) 1.86×10^{13} atoms/mm^2

3.9.5 1.84×10^{13} atoms/mm^2

3.9.6   (*a*) 3.03×10^6 atoms/mm, (*b*) 2.14×10^6 atoms/mm, (*c*) 3.50×10^6 atoms/mm

3.9.7   (*a*) 2.55×10^6 atoms/mm, (*b*) 3.60×10^6 atoms/mm, (*c*) 1.47×10^6 atoms/mm

3.10.2   −3.8%

3.10.3   −4.9%

3.11.6   0.3146 nm

3.11.7   0.19375 nm

3.11.8   (*a*) FCC, (*b*) 0.4080 nm (*c*) Au

3.11.9   (*a*) BCC (*b*) 0.5021 nm (*c*) Ba

3.11.10 (*a*) FCC; (*b*) 0.4087 nm (*c*) Ag

3.11.11 (*a*) FCC; (*b*) 0.3892 nm (*c*) Pd

Chapter 4

4.1.7 1.15×10^{-7} cm

4.1.8   374 atoms

4.1.9 1.19×10^{-7} cm

4.1.10   233 atoms

4.3.6   0.036 nm

4.4.7   10.65

4.4.8   12.75

4.4.9   6.0

4.4.10   9.64

4.5.3   (a) 1.34×10^{24} vacancies/m^3, (b) 4.14×10^{-5}
4.5.4   (a) 3.13×10^{22} vacancies/m^3, (b) 3.70×10^{-6}
4.7.2   6,553 s
4.7.3   13,590 s
4.7.4   0.262 wt %
4.7.5   0.34 wt %
4.7.6   7.06 h
4.7.7   0.354 wt %
4.7.8   1.10 mm
4.7.9 2.34×10^{-4} cm
4.7.10 8.76×10^{-4} cm
4.7.11   1509 S
4.7.12 7.07×10^{-5} cm
4.7.13   7.42 h
4.8.1 9.06×10^{-15} m^2/s
4.8.2 7.04×10^{-13} m^2/s
4.8.3 1.32×10^{-22} m^2/s
4.8.4   22.5 kJ/mol
4.8.5   143 kJ/mol
4.8.6   246 kJ/mol

Chapter 5

5.1.6 $5.88 \times 10^{-4}\ \Omega$
5.1.7   0.466 m
5.1.8   (a) $3.64 \times 10^7\ (\Omega \cdot m)^{-1}$; (b) 3.64×10^6 A/m^2
5.1.9 7.79×10^{-4} m
5.1.14 $2.72 \times 10^{-8}\ \Omega \cdot m$
5.1.15   305°C
5.1.16   543°C
5.3.7   4.28 atoms/m^3
5.3.8 $0.045\ \Omega \cdot m$
5.3.10   2855 $(\Omega \cdot m)^{-1}$
5.3.11   10.94 $(\Omega \cdot m)^{-1}$
5.4.7   (a) $5.0 \times 10^{21}\ e^-$/m^3; 4.5×10^{10} holes/m^3 (b) $9.26 \times 10^{-3}\ \Omega \cdot m$
5.4.8 $9.26 \times 10^{21}\ e^-$/m^3
5.4.9 7.66×10^{19} holes/m^3
5.4.10   (a) 5.0×10^{15} holes/cm^3; $4.5 \times 10^4\ e^-$/cm^3
         (b) $\mu_n = 1000$ cm^2/V $\cdot$ s, $\mu_p = 300$ cm^2/V $\cdot$ s; (c) $\rho = 4.17\ \Omega \cdot$ cm
5.4.11   (a) 8.0×10^{15} holes/cm^3; $2.8 \times 10^4\ e^-$/cm^3
         (b) $\mu_n = 800$ cm^2/V $\cdot$ s, $\mu_p = 250$ cm^2/V $\cdot$ s; (c) $3.12\ \Omega \cdot$ cm
5.4.12   (a) $5.79 \times 10^{18}\ e^-$/cm^3; (b) 1.16×10^{-4} As/Si atoms
5.4.13   (a) 2.07×10^{19} h/cm^{-3}; (b) 4.13×10^{-4} B/Si atoms
5.7.1 $8.36 \times 10^{-6}\ (\Omega \cdot$ cm$)^{-1}$
5.7.2   (a) $1.8 \times 10^4\ (\Omega \cdot m)^{-1}$ (b) $2.2 \times 10^4\ (\Omega \cdot m)^{-1}$
5.7.3   (a) $2.3 \times 10^{-5}\ (\Omega \cdot m)^{-1}$ (b) $3.08 \times 10^4\ (\Omega \cdot m)^{-1}$
5.7.4   (i) (a) 0.833 (b) 0.167; (ii) (a) 0.987 (b) 0.0135
5.7.5   (i) (a) 0.964 (b) 0.036; (ii) (a) 0.371 (b) 0.629

Chapter 6

6.1.5   30.0%
6.1.6   0.0638 cm

6.1.7	54.3%
6.1.13	57.9%
6.1.14	0.095 mm
6.1.15	47.8%
6.1.16	60.3%
6.2.3	31.4 MPa
6.2.4	1.27 GPa
6.2.5	2.35 GPa
6.2.6	13.7 ksi
6.2.8	0.425
6.3.1	25.8%
6.3.2	(a) U.T.S. = 76 ksi; (b) Elongation = 19%
6.3.3	(a) U.T.S. = 524 MPa; (b) Elongation = 19%
6.3.4	(b) Y.S. = 64 ksi, (c) $E = 28 \times 10^6$ psi
6.3.5	(b) Y.S. = 441 MPa
6.3.6	(a) $\sigma_E = 855$ MPa, $\epsilon_E = 0.066$; (b) $\sigma_T = 911$ MPa, $\epsilon_T = 0.064$
6.3.7	(a) $\sigma_E = 107$ ksi, $\epsilon_E = 0.235$ (b) $\sigma_T = 132$ ksi, $\epsilon_T = 0.21$
6.3.8	(a) $\sigma_E = 955$ MPa, $\epsilon_E = 0.384$ (b) $\sigma_T = 1320$ MPa, $\epsilon_T = 0.325$
6.5.13	(a) 24.5 MPa; (b) 0
6.5.14	(a) 20.4 MPa; (b) 0
6.5.15	1.715 MPa
6.6.6	(a) 25% (b) 0.277 in.
6.6.7	(a) 18.7% (b) T.S. $\cong$ 60 ksi, Y.S. $\cong$ 51 ksi, Elong. $\cong$ 26%
6.6.8	(a) 42.6%; (b) T.S. $\cong$ 78 ksi, Y.S. $\cong$ 62.5 ksi, Elongation $\cong$ 6.5
6.8.10	121.1 kJ/mol
6.8.11	72.1 kJ/mol, 84 min
6.8.12	98.5 kJ/mol
6.9.7	0.0434 in
6.9.8	1.70 mm
6.9.9	51.8 ksi
6.9.10	2.33 mm
6.9.11	(a) 0.072 in (b) 0.163 in
6.9.12	936 MPa
6.9.13	0.76 mm^7
6.11.6	4.4×10^{-7} in/(in)(h)

Chapter 7

C7.2.1	16,100 mer/mol
C7.2.2	196,000 g/mol
C7.2.3	44.2 mers
C7.2.4	106.3 mers
C7.2.5	22,750 g/mol
C7.2.6	PS = 0.604, PAN = 0.396
C7.2.7	PAN = 0.314 PB = 0.430 PS = 0.256
C7.2.8	$f_{VC} = 0.596 f_{VA} = 0.404$
C7.9.1	1.78 g S
C7.9.2	0.169
C7.9.3	4.70 g S
C7.9.4	28.2 kg S

C7.9.5 0.0211
C7.9.6 7.35 wt %
C7.9.7 8.16 wt %
C7.9.8 10.2 kg
C7.9.9 7.41 wt %
C7.10.1 (*a*) 109.6 days; (*b*) 4.15 MPa
C7.10.2 13.3 days
C7.10.3 17.5 kJ/mol
C7.10.4 (*a*) 39.1 days; (*b*) (i) 3.87 MPa, (ii) 1.39 MPa
C7.10.5 (*a*) 24.5 days; (*b*) 68.3 days
7.12.1 (*e*)
7.12.2 (*a*)
7.12.3 (*c*)
7.12.4 (*c*)
7.12.5 (*d*)
7.12.6 (*b*)
7.12.7 (*c*)
7.12.8 (*d*)

Chapter 8

All percentages are weight percents

8.4.2 (*a*) (i) $L + \alpha$, (ii) L(62% Ni), $\alpha = 73\%$ Ni, (iii) % $\alpha = 72.7\%$, % $L = 27.3\%$; (*b*) 100% liquid only

8.6.1 (*a*) All liquid; (*b*) (i) $L + \alpha$, (ii) $L = 65\%$ Ag, $\alpha = 7.9\%$ Ag, (iii) % $L = 30.0\%$, % $\alpha = 70.0\%$; (*c*) (i) $L + \alpha$ (ii) $L = 71.9\%$ Ag, $\alpha = 7.9\%$ Ag, (iii) % $L = 26.7\%$, % $\alpha = 73.3\%$; (*d*) (i) $\alpha + \beta$, (ii) $\alpha = 7.9\%$ Ag, $\beta = 91.2\%$ Ag, (iii) % $\alpha = 79.5\%$, % $\beta = 20.5\%$

8.6.2 (*a*) $L = 692$ g, $\beta = 57.7$ g; (*b*) $L = 435$ g, $\beta = 315$ g; (*c*) 101 g (*d*) 334 g

8.6.3 88.6% Sn; 11.4% Pb

8.6.4 31.4% Sn, 68.6% Pb

8.6.5 (*a*) Hypoeutectoid; (*b*) ≈10% Sn; (*c*) % $L = 13.6\%$, $L = 61.9\%$ Sn, % $\alpha = 86.4\%$, $\alpha = 19.2\%$ Sn; (*d*) % $\alpha = 92.6\%$, $\alpha = 19.2\%$ Sn, % $\beta = 7.4\%$, $\beta = 97.5\%$ Sn; (*e*) % $\alpha = 75.8\%$, % $\beta = 24.2\%$

8.7.1 (*a*) (i) $\beta + L$, (ii) $\beta = 63\%$ Re, $L = 38\%$ Re, (iii) $\beta = 8\%$, $L = 92\%$; (*b*) (i) $\beta + L$, (ii) $\beta = 55.0\%$ Re, $L = 30\%$ Re, (iii) $\beta = 40.0\%$, $L = 60.0\%$; (*c*) (i) $\alpha + \beta$, (ii) $\alpha = 35\%$ Re, $\beta = 55.0\%$ Re, (iii) $\alpha = 75.0\%$, $\beta = 25.0\%$

8.7.2 (*a*) All liquid phase; (*b*) (i) $\beta + L$, (ii) $\beta = 55.0\%$ Re, $L = 30\%$ Re, (iii) % $\beta = 12.0\%$, % $L = 88.0\%$; (*c*) (i) $L + \alpha$, (ii) $L = 30\%$ Re $\alpha = 35\%$ Re, (iii) % $L = 40.0\%$, % $\alpha = 60.0\%$

8.7.5 % $\gamma = 33\%$

8.7.6 % $\gamma = 18.2\%$

8.7.7 % $L = 7.1\%$, % $\delta = 92.9\%$

8.7.8 4.16% Ni

8.8.2 (*a*) (i) $\alpha = 100\%$ Cu, $L = 18\%$ Pb, (ii) % $\alpha = 44.4\%$; % $L = 55.6\%$; (*b*) (i) $\alpha = 100\%$ Cu, $L = 64\%$ Cu, 36% Pb, (ii) % $\alpha = 72.2\%$, % $L = 27.8\%$; (*c*) (i) $\alpha = 100\%$ Cu; $L_2 = 87\%$ Pb, (ii) % $\alpha = 88.5\%$, % $L_2 = 11.5\%$; (*d*) (i) $\alpha = 100\%$ Cu, $\beta = 100\%$ Pb, (ii) % $\alpha = 90\%$, % $\beta = 10\%$

8.8.3 (*a*) (i) $L_1 = 36\%$ Pb, $L_2 = 87\%$ Pb, (ii) % $L_1 = 33\%$, % $L_2 = 67\%$;

(b) (i) α = 100% Cu, L_2 = 87% Pb, (ii) % α = 19.5%; % L_2 = 80.5%;
(c) (i) α = 100% Cu, β = 100% Pb, (ii) % α = 30%, % β = 70%

8.8.4   10.8% Pb, 88.2% Cu

8.10.4 Peritectic reaction at 903°C: α(32.5% Zn) + L(37.5% Zn) → β(36.8% Zn)
Peritectic reaction at 835°C: β(56.5% Zn) + L(59.8% Zn) → γ(59.8% Zn)
Peritectic reaction at 700°C; γ(69.8% Zn) + L(80.5% Zn) → δ(73.0% Zn)
Peritectic reaction at 598°C; δ(76.5% Zn) + L(89% Zn) → ϵ(78.6% Zn)
Eutectoid reaction at 558°C; δ(75% Zn) → γ(70% Zn) + ϵ(78.6% Zn)
Peritectic reaction at 424°C: ϵ(87.5% Zn) + L(98.3% Zn) → η(97.3% Zn)
Eutectoid reaction at 250°C: β'(47% Zn) → α(37% Zn) + γ(59% Zn)

8.10.5 Eutectic reaction at 639.9°C: L(0.1% Ni) → (Al)(0.05% Ni) + $NiAl_3$(42% Ni)
Peritectic reaction at 854°C: L(28% Ni) + Ni_2Al_3(55% Ni) → $NiAl_3$ (42%)
Peritectic reaction at 1133°C: L(44% Ni) + $NiAl$(62% Ni) → Ni_2Al_3(59% Ni)
Peritectoid reaction at 700°C: $NiAl$(77% Ni) + Ni_3Al(86% Ni) → Ni_5Al_3(81% Ni)
Peritectic reaction at 1395°C: L(87% Ni) + $NiAl$(83% Ni) → Ni_3Al(86% Ni)
Eutectic reaction at 1385°C: L(86.7% Ni) → (Ni)(88.8% Ni) + Ni_3Al(86% Ni)

8.10.6 Eutectoid reaction at 906°C: (Ni)(29.1% Ni) → Ni_3V(25% V) + Ni_2V(30.5% V)
Eutectoid reaction at 890°C: (Ni)(35.2% V) → Ni_2V(31.5% V) + σ'(51.5% V)
Eutectic reaction at 1202°C: L(47.5% V) → (Ni)(39.6% V) + σ'(51.1% V)
Peritectic reaction at 1280°C: L(63.8% V) + (V)(38.1% V) → σ'(73.3% V)
Peritectoid reaction at 900°C: (V) (75.3% V) + σ'(70.7% V) → NiV_3(74.3% V)

8.11.1 A = 20%; B = 30%, C = 50%

Chapter 9

All percentages are weight percents.

9.2.9   % Austenite = 61.5%; % ferrite = 38.5%

9.2.10   (a) 38.5%; (b) % eutectoid ferrite = 54.3%; % eutectoid cementite = 7.2%

9.2.11   0.55% C

9.2.12   0.424% C

9.2.13   0.077% C

9.2.14   % proeutectoid cementite = 4.3%; % austenite = 95.7%

9.2.15   (a) 4.25%; (b) eutectoid cementite = 11.2%, eutectoid ferrite = 84.5%

9.2.16   1.08% C

9.2.17   1.41% C

9.2.18   0.585% C

9.2.19   (a) 12.8%; (b) eutectoid ferrite = 77.0%, eutectoid cementite = 10.2%

9.2.20   0.391% C

9.2.21   0.246% C

9.2.22   (a) 1.70% Fe_3C; (b) eutectoid ferrite = 86.8%, eutectoid cementite = 11.5%

9.3.27   (a) Martensite; (b) tempered martensite; (c) coarse pearlite; (d) martensite, marquenching; (e) bainite, austempering; (f) spheroidite

9.4.14 $\frac{3}{4}$R: RC = 35; center: RC = 32

9.4.15   Center: RC = 52; surface: RC = 53

9.4.16   surface: RC = 53; M-R: RC = 51

9.4.17   (a) 53RC; (b) 52RC

9.4.18   (a) 36RC; (b) 31RC

9.4.19   5.4°C/s

9.4.20   7°C/s

9.4.21   4°C/s
9.4.22 $S = 53RC$, $\frac{3}{4}R = 51RC$, $MR = 45RC$, $C = 38RC$
9.4.23 $S = 41RC$, $\frac{3}{4}R = 36RC$, $MR = 32RC$, $C = 30RC$
9.4.24 $S = 52RC$, $\frac{3}{4}R = 48RC$, $MR = 40RC$, $C = 37RC$
9.4.25 $S = 53RC$, $\frac{3}{4}R = 51RC$, $MR = 50RC$, $C = 48RC$
9.4.26   Bainite + martensite + austenite
9.4.27   Bainite + martensite + austenite
9.4.28   Ferrite + bainite + martensite + austenite
9.5.15   8.3%
9.5.16   (*a*) 87.8% (*b*) 5.0% (*c*) 7.2%
9.10.1   (*b*)
9.10.2   (*d*)
9.10.3   (*b*)
9.10.4   (*e*)
9.10.5   (*b*)
9.10.6   (*d*)
9.10.7   (*b*)
9.10.8   (*c*)
9.10.9   (*c*)
9.10.10 (*a*)
9.10.11 (*c*)

Chapter 10

10.2.4   0.414
10.2.5   (*a*) 8, cubic coordination; (*b*) 8, cubic coordination
10.2.6   4.91 g/cm³
10.2.7   4.88 g/cm³
10.2.8   (*a*) [110] = 2.57 Ba^{2+} and O^{2-}/nm, [111] = 1.05 Ba^{2+} and O^{2-}/nm;
         (*b*) [110] = 3.30 Co^{2+} and O^{2-}/nm, [111] = 1.35 Co^{2+} and O^{2-}/nm
10.2.9   (*a*) (111) = 7.37 Fe^{2+} and O^{2-}/nm², (110) = 12.0 Fe^{2+} and O^{2-}/nm²;
         (*b*) (111) = 5.33 K^+ and Br^-/nm², (110) = 3.27 K^+ and Br^-/nm²
10.2.10 (*a*) 6.70 g/cm³; (*b*) 2.82 g/cm³
10.2.11 (*a*) 0.542; (*b*) 0.525
10.2.12 4.06 g/cm³
10.2.13 4.17 g/cm³
10.2.14 F^- occupy all tetrahedral sites
10.2.15 6.35 g/cm³
10.2.16 Zero
10.2.17 (*a*) [111]: (1.14 Zr^{4+} and 2.28O^{2-})/nm; (*b*) [110]: 2.79 Zr^{4+}/nm
10.2.18 (*a*) (111) = 9.28 Hf^{4+}/nm²; (*b*) (110) = 5.68 Hf^{4+}/nm²; (*c*) (110) = 11.4 O^{2-}/
         nm²
10.2.19 0.701
10.2.20 1.0
10.2.23 0.560
10.2.24 6.47 g/cm³
10.6.6 1.59×10^{-4} m
10.6.7   3674
10.7.5   89.8 μm

10.7.6   719 MPa
10.7.7   424 μm
10.7.8   (a) 14.6 μm; (b) 1080 MPa
10.9.16 374 kJ/mol
10.9.17 (a) 314 kJ/mol (b) $10^{11.44p}$
10.9.18 (a) 838 kJ/mol (b) 844°C
10.9.19 334 kJ/mol

Chapter 11

11.2.10 $M_s = 1.70 \times 10^6$ A/m, $B_s = 2.14$ T
11.2.11 $M_s = 2.51 \times 10^6$ A/m
11.2.12 $M_s = 1.96 \times 10^6$ A/m
11.2.13 $M_s = 1.69$ μ_B/atom
11.2.14 $B_s = 0.645$ T
11.2.15 $M_s = 1.98 \times 10^6$ A/m
11.7.16 Wt % Fe = 91.72; wt % B = 2.96%; wt % Si = 5.32%
11.8.3   ~153 kJ/m^3
11.8.4   ~48 kJ/m^3
11.8.5   ~0.08 J
11.9.5   (a) 4; (b) 2; (c) 5
11.9.6 2.56×10^5 A/m, $B_s = 0.32$ T

Chapter 12

12.3.2   −0.627 V
12.3.3   −0.314 V
12.3.4   −0.403 V
12.3.5   −0.444 V
12.3.6   −0.596 V
12.3.7   −1.109 V
12.3.8   0.071 M
12.3.9   0.050 M
12.3.10 −0.046 V
12.3.11 −0.0224 V
12.4.1   17.8 g/h
12.4.2   1408 s
12.4.3   (a) 0.50 A (b) 7.9×10^{-5} A/cm^2
12.4.4 1.06×10^{-2} g/min
12.4.5   350 days
12.4.6   13.8 days
12.4.7 4.52×10^{-7} A/cm^2
12.4.8   6.54 mdd
12.4.9 2.97×10^{-3} mm
12.4.10 8.36×10^{-8} A/cm^2
12.4.11 15.2 years
12.4.12 (a) 2.17×10^{-8} A; (b) 9.56×10^{-5} A/cm^2
12.4.13 231 days
12.6.2   W = 2.01; Na = 0.574; Hf = 1.62; Cu = 1.74; Mn = 1.70; Sn = 1.15
12.6.5   101 μg/cm^2

12.6.6   23.6 h
12.7.10 0.652 A
12.7.11 0.870 kg

Chapter 13

13.3.5   (a) 72.4 wt % C, 27.6 wt % epoxy; (b) 1.59 g/cm^3
13.3.6   (a) 75 volume % C, 25 volume % epoxy; (b) 81.0 wt % C, 19.0 wt % epoxy
13.3.8   35.6 × 10^6 psi
13.3.9   (a) 2.18 × 10^5 psi; (b) 0.994
13.3.10 17.8 × 10^6 psi
13.3.11 (a) 388 × 10^3 psi; (b) 0.99
13.3.13 12.5 GPa
13.3.14 9.42 GPa
13.6.19 22.4 sacks cement, 4120-lb wet sand, 8000-lb gravel, 109-gal water
13.6.20 12.1 sacks cement, 2255-lb wet sand, 3670-lb wet gravel, 54.2-gal water
13.8.10 53.8%
13.8.11 156.0 g
13.8.12 107.4 g
13.10.1 229 GPa
13.10.2 f(SiC) = 0.265
13.10.3 E_c = 213 GPa
13.10.4 ρ(Al) = 2.92 g/cm^3
13.10.5 f(SiC) = 0.417
13.10.6 RBSN σ_c = 1556 MPa, SiC σ_c = 718 MPa, SiC cracks first
13.10.7 (a) E_c = 356 GPa; (b) ϵ_c = 3.7 × 10^{-3}
13.10.8 Al$_2$O$_3$ cracks at 953 MPa

Chapter 14

14.2.2   (a) 0.39 → 0.77 μm; (b) 0.01 → 0.39 μm; (c) 0.77 → 100 μm
14.2.3   539.8 nm, green
14.2.4   2.39 eV
14.4.3   (a) 3.6%; (b) 4.4%
14.4.4   0.0315
14.4.5   0.0684 cm^{-1}
14.4.6   0.903
14.4.8   (a) 551 nm; (b) 1823 nm; (c) 976 nm
14.4.9   (a) 479 nm (transmitted); (b) 548 nm (absorbed); (c) 300 nm (transmitted)
14.5.5   (a) 2.95 × 10^{-3} s; (b) 4.36 × 10^{-6}%
14.5.6   0.0379 s
14.7.4   −0.973 dB/km
14.7.5   55 km
14.7.6   5.67%
14.7.11 87.23°
14.8.9   0.0462 T
14.8.10 3.40 K

INDEX